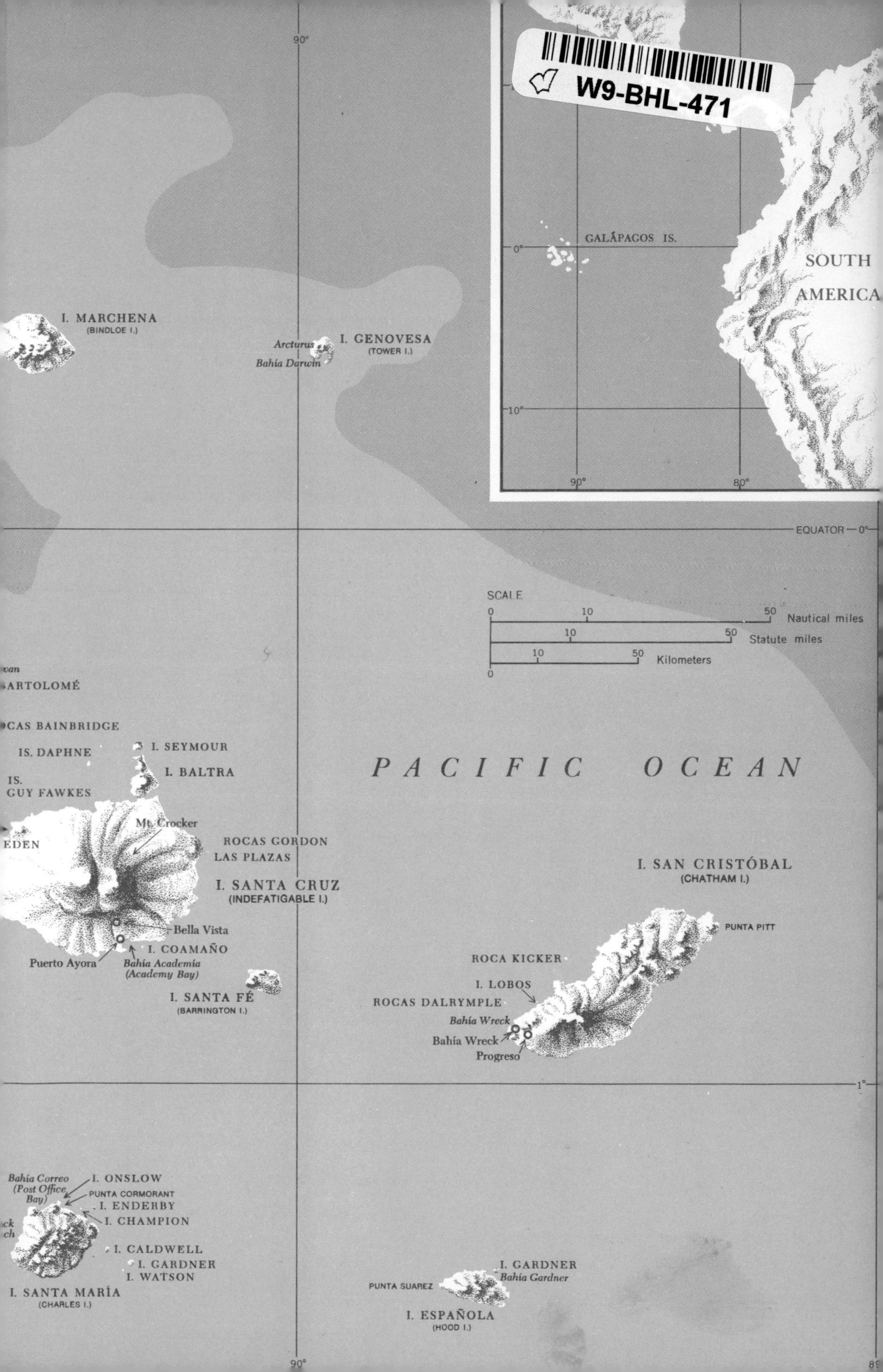

W9-BHL-471
GALÁPAGOS IS.
SOUTH AMERICA
90°
80°
0°
10°
I. MARCHENA
(BINDLOE I.)
I. GENOVESA
(TOWER I.)
Arcturus
Bahía Darwin
EQUATOR — 0°
SCALE
0 10 50 Nautical miles
10 50 Statute miles
0 10 50 Kilometers
ARTOLOMÉ
CAS BAINBRIDGE
IS. DAPHNE
I. SEYMOUR
I. BALTRA
IS.
GUY FAWKES
PACIFIC OCEAN
EDEN
Mt. Crocker
ROCAS GORDON
LAS PLAZAS
I. SANTA CRUZ
(INDEFATIGABLE I.)
Bella Vista
I. COAMAÑO
Puerto Ayora
Bahía Academía
(Academy Bay)
I. SANTA FÉ
(BARRINGTON I.)
I. SAN CRISTÓBAL
(CHATHAM I.)
PUNTA PITT
ROCA KICKER
I. LOBOS
ROCAS DALRYMPLE
Bahía Wreck
Bahía Wreck
Progreso
1°
Bahía Correo
(Post Office
Bay)
I. ONSLOW
PUNTA CORMORANT
I. ENDERBY
I. CHAMPION
I. CALDWELL
I. GARDNER
I. WATSON
I. SANTA MARÍA
(CHARLES I.)
I. GARDNER
Bahía Gardner
PUNTA SUAREZ
I. ESPAÑOLA
(HOOD I.)
90°

Flora of the Galápagos Islands

Contributors

Edward F. Anderson
David M. Bates
Derek Burch
Lincoln Constance
Richard S. Cowan
Arthur Cronquist
Uno Eliasson
Leslie A. Garay
Amy Jean Gilmartin
John Thomas Howell
Tetsuo Koyama
Mildred Mathias
Harold N. Moldenke
Reid Moran
Conrad V. Morton
Peter H. Raven
Charlotte Reeder
John R. Reeder
Charles M. Rick, Jr.
Keith Roe
Reed C. Rollins
Velva E. Rudd
Henry J. Thompson
David L. Walkington
Dieter C. Wasshausen
U. T. Waterfall
Grady L. Webster
Delbert Wiens

Ira L. Wiggins
and Duncan M. Porter

Flora of the Galápagos Islands

1971
Stanford University Press
Stanford, California

Stanford University Press
Stanford, California

Printed in the United States of America
ISBN 0-8047-0732-4
LC 78-97917

To the memory of E. Yale Dawson

Preface

In 1964 some 65 scientists from nearly a dozen countries took part in the field operations of the Galápagos Islands Scientific Project. During these six weeks, a plan to produce a modern flora of the islands began to take shape in talks between the late E. Yale Dawson and the senior author. It was agreed that Dawson should take primary responsibility for the book and should guide the project from his position as Director of the San Diego Natural History Museum.

We were of course encouraged in these plans by the special character of the islands—they hold a particular interest for biologists because of their isolation from excessive interchange of biotic elements with other regions, and have thus yielded profound and dramatically evident evolutionary developments among both plants and animals. Ecologists, environmentalists, conservationists, and biologists would perhaps welcome a manual wherein they might readily identify the vascular plants they would be likely to find in the islands. It would serve also to put on record the present makeup of the native flora, and it would furnish, as well, information about the plants known to have been introduced by man. Detailed descriptions, carefully constructed keys, and line drawings of representatives of each recognized genus would be found particularly useful; and distribution maps would provide reference foci for investigations of the biogeography of the islands, as well as making possible critical comparisons with floras of other areas. And of course such a work would hold especial interest for biologists who would be working at the Charles Darwin Foundation's research station on Isla Santa Cruz, as it would for the people joining the tours of the islands that were being inaugurated by several agencies.

A few months after Dr. Dawson's return to San Diego from the Galápagos Islands, he accepted a curatorship in the U.S. National Museum of Natural History and withdrew from active participation in the work on the flora, since his new position would demand full-time attention to his major field of interest, marine algology. But until his tragic death in the surf of the Red Sea in June 1967, he continued to tender advice and suggestions on numerous aspects of the project. We who have completed the task can only regret that he did not live to see this particular dream become a reality.

Prior to Dr. Dawson's departure from San Diego, he and Wiggins planned the broad outlines of the book and began enlisting help from specialists at home and abroad. Wiggins would take over prime responsibility for the program of research and writing, under the sponsorship of the California Academy of Sciences, where he was a Research Associate in botany. And because expeditious completion of the manuscript for the book demanded the full-time assistance of a trained botanist, Porter was engaged as a post-doctoral fellow, serving in that capacity from July 1966 through August 1967. Porter then took over responsibilities of junior author and continued to work on several families while on the faculty of the University of San Francisco in 1967–68, and subsequently as Curator of Central American Collections at the Missouri Botanical Gardens in St. Louis.

Constraints on island-to-island transportation at the time of the 1964 expedition had forced the senior author to restrict his collecting of botanical specimens to a single island, Isla Santa Cruz. Thus, to obtain some familiarity with the vegetative profile elsewhere in the archipelago, both authors spent several weeks collecting specimens, obtaining photographs, and recording their observations on the flora of thirteen separate islands in January and February 1967 (see p. 44). Preparation of a final draft then began in earnest, for both authors and collaborators.

During the field season of 1964 (see pp. 43f), the Charles Darwin Foundation for the Galápagos Islands, through Dr. David Snow, the director of its research station on Isla Santa Cruz, was extremely helpful to Dr. Dawson and the senior author. In particular, Dr. Snow gave the senior author many duplicates from his Galápagos collections for the Dudley Herbarium at Stanford University, where they were very useful during the preparation of this work. We are grateful also to Dr. Snow's successor, Roger Perry, for his courtesies during our brief stay on Isla Santa Cruz in 1967, and to other officers of the Foundation for their help.

Donald Kyhos, now on the faculty of the University of California, Davis, painstakingly examined over a hundred bud fixations of Galápagos plants that we collected on Isla Santa Cruz in 1964. Despite imperfections in the fixations, Dr. Kyhos succeeded in obtaining chromosome counts for approximately half of them; these counts are given in the appropriate species descriptions. We are most grateful not only for his skill, but for his patience and generosity.

Special thanks are also due a number of people who collected specimens for us in 1964, in some cases from islands to which we had no access: Allyn G.

Smith, John R. Hendrickson, Luis Fournier, Jacqueline and André DeRoy, E. Gorton Linsley, Carl B. Koford, Sigurd Horneman, Syuzo Itow, Paul E. Silva, and the late Robert E. Usinger.

We are grateful to S. M. Walters of Cambridge University for the loan of several types and other specimens, and to scientists at Kew, the Gray Herbarium, the New York Botanical Garden, the Brooklyn Botanic Garden, the University of California, Davis, the California Academy of Sciences, the Missouri Botanical Garden, the Smithsonian Institution, and the Dudley Herbarium at Stanford University for providing our group with hundreds of sheets for examination, thus enabling us to get a more accurate idea of the range of variation among the many species than would have been possible without their help. We would like to thank also Irene Manton of Leeds University, F. M. Jarret of the Royal Botanical Gardens at Kew, and Rolla Tryon of the Gray Herbarium for their helpful advice about certain ferns.

Among many other courtesies, the officers and staff of the California Academy of Sciences kindly allowed us to borrow specimens from the herbarium and gave us working space. Similar courtesies were accorded us by the director and personnel of the Division of Systematic Biology at Stanford University. We are deeply grateful also to Kenneth K. Bechtel for his support of the Galápagos International Scientific Project in 1964, for his continued interest in the work carried forward under the sponsorship of the California Academy, and for a grant that made possible the printing of the color plates.

Color transparencies from which the color plates were made were generously loaned by David Cavagnaro, Paul A. Colinvaux, George E. Lindsay, Robert T. Orr, and Allyn G. Smith. We appreciate greatly their interest and help.

Most of the drawings are the work of Jeanne Russell Janish and the senior author, and are reproduced here for the first time. The others, in most cases identified by initials, are by Phyllis Plattner, Leslie A. Garay, Mary Benson, Charles C. Clare, Jr., Charlotte Reeder, and a number of other professional illustrators working over the past seventy years. To Mrs. Janish and the others, and to Nancy Fouquet and Deborah Daymon for the basic internal maps and the endsheet map, we are most grateful. Other illustrations are reproduced from books and journals by permission of the director of the Missouri Botanical Garden, the Hunt Botanical Library, the director of the U.S. National Museum of Natural History (most of these are by Agnes Chase), the director of the Brooklyn Botanic Garden, the managing editor of the *American Journal of Botany*, and the Regents of the University of California.

Jane Bavelas and Jean McIntosh examined the entire text for such matters as technical and nomenclatural consistency. And at Stanford University Press,

William W. Carver, Barbara Mnookin, and Humphrey Stone devoted endless attention to the myriad editorial and design details that contribute to the production of an acceptable book. To them, and to all the others at the Press who have given expert attention and effort to the project, we extend our heartfelt appreciation.

Finally, we must make clear that this project would have been entirely beyond our means without the financial aid of the National Science Foundation, provided through Grant GB-5254. We extend our sincere thanks not only for this aid but for the interest, understanding, and patience of the Foundation's officers during the course of the project.

Imperfections in a book of this sort being unavoidable, we can only hope that serious errors are few. The individual collaborators are responsible only for their own taxonomic decisions and for the technical treatment of their plant groups; the responsibility for all other decisions is ours.

I.L.W.
D.M.P.

February 8, 1971

Contents

Detailed Description of Color Plates

(*The plates follow p. 42*)

1. Arcturus Lake, Isla Genovesa (Tower Island). Genovesa is in fact a low-lying island, though very likely the shell of an earlier and taller peak; the lake, which occupies its crater, lies at sea level and rises and falls with the tide, percolating through its lava basin. Typical Littoral Zone vegetation: *Bursera graveolens* in the foreground, *Rhizophora mangle* along water's edge. (Paul A. Colinvaux)

2. Cumulo-nimbus clouds over Isla Santa Cruz, the characteristic view from the sea during the rainy season, January to May. (Ira L. Wiggins)

3. Southwest end of Wreck Bay, Isla San Cristóbal, from the harbor. Government buildings stand near the palms and at extreme right; the palms and *Ficus indicus* (dark green tree) are introduced. The town of Progreso lies inland to the left, hidden by clouds. (George E. Lindsay)

4. Main waterfront street of Wreck Bay, Isla San Cristóbal, chief port of the archipelago. This is the business district, its buildings facing northwest; there are no pavements and few conveniences. (George E. Lindsay)

5. Isla Fernandina, seen from a point near Tagus Cove, Isla Isabela. The volcanic rock is typical of these shores. (Paul A. Colinvaux)

6. Caldera on Isla Fernandina, after the explosive eruption of 1968. The level of the lake dropped 100 m, and ash and cinders covered vegetation to a depth of 3 to 5 m on the southwest flank of the volcano, almost to sea level. (Paul A. Colinvaux)

7. Caldera on Isla Fernandina, 1966. The salt lake lies 900 m below the rim of the crater. The island at the left is a parasitic crater. (Paul A. Colinvaux)

8. West end of Isla Bartolomé. Plants of a *Scalesia* species are among the lava boulders in the foreground; at the right are mangroves (*Rhizophora mangle* and *Laguncularia racemosa*). (Ira L. Wiggins)

9. Isla Bartolomé, with Isla San Salvador in the background, Sullivan Bay between the islands. The rock spire is the edge of a volcanic crater that has been blown apart; the parallel ridges are successive volcanic flows alternated with ash deposits. Sand dunes are built up by prevailing winds from the left. Typical Littoral Zone vegetation (see legend for color plate 8, above). (Ira L. Wiggins)

10. Pelicans at Punta Espinosa, Isla Fernandina. These awkward birds nest either on the ground or on loose platforms of sticks and twigs atop shrubs—here, red mangrove (*Rhizophora mangle*) in the immediate foreground, white mangrove (*Laguncularia racemosa*) just behind. (Ira L. Wiggins)

11. Pink-footed booby on shore of Darwin Bay, Isla Genovesa. The booby is a clumsy bird when on land or perched in shrubs, but is a splendid aviator and diver.

This species nests in a shallow cup, lined only with sand, packed soil, or pebbles. The bush here is a battered red mangrove. (Ira L. Wiggins)

12. Masked booby with newly hatched chick, Punta Suárez, Isla Española. The hatchlings, at first almost naked, soon sport whitish pinfeathers and, within a few days, a dense coat of white down. An unguarded hatchling quickly falls prey to a gull or skua. (Ira L. Wiggins)

13. Blue-footed booby and eggs, Punta Suárez. The rougher lava pebbles have been kicked and scraped from a circle about a foot in diameter. Male and female alternate egg-tending during incubation, the absent bird rarely returning in less than 24 hours. Several acres of this rookery were almost devoid of vegetation. (Ira L. Wiggins)

14. Basking marine iguanas, Punta Espinosa, Isla Fernandina. The animals sprawl about on sandy beaches for much of the middle part of the day. Mixed white and red mangrove. (Ira L. Wiggins)

15. Fledgling masked booby, Darwin Bay, Isla Genovesa. This chick is about three-fourths grown, the down nearly an inch thick. The shrub, *Cryptocarpus pyriformis*, is abundant at low elevations throughout the archipelago. (Ira L. Wiggins)

16. Frigate birds, Darwin Bay. These females had not yet chosen mates and were awaiting the displaying males. Their perch is a tangled mass of *Cryptocarpus pyriformis*, about 50 m from the water's edge. (Ira L. Wiggins)

17. Frigate bird, Darwin Bay. This was one of several males observed sitting quietly on the *Cryptocarpus* shrubs, gular pouches distended; later they would begin aerial displays and gather sticks for nesting platforms, each male pilfering such sticks as he could from his neighbors' nests. (Ira L. Wiggins)

18. Galápagos tortoise at the Charles Darwin Research Station, Isla Santa Cruz. The tortoises eat many kinds of native vegetation, including fallen *Opuntia* pads, water fern (*Azolla*), and several herbs and grasses. In captivity, they do well on pumpkins, *Opuntia* pads, and vegetable trimmings. (Ira L. Wiggins)

19. Flightless cormorant at Punta Espinosa, Isla Fernandina. These birds are superb divers and swimmers, easily overtaking fish in an open chase; they spend several hours a day perched on the rocks, drying out their feathers. (Ira L. Wiggins)

20. Marine iguana, Isla Española. The iguanas on this island are more brightly colored than those on the other islands. They dive to 20 m or more when the algae on which they feed are depleted at shallower depths. (Ira L. Wiggins)

21. Southern shore of Isla Pinta. Here we see typical Littoral Zone vegetation, with low *Cryptocarpus pyriformis* bushes in the foreground, red mangrove (*Rhizophora mangle*) beyond the beach. A young bull sea lion, at the extreme right, defends his harem. A strip of black lava surrounds the base of the island's volcanic crater, with stringers running to the sea from the sandy beach. (David Cavagnaro)

22. Red mangrove seedling, Conway Bay, Isla Santa Cruz. The seed germinates while still attached to the parent plant, forming a spindle-shaped seed-root; when this becomes too heavy for its twig attachment, it plunges into the mud or sand and plants itself. (Ira L. Wiggins)

23. Prop roots of the red mangrove, low tide at Conway Bay. The roots descend from the undersides of spreading branches, forming intricately gnarled and branched attachment structures. Sand may often be washed away to expose the holdfasts; and on bare rock, wave action causes the prop tips to become clubbed. (Ira L. Wiggins)

24. Red mangrove stand at Conway Bay. Each established shrub spreads radially until it fills the area available or meets another colony. Mangroves are confined to

soil or rocks washed by salt water at frequent intervals; the interlaced branches and prop roots are almost impenetrable. (Ira L. Wiggins)

25. The shore at Conway Bay. A young red mangrove stands in the shallow water. In the right foreground, *Cryptocarpus pyriformis*; on the spit in the middle distance, *Opuntia echios* var. *gigantea, Jasminocereus thouarsii* var. *delicatus*, and more mangroves and *Cryptocarpus*. The greenish foam among the mangrove props is rich in diatoms: a single tablespoonful from Punta Espinosa, on Isla Fernandina, contained over 120 species of diatoms. (Duncan M. Porter)

26. Salt marsh, Conway Bay, Isla Santa Cruz. The dull gray-green vegetation in the foreground is a mixed stand of *Batis maritima, Sporobolus virginicus*, and stunted *Cryptocarpus pyriformis*. Just beyond the man are straggly shrubs of *Avicennia germinans*, and the Arid Zone vegetation beyond the pond includes *Bursera graveolens, Maytenus octogona, Lycium minimum*, and *Croton scouleri*. (Ira L. Wiggins)

27. Trunk of mature *Opuntia echios* var. *gigantea*, Academy Bay, Isla Santa Cruz. As the trunk grows larger, the spine clusters are gradually crowded off. The thin, papery plates are at first tightly laminated, but the outer ones steadily loosen and fall away; insects, scorpions, and geckos often hide under the plates and in the vertical fissures. (Ira L. Wiggins)

28. *Opuntia echios* var. *barringtonensis*, Barrington Bay, Isla Santa Fé. Mature trees of this variety are characteristic of the stands near the coast of Santa Fé. The young plant in the left foreground displays the heavy spine cover of immature specimens. The stiff, dark green shrubs in the middle distance are chiefly *Scutia pauciflora*. (Ira L. Wiggins)

29. Arid Zone vegetation as seen from atop a 20-meter escarpment about one-half km inland from the Charles Darwin Research Station, Academy Bay, Isla Santa Cruz; the view is to the southwest. Predominant here are *Opuntia echios* var. *gigantea, Jasminocereus thouarsii* var. *delicatus, Croton scouleri, Bursera graveolens, Cryptocarpus pyriformis, Maytenus octogona, Alternanthera echinocephala*, and *Castela galapageia*; mangroves (*Rhizophora mangle*) form dense stands along the shore at Puerto Ayora (middle distance at right), and patches of *Hippomane mancinella* are just inland from salt water. (George E. Lindsay)

30. Spines on the trunk of *Opuntia megasperma* var. *orientalis*, Isla Gardner (near Española). In young plants the spines of adjacent clusters overlap, but as the trunk grows the spines are gradually separated. The plates are generally gray in this species, in contrast to the bright red of *O. echios*. (Ira L. Wiggins)

31. *Opuntia insularis* near Tagus Cove, Isla Isabela. This species does not attain the majestic proportions of *O. echios* and *O. galapageia*. On young plants the spine armament is extremely dense. (Ira L. Wiggins)

32. *Opuntia galapageia* var. *galapageia*, Isla Pinta. Though this *Opuntia* is less of a tree than *O. echios*, the trunk is relatively stouter. The nearly leafless, lower shrubs are *Croton scouleri* var. *scouleri*; the few small patches of green, *Scutia pauciflora*. This photograph, taken near the end of the dry season, illustrates the extreme aridity of the lower stretches of smaller islands. (David Cavagnaro)

33. Typical Arid Zone vegetation on cliff face, growing in crevices in lava blocks, Academy Bay, Isla Santa Cruz. The cacti are *Opuntia echios* var. *gigantea* and *Jasminocereus thouarsii* var. *delicatus*; the slender, small, white-trunked trees, *Croton scouleri*; the trees with pale green leaves at tip, *Bursera graveolens*; the intricately branched, apparently leafless shrub in center, *Lycium minimum*. Boobies, cormorants, and gulls perch on the rocks. (Allyn G. Smith)

34. *Opuntia megasperma* var. *megasperma*, Isla Champion (near María). This flower had opened for the first time the previous day and had been clear, pale yellow, like the flower in color plate 35 (below). On this second day the petals were flushed with rose, the anthers had shed their pollen, and the stigma lobes were expanded and receptive. By the end of the third day the petals would be dingy red and tightly inrolled, and would not open again. (Ira L. Wiggins)

35. *Opuntia megasperma* var. *megasperma*, Isla Champion. A freshly opened flower, pale, waxy yellow; the stigma lobes are still tightly clamped. The black specks are flies and tiny beetles, which swarm over each new blossom; some become trapped when the petals wither and coalesce at their tips. (Ira L. Wiggins)

36. *Opuntia helleri*, Isla Genovesa. This cactus is a low-growing, bushy plant rarely over 1.5 m tall; the spines are bristle-like and fairly flexible—less formidable than they appear. The twigs just behind the flower are of *Chamaesyce viminea*. (Ira L. Wiggins)

37. Arid Zone, Isla Isabela. The bright green, large-leaved shrubs are *Scalesia affinis*; the tall cacti, *Jasminocereus thouarsii* var. *sclerocarpus*; the gray, low, apparently leafless plants, probably *Borreria ericaefolia* or *B. dispersa*. The terrain is characteristic of the volcanic scoria that covers many square miles of Islas Isabela and Fernandina. (David Cavagnaro)

38. *Opuntia echios* var. *gigantea*, Academy Bay, Isla Santa Cruz. An open flower is often as much as 9 cm across. The anthers dehisce and shed nearly all their pollen before the stigma lobes become receptive; the latter still form a tight ball at the style apex. Finches, doves, and mockingbirds feed on the stamens, and are doubtless effective pollinators. (David Cavagnaro)

39. Dry season, Isla Genovesa (Tower Island). The island reaches no more than 70 m and lies completely in the Arid Zone; its surface is intricately fissured by crevasses. The trees are chiefly *Bursera graveolens*, with scattered *Croton scouleri* and shrubs of *Chamaesyce viminea*, *Heliotropium angiospermum*, and *Maytenus octogona*. (Allyn G. Smith)

40. Fine specimens of *Opuntia echios* var. *gigantea* and *Jasminocereus thousarsii* var. *delicatus*, Academy Bay, Isla Santa Cruz. *Scalesia affinis* stands in the center, middle ground; in the distance, mixed *Alternanthera echinocephala*, *Cryptocarpus pyriformis*, *Maytenus octogona*, *Castela galapageia*, *Croton scouleri*, and *Bursera graveolens*. (Allyn G. Smith)

41. *Jasminocereus thouarsii* var. *delicatus*, Conway Bay, Isla Santa Cruz. Here we see the characteristic branching and spination of a healthy, half-grown plant. The small spray of leaves at upper left is of *Bursera graveolens*. (Ira L. Wiggins)

42. Arid Zone vegetation near Tortuga Bay, Isla Santa Cruz. The feather-duster bushes in the foreground are *Scalesia helleri*; behind them, *Opuntia echios* var. *gigantea* and *Jasminocereus thouarsii* var. *delicatus*. Light patches on the lava are lichens, and the leafless shrub at left is *Castela galapageia*. (David Cavagnaro)

43. Transition Zone forest, Isla Santa María, an almost pure stand of *Bursera graveolens* inland from the east end of Post Office Bay. Various shrubs and herbs form an open understory. (Ira L. Wiggins)

44. Transition Zone, Isla Santa Cruz: the trail to Bella Vista, 130 m elevation. The trees along the trail include *Pisonia floribunda*, *Bursera graveolens*, *Psidium galapageium*, and *Piscidia carthagenensis*. Herbaceous vegetation includes several legumes (with trifoliate leaves), *Abutilon depauperatum*, *Ipomoea triloba*, and *Alternanthera filifolia*. The gray streamers are a lichen, *Ramalina* sp. (Ira L. Wiggins)

45. Junction of old and new trails, Isla Santa Cruz. The main shrubby growth along the trail is *Tournefortia pubescens*; other plants in the margins are *Trema micrantha* (one or two, with spreading-ascending branches and drooping leaves), *Piscidia carthagenensis* (dark green leaves, back from trail), and *Psidium galapageium* (tallest tree right of center). Herbaceous plants in the trail include *Sida spinosa, Mentzelia aspera, Physalis galapagoënsis*, and *Capraria biflora*. (Robert T. Orr)

46. Transition Zone, en route to Bella Vista, Isla Santa Cruz. Visible are a small *Opuntia echios* var. *gigantea* (right foreground), *Tournefortia pubescens* (dark green-leaved shrub, left foreground), *Bursera graveolens* (white-trunked trees), and *Acacia rorudiana* (feathery pale green tree with flattish top, right middle ground); the low gray-green herbs along the trail are chiefly *Trianthema portulacastrum*. (Robert T. Orr)

47. Transition Zone, west slopes near Tagus Cove, Isla Isabela. Bare lava flows alternate with strips of cinder slope covered with *Bursera graveolens, Pisonia floribunda, Psidium galapageium*, and various lower shrubs. The cover on Isla Santa Cruz at the same altitude is more open. (Ira L. Wiggins)

48. *Bursera graveolens*, inland from Caleta Black, Isla Isabela. Individual blossoms are less than 1 cm broad and have very little fragrance; the volatile oils in the leaves, bark, and fruit are, by contrast, pleasantly aromatic. (Ira L. Wiggins)

49. *Bursera* tree near Caleta Black, Isla Isabela. The soil is so thin that the root has followed fissures in the lava, while remaining above the surface. The slender stems behind the tree are of *Croton scouleri* var. *scouleri*. (Ira L. Wiggins)

50. Transition Zone, southwestern slopes of Isla Fernandina. This area was partially covered with ash and cinders following the explosive eruption of the volcano on Fernandina in June–July 1968. The cactus is *Opuntia insularis*; the yellow-flowered bush in the center, *Cassia occidentalis*. The trees include *Croton scouleri* and *Bursera graveolens*, and grasses and sedges are abundant. (David Cavagnaro)

51. A fine stand of *Scalesia pedunculata* on the north slope of Mount Crocker, Isla Santa Cruz. The understory is dominated by various moderately large ferns, and many *Scalesia* seedlings appear at the margins of the forest. (David Cavagnaro)

52. Dense growth of *Scalesia pedunculata* seedlings about 1 km south of Bella Vista, south slope of Isla Santa Cruz. The trifoliate vines climbing over the *Scalesia* plants are probably *Vigna luteola*. (Robert T. Orr)

53. *Scalesia pedunculata* forest: the delicate canopy overhead. (David Cavagnaro)

54. Open stand of *Scalesia incisa* on the flat top of Isla Wolf. Many low, shrubby plants of *Opuntia helleri* grow in the spaces between the dominant shrubs; several sedges and grasses and *Trianthema portulacastrum* form herbaceous cover. The island is wind-swept and fog-drenched much of the time. (David Cavagnaro)

55. Forest of *Scalesia pedunculata*, north slope of Isla Santa Cruz. The low light intensity under the dense *Scalesia* canopy provides optimum conditions for epiphytic mosses, liverworts, and ferns, including several species of *Elaphoglossum* and *Nephrolepis*. (David Cavagnaro)

56. *Zanthoxylum fagara* forest near summit of Isla Pinta, just above the *Scalesia* Zone. On this island, *Zanthoxylum* occupies the niche elsewhere characterized as the *Miconia* Zone. Dense growths of mosses, liverworts, and ferns occupy the trunks and branches of trees; the large terrestrial ferns include *Pteridium aquilinum* var. *arachnoideum* and *Diplazium subobtusum*. (David Cavagnaro)

57. Stand of *Croton scouleri* var. *grandifolius* with scattered shrubs of *Scalesia incisa* in the lichen-rich fog belt on Isla Pinzón. (David Cavagnaro)

58. Fern-Sedge Zone, Isla Santa Cruz. The endemic tree fern, *Cyathea weatherbyana*, thrives in the fog-swathed depressions and slopes it shares with spongy masses of *Sphagnum*; strap-shaped fronds of *Elaphoglossum engelii* are also abundant at the edges of the *Sphagnum* blankets. (David Cavagnaro)

59. Fern-Sedge Zone, Isla Santa Cruz. The sparsely covered lava ridges in the foreground support lichens, low shrubs of *Pernettya howellii*, and several weedy herbs. A band of *Scalesia pedunculata* trees stands on the ridge at the right of the mountain in the center, and an extensive mixed growth of *Scalesia* and *Miconia robinsoniana* encircles the base of the volcanic cone beyond. (David Cavagnaro)

60. *Miconia* Zone, Isla Santa Cruz. A dense tangle is formed from the intergrown broad-leaved *Miconia robinsoniana* shrubs, tree ferns (*Cyathea weatherbyana*), and the dry and brownish fronds of *Pteridium aquilinum* var. *arachnoideum*. The bare branches at the left are of *Pisonia floribunda*. (Ira L. Wiggins)

61. An almost impenetrable stand of *Miconia robinsoniana* along the trail to Mount Crocker, Isla Santa Cruz. The ferns in the lower left corner and at the base of the shrub at right are *Blechnum occidentale* var. *puberulum*. Beyond are strips of darker green *Scalesia* forest with Transition Zone and Arid Zone vegetation stretching to the sea at far left. (David Cavagnaro)

62. Contact area between the *Miconia* Zone and the Fern-Sedge Zone above Bella Vista, Isla Santa Cruz. The tree fern, *Cyathea weatherbyana*, with bright green fronds, arches above the dry, brownish fronds of fern brake, *Pteridium aquilinum* var. *arachnoideum*, with leaves of *Miconia robinsoniana* between the two; a similar mixture clothes the hillside in the background. (Allyn G. Smith)

63. *Passiflora foetida* var. *galapagensis* near the Charles Darwin Research Station, Academy Bay, Isla Santa Cruz. The flowers wither and close before midday; glands dotting the herbage emit an unpleasant odor. (Allyn G. Smith)

64. *Habenaria monorrhiza* on the southwesterly slope of Mount Crocker, Isla Santa Cruz. This terrestrial orchid grows in pockets of soil on lava ridges; fleshy tuber-like roots store water to sustain the plant in dry periods. (Ira L. Wiggins)

65. *Tillandsia insularis*, epiphytic on the trunk of a small tree of *Pisonia floribunda*, between Academy Bay and Bella Vista, Isla Santa Cruz. The shining green leaves at the lower left are of *Psychotria rufipes*; lichens and minute liverworts share the tree trunk with the *Tillandsia*, whose leaf bases often hold rainwater that becomes a habitat for mosquito larvae. (Ira L. Wiggins)

66. *Lantana peduncularis*, Isla Champion, a common and widespread small shrub on most of the islands in the archipelago. The endemic carpenter bee works over the flowers assiduously during the early hours of the day. (Ira L. Wiggins)

67. *Adiantum macrophyllum* in the *Miconia* Zone above Bella Vista, Isla Santa Cruz. The bright red of the freshly expanded fronds is soon displaced by deep green. The three-veined leaves are of *Fleurya aestuans*. (David Cavagnaro)

68. *Cordia leucophlyctis*, Conway Bay, Isla Santa Cruz. Straggly shrubs of this plant are usually scattered irregularly in partial shade in the Arid and Transition Zones; by midseason the scarlet berries contrast attractively with the white flowers. (Ira L. Wiggins)

69. *Sesuvium portulacastrum*, along the beach near the Charles Darwin Research Station, Isla Santa Cruz. This succulent weed is nearly worldwide in distribution in warm temperate and tropical regions, growing readily in salt-impregnated soils; the pink flowers open in the early morning and usually close before noon. (Allyn G. Smith)

70. *Capsicum frutescens*, along the trail between Academy Bay and Bella Vista,

Isla Santa Cruz. These chili peppers, popular as a condiment among the residents, are scattered widely along trails. (Allyn G. Smith)

71. *Lecocarpus pinnatifidus* on ash and lava scoria near the eastern end of Post Office Bay, Isla Santa María. The small shrubs with feather-duster clusters of leaves at the tips of slender stems are *Scalesia villosa*; *Bursera graveolens* covers the slope in the background. (Duncan M. Porter)

72. *Notholaena galapagensis* near Academy Bay, Isla Santa Cruz. Large numbers of dust-like brown spores are borne on the underside of the margins of leaf segments. This fern inhabits shaded niches in the bare lava of the Arid Zone. (Allyn G. Smith)

73. *Hippomane mancinella*, Caleta Black, Isla Isabela. The fruit becomes pale yellow and resembles a small crabapple when ripe, but is lethally poisonous if eaten; juice from a cut or broken twig causes painful blisters if allowed to come in contact with the skin. *Hippomane* is common near the strand on most of the islands. (Duncan M. Porter)

74. *Acacia rorudiana*, Isla Española. The minute flowers, borne in dense heads, are delicately fragrant, and the flowers as well as the green fruits are eaten by land iguanas and feral goats. This is the commonest *Acacia* in the archipelago. (Ira L. Wiggins)

75. Shrub of *Croton scouleri* var. *scouleri* in full flower on Isla Champion. The staminate flowers greatly outnumber the carpellate. Sap in the twigs and stems is yellowish to red when fresh but turns black on exposure to the air and produces permanent brown stains on fabrics it touches. (Ira L. Wiggins)

76. *Scalesia affinis*, northern coast of Isla Isabela. This species is a low, dense shrub that grows typically in fissures in lava flows, occasionally in sand or scoria at the upper margins of beaches. The white patches on the rocks are plaques of crustose lichens. (David Cavagnaro)

77. *Datura arborea*, along trail near Bella Vista, Isla Santa Cruz. The pendent flowers are nearly a foot long and emit a cloyingly heavy fragrance. The cordate leaves belong to a vine of *Ipomoea nil*. (Ira L. Wiggins)

78. Lush growth of *Portulaca oleracea* and tufts of the sandbur grass, *Cenchrus platyacanthus*, on lava near the Charles Darwin Research Station, Academy Bay, Isla Santa Cruz. *Portulaca* seeds are a staple food for some of the island's smaller finches and for the small native dove. (Allyn G. Smith)

79. *Portulaca howellii*, Isla Champion. The leaves and flowers last only a few weeks, the plants remaining during most of the year as bare, dropsical gray branches; the Galápagos plants have in fact only recently been recognized as a distinct species. (Duncan M. Porter)

80. *Urera caracasana* along the trail near Bella Vista, Isla Santa Cruz. The flowers and fruits of this nettle (it has stinging hairs) are borne in clumps along the stems. The two yellow flowers above and behind the fruits are of *Momordica charantia*, a vine of the cucumber family (see legend for color plate 96). (Ira L. Wiggins)

81. *Parkinsonia aculeata* flowers and young fruit on Isla Española. The buds, flowers, leaves, and young fruits of this shrub are eaten voraciously by land iguanas. (Ira L. Wiggins)

82. Flower of *Passiflora colinvauxii* (an endemic plant first discovered in 1964) among leaves of *Alternanthera halimifolia*, along the trail through the Transition Zone west of Bella Vista, Isla Santa Cruz. Three or four of the lunate leaves of *P. colinvauxii* can be discerned. The small white flowers in the upper left corner are of *Borreria laevis*, a common weed. (Paul A. Colinvaux)

83. *Zanthoxylum fagara* near Bella Vista, Isla Santa Cruz. This member of the

Rutaceae (Orange Family) is ubiquitous in the Galápagos, growing from sea level upward into the *Miconia* Zone; the twigs are armed with "cat's claw" spines. (Ira L. Wiggins)

84. *Clerodendrum molle* in the Transition Zone inland from Academy Bay, Isla Santa Cruz. The flowers are pleasantly fragrant and often produced in profusion. (Allyn G. Smith)

85. *Prosopis juliflora*, Isla Española, the mesquite tree of the Galápagos. The herbage, flowers, buds, and green fruit of this legume are eaten by finches, iguanas, goats, burros, and cattle. (Ira L. Wiggins)

86. *Cassia bicapsularis* in the Transition Zone north of Academy Bay, Isla Santa Cruz. The white-barked trees at the left are *Bursera graveolens*; the feathery leaves in the upper right corner, *Acacia rorudiana*. (Robert T. Orr)

87. New growth of *Tribulus cistoides* in cinder bed on Isla Española. The dead, gray stems are from the previous year's growth, the plant dying back to the root crown each dry season. (Duncan M. Porter)

88. Endemic *Darwiniothamnus tenuifolius* var. *glandulosus* at the upper margin of the beach at Caleta Black, Isla Isabela. The gland-tipped hairs on the leaves and stems trap small insects. The bright green, leafless stems crossing the picture are of *Sarcostemma angustissima*, an endemic milkweed. (Ira L. Wiggins)

89. *Doryopteris pedata* var. *palmata* in a humus-filled crack in the lava along the trail near Bella Vista, Isla Santa Cruz. The leaf blades are somewhat dimorphic, the sterile ones being shorter and less narrowly lobed than the fertile ones. This fern is common on several islands in the archipelago, growing only where it is shaded during most of the day. (Ira L. Wiggins)

90. *Cardiospermum galapageium* in the Transition Zone between Academy Bay and Bella Vista, Isla Santa Cruz. A scrambling and climbing endemic vine, sometimes reaching heights of 6 or 8 meters; its fruits are inflated and about an inch in diameter, resembling small Chinese lanterns. (Allyn G. Smith)

91. Buds and open flower of *Cordia lutea* near Puerto Ayora, Isla Santa Cruz. This plant varies in habit from a low, sprawling shrub to a tree as much as 10 m tall; the flowers are produced in great quantities and harbor many small insects. (Allyn G. Smith)

92. *Psilotum nudum* on a moist cliff well up toward the rim of the crater on Isla Fernandina. This primitive fern is relatively rare in the Galápagos. Fronds of *Polypodium tridens* are at the extreme top, flanked by fronds of an undetermined *Blechnum* or *Polypodium* at left and right just above the broomlike *Psilotum*. (David Cavagnaro)

93. Flowering spray of *Grabowskia boerhaaviaefolia*, Isla Española. The leaves are moderately fleshy and coated with a thin, whitish bloom; the fruit is blue-black and the size of a small pea. (Ira L. Wiggins)

94. *Lycium minimum*, Punta Suárez, Isla Española. The twigs are stiff and spine-tipped; finches strip the small red berries from the plants virtually as fast as they ripen. (Ira L. Wiggins)

95. Lichens—fruticose at left, crustose at right—on rough lava on Islas Las Plazas. The orchilla lichen once was harvested extensively in the Galápagos for sale to the dye industry, before the advent of aniline dyes. (Ira L. Wiggins)

96. Ripe fruit of *Momordica charantia* near Puerto Ayora, Isla Santa Cruz. Each seed is surrounded by a bright red, juicy aril; the arils (but not the seeds, which are moderately toxic) are often eaten by birds or children. The yellow-orange capsule splits irregularly, exposing the arillate seeds. (Allyn G. Smith)

Flora of the Galápagos Islands

Introduction

The Galápagos Islands are a territory of the Republic of Ecuador. And though they are officially listed by that government as El Archipiélago de Colón, most local inhabitants and many of the Ecuadorian nationals who visit the islands persist in calling them Las Islas de los Galápagos, the Islands of the Tortoises. Elsewhere in the world, of course, the name Galápagos has long been current.

The archipelago lies astride the Equator roughly 600 miles from the port of Guayaquil, or about 500 miles west of Punta Arena (La Puntilla), the westernmost point of continental Ecuador. It extends from approximately 89°14′ to 92°01′ West Longitude and from 1°25′ South to 1°40′ North Latitude; its span is thus about 190 statute miles (304 km) east to west, or 213 miles (341 km) northwest to southeast, i.e. from Isla Darwin to the southeastern tip of Isla Española. Though the islands appear to have no close geologic affinities with the mainland of North or South America, their native fauna and flora derive in the main from the American tropics.

THE ISLANDS

Forty-five islands, rocks, and groups of islands taken as units—all those of appreciable size or importance—are listed in Table 1. From most of these at least a few biological specimens have been obtained during the past 145 years. There are, in addition, scores of smaller rocks and islets that bear no official names. Most of these are inaccessible—their precipitous cliffs, rising directly out of the water, present such hazards to landing parties that no one has attempted collections of plants or animals on them. Even some of the larger islets rise so abruptly from the sea that no landing has been made on them. Among these named but unexplored islets are Isla Wolf, Isla Daphne Chica, Isla Sin Nombre, and Roca Redonda. On such islets the cliffs have been undercut by wave action to such a degree that a pronounced overhang makes landing impossible. Isla Wolf, which rises to 830 ft (253 m), has defied attempts to reach its plateau. In a few places, huge blocks of lava have tumbled from the cliffs and built up talus heaps on which landings can be

TABLE 1. ISLANDS AND ROCKS OF THE GALÁPAGOS ISLANDS

Ecuadorian Name	English Name	Approximate Area*		Altitude of Highest Peak		Longitude and Latitude
		Square statute miles	Square kilometers	Feet	Meters	
Isla Albany	Albany Island	0.04	0.10	197	60	90°50′32″ 0°10′11″ 90°50′52″ W 0°10′22″ S
Isla Baltra	South Seymour Island	9.69	25.10	?	?	90°14′49″ 0°24′31″ 90°17′52″ W 0°28′53″ S
Isla Bartolomé	Bartholomew Island	0.48	1.24	359	109	90°32′23″ 0°16′40″ 90°33′35″ W 0°17′13″ S
Islas Beagle	Beagle Islands	0.03	0.08	?	?	90°37′36″ 0°28′38″ 90°37′55″ W 0°29′00″ S
Isla Caldwell	Caldwell Island	0.08	0.21	374	114	90°20′00″ 1°17′08″ 90°20′24″ W 1°17′37″ S
Isla Champion	Champion Island	0.04	0.10	152	46	90°23′00″ 1°14′03″ 90°23′12″ W 1°14′15″ S
Isla Coamaño	———	0.02	0.05	?	?	90°16′39″ W 0°45′21″ S
Isla Cowley	Cowley Island	0.01	0.03	?	?	90°57′42″ W 0°22′57″ S
Isla Daphne Major	Daphne Major	0.13	0.34	?	?	90°22′10″ 0°25′00″ 90°22′27″ W 0°25′26″ S
Isla Daphne Minor (Daphne Chica)	Daphne Minor	0.03	0.08	?	?	90°21′00″ 0°23′28″ 90°21′10″ W 0°23′40″ S
Isla Darwin (Guerra)	Culpepper Island	0.90	2.33	550	168	91°59′50″ 1°38′42″ 92°00′40″ W 1°39′50″ N
Isla Eden	Eden Island	0.01	0.03	?	?	90°32′00″ 0°33′17″ 90°32′18″ W 0°33′37″ S
Isla Enderby	Enderby Island	0.07	0.18	368	112	90°21′32″ 1°13′42″ 90°21′46″ W 1°13′57″ S
Isla Española	Hood Island	22.5	58.00	650	198	89°33′45″ 1°20′30″ 89°44′40″ W 1°24′32″ S
Isla Fernandina	Narborough Island	245.00	635.00	4,900	1,494	91°22′35″ 0°15′33″ 91°39′40″ W 0°30′15″ S
Isla Gardner (near Española)	Gardner Island (near Hood)	0.22	0.57	160	49	89°38′17″ 1°20′15″ 89°38′53″ W 1°20′44″ S
Isla Gardner (near Santa María)	Gardner Island (near Charles)	0.30	0.78	746	227	90°17′16″ 1°19′37″ 90°18′00″ W 1°20′12″ S
Isla Genovesa	Tower Island (Ewres)	6.70	17.35	250	76	89°56′11″ 0°17′49″ 89°58′46″ W 0°20′40″ S
Islas Guy Fawkes	Guy Fawkes Islands	0.02	0.05	55	17	90°31′25″ 0°30′36″ 90°31′45″ W 0°30′54″ S
Isla Isabela (Santa Gertrudis)	Albemarle Island	1,803.00	4,670.00	5,600	1,707	90°47′00″ 0°10′00″ N 91°35′42″ W 1°03′02″ S

Isla Lobos	———	0.04	0.10	?	?	89°34′03″ W 0°51′13″ S
Los Hermanos (Islas Los Hermanos)	Crossman Islands	0.33	0.85	550	168	90°44′55″ 0°50′30″ 90°48′50″ W 0°51′10″ S
Isla Marchena (Torres)	Bindloe Island	50.00	130.00	1,125	343	90°24′00″ 0°16′45″ 90°32′31″ W 0°23′25″ N
Isla Mosquera	———	0.03	0.08	?	?	90°16′35″ W 0°24′10″ S
Isla Onslow	Onslow Island	0.005	0.01	82	25	90°25′16″ W 1°12′50″ S
Isla Pinta (Geraldino)	Abingdon Island	23.00	60.00	2,550	777	90°43′03″ 0°32′32″ 90°47′23″ W 0°38′47″ S
Isla Pinzón	Duncan Island	6.93	13.05	1,502	458	90°38′28″ 0°35′05″ 90°41′16″ W 0°37′54″ S
Las Plazas (Islas Plaza)	———	0.09	0.23	?	?	90°09′26″ 0°34′35″ 90°10′03″ W 0°34′57″ S
Isla Rábida	Jervis Island	1.89	4.90	1,203	367	90°41′45″ 0°23′45″ 90°43′12″ W 0°25′30″ S
Isla San Cristóbal (Grande)	Chatham Island (Dassigney)	213.00	552.00	2,350	715	89°14′27″ 0°40′55″ 89°37′38″ W 0°57′13″ S
Isla San Salvador (Santiago, Olmedo)	James Island (Gil, York)	221.00	572.00	2,974	905	90°32′42″ 0°07′36″ 90°52′18″ W 0°22′21″ S
Isla Santa Cruz (Bolivia, Chavez, Valdez, San Clemente)	Indefatigable Island (Norfolk, Porter)	349.00	904.00	2,835	864	90°10′05″ 0°28′43″ 90°32′55″ W 0°46′23″ S
Isla Santa Fé	Barrington Island	9.30	24.00	850	259	90°01′39″ 0°47′42″ 90°05′27″ W 0°50′14″ S
Isla Santa María (Floreana)	Charles Island	66.00	171.00	2,100	640	90°21′07″ 1°13′04″ 90°30′37″ W 1°21′30″ S
Isla Seymour	North Seymour Island	0.71	1.84	?	?	90°16′32″ 0°23′10″ 90°17′37″ W 0°23′55″ S
Isla Sin Nombre	Nameless Island	0.03	0.08	369	111	90°35′07″ W 0°40′05″ S
Isla Tortuga	Brattle Island	0.48	1.24	610	186	90°52′28″ W 1°00′15″ S
Isla Watson	Watson Island	0.01	0.03	185	56	90°18′30″ W 1°20′30″ S
Isla Wolf (Gasna, Genovesa, Nuñez)	Wenman Island	1.10	2.85	830	253	91°49′15″ W 1°22′30″ N
Rocas Bainbridge	Bainbridge Rocks	0.28	0.73	160	49	90°33′22″ 0°20′20″ 90°35′11″ W 0°22′12″ S
Roca Blanca	White Rock	0.001	0.003	31	9	90°51′28″ W 0°32′55″ S
Rocas Dalrymple	Dalrymple Rocks	0.003	0.008	?	?	89°37′30″ W 0°51′00″ S
Rocas Gordon	Gordon Rocks	0.013	0.034	?	?	90°08′25″ 0°37′37″ 90°08′35″ W 0°38′52″ S
Roca Kicker (León Dormiente)	Kicker Rock	0.04	0.10	486	148	89°31′05″ W 0°46′30″ S
Roca Redonda	Round Rock	0.003	0.008	220	67	91°37′30″ W 0°16′32″ N

* As measured with a Keuffel and Esser Polar Planimeter on United States Hydrographic Office marine charts.

made in calm weather; cited specimens from Wolf collected prior to 1964 were obtained from these talus slopes, not from the main mass of the island. In January and February 1964, a group of biologists was landed on the summit of Wolf by helicopter, through the good offices of the United States Navy and the Government of Ecuador. We have examined a few of the specimens obtained by members of this party.

All of the islands of appreciable size, and many of the smaller rocks, were originally named by English sea captains and buccaneers. Since most of the early visits to the islands were made by nationals from England, France, and the eastern seaports of what is now the United States, Spanish names were rarely applied, and these were largely ignored by the crews of the whalers, privateers, and men-of-war that called at the islands to replenish fresh-meat supplies with the native tortoise. When the people of Ecuador began attempts to colonize the larger islands, Ecuadorian names were applied and officially adopted. At present, the charts issued by the United States Hydrographic Office carry both the official names and the displaced, English ones. We have taken the position that names used by nationals of a country for their own territory deserve recognition, even when one is not writing in the language of that country. Thus throughout this book, in giving distribution data, we use the official Ecuadorian names for the individual islands. And though we use the English names for many of the other place names (bays and the like), we have, in many cases, given the Ecuadorian names as well.

Table 1, then, which is similar to a table prepared by the late Joseph R. Slevin (1959, 25–26), should help familiarize English-speaking people with the Ecuadorian names for the islands. In the table, names placed in parentheses are those that have been, or may now be, used in place of the more recognized names. Since some islands have been rechristened rather frequently, there are no doubt still other names that have escaped our attention. Very early names in either language that have not been in use during the past century and a half are ignored—though one such name deserves mention. After the discovery of the islands in 1535, by a party bound for Puerto Viejo from Panama and caught in a westerly-setting current when becalmed, they were often called Las Islas Encantadas, the Enchanted Islands. This appellation occurs rarely now, and never officially.

The total land area of the islands and rocks above sea level included in the table is approximately 3,033 mi^2 (7,856 km^2). The remaining land area—that in the unlisted islets and rocks—could be no more than a few square miles. The areas given have been derived by applying a planimeter to the Hydrographic Office charts and calculating the number of square statute miles from the nautical mileage used on the charts.

Isla Isabela (Albemarle Island) is by far the largest of the Galápagos

Islands, stretching 84 mi (134 km) and varying from 7.5 to 21 mi (12 to 34 km) in width in the part north of Istmo Perry and to 26 mi (42 km) north-south across the broad basal "foot" of the island. Isabela's total land surface is about 1,803 mi^2 (4,670 km^2), which is more than the total area of all the other islands combined, and over four times as large as the second largest, Isla Santa Cruz. The next five larger islands, in order of size, are Islas Fernandina, San Salvador, San Cristóbal, Santa María, and Marchena. The rest of the islands in the archipelago range in area from 23 mi^2 (60 km^2) downward to mere specks on the charts, with less than 0.1 mi^2 (0.26 km^2) of land above the sea. Among the smaller islands are Islas Wolf and Darwin, the extreme outposts of the Galápagos group; these are located 116 mi (186 km) and 143 mi (229 km), respectively, northwest of the northern tip of Isabela.

The elevations above sea level attained by the rocks and islands in the archipelago vary from barely above the wash of the sea to an extreme of 5,600 ft (1,707 m), attained by Volcán Wolf near the northern end of Isla Isabela. Among all the others, only Islas Fernandina and San Salvador reach elevations above 2,900 ft (884 m). Slopes are generally substantial, often precipitous. One of the more spectacular examples is Roca Kicker; this little islet, just north of Isla San Cristóbal, is only 0.04 mi^2 (0.1 km^2) but rises to 486 ft (148 m). The elevations given in the table are those shown on the Hydrographic Office charts, and some are only tentative, the consequence of insufficient survey work in the archipelago (various authors have in fact given appreciably different elevations for some of the islands and volcanic peaks). A question mark instead of an altitude for a given island means that no elevation for that island has been published to date.

The latitude and longitude given for each of the smaller islets indicate the approximate position of the highest point on the islet, as shown on the Hydrographic Office charts, and thus do not necessarily locate the geographical center of the islet. For the larger islands, and for some small ones, readings are given for the extremes of their east-west and north-south limits. The positions are approximate only, having been scaled from the charts with rule and dividers.

The map of the archipelago reproduced on the endpapers of this book shows all except the smallest of the islets and rocks in their geographical relationship to one another. General topography and the more important place names are also indicated. But at this scale, detailed delineation of coastlines and minor topographical refinements is impossible.

SETTLEMENT PATTERN

Five islands in the archipelago are now inhabited on a permanent basis. A sixth, Isla Baltra, will be continuously occupied in the near future, or may be already. The only airstrip in the archipelago, built on Baltra during World War II, has been used at irregular intervals. If a planned weekly air shuttle from the mainland to the islands proves successful, airline personnel and their families will live on the islet. In that event, water will have to be brought in, or cisterns built to store rainwater, for Baltra has no underground water resources.

The settlements on the three main inhabited islands are rather dispersed and in each case consist largely of two centers of activity—a small village on the coast and a second a few kilometers inland and from 125 to 400 m higher, where the soil provides a living of sorts. The coastal village typically consists of a meagerly stocked store or two, moorings for the small local fishing boats, a marine fuel depot servicing the fishing boats and occasionally selling diesel oil to commercial fishermen from distant ports or to yachtsmen, an unpretentious boat-repair and boat-building shop, two or three government buildings, a small schoolhouse, and the scattered dwellings of the families engaged in these enterprises. The inland village usually is even more scattered, with fewer conveniences, and its inhabitants depend chiefly on the coastal village for whatever their cultivated fields do not supply. All these villages seem to be rather loosely organized; the pair on San Cristóbal are Wreck Bay (Bahía Wreck) and Progreso; those on Santa Cruz, Puerto Ayora at Academy Bay (Bahía de la Académia), and Bella Vista, about 4 km inland; those on Isabela, Villamil, at the coast, and Santo Tomás, a few kilometers inland. Most family plots and homes are situated well back from the coast and at a higher elevation than the center of the little trading communities, near good soil and in the rainy belt.

San Cristóbal has been occupied continuously since 1869. In that year Manuel J. Cobos departed Guayaquil to establish the settlement of Progreso near the western end of the island. Cobos was a cruel, tyrannical master, and the slaves who worked the new hacienda suffered inhuman treatment before breaking into his bedroom one night and butchering him with their machetes. The following years saw a good deal of lawlessness, but Progreso survived those episodes and is now the center of a reasonably successful agricultural development. The village, surrounded by farmland, is on a plateau from 5 to 9 km from the port of Wreck Bay (Slevin, 1959, 105–6). Estimates of the population of the island vary, but in 1967 the local people said there were about 1,200, all told, at Wreck Bay and Progreso. Water for domestic use is piped from a lake above Progreso to both villages, and irrigation of farmland depends in part on the same lake. A small diesel plant generates enough electricity to light a few low-wattage lamps along the streets and in some buildings, but most of the inhabitants must depend on oil lamps or retire soon after dark. The central part of the streets at Wreck Bay and the road to Progreso are paved with partially trimmed blocks of lava, which allow some passage during heavy rains but make for a rough ride in a wheeled vehicle. Most of the trails and roads are unpaved and become quagmires during protracted periods of rainfall.

This island has perhaps the greatest agricultural potential of all the islands in the group, for the soil is good and the lake at El Junco (the Place of the Sedges) might support limited irrigation if careful schedules for its use are developed. Crops grown in the milpas (fields cultivated by plowing or by hoe) in the vicinity of Progreso include the field crops coffee, maize, peanuts, sugar cane, tobacco, and some elephant grass for stock feed; such vegetables as avocados, beans, breadfruit, cabbage, cauliflower, lettuce, manihot, onion, otoy, peppers, pumpkins, potatoes, papa china (a root plant of the *Arum* family and a rich source of starch), squash, sweet potatoes, tomatoes, and probably a few others in small garden patches. Among the fruits produced in moderate quantities are acerolas (one of the hawthorns), bananas, chirimoyas, grapefruit, guavas, lemons, limes, mangoes, oranges (though all citrus trees are heavily infested with scale insects), papayas, pineapples (small but of excellent flavor), plantains, sauersops, tamarinds, and watermelons. The chief export crops are coffee and a limited amount of sugar produced by a small mill at Progreso. Most of the other produce is consumed locally. Cattle graze on the grassy and sedge-covered areas above and east of Progreso, and beef is shipped to Guayaquil on the hoof (Koford, 1966, 285).

The southeastern part of Isabela has been inhabited since 1893 (Slevin, 1959, 107–8), Santa María intermittently between 1902 and 1929 and continuously since. Santa Cruz was colonized in the middle 1920's, though precariously in the early years. William Beebe (1924; 1926) found only three or four young men clinging tenaciously to their settlement near Academy Bay when he called there, the rest of the settlers having returned to Scandinavia or moved on to San Cristóbal, to the mainland of Ecuador, or to other South American countries. These earlier colonists had been the victims of a swindle that promised much but delivered virtually nothing, for the island could not support even the first group that landed. Today, Academy Bay (more precisely, the outskirts of Puerto Ayora) is the home of the Charles Darwin Research Station, operated under the auspices of the Charles Darwin Foundation for the Galápagos. The Foundation was conceived on an international basis in the 1950's, and the Research Station was dedicated during the Galápagos International Scientific Project, on January 21, 1964. The Foundation enjoyed support from UNESCO, financial aid from several national organizations, and the cooperation of the Ecuadorian government. A full-time Director is in residence throughout the year, and modest facilities are available for a score or so of scientific personnel.

A settlement at James Bay, in the southwestern part of Isla San Salvador, has fluctuated markedly in number of inhabitants and even today is occupied only because the salt deposits in an old crater are being worked. Some fishing is also carried on from this port, but little of the catch is exported. Agriculture has proved impractical, for the only land that might support crops lies several kilometers inland, near the summit of the mountain, and is small in area. Moreover, the island's water is inadequate for crop irrigation. Goats, pigs, burros, a few wild cattle, and feral dogs and cats roam the countryside, as they do on several other islands. In 1967 several large herds of goats were seen on San Salvador. At present they seem to be no particular nuisance, but a few dry years could easily tip the balance, and the goats could then do tremendous damage to the native vegetation in a short time.

The total population of the archipelago is between 2,000 and 3,000, most of the people living in the pairs of villages on San Cristóbal, Santa Cruz, and Isabela. A few live on Santa María, at Black Beach and in the vicinity of Wittmers's farm, and another small group, possibly numbering a little over 100, lives at James Bay on San Salvador. The smaller islands are visited occasionally by fishermen, yachtsmen, or scientists, but none has a reliable source of fresh water, and one can only wonder how many people have died of thirst on their desolate slopes since the archipelago was discovered in 1535.

A territorial governor has headquarters at Wreck Bay, on San Cristóbal. The Customs Office, where all craft entering territorial waters must clear before going on to other islands, is also at Wreck Bay. There is a small garrison at Academy Bay, on Santa Cruz, and similar stations on San Salvador and at Villamil on Isabela. Northeast of Villamil is the one small penal colony. A naval patrol boat operates among the islands to discourage poaching by foreign fishing vessels and to prevent capture of the Galápagos tortoise and other native animals.

According to recent reports from the Director of the Charles Darwin Research Station, the islanders are encroaching steadily on the forested part of several of the larger islands, generally with grave effects on the native flora and fauna. The trend is hard to combat, owing to ignorance of ecological hazards or to outright antipathy toward conservation efforts. Government controls, however, including a total ban on the exportation of native animals, are being stiffened. Moreover, most of Isla Santa Cruz has been set aside as a wildlife preserve by edict of the Ecuadorian government, and there is a movement afoot to have the entire archipelago declared a national park.

PHYSIOGRAPHY

The entire archipelago is the result of volcanic activity of uncertain age, and some eruptive and explosive phenomena continue. The caldera that occupies the central part of Isla Fernandina was considerably altered by a violently explosive disturbance in June and July 1968. The presence of one or more volcanic cones is characteristic of all the larger islands, and of many of the smaller ones as well. Extensive lava flows from secondary or "parasitic" cones on the flanks of major volcanoes have radiated in all directions, producing vast basaltic lava fields around their bases. The slopes near the sea often are no more than 2° on the larger islands, increasing gradually to 10° or 15° midway, and becoming as steep as 25° near the rims of major craters. The general appearance is that of rather symmetrical cones set on broad, gently sloping bases, becoming steeper toward the rims of the craters, which are often distinctly truncated at their summits.

Because shore lines are relatively free of significant indentations, good harbors or protected bays are few. Furthermore, since the islands are volcanic in origin and situated beyond the continental shelf, they rise abruptly from depths in excess of 2,000 meters and are sloped more steeply below sea level than above it. As a result, anchorage is poor and unreliable, even in coves that would appear to provide adequate protection from storms.

Owing to the porosity of the country rock and to the extensive deep fissures in the lava, water percolates rapidly to considerable depths—well beyond the reach of the roots of most plants—and returns to the surface in the form of springs or seeps in only a few places. There are virtually no permanently flowing streams, although temporary watercourses usually follow heavy rains for a short time. Heavy downpours occur from time to time, but usually on an irregular cycle of about ten years' length, making their deluges at present unpredictable. A few minor springs occur in or just below the belt of heaviest forest, which coincides with the belt of heaviest rainfall. There is one known exception to this situation: on San Cristóbal there is a small but permanent stream; but because it flows to the sea through rough terrain its flow has not been used to water arable land.

Many of the lava flows on Fernandina, Isabela, and San Salvador, and some lesser flows on other islands, are so recent that vegetation has not yet colonized them, and soil is practically nonexistent within their limits. The bare rock of these flows is extremely rough and rugged; in some parts solid and stable underfoot, the rock may in other parts of the same flow be broken into innumerable slabs, with projecting knobs and ridges, into small-to-huge angular blocks, or into thin, knifelike shards. Most of these flows, regardless of the major outlines of their surfaces, consist of material with abrasive and cutting edges and are jumbled together in huge barricades, ridges, and "malpais," making travel over them slow, painful, and dangerous. Beebe (1926) described some of the hazards presented by loose slabs lying on steep slopes on Isabela, and by the jumbled blocks on the northeastern slopes of Santa Cruz.

The interiors of the volcanic cones are marked by steep, precipitous cliffs that drop to a relatively flat floor several score to several hundred meters below the rim. Some craters contain permanent lakes, e.g. the western crater on San Cristóbal, the craters on Fernandina and Genovesa, and the small crater at Tagus Cove on Isabela. The water in the San Cristóbal lake is fresh and supplies water for the villages of Progreso and Wreck Bay. That in the others is salty.

Interdigitated with, or above, the basaltic lava flows are areas covered by volcanic ash, clinkers, and scoria. Plants often are present on at least some of this kind of substrate and absent on the lava flows, thus presenting alternating streaks or strips of bare rock and areas that are well colonized by shrubs.

The strand lines vary greatly. Many kilometers of shoreline are marked by vertical or even undercut cliffs, caverns, and grottoes into which the Pacific swells crash with considerable violence during calm weather and with tremendous, roaring force during storms. Along stretches of lower relief, the waves pound upon jagged, angular rocks of sizes running the gamut from pebbles to boulders as large as a city skyscraper. Both types of shoreline are dangerous and much of the time prohibit landings from the sea.

At irregular intervals are beaches, with sand and pebbles either made up entirely of fragmented volcanic ejecta and ash or containing high percentages of organic matter in the form of broken mollusk shells, spines of sea urchins, and other calcareous materials. Some coral is present, but not in reef-building quantities. Blocks of lava broken from nearby flows often are scattered at random along these beaches and in the adjacent shallow water, where they present further hazards to small boats sailing close inshore or attempting landings. Such conditions are the rule throughout the archipelago.

GEOLOGY

The Galápagos Islands occur along several fault zones, the major one of which runs approximately through the craters of Islas Darwin, Wolf, San Salvador, Santa Cruz, and Española. The others, somewhat curved, cut across the principal fault line at nearly right angles. Islas Santa María and San Cristóbal mark the most southerly of these transverse fault zones; the southerly part of Isabela and Isla Santa Cruz indicate another; and one of the middle craters on Isabela and the craters of Fernandina, San Salvador, and Genovesa indicate still a third. These transverse fault lines are more or less curved, and very likely lesser lines of weakness in the earth's crust occur in the region to form a pattern still incompletely known. A secondary fault line roughly parallel to the main northwest-southeast one appears to run through Pinta, Marchena, and San Cristóbal, or it may be that Pinta, Marchena, and Genovesa lie along a curved fault that does not extend farther southeast toward San Cristóbal at all. All of these known and postulated fault zones, marked by the presence of volcanic and seismic activity, lie within the broad outlines of the Pacific "Ring of Fire" that has been marked by similar volcanic and crustal activity over an extended period.

Whatever the exact position and directional trends of the faults associated with the islands, it is clear that all of the islands are of volcanic origin. There is at least one major crater on each of the larger islands—San Cristóbal and San Salvador each have two, and Isabela six. And because none of the land has undergone appreciable erosion, the craters are clearly defined and the rocks are reasonably fresh and unweathered (Chubb, 1933).

The age of the archipelago has not been determined accurately. Preliminary investigations, based on shifts in the magnetic polarity of lava that cooled in place, indicate that the islands east of the Darwin–Española fault line are somewhat more than one million years old, those west of that line less than a million years old. The use of other methods of determining age may put a more precise dating on various islands, but such determinations are yet to be made (personal communication from Allan Cox). Only scanty and imperfectly preserved fossil material has been found on the islands, none of it older than Pliocene, and most or all of it probably of Pleistocene age.

But as one would expect among islands of oceanic, volcanic origin, the character of the vulcanism and the islands produced by it remain the chief focus of interest to the geologist or naturalist investigating the archipelago and its biota. There are upward of 2,000 secondary or parasitic craters scattered thickly over the flanks of the 20 major volcanoes. Many of these have produced lava flows, and a few have given rise to finer ejecta. Most of the parasitic vents arose as a result of solidification of the central core of the main crater: with the formation of plugs too massive to be pushed upward by pressure from below, the molten materials vented more or less laterally or obliquely upward, coming to the surface on the flanks at about the level of or slightly below the floor of the principal crater.

All of the major volcanoes are of one type, of which those on Fernandina and Isabela are good examples. They are circular or oval in outline at the rim and around their sea-level bases; and they are 25 to 30 km in diameter at sea level, but considerably larger where they rest on the ocean floor. Each possesses a deep crater: the Fernandina crater descends to about 900 m below the highest part of the rim; most of the others are from 150 to 200 m in depth. Rims are commonly 400 to 900 m above the level of the sea, though that on Fernandina exceeds 1,450 m, Cerro Azul on Isabela reaches almost 1,690 m, and Volcán Wolf on the same island towers a little over 1,700 m above the sea. Calderas are 5 to 9 km across, the inner walls are steep, and the floors are the apices of the old volcanic plugs that formed when the major mountain-building outflow of basaltic lava occurred. Benches, terraces, and hanging "platforms" occur at various elevations inside several of the larger calderas. Subsidence of at least some of the crater floors has taken place from time to time, the floor in Fernandina having dropped over 100 m during the activities in June and July 1968 (Colinvaux, 1968). Early in 1954 there was an uplift of about 4 m along the shore of Urvina Bay on Isabela, which was followed by an eruption of nearby Volcán Alcedo in November of the same year. "Many such uplifts of volcanic origin must have accompanied the growth of the Archipelago, but regions of broader, regional uplifts have yet to be discovered. . . . The principal volcanoes are huge basaltic shields with large summit calderas formed by collapse. Spectacular circumferential fissures and lines of eruptive conelets of cinders and spatter border several of the calderas, and radial fissures cut the flanks of most of the major cones" (Williams, 1966, 65).

The major volcanoes on Isabela, beginning in the southwest and progressing eastward, then northward, are Cerro Azul, altitude 1,689 m, Sierra Negra (Volcán Grande, Volcán Santo Tomás), 1,486 m, Volcán Alcedo, 1,113 m, Volcán Darwin, 1,311 m, and Volcán Wolf (Mount Whiton), 1,707 m. Volcán Darwin (called Mount Williams by Beebe) burst into activity through a series of conelets on its easterly flank in 1923; Beebe (1925) gives a vivid account of the event.

Each major volcano is surrounded by a series of lava flows; the younger flows partially overlie earlier ones, although all of them emerge on the flanks of the main cones some distance below the rim. The lava flows from the major volcanoes on Isabela have built up cols between their respective cones, and the major cones and their lava fields and beds of cinders thus form an undulating ridge from one end of the island to the other. The northwesterly tip of this ridge, which at present extends approximately 116 km, has been partially removed by erosion.

The upper flanks of the major volcanoes are grayish or light fawn in color, with radiating streaks of black lava flows covering the paler ash deposits. The lighter colored slopes are composed mainly of loosely compacted, friable tuff, containing numerous broken fragments of lava. In contrast, the lava flows themselves are made up of columnar structures along the faces of numerous fissures. Their surfaces may be ropy, corded, wrinkled, festooned, or covered with spires and jagged fragments that snapped off during the cooling and while the underlying lava was still molten. Rarely, the surface is smooth. In some instances molten lava flowed from under the solidified surficial layer, leaving tubes of varying dimensions; the roofs of these tubes often collapsed or formed a thin shell that may or may not bear the weight of a person walking over it. Andesitic and rhyolitic lavas are unknown in and around the Galápagos volcanoes, in sharp contrast with the volcanic areas in the Ecuadorian Andes.

The larger islands are composed chiefly of ejecta from the major volcanoes (Banfield, Behre & St. Clair, 1956). These authors observe also that the bottom topography of the ocean between the islands and the mainland is not fully known. But the Cocos and Carnegie Ridges notwithstanding, there is no indication that the Galápagos Islands have ever been connected with the continent by an above-water "land bridge." Some of the smaller islands, such as Baltra, Santa Fé, and Las Plazas, are fault blocks without central volcanic vents (McBirney & Aoki, 1966), as are scores of the small rocks that rim the shores of the larger islands.

Yet among all this volcanic activity a few small deposits of sedimentary rocks containing poorly preserved fossils of marine mollusks have been found. They are believed to be of Pliocene and/or Pleistocene age and, so far as is known at present, are restricted to a small area across the channel from the tiny islets of Las Plazas on the eastern end of Santa Cruz, to three or four areas on Baltra, and to one area on Isabela 10 to 15 km northeast of the village of Villamil. The fossils in these deposits are imperfectly preserved and many of them are unidentifiable, but their affinities seem to be chiefly with the mollusks of the American tropical seas, particularly with those of the lower Gulf of California; a few appear to be linked to Antillean forms (Dall & Ochsner, 1928).

CLIMATE

Diurnal temperature ranges in the Galápagos Islands are small, averaging about 9°F (5°C) on the windward sides of the larger islands, and increasing to 14–18°F (8–10°C) on the leeward sides (Palmer & Pyle, 1966). The annual range is also low, somewhere in the neighborhood of 20–22°F (11–12°C). From December through June the average peak daytime temperature is about 84°F (29°C), and in the month of August the average upper limit is about 66°F (19°C).

Two, rather than four, seasons are readily discernible, a "rainy" season that coincides with the "warm" season and a dry, cooler season. In spite of the relatively moderate temperatures throughout the year, the cool July–November period is marked by fog or heavy overcast (without rainfall) and steady winds that increase the chill factor, making existence outside a shelter uncomfortable at best.

The Galápagos Islands are in the "dry zone" of the equatorial Pacific and receive much less rain than areas to the north and south. On the average, the archipelago receives less than 75 cm of rainfall per year, though the total annual precipitation fluctuates markedly from one year to the next. For example, the records for Wreck Bay, at the western end of San Cristóbal, show that in 1950 only 3.55 cm of rain fell during the entire year, whereas during the "El Niño" year of 1953, that station received a total of 141.9 cm (Palmer & Pyle, 1966, 94).

Other factors complicate the situation, among them the shifts in ocean currents, steepness and direction of slopes (which may vary on the same island as well as from island to island), elevation of a particular area or zone, and the "rain shadows" cast by major peaks. When moist air masses accumulate around the islands, the clouds generally occupy a layer between a lower limit of 200 to 300 m and an upper limit of 400 to 500 m, and it is in this zone that most of the precipitation occurs. Below it, especially in gently sloping terrain and near the coast, there may be an overcast, but no rainfall, for weeks; yet only two or three kilometers away and no more than 250 m higher, the ground may be soaked by repeated downpours. Above this relatively narrower zone of heavy rainfall is an area of diminishing precipitation; here, only drizzles and early morning "fog-drip" bring small amounts of moisture. This zone has an indefinite lower limit, since changes in surface topography and in the air currents that swirl around the peaks and ridges tend to push the rain upward along valleys and downward along some ridges. In general, however, this region of fog and nonprecipitating clouds begins at or near the lower limit of the *Miconia* Zone (see the discussion of vegetation zones below) and extends upward to the summits of the peaks. Fog-drip is more pronounced in the shrubby *Miconia* Zone than in the Fern-Sedge Zone above it, for condensation is directly correlated with the leaf surface area of the vegetation, and with leaf shape and size.

An example of the differences in rainfall occurring at areas near sea level and at those a few hundred meters higher and less than ten kilometers away can be seen in the records of two stations in one 24-hour period. In February 1964 a brisk shower brought 7 mm of rain to the Charles Darwin Research Station, which is situated on Santa Cruz at an elevation of less than 10 m. During the same period heavy rains occurred nearby throughout the day and much of the night, producing a total of 65 mm at Bella Vista, approximately 220 m higher than the coastal station and not over 10 km distant. In other periods of several consecutive days, the Research Station received no precipitation at all, while Bella Vista had daily rains of from 5 to 50 mm during each downpour.

Variations between the precipitation on southerly-facing slopes and that on the north are equally striking. The south slopes always get more rain, and always at lower elevations, than the northerly and northeasterly areas. These differences in rainfall are graphically shown by the vegetative cover on the islands. *Scalesia* forests extend from about 180 m above sea level to about 400 m on the southerly faces of the larger islands (they are particularly evident on the slopes of Santa Cruz), but on the north slopes of the same peaks the upper limit reaches nearly 700 m, and the lower margin of the forest is correspondingly higher.

Lower islands, such as Baltra, Eden, the two Daphnes, Española, Santa Fé, and Genovesa, rarely receive drenching rains and are wholly desert in climate and vegetative cover. They are neither high enough nor large enough to force moisture-bearing air masses upward sufficiently to bring about condensation and rainfall. Their infrequent showers are in the nature of fringe benefits from the precipitating masses that accumulate around the neighboring, larger islands, or are the usual vagrant showers that occur over tropical seas.

SOIL ZONES

Laruelle (1966) recognized five pedological zones on Santa Cruz, the only island in the archipelago whose soils have been investigated. But for these purposes, one island may be enough; in view of the homogeneity of the volcanic materials from island to island and the uniformity of climatic regimes of the whole area, there is probably little departure from the Santa Cruz soil pattern on the other larger islands.

Laruelle's Zone I extended from sea level at Academy Bay northward to an altitude of 100 to 120 m, and its soil is characterized as being thin, made up of surficial or interstitial soil materials having a high percentage (63 to 90 per cent) of saturated bases. Usually reddish in color, the soil in Zone I has a high (about 28 per cent) clay content, is rarely more than 5 cm in depth, and is a lithosol. A few scattered, small pockets of deeper residual soil occur in this area. The zone coincides with the Arid Zone of vegetation.

Zone II, which occupies the same area as the Transition Zone of plants, begins at an elevation of 100 to 120 m, and rarely has soil depths exceeding 70 cm. It exhibits an "AC" profile, developed mainly from parent basaltic rock to clay, and has a base saturation ranging roughly from 64 to 89 per cent. There are frequent and extensive outcrops of lava rock on which soil has not formed, or from which the small amount of soil once present has been removed by erosion.

Soil Zone III begins at the lower limits of the *Scalesia* forest at an altitude of about 180 to 200 m, and thus is coextensive with the *Scalesia* Zone of vegetation. The soil is still less than 1 m in depth, contains an admixture of ash and clay, has a reddish brown hue, darkened locally by humification, and exhibits base saturations of 57 to 74 per cent. The soil in this zone has an "ABC" profile, with humus demonstrable in the upper 5 to 13 cm, and with a certain amount of tufflike material, ash, and interspersed layers of fragmented basaltic and pyroclastic materials in the "C" horizon.

The boundary between Zones III and IV is poorly marked, but soil in the fourth zone is brownish rather than reddish in color, is deeper, has less influence from basaltic rocks and more from pyroclastics, and shows a dramatic drop in the base-saturation percentage, to less than 10 per cent. On Santa Cruz, Soil Zone IV approximates the *Miconia* Zone, but because our knowledge of the soils on the other islands is so meager, we can only guess what the soil/shrub correlations might be there.

The soils in Zone V, which coincides with the upper or Fern-Sedge vegetational zone, have a distinctly acid reaction. The acidity is reflected in the dense growths of *Sphagnum* that occupy the potholes and swales of that area.

According to Laruelle (1966), a climatic sequence has been important in the development of the first three soil zones. A lithosequence, shared in the development of Zone II to a limited extent, became the dominant factor in Soil Zones IV and V. Accordingly, the *Scalesia* Zone, delimited on the basis of vegetation types, is a "transition zone" insofar as soil is concerned. This relationship may or may not hold for other islands of the archipelago. One feature, however, is quite obvious: soils of sufficient depth to permit even moderate manipulation for agricultural pursuits exist only in and above the areas occupied by *Scalesia* forests.

VEGETATION ZONES

The vegetative cover of the Galápagos Islands has been discussed at some length by Robinson (1902), Stewart (1911; 1915), and Bowman (1961). But though their observations paint the picture in bold strokes, they by no means treat comprehensively the details of microecological variations. Not even a respectable beginning has been made on controlled, carefully instrumented studies of the plant zones of the archipelago.

Owing to the total absence of potable water on the smaller islets and its infrequency and unreliability during dry years on most parts of the larger ones, no one has spent extended periods in the field to make detailed ecological studies. Preliminary observations have been made at the more usual ports of call, such as Wreck Bay, Black Beach (Santa María), Academy Bay, and Villamil, and at farming centers back of these ports. In contrast, there are hundreds of square miles on Isabela that have never been so much as visited by biologists. Consequently, little is known about the similarities or differences in the ecological communities of similar habitats on different islands, or even on different parts of the same island. *Miconia*, for example, has not been found, and may not occur, on Fernandina, Isabela, or San Salvador; there is a shrubby zone on these islands, but the exact composition of its plant cover is still to be determined. Thus there is no dearth of fascinating problems for an ecologist in the Galápagos Islands. Comparative, intra- and inter-island studies of soil and plant zones, faunal zones, microecological niches, and peculiarities of life cycles among the more exotic species are still to be made. Refinements in delimitation of the life zones postulated thus far, as well as recognition of subdivisions or modifications of these zones, will emerge only from further studies in the field and laboratory.

Withal, six fairly recognizable vegetation zones can be distinguished in the Galápagos. Photographs generally characteristic of these zones are included in the section of full-color illustrations.

Littoral Zone. Beginning at the sea's edge, the first zone is the Littoral Zone, in which the influence of salt is direct and limits the plant habitat to a narrow strip, the individual members of the community all growing within a few meters of the sea or of a salt lagoon. The dominant plants of this zone are shrubs or small trees, a few shrubs that never become arborescent, and several salt-tolerant herbaceous species. Among the larger shrubs and small trees are the red mangrove (*Rhizophora mangle*), the white mangrove (*Laguncularia racemosa*), the button mangrove (*Conocarpus erecta*), and the black mangrove (*Avicennia germinans*). In favorable habitats the white mangrove sometimes becomes a tree as much as 25 m tall, with a trunk 2 dm or more in diameter. It reaches these proportions along a small lagoon behind the beach at Black Cove (Caleta Black) on the west shore of Isabela, and at a few other spots on this and other islands. *Lycium minimum* and *Grabowskia boerhaaviaefolia* are common halophytes, forming shrubs 2–3 m high, and are fairly widespread throughout the islands. Less common, but known to be present in several areas, is a close relative, *Nolana galapagensis.* It is abundant on Isla Coamaño, a low-lying islet that bears the navigational light just outside Academy Bay, and at Cormorant Bay in the northeastern sector of Santa María. This plant forms dense thickets 1–2 m high and grows either on sand or among jumbled rocks. Where the strand is sandy, *Ipomoea pes-caprae* is common, and is often accompanied by the perennial grass *Sporobolus virginicus* and by such succulent herbs as *Sesuvium, Trianthema,* and *Heliotropium curassavicum.*

The list of plants confined to, or occurring chiefly in, the Littoral Zone is small and differs less from a list of plants from a comparable mainland zone than do lists covering other zones from the higher elevations. The species marked with an asterisk in each of the following zone lists are those that either are confined to that zone or grow chiefly in it. Those without such a designation are adaptable to this and to adjacent zones, sometimes occupying certain niches in several plant zones. Those of the Littoral Zone are:

**Atriplex peruviana*
**Avicennia germinans*
**Batis maritima*
**Conocarpus erecta*
Cryptocarpus pyriformis
Cyperus brevifolius
Cyperus elegans
Cyperus ėsculentus
Cyperus laevigatus
Cyperus ligularis
Cyperus polystachyos
Eleocharis atropurpurea
Grabowskia boerhaaviaefolia
**Heliotropium curassavicum*
**Ipomoea pes-caprae*
**Laguncularia racemosa*
**Lycium minimum*
Maytenus octogona
Mollugo cerviana
**Nolana galapagensis*
**Rhizophora mangle*
**Salicornia fruticosa*
**Sesuvium edmonstonei*
Sesuvium portulacastrum
**Sporobolus virginicus*
Trianthema portulacastrum

Littoral Zone vegetation is lacking along the parts of the coast where sheer cliffs of basalt rise directly from the sea to heights of 10 m or more, but is well represented wherever there are beaches or lower, broken and tumbled lava boulders. The greatest development occurs where the sea has invaded shallow valleys or rifts, as it has at the landward side of Darwin Bay on Genovesa; at Tortuga, Conway, and Academy Bays on Santa Cruz; at Post Office Bay and Cormorant Bay on Santa María; near James Bay on San Salvador; and at numerous places on Isabela and Fernandina. Smaller areas of Littoral Zone vegetation are scattered along the coasts of almost all of the large and moderately sized islands, and occasionally on some of the very small ones. Relatively few species occupy the zone, and its relatively simple profile would seem to make a critical, detailed study of the habitat uncomplicated and possibly very rewarding.

One variation of this habitat, bare rock with almost no sand present, occurs most notably at Punta Espinosa at the northeastern quarter of Fernandina. Here the *Rhizophora mangle* shrubs have become established in narrow crevices in the lava. Many of their prop roots impinge on bare, impenetrable rock, where they rub back and forth in the wash of the breakers. These prop roots develop grotesque "club-footed" swellings at their tips, and of course never obtain a foothold on the rock.

The type of habitat usually associated with mangrove "swamps," soft mud and sand partially or wholly exposed at low tide and inundated at high water, also occurs in the Galápagos. Such localities are at Tortuga Bay and Conway Bay on Santa Cruz; at James Bay on San Salvador; in a tiny cove at the westerly end of Bartolomé; around a lagoon between Post Office Bay and Cormorant Bay on Santa María; and at other isolated spots.

Arid Zone. Immediately inland from the Littoral Zone, the vegetation is clearly xerophytic and remains so inland to an elevation of 80 to 120 m (and sometimes much higher) on the southerly faces of the islands, and upward to 200 or 300 m on the lee side of the larger islands. This zone is dominated by arborescent and shrubby species of *Opuntia* and by two cerioid genera, *Jasminocereus* and *Brachycereus.* Individual specimens of *Opuntia* and *Jasminocereus* often become quite tall and quite majestic. The zone is frequently occupied also by a number of small-leaved, often spiny, shrubs and small trees. A number of annual herbs appear during the wet season, providing a temporary blanket of rich green for a few weeks; some of the perennial herbs also die back at the end of the short growing season and are not apparent in the dry season. The principal species of this zone—not all present in all parts, or even on all islands—include the following:

Acacia macracantha
Acacia rorudiana
Alternanthera echinocephala
Aristida repens
Aristida subspicata
Borreria dispersa
Borreria ericaefolia
Borreria linearifolia
Brachycereus nesioticus
Bursera graveolens
Castela galapageia
Cenchrus platyacanthus
Chamaesyce amplexicaulis
Chamaesyce nummularia
Chamaesyce punctulata
Chamaesyce viminea
Clerodendrum molle
Coldenia darwinii
Coldenia fusca
Coldenia nesiotica
Cordia leucophlyctis
Cordia lutea
Cordia revoluta
Croton scouleri
Cryptocarpus pyriformis
Cyperus anderssonii
Desmanthus virgatus
Erythrina velutina
Gossypium barbadense var. *darwinii*
Gossypium klotzschianum
Jasminocereus thouarsii
Lantana peduncularis
Maytenus octogona
Mentzelia aspera
Opuntia echios
Opuntia galapageia
Opuntia helleri
Opuntia megasperma
Parkinsonia aculeata
Piscidia carthagenensis
Prosopis juliflora
Scalesia affinis
Scutia pauciflora
Waltheria ovata

Our list differs from that published by Stewart (1911, 207) in minor details only; it includes some species not known to him or omitted because he had less opportunity to penetrate far inland on some islands; also, there have been nomenclatural changes since his time, with the result that different names in the two lists may apply to the same plant.

Much of the cover in the Arid Zone is quite open, exhibiting the regular spacing that is so well known to students of the deserts of the southwestern United States and adjacent Mexico. However, the spacing in the Galápagos is not as regular as in the North American deserts, owing to differences in the extent of fracturing and in the development of crevices in the lava, tending to force plants to follow lines of faulting and fissuring and thus concentrating growth in some areas and reducing or eliminating it in others. Some species form dense thickets where the fracturing is extensive, notably *Cryptocarpus pyriformis, Scutia pauciflora,* and *Alternanthera echinocephala;* all form thickets so dense as to make passage difficult. *Croton scouleri* also forms dense stands in occasional swales, but though the crowns are interlaced the trunks are far enough apart to permit movement among them.

Lianas are few, *Sarcostemma angustissima* and *Passiflora foetida* var. *galapageia* being the chief vines, with two or three leguminous plants playing a minor role in this form of growth. None of these is included in the list because they extend into other zones also, and are too insignificant a part of the total cover.

Transition Zone. Many of the species in this plant community are evergreen. Their leaves are ample and in most cases bear little or no pubescence. Thus they provide a deeper hue of green to the landscape than do the narrower, grayish leaves of most plants in the Arid Zone. The vegetative cover is made up of some xerophytic plants that have pushed upward from the Arid Zone, intermingling with more mesophytic representatives from the moister *Scalesia* Zone just above. The trees here are taller and more closely spaced than those of the Arid Zone, and the understory is more fully developed, containing many shrubs not found in the drier habitats at lower levels. Some epiphytic plants, such as *Tillandsia insularis,* a *Peperomia* or two, one orchid (*Ionopsis utricularioides*), and a number of lichens and bryophytes live on the trees and on some of the larger shrubs. A few ferns, such as *Adiantum concinnum, Trachypteris pinnata,* and *Polypodium tridens,* are able to survive extended periods of drought in shaded, rocky areas. Annual herbs are less numerous than in the Arid Zone, but grasses become more abundant than in the lower region. Extensive open forests consisting of nearly pure stands of *Bursera graveolens* are conspicuous in the Transition Zone on several of the larger islands, and are particularly apparent on Santa María, Isabela, and San Salvador. They are also a dominant feature of the vegetative cover along the northern slopes on Santa Cruz. This tree, when in new leaf, imparts a distinctive light yellow-green cast to the landscape.

Momordica charantia, Rhynchosia minima, Anredera ramosa, Cissampelos galapagensis, and two species of *Cardiospermum* increase the incidence of lianas, though they by no means approach the conspicuous role this growth form plays in the rain forest of the mainland.

Since all of the plants known to occur in the Transition Zone occur also in the Arid Zone below it or in the *Scalesia* Zone just above, none can be singled out as predominantly an inhabitant of this zone. Those listed below make up the bulk of the vegetation in the Transition Zone:

Adiantum concinnum
Anredera ramosa
Bursera graveolens
Cardiospermum corindum
Cardiospermum galapageium
Cassia occidentalis
Castela galapageia
Chamaesyce viminea
Cissampelos galapagensis
Cissampelos pareira
Chiococca alba
Clerodendrum molle
Cordia lutea
Cordia revoluta
Croton scouleri
Dalea tenuicaulis
Darwiniothamnus tenuifolius
Doryopteris pedata var. *palmata*
Elytraria imbricata
Galactia striata
Gossypium barbadense var. *darwinii*
Heliotropium angiospermum
Ionopsis utricularioides
Ipomoea linearifolia
Lantana peduncularis
Maytenus octogona
Merremia aegyptica
Momordica charantia
Opuntia echios
Peperomia petiolata
Pisonia floribunda
Polypodium dispersum
Polypodium steirolepis
Polypodium tridens
Psidium galapageium
Psychotria rufipes
Rhynchosia minima
Tillandsia insularis
Tournefortia pubescens
Trachypteris pinnata
Waltheria ovata
Zanthoxylum fagara

This zone covers a large area of the Galápagos Islands and has been inadequately explored by botanists. On Santa Cruz it begins at the top of the escarpment back of Academy Bay at an elevation of about 80 m and continues upward to the lower edge of the *Scalesia* Zone at between 180 and 200 m. On the northern and northeastern sides of all of the larger islands, both its lower and upper limits occur somewhat (often as much as 120 m) higher, owing to the decreased rainfall on these slopes. On the other hand, Transition Zone vegetation is encountered only a few meters above sea level on some of the extreme southwestern parts of Isabela and Fernandina where moisture-laden air currents first strike large land surfaces and dump much more water than can be squeezed from clouds passing inland only a few kilometers away. In these two areas on Isabela and Fernandina the Arid Zone is almost lacking. Aside from these exceptions, the altitudinal limits of the Transition Zone seem to fit the pattern exhibited on Santa Cruz fairly well, but because less attention has been given San Cristóbal, San Salvador, Santa

María, and some of the higher small islands, accurate delimitation of this zone, as of other zones, must await further field studies.

Scalesia Zone. On the southerly slopes of Santa Cruz, this zone begins at about 180 m above sea level. It extends upward to an elevation of about 400 m, and in places to as high as 550 m, then gradually surrenders to the *Miconia* Zone. On San Salvador, San Cristóbal, and Santa María, the lower edge of the zone is encountered at about the same level as on Santa Cruz, although variations occur from place to place owing to differences in direction of slope and depth of gulleys. On Fernandina and on Isabela south of Istmo Perry (the narrow region between Elizabeth Bay and Cartago Bay), the *Scalesia* Zone is characterized by *Scalesia cordata*; and on Fernandina and on Isabela northward from Istmo Perry by *S. microcephala*. These two species occupy about the same habitats filled by *S. pedunculata* on Santa Cruz, San Cristóbal, San Salvador, and Santa María. On these four islands *S. pedunculata* commonly reaches heights of 15 m, and under favorable environmental conditions towers to 20 m or more. Two important trees in the zone, on Fernandina and Isabela as well as on the four other large islands, are *Psidium galapageium* and *Pisonia floribunda*. Both produce trunks of sufficient size and sufficiently hard wood to make them suitable for building purposes, ship and boat repairs, and local cabinet work. At one time timber was exported to Guayaquil. As a result, the larger trees of these species have been nearly eliminated from areas readily accessible to the inhabitants, or, before them, to seamen needing timber to repair their ships.

The understory of shrubs is prominent and fairly uniformly developed over large tracts in the *Scalesia* Zone. Heavy masses of several leguminous vines, two or three members of the Cucurbitaceae, and occasional vines of *Passiflora, Ipomoea,* and some epiphytes increase the density of this part of the forest. Several species of ferns, orchids, *Tillandsia,* and *Peperomia* are common epiphytes and occasional inhabitants of fallen logs. The general appearance of the *Scalesia* Zone, whether its dominant tree sunflower is *S. pedunculata, S. cordata,* or *S. microcephala,* is that of the true rain forest of the mainland, although the lianas are far less abundant and the understory of shrubs boasts fewer species. Too little is known about the interrelations among the various genera and species that occupy the *Scalesia* Zone on Fernandina and Isabela for us to explain why there should be such a marked difference between the vegetative cover on these two islands and that of the other four principal islands. Perhaps Fernandina and Isabela, the younger islands, have not been above the sea long enough for soils as complex as those on Santa Cruz to have developed. Further, their having a larger mass of mountain peaks stretching across a broader front may have been important in determining the climatic factors that impinge upon and shape the character of the vegetation. At any rate, both islands need much more field work—an assignment calling for several hardy botanists who can spend weeks in the field during all seasons of the year, and preferably during dry years and through one of the El Niño years of bountiful rainfall.

The common and more conspicuous plants of the *Scalesia* Zone are the following (those known to occur only or chiefly on Isabela and Fernandina but not at all on Santa Cruz are marked with a dagger):

†*Acalypha parvula* var. *parvula*
**Adenostemma lavenia*
Adiantum concinnum
**Adiantum henslovianum*
**Adiantum macrophyllum*
†*Ambrosia artemisifolia*
**Asplenium auritum*
**Asplenium cristatum*
Asplenium feei
**Asplenium formosum*
Asplenium praemorsum
Asplenium serra
†*Baccharis gnidifolia*
**Blechnum occidentale* var. *puberulum*
**Blechnum polypodioides*
Borreria laevis
†*Bulbostylis hirtella*
Cheilanthes microphylla
Chiococca alba
Cissampelos pareira
**Conyza bonariensis*
Ctenitis pleiosoros
Ctenitis sloanei
Darwiniothamnus tenuifolius var. *tenuifolius*
†*Darwiniothamnus tenuifolius* var. *glabriusculus*
Dennstaedtia cicutaria
**Desmanthus virgatus*
Doryopteris pedata var. *palmata*
†*Drymaria rotundifolia*
†*Duranta dombeyana*
**Epidendrum spicatum*
†*Eupatorium solidaginoides*
†*Froelichia juncea* var. *juncea*
†*Gnaphalium vira-vira*
†*Hyptis spicigera*
Ionopsis utricularioides
†*Ipomoea alba*
**Justicia galapagana*
†*Linum harlingii*
**Lycopodium dichotomum*
**Lycopodium passerinoides*
†*Mentha piperita*
Nephrolepis cordifolia
Passiflora suberosa
Peperomia galapagensis
Peperomia galioides
**Pisonia floribunda*
Plumbago scandens
**Polypodium aureum*
Polypodium dispersum
Polypodium lanceolatum
**Polypodium phyllitidis*
Psidium galapageium
**Psychotria rufipes*
†*Scalesia cordata*
†*Scalesia microcephala*
**Scalesia pedunculata*
†*Scoparia dulcis*
Tillandsia insularis
**Tournefortia rufo-sericea*
†*Triumfetta semitriloba*
Zanthoxylum fagara

In this zone the trunks, branches, twigs, and in some species even the leaves bear large numbers of epiphytic liverworts and mosses. Saprophytic fungi are common to abundant during the early part of the rainy season, varying in size from delicate little plants with caps no more than 3 or 4 mm in diameter to large boletes, *Amanita-* and *Agaricus*-like species with caps as much as 15 cm across. Several different types of puffballs also are fairly common.

Margins of cleared areas and the sides of trails are almost smothered by heavy masses of intertangled vegetation through which one forces his way with difficulty and often with considerable discomfort, owing to the fire ants that seem to inhabit every twig and leaf. Sharp, hooked spines on some of the plants also tend to impede one's progress in breaking through these peripheral blankets of vegetation. Beyond this barrier the forest is fairly open, and passage is easy. The intensity of the light at ground level is low, for the crown of the forest canopy is extremely dense.

During the dry season, there is an accumulation of thick fog through most of the night and well into the forenoon of the next day. Moisture condenses in appreciable amounts on the leaves and twigs of plants, and drips to the ground from the leaf tips or trickles down the branches and trunks of the trees, lianas, and shrubs. The coatings of mosses, lichens, and liverworts on many of the trees and shrubs also are condensation stations during the foggy periods, but much of the water forming on them is absorbed; only a small portion reaches the ground. No records have been found, if indeed any exist, concerning the total amount of moisture made available to plants and their insect and avian inhabitants through fog condensation. An investigation of this phenomenon might yield illuminating results.

Miconia Zone. Above the maximum expression of the *Scalesia* Zone there is a broad band of gradually diminishing rainfall, where the larger trees thin out and finally disappear. This change in the vegetative cover is brought about not by reduction of rainfall alone, but also by pedological transition in which the base-saturation percentage drops from above 50 per cent to between 1 and 6 per cent. Thus, the soil in the *Miconia* Zone is slightly acid instead of neutral or moderately basic. The soil on Fernandina and Isabela has not been analyzed, but it is known that the upper slopes around the major volcanic peaks are covered with a deeper and more extensive accumulation of ash and cinders than occurs on the older islands east of them. This undoubtedly has an important effect upon the soil reactions and thereby on the character of the plants that inhabit the areas so affected.

The irregularities in the outline of the upper and lower altitudinal limits of the *Miconia* Zone are quite marked. In general, the lower edge of this shrub-covered zone is reached at between 400 and 550 m, depending upon direction of slope, extent of exposure to rain-bearing air currents or of blanketing by adjacent ridges or peaks, and perhaps local variations in soil characteristics. On Isabela and Fernandina the fluctuations are even more pronounced, with the relative amounts of ash constituting the substrate having an important role in providing favorable or unfavorable conditions for the shrub cover. On these two islands *Miconia* is, so far as is kown at present, wholly lacking, and its place taken by such shrubs as *Baccharis, Darwiniothamnus, Hyptis,* and *Dodonaea.* The zone is developed most extensively on Santa Cruz, but is represented also on San Cristóbal, San Salvador, and Santa María by a more open stand of *Miconia* and an admixture of other shrubs. On Santa Cruz it is an area of low, dense shrub cover, in which the woody element is almost a pure stand of *Miconia robinsoniana* with closely intertangled branches reaching a height of 3 or 4 m, most of the branches heavily ensheathed with lichens. Contiguous shrubs are usually tightly interlocked, and it is difficult to work one's way through except along trails opened by wild cattle and burros that graze in the zone just above and find good refuge from hunters in the dense shrubby cover. On Santa María the shrub cover is less dense, but is readily recognized. On this island burros are more numerous than feral cattle, and their trails and tracks are conspicuous over most of the island. Cattle are present on San Cristóbal, also; and both burros and goats range from sea level to the tops of the peaks on San Salvador, where the shrub cover is intermediate in density between that on Santa Cruz and that on the more open slopes of Isabela and Fernandina.

Bowman (1961) applied the name Brown Zone to the lower part of the *Miconia* Zone on Santa Cruz. Here, a dense growth of epiphytic liverworts, mosses, and lichens on the shrubs, aided by an intermixture of shrubs that are less common farther above (*Tournefortia, Macraea, Darwiniothamnus,* etc.), tends to dilute the deep green of the *Miconia* leaves; moreover, during the dry season the epiphytes turn brown and many of the mature leaves on the shrubs fall. Thus at any time, but especially during the dry season, the brownish tinge of the lichens and liverworts appears pronounced at this level, and a distinct brownish tinge is apparent from a considerable distance. But inasmuch as this "zone" is no more than a sort of transition from the *Scalesia* forest to the lower shrubby cover, and since the *Miconia* shrubs enter into this area, it seems advisable to include it within the *Miconia* Zone, rather than to distinguish it as a separate zone. The brown aspect is indeed less apparent on other islands and not at all visible on some parts of them.

None of the plants listed for the *Miconia* Zone is exclusively limited to the zone; they occur also either in the *Scalesia* Zone below it or in the Fern-Sedge Zone above it. Most of the secondary members of the community grow along trails, around fire-created openings in the shrubby cover, in the bottoms of ravines, and on occasional ridges devoid of the dominant shrubs. Ferns and herbaceous plants are important members of the community in the openings, with *Lycopodium cernuum, L. passerinoides*, and *L. thyoides* quite conspicuous in such habitats. *Adiantum macrophyllum*, with its young fronds providing splashes of bright red among the deep green masses of shrubby foliage, is particularly showy.

In the following list of plants common in the *Miconia* Zone, those marked with a dagger are present on Fernandina or Isabela (or both) but are lacking or nearly so on Santa Cruz, San Cristóbal, Santa María, and San Salvador. The plants listed do not, of course, represent a complete catalog of the plants known from the zone, but are rather those that give the zone the characteristic aspect seen by one entering the area.

†*Acalypha parvula* var. *parvula*
Adiantum macrophyllum
†*Ambrosia artemisifolia*
Apium leptophyllum
†*Baccharis gnidifolia*
†*Calceolaria meistantha*
Centella asiatica
Cuphea carthagenensis
Cyathea weatherbyana
Cyperus brevifolius
Cyperus rivularis
Cyperus virens
†*Darwiniothamnus tenuifolius* var. *glabriusculus*
†*Darwiniothamnus tenuifolius* var. *glandulosus*
Dicranopteris flexuosus
†*Dodonaea viscosa* var. *galapagensis*
†*Drymaria rotundifolia*
†*Duranta dombeyana*
Eleocharis maculosa
†*Eragrostis mexicana*
†*Eupatorium solidaginoides*
†*Euphorbia equisetiformis*
†*Froelichia juncea* var. *juncea*
Galium galapagoense
†*Gnaphalium vira-vira*
†*Hyptis gymnocaulos*
Jaegeria gracilis
†*Linum harlingii*
Lobelia xalapensis
Lycopodium cernuum
Lycopodium clavatum
Lycopodium dichotomum
Lycopodium passerinoides
Lycopodium thyoides
Macraea laricifolia
Miconia robinsoniana
Ophioglossum reticulatum
Panicum dichotomiflorum
Panicum glutinosum
Panicum hirticaule
Panicum laxum
Paspalum conjugatum
Pennisetum purpureum
†*Pilea peploides*
Pityrogramma calomelanos
†*Polygonum hydropiperoides*
Polygonum opelousanum
Pteridium aquilinum var. *arachnoideum*
Rhynchospora rugosa
†*Scalesia cordata*
†*Scalesia microcephala*
Scleria pterota
Setaria geniculata
Tournefortia pubescens
†*Verbena glabrata* var. *tenuispicata*
Zanthoxylum fagara

Stewart (1911; 1915a) did not recognize this shrubby zone, probably because at the time of his visit (1905–6) there were no trails leading into the interior of Santa Cruz, where this zone is most prominent. Neither did he report *Miconia* from Santa Cruz in his papers.

Fern-Sedge Zone. This is the vegetation complex that Bowman (1961) termed the Grassy Zone. Careful consideration of the genera and species most prevalent above the *Miconia* Zone reveals that many of the narrow-leaved plants appearing at first glance to be grasses are in fact sedges of the genera *Cyperus, Eleocharis, Rhynchospora,* and *Scleria.* Moreover, a prominent role is played by a number of ferns. Hence our choice of name for the zone.

The Fern-Sedge Zone occurs as low as 500 m in a few areas, possibly where fire has destroyed the normal shrubby cover, and may have its lower margin pushed upward as high as 625 m—rarely even to 700 m—where conditions have permitted the shrubby cover to encroach along a slope unusually well soaked by rain from clouds forced through a particular gap or up a critical canyon. Mostly, however, the lower edge of the Fern-Sedge Zone occurs at about 525 to 550 m above sea level. The zone reaches to the tops of the peaks, and is present on all larger islands. This region lies above the upper limits of most of the cumulonimbus clouds that surround the islands much of the time during the wet season, and consequently gets moisture mostly in the form of a fine drizzle or fog. Many times, while a heavy rain drenches the *Scalesia* forests and spasmodic showers straggle into the *Miconia* Zone, the Fern-Sedge Zone is left with only the haphazard "blow-over" wisps of clouds that have already lost most of their moisture and can drop only small amounts, if any, on the higher slopes.

The soil is fairly deep in this zone, if the conditions on Santa Cruz can be accepted as typical, and has a slightly acid reaction. It retains moisture much longer than the soils in lower zones, lining many swales with a relatively impervious blanket that holds water within their confines for considerable periods. Since the plants that grow in the Fern-Sedge Zone are with one exception low-growing forms with narrow leaves or with other mechanisms for reducing water loss, they do not require as much water per plant as do those in the wetter areas. Very few annuals occur here, most plants being perennial sedges, ferns, forbs, and a few low bushes. The notable exception is the tree fern, *Cyathea weatherbyana,* which has fronds 2 to 3 m long, these supported on a trunk 1 dm or more in diameter and as much as 3 m high.

The centers of swales and small potholes, some of which were formed where lava tubes collapsed, hold water for a short time following the irregularly spaced showers. Such habitats support striking assemblages of sedges, grasses, and two species of *Sphagnum* (peat moss, a nonvascular plant). The margins of such pools often are rimmed by ferns and two or three species of *Lycopodium.* Some of the smaller, ephemeral pools, or their beds after the water disappears, often contain annual forbs, such as *Apium leptophyllum, Mecardonia dianthera, Lindernia anagallidea,* and similar inconspicuous herbs. *Ludwigia peploides* subsp. *peploides* and *L. leptocarpa* occur sparingly around the margins of such pools and along some of the intermittent drainage courses.

Four terrestrial orchids occur in this area, none of which was known from the Galápagos Islands until Svenson collected *Liparis nervosa* in 1930. Then, in 1957, Bowman found *Habenaria monorrhiza* in the area approaching the summit of Mount Crocker, on Santa Cruz. The latter is the commonest orchid of the four, and grows in deep soil at elevations ranging from about 500 m to the summit of Mount Crocker (835 m), and to at least 1,000 m on Isabela. The other two were collected first by Inga and Uno Eliasson in 1967: *Habenaria alata* occurs on the floor of the crater of Volcán Santo Tomás, on Isabela, at an altitude of 870 m; *Prescottia oligantha* occupies shaded habitats near the lower limits of the Fern-Sedge Zone on Santa Cruz (at about 500 m above sea level) along the trail from Bella Vista to Mount Crocker.

The diminutive *Ophioglossum reticulatum* occurs over most of the higher area, some plants within a few meters of the highest point on Mount Crocker; and it has been taken on San Salvador and Isabela by Eliasson at elevations ranging from about 800 to over 900 m. This tiny fern is rather easily overlooked, since it is usually hidden among other types of vegetation.

Lichens, closely crowded and some as much as 15 to 20 cm in height, cover extensive areas on barren cliffs and exposed rocks, offering pastel shades and hues of green, yellow, orange, gray, and red.

Areas that have been swept by dry-season fires often are closely inhabited by huge plaques of the liverworts *Marchantia chenopoda* and *Anthoceros simulans.* The latter is particularly abundant in both the Fern-Sedge and *Miconia* zones on Santa Cruz, some of the sporophytes being so crowded as to resemble a green brush with "bristles" up to 20 cm high.

Jaegeria crassa and *J. gracilis* are fairly common in the Fern-Sedge Zone, the two species growing together at the summit of Mount Crocker. *Gnaphalium purpureum,* though not common, occurs on the more open ridges. On the flanks of these rocky strips, one often finds numerous plants of *Lobelia xalapensis, Cuphea carthagenensis, Hypericum uliginosum, Centella asiatica,* and *Pernettya howellii.*

But again, the outstandingly spectacular plant of the Fern-Sedge Zone is the endemic tree fern, *Cyathea weatherbyana,* which occupies ravines, deep swales, pockets at the bases of slumped cliffs, and large potholes. In such habitats, often inaccessible without ropes, the plants are all the more tantalizing. In all of these areas the lacelike patterns of the immense fronds produce graceful arches, casting shadows that are at once a delight and a despair to the photographer or artist.

The following are the more important species of the Fern-Sedge Zone (a dagger marks those that are found solely or chiefly on Fernandina and Isabela, but scarcely at all on the other islands):

Apium leptophyllum
†*Botrychium underwoodianum*
†*Calceolaria meistantha*
Centenella asiatica
Cuphea carthagenensis
**Cyathea weatherbyana*
**Cyperus grandifolius*
Cyperus rivularis
Cyperus virens
**Dicranopteris flexuosa*
†*Elaphoglossum tenuiculum*
†*Elaphoglossum yarumalense*
Eleocharis fistulosa
Eleocharis maculosa
Eleocharis mutata
Eleocharis nodulosa
Eleocharis sellowiana
†*Equisetum bogotense*
Gnaphalium purpureum
Grammitis serrulata
**Habenaria alata*
**Habenaria monorrhiza*
**Hypericum uliginosum*
Hypolepis hostilis
**Jaegeria crassa*
Jaegeria gracilis
†*Jungia hirsuta*
**Lindernia anagallidea*
Liparis nervosa
Lobelia xalapensis
Ludwigia leptocarpa
Ludwigia peploides subsp. *peploides*
Lycopodium cernuum
Lycopodium clavatum
Lycopodium thyoides
**Mecardonia dianthera*
**Mildella intramarginalis*
†*Nama dichotomum*
†*Notholaena aurea*
**Ophioglossum reticulatum*
†*Pellaea sagittata*
**Pernettya howellii*
Pityrogramma calomelanos
†*Plumbago coerulea*
†*Polystichum muricatum*
**Prescottia oligantha*
Pteridium aquilinum var. *arachnoideum*
**Rhynchospora corymbosa*
Rhynchospora rugosa
**Scleria hirtella*
Scleria pterota
†*Stellaria media*

The native plants in the Fern-Sedge Zone deserve a great deal more critical field and herbarium work. Owing to the difficult logistic problems attendant upon extended encampments in the areas where this zone occurs, no one has devoted as much time to it as to those nearer the coast. A party of three or four men, experienced in carrying on field investigations under rather primitive conditions, could profitably devote several weeks to making collections, taking photographs, and recording relative abundance of the various species in the Fern-Sedge Zone on each of the islands having elevations high enough for the formation to have developed.

Similar efforts should be undertaken in the *Miconia* and *Scalesia* zones, for they also have been investigated in only superficial fashion. A full year in the field, during which a party could observe the changing vegetation and make collections in all seasons, would be extraordinarily rewarding. Such a field party should make several circuits of the islands, comparing the vegetation of one exposure with that of another, and making similar comparisons of vegetation from one island to the next. There undoubtedly are many niches that have not yet been seen or recorded; Isabela especially, owing to its great area, is only minimally known.

THE FAUNA

In most phyla of the animal kingdom, the Galápagos Islands fauna can offer some unusual species but is poor in total species. Only a cursory accounting of the fauna will be attempted here.

Arachnida and Chilopoda. Both spiders and scorpions are fairly numerous as individuals, but are represented by comparatively few species in the Galápagos Islands. Large spider webs are encountered among the shrubs and trees in all of the vegetation zones, though less commonly in the Fern-Sedge Zone than in those below it. The scorpions seen during the 1964 expedition were moderate in size and fairly common under stones, in crevices in the lava, and under the exfoliating bark of several trees. The local inhabitants genuinely fear them, for their sting is very painful even to a mature man or woman and perhaps fatal to children. Centipedes are nearly as numerous as scorpions and are also looked upon with a certain amount of fear, although the villagers concede that the centipede bite is less troublesome than the sting of a scorpion. Some of the centipedes are fully 20 cm in length and the articulated body about 1 cm wide. The mockingbirds include them regularly in their diet, and seem to have little trouble in dispatching them with a few well-aimed pecks at the anterior end.

Insects. Entomologists report that the insect fauna of the archipelago is relatively poor in genera and species (Williams, 1907; Van Dyke, 1953). Insects that function as pollinators of the native plants are particularly few, there being only a single bee, an endemic carpenter bee (*Xylocopa darwinii*), in contrast to the dozens of genera and species of solitary bees (and some colonial ones) in arid regions elsewhere in the world. Likewise, the butterflies are poorly represented, only seven being known to occur in the Galápagos, and only two of these, *Agraulis vanillae galapagensis* and *Urbanus dorantes galapagensis,* believed to be endemic to the islands. Hawkmoths are slightly better represented, for there are 12 species distributed among seven genera, several of them having been observed visiting flowers of several different plants early in the evening and at about daybreak. One of them, *Epistor lugubris delanoi,* is diurnal in its activities and has been seen foraging among flowers in midday (Linsley, 1966; Rick, 1966). Beetles are more numerous, occurring in several, if not all, of the vegetation zones, and some are regularly found in and around the flowers of the cacti, but their role as pollinators is incompletely known. Ants, mosquitoes, and flies are present in large numbers, but among these groups also the number of genera and species represented is fairly low as compared with similar groups in the mainland habitats. The Hemiptera are represented by four genera and five or six species in the Orsillinae, only one of which is definitely known to be endemic (Usinger & Ashlock, 1966).

Vercammen-Grandjean (1966) reported that there is no record of any parasitic Acarina from the Galápagos but predicted that a close search would reveal some among the birds. He must have overlooked the ectoparasites on the iguanas, which are often quite liberally infested with ticks.

Amphibians. None are known from the Galápagos Islands.

Reptiles. Sea turtles come ashore on a number of beaches to dig nesting holes and lay their eggs, but of course spend the major part of each year at sea. This group, although said to be less abundant than half a century ago, seems in no immediate danger of extinction.

The giant tortoises of the Galápagos, the animals for which the islands were named, have already suffered the extinction of several subspecies, and populations on some of the other islands have been reduced to such a point that they are facing total disappearance—and may already have gone the way of those once present on Santa Fé and one or two other islands. Van Denburgh (1914) described 15 species among the Galápagos tortoises, but all have subsequently been reduced to subspecies under the single species, *Geochelone elephantopus.* The total population of this magnificent creature has been reduced from an estimated 2 to 3 million at the end of the eighteenth century to probably 3,000 or less. Two different colonies are now known on Santa Cruz, and two or three on Isabela. None have been seen on Fernandina for some time, and those on Española have been reduced to such low numbers that their return to normal colonies seems doubtful. All Galápagos native animals are now protected by government edict, but, sad to say, occasional poaching still occurs. A graver threat are the feral dogs, cats, and pigs, which destroy many newly hatched animals, and the wardens report that pigs, and possibly dogs, have been known to kill mature individuals. Hendrickson (1966) summarized the present state of the tortoises as observed by him in 1964.

Marine iguanas (*Amblyrhynchus cristatus*) are particularly numerous on Fernandina and common on Isabela; and there are large colonies on Santa Cruz, San Salvador, Santa María, Española, and San Cristóbal. Those on Española show rather bright markings of green and red, in contrast to a uniform dull gray (when dry) to black on the other islands visited in 1967. There is some variation in size of individuals on some islands as compared with mature animals on other islands, those on San Salvador and Santa Fé appearing to be smaller than the ones on Santa Cruz, Fernandina, Isabela, and Española. Among the animals on the last four islands, mature individuals may approach, or even attain, overall lengths of a meter or slightly more. All of them feed on seaweed, either by stripping rocks exposed at low tide or by diving to depths of 10 m or more.

Land iguanas (*Conolophus* spp.) are dwindling in numbers owing to depredations by feral dogs and pigs and to illegal hunting for human food. Small colonies of these huge endemic lizards occupy some of the less accessible parts of the Arid and Transition zones on Santa Cruz, in the northern part of Isabela and adjacent Fernandina, and on San Salvador and Española. They are fairly well represented on Santa Fé also, but less numerous on Santa María and San Cristóbal. In 1964 there were several mature animals on Las Plazas, where they were observed high in the spiny shrubs eating the tender twigs and buds of *Maytenus, Castela,* and one or two other shrubs. The land iguana is less heavily crested than the marine animal, is somewhat stubbier and heavier with respect to length, and, with the exception of the marine iguanas on Española, is more colorful, being tan to orange in general hue, with some added deeper red or orange on the throat, and with a lighter ventral surface. They move more rapidly than the marine iguanas, and were not observed swimming by any of our groups in either 1964 or 1967. *Conolophus,* like *Amblyrhynchus,* is endemic to the Galápagos Islands.

There is only one snake in the Galápagos—*Dromicus biserialis,* a slender, racer-like creature, endemic to the islands, with an attractive color scheme of stripes and spots of green, yellow, gray, and black, and in some specimens very small flecks of red. They are shy snakes, moving rapidly when disturbed and quickly taking cover under stones, in crevices in the lava, or in dense vegetation. The one species is segregated into several island-limited subspecies.

The lava lizard (*Tropidurus*) is represented by seven endemic species, with no island having more than a single form within its shores. *T. albemarlensis* has been taken on Fernandina, Isabela, Baltra, Santa Cruz, Santa Fé, Rábida, and San Salvador. Each of the other six species is confined to a single island. The males are often twice as heavy as the females, and weights of adults range from 9.88 gms for small females to 69.67 gms for large males. The females are slightly more brightly colored than the males, with more red on the throat and the sides of the head than in the male. Carpenter (1966) presents a fascinating account of their display behavior in the breeding season.

Geckos are common around most buildings and probably occur on all of the larger and moderate-sized islands. At the Charles Darwin Research Station at Academy Bay, they appeared all over the walls, ceilings, and outer surfaces of the dormitory and laboratory buildings every evening as soon as darkness set in.

Birds. According to Swarth (1931) there are 108 species and subspecies of birds known to occur in the Galápagos Islands; of these, 89 reside and breed in the archipelago and 77 of them are endemic. The commonest and most thoroughly studied group of Galápagos birds is the Geospizidae, or Galápagos finches, whose affinities are uncertain. Swarth recognized 37 species and subspecies, and Lack (1945), who was mainly interested in variation among the species, recognized 29 taxa, in the main following Swarth's classification. These birds are very numerous on some islands, and present to some degree on virtually every one, the larger birds living toward the northern limits of the archipelago and the smaller, more slender birds inhabiting southern areas. Most species subsist, as one would suppose, on routine diets of seeds, fruits, and/or insects, but three are singularly divergent in diet. One, *Cactospiza pallida*, regularly uses cactus spines or a short twig plucked from a nearby bush to impale larvae of insects buried too deeply in their burrows to permit access of the birds' beaks! Another, *Geospiza difficilis,* pecks at the base of the emerging secondary feathers of boobies until the follicles bleed, then sips the oozing blood. The third, *G. fuliginosa*, perches on the marine iguanas and assiduously grooms them for ectoparasites (Amadon, 1966; Lack, 1945). Bowman (1963) draws some very apt comparisons between the beaks of the Geospizidae and certain handtools used by man, and correlates the forms of these organs with the food-gathering habits of their owners.

It would be inappropriate to list all of the Galápagos birds, but passing mention of a few of the more intriguing ones may be forgiven. Among the oddities are the flightless cormorant (*Nannopterum harrisi*), present in considerable numbers around Fernandina, Isabela, Wolf, and other areas; the small Galápagos penguin (*Spheniscus mendiculus*), seen in the same area; the small Galápagos green heron (*Butorides sundevalli*), that sits on branches of mangroves and dives into the water below for fish, very much in the manner of a kingfisher; the very tame and beautiful small endemic dove (*Nesopelia galapagoensis*); the Galápagos hawk (*Buteo galapagoensis*); an endemic duck (*Paecilonitta galapagensis*); several subspecies of mockingbirds (*Nesomimus* spp.), so tame that they often alight on one's head, shoulders, or even boots; and the fork-tailed gull (*Creagrus furcatus*), which does its fishing at night, to name only a few. Bryan Nelson (1968) gives a very readable chronicle of a year spent on two of the islands, Genovesa and Española, with his wife, studying the seabirds of the archipelago. His book paints a fascinating picture of the vicissitudes the avifauna of the area faces from year to year—even from month to month.

Mammals. The known endemic mammal fauna of the Galápagos is limited to ten species belonging to three orders. The Chiroptera are represented by two insectivorous bats, *Lasiurus brachyotis* and *L. cinereus.* The first bat collected in the Galápagos was taken in 1891 by Baur and Adams on San Cristóbal. It was *L. brachyotis* and the only known species based on collected specimens until 1952, when Bowman collected two specimens of *L. cinereus* on Santa Cruz.

The pinnipeds also are represented by two species, the Galápagos fur seal (*Arctocephalus australis galapagoensis*) and a sea lion, *Zalophus californianus wollebaeki.* The fur seal has its closest affinities with fur seals of the Southern Hemisphere, whereas the sea lion is a subspecies of the California sea lion (Orr, 1966). We saw small numbers of fur seals and sea lions at Isla Wolf in 1967, and many sea lions and a few fur seals around Punta Espinosa and in Tagus Cove during our calls there. Sea lions are numerous throughout the archipelago and fishermen blame them for taking large numbers of fish. We saw sea lion bulls repel a large shark on the south side of Isla Champion in 1967, by swimming to meet it while the cows and pups of the harems dashed frantically for shore. The bulls swam between the shark and the rookery, actually nudging it as they pursued a semicircular pattern, until the shark made off for deeper water. Three such encounters were seen during the course of a two-hour visit to the island in late February.

All native land mammals known in the Islands are rodents. Six species of rice rats are recognized, two in the subgenus *Oryzomys* (*O. galápagoensis* on San Cristóbal; *O. bauri* on Santa Fé); and four in the subgenus *Nesoryzomys* (*N. indefessus* on Santa Cruz and Baltra, *N. darwini* on Santa Cruz, *N. swarthi* on San Salvador, and *N. narboroughi* on Fernandina). It seems strange that none has been found on Isabela, but this may be the result of insufficient field trapping on that island. These native rodents may face severe, even lethal competition from the introduced house rat.

Introduced mammals include both the house rat, *Rattus rattus,* and the house mouse, *Mus musculus;* both are present around all settlements.

Goats were introduced by sea captains before the end of the eighteenth century, with the intent of providing food for shipwrecked seamen. They have survived and multiplied, although sheep introduced at about the same time soon disappeared. Goats have done great damage to vegetation on Santa Fé and Española, although less on the latter island than on Santa Fé. They are present in large numbers on several of the large and moderate-sized islands. Efforts to reduce their numbers by hunting have been made during the past few years, but it is too early to assess the long-term effects of these efforts.

Cattle and burros have gone wild on Isabela, San Cristóbal, Santa Cruz, and Santa María. Both animals occupy the higher parts of the islands, but the burros range farther than the cows and frequently come down to sea level in uninhabited areas. Neither seems to have done serious damage to the native vegetation, although wide areas have suffered from brush fires set by hunters during dry seasons. Burned-over land remains unproductive—and subject to erosion—for several years.

Pigs have gone wild on several islands, and constitute a grave menace to the tortoises on islands where both occur. The pigs not only eat hatchling tortoises but kill mature animals by grasping the edge of the carapace near the tortoise's front leg and violently shaking their heads while pulling backward with all their strength until a piece of the shell breaks away. This process is repeated until the vitals of the tortoise are exposed and the animal is killed. The pig (or pigs) then devour the luckless tortoise. A campaign of shooting pigs has been waged for several years by government wardens, with some success in reducing their numbers, but in this case also it is too early to be assured of the tortoise's ability to survive the advent of man and his feral animals.

Dogs and cats have also gone wild and menace the hatchling tortoises, the land iguanas, and the avifauna of the larger islands. There have been reports of dogs employing much the same tactics as the pigs in destroying mature tortoises, but these reports may or may not be accurate. The cats certainly make inroads on the ranks of the extraordinarily tame land birds, on the lava lizard, and probably on the populations of very young iguanas. They probably kill at least a few of the newly hatched tortoises.

We have seen no reports of guinea pigs escaping and becoming established in the wild in the Galápagos. But the potential for such a development exists, for Ecuadorians raise this small rodent for food, and it would be no more than natural for some of them moving to the archipelago to take a stock with them. If it is once introduced, escapes surely will occur, though to what extent the animal could maintain itself in the wild is uncertain. Perhaps it would be eliminated quickly by hawks, owls, snakes, and the feral cats, but wild stocks of this group are widespread in South America, and there is no particular reason to suppose they would be less successful in the islands than on the mainland, where they are native.

DISCOVERY AND EARLY HISTORY

Inca legend says that about 60 years prior to the arrival of the Spaniards, a powerful Incan named Antarqui sailed away from the South American coast from near the present city of Manta with an armada of balsa sailing rafts, and that during his voyages he discovered two islands, which he called Nina-Chumbi (Island of Fire) and Hahua-Chumbi (Outer Island). Some early historians believe these islands may have been two of the Galápagos group. Heyerdahl and Skjölsvold (1956) found pottery fragments they believe to have been left by settlers from the Wari civilization of Peru about 700 A.D., but the suggestion is doubtful and no other traces of pre-Columbian man have been uncovered in the archipelago.

In 1534 Emperor Carlos V of Spain granted authority to Fray Tomás de Berlanga, Fourth Bishop of Panamá, to arbitrate disputes among certain South American conquistadors. Accordingly the Bishop set out from Panama in February 1535 to sail to Puerto Viejo in what is now Ecuador. After seven days' sailing, the ship was becalmed, then drifted without wind for six days in a westerly-setting current, when at last an island was sighted. The party's water supply had by that time been reduced to barely enough for two more days, and boats were accordingly sent ashore in search of water. None was found. The returning seamen reported seeing the huge tortoises, the first record of these animals. The little ship then drifted for three more days, without water, and much hardship was endured by the crew and its passengers. Two more days passed in digging wells from which only water saltier than that in the sea was obtained, and the men chewed pads of the cactus to extract small amounts of moisture. Finally, on Passion Sunday, they found a ravine containing about a hogshead of potable water, the small basin refilling slowly as they dipped out the precious fluid. Eight hogsheads were filled and the ship again set sail for the mainland, since a light breeze had arisen. One man had died during the days of their frantic search for drinkable water, and two days after they left the islands a second man died. Ten horses were also lost to thirst.

Fray Tomás reported that there were many birds on the islands, along with seals, turtles, iguanas, and tortoises. He mentioned also that the birds were so tame they were easily caught by hand. In his letter to Emperor Carlos he wrote, "On the sands of the shore, there were small stones, that we stepped on as we landed, and they were diamond-like stones and others amber-colored; but on this whole island, I do not think that there is a place where one might sow a bushel of corn, because most of it is full of very big stones." The ship was becalmed a second time just within sight of land, and when all water was again exhausted the party suffered once more the tortures of thirst. But the ship managed to gain the harbor before further deaths occurred.

Following the discovery of the islands by the Bishop, no other ships approached the islands for many years, and the archipelago was nearly forgotten, or at least shunned by seafarers. The islands appear on maps and charts as early as 1754 (Slevin, 1959), where they were named "Isolas de Galápagas." Individual islands went unnamed until 1684, when William Ambrose Cowley assigned names to some of them; more were assigned in 1793–94 by

Capt. Colnett. For the most part these early names honored English kings, admirals, noblemen, buccaneers, and early visitors to the archipelago. For many years the islands were used chiefly as a base of operations for the pirates who preyed on Spanish galleons bound for Panama from the Philippines. Later they were used as rendezvous points by whalers of several nations. From time to time ships were careened on sandy beaches for repairs, or landings were made to capture tortoises for fresh meat, but no scientific collecting was done in the islands until the first quarter of the nineteenth century.

HISTORY OF BOTANICAL COLLECTING

The first botanical specimens of record taken in the Galápagos Islands were collected by the naturalists David Douglas and Dr. John Scouler. Douglas had been commissioned by the Royal Horticultural Society to collect plant specimens and seeds in northwestern North America and to send the botanical material back to England for study and for the introduction of exotic plants into English gardens. Scouler, a medical doctor, had signed on as ship's surgeon, expressly to gain passage to that part of the world; he, also, hoped to collect specimens of plants and animals, for he was as keenly interested in natural history as in medicine. They sailed in the Hudson's Bay Company's brig *William and Ann*, Captain Henry Hanwell in command, on July 26, 1824. The ship touched at Rio de Janeiro, sailed through the Straits of Magellan, stopped briefly at San Juan Fernando Island to fill the water casks, then struck out for the Galápagos Islands. Douglas's diary contains the following brief account of their arrival and activities:

"At noon on Sunday, 9th [January 1825], Chatham Island was seen; we passed along the east side at 4 P.M. of the same day, fifteen miles from shore. It is not mountainous and apparently but little herbage on it. On the morning of 10th (Monday) I went ashore on James Island, thirty-seven miles west of Chatham Island. It is volcanic, mountainous, and very rugged, with some fine vestiges of volcano craters and vitrified lava; the hills are not high, the highest being about 2000 feet above the level of the sea. The verdure is scanty in comparison with most tropical climates, arising, no doubt, from the scarcity of fresh water, although at the same time some of the trees in the valleys are large, but very little variety; few of them were known to me. My stay was three days, two hours on shore each time." Douglas mentioned the tameness and variety of the birds, noting that he collected and skinned 45 of 19 genera, "and had the mortification to lose them all except one species of *Sula*; by the almost constant rain of twelve days after leaving the islands I could not expose them on deck and no room for them below." With equal chagrin he reported: "I was nearly as unfortunate with plants, my collection amounting to 175 specimens, many of them, no doubt, interesting. I was able to save only forty. Never in my life was I so mortified, touching a place where everything, indeed the most trifling particle, becomes of interest in England, and to have such a miserable collection to show I have been there. . . . With no small labor I dried the few now sent home; what they may be I cannot say."

Douglas briefly mentioned the tree *Opuntia* (*O. galapageia,* probably); a *Gossypium* "with large yellow flowers and yellow cotton" (*G. barbadense* var. *darwinii*); and a seed that he thought, mistakenly, might belong to the Coniferae (Wilks, 1914, 100).

Sir Joseph Hooker described or listed 13 plants collected in the Galápagos by Scouler and five collected by Douglas, in the paper in which he reported on Darwin's collections (Hooker, 1847). Robinson (1902, 221) credited the two men together with having collected 23 numbers. What happened to the other 17 specimens Douglas mentioned in his *Journal* is not known. Nor is there a record of the number of specimens collected by Scouler; his collections apparently were made independently of Douglas's efforts.

Shortly after the visit of Douglas and Scouler, James Macrae collected 41 numbers on Isabela. Only seven of Macrae's plants, gathered between March 26 and April 2, 1825, duplicated specimens Douglas and Scouler had obtained in January; 34 were new records for the islands, and 20 were previously undescribed. Thirty-seven of them were identified by Hooker as incidental additions to the report on the Darwin collections (Hooker, 1847).

Hugh Cuming, in the course of an extensive cruise on his yacht, put in at the Galápagos Islands sometime during 1829 and is believed to have gone ashore on Española, Isabela, San Cristóbal, San Salvador, and Santa María. Four of the nine numbers he is reported to have collected in the Galápagos were identified by Hooker (1847).

Charles Darwin, during his famous voyage on H.M.S. *Beagle,* spent from September 15 to October 20, 1835, in the Galápagos, and collected vascular plants on Isabela, San Cristóbal, San Salvador, and Santa María. About 74 per cent of his 209 numbers were new records for the archipelago, and 78 of them were described as new by Hooker (1847). Darwin's collection, the largest obtained up to that time, was the primary foundation for subsequent work on the vascular plants of the islands. Several of the Darwin specimens, which are housed chiefly in the herbarium of Cambridge University, were borrowed by various specialists for this study.

Three years after Darwins' visit, Adm. Abel du Petit-Thouars and Adolphe-Simon Neboux, sailing on the *Venus,* called at the Galápagos. Between June 21 and July 15, 1838, they collected five numbers from Santa María, three of them new taxa. One of these, *Jasminocereus thouarsii,* commemorates the Admiral's name. However, Petit-Thouars and Neboux, under orders to carry on surveys of several possible harbors, had little opportunity to collect plants ashore.

H.M.S. *Herald*, commanded by Capt. Beechey, stopped at San Cristóbal, San Salvador, and Santa María between January 6 and January 16, 1846. Aboard were two naturalists, Thomas Edmonstone (spelled Edmonston in some sources) and Dr. John Goodridge, who collected 41 numbers from the three islands. The records seem to indicate that most of the collections were made by Edmonstone, for Goodridge's name appears infrequently in Hooker's 1847 paper on the plants of the Galápagos. Hooker named five species in honor of Edmonstone (*Iresine edmonstonei*, p. 190; *Phaca edmonstonei*, p. 227; *Sesuvium edmonstonei*, p. 221; *Solanum edmonstonei*, p. 201; and *Spondias edmonstonei*, p. 230). Edmonstone collected the type specimens of three of these, Darwin one, and Goodridge one. Twenty-one of the Edmonstone and Goodridge collections constituted new records for the archipelago. Capt. James Wood, commanding H.M.S. *Pandora*, which accompanied the *Herald*, collected 27 numbers, 15 of which were new records for the Galápagos; none, however, was new to science.

Professor N. J. Andersson, sailing aboard the Swedish frigate *Eugenies*, spent ten days in the Galápagos (July 11–20, 1852) and collected plants on four islands, among them Santa Cruz, which had not been explored by the earlier naturalists. He obtained a remarkably large collection, considering the briefness of his stay, returning to Sweden with 325 numbers. Thirty-one per cent of this collection represented new records, and an even 50 were published as new species (Andersson, 1855). Andersson reported on the vascular plants; the mosses were worked up by J. Ängström.

Sixteen years later, between July 22, 1868, and sometime in January 1869, Dr. A. Habel visited the islands, making perhaps three trips from Guayaquil, on the mainland, aboard a small sailing craft. He visited four islands—Española, Marchena, Pinta, and Santa Cruz—and Kew botanists, who reported on at least a part of his material, credited him with 69 numbers. Additional specimens may have been deposited in Vienna, where he took a large collection of birds and animals in connection with his own special studies. Robinson (1902) credits eight of Habel's collections as new records and two as new species.

The next naturalists to visit the islands were Drs. Thomas Hill and Franz Steindachner. They spent June 10 to June 19, 1872, collecting on Isabela, Rábida, San Salvador, Santa Cruz, and Santa María, taking 96 numbers, two of them new records, and obtained only a single undescribed species. It was becoming more difficult to find novelties!

In August 1875, Dr. Theodor Wolf visited Isabela, San Cristóbal, Santa Cruz, and Santa María. Because his specimens were lost while in storage in Guayaquil, only seven of his collections were reported on. Some of his sight records, for example that of *Salvinia* "in a brook near [a] hacienda" on Santa María, have not been confirmed by later field workers, and remain uncertain.

Two Italians, Gaetano Chierchia and Cesare Marcacci, collected on San Cristóbal and Santa María between March 21 and March 31, 1884. Marcacci collected only algae, but Chierchia obtained 44 numbers of vascular plants. Twenty-two of these were new to the islands, insofar as records then covered the flora, and two had not been described previously.

Leslie A. Lee, an American, spent a week (April 5–11, 1888) collecting plants on San Cristóbal, San Salvador, Isabela, Española, and Santa María, and secured 42 numbers. Only four of this lot were new records for the archipelago, and none was new to science. Most of his specimens are in the U.S. National Herbarium, but a few duplicates were sent elsewhere.

Three years later (March 28 to April 4, 1891), Alexander Agassiz collected 41 numbers of flowering plants and ferns on Pinzón, San Cristóbal, and Santa María. One of these was described by Rose as a new species (*Oxalis agassizii* Rose), but subsequent workers have relegated it to synonymy under *O. cornellii* Anderss. One other species was a new find for the islands. One reason fewer new records and new taxa were being found was that collectors were mainly interested in gathering and studying other organisms, and plant materials were a secondary consideration. Another was that they did not stay long enough to penetrate to the interior; most of their collecting was carried on within a short distance of the shore or along trails leading to small farm plots, where water was available.

Dr. Georg Baur spent three months (June 6 to September 6, 1891) visiting 13 of the islands, but missing Fernandina. He was a keen observer and collector, and obtained many plants that had not been seen since Darwin's day. A number of his specimens still constitute the only record for the species from several islands. His collection of flowering plants and ferns, 385 sets in all, was reported on by B. L. Robinson and J. M. Greenman (1895), who described 37 new taxa (some of them varieties or formas) from among his material. A set at the Gray Herbarium of Harvard University is almost complete and is probably the best existing representation of Baur's collection.

At the end of the nineteenth century, two young naturalists, R. E. Snodgrass and Edmund Heller, participated in the Hopkins Expedition to the Galápagos sent out from Stanford University. The two spent from December 10, 1898, until mid-June 1899 in the archipelago and collected 949 numbers of vascular plants, taken from 16 of the islands. They seem to have been the first botanical collectors to land on, or at least collect plant specimens from, Baltra, Darwin, Fernandina, and Wolf. Very likely they landed on Seymour also, for at that time Baltra (or South Seymour) and Seymour (North Seymour) were often lumped simply as "Seymour." A complete set of the Snodgrass and Heller collection of vascular plants is in the Gray Herbarium, and a partial set is in the Dudley Herbarium at Stanford. Their collection, by far the largest taken during a continuous period up to that date, was reported on by B. L. Robinson (1902). In addition to naming several new species, varieties, and formas, Robinson presented a penetrating discussion of the vegetation and its relationship to the flora of the adjacent mainland of continental South America and speculated about the origin of the islands.

An ambitious expedition was undertaken by the California Academy of Sciences in 1905–6. Eleven men sailed from San Francisco aboard the schooner-rigged *Academy* in May 1905 and landed on Española on September 24. They remained in the archipelago just over a year, departing from Darwin on September 25, 1906. During the year, collections of reptiles, birds, marine organisms, land shells, insects, and plants were amassed. Alban Stewart, a graduate student in systematic botany who had been given the responsibility for collecting plants, went ashore at several different points on most of the larger islands and succeeded in taking over 3,000 numbers, many of them in sets of three or four duplicates. He collected on every one of the larger islands at least once and on virtually all of the smaller ones that are at all accessible. And though the summit of the main crater on Santa Cruz defied his best attempts, owing to dense brush on the south side and to extremely rugged volcanic detritus on the northern and northeastern sides, he did reach the tops of a number of peaks on other islands. Stewart drew upon his collections for his Ph.D. dissertation at Harvard University, where he worked under the supervision of B. L. Robinson. In his dissertation, which was published in the *Proceedings of the California Academy of Sciences* (Stewart, 1911), he reported 615 specific and subspecific taxa as occurring in the Galápagos, describing ten of them as new. A substantial number of new records, new either for the archipelago or for individual islands, were established by his work. An excellent set of his collections is in the Gray Herbarium; and a set that compares favorably is in the herbarium of the California Academy of Sciences. Scattered specimens, representing duplicates of some of his more copious collecting, can be found at the U.S. National Herbarium, at the Missouri Botanical Garden, and in one or more herbaria in Europe.

We have found no records of additional collections of plants obtained in the Galápagos Islands between the departure of the California Academy of Sciences Expedition and the arrival of the Harrison Williams Galápagos Expedition in 1923. The stresses of World War I were too far-reaching and intense to permit much field work there or anywhere else during its course. William Beebe wrote graphically of the discoveries made by the 1923 expedition and about some of the frustrations that accompanied it (Beebe, 1924). One of the most interesting episodes he recorded was an eyewitness account of a sudden increase in volcanic activity from subsidiary vents on the northeasterly slopes of Volcán Darwin (called Mount Williams in his account). William Morton Wheeler, Ruth Rose, and Beebe worked as a team to collect plants when more exciting enterprises slacked off a bit, and obtained specimens from Baltra and the northern part of Santa Cruz. Sand burs were mentioned as a source of major discomfort to the field workers. There are of course no native mammals in whose wool these fruiting structures could be transported from one part of the island to another, but goats have been running wild on many of the larger islands since early in the 1800's, and have spread the burs from two grasses widely over the terrain, often into areas that seem too rugged even for these sure-footed creatures. The Wheeler, Rose, and Beebe collections were not reported on as a unit, but some of them have been cited in several papers. The main collections are in the New York Botanical Garden's herbarium; a few duplicates are housed elsewhere.

The following year two more expeditions visited the Galápagos. One was the Norwegian Zoological Expedition under the leadership of Alf Wollebaek, the other a British group aboard the yacht *St. George*, with James Hornell as scientific director. We have found only a few notes on the plant specimens collected by Wollebaek, which were given incidentally in a report on the collections of Miss Borghild Rorud, and these introduced no new records or new taxa (Christophersen, 1932). During the period July 25–August 7, 1924, personnel of the *St. George* expedition collected plants on Isabela, San Salvador, Santa Cruz, and Santa María. In reporting on these specimens, L. A. M. Riley (1925) described three new species and published the first record of *Psilotum nudum* from the Galápagos Islands (on San Salvador).

From July 1926 through February 1927, Miss Rorud (later Mrs. B. Rambech), one of a group of Norwegian settlers on the island, collected 262 numbers of vascular plants on Santa Cruz. These were reported on by Erling Christophersen (1932); he described two new species (*Acacia rorudiana* and *Periloba galapagensis* = *Nolana galapagensis* [Christoph.] I. M. Johnston) and two varieties (*Mollugo snodgrassii* var. *santacruziana* and *Froelichia nudicaulis* var. *longispicata*).

Henry K. Svenson, a botanist with the Brooklyn Botanic Garden, was on the Vincent Astor Expedition to the Galápagos in 1930 and spent April 1 to April 15 on the islands. He obtained 300 numbers of vascular plants, described two new species (*Luffa astorii* and *Verbena townsendii*) and three new varieties (in *Clerodendrum, Tribulus*, and *Croton*), and reported a number of new records for the archipelago. He landed on only four of the islands —Baltra, Genovesa, Santa Cruz, and Santa María.

The Templeton Crocker Expedition of 1932 spent the period April 15–June 16 in the islands. John Thomas Howell, representing the California Academy of Sciences, explored several previously untouched areas in the interior of two or three of the larger islands, and was a member of the party that made the first ascent of Mount Crocker, the main volcano on Santa Cruz. He collected on 13 of the islands and obtained a great many specimens represent-

Arcturus Lake, Genovesa

2. Santa Cruz, as so often seen from the sea

Wreck Bay, San Cristóbal

4. Main street of Puerto Ayora, Wreck Bay

Fernandina, as seen from a point near Tagus Cove, Isabela

6. Fernandina caldera after eruption, 1968

7. Fernandina caldera before eruption

8. Mangroves and lava fragments, Bartolomé

9. Broken terrain of Bartolomé; San Salvador beyond

10. Pelicans, Fernandina

11. Pink-footed booby

12. Masked booby, with hatchling

13. Blue-footed booby

14. Marine iguanas, Punta Espinosa

15. Fledgling masked booby

16. Frigate birds, female

17. Frigate bird, male

18. Galápagos tortoise

19. Flightless cormorant

20. Marine iguana, Española

21. Sea lions, Pinta; typical Littoral Zone vegetation

22. Mangrove seedling

23. Mangrove prop roots

24. Mangrove stand (*Rhizophora mangle*)

25. Young red mangroves, Conway Bay

26. Salt marsh, Santa Cruz

7. Mature *Opuntia*

28. *Opuntia echios* var. *barringtonensis*, Santa Fé

). Arid Zone, higher reaches, Santa Cruz

30. *Opuntia megasperma* var. *orientalis*

. *Opuntia insularis*

32. *Opuntia galapageia* var. *galapageia*, Pinta

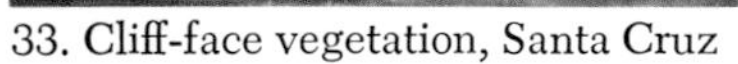
33. Cliff-face vegetation, Santa Cruz

37. Arid Zone vegetation, Isabela

34. *Opuntia megasperma*, second day

35. *Opuntia megasperma*, newly opened

36. *Opuntia helleri*, Genovesa

38. *Opuntia echios* var. *gigantea*

39. Arid Zone in dry season, Genovesa

40. *Opuntia* and *Jasminocereus*, Santa Cruz

41. *Jasminocereus thouarsii* var. *delicatus*

42. *Scalesia helleri*, Tortuga Bay

43. *Bursera* forest, Santa María

44. Trail to Bella Vista

45. Trail junction, Santa Cruz

46. Mixed growth of Transition Zone, Santa Cruz

47. Cinder slopes and Transition Zone forest, Isabela

8. *Bursera graveolens*, dominant in Transition Zone

49. *Bursera*, root following lava fissure

50. Transition Zone vegetation, Fernandina

51. *Scalesia pedunculata* forest, north slope of Mount Crocker, Santa Cruz

. *Scalesia* seedlings

Zanthoxylum fagara and ferns, Pinta

53. *Scalesia pedunculata*

54. *Scalesia* Zone, Isla Wolf

55. Under the canopy, *Scalesia* forest

57. *Croton*, festooned with lichens; a lone *Scalesia*

58. Ferns and *Sphagnum*, Santa Cruz

59. Grassland, Santa Cruz

60. Tree fern (*Cyathea weatherbyana*)

61. Dense stand of *Miconia robinsoniana*, Santa Cruz

62. Border between zones, Santa Cruz

3. *Passiflora foetida*

64. *Habenaria monorrhiza*

65. *Tillandsia insularis*

3. *Lantana peduncularis*

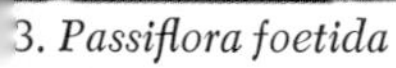

67. *Adiantum macrophyllum*

. *Cordia leucophlyctis*

. *Sesuvium portulacastrum*

70. *Capsicum frutescens*

71. *Lecocarpus pinnatifidus*

72. *Notholaena galapagensis*

73. *Hippomane mancinella*

74. *Acacia rorudiana*

75. *Croton scouleri* var. *scouleri*

76. *Scalesia affinis*

77. *Datura arborea*

78. *Portulaca oleracea*

79. *Portulaca howellii*

). *Urera caracasana*

81. *Parkinsonia aculeata*

. *Passiflora colinvauxii*

83. *Zanthoxylum fagara*

. *Clerodendrum molle*

85. *Prosopis juliflora*

Cassia bicapsularis

87. *Tribulus cistoides*

88. *Darwiniothamnus tenuifolius*

89. *Doryopteris pedata* var. *palmata*

90. *Cardiospermum galapageium*

91. *Cordia lutea*

92. *Psilotum nudum*

93. *Grabowskia boerhaaviaefolia*

94. *Lycium minimum*

95. Lichens, Las Plazas

96. *Momordica charantia*

ing species that had been sampled inadequately by earlier collectors, including several that had been taken only by Darwin. Several of his papers were based on parts of this collection (1933a, b, c; 1934a, b; 1937; 1941a, b).

Professor Folke Fagerlind and his assistant, Göran Wibom, stayed in the Galápagos Islands from April through June 1953, collecting chiefly on Baltra, Española, San Cristóbal, San Salvador, Santa Cruz, and Santa Fé. No unified account of their collections has been published to date but some of their specimens have been cited by Scandinavian writers (Harling, 1962; and Eliasson, 1965a, b; 1966; 1967).

In 1959 Dr. Gunnar Harling and his Ecuadorian assistants, Manuel Lugo R. and Holguer Lugo S., collected plants on ten different islands. The combined efforts of the Fagerlind-Wibom and Harling-Lugo teams yielded about 1,600 numbers. Both collections are housed in the herbarium of the Botany Department of the Naturhistoriska Riksmuseum, Stockholm. The references cited just above to papers by Eliasson apply to certain selected groups represented in the Harling and Lugo collections.

Robert I. Bowman, of the Department of Biology at San Francisco State College, made ornithological studies in the Galápagos in 1957 and again in 1959. During both visits he collected plant specimens from the ecological niches occupied by the finches he was studying. These collections were deposited at the University of California at Berkeley and at the California Academy of Sciences. Bowman obtained over 200 numbers of vascular plants, many of which are cited in the present work.

E. Yale Dawson collected some seed plants, mainly cacti, in the Galápagos Islands during January 1962. His specimens were deposited in the herbarium of the Allan Hancock Foundation, University of Southern California, Los Angeles.

During 1964 more than 60 scientists spent some six weeks in the Galápagos Islands in connection with the dedication of the Charles Darwin Research Station, at Academy Bay on Santa Cruz, and in furthering the field work done under the Galápagos Islands Scientific Program. The senior author was one of the botanists included in this group and collected nearly 1,000 numbers, all taken on Santa Cruz. In addition to his own collections, now at the Dudley Herbarium, Stanford University, he was privileged to extract from the collections made by Dr. David Snow about 200 duplicates from San Cristóbal, Isabela, Santa Cruz, and Santa Fé, and a few from other areas. During the period of field work—January and February 1964—F. Raymond Fosberg collected specimens, which are now deposited in the U.S. National Herbarium, as did the late E. Yale Dawson, whose small set, some of it from the interior of the previously unvisited and normally inaccessible Isla Wolf, was presented to the Dudley Herbarium. Dr. C. B. Koford, Dr. Gorton E. Linsley, the late Dr. Robert Usinger, Dr. John R. Hendrickson, and Allyn G. Smith collected vascular plants on several islands to which the senior author had no access, and presented their materials to him. Another expedition member, Luis A. Fournier, then a graduate student at the University of California at Davis, collected nearly 300 numbers on several different islands. His entire collection was loaned to us through the courtesy of Dr. John M. Tucker, and Fournier's specimens were included in the materials studied by the authors and collaborating specialists.

During the activities of the 1964 Galápagos Islands Scientific Program, Dr. Syuzo Itow, now at the Institute of Biology, Nagasaki University, made collections on Española, San Cristóbal, Santa Cruz, Santa Fé, and Santa María. Of his specimens, some 269 were loaned to the senior author and were also included in the materials studied.

In the summer of 1966 Dr. Paul A. Colinvaux, of Ohio State University, spent nearly two months in the Galápagos obtaining cores from bogs and lakes as part of his work on the pollen sequence in sediments. He collected over 400 specimens of vascular plants from the environs of his stations, referring most of them to the senior author for determination. The duplicates are deposited in the Dudley Herbarium. Several of his collections from the inner slopes of the crater on Fernandina were new records for that island (Colinvaux, Schofield & Wiggins, 1968).

In November of the same year, Dr. Uno Eliasson and his wife, Inga, both representing the University of Göteborg, landed in the Galápagos Islands, remaining until the following June. They visited the larger islands, made ascents of the five major volcanoes on Isabela, and collected valuable material in areas difficult of access or not covered previously. In addition to the specimens cited by Eliasson in the families constituting the Centrospermae, others are cited by several workers who have had access to some of his material, notably in the treatments of the Euphorbiaceae, the Gramineae, the Orchidaceae, and the ferns and their allies. Eliasson himself has reported on some of the more striking finds that resulted from his 1966–67 field work (Eliasson, 1968a, b, c).

Both authors were members of a group sponsored by the University of California Extension Service, Berkeley; by the Departments of Botany and of Zoology at the University of California, Los Angeles; and by the California Academy of Sciences, that visited the archipelago in January and February 1967 to study and photograph both fauna and flora. We collected at 14 separate landings on 13 different islands, and obtained 571 numbers of ferns and flowering plants. The first set of this collection is deposited in the herbarium of the California Academy of Sciences; duplicates are to be distributed to other herbaria.

Mr. and Mrs. André DeRoy, who live at Academy Bay, visited Isla Rábida late in the spring of 1967 and obtained about 30 numbers, mostly from high on the slopes of the main crater. Their specimens were sent to us for inclusion in the present study and are cited where appropriate.

No doubt other individuals and small groups, not known to us, have collected plant specimens in various parts of the islands. Their omission from this account should not detract from any credit due them; rather, it springs solely from the incomplete nature of our data. Indeed, we welcome the opportunity to examine and, insofar as we are able, to provide determinations for any specimens that other visitors to the Galápagos Islands may wish to refer to us.

CONTRIBUTORS

The botanists in the list that follows have devoted considerable time and effort to the study of plant specimens and the preparation of manuscripts. Most of them have given particular attention to certain genera or families over an extended period and are generally considered experts in the taxonomy of the groups assigned them. Their contributions, noted here, are also indicated by footnotes to the text.

Edward F. Anderson (in cooperation with David L. Walkington): Cactaceae

David M. Bates: Malvaceae

Derek Burch: *Chamaesyce* and *Euphorbia*

Lincoln Constance (in cooperation with Mildred Mathias): Umbelliferae

Richard S. Cowan: Caesalpinioideae

Arthur Cronquist: Compositae

Uno Eliasson: Centrospermae (families Chenopodiaceae through Caryophyllaceae in the Engler and Prantl System)

Leslie A. Garay: Orchidaceae

Amy Jean Gilmartin: Bromeliaceae

John Thomas Howell (in cooperation with Duncan M. Porter): Polygalaceae

Tetsuo Koyama: Cyperaceae

Mildred Mathias (in cooperation with Lincoln Constance): Umbelliferae

Harold N. Moldenke: Avicenniaceae and Verbenaceae

Reid Moran: Crassulaceae

Conrad V. Morton (in cooperation with Ira L. Wiggins): Ferns and allies

Peter H. Raven: Onagraceae

John R. and Charlotte Reeder: Gramineae

Charles M. Rick, Jr.: *Lycopersicon*

Keith Roe: *Solanum*

Reed C. Rollins: Cruciferae

Velva E. Rudd: Mimosoideae and Faboideae

Henry J. Thompson: Loasaceae

David L. Walkington (in cooperation with Edward F. Anderson): Cactaceae

Dieter C. Wasshausen: Acanthaceae

U. T. Waterfall: *Cacabus* and *Physalis*

Grady L. Webster: Euphorbiaceae (exclusive of *Chamaesyce* and *Euphorbia*)

Delbert Wiens: Viscaceae

We extend our hearty thanks to each, for aid so freely given. Without their help the book would have been considerably longer in the making—and would have achieved less.

HERBARIUM MATERIAL UTILIZED

The major collections of Galápagos plants in the United States are in the herbarium of the California Academy of Sciences in San Francisco, at the Gray Herbarium of Harvard University, and at the Dudley Herbarium of Stanford University, but there are also important collections at the Brooklyn Botanic Garden, the New York Botanical Garden, and the U.S. National Herbarium in the Smithsonian Institution, Washington, D.C. Through the courtesy of the curators of these herbaria, most of the valuable collections made by United States citizens were available for study, either through loans to the various specialists or on request at the premises.

All specimens collected by the two authors during the 1964 and 1967 field expeditions were made available to contributing specialists through loan, and in each case where duplicate specimens were on hand, the specialist or his institution was permitted to retain one. All earlier collections from the Galápagos Islands at the California Academy of Sciences and in the Dudley Herbarium were sent on loan to collaborators. Thus, a large percentage of the existing material was available for critical examination during preparation of the book.

The keepers and curators at the Royal Botanical Gardens at Kew and at the herbarium of Cambridge University generously allowed several of us to borrow critical specimens from the Darwin gatherings of 1835, so that we could establish limits of poorly known species. Those that were too fragile to entrust to shipment they examined for us, sending precise notes on characters pertinent to certain puzzling situations. Their aid is greatly appreciated.

SCOPE OF THE BOOK AND METHODS OF TREATMENT

For one or two botanists to have dealt with all of the plant groups at hand would have been impractical, if not unwise, even in as limited an area as the

Galápagos Islands. Furthermore, to have included the algae, fungi, bacteria, liverworts, mosses, and lichens would have produced an inconveniently bulky work. Thus we deal only with the Tracheophyta, and have divided the work among 30 specialists. Within the Tracheophyta are the groups that Eames (1936) called Pteropsida, Sphenopsida, Lycopsida, and Psilopsida. All of their representatives have a definite vascular system for the transportation, first, of water and dissolved mineral salts to the plant body and, second, of elaborated foods manufactured in the leaves or other photosynthetic organs to other parts of the plants' systems. All four of these groups are represented in the Galápagos, and each is included in the keys at the beginning of the taxonomic treatment. Not all subdivisions of these groups are present in the islands, however—for example, there are no conifers (Gymnospermae)—and indeed many families common on the mainland of South America are totally lacking in the archipelago.

In order to make the book comprehensive and as useful as possible to anyone interested in the identity of plants found in the archipelago, we have included every native vascular plant for which we have been able to find an authenticating specimen, and we have used the records covering them to the fullest degree. In addition, we have accorded equal treatment to the introduced weeds that have persisted long enough to have been collected by field workers, and to such escapes from cultivation as have been able to reproduce themselves spontaneously without aid from man. The escapes include several large trees and shrubs, some perhaps on rather flimsy claims to permanence, but we have excluded the obviously cultivated vegetation that has not met the criterion of persisting without help from agriculturists or gardeners.

At a number of points in the text are statements indicating skepticism about the actual occurrence of certain species reported from the Galápagos Islands by earlier writers. Often this skepticism arises because present knowledge of the range and distribution of the species in question is such that we are quite sure the species fails to occur anywhere near our area. And there are other bases for doubt. Theodor Wolf reported seeing *Salvinia* in a stream on Santa María, but he could produce no specimens to document the report; and no one since has been able to find the plant, either on Santa María or elsewhere in the Galápagos Islands. Several dubious records of other species were based on collections made by Capt. Wood, and although specimens exist, it is practically certain that a mixing of labels occurred between the time they were collected and the time they were first studied by botanists. This is obvious in a number of instances, for plants known to occur in abundance at his other ports of call, but not in the Galápagos, bear labels giving the archipelago as their source.

From family to family occur notes mentioning genera or species excluded from the flora. These exclusions have been made for various reasons: for such as those just mentioned; because subsequent study has revealed that a specimen (perhaps inadequately representing the plant) had been incorrectly determined by the first worker reporting it from the Galápagos; because a change in nomenclatural interpretation has removed a taxon from the currently accepted legitimate names; or for other reasons, also explained in the text. There are about 40 such excluded taxa scattered through the book.

Table 2 gives the numerical breakdown—by taxonomic categories and major plant groups—of the taxa embraced in the flora of the Galápagos Islands. When we add the subspecific taxa (subspecies and varieties) to the number of species now known to grow spontaneously in the Galápagos Islands, the grand total of such taxa is 702, as compared with Stewart's total list of 682. Stewart, however, included 15 unidentified species in his list, whereas we have included none in this category. Only three of our 702 taxa—the Wolf reports of *Salvinia*, *Callitriche*, and *Turnera* are given recognition without supporting specimens, and only with considerable reluctance.

The number of accepted endemic taxa has dropped from Stewart's 252 to only 228 in our coverage. The drop is attributable chiefly to two factors: several plants believed to be endemic when Stewart wrote his paper in 1911 have subsequently been found to occur on the adjacent mainland of South America; and a more conservative interpretation of the taxonomic status of several others has resulted in their assignment to synonymy, thus removing them from the tabulation. Thus Stewart recognized about 37 per cent of his

TABLE 2. NUMBERS OF FAMILIES, GENERA, SPECIES, AND SUBSPECIFIC TAXA INCLUDED IN THE TAXONOMIC TREATMENT, BY MAJOR PLANT GROUPS

	Major Plant Group					
Category	Ferns and Allies	Apetalae	Gamopetalae	Polypetalae	Monocotyledonae	Total
Families	10	15	23	44	15	107
Genera	39	32	107	114	56	348
Species	89	68	186	185	114	642
Subspecies and varieties*	1	15	15	25	4	60
Endemics (species or subspecific taxa)	10	43	88	70	17	228
Recently introduced taxa	0	6	23	39	9	77
New records (first report of taxon for Galápagos)	15	2	12	7	6	42

* A species widely distributed elsewhere, but represented in the Galápagos only by one of its subspecies or varieties, is tabulated in the "species" row. Where two or more races of a given species occur in the Islands, one is accounted for in the "species" row, the others in the "subspecies and varieties" row.

total flora as being endemic, whereas our treatment counts only 32.5 per cent as endemic to the Galápagos Islands.

Careful searches have been made in an extensive gamut of botanical literature, and through a considerable number of books and articles of a general nature that deal with the Galápagos. Our searches have revealed a number of records, easily overlooked in such scattered sources, that are apparently new to the mainstream of botany; we have incorporated these in the text. Literature cited in connection with papers published by earlier workers has been a fruitful source of important information, and has provided clues to obscure papers of some significance.

The scattered papers involved in this type of search have furnished information helpful in the disposition of many names, including the relegation of some of them to synonymy. Admittedly, the lists of synonyms are in some cases incomplete, owing to the absence of critical treatments of the genera involved. In other cases, the synonymy of a taxon is so voluminous that a reference to a monograph has been given in lieu of the total, unwieldy list. And in still other cases, work verging on the monographic has been necessary in order to settle questioned taxonomic points; here, since full listings of synonyms are generally lacking elsewhere, we have included them. We have tried to give accurate references for all taxa treated, whether as accepted in our treatment or as relegated to synonymy.

Scattered throughout the text are references to papers and books that bear on particular points. Each such reference is given in a footnote or as a parenthetical inclusion (of the date an author's paper was printed) in text. In some instances the page on which the particular point is treated is also given. In those cases in which an author produced two or more papers cited in our work within the compass of a single year, the separate publications are indicated by the letters a, b, etc., following the date. All such references, and those included in this Introduction, are combined in the section headed "Literature Cited," following the taxonomic treatment. Insofar as possible, we have used abbreviations recommended in *Botanico-Periodicum-Huntianum*, published by the Hunt Botanical Library, Pittsburgh, Pennsylvania (Lawrence *et al.*, 1968).

Our contributing specialists were invited to prepare descriptions, keys, synonymy, statements of range, citations of specimens examined, and such added notes as were considered desirable for the groups in their particular assignments. (Only such minor changes as were necessary to ensure uniform treatment and form have been made by the authors.) The sizes of the groups delegated to specialists ranged from a single genus containing a single species to a group of families including over 60 species. All groups not thus assigned were the responsibility of the authors. The citations of specimens, which would have run in sum to hundreds of pages in type, were felt not to justify the considerable cost of typesetting; they are, however, included in the original manuscript, which will be preserved at the Dudley Herbarium for consultation in such matters.

Keys to the major groups of vascular plants known to exist in the Galápagos Islands begin the taxonomic treatment. We have departed from a wholly alphabetical arrangement of the families taken up in order to keep together certain major groups of plants. The first group comprises all plants that reproduce by spores instead of by seeds and require the presence of water (though only a minute film on the gametophyte of many is sufficient) to bring about transfer of the motile antherozoid to the female archegonium and subsequent fertilization of the egg. Thus families of spore-bearing plants, consisting of those belonging in the Lycopsida, Psilopsida, Sphenopsida, and the Filicineae of the Pteropsida, are arranged in a single alphabetical sequence in the first segment of the text. Each family description is followed by a key to the genera, if necessary, and by genera in alphabetical sequence; each genus is treated similarly (i.e. description, key to the species if necessary, and species in alphabetical sequence). Following this first group, and with the same treatment, the seed plants (Angiospermae) are arranged in four major text segments. The first three of these are the three series of the Dicotyledones: the Apetalae (flowers without petals), the Gamopetalae (petals fused to one another at least at the base to form a common structure that falls or withers as a unit), and the Polypetalae (petals present and free from one another, falling as separate units or at least not as an organic unit). The fifth group, finally, is the Monocotyledones. As noted earlier, the class Gymnospermae is not represented in the Galápagos.

Identifying keys are provided for all nonsingle entities recognized. It is of course impossible to construct accurate, workable keys for a family (or genus) containing several to many genera (or species) in such a manner as to keep the generic or specific epithets in alphabetical sequence within the keys. But since each such group of entities is presented in alphabetical sequence in its appropriate part of the text, it was felt unnecessary to number the epithets, either in the keys or in the text. The page running heads, moreover, afford a further aid to text location.

We have supplied the legitimate name of, and a reference to the place of publication of, each genus, species, or subspecific taxon recognized; provided synonymy when appropriate; given a description for each formal category of classification; given a short statement concerning the size and geographic distribution of families and genera; and indicated the habitat and distribution (worldwide and interisland) of each species and subspecific taxon. The interisland distribution is given under "Specimens examined": in nearly all cases (exceptions are indicated), the list of islands from which collections have been examined can be taken to coincide with the known distribution; the number of specimens examined from a given island, for a given species, in fact varied from one to several dozen.

Where the presentation of additional information about the history, ecology, or utilization of a taxon has seemed advisable, such data have been included, following the description and statement of range. It has often been necessary to append, as well, a brief discussion relative to taxonomic status or interpretation; to systematic interpretations of earlier workers; to details of unusual distribution patterns; to relationships of taxa under consideration to others within the genus or family; etc. This kind of information is, of course, highly variable in amount and reliability, and the comments accordingly fluctuate from none at all to rather lengthy paragraphs.

Maps showing the general distribution of each species within the archipelago have been prepared, but no attempt has been made to provide a spot on these maps to match each specimen examined or recorded; instead, each known collection site is indicated, as far as is practical on small-scale maps. Maps showing generic distribution in tropical and subtropical North and South America and in the Caribbean had been contemplated, but it was found that data for the mainland are just too spotty to provide reliable range information. Worldwide range is, however, covered synoptically in concise verbal statements.

Each genus represented in the archipelago is illustrated by a pen-and-ink drawing showing all or part of a plant, often accompanied by detailed drawings of one or more parts important in differentiating taxa. These illustrations should be useful to botanists unfamiliar with the native plants of the Galápagos Islands. A few drawings are reproduced, by permission, from books that do not deal directly with the plants of the archipelago; most, however, are of Galápagos specimens and are presented here for the first time. Photographs by several hands have been used to assemble the section of color illustrations. These show characteristic habitats and individual plants and flowers; we believe they constitute a useful base for grasping the essential nature of the Galápagos flora.

In the glossary, which follows the text, we have tried to include all technical terms used in the text. The definitions are as succinct as possible, but we hope sufficient for all ordinary needs.

The index includes every taxonomic name occurring in the main body of the text, both those accepted and those relegated to synonymy. Page references are also given for species in the lists of those characteristic of the vegetation zones, as well as for taxa otherwise referred to in the Introduction. No typographic distinction has been made between epithets of long standing (and still recognized) and those applied as new taxa, new status, or new combinations by ourselves or our collaborators in the preparation of this book. Synonyms, however, appear in the index in italic type. All new taxa, new combinations, and necessary changes in status—there are about 30 of them—have been published in separate papers prior to the publication of this book (references for these are given directly after the names or changes involved). We have not provided a list of the authors, numbering several hundred, who are responsible for the legitimate taxonomic names and the synonyms.

Darwin found much in the Islands to stimulate a theoretical bent. We would hope that this account of the plant life of the Islands will, in some similar fashion, be found challenging beyond its basic purposes.

Major Plant Groups

Major Plant Groups

In the four pages that follow, synopses and keys are provided for all major vascular plant groups represented in the Galápagos Islands. The families making up these groups are then taken up in the succeeding text (the arrangement of the text into five sections is explained on p. 50).

Division TRACHEOPHYTA*

Plants with a well-developed system of vessels and/or tracheids through which water containing mineral salts in solution is transported from the roots to the leaves and branches of the plant body, and a system of sieve tubes and other elements (constituting the phloem) through which foods manufactured in the photosynthetic areas (usually leaves) are moved to living cells (and places of storage) in other parts of the organism. [Some parasitic plants draw nutrients from their hosts and do not carry on photosynthesis; these nevertheless possess transportation systems and belong in this division.] Sporophytes are independent of the gametophyte at maturity, and constitute the dominant element in the phase of the life history involving alternation of generations.

Key to the Subdivisions of the Tracheophyta

Plant body consisting of naked, dichotomously branched stems lacking true roots and true leaves (the axes with some leaflike lobes); sporangia terminal on short lateral branches, large, 3-loculed, with walls several cells thick.........PSILOPSIDA, p. 56

Plant body consisting of roots, stems, and true leaves; sporangia not trilocular, mostly not terminal, usually small, mostly with walls 1 cell thick:

- Leaves minute, whorled, connate into a sheath; sporangia borne on sporangiophores grouped into a terminal cone; stems articulate at the nodes, heavily coated with bands and tubercles of silica..........................SPHENOPSIDA, p. 56
- Leaves larger (or if minute not connate into a sheath); sporangia not borne on sporangiophores (sporophylls forming terminal cones in some species of *Lycopodium* but without sporangiophores); stems not articulate at the nodes, not coated with silica:
  - Sporangia large, cauline, adaxial in relation to sporophylls, lacking an annulus; leaves small in relation to entire plant, 1-veined (never net-veined), numerous, entire or rarely denticulate or serrulate but not divided into segments nor dissected; leaf-gaps not present in central stele; stems normally protostelic............. .... LYCOPSIDA, p. 56

* Keys and text through Filicales contributed by C. V. Morton and Ira L. Wiggins.

Sporangia smaller, not cauline, marginal or abaxial in position, with an annulus in most Filicineae; leaves megaphyllous (large in relation to entire plant), few, usually net-veined and often divided to dissected into segments in many Filicineae, or if minute (in *Azolla*) then bilobed with one lobe submerged; leaf-gaps normally present in central stele; stems siphonostelic or dictyostelic. . . . . PTEROPSIDA (below)

Subdivision PSILOPSIDA

Sporophytes without true leaves (leaves represented by scale-like flaps of tissue on stems) and without true roots, irregularly enlarged processes at base of stem serving as organs of absorption. No leaf-gaps in stele of stem. Stems dichotomously branched, bearing terminal homosporous sporangia. Water necessary for movement of antherozoid to archegonium to bring about fertilization of egg cell. Gametophyte and sporophyte generations nutritionally independent.

Represented in our area by a single genus (*Psilotum*) . . . . . PSILOTACEAE, p. 174

Subdivision SPHENOPSIDA

Horsetails. Sporophyte usually erect, with roots, stem(s) and small leaves in whorls at the nodes of the stem and connate into a short sheath. Stems hollow but with transverse diaphragm at each node, articulated at the nodes, the diaphragm thus remaining in one internode when disarticulated; outer surface ribbed and coated with silica. Groups of sporangia borne on underside of an umbrella-like sporangiophore, closely crowded in terminal strobili. Sporophyte dominant. Gametophytes small, mostly shallowly subterranean, the two generations nutritionally independent. Spores each equipped with 4 hygroscopic elaters.

Represented in our area by a single genus (*Equisetum*) . . . . EQUISETACEAE, p. 66

Subdivision LYCOPSIDA

Commonly called "Club Mosses." Sporophytes with roots, stems, and small but true leaves, but without leaf-gaps in the vascular cylinder of stem. Sporophylls usually arranged in a cone or strobilus at tips of branches, bearing sporangia on upper surface (adaxial side). Sporophyte and gametophyte nutritionally independent very soon after development of root and stem of the dominant sporophyte.

Represented in our area by a single genus (*Lycopodium*) . . . . LYCOPODIACEAE, p. 75

Subdivision PTEROPSIDA

Ferns and seed plants. Sporophyte conspicuously dominant, with well-developed roots and stems and large leaves. Leaf-gaps present in stele of stem. Sporangia borne on lower (abaxial) surface of leaves or, among seed plants (in all of ours), enclosed by modified leaflike structure(s) called carpel(s).

KEY TO THE CLASSES OF THE PTEROPSIDA

Plant not producing seeds, reproducing by spores; gametophyte free-living, usually green; motile, flagellate antherozoids produced by antheridia; plants homosporous (except in Azollaceae); sporangia mostly neither totally covered nor concealed... FILICINEAE (see key)

Plant producing seeds; gametophyte not free-living, closely attached to sporophyte, not green; motile, flagellate antherozoids not present; plants heterosporous; sporangia (in ours) wholly concealed within anthers or pistil(s)..........ANGIOSPERMAE, p. 179

KEY TO THE ORDERS OF THE FILICINEAE

Plant body fleshy, lacking sclerenchyma, consisting of a fleshy rootstock and a single leaf (young plants sometimes with 2–3 sterile leaves); sporangia large, partially embedded in sporophyll, borne on an erect fertile segment above a sterile blade; each sporangium normally producing several thousand spores; gametophyte fleshy, a subterranean tuber-like body, colorless...........................OPHIOGLOSSALES (below)

Plant body containing sclerenchyma, normally with more than 1 leaf; sporangia small, borne on surface of abaxial side of leaf or at its margin, not embedded in or coalesced to the tissue of the sporophyll axis or its branches; each sporangium containing relatively few spores (usually 64 in Polypodiaceae); gametophyte flat, thin, not fleshy, growing on surface of substrate, green................FILICALES (below)

Order OPHIOGLOSSALES

With the characteristics set forth in the key. A primitive order consisting of a single family of 4 or possibly 5 genera.

Represented in our area by 2 genera...............OPHIOGLOSSACEAE, p. 81

Order FILICALES

Stem (or rhizome) either subterranean or exposed on surface of soil, rocks, or trunks and branches of trees, or in tree ferns erect and often reaching height of 10 m (or more). Sporangia borne on abaxial surface of leaves or at their margins, not in or on tissues of axis or branches of axis of fertile part of sporophyll. Each sporangium normally producing less than 100 spores (usually 64 in Polypodiaceae). Gametophyte developing on surface of substrate, rarely subterranean, green.

KEY TO THE FAMILIES OF THE FILICALES*

Plants small (mostly 1–2 cm long) floating aquatics; stems dichotomous; leaves minute, bilobed, the smaller lobe submerged; sporangia of 2 kinds, 1 containing 64 microspores, the other a single megaspore, each kind in a separate sporocarp...AZOLLACEAE, p. 61

Plants larger (*Grammitis* sometimes only 2 cm tall), not floating; stems rarely dichotomous; leaves not bilobed; sporangia homosporous:

Leaves filmy, only 1 cell thick (except along veins), scattered on filiform, creeping rhizomes; sporangia borne in basipetal succession on an elongate receptacle or in an elongate indusial tube; plant often with hairs but without scales...HYMENOPHYLLACEAE, p. 69

Leaves not filmy, much thicker, more than 1 cell thick, fasciculate or scattered on rhizome, this rarely filiform; sporangia not borne in basipetal succession; sporophyte mostly with scales on rhizome, often on stipes and rachises also; hairs present or absent:

* Salviniaceae (see p. 176) is not included in the key; Wolf reported seeing a *Salvinia* in the Galápagos in 1879, but none has been seen since.

Rachis appearing dichotomous, the main axis terminated by a dormant bud; only ultimate axes with blades, these pectinately pinnatifid; sporangia 3–4 in a sorus, developing simultaneously, with a complete oblique annulus, dehiscing longitudinally; plants with hairs but no scales. . . . . GLEICHENIACEAE, p. 67

Rachis not appearing dichotomous, not terminated by a dormant bud; axes with blades; sporangia usually numerous in a sorus, developing at different times, with a vertical incomplete annulus (vertical-oblique complete annulus in Cyatheaceae), dehiscing (except in Cyatheaceae) transversely; plants with hairs or scales or both:

Plants becoming trees with a trunk 8–10 cm in diameter, to 5 m high or more; trunks with numerous vascular bundles arranged in a circle; sporangia with oblique-vertical annulus; indusium hemispherical. . . . . . CYATHEACEAE, p. 63

Plants small or large, but not trees; caudices with 6 or fewer vascular bundles variously arranged; sporangia with vertical incomplete annulus; indusium present or absent, but never hemispherical. . . . . POLYPODIACEAE, p. 83

Ferns and Allies

Ferns and Allies

The foregoing has supplied synopses and keys to all major vascular plant groups represented in the Galápagos Islands. And, with the exception of the Angiospermae, which are taken up in four subsequent sections of the text, all of these groups have been keyed to the family level. The families thus represented—i.e., all but those of the Angiospermae—are taken up here in a single alphabetic sequence.

AZOLLACEAE. Mosquito Fern Family*

Small or minute annual floating, aquatic, heterosporous plants with dorsiventral, pinnately or dichotomously branching stems and simple roots hanging free on underside or occasionally penetrating mud. Leaves alternate, in 2 rows, each divided into 2 lobes; upper lobe with photosynthetic tissue and stomata, imbricate with adjacent leaves when young, its upper surface minutely papillate, forward margin eventually more or less erect; lower lobe thinner, with cavity filled with mucilage containing filaments of *Anabaena*, its lower surface immersed in water. Sori borne in pairs on submerged leaf lobes, each pair consisting of 2 microsporocarps, or 2 megasporocarps, or 1 of each, each sorus completely surrounded by a saclike indusium; microsporangium-sorus globose, containing several to many long-stalked microsporangia, each of these containing 4–10 spore masses (massulae), each massula furnished with anchor-shaped, septate or nonseptate glochidia on outer surface; megasporangium-sorus acorn-shaped, smaller than microsporangium sorus, containing a single smooth or warty megaspore at its rounded base, and having 3(4) massulae serving as floating apparatus. Microgametophytes developing within the free-floating massulae, becoming attached to megaspore by their glochidia, each microgametophyte producing 8 antherozoid mother cells. Megagametophyte giving rise to small, green prothallium with single archegonium or, if this is not fertilized, to 1 or more additional archegonia; young sporophyte attached to prothallium for varying periods after development of embryo.

A monogeneric family with 8 living species widely distributed in tropical and warm temperate regions.

* Contributed by C. V. Morton and Ira L. Wiggins. See also C. Christensen (1938); H. K. Svenson (1944); Arthur J. Eames (1936, 242–55).

Azolla Lam., Encycl. Méth. 1: 343. 1783

Characters of the family.

Azolla microphylla Kaulf., Enum. Fil. 273. 1824

Azolla caroliniana sensu Robins., Proc. Amer. Acad. 38: 115. 1902, and sensu Stewart, Proc. Calif. Acad. Sci. IV, 1: 27. 1911, non Willd.
Azolla filiculoides sensu Svens., Bull. Torrey Bot. Club 65: 329. 1938, non Lam.

Plant pinnately branching, 1–2.5 cm wide, 1.5–4 cm long; leaves closely imbricate on young growth, later separated nearly their own width, the upper lobe broadly ovate to nearly orbicular, 1–1.5 mm long, acute or obtuse, entire, the margin 1 cell thick and devoid of chlorophyll, lower lobe acute-acuminate, slightly longer than upper lobe, base asymmetrical, abaxial side somewhat hastate; sporocarps usually borne on abaxial side of branch, pairs of megasporocarps usually alternating with a pair of microsporocarps or with pair made up of both types; microsporocarp-sorus globose, 1–1.3 mm in diameter, the indusium hyaline, slightly apiculate at apex; megasporocarps 0.4–0.6 mm long, the conical apical part shallowly and finely pitted in lines converging toward apex, reddish brown; megaspore smooth or very faintly pitted in lines from base to apex, greenish; glochidia 40–60 μ long, mostly 3–5-septate, but some only once or twice septate.

On freshwater ponds and slowly moving streams, in some of the West Indies, and in Central and northern South America, including Brazil and Peru. Specimens examined: Isabela, San Cristóbal, San Salvador, Santa Cruz, Santa María.

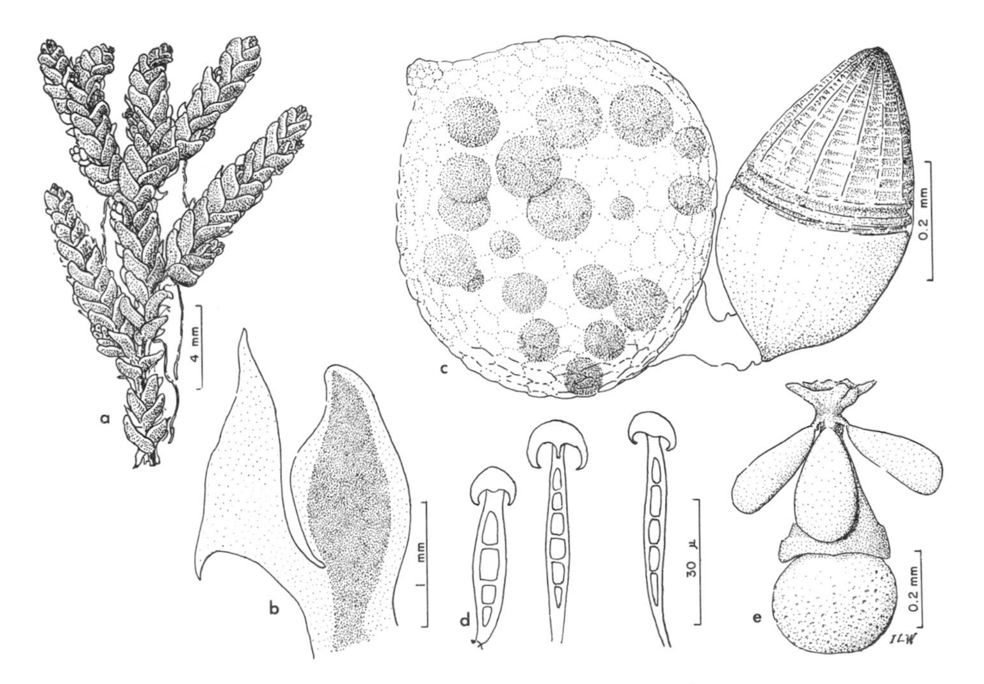

Fig. 1. *Azolla microphylla*: *a*, frond from above; *b*, lower leaf (left) and upper leaf removed from stem; *c*, sori of microsporangia (left) and megasporangium; *d*, glochidia; *e*, megasporangium with attached "flotation" massulae.

Until Chamisso's type collection can be examined, the identity of this species must remain somewhat uncertain. It was reported as collected "in California" but no such species has been found in that area since, and it is probable that if the Chamisso material proves to be this, he collected it in Brazil, for that is the only area touched by him where it is now known to occur. On the basis of the original description of the "smooth" megaspore, the pinnate branching, and the septate glochidia, we interpret our material as belonging here.

Fossil megaspores and massulae containing microspores of *Azolla filiculoides* Lam. have been reported by Schofield and Colinvaux (1969, 623–28) from sediments about 8 m below the present bottom of the volcanic lake El Junco, on Isla San Cristóbal. The stratum of black organic material in which the fossils were found is about 0.5 m thick, is sandwiched between inorganic, nonfossiliferous sediments, and has been radiocarbon-dated at "more than 48,000 years [before present]." Positively identified material of living *A. filiculoides* in the Galápagos Islands is unknown.

See Fig. 1.

CYATHEACEAE. Tree Fern Family*

Robust tree ferns with erect trunks more or less covered with subpermanent bases of old fronds, and in some, by many pendent roots, the apex and bases of stipes covered with slender scales. Stipes closely arranged around apex of trunk, scaly at least at base, often bearing conical, firm spines or prickles. Fronds large, usually bipinnate and more or less deeply tripinnatifid, more or less scaly on lower surfaces of rachises and costae, the scales often bullate; costules of pinnule lobes nearly at right angles to the costae, their veins pinnate, simple or forked. Sori dorsal, large, borne just above middle of veins or at a fork, the receptacle large and often elongate. Indusium more or less globose, cupulate, hemispherical (in ours), reduced to a scale or absent. Sporangia borne in succession, each with oblique annulus completely encircling sporangium and with a poorly defined stomium. Hairlike paraphyses usually present. Spores trilete, usually 64 to a sporangium.

About 700 species in a single genus, occurring in humid areas of all tropics, particularly well represented in moist, tropical mountainous areas.

Cyathea Smith, Mem. Acad. Turin 5: 416. 1793

Characters of the family.

Cyathea weatherbyana (Morton) Morton, Amer. Fern Jour. 59: 65. 1969

Hemitelia multiflora sensu Stewart, Proc. Calif. Acad. Sci. IV, 1: 20. 1911, non R. Br.
Hemitelia aff. *subcaesia* sensu Svens., Bull. Torrey Bot. Club 65: 318. 1938, non Sodiro.
Hemitelia weatherbyana Morton, Leafl. West. Bot. 8: 188. 1957.

Showy tree fern with sturdy trunk to 6 m tall and 2.5–3 dm thick, its upper part and tip covered with linear-lanceolate scales 4–5 mm wide at base and 2–2.5 cm long, the scales markedly attenuate at apex, pale and erose along margins, darker brown to nearly black near base and in lower part of central section; fronds to 3 m

* Contributed by C. V. Morton and Ira L. Wiggins.

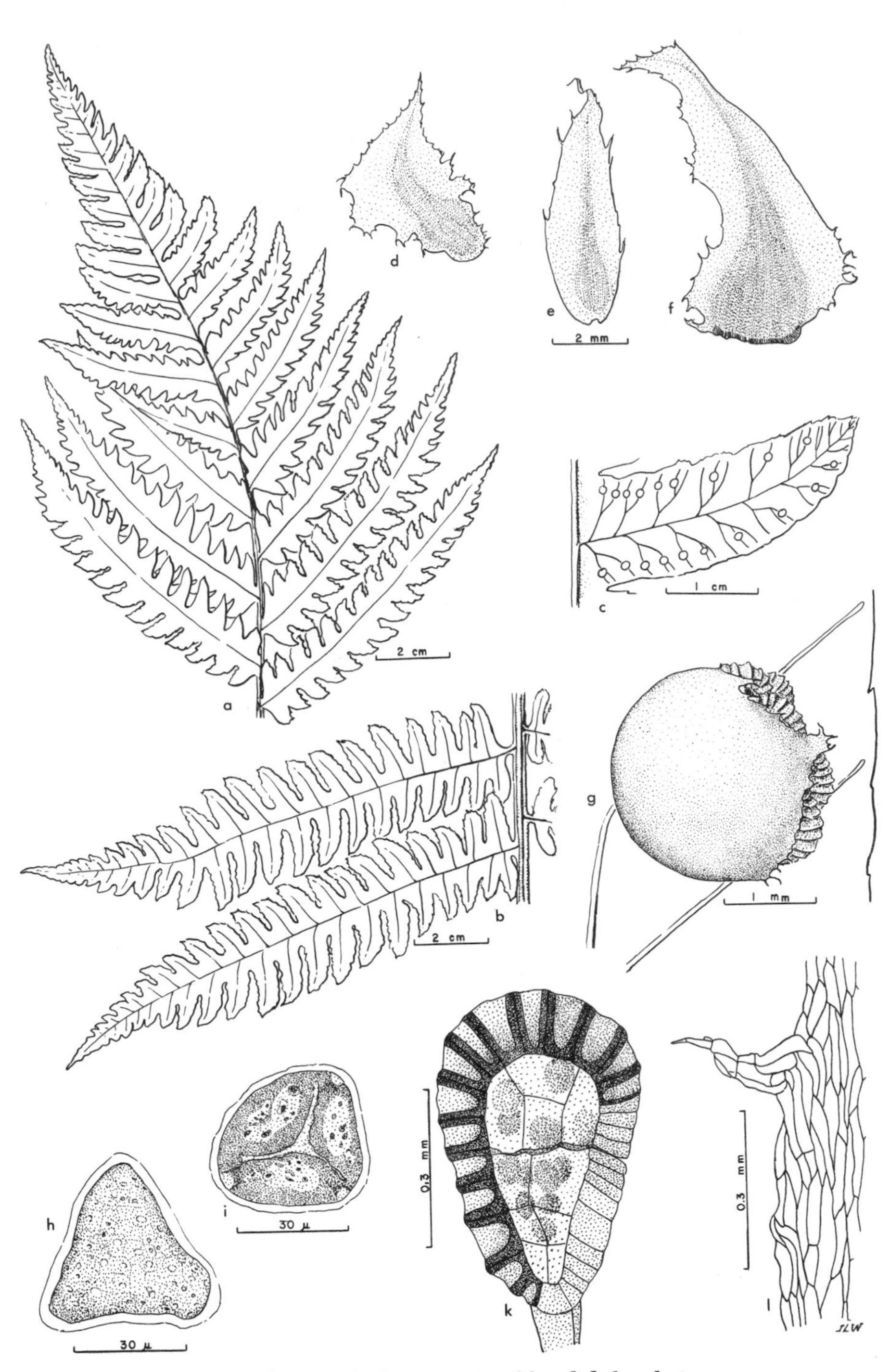

Fig. 2. *Cyathea weatherbyana*: *a*, tip of frond; *b*, basal pinnae; *c*, pinnule; *d–f*, scales from stipe and rachis; *g*, sorus; *h, i*, spores; *k*, sporangium; *l*, cell arrangement along margin of rachis scale.

long, the stipes castaneous, more or less scaly and arachnoid-pubescent with pale rufous to yellowish, slender multicellular hairs, often sparingly conical-subspinose near base, the rachis stramineous, unarmed, more or less furfuraceous, distinctly canaliculate along upper side; lamina bipinnate-pinnatifid, the pinnae on petiolules 6–15 mm long, oblong-lanceolate, 3.5–4.5 dm long, 1–2 dm wide, rachilla brownish, furfuraceous, canaliculate along upper side; pinnules 15–20 pairs on each pinna, 10–15 cm long, 2–3.5 cm wide, acuminate, alternate and sessile; segments about 15–24 pairs, oblong, 16–22 mm long, 5–8 mm wide, slightly falcate, acute to obtuse at apex, the margins subdenticulate to crenulate, attached to a common wing 1–2.5 mm wide along rachilla, scaly along costae, finally becoming more or less scaleless, with slender, multicellular hairs prominent along costae beneath and sparsely present on lamina between them; veins free, 7–9 pairs, all forked once or twice; sori 1–2 mm broad, arranged submedianly along forks of veins, often present on each of 2 or 3 branches of veins near bases of segments; indusium nearly hemispherical, partially enclosing sporangia, thin, pale yellowish, the margins irregularly erosulate to subfimbriate, attached at proximal base of receptacle; sporangia broadly ovoid to subglobose, 0.4–0.5 mm wide, slightly longer, annulus prominent, dark brown, about 0.2 mm thick, of 15–20 thick-walled cells; spores slightly triquetrous, rounded, the exine appearing perforate on outer, convex face, 45–55 μ in diameter, without spines, crests, or ridges.

Endemic; in craters, gullies, sinkholes, and occasionally on less rugged slopes, chiefly above 400 m. Specimens examined: Isabela, San Cristóbal, San Salvador, Santa Cruz.

This tree fern is abundant above the *Miconia* Zone at about 375 to 400 m. It is the dominant plant in many of the craters, in deep gullies, and in potholes formed where lava tubes have caved in. Some grow on nearly vertical walls of craters and gullies, and form a dense cover of interlacing fronds. Sporulating fronds produce tremendous quantities of yellowish spores. J. T. Howell gives a good account of the fern's habitat in "Up under the Equator," in Sierra Club Bull. 27: 79–82. 1942.

See Fig. 2 and color plates 58, 60, and 62.

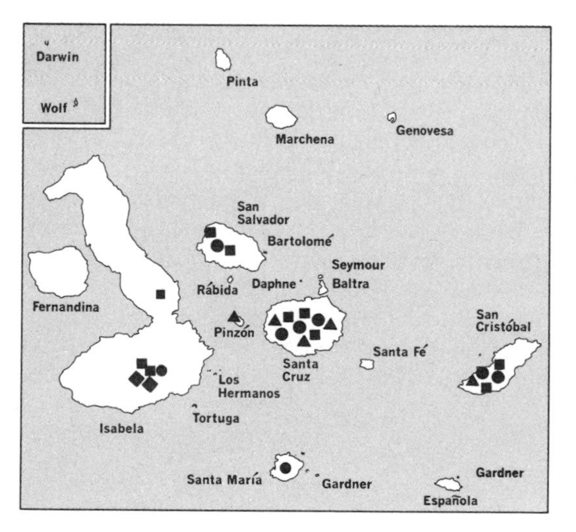

● *Azolla microphylla*
■ *Cyathea weatherbyana*
◆ *Equisetum bogotense*
▲ *Dicranopteris flexuosa*

EQUISETACEAE. HORSETAIL FAMILY*

Rushlike perennial plants with articulated aerial stems from brown or black, wide-creeping, branching rhizomes. Aerial stems annual or perennial, erect, fluted; branches in whorls at nodes or the stems unbranched; internodes hollow or in smaller branches containing thin-walled cells without chlorophyll; harsh deposits of silica on surface in the form of tubercles, transverse ridges, rosette-like aggregations, or uniformly distributed. Stomata in grooves between longitudinal ridges. Leaves reduced to small scales more or less coalesced into cylindrical sheath at each node, tips of teeth free, hyaline or membranous. Strobili terminal, of peltate crowded sporangiophores (sporophylls) each bearing several oblong to subglobose, thin-walled sporangia on its under surface near margins of shield. Spores greenish, spherical, equipped with 4 narrowly spatulate hygroscopic elaters. Gametophyte subterranean, small, lobed.

A monogeneric family with about 25 species distributed widely over all of the major land masses of the world except New Zealand, Australia, and Antarctica.

Equisetum L., Sp. Pl. 1061. 1753; Gen. Pl. ed. 5, 484. 1754

Characters of the family.

Equisetum bogotense HBK., Nov. Gen. & Sp. Pl. 1: 42. 1816

Rhizome dark reddish brown, creeping and much branched, nodose, bearing teeth-sheaths 3–4 mm long, these dark brown to nearly black, internodes 1.5–2.5 mm broad, 5–15 mm long; slender rootlets arising at nodes, profusely covered with crinkled, dark brown hairs; aerial stems cespitose, erect or nearly so, simple or moderately branched, 1–2 mm thick, 8–40 cm tall, sometimes bearing whorls of 2–6 lateral, spreading-ascending branchlets at 1 or more of lower nodes; internodes of main branches 5–35 mm long, deeply 4- to 6-grooved, the ridges nearly flat on outer surface, about 0.2–0.4 mm wide, closely set transversely with low, irregularly forked ridges of silica, more or less glistening, the grooves finely tuberculate with tubercles arranged in transverse rows; leaf sheaths of main stems narrowly campanulate, 3–5 mm long, about 2–3 mm wide, pale green with brownish, membranous, lance-ovate free teeth 1–2.5 mm long; lateral branches much shorter and thinner than main stems, mostly 2–6 cm long, about 1 mm thick, the internodes 5–13 mm long; strobili ellipsoidal to oblong, 6–12 mm long, 3–4 mm broad, pale brown to yellow-brown, exserted 3–8 mm from subtending leaf sheath; sporangiophores peltate, polygonal, 1.2–1.8 mm broad, more or less spongiose, the outer surface minutely reticulate with whitish, relatively large cells; spores spherical, pale yellowish green, about 55–60 μ in diameter, finely and faintly rugulose to nearly smooth; elaters about two-thirds as long as circumference of spores.

Margins of water courses, temporary pools, lakes, and in thickets in moist habitats, Venezuela to Argentina and Chile. Rare in the Galápagos Islands. Specimens examined: Isabela.

The presence of this species on Isla Isabela at elevations of 770 and 960 m is

* Contributed by C. V. Morton and Ira L. Wiggins.

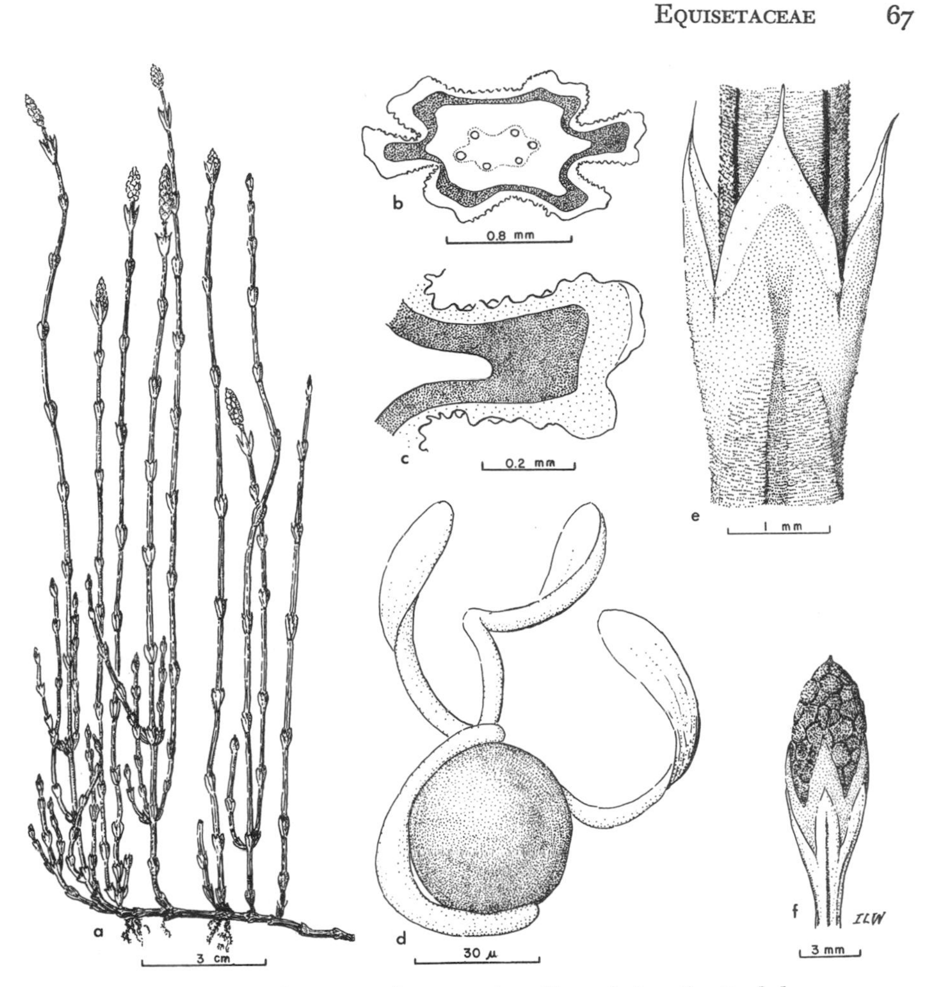

Fig. 3. *Equisetum bogotense*: *b*, transection of branch, heavily stippled area representing chlorenchyma; *c*, transection of ridge; *d*, spore, with elaters; *e*, sheathing leaves at apex of joint; *f*, strobilus.

unexpected, for on the mainland, where it is common in the Andes from Venezuela to Chile, it never occurs lower than 1,400 m, and usually is found between 2,500 and 3,800 m.

See Fig. 3.

GLEICHENIACEAE. CORAL FERN FAMILY*

Rhizome long-creeping, producing fronds at rather long intervals, the younger parts closely invested with stiff, dark brown, septate hairs or narrow scales; vascular strand simple. Fronds usually long, scrambling or climbing, often forming dense

* Contributed by C. V. Morton and Ira L. Wiggins.

thickets. Main rachis bearing opposite pairs of lateral branches, growth of rachis arrested by growth of each pair, apical bud enclosed in hairs, scales, or stipule-like leaflets; rachis branches unbranched and bearing pinnately arranged pinnules, or short, leafless, and bearing a pair of branches in same manner as main rachis with its terminal bud arrested, thus only ultimate branches leafy, or repeatedly branched with branches below terminal branches also leafy. Ultimate branches ("pinnae") pectinately lobed almost to costa, segments short and round to long and narrow, costule and veins free, veins forked. Sori dorsal on veins, variously arranged at different points between costa and tip of vein; sporangia few, large, all maturing at same time. Indusium none. Sporangia obovoid, with complete, oblique annulus, dehiscing by vertical slit. Spores monolete or trilete, smooth, translucent, numerous (usually more than 200 to a sporangium).

A family of 3 genera and about 120 species. Widely distributed in the tropics and subtropics of the world, excluding North Africa, and extending far into the South Temperate Zone. Represented in the Galápagos Islands by a single species.

Dicranopteris Bernh., Neues Jour. Bot. Schrad. 1^2: 26, 28. 1806

Indument on rhizome and fronds of multicellular hairs; scales absent. Apex of main rachis arrested during development of lateral branches but later continuing growth, the fronds thus indefinitely elongate; primary branches several times forked, included buds usually dormant, a short, pectinately lobed accessory branch present at base of each primary branch and similar deflexed, accessory branches usually present at base of forks of lateral branch systems; only ultimate branches with pectinate laminae; segments entire, usually glaucous beneath. Veins twice-forked. Sori 1 to a vein-group on the acroscopic branch, often containing 8 or more sporangia. Paraphyses absent.

About 10 species in the tropics of the Old and New Worlds, extending beyond Tropic of Cancer into New Zealand.

Dicranopteris flexuosa (Schrad.) Underw., Bull. Torrey Bot. Club 34: 254. 1907

Mertensia flexuosa Schrad., Goett. Gel. Anz. 1824: 864. 1824.
Gleichenia flexuosa Mett., Ann. Mus. Lugd. Bat. 1: 50. 1863.
Gleichenia dichotoma sensu Robins., Proc. Amer. Acad. 38: 109. 1902, non Hook. 1846.
Gleichenia linearis sensu Stewart, Proc. Calif. Acad. Sci. IV, 1: 20. 1911, non Clarke. 1880.

Rhizomes long-creeping, extensively branched, slender, 1.5–3 mm thick, dark reddish brown, spreading septate hairs 1–1.5 mm long clothing younger parts; fronds distantly spaced, erect or scrambling, 3–15 dm long or possibly longer; stipes terete, firm to wiry, smooth and glabrous, 1.5–2 mm thick, rusty brown below, yellowish green above; arrested buds enclosed by stipule-like brown segments 3–5 mm long; basal accessory branches ovate-lanceolate to broadly lanceolate in outline, 3–10 cm long, 8–35 mm wide, deeply pinnatifid, the segments linear to oblong or deltoid, 1.5–4 mm wide, to 10 mm long, acute to obtuse, alternate, entire, yellow-green above, glaucous beneath, glabrous; terminal leafy branches oblong in outline, 8–25 cm long, 3–4 cm wide, pinnatifid, the segments numerous, alternate, oblong to lanceolate, 5–40 mm long, 2–4 mm wide, unequal, acute to obtuse, glaucous beneath; veins prominulous beneath, free; sori circular, usually consisting of 5–9 sporangia, forming single row on each side of costa about one-third of the way from costa to margin, naked; sporangia reddish brown, about 0.4 mm long, two-thirds as wide,

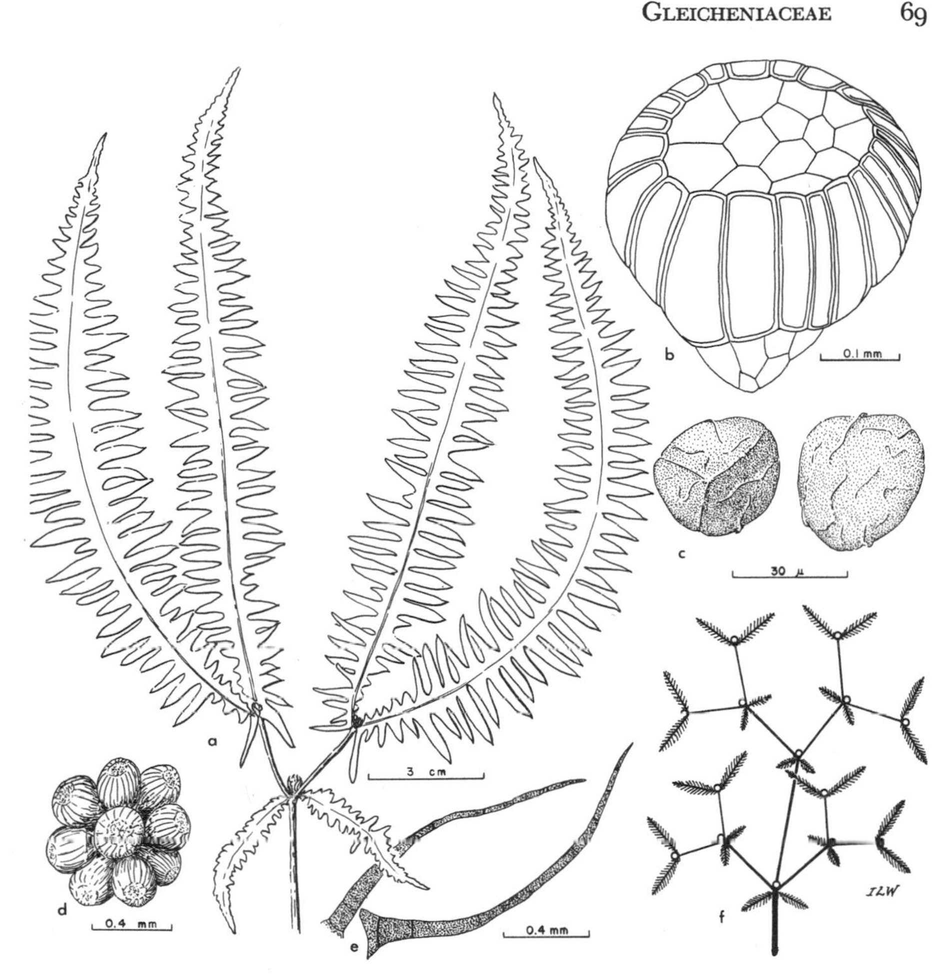

Fig. 4. *Dicranopteris flexuosa*: *a*, terminal part of branch system; *b*, sporangium; *c*, spores; *d*, sorus; *e*, septate hairs from rhizome; *f*, schematic plan of branching in typical plant.

broadly obovoid, dehiscing vertically, the stalks very short or sporangia sessile; spores trilete, very thin-walled, yellowish brown, faintly rugulose to nearly smooth, 40–45 μ in diameter.

Open areas, margins of forests, grassy swales and slopes, West Indies to Brazil and Peru. Specimens examined: Pinzón, San Cristóbal, Santa Cruz.

See Fig. 4.

HYMENOPHYLLACEAE. Filmy Fern Family*

Epiphytic or terrestrial ferns, usually with long-creeping, slender, scaleless, densely pubescent rhizomes. Fronds distantly placed to rather crowded, extremely variable

* Contributed by C. V. Morton and Ira L. Wiggins. For a recent study of the genera and sections, see Morton (1968, 153–214).

in size, shape, and color, from very small and simple to fairly large and greatly divided with many ultimate divisions, usually with only 1 vein to an ultimate segment; blade (except at veins) mostly 1 cell in thickness, yellow-green to deep green, or bronze-iridescent. Veins delicate, forking, free. Sori terminal on ultimate, 1-veined segments, or marginal at vein endings on multiveined leaflets; receptacles columnar, more or less elongated, the base enclosd in a tubular- to conical-based indusium, distal edges of indusium more or less dilated, narrowly conical, or deeply divided into 2 lips. Sporangia sessile, obovoid to turbinate, with oblique annulus, dehiscing vertically without definite stomium. Spores trilete, rounded-tetrahedral to globose, the laesurae present as ridges, without a perispore, granulate to papillate. Prothallia green, chiefly filamentous, monoecious.

A family of 6 genera and about 625 species, widely distributed in wet or humid habitats in tropical and subtropical regions.

Indusium nearly lacking a tube, the valves semicircular, not black-margined; rhizome about 0.3 mm thick, with only a few pale, inconspicuous rhizoids; blades glabrous, or, if hairy, the marginal hairs slender, stalked, pale, not located mostly in the sinuses . . . . *Hymenophyllum*

Indusium long-tubular at base, tipped with 2 semicircular, black-margined valves; rhizomes thicker, 0.5–0.6 mm thick, densely covered with coarse black rhizoids; marginal stellate hairs coarse, black, sessile, located mostly in the sinuses between segments . . . . *Trichomanes*

Hymenophyllum Smith, Mem. Acad. Turin 5: 418. 1793

Delicate epiphytic ferns with very slender wide-creeping, reddish brown, flexuous rhizomes producing relatively few rhizoids and distant, small, upright, spreading or drooping fronds. Fronds short-stipitate, with suborbicular, linear, or deltoid-ovate blades, these green or often reddish, elastic, once to several times pinnatifid or pinnate, the primary rachis usually more or less alate, the wings thin, pale green; ultimate segments plane or undulate-crisped, linear or oblong, entire or finely serrulate, glabrous or sparsely hirsute, each with 1 veinlet. Sori terminal on segments, mostly on basal acroscopic segments of upper pinnae. Indusium bivalvate, attached at base of receptacle, base nearly free or somewhat immersed in leaf tissue, lobes orbicular to triangular-ovate, subequal, cucullate, longer than the hollow, conical basal part. Receptacle columnar or rarely subglobose, of limited growth, usually shorter than valves. Sporangia obovoid, sessile or nearly so, reddish brown.

About 300 species, widely distributed throughout the wet tropics of the world.

Blades with obviously stalked, stellate hairs on the margins; rachis wing usually 0.6–1 mm wide on each side of rachis; segments mostly more than 1 mm wide (Subg. *Sphaerocionium*) . . . . *H. hirsutum*

Blades glabrous; rachis wing 0.2–0.35 mm wide on each side of rachis; segments 0.6–1 mm wide (Subg. *Mecodium*):

Pinnules and segments undulate, pinnules often drooping; segments shallowly or not at all emarginate; indusial valves unequal, often slightly toothed; blades not gradually narrower at base . . . . *H. lehmannii*

Pinnules and segments flat, not undulate, pinnules not drooping; segments deeply and narrowly notched at apex; indusial valves equal, entire; blades (in ours) gradually narrowed toward base . . . . *H. polyanthos*

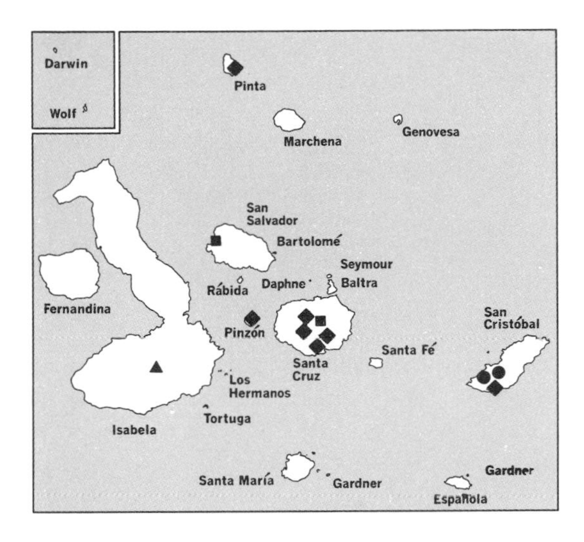

● *Hymenophyllum hirsutum*
■ *Hymenophyllum lehmannii*
▲ *Hymenophyllum polyanthos*
◆ *Trichomanes reptans*

Hymenophyllum hirsutum (L.) Sw., Jour. Bot. Schrad. 1800²: 99. 1802 (as to type)*

Trichomanes hirsutum L., Sp. Pl. 1098. 1753.
Trichomanes ciliatum Sw., Nov. Gen. Sp. Pl. Prodr. 136. 1788.
Hymenophyllum ciliatum Sw., Jour. Bot. Schrad. 1800²: 100. 1802.

Rhizomes filiform, wide-creeping, much branched and in tangled mats, bearing few, distant, pale reddish to brown to tawny, very slender rhizoids; fronds many, suberect to spreading, distant or subdistant, 2–10 cm long or more, the stipes short, broadly alate, the lower part and margins of fronds conspicuously pubescent with coarse, stipitate, forked or stellate, rusty brown hairs; blades narrowly oblong to subovate or deltoid, obtuse to acute at apex, 1–3 cm wide or more, 2–8 cm long or more, deeply bipinnatifid, with wiry, rusty red–pubescent rachises, similar hairs on margins and veins of blade, copiously so on costae and costules; pinnae approximate, spreading to laxly ascending, oblong to rhombic-deltoid, rounded-obtuse, obliquely pinnatifid; pinnules 3–5 pairs, simple, or some larger ones acutely bifid, segments linear to linear-oblong, 0.5–1 mm wide; sori occupying apical part of blade, solitary, with orbicular, convex valves, these free to the rounded base, stellate-ciliate like blade segments; receptacle included.

On tree trunks and fallen logs in humid, deeply shaded habitats, West Indies and southern Mexico to southern Brazil; reported from the tropics of the Old World. This complex needs critical study. Specimens examined: San Cristóbal.

Hymenophyllum lehmannii Hieron., Bot. Jahrb. Engler 34: 428. 1904

Trichomanes pusillum sensu Stewart, Proc. Calif. Acad. Sci. IV, 1: 27. 1911, non Sw. 1788.

Rhizome filiform, bearing very few reddish brown rhizoids near insertion of stipes, creeping extensively along branches of trees and large shrubs, often 5 dm long or more; fronds distant, 1.4–3 cm apart; stipes 8–14 mm long, castaneous, narrowly alate from base upward and bearing a few simple or once-forked rusty brown hairs

* Concerning the typification of *H. hirsutum*, see Morton (1947, 174).

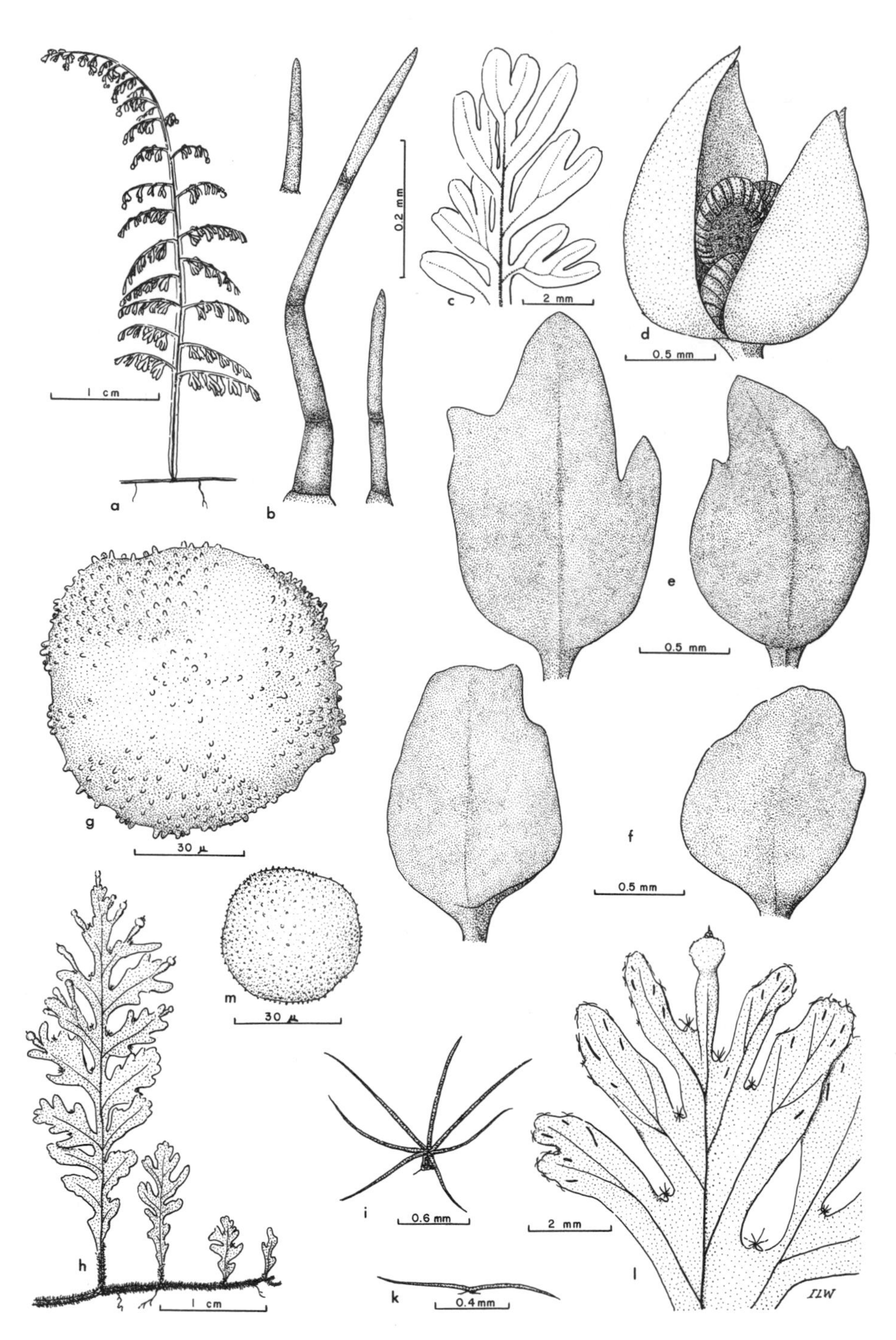

Fig. 5. *Hymenophyllum lehmannii* (*a–g*): *b*, hairs from rhizome; *c*, terminal part of pinna; *d*, indusium; *e, f*, pairs of valves of indusia; *g*, spore.
Trichomanes reptans (*h–m*): *i*, stellate hair from sinus in leaf blade; *k*, bifid hair from margin of frond; *l*, terminal part of pinna; *m*, spore.

near base, nearly or fully glabrous above; blades oblong, 1–1.5(2) cm wide, 2.5–6 cm long, pinnatifid, the pinnules undulate, again lobed or emarginate, drooping; ultimate segments linear-oblong, about 1 mm wide, 2–3 mm long, wings of rachis plane, not undulate, 0.2–0.35 mm wide on each side of rachis, glabrous, light green but turning red in age; veins relatively coarse, only a single branch to each segment, often terminating in slight emargination; sori at tips of sharply drooping or spreading-drooping ultimate segments on upper one-third to half of frond; indusial valves broadly ovate, 1–1.6 mm long, 1–1.2 mm wide, the members of a pair usually unequal, entire to shallowly once- or twice-lobed; receptacle short, stoutish, about 0.2 mm long; sporangia tumidly obovoid, relatively large, 0.5–0.6 mm long, 0.4 mm wide, 0.2–0.3 mm thick, annulus surrounding almost five-sixths of sporangium, bright reddish brown, lustrous, consisting of 20–28 thickened cells, the side walls nearly hyaline; spores trilete, deep green, apparently 128 in each sporangium, about 80 μ in diameter, minutely papillate, the papillae irregularly grouped.

In rain forests in Colombia, on trees and shrubs in fog zone in the Galápagos Islands; distribution elsewhere uncertain. Specimens examined: San Salvador, Santa Cruz.

See Fig. 5.

Hymenophyllum polyanthos (Sw.) Sw., Jour. Bot. Schrad. 1800²: 102. 1802

Trichomanes polyanthos Sw., Nov. Gen. Sp. Pl. Prodr. 137. 1788.

Rhizome long-creeping, dark brown, 0.3–0.4 mm thick; fronds erect, glabrous, subdistant to distant (0.5–6 cm apart), 7–30 cm tall; stipes slender, 1–3.5 cm long, very narrowly alate, reddish brown, with a few widely scattered red-brown hairs near base; blades elliptic-lanceolate, narrowly oblong-lanceolate, or oblanceolate, in ours 6–13 cm long, 1.2–2.2 cm wide at broadest point, gradually narrowed toward base; rachis narrowly alate, the wing 0.2–0.35 mm wide on each side of the dark brown rachis, plane, light yellow-green or reddish near base; pinnae about 10–20 pairs, ascending (not at all drooping), narrowly ovate to oblong, middle ones 9–14 mm long, 4–6 mm wide, pinnatifid or a few inferior segments again lobed, the inferior basal segments distinctly larger than the superior; segments 0.6–1.1 mm wide, 1.5–3(4) mm long, plane, the margins not at all undulate, the tip deeply emarginate (notch 0.2–0.4 mm deep), bright yellow-green, the nerves brownish, solitary in each segment, often ending about 0.2 mm below sinus in sterile segments; indusial valves suborbicular, 0.8–1.2 mm long and wide, entire.

Widespread in tropical regions of both the Old and the New Worlds. Specimen examined: Isabela.

The Galápagos specimen referred to here is not typical, the blades being linear and narrowed toward the base. This species is an aggregate that needs monographic study. Indeed, the whole *Hymenophyllum polyanthos* group needs exhaustive study. The specimens cited under *H. lehmannii* seem to differ from *H. polyanthos*, sensu stricto, in having indusial valves narrower than broad and often shallowly lobulate, in contrast to that of *H. polyanthos*, which has valves nearly circular and strictly entire. Also, the segments in *H. lehmannii* are undulate-crispate, whereas those in *H. polyanthos* are flat and plane. *H. lehmannii* probably is closely related to *H. nigrescens* Liebm.

Trichomanes L., Sp. Pl. 1097. 1753; Gen. Pl. ed. 5, 485. 1754

Rhizome filiform or nearly so, closely attached to trunks and branches of trees, to wet rocks, or to very rocky soil, the rhizomes short-creeping, with erect subfascicu-late or scattered fronds and wiry, numerous rhizoids. Fronds many, erect, spreading or pendent, subsessile to long-stipitate; blades green, small, orbicular, entire and flabellate-veined to elongate and several times pinnatifid to pinnate, glabrous, hir-sute, or glandular, when pubescent the hairs simple, forked, or stellate, borne chiefly to exclusively on margins and along veins; ultimate segments oblong to linear, usu-ally adnate and decurrently joined, entire, truncate to emarginate at apex. Indusium tubular, cylindrical-cuneate, to salverform, truncate and entire or with flaring, en-tire limb or rarely somewhat bivalvate, valves shorter than tube. Receptacle filiform, usually elongate in age, usually exserted.

About 330 species distributed widely in humid habitats, mostly in tropical and subtropical areas.

Trichomanes reptans Sw., Nov. Gen. Sp. Pl. Prodr. 136. 1788*

Hymenophyllum polyanthos sensu Stewart, Proc. Calif. Acad. Sci. IV, 1: 21. 1911, non Sw.
Hymenophyllum hirsutum sensu Stewart, loc. cit., non Sw.
Trichomanes krausii sensu Svens., Bull. Torrey Bot. Club 65: 329. 1938, non Grev. and Hook.

Plants often forming dense mats on rocks or on trunks and branches of trees; fronds greatly variable, broadly oblong to lanceolate-oblong, 2.5–4(9) cm long, 1.5–2.5 cm wide, bipinnate or pinnatifid almost to rachis, rachis wing 1 mm wide or less; stipe 0.5–1.5 cm long, densely tomentose with black hairs or the pubescence sparse; pinnae linear, about 1–1.8 mm wide, unequal, usually with dark stellate hair on short stalk at bottom of each sinus and with a few hairs varying from simple to 3- or 4-branched along margins of segments, in ours when hairs bifid the 2 arms almost parallel to and in same plane with subtending margin of the leaf; costules almost devoid of connected lateral veinlets, false veins often present, but often few and distant from costule, mainly at slight angle to or almost parallel with main costular vein; sori several, on upper and terminal lobes; indusia 3–4.5 mm long, 0.6–0.8 mm wide, partly immersed, slightly winged, the lips semiorbicular, broader than tube, 0.8–1.2 mm wide, narrowly but distinctly dark-margined, erect or spreading-ascending; receptacle bristle-like, exserted in age; sporangia sessile, crowded, light to dark brown, obovoid-discoid, 0.3–0.4 mm long, 0.2–0.3 mm wide, the oblique annulus consisting of 18–24 thickened cells, splitting at right angles to and near end of annulus; spores tetrahedral-spherical, pale greenish yellow, about 35–40 μ in diameter, thin-walled, granulate, the granules short, rounded or blunt.

In humid habitats on trunks of trees, on dead logs, and on bare rocks in deep

* Wessels Boer, in his revision of the American species of the section *Didymoglossum,* listed *Stewart 899* as *T. krausii* and *Howell 9200* as *T. reptans,* but distribution maps 13 and 14 show only *T. reptans* in the Galápagos Islands. The plants represented by the Stewart and Howell collections do not appear different, and it is likely that only one spe-cies, *T. reptans,* occurs in the islands. The differences mentioned in the key by Wessels Boer are rather obscure and may not hold: *T. reptans* is in general larger and coarser, with broader, less lobed or unlobed segments; the segments of *T. krausii* are usually quite narrow and deeply lobed; the stipe in *T. krausii* is short and densely black-hairy, and the blades usually are long-decurrent at the base.

shade, West Indies and Guatemala, southward to Brazil and Bolivia; common in wetter forests of the Galápagos Islands. Specimens examined: Pinta, Pinzón, San Cristóbal, Santa Cruz.

See Fig. 5.

LYCOPODIACEAE. CLUBMOSS FAMILY*

Terrestrial or epiphytic perennial plants of erect, trailing, decumbent, or pendent habit, the stems usually with numerous alternate or repeatedly dichotomous branches, leafy nearly throughout. Leaves in 4 to many ranks, opposite or spiral, simple, small, 1-nerved, more or less continuous with axis, persistent, mostly uniform, usually imbricate. Sporophylls similar to vegetative leaves and not borne in a spike, or bractlike, much modified and compactly imbricate in cylindrical, terminal strobili, the leaves of lower part of stalk of a strobilus reduced or similar to vegetative leaves. Sporangia uniform, 1-celled, 2-valved, mostly reniform or spherical, more or less compressed. Spores uniform, minute, trilete, globose or subglobose, numerous. Gametophyte a fleshy, subterranean, monoecious, tuber-like structure.

Two genera: *Phylloglossum*, consisting of a single species in New Zealand and Australia; and *Lycopodium*, with about 500 species widely distributed from tropical to boreal zones in both the Old and the New Worlds.

Lycopodium L., Sp. Pl. 1100. 1753; Gen. Pl. ed. 5, 486. 1754

Characters of the family. Leaves in 4–24 ranks, usually closely imbricate, from rigidly ascending and appressed to branches to spreading or reflexed, usually decurrent, the members of some ranks in a few species differing in size and shape from those of other ranks. Spores very numerous, sulphur yellow.

Mainly in the temperate regions and at high elevations in the mountains of the tropics.

Plants always terrestrial, the main stems prostrate or arching, bearing rhizophores at intervals, giving off erect branches, these with the dichotomies unequal (apparently monopodial branching); sporophylls unlike foliage leaves in shape, borne in terminal strobili:
- Foliage leaves in only 4 ranks, the 2 of the upper and lower sides much smaller than those of lateral ranks . . . . *L. thyoides*
- Foliage leaves in many ranks, all alike:
  - Strobili short, ellipsoidal, about 1 cm long, sessile; foliage leaves arcuate, not setiferous at apex . . . . *L. cernuum*
  - Strobili long-cylindric, several cm long, 2 to 4 borne on an elongate, bracteate peduncle; foliage leaves straight, bearing a long, apical seta . . . . *L. clavatum*

Plants usually epiphytic (terrestrial in *L. reflexum*), the main stems erect or pendent, not creeping, rooting at base only, the dichotomies subequal; sporophylls similar in shape to leaves of vegetative areas, though sometimes smaller:
- Plants terrestrial or saxicolous; leaves reflexed above middle in age, subulate, 4.5–6 mm long, broadest at base (about 1 mm wide), sides straight, not bisulcate, with several minute marginal teeth, especially near base; almost all leaves fertile, sporophylls exactly similar in shape to sterile leaves . . . . *L. reflexum*

---

* Contributed by C. V. Morton and Ira L. Wiggins.

Plants epiphytic, pendent; leaves erect or spreading, not reflexed, 12 mm long or more, not toothed; spore-bearing leaves confined to ultimate or penultimate branches:
Leaves narrowly linear or almost filiform, 14–25 mm long or more, 0.5–0.7 mm wide and nearly same width throughout, with straight sides, slightly bisulcate abaxially, in 8 (or sometimes 10?) ranks; sporophylls similar to sterile leaves or only slightly smaller .... *L. dichotomum*
Leaves broadly lanceolate, 12–18 mm long, 2.5–3.5 mm wide, broadest near middle, flat, not bisulcate, in 6 ranks, the sides curved; sporophylls obviously smaller than sterile leaves.. *L. passerinoides*

Lycopodium cernuum L., Sp. Pl. 1103. 1753
Lycopodium capillosum Willd. in Spring, Flora 21[1]: 165. 1838.

Plants terrestrial, the main stems woody, terete, 2–3 mm thick, arching and rooting at intervals, producing erect or strongly ascending branches, the secondary branches decurved, spreading and again branching to form a narrowly pyramidal, erect branch system to 6 dm tall (in ours usually 2–4 dm), the ultimate branches ending in sessile cylindric strobili 4–8 mm long; leaves arranged spirally in 16–24 ranks, subulate-attenuate, 2–4.5 mm long, 0.2–0.4 mm wide at base, acicular, herbaceous, strongly convex abaxially, slightly concave to plane adaxially, carinate and unevenly decurrent at base; lateral branches 2–10(15) cm long, bearing crowded leaves like main stem; strobili many, subpendent, mostly 3–8 mm long, 2–3 mm thick; sporophylls ovate-acuminate, 1.5–2 mm long, constricted at base, obliquely ciliolate, closely appressed-imbricate in about 10 ranks, thinner and more yellowish than vegetative leaves; sporangia immersed, 0.5–0.7 mm in diameter, subglobose, the valves unequal; spores spherical, thin-walled, sulphur yellow, subhyaline in transmitted light, their contents more or less granular, about 40 μ in diameter.

Florida to tropical South America and in the Old World; in the Galápagos Islands, common in moderately moist habitats in the *Miconia* scrub and on the slopes and ridges in the Fern-Sedge Zone. Specimens examined: Isabela, San Cristóbal, Santa Cruz.

See Fig. 6.

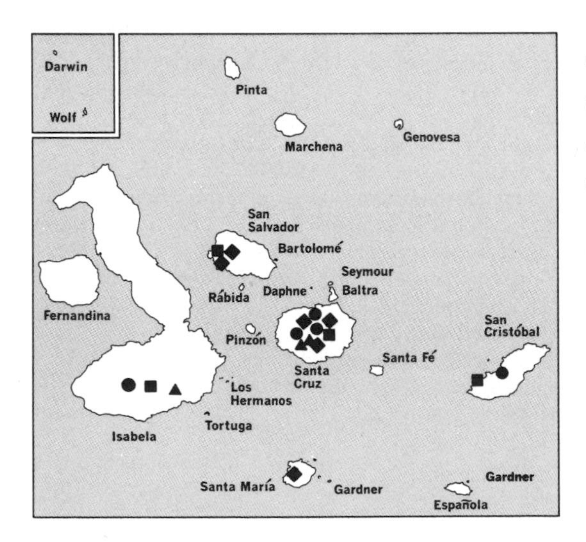

● *Lycopodium cernuum*
■ *Lycopodium clavatum*
▲ *Lycopodium dichotomum*
◆ *Lycopodium passerinoides*

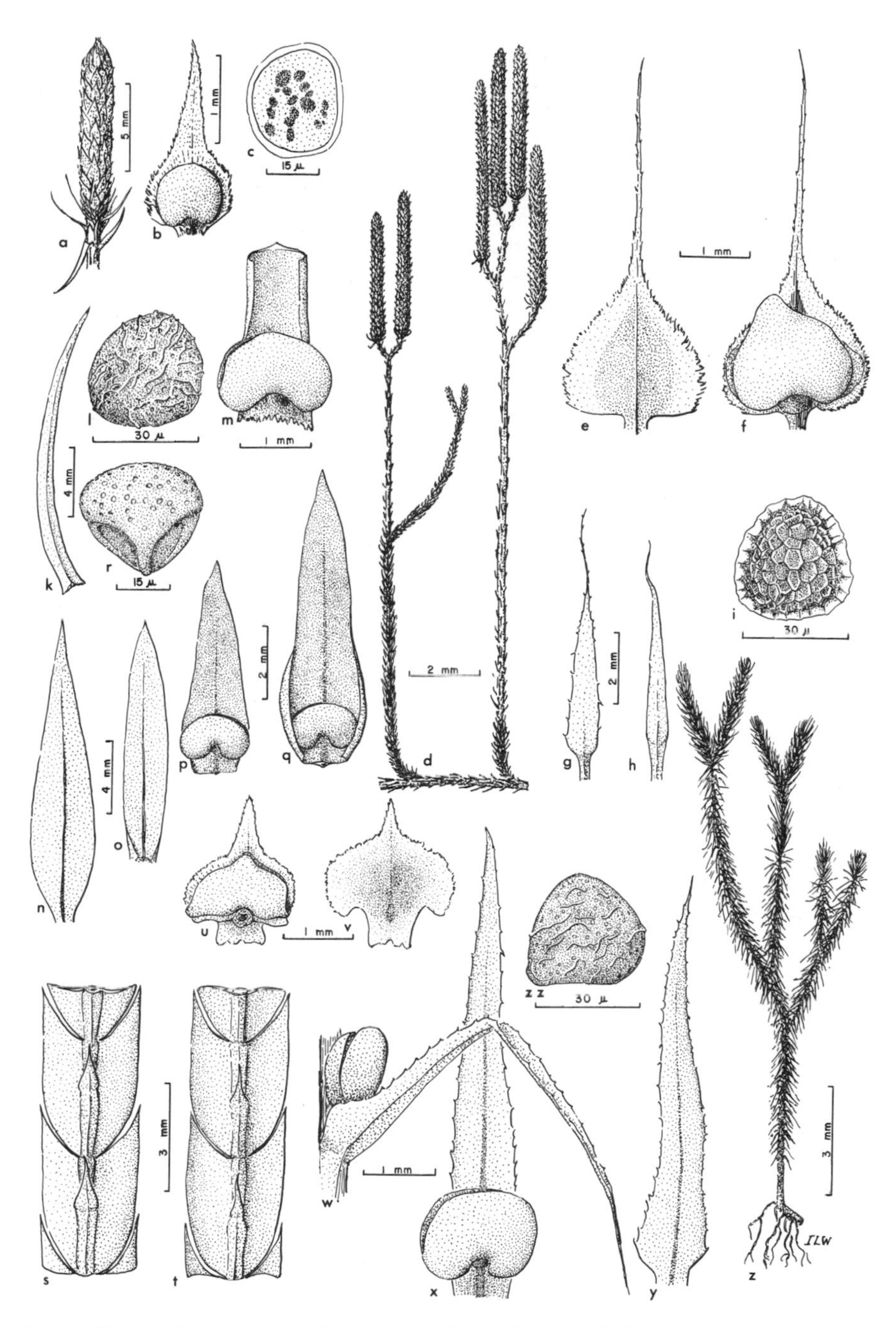

Fig. 6. *Lycopodium cernuum* (*a–c*): *a*, strobilus; *b*, sporophyll and sporangium; *c*, spore. *L. clavatum* (*d–i*): *g*, leaf from main stem; *h*, leaf from aerial branch. *L. dichotomum* (*k–m*): *m*, base of sporophyll and sporangium. *L. passerinoides* (*n–r*): *n, o*, leaves from lower two-thirds of branch. *L. thyoides* (*s–v*): *v*, abaxial side of sporophyll. *L. reflexum* (*w–zz*).

Lycopodium clavatum L., Sp. Pl. 1101. 1753

Plant terrestrial; main stem creeping to decumbent, 1.5–2.5 mm thick, to 1 m long or more, distantly rooted, sparingly branched laterally but giving rise to many erect aerial branches 1–5 dm tall, these again branched freely, the tertiary and subsequent branchlets 1–15 cm long, spreading to ascending; leaves spirally arranged in about 10 ranks, acicular-subulate, those on horizontal main stem 5–8 mm long, 0.6–0.8 mm wide at base, concave adaxially, convex to subcarinate abaxially, decurrent, bearing a white- to horn-colored, slender, somewhat crinkled seta 2–3.5 mm long at apex and 3–8 fragile, slender, spreading teeth 0.1–0.2 mm long along margins; leaves on upright and ascending branches more slender, 0.2–0.6 mm wide at base, their terminal setae 1–2 mm long, marginal teeth few or lacking, all leaves glossy, yellow-green, closely spaced, ascending-spreading, imbricate; stalks of strobili (borne at apices of few to many erect or ascending branches) 3–15 cm long, about 1 mm thick, their leaves 3–4 mm long, including the 2-mm-long, whitish, antrorsely denticulate setae, 0.4–0.6 mm wide at base, closely appressed to stem; strobili mostly 2 on a stalk, but sometimes 3 or 4, each strobilus slenderly cylindric, 3–6 cm long, 3–4 mm thick, the sporophylls closely imbricate in about 8 ranks, broadly ovate-aristate, constricted at base, pale brown in basal central portion, gradually paler to nearly hyaline along fimbriolate-ciliolate margin, 2–2.4 mm long, exclusive of terminal seta, this nearly white or pale yellowish, 0.6–0.8 mm long, smooth, somewhat sinuous or contorted, 1.4–1.6 mm wide just below midpoint; sporangia globose-reniform, about 1 mm wide, smooth; spores pale yellow in masses, pale greenish by transmitted light, spherical to nearly so, about 45 μ in diameter, thin, hyaline reticulations about 5–6 μ high, edges slightly and irregularly scalloped.

In moist to slightly dry habitats, usually in shaded areas, circumboreal in the Northern Hemisphere; Alaska and northern Canada southward to tropical South America in mountainous areas. Specimens examined: Isabela (new record), San Cristóbal, San Salvador, Santa Cruz.

See Fig. 6.

Lycopodium dichotomum Jacq., Enum. Stirp. Vindob. 314. 1762

Lycopodium barbatum Christ, Bull. Herb. Boiss. II, 5: 254. 1905.
Urostachys dichotomum Herter, Beih. Bot. Centralbl. 39: 249. 1922.

Plant epiphytic, 1–6 dm tall, usually pendent from an arcuate base, 2- to 5-times dichotomous, the divisions of about equal thickness, in ours 1–1.25 cm thick except at the slightly narrower apical, sporophyll-bearing region; stems and main branches 2–4 mm thick; leaves closely crowded, imbricate, spirally arranged in 8–12 ranks, spreading to ascending-spreading, linear-subulate, 1.5–2 cm long, 0.6–1.2 mm wide at base, decurrent, rigid, subfalcate, herbaceous, entire, subconvex to nearly flat abaxially, slightly concave adaxially, low-carinate at base, the tip acicular, whitish, straight; sporophylls similar to leaves, or slightly shorter and narrower, sometimes involute; sporangia up to 2 mm wide, much wider than sporophylls, reniform with a deep sinus; spores trilete-spherical, pale yellow in masses, nearly colorless by transmitted light, about 40–45 μ in diameter, finely low-rugulose on the outer, convex surface, nearly smooth on commissural faces.

On trunks and larger branches of forest trees, West Indies to Brazil. Specimens examined: Isabela, Santa Cruz.

See Fig. 6.

Lycopodium passerinoides HBK., Nov. Gen. & Sp. Pl. 1: 41 (folio ed. 33). 1816

Lycopodium taxifolium sensu Stewart, Proc. Calif. Acad. Sci. II, 1: 28. 1911, non Sw.

Plant epiphytic, pendent from larger branches of trees and large shrubs, often forming clumps 2–5 dm across; branches lax, dichotomously branched 2–5 times, 1.5–6 dm long, about 2.5–4 mm thick (exclusive of leaves), ridged by the decurrent bases of leaves, the leaves in 6 ranks, yellow-green, ascending, subappressed, or slightly spreading, nearly straight, 10–18 mm long, 2.5–3 mm wide, broadest near middle, flat but not sulcate, the margins curved, entire, acute at apex; sporophylls similar in shape and texture to vegetative leaves, 5–6 mm long, imbricate, ovate-subulate, the broad base 2–2.5 mm wide, somewhat clasping sporangium, abruptly narrowed to subulate apical portion; sporangia reniform, about 2 mm wide, the sinus fairly deep; spores pale yellow in masses, nearly colorless in transmitted light, 40–48 μ in diameter, verrucose-reticulate, punctate within reticulations on outer, curved face, smooth on commissural faces, trilete ridges sharply marked.

In humid habitats on trees and large shrubs, Venezuela to Peru. Specimens examined: San Salvador, Santa Cruz, Santa María.

See Fig. 6.

Lycopodium reflexum Lam., Encycl. Méth. 3: 653. 1789

Urostachys reflexus Herter, Beih. Bot. Centralbl. 39: 249. 1922.

Plant terrestrial or saxicolous, rigidly erect or sometimes decumbent, 0.6–7 dm tall, simple to several times dichotomous (ours mostly 0.6–3.5 dm tall, simple to twice dichotomous), in larger plants with decumbent bases, the divisions rigidly erect

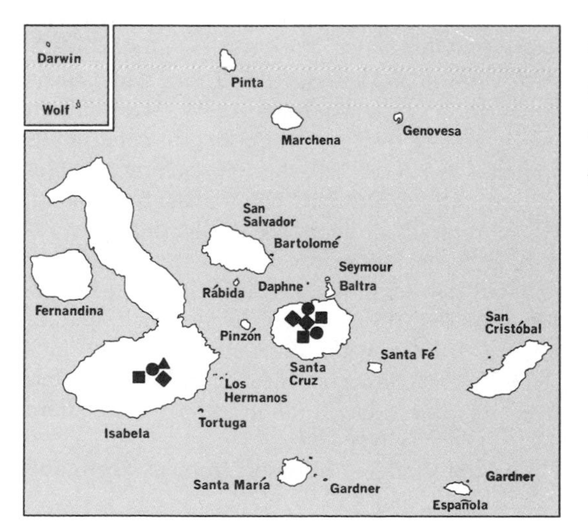

● *Lycopodium reflexum*

■ *Lycopodium thyoides*

▲ *Botrychium underwoodianum*

◆ *Ophioglossum reticulatum*

and coarctate; stems (exclusive of leaves) 1–2.5 mm thick, when including leaves about 10–12 mm across throughout, bearing sporangia almost to base of plant; leaves yellow-green, crowded, spirally arranged in 12–16 crowded ranks, subulate, 4.5–6 mm long, about 1 mm wide at base, tapering gradually to narrowly acute to attenuate, cartilaginous apex, decurved to sharply reflexed from base in age, rigidly herbaceous, plane and lustrous on both faces, stiffly hyaline-denticulate or denticulate-ciliolate along margins, the costa apparent only at very base, short-decurrent; sporophylls like leaves; sporangia transversely oblong-reniform, 1–1.2 mm wide, with small, shallow sinus; spores verrucose-punctate as in *L. passerinoides*, 40–44 μ in diameter, sharply triradiate-ridged on inner face, commissural planes smooth.

Cliffs, steep banks, margins of thickets, at low and moderate elevations, West Indies and Mexico to Bolivia and Paraguay. Specimens examined: Isabela, Santa Cruz.

Possibly often overlooked in the Galápagos Islands, owing to its relatively low stature and its tendency to grow under shrubs and on inaccessible faces of steep ravines and canyons.

See Fig. 6.

Lycopodium thyoides Humb. & Bonpl. ex Willd. in L., Sp. Pl. ed. 4, 5: 18. 1810

Lycopodium complanatum sensu Stewart, Proc. Calif. Acad. Sci. IV, 1: 28. 1911, and sensu Svens., Bull. Torrey Bot. Club 65: 329. 1938, non L.

Plant terrestrial, the main stems creeping or arching, rooting at intervals, 2–3 mm thick, terete, branching freely and spreading over several square meters in favorable situations, their leaves virtually all alike, in 6–8 rows, subappressed but scarcely overlapping, lance-subulate, about 2–4 mm long, low-carinate toward base abaxially, concave adaxially, entire, acute at apex, decurrent; aerial branches erect or ascending, 2–4 dm tall, branching freely, the lateral branches ascending, spreading, or arching-drooping, often somewhat flabelliform in outline; leaves of ultimate branches of 2 kinds, in 4 ranks, all strongly decurrent, the leaves of lateral ranks much wider than those of dorsal and ventral ranks, thus forming flattened sprays, branchlets 2–2.5 mm wide, about half as thick, lateral leaves 3–4 mm long, carinate and convex abaxially, deeply concave adaxially, entire, the tips ascending-spreading, cartilaginous; leaves of dorsal and ventral ranks about 1.5–2 mm wide, those of abaxial side same length as laterals, those of adaxial side only about half to three-fourths as long, narrowly subulate, adnate over half their length; sporophylls borne in terminal strobili on erect branches to 2.5 dm tall, these branching dichotomously 2 to 5 times, strobili erect, coarctate, 2–2.6 mm across, 2–3.5 cm long, stalks terete, ridged, 1–1.5 mm thick; sporophylls brownish centrally, paler toward margins, broadly ovate-attenuate to broadly hastate-ovate, 1.8–2.2 mm long, 1.4–1.8 mm wide, subpeltate on stalks 0.3–0.5 mm long, margins undulate and irregularly lacerate-denticulate, tips ascending or slightly spreading to recurved; sporangia cordate-reniform, 1–1.2 mm high, 1.4–1.6 mm wide, distinctly apiculate, sinus 0.3–0.4 mm deep; spores sharply triradiate on inner surface, rounded on outer. 38–42 μ in diameter, reticulate-foveolate, the ridges around cells of reticulum hyaline, very thin, irregularly crenulate, about 5–6 μ high.

In moist swales and ravines, and in partial shade on hillsides, tropical North and South America. Specimens examined: Isabela, Santa Cruz.

See Fig. 6.

OPHIOGLOSSACEAE. ADDER'S TONGUE FAMILY*

Sporophytes terrestrial herbs (or epiphytic in 1 or 2 species), with short, fleshy, non-scaly subterranean rhizome, and solitary or few fronds straight or slightly bent in vernation (never circinate). Fronds dimorphic or usually consisting of sterile blade similar to sterile frond and an erect, spicate or paniculate, long-stipitate fertile segment, both on same stipe; blade simple, lobed, or variously compound. Sporangia eusporangiate, large, naked, the walls more than 1 cell thick, without annulus, bivalvate and dehiscing by a slit, arranged along nonfoliar pinnae above the photosynthetic, sterile blade, or fused into a continuing spikelike structure with adjacent sporangia separated by nonsporogenous cells, each sporangium producing several hundred thick-walled spores. Prothallia of terrestrial species hypogeous, tuberous, with endophytic mycorhiza, without chlorophyll.

A family of 4 or possibly 5 genera and 50 to 90 species, depending upon one's systematic viewpoint. Widely distributed from the tropics to the arctic regions; represented in the Galápagos Islands by 2 genera, each with a single species.

Sterile blade compound; veins free; sporangia free, borne along both sides of one or more slender pinnae above sterile blade . . . . *Botrychium*
Sterile blade simple; veins anastomosing; sporangia fused in an erect, spikelike segment above sterile blade . . . . *Ophioglossum*

Botrychium Sw., Jour. Bot. Schrad. 1800²: 8, 110. 1802

Fleshy terrestrial plants from erect, small, hypogeous rhizome bearing 2 to many more-or-less corrugated and fleshy roots. Fronds usually 1–3, erect, short to elongate, basal part of stipe hypogeous; sterile blade erect or somewhat curved to bent down, sessile to long-stalked, 1 to 4 times pinnately or ternately divided or compound, free-veined. Sporophylls long-stalked, the basal stipe common with that of sterile blade, the sporangiophore spikelike or 1–5 times pinnate-paniculate, erect, its large globose sporangia distinct, sessile or subsessile, borne in 2 rows on opposite sides of ultimate segments of sporophyll. Spores many, sulphur yellow. Bud for frond of following season enclosed in base of common stipe at apex of rhizome.

About 25 species, native chiefly in temperate regions of both hemispheres, particularly well represented in North America.

Botrychium underwoodianum Maxon, Bull. Torrey Bot. Club 32: 220, *pl. 6.* 1905

Fronds to 32 cm high; sterile stalk to 8 cm long, diverging from fertile stalk 2–3 cm from base of plant; sterile leaf blade pentagonal, to 12 cm long, 16 cm wide; ultimate segments suboval to spatulate, crenate, 5–12 mm long, 3–7 mm wide; venation flabellate; fertile stalk to 25 cm long; fertile leaf blade (fruiting spike) to 7 cm long and 5 cm wide; sporangia yellowish brown, 0.7–1.0 mm in diameter. (Description adapted from Eliasson, 1968, 630.)

Native in Jamaica, Costa Rica, Colombia, and Venezuela, and now known from the Galápagos Islands (Isabela), a new record for the Galápagos Islands.

See Fig. 7.

* Contributed by C. V. Morton and Ira L. Wiggins.

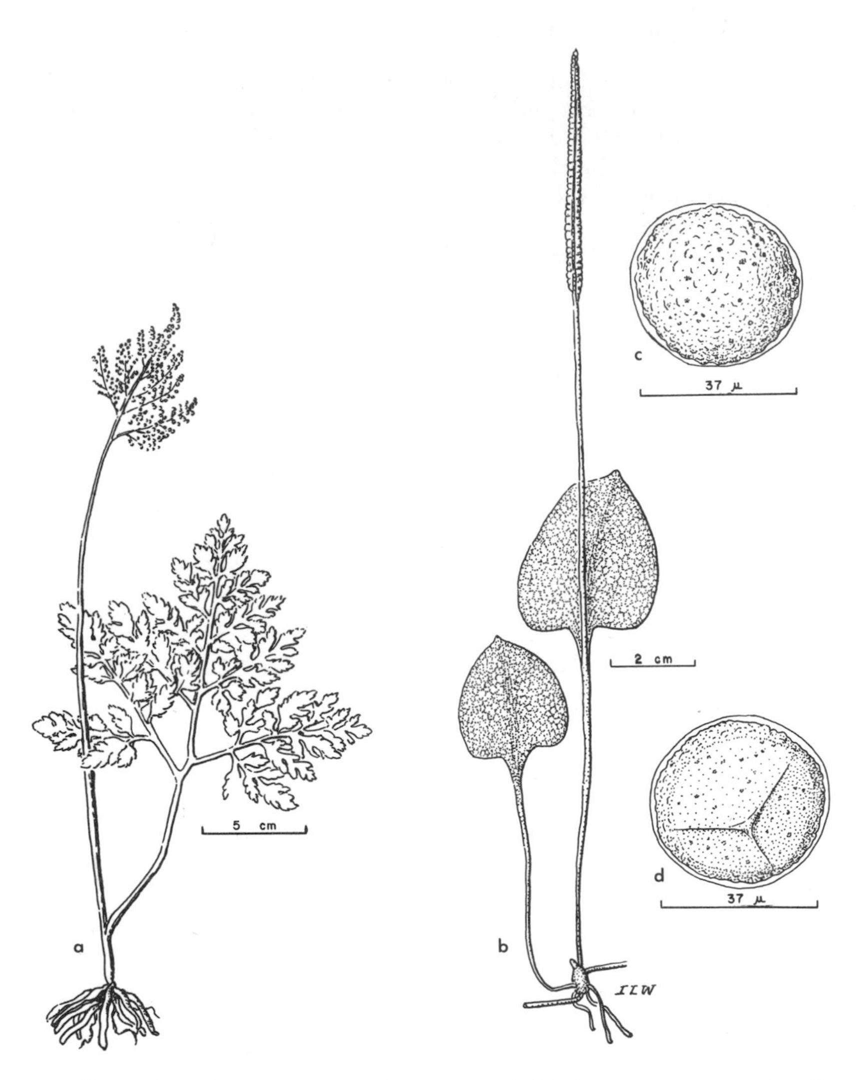

Fig. 7. *Botrychium underwoodianum* (*a*). *Ophioglossum reticulatum* (*b–d*). (Fig. 7*a* drawn from halftone in Bot. Notisser 121: 630–32. 1968.)

Ophioglossum L., Sp. Pl. 1062. 1753; Gen. Pl. ed. 5, 484. 1754

Rhizome usually terminating in erect, exposed bud that functions the following season. Leaves erect in vernation, 1–4 from each rhizome, fleshy, glabrous, arising at side of apical bud. Sterile blade simple, sessile or short-petiolate, linear-lanceolate, ovate, cordate, or reniform, with reticulate venation, the areoles simple or with free and anastomosing veinlets. Common stalk of fertile fronds short to elongate, slender. Sporophylls simple, slender, long-stalked, spicate, the large globose sporangia marginal, coalesced into 2 ranks, transversely dehiscent. Spores spherical, trilete, smooth or essentially so.

A widespread genus of 50 or more rather poorly defined species and subspecies.

Ophioglossum reticulatum L., Sp. Pl. 1063. 1753

Yellowish green plant 5–25 cm tall from short fleshy rhizomes 5–12 mm long, 3–6 mm thick; roots fleshy, 1–2.5 mm thick, sterile fronds usually solitary, the blade smooth, glabrous, ovate to ovate-cordate, 12–16 mm long, 10–22 mm broad, entire, obtuse to shallowly cordate at base, mostly obtuse at apex, the stipe 2–10 cm long, gradually widening near blade; stipe of fertile spike a continuation of that blade, exceeding lower part and rising to 18 cm above sterile lamina; fertile spike to 6 cm long, pale greenish at first, later yellowish, about 4–5 mm wide; spores spherical, pale cream when shed, smooth or nearly so, containing conspicuous oil droplets.

In grassy areas, in forests under fairly dense shade, along trails, and among low herbaceous vegetation; pantropical in distribution. Specimens examined: Isabela, Santa Cruz. Found on San Salvador by Eliasson.

See Fig. 7.

POLYPODIACEAE. FERN FAMILY*

Leafy ferns of various habit with hairy or paleaceous, creeping to erect rhizomes. Fronds spreading to erect, usually with naked, paleaceous or pubescent stipes; blades simple to several times pinnate or pinnatifid, vernation circinate. Sporangia borne on under (abaxial) side of leaf blades, or on special nonfoliose blades, mostly short- to long-stalked, provided with incomplete vertical ring of thickened hygroscopic cells called the annulus, dehiscing transversely. Sori naked or with membranous indusium. Spores smooth to variously spinulose, ridged, or winged, monolete or trilete. Prothallia green, often cordate, readily visible without a lens, at least at maturity.

Widely distributed in all continents, and extending into arctic regions and to snow line in high mountains in the northern hemisphere. Occupying habitats from saturated wet bogs or open water to dry, desert washes and hillsides.†

Rhizomes clothed with hairs only; fronds large, compound, the primary pinnae borne at an acute angle; groove of central rachis continuous above, not opening to receive grooves from rachises of pinnae; sori submarginal (Subfam. Dennstaedtioideae):

- Sporangia borne in an elongate coenosorus on a submarginal vein, protected by incurved margin (false indusium) and an inconspicuous hyaline inner indusium; blades thick in texture; segments large and elongate, tomentose beneath, some connected by low, lunate, costal lobes . . . . *Pteridium*
- Sporangia borne in discrete, small sori, these with or without an inner indusium; blade thin in texture; segments smaller, not elongate, not tomentose beneath, not connected by low, costal lobes:
  - Indusium cuplike, formed from modified margin fused with an inner indusium; blades anadromous throughout; axes not spiny . . . . *Dennstaedtia*
  - Indusium not cuplike, consisting of small, recurved, marginal flap, an inner indusium

---

* Contributed by C. V. Morton and Ira L. Wiggins. Species of Polypodiaceae excluded from those listed from the Galápagos Islands are discussed on pp. 173–74.

† NOTE ADDED IN PROOF: We had assumed that Wood's collection of *Histiopteris incisa* was in fact collected from the mainland, and *Histiopteris* is therefore excluded from the key. New information, however—see the added note on p. 173—makes it clear that *H. incisa* does indeed occur in the islands.

lacking; blades catadromous above basal pair of pinnae; axes sometimes spiny .... *Hypolepis*

Rhizomes clothed with flattened, thin scales, sometimes hairy also:

Pinnae articulate and deciduous individually from rachis, this indicated by an obvious rim around base of each petiolule (Subfam. Nephrolepidioideae).... *Nephrolepis*

Pinnae not articulate:

Stipes of fronds articulate to rhizome, leaving large round scars on rhizome when falling; lamina of blades sometimes with scales on undersurface; veins complexly reticulate (free and once-forked in *Polypodium dispersum*); sori rotund, exindusiate (Subfam. Polypodioideae) .... *Polypodium*

Stipes of fronds not articulate to rhizome; lamina of blades not scaly on surfaces (except in some species of *Elaphoglossum*); veins free (except in *Hemionitis, Doryopteris, Tectaria,* and *Antrophyum*); sori round or elongate; indusia present or absent:

Sterile blades prostrate-spreading, forming a basal rosette, simple, entire, oblong or spatulate; fertile blades erect, pinnate or pinnatifid, much longer than sterile .... *Trachypteris*

Sterile blades erect or ascending, not forming a basal rosette; fertile blades conform, only slightly dimorphic, or strongly dimorphic (in some species of *Blechnum*):

Sporangia not in sori, borne all over dorsal surface of fertile blades; blades simple, entire, coriaceous, bearing scales on lower surface (scales often few and submicroscopic), lacking true hairs (Subfam. Elaphoglossoideae).... .... *Elaphoglossum*

Sporangia not borne all over dorsal surface, borne on veins in lines or in distinct sori but not between veins:

Veins all simple and unbranched; blades simple, minute, merely toothed, sori confluent in contracted terminal part of blade, exindusiate (Subfam. Grammitidioideae) .... *Grammitis*

Veins branched or else the blades compound:

Sori exindusiate, dorsal or submarginal, the sporangia sometimes covered by a reflexed, modified margin (false indusium); spores trilete:

Blades simple and entire; sporangia borne in lines in deep grooves (Subfam. Vittarioideae) .... *Antrophyum*

Blades compound, or if simple then not entire; sporangia not borne in grooves:

Venation catadromous above basal pair of pinnae; sporangia borne on continuous submarginal commissure, covered by reflexed margin; blades pinnate-pinnatifid, basal pair of pinnae with a large, pinnatifid, basiscopic lobe (Subfam. Pteridoideae).......... *Pteris*

Venation anadromous throughout; sporangia not borne on continuous marginal commissure (except sometimes in *Doryopteris*, with simple lobed blades) (Subfam. Adiantoideae):

Sporangia borne on veins on underside of a reflexed modified marginal flap .... *Adiantum*

Sporangia not borne under a reflexed marginal flap:

Sporangia borne at or near ends of veins, often covered by a modified margin:

Blades simple, palmately lobed; venation free or reticulate; sporangia borne in discrete sori or on a more or less continuous submarginal commissure .... *Doryopteris*

Blades pinnate or bipinnate; venation free:

Stipes and rachises grooved above, the edges of grooves with short, stiff, bicellular hairs.................. *Mildella*

Stipes and rachises sulcate to terete, the indument otherwise, if present:
Margins not reflexed, not modified, hardly covering sporangia; blades hairy or yellow-waxy beneath........ .... *Notholaena*
Margins reflexed, modified, whitish; blades not yellow-waxy beneath:
Blades fully bipinnate, the pinnules stalked; pinnules cordate at base, glabrous; axes glabrous, sparingly scaly .... *Pellaea*
Blades bipinnate, the pinnules sessile; pinnules not cordate at base, pilose; axes densely pilose.......... .... *Cheilanthes*
Sporangia borne dorsally along veins, not submarginal:
Blades simple, palmately lobed, pilose beneath; venation reticulate .... *Hemionitis*
Blades bipinnate or tripinnate, glabrous beneath or waxy, not pilose; venation free:
Blades with white or yellow wax beneath (except in juveniles), large and coarse, perennial from large rhizome .... *Pityrogramma*
Blades lacking wax beneath, small and delicate, apparently annual from minute rhizome.............. *Anogramma*
Sori indusiate (indusium sometimes lacking or minute in *Thelypteris* and *Ctenitis*), dorsal, never submarginal, margin never reflexed nor indusium-like; spores monolete:
Sporangia borne in a coenosorus on a continuous, subcostal commissure (Subfam. Blechnoideae) .... *Blechnum*
Sporangia not in a coenosorus, not subcostal:
Sori and indusia elongate along veins; venation anadromous:
Sporangia borne on veins; rhizome scales clathrate, i.e. with broad cells, dark walls, and clear lumina; steles of stipe 2, uniting below base of blade into X-shaped bundle (Subfam. Aspleni-oideae) .... *Asplenium*
Sporangia borne in a line adjacent to veins; rhizome scales fibrous, i.e. with narrow, elongate cells with brown walls, narrow lumina, and pale brown outside walls; steles of stipe 2, uniting below blade into U-shaped bundle (Subfam. Athyrioideae)......... .... *Diplazium*
Sori round, or if elongate (in *Thelypteris linkiana*) then exindusiate:
Rachis above (and blades throughout) lacking hairs, but sometimes with glands; venation anadromous:
Plants usually climbing trees; rhizomes elongate, thick, densely scaly, with distant, distichous fronds; blades not scaly..... .... *Rumohra*
Plants terrestrial; rhizome erect, with fasciculate, polystichous fronds; blades scaly (Subfam. Dryopteridoideae):
Blade sparingly scaly, densely glandular; segments not spinulose; indusia reniform..................... *Dryopteris*
Blade densely scaly, not glandular; segments spinulose; indusia absent .... *Polystichum*
Rachis above pubescent, and blades usually more or less pubescent elsewhere also; venation catadromous above basal pair of pinnae:

Hairs of rachis and costae above acicular, 1-celled; veins reaching margin; sporangia sometimes setose (Subfam. Thelypteridoideae) .... *Thelypteris*

Hairs of rachis and costae above not acicular, jointed, many-celled; rhizome scales not ciliate; veins not reaching margin; sporangia not setose (Subfam. Tectarioideae):

Veins all free; blades bipinnate or tripinnate, hairy and scaly .... *Ctenitis*

Veins complexly anastomosing; blades once-pinnate, basal pair of pinnae with 1 or several basiscopic lobes, nearly glabrous, scaleless .... *Tectaria*

Adiantum L., Sp. Pl. 1094. 1753; Gen. Pl. ed. 5, 485. 1754

Terrestrial ferns of moist, shaded habitats; rhizomes thickish and nearly erect to slender and wide-creeping, paleaceous. Fronds distichous or borne in several ranks, few to many, suberect to drooping; stipes firm, often nearly wiry, dark brown to black, usually lustrous. Blades pinnate to quintipinnate or decompound, rachises usually dark, polished, dissection of pinnae or segments extremely variable; pinnules glabrous, minutely puberulent or distantly stellate-paleaceous, often more or less glaucous beneath, sessile or petiolulate, membranous or coriaceous, sometimes articulate and in some species deciduous. Veins free or casually anastomosing, their ultimate branches terminating in tooth or sinus. Sori marginal, sporangia borne along and sometimes between ends of ultimate veinlets, extending into the membranous, sharply reflexed lobes of margins or false indusia, sometimes whole margin so modified. Sporangia with pedicel 2 or 3 cells thick, the annulus usually of about 18–22 thickened cells. Spores dark, smooth to shallowly foveolate or irregularly rugulose or both, trilete, tetrahedral or globose.

About 200 species, widely distributed, represented particularly well in the American tropics, and less extensively far into the temperate zones.*

Blades simple-pinnate; pinnae usually 4–6 pairs, 4–8 cm long, 2–5 cm wide, subsessile; sori elongate, 2–5 cm long, continuous on upper or upper and lower margins; rachis glabrous; veins running to teeth in sterile pinnae .... *A. macrophyllum*

Blades pedate or bipinnate to decompound; ultimate segments rarely over 2 cm long, usually shorter, mostly petiolulate; sori continuous or discontinuous, usually orbicular-reniform and separated along margin of segments; rachis often glabrous, glandular-pilosulous, or rusty-pilose:

Blades pedate, basiscopically developed, the lowest pinna forked 2 or 3 times, with solitary pinnule below each fork; rachis glabrous; veins run to teeth..... *A. patens*

Blades bipinnate, not pedate; rachis glabrous or variously pubescent:

Veins in sterile pinnae terminating in the teeth; rachises and petiolules obviously rusty-pilose .... *A. villosum*

Veins in sterile pinnae terminating in the sinuses; rachises and petiolules glabrous or glandular-pilosulous, not rusty-pilose:

Rachis and petiolules glabrous; lowest superior pinnule equally subtruncate at base, not dimidiate .... *A. concinnum*

Rachis and petiolules at least slightly glandular-pilosulous; lowest superior pinnule usually unequal at base, obviously dimidiate.............. *A. henslovianum*

---

* Three species reported from the Galápagos Islands but not collected recently are not included in the key. They are briefly described following the description of *A. villosum*.

Adiantum concinnum Humb. & Bonpl. ex Willd. in L., Sp. Pl. ed. 4, 5: 451. 1810

Adiantum cuneatum sensu Hook. f., Trans. Linn. Soc. Lond. 20: 168. 1847, non Langsd. & Fisch.
Adiantum aethiopicum sensu Moore, Ind. Fil. 19. 1857, non L.
Adiantum henslovianum sensu Stewart, Proc. Calif. Acad. Sci. IV, 1: 11. 1911, pro parte, non Hook. f.

Rhizome stout, short-creeping, to 5 mm thick, closely invested with uniformly bright brown, lance-acicular, entire scales; fronds few to several, sometimes fasciculate, usually more or less drooping, 2–10 dm long, the stipes polished, glabrous, castaneous to atropurpureous, 0.5–2 mm thick, rather fragile, firmly terete at maturity; blades lanceolate, lance-oblong, or triangular-oblong, 1.5–7 dm long, 6–25 cm wide, bi- to tripinnate, usually attenuate at apex, the rachis straight or faintly zigzag, glabrous, polished like stipe; pinnae many, laxly ascending, subsessile, alternate, rather closely arranged, mostly triangular-oblong, 6–15 cm long, 2–5 cm wide, the basal pinnule of each often reduced and overlapping rachis; segments membrano-papyraceous, yellowish green to dark green, glabrous, rhombic-obovate to broadly trapeziform, short-petiolulate, not articulate, 5–15 mm long, incised-lobed, the fertile lobes deeply emarginate, bearing orbicular-reniform sorus at base of sinus between adjacent lobes; indusia membrano-papyraceous, dull, glabrous, pale brown around center, margins nearly white, irregularly crenulate to entire; sporangia rufous-brown, dull, 8–40 sporangia in sorus; spores subspherical, 50–60 μ in diameter, smooth, pale reddish brown.

Moist to dryish rocky banks, under shrubs, and around shading rocks, West Indies and Mexico to tropical South America. Specimens examined: Fernandina, Isabela, Pinta, San Salvador, Santa Cruz, Santa María.

In the juvenile stage *A. concinnum* is sometimes difficult to distinguish from *A. henslovianum*. However, if specimens are examined under a lens of high magnification, *A. henslovianum* is usually set apart by its slightly glandular-pilosulous rachis and petiolules.

Adiantum henslovianum Hook. f., Trans. Linn. Soc. Lond. 20: 169. 1847

Adiantum parvulum Hook. f., op. cit. 168.
Adiantum concinnum sensu Stewart, Proc. Calif. Acad. Sci. IV, 1: 11. 1911, pro parte, non Humb. & Bonpl.
Adiantum diaphanum sensu Stewart, loc. cit., non Blume.

Rhizomes 3–6 mm thick, short-creeping to subascending, densely clothed with dark brown, sublustrous, lance-acicular scales 1.5–3 mm long, their margins entire; fronds few to several, more or less fasciculate, 1–6 dm tall, stipes sometimes about equaling blades, firm, 1.5–2.5 mm thick, deep brownish red to atropurpureous, canaliculate along upper side, glabrous to faintly and sparsely glandular-puberulent; blades ovate-triangular, 10–35 cm long, 10–30 cm wide, bi- or tripinnate; primary pinnae 6–30, gradually reduced upward, the middle and upper ones simple-pinnate, the basal triangular, the others linear-oblong, 10–20 cm long, mostly 1.5–5 cm wide at base, acute, the secondary pinnules lanceolate, 4–7 cm long, about 2 cm wide; rachises more or less finely glandular-puberulent, or sometimes almost glabrous; segments membranous-papyraceous, pale yellow-green to deep green, rhombic-lunate or some of the lower suborbicular, delicately short-petiolulate, the upper base often overlapping rachis, lower margins entire, upper denticulate in sterile fronds, crenulate in fruiting ones, all glabrous except at extreme base and on petiolules where minutely glandular-puberulent, or in some very sparsely glandular-pilosulous on undersurface; veins free, forking 1–4 times, termi-

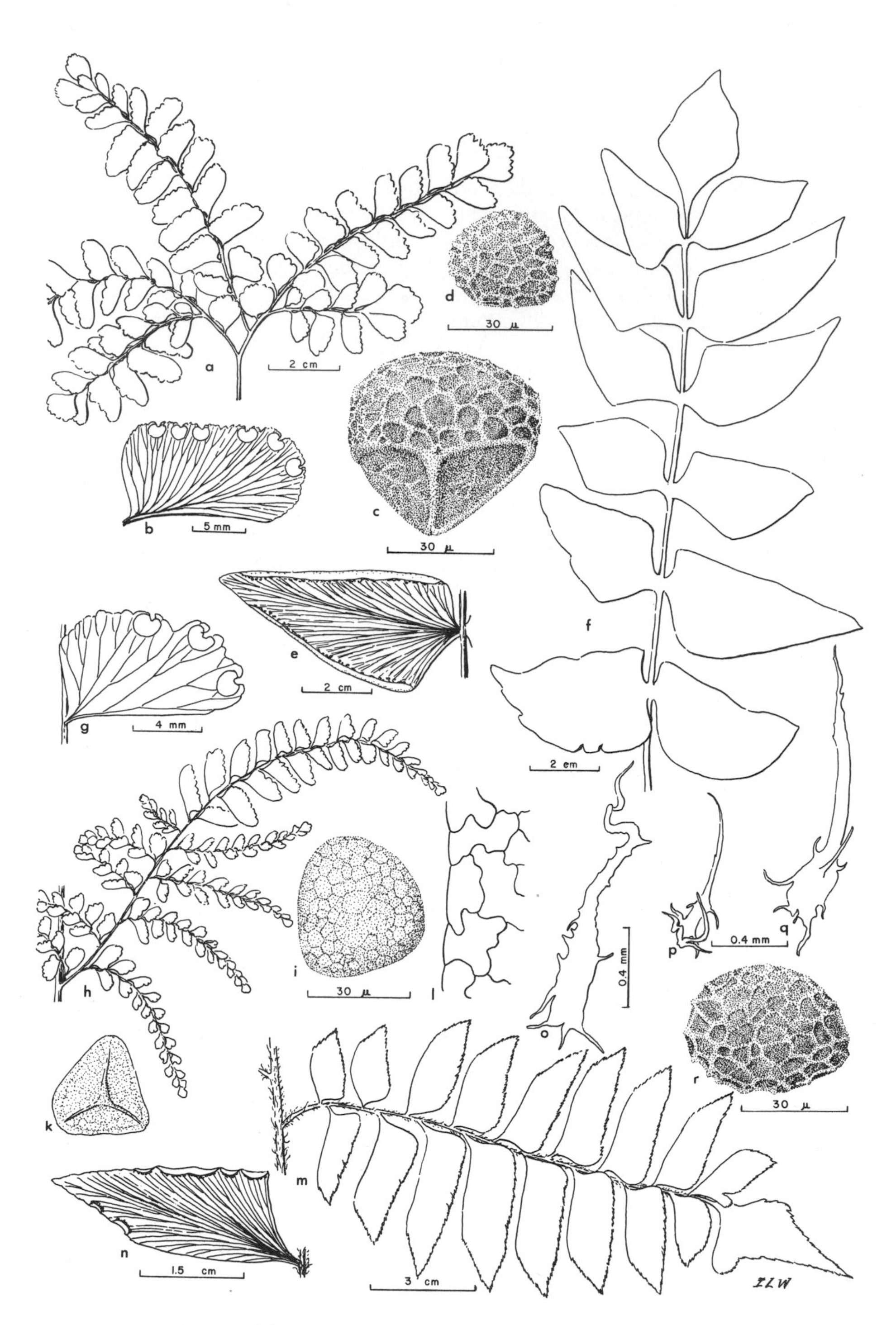

Fig. 8. *Adiantum patens* (*a–c*). *A. macrophyllum* (*d–f*). *A. henslovianum* (*g–l*). *A. villosum* (*m–r*).

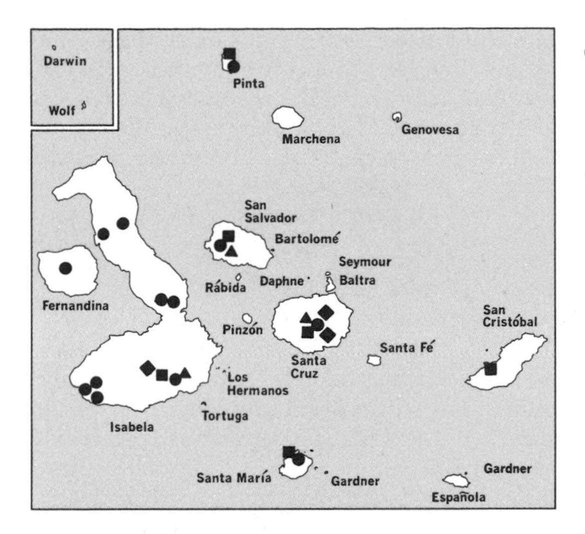

nating in minute sinuses between denticles of margin; sori lunate, 1.5–2 mm long, mostly 2–5 on a segment; indusia membranous, all except margin sordid white to greenish white, moderately invested with short, gland-tipped hairs and a few scattered, spreading, much longer nonglandular hairs, the margin pale brown, tightly curled under, very thin (1 cell thick); sporangia 0.3–0.4 mm long, annulus pale reddish brown, sublustrous, sides pale, sordid white, dull; spores trigonous-orbicular, 35–60 μ in diameter, reddish brown by transmitted light, shallowly reticulate-foveolate.

In moist habitats in shade of trees, shrubs, and rocks, and in caves; rarely on the mainland, in Venezuela and Ecuador. Specimens examined: Isabela, Pinta, San Cristóbal, San Salvador, Santa Cruz, Santa María.

See Fig. 8.

Adiantum macrophyllum Sw., Nov. Gen. Sp. Pl. Prodr. 135. 1788

Rhizomes stout, short-creeping, 4–10 mm thick, 2–8 cm long, closely paleaceous with dark brown, dull scales 4–8 mm long; fronds few, closely distichous, erect-spreading, 1.5–8 dm tall, the stipes shining black, 8–45 cm long, 1.5–2.5 mm thick, glabrous; blades ovate to oblong, 10–30 cm long, 8–20 cm wide, simple-pinnate; pinnae 3–8 pairs, plus terminal pinna, firmly papyraceous, the young bright red, broadly ovate to suborbicular and entire when first expanded, subglabrous beneath, mostly opposite, the upper of mature fertile blades narrowly triangular from inequilateral rectangular base, lower base straight and more or less perpendicular to rachis, upper base often overlapping rachis, sessile, 4–10 cm long, 2–5 cm wide, acute to acuminate at apex, the sterile pinnae deltoid-cordate, often coarsely toothed or incised along upper side or along both sides on distal half to two-thirds, usually acuminate, the basal 1–3 pairs broadly triangular from rounded to subtruncate base, short-petiolulate or sessile; costae obscure, subcentral, all veins closely crowded, free, repeatedly dichotomous; sori continuous in heavy line along upper margin or both margins, rarely reaching apex of pinnule; modified, reflexed margins 1–2.5 mm wide, soon pushed outward by expanding sporangia, these tumidly obovoid, 0.25–

0.3 mm long, annulus reddish brown, sublustrous, sides gray-brown, dull; spores subspherical to trigonous, coarsely and irregularly reticulate-rugulose with low, broad ridges, and microscopically foveolate within reticulae, about 40 μ in diameter.

Moist, shady ravines, forests, and margins of densely shrubby areas, West Indies and Mexico to northern South America. Specimens examined: Isabela, San Salvador, Santa Cruz.

See Fig. 8 and color plate 67.

Adiantum patens Willd. in L., Sp. Pl. ed. 4, 5: 439. 1810

Adiantum diaphanum sensu Stewart, Proc. Calif. Acad. Sci. IV, 1: 11. 1911, pro parte, non Blume.
Adiantum henslovianum sensu Stewart, loc. cit., pro parte, non Hook. f.

Rhizome slender, 1–1.5 mm thick, long-creeping, closely clothed with reddish brown to castaneous, lanceolate, sublustrous scales 1–2.5 mm long, base usually cordate to hastate-caudate; fronds few, 1.5–4 dm tall, stipes brownish red, shining, glabrous except for a few scales on basal 1–2 cm, about 0.75–1 mm thick; blades when young often merely ternate instead of pedate, lower pinnae nearly as long as rest of blade, basiscopically developed, the lowest pinna forked 1 to 3 times and bearing solitary pinnule on distal side just below each fork, the main pinnae oblong, 5–15 cm long, 2.5–4 cm wide; pinnules thin-papyraceous, short-petiolulate, rhombic-rectangular to obovate, nearly horizontal, 13–20 mm long, 7–10 mm wide, glabrous, upper base usually overlapping rachis, apex rounded to subtruncate, proximal margin entire, upper margin crenate, terminal pinnule obovate-cuneate, truncate, rounded or shallowly lobed at apex; veins forking 3–6 times, each ultimate veinlet ending in minute tooth; sori round-reniform, situated in sinuses of crenulations, 1.2–1.8 mm in diameter, the reflexed, modified margin membranous, brownish around sinus, the rest pale to nearly white; sporangia yellowish brown, relatively few in a sorus; spores orbicular-trigonous, 60–65 μ in diameter, faintly rugulose-foveolate, with scattered areas of minute, blunt but slender papillae near commissural planes.

In moist habitats under shrubs, trees, and rocks, and on seepy slopes, Mexico to northern South America. Specimens examined: Isabela, Santa Cruz.

See Fig. 8.

Adiantum villosum L., Syst. Nat. ed. 10, 2: 1328. 1759

Adiantum petiolatum sensu Stewart, Proc. Calif. Acad. Sci. IV, 1: 12. 1911, non Desv.

Rhizomes markedly nodose, short-creeping, 5–8 mm thick, paleaceous with dull to sublustrous, dark brown acicular-lanceolate scales; fronds rather crowded, mainly in 2 rows, rigidly ascending, 5–10 dm tall, the stipes stout, angulate, deeply atropurpureous, polished, 3.5–6 dm long, deciduously rusty-furfuraceous or invested with rusty lance-acicular scales, these attenuate at apex, irregularly long-toothed along margins and on upper surface near base; blades ovate to transversely oblong, except for the elongate terminal pinna, 25–50 cm long, 25–40 cm wide, bipinnate; rachises closely covered with same type of narrow scales as stipes, these less deciduous than those on stipe; lateral pinnae 2–6 pairs; chiefly alternate, spreading to laxly ascending, linear to lance-oblong, up to 30 cm long; sterile pinnules oblong-trapeziform, 2–4 cm long, 8–15 mm wide, spreading, more or less imbricate, upper margin sharply biserrulate, lower margin entire, slightly concave-curved, short-petiolulate to subsessile, the terminal pinnule unequally ovate, usually rounded-auriculate on one or both sides, attenuate, to 5 cm long; fertile pinnules 15–20 mm long, 5–7 mm wide, thinner and more lustrous than the sterile, closely crowded, short-petiolulate, dimidiate-oblong to narrowly rhombic, lower margin nearly straight and entire, apex abruptly acute to acuminate, upper margin shallowly crenulate to subentire, often the segments bearing sori somewhat arcuate, some pinnules with superior auricle; sori along upper margin and around apex, continuous or more or less interrupted and arcuate, about 1 mm wide when mature; indusia thin but rather firm, pale brown, margins irregularly and minutely denticulate or merely sinuate; sporangia pale brown, dull, the stalks as long as body or slightly longer; spores subspherical, 40–60 μ in diameter, very shallowly and irregularly reticulate-foveolate, reddish brown by transmitted light.

Shaded rocky banks, around boulders, and under shrubs and trees, West Indies and Mexico to northern South America. Specimens examined: Santa Cruz.

Adiantum villosum typically has the indusium continuous along the upper edge of the segments and around the apex. But though the Galápagos specimens sometimes show this character, they more often have the sorus broken up into more or less discrete sori, thus approaching *A. tetraphyllum* Humb. & Bonpl.

See Fig. 8.

ADIANTUM SPECIES EXCLUDED FROM KEY

Adiantum alarconianum Gaud., Voy. Bonite Bot. *t.* 99. 1846

Adiantum incisum Presl, Rel. Haenk. 1: 61, *pl. 10, f. 3.* 1825, non Forsk. 1775.

Reported from the Galápagos Islands by Moore, Ind. Fil. 28, 1857, without a collector's name. The species is rare on the mainland, in Ecuador. It is simply pinnate like *A. macrophyllum*, but the pinnae are deeply incised.

Adiantum petiolatum Desv., Mag. Ges. Naturforsch. Freund. Berlin 5: 326. 1811

Reported from the Galápagos Islands by Moore (op. cit. 29), from the collection of Wood. The species will run to *A. macrophyllum* in the key presented here, but the pinnae are smaller and relatively narrower and more finely toothed, glaucescent beneath, and the sori small and discrete.

Adiantum tetraphyllum Humb. & Bonpl. ex Willd. in L., Sp. Pl. ed. 4, 5: 441. 1810

Reported by Moore (op. cit. 34), from the Galápagos Islands, collected by Wood. It is common on the mainland but not known otherwise from the Galápagos Islands. It is near *A. villosum*, from which it differs especially in having the sori discrete rather than almost or entirely continuous.

Anogramma Link, Fil. Spec. 137. 1841

Small, terrestrial ferns with very small, inconspicuous rhizomes bearing reddish, slender scales, these gradually grading into hairs at base of stipe. Fronds few, tufted; lamina small, bi- or tripinnate. Pinnules slightly decurrent, incised, glabrous or sparsely villous, membranous, with forked, free veins. Sporangia borne in lines along the forking veins; paraphyses and indusia none; annulus of 20–22 thickened cells. Spores tetrahedral, faintly ribbed to nearly smooth.

About 6 species with wide distribution in tropical and subtropical, humid areas; circumequatorial.

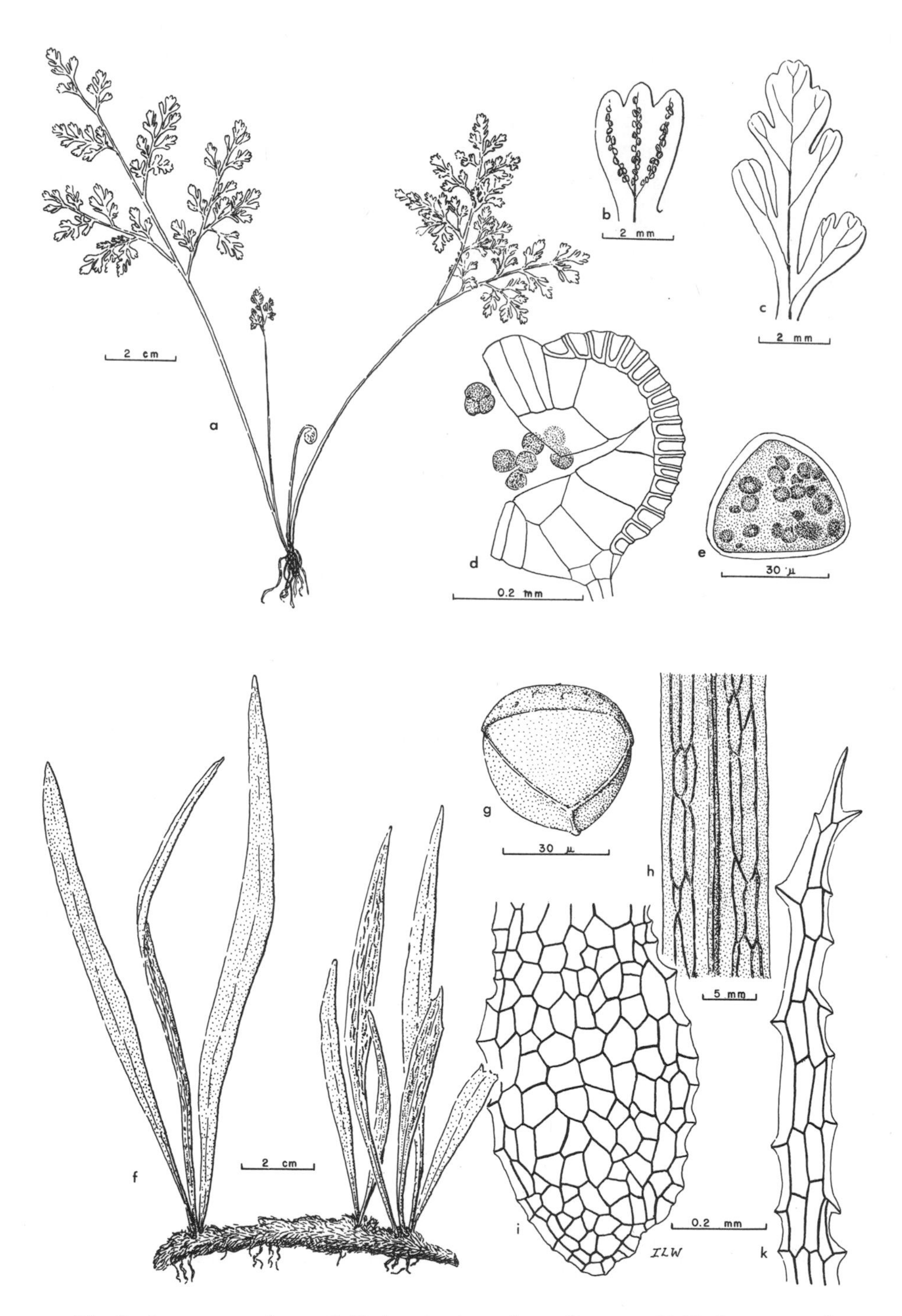

Fig. 9. *Anogramma chaerophylla* (*a–e*). *Antrophyum lineatum* (*f–k*): *h*, pattern of sporangia on blade; *i*, cell arrangement in base of rhizome scale; *k*, tip of same scale.

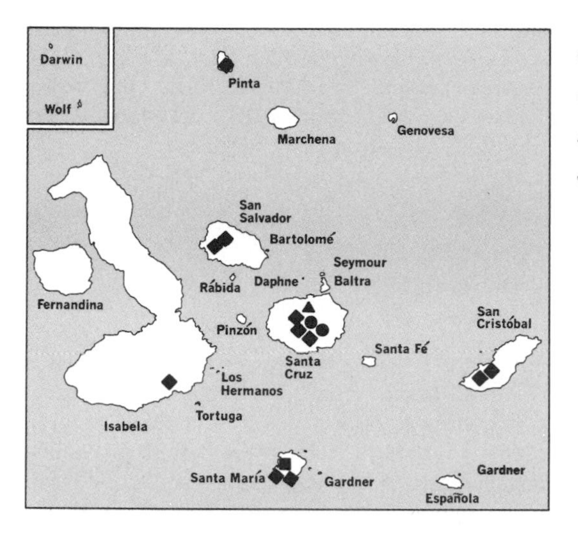

● *Adiantum villosum*
■ *Anogramma chaerophylla*
▲ *Antrophyum lineatum*
◆ *Asplenium auritum* var. *auriculatum*

Anogramma chaerophylla (Desv.) Link, Fil. Spec. 137. 1841

Gymnogramma chaerophylla Desv., Mag. Ges. Naturforsch. Freund. Berlin 5: 305. 1811.
Gymnogramma leptophylla sensu Robins., Proc. Amer. Acad. 38: 109. 1902, non Desv.
Anogramma leptophylla sensu Svens., Bull. Torrey Bot. Club 65: 308. 1938, non Link.

Small, delicate, apparently annual fern with very short, slender, erect or ascending rhizomes 2–3 mm long, these bearing very fragile, slender scales 1–2 mm long, and numerous delicate rootlets; fronds erect, 10–15 cm tall, stipes slender, at first greenish, later stramineous, finally reddish brown, at least at base, sparingly villous with slender, few-celled simple hairs, soon glabrate; blades broadly ovate to deltoid, 4–8 cm long, 2.5–6 cm wide at base, tripinnate, the pinnules cuneate-obovate, 2–3 mm long and 0.6–2 mm wide, thin, entire, emarginate or 3- or 4-lobed at apex, the ultimate segments obtuse to rounded or rarely subacute at apex; veins branching to provide single vein to each segment, free; sporangia nearly globose, 0.1–0.2 mm in diameter, pale rufous brown, arranged in dichotomous lines along veins of segments; indusium none; spores often more or less triangular, about 50–55 μ in diameter, containing conspicuous oil droplets or granules, the surface smooth.

Widely distributed in semi-arid habitats in tropical America, but inconspicuous, often under the overhang of rocks or in caves and potholes, and easily overlooked. Specimens examined: "Galápagos Islands" without further locality data and without a collecting number, and Santa María.

See Fig. 9.

Antrophyum Kaulf., Enum. Fil. 197. 1824

Epiphytic or saxicolous plants, very rarely on soil; rhizomes relatively stout, short-creeping to suberect, usually covered with a dense mass of slender, hairy roots and bases of old fronds, their tips bearing closely crowded, delicate, deciduous, clathrate, brown scales, their cell walls smooth or papillose. Fronds more or less cespitose, glabrous, membranous, coriaceous, or fleshy, sessile or on alate stipes; costae complete or vestigial; secondary venation reticulate, forming long, costal areolae and shorter, divergent lateral ones, closed along margin or open, the closed without included veinlets. Sporangia in simple to branching lines, mostly along the longitudinal veins, free or somewhat interconnected to completely reticulate on all veins in mature fronds, superficial or immersed in deep, narrow grooves. Indusia none. Paraphyses present and variable in shape, or absent (wanting in ours). Sporangia tumidly obovoid to subglobose, dark brown, annulus with 14–18 thick-walled cells. Spores trilete or monolete, smooth or slightly and irregularly rugulose.

About 30 species widely distributed in the tropics of the Old and New Worlds.

Antrophyum lineatum (Sw.) Kaulf., Enum. Fil. 199. 1824

Hemionitis lineata Sw., Nov. Gen. Sp. Pl. Prodr. 129. 1788.
Loxogramme lineata Presl, Tent. 215. 1836.
Polytaenium lineatum J. Smith, Jour. Bot. Hook. 4: 68. 1841.

Rhizomes 4–5 mm thick, but appearing larger, owing to persistent roots and bases of old fronds; scales narrowly lanceolate, 2–4 mm long, 0.6–1.2 mm wide at base, acute to slightly attenuate apically, gray-brown, sublustrous, the clathrae running parallel to axis, margins slightly irregular; fronds in fascicles of 3–6 or more, fascicles 3–8 cm apart, erect, blades linear-elliptic to narrowly lance-elliptic, 5–18 cm long, 5–12 mm wide, acute or sometimes shallowly bifurcate at apex, gradually narrowed to subsessile base, subcoriaceous, margins entire, deep green above, slightly yellowish brown beneath, costa visible but not prominent, impressed above, more evident and pale brownish beneath, lateral veins obscure except as indicated by soral lines beneath, tissue firm, sublustrous and bearing scattered, thin, white, appressed, peltate squamellae on upper surface; soral lines obvious on undersurface, 2–4 on each side of costa, forming narrow reticulae 3–24 mm long on larger fronds, darker brown than adjacent tissue; sporangia completely hidden in grooves until nearly mature, slightly compressed laterally, 0.25–0.3 mm long, on very short stalks less than 0.1 mm long; annulus of 14–18 heavily thickened cells; spores dark brown, about 40–50 μ in diameter, outer surface irregularly rugulose with very fine, low ridges.

On trees and shaded moist rocks, West Indies and Mexico, south to Bolivia at low and medium elevations. Specimen examined: Santa Cruz, the first record of this genus from the Galápagos Islands.

See Fig. 9.

Asplenium L., Sp. Pl. 1078. 1753; Gen. Pl. ed. 5, 485. 1754

Small, delicate to large terrestrial, rock-inhabiting or epiphytic ferns of moist habitats with short-creeping to erect, scaly rhizomes, the scales with dark cross walls between cells. Fronds more or less clustered, erect or erect-spreading, not articulated to rhizomes, all alike or rarely slightly dimorphic. Blades simple to quadripinnate, or variously pinnatifid; rachises dark and lustrous to green and succulent, often narrowly winged or margined. Pinnae or segments toothed, incised, or less commonly entire, glabrous or rarely puberulous or pilose. Veins free and forked, or rarely somewhat anastomosing. Sori oval to narrowly linear, distinct, nearly straight, borne on the free, usually oblique, ultimate veins. Indusia present, lateral, as long as sori, mostly membranous. Sporangia ovoid to subglobose, with 18–28 thickened cells in annulus. Spores monolete, perisporiate, bilateral, generally rugulose or cristate, sometimes spinulose or granulose between crests.

About 650 species widely distributed, chiefly in tropical and subtropical regions, in moist wooded or rocky areas, many species mainly epiphytic.*

Blades membranous to chartaceous; veins and sori not at a very acute angle to the costae; rachises scaleless and glabrous or nearly so (Sect. *Asplenium*):

Stipe and base of rachis dark or atropurpureous and shining; blades simple-pinnate, less than 2.5 cm wide, the upper margin of the pinnae more or less incised. . . . . *A. formosum* var. *carolinum*

Stipe and base of rachis green, stramineous, or reddish, not dark; blades simple-pinnate to tripinnate, more than 2.5 cm wide:

Blades simple-pinnate with a conform terminal pinna; indusia rather short and broad, 2–4 times as long as broad, whitish; plants terrestrial. . . . . *A. feei*

Blades simple-pinnate, terminating in a pinnatifid apex rather than a conform terminal pinna, or bi- or tripinnate:

Blades bipinnate or tripinnate; rachis green-alate; blades thin, bright green, the pinnae subsessile, the ultimate segments linear or oblong. . . . . *A. cristatum*

Blades simple-pinnate:

Plants epiphytic; indusia thick, greenish white. . . . . *A. auritum* var. *auriculatum*

Plants terrestrial; indusia thin:

Blades oblong to linear, the stipes much shorter than the blades; lower pinnae not larger or basiscopically developed; blades glabrous. . . . . *A. otites*

Blades triangular in outline, the stipes much longer than the blades; lower pinnae (or segments) the largest, basiscopically developed; blades sep-tate-hairy . . . . *A. pumilum*

Blades chartaceous to subcoriaceous; veins and sori at a very acute angle to the costae or sometimes parallel to them; rachises sometimes scaly; blades simple-pinnate or subbipinnate (Sect. *Sphenopteris*):

Rhizomes erect; sori decidedly spreading; stipes and rachises scaly; pinnae obviously scaly beneath; blades subbipinnate. . . . . *A. praemorsum*

Rhizomes creeping; sori costal or slightly spreading; stipes and rachises only slightly scaly; pinnae not scaly beneath; blades simple-pinnate. . . . *A. serra* var. *imrayanum*

Asplenium auritum Sw. var. **auriculatum** (Hook. f.) Morton & Lellinger, Mem. N.Y. Bot. Gard. 15: 19. 1966

Asplenium marinum L. var. *auriculatum* Hook. f., Trans. Linn. Soc. Lond. 20: 170. 1847.
Asplenium lunulatum sensu Robins., Proc. Amer. Acad. 38: 107. 1902, non Sw.
Asplenium sulcatum sensu Stewart, Proc. Calif. Acad. Sci. IV, 1:15. 1911, non Lam.

Epiphyte; rhizomes short-creeping, often 1–2 cm long, closely enmeshed in tangle of dark brown, slender rootlets and feltlike mass of dark brown, sparingly septate slender hairs; fronds fasciculate, erect or ascending, 1.5–4 dm long, the stipe one-half to two-thirds as long as blade, canaliculate adaxially, the channel flanked by narrow green line on each side almost to base of stipe, with very narrow lance-subulate to nearly hairlike crinkled scales when young, later glabrate; blade oblong-lanceolate, acuminate, not reduced at base, 1.5–3 dm long, mostly 5–12 cm wide, the rachis canaliculate and green on upper side, darker (to nearly black) on abaxial side, glabrous or with a few scattered, hairlike scales when young; pinnae thick-herbaceous, bluish green, opaque, 10–20 pairs, subsessile or those on basal third of frond short-

* Three species reported from the Galápagos but not collected recently are not included in the key. They are briefly described following the description of *A. serra*.

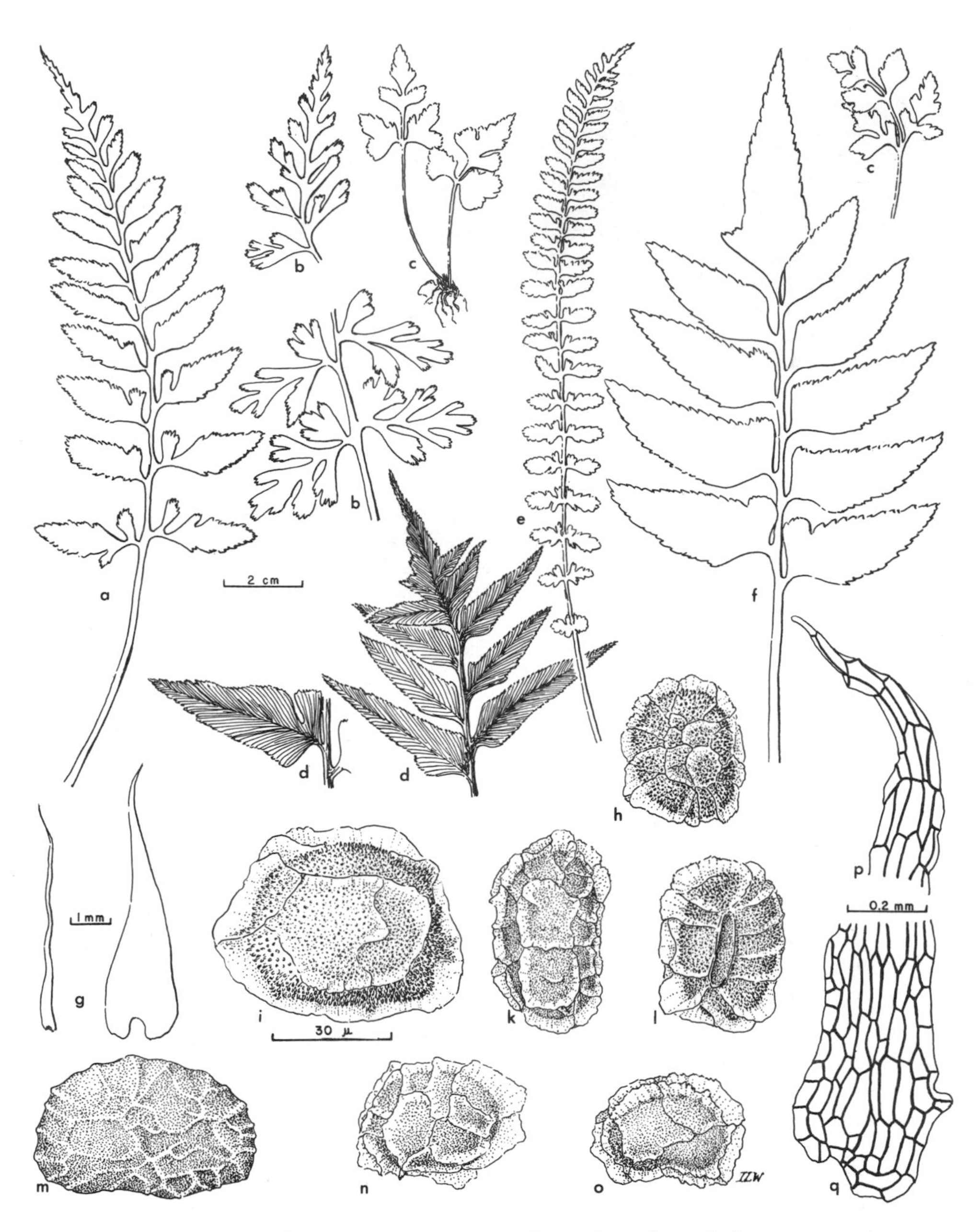

Fig. 10. *Asplenium auritum* var. *auriculatum* (*a, g, k, p, q*): *k*, spore; *p, q*, apical and basal parts of scale. *A. cristatum* (*o*). *A. feei* (*f, i*). *A. formosum* var. *carolinum* (*e, n*). *A. praemorsum* (*b, m*). *A. pumilum* (*c, h*). *A. serra* (*d, l*).

petiolulate, rounded to acute at apex, unequal at base, the proximal side usually cuneate, distal side with simple, basal, adnate auricle toothed at its apex, rest of pinnae faintly toothed or sinuate as well as toothed; veins 1-forked near costa, inconspicuous, the tip of each branch appearing as a narrowly elliptical or oblanceolate-linear, white ridge on upper (adaxial) surface of pinnae, this ridge turning black in age; sori oblique, 2–3 mm long, 1.5–2 mm wide at maturity, reddish brown, rather closely crowded on many fronds; indusia 0.6–0.8 mm wide, thickish, greenish white, entire; sporangia numerous and crowded, 0.3–0.4 mm long, about half as wide, pale reddish brown, annulus of 18–22 thickened cells; spores dark brown, oblong to subreniform, about 40 μ thick, 60–70 μ long, coarsely reticulate with low, relatively thick ridges, the inter-ridge areas not spinulose.

In forests of tropical America, the exact distribution undetermined, but known from Colombia, Venezuela and Ecuador. Specimens examined: Isabela, Pinta, San Cristóbal, San Salvador, Santa Cruz, Santa María.

Although predominantly epiphytic, plants of this species occasionally occur on moist, moss-covered rocks in shaded areas and in caverns in the lava.

See Fig. 10.

Asplenium cristatum Lam., Encycl. Méth. 2: 310. 1786

Asplenium cicutarium Sw., Nov. Gen. Sp. Pl. Prodr. 130. 1788.
Asplenium macraei sensu Hook. f., Trans. Linn. Soc. Lond. 20: 169. 1847, non Grev. & Hook.
Asplenium myriophyllum sensu Stewart, Proc. Calif. Acad. Sci. IV, 1: 14. 1911, and sensu Christens. in Christoph., Nyt Mag. Naturvid. 70: 69. 1932, non (Sw.) Presl.

Terrestrial, ours mostly in caverns; rhizomes decumbent to erect, 5–10 mm thick, 2–8 cm long, usually encased by numerous remnants of stipes from previous seasons, the tip covered with dark castaneous, lance-attenuate, cordate-based, rigid scales; fronds 5–20, cespitose, suberect, 2–7.5 dm tall, the stipes from one-half to as long as blades, olive-gray, or reddish brown to very drak brown, narrowly green-margined near blades, these dark green, herbaceous-membranous, lanceolate to deltoid-oblong, 1–4 dm long, 5–16 cm wide, acuminate at apex, acuminate to abruptly truncate at base, tripinnate or the pinnules pinnatifid; rachis greenish, fuscous, or olive-brown, slightly greenish-alate toward apex; pinnae up to 20 pairs or more, alternate, close to subdistant, straight or slightly arcuate, oblong-lanceolate, 3–9 cm long, 1–2.5 cm wide, acute to acuminate, sessile, the superior basal lobe usually overlapping rachis; pinnules 6–15 pairs, alternate, spreading, obliquely deltoid-oblong, ovate-oblong, or oblong, obtuse at apex, coarsely pinnatifid to pinnatisect, variable in dentation or lobing, degree of adnation, and depth of incising; veinlets solitary in ultimate segments, fairly evident, free; sori short, oval to elliptical, 1–2 mm long; indusia membranous, whitish, dull; spores dark brown, broadly ellipsoidal or oblong, 42–50 μ thick, 55–65 μ long, coarsely reticulate-winged, the surfaces within reticulations smooth to microscopically papillose-spinulose.

In pockets of rich soil on moist slopes among rocks or along water courses and in caverns in dense shade, West Indies and Mexico to Brazil and Bolivia. Specimens examined: Isabela, Pinta, San Cristóbal, San Salvador, Santa Cruz.

Among plants collected on different islands, there is considerable variation in the size of the spores, in the relative coarseness or fineness of the wings, and in the reticulations.

See Fig. 10.

Asplenium feei Kunze ex Fée, Gen. Fil. 194. 1852

Asplenium nigrescens Hook. f., Trans. Linn. Soc. Lond. 20: 170. 1847, non Blume.
Asplenium sanguinolentum Kunze ex Mett., Abhandl. Senckenb. Naturf. Gesell. 3: 192, *t. 4, f. 10.* 1858.
Asplenium nubilum Moore, Ind. Fil. 150. 1859. A new name for *A. nigrescens* Hook. f., non Blume.
Asplenium anisophyllum var. *latifolium* Hook., Sp. Fil. 3: 111. 1860, pro parte with respect to Galápagos specimens.
Asplenium serra sensu Stewart, Proc. Calif. Acad. Sci. IV, 1: 15. 1911, pro parte.

Terrestrial; rhizomes short and stout to rather elongate, densely paleaceous near apex with castaneous lanceolate-acuminate scales 3–5 mm long; fronds 5–15, fasciculate, 2–9 dm tall, the stipes about half as long as blades, greenish brown, scaly at base, somewhat fibrillose upward; blades oblong, 8–17 cm wide, simple-pinnate, the rachis green, narrowly winged; pinnae 6–14 pairs, petiolulate, broadly lanceolate, 5–7 cm long, usually 1.5–2 cm wide, attenuate or acute at apex, the base unequal, narrowly cuneate on lower side, roundish-auriculate on upper, crenate-serrate to biserrate; veins rather inconspicuous, usually forked once near costa, sometimes again forked toward margin; sori oblong to elliptic, 3–6 mm long, 2–3 mm wide; indusia broad, membranous, pale green to nearly white, vaulted over sporangia, margins entire; sporangia numerous, crowded, ovoid, 0.4–0.5 mm long, half as wide, slightly compressed laterally, shining, the annulus dark brown, of 20–24 thickened cells, lateral walls much paler, stalks 1 cell in thickness, 2–4 times as long as body; spores dark brown, oblong, 55–60 μ thick, 75–85 μ long, coarsely reticulate with thin, nearly hyaline wings 6–10 μ high, the areas within reticulations minutely spinulose with slender papillae 2–4 μ long.

In shaded areas among shrubs and trees or sometimes in nearly full sun, Mexico and Cuba to Brazil and Bolivia. Specimens examined: Isabela, San Cristóbal, San Salvador, Santa Cruz.

See Fig. 10.

Asplenium formosum Willd. var. **carolinum** (Maxon) Morton, Amer. Fern Jour. 59: 66. 1969

Asplenium "subulatum" sensu Hook. f., Trans. Linn. Soc. Lond. 20: 169. 1847, non *A. subalatum* Grev. & Hook.
Asplenium "farinosum" sensu Robins. & Greenm., Amer. Jour. Sci. III, 1: 149. 1895, non *A. formosum* Willd.
Asplenium carolinum Maxon, Contr. U.S. Nat. Herb. 17: 148. 1913.

Terrestrial; rhizomes ascending to erect, woody, 4–5 mm thick, 1–3 cm long, more or less sheathed by bases of old stipes, the tips paleaceous with linear-lanceolate scales 1.5–2.5 mm long, their margins pale yellowish brown, median stripe dark brown and lustrous; fronds few to many, 5–40 cm long, ascending, leaf tissue firm to chartaceous, dull green, minutely puberulous beneath; stipes short, to 1 mm thick, subterete, narrowly sulcate with narrowly alate margins on adaxial side; blades linear-lanceolate, 4–22 cm long, 1.2–2 cm wide, pinnate, acute at apex, gradually reduced below; pinnae numerous, spreading, adjacent to distant, the lowermost minute, the next deltoid, deeply parted, the middle 6–12 mm long, 3–5 mm wide, oblong, the base cuneate to subrectangular, subauriculate at superior basal margin, auricle bi- to quadridentate, the rest of upper margin crenate-serrate to obliquely incised, crenations sometimes faintly bidentate, lower margin entire to obliquely crenate; veins 4–6 pairs per pinna, strongly oblique, the superior basal vein 1- to 3-forked, the others usually simple, all nearly concealed; sori 2 or 3 per pinna, borne mostly on outer half to two-thirds of pinnae, elliptical, oblique, tumid,

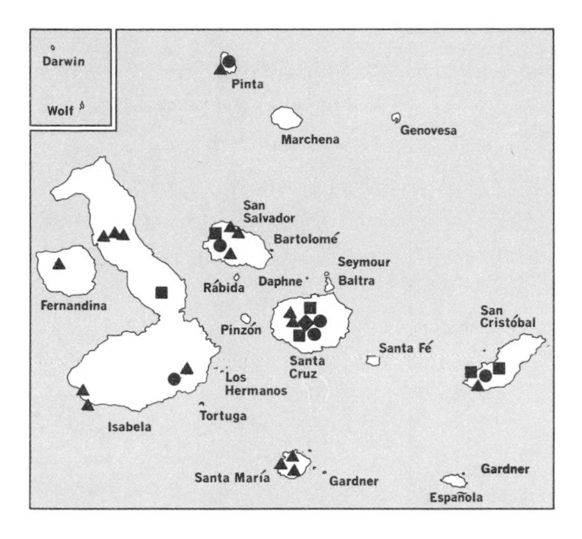

- ● *Asplenium cristatum*
- ■ *Asplenium feei*
- ▲ *Asplenium formosum* var. *carolinum*
- ◆ *Asplenium otites*

1.2–2.5 mm long; indusia firm-chartaceous, whitish; sporangia crowded, deep reddish brown, sublustrous; spores reddish brown, about 40 μ thick, 53–60 μ long, coarsely reticulate-alate, wings hyaline, the areas between wings smooth to irregularly and microscopically spinulose.

Endemic; among rocks, in crevices in the lava, on moist cliffs, and in caverns in moist habitats. Specimens examined: Fernandina, Isabela, Pinta, San Cristóbal, San Salvador, Santa Cruz, Santa María.

See Fig. 10.

Asplenium otites Link, Hort. Reg. Bot. Berol. Descr. 2: 00. 1833

Asplenium poloense Rosenst., Rep. Spec. Nov. Fedde 12: 469. 1913.

In rocky soil and on mossy rocks; rhizomes short, erect; fronds few, fasciculate, to 16 cm tall, usually about 3 cm broad, often smaller; stipes brownish, narrowly ridged, about one-fourth as long as blade; rachis brownish, narrowly ridged, nonproliferous; blades pinnate, with 12–14 pairs of pinnae, these ascending and more or less falcate, oblong-subrhombic, obtuse, distinctly dimidiate, lower margin straight and entire, upper margin sharply and irregularly few-dentate, teeth 6–8, somewhat hooked, blade reduced upwardly to pinnatifid apex; veins 5–7 on upper side, the lowermost 2–4 1-forked or occasionally 2-forked, the 3–5 veins of lower side all simple; sori medial, few, 3–7 per pinna.

In shaded, moist habitats from Honduras to Peru. Specimens examined: Santa Cruz.

Asplenium praemorsum Sw., Nov. Gen. Sp. Pl. Prodr. 130. 1788

Asplenium nigricans Kunze, Linnaea 9: 69. 1834.
Asplenium furcatum sensu Robins., Proc. Amer. Acad. 38: 107. 1902, non Thunb.

Epiphytic or terrestrial; rhizomes erect or nearly so, 5–10 mm thick, closely invested by lance-attenuate, minutely and irregularly denticulate scales, these dark brown to black, lustrous, the margins sometimes narrowly and slightly paler brown than median area, base usually cordate-caudate; fronds few to several, ascending to stiffly errect; stipes and rachises dark brown to black, with tan to dark brown, nar-

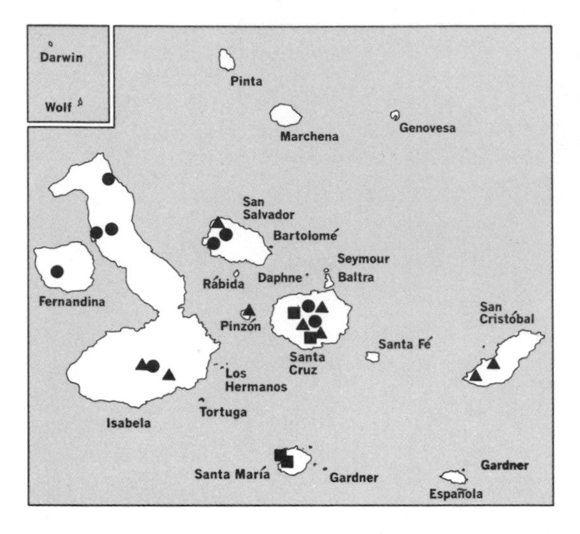

row wings, copiously supplied with spreading dark scales with filiform-attenuate apices; stipes 5–15 cm long, 1–1.5 mm thick; blades triangular-ovate, 5–35 cm long, 2–7 cm wide, bipinnate to bipinnate-pinnatifid below, pinnate above; pinnae 16–30, opposite to alternate, to 3.5 cm long, to 2.5 cm wide, inequilaterally quadrangular, incised into linear, obtriangular to obovate segments, these irregularly dentate at apex; veins 2- or 3-forked, free, each veinlet terminating in blunt, short tooth; sori linear to narrowly oblong, 3–7 mm long, arranged in narrowly palmate pattern, 3–5 per segment; indusia thin, nearly membranous, pale brown, dull, the margins entire to faintly sinuous; sporangia tumidly obovoid, about 0.4 mm long, 0.2–0.3 mm wide, annulus dark reddish brown and lustrous, side walls pale gray-brown, dull to sublustrous; spores reddish brown, broadly subreniform, 75–82 μ long, 48–55 μ thick, coarsely reticulate with very low, narrow ridges, smooth within reticulations.

In shaded, moist habitats in the forest and upper shrubby zones, West Indies and Mexico to Brazil and Bolivia. Specimens examined: Fernandina, Isabela, San Salvador, Santa Cruz.

See Fig. 10.

Asplenium pumilum Sw., Nov. Gen. Sp. Pl. Prodr. 129. 1788

Asplenium anthriscifolium Jacq., Coll. Bot. 2: 103, *pl. 2, f. 3, 4.* 1788.
Asplenium humile Spreng., Neu. Entd. 3: 6. 1822.

Terrestrial, usually in leaf mold; rhizomes 2–5 mm thick, 3–10 mm long, erect, bearing tuft of small, brown-costate, filiform to acicular scales at apex; fronds few to several, cespitose, 5–20 cm tall, the stipes nearly always longer than blades, deciduously dark-fibrillose, castaneous abaxially, greenish and canaliculate adaxially, the channel with thin, nearly membranous greenish margins; blades soft-herbaceous, whitish hirsutulous on both sides, variable, in smaller forms 3-lobed or 3-parted, deltoid-ovate, acute, 3–5 cm long, 2.5–4 cm wide, basal divisions obliquely ovate and subentire to irregularly dentate, the apical deltoid-ovate from cuneate base, subentire, dentate to lobed about base; larger blades broadly pentagonal, 5–12 cm long, 3–5 cm wide, with 2–4 pairs of oblique-spreading pinnae, these acute to acuminate at apex, obliquely incised, the basal ones petiolulate, unequally deltoid; sori rather numerous, linear, nearly medial, to 1 cm long; indusia thin, whitish, deciduously ciliate to shallowly fimbriate-lacerate; spores broadly ellipsoidal, reddish brown, about 55 μ long, 40 μ thick, coarsely reticulate-alate, the areas between wings inconspicuously papillate-spinulose.

On dryish to moist, shaded, rocky banks, Florida, West Indies and Mexico to Brazil. Specimens examined: Santa Cruz, Santa María.

See Fig. 10.

Asplenium serra Langsd. & Fisch. var. **imrayanum** Hook., Sp. Fil. 3: 156. 1866

Epiphytic and terrestrial in rich humus; rhizomes creeping, woody, 1–2 cm thick, 5–20 cm long, invested with tufted brown, iridescent, strongly clathrate, lance-attenuate to filiform scales 5–15 mm long; fronds distichous, erect-spreading, 1.5–6 dm long; stipes subquadrangular, brown to atropurpureous, deciduously paleaceous, 10–25 cm long, 1–2 mm thick; blades oblong-acuminate, 1–3 dm long, 5–12 cm wide, pinnate, the stipe rachis-like; pinnae few to many, horizontal or oblique, alternate or subopposite, short-petiolulate, subdistant, lance-ovate to lance-attenuate, 2–8 cm long, 1–2.5 cm wide, unequally cuneate at base, distal side usually slightly broader; costae fibrillose-paleaceous to nearly glabrous beneath; veins oblique, 2- to 4-forked, glabrescent, brownish; margins of pinnae finely serrate to triply serrate or inciso-serrate; sori 3–12 mm long, at sharp angle to costa, inframedial, imbricate or forming single row along each side of costa; indusia firm-membranous, pale brown, margin more or less crenulate; sporangia dark brown, crowded, the annulus of 18–24 thickened cells, lustrous; spores broadly ellipsoidal, 45–50 μ long, 30–35 μ thick, coarsely reticulate-alate, dark brown, microscopically papillate-spinulose within reticulations, wings pale brown to subhyaline, irregular.

On trunks of trees and in rich humus on forest floor, at middle and higher elevations, Lesser Antilles and possibly on the mainland adjacent to the Galápagos Islands. Specimens examined: Isabela, Pinzón, San Cristóbal, San Salvador, Santa Cruz.

Typical *Asplenium serra* is widespread and variable in continental North and South America. Hooker's var. *imrayanum* differs in having the mature pinnae larger and broader, with the lower base more strongly curved and excised, and the sori diverging at a broader angle from the costae. It occurs primarily in the Lesser Antilles and on the Galápagos Islands, but may also occur on the continent. The largest and most typical specimen from the Galápagos Islands is *Stewart 856* (CAS), which is quite immature. Most of the Galápagos specimens represent small plants that are evidently depauperate; they were probably taken from relatively xeric, unfavorable habitats.

See Fig. 10.

ASPLENIUM SPECIES EXCLUDED FROM KEY

Asplenium laetum Sw., Syn. Fil. 79, 271. 1806

Reported from the Galápagos Islands from the Captain Wood collection, but not found since. It is somewhat similar to *A. otites,* but is larger, with dark rachis and stipe, and the pinnae have microscopic white hairs beneath when young.

Asplenium rutaceum (Willd.) Mett., Abhandl. Senckenb. Naturf. Gesell. 3: 129, *t. 5, f. 32, 33.* 1858

Aspidium rutaceum Willd. in L., Sp. Pl. ed. 4, 5: 266. 1810.

Reported from the Galápagos Islands by Hooker and Baker (Synopsis Filicum 220. 1867), but not found by recent collectors. It differs from all others of the Galápagos *Asplenium* specimens in having the apex of the frond flagelliform and radicant.

Asplenium serratum L., Sp. Pl. 1079. 1753

Reported from the Galápagos Islands on the basis of a Wood collection, but not collected recently. This differs from the other Galápagos species in having the blades quite simple and not at all pinnate or lobed.

Blechnum L., Sp. Pl. 1077. 1753; Gen. Pl. ed. 5, 485. 1754

Terrestrial or climbing ferns, mainly in open areas or among open stands of shrubs or in open forest, with paleaceous erect or creeping rhizomes, often stoloniferous. Fronds small to large, mostly cespitose, not articulate to rhizome, the blades simple and entire to simple-pinnate. Pinnae distinct or confluent at base, sometimes articulate, nearly uniform, or the fertile slightly to much narrower than the sterile. Veins forked, branches in sterile pinnules free (in ours), parallel, excurrent. Sori elongate-linear, usually continuous, borne near or against costa on elongate transverse veinlet connecting adjacent main veins. Indusia linear, continuous, facing costa, at first closely applied to costa, at length pushed outward by the enlarging sporangia. Sporangia forming heavy double line near costa. Spores monolete.

About 190 species, with wide distribution throughout the New World and the Old.

Sterile and fertile fronds almost alike, the fertile with a broad marginal sterile band; pinnae not more than 6 cm long (Subg. *Blechnum*):
- Blades pinnate at base, the lower pinnae nearly or quite free from rachis; largest pinnae usually 3–6 cm long; lowest pinnae not or scarcely smaller than the middle; rachis usually obviously glandular-pilose beneath........*B. occidentale* var. *puberulum*
- Blades pinnatisect, all segments fully adnate at base to rachis, without a superior auricle; largest segments mostly not over 2.5 cm long; lowest segments obviously smaller than the middle; rachis glabrous or sometimes minutely glandular-pilosulous..... .... *B. polypodioides*

Sterile and fertile fronds dissimilar, the fertile with pinnae contracted and almost fully fertile, lacking any sterile marginal band (Subg. *Lomaria*):
- Sterile blades 35–75 cm long, fully pinnate at base, the pinnae not at all adnate; pinnae elongate (10–23 cm), acuminate, serrate above middle...........*B. falciforme*
- Sterile blades 20–35 cm long, deeply pinnatisect only, all segments fully adnate throughout; segments short (4 cm long or less) and rounded at apex, entire..*B. lehmannii*

Blechnum falciforme (Liebm.) C. Chr., Ind. Fil. 154. 1905

Lomaria falciformis Liebm., Vidensk. Selsk. Skr. V, 1: 11. 1849.
Acrostichum aureum sensu Stewart, Proc. Calif. Acad. Sci. IV, 1: 11. 1911, non L.
Struthiopteris falciformis Broadh., Bull. Torrey Bot. Club 39: 365. 1912.

Terrestrial fern with erect or ascending rhizome invested with lanceolate to ovate-lanceolate scales 1–3 cm long; stipes cespitose, 30–80 cm long, often angulate, stramineous to reddish brown, bearing varying number of stramineous to reddish scales; blades 35–80 cm long, 15–32 cm wide, oblong, abruptly reduced at base, without vestigial pinnae, gradually reduced toward apex, the rachis with scales like those on stipe; pinnae 18–32 pairs, linear to lanceolate, falcate to nearly straight, the lower more strongly curved than the upper, apex acuminate, base unequally and usually decidedly cordate; lower pinnae distinctly petiolulate, 12–23 cm long, 14–25 mm wide, margins finely, sharply, and regularly cartilaginously toothed, the teeth usually pointing forward, sometimes incurved; laminar tissues membranous to herbaceous; veins distinct, rather delicate, 12–15 per cm; fertile blades about size of sterile, but pinnae 11–22 cm long, 3–5 mm wide, with sterile apex 3–12 mm long, base cordate, usually petiolate; sporangia dark brown; indusium lacerate.

In damp soil rich in humus, Mexico to Colombia. Specimens examined: Isabela, Santa Cruz.

Blechnum lehmannii Hieron., Bot. Jahrb. Syst. 34: 473. 1904

Struthiopteris maxonii Broadh., Bull. Torrey Bot. Club 39: 268. 1912.
Blechnum maxonii C. Chr., Ind. Fil. Suppl. 1: 16. 1913.
Blechnum mexiae Copeland, Univ. Calif. Publ. Bot. 17: 32, *t.* 7. 1932.

Terrestrial; rhizome creeping along surface but tip ascending or erect, whole rhizome 5–7 mm thick, 5–45 cm long, closely invested with bases of old fronds, roots densely black-hairy, terminal part of rhizome scaly with linear-lanceolate, entire, dark purplish brown to black, lustrous scales 4–10 mm long, 1–1.5 mm wide at base; sterile fronds 15–40 cm tall; stipes stramineous to pale green, 1–6 cm long, 0.6–1.2 mm thick, rounded abaxially, prominently grooved adaxially, margins of groove abruptly rimmed, closely beset with minute papillae, sublustrous, bearing a few paler, broader scales than those of rhizome near bases of stipes; lamina narrowly elliptic to narrowly oblanceolate, usually gradually reduced toward each end, 12–35 cm long, 2.5–4.3 cm wide, pinnatisect, 8–26-jugate, lower pinnae sometimes opposite, others alternate, deep green, sublustrous, with submarginal whitish hydathodes 0.4–0.8 mm across on upper surface, paler beneath, terminal pinna 1–4 cm long (1–2.5 cm in ours), the lower 2–8 mm long, 7–9 mm wide, the middle rounded at apex, somewhat falcate, 1–2.2 cm long, 9–14 mm wide, margins plain to faintly and irregularly revolute; rachis grooved and conic-papillate like stipe; veins obscure, mostly 1-forked, midvein slightly elevated on upper surface; fertile fronds usually somewhat exceeding the sterile, their stipes 10–25 cm long, papillate like stipes of sterile fronds; fertile laminae mostly broader than sterile; pinnae linear, 1–3.5 cm long, 1.5–2.5 mm wide; indusium delicate, entire, papery, pale brown to stramineous; sporangia densely crowded, obovoid, about 0.4 mm long; spores broadly ellipsoidal, 50–55 μ long, 30–40 μ thick, faintly reticulate with low, irregular ridges, the reticulae 4–6 μ across, broadly to narrowly rectangular.

In heavily shaded forest or shrubby cover at moderate elevations, Costa Rica to Colombia, Venezuela, and Brazil. Specimen examined: Santa Cruz, a new record for the Galápagos Islands.

Blechnum occidentale L. var. **puberulum** Sodiro, Recens. Crypt. Vasc. 32. 1883

Blechnum occidentale var. *caudatum* sensu Hook., Sp. Fil. 3: 51. 1860, pro parte, with respect to Galápagos specimens.

Rhizomes erect, sometimes elongate and to 12 cm long, 8–20 mm thick, closely invested with dark brown, lanceolate scales 4–8 mm long, their central line darker

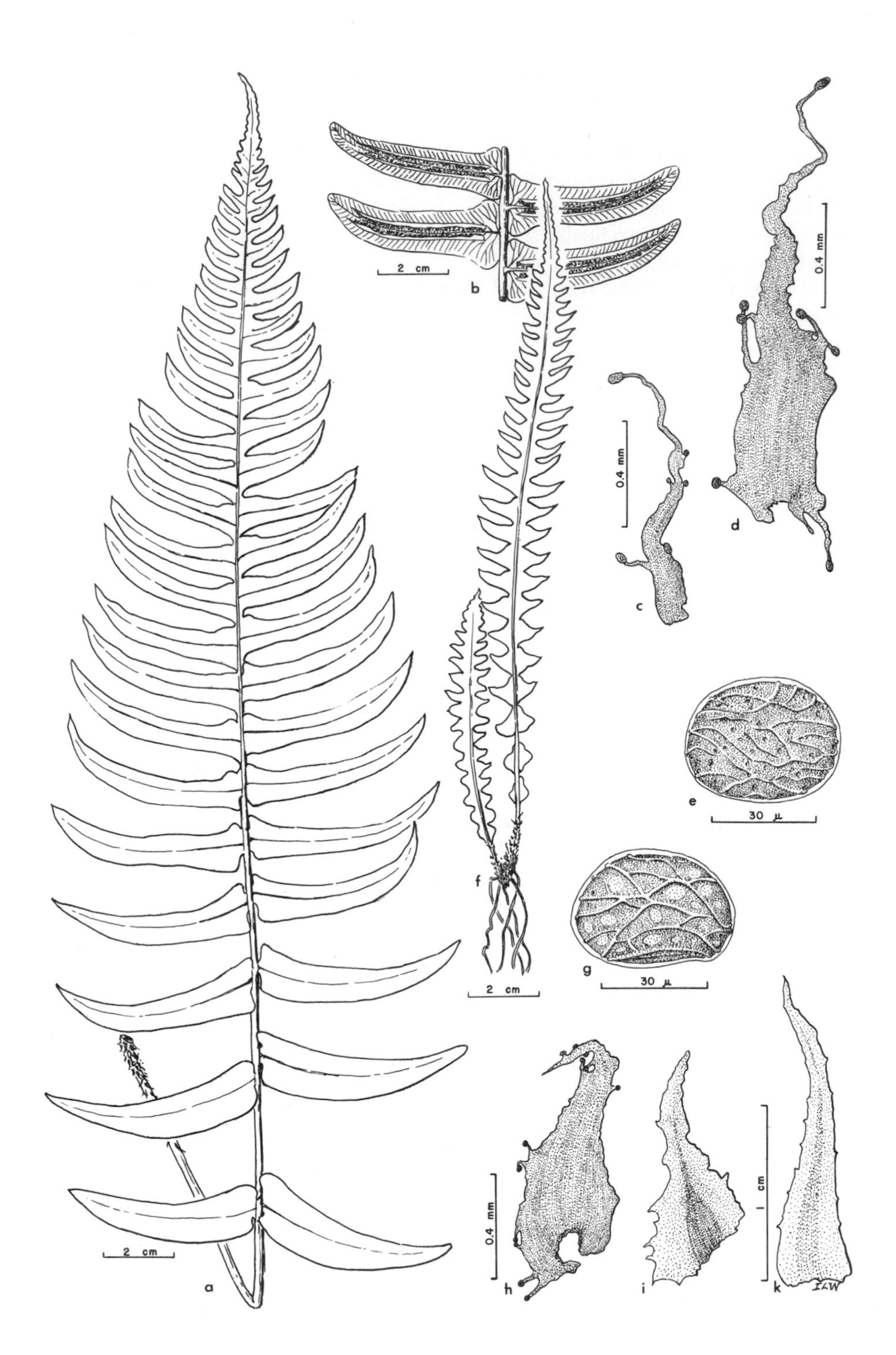

Fig. 11. *Blechnum occidentale* var. *puberulum* (*a–e*). *B. polypodioides* (*f–i*).

brown and more lustrous than marginal bands; stolons usually present, 1.3–2 mm thick, dark brown; stipes fasciculate, 6–16 cm long, shorter than blades, stramineous, 0.7–2.5 mm thick, glabrous but sparsely scaly on basal 2–4 cm; blades ovate to lanceolate, mostly 16–50 cm long, 5–25 cm broad, pinnate at base, pinnatisect above and gradually reduced toward the elongate, pinnatifid, more or less acuminate apex; rachis puberulous, pilosulous or glandular-pilosulous, roughened by minute conical pustules or papillae visible under high magnification; pinnae and segments horizontal or nearly so, numerous, (10) 15–40 pairs, the middle 2.5–7 cm long, 7–13 mm wide, subobtuse to acuminate, sometimes slightly apiculate, margins entire and smooth to shallowly crenulate-undulate and bearing microscopic crystalline teeth, the lower 3–10 pairs free from rachis, sessile, cordate at base, rounded-biauriculate, the upper auricle rounded, overlapping rachis, lowest pinnae slightly smaller than middle ones but not strongly reduced or minute, all subcoriaceous, light green, glabrous except sometimes immediately adjacent to rachis beneath; fertile blades essentially like sterile but with somewhat longer stipes; sori costal, extending from near base to slightly below apex of pinna, sterile band between sorus and margin half again to 3 times as wide as sporulating band; indusia thin, entire, glabrous, 0.6–1 mm wide before rupture by enlarging sporangia, then somewhat obscure; sporangia very numerous, pale rufous brown, about 0.15–0.2 mm long, half as thick; annulus of 12–16 thickened cells; spores bilateral, broadly ellipsoidal, 40–45 μ thick, 55–60 μ long, surfaces smooth to finely and sparingly ridged with minute, irregularly spaced, rounded ridges, brownish gold by transmitted light.

In rather open and sometimes dry situations, Florida, the West Indies and Mexico to Argentina. Specimens examined: Isabela, Pinta, Pinzón, San Cristóbal, San Salvador, Santa Cruz, Santa María.

See Fig. 11 and color plate 61.

Blechnum polypodioides Raddi, Opusc. Sci. Bologna 3: 294. 1819*

Asplenium blechnoides Lag. ex Sw., Syn. Fil. 76. 1806.
Blechnum unilaterale Sw., Mag. Ges. Naturforsch. Freund. Berlin 4: 79, *t. 3, f. 1.* 1810.
Blechnum blechnoides C. Chr., Ind. Fil. 151. 1905, non Keys. 1873.

Rhizomes usually erect, to 6 cm long, stoloniferous, paleaceous; scales lanceolate-subulate, dark brown; fronds fasciculate, few, erect, 20–70 cm long; stipes of sterile fronds much shorter than blades, 2–8 cm long, 0.5–1.3 mm thick, stramineous, sparingly scaly toward base, scales thin, pale brown, with conspicuous, elongate, hair-like teeth, sparingly or rather densely pilosulous with flaccid, septate, pale hairs; stipes of fertile blades similar but more elongate, to 30 cm long; blades linear, attenuate and pinnatifid toward apex, gradually narrowed downward, pinnatifid at middle, here 1.5–5(6) cm wide, middle segments lanceolate, widest at base, horizontal, acuminate, subfalcate, the apex cuspidate, entire, pinnatisect at base, lower

* It is regrettable that the well-known name *Blechnum unilaterale* Swartz must be abandoned, but there is no help for it, since it is superfluous and therefore an illegitimate name. When transferring *Asplenium blechnoides* Lagasca ex Swartz to *Blechnum*, Swartz provided the new epithet *unilaterale*, since in his time an epithet like *B. blechnoides* would have been inappropriate; nevertheless, according to our present Code, an epithet may not be rejected merely because it is inappropriate. It is also unfortunate that the earliest available subsequent epithet should have been *B. polypodioides* Raddi, because the later, illegitimate name *B. polypodioides* (Swartz) Kuhn has been much used for an entirely different species because of its adoption in the Index Filicum.

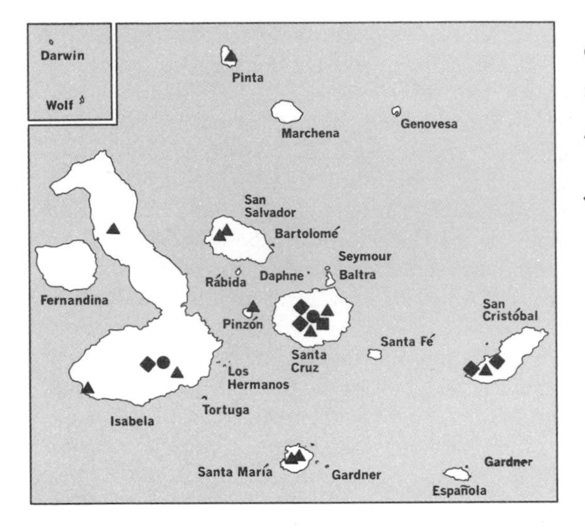

● *Blechnum falciforme*
■ *Blechnum lehmannii*
▲ *Blechnum occidentale* var. *puberulum*
◆ *Blechnum polypodioides*

segments entirely free from one another and often distant but entirely adnate to rachis, rounded at lower base, decurrent, the lowest triangular to semicircular, much broader than long, all pale green, subpapyraceous, rachis and costae glabrous or minutely glandular-pilosulous, undersurface sparsely so, margins not ciliate, minutely scaberulous-toothed with crystalline denticles (visible under high magnification); fertile blades exceeding sterile but essentially like them; sori costal, often not reaching base or apex of segments, usually leaving broad, sterile band between sorus and margins; spores oblong-reniform, about 35 μ thick, 60–65 μ long, irregularly and faintly ridged longitudinally.

In forests, among shrubs, and on grassy, open ground in moist parts of tropical America; both epiphytic and terrestrial. Specimens examined: Isabela (a new record for this island), San Cristóbal, Santa Cruz.

See Fig. 11.

Cheilanthes Sw., Syn. Fil. 5, 126. 1806

Small terrestrial ferns mostly in arid habitats. Often on or among rocks and boulders, with massive, nodose-multicipital or creeping rhizomes closely covered with thin or sclerotic scales, and with tomentose, glandular-pubescent, paleaceous, ceraceous, or naked stipes and blades. Fronds usually clustered, pinnate to decompound, the segments often small and beadlike. Sori submarginal at the enlarged tips of veins, often in contact but scarcely confluent laterally. Indusia pouchlike, introrse, formed by the more or less modified, inflexed margins, discrete or confluent, sometimes obsolescent or lacking. Sporangia with 14–24 thickened cells forming the annulus and a many-celled stomium. Spores trilete, globose-tetrahedral, smooth, granulose, or corrugated.

About 125 species widely distributed in mainly arid habitats in tropical to warm temperate regions.

Segments larger, longer than broad, not beadlike; axes hairy but not scaly. . *C. microphylla*
Segments minute and beadlike; axes densely scaly. . . . . *C. myriophylla*

Cheilanthes microphylla (Sw.) Sw., Syn. Fil. 122. 1806.

Adiantum microphyllum Sw., Nov. Gen. Sp. Pl. Prodr. 135. 1788.
Adiantum pubescens Poir. in Lam., Encycl. Suppl. 1: 141. 1810, non Schkuhr. 1809.
Cassebeera microphylla J. Smith, Jour. Bot. (Hook.) 4: 159. 1841.
Allosorus microphyllus Liebm., Vidensk. Selsk. Skr. V, 1: 219. 1849.
Cheilanthes heterotricha Anderss., Kongl. Svensk. Vet.-Akad. Handl. 1853: 129. 1855.
Notholaena microphylla Keys., Polyp. Cyath. Herb. Bung. 28. 1873 (as *Nothchlaena*).

Rhizome multicipital, short- or long-creeping, clothed with minute linear-lanceolate to attenuate-filiform, concolorous, somewhat tortuous, light brown scales 1–1.5 mm long, or sometimes the larger scales with darker center below middle; fronds approximate or somewhat clustered, 15–40 cm tall; stipes terete, slender, dark purple-brown to nearly black, irregularly puberulent with reddish brown hairs, a few lance-ovate scales near base, usually slightly shorter than blades; blades oblong to elliptic-lanceolate, 10–18 cm long, 2.5–4 cm wide, bipinnate-pinnatifid to tripinnate below, rachises similar to stipes; pinnae distant, 10–15 pairs, usually alternate, the upper merely dentate; ultimate segments ovate to oblong, obtuse to subacute, sessile, the base oblique, to 3.5 mm wide but usually smaller, green on both surfaces, sparingly pubescent with short, coarse, somewhat crinkly hairs, margins shallowly crenulate to coarsely denticulate; sori almost continuous along margins; indusia narrow, dentate-crenulate, formed by the weakly revolute margins, rarely continuous around apex; sporangia reddish brown, dull.

In rocky, usually arid habitats from Florida, Alabama, Texas, and Sonora south-

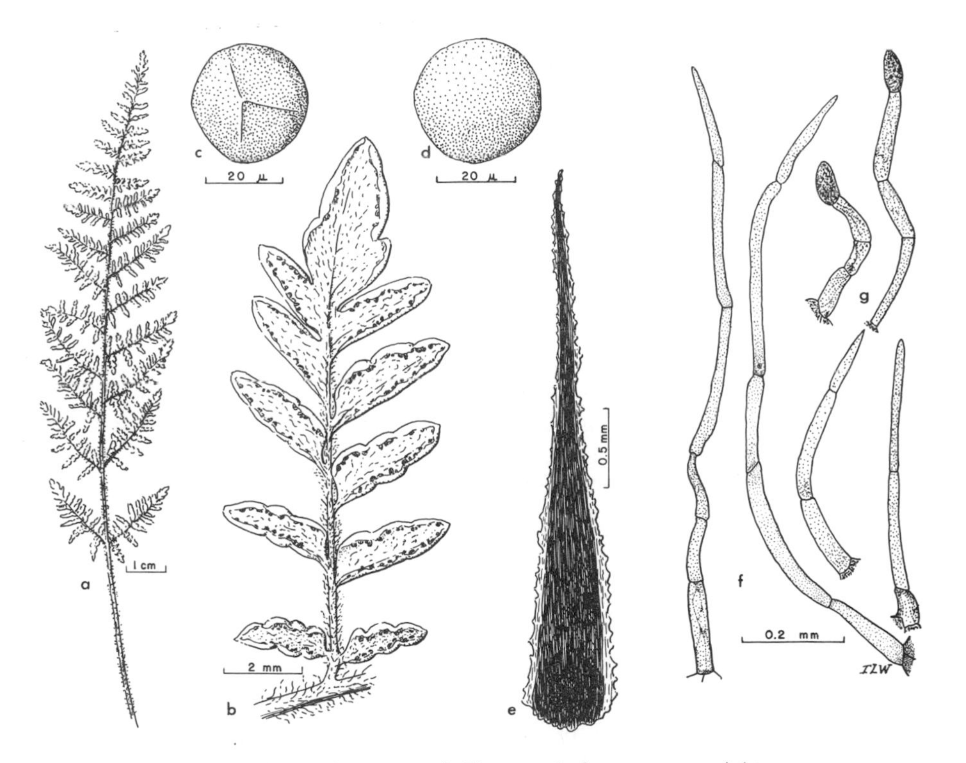

Fig. 12. *Cheilanthes microphylla*: *e*, scale from near tip of rhizome; *f*, nonglandular hairs from near base of stipe; *g*, glandular hairs from same area.

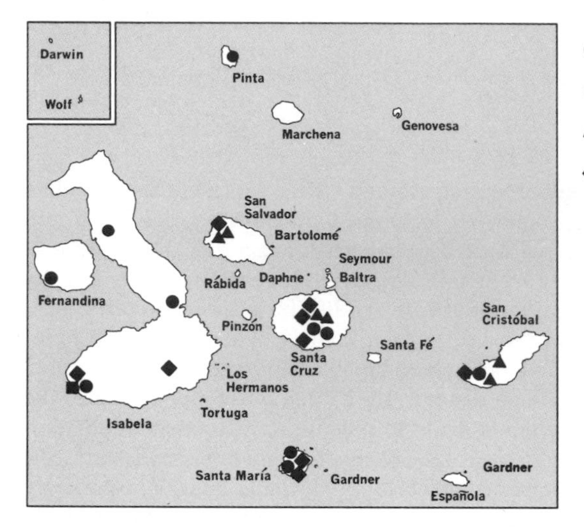

● *Cheilanthes microphylla*
■ *Cheilanthes myriophylla*
▲ *Ctenitis pleiosoros*
◆ *Ctenitis sloanei*

ward through the West Indies and Central America to Ecuador. Specimens examined: Fernandina, Isabela, Pinta, San Cristóbal, Santa Cruz, Santa María.
See Fig. 12.

Cheilanthes myriophylla Desv., Mag. Ges. Naturforsch. Freund. Berlin 5: 328. 1811

Rhizome massive, erect or decumbent, of several to many erect or ascending branches 3–5 mm thick, 5–12 mm long, densely clothed with linear-subulate to lanceolate, acuminate-attenuate, tortuous, dark brown to nigrescent scales 0.3–1 mm wide, 3–4 mm long, their margins thin, somewhat lighter than central parts, scales at tips of rhizome branches lighter and long-attenuate; fronds erect, closely crowded, 2–3.5(5) dm tall; stipes about 1 mm thick, cinnamomeous to purplish brown at maturity under scales, invested with appressed persistent lanceolate to filiform grayish or ferruginous scales, the narrower scales often tortuous; blades narrowly lanceolate, short-acuminate, tripinnate or quadripinnate, to about 2 dm long, rachises clothed similarly to stipes; pinnae somewhat distant, short-stalked, alternate, ovate to subdeltoid-lanceolate; pinnules up to 12 pairs, alternate, bipinnate; ultimate segments beadlike, orbicular to narrowly obovate, frequently arranged ternately with terminal division larger than lateral ones and cuneate at base, with a few long hairs on upper surface, densely paleaceous on all rachises and costae on lower surface; scales of costae and rachises brown, ovate with cordate base and subentire to strongly erose-denticulate along margins or filiform and tortuous, overlapping and extending beyond margins of segments; indusia formed by the closely reflexed margins of segments, obscuring sporangia.

In rocky situations, usually in shaded spots, northern Mexico to Chile and Argentina. Specimen examined: Isabela.

Ctenitis (C. Chr.) C. Chr. in Verdoorn, Man. Pterid. 544. 1938

Terrestrial ferns of moderate to large size. Rhizomes short, ascending or erect, rarely short-creeping, invested with broad, dark brown scales, these becoming

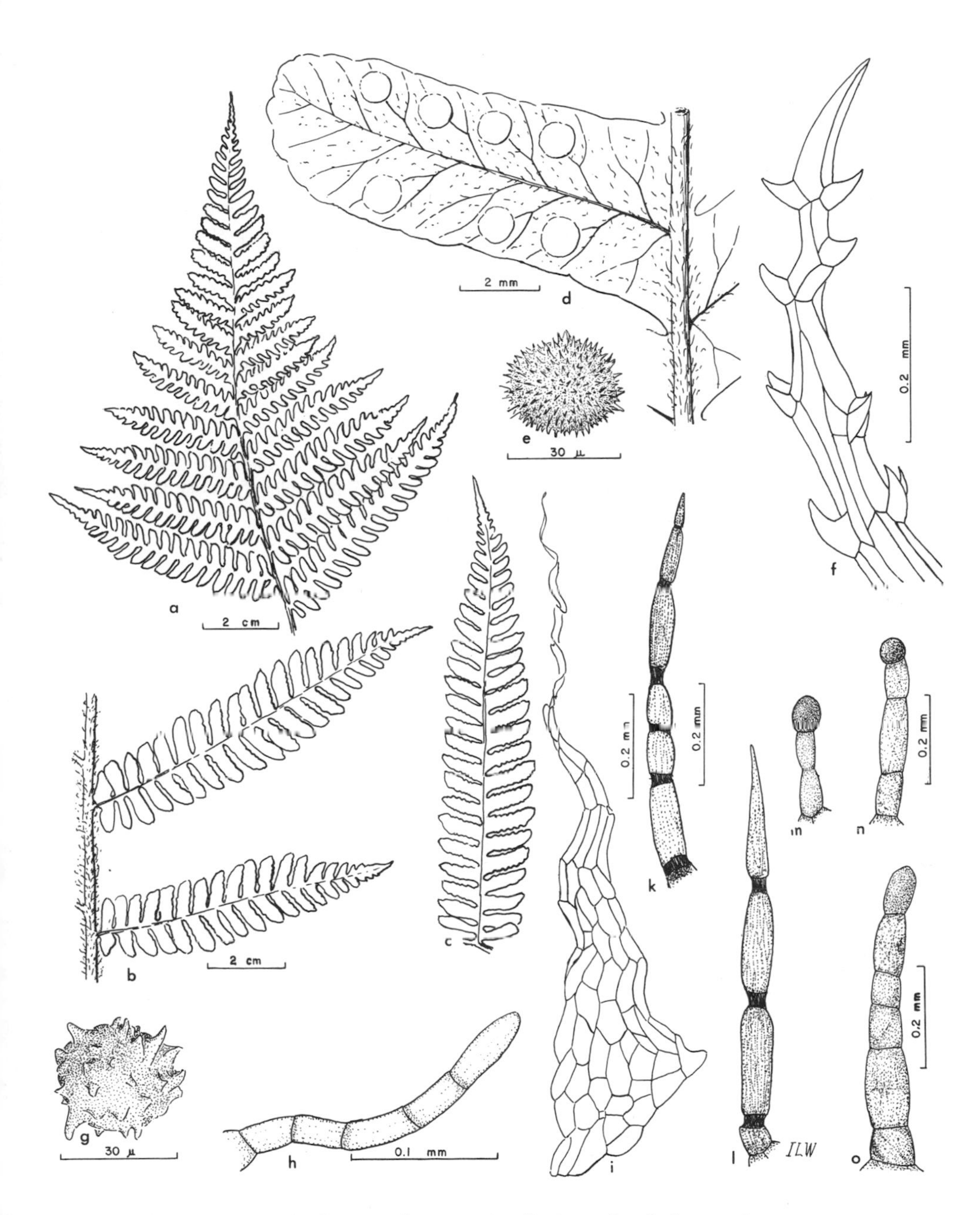

Fig. 13. *Ctenitis pleiosoros* (*a–f*) : *f*, tip of scale from rachis. *C. sloanei* (*g–o*) : *h, k, l*, nonglandular hairs; *m–o*, glandular hairs.

narrower and less crowded on stipes and rachises of many specimens. Fronds ovate-oblong to deltoid, pinnate-pinnatifid to decompound, broad at base, the pinnae catadromous (except the basal), usually herbaceous, the axes bearing toothed scales beneath, multicellular hairs above; ultimate segments or divisions usually rounded or obtuse, costae mostly tomentose above with septate hairs, not decurrent. Veins free, simple or branched. Sori mostly small, dorsal on veins; indusia round, reniform, or rarely lacking. Sporangium smooth, rounded, somewhat compressed laterally; annulus of 14–16 thick-walled cells. Spores monolete, minutely echinate.

About 150 species widely distributed in the warmer parts of the Old and New Worlds.

Blades bipinnate, the lowest pair of pinnae similar to the middle, not unequal-sided, not basiscopically developed; primary pinnae nearly sessile; hairs all or mostly elongate, hyaline, with colorless cross walls, many-celled with elongate cells, not glandular.... .... *C. pleiosoros*

Blades tripinnatifid to quadripinnate, the lowest pair of pinnae unequal-sided, the lower inferior pinnules longer than the superior; primary pinnae obviously stalked; hairs of axes and segments all short, many-celled, with dark cross walls and short cells, mostly glandular.. *C. sloanei*

Ctenitis pleiosoros (Hook. f.) Morton, Leafl. West. Bot. 8: 190. 1957 (as "*pleiosora*")

Polypodium pleiosoros Hook. f., Trans. Linn. Soc. Lond. 20: 166. 1847.
Nephrodium villosum sensu Robins., Proc. Amer. Acad. 38: 110. 1902, non Presl.
Dryopteris villosa sensu Stewart, Proc. Calif. Acad. Sci. IV, 1: 20. 1911, non Kuntze.
Dryopteris tricholepis sensu Stewart, op. cit. 19, non C. Chr.
Dryopteris pleiosoros Svens., Bull. Torrey Bot. Club 65: 316. 1938 (as "*pleiosora*").

Rhizome short, ascending or very short, closely covered with slenderly lance-linear, chestnut brown scales 2–3 cm long, 1.5–2 mm wide, these concolorous, lustrous, faintly denticulate with subremote, spreading to slightly antrorse, simple or subapically forked teeth; fronds 4–8 dm tall; stipe 1.5–3 dm long, canaliculate-striate, about 3–4 mm thick, rufous or tawny, sparsely to moderately clothed with slender scales similar to those of rhizome but of darker hue and gradually reduced in size upwardly, with several-celled, slender reddish or brownish nonglandular hairs intermingled with scales; blades narrowly ovate-lanceolate, 4–7 dm long, 2–3 dm wide, bipinnate; lowest pair of pinnae similar to middle ones, 8–12 cm long, 2–2.5 cm wide, equal-sided; primary pinnae nearly sessile, rachises rather densely villous with tawny to reddish, nonglandular hairs; pinnules sessile, thin, those near middle of primary pinnae about 9–13 mm long, 4 mm wide, obscurely and shallowly crenate-denticulate, obtuse at apex, sparsely villous on upper surface, more copiously so beneath; costae more densely villous than lamina; veins free, 1(2)-branched; sori 4–8 on each side of costa, 0.6–1 mm in diameter, dark brown; sporangia subglobose, 0.2–0.3 mm long, annulus prominent, lustrous, of 12–14 thick-walled cells; spores broadly ellipsoidal to subglobose, 35–40 μ long, closely echinate, the spicules linearly arranged, body dark brown, nearly opaque.

Endemic; in shade and in moist gullies and potholes in lava, upper *Scalesia* Zone and into Fern-Sedge Zone. Specimens examined: San Cristóbal, San Salvador, Santa Cruz.

See Fig. 13.

Ctenitis sloanei (Poeppig) Morton, Amer. Fern Jour. 59: 66. 1969*

Polypodium sloanei Poeppig ex Spreng. in L., Syst. Nat. ed. 16, 4: 59. 1827.
Polypodium paleaceum Hook. f., Trans. Linn. Soc. Lond. 20: 166. 1847.
Alsophila sp. Robins. & Greenm., Amer. Jour. Sci. 50: 149. 1895.
Dryopteris furcata sensu Stewart, Proc. Calif. Acad. Sci. IV, 1: 18. 1911, non Kuntze.
Dryopteris ampla sensu Svens., Bull. Torrey Bot. Club 65: 314. 1938, non Kuntze.

Graceful fern 6–15 dm tall; rhizome and lower part of stipes densely clothed with reddish brown, sinuous scales 2–3.5 cm long, 1.5–3 mm wide, these long-attenuate; stipes distinctly canaliculate on upper side, to 6 dm long, scaly but less densely so than rhizome, scales broadly ovate to ovate-lanceolate, 10 mm long or less, irregularly 2- or 3-toothed near apex or entire; blades yellowish green, 3–6 dm broad at base, broadly ovate to deltoid, tripinnatifid; pinnae oblong, 5–15 cm long, 1–5 cm wide, lowermost pair with inferior pinnules longer than superior, upper pinnules gradually smaller, primary pinnae obviously stalked; rachis scaly with small, narrow scales and more or less pubescent with slender, reddish, several-celled hairs; pinnules about 10–15 pairs on lower pinnae, acute to attenuate at apex, pinnatifid halfway to center or deeper, the segments acute to obtuse, 2–6 mm long, the smaller entire, the larger undulate-crenulate, sparingly hairy-scaly on upper surface, slightly more densely so beneath, veins and margins bearing a few to many short, stoutish, glandular hairs on undersurface of pinnules; veins free, 1- or 2-branched between costa and margin; sori circular, about 0.6 mm across, borne in 2 lines, each usually on more distal branch of primary vein; indusium lacking or minute; sporangia reddish, crowded, about 0.2 mm wide, lustrous; stout, round-tipped paraphyses about 0.2–0.3 mm long, numerous among sporangia; spores subspherical, coarsely echinate, dark brown, processes often curved to sub-hooked, about 40–45 μ long.

In humid habitats in forests or among shrubs, Florida, the West Indies, and Guatemala to Bolivia. Specimens examined: Isabela, San Cristóbal, San Salvador, Santa Cruz, Santa María.

See Fig. 13.

* It is unfortunate that the application of the name *Ctenitis ampla* (Humb. & Bonpl.) Ching must be changed. It has been applied since the time of Mettenius to the widely distributed species described above. However, a photograph of the holotype of *Polypodium amplum* Humb. & Bonpl. ex Willd. in the Willdenow Herbarium (No. 19722), Berlin, obtained by Dr. Tryon, shows that the epithet *amplum* properly belongs to the species called *Dryopteris nemophila* (Kunze) C. Chr. in Christensen's monograph of *Dryopteris* (Dansk Vidensk. Selsk. Skr. ser. 8, 6: 57, *fig. 11.* 1920). Christensen had some doubt that he was applying the name *ampla* correctly (op. cit. 49). The type specimen of *P. amplum* is identical to *Fendler 204* from Venezuela, which Christensen has identified as *Dryopteris nemophila*, and this in turn does not seem different from Peruvian material of the species, which is doubtless the same as the types of *Aspidium nemophilum* Kunze and *A. catocarpum* Kunze, both of which came from Peru. In addition to the characters pointed out by Christensen (loc. cit., pt. 2, p. 49), it may be mentioned that the pinnae in *Ctenitis sloanei* (*Dryopteris ampla* sensu Christensen) are long-stalked and the pinnules nearly sessile. The most obvious character distinguishing the species is the apex of the pinnules, which is attenuate in *C. sloanei* and somewhat obtuse in *C. ampla*. The rhizome scales are somewhat different, those of *C. sloanei* forming a dense light brown mass and those of *C. ampla* being larger and darker brown, though similar in structure. The veins beneath in *C. sloanei* always bear minute glandular hairs, whereas those of *C. ampla* are usually eglandular.

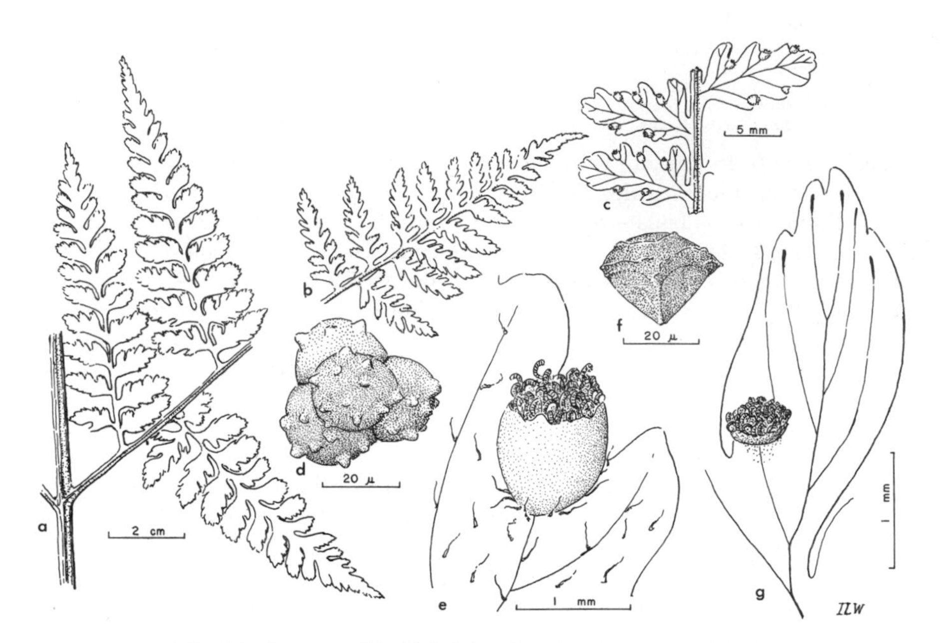

Fig. 14. *Dennstaedtia globulifera* (*a–e*). *D. cicutaria* (*f, g*).

Dennstaedtia Bernh., Jour. Bot. Schrad. 1800²: 124. 1802

Medium-sized to large terrestrial ferns of erect, ascending or spreading habit on wet, shaded slopes and forest margins, with slender, creeping, densely hairy rhizomes. Fronds not articulate to rhizome, distichous, subdistant, long-stipitate, without scales; blades ample, mostly narrowly deltoid, bi- to quadripinnate, the ultimate segments incised to sinuate-lobate. Veins anadromous, simple or pinnately branched, free, often prominent. Sori numerous, marginal, terminal, solitary and mostly distant. Indusium extrorse, convex, adnate at sides to margin of modified, usually recurved leaf lobule, the soral pouch so formed shallowly to deeply cyathiform, shallowly 2-lobed or entire; receptacle punctiform or transversely oblong. Spores trilete.

About 75–80 species, mostly in tropical regions but a few species extending into the temperate zones.

Indusial cup small and shallow; axes of pinnules very narrowly alate or merely lined. . . . . *D. cicutaria*
Indusial cup large and deep; axes of pinnules obviously green-alate. . . . . . . *D. globulifera*

Dennstaedtia cicutaria (Sw.) Moore, Ind. Fil. 97. 1857

Dicksonia cicutaria Sw., Jour. Bot. Schrad. 1800²: 91. 1802.
Hypolepis repens sensu Svens., Bull. Torrey Bot. Club 65: 332. 1938, non Presl.

Rhizome long-creeping, dark brown to nearly black, 5–8 mm thick, naked or nearly so; stipes distant, erect, to 1 m tall, 8–10 mm thick, dull yellowish brown (or darker near base), minutely puberulent, glabrescent, with many wiry, nearly black roots; blades deltoid, acuminate, 1–2.5 m long (whole frond often 3 m tall), 1–2.5 m broad, quadripinnate, with pinnules again pinnately incised, ultimate lobules small and delicate; basal pinnae subopposite, oblong-ovate, 6–10 dm long, 3–5 cm wide, short-stalked, acuminate; secondary pinnae strongly anadromous, numerous, alternate, spreading, linear-oblong, abruptly acuminate, 15–30 cm long, 4–9 cm wide; pinnules close, horizontal, oblong, acute to acuminate or sometimes obtuse, subsessile, often pinnate at base and pinnatisect or pinnatifid above; ultimate segments oblong from decurrent-cuneate base, 3–7 mm long, 1–3 mm wide, obliquely dentate-incised; leaf tissue light green, herbaceous, lustrous above, glabrous or hirsutulous in proximity of veins; tertiary rachises and costae laxly yellowish-hirsute above, thinly hirtellous beneath; veins markedly oblique, translucent, mostly 1-forked, impressed above, elevated beneath; sori small (about 0.3–0.4 mm in diameter), seated at distal sinuses of lobes, pouch usually vertical, shallowly cyathiform; spores about 25–30 μ in diameter, slightly verrucose on outer surface, golden brown.

Wet forests, stream banks, wooded hillsides, margins of forests and thickets from 100 to 2,600 meters altitude, central Mexico and the Greater Antilles to Bolivia and southern Brazil. Specimens examined: Santa Cruz.

No fruiting plants of this fern could be found during January and February in 1964 or 1967. Howell's specimen, collected in May, bore a few scattered sori but many of the included spores were immature.

See Fig. 14.

Dennstaedtia globulifera (Poir.) Hieron., Bot. Jahrb. Engler 34: 455. 1904

Polypodium globuliferum Poir. in Lam., Encycl. Méth. 5: 554. 1804.
Dicksonia altissima Smith in Rees, Cycl. 11. 1808.
Dicksonia punctulata Poir. in Lam., Encycl. Suppl. 2: 475. 1811.
Dicksonia tenera Presl, Del. Prag. 1: 189. 1822.
Dicksonia exaltata Kunze, Bot. Zeit. 8: 59. 1850.
Dennstaedtia tenera Mett., Ann. Sci. Nat. V, 2: 261. 1864.
Dicksonia lagerheimii Sodiro, Crypt. Vasc. Quit. 50. 1893.
Dicksonia globulifera Kuntze, Rev. Gen. Pl. 3²: 378. 1898.
Dennstaedtia lagerheimii C. Chr., Ind. Fil. 217. 1905.
Polystichum apiifolium sensu Stewart, Proc. Calif. Acad. Sci. IV, 1: 26. 1911, non C. Chr.
Dryopteris furcata sensu Stewart, op. cit. 18, pro parte, non Kuntze.
Dennstaedtia tamandarei Rosenst., Hedwigia 56: 359. 1915.
Dennstaedtia bradeorum Rosenst., Rep. Spec. Nov. Fedde 22: 3. 1925.

Stipes stout, 1 m long or more, bright yellow-brown, densely pilose with reddish, septate hairs 1–5 mm long on basal part, finely puberulent to scaberulous and with minute conical processes above; blades deltoid, 1–2 m long, nearly as broad as long at base, coarsely tripinnate, the ultimate segments crenately lobed or parted; basal pinnae petiolulate, subovate, acuminate, 7–10 dm long, 2–4 dm wide, the rachis yellowish brown, obtusely ridged on upper side, sparingly to moderately pilose with brownish septate hairs, glabrescent and scabrous beneath; secondary pinnae linear-oblong to subdeltoid-oblong, petiolulate, long-acuminate, the larger 12–20 cm long, 3–6 cm wide, pinnules at base of pinnae subdistant, laxly spreading, 1.5–3.5 cm long, 7–15 mm wide, subauriculate, obliquely oblong to ovate-oblong, mostly rounded at apex, the subterminal ones sometimes acute, subsessile or those toward tips of pinnae joined by broad foliaceous wing; larger lobes of segments 3–5 pairs, rounded-obtuse to subtruncate, lightly and broadly dentate; leaf tissue soft-herbaceous, yellow-green, glabrous above, setose beneath; veins pinnately branched, prominent beneath; sori globose, 1.2–1.6 mm in diameter, vertical, soral pouch

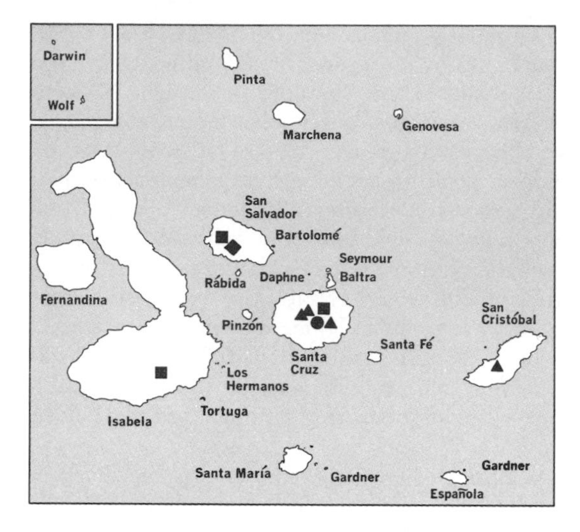

deeply cyathiform, 0.8–1.6 mm deep, margins irregularly crenulate to subentire; spores trilete, about 25–50 μ in diameter, coarsely verrucose on the rounded, outer surface, brownish by transmitted light.

Moist, shaded hillsides, seepy ravines, borders of forests, and along streams, eastern and central Mexico to Jamaica, Argentina, Uruguay, and Paraguay. Specimens examined: Isabela, San Salvador, Santa Cruz.

See Fig. 14.

Diplazium Sw., Jour. Bot. Schrad. 1800[2]: 61. 1802

Medium-sized to larger ferns of upright habit, from creeping to erect, massive rhizomes paleaceous at and near apex, the cells of scales dark-colored, dense, the walls equally thickened. Stipes not articulate to rhizomes. Blades ample, simple to quadripinnate, linear to deltoid, the divisions, when present, small to quite large; texture papyraceous, coriaceous, or spongiose-herbaceous and often succulent. Veins free (in ours) or obliquely areolate without included veinlets. Sori elliptic to elongate-linear, laterally attached, at least those of basal veins on distal branches partly double, the 2 parts opposed on common receptacle and facing away from each other. Indusia membranous, following sori, attached laterally throughout and facing outward. Sporangia with stalks consisting of 3 rows of cells above, sometimes only 2 rows at base; annulus of 12–18 thickened cells. Spores bilateral, slightly reniform, relatively large, in ours prominently reticulate-alate, rusty in transmitted light, the wings nearly hyaline.

About 350 species widely distributed in the tropics of the Old and New Worlds.

Diplazium subobtusum Rosenst., Rep. Spec. Nov. Fedde 7: 296. 1909

Diplazium expansum sensu Morton, Leafl. West. Bot. 8: 190. 1957, non Willd.

Rhizome erect or nearly so, 1.5–2 cm thick but appearing thicker, owing to adherence of bases of old stipes, the tip densely clothed with dark brown, dull to sublustrous, oblong-lanceolate scales 10–15 mm long; fronds several, cespitose, arching-ascending, 1–2 m long, the stipes 4–8 mm thick, markedly ridged-canaliculate, dark

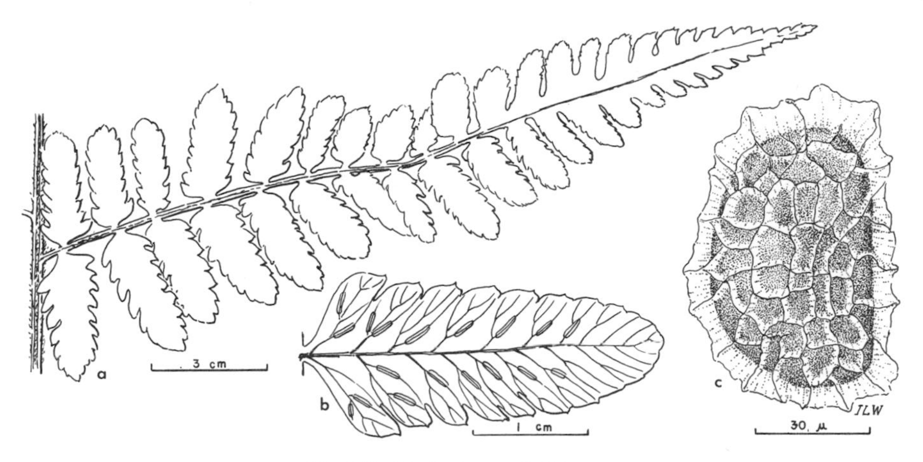

Fig. 15. *Diplazium subobtusum.*

brown at base, paler above, shorter than blades, pubescent, at least within sulcus; blades narrowly ovate to lanceolate, long-acuminate, 5–15 dm long, 4–8 cm wide, bipinnate, the rachis deeply sulcate above, pubescent; pinnae 8–15 pairs, mostly alternate but lower 1–3 pairs often opposite or subopposite, spreading, distinctly petiolulate, the stalks 1–1.5 cm long, pinnae oblong-acuminate, the larger 2–4 dm long, 5–15 cm wide, the lower half to two-thirds of each pinna fully pinnate, the remaining apical part coarsely pinnatifid, rachis septate-hirtellous with brownish hairs; pinnules spreading, approximate to distant, narrowly deltoid-oblong, obtuse, acute or acuminate at apex, subtruncate at base, the larger petiolulate, 4–8 cm long, 8–20 mm wide, pinnatifid halfway to costa or less, segments broadly oblong, acutish to rounded at apex, coarsely and shallowly dentate; costules tawny-hirtellous to glabrate beneath, leaf tissues adjacent nearly glabrous; veins 4–8 pairs, oblique, 2- or 3-forked or those near tip of segments simple, all free; sori 2–10 pairs, 2–4 mm long, mostly double; indusia yellowish brown, membranous, about 1 mm wide, quickly pushed aside by enlarging sporangia, entire but often ruptured as sporangia expand; sporangia about 0.6 mm long, 0.4 mm wide, annulus dark reddish brown, lustrous, side walls yellowish to nearly white, shining; spores slightly reniform, reddish brown, 70–80 μ long, 50–60 μ thick, prominently reticulate-alate, the wings very thin, 6–15 μ high, irregularly crenulate.

In humid habitats, Ecuador. Specimens examined: San Cristóbal, Santa Cruz.

Not common; many of the plants showed no fruiting blades during January and February in 1964 and 1967.

See Fig. 15 and color plate 56.

Doryopteris J. Smith, Jour. Bot. (Hooker) 4: 462. 1841

Rhizome erect, usually ascending or decumbent, bearing relatively few narrowly lanceolate, sclerotic, dark brown to nearly black scales 2–5 mm long. Stipes cespitose, atropurpureous to black. Blades usually pedately parted, the basal pinnae or primary segments longer than second pair and lobed on lower side with inner lobe longest, often also lobed on upper side, some fronds entire or trilobate in young plants. Venation areolate or free. Indusium formed by the reflexed, modified margins of segments, continuous. Sori marginal or intramarginal, contiguous or usually continuous. Sporangia usually long-stalked, but short-stalked in a few species. Spores tetrahedral-globose, yellow to brown, with smooth or roughened surface.

About 25 species in the tropics of the Old and New Worlds, the chief center of distribution in Brazil.

Venation free; blades lacking a bud at apex of stipe . . . . *D. concolor*
Venation markedly areolate; proliferous bud at apex of stipe . . . . . *D. pedata* var. *palmata*

Doryopteris concolor (Langsd. & Fisch.) Kuhn in von Decken, Reisen Öst.-Afr. 3[3]: 19. 1879

Pteris concolor Langsd. & Fisch., Icon. Fil. 19, *t. 21.* 1810.

Rhizome small to stout, more or less short; scales of the stipe bases ovate-lanceolate to attenuate, with a dark reddish to blackish sclerotic central stripe and a broader, lighter to hyaline marginal band along each margin, entire or finely toothed; stipes reddish brown to black, scaly toward base or with a few short, brown hairs and often faintly alate-angled on upper side, otherwise glabrous; fertile and sterile blades similar, without proliferating buds, sterile fronds 4–15 cm long, their blades 2–5 cm long, pentagonal, bi- or tripinnatifid, their numerous ultimate segments deltoid, oblong, or ovate-oblong, acute to broadly rounded, entire to shallowly crenate, margins with white to brown cartilaginous border; veins all free, a hydathode on upper surface of frond at tip of veinlet; fertile fronds 5–35 cm tall, blades 3–15 cm long, usually tripinnatifid, pentagonal to suborbicular, the ultimate segments deltoid to narrowly lanceolate, usually acute to acuminate, entire, or sometimes weakly crenulate; soral lines usually broken at sinuses; sporangia nearly sessile; spores yellow, exospore smooth, whitish, poorly to moderately developed.

In open forest, along margins of thickets, and around rocks and outcrops, West Indies, South America, Africa, India, China, Malaysia, Australia, and Oceania. Specimens examined: San Salvador.

Probably more common on the islands than the paucity of specimens indicates, for this species can easily be confused with *D. pedata* var. *palmata* in the field.

Doryopteris pedata (L.) Fée var. **palmata** (Willd.) Hicken, Revist. Mus. la Plata 15: 253. 1908

Pteris palmata Willd. in L., Sp. Pl. ed. 4, 5: 357. 1810.
Doryopteris palmata J. Smith, Jour. Bot. (Hooker) 4: 163. 1841.
Litobrochia palmata Moore, Ind. Fil. 342. 1862.
Pteris pedata var. *palmata* Baker in Mart., Fl. Bras. 1[2]: 408. 1870.
Pteris pedata gemmipara Sodiro, Anal. Univ. Quito 8: 68. 1893.
Doryopteris pedata subsp. *palmata* Hassler, Trav. Mus. Farm. Fac. Cienc. Med. Buenos Aires 21: 20. 1919 (as *Dryopteris*).
Doryopteris mayoris Rosenst., Mém. Soc. Neuchatloise 5: 51, *t. 2, f. 2.* 1912.

Stipes cespitose, castaneous to nigrescent, usually plane on upper side, at least below middle, rarely sulcate or subterete; fertile and sterile blades producing buds at bases just at apex of stipes, very rarely also from some blades of plant at marginal positions; sterile fronds to 37 cm tall, blades 1.5–14 cm long, deltoid, hastate and shallowly 3-lobed in smaller blades, narrowly pentagonal and deeply 5-lobed in larger blades, the lateral lobes shorter than the central, ultimate segments 2 to several on each lobe, these deltoid to ovate-lanceolate, margins often minutely denticulate, acute or rounded; fertile fronds 11–40 cm long; blade 4–15 cm long, usu-

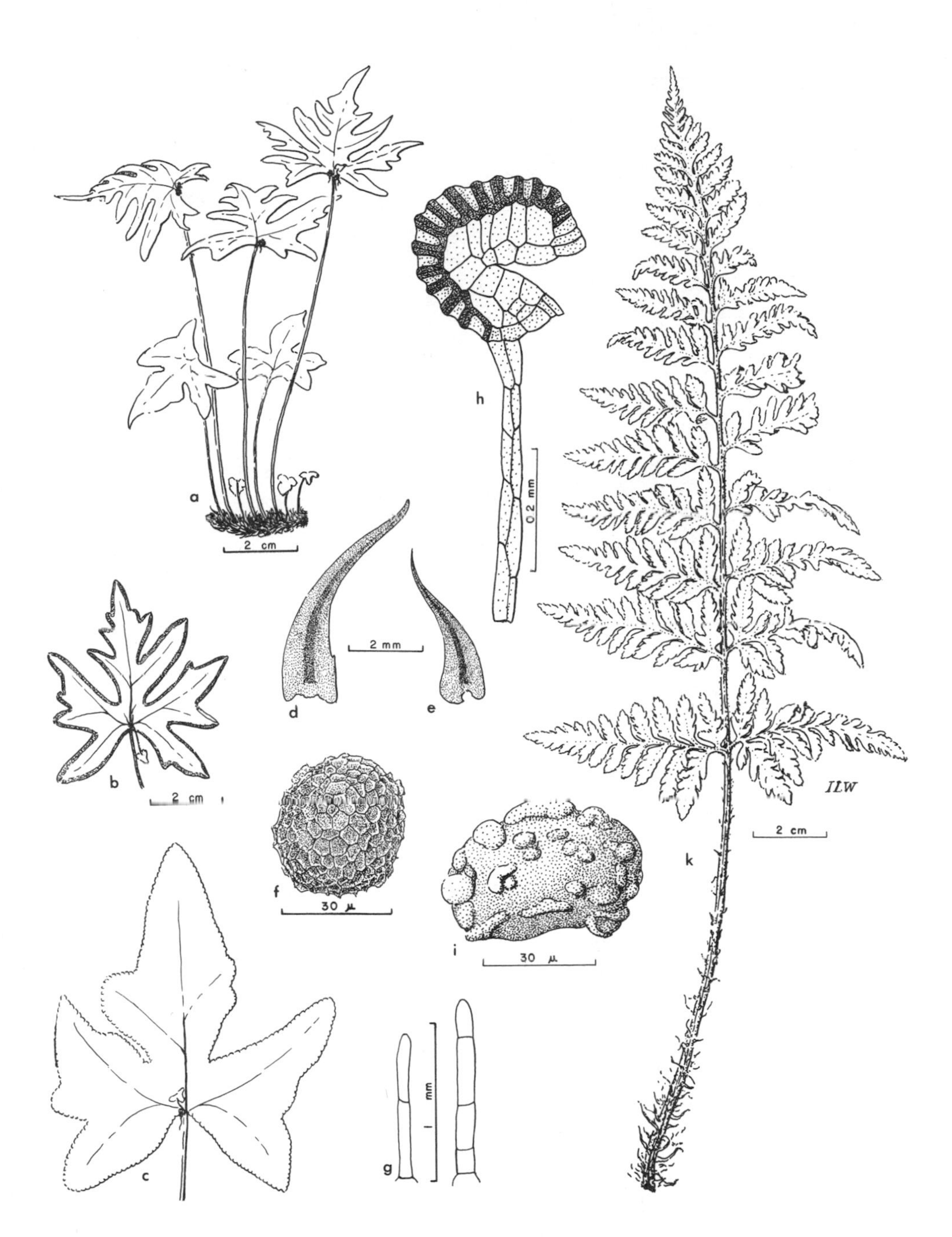

Fig. 16. *Doryopteris pedata* var. *palmata* (*a–g*). *Dryopteris patula* (*h–k*).

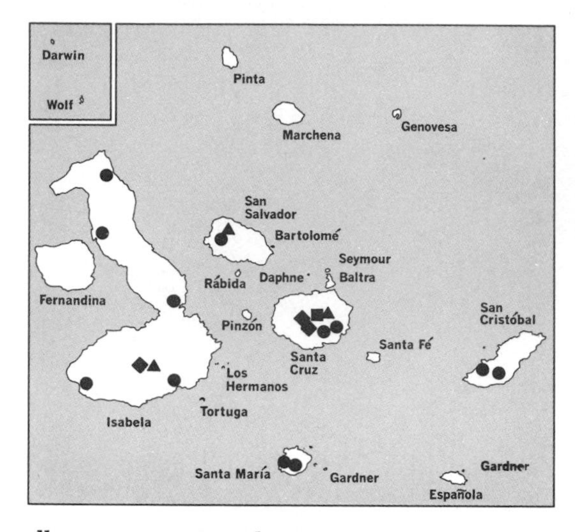

ally ovate-pentagonal, rarely suborbicular, sometimes deeply 5-parted, all lobes about same length, ultimate segments deltoid to linear; soral lines continuous on some segments; spores with whitish, rugulose perispore.

In moderately moist habitats in partial shade, or rarely in full sun, Mexico to Bolivia. Specimens examined: Isabela, San Cristóbal, San Salvador, Santa Cruz, Santa María.

See Fig. 16 and color plate 89.

Dryopteris Adans., Fam. Pl. 2: 20, 551. 1763

Terrestrial ferns of small to large stature, with short, stout, short-creeping, ascending, or erect paleate rhizomes; scales broad, entire or glandular-margined, rarely lacerate or fimbriate, the cells elongate and often sinuate. Fronds borne singly or usually in a cluster; stipes elongate, commonly scaly; blade bipinnatifid to decompound, catadromous in development, axes glabrous or scaly and often short-stipitate-glandular, but not pubescent; minor axes decurrent on major axes, forming sides of dorsal groove; ultimate segments herbaceous to coriaceous, almost entire to coarsely toothed; veins free, forked 1–4 times. Sori normally dorsal on veins, rounded; indusia round-reniform, attached by inner edge of sinus or sometimes more extensively affixed, rarely wanting. Sporangia with 14–22 thick-walled cells; spores bilateral, the epispore shrunken to make surface echinate to tuberculate or coarsely and irregularly rugose.

About 150 species, widespread in distribution; particularly well represented in the Northern Hemisphere but also well entrenched south of the Equator.

Dryopteris patula (Sw.) Underw., Our Nat. Ferns, ed. 4, 117. 1893

Aspidium patulum Sw., Kongl. Svensk. Vet.-Akad. Handl. 1817: 64. 1817.

Rhizome stout, erect or ascending, clothed by closely crowded, thin, somewhat sinuate-margined, pale ferruginous, lanceolate scales 4–10 mm long; fronds several, rather closely set near apex of rhizome, 2.5–10 dm tall, stipes erect, half to two-thirds as long as blade, densely paleaceous at and near base, gradually less so upward, narrowly and rather deeply grooved adaxially, bearing few to many scattered or more or less densely arranged short-stipitate to subsessile glands in groove and/or along margins of lateral side of stipe; rachis grooved adaxially, the ridges not interrupted by incoming ridges of pinnae, more or less glandular like stipe; blades (in ours 15–22 cm long, 7–10.5 cm wide), triangular-ovate to triangular-lanceolate, pinnae pinnately arranged, anadromous, ovate-triangular, the lowermost 4–6 cm long, about 3.5 cm wide at base, bipinnate-pinnatifid, thin to subcoriaceous; pinnules glabrous, lance-ovate, the larger 18–22 mm long, 6–9 mm wide, pinnatifid bases of ultimate segments decurrent-adnate, the distal side of larger ones crenate-dentate, progressively less so toward tips of pinnae; all pinnae gradually smaller toward apex of blade, this pinnatifid-crenate; veins forked (1)2–3(4) times, departing from costule at sharp angle; upper surface of segments marked by linear-elliptic, minute brownish hydathode-like structure immediately above each sorus; sori in 2 rows midway between costule and margins of pinnule, usually 1 sorus per segment; indusium round, with deep sinus, pale rufous, puberulent, especially around margins, these finely sinuate; sporangia dark brown, tumidly obovoid to nearly globose, 0.3–0.4 mm long, dull, annulus of about 14 thick-walled cells, stalk equaling or longer than body, 2 cells thick; spores ellipsoid-reniform, dark brown, about 50–55 μ long, 30–35 μ thick, coarsely and irregularly rugose-tuberculate.

Widely distributed in the West Indies and Mexico, south into Ecuador and Brazil. Specimen examined: Santa Cruz.

This fern superficially resembles *Rumohra adiantiformis*, but can be distinguished by the copious glands on the axes, which are lacking in *Rumohra*.

See Fig. 16.

Elaphoglossum Schott ex J. Smith, Jour. Bot. (Hooker) 4: 148. 1841

Epiphytic and terrestrial ferns of small or moderate size. Rhizomes short-creeping, dictyostelic, with few vascular bundles and without sclerenchymatous bands, the surface densely scaly. Stipes subcespitose or approximate (rarely distant), articulate or nonarticulate with rhizome. Laminae erect, simple and entire, sometimes with pronounced cartilaginous border, firm to coriaceous, scaly or glabrous, the fertile fronds usually slightly smaller, longer-stiped, and narrower than the sterile. Veins evident or immersed and obscure, mostly forked at or near base, then straight and parallel, usually free but sometimes anastomosing sparingly and irregularly, sometimes connected at tips. Sporangia covering all or most of undersurface of fertile fronds; annulus of 10–14 thickened cells. Paraphyses none. Spores trilete, elliptic, brownish, the thick epispore shrinking, the spore then angular-reticulate, reticulate-flaky, or tuberculate.

Over 400 species in all tropical and some subtropical countries; especially numerous in the Andes.

See color plate 55.

Blades persistently and densely scaly beneath, the scales imbricate or not:
- Scales beneath imbricate, broad .... *E. engelii*
- Scales beneath not imbricate, some of them deeply stellately dissected and leaving some leaf surface visible .... *E. yarumalense*

Blades scaleless or sparsely scaly beneath, the scales not imbricate:
- Rhizome thick, 5–6 mm in diameter, short-creeping; rhizome scales concolorous, dark brown, large; blades large, 30 cm long or more, 4 cm wide or more, short-stipitate .... *E. firmum*

Rhizome slender, 2–2.5 mm in diameter, long-creeping:
- Rhizome scales brown, concolorous; stipe scales stellately dissected; blade scales obviously dissected . . . . *E. tenuiculum*
- Rhizome scales bicolorous, with a dark, shining central band above the middle; stipe scales broad, not stellately dissected; blade scales minute but not hairlike:
  - Blades 3.5–4.5 cm broad; stipes 1.5 mm across or more; rhizomes very widely creeping, the fronds distant . . . . *E. glossophyllum*
  - Blades less than 1 cm broad; stipes slender, 0.4–0.8 mm across; rhizomes shorter, the fronds less distant . . . . *E. minutum*

Elaphoglossum engelii (Karst.) Christ, Monogr. Elaph. 31. 1899*

Acrostichum engelii Karst., Fl. Colomb. 1: 118, *t.* 59. 1860.
Elaphoglossum petiolatum sensu Svens., Bull. Torrey Bot. Club 65: 318. 1938, pro parte, non Urban.

Rhizome short-creeping, 5–6(8) mm thick, usually branching and adjacent branches more or less parallel and bound together by interlacing roots and bases of petioles, densely scaly with dark brown, lanceolate to ovate-lanceolate, lustrous scales 2.5–5 mm long, their margins entire to short-denticulate, not ciliate like scales of fronds; stipes subdistichous, 2.5–3.5 mm thick, 7–30 cm long, densely scaly with dark scales similar to but slightly smaller than those of rhizome, gradually replaced upward by paler, long-ciliate scales like those on lamina; laminae thick-leathery, 1–3.5(4) cm wide, 15–30 cm long, usually strongly revolute, densely imbricate-scaly on upper surface with whitish to silvery, papyraceous, ovate to nearly orbicular long-ciliate scales 1–1.5(2) mm long, these deciduous in irregular patches with age, lower surface more densely imbricate-scaly with pale reddish brown, long-ciliate, lanceolate to oblong-lanceolate scales 1.2–2.5 mm long; veins immersed and totally hidden by scales, simple and free or some forking once near base, thence parallel and free to margin, not joining; sporangia densely crowded, dark reddish brown, obovoid to nearly globose, about 0.3 mm long, on slender stalks 0.4–0.6 mm long; annulus of 11–13 thickened cells; spores dark reddish brown, broadly ellipsoidal, about 40–45 μ long, about 25 μ broad, the epispore distinctly and regularly angled-reticulate, elevated walls of reticulations about 2–5 μ high, thin.

On rocks in partial shade, on walls of ravines, and in lava caverns, Colombia to Bolivia. Specimens examined: Isabela, San Salvador, Santa Cruz.

See color plate 58.

Elaphoglossum firmum (Mett.) Urban, Symb. Antill. 4: 59. 1903†

?*Oflersia langsdorffii* sensu Hook. f., Trans. Linn. Soc. Lond. 20: 161. 1847, non Presl.
Acrostichum firmum Mett. ex Kuhn, Linnaea 36: 55. 1869.
?*Acrostichum muscosum* sensu Robins., Proc. Amer. Acad. 38: 104. 1902, non Swartz.
Elaphoglossum petiolatum sensu Svens., Bull. Torrey Bot. Club 65: 318. 1938, pro parte, non Urban.
Elaphoglossum longifolium sensu Morton, Leafl. West. Bot. 8: 192. 1957, probably non J. Smith.

Mainly terrestrial with short, branched, short-creeping and ascending rhizomes 5–6(9) mm thick, densely clothed with dark brown, concolorous, somewhat lacer-

* The identification of the Galápagos plants with the Colombian and Venezuelan *E. engelii* is somewhat doubtful, like many *Elaphoglossum* identifications. See comments by Morton (1957, 191–92).

† The Santa Cruz specimens Morton referred to as *E. longifolium* (1957, 192) probably are not of that species, differing in having cartilaginous margins on the sterile blades and in having veins not marginally connected. They agree fairly well with *E. firmum* from Cuba.

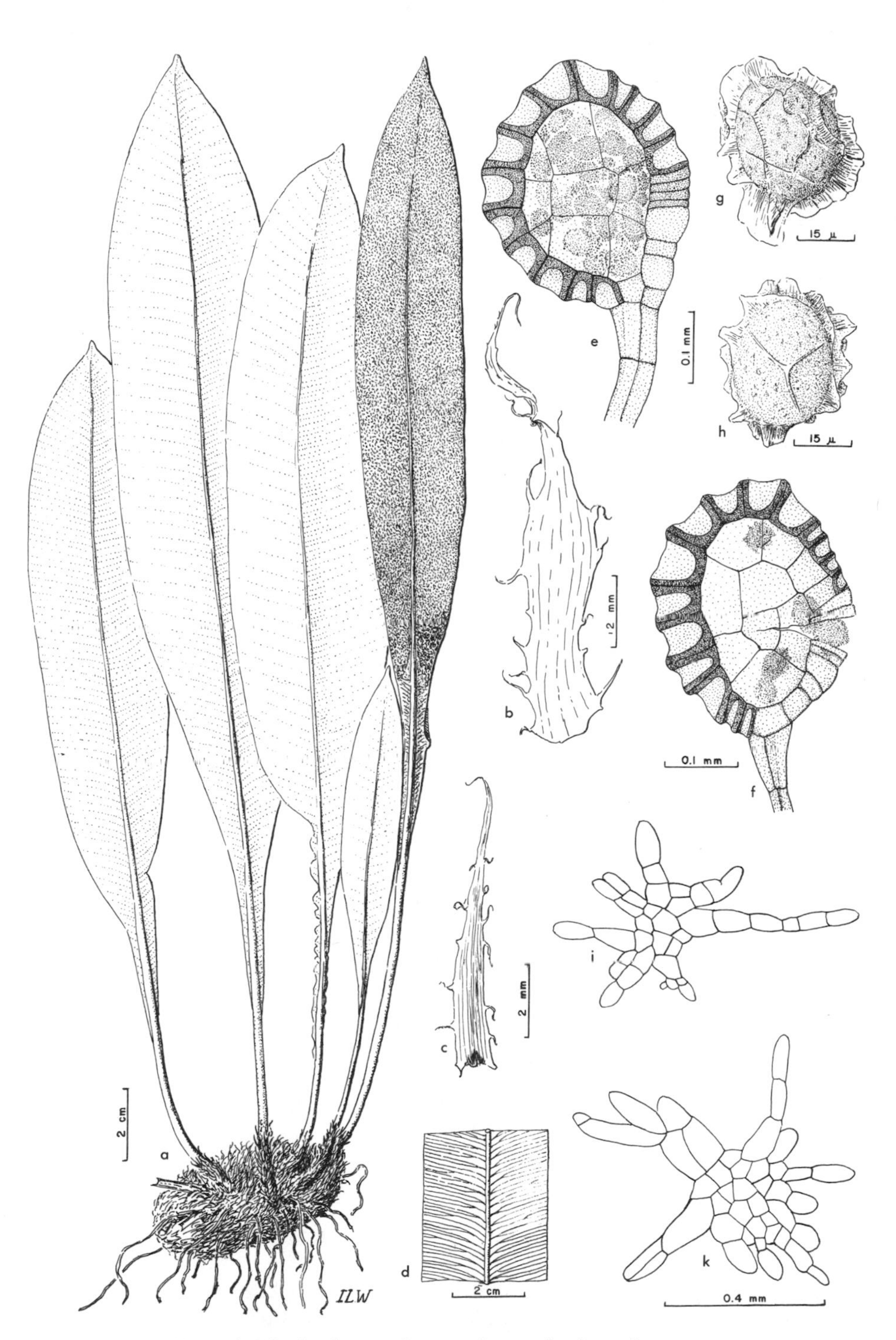

Fig. 17. *Elaphoglossum firmum*: *b, c,* scales from rhizome; *i, k,* stellate scales from lower surface of sterile blade.

ate-margined scales 4–10 mm long; fronds erect, approximate, 5–12 on each massive rhizome; sterile fronds on stipes 2–9 cm long, 1.5–2 mm thick; laminae narrowly cuneate-decurrent at base, acute to slightly acuminate at apex, 15–35 cm long, 3–5 cm broad, glabrous, scaleless or younger fronds bearing a few widely scattered, irregularly stellate-few-rayed, dark brown scales 0.2–0.5 mm across; margins of blade thickened and cartilaginous; costa prominent, pale brown; veins numerous, forked once at or near base, occasionally anastomosing part way to margin, then forking again, rarely simple, branches nearly parallel, obvious to faintly visible on 1 or both sides; fertile fronds few, in ratio of about 1 to 7 sterile fronds, their stipes usually about twice as long as or slightly longer than laminae of sterile fronds, wholly covered beneath by crowded sporangia except at naked cuneate-decurrent base; sporangia deep reddish brown, ovoid to subglobose, slightly flattened laterally, about 0.5 mm long, 0.3 mm wide, 0.2 mm thick, the annulus broad, conspicuous, of 12–14 thickened cells; spores broadly ellipsoid, 45–50 μ long, 35–40 μ wide, irregularly cristate-ridged in coarse reticulum on convex surface, reddish brown.

A common inhabitant of moist habitats in Colombia and the West Indies. Specimens examined: Isabela, Santa Cruz.

See Fig. 17.

Elaphoglossum glossophyllum Hieron., Hedwigia 44: 180. 1905

Elaphoglossum linguiforme Hieron., Bot. Jahrb. Engler 34: 542. 1904, non Moore 1857.
Elaphoglossum petiolatum sensu Svens., Bull. Torrey Bot. Club 65: 318. 1938, pro parte, non Urban.

Rhizome long-creeping, 2–2.5 mm thick, 5–10 cm long or more, brownish, nearly naked except near apex, here usually densely clothed with ovate scales 1–2 mm wide, 2–5 mm long, their median stripe lustrous, dark brown to nearly black near apex of each, otherwise reddish brown, dullish; fronds 5–20 mm apart, erect; stipes slender, 1–2(3) mm across, 10–20 cm long, glabrous and naked except for a few scales at extreme base; laminae lanceolate-acuminate, 1–2 dm long, 3–5 cm wide, glabrous above, sparsely scaly with minute deltoid-linear, peltately affixed scales, these usually short-ciliolate at base; veins slender, 0.6–1 mm apart, immersed and inconspicuous above, more obvious beneath, parallel, free at tips; fertile fronds longer and narrower than the sterile, their laminae 12–17 cm long, 2–3.5 cm wide; sporangia and spores light reddish brown, sporangia glossy, about 0.4 mm long, 0.3 mm wide, distinctly flattened laterally; annulus of 12–14 thickened cells, rather broad; spores ellipsoidal, 38–45 μ long, 28–33 μ broad, with narrow, rounded ridges forming irregular, sinuate-outlined, coarsely meshed reticulae on outer surface, the ridges 2–3 μ high.

Moist, shaded places and precipitous slopes and sides of ravines, Colombia. Specimens examined: Isabela, Santa Cruz.

Elaphoglossum minutum (Pohl) Moore, Ind. Fil. 12. 1857

Acrostichum minutum Pohl ex Fée, Mém. Foug. 2: 39, *t. 10, f. 3.* 1845.
Acrostichum leptophyllum Fée, op. cit. 45, *t. 17, f. 1,* non Lam. 1805.
Elaphoglossum leptophyllum Moore, Ind. Fil. 11. 1857.*

Rhizome short-creeping, 2–2.5 mm thick, reddish brown, closely covered near apex with reddish brown, lance-deltoid scales 1.5–2 mm long, their median stripe usually

* Additional synonymy: see Alston, Bol. Soc. Broteriana II, 32: 16. 1958.

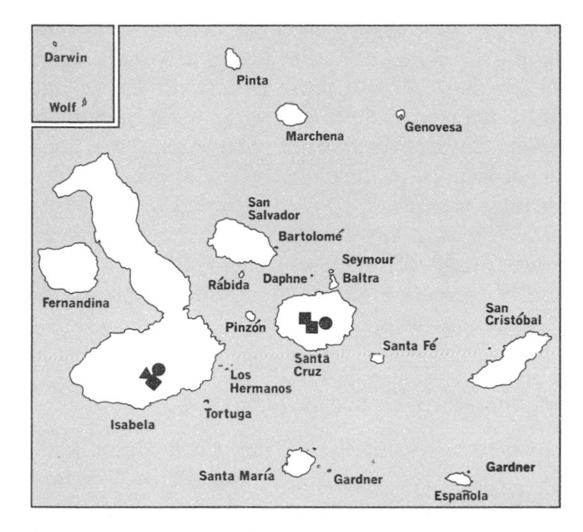

● *Elaphoglossum glossophyllum*
■ *Elaphoglossum minutum*
▲ *Elaphoglossum tenuiculum*
◆ *Elaphoglossum yarumalense*

lustrous and nearly black, margins very thin, often ciliolate but cilia soon deciduous; stipes more or less distichous, 3–6 mm apart, 4–8 cm long, 0.4–0.8 mm broad, slender, rounded abaxially, shallowly and broadly canaliculate adaxially, laterally minutely subalate or lined, glabrous or sparsely scaly at extreme base, the scales ovate, broad, not dissected, bicolorous; laminae linear-oblanceolate, acute to short-acuminate at apex, gradually attenuate toward base, 8–20 cm long, 7–15 mm wide, bright green and glabrous above, paler and sparsely sprinkled below with minute, dark brown, orbicular, ciliolate scales about 0.2 mm broad, and deltoid-lanceolate, fimbriate-ciliolate, brown scales 0.4–1 mm long, these completely random in directional orientation; sporangia very closely crowded over entire abaxial surface of fertile lamina, obovoid-subglobose, 0.3–0.4 mm long, about three-fourths as wide, turgid, cinnamon brown, stalks equaling or exceeding body; annulus dull to sublustrous, of 12–14 slightly thickened cells; spores oblong or broadly ellipsoidal, 40–55 μ long, 28–38 μ across, dark reddish brown, irregularly and moderately coarsely reticulate with low, rounded ridges.

On rotting logs, moist banks, and sides of ravines in shrubby or forested areas, Costa Rica and northern South America. Specimens examined: Santa Cruz.

Elaphoglossum tenuiculum (Fée) Moore ex Christ, Monogr. Elaph. 93. 1899

Acrostichum tenuiculum Fée, Mém. Foug. 10: 6, *t. 29, f. 2.* 1866.

Rhizome short-creeping, sparingly branched, 1.5–3 mm thick, densely scaly with dark reddish brown, attenuate, more or less sinuate scales 2–3.5 mm long, 0.2–0.4 mm wide at base, appressed or tips slightly spreading; fronds subdistichous, 4–8 mm apart, erect, 5–30 cm tall, stipes very slender, greenish to stramineous, 7–18 cm long, 0.3–0.8(1) mm across, terete but canaliculate adaxially, squamose and slightly to moderately beset with slender-rayed, stellate trichomes arising from umbonate sessile processes; laminae narrowly oblanceolate to linear-lanceolate, sterile ones 4.5–15 cm long, 3–11 mm wide, gradually attenuate at base, acute or attenuate at apex, moderately squamose on upper surface with very few lanceolate and many orbicular, sessile scales on bosses bearing 2–8 pale rufous to whitish stellately radiating trichomes, these varying from subappressed to ascending-spreading, 0.2–0.8 mm long, lower surface somewhat more densely invested; midnerve visible but inconspicuous above, moderately elevated and prominulent beneath, more densely stellate beneath than above; fertile fronds similar to but slightly smaller than the sterile; veins immersed and obscure, branching once near base or simple, free; sporangia densely crowded over whole abaxial surface (obscuring midnerve) of fertile lamina, obovoid, 0.4 mm long, about 0.2 mm wide, olivaceous to reddish brown, sublustrous, annulus of 10–12 thickened cells; spores broadly ellipsoidal, 35–45 μ long, 25–30 μ across, pale reddish brown, irregularly and coarsely reticulate on outer surface, walls of reticulae smooth, 3–5 μ high.

In moist habitats, usually in shade, Venezuela and Colombia. Specimens examined: Isabela, a new record for the Galápagos Islands.

Elaphoglossum yarumalense Hieron., Bot. Jahrb. Engler 34: 556. 1904

Rhizome rather massive, short-creeping to nearly erect, 5–8 mm thick, densely invested with dark brown, sublustrous, lanceolate or lance-deltoid scales 3–5 mm long, these rounded to subcordate at base, attenuate, irregularly denticulate, often somewhat sinuous; fronds approximate, 15–20(30) cm tall, erect, stipes terete or slightly compressed, 1–1.5 mm thick, rather closely clothed with pale rufous lanceolate to lance-ovate scales 2–5 mm long, these cordate at base, margins denticulate and dissected-ciliolate on lower half or more, laminae of sterile fronds oblong, 9–13 cm long, 2.5–3.2 cm wide, obtuse to rounded at apex, obtuse to broadly cuneate at base, bearing almost imbricate orbicular to broadly deltoid-ovate, stellately dissected scales on both surfaces, these mostly about 1 mm across or 1.5–2 mm long, those on upper surface whitish, those below rufous, scales along midnerve larger and darker, midnerve pale, prominent; veins visible but inconspicuous, branching once about one-fourth to one-third of way from costa, free, parallel, each branch terminating in a translucent, obovoid hydathode near margin; fertile fronds with stipes equaling to slightly surpassing entire sterile fronds, blades 5–9 cm long, 8–9 mm wide, densely scaly with imbricate scales on upper surface and along midnerve beneath, rest of abaxial surface wholly covered with crowded sporangia, these minute, about 0.2 mm long, half as wide, pale reddish brown, annulus of 10–13 moderately thickened cells; spores broadly ellipsoidal, about 55 μ long, 35 μ across, thin-walled, nearly smooth (possibly immature, but no reticulations visible).

Trunks of trees and rocky slopes in shady places, known elsewhere only from the type locality in Colombia. Specimen examined: Isabela, a new record for the Galápagos Islands.

Grammitis Sw., Jour. Bot. Schrad. 1800[2]: 17. 1802

Small to minute epiphytic ferns, or occasionally terrestrial. Rhizomes short-creeping to erect, dictyostelic, closely invested by small scales. Stipes hirsute with 1-celled, spreading hairs, scaleless, approximate, congested or rarely remote, firmly attached to rhizome, not articulate. Lamina simple, linear to lanceolate, entire, crenate, or shallowly to deeply pinnatifid to pinnatisect, setose or glabrate, scaleless, membranous to coriaceous or fleshy, the costa usually prominent, veins usually free and forked, readily visible or immersed and obscure. Sori usually on lowest acropetal veinlet of forked veins, round to elongate, sometimes confluent, exindusiate, with or without setae. Sporangia on slender stalks consisting of single chain of cells except at apex. Annulus of 8–16 thickened cells. Spores trilete, greenish.

About 400 species, sensu lato, widely distributed in the mountain forests of the tropics of the New and Old Worlds, but rare in Africa.

Grammitis serrulata (Sw.) Sw., Jour. Bot. Schrad. 1800[2]: 18. 1802

Acrostichum serrulatum Sw., Nov. Gen. Sp. Pl. Prodr. 128. 1788.
Xiphopteris serrulata Kaulf., Enum. Fil. 85. 1824.
Polypodium serrulatum Mett., Fil. Hort. Lips. 30. 1856.
Xiphopteris extensa Fée, Mém. Foug. 11: 14. 1866.
Polypodium duale Maxon, Contr. U.S. Nat. Herb. 16: 61. 1912.

Minute fern most commonly epiphytic on branches of trees or shrubs closely covered with mosses or liverworts, but sometimes growing on mossy rocks or soil; rhizomes slender, short, short-creeping or nearly erect, about 0.5 mm thick, rather closely invested by lanceolate to lance-ovate, brown, entire scales 0.6–1.5 mm long, these rounded to round-emarginate at base, acute to obtuse at apex; fronds 1–5 cm tall, all erect or the lower, vegetative ones somewhat spreading; sterile fronds about 2 mm wide, serrulately toothed about halfway to central costa, glabrous, teeth about 1 mm wide at bases, acute, a single brown, unbranched vein running to tip of each tooth; fertile fronds usually 1.5 to 2 times as long as the sterile, the entire stipe nearly as long as the slightly more remotely toothed sterile part of blade, this only 0.5–0.8 mm wide, the fertile part terminal, 3–13 mm long, scarcely toothed, margins merely sinuate, thin, incurved, fertile veins subparallel to costa at base, then ascending, sori elongate along base of vein and thus subcostal, confluent; sporangia pale brown, about 0.2 mm long, numerous, closely crowded the full length of fertile part of frond; spores globose or slightly ovoid, slightly greenish in transmitted light, trilete but outlines of faces often indistinct, surface smooth or only faintly roughened.

Common in wetter parts of the American tropics, occurring also in Mauritius and tropical Africa; often difficult to find, owing to its habit of growing among dense mats of mosses and liverworts. Specimens examined: Pinta, Santa Cruz, a new record for the Galápagos Islands.

The 2 specimens examined are the only known specimens of the species and

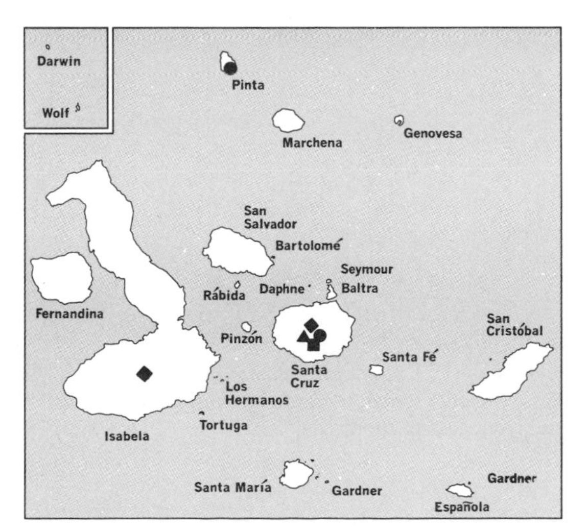

● *Grammitis serrulata*
■ *Hemionitis palmata*
▲ *Hypolepis hostilis*
◆ *Mildella intramarginalis*

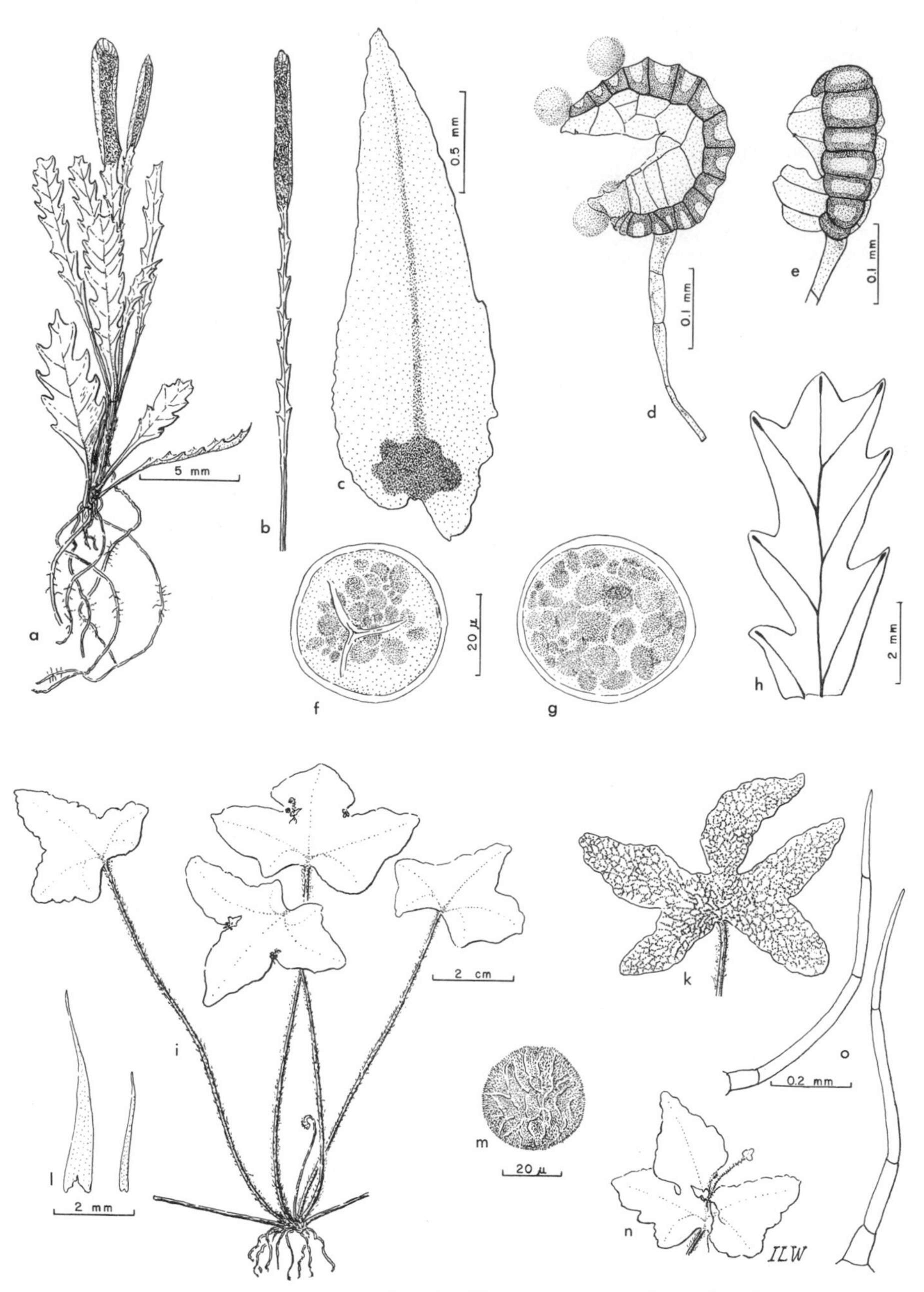

Fig. 18. *Grammitis serrulata* (*a–h*). *Hemionitis palmata* (*i–o*): *k*, pattern of sporangia on blade; *l*, rhizome scales; *n*, blade bearing plantlet; *o*, paraphyses from among sporangia.

genus from our area. Probably it is far more abundant than the paucity of collections would indicate, for it is very inconspicuous.

See Fig. 18.

Hemionitis L., Sp. Pl. 1077. 1753; Gen. Pl. ed. 5, 485. 1754

Terrestrial ferns with a dictyostelic, suberect to short-creeping rhizome bearing thin, brown, slender scales, the smaller ones becoming hairlike, the broader ones dark-costate in age. Fronds tufted, small to moderate in size, the maroon to dark stipes hairy at least at base; lamina somewhat dimorphic, the sterile fronds shorter-stiped than the fertile, the blades simple, herbaceous, roundish, cordate, pentagonal or palmately lobed, not elongate, the margins entire or lobed. Venation reticulate, without included veinlets. Sporangia borne along veins, distinctly reticulate; annulus of 14–20 thickened cells; stomium large. Paraphyses present. Spores trilete, globose, reticulate-spinulose.

A genus of 7 American and a single Asiatic species, growing in moist, tropical habitats.*

Hemionitis palmata L., Sp. Pl. 1077. 1753

Rhizomes ascending or sometimes short-creeping, 1–3 cm long, 3–4 mm thick, closely invested by pale brown, lanceolate-attenuate scales, bearing fragments of old stipes and many dark brown to nearly black roots; fronds erect or nearly so, in clusters of 5–15; stipes of sterile fronds 5–10 cm tall, reddish brown to chestnut brown, shiny, more or less scaly and hairy at base, hairy above, the lamina palmately 3- to 5-lobed, 2–4 cm long, 3–5 cm wide, the lobes often unequal, obtuse at apex, crenate, 2 often considerably fused to produce an asymmetrical blade, deep green above, slightly paler beneath and with conspicuous, brown midveins in each lobe, arachnoid-pubescent on both surfaces or glabrate above; fertile fronds on stipes to 3 dm tall, scaly and hairy like those of the sterile, the fertile lamina up to 5 cm long, 3–7 cm wide, the lobes more symmetrical and narrower and more acute than those of sterile fronds, often producing vegetative plantlets near base of central lobe or in the sinuses between lateral and terminal lobes; sporangia numerous, rufous, closely crowded in reticulate patterns over more or less the whole lower surface, but failing to develop over irregular areas in some fronds, intermingled with slender, bristle-like paraphyses; spores slightly trigonous to round in outline, trilete, about 30 μ in diameter, closely but bluntly fine-spinulose over whole surface, with irregular, low ridges over much of outer surface.

In shaded, moist habitats from Mexico and the West Indies to Bolivia. Specimens examined: Santa Cruz.

The specimens examined constitute the only collection of the species made in the Galápagos Islands. The plants were in full fruiting condition when collected, and several fronds bore plantlets in the sinuses.

See Fig. 18.

Hypolepis Bernh., Neues Jour. Bot. Schrad. 1²: 34. 1806

Moderate-sized to large, erect or semi-erect ferns with slender, wide-creeping, sparsely hairy, nonpaleaceous rhizomes. Fronds distant, the stipes stramineous to

* According to Copeland (*Genera Filicum* 73. 1947).

atropurpureous, elongate, woody or subwoody, not articulate with rhizome, glabrescent, smooth or somewhat spinulose, canaliculate on upper side. Blades ample, oblong-ovate, bi- to quadripinnate, with relatively few subopposite, ascending pinnae; ultimate segments lobed or pinnatifid, glabrous to densely pubescent or viscid-glandular. Veins free. Sori circular, relatively small, solitary at end of anterior veinlet of each group at sinuses of ultimate segments, marginal or nearly so, indusiate by the turning inward of a rounded or oblong marginal flap of modified tissue, or rarely inframarginal and naked, the margin then unmodified. Receptacle punctiform or transversely oblong, slightly elevated. Sporangia obovoid, only moderately tumid, the annulus of 12–16 thickened cells. Spores monolete or trilete, oblong, usually spinulose or tuberculate-warty, rarely smooth.

About 45 species, pantropical and ranging from Japan to New Zealand and in South Africa.

Hypolepis hostilis (Kunze) Presl, Tent. Pterid. 162. 1836

Cheilanthes hostilis Kunze, Linnaea 9: 86. 1834.
Hypolepis parviloba Fée, Crypt. Vasc. Brésil. 1: 53, *t. 20, f. 1.* 1869.
Hypolepis buchtienii Rosenst., Rep. Spec. Nov. Fedde 25: 58. 1928.
Hypolepis repens sensu Svens., Bull. Torrey Bot. Club 65: 319. 1938, non Presl.

Rhizomes (in ours) about 3 mm thick, ebenaceous, sublustrous, sparsely hirsute with very coarse, simple, dark reddish brown hairs near apex, soon glabrescent,

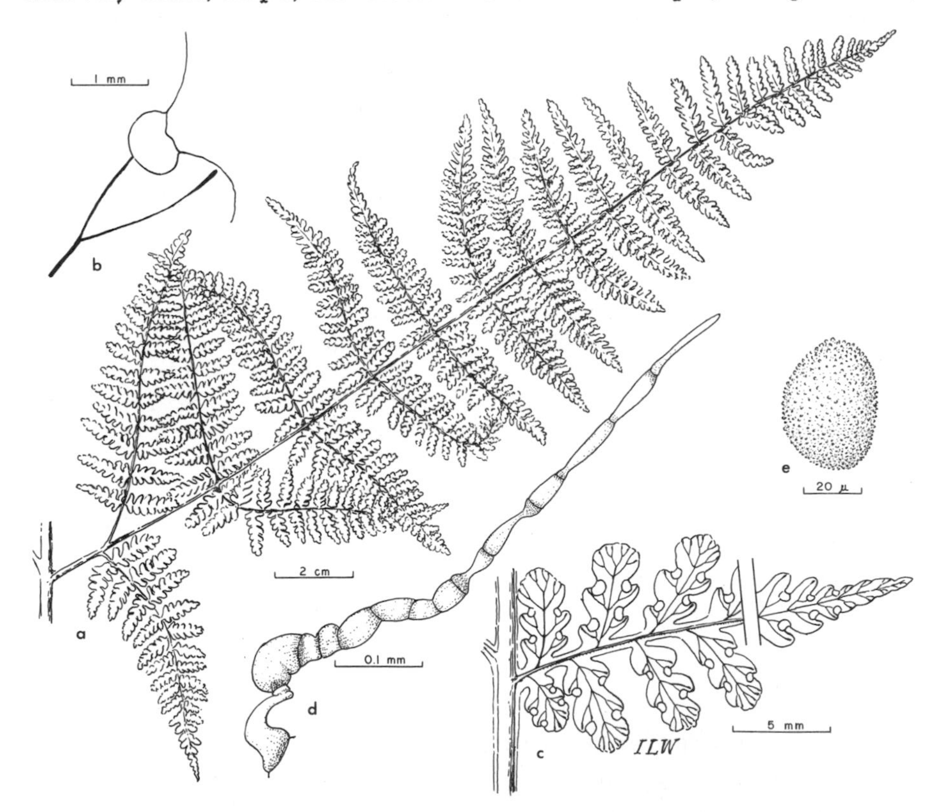

Fig. 19. *Hypolepis hostilis*: *a*, one of lowermost pinnae; *b*, relationship of vein and branch to sorus; *c*, pinnule; *d*, hair from tip of rhizome; *e*, spore.

minutely and sparingly aculeate; fronds to 2 m tall, stipe ebenaceous to atropurpureous at base, gradually lighter upward and rusty-stramineous at base of blade, lustrous, deeply canaliculate on dorsal side; blade quadripinnate to quadripinnatifid-pinnatifid, about equaling stipe, the rachis and rachillae of pinnae straight, pale stramineous, moderately to sparsely aculeate, glabrous or in some very sparsely hirsute with spreading, coarse, simple hairs; pinnae ascending, equilateral or essentially so, to 2 dm long, 8–12 cm wide at base, lower ones only slightly larger than the upper; ultimate segments oblong, rounded to subacute at apex, 7–10 mm long, 2–3.5 mm wide, rather deeply crenulate along margins save on smallest ones near tip of pinna, in ours sparsely pubescent beneath, especially along costules, with short, stoutish, nonglandular, sharp-pointed hairs intermingled with more numerous few-celled, gland-tipped trichomes; sori about 1 mm in diameter, usually 3–4 on an ultimate segment, pale ferruginous; spores pale yellow, broadly elliptic, 40–45 μ long, 30–35 μ broad, irregularly and roughly tuberculate-warty over whole surface.

In forests and shrubby cover or along their margins and in cut-over areas, Colombia and Brazil to Bolivia. Specimens examined: Santa Cruz.

Very abundant on Santa Cruz.

See Fig. 19.

HYPOLEPIS SPECIES EXCLUDED FROM THE FLORA

Hypolepis repens (L.) Presl, Tent. Pterid. 162. 1836

Lonchitis repens L., Sp. Pl. 1078. 1753.
Dicksonia aculeata Spreng., Neue Entdeck. 3: 7. 1822.
Cheilanthes repens Kaulf., Enum. Fil. 215. 1824.

This species was reported from the Galápagos Islands by Hooker (1858, 2: 64) on the basis of a Wood collection, which has not been seen. If in fact it came from the Galápagos, it probably represents *H. hostilis* rather than *H. repens.*

Mildella Trev., Rendic. Ist. Lombardo, Milano II, 9: 810. 1876

Small terrestrial or epiphytic ferns to about 4 dm tall, the fronds rather stiff, clustered near tips of short-ascending to much-creeping, multicipital rhizomes. Rhizome scales closely crowded, linear-lanceolate, markedly bicolorous. Stipe and rachis stramineous, light brown, castaneous or atropurpureous to black, adaxially flattened or grooved, smooth, lustrous, bearing many, often closely crowded, 1- to few-celled, coarse, stiff to lax, cylindric to slightly clavate hairs on adaxial face. Blades pinnate-pinnatifid to bipinnate, usually shorter than stipes. Basal pinnae opposite, inequilateral, basiscopically developed with broadly adnate, triangular, oblong or linear, subcoriaceous segments. Inferior basal segment of basal pinna pair entire to more or less pinnatifid. Upper pinnae gradually simpler and smaller, less strongly inequilateral. Upper segments simple or 1-lobed, usually ascending, broadly adnate and decurrent on rachis, and often mutually adnate, especially near apex of frond. Veins free, 1-forked or simple, the tips usually almost reaching margins of segments and flaring to bear (1)2 short-stalked sporangia. Indusium thin, inframarginal, entire, erose or ciliolate, closely overarching sporangia. Annulus of 18–22 thick-walled cells. Spores tetrahedral, microscopically roughened, light gray, pale brown, or dark brown.

About 8 species, ranging from Mexico and Panama, usually at higher elevations, to Haiti, and from Pakistan and India into western China.

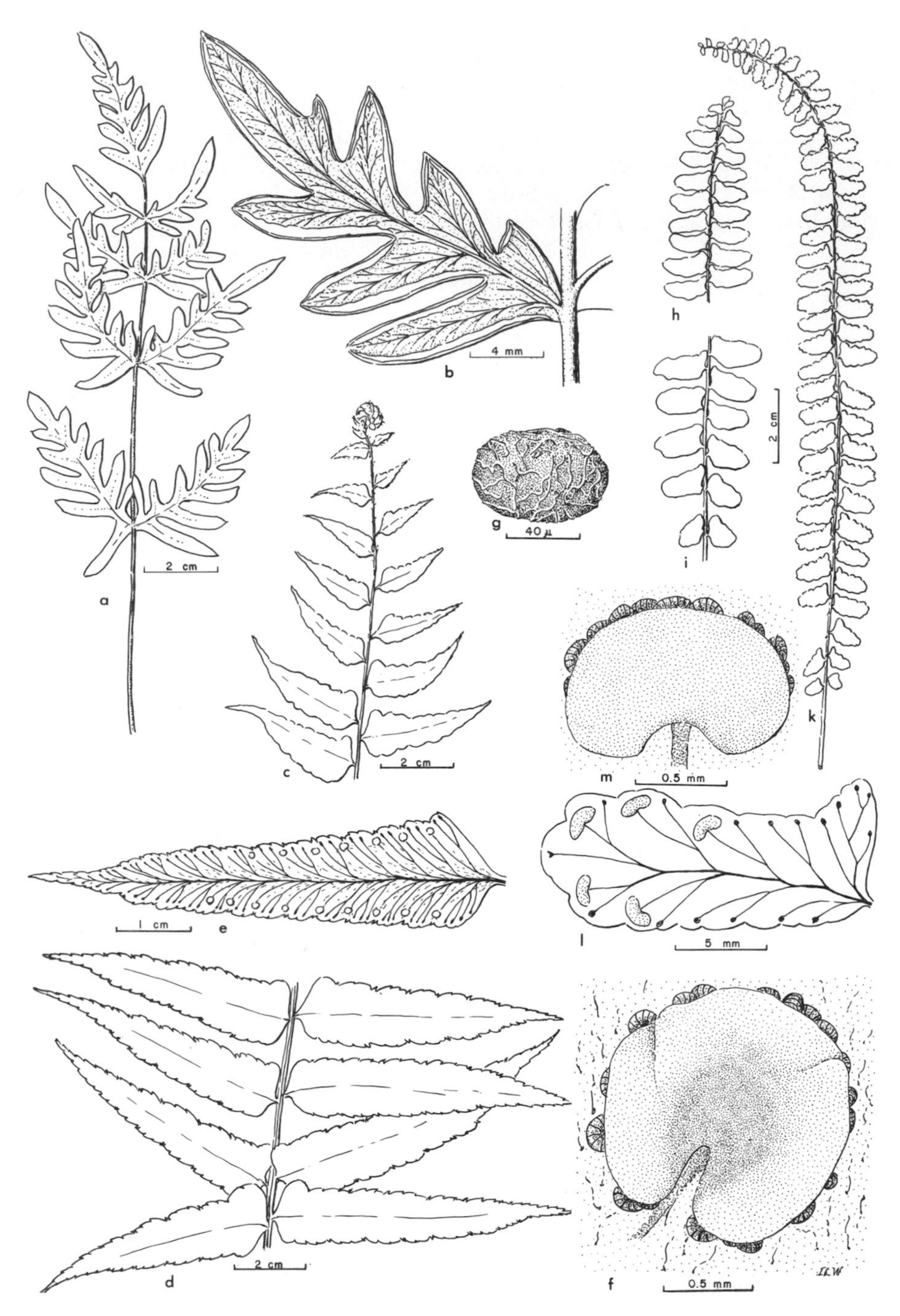

Fig. 20. *Mildella intramarginalis* var. *intramarginalis* (*a, b*). *Nephrolepis biserrata* (*c–g*). *N. cordifolia* (*h–m*).

Mildella intramarginalis (Kaulf. ex Link) Trev., Rendic. Ist. Lombardo, Milano II, 9: 810. 1876, var. **intramarginalis**

Pteris intramarginalis Kaulf. ex Schlecht. & Cham., Linnaea 5: 613. 1830, nomen nudum.
Pteris intramarginalis Kaulf. ex Link, Hort. Reg. Bot. Berol. ed. 2, 2: 34. 1833.
Cheilanthes intramarginalis Hook., Sp. Fil. 2: 112. 1852.
Pellaea intramarginalis J. Smith, Cat. Kew Ferns 4. 1856.

Rhizomes at first simple and erect, soon multicipital with ascending branches, short-creeping or at length long-creeping, 5–8 cm long, 3.5–5 mm thick, densely scaly at and near tip, eventually nearly naked, dark reddish brown; rhizome scales narrowly lanceolate to almost linear, the larger 3–4.5 mm long, 0.4–0.8 mm wide near base, attenuate toward apex, strongly bicolorous, the central stripe lustrous, atropurpureous to nearly black, this sclerotic band in ours one-half to three-quarters the width of scale, margins very thin, pale brown to castaneous, minutely and irregularly ciliolate or papillate; fronds 12–40 cm tall, erect or ascending, rather stiff; stipes and rachises shallowly and broadly grooved adaxially, castaneous to dark brown, closely covered with short, stiff, mostly bicellular hairs along groove, the rest glabrous and lustrous; blades long-lanceolate to ovate-triangular, 7–15 cm long, 3–6(7) cm wide; lower pinnae inequilateral-triangular, to 5 cm long, spreading-ascending, pinnatifid or pinnate, basiscopic, the segments adnate, to 2 cm long, 4 mm wide, diminishing toward tip of pinna; upper adnate segments simple, gradually becoming smaller and producing pinnatifid apex; indusium papery, nearly hyaline, about 0.6 mm wide, continuous except for small gap at apex of each segment, dull, glabrous, entire; sporangia slightly exserted from indusium at maturity, tumidly obovoid, 0.3–0.35 mm long, the annulus dark brown, dull; spores dark gray (deep brown by transmitted light), about 70 μ in diameter, irregularly and relatively coarsely scabrous-roughened on outer surface.

Among rocks and on cliffs, Mexico to Panama. Specimens examined: Isabela, Santa Cruz.

A notable extension of range, the previously southernmost station being in the region of Volcán Chiriquí, Panama.

See Fig. 20.

Nephrolepis Schott, Gen. Fil. *pl.* 3. 1834

Terrestrial or epiphytic ferns with erect, scandent, or creeping rhizomes and usually stoloniferous, the rhizomes paleaceous and bearing tubers in some species. Fronds erect to laxly pendent, firmly attached to rhizome, not articulate, slowly determinate, narrow, elongate, the stipe and rachis both firm, lustrous; blades mostly linear, uniform nearly full length, simple-pinnate with many spreading, sessile or petiolulate pinnae, these articulate to rachis and deciduous. Pinnae linear to narrowly oblong, subentire, denticulate, or crenate, usually more or less auriculate on upper side at base. Veins free, oblique, 1–4-forked, the ultimate branches markedly clavate at apex, sterile ones excurrent nearly to margin. Sori terminal on first distal veinlet, supramedial to submarginal, in single row on each side of costa. Indusia lunate to orbicular, firmly attached at sinus, oblique or not, persistent but sometimes obscured in age. Sporangia numerous, rusty brown, stalked, the stalks slightly shorter than, equaling, or slightly exceeding body. Spores monolete.

Possibly 20 species, widely distributed in the Old and New Worlds.

See color plates 55 and 56.

Indusia orbicular, the sporangia radiating in all directions; pinnae large, 6–12 cm long or more, pilose along midrib beneath, margins biserrate . . . . *N. biserrata*
Indusia semicircular or lunate-reniform, the sporangia pointing toward apex and upper margins, not past the sinus of the indusium; pinnae small, 1–3 cm long, glabrous beneath, margins crenate . . . . *N. cordifolia*

Nephrolepis biserrata (Sw.) Schott, Gen. Fil. *sub pl.* 3. 1834.

Aspidium biserratum Sw., Jour. Bot. Schrad. 1800²: 32. 1802.
Nephrolepis acuta sensu Robins., Proc. Amer. Acad. 38: 110. 1902, non Presl.

Terrestrial or epiphytic; rhizomes erect or ascending, 10–25 cm long, stout, densely covered with many roots to 1.5 mm broad and with lance-attenuate, pale castaneous, lustrous scales to 8 mm long, freely stoloniferous; fronds wide-spreading, ascending-spreading, or the largest often recumbent, 1–5 m long, the stipes short (1–4 dm long), stout, angulate, pale brown, lustrous, densely clothed with slender, paleaceous, pale brown scales basally, becoming scaleless or nearly so above; blades linear-oblong to broadly linear, 0.7–4 m long, 15–40 cm wide, only slightly narrowed at base, the rachis 2–3.5 mm broad, pale brown, lustrous, finely fibrillose-paleaceous, canaliculate along upper side; pinnae many, alternate, 1–2 cm apart, spreading, linear-attenuate, 5–15 cm long, 1–2.5 cm wide, subcaudate at apex, roundish to rounded-auriculate on 1 or both sides at base, the margins variable, finely dentate-serrulate in sterile pinnae, curved-crenate with finer serrulate teeth on crenations in fertile pinnae, the petiolules 1–2.5 mm long, surrounded by dense rings of hairs or minute, short scales, finely rusty-hirsutulous beneath; leaf tissues membrano-herbaceous, translucent; veins 2–4-forked, 2–4 sterile branches running to each crenation; sori intramarginal, about 2 mm in diameter, the contiguous ones sometimes touching but usually discrete; indusia orbicular, attached at base of very narrow sinus; sporangia many, radiating in all directions from attachment of indusium, tumidly obovoid, about 0.2 mm wide, two-thirds as thick, relatively long-stalked, rusty brown, lustrous, the annulus of 16–20 thickened cells; spores ellipsoidal-oblong, rusty red, irregularly verrucose-reticulate, 44–48 μ long, 26–34 μ wide.

In shaded, moist places, West Indies, southern Florida, and Mexico to Peru and

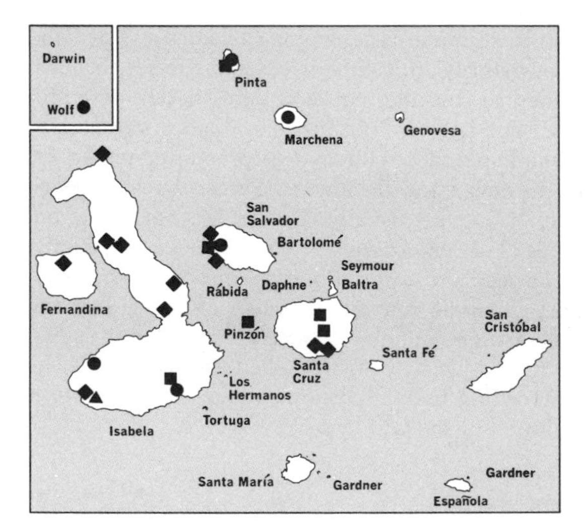

● *Nephrolepis biserrata*
■ *Nephrolepis cordifolia*
▲ *Notholaena aurea*
◆ *Notholaena galapagensis*

southern Brazil; widely distributed in the tropics of the Old World. Specimens examined: Isabela, Marchena, Pinta, San Salvador, Wolf.

Strangely, no specimens of this species from Isla Santa Cruz have come to our attention. However, since it forms extensive brakes on Islas Isabela and Pinta at elevations no greater than those attained on Isla Santa Cruz, it reasonably can be expected to occur on Santa Cruz. It should be sought in caverns and on shaded cliffs from the denser forests of the *Scalesia* Zone to above the *Miconia* Zone.

See Fig. 20.

Nephrolepis cordifolia (L.) Presl, Tent. Pterid. 79. 1836

Polypodium cordifolium L., Sp. Pl. 1089. 1753.
Nephrodium pectinatum sensu Hook. f., Trans. Linn. Soc. Lond. 20: 170. 1847, non Link.
Nephrolepis pectinata sensu Robins., Proc. Amer. Acad. 38: 110. 1902, and sensu Stewart, Proc. Calif. Acad. Sci. IV, 1: 22. 1911, non Schott.

Mostly epiphytic on trunks of trees in ours, but also on ground and on shaded cliffs; rhizomes suberect, 1–12 cm long, densely invested by slender, thin, ferruginous scales and bearing numerous wiry stolons 0.4–0.8 mm thick, these bearing ovoid, paleaceous tubers 1–1.5 cm long from 1 to 10 cm apart; fronds numerous, erect, spreading, or subpendent, 2–8 dm long, the stipes 5–15 cm long, 0.5–1.5 mm thick, light brown, lustrous, densely fibrillose-paleaceous in youth but soon nearly glabrous; blades narrowly linear, 2–6 dm long, 2–6 cm wide, attenuate toward base, the rachis deciduously fibrillose-paleaceous like stipes; pinnae numerous, alternate, horizontal, contiguous to imbricate, oblong or linear-oblong, from an unequal base, larger ones 1.5–3 cm long, 4–8 mm wide, rounded to subacute at apex, subcordate on lower side at base, the auricle variously developed and often overlying rachis, margins cartilaginous, lightly crenate; leaf tissue coriaceous or thick-herbaceous, deciduously rusty-fibrillose beneath; veins oblique, 1-forked, immersed; sori slightly nearer margins than costa, in single row on each side of costa, 1.5–2 mm wide; indusia broadly reniform to lunate, broadly attached in the wide sinus, firm, thin, persistent, the distal edge facing apex of pinna; sporangia about the same as those of *N. biserrata*, but all facing toward apex of indusium, reddish brown, lustrous; spores broadly ellipsoidal, 40–50 μ long, 25–30 μ wide, rusty brown, verrucose-reticulate and slightly warty.

In shaded, damp situations on soil, or epiphytic, Mexico and the West Indies to southern Brazil and Bolivia; widely distributed in the Old World tropics. Specimens examined: Isabela, Pinta, Pinzón, San Salvador, Santa Cruz.

See Fig. 20.

Notholaena R. Br., Prodr. Fl. Nov. Holl. 145. 1810

Small, xerophytic ferns of rocky habitats with short, multicipital, hairy or scaly rhizomes. Fronds erect or ascending, rather rigid, often borne in tufts. Stipes usually dark and rigid, often longer than blades; blades linear to pentagonal, pinnate to quadripinnate, more or less lobed or pinnatifid, firm-herbaceous to coriaceous, sometimes white- or yellow-ceraceous beneath. Veins all free, usually enlarged at tips. Sori submarginal, orbicular to oblong, borne at tips of veins. Indusium none, leaf margins not modified, sometimes revolute and partially concealing sporangia, in some the sporangia more or less protected and obscured by hairs, scales, or the waxy coating of undersurface of blade.

About 60 New World species and some in the Old World, mainly in arid regions.

Blades rusty-tomentose beneath, not ceraceous, linear, reduced at base, longer than stipe; pinnae not basiscopically developed .... *N. aurea*
Blades yellow-ceraceous beneath, not hairy, broadest at base, shorter than stipe; pinnae basiscopically enlarged .... *N. galapagensis*

Notholaena aurea (Poir.) Desv., Mém. Soc. Linn. Paris 6: 219. 1827

Pteris aurea Poir., Encycl. Méth. 5: 710. 1810.
Notholaena ferruginea Hook., Sec. Cent. of Ferns *sub t.* 52. 1861, non Desv. 1813.
Notholaena bonariensis C. Chr., Ind. Fil. 6. 1905; 459. 1906.*

Rhizome short-creeping, 3–4 mm thick, dark brown, scales lance-linear, 1–1.8 mm long, the center line castaneous, a lustrous sclerotic, black band, the margins pale brown, entire, thin, hyaline; stipes rather closely spaced, terete, 1.5–4(10) cm long in our material, castaneous to nearly black, invested with coarse, nearly straight, subappressed and some laxly spreading, whitish septate hairs 0.4–1 mm long; fronds 10–20(60) cm tall; lamina linear-elliptic, long-attenuate at base, pinnate-pinnatifid almost to apex, this gradually or abruptly narrowed, acute to obtuse; pinnae numerous, 25–40 pairs, oblong or deltoid-oblong, in ours mostly 5–8 mm long, obtuse or acute, cut half to three-fourths of distance to costa into linear-oblong, entire, obtuse lobes, sparsely villosulous with whitish to golden-tinged, several-celled, straight hairs on upper surface, more densely so beneath, densely white- to tawny-tomentose when young; veins oblique, 1–2-forked; sporangia borne on clavate to subpunctiform ends of veins, margin thin and subhyaline, narrowly and rather closely revolute; sporangia obovoid, tumid, 0.4–0.5 mm long, annulus dark brown to nearly black, more or less lustrous, of 22–24 thickened cells; spores trilete, dark brown, mostly 60–65(75) μ in diameter, outer surface faintly and irregularly reticulate, reticulations very low and indistinct.

In a variety of habitats, from moist to dry, usually on or among rocks, southwestern United States, through Mexico, West Indies, and Central America, south along the Andean regions to Argentina and Chile, occurring mainly between 700 and 4,000 m. Specimens examined: Isabela, a new fern record for the Galápagos Islands.

See Fig. 21.

Notholaena galapagensis Weatherby & Svens., Bull. Torrey Bot. Club 65: 319, *pl. 1, fig.* 2. 1938.

Notholaena candida sensu Hook., Sp. Fil. 5: 110. 1864, pro parte, with respect to Galápagos specimens.
Notholaena sulphurea sensu Stewart, Proc. Calif. Acad. Sci. IV, 1: 22. 1911, non J. Smith.

Rhizome short and oblique, invested with castaneous to brownish, lustrous, lanceolate-subulate, acuminate scales with a very narrow, paler, irregularly serrulate margin, sometimes with a few cilia and an ephemeral trichome at tip, the walls of its cells thick; fronds 7–26 cm tall; stipes slender, castaneous, shiny, terete, with lance-ovate, cordate-based scales near base, these nearly concolorous; blades ovate to deltoid-ovate, lowest pair of pinnae largest; pinnate-pinnatifid or bipinnate-pinnatifid at base, acute or slightly acuminate at pinnatifid apex, texture firm-herbaceous to coriaceous, the veins hidden; upper surface glabrous, lower surface pale yellow- to white-ceraceous; pinnae 5–7 pairs, lowest pair distant, ovate-lanceolate to deltoid-

* For complete synonymy, see Tryon (1956, 35–36).

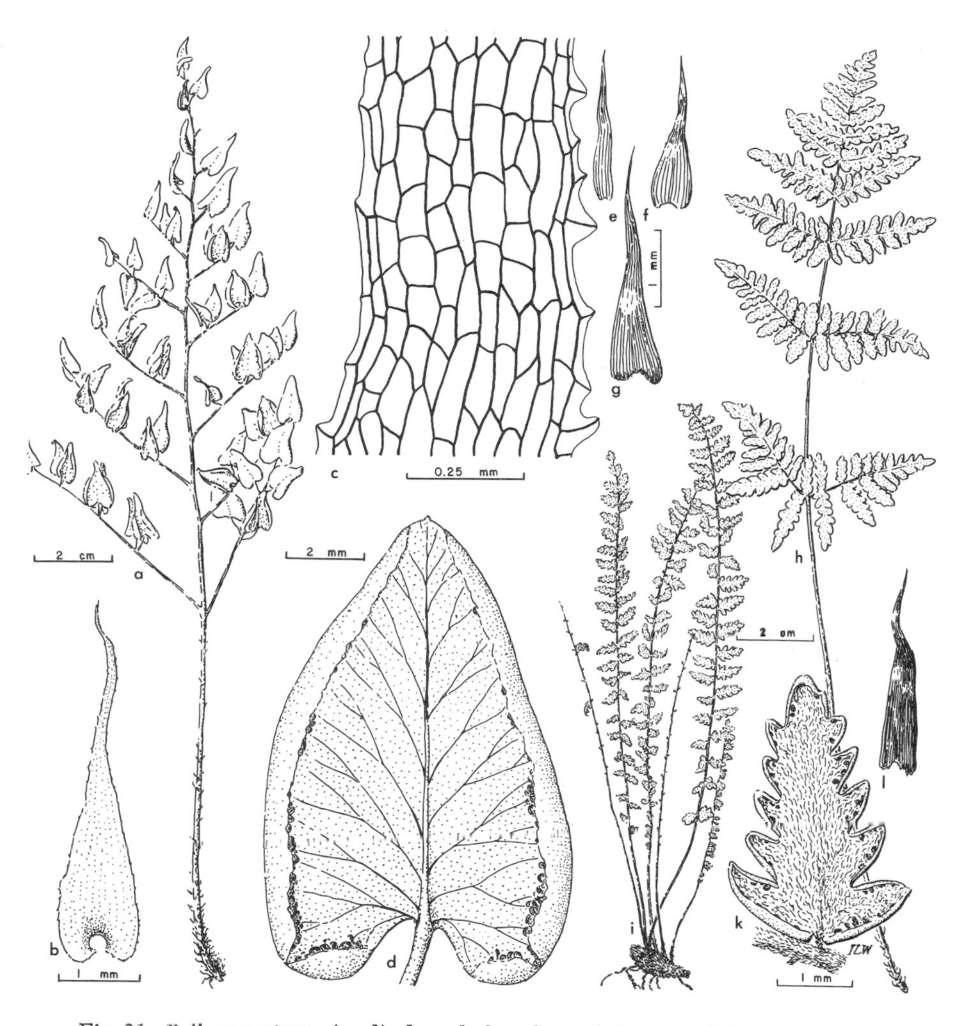

Fig. 21. *Pellaea sagittata* (*a–d*): *b*, scale from base of stipe; *c*, cellular arrangement of scale; *d*, pinnule, lower surface. *Notholaena galapagensis* (*e–h*): *e–g*, scales from stipe. *N. aurea* (*i–l*): *k*, lower surface of single pinna; *l*, scale from rhizome.

ovate, inequilateral, with variously lobed, toothed, or entire pinnules; upper pinnae linear to narrowly lanceolate, acute but with obtuse lobules; veins simple or 1-forked; sporangia borne on dilated tips of veins, each normally containing 64 spores; margins of leaf slightly revolute, unmodified.

Endemic; a plant of rocky habitats, from sea level upward to about 900 m. Specimens examined: Fernandina, Isabela, San Salvador, Santa Cruz.

See Fig. 21 and color plate 72.

Pellaea Link, Fil. Sp. Hort. Bot. Berol. 59. 1841, nom. conserv.

Small or rarely large, rigid, rock-inhabiting ferns with slender, long-creeping to short and stout, multicipital rhizomes. Scales narrowly linear-lanceolate, pale brown, ferruginous or brown to nearly black, usually concolorous and dull. Fronds erect to somewhat recurved; stipes and rachises wiry, slender, tan, pinkish, pale brown to purplish black, without wings or conspicuous glands, often more or less scaly near base, glabrous or less commonly somewhat pubescent; blades pinnate to quadripinnate, lanceolate, ovate or oblong; segments thick, leathery, articulate and deciduous, their costae often obscure. Veins free, 1–3-forked, tips usually broadened slightly. Sori intramarginal, elongate, terminal on the enlarged tips of veins, more or less contiguous and often mutually touching (not confluent), forming marginal line overarched by reflexed margins of fertile segments, margin often modified to a thin, membranous, nearly hyaline tissue. Paraphyses rarely present. Sporangia tumidly obovoid, dark brown or ferruginous, annulus of 12–18(20) thick-walled cells. Spores tetrahedral, 32 or 64 per sporangium, smooth to reticulate-spinulose.

A genus of 70 to 80 species occurring mostly in dry habitats from southern Canada to Chile and New Zealand and in South Africa.

Pellaea sagittata (Cav.) Link, Fil. Sp. Hort. Bot. Berol. 60. 1841, var. **sagittata**

Pteris sagittata Cav., Descr. Pl. 267. 1802.
Allosorus sagittatus Presl, Tent. Pterid. 153. 1836.
Platyloma sagittata J. Smith, Jour. Bot. Hook. 4: 160. 1841.
Pellaea cordata f. *sagittata* Davenp., Bot. Gaz. 21: 261. 1896.

Rhizome stoutish, compactly multicipital, decumbent; rhizome scales matted, tan to ferruginous, concolorous, dull, elongate, lanceolate-triangular, usually slightly to markedly cordate basally, straight or slightly curved or sinuous apically, margins minutely and irregularly denticulate, tip filiform; scales of stipe appressed, tawny to sordid white, concolorous, dull, broadly ovate-lanceolate, cordate to pseudopeltate, somewhat tortuous, clathrate with elongate cells parallel to axis; fronds 15–75 cm tall, erect, straight or nearly so, stiff, approximate, dimorphic, the sterile fronds shorter and usually with larger segments than the fertile; stipe and rachis glabrous or slightly puberulous, stramineous to tan or mottled, darker in age, stipe bearing sordid white scales nearly or completely its full length, breaking irregularly, rachis often somewhat zigzag-flexuous; blade 10–45 cm long, 5.5–32 cm wide, ovate-triangular to rhomboid, pinnate, more commonly bipinnate, rarely tripinnate, pale

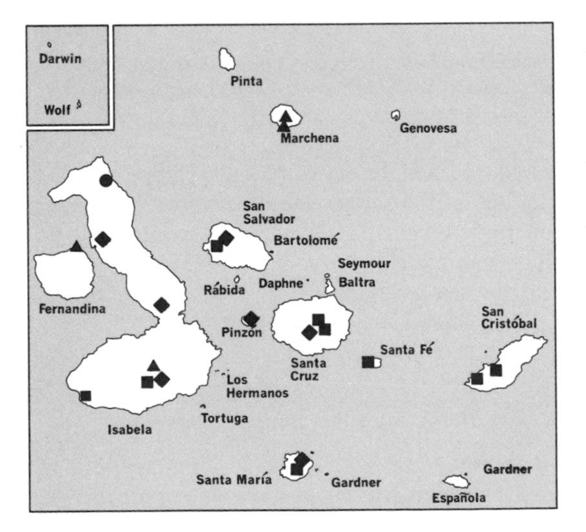

● *Pellaea sagittata*

■ *Pityrogramma calomelanos* var. *calomelanos*

▲ *Pityrogramma calomelanos* var. *aureoflava*

◆ *Pityrogramma tartarea*

green, pinnae broadly ascending; segments 3–18 per pinna, 5–55 mm long, 3–27 mm wide (in ours to 12 mm long, 9 mm wide), broadly ovate-cordate to oblong-sagittate, acute to obtuse, herbaceous to coriaceous, glabrous to puberulent, especially along the enrolled margins (mostly glabrous in ours); margin white or tawny, the membrane 0.6–0.8 mm wide, crenulate; sporangia with stalks about half as long as body, mostly pale reddish brown or ferruginous; spores ellipsoid to tetrahedral-subglobose, yellow to pale brown, sparsely marked with prominent rugae.

On dry banks, among rocks (mostly in limestone), on adobe and rock walls, in shade or full sun, northern Mexico and Guatemala to Colombia and Bolivia. Specimen examined: Isabela, a new record for the Galápagos Islands, the nearest known locality heretofore being in Imbabura and Pichincha provinces in Ecuador.

See Fig. 21.

Pityrogramma Link, Handb. Gewächse 3: 19. 1833

Small to moderate-sized terrestrial ferns of dryish banks and slopes in more or less open habitats; rhizomes dictyostelic, short-creeping to erect, densely invested with narrow, concolorous, brownish, attenuate scales. Fronds cespitose, uniform, not articulate to rhizome, erect or ascending; stipes often scaly at base, otherwise glabrous, firm, dark colored, lustrous. Blades herbaceous to subcoriaceous, pinnate to tripinnate, linear to deltoid, pentagonal, the undersurface covered with white, ochraceous, or bright yellow powder, or in a few species pubescent, sometimes glandular on upper surface, rarely bearing small scales. Veins free. Sori distributed along veins, naked, usually confluent at maturity. Spores trilete, globose-tetrahedral, with an irregularly reticulate and ribbed epispore.

About 12 species, mostly in tropical America but extending well into the temperate zones, a few present in Africa and Madagascar.*

Pinnae inequilateral; lamina long-triangular to deltoid, the apex evenly long-acuminate; apical pinnae nearly or quite at right angles to the rachis, obtuse or rounded, obtusely lobed or pinnatifid . . . . *P. tartarea*

Pinnae equilateral; lamina lanceolate to ovate-lanceolate, the apex often acute or abruptly acuminate; apical pinnae strongly ascending; pinnules ascending, those on basiscopic side more strongly so than those on acroscopic side, all acute, serrate or acutely pinnatifid:

- Wax on lower surface of blades white . . . . *P. calomelanos* var. *calomelanos*
- Wax on lower surface of blades yellow . . . . *P. calomelanos* var. *aureoflava*

Pityrogramma calomelanos (L.) Link, Handb. Gewächse 3: 20. 1833, var. **calomelanos**

Acrostichum calomelanos L., Sp. Pl. 1072. 1753.†

Rhizome decumbent to erect, 1–2 cm thick, to 7 cm long, mostly obscured by many slender, crowded roots, apex densely clothed with yellowish brown, lanceolate scales 2–3.5 mm long; fronds fasciculate, erect, 1–10 dm tall, stipes to 2.5 mm thick, nearly as long as blades, sparsely to moderately scaly basally, atropurpureous, lustrous, glabrous, conspicuously grooved on upper side; rachis similar to upper part of stipe; blades lanceolate to deltoid-ovate, acuminate, 2–6 dm long, 5–25 cm wide,

* Key adapted from R. M. Tryon (1962, 58–65).

† For additional synonymy, see Tryon (1962, 60).

Fig. 22. *Pityrogramma tartarea* (*a, b*). *P. calomelanos* var. *calomelanos* (*c–f*).

bipinnate; pinnae numerous, erect-spreading, sometimes moderately arcuate, adjacent or lower and middle ones subdistant, middle ones largest, triangular-lanceolate, attenuate, ours 1–5 cm wide, 2–18 cm long, stalked; pinnules on basiscopic side of pinnae slightly more ascending than those on acroscopic side, all more or less oblique, lance-elliptic to oblong, acute or terminal one acuminate, acutely serrate or largest ones obliquely and often only partially pinnatifid, lobes acute and serrulate; leaf tissue chartaceous to membrano-herbaceous, lustrous, dark green above, closely white-ceraceous beneath, the wax obscuring young sporangia but these soon extending above wax along the rather closely arranged veins; mature sporangia subglobose-obovoid, about 0.4 mm long, 0.3 mm wide, nearly as thick, annulus rusty brown, lustrous; spores reddish brown, 70–85 μ in diameter, irregularly and coarsely ridged around equatorial region, elsewhere cerebriform-reticulate, areas between ridges thin and nearly hyaline.

In dryish to moderately damp habitats in open or partial shade, southern Florida,

West Indies, and Mexico to Brazil, Argentina, Bolivia, and Paraguay; widely introduced into the Old World tropics. Specimens examined: Isabela, San Cristóbal, San Salvador, Santa Cruz, Santa Fé, Santa María.

See Fig. 22.

Pityrogramma calomelanos var. **aureoflava** (Hook.) Weatherby ex Bailey, Man. Cult. Pl. 64. 1926

Gymnogramma calomelanos var. *aureoflava* Hook., Gard. Ferns, *t. 50 text.* 1826.
Pityrogramma austroamericana Domin, Publ. Fac. Sci. Univ. Charles 88: 7. 1928.
Pityrogramma calomelanos var. *austroamericana* Farw., Am. Midl. Nat. 12: 280. 1931.

Differing from var. *calomelanos* only in bright yellow color of the wax on undersurface of blades.

Costa Rica to Bolivia, Argentina, and Brazil; introduced into the Hawaiian Islands, Java, and Tahiti, and probably into Africa. Specimens examined: Fernandina, Isabela, Marchena.

Pityrogramma tartarea (Cav.) Maxon, Contr. U. S. Nat. Herb. 17: 173. 1913

Acrostichum tartareum Cav., Descr. Pl. 2: 42. 1802.*

Rhizome obliquely ascending to erect, 1–2 cm thick, 1–8 cm long, clothed with numerous slender, hairy, reddish brown roots, the apex by stiff, acicular-lanceolate, dark brown, lustrous scales 2–8 mm long; fronds several, spreading-ascending, 5–15 dm tall, stipes as long as to half again as long as blade, 1.5–3 mm thick, dark brown to atropurpureous, lustrous, deeply canaliculate on upper side, brown-paleaceous with slenderly lanceolate scales to 1 cm long at base, glabrous above; rachis like upper part of stipe; blades elongate-deltoid, acuminate, bipinnate, in ours 1–6 dm long, 1–5 dm wide; pinnae at right angles to rachis to slightly arcuate-ascending, lower and middle ones stalked, distant, upper ones sessile and adjacent, lowermost the largest, to 15 cm long, 6 cm wide; pinnules distant, spreading, oblong, rounded-obtuse, or those near tips of pinnae often subacute, smaller ones entire, larger ones crenate to coarsely pinnatifid, the basiscopic about half again as long as the acroscopic, especially on lower pinnae; leaf tissue coriaceous, dark glossy green above, densely white-ceraceous beneath; sporangia distributed evenly along all veins, becoming nearly or quite confluent in early maturity, like those of *N. calomelanos*; spores dark reddish brown, 63–75 μ in diameter, equatorial band coarsely ridged, convex surface and most of commissural faces cerebriform-rugulose, areas between ridges thin, subhyaline.

Open to moderately shaded slopes, in lava caverns, and on partially shaded cliffs, West Indies and Mexico to Brazil and Bolivia. Specimens examined: Isabela, Pinzón, San Salvador, Santa Cruz, Santa María.

See Fig. 22.

Polypodium L., Sp. Pl. 1082. 1753; Gen. Pl. ed. 5, 485. 1754

Epiphytic or terrestrial ferns ranging from very small to large. Rhizome usually stout, erect, short-creeping to long-creeping, invested with paleaceous, entire to irregularly ciliate scales, these often more or less basally peltate. Stipes approxi-

* For additional synonymy, see Tryon (1962, 65).

mate to distant, articulate with rhizome and leaving round scars or short knobs on it, shorter than or equaling blade. Lamina simple and entire to pinnatisect or deeply pinnatifid, rachis often more or less alate, sinuses usually rounded but sometimes acute and rather narrow, tissues glabrous, squamose, or pubescent, or in some both scales and hairs present on 1 or both sides, glaucous or not, the tissue membranous to chartaceous. Veins anastomosing to form various sizes, shapes, and combinations of large and small areoles, or rarely free and simple or 1-forked, in those forming areoles 1 or more free veinlets usually present within at least some areoles (in the subgenus *Pleopeltis* veinlets forming a plexus within an areole, several of them joining to supply sorus). Sori usually circular, small to large, terminal or subterminal on free veinlets, in 1 to many series on a blade, superficial or partially immersed in mesophyll, with or without paraphyses. Indusium lacking. Sporangia few to many, annulus of 12–18 thick-walled cells, stalk short to moderate in length. Spores monolete, ellipsoidal, oblong, or oblong-reniform, smooth or more commonly verrucose-reticulate or strongly tuberculate.

A genus, sensu lato, of about 320 species, abundantly represented in the tropics and subtropics of the Old and New Worlds, with some species ranging well into the cold temperate regions of both the southern and the northern hemispheres.

See color plate 56.

Fronds simple:

Blades with obvious scales on lower surface; sori borne at confluence of several veins; stalked paraphyses with an enlarged, scalelike tip mixed with sporangia (Subg. *Pleopeltis*) .... *P. lanceolatum*

Blades scaleless on lower surface, smooth and shining; sori borne at tip of free, included veinlet, mostly in 2 rows between lateral veins; paraphyses none (Subg. *Campyloneurum*):

Blades less than 4 cm broad; lateral veins irregular and often obscure; primary areoles irregular, mostly elongate parallel with or at a low angle to the costa; venation highly irregular .... *P. angustifolium* var. *amphostenon*

Blades normally 4–8 cm broad; lateral veins straight and well developed; primary areoles transversely elongate, typically with 3 included veinlets, the 2 lateral soriferous at apex, the central sterile, often joining with the next outer, thus forming 2 minor areoles .... *P. phyllitidis*

Fronds deeply pectinately pinnatifid:

Venation copiously areolate, the costal areole sterile, the second areole fertile, the sorus at juncture of usually 2 included veinlets (or at apex of a solitary veinlet in juveniles); blades glaucous beneath, not scaly, with prominently raised venation; rhizome widely creeping, thick, epiphytic, with very large, bright brown scales (Subg. *Phlebodium*) .... *P. aureum* var. *areolatum*

Veins free or with 1 series of areoles, if areolate the costal areole fertile, the sori borne at tip of a solitary included veinlet; blades not glaucous beneath, the veins not raised; rhizome thinner, terrestrial or epiphytic, the scales smaller, usually darker (Subg. *Polypodium*):

Blades not scaly beneath on the surface, some few scales present on the costa beneath; veins free, 1- or 2-forked; plants terrestrial .... *P. dispersum*

Blades obviously scaly beneath and more or less above also; veins immersed, not obvious, usually partly anastomosing to form some costal areoles; plants epiphytic or terrestrial:

Rhizomes widely creeping, epiphytic, slender, 1–1.6 mm thick, the scales often pale brown with a darker central stripe; fronds distant, small, usually 5–9 cm long and 3–4 cm wide, not reduced at base, on stipes almost as long as the blades .... *P. polypodioides* var. *burchellii*

Rhizomes short-creeping, thicker, the scales often darker; fronds subfasciculate, larger:
  Scales on lower surface rather distant, not at all imbricate or completely covering the surface; blades not reduced at base; stipes almost or as long as blades; plants epiphytic .... *P. steirolepis*
  Scales on lower surface abundant, imbricate, completely covering the surface; blades reduced at base or not; stipes shorter than blades:
    Blades obviously reduced at base; stipes short, 3–4 cm long; segments never lobed .... *P. insularum*
    Blades not or scarcely reduced at base; stipes usually much longer, up to 14 cm long or more; segments entire, or often with 2 or 3 prominent lobes .... *P. tridens*

Polypodium angustifolium Sw. var. **amphostenon*** (Kunze) Hieron., Bot. Jahrb. Engler 34: 532. 1904

Polypodium amphostenon Kunze ex Klotzsch, Linnaea 20: 399. 1847.

Terrestrial and epiphytic; rhizomes short-creeping, thickish, 3–7 mm thick, firm, nodose, usually pruinose, deciduously paleaceous with dark brown, nearly concolorous, sparsely ciliolate, lance-ovate, attenuate scales 2–3.5 mm long, sometimes branched and multicipital; fronds numerous, approximate to subdistant (1 mm to 6 cm apart), dorsal on rhizome, erect, ascending or subpendent, 1–7 dm tall, the stipes very short, to 1 dm long, upper part bearing long-decurrent blade tissue as narrow wing; blades narrowly linear to oblong, 1–3.5 dm long or more, 3–25 mm wide, evenly long-attenuate downward, usually more abruptly so to a rather rigid, subalate-caudiform tip, simple, often falcate, plane or often somewhat repand, glossy green above, paler and duller beneath, glabrous, the margin entire, revolute, tissue firm-coriaceous in age or in drying; costa elevated beneath, prominent above, yellowish; veins deeply immersed, oblique, subequal, variously joined in 1–2 irregular rows of areoles, these mostly with 1–2 included veinlets in each; sori 1–5-seriate, dorsal or subterminal on free veinlets, 1.5–2 mm in diameter; sporangia pale reddish brown, obovoid, 0.4–0.5 mm long, about 0.3 mm wide, stalk about equaling body, annulus of 12–14 thick-walled cells, dark reddish brown, lustrous, the side walls many-celled, subhyaline; spores broadly ellipsoidal to weakly reniform, 75–80 μ long, 45–60 μ wide, coarsely verrucose-reticulate, deep brown, opaque.

On cliffs, walls, rocky banks, and trunks of trees, southern Florida, the West Indies, and Mexico to Argentina. Specimens examined: San Salvador, Santa Cruz. See Fig. 23.

Polypodium aureum L. var. **areolatum** (Humb. & Bonpl.) Baker in Mart., Fl. Bras. 1²: 528, *pl. 32, f. 12, pl. 33, f. 18.* 1870

Polypodium areolatum Humb. & Bonpl. ex Willd. in L., Sp. Pl. ed. 4, 5: 172. 1810.
Pleopeltis areolata Presl, Tent. Pterid. 193. 1836.
Goniophlebium areolatum Presl, op. cit. 186.
Phlebodium areolatum J. Smith, Jour. Bot. Hook. 4: 59. 1841.
Chrysopteris areolata Fée, Mém. Foug. 9: 27. 1857.

Rhizome to 1.5 cm thick, multicipital, densely clothed in reddish brown, ovate-deltoid to lanceolate-acuminate scales 5–9 mm long, 0.5–3 mm wide, the bases

* Corrected to "*amphistemon*" by Hieronymus. Though the original "*amphostenon*" is rather obscure etymologically, it should stand; "*amphistemon*" is hardly more meaningful.

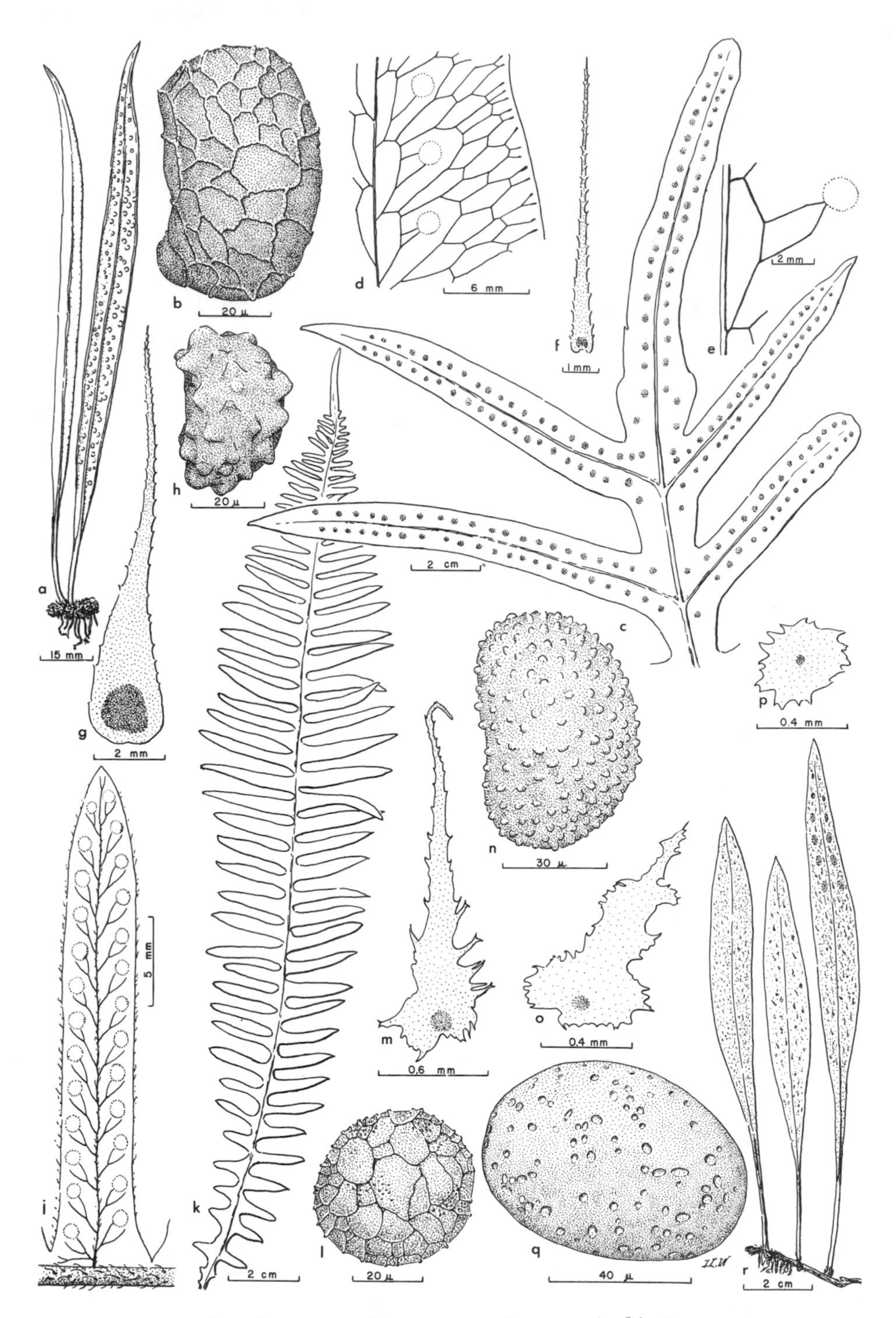

Fig. 23. *Polypodium angustifolium* var. *amphostenon* (*a, b*). *P. aureum* var. *areolatum* (*c–h*). *P. dispersum* (*i–l*). *P. insularum* (*m, n*). *P. lanceolatum* (*o–r*).

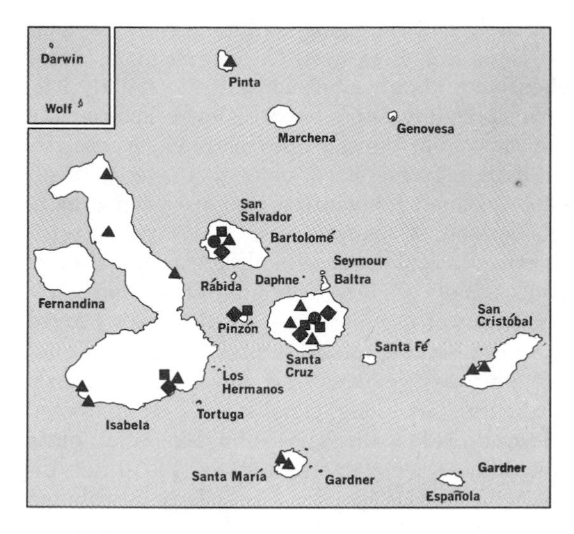

● *Polypodium angustifolium* var. *amphostenon*

■ *Polypodium aureum* var. *areolatum*

▲ *Polypodium dispersum*

◆ *Polypodium insularum*

rounded to shallowly cordate, peltately attached, attenuate, minutely denticulate with numerous teeth; stipes 2–10 dm long, 1–4 mm thick, terete or more commonly canaliculate along upper side, reddish brown, lustrous, glabrous except for narrow ring of scales just above attachment to rhizome; blades deltoid to broadly ovate, 1.5–10 dm long, 1 dm wide at base, deeply pinnatifid, coriaceous, bright yellow-green above, glaucous beneath, segments opposite to alternate, 7–35 or more, linear-oblong, nearly same width the full length or sometimes slightly constricted basally and narrowed abruptly to acute or obtuse apex, slightly undulate to plane, often shallowly crenulate, connate at base to form wing 2.5–12 mm wide along rachis, sinuses rounded or truncate, distant as much as 2 cm; venation variable, essentially of anastomosing veins with single series of areolae along each side of and parallel to costa, several irregular series of smaller areoles between costal series and margins of segment, only major areoles bearing sori; sori in single row along each side of costa about halfway between it and margin, each borne at tip of a free veinlet within a costalar areole, about 1.5–2 mm in diameter; sporangia yellowish brown to reddish brown, about 0.4 mm long, stalks slender, equaling or slightly exceeding length of body; annulus of 12–16 dark brown, thick-walled cells, lustrous; lateral walls many-celled, nearly hyaline; spores reniform, pale yellowish brown, 45–55 μ long, about 30 μ broad, coarsely tuberculate with rounded, blunt tubercles.

On tree trunks, rocks, cliffs, and rocky banks, Florida, West Indies, and Mexico to Argentina. Specimens examined: Isabela, Pinzón, San Salvador, Santa Cruz.

See Fig. 23.

Polypodium dispersum A. M. Evans, Amer. Fern Jour. 58: 173, *pl. 27.* 1968

Polypodium paradiseae sensu Hook. f., Trans. Linn. Soc. Lond. 20: 165. 1847, and sensu Anderss., 1855, non Langsd. & Fisch.
Polypodium pectinatum sensu Stewart, Proc. Calif. Acad. Sci. IV, 1: 24. 1911, non L.

Rhizome usually black, short-creeping to ascending, 1–5 cm long, 2–3.5 mm thick, somewhat nodose, bearing closely crowded, circular scars about 1 mm broad, younger parts invested by closely crowded, narrowly triangular, long-acuminate, dark reddish scales, the margins minutely and irregularly rather sparsely denticulate; fronds approximate, erect, 1–6.3 dm tall, stipe 2–10 cm long or more, about 1 mm thick, atropurpureous to ebeneous, subterete but adaxial side slightly flattened and closely hirsute with erect-spreading, stiff, grayish, few-celled, simple hairs, abaxial side usually glabrate and sublustrous; blades narrowly lanceolate, 10–48 cm long, abruptly reduced at base; segments linear-oblong, attenuate, 15–45 mm long, 2–5 mm wide, obtuse at apex, connate into narrow, irregular wing at base, the black costae and main rachis moderately to sparsely scaly, scales narrowly triangular to linear-subulate, dark brown, denticulate, short-fimbriate at base, sometimes slightly hastate; basal segments usually slightly deflexed and sometimes reduced to auricles; veins free, immersed, inconspicuous, running diagonally toward margins of segments, 1(2)-forked; sori uniseriate, medial, round or sometimes oblong, borne at tip of distal branch of a primary veinlet, 1–1.5 mm in diameter, pale reddish brown; sporangia obovoid, about 0.3 mm long, annulus of 12–14 thickened cells, pale reddish brown, slightly lustrous, side walls ivory, subhyaline, their outer surfaces minutely papillate, usually with 1–4 setae; spores globose, 43–50 μ in diameter, finely verrucose-reticulate, with minute papillae irregularly distributed along the low ridges, 32 in each sporangium.

In leaf mold, on fallen logs, on rocks in pockets of soil in lava, and epiphytic on lower trunks and larger branches of trees; widespread in tropical America. Specimens examined: Isabela, Pinta, San Cristóbal, San Salvador, Santa Cruz, Santa María.

See Fig. 23.

Polypodium insularum (Morton) de la Sota, Revist. Mus. de la Plata n.s. Bot. 11: 152. 1967.

Polypodium lepidopteris sensu Robins., Proc. Amer. Acad. 38: 112. 1902, non Kunze.
Polypodium bombycinum var. *insularum* Morton, Leafl. West. Bot. 8: 193. 1957.

Rhizome markedly nodose, short-creeping, 4–5 mm thick, densely clothed with rusty red-brown, lanceolate-attenuate, slender scales 3–6 mm long, their bases 0.5–1 mm wide, abruptly narrowing above base, irregularly denticulate-fimbriate, basal part darker brown than margins and tips; fronds 3–10 mm apart, erect, stipes short and stout, 1–3 cm long, 1–2 mm broad, moderately paleaceous with ovate-lanceolate, attenuate scales, these with dark brown centers, closely fimbriate-denticulate with pale to nearly white teeth; blades lance-oblong, broadest near middle, 1–3 dm long, 2.5–6 cm wide, dull green and invested by closely imbricate, sordid-gray, fimbriate-denticulate scales similar to those of stipe on upper surface, and by rusty red-brown, bicolorous scales beneath; segments subopposite to alternate, 20–50 pairs, lance-oblong, 1–3 cm long, 3–7 mm wide, sinuses broadly rounded, apex acute to rounded, basal 1 to 3 pairs reduced to deltoid auricles; leaf tissue coriaceous, thick; sori uniseriate, 1–1.5 mm in diameter, almost wholly obscured by scales until maturity; sporangia obovoid, about 0.4 mm long, annulus of 12–14 thickened, red-brown, lustrous cells; spores reniform to ellipsoidal, 60–70 μ long, about 45 μ broad, minutely papillate.

Endemic; on tree trunks and branches in deeply shaded forests. Specimens examined: Isabela, Pinzón, San Salvador, Santa Cruz.

See Fig. 23.

Polypodium lanceolatum L., Sp. Pl. 1082. 1753

Pleopeltis macrocarpa Kaulf., Berlin Jahrb. Pharm. 21: 41. 1820.
Pleopeltis lanceolata Presl, Tent. Pterid. 193. 1836, non Kaulf. 1824.
Lepicystis lanceolata Diels in Engl. & Prantl, Nat. Pflanzenfam. 1[4]: 323. 1899.

Rhizomes slender, long-creeping, 1.2–1.8 mm thick, the terminal 1–5 cm clothed with peltate-orbicular, ovate or lanceolate, sometimes abruptly attenuate scales 1.4–2.4 mm long, 0.8–1.2 mm wide, all except the narrow marginal portion black, dull to lustrous, margins pale to dark brown, erose-denticulate to short-fimbriolate, sometimes whole scale uniformly dark brown, the central areas of elongate-quadrate cells with thick walls, dark brown, prominent, the lumina evident; orbicular scales attached near middle, lanceolate ones near base, some scales bearing a few reddish brown, multicellular simple hairs, but these not obscuring body; older parts of rhizome black, dull, naked or nearly so; stipes slender, 1–8 cm long, 0.4–1.2 mm broad, green to stramineous, or nearly black at base, naked or with a few orbicular scales; blades simple, narrowly lanceolate, 4–25 cm long, 5–25 mm wide, tapering gradually toward both ends, bright green above, pale yellow-green beneath, bearing a few scattered, circular, dark-centered, erose-serrulate, pale-margined scales interspersed with fewer lance-ovate scales 0.8–1.2 mm long on upper surface, more numerous, paler scales beneath; veins deeply immersed, anastomosing, especially in vicinity of sori, these furnished by plexus of veinlets from adjacent veins of areole; sori relatively large, usually ellipsoidal, 3–4 mm long, about 2 mm wide, the long axis parallel with costa; scale-like paraphyses obscuring sporangia until nearly mature; sporangia reddish brown, obovoid, 0.3–0.4 mm long, annulus of 12–16 thickened cells, sublustrous; spores yellowish green, ellipsoidal, 70–75 μ long, 55–60 μ broad, nearly smooth but faintly and irregularly punctate.

Mostly epiphytic but occasionally on moist, shaded lava blocks, tropical America, Africa, and Asia. Specimens examined: Isabela, Pinta, Pinzón, San Cristóbal, San Salvador, Santa Cruz, Santa María.

This small epiphytic fern with simple fronds and large sori is common in the moister, well-shaded forests of the Galápagos Islands. Its peltate scales and peltate paraphyses are very distinctive.

See Fig. 23.

Polypodium phyllitidis L., Sp. Pl. 1083. 1753

Campyloneurum phyllitidis Presl, Tent. Pterid. 190. 1836.
Polypodium crassifolium sensu Stewart, Proc. Calif. Acad. Sci. IV, 1: 23. 1911, non L.

Epiphytic and terrestrial; rhizome creeping, stout, 5–15 mm thick, including the closely crowded, stout knobs formed by disarticulating fronds, only apical parts scaly, bearing masses of brown-tomentose rootlets; scales of rhizomes appressed-imbricate, pale brown with paler margins, broadly ovate to oblong, acute, subpeltate at or near base, clathrate, the margins erose-fimbriolate; fronds few or several, rather crowded, erect; stipes slender, 1–10 cm long, narrowly alate, the wing widening upward; blades linear to lanceolate, tapering equally in both directions, 2–8(12) dm long, 3–10 cm wide at middle, simple, acute, long-acuminate, or attenuate at apex, rigidly chartaceous, straight, yellowish green, glabrous, lustrous on both surfaces, entire to irregularly repand, often narrowly revolute, cartilaginous; costa stoutish, prominent above, markedly elevated beneath, yellowish; lateral veins oblique, elevated beneath, straight, connected by many arcuate transverse veinlets

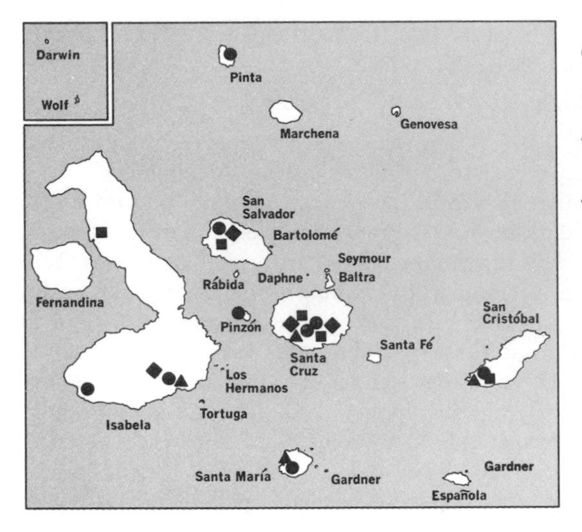

forming short areoles with mostly 3 excurrent included free veinlets, the central veinlet often uniting with next outer one; sori numerous, golden brown, biseriate between main lateral veinlets, superficial, terminal or subterminal on veinlets; sporangia tumidly obovoid, 0.4–0.5 mm long, the hyaline lateral walls glistening, stalks about as long as body; annulus of 16–20 thickened cells; spores broadly ellipsoidal, 65–70 μ long, about 45 μ broad, pale greenish yellow, irregularly and shallowly punctate to nearly smooth.

On trees, bushes, and rocky slopes, sometimes in rather dry situations (from 150 to 600 m altitude in the Galápagos Islands), Florida, West Indies, and Mexico to Uruguay. Specimens examined: Isabela, San Cristóbal, San Salvador, Santa Cruz.

Reported from Pinta on the basis of a sight identification by Williams, but no specimen has been found in herbaria.

Polypodium polypodioides (L.) Watt. var. **burchellii** (Baker) Weatherby, Contr. Gray Herb. 124: 29. 1939

Polypodium incanum Sw. var. *burchellii* Baker in Mart., Fl. Bras. 1[2]: 526. 1870.

Rhizome slender, widely creeping, 1–1.6 mm thick, the terminal few cm densely clothed with dark reddish brown, linear-lanceolate to acicular scales 2–3.5 mm long, about 0.4–0.6 mm wide, older ones with an obvious dull black-sclerotic central band occupying about two-thirds to three-fourths of width, margins paler brown, very thin, fimbriate-denticulate; fronds distant (1–4 cm apart), erect, 5–10(15) cm long (in ours), 2–3.5 cm wide; stipes shorter than blades, imbricately scaly with ovate to lanceolate-ovate scales 1.5–2.5 mm long, abaxial scales ascending-appressed with dark brown to nearly black areas near base, margins paler, adaxial scales spreading at right angles to stipe, longer and narrower and less markedly dark-centered than abaxial scales; blades widest at base, pinnatifid, the segments oblong-oblanceolate to oblong, 2–5 mm wide, bright green and sparsely scaly above, densely imbricate-scaly beneath, scales ovate-lanceolate, 0.6–0.8 mm wide, tapering gradually to attenuate apex, margins pale brown to silvery gray; leaf tissue thick, opaque; segments more or less involute in drying; sori few, 1–1.3 mm in

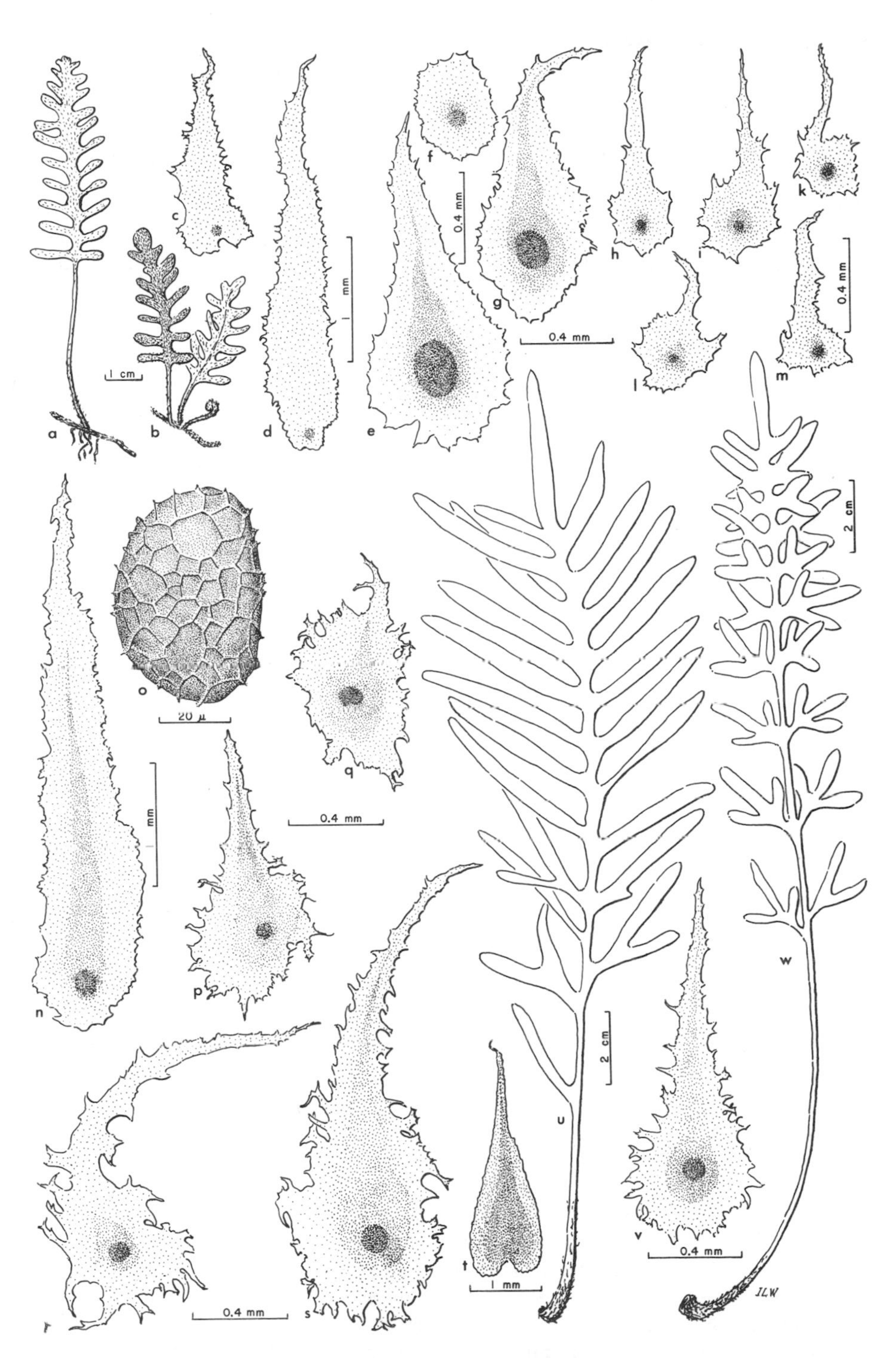

Fig. 24. *Polypodium polypodioides* var. *burchellii* (*a–o*): *c, d,* scales from rhizome; *e–n,* scales from frond. *P. tridens* (*p–w*): *s, v,* scales from frond; *t,* scale from rhizome.

diameter, obscured by scales until nearly mature; sporangia red-brown, tumid, obovoid, about 0.4 mm long, more or less lustrous, annulus with 12–14 thickened cells; stalks slender, equaling or slightly exceeding body; spores broadly ellipsoidal, greenish and translucent, very shallowly and irregularly faveolate, 55–60 μ long, 40–45 μ broad.

On bushes, trees, and rocks throughout much of tropical and subtropical South America, from near sea level to about 2,000 m. Specimens examined: Isabela, San Cristóbal, Santa Cruz, Santa María.

See Fig. 24.

Polypodium steirolepis C. Chr., Amer. Fern Jour. 7: 33. 1917

?*Polypodium nigripes* Hook., Sp. Fil. 5: 17. 1863, non Hassk. 1844.
Polypodium thyssanolepis sensu Stewart, Proc. Calif. Acad. Sci. IV, 1: 26. 1911, non A. Braun.

Rhizome creeping, branching, 3–4 mm thick, closely invested by lanceolate-attenuate scales about 1 mm wide at base, 2–3.5 mm long, the central band black-sclerotic, clathrate, margins pale rufous brown, 4–10 cm long, sparsely scaly to naked and glabrous; sterile fronds smaller and with shorter, broader segments than the fertile; blades deltoid-oblong to broadly lanceolate-oblong, pinnatifid, 5–9(12) cm long, 3–4(8) cm wide at base, segments oblong, round and slightly serrulate or acute near apex, 2–4 cm long, 5–10 mm wide, margins slightly undulate, the terminal segment lanceolate, 2–6 cm long, 8–15 mm wide, acute, all green and bearing scattered, irregularly fimbriate-denticulate, circular to lanceolate scales or almost naked on upper surface, densely scaly beneath, the scales here with darker centers, scales never wholly obscuring paler green undersurface; sori somewhat impressed into leaf tissue, circular to ellipsoidal, 2–3 mm in diameter, dull reddish brown, arranged in single median series along each side of costa, usually 5–10 pairs on a segment, sporangia about 0.4 mm long, annulus of 12–16 thickened cells; spores greenish, ellipsoidal, 65–75 μ long, 40–50 μ wide, finely and faintly reticulate-ridged and papillate in irregular patches.

Epiphytic on trees and large shrubs, occasionally on rocks or rocky soil, Venezuela and Ecuador. Specimens examined: Isabela, San Salvador, Santa Cruz.

Polypodium tridens Kunze, Farnkr. 1: 23, *t. 13. f. 1.* 1840

Marginaria incana sensu Hook. f., Trans. Linn. Soc. Lond. 20: 166. 1847, non Presl.
Polypodium squamatum sensu Robins., Proc. Amer. Acad. 38: 113. 1902, and sensu Stewart, Proc. Calif. Acad. Sci. IV, 1: 25. 1911, non L.

Rhizome creeping and nodose-branching, 3–6 mm thick, closely invested with lanceolate scales 2–4 mm long, these with black-sclerotic central bands and reddish brown, almost entire margins; fronds usually fasciculate, 2–6 dm tall; stipes 1–4 dm long, stoutish, 0.6–4 mm thick, more or less scaly with fimbriate-margined scales or eventually naked; blades pinnate or pinnatifid, oblong to oblong-lanceolate, 3–20 cm wide, stiffly erect, the lowermost segments usually bifid or trifid one-half to four-fifths of way to base, the upper simple, the fertile usually slightly narrower than the sterile, mostly acute at apex, often petiolulate near base of blade, entire, faintly undulate, the upper surface closely to sparsely scaly with lanceolate-attenuate scales with pale gray fimbriae around base and part way to the slender apex, the scales not completely covering or obscuring the bright green surface of blade; undersurface completely hidden by broader, ovate-lanceolate scales with darker brown margins and black centers, the margins more conspicuously fimbriate-denticulate; sori circular, 2–3 mm in diameter, medianly arranged in single row along each side of costa, reddish brown, not impressed; sporangia closely crowded, lustrous, red-brown, about 0.4 mm long; spores ellipsoidal, 50–65 μ long, 35–45 μ broad, greenish, translucent, minutely papillate, without faveolate ridges.

Endemic. Both epiphytic and terrestrial; possibly sparingly present on the mainland in Ecuador. Specimens examined: Española, Fernandina, Isabela, Marchena, Pinta, Pinzón, Rábida, San Cristóbal, San Salvador, Santa Cruz, Santa Fé, Santa María.

See Fig. 24.

Polystichum Roth, Rom. Arch. 2^1: 106. 1799

Coarse, rigid ferns of terrestrial habitats. Rhizomes erect or decumbent, usually short, densely covered with scales, these entire to lacerate. Fronds stiffly ascending, not articulate to rhizomes. Stipes densely fasciculate, scaly at least at base. Blades uniform, simple to tripinnate (bipinnate or bipinnate-pinnatifid in ours), often proliferous, more or less scaly along rachises and costae, and sometimes on the harsh segments, these usually auriculate, spinulose. Veins free. Sori dorsal to subterminal on veins, circular; indusia superior, orbicular, centrally peltate, persistent or caducous (lacking in ours). Sporangia numerous, the annulus usually consisting of 18 cells or more. Spores monolete, oblong to subglobose, nearly always spinulose or tuberculate.

About 175 species in tropical and cool temperate regions of the world.

Blades strictly bipinnate, the pinnules toothed but not pinnatifid or pinnatisect; largest pinnae about 14 cm long; basal stipe scales clear brown. . . . . *P. gelidum*

Blades subtripinnate, the basal pinnules deeply pinnatifid or pinnatisect at base; largest pinnae more than 20 cm long; basal stipe scales dark castaneous. . . . . . *P. muricatum*

Polystichum gelidum (Kunze) Fée, Gen. Fil. 278. 1852

Aspidium gelidum Kunze, Linnaea 20: 365. 1847.
Polystichum aculeatum sensu Stewart, Proc. Calif. Acad. Sci. IV, 1: 26. 1911, non Schott.
Polystichum pycnolepis sensu Svens., Bull. Torrey Bot. Club 65: 327. 1938, non Moore.

Rhizome erect or ascending, nodose, 1–3 cm thick (including woody bases of old stipes), densely clothed with reddish brown, thin, narrowly lanceolate, entire to minutely denticulate scales 1–3.5 cm long, 3–5 mm wide, these more or less tortuous and finely wrinkled; fronds fasciculate, erect or ascending, 4–12 dm tall; stipes about as long as blades, reddish, 3–5 mm broad, sulcate, densely clothed with thin, bright brown scales like those of rhizome but smaller; blades narrowly ovate-lanceolate, 3–6 dm long, 12–30 cm wide, bipinnate, the rachis and rachillae densely scaly with linear or lance-linear, spreading, reddish scales 3–12 mm long; pinnae alternate, 20–30 pairs, narrowly lanceolate, 8–14 cm long, 1.5–2.5 cm wide, pinnate or the uppermost pinnatifid; segments broadly ovate to somewhat trapezoidal, 3–12 mm long, 3–6 mm wide, spinulose-serrulate, apiculate, those on basal part of pinna usually slightly to markedly auriculate on distal side, firm, coriaceous, bright glossy green and glabrous on upper surface, distinctly paler, dull, and sparingly scaly beneath, the scales arachnoid-fimbriate; veins free, inconspicuous, 1–5-forked; sori 0.6–1 mm in diameter, about median, 1–4 borne on auricular part, mostly in 1 row

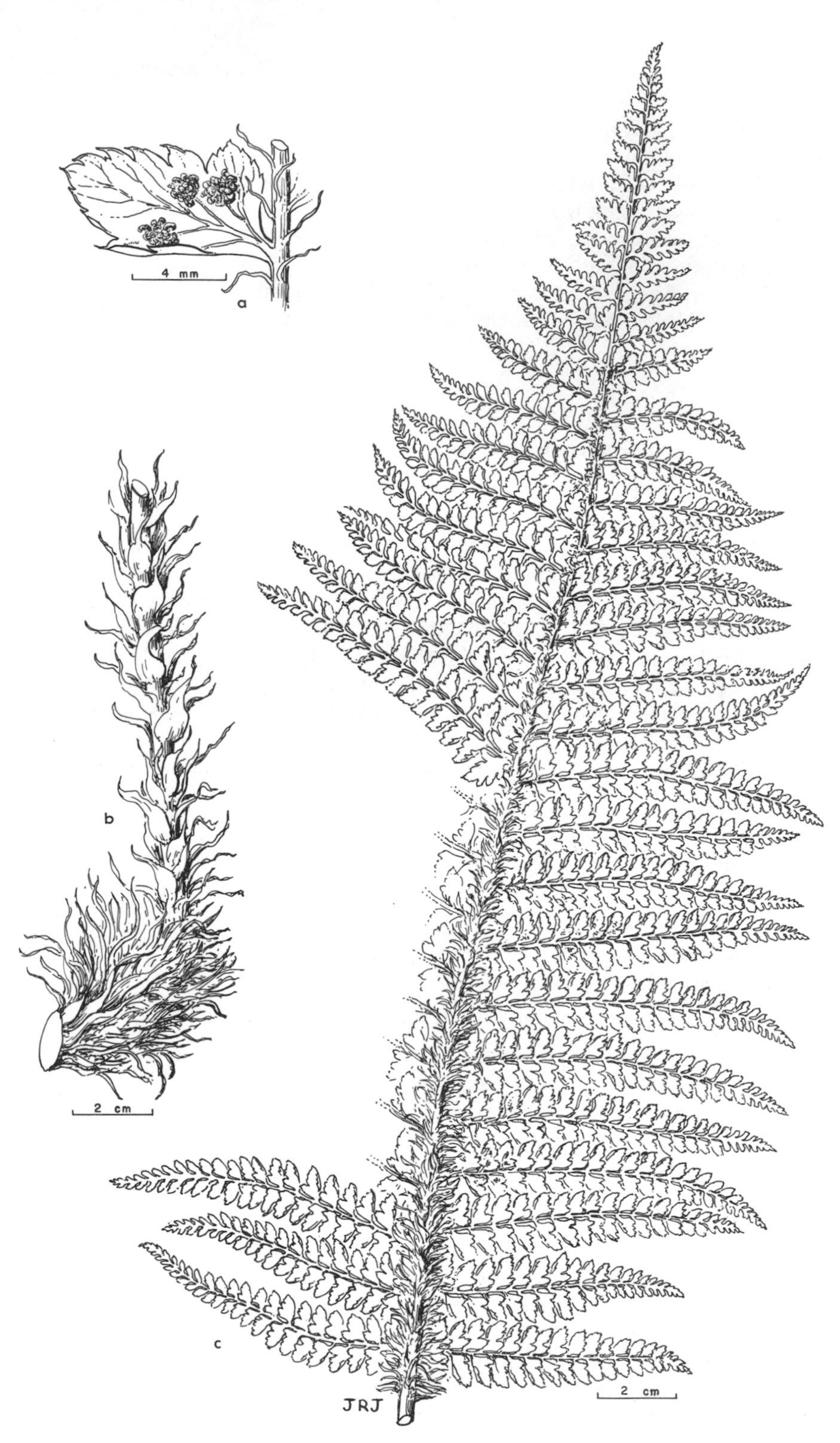

Fig. 25. *Polystichum gelidum.*

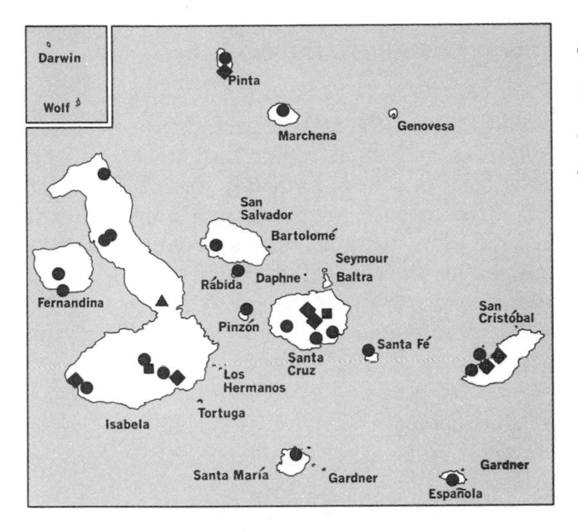

● *Polypodium tridens*
■ *Polystichum gelidum*
▲ *Polystichum muricatum*
◆ *Pteridium aquilinum* var. *arachnoideum*

along each side of rest of costule, or on smaller segments only 1 row along distal side of costa; sporangia reddish, about 0.3 mm long, annulus dullish; spores subglobose to broadly oblong, 55–60 μ long, 35–45 μ broad, irregularly and coarsely reticulate-alate, with some tubercles, minutely spinulose between wings, these about 2–10 μ high, erose-serrulate.

In moist, often shady places, Venezuela and probably northern Ecuador. Specimens examined: Isabela, Santa Cruz (identification of these specimens with the Venezuelan *P. gelidum* is problematical).

See Fig. 25.

Polystichum muricatum (L.) Fée, Gen. Fil. 278. 1852

Polypodium muricatum L., Sp. Pl. 1093. 1753.

Rhizome erect, massive, 4–6 cm thick (including persistent bases of old stipes and tangled masses of dark brown roots), apical part ensheathed in dark brown, entire, sublustrous, lanceolate-attenuate scales 1.5–3 cm long, 2–4 mm wide; fronds fasciculate, 6–20 dm tall; stipes 1–6 dm long, 4–6 mm thick, deeply sulcate adaxially, densely scaly, the basal scales dark castaneous, 1–2 cm long, 4–6 mm wide, thin, marginal bands paler castaneous than major central part, scales higher on stipe lighter, reddish brown, intermingled with very narrow, hairlike scales; blade subtripinnate, broadly lanceolate, 2.5–12 dm long, to 4.5 dm wide at base, the basal pinnae largest, to 3 dm long, 2–3 cm wide, pinnate, 30–50-jugate; rachillae scaly and hairy with slender, attenuate, pale reddish hairs and scales beneath, less densely so above; segments ovate-oblong, 9–15 mm long, 3–5 mm wide exclusive of auricles, 1–2 mm long on distal side of base, spinulose-serrate, apiculate, dark green, glossy above, paler and sublustrous to dull beneath, rachis and rachillae sulcate above; veins free, inconspicuous, 1- or 2-forked; sori 1–1.4 mm in diameter, pale castaneous; sporangia scarcely sublustrous, crowded, annulus with 18–22 thick-walled cells; spores globose to broadly oblong-ellipsoid, dark brown, 50–55 μ long, 35–45 μ wide, coarsely tuberculate-spinulose, the processes blunt to truncate, 4–8 μ high.

Widespread in tropical America, from West Indies to Venezuela and Colombia. Specimen examined: Isabela, a new fern record for the Galápagos Islands.

Pteridium Gled. ex Scop., Fl. Carn. ed. 1, 169. 1760

Rhizome extensively creeping, much-branched, usually well below surface of soil, closely invested with hairs but devoid of scales, the stele a perforated solenostele containing vessels. Fronds large, alternate and often rather distant; stipes long, with several to many vascular bundles; blades coarse, pinnately dissected, usually tripinnate, the lower pinnae with nectaries at base; segments numerous, ovate to linear, with revolute margins. Veins free. Sori marginal, usually continuous. Sporangia borne between outer indusium and an inner indusium. Spores trilete, brown, minutely spinulose, tetrahedral-globose.

A single, worldwide species that has proliferated into many subspecific groups varying greatly in their characteristics.

Pteridium aquilinum (L.) Kuhn var. **arachnoideum** (Kaulf.) Herter, Revist. Sudamer. Bot. 5: 21. 1937

Pteris arachnoidea Kaulf., Enum. Fil. 190. 1824.
Pteris aquilina var. *caudata* sensu Robins. & Greenm., Amer. Jour. Sci. III, 1: 149. 1895, non Link.
Pteris aquilina var. *esculenta* sensu Robins., Proc. Amer. Acad. 38: 114. 1902, non Hook. f.

Fronds coarse, 1–3 m tall; stipe usually shorter than blade, dark reddish brown to nearly black and densely pilose near base, glabrous and sublustrous higher; blades

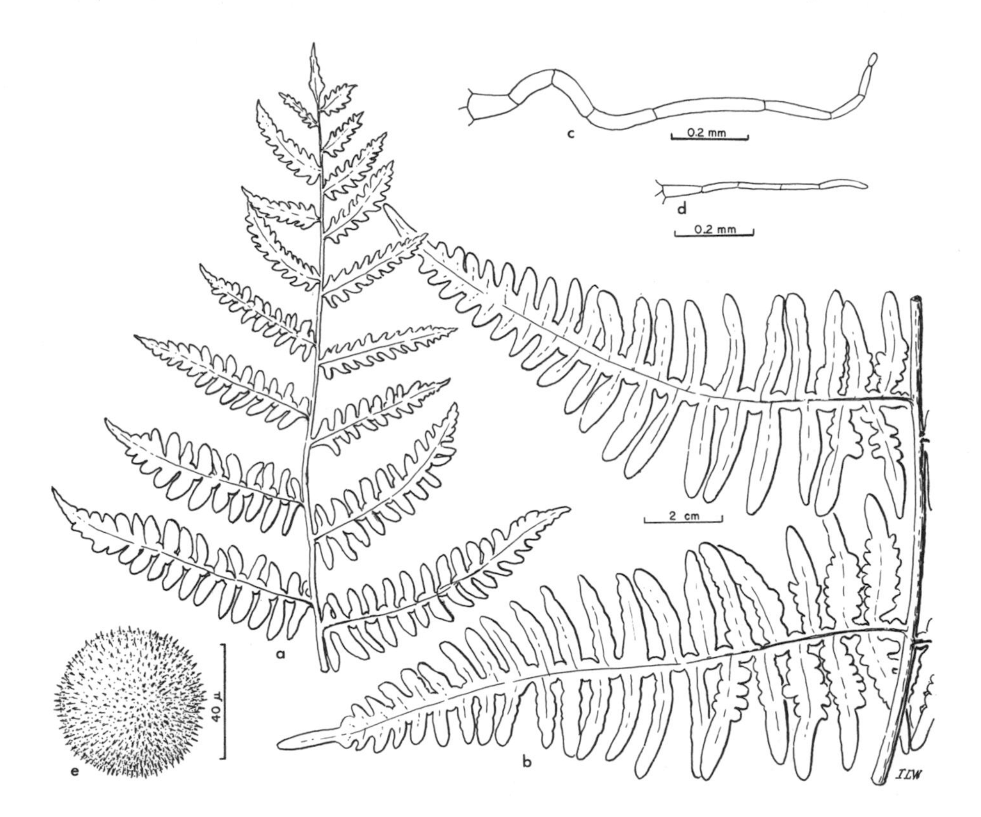

Fig. 26. *Pteridium aquilinum* var. *arachnoideum.*

0.5–2 m long, ovate-triangular to narrowly triangular, tripinnate or quadripinnate; primary pinnae narrowly triangular to lanceolate, 1–6 dm long; costules of penultimate segments usually pubescent beneath with arachnoidly tangled hairs, less densely so to glabrate on upper surface; free, semilunate lobes present along rachises, costae, and costules; ultimate segments ovate to linear, margins spreading-pubescent, or some segments glabrous, usually with farinaceous appearance, midnerves usually pubescent with dark or bicolorous hairs; membranous wings usually present along veins and midnerves; indusium ciliate, glabrous to pubescent; sporangia about 0.13–0.15 mm long, about 0.1 mm wide, annulus with 14–16 thickened cells; spores about 50 μ in diameter, closely and minutely spinulose, spinules about 2 μ long.

Open slopes, rocky places, in forests, grasslands, and thickets, and sometimes in cleared fields, from near sea level to 3,000 m, in the West Indies and Mexico, to the southerly parts of South America. Specimens examined: Isabela, Pinta, San Cristóbal, Santa Cruz.

This fern is more abundant in the Galápagos Islands than the relatively few collections in herbaria indicate. It forms extensive brakes in the upper parts (between 300 and 900 m or somewhat higher) of several of the larger islands.

See Fig. 26 and color plates 56, 60, and 62.

Pteris L., Sp. Pl. 1073. 1753; Gen. Pl. ed. 5, 484. 1754

Coarse ferns of moist forests and rocky slopes, with stout, creeping to erect rhizomes copiously invested with thin scales. Fronds erect or ascending, fasciculate, with firm, terete to angulate, stramineous or green stipes, these rarely atropurpureous, usually lustrous and glabrescent. Blades pinnate to quadripinnate, elongate, triangular or broadly pentagonal, often only the basal parts decompound, then of complex structure. Segments small to large, not articulate in most species. Veins lax and distinct to close, often rather prominent, free (in ours) or joined in angular areoles, or only the basal anastomosing. Sori mostly linear. Indusia linear, introrse, membranous, formed by the highly modified, reflexed leaf margins. Sporangia borne in continuous row along each margin of a filiform receptacle connecting ends of veinlets, usually absent from extreme tips of segments and from sinuses. Paraphyses present or lacking. Spores trilete.

About 250 species of tropical and subtropical distribution in the Old and New Worlds.

Pteris quadriaurita Retz., Obs. Bot. 6: 38. 1791

Pteris repandula Link, Fil. Hort. Berol. 56. 1841.
Pteris biaurita var. *repandula* Kuhn, Bot. Jahrb. Engler 24: 99. 1897.
Histiopteris incisa sensu Stewart, Proc. Calif. Acad. Sci. IV, 1: 21. 1911, non J. Smith.*

Rhizome woody, erect or obliquely ascending, bearing small, acicular, castaneous, pale-margined scales near apex; fronds to 1.5 m tall, the stipe often nearly equaling blade, deeply sulcate, brownish and scaly at base, stramineous and glabrous above; blades oblong, to 1 m long, 3–5 dm broad, pinnate; pinnae 7–15 pairs, all except basal pairs simple, narrowly deltoid-oblong to broadly lanceolate, 10–80 cm long,

* The true *Histiopteris incisa* (Thunb.) J. Smith was reported from the Galápagos but had not been collected again until recently. See p. 173 for a brief discussion.

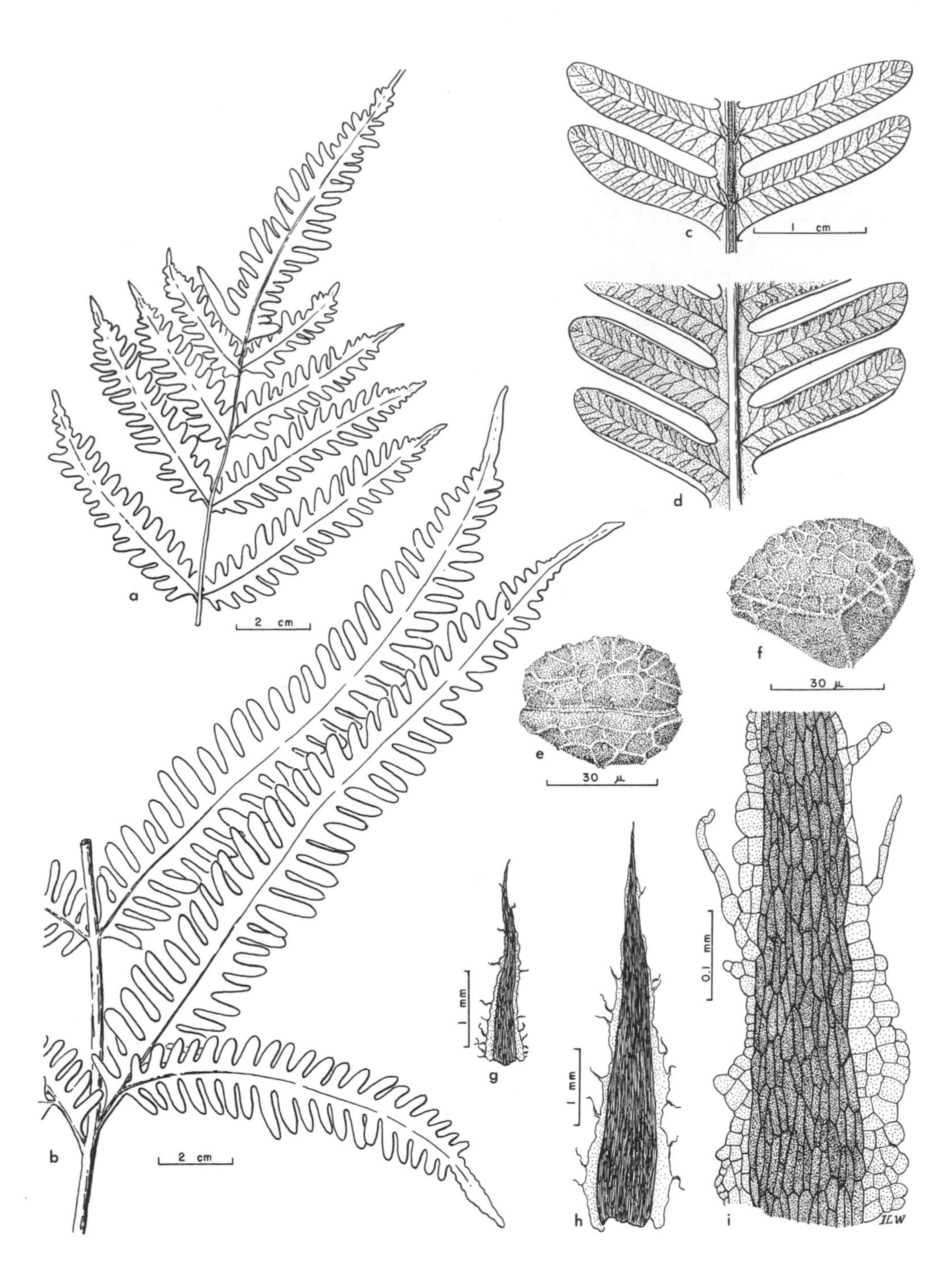

Fig. 27. *Pteris quadriaurita.*

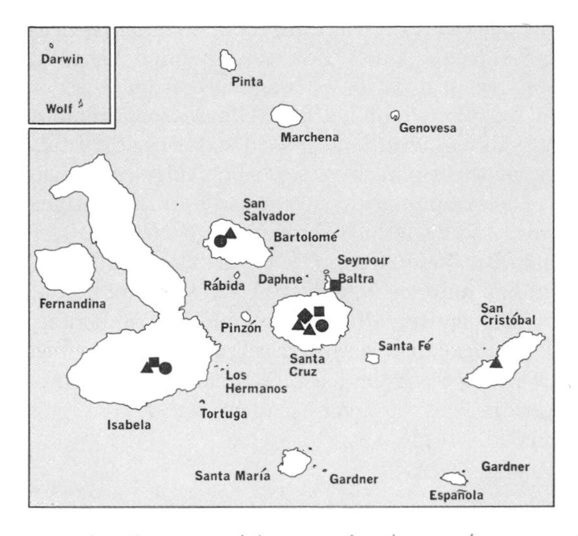

● *Pteris quadriaurita*
■ *Rumohra adiantiformis*
▲ *Tectaria aequatoriensis*
◆ *Thelypteris balbisii*

very deeply pinnatifid except for the caudate-acuminate, terminal 1–3 cm, segments on lower side of pinnae slightly longer than those on upper side; basal pinnae with elongate, pinnatifid basal inferior pinnule; rachises strongly elevated, costae bearing a pale, acicular, stout, spinose process at junction of costa and costule, these protuberances 1–1.5 mm long, ascending; ultimate segments linear to oblong, 1–3 cm long, 3–6 mm wide, straight to subfalcate, entire, rounded-obtuse to acute, the margins lightly callose, sinuses narrow and acute; veins numerous, oblique, all free, mostly 1-forked at or near base; sori not quite reaching tips of segments or the sinuses between them; indusia pale, thin, stramineous to pale brown, about 1 mm wide, entire; spores 60–65 μ long, coarsely reticulate-rugose on outer surface, brownish.

On shady slopes and in ravines and openings in the forest and thickets, throughout a considerable altitudinal range, but always in moist habitats, from Florida into the tropics of South America and in the Old World. Doubtless in the sense circumscribed here a species aggregate. Specimens examined: Isabela, San Salvador, Santa Cruz.

See Fig. 27.

PTERIS SPECIES EXCLUDED FROM THE FLORA

Pteris propinqua Agardh, Rec. Sp. Gen. Pter. 65. 1839

Pteris biformis Splitgb., Tijdschr. Nat. Gesch. 7: 422. 1840.
Pteris hostmanniana Ettingsh., Denkschr. Akad. Wien 23: 55. 1864.
Pteris blanchetiana Presl in Ettingsh., Farnkr. 103, *pl. 36, f. 7.* 1865.

This species was reported from the Galápagos Islands by Hooker f. on the basis of a Wood specimen that was probably obtained on the mainland. It differs, among other characters, in having areolate venation.

Rumohra Raddi, Opusc. Sci. Bologna 3: 290. 1819

Rhizome epiphytic, long-creeping, thick, densely paleaceous, with large, reddish brown, sparsely long-fimbriate, ovate-lanceolate scales, the cells small, not much longer than broad, in many series, the cell walls brown, thin, the lumina large, hyaline. Fronds scattered, distant. Stipe elongate, thick, glabrous, rounded beneath, 2-ridged and conspicuously sulcate above, scaly at base, roughened from bases of fallen scales; blades large, deltoid, tripinnate-pinnatifid at base, anadromous throughout, the rachis deeply sulcate above, with large raised median ridge, this unbroken and not opening to receive ridges of pinnae axes, median ridge of pinnae similarly continuous, margins of petiolules of pinnae decurrent and forming margins of rachillae; blades coriaceous, glabrous and eglandular throughout, even in grooves of rachis, completely scaleless at maturity. Veins free, obscure or barely prominulous, not reaching margins. Sori dorsal on anterior branchlet of a veinlet, indusiate; indusium large, flat, perfectly circular, entire, attached at middle, umbonate, promptly deciduous. Spores monolete, anisopolar, narrowly elliptic in polar view, concavo-convex in profile, the perine with a few rounded prominences, these sometimes partly fused to form low wings, the laesurae long, slender ridges.*

A monotypic genus in the New and Old Worlds.

Rumohra adiantiformis (Forst.) Ching, Sinensia 5: 70. 1934

Polypodium adiantiforme Forst., Prodr. 82. 1786.
Rumohra aspidioides Raddi, Opusc. Sci. Bologna 3: 290, *t. 12, f. 1.* 1819.
Polystichum coriaceum Schott, Gen. Fil. *t.* 9. 1834.
Polystichum adiantiforme J. Smith, Hist. Fil. 220. 1875.

Rhizome stout, long-creeping, rather tortuous, more or less flattened, 1–1.5 cm thick, densely clothed with soft, imbricate-spreading, reddish brown, ovate-lanceolate, attenuate, erose-denticulate scales 2–5 mm wide, 6–20 mm long; fronds few, 3–10 cm apart, erect, 5–15 dm tall; stipes shorter than blades or rarely equaling them, sulcate adaxially, reddish brown, scaly at base, or sometimes sparsely so to base of blade; blades deltoid, 2.5–9 cm long, 2–7.5 dm wide at base, tripinnate, rachis reddish, glabrous, with scattered spots of darker red over whole surface; pinnae few, oblique, rigid, straight, petiolulate, the basal largest and asymmetrically deltoid, long-acuminate, with 8–12 pairs of petiolulate, narrowly triangular, acuminate pinnules; upper pinnae oblong-acuminate to lance-deltoid, gradually smaller and simpler than the lower; segments lance-elliptic to oblong-ovate, asymmetrically cuneate at base, acute, the larger obliquely incised, the smaller bluntly serrate; veins immersed, oblique, 1–3 branched, tips free; sori large (1.5–2 mm in diameter), slightly inframedial; indusia circular-peltate, 0.8–1.2 mm across, glabrous, caducous, margins entire or faintly crenulate; sporangia numerous, dark brown to nearly black, lustrous, about 0.2 mm in diameter; annulus of 14–16 thickened cells; spores terete in polar view, slightly concavo-convex in profile, dark reddish brown, 35–40 μ long, about 25 μ broad, the surface bearing rounded prominences 2–4 μ high, and with low, narrow ridges.

On tree trunks, rotting logs, moist rocks, and rocky soil, tropics of the Old and New Worlds. Specimens examined: Isabela, Santa Cruz.

See Fig. 28.

Tectaria Cav., An. Hist. Nat. (Madrid) 1: 115. 1799

Large ferns of shady banks and wet forests, with a stout, woody, short-creeping to erect, paleaceous rhizome. Fronds few, suberect to recurved toward tips; stipes

* Description of spore from W. F. Harris (1955).

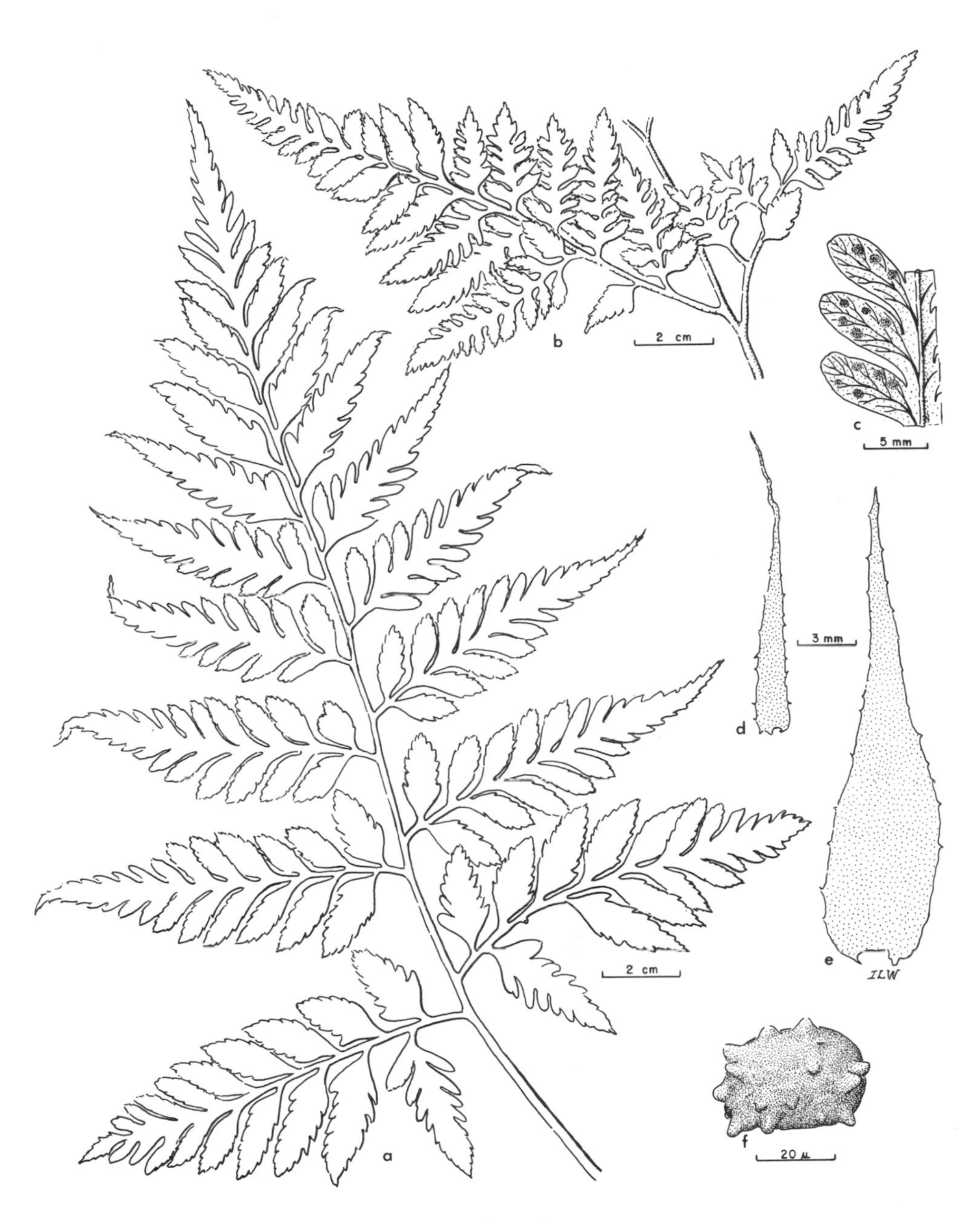

Fig. 28. *Rumohra adiantiformis.*

stramineous to atropurpureous, usually firm and shining, not articulate with rhizome. Blades narrowly elongate and simple, broader and pinnatifid to bipinnatifid, to deltoid and pinnately decompound, ultimate segments coarse. Veins anastomosing intricately, the areoles with (in ours) or without included free veinlets. Sori many, irregularly disposed or in open rows near ultimate costae, but sometimes on included veinlets. Indusia persistent or caducous (sometimes apparently lacking), orbicular and centrally peltate to round-reniform.

About 125 species widely distributed in the tropics.

Tectaria aequatoriensis (Hieron.) C. Chr., Ind. Fil. Suppl. 3: 176. 1934

Nephrodium macrophyllum sensu Robins., Proc. Amer. Acad. 38: 110. 1902, non Baker.
Aspidium aequatoriense Hieron., Hedwigia 46: 353. 1907.
Aspidium martinicense sensu Stewart, Proc. Calif. Acad. Sci. IV, 1: 13. 1911, non Spreng.

Rhizome erect, 1.5–2 cm thick, closely invested with castaneous, lanceolate, fimbriate scales 3–6 mm long, 1–2 mm wide; fronds fasciculate, erect, 4–15 cm tall, the stipes about equaling or slightly longer than blades, stramineous, sulcate adaxially, rounded abaxially, scaly at base but glabrate above; blades submembranous, oblong to ovate-oblong, 1.5–7.5 dm long, 1.2–5 dm wide, simple-pinnate; rachis sulcate adaxially, puberulent with septate hairs in groove; pinnae 2–10 pairs, oblique, basal pair petiolulate, with 2 or 3 more or less elongate basal lobes and a smaller, shorter, distal lobe, or this lacking, otherwise the pinnae shallowly lobed or nearly entire, the lateral pinnae oblong, acuminate, 12–25 cm long, 2–5 cm wide, the lower sessile, the upper adunate throughout, subentire to shallowly lobed, puberulous with septate hairs along costae and costules above, apex composed of coadunate, acuminate, decurrent pinnae, margins glabrous or subciliate, sinuses glabrous; areoles in numerous rows, the costal solitary or 2 between adjacent costules, these elongate parallel to costae, the proximal at right angles and parallel to costules, others pentagonal or hexagonal, in 17–19 rows between costa and margin, major ones in 2–4 rows, subdivided into numerous minor areoles, often with free, included veinlet; sori large, about 2 mm in diameter, in 2 rows between costules, borne on outer margin of costal areole at base of an outwardly extending veinlet; indusia persistent, reniform to circular, the basal sinus lobes overlapping and closing sinus, the indusia thus appearing peltate, their margins usually curled upward and inward at maturity; sporangia numerous, obovoid, about 0.3 mm long, slightly compressed laterally, annulus prominent, dark reddish brown, lustrous, the sides paler, stalks 2–3 times as long as body; spores broadly ellipsoidal, irregularly reticulate with thin, winglike ridges, the body about 50 μ long, two-thirds as wide.

In moist habitats, northern Ecuador. Specimens examined: Isabela, San Cristóbal, San Salvador, Santa Cruz.

See Fig. 29.

Thelypteris Schmidel, Icon. Pl. ed. Keller, 45, *pl. 11.* 1763

Terrestrial ferns of moderate, or rarely large, size. Rhizome short-creeping, long-creeping, ascending, or erect, dictyostelic, rather densely scaly and often bearing simple, branched, or stellate hairs among scales. Stipe and rachis often more or less pubescent, usually shorter than blade. Lamina simple to pinnate-pinnatifid, or rarely more intricately compound, pubescent on costae and costules with simple, unicellular hairs and often densely pubescent elsewhere with simple or stellate hairs; rachis sometimes proliferous in axils or at apex. Veins free and reaching margins

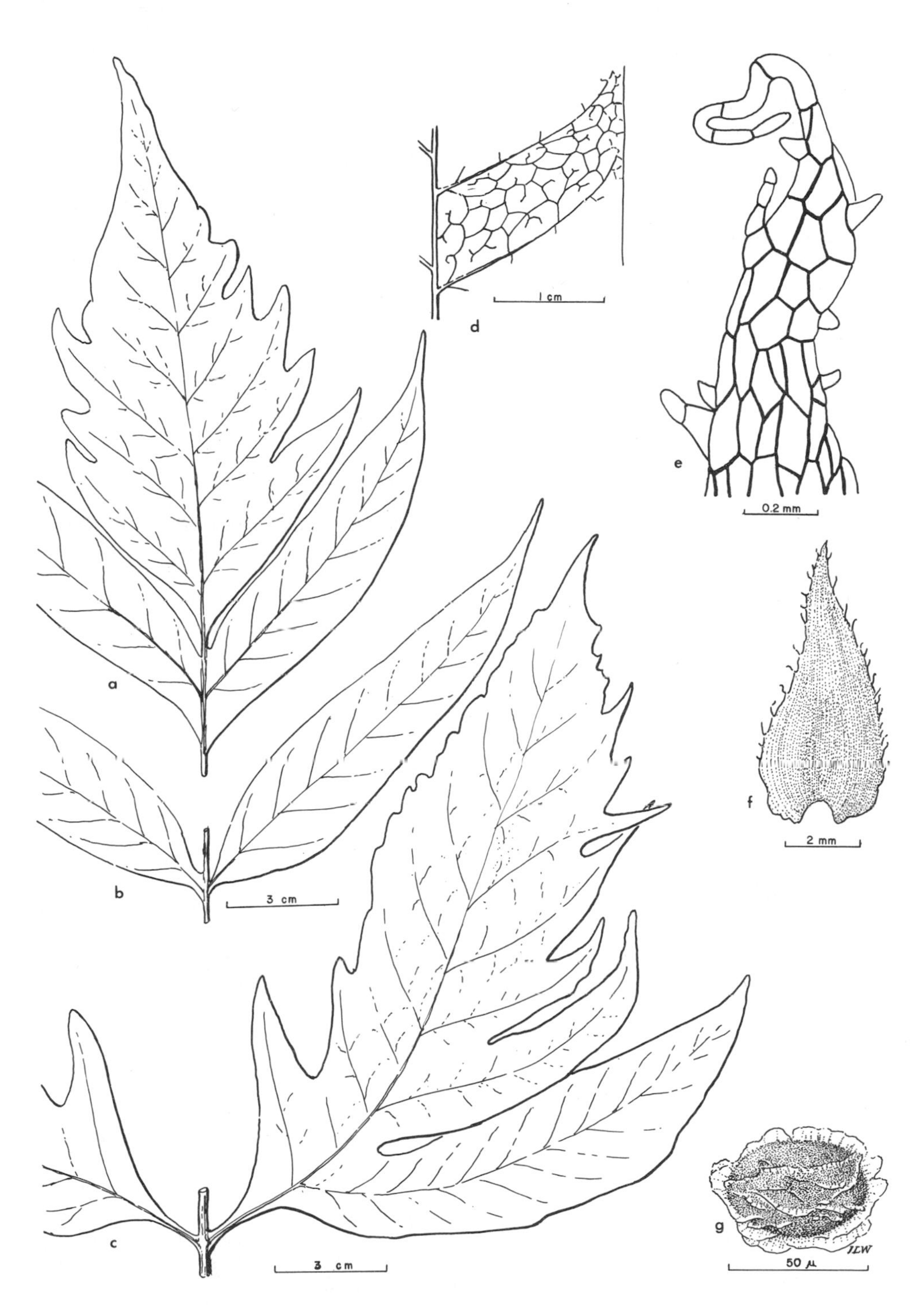

Fig. 29. *Tectaria aequatoriensis.*

of segments or anastomosing in pairs with secondary excurrent veinlets, or only the lowest veins of adjacent segments meeting and uniting at or below the intervening sinus. Sori dorsal, or rarely terminal on veins, circular or rarely ellipsoidal-elongate. Indusium round-reniform or lacking. Sporangia smooth or setulose, annulus of 12–14(18) thickened cells. Spores monolete, bilateral, usually oblong, tuberculate, faveolate or spinulose.

Nearly 900 species in all parts of the world in which ferns thrive, from the tropics well into the cooler temperate zones.*

Lowest pair of veins reaching margin above sinus between segments, or one vein running to sinus and not meeting other; blades narrowed at base (Subg. *Lastrea*):
  Sori obviously elongate along the veins, exindusiate; lowest vein running to sinus; auricle-like basal pinnae absent. . . . . *T. linkiana*
  Sori round, indusiate; lowest pair of veins reaching margin well above sinus; blades very gradually reduced at base, the lowest pinnae minute and auricle-like:
    Veins in segments more than 12 pairs; surfaces beneath with obvious red or orange glands; indusia obvious, with red, sessile glands and few setae. . . . . . . *T. balbisii*
    Veins in segments 6–8 pairs; surfaces beneath not glandular; indusia minute, with elongate setae, not glandular. . . . . *T. pilosula*
Lowest pair of veins meeting at sinus or united within the leaf tissue, or several pairs connivent or united; blades not narrowed at base or only slightly so, with no minute auricle-like pinnae:
  Rachis hairs simple (or very rarely 1-forked?); blades pinnatifid at apex, without a conform terminal pinna; sporangia without setae (Subg. *Cyclosorus*):
    Basal veins of adjacent costules united within leaf tissue into an excurrent veinlet running to sinus; rhizome erect or suberect, never creeping. . . . . . *T. quadrangularis*
    Basal veins of adjacent costules connivent and running into sinus (except toward apex of blades and in juveniles), not definitely united with one another below sinus to form an excurrent veinlet; rhizome erect or creeping; blades membranous; pinnae nearly sessile:
      Pinnae incised somewhat less than three-fourths the way to costa, with an obvious elongate, yellowish cartilaginous membrane extending down from sinus; usually 2 pairs of connivent veins; basal inferior segments of lower pinnae obviously reduced in size; dark ciliate scales present on costae beneath; rhizome creeping and fronds scattered. . . . . *T. invisa* var. *aequatorialis*
      Pinnae incised about three-fourths the way to costa, without an elongate, yellowish cartilaginous membrane extending down from sinus; only 1 pair of connivent veins; basal segments of lower pinnae not reduced in size; scales lacking on costae beneath; rhizomes erect or creeping:
        Rhizome erect, with broad, thin eciliate scales; superior basal segments elongated; veins on upper surface not hairy. . . . . *T. patens*
        Rhizome creeping, with narrow, ciliate scales; superior basal segments usually not elongated; veins on upper surface slightly hairy. . . . . *T. kunthii*
  Rachis hairs (at least those on lower side) partly 2- to 4-branched at apex; blades with the terminal pinnae conform or nearly so; sporangia (or at least some of them) with elongate setae; lowest veins anastomosing into an excurrent veinlet running to sinus or with several pairs of veins uniting (Subg. *Goniopteris*):
    Pinnae 5–12 pairs, lobed at least one-third the way to costa; lowest 1 or 2 pairs of veins uniting and giving off an excurrent veinlet. . . *T. tetragona* subsp. *aberrans*
    Pinnae 4 or 5 pairs, not lobed, merely crenate; 5 or 6 pairs of uniting veins. . . . . *T. poiteana*

* Two species reported from the Galápagos Islands, but either not collected recently or of doubtful origin, are not included in the key. They are briefly described on p. 170.

Thelypteris balbisii (Spreng.) Ching, Bull. Fan Mem. Inst. Bot. 10: 250. 1941

Polypodium balbisii Spreng., Nov. Act. Acad. Caes. Leop. Carol. 10: 228. 1821.
Aspidium sprengelii Kaulf., Flora 6: 365. 1823. (Based on *Polypodium balbisii* Spreng.) (Illegit.)
Dryopteris sprengelii Kuntze, Rev. Gen. Pl. 2: 813. 1891. (Illegit.)
Dryopteris balbisii Urban, Syst. Antill. 4: 14. 1903.
Thelypteris sprengelii Proctor, Bull. Inst. Jam. Sci. Ser. 5: 65. 1953. (Illegit.)*

Rhizome erect or ascending, 1.5–2 cm thick, 5–12 cm long, apical parts clothed with lance-attenuate, glossy brown scales 5–12 mm long, 1–2 mm wide, fronds several, closely fasciculate, strongly ascending, 6–8(20) dm tall, the stipe relatively short, 1–4 dm long, stramineous, markedly sulcate, scaly and brownish at base; blades oblong to lanceolate, 5–17 dm long, 20–40 cm wide, pinnate, acuminate, abruptly reduced at base, basal 4–6 pairs of pinnae distant, greatly reduced; rachis stramineous, glabrate, sulcate along upper side; pinnae many, opposite or subopposite, horizontal, lanceolate-attenuate, 10–20 cm long, 1–3 cm wide, deeply pinnatifid, the costa strigillose above, thinly pilosulous to glabrate beneath; segments close, spreading, linear, acute, entire, membranous, sparingly and deciduously strigillose above, glandular and thinly short-puberulous to glabrate along costae and the simple veins, these 13–18 pairs, oblique, red-glandular on surface beneath; sori circular, about 1 mm in diameter, often partially concealed by the revolute margins, slightly supramedial; indusium circular-reniform, sinus running almost to center, subpeltate, 0.4–0.6 mm in diameter, bearing several bright reddish or amber globose sessile glands around margins and nearby, a few simple, few-celled setae on upper surface among glands, persistent put pushed upward by developing sporangia, the latter subglobose, slightly compressed laterally, reddish brown, the annulus of 12–15 thickened cells; spores subglobose-ellipsoidal, 35–42 μ long, 28–30 μ broad, reddish amber, minutely papillate.

In shaded places among shrubs and trees, on moist rocky slopes, in ravines, and along trails, West Indies, Mexico, and northern South America. Specimens examined: Santa Cruz, a new record for the Galápagos Islands; many plants had fronds 2 m tall, and were producing an abundance of spores early in February, 1964.

The Galápagos material has the habit, venation, and sessile orange or reddish glands of typical *T. balbisii*; the hairs along the costae of the pinnae beneath are somewhat longer (as much as 2 mm long) and are made up of 5–8 slender, cylindrical cells. *T. balbisii* has been described as having indusia devoid of setae, whereas our material has several rigid, unicellular marginal setae. In spite of this divergence from typical *T. balbisii*, the differences scarcely seem sufficient to separate the Galápagos population as a different species.

See Fig. 30.

Thelypteris invisa (Sw.) Proctor var. **aequatorialis** (C. Chr.) Morton, Contr. U. S. Nat. Herb. 38: 61. 1967

Dryopteris oligophylla Maxon var. *aequatorialis* C. Chr., Dansk. Vidensk. Selsk. Skr. VII, 10^2: 189. 1913.

Rhizome creeping, stout and woody, 8–12 mm thick, apical part scurfy with thin, dark reddish brown, setulose-ciliolate, more or less tortuous scales; fronds few, distant, erect-arching, rigid, 0.5–3 m tall; stipes usually as long as blades, stoutish, brown and scaly at base, stramineous, glabrate, terete but sulcate along upper side; blade oblong, in ours 40–120 cm long, 20–70 cm wide, pinnate-pinnatifid, gradu-

* For more on the nomenclature of this species, see Morton (1963, 62–64).

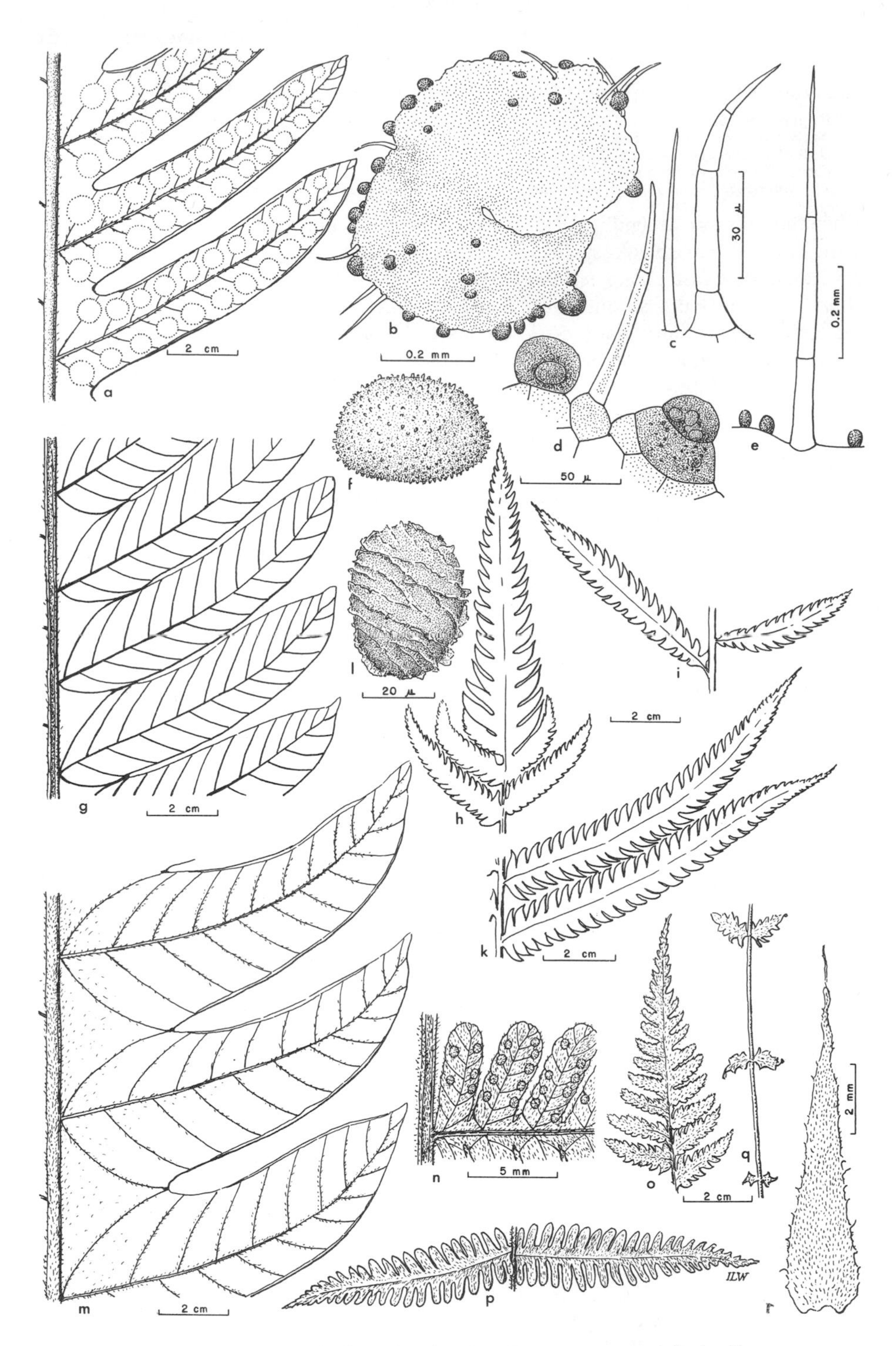

Fig. 30. *Thelypteris balbisii* (*a–f*). *T. invisa* var. *aequatorialis* (g–*l*). *T. patens* (*m*). *T. pilosula* (*n–r*).

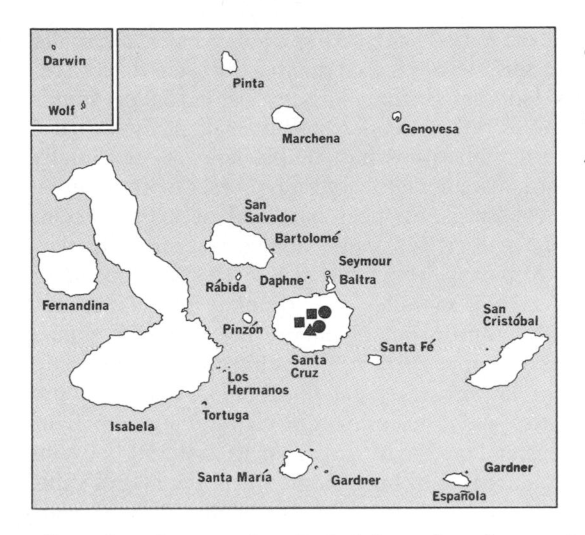

● *Thelypteris invisa* var. *aequatorialis*

■ *Thelypteris kunthii*

▲ *Thelypteris linkiana*

ally reduced upward and slightly reduced toward base; rachis stramineous, finely puberulous or pilosulous with slender, simple hairs; pinnae numerous, alternate, linear-attenuate or narrowly lanceolate-attenuate, central ones 10–35 cm long, 1.5–3.5 cm wide, deeply pinnatifid, the lower 1–5 pairs about one-half as long as the middle, slightly narrowed toward bases, usually short-petiolulate; middle and upper pairs of pinnae broadest at base, sessile, all pinnae dark glossy green above, paler and dull beneath; costae prominent, these, the costules, veins, and indusia densely pilose beneath with simple hairs 0.3–1 mm long, the costae deeply sulcate and moderately pilosulous with hairs 0.2–0.4 mm long on upper surface, or in some plants nearly glabrous above; segments oblique, subfalcate, acute (often minutely apiculate), entire, 2–5 mm wide at base, in ours 1–1.5 cm long, the sinuses narrowly open, acute, lowermost pinnae with equally reduced distal and proximal basal segments, but none vestigial or abortive; leaf texture herbaceous-coriaceous; veins 10–14 pairs; sori many, 1 mm in diameter, mostly on basal half to two-thirds of each fertile segment, contiguous at maturity; indusia circular-reniform, 0.6–0.8 mm in diameter, firm, pale reddish, densely setulose-pilose with stiff hairs 0.2–0.5 mm long; sporangia dark brown, lustrous, about 0.4 mm long, stalks about equaling body, annulus of 14–20 thickened cells; spores broadly oblong, dark brown, 40–45 μ long, 30–35 μ broad, minutely spinulose-alate, the wings often in diagonal, branching rows, 2–4 μ high.

In shaded or semi-shaded habitats, Ecuador to Bolivia. Specimens examined: Santa Cruz.

See Fig. 30.

Thelypteris kunthii (Desv.) Morton, Contr. U. S. Nat. Herb. 38: 53. 1967

Nephrodium kunthii Desv., Mém. Soc. Linn. Paris 6: 258. 1827.
Dryopteris normalis C. Chr., Arkiv. för Bot. 9[11]: 31. 1910.
Thelypteris normalis Moxley, Bull. So. Calif. Acad. Sci. 19: 57. 1920.
Filix-mas augescens var. *normalis* Farw., Amer. Midl. Nat. 12: 253. 1931.
Lastrea normalis Copeland, Gen. Fil. 139. 1947.

Rhizome short- or long-creeping or sometimes ascending, dark brown or rusty brown, 4–6 mm thick, the apex closely invested with rusty brown, lanceolate scales

2–3 mm long, 0.4–0.8 mm wide, intermingled with stiff spreading hairs, the scales ciliolate and sparsely setose on outer surface when young, later the setae deciduous, the scales falling and leaving only hairs on rhizome slightly behind apex; fronds erect or ascending, 6–12 dm tall, stipes rather slender, 2–8 dm tall, scaly on basal 1–5 cm or merely sparsely pubescent with spreading, simple hairs, or essentially glabrate, stramineous, often lustrous, canaliculate abaxially, beset with minute, globose, sessile or subsessile, pale yellow glands near base and on central rachis among pinnae; pinnae alternate above first 3–6 pairs, oblong-lanceolate to narrowly-lanceolate, 5–12 cm long, 8–20 mm wide, pinnatifid into 15–30 segments or more, these distinct to within 1–3 mm of rachilla, then connate into continuous medial wing 2–6 mm wide, their tips obtuse but apiculate, the tissue of pinnae firm, dark green and sparsely pilose on upper surface, the veins and costules more densely pilose than intervein surfaces, lower surface paler, dull, more densely pilose and bearing many yellowish, globose, sessile or subsessile glands; lowest pair of veins entering sinus at its base, with small wedge of cartilaginous material between their tips at base of sinus, the veins not fusing to form single trace, the lateral veins 7–9(11) pairs, slightly clavate-swollen at tips and ending at or just short of margin of segment, usually brownish, especially at tips; sori orbicular, borne on 5–8 pairs of veins, about medial, 0.6–0.8 mm in diameter, yellowish brown; indusium circular-reniform, conspicuously setose-hirsute with unicellular hairs 0.2–0.5 mm long, and stipitate-glandular on margins and surface, attached centrally (at base of sinus); sporangia obovoid-compressed, about 0.3 mm long, 0.2 mm wide, and half as thick, the stalk equaling body, annulus pale brown, of 16–18 thickened cells.

Among shrubs (in *Miconia* Zone at about 450 to 500 meters in the Galápagos Islands), Florida, Greater Antilles, tropical Mexico, and South America. Specimens examined: Santa Cruz.

Thelypteris linkiana (Presl) Tryon, Rhodora 69: 6. 1967

Gymnogramma diplazioides Desv., Mém. Soc. Linn. Paris 6: 214. 1827.
Gymnogramma polypodioides Link, Hort. Berol. 2: 50. 1833, non Spreng. 1827.
Grammitis linkiana Presl, Tent. Pterid. 209. 1838.
Dryopteris diplazioides Urban, Symb. Antill. 4: 21. 1903, non Kuntze 1891.
Dryopteris linkiana Maxon, Jour. Wash. Acad. Sci. 14: 199. 1924.
Thelypteris diplazioides Proctor, Bull. Inst. Jam. Sci. Ser. 5: 59. 1953, non Ching 1941.

Rhizome erect or nearly so, 4–8 mm thick, to 15 cm long or more, closely ensheathed by bases of old stipes and plunging dark brown roots; rhizome scales thin, concolorous, pale yellowish brown, lanceolate to oblong-lanceolate, 3–10 mm long, blunt to attenuate at apex, rather closely appressed-ascending; fronds closely cespitose, erect; stipes about one-fourth to one-third as long as lamina, sparingly scaly at extreme base, often minutely brown-hirtellous, at least at base, canaliculate adaxially, pale green to stramineous, to 5 mm broad, blades lance-elliptic, pinnate-pinnatifid, 3–12 dm long, 13–28 cm wide, rachis more or less tetragonal, puberulent to glabrate, stramineous; pinnae 25–40 pairs, attenuate, narrowly lanceolate, to 13 cm long, 1–2.5 cm wide, pinnatifid halfway to costa, merely serrate at apex, lower pinnae gradually reduced, distinctly reflexed, often only 1–2.5 cm long, dark green and sparsely puberulent to glabrate above, paler and with few minute, simple hairs beneath, segments entire; rachis and costae pale green, hispidulous with spreading, simple, whitish or yellowish hairs 0.1–0.6 mm long; veins 5–7 pairs in each segment, lowermost veins from adjacent segments reaching sinus at different levels, one at base of sinus, other just above it but the two not meeting; sori orbicular to oblong, the larger elongate along veins, exindusiate; sporangia tumidly obovoid-subglobose, about 0.3 mm long, stalk shorter than body, reddish brown, annulus sublustrous, of 12–16 thickened cells; spores oblong-reniform, pale reddish brown, 38–44 μ long, 22–30 μ broad, minutely papillate, the papillae 1–2 μ high, rounded, irregularly distributed.

In moist habitats, margins of and in shrubby or forested areas, Mexico to Brazil and Ecuador. Specimen examined: Santa Cruz, new to the Galápagos Islands.

Thelypteris patens (Sw.) Small, Ferns SE. States 476. 1938

Polypodium patens Sw., Nov. Gen. Sp. Pl. Prodr. 133. 1788.
Nephrodium molle sensu Hook. f., Trans. Linn. Soc. Lond. 20: 171. 1847, non Schott.
Dryopteris patens Kuntze, Rev. Gen. Pl. 2: 813. 1891.
Dryopteris parasitica sensu Stewart, Proc. Calif. Acad. Sci. IV, 1: 19. 1911, pro parte, non Kuntze.
Dryopteris dentata sensu Svens., Bull. Torrey Bot. Club 65: 315. 1938, non C. Chr.

Rhizome erect, stoutish, 1–4 cm thick, 3–15 cm long, bearing many brown, pubescent roots, apex closely clothed with pale brown, ovate to ovate-lanceolate scales 5–12 mm long, these sparingly hirsute with straight, erect, colorless hairs in youth but soon glabrate; fronds several, fasciculate, 0.5–2 m tall, erect; stipe about one-half to nearly as long as blade, quadrangular, stramineous, scaly at base, scales broad, not ciliate, sparsely pubescent in youth, later glabrate; blade oblong, abruptly short-acuminate, 3–12 dm long, 2–4.5 dm wide, pinnate, the rachis pale-stramineous, pubescent, canaliculate along upper side; pinnae many, alternate, usually adjacent, narrowly oblong-attenuate, 10–25 cm long, 1–3 cm wide, the lower slightly shorter and moderately to strongly reflexed, all broadest at base, pinnatifid about three-fourths the way to costa, without cartilaginous membrane below sinus, the basal segments of some lower pinnae enlarged, dentate on 1 or both sides, pilosulous above costules but veins glabrous, both costules and veins pilosulous beneath, surface minutely glandular, glands pale yellow, sessile or short-stalked; segments oblong to linear-oblong, close, slightly to markedly falcate, acute, the sinuses open but narrow and acute, margins narrowly revolute; veins 8–15 pairs, simple basal ones arcuate, ending in small cartilaginous membrane occupying sinus but

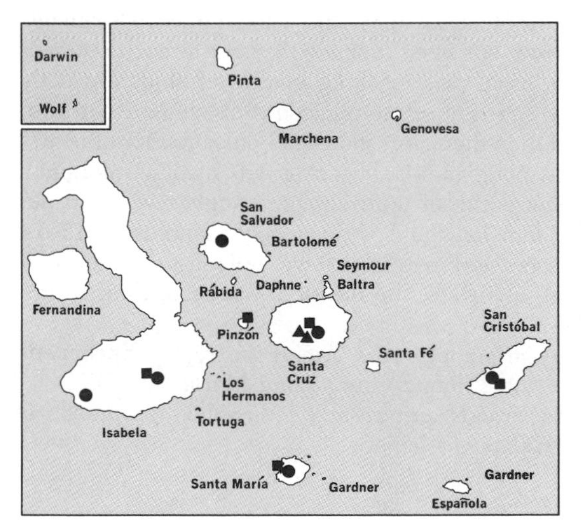

● *Thelypteris patens*
■ *Thelypteris pilosula*
▲ *Thelypteris poiteana*

not extending toward costa, only 1 pair of veins connivent and these often barely so in immature plants; sori supramedial, close, about 1–1.3 mm in diameter; indusia pale brown, circular-reniform, persistent, 0.8–1 mm in diameter, conspicuously white-pilose; sporangia obovoid-subglobose, about 0.4 mm long, rather dull, annulus of 14–18 thickened cells; spores black in masses, dark amber by transmitted light, subglobose to broadly oblong, 40–55 μ long, 33–40 μ broad, minutely and sharply spinulose-alate, i.e., spines forming outer margins of thin wings, about 2 μ high.

In margins of forests and shrubby areas in moist, semi-shaded places, Florida, West Indies, and Mexico to Argentina and Chile. Specimens examined: Isabela, San Cristóbal, San Salvador, Santa Cruz, Santa María.

Thelypteris pilosula (Klotzsch & Karst.) Tryon, Rhodora 69: 7. 1967

Aspidium pilosulum Klotzsch & Karst. ex Mett., Fil. Hort. Bot. Lips. 90 (error as "*lasiesthes*") 130 (corrected). 1856.
?*Polypodium rude* sensu Hook., Syn. Fil. 4: 243. 1864, non Kunze.
Dryopteris parasitica sensu Stewart, Proc. Calif. Acad. Sci. IV, 1: 19. 1911, pro parte, non Kuntze.
?*Dryopteris rudis* sensu Svens., Bull. Torrey Bot. Club 65: 317. 1938, non C. Chr.
Dryopteris colombiana sensu Svens., op. cit. 314, non C. Chr.

Rhizome erect, slender, 3–6 mm thick, but appearing much larger owing to heavy sheathing of old stipe bases and masses of dark brown roots, terminal parts clothed by reddish brown, concolorous, lanceolate scales 4–7 mm long, these hirsute on outer surface and setose-ciliolate along margins; fronds several, fasciculate, erect or slightly arcuate, 3–10 dm tall; stipes slender, much shorter than blades, stramineous to dark brown, scaly basally, closely puberulent with somewhat retrorse hairs 0.3–1 mm long; rachis brownish or stramineous, pubescent; blades lance-oblong, narrowing gradually to apex and base, pinnate-pinnatifid, 18–70 cm long, 7–30 cm wide at about middle, pinnae 15–30 pairs, sessile, spreading, opposite or alternate, oblong-attenuate, 4–15 cm long, 10–18 mm wide, deeply pinnatifid except at attenuate tip, this shallowly dentate, rather harshly pilosulous on both surfaces, dark green and glossy above, slightly paler, dull, and more densely pubescent beneath, lower 3–7 pairs greatly reduced to incised to merely dentate, the deltoid vestiges 3–12 mm long; ultimate segments oblong, 5–9 mm long, 2–3 mm wide, faintly crenulate, rounded or obtuse at apex, the wing between segments 0.5–1 mm wide, sinuses acute, narrow but open; veins 6–8 pairs in each segment of middle pinnae, whitish, oblique, basal pair reaching margin of sinus distinctly (0.4–1.3 mm) above base of sinus, sparsely setose-pilosulous above and beneath; sori supramedial, circular, 0.5–0.8 mm in diameter; indusia circular-reniform, nearly hyaline-whitish, 0.3–0.4 mm across, nonglandular, bearing 3–9 hyaline, marginal, radiating setae equaling or exceeding width of indusium; sporangia few, 5–10 per sorus, tumidly obovoid, about 0.3 mm long and wide, glossy, annulus of 12–14 thickened, dark red-brown cells; spores dark reddish brown, oblong to faintly reniform, 45–55 μ long, 33–40 μ broad, faveolate, the ridges 2–7 μ high, thin, microscopically erose-denticulate.

In moist, shady places in forests and among rocks, American tropics. Specimens examined: Isabela, Pinzón, San Cristóbal, Santa Cruz, Santa María.

The combination of nonglandular fronds and greatly reduced lower pinnae is distinctive among the ferns of the Galápagos Islands.

See Fig. 30.

Thelypteris poiteana (Bory) Proctor, Bull. Inst. Jam. Sci. Ser. 5: 63. 1953

Polypodium crenatum Sw., Nov. Gen. Sp. Pl. Prodr. 132. 1788, non Forsk.
Lastrea poiteana Bory, Dict. Class. 9: 233. 1826.
Dryopteris poiteana Urban, Symb. Antill. 4: 20. 1903.
Dryopteris reticulata sensu Stewart, Proc. Calif. Acad. Sci. IV, 1: 19. 1911, non Urban.

Rhizome short-creeping, 8–15 mm thick, the knobby bases of old stipes persisting, thus appearing 1–3 cm thick, apex scantily paleaceous with stellate-pubescent scales; fronds few, close, erect or ascending-spreading, 4–10 dm long; stipes pale reddish to stramineous, quadrangular, markedly sulcate, scaly at base, puberulent with 2- or 3-forked hairs when young, soon glabrate; blades oval to broadly deltoid-oblong, 20–40 cm long, 18–30 cm wide, simple-pinnate; pinnae 2–6 pairs, plus a terminal conform one, distant, subopposite or alternate, usually oblique, oblong to linear-elliptic or oblanceolate, abruptly acuminate, rounded to acute basally, subentire to coarsely serrate-crenate, 10–25 cm long, 2–6 cm wide, often gemmiparous in axils, the basal pinnae short-petiolulate, the middle and upper sessile, dark green and sublustrous above, slightly paler and dullish beneath, pilose with stiff, spreading, simple hairs on both surfaces, a few on costa of lower surface sometimes bifurcate; costae elevated, stramineous, densely hirsute; primary veins 20 pairs or more, oblique, terminating in the shallow sinuses between crenations of margins, secondary veins 6–8 pairs, all except outer 2 or 3 pairs connivent medianly between primary veins and forming excurrent veinlets, these almost to obviously joining next connivent pair above, occasionally an extra branching vein running from a primary vein to an excurrent veinlet or extending only part way to it and stopping in leaf tissue; sori circular, about 2 mm in diameter, in 2 rows between adjacent primary veins, borne on the diagonally arranged connivent veins or on free veinlets stopping in leaf tissue; indusia none; sporangia reddish brown, crowded, relatively small (0.2–0.3 mm long), at least some short-setose with 2–6 hyaline unicellular hairs; spores bright amber, broadly oblong to subreniform, 40–46 μ long, 30–35 μ broad, faveolate, the ridges 4–7 μ high, microscopically crenate-erose.

In moist, shady places, West Indies to southern Brazil and Peru. Specimens examined: Santa Cruz.

See Fig. 31.

Thelypteris quadrangularis (Fée) Schelpe, Jour. So. Afr. Bot. 30: 196. 1964

Polypodium molle Jacq., Coll. Bot. 3: 188. 1789, non Schreb. 1771, non All. 1785.
Nephrodium quadrangulare Fée, Mém. Foug. 5: 308. 1852.
Dryopteris mollis Hieron., Hedwigia 46: 348. 1907.
Dryopteris quadrangularis Alston, Jour. Bot. 75: 253. 1937.
Cyclosorus quadrangularis Tardieu, Notul. Syst. 14: 345. 1952.

Rhizome erect or nearly so, reddish brown, ensheathed by bases of old stipes and roots, densely scaly near apex; rhizome scales and those of basal parts of stipes concolorous, thin, pale rufous, minutely and irregularly denticulate, lanceolate to lance-attenuate, 3–5 mm long or more, cells clathrately arranged; fronds clustered, erect, 6–12 dm tall; stipes pale green to stramineous, broadly sulcate adaxially, sparingly to moderately hirsute with spreading, slender, white hairs; laminae lanceolate to elliptic-lanceolate, 30–60 cm long, 15–20 cm wide; rachis and costae moderately to densely hirsute with spreading white hairs to 1 mm long; pinnae narrowly lanceolate, about 30–40 pairs, the largest 8–11 cm long, to 15 mm wide, sessile or subsessile, pinnatifid half to two-thirds the way to costa, pinnules

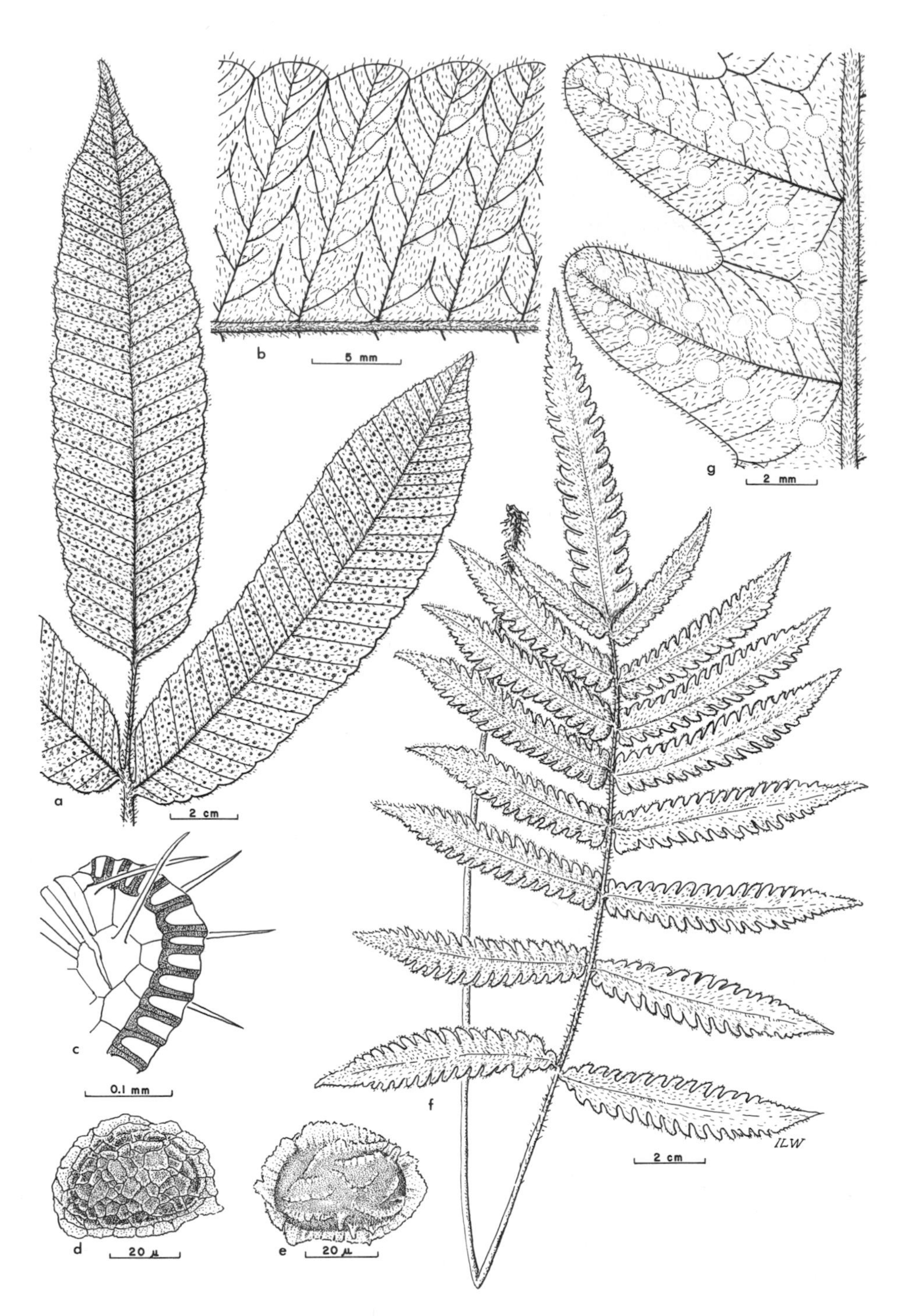

Fig. 31. *Thelypteris poiteana* (*a–d*). *T. tetragona* subsp. *aberrans* (*e–g*).

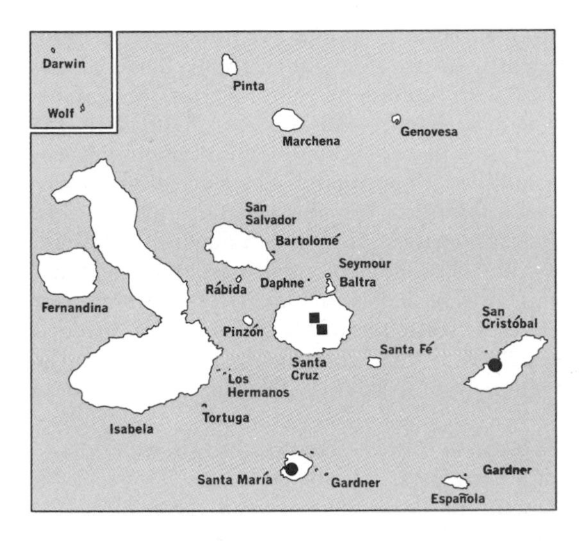

● *Thelypteris quadrangularis*

■ *Thelypteris tetragona* subsp. *aberrans*

slightly antrorse-arcuate, obtuse or rounded, 2–3 mm wide at base, superior basal one slightly longer than inferior segment of same pinna; veins 5–6 pairs in each pinnule, the lowermost from adjacent segments meeting in leaf tissue below sinus and sending an excurrent veinlet to base of sinus, next pair reaching sinus 0.5–1 mm above its base, both surfaces of leaf tissue more or less hirsutulous; sori orbicular, about 1 mm in diameter, borne on veins on both sides of costa about two-thirds the way to margin; indusia round-reniform, bearing several fine, white, stiffish setae on upper surface, sinus relatively wide, reaching center of indusial scale; sporangia about 0.3 mm long, turgid, reddish brown, side walls thin, hyaline; annulus of 12–16 thickened cells, reddish brown; spores reniform, 30–38 μ long, 22–25 μ broad, sparingly papillose, the papillae rounded, about 2 μ high, sometimes several confluent to form short ridges.

In shaded, damp habitats, West Indies and Mexico to Colombia; also in Africa. Specimens examined: San Cristóbal, Santa María, a new record for the Galápagos Islands.

Thelypteris tetragona (Sw.) Small subsp. **aberrans** Morton, Leafl. West. Bot. 8: 194. 1957

Dryopteris pseudotetragona sensu Stewart, Proc. Calif. Acad. Sci. IV, 1: 19. 1911, non Urban.
Dryopteris tetragona sensu Svens., Bull. Torrey Bot. Club 65: 317. 1938, non Urban.

Rhizome creeping, 5–7 mm thick, sparsely paleaceous; fronds numerous, subfasciculate; stipes erect from curved bases, usually longer than blade, 30–50 cm long, 2–3 mm broad at base, deeply sulcate, stramineous, densely paleaceous at base, scales narrowly lanceolate, 3–6 mm long, 0.5–1 mm broad, long-attenuate, entire, pale brown, glabrous within, with minute sessile stellate hairs externally, nearly epaleaceous upward, provided sparingly with minute stellate (2–6-radiate) hairs; blades pinnate, subdeltoid, to 35 cm long, 30 cm broad at base, not reduced basally, rachis pilose with hyaline, spreading, 1- to few-celled, subrigid hairs about 1 mm long, these simple or some forked at apex; pinnae herbaceous, 5–12 pairs with a subconform terminal pinna, subopposite, subhorizontally spreading, sessile, linear-lanceolate, 10–17 cm long, 1.5–3.2 cm wide, truncate or scarcely narrowed at base, acuminate, lobed one-third to halfway to costa, pilosulous on costa, veins, and surface beneath with simple, unicellular, subrigid, spreading hairs 0.3–0.5 mm long, above antrorsely strigose on costa, with setiform hairs on costules, the surface glabrous; segments broad, 8–9 mm long, 5–6 mm wide, obtuse, slightly falcate, entire, ciliate; veins 8–11 pairs, basal 1 or 2 pairs anastomosing and giving off excurrent veinlets, the lowest free (if more than 1 pair producing excurrent veinlets), the upper running to sinus, sometimes a third pair connivent to sinus, other veins free; sori small, circular, about 0.8 mm in diameter, slightly inframedial; indusium lacking; sporangia few, intermingled with subrigid hairs, upper part of sporangium often setose with 1–3 rigid, deciduous, hyaline hairs, annulus of 14–18 thickened cells; spores broadly oblong, terete in cross section, reddish amber, 50–55 μ long, 30–40 μ broad, thin-alate, the wings 2–15 μ high, their outer edges irregularly erose-undulate, radially striate, and minutely tuberculate with pale brown thickenings.

Endemic. In dense shade in moist habitats in the *Scalesia* and *Miconia* zones, at present known only from Isla Santa Cruz.

There is considerable variability among the lower pairs of veins involved in the anastomosing and formation of excurrent veinlets, even on the same frond. Juvenile fronds tend to have only one pair of anastomosing veins and to be without an excurrent veinlet, the anastomosis occurring almost at the sinus. Larger fronds of more mature plants frequently have two pairs anastomosing and forming excurrent veinlets, the third pair joining just below or at the base of the sinus. The variation in venation is unusual in the subgenus *Goniopteris* and suggests possible hybridity.

See Fig. 31.

THELYPTERIS SPECIES EXCLUDED FROM KEY

Thelypteris brachyodus (Kunze) Ching, Bull. Fan Mem. Inst. Biol. Bot. 6: 286. 1936

Polypodium brachyodus Kunze, Linnaea 9: 48. 1834.
Dryopteris glandulosa var. *brachyodus* C. Chr., Dansk. Vid. Selsk. Skr. VII, 10^2: 172. 1913.
Dryopteris glandulosa (Desv.) C. Chr. var. *brachyodus* (Kunze) C. Chr., Bull. Fan Mem. Inst. Biol. Bot. 6: 286. 1936.

This species has been reported from the Galápagos Islands on the basis of a Wood collection. Probably it came from the mainland. It is a very large species, with the pinnae lobed one-fourth to one-third the way to the costae, and with 4 to 6 pairs of anastomosing veins. It has a prominent recurrent cartilaginous membrane coming back from the sinus. It lacks forked or branched hairs.

Thelypteris totta (Thunb.) Schelpe, Jour. So. Afr. Bot. 29: 91. 1963

Polypodium tottum Thunb., Prodr. Pl. Cap. 172. 1800.
Aspidium gongylodes Schkuhr, Krypt. Gew. 1: 193, *t. 33c*. 1809 (as "*goggilodus*").
Nephrodium unitum R. Br., Prodr. Fl. Nov. Holl. 148. 1810.
Dryopteris gongylodes Kuntze, Rev. Gen. Pl. 2: 811. 1891.
Thelypteris gongylodes Small, Ferns SE. States 248. 1938.

This species, differing from *T. quadrangularis* in having the rhizome elongate, creeping, black, and nearly scaleless, the fronds scattered, the blades subcoriaceous, not reduced at base, and the pinnae short-stalked, was reported as having been collected on San Cristóbal by Chierchia (Caruel, Rendic. Acad. Lincei 5: 625. 1889) as *Nephrodium unitum*. But it is likely that the specimen was misidentified and really belongs to one of the other species treated above.

Trachypteris André ex Christ, Denkschr. Schweiz. Naturf. Ges. 36: 150. 1899

Terrestrial ferns of small stature from small, erect to decumbent, densely scaly rhizome. Fronds dimorphic; sterile ones few, in basal rosette, spreading; fertile ones 1 to few, in erect or ascending clusters; stipes scaly at least at base, the laminae deeply pinnatifid to simple-pinnate, densely scaly beneath, glabrate above. Veins deeply immersed and obscure, anastomosing. Sporangia borne along veins nearly to unmodified margins of segments. Indusia and paraphyses none. Spores reticulate-spinulose.

One species in northern South America, another in Madagascar.

Trachypteris pinnata (Hook. f.) C. Chr., Ind. Fil. 634. 1906

Hemionitis pinnata Hook. f., Trans. Linn. Soc. Lond. 20: 167. 1847.
Acrostichum aureonitens Hook. f., Icon. Pl. 10: *pl. 933.* 1854.
Trachypteris aureonitens André ex Christ, Denkschr. Schweiz. Naturfors. Gesell. 36: 151. 1899.

Rhizome erect or nearly so, 1–1.5 cm long, about 5 mm thick, densely clothed with closely packed sheath of dark brown, dull roots, apex paleaceous with dark reddish brown, lance-linear scales 1.5–2.5 mm long, their margins minutely and fairly closely denticulate; sterile leaves 5–25, spreading in rosette 5–25 cm across, curling upward into subglobose ball when subjected to protracted drought, spatulate to obovate, to 12 cm long, 5–30 mm wide near apex, 2–10 mm wide at the narrowed base, margins entire, undulate or rarely slightly lobed, upper surface dark green, smooth, subtomentose with narrow, nearly capillary, white to rusty scales about 1 mm long, lower surface densely covered with bright rust-red lanceolate scales 0.5–3.5 mm long, their cells clathrately arranged; fertile fronds few, erect, 0.5–3.5 dm tall, pinnate or pinnatifid; stipes usually longer than blades, broadly and shallowly sulcate along upper surface, dark reddish to atropurpureous, the blade broadly ovate to deltoid, 4–12 cm long, 3–15 cm wide; pinnae 1–6 pairs, linear, 1–8 cm long, 3–6 mm wide, dark green and glabrate above, densely paleaceous beneath; veins immersed, obscure or invisible; sporangia closely crowded on veins over most of undersurface, tumid, about 0.2 mm long and wide, annulus of 20–24 thickened

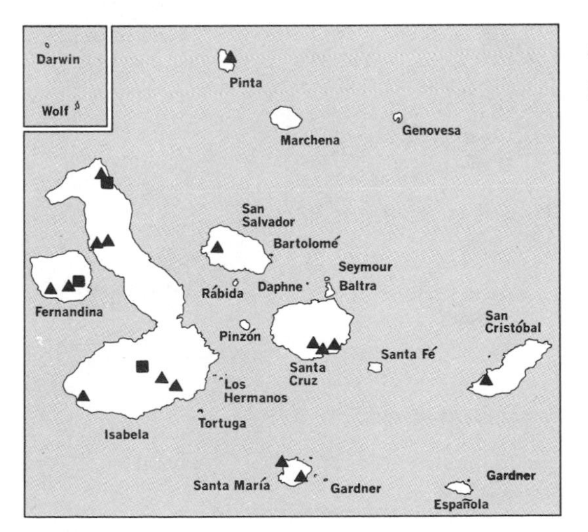

▲ *Trachypteris pinnata*

■ *Psilotum nudum*

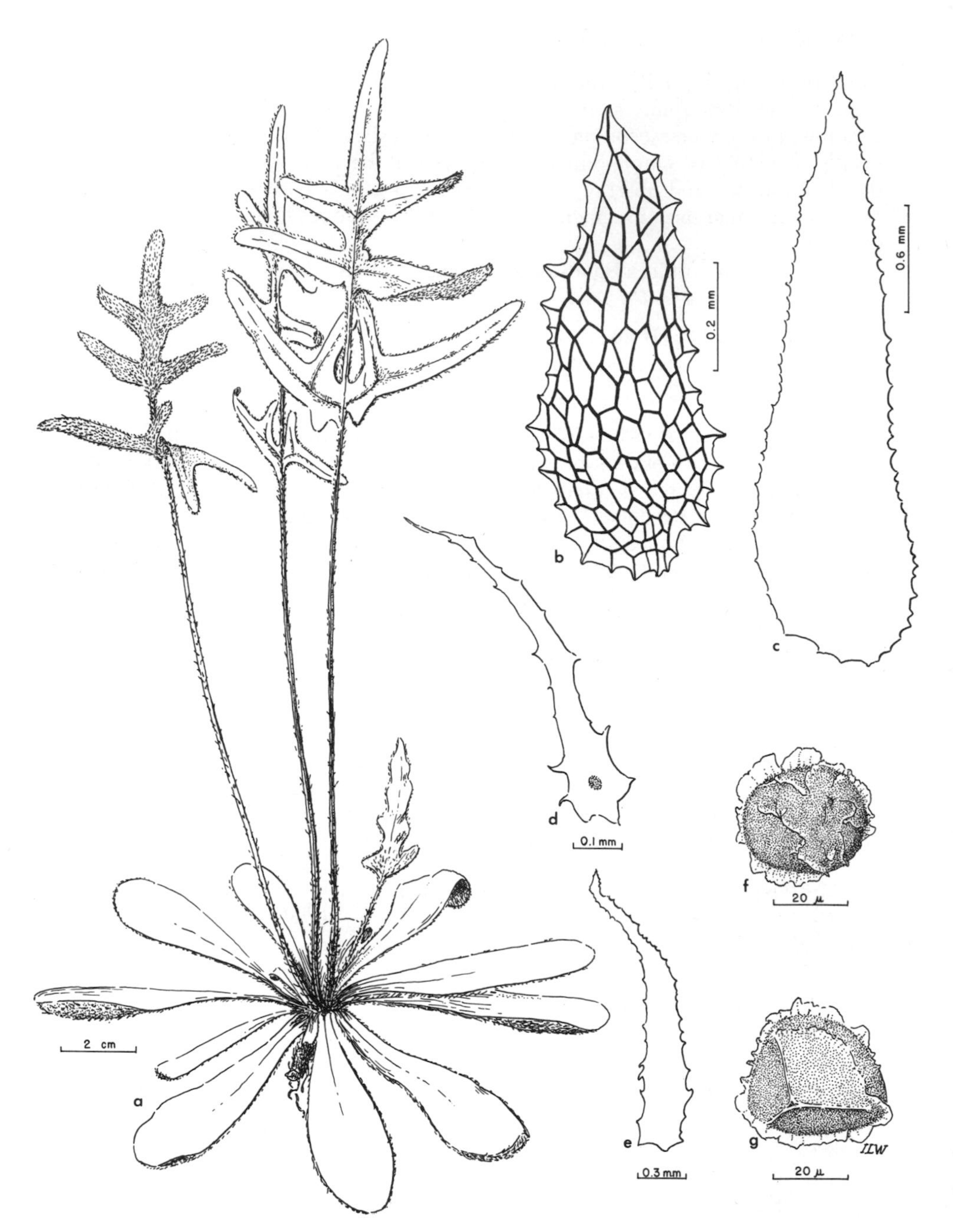

Fig. 32. *Trachypteris pinnata.*

cells, golden brown; spores trilete, 35–40 μ in diameter, irregularly and moderately alate, the wings very thin and appearing like spinulose processes in narrowest profile, 1–5 μ high, the margins slightly serrulate.

On rocks and rocky soil in moderately dry to fairly moist habitats, in open areas, and in woodlands, Brazil, Argentina, Bolivia, and Peru. Specimens examined: Fernandina, Isabela, Pinta, San Cristóbal, San Salvador, Santa Cruz, Santa María.

See Fig. 32.

POLYPODIACEAE SPECIES EXCLUDED FROM THE FLORA

Acrostichum aureum L., Sp. Pl. 1069. 1753

Reported from the Galápagos Islands from a Wood collection, but not found since. The specimens so identified by Stewart are *Blechnum falciforme* Broadh.

Adiantum (Three species; see p. 91 for a brief discussion.)

Ananthacorus angustifolius (Sw.) Underw. & Maxon, Contr. U.S. Nat. Herb. 10: 487. 1908

Pteris angustifolius Sw., Nov. Gen. Sp. Pl. Prodr. 129. 1788.
Taenitis angustifolia Spreng. in L., Syst. Nat., ed. 16, 4: 42. 1827.
Vittaria angustifolia Baker in Mart., Fl. Bras. 1[2]: 544. 1870.

Reported from the Galápagos Islands on the basis of a Wood collection that probably came from the mainland.

Asplenium (Three species; see pp. 101–2 for a brief discussion.)

Cyclopeltis semicordata (Sw.) J. Smith, Bot. Mag. Curtis 72: Comp. 36. 1846

Polypodium semicordatum Sw., Nov. Gen. Sp. Pl. Prodr. 132. 1788.
Aspidium semicordatum Sw., Jour. Bot. Schrad. 1800[2]: 31. 1802.

Reported from the Galápagos Islands on the basis of a Wood collection, probably in error with respect to locality.

Cystopteris fragilis (L.) Bernh., Schrad. Neues Jour. Bot. 1[2]: 27, *pl. 2*, *f.* 9. 1806

Polypodium fragile L., Sp. Pl. 1091. 1753.

Reported from Santa María by Theodor Wolf from a field identification. An extremely doubtful record in the absence of an extant specimen.

Histiopteris incisa (Thunb.) J. Smith, Hist. Fil. 295. 1875

Pteris incisa Thunb., Prodr. Pl. Cap. 171. 1800.

This species, reported from the Galápagos Islands by Hooker f. on the basis of a Wood specimen that probably was collected on the mainland, differs from *Pteris* in having an elongate, widely creeping rhizome with distant fronds, in having the basal pinnules enlarged and stipuliform, and in having the spores monolete, among other characters. [N.B. After pagination had been set, a specimen of *H. incisa* collected in a sphagnum bog between El Camote and Mount Crocker, at 680 m altitude, on Isla Santa Cruz (*Itow 40403-6*, 4 April 1970), was submitted by Dr. Itow for identification; his specimen dispels uncertainty about the Wood collection and the presence of this species in the Galápagos Islands.]

Hypolepis repens (L.) Presl, Tent. Pterid. 162. 1836 (See p. 129 for a brief description.)

Lonchitis repens L., Sp. Pl. 1078. 1753.
Dicksonia aculeata Spreng., Neue Entdeck. 3: 7. 1822.
Cheilanthes repens Kaulf., Enum. Fil. 215. 1824.

Thelypteris (Two species; see p. 170 for a brief discussion.)

PSILOTACEAE. WHISK-PLANT FAMILY*

Erect or pendulous, epiphytic or terrestrial plants superficially resembling some species of *Lycopodium*. Gametophytes hypogeous or in fissures in bark of trees, small, brown, elongate-cylindric, bearing antheridia and archegonia from subepidermal cells. Sporophytes with branching, slenderly coralloid rhizomes devoid of roots; aerial stems branching dichotomously, photosynthetic, angled or flattened. Leaves alternate, small, elliptic to broadly oblanceolate or reduced to minute scales. Sporangia 3-lobed, rather large, borne near apex of branches on adaxial sides of scales or appendages, the fertile branches erect in *Psilotum*, horizontal in *Tmesipteris*. Spores numerous, in tetrads, monolete, all alike, the walls smooth to finely reticulate.

A family of 2 genera and a few species, widely distributed in humid, tropical and subtropical areas of the Old and New Worlds.

Psilotum Sw., Jour. Bot. Schrad. 1800[2]: 109. 1802

Stems erect and plants terrestrial or stems pendent and plants epiphytic. Appendages reduced to minute scales. Otherwise, with characters of the family.

A genus of 2–5 species, in tropical and subtropical humid areas.

Psilotum nudum (L.) Palisot,† Prodr. Fam. Aethog. 106, 112. 1805

Lycopodium nudum L., Sp. Pl. 1100. 1753.
Psilotum triquetrum Sw., Jour. Bot. Schrad. 1800[2]: 109. 1802. (*Psilotum triquetrum* Sw. is a superfluous, illegitimate name based on *Lycopodium nudum* L.)

Plant terrestrial, perennial, from intricately branching and creeping coralloid rhizomes; aerial stems often crowded in clumps, rigidly erect, dichotomously 3–5-branched, thus broomlike, 1–5 dm tall (mostly under 2.5 dm in ours), the main stalks simple, triangular in cross section, 3–4 mm thick; branches faintly winged along angles, 0.5–1 mm thick; leaves reduced to minute, obliquely spreading, subulate scales about 1 mm long; sporophylls rudimentary, more or less bilobed; sporangia yellowish to light brown, shining, depressed-globose, 3-celled and 3-lobed,

* Contributed by C. V. Morton and Ira L. Wiggins.

† Palisot de Beauvois, usually cited as "Beauv." ("Beauvois" was the place where he lived.) The combination *P. nudum* is often erroneously attributed to Grisebach.

about 2 mm in diameter; spores oblong or bluntly ellipsoidal, 35–40 μ long, 20–25 μ wide, with 2 parallel lines along inner face, the surface evenly and minutely reticulate-pitted.

In moist habitats in forests or among dense shrubs, along banks of water courses, on dripping cliffs, and about bases of trees in humus and dense shade, southern United States to southern South America, widely distributed in the tropics and subtropics of both the Northern and the Southern Hemispheres. Specimens examined: Fernandina, Isabela.

Riley (1925, 231) reported this species (as *P. triquetrum* Sw.) from Isla San Salvador, based on collections made in 1924; we have not seen his specimens. It probably occurs on Isla Santa Cruz, and should be sought there in the moister, deeply shaded areas.

See Fig. 33 and color plate 92.

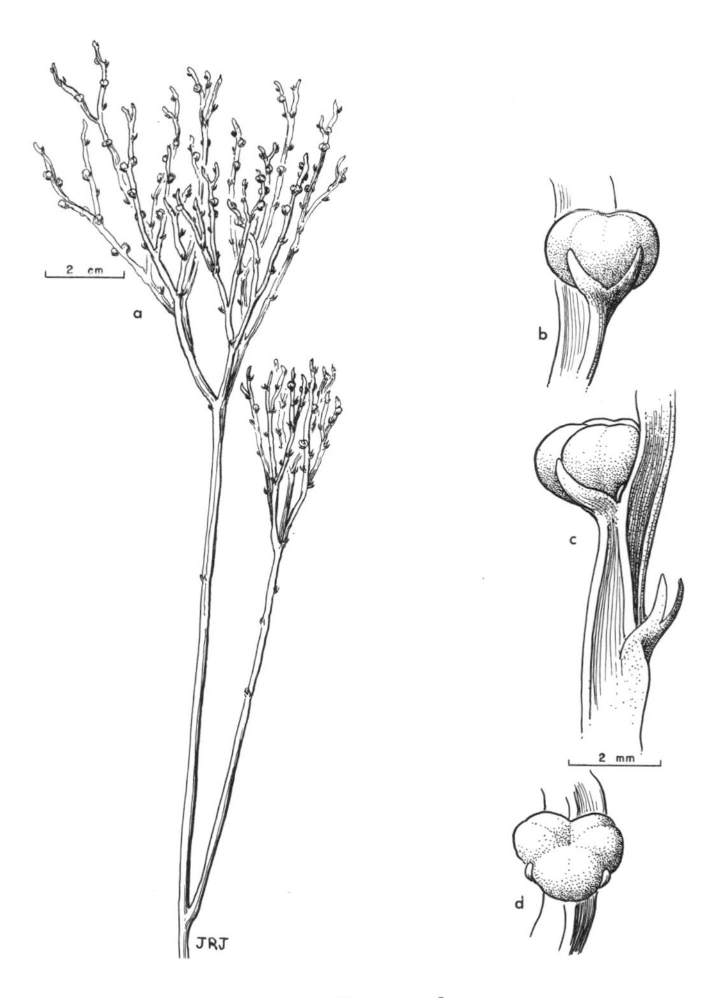

Fig. 33. *Psilotum nudum.*

SALVINIACEAE. WATER FERN FAMILY*

Small floating plants with stems simple or very sparingly branched, and leaves borne in 2 ranks, horizontally spreading, with papillae or short hairs on upper surface, and brown-hairy beneath. Spores heterosporous, borne in megasporangia and microsporangia.

A monogeneric family containing about a dozen species, of wide geographic distribution.

Salvinia Adans., Fam. Pl. 2: 15. 1763

Characters of the family. Leaves cordate, 1–1.5 cm long, not appreciably imbricate, rough-papillose above, densely brown-hairy beneath. Sporocarps borne in groups of 4's or 8's, the upper containing about 10 megasporangia, each of latter usually containing a single megaspore, the other sporocarps each containing numerous microsporangia, with many microspores in each of these.

Floating on quiet water or in slowly moving streams and ditches.

Theodor Wolf reported (1879) seeing *Salvinia* in a brook near the hacienda on Isla Santa María, but no material was collected. Although diligently sought, it has not been found since.

Some of Wolf's sight identifications are not to be trusted (e.g., *Cystopteris fragilis*), but *Salvinia* is such a distinctive plant that it could hardly be confused with anything else, though the flowering plant *Lemna* somewhat simulates it. If Wolf actually saw *Salvinia*, the species probably was *S. auriculata* Aublet (Hist. Pl. Guian. 2: 969, *t. 367*. 1775), which is the commonest species in South America. *Salvinia* can be distinguished from *Azolla* by its much larger, round to oblong, hairy leaves, and by the different arrangement of the leaves.

* Contributed by C. V. Morton and Ira L. Wiggins.

Dicotyledones
Series I: Apetalae

Dicotyledones
Series I: Apetalae

For convenience we have arranged the Angiospermae in four sections of text. In this section, following a synopsis to the class and a key to its subclasses and series, the apetalous families are taken up in a single alphabetical sequence.

Class ANGIOSPERMAE*

Herbs, shrubs, trees, or vines with vessels, fiber cells, and often tracheids in xylem of vascular bundles. Microspore nucleus usually dividing once before dispersal of pollen grains, microgametophyte thus binucleate or 2-celled at time of dispersal. Megasporophylls (carpels) often elongate at apex to form a style, this bearing a receptive area (stigma) at or near its summit. Carpels 1 to several in each perfect or pistillate flower, if more than 1 either all free and distinct or all more or less coalesced and pistil then made up of more than 1 carpel. Styles and stigmas in multicarpellate pistils distinct or more or less coalesced. Ovules borne inside carpels on placental tissue of ovary, the placenta central, peripheral, or basal; each ovule attached to placenta by slender filament, the funiculus. Microgametophytes carried from anthers to stigmas by gravity, wind, water, or insects (rarely by birds or bats), or transferred by mechanical contact owing to growth movements of flower parts, the pollen grains germinating on stigma and producing a pollen tube that grows to ovules in ovary. One male gamete nucleus uniting with nucleus of egg cell, the other uniting with endosperm nucleus (or nuclei) to initiate growth of food-supplying endosperm of seed. Seeds with 1 or 2 coats. Embryo minute to large, containing much stored food, or food occurring in endosperm, or some stored in each tissue. Endosperm copious, scanty, or none. Cotyledons 1 or 2.

Key to the Subclasses and Series of the Angiospermae

Cotyledons 2; leaves usually netted-veined; vascular bundles "open," with a meristem (cambium) between xylem and phloem in bundles of stem (Subclass Dicotyledones, p. 180):
- Petals none; perianth segments all alike in texture and color, or represented by bracts, or none . . . . Series I. Apetalae, p. 180
- Petals present; inner whorl of perianth segments (corolla) usually differing markedly in color, or texture, or size, or in all these characters from outer whorl (calyx):
  - Petals united into a tube or cup, at least at the base and often for much of their

---

* Descriptions of groups above families and keys to these groups by Ira L. Wiggins.

length, the corolla lobes either shorter or longer than the tube...Series II. GAMOPETALAE, p. 257
Petals distinct, free from one another at least at the base and usually for their full length.................................Series III. POLYPETALAE, p. 513
Cotyledon 1; leaves usually parallel-veined; vascular bundles "closed," without a meristem (cambium) between xylem and phloem in stem..Subclass MONOCOTYLEDONES, p. 781

Subclass DICOTYLEDONES

Stems of pith, wood, and cortex (bark), a layer of thin-walled, meristematic cells (cambium) lying between xylem and phloem and adding new cells to both these tissues; cambium functional throughout life of plant. Medullary rays and xylem rays traversing xylem of stem. Leaves usually pinnately or palmately veined, the veinlets in a reticulate pattern, rarely parallel. Floral parts in 4's or 5's, rarely in 3's or 6's. Embryo of seed mostly with 2 cotyledons. First leaves of young seedling opposite.

Series I. APETALAE

Dicotyledonous plants with floral envelope consisting of outer whorl (calyx) only, or sometimes this also lacking; inner whorl (corolla) completely lacking. Otherwise, with characters of the subclass.

[A few genera in some families chiefly polypetalous or gamopetalous have become apetalous. The two families to which such genera belong (the Euphorbiaceae and the Sapindaceae) have been included in the key, in each case with a page reference to the position of the family in the text.]

KEY TO THE FAMILIES OF THE APETALAE*

Flowers borne in dense, often fleshy spikes, often more or less embedded in tissues of axis:
  Leaves leathery; plant parasitic on shrubs, trees, and woody vines; ovary inferior; fruit a berry, its flesh viscid-sticky............................VISCACEAE, p. 251
  Leaves herbaceous to fleshy and juicy; plant autotrophic, not parasitic on other plants; ovary superior; flesh of fruits (when present) not viscid-sticky:
    Flowers bisexual; perianth segments 5, free or connate at their bases; plants scandent vines; leaf blades thin-herbaceous, broadly ovate........BASELLACEAE, p. 208
    Flowers unisexual, usually naked or perianth reduced to a bract; plants ascending, erect, or pendent, but not long-scandent vines; leaf blades linear, elliptic, or reduced to minute scales:
      Leaves alternate, blades elliptic, petiolate; plants often epiphytic on shrubs and trees ....PIPERACEAE, p. 231
      Leaves opposite, blades linear to spatulate and sessile, or reduced to minute scales; plants not epiphytic:
        Ovary 4-celled; leaves linear to spatulate; fruit a berry....BATIDACEAE, p. 208
        Ovary 1-celled; leaves reduced to a pair of minute scales at apex of each fleshy joint of stem; fruit a utricle (*Salicornia*).......CHENOPODIACEAE, p. 210
Flowers variously arranged, but not in fleshy spikes nor embedded in axis:
  Ovary inferior; plant parasitic on shrubs, trees, and vines; fruit with viscid-sticky pulp .... VISCACEAE, p. 251
  Ovary superior; plants not parasitic; fruit dry or fleshy, but not viscid-sticky:
    Leaves alternate (basal ones sometimes opposite):

* Key by Ira L. Wiggins.

Perianth segments usually 6, in 2 whorls:
Fruit a large berry or drupe, 5–20 cm long; stipules absent; outer perianth segments smaller than inner; plant a medium to large tree. . LAURACEAE, p. 213
Fruit an achene, less than 5 mm long; stipules present, sheathing the stem; inner and outer perianth whorls similar; plants (in ours) herbaceous. . . . . POLYGONACEAE, p. 236
Perianth segments 3–5, in a single whorl, or sometimes only 1 and bractlike:
Ovary 1-celled, fruit 1-seeded:
Leaves bearing elliptic to short-linear cystoliths, often with stinging hairs also; stamens inflexed in bud and erupting explosively. . . . URTICACEAE, p. 243
Leaves without cystoliths and/or stinging hairs; stamens not markedly inflexed, nor bursting explosively at anthesis:
Fruit juicy or pulpy, a berry or a drupe:
Leaves not distichous, not asymmetrical at base; fruit a berry. . . . . PHYTOLACCACEAE, p. 229
Leaves distichous, asymmeterical at base; fruit a small drupe. . . . . ULMACEAE, p. 240
Fruit dry, a utricle or an anthocarp:
Calyx tubular, often brightly colored; ovary apparently inferior, but not adnate to calyx tube; fruit an anthocarp. . . . . NYCTAGINACEAE, p. 222
Calyx not tubular, not brightly colored; ovary obviously superior:
Bracts and sepals often scarious; stamens connate into a ring at base; leaves rarely scurfy. . . . . AMARANTHACEAE, p. 184
Bracts and sepals herbaceous or fleshy, not scarious; stamens free at base; leaves usually scurfy. . . . . CHENOPODIACEAE, p. 210
Ovary 2- to several-celled; fruit 2- to several-seeded:
Stipules usually present; sap often milky or bright-colored, more or less corrosive and irritating to skin; marginal capitate or cupulate leaf glands often present; seeds usually carunculate. . . . . EUPHORBIACEAE (Series III), p. 568
Stipules none; sap not milky, not bright-colored; marginal capitate or cupulate leaf glands none; seeds not carunculate:
Fruit a juicy, usually 8-celled berry; leaves not appearing varnished, to 20 cm long; embryo annular. . . . . PHYTOLACCACEAE, p. 229
Fruit a winged capsule, 1–2-celled; leaves appearing varnished, 3–6 cm long; embryo curved but not annular (*Dodonaea*). . . . . SAPINDACEAE (Series III), p. 747
Leaves opposite (rarely a few alternate):
Stamens strongly inflexed in bud, straightening explosively at anthesis; cystoliths on both surfaces of leaves. . . . . URTICACEAE, p. 243
Stamens not markedly inflexed, not straightening explosively; leaves without cystoliths:
Ovary 1-celled; fruit 1-seeded; leaves exstipulate:
Calyx tubular and often brightly colored; ovary usually appearing inferior, but not adnate to calyx tube. . . . . NYCTAGINACEAE, p. 222
Calyx not tubular, not brightly colored; ovary obviously superior:
Bracts and sepals scarious; stamens connate into ring at base; leaves not reduced to scales. . . . . AMARANTHACEAE, p. 184
Bracts and sepals fleshy or herbaceous, not scarious; stamens free; leaves reduced to minute scales (*Salicornia*). . . . . . CHENOPODIACEAE, p. 210
Ovary 2-celled to several-celled; fruit usually several- to many-seeded; leaves stipulate or exstipulate:
Fruit circumscissile; herbage more or less succulent; calyx lobes appendaged just below apex on outer surface. . . . . AIZOACEAE, p. 182

Fruit not circumscissile; herbage not succulent; calyx lobes not dorsally appendaged:
Leaves usually in whorls; seeds ridged or tuberculate, moderately compressed laterally; capsule membranous, splitting longitudinally...... .... MOLLUGINACEAE, p. 217
Leaves opposite, not in whorls; seeds otherwise than above; fruit a chartaceous to woody capsule or a schizocarp:
Styles 2; fruit 2-loculed (or 4-loculed by a false partition); stipules none; flowers mostly solitary in axils of leaves... CALLITRICHACEAE, p. 210
Styles 3; fruit usually 3-loculed, capsular; stipules mostly present, often conspicuous; flowers sometimes in cyathia...................... ..........................EUPHORBIACEAE (Series III), p. 568

AIZOACEAE. AIZOON FAMILY*

Herbs or shrubs. Leaves simple, alternate or opposite, often fleshy, sometimes small and scalelike. Flowers usually bisexual, actinomorphic. Calyx free or tube adnate to ovary; calyx lobes 5–8. Petals (staminodia) numerous or absent. Stamens few to many, perigynous, free or united into bundles at base; anthers 2-celled. Ovary superior or inferior, 1- to several-loculed. Ovules solitary to many in a locule. Fruit a capsule, nutlike or berry-like, often clasped by the persistent calyx. Embryo strongly curved.

About 130 genera and 1,200 species, widely distributed in tropical and subtropical regions, most strongly represented in South Africa.

Leaves linear to oblanceolate, at least 4 times as long as wide, very succulent; ovary 3- to 5-loculed .... *Sesuvium*
Leaves obovate, spatulate or subelliptical, 1.5–2 times as long as wide, flat; ovary 1- or 2-loculed .... *Trianthema*

Sesuvium L., Syst. Nat. ed. 10, 1058. 1759

Perennial succulent herbs or low shrubs. Leaves opposite, fleshy, linear to oblong or oblanceolate. Flowers seemingly axillary. Calyx lobes 5, united below, mucronate near apex. Petals none. Stamens many. Ovary free, 3- to 5-loculed; styles 3–5. Ovules numerous in each locule. Capsule circumscissile. Testa of seeds smooth or wrinkled.

About 8 species on tropical and subtropical coasts.

Seeds with brainlike furrows; stems and branches densely covered with scales; calyx lobes white .... *S. edmonstonei*
Seeds smooth; stems and branches glabrous, not covered with scales; calyx lobes purplish .... *S. portulacastrum*

Sesuvium edmonstonei Hook. f., Trans. Linn. Soc. Lond. 20: 221. 1847
Sesuvium eastwoodianum Howell, Leafl. West. Bot. 10: 352. 1966.

Ascending, freely branched; stems and branches densely covered with rounded scales, margins of scales bent upward; leaves sessile or nearly so, oblanceolate, obtuse at apex, narrowed toward base, terete, 1.5–3(4) cm long, 0.3–0.5 cm wide,

* Contributed by Uno Eliasson.

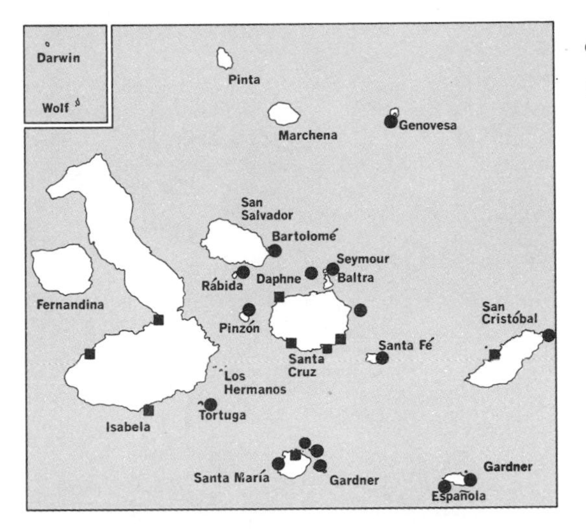

● *Sesuvium edmonstonei*

■ *Sesuvium portulacastrum*

often covered on both sides with a layer of scales, dilated into open sheath at base; peduncles 5–15(20) mm long; calyx 7–8(10) mm long, lobes 4–5(6) mm long, triangular, white within; seeds about 1 mm in diameter, black, round-reniform, with brainlike furrows.

Endemic; found over most of the archipelago on rocky shores and shell-sand banks. Specimens examined: Bartolomé, Caldwell, Champion, Daphne Major, Enderby, Española, Gardner (near Española), Genovesa, Pinzón, Las Plazas, Rábida, San Cristóbal, Santa Fé, Santa María, Seymour, Tortuga.

The color of the whole plant varies from grayish green to orange or bright red. See Fig. 34.

Sesuvium portulacastrum L., Syst. Nat. ed. 10, 1058. 1759

Stems glabrous, procumbent or ascending, often rooting at nodes; leaves sessile or nearly so, oblanceolate or oblong, obtuse to acutish at apex, narrowed toward base, 3–5 cm long, 0.5–1 cm wide, dilated into membranous open sheath at base; peduncles 15–25 mm long; calyx about 10 mm long, lobes 6–7 mm long, triangular, purplish within; seeds about 1.2 mm in diameter, shiny black, smooth, rounded; funiculus prominent.

Widely distributed in the tropics of the world; in the archipelago, on sandy shores. Specimens examined: Isabela, San Cristóbal, Santa Cruz, Santa María.

See color plate 69.

Trianthema L., Sp. Pl. 223. 1753; Gen. Pl. ed. 5, 105. 1754

Herbs, sometimes suffrutescent. Leaves opposite, the blades of each pair of different size. Bases of 2 opposite petioles united by thin membrane to form sheath. Flowers axillary. Calyx lobes 5, united below, mucronate near apex. Petals none. Stamens 5 to many. Styles 1 or 2. Ovary free, 1- or 2-loculed, with 1 to many ovules in each locule. Capsule circumscissile, the lid incompletely closed at its base by transverse, membranous dissepiment enclosing 1 or 2 seeds.

About 20 species in tropical and subtropical regions.

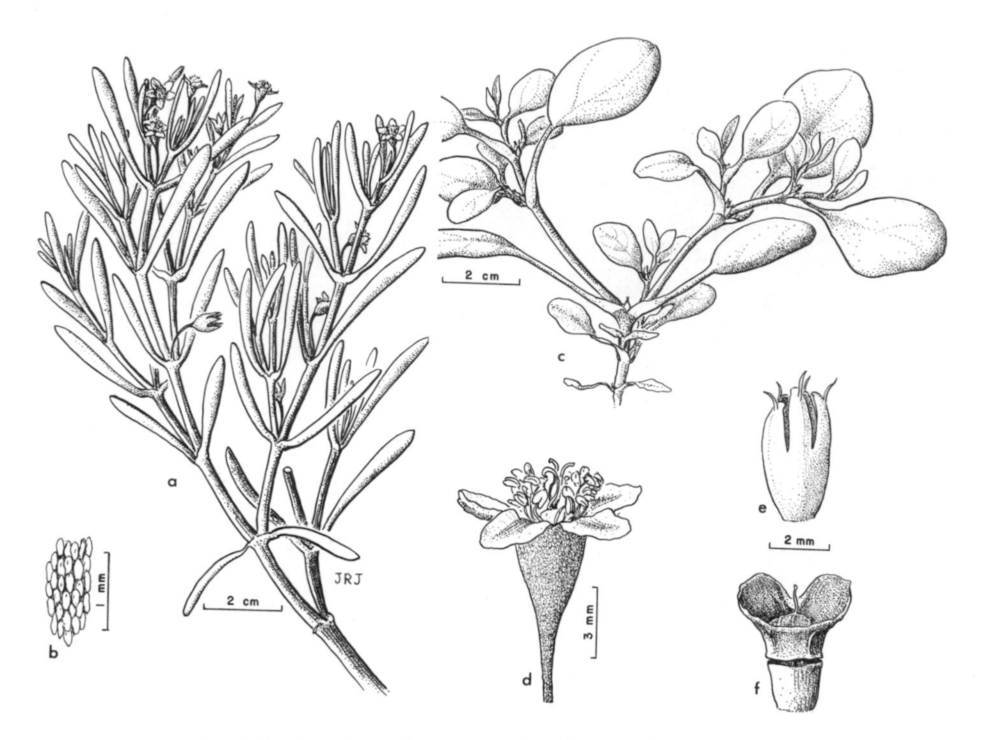

Fig. 34. *Sesuvium edmonstonei* (*a, b*): *b*, epidermal vesicles on stems. *Trianthema portulacastrum* (*c–f*).

Trianthema portulacastrum L., Sp. Pl. 223. 1753

Stem prostrate, ascending or erect, glabrous, often reddish; leaves somewhat succulent but flat; leaf blades obovate, spatulate, or subelliptic, obtuse or acute at base, rounded to truncate at apex, usually mucronulate, 1–2(3) cm long, 0.7–2 cm wide; petioles 5–25 mm long; flowers solitary, partly concealed within petiolar sheath; calyx purplish within; calyx lobes about 5 mm long; style 1; seeds black, 1.3–1.5 mm in diameter, reniform, the testa wrinkled.

A species of wide distribution in the tropics of the world; widely distributed in the Galápagos Islands. Specimens examined: Baltra, Coamaño, Española, Gardner (near Española), Isabela, Pinzón, Rábida, San Cristóbal, Santa Cruz, Santa Fé, Santa María, Seymour, Tortuga.

See Fig. 34 and color plate 46.

AMARANTHACEAE. Pigweed Family*

Herbs or shrubs, rarely small trees. Leaves opposite or alternate, entire, exstipulate. Flowers actinomorphic, unisexual or bisexual, generally small, arranged in spikes, heads, racemes, or panicles, often with scarious bracts and bractlets. Perianth segments (sepals) generally 3–5, free or united below into a tube. Petals none. Sta-

* Contributed by Uno Eliasson.

mens generally 5, hypogynous, opposite sepals, often alternating with staminodia or pseudostaminodia; filaments united below into a cup or short tube; anthers with 2 or 4 pollen sacs, dehiscing by longitudinal slits. Ovary 1, superior, 1-loculed; stigma capitate or 2–3(6)-fid. Ovules 1 to several. Embryo curved. Fruit indehiscent or transversely dehiscent.

About 65 genera and 850 species in tropical, subtropical, and temperate regions.

Leaves alternate; anthers with 4 pollen sacs:
- Annual herbs; fruit 1-seeded . . . . . *Amaranthus*
- Shrubs 1–2 m high; fruit a many-seeded berry-like capsule . . . . . *Pleuropetalum*

Leaves opposite; anthers with 2 pollen sacs:
- Perianth-segments united at least to middle into a tube . . . . . *Froelichia*
- Perianth-segments free to base:
  - Stigma capitate; anthers 5, alternating with laciniate pseudostaminodia . . . . . *Alternanthera*
  - Stigma 2- or 3-fid; anthers 2 or 5; pseudostaminodia present or absent:
    - Anthers 2; staminodia 3; pseudostaminodia lacking; leaves mainly crowded in basal rosette . . . . . *Lithophila*
    - Anthers 5:
      - Heads few-flowered (6- to 8-flowered); leaves ovate-lanceolate; anthers alternating with short pseudostaminodia . . . . . *Iresine*
      - Heads many-flowered (20 flowers or more); leaves subulate; pseudostaminodia none . . . . . *Philoxerus*

Alternanthera Forsk., Fl. Aegypt. Arab. 28. 1775

Herbs, shrubs, or rarely small trees. Leaves opposite, entire, linear to suborbicular, often mucronulate. Inflorescences of terminal or axillary rounded or cylindrical heads, solitary or to several-glomerate. Flowers bisexual, each flower subtended by 1 ventral bract and 2 lateral bractlets, the bractlets generally keeled. Sepals 5, free, lanceolate or triangular-lanceolate, acute, glabrous or hairy, often of different lengths. Petals absent. Stamens usually 5, filaments at least basally united, forming

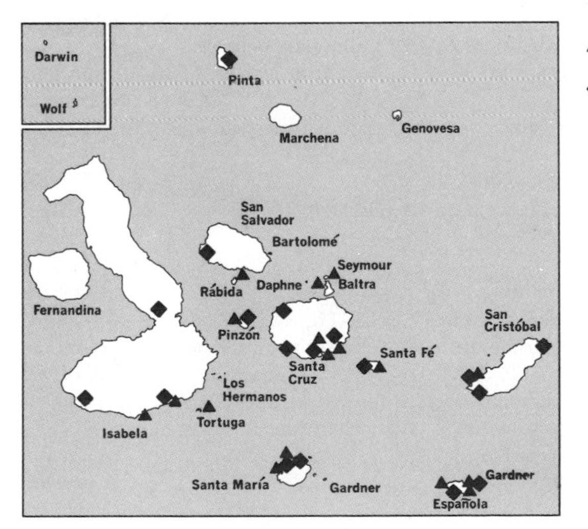

▲ *Trianthema portulacastrum*

◆ *Alternanthera echinocephala*

tube where fertile stamens alternate with laciniate pseudostaminodia of varying lengths. Anthers with 2 pollen sacs, introrse, dehiscing longitudinally. Pistil 1; ovary rounded or ovoid, 1-loculed with 1 ovule. Stigma capitate.

About 200 species in tropical and subtropical regions.

Leaves narrowly linear to linear-lanceolate, generally 1–4(7) mm wide:
  Leaves densely pubescent (especially beneath), linear, 1–2 mm wide; heads generally terminal and solitary . . . . *A. flavicoma*
  Leaves glabrous or subglabrous, linear to linear-lanceolate, 2–4(7) mm wide; heads terminal or axillary, solitary or 2- to 7-glomerate (leaves oblanceolate, 4–7(15) mm wide in *A. filifolia* subsp. *pintensis* and subsp. *rabidensis*):
    Stems and leaves not glaucous; leaves not succulent:
      Leaves linear or linear-oblanceolate:
        Heads minute, rounded, 3–4 mm in diameter; stems sparsely branched . . . . *A. filifolia* subsp. *microcephala*
        Heads rounded or oblong, generally 5–15 mm long; stems generally profusely branched:
          Flowers loosely imbricate to subsquarrose, varying from subglabrous to rather densely covered with yellow trichomes; flower tips generally straight . . . . *A. filifolia* subsp. *filifolia*
          Flowers closely and smoothly imbricate, densely covered with yellow trichomes; flower tips often incurved . . . . *A. filifolia* subsp. *nudicaulis*
      Leaves oblanceolate or oblong-oblanceolate:
        Leaves with stellate trichomes; flowers closely imbricate . . . . *A. filifolia* subsp. *pintensis*
        Leaves with simple trichomes; flowers loosely imbricate . . . . *A. filifolia* subsp. *rabidensis*
    Stems and leaves glaucous; leaves somewhat succulent:
      Heads generally globose or subglobose, 3–4 mm in diameter; subshrubs about 3 dm high; most of the branches diverging near base of plant . . . . *A. filifolia* subsp. *glauca*
      Heads generally cylindrical, 7–10 mm long, 3.5–4.5 mm in diameter; bushes about 5 dm high; rather uniformly branched . . . . *A. filifolia* subsp. *glaucescens*
Leaves oblanceolate, elliptic, ovate, or suborbicular:
  Stems and leaves glabrous and glaucous . . . . *A. galapagensis*
  Stems and leaves not glaucous; leaves more or less pubescent beneath:
    Flowers arcuate-spreading or squarrose; sepals subequal in length, (5)6–10 mm long . . . . *A. echinocephala*
    Flowers loosely to closely imbricated; sepals distinctly unequal; outer sepal never exceeding 4 mm in length:
      Leaves at least twice as long as wide:
        Leaves elliptic-lanceolate, oblanceolate or narrowly obovate, generally 0.5–1.5 cm wide:
          Hairs on stems simple . . . . *A. snodgrassii*
          Hairs on stems much-branched . . . . *A. vestita*
        Leaves elliptic, ovate, or oblong-ovate, (1)1.5–4(6) cm wide:
          Plant a trailing or bushy perennial; leaves elliptic to ovate, acute at apex . . . . *A. halimifolia*
          Plant a small tree; leaves oblong-ovate, rounded at apex . . . . *A. rugulosa*
      Leaves nearly as wide as long, sometimes wider than long:
        Plant shrubby; leaves 1.5–5 cm long . . . . *A. helleri*
        Plant prostrate; leaves 0.4–0.9 cm long . . . . *A. nesiotes*

Alternanthera echinocephala (Hook. f.) Christoph., Nyt Mag. Naturvid. 70: 73. 1932 (1931).

Brandesia echinocephala Hook. f., Trans. Linn. Soc. Lond. 20: 189. 1847.
Telanthera echinocephala Moq. in DC., Prodr. 13: 373. 1849.
Telanthera argentea Anderss., Kongl. Svensk. Vet.-Akad. Handl. 1853: 168. 1855.
Telanthera argentea var. *robustior* Anderss., loc. cit.
Telanthera argentea var. *nudiflora* Anderss., op. cit. 169.
Telanthera argentea var. *bracteata* Anderss., loc. cit.
Telanthera echinocephala var. *robustior* Anderss., Kongl. Svensk. Vet.-Akad. Handl. 1859: 63. 1861.
Telanthera echinocephala var. *nudiflora* Anderss., loc. cit.
Telanthera echinocephala var. *bracteata* Anderss., loc. cit.
Achyranthes echinocephala Standl., Jour. Wash. Acad. Sci. 5: 74. 1915.
Alternanthera tubulosa Suesseng., Rep. Nov. Spec. Regn. Veg. 42: 56. 1937.

Shrub to 2 m high; stems erect, branched, glabrous or with appressed to spreading, unbranched, many-celled trichomes; leaves lanceolate to elliptic, 4–9 cm long, 1–3 cm wide, mucronulate, with sparse to dense, unbranched hairs on both surfaces; flower heads terminal and axillary, rounded, 1–1.5 cm across; flowers loosely imbricate to arcuate-spreading; bracts and bractlets ovate, mucronate, 2–4.5 mm long, glabrous or with scattered hairs; sepals nearly equal, lanceolate, acute, 5–10 mm long, glabrous or slightly hairy; stamen tube 4–5 mm long; anthers about 2 mm long, hiding the short, free filaments; pseudostaminodia laciniate, somewhat shorter to somewhat longer than stamens.

A common coastal plant in the Galápagos Islands, distributed widely; also in Peru. Specimens examined: Española, Gardner (near Española), Isabela, Pinta, Pinzón, San Cristóbal, San Salvador, Santa Cruz, Santa Fé, Santa María.

See Fig. 35 and color plates 29 and 40.

Alternanthera filifolia (Hook. f.) Howell, Proc. Calif. Acad. Sci. IV, 21: 102. 1933 (sensu lato, the description covering all subspecies)

Extremely variable, forming shrubs 0.5–1.5 m high; leaves linear, linear-oblanceolate, oblanceolate, or oblong-oblanceolate, 2–5 cm long, 2–7 mm wide, green or glaucous, subglabrous or with simple or stellate trichomes; flower heads terminal or axillary, solitary or 2–7 glomerate, rounded or cylindrical, 3–15 mm long, 3–5 mm wide; flowers loosely or closely imbricate, sometimes subsquarrose, either subglabrous or more or less densely covered with white or yellowish simple trichomes, flower tips incurved, straight, or divergent; bracts and bractlets ovate, acute, 1.5–2 mm long; sepals lanceolate-triangular, outer one 1.7–3 mm long, inner ones 1.1–2.5 mm long; anthers about 0.6 mm long; pseudostaminodia laciniate, shorter than or equaling stamens.

Occurring over a large part of the archipelago. The distribution of *A. filifolia*, sensu lato, in areas outside the Galápagos is somewhat uncertain. It was long considered to be endemic, but plants collected in Chile have been referred to *A. filifolia* subsp. *nudicaulis* in the literature.

In this treatment 7 subspecies have been distinguished. Especially within the subspecies *filifolia* it is possible to recognize several morphologically different populations, but the differences among these entities are generally very small and hardly worth distinguishing. In several cases intermediate specimens can be found. Howell (Proc. Calif. Acad. Sci. IV, 21: 103–4. 1933) recognized *A. filifolia* subsp. *subsquarrosa*, subsp. *margaritacea*, and subsp. *sylvatica* from Islas San Salvador, Isabela, and Santa Cruz, respectively. Because a series of intergrading specimens exists

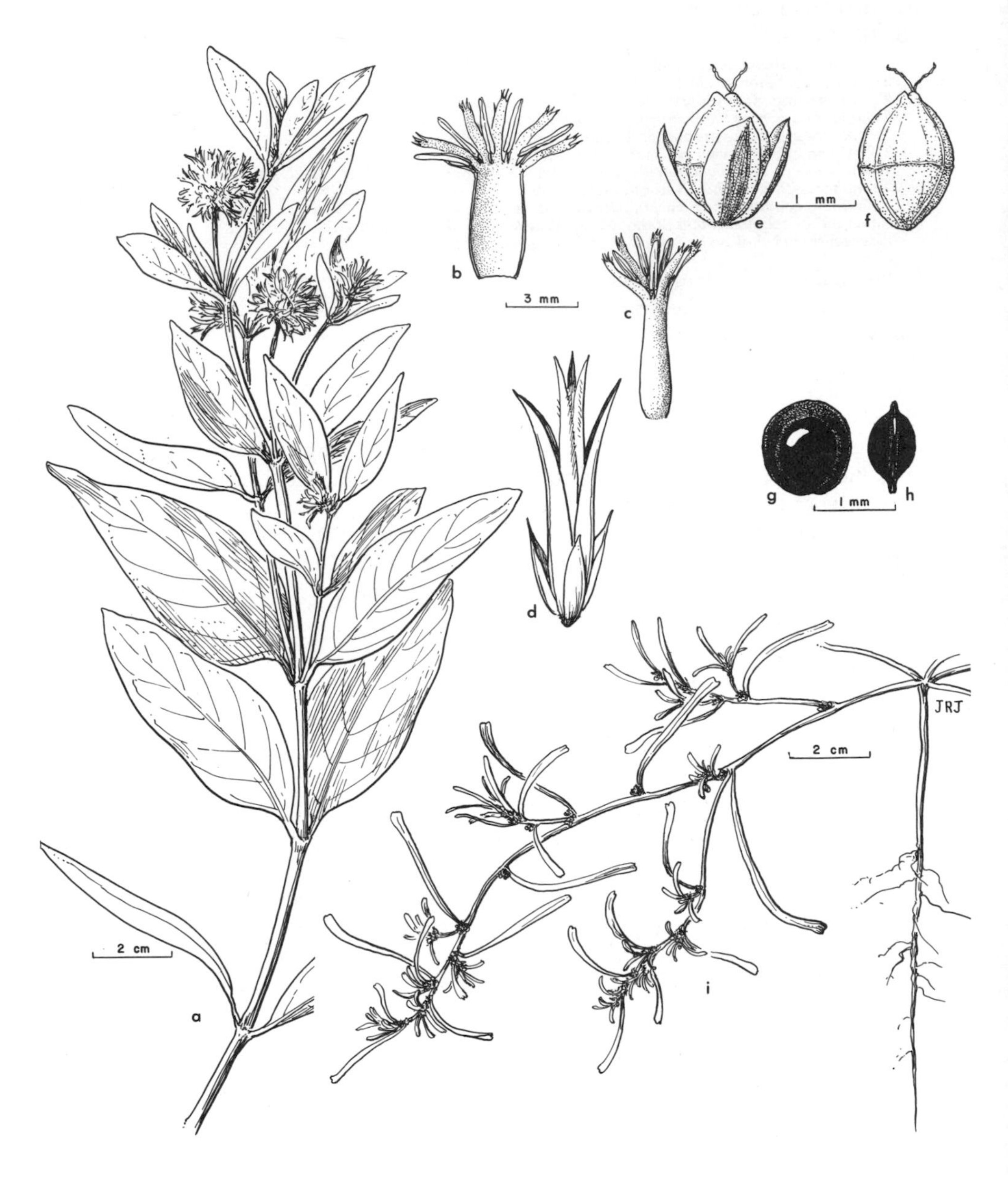

Fig. 35. *Alternanthera echinocephala* (*a–d*): *b*, flower, with tube split; *c*, single flower from exterior; *d*, fruit encased in bracts. *Amaranthus sclerantoides* (*e–i*).

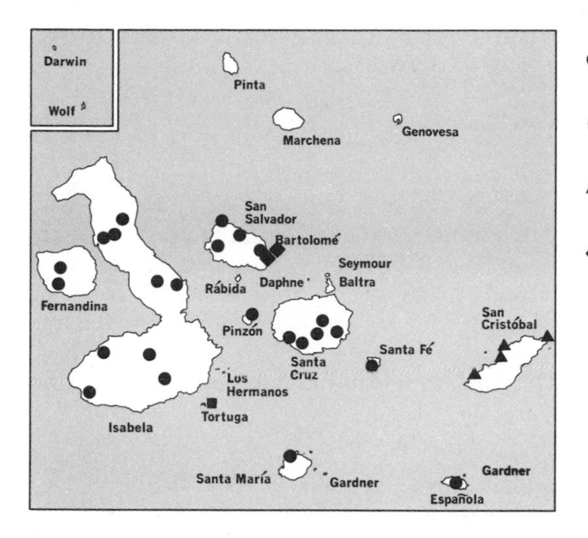

● *Alternanthera filifolia* subsp. *filifolia*
■ *Alternanthera filifolia* subsp. *glauca*
▲ *Alternanthera filifolia* subsp. *glaucescens*
◆ *Alternanthera filifolia* subsp. *microcephala*

between each of these "subspecies" and subsp. *filifolia* sensu stricto, they have been referred to synonymy under subsp. *filifolia*, which has given that subspecies rather wide limits of variation. Other specimens examined, from several islands, were more or less intermediate in character and could not be assigned readily to any particular subspecies.

Alternanthera filifolia subsp. **filifolia**

Bucholtzia filifolia Hook. f., Trans. Linn. Soc. Lond. 20: 192. 1847.
Telanthera filifolia Moq. in DC., Prodr. 13: 368. 1849.
Achyranthes hookeri Standl., Jour. Wash. Acad. Sci. 5: 74. 1915.
Alternanthera filifolia subsp. *subsquarrosa* Howell, Proc. Calif. Acad. Sci. IV, 21: 103. 1933.
Alternanthera filifolia subsp. *margaritacea* Howell, op. cit. 104.
Alternanthera filifolia subsp. *sylvatica* Howell, loc. cit.

Leaves linear to linear-oblong or linear-lanceolate, (1)2–5(7) cm long, 2–6 mm wide, green, with sparse, almost unbranched trichomes on both surfaces, but much denser beneath; heads 3–7-glomerate, rarely solitary, rounded to oblong, 7–10(15) mm long, 3–5 mm in diameter, flowers loosely appressed, glabrous or pubescent, the tips straight, not divergent.

Widespread in the Galápagos and including several recognizable populations. Specimens examined: Española, Fernandina, Isabela, Pinzón, San Salvador, Santa Cruz, Santa Fé, Santa María.

Alternanthera filifolia subsp. **glauca** Howell, Proc. Calif. Acad. Sci. IV, 21: 103. 1933

Stems profusely branched, glaucous, most of the branches diverging near base of plant; internodes 2.5–4(5) cm long; leaves linear-lanceolate, 2.5–3.5 cm long, 3–5(8) mm wide, thickish, glabrous or nearly so, glaucous, lowest leaves sometimes elliptic-lanceolate, to 4 cm long and 1 cm wide; heads generally 2–4 glomerate, generally rounded and 3–4 mm in diameter, sometimes oblong and to 8 mm long and 4 mm wide; flowers with straight, not appressed tips.

Known only from Isla Tortuga.

Alternanthera filifolia subsp. **glaucescens** (Hook. f.) Eliass., Madroño 20: 264. 1970

Bucholtzia glaucescens Hook. f., Trans. Linn. Soc. Lond. 20: 191. 1847.
Telanthera glaucescens Moq. in DC., Prodr. 13: 369. 1849.
Telanthera strictiuscula Anderss., Kongl. Svensk. Vet.-Akad. Handl. 1853: 166. 1855.
Telanthera angustata Anderss., Kongl. Svensk. Vet.-Akad. Handl. 1857: 61. 1857.
Achyranthes glaucescens Standl., Jour. Wash. Acad. Sci. 5: 74. 1915.
Achyranthes strictiuscula Standl., op. cit. 75.
Alternanthera glaucescens Howell, Proc. Calif. Acad. Sci. IV, 21: 104. 1933.
Alternanthera glaucescens f. *strictiuscula* Howell, op. cit. 105.

Stems generally profusely branched, glaucous; internodes generally 2.5–5 cm long; leaves linear-oblanceolate, 1.5–3(6) cm long, 3–5(6) mm wide, glabrous or nearly so, glaucous, somewhat succulent; heads solitary or 2- or 3-glomerate, ovoid or oblong, (4)6–8(10) mm long, 3.5–4.5 mm wide; flowers closely imbricate, the tips incurved at top of head.

Specimens examined from San Cristóbal, to which this subspecies seems to be restricted.

Alternanthera filifolia subsp. **microcephala** Eliass., Madroño 20: 264. 1970

Alternanthera glaucescens f. *strictiuscula* sensu Eliass., Svensk. Bot. Tidskr. 60: 413. 1966, non Howell 1933.

Stems slender, little branched, sometimes slightly glaucous; internodes 5–8 cm long; lower leaves linear to linear-oblanceolate, 3–4 cm long, 2–7 mm wide, glabrous; leaves higher on stem filiform or linear, 2–4 cm long, 0.5–1 mm wide, glabrous; heads few, in upper part of plant, terminal or on short axillary peduncles, rounded, 3–4 mm across; flowers loosely imbricate, with straight tips.

Specimens examined: Bartolomé, San Salvador.

Alternanthera filifolia subsp. **nudicaulis** (Hook. f.) Eliass., Madroño 20: 264. 1970

Bucholtzia nudicaulis Hook. f., Trans. Linn. Soc. Lond. 20: 191. 1847.
Telanthera nudicaulis Moq. in DC., Prodr. 13: 369. 1849.
Achyranthes nudicaulis Standl., Jour. Wash. Acad. Sci. 5: 74. 1915.
Alternanthera nudicaulis Christoph., Nyt Mag. Naturvid. 70: 73. 1931.

Leaves linear to linear-oblanceolate, 2–5 cm long, 2–3 mm wide, green, with somewhat branched trichomes beneath, very sparsely hairy above with simple or slightly

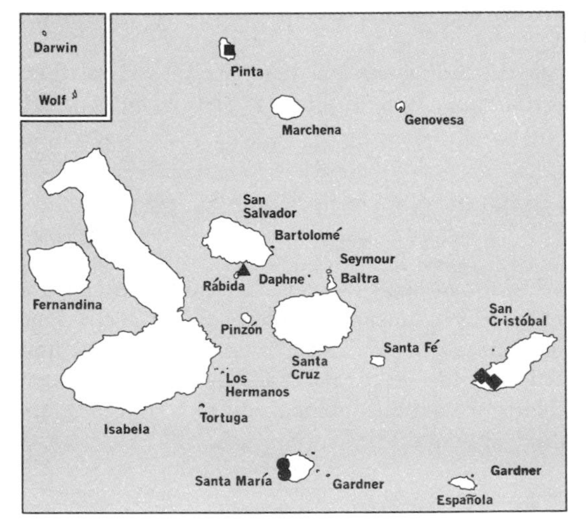

● *Alternanthera filifolia* subsp. *nudicaulis*
■ *Alternanthera filifolia* subsp. *pintensis*
▲ *Alternanthera filifolia* subsp. *rabidensis*
◆ *Alternanthera flavicoma*

branched trichomes; heads solitary or 2–7-glomerate, rounded or oblong, 5–10 mm long, 4–5 mm wide; flowers loosely imbricate, densely covered with yellow trichomes.

In its typical form, restricted to Isla Santa María.

Alternanthera filifolia subsp. **pintensis** Eliass., Madroño 20: 264. 1970

Stems rather profusely branched, with, at least in upper parts, white, somewhat branched trichomes; internodes 3–5(7) cm long; leaves oblanceolate or oblong-lanceolate, 2–3(6) cm long, 4–7(15) mm wide, with white, stellate hairs especially prominent beneath; heads solitary or in glomerules of 2–7 heads, the heads cylindrical to ovoid, 5–8 mm long, 3–4 mm wide; flowers rather closely imbricate but tips free, straight; flowers bearing dense, yellow, simple trichomes.

Known only from Isla Pinta.

Intermediate between *A. filifolia* subsp. *nudicaulis* and *A. snodgrassii.*

Alternanthera filifolia subsp. **rabidensis** Eliass., Madroño 20: 265. 1970

Stems and branches with white, simple trichomes in upper parts; internodes 2–5 cm long; leaves oblanceolate, 2–3 cm long, 4–7(10) mm wide, with white, simple trichomes prominent beneath; heads solitary or 2–4-glomerate, rounded or cylindrical, 4–10 mm long, 3–4 mm wide; flowers loosely imbricate, with slightly incurved tips; flowers with yellowish, simple trichomes.

Known only from Isla Rábida.

Alternanthera flavicoma (Anderss.) Howell, Proc. Calif. Acad. Sci. IV, 21: 107. 1933

Telanthera flavicoma Anderss., Kongl. Svensk. Vet.-Akad. Handl. 1853: 166. 1855.
Achyranthes flavicoma Standl., Jour. Wash. Acad. Sci. 5: 74. 1915.
Alternanthera flosculosa Howell, Proc. Calif. Acad. Sci. IV, 21: 107. 1933.

Erect shrub 1–2 m high; branches ascending or semi-erect, striate, at least in upper parts pubescent with branched hairs; internodes (3.5)4–6 cm long; leaves linear, (1.5)2–3(4.5) cm long, 1–2 mm wide, 1-nerved, bearing yellow-white stellate hairs on both surfaces; flower heads generally terminal and solitary, rarely 2- or 3-glomerate, sometimes axillary, sessile, rounded or ovoid, 3–5(7) mm long, 2.5–3 mm wide, densely pubescent with white, almost unbranched hairs; flowers rather closely imbricate, with tips subappressed; bract and bractlets ovate, acute, about 1.5 mm long; sepals lanceolate-triangular, acute, outer one about 2 mm long, inner ones about 1.7 mm long; anthers about 0.7 mm long; pseudostaminodia laciniate, distinctly shorter than stamens.

Endemic. According to the label on the type specimen (*Andersson s.n.*), it was found on Isla Santa María (Floreana), but it might have been mislabeled. All later collections of the species have been made near Wreck Bay on Isla San Cristóbal.

Alternanthera galapagensis (Stewart) Howell, Proc. Calif. Acad. Sci. IV, 21: 108. 1933

Telanthera galapagensis Stewart, Proc. Calif. Acad. Sci. IV, 1: 57. 1911.
Achyranthes galapagensis Standl., Jour. Wash. Acad. Sci. 5: 74. 1915.

Suffrutescent plant to 3 dm high, with glabrous and glaucous stems and leaves; stems and branches striate; internodes 1.5–2.5 cm long; nodes bearing white or

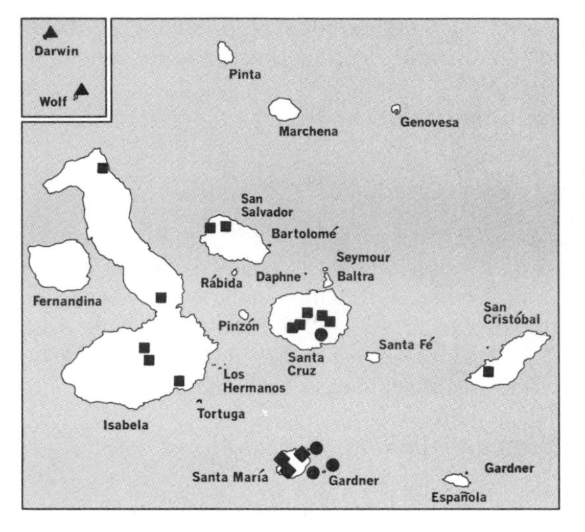

● *Alternanthera galapagensis*
■ *Alternanthera halimifolia*
▲ *Alternanthera helleri*
◆ *Alternanthera nesiotes*

yellowish, more or less simple trichomes; leaves spatulate, subobovate, or elliptic, attenuate at base, obtuse and often rounded at apex, 1.5–2.5(3) cm long, 0.7–1(1.5) cm wide; flower heads 3 or 4 together in terminal clusters or solitary in axils of upper leaves, 5–10 mm long, about 4 mm broad; flowers loosely imbricate; bracts about 1.5 mm long, the ventral especially very broad, suborbicular, with midnerve projecting into a tip; sepals ovate-triangular, hairy with simple trichomes between nerves and toward tip, outer sepal about 2.5 mm long, inner ones a bit less than 2 mm long; anthers about 1 mm long; pseudostaminodia laciniate, shorter than stamens.

Endemic, with a peculiar distribution. Specimens examined: Caldwell, Enderby, Gardner (near Isla Santa María), Santa Cruz.

Alternanthera halimifolia (Lam.) Standl. in Pittier, Man. Plant. Usual. Venez. 145. 1926

Achyranthes halimifolia Lam., Encycl. Méth. 1: 547. 1783.
Alternanthera halimifolia subsp. *macrophylla* Howell, Proc. Calif. Acad. Sci. IV, 21: 112. 1933.

Stems either spreading and rooting or ascending and forming bushes to 1 m high; stems branched, glabrous or more or less hirsute with stellate hairs; leaves elliptic to ovate, 2–10 cm long, 1–6 cm wide, acute and mucronulate at apex, cuneate or attenuate at base; stellate hairs more numerous beneath than above; flower heads axillary, rounded or subcylindrical, 4–8 mm long, to 6 mm wide; flowers loosely imbricate or subsquarrose, glabrous or pubescent with simple trichomes; bract and bractlets ovate, acuminate, about 3 mm long; sepals lanceolate-triangular, acute to acuminate, outer sepal 3.5–4 mm long, inner ones 2.5–3 mm long; anthers 1–1.5 mm long; pseudostaminodia laciniate, somewhat shorter to somewhat longer than stamens.

From Chile northward along the coast to Ecuador, but found also as an occasional weed far from this area; occurs over most of the archipelago, from the arid lowlands to the humid mountaintops, sometimes constituting an important part of the ground vegetation in the *Scalesia* forest. Specimens examined: Isabela, San Cristóbal, San Salvador, Santa Cruz.

The difference in leaf size is great and seems to be in direct response to the environment. The largest plants with large leaves are found in moist regions, whereas in the arid lowlands the plants become low and bear smaller leaves.

Alternanthera helleri (Robins.) Howell, Proc. Calif. Acad. Sci. IV, 21: 109. 1933

Telanthera helleri Robins., Proc. Amer. Acad. 38: 138. 1902.
Achyranthes helleri Standl., Jour. Wash. Acad. Sci. 5: 74. 1915.
Telanthera helleri var. *obtusior* Robins., op. cit. 139.
Alternanthera helleri f. *obtusior* Howell, Proc. Calif. Acad. Sci. IV, 21: 110. 1933.

Subshrub to 6 dm tall; stems and branches at least in upper parts more or less covered with white or yellowish, branched, often stellate hairs; petioles 3–8 mm long, more or less stellate-pubescent; leaf blades broadly ovate, oval, or suborbicular, (1)1.5–5 cm long, (1)1.5–3 cm wide, obtuse or truncate at base, generally rounded, rarely acute at apex, stellate-pubescent on both surfaces, more densely so beneath than above; flower heads 4–9 mm long, 3.5–4 mm broad; flowers closely imbricate; bracts ovate, acute at apex, about 2 mm long, slightly hairy to almost glabrous; sepals triangular, with simple or somewhat branched hairs between nerves; outer sepal about 3 mm long, inner ones about 2 mm long; anthers 0.8–1 mm long; pseudo-staminodia laciniate, shorter than stamens.

Endemic. Specimens examined: Darwin, Wolf.*

Chilean material (*Jaffuel 1620* [GH]) resembles this species but has larger bracts and bractlets with sharper and more elongate tips, and leaves that are more acute at the base. This Chilean material seems closer to *A. halimifolia*.

Alternanthera nesiotes Johnston, Contr. Gray Herb. 68: 83. 1923

Root solid, vertical; stems numerous, prostrate, with silvery, unbranched hairs; internodes short, 0.2–1(1.5) cm long; leaves suborbicular, 4–9 mm broad, grayish brown, pubescent with unbranched, white trichomes on both surfaces, denser below than on upper surface, young leaves and shoots densely pubescent throughout, thus having silvery appearance; flower heads solitary in axils of leaves, rounded, about 3 mm broad, 4–6-flowered; bracts ovate, acute, hyaline, with a brownish midnerve, somewhat hairy, about 2 mm long, the tips recurved; sepals ovate-triangular, hairy, outer one about 2 mm long, inner ones about 1.7 mm long; anthers about 0.6 mm long; pseudostaminodia laciniate, about as long as stamens.

Endemic. Known only from Isla Santa María; growing near the shore on heavily weathered lava.

Alternanthera rugulosa (Robins.) Howell, Proc. Calif. Acad. Sci. IV, 21: 111. 1933

Telanthera rugulosa Robins., Proc. Amer. Acad. 38: 139. 1902.
Achyranthes rugulosa Standl., Jour. Wash. Acad. Sci. 5: 74. 1915.

Low tree, 3–4 m tall; branches and leaves pubescent with branched, generally stellate hairs; petioles 3–5 mm long, stellate-pubescent; leaves oblong-ovate, rounded

* Isla Darwin rises precipitously out of the sea, with vertical or nearly vertical cliffs unbroken by gullies or canyons; only experienced climbers can scale its walls. Until January 1964 no botanists had had access to the summit. At that time, several scientists were set on top of the island by helicopter. Among them were John Hendrickson, E. Yale Dawson, and F. R. Fosberg, each of whom collected a few plants. The Hendrickson and Dawson collections were given to Ira Wiggins; the specimens taken by Fosberg were deposited in the United States National Herbarium.

at apex, 2.5–5 cm long, 1.5–2.5 cm wide, with subpersistent stellate hairs above and densely stellate-pubescent beneath; flower heads 6–11 mm long, 6–9 mm broad; flowers loosely imbricate, to arcuate-spreading; bracts ovate, acute at apex, about 2.5 mm long; sepals triangular, acute, glabrous, outer one 4.5–5 mm long, inner ones about 3 mm long; anthers about 1.5 mm long; pseudostaminodia laciniate, equal to or shorter than stamens.

In the Galápagos Islands known only from Isla San Cristóbal. Previously regarded as endemic, but Peruvian material (*Worth & Harrison 15681* [GH]) corresponds well with this species.

Alternanthera snodgrassii (Robins.) Howell, Proc. Calif. Acad. Sci. IV, 21: 109. 1933

Telanthera snodgrassii Robins., Proc. Amer. Acad. 38: 140. 1902.
Achyranthes snodgrassii Standl., Jour. Wash. Acad. Sci. 5: 75. 1915.

Shrub to 6 dm high; stems and branches striate, covered with white, simple (or nearly simple) trichomes; internodes (2.5)4–6(8) cm long; leaves elliptic-lanceolate to narrowly obovate, (2)2.5–4 cm long, (0.5)0.7–1.5 cm wide, mucronulate, sparsely hairy with simple or few-branched hairs on both surfaces, young leaves more densely hairy; flower heads terminal or axillary, solitary or in 2–4-headed glomerules, rounded or oblong, 4–8 mm long, about 4 mm wide; flowers rather loosely imbricate, their outer parts with simple white trichomes, tips straight; bract and bractlets ovate, acute, about 2 mm long; sepals lanceolate-triangular, acute, outer one 3.5–4 mm long, inner ones about 3 mm long; anthers about 1 mm long; pseudostaminodia laciniate, about equaling stamens.

Endemic; apparently restricted to the islands of Baltra and Seymour, and to the adjacent northern part of Isla Santa Cruz. Specimens examined: Santa Cruz, Seymour.

Closely related to *A. vestita*, which differs mainly in having stellate pubescence.

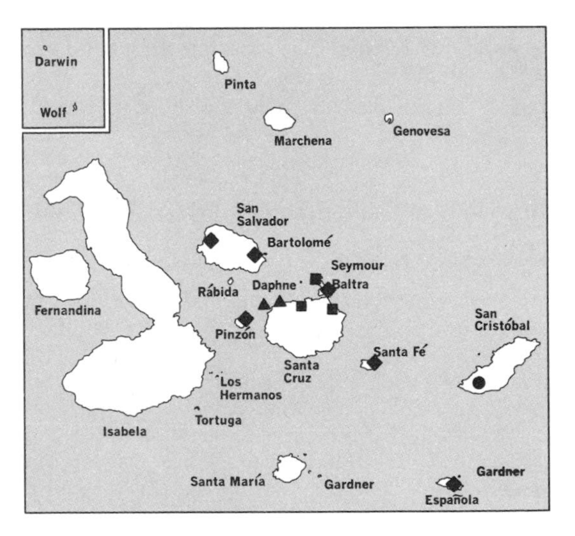

● *Alternanthera rugulosa*
■ *Alternanthera snodgrassii*
▲ *Alternanthera vestita*
◆ *Amaranthus anderssonii*

Alternanthera vestita (Anderss.) Howell, Proc. Calif. Acad. Sci. IV, 21: 108. 1933

Telanthera vestita Anderss., Kongl. Svensk. Vet.-Akad. Handl. 1853: 169. 1855.
Achyranthes vestita Standl., Jour. Wash. Acad. Sci. 5: 75. 1915.

Much-branched shrub to 1 m high; branches striate, generally densely covered with white or yellowish white stellate hairs; internodes 2.5–6(10) cm long; leaves elliptic, lanceolate or oblanceolate, mucronulate, (1.5)2–4(6) cm long, (0.4)0.5–1.5(2) cm wide, covered with stellate hairs on both surfaces, more densely so beneath than above, young leaves often very densely pubescent; flower heads terminal or axillary, solitary or 2- or 3-glomerate, rounded or oval, 5–7 mm long, 4–5 mm wide; flowers rather loosely imbricate, covered with white, unbranched hairs, the tips straight or slightly bent outward, not appressed; bract and bractlets ovate, acute, about 2.5 mm long; sepals lanceolate-triangular, acute, outer one (in mature flowers) 3.5–4 mm long, inner ones 3.2–3.5 mm long; anthers about 1 mm long; pseudostaminodia laciniate, about equaling stamens.

Previously thought to be endemic to the Galápagos, but Chilean material (*Jaffuel 1650* [GH]) matches the Galápagos specimens almost perfectly. In the Galápagos probably restricted to Isla Santa Cruz and the nearby islet Eden.

Amaranthus L., Sp. Pl. 989. 1753; Gen. Pl. ed. 5, 427. 1754

Mostly annual herbs, rarely perennials. Leaves alternate. Flowers unisexual or bisexual, monoecious, dioecious or polygamous, arranged in axillary or terminal clusters, the clusters often forming simple or branched spikes. Sepals (1)3–5, sometimes hardening at base in fruit. Stamens (1)2–5, anthers with 4 pollen sacs; filaments free nearly to base. Stigmas 2–4; ovary 1-celled, 1-ovuled. Fruit a utricle, indehiscent or transversely dehiscent. Seeds lenticular, shining, generally dark.

About 60 species, mainly in tropical and subtropical regions; some species are widespread weeds. In the Galápagos 9 species are known, 4 of them endemic.

Sepals 5:
  Plant armed with spines at nodes . . . . *A. spinosus*
  Plant without spines:
    Perianth-segments oblong to oblong-oblanceolate, not spatulate-expanded; base of segments generally not swollen in fruit:
      Sepals all alike, subequal, generally shorter than utricle . . . . *A. dubius*
      Sepals dissimilar, outer one bractlike, much longer than inner ones, at least the longest sepal exceeding utricle . . . . *A. quitensis*
    Perianth-segments spatulate, their bases indurated and swollen in fruit:
      Flowers in condensed, axillary cymules; seeds roundly ovate, about 0.6 × 0.8 mm; leaves elliptic to obovate . . . . *A. anderssonii*
      Flowers in clusters forming elongate spikes; seeds rounded or roundly ovate, 0.9–1 mm in diameter; leaves linear to ovate-lanceolate . . . . *A. squamulatus*
Sepals 1–3 (sometimes 4):
  Leaves ovate or rhombic-ovate; sepals 3:
    Utricle rugulose; leaves not emarginate at apex . . . . *A. gracilis*
    Utricle smooth; leaves emarginate at apex . . . . *A. lividus*
  Leaves linear to elongate-obcordate; sepals 1–4:
    Sepal 1; stamen 1; axis of flower clusters thickened and curved in fruit . . . . *A. furcatus*
    Sepals 2–4; stamens 2 or 3; axis of flower clusters not thickened in fruit . . . . *A. sclerantoides*

Amaranthus anderssonii Howell, Proc. Calif. Acad. Sci. IV, 21: 95. 1933

Scleropus urceolatus Anderss., Kongl. Svensk. Vet.-Akad. Handl. 1853: 162. 1855, non *Amaranthus urceolatus* Benth. 1844.
Amblogyne urceolata Anderss., op. cit. 1855: 59. 1857.
Amaranthus anderssonii f. *erectus* Howell, Proc. Calif. Acad. Sci. IV, 21: 96. 1933.

Stems spreading or erect, often profusely branched, subglabrous or slightly villous; leaves elliptic to obovate, mucronulate, (0.5)1–3 cm long, (0.3)0.5–1.5 cm wide; flowers monoecious in condensed, axillary cymules; sepals 5; stamens normally 2; stigmas 3; sepals of staminate flowers lanceolate to narrowly elliptic, about 1.5 mm long; sepals of pistillate flowers spatulate, 2–2.5 mm long; stigmas generally protruding from the pistillate flower; each flower generally subtended by a narrowly ovate bractlet; middle part of sepals darkening and striate in fruit, the bases then indurated, swollen, and coalescent; the seed lenticular, roundly ovate, about 0.6 mm wide, 0.8 mm long when viewed from the flattened side, reddish brown or chestnut brown.

Endemic; rather widespread at low elevations. Specimens examined: Baltra, Española, Pinzón, San Salvador, Santa Fé.

Amaranthus dubius Mart., Pl. Hort. Erlang. 197. 1814

Stem erect, generally branched; leaves ovate or rhombic-ovate, 4–7 cm long, 2.5–5 cm wide, mucronulate; inflorescences terminal and axillary spikes (clusters forming spikes), often branched; bracts short, generally about 1 mm long, pointed, usually distinctly shorter than sepals; flowers monoecious; sepals 5, subequal, about 1.5 mm long; stamens 5; stigmas 3; utricle usually exceeding sepals, circumscissile, upper half somewhat rugulose, the 3 stigmas persisting as short horns, lower half smooth or nearly so; seed lenticular, about 1 mm in diameter, dark reddish brown to black, shining.

Widespread in tropical regions, sometimes occurring as a weed in temperate regions. Specimens examined: Española, Isabela, San Cristóbal, San Salvador, Santa Cruz, Santa María.

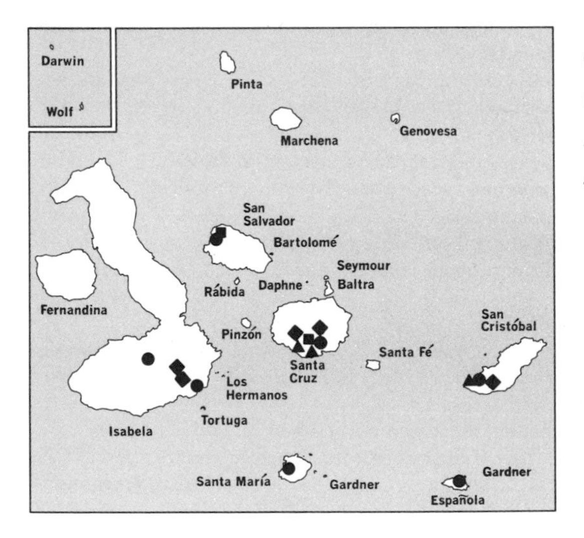

● *Amaranthus dubius*
■ *Amaranthus furcatus*
▲ *Amaranthus gracilis*
♦ *Amaranthus lividus*

Amaranthus furcatus Howell, Proc. Calif. Acad. Sci. IV, 21: 94. 1933

Annual, to 15 cm tall; stems prostrate; leaves shallowly cleft with lobes somewhat divergent, gradually attenuate at base, 10–35 mm long, to 3 mm wide at apex; flowers monoecious in short axillary clusters or spikes, the axes of which become corky-thickened and curved in fruit; bracts linear-lanceolate; sepal 1; staminate flowers with single stamen, the sepal about 1 mm long; pistillate flowers with 3 stigmas, the sepal 1.1–1.4 mm long; utricle rugulose, indehiscent; seed shining black, about 1 mm in diameter.

Endemic. Specimens examined: San Salvador, Santa Cruz.

Closely related to *A. scleranthoides* but differing from it in having flowers with a single sepal and stamen, in the divergent lobes at the ends of the leaves, and in the thickened, curved axis of the flower clusters.

Amaranthus gracilis Desf., Tabl. École Bot. 43. 1804

Stem erect, often branched, to 6 dm high, glabrous or slightly hairy on upper parts; leaves on petioles 1–3(8) cm long; leaf blades ovate or roundly triangular, truncate or obtuse at base, obtuse and mucronulate at apex, usually 4–6 cm long, 2.5–4 cm wide; flowers monoecious, mainly in elongate, terminal inflorescences (clusters forming spikes); sepals 3, whitish with green midnerve, those of staminate flowers lanceolate, about 1 mm long, those of pistillate flowers oblong-spatulate, subacute to acuminate, 1–1.3 mm long; stamens (2)3; stigmas 3; utricle rugulose, indehiscent; seed about 1 mm in diameter, black, shining.

Of nearly worldwide distribution; common in tropical regions and occurring as a weed in temperate regions. Specimens examined: San Cristóbal, Santa Cruz.

Amaranthus lividus L., Sp. Pl. 990. 1753

Amaranthus viridis L., Sp. Pl. ed. 2, 1405. 1763.
Amaranthus polygonoides Zoll., Syst. Verz. Java 73. 1845–46, non L. 1760.
Euxolus viridis Moq. in DC., Prodr. 13: 273. 1849.
Amaranthus lividus var. *polygonoides* Thell. in Aschers. & Graebn. 5: 1. 1919.

Stems procumbent to ascending, generally branched, glabrous, 1–5 dm high; leaves on petioles (0.5)1–3(5) cm long; leaf blades ovate or rhombic-ovate, 1–5 cm long, 0.7–4 cm wide, acute at base, generally emarginate at apex; flowers monoecious in axillary clusters and in more or less well-developed, elongate, terminal inflorescences (clusters forming spikes); sepals normally 3, whitish with green midnerve, those of staminate flowers oblong or oblong-elliptic, subacute, about 1 mm long, those of pistillate flowers oblong or oblong-spatulate, obtuse or somewhat acute, about 1.5 mm long; stamens 2 or 3; stigmas 3; utricle smooth, sometimes minutely wrinkled, green, indehiscent; seed 1–1.2 mm in diameter, dark reddish brown, shining.

Widely distributed in the tropics of the world and occurring as a weed in temperate regions. Specimens examined: Isabela, San Cristóbal, Santa Cruz.

Probably introduced into the Galápagos Islands in recent times.

Amaranthus quitensis HBK., Nov. Gen. & Sp. Pl. 2: 194. 1817

Stems erect, often branched; leaves ovate, mucronulate, (3)5–10(12) cm long, 2.5–6 cm wide; inflorescences mainly terminal, compound (of clusters in branched spikes), but also in small axillary clusters; clusters each subtended by an ovate, acuminate bract 2–3 mm long; flowers monoecious; sepals 5, the outer generally bractlike, 2.5–3 mm long, with midnerve excurrent, forming apiculation, inner sepal about 1.5 mm long, mucronate, tips of all sepals straight or spreading away from utricle; stamens 5; stigmas 3; utricle shorter than longest sepal, circumscissile, upper half strongly rugulose; persistent stigmas forming 3 prominent horns; seed about 1 mm in diameter, lenticular, dark reddish brown to black, shining.

A South American species introduced as a weed into Europe. Specimens examined: Isabela, Santa Cruz, Santa María.

The Galápagos specimens belong to a form with bracts somewhat shorter than those of the type. Young individuals often are difficult to distinguish from *A. dubius*. The seeds are used extensively as human food in the high Andes.

Amaranthus sclerantoides (Anderss.) Anderss., Kongl. Svensk. Vet.-Akad. Handl. 1859: 59. 1861

Euxolus sclerantoides Anderss., Kongl. Svensk. Vet.-Akad. Handl. 1853: 163. 1855.
Amaranthus sclerantoides f. *chathamensis* Robins. & Greenm., Amer. Jour. Sci. 50: 140. 1895.
Amaranthus sclerantoides f. *hoodensis* Robins. & Greenm., loc. cit.
Amaranthus sclerantoides f. *albemarlensis* Stewart, Proc. Calif. Acad. Sci. IV, 1: 55. 1911.
Amaranthus sclerantoides f. *abingdonensis* Stewart, loc. cit.
Amaranthus sclerantoides f. *rugulosus* Howell, Proc. Calif. Acad. Sci. IV, 21: 94. 1933.

Annual, to 15 cm high; stems prostrate or ascending; leaves greatly variable in shape, linear, linear-obcordate, or obcordate, 3–20(30) mm long, 1–5(10) mm wide, mucronate, gradually attenuate at base; flowers monoecious in small axillary clusters; bracts about half as long as sepals; staminate flowers with 3 or sometimes 2 ovate-elliptic sepals, 1.4–1.7 mm long; stamens generally 3, rarely 2; pistillate flowers with 3 or 4 lanceolate or ovate-lanceolate sepals slightly unequal in length, the longest 1.6–2 mm long; stigmas 3; utricle smooth or wrinkled, dehiscent or indehiscent; seed about 1 mm in diameter, black, shiny.

Endemic; widespread in the Galápagos Islands, mainly as a coastal plant, especially common on sand dunes. Specimens examined: Baltra, Coamaño, Daphne Major, Darwin, Española, Gardner (near Española), Genovesa, Isabela, Las Plazas, Pinta, San Cristóbal, San Salvador, Santa Cruz, Santa Fé, Santa María, Wolf.

See Fig. 35.

Amaranthus spinosus L., Sp. Pl. 991. 1753

Stem erect, generally branched, to 1 m high, normally glabrous; petioles 1–3 cm long; leaf blades rhombic-ovate or ovate-lanceolate, attenuate at base, obtuse and mucronulate at apex, 4–7 cm long, 1–3 cm wide; most of leaf axils bearing 2 stout spines about 1 cm long; flowers monoecious; staminate flowers in upper part of elongate, terminal inflorescences (clusters forming spikes); pistillate flowers in lower parts of spikes and in axillary clusters; sepals 5, whitish with green midnerve, those of staminate flowers ovate, acute, about 1.5 mm long, those of pistillate flowers ovate, oblong, or oblong-ovate, acute, about 1.5 mm long; stamens 5; stigmas 2 or 3; utricle often, but not always, circumscissile; seed about 0.7 in diameter, dark reddish brown to nearly black, shining.

Widespread in tropical and subtropical regions of the world; occurring generally as a weed in temperate regions. Specimens examined: Isabela, San Cristóbal, Santa Cruz.

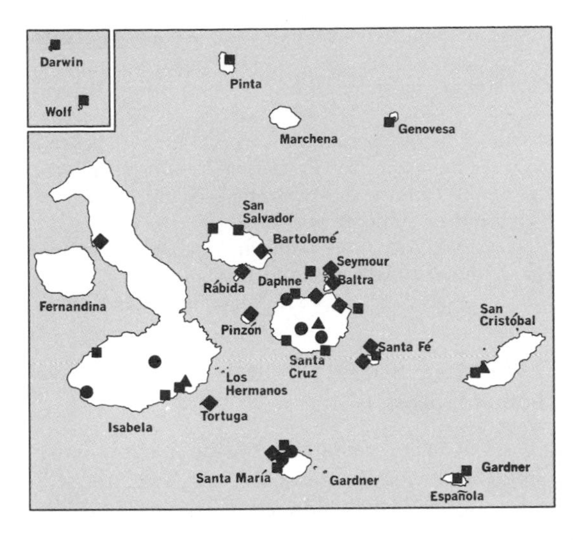

● *Amaranthus quitensis*
■ *Amaranthus sclerantoides*
▲ *Amaranthus spinosus*
◆ *Amaranthus squamulatus*

Amaranthus squamulatus (Anderss.) Robins., Proc. Amer. Acad. 43: 22. 1907

Scleropus squamulatus Anderss., Kongl. Svensk. Vet.-Akad. Handl. 1853: 162. 1855.
Amblogyne squarrulosa A. Gray, Proc. Amer. Acad. 5: 169. 1861.
Amaranthus squarrulosus Uline & Bray, Bot. Gaz. 19: 270. 1894.

Stem erect, 4–10 dm high, often branched, glabrous or slightly villous in upper parts; leaves linear, lanceolate or ovate-lanceolate, mucronulate, 3–5 cm long, 0.2–1.2 cm wide; flowers monoecious, glomerate, the clusters forming axillary and terminal spikes; sepals 5; stamens and stigmas 3; sepals of staminate flowers narrowly elliptic to oblong-oblanceolate, acute, about 2 mm long; sepals of pistillate flowers broadly spatulate, apiculate, about 2 mm long; each flower subtended by 1–3 bracts; bracts in outer part of cluster ovate, with sharp-pointed tips, generally exceeding sepals; bracts in inner part of cluster usually shorter than sepals, ovate, acute or nearly so; base of sepals swollen and indurate in fruit; seed lenticular, rounded or roundly ovate, 0.9–1.1 mm in diameter, dark purplish to black, shining.

Endemic; widespread at low elevations. Specimens examined: Baltra, Isabela, Pinzón, Rábida, San Salvador, Santa Cruz, Santa Fé, Santa María, Seymour, Tortuga.

Froelichia Moench, Meth. 50. 1794

Generally low, often broomlike, shrubs. Leaves opposite, the upper often reduced to scales. Flowers bisexual, slightly and obscurely bilabiate, arranged in terminal spikes or heads. Each flower subtended by abaxial bract and 2 lateral bractlets. Perianth-segments 5, united at least to middle to form a tube, lobes glabrous, tube often woolly, hardening and sometimes developing wings in fruit. Stamens 5, filaments forming 5-lobed tube; anthers with 2 pollen sacs, attached at incisions of tube. Stigma capitate or obscurely 2-lobed. Ovary 1-celled, with 1 ovule.

About 20 species in tropical America, 2 of them endemic in the Galápagos.

Inflorescences elongate spikes, the spikes acute at apex; shrub about 1 mm tall:
  Perianth-tube without wings or with thick, narrow wings about 0.5 mm wide in fruit; perianth-tube glabrous or nearly so. . . . . *F. juncea* subsp. *juncea*

Perianth-tube conspicuously winged with thin wings about 1 mm wide in fruit; perianth-tube often tomentulose . . . . *F. juncea* subsp. *alata*

Inflorescences rounded heads or cylindrical spikes, the spikes obtuse at apex; shrubs broomlike, 3–5 dm tall:

Perianth-tube in fruit almost or quite without wings; perianth 3–3.5 mm long . . . . *F. nudicaulis* subsp. *lanigera*

Perianth-tube in fruit developing wings about 1 mm wide; perianth 4–5 mm long:

Spikes oblong; upper part of stem glabrous or subglabrous . . . . *F. nudicaulis* subsp. *nudicaulis*

Spikes rounded, capitate; upper part of stem subglabrous or silvery-lanate . . . . *F. nudicaulis* subsp. *curta*

Froelichia juncea Robins. & Greenm., Amer. Jour. Sci. 50: 143. 1895
(sensu lato, the description covering both subspecies)

Shrub to 1 m high, woody and 5 cm thick at base; stems glabrous or sparsely hairy toward apices; internodes 10–20 cm long; lower leaves linear, 2–8 cm long, about 2 mm wide, 1-nerved, pilose on both sides, upper leaves reduced to triangular scales about 2 mm long; inflorescences of terminal, acute spikes, often branched, 1.5–5 cm long, 6–7 mm wide, bract 1–1.3 mm long; bractlets 2–2.5 mm long, the inner generally exceeding the outer; perianth glabrous or nearly so, 3.5–4.5 mm long, free lobes oblong or oblong-elliptic, obtuse at apex, about 1.5 mm long, 0.5 mm wide, perianth-tube becoming hardened, darkened, striated, often winged in fruit; stamen-tube 2.5–3 mm long, its lobes about half as broad as long, slightly emarginate at apex.

Endemic; weathered lava fields on Isabela and Santa Cruz.

See Fig. 36.

Froelichia juncea subsp. **juncea**

Stems and twigs glabrous or sparsely hairy in distal parts; perianth glabrous or nearly so, the tube wingless in fruit or with thick, narrow wings about 0.5 mm wide.

Froelichia juncea subsp. **alata** Howell, Proc. Calif. Acad. Sci. IV, 21: 116. 1933

Froelichia nudicaulis Hook. f. var. *longispicata* Christoph., Nyt Mag. Naturvid. 70: 74. 1932.
Froelichia nudicaulis subsp. *longispicata* Howell, Proc. Calif. Acad. Sci. IV, 21: 115. 1933.

Distal parts of stems and twigs generally with subappressed, silvery hairs; perianth often tomentulose; perianth-tube about 1 mm wide in fruit, with thin wings. n = 16.*

Known only from Santa Cruz. Floral characters intermediate between those of *F. juncea* subsp. *juncea* and *F. nudicaulis.*

Froelichia nudicaulis Hook. f., Trans. Linn. Soc. Lond. 20: 192. 1847
(sensu lato, the description covering all three subspecies)

Low, more or less broomlike shrubs 3–5 dm high; stems woody basally, subglabrous to pilose or silvery-lanate; internodes 5–10 cm long; inflorescences terminal, either rounded heads 6–14 mm broad, or oblong-cylindrical spikes 10–25 mm long, 7–10

* Count by D. Kyhos from buds from same plant as herbarium specimens *Wiggins 18428* (DS, GB) from Santa Cruz.

Fig. 36. *Froelichia juncea* (*a–c*). *Iresine edmonstonei* (*d–g*).

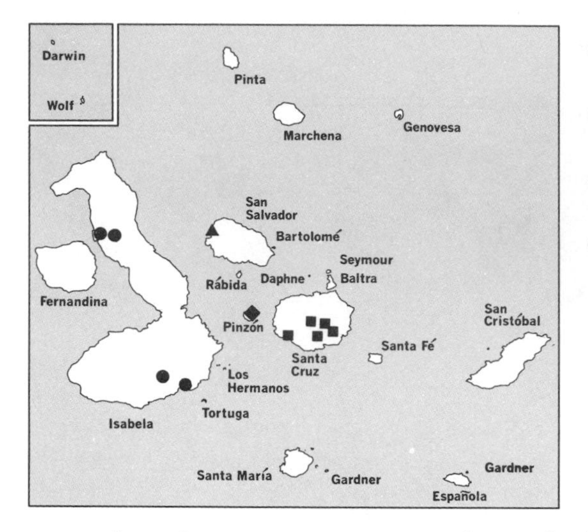

mm wide; spikes sometimes interrupted; tip of inflorescence rounded or abruptly obtuse; rachis long-woolly; each flower subtended by a scarious bract 1.5–2 mm long; bractlets 3–4 mm long; perianth 3–5 mm long, the free lobes oblong, obtuse apically, about 1.5 mm long, 0.4 mm wide; perianth-tube woolly to subglabrous, becoming indurated and darkened in fruit, sometimes with wings to 1 mm wide; stamen-tube 2.5–3.5 mm long, the lobes rounded or emarginate apically.

Endemic. Three subspecies distinguishable.

Froelichia nudicaulis subsp. **nudicaulis**

Stems slender, divergently branched, glabrous to subglabrous; spikes oblong, cylindrical, to 25 mm long and 7 mm wide in the single specimen seen; perianth about 4 mm long, the tube woolly, indurated and with wings about 1 mm wide in fruit.

Endemic. Specimen examined: San Salvador. Reported also from San Cristóbal and Santa María (Howell, 1933).

Froelichia nudicaulis subsp. **curta** Howell, Proc. Calif. Acad. Sci. IV, 21: 115. 1933

Stems low, erect, and broomlike, subglabrous or with silvery, appressed hairs; spike capitate, rounded, 10–14 mm across; perianth 4–5 mm long, the tube woolly, hardening in fruit and then developing wings about 0.7 mm wide.

Restricted to Isla Pinzón.

Froelichia nudicaulis subsp. **lanigera** (Anderss.) Eliass., Madroño 20: 265. 1970

Froelichia lanigera Anderss., Kongl. Svensk. Vet.-Akad. Handl. 1857: 63. 1859.
Froelichia lanata Anderss., op. cit. *pl. 3, fig. 1.*
Froelichia scoparia Robins., Proc. Amer. Acad. 38: 136. 1902.
Froelichia lanigera subsp. *scoparia* Howell, Proc. Calif. Acad. Sci. IV, 21: 116. 1933.

Stems low, erect and broomlike, silvery-pubescent with subappressed hairs; spike either capitate and (6)7–10 mm in diameter, or oblong-cylindric, to 20 mm long and 10 mm wide, often interrupted and somewhat branched; perianth 3–3.5 mm long, the tube woolly, in fruit becoming indurated but nearly or quite without wings.

Occurring on fresh lava fields, especially at high altitudes; sometimes forming patches several meters in diameter on otherwise sterile and barren lava fields. Specimens examined: Fernandina, Isabela.

Iresine P. Br., Nat. Hist. Jamaica 358. 1756

Herbs or shrubs. Leaves opposite, entire, sometimes fleshy. Inflorescences of small and generally numerous heads. Flowers unisexual or bisexual. Each flower subtended by 1 ventral bract and 2 lateral bractlets. Perianth-segments (sepals) 5. Stamens 5, united at base into low cup. Pseudostaminodia short or absent. Stigma lobes generally 2.

About 80 species in America and Australia. The one collection reported from the Galápagos is of doubtful origin.

Iresine edmonstonei Hook. f., Trans. Linn. Soc. Lond. 190. 1847

Erect herb about 1 m tall; branches glabrous, somewhat striate; leaves ovate-lanceolate, glabrous, 2–4 cm long, 1–1.5 cm wide, on petioles 5–10 mm long; flower heads dispersed in open panicle; heads rounded or oblong, 3–5 mm long, 2–4 mm thick, few (6–8)-flowered; flowers small, about 2 mm long; bract roundly cordate, about 0.7 mm long; bractlets suborbicular, about 1 mm long, apiculate at apex, not keeled; perianth-segments oblong-lanceolate, about 2 mm long, obtuse at apex, strongly woolly at base with simple trichomes 1.5–2 mm long; pseudostaminodia lacking or present only as a minute tooth between stamens; anthers about 0.6 mm long; stigma-lobes 2, about 0.2 mm long.

In the Galápagos Islands this species is known from a single collection (*Edmonstone* s.n. [K]) from Isla Santa María (Floreana), but there is a possibility the specimen was mislabeled. The Galápagos specimen agrees very well with material from the mainland (*Hitchcock 20118*, between Guayaquil and Salinas, June 1923 [GH]). An introduction of this plant to Isla Santa María by early settlers is quite possible.

See Fig. 36.

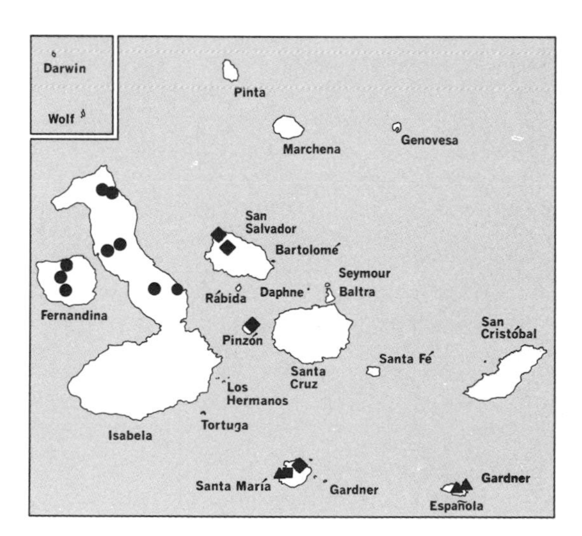

● *Froelichia nudicaulis* subsp. *lanigera*

■ *Iresine edmonstonei*

▲ *Lithophila radicata*

◆ *Lithophila subscaposa*

Lithophila Sw., Prodr. Veg. Ind. Occ. 14. 1788

Perennial plants with ascending to erect stems and basal rosette of leaves. Stems often branched, naked or with small leaves. Flowers bisexual, in terminal, rounded or cylindrical heads or spikes. Each flower subtended by ventral bract and 2 lateral bractlets. Perianth-segments 5. Fertile stamens 2; staminodia 3; pseudostaminodia lacking. Stigma 2-lobed. Ovary 1-celled, 1-ovuled.

About 15 neotropical species, 2 of them endemic to the Galápagos Islands.

Basal leaves subulate, pointed, 1.5–2 mm wide; coastal plant. . . . . *L. radicata*
Basal leaves oblong-lanceolate, obtuse, 5–10 mm wide; highland plant. . . . . *L. subscaposa*

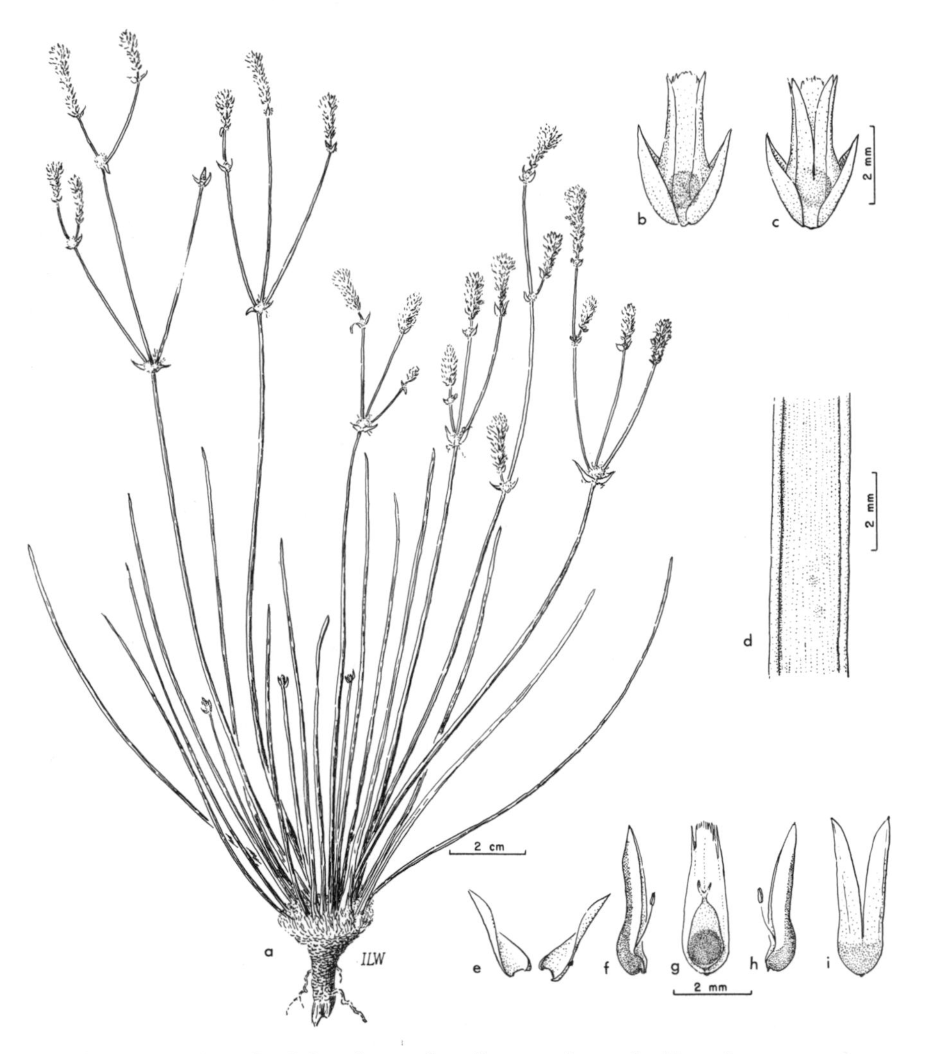

Fig. 37. *Lithophila radicata*: *b, c,* flower with attached bractlets; *d,* inner surface of leaf; *e,* bractlets; *f, h,* lateral tepals with accompanying stamens; *g,* dorsal tepal and pistil; *i,* ventral tepals, connate near base.

Lithophila radicata (Hook. f.) Standl., Jour. Wash. Acad. Sci. 5: 396. 1915

Alternanthera radicata Hook. f., Trans. Linn. Soc. Lond. 20: 262. 1847.
Alternanthera acaulis Anderss., Kongl. Svensk. Vet.-Akad. Handl. 1853: 164. 1855.
Iresine radicata Kuntze, Rev. Gen. Pl. 542. 1891.

Leaves crowded at apex of thick, vertical root, rushlike, pointed, 5–15 cm long, 1.5–2 mm wide, densely silky-hairy at base; stems several, (10)15–20(25) cm tall, often branched above; flower-heads (6)8–20 mm long, about 8 mm wide; bract and bractlets 2.5–3 mm long, whitish, hyaline, bractlets keeled; perianth-segments oblong, obtuse, about 3 mm long, 0.8 mm wide, whitish, hyaline, the 2 lateral segments keeled; filaments 1.5–2 mm long, anthers about 0.8 mm long; pistil about 3 mm long, ovary 1–1.5 mm long, gradually narrowing to a short style; stigma-lobes about 0.3 mm long; seed about 0.8 mm broad, round-reniform, reddish brown.

Endemic; a coastal plant growing in open, sunny places among lava and shell fragments along beaches. Specimens examined: Española, Santa María. Known also from San Cristóbal.

See Fig. 37.

Lithophila subscaposa (Hook. f.) Standl., Jour. Wash. Acad. Sci. 5: 396. 1915

Alternanthera subscaposa Hook. f., Trans. Linn. Soc. Lond. 20: 189. 1847.
Iresine subscaposa Kuntze, Rev. Gen. Pl. 542. 1891.

Basal rosette of leaves at apex of vertical root; leaves oblong-lanceolate, (2)5–8(12) cm long, 6–8 mm wide, obtuse, often rounded at apex, hairy at base; nervature characteristic, with short, profusely branched nerves diverging from midnerve at right angles; stems several, upper part generally somewhat branched, bearing a few oblong-lanceolate leaves 3–10 mm long, 2–3 mm wide; flower-heads 3–6(10) mm long, 5–6(7) mm wide; bract and bractlets 2–2.5 mm long, whitish, hyaline, bractlets keeled; perianth-segments oblong-oblanceolate, obtuse, about 3 mm long, 0.8 mm wide, whitish, hyaline, sometimes greenish in central part; filaments about 2 mm long, anthers about 1 mm long; ovary about 1.8 mm long, ovoid, style distinct, about 0.3 mm long, stigma-lobes minute, about 0.15 mm long; seed about 1 mm in diameter, round-reniform, reddish brown.

Endemic; in rocky habitats in the highlands. Specimens examined: Pinzón, San Salvador, Santa María.

Philoxerus R. Br., Prodr. 416. 1810

Low, shrubby plants with narrow, opposite leaves. Flowers bisexual, imbricate, in short to elongate spikes. Perianth-segments (sepals) 5. Each flower subtended by 1 bract and 2 lateral bractlets. Stamens 5; filaments connate basally into a cup. Pseudostaminodia none. Ovary with 2 or 3 stigmas.

Some 10 species, arid lowlands of tropical America, West Africa, and Australia.

Philoxerus rigidus (Robins. & Greenm.) Howell, Proc. Calif. Acad. Sci. IV, 21: 98. 1933

Alternanthera rigida Robins. & Greenm., Amer. Jour. Sci. 50: 143. 1895.
Lithophila rigida Standl., Jour. Wash. Acad. Sci. 5: 396. 1915.

Low shrub, profusely and compactly branched; internodes 5–12 mm long, glabrous; leaves subulate, pungent, 3–4 mm long, 1 mm wide or less, with broad bases completely surrounding stem; axils of leaves lanate, proliferous; inflorescences elongate,

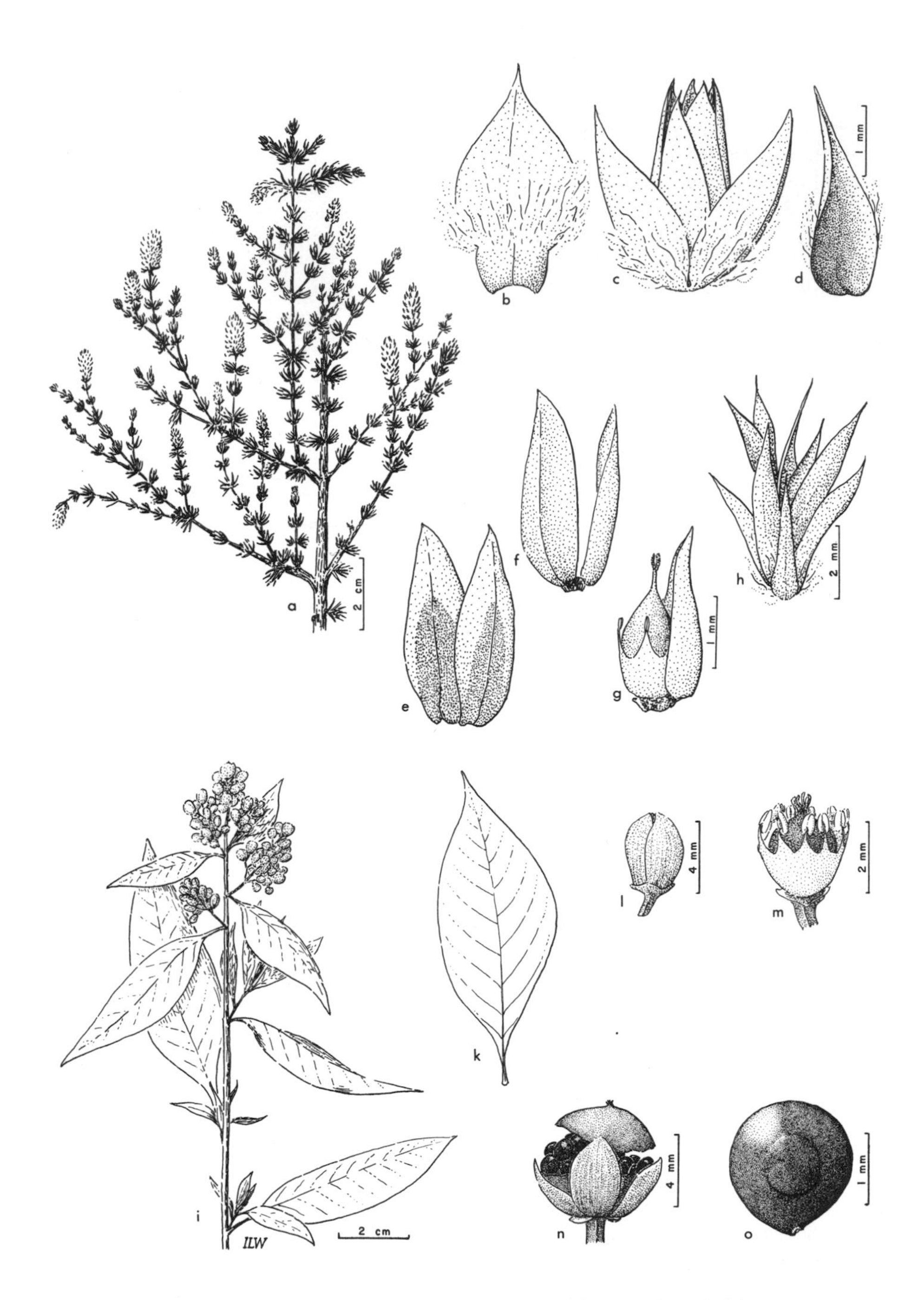

Fig. 38. *Philoxerus rigidus* (*a–h*): *b, d,* bracts; *c,* flower subtended by bracts; *e, f,* tepals; *g,* inner tepal attached to receptacle, stamens and pistil intact; *h,* fascicle of leaves. *Pleuropetalum darwinii* (*i–o*).

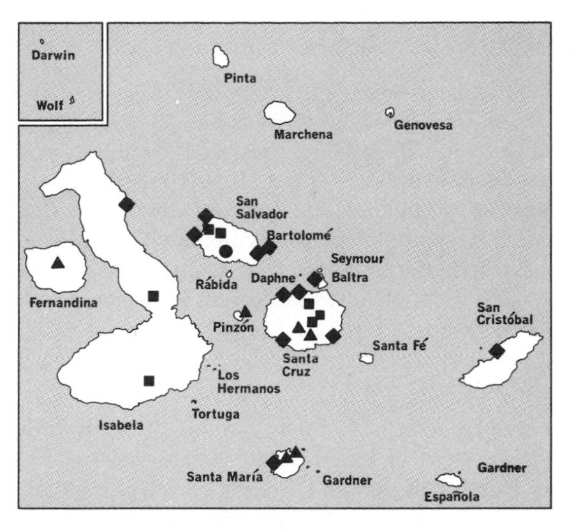

terminal spikes (6–17 mm long, 5–6.5 mm wide in single specimen seen); bract, bractlets, and perianth white; ventral bract ovate-cordate, about 2 mm long; bractlets ovate-lanceolate, keeled, 2.5–2.7 mm long, apex often bent toward axis of inflorescence; perianth-segments oblong-ovate, obtuse, 2.6–3 mm long, the 2 lateral and inner ones obtusely keeled; stamens about 2.3 mm long, anthers about 0.8 mm long; pistil about 1.7 mm long; stigmas 2, elongate-triangular, 0.3 mm long.

Endemic; known from only 2 collections on Isla San Salvador.

See Fig. 38.

Pleuropetalum Hook. f., Proc. Linn. Soc. Lond. 1: 278. 1845

Shrubs with alternate, entire leaves. Flowers bisexual, on short pedicels, racemose or paniculate. Each flower with 2 bractlets (sometimes interpreted as sepals). Perianth-segments 5, obtuse. Stamens 5–10, the filaments connate at base into low cup. Ovary with 2–6 stigma lobes. Fruit a berry-like capsule, few- to many-seeded.

Five species in Central and South America.

Pleuropetalum darwinii Hook. f., Lond. Jour. Bot. 5: 109. 1846

Shrub 1–2 m high; stems glabrous except in upper parts near inflorescence where invested by brown, papilla-like trichomes; leaves elliptic to ovate, acute at base, acuminate or cuspidate at apex, 5–10 cm long, 2–5 cm wide, petioles 0.5–2 cm long; inflorescences terminal panicles; bractlets broadly ovate or cordate, about 1 mm long, 1.5 mm wide, with prominent midnerve; perianth-segments oval, 4–5 mm long, 2.5–3 mm wide, orange-colored on inner surface, with marked, parallel nerves on outer, convex surface; stamens about 8; stigma-lobes 4–5(6); fruit a many-seeded, berry-like capsule, black or nearly so, circumscissile; seeds reniform-lenticular, 1–1.5 mm in diameter, black, shining. n = 17.*

Endemic. Specimens examined: Isabela, San Salvador, Santa Cruz.

See Fig. 38.

* On plant grown from seeds taken on Isabela (Eliasson, Bot. Notiser 123: 150. 1970).

BASELLACEAE. BASELLA FAMILY*

Twining herbs, generally perennials, often somewhat fleshy, glabrous. Leaves alternate, entire, exstipulate. Flowers usually bisexual, actinomorphic, in axillary or terminal spikes or racemes. Each flower with 2 or 4 bractlets, if 4 arranged in 2 decussate pairs. Lower bractlets often connate at base, upper often tepaloid. Tepals 5, free or united near base. Stamens 5, epipetalous, extrorse, adnate to base of tepals. Ovary superior, 1-loculed, with 1 campylotropous ovule. Style simple or cleft into 3 segments. Seed globose. Embryo spirally twisted or semicircular.

About 4 genera and 25 species in warm parts of America, Africa, and Asia.

Anredera Jussieu, Gen. 84. 1789

Vines with shoots from rhizome. Leaves lanceolate, ovate or cordate, often slightly fleshy. Flowers usually bisexual. Pedicels short or lacking, with persistent or caducous bracts. Apex of pedicels with 2 pairs of decussate bractlets, lower pair small, connate at base, upper pair tepaloid. Tepals connate at base. Filaments broadened toward base. Ovary free, more or less concealed by connate bases of tepals. Style 1. Stigma entire or split into 3 branches. Fruit globose, enclosed by perianth.

A genus of 5–10 species, ranging from southern United States and the West Indies southward to Argentina. One species in the Galápagos Islands.

Anredera ramosa (Moq.) Eliass., Madroño 20: 266. 1970

Tandonia ramosa Moq. in DC., Prodr. 13: 227. 1849.
Boussingaultia ramosa Hemsley, Biol. Centr.-Amer. 3: 27. 1882.
Boussingaultia baselloides in Galápagos literature, non HBK. 1825.

Stems twining, glabrous; leaves lanceolate to ovate or subcordate, acute, acuminate or cuspidate at apex, 2–5 cm long, 0.5–2 cm wide; flowers in spikes or spikelike racemes; pedicels 0.5–1 mm long; each flower subtended by a subulate bract 1.5–2 mm long; lower bractlets ovate-triangular, about 1 mm long; upper bractlets suborbicular, about 1.5 mm long, swollen at base; tepals broadly oval, about 1.5 mm long, white, distinctly exceeding upper bractlets; filaments slender, about 1.5 mm long, anthers about 0.4 mm long; style about 1.2 mm long; stigma entire, not lobed, papillate.

Known from Mexico, Central America, and the Gálapagos Islands. Specimens examined: Fernandina, Pinzón, Santa Cruz, Santa María.

An incompletely known species, the delimitation of which needs further research. See Fig. 39.

BATIDACEAE. BATIS FAMILY*

Glabrous, succulent, low, dioecious or monoecious shrubs. Leaves opposite, entire. Flowers bracteate, solitary or in axillary strobili. Staminate flowers with cup-shaped perianth and 4 stamens alternating with 4 staminodia. Anthers 2-celled, introrse. Pistillate flowers without perianths, each consisting of a single pistil. Ovary 4-loculed, with 1 ovule in each locule. Stigma 2-lobed, sessile. Seeds without endosperm.

* Basellaceae and Batidaceae contributed by Uno Eliasson.

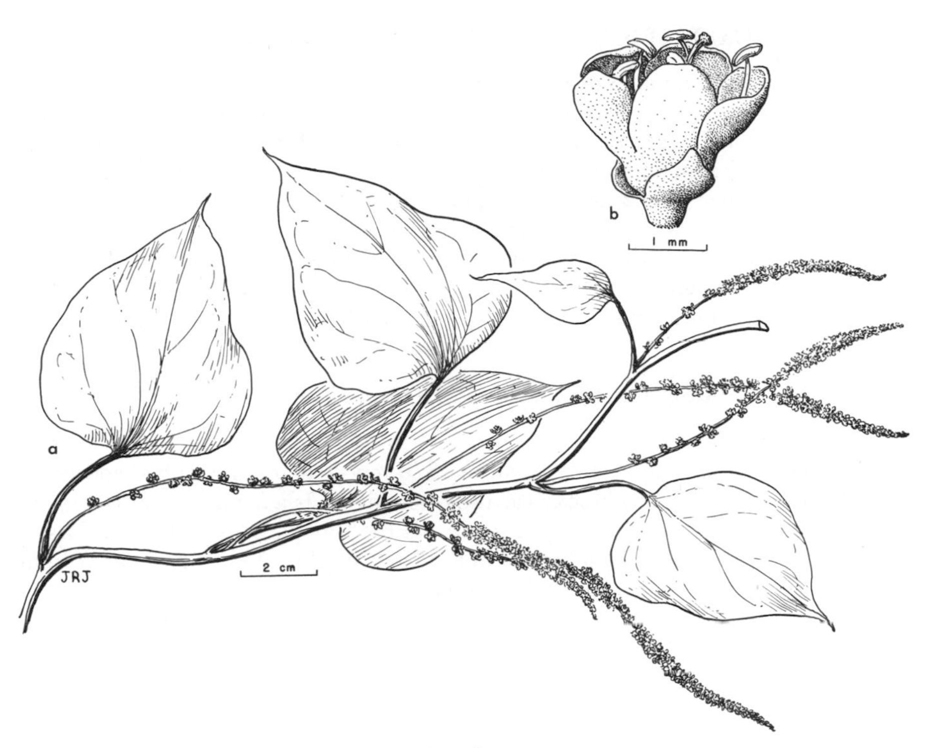

Fig. 39. *Anredera ramosa.*

A bitypic tropical family of uncertain phylogenetic relationship. One species along the coasts of tropical America, the Hawaiian and the Galápagos Islands; the second hitherto found only in south New Guinea.

Batis L., Syst. Nat. ed. 10, 2: 1380. 1759

Characters of the family.

Batis maritima L., Syst. Nat. ed. 10, 2: 1380. 1759

Dioecious shrub; stems sprawling to ascending, much branched, branches mostly opposite; leaves narrowly oblanceolate, fleshy, 1–2 cm long, 0.2–0.6 cm broad; staminate strobili ovoid to oval, 3–8 mm long; staminate flowers with shallowly cup-shaped perianth; stamens 4, alternating with rhomboid-spatulate staminodia, each flower in axil of a rounded bract; pistillate strobili more or less cylindrical, 5–20 mm long; pistillate flowers each consisting of pistil in axil of roundish, deciduous bract, without perianth; ovaries ovoid-oblong, coalescent to form fleshy fruit.

Ranging from California to Florida and south to Brazil and Peru, occurring also in the Hawaiian Islands; on sandy shores, often around salt lagoons, on several of the Galápagos Islands. Specimens examined: Baltra, Bartolomé, Isabela, San Cristóbal, San Salvador, Santa Cruz, Santa María.

See Fig. 40 and color plate 26.

CALLITRICHACEAE. WATER-STARWORT FAMILY*

Monoecious annual, aquatic or terrestrial herbs with slender stems and entire, exstipulate, mostly opposite leaves. Flowers unisexual, regular, solitary in leaf axils, the staminate and carpellate flowers usually in separate axils; perianth none, 2 hornlike bracteoles subtending each flower; staminate flower consisting of 1 stamen; carpellate flower a 2-lobed, 2-carpeled, 2-loculed pistil, the ovary superior. Ovule 1 in each locule. Styles 2, filiform, papillose. Fruit a schizocarp, separating at maturity into 2 or 4, usually dorsally winged or keeled, mericarps. Endosperm fleshy. Embryo straight.

A monogeneric family with 25 to 30 species nearly worldwide in distribution.

Callitriche L., Sp. Pl. 969. 1753; Gen. Pl. ed. 5, 5. 1754

Characters of the family.

According to Robinson (1902), Wolf in 1879 reported an unidentified species of *Callitriche* as occurring in a brook near the hacienda on Isla Santa María. However, no collections of the genus have been seen from the Galápagos Islands, and Fassett (1951), in his monograph of *Callitriche* in the New World, cited no specimens from the archipelago. Consequently the determination by Wolf is suspect.

CHENOPODIACEAE. GOOSEFOOT FAMILY†

Herbs or shrubs, rarely small trees. Leaves generally alternate, rarely opposite, entire or irregularly lobed, exstipulate. Flowers unisexual or bisexual, normally small and greenish, solitary or glomerate, often arranged in cymes or spikes. Perianth simple, of 2–5 segments, these united below, persistent after flowering. Stamens as many as perianth segments or fewer, opposite segments. Filaments slender. Anthers inflexed in bud, generally 4-celled. Ovary superior, 1-celled. Stigmas generally 2, rarely 3–5. Ovule solitary. Fruit a utricle or nutlet, often enclosed in perianth. Embryo generally surrounding endosperm, either simply bent or spirally twisted.

About 100 genera and 1,400 species, in nearly all parts of the world. The majority are halophytic plants.

Stems and branches jointed, succulent; leaves reduced to minute scales........*Salicornia*
Stems and branches not jointed; leaves of normal appearance:
  Flowers unisexual; pistillate flowers with 2 bractlets that enlarge and enclose the fruit .... *Atriplex*
  Flowers bisexual; without bracts ....*Chenopodium*

Atriplex L., Sp. Pl. 1052. 1753; Gen. Pl. ed. 5, 472. 1754

Annual or perennial herbs or bushes, often with a mealy pubescence of inflated hairs. Leaves alternate, rarely opposite, entire or irregularly lobed. Flowers monoecious or dioecious, sometimes also a few bisexual flowers present, clustered; clusters axillary or in solitary or panicled spikes. Staminate flowers with 3–5-parted perianth

* Contributed by Duncan M. Porter.
† Contributed by Uno Eliasson.

Fig. 40. *Batis maritima* (*a*). *Atriplex peruviana* (*b, c*).

and 3–5 stamens, bractlets none. Pistillate flowers generally without perianth but with 2 bractlets that enlarge with age and enclose fruit. Stigmas 2, rarely 3. Pericarp very thin. Seed erect or inverted, rarely horizontal. Embryo annular.

About 200 species in temperate and tropical regions.

Atriplex peruviana Moq. in DC., Prodr. 13: 102. 1849

Shrub to 1 m high, richly branched; stems woody at base, more or less scurfy in upper parts; leaves slenderly petiolate, blades ovate, ovate-rhombic, or rounded, 1.5–4 cm long, 1–2 cm wide, attenuate at base or this cuneate to obtuse, acute, or mucronulate at apex, scurfy on both sides; flowers monoecious or dioecious; staminate flowers densely clustered in terminal or axillary panicles or branched spikes, perianth 5-parted, stamens 5; pistillate flowers axillary or in clusters in lower part

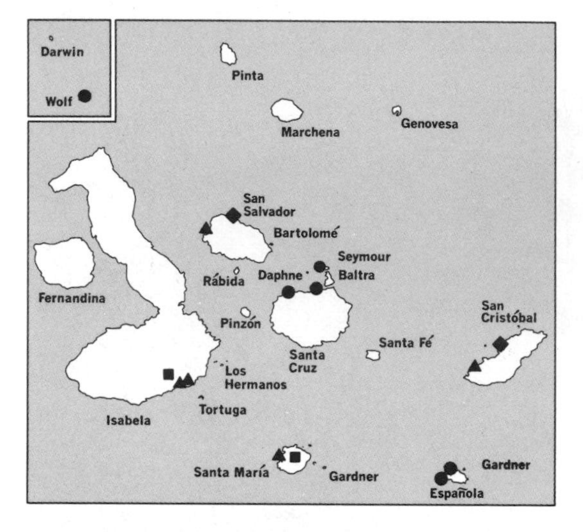

of staminate inflorescence, fruiting bracts united to middle, rhombic, 4–6(8) mm long, 3–4(6) mm wide, entire or with pair of teeth near middle.

Peru and Chile and on the shores of several islands in the Galápagos. Specimens examined: Española, Santa Cruz, Seymour, Wolf.

The affinity of the Galápagos plant to that of Peru and Chile is in need of further study.

See Fig. 40.

Chenopodium L., Sp. Pl. 218. 1753; Gen. Pl. ed. 5, 103. 1754

Annual or perennial herbs or bushes, rarely small trees, often with mealy pubescence of minute, inflated hairs, sometimes glandular. Leaves alternate, entire or irregularly lobed. Flowers perfect, rarely unisexual, generally in small clusters in simple or compound spikes. Perianth segments generally 5. Stamens usually 5, sometimes 4, rarely fewer. Stigmas generally 2. Utricle enclosed in perianth. Seeds lenticular, horizontal or vertical. Embryo annular.

A cosmopolitan genus with about 150 species, many of them widely distributed weeds. Two species have been found in the Galápagos Islands, both of which are probably of recent introduction.

Leaves lanceolate or ovate-lanceolate, gland-dotted beneath; seed about 0.8 mm in diameter . . . . *C. ambrosioides*
Leaves ovate-rhombic, not gland-dotted; seed about 1.5 mm in diameter . . . . *C. murale*

Chenopodium ambrosioides L., Sp. Pl. 219. 1753

Annual or perennial, to 1 m tall; stem simple or branched, glabrous or puberulent; leaves short-petioled, (3)5–8(12) cm long, (0.5)1–2(3) cm wide, gland-dotted beneath, lower leaf blades ovate-lanceolate, sinuate-toothed, upper lanceolate, entire or shallowly toothed; flower clusters arranged in spikes; perianth glandular; flowers polygamous; bisexual flowers with 5 stamens and 2–3 short stigmas, female

Fig. 41. *Chenopodium murale* (*a–c*). *Salicornia fruticosa* (*d, e*).

flowers with 3 long stigmas; seed horizontal, rarely vertical, 0.7–0.8 mm in diameter.

In temperate and tropical regions all over the world. Specimens examined: Isabela, Santa María.

Chenopodium murale L., Sp. Pl. 219. 1753

Annual, to 6 dm tall, often profusely branched; stems and branches glabrous or sparsely mealy; petioles equaling or shorter than blades, the latter ovate-rhombic to triangular with irregularly dentate margins, 2–8 cm long, 1–7 cm wide, glabrous above, often mealy beneath; flowers in clusters, arranged in dense paniculate inflorescences; perianth mealy; flowers bisexual, with 5 stamens and 2 stigmas; seed horizontal, 1.3–1.5 mm in diameter.

In temperate and tropical regions of the world. Specimens examined: Isabela, San Cristóbal, San Salvador, Santa María.

See Fig. 41.

Salicornia L., Sp. Pl. 3. 1753; Gen. Pl. ed. 5, 4. 1754

Annuals or perennials, often suffrutescent. Stems and branches jointed, glabrous, fleshy, the joints dilated at apex into a short sheath. Leaves rudimentary, reduced to minute scales. Flowers in groups of 3–7, sunken in joints of stems; flowering areas decussately opposite. Fertile joints forming terminal, cylindrical spikes. Flowers perfect or polygamous. Perianth obpyramidal, with small orifice at top through which stamens and stigmas protrude. Stamens 1 or 2. Stigmas 2. Fruit a utricle, enclosed by the spongy perianth at maturity. Seed compressed.

About 35 species on seashores and salt-pans in temperate and tropical regions.

Salicornia fruticosa L., Sp. Pl. ed. 2, 5. 1762

Salicornia peruviana HBK., Nov. Gen. & Sp. Pl. 2: 193. 1817.

Perennial, suffrutescent, to 3 dm tall, profusely branched with ascending or erect branches; sterile joints 7–12 mm long; sheaths with very obtuse-angled lobes; flowering joints 2–4 mm long; flowers perfect or polygamous; perianth obpyramidal, outer part rhomboid or rounded, about 2 mm across; stamens 1–2; anthers 0.8–1 mm long; stigmas about 0.5 mm long; seed 1.3–1.5 mm long, 0.9–1 mm wide, pale brown, hispidulous.

Widely distributed on seashores throughout the world. Specimens examined: San Cristóbal, San Salvador, a new record for the Galápagos Islands.

See Fig. 41.

LAURACEAE. LAUREL FAMILY*

Shrubs or trees (*Cassythea* a parasitic, twining perennial herb), mostly with evergreen leaves and aromatic foliage and bark. Leaves alternate, exstipulate, simple, usually coriaceous and punctate with droplets of aromatic oils, entire, prominently

* Contributed by Ira L. Wiggins.

nerved. Flowers axillary or subterminal in panicles, spikes, racemes, or umbels, usually perfect but sometimes unisexual, regular, mostly 3-merous, small, white, yellowish, or greenish. Stamens usually of 4 whorls of 3 stamens each, adnate to perianth tube, the innermost often reduced to staminodia, 1 (usually the 3d) whorl bearing a pair of sessile, glandular protuberances basally; anthers basifixed, 2- or 4-celled, dehiscing by subterminal valves opening upward. Pistil 1; ovary superior or nearly so, 1-celled; ovule solitary, pendulous from a parietal placenta. Style and stigma 1, stigma capitate to 2- or 3-lobed. Fruit a berry or drupe. Seed with large, straight embryo. Endosperm none.

A family of 45 genera and over 1,100 species in tropical and subtropical regions of Asia and America, with a few in Africa and the Mediterranean region; a few genera ranging into temperate zones.

Persea Gaertn. f., Fruct. & Sem. 3: 222. 1805

Trees or shrubs, with large leathery leaves and flowers in axillary or subterminal pedunculate panicles. Perianth 6-lobed, in 2 series, outer whorl usually smaller than inner. Functional stamens 9, 1st and 2d series eglandular, 3d whorl with gland on each side at base of broad, short filament. Anthers extrorsely 4-celled. Fruit small to quite large, ovoid to ellipsoid.

About 60 species, native chiefly in the New World and occurring as far north as Virginia along the Atlantic coast; introduced and cultivated in many tropical and subtropical areas.

Persea americana Mill., Gard. Dict. ed. 8. 1768

Laurus persea L., Sp. Pl. 370. 1753.
Persea gratissima Gaertn. f., Fruct. & Sem. 3: 222, *pl. 221*. 1807.
Persea edulis Raf., Sylva Tellur. 134. 1838.
Persea gratissima var. *vulgaris* Meisn. in DC., Prodr. 15^1: 53. 1864.
Persea gratissima var. *macrophylla* Meisn., loc. cit.
Persea gratissima var. *oblonga* Meisn., loc. cit.
Persea persea Cockerell, Bull. Torrey Bot. Club 19: 95. 1892.

Sturdy tree with rounded crown, often 6–12 m high and sometimes to 30 m tall; petioles 1.5–5 cm long, pubescent or at length glabrate, leaf blades coriaceous, dark green, ovate-oblong, 10–20(30) cm long, 3–10 cm wide, crowded at tips of branches and twigs, shining above, dull and glaucous beneath, glabrous to pubescent throughout, veins prominent and yellow-pubescent beneath; flowers many, in axillary and subterminal panicles, peduncles and pedicels short-pubescent with yellowish to brownish hairs; bracts tomentose, fugacious; pedicels 2–5 mm long, densely brown-pubescent; perianth 5–15 mm broad, lobes yellowish or pale yellow, silky-tomentose on both surfaces, 3–6 mm long, 2–3 mm wide, outer lobes smaller than the ovate-oblong to lanceolate inner ones; stamens of 2 outer whorls with ovoid-oblong, introrse anthers, their filaments flat, pilose, 2.5–3.5 mm long; stamens of inner whorl longer, with extrorse anthers, their filaments bearing a pair of short-stiped, ovoid glands at base of each; staminodia short-stiped; style filiform; stigma discoid; ovary ovoid, 4–5.5 mm long at anthesis; fruit usually ovoid, yellowish green tinged with purple when ripe, shining and glabrous, 7–10 cm thick, 7–20 cm long, rind coriaceous, flesh thick, oily and pulpy, 1.5–2 cm thick, edible.

Native from Mexico to Central America and possibly in the West Indies; now

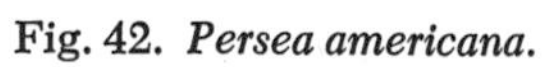

Fig. 42. *Persea americana.*

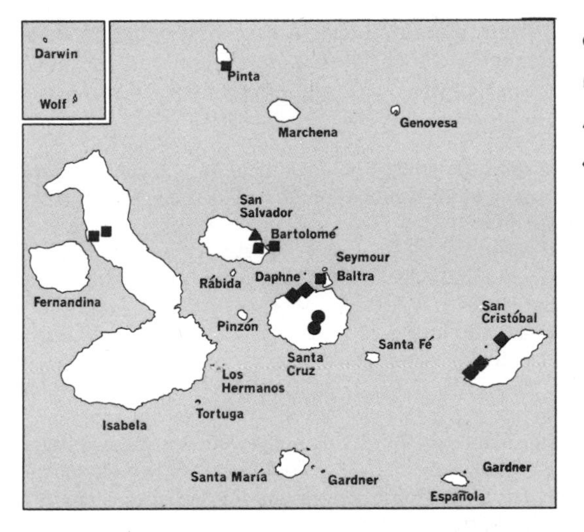

● *Persea americana*

■ *Mollugo cerviana*

▲ *Mollugo crockeri*

◆ *Mollugo flavescens* subsp. *flavescens*

cultivated widely in these regions and in nearly all tropical and subtropical countries. Introduced into the Galápagos Islands and growing wild in and above the forest zones on several of the larger islands. Specimens examined: Santa Cruz, a new record for the Galápagos Islands.

One specimen examined was growing above the *Miconia* Zone at an altitude of nearly 600 m. Trees around dwellings at Bella Vista planted as seedlings in the middle 1930's are now over 20 m tall and produce large crops of fruits 15–20 cm long each year. Standley (1922, 290–91) gives an interesting synopsis of the characteristics of the tree and the uses to which the fruits, seeds, and wood have been put by natives and inhabitants of Mexico and other tropical American areas.

See Fig. 42.

MOLLUGINACEAE. MOLLUGO FAMILY*

Annual or perennial herbs with opposite, alternate, or verticillate leaves. Flowers actinomorphic, bisexual, solitary or cymose. Sepals free or united at base. Petals small or lacking. Stamens hypogynous or slightly perigynous. Anthers 2-celled. Ovary usually several-loculed, several ovules in each locule. Fruit a membranous, loculicidally dehiscent capsule, usually surrounded by the persistent calyx. Embryo curved.

A single genus of about 20 species in the tropics and subtropics.

Mollugo L., Sp. Pl. 89. 1753; Gen. Pl. ed. 5, 39. 1754

Annual or perennial herbs. Leaves linear to narrowly obovate, often subverticillate. Sepals 5. Petals none. Stamens 3–9. Stigmas 3. Ovary 3–5-loculed. Seeds many, angularly semicircular to turgidly reniform, the surface usually tuberculate or reticulate.

* Contributed by Uno Eliasson.

About 20 species in tropical and subtropical regions. In this treatment 5 species have been recognized from the Galápagos Islands, 4 of them endemic to the area. The genus is widely distributed in the archipelago, though some taxa are restricted to 1 or to only a few of the islands.

Seeds angularly semicircular, 0.2 × 0.3 mm, the surface finely reticulate. . . . . *M. cerviana*
Seeds turgidly reniform or ovate-reniform, 0.4–0.6 mm in diameter, the surface tuberculate:
  Stems, leaves, pedicels, and sepals glandular:
    Cauline leaves linear, 0.6–0.8 mm wide. . . . . *M. crockeri*
    Cauline leaves oblanceolate, 1.3–2(3) mm wide. . . . *M. floriana* subsp. *gypsophiloides*
  Stems, leaves, pedicels, and sepals glabrous:
    Sepals 3–3.5 mm long; pedicels 15–30 mm long. *M. snodgrassii*
    Sepals 1.5–2.7 mm long; pedicels 1.5–8 mm long:
      Sepals 2.5–2.7 mm long. . . . . *M. floriana* subsp. *floriana*
      Sepals 1.5–2 mm long:
        Leaves linear, 20–30 mm long; plant perennial, distinctly woody at base. . . . . *M. floriana* subsp. *santacruziana*
        Leaves linear to oblanceolate, not over 20 mm long; plant an herbaceous annual:
          Seeds not deeply ridged on back; stamens 3–4:
            Seeds slightly over 0.5 mm in diameter; funiculus prominent; leaves oblanceolate, spatulate, or obovate. . . . . . *M. flavescens* subsp. *flavescens*
            Seeds slightly less than 0.5 mm in diameter; funiculus not prominent; leaves linear, oblanceolate, or spatulate. . . . . . *M. flavescens* subsp. *gracillima*
          Seeds deeply ridged on back; stamens 3–8:
            Stamens (5)6(8); funiculus prominent. . . . . . *M. flavescens* subsp. *insularis*
            Stamens usually 3; funiculus not prominent. . . . . *M. flavescens* subsp. *striata*

Mollugo cerviana (L.) Ser. in DC., Prodr. 1: 392. 1824
Pharnaceum cerviana L., Sp. Pl. 272. 1753.

Delicate annual to 10 cm high; stems filiform, glabrous; lowest internode 20–35 mm long, upper internodes 12–20 mm long; leaves glaucous, linear, 4–8 mm long, about 0.5 mm wide; pedicels divaricate, finely filiform or capillary, 3–6(10) mm long; sepals 4, elliptic, obtuse, 1–1.5 mm long, green-veined, white-margined; stamens 4; seeds angularly semicircular, 0.2 × 0.3 mm, surface finely reticulate.

A species of wide distribution in warm regions of the world. In the Galápagos Islands known from scattered localities on several islands. Specimens examined: Baltra, Bartolomé, Isabela, Pinta, San Salvador.

Mollugo crockeri Howell, Proc. Calif. Acad. Sci. IV, 21: 20. 1933

Perennial; stems lignified at base, to 2 dm tall, glandular-puberulent; internodes 1–3 cm long; leaves linear, (5)7–20 mm long, (0.5)0.6–0.8 mm wide, glandular-puberulent, the lowest sometimes linear-oblanceolate, to 2.5 mm wide; pedicels 5–8 mm long, filiform, glandular-puberulent, normally 3 from same node; sepals 5, oblong-elliptic, 2.5–2.8 mm long, whitish with 3 green veins, glabrous or glandular-puberulent; stamens (4)6–7; seeds turgid-reniform, 0.5–0.6 mm in diameter, black, shiny, the surface with low tubercles arranged in rows dorsally; funiculus prominent.

Endemic; known only from San Salvador.

Mollugo flavescens Anderss., Kongl. Svensk. Vet.-Akad. Handl. 1853: 226. 1855 (sensu lato, the description covering all subspecies)

Usually annual, sometimes perennial (in subsp. *insularis*); stems prostrate, decumbent, or erect, to 4 dm tall, glabrous; internodes 0.7–6 cm long; leaves linear, oblanceolate, spatulate, or obovate, 5–20 mm long, 0.5–6 mm wide; pedicels 1.5–8 mm long, filiform, glabrous, (1)3–8 from same node; sepals 4–5, oblong-elliptical, 1.5–2 mm long, whitish, with green veins; stamens 3–6(8); seeds turgid-reniform, 0.4–0.6 mm in diameter, dark reddish brown or black, the surface shiny, tuberculate, sometimes ridged dorsally.

Endemic; occurs on practically all of the islands in the archipelago.

In this treatment *M. flavescens* has been given a rather wide circumscription. It is a highly variable species. A large number of different populations occur in different localities, but the differences are small, and it seems meaningless to attempt to distinguish these populations as separate taxa, since intergrading occurs among them. Four subspecies, previously recognized as species, have been distinguished, but some intergradations occur even among these.

Mollugo flavescens subsp. **flavescens**

Internodes 1.5–6 cm long; leaves oblanceolate, spatulate, or obovate, 8–15(20) mm long, (1.5)2–6 mm wide; pedicels 1.5–8 mm long, sepals 4 or 5; stamens 3–4(6); seeds 0.5–0.6 mm in diameter, not ridged on back; funiculus prominent.

Apparently restricted to Isla San Cristóbal and to the northern parts of Isla Santa Cruz.

Mollugo flavescens subsp. **gracillima** (Anderss.) Eliass., Madroño 20: 265. 1970

Mollugo gracillima Anderss., Kongl. Svensk. Vet.-Akad. Handl. 1853: 226. 1855.
Mollugo gracilis Anderss., op. cit. 1857: *t. 15, fig.* 3. 1861.
Mollugo gracillima Anderss. subsp. *latifolia* Howell, Proc. Calif. Acad. Sci. IV, 21: 17. 1933.
Mollugo flavescens Anderss. subsp. *angustifolia* Howell, op. cit. 18.
Mollugo flavescens Anderss. subsp. *intermedia* Howell, loc. cit.

Internodes (1)1.5–3.5 cm long; leaves linear, oblanceolate, or spatulate, 5–12(18) mm long, 0.5–3 mm wide; pedicels 2–8 mm long; sepals 4–5; stamens 3, rarely 4; seeds 0.4–0.5 mm in diameter, not ridged on back but sometimes lineate; funiculus not prominent.

Widely distributed over much of the archipelago. Specimens examined: Baltra, Bartolomé, Fernandina, Isabela, Marchena, Pinta, Pinzón, Rábida, San Cristóbal, San Salvador, Santa Cruz, Santa Fé, Santa María, Tortuga.

Mollugo flavescens subsp. **insularis** (Howell) Eliass., Madroño 20: 265. 1970

Mollugo insularis Howell, Proc. Calif. Acad. Sci. IV, 21: 19. 1933.

Internodes 0.7–2.5 cm long; leaves spatulate or narrowly oblanceolate, rarely linear, 5–12(20) mm long, (0.5)1–3(6) mm wide; pedicels 1.5–3(8) mm long; sepals 5; stamens (3)6(8); seeds 0.5–0.6 mm in diameter, deeply ridged on back; funiculus prominent.

Known only from Islas San Cristóbal and Santa María.

Either annual or perennial.

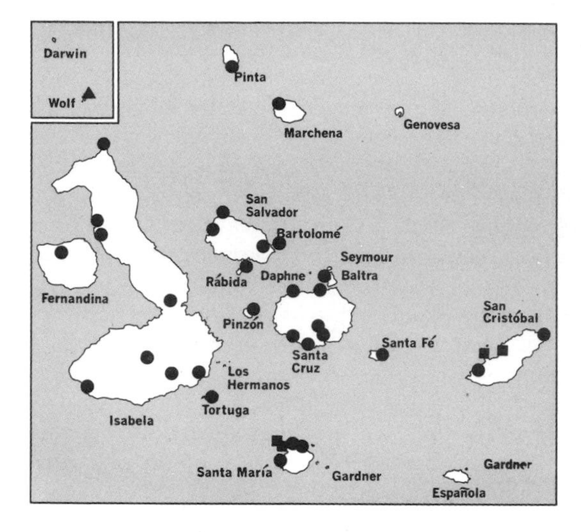

Mollugo flavescens subsp. **striata** (Howell) Eliass., Madroño 20: 265. 1970

Mollugo striata Howell, Proc. Calif. Acad. Sci. IV, 21: 19. 1933.

Internodes (1)1.5–3 cm long; leaves linear-oblanceolate, 8–18 mm long, 1–1.5 mm wide; pedicels 1–4 mm long; sepals 5; stamens generally 3; seeds 0.5–0.6 mm in diameter, ovate-reniform, deeply ridged on back; funiculus not prominent.

Known only from the type locality on Isla Wolf.

Additional collections are needed to determine the range of variation within this taxon and to indicate the relationship to other subspecies in the complex.

Mollugo floriana (Robins.) Howell, Proc. Calif. Acad. Sci. IV, 21: 21. 1933 (sensu lato, the description covering all subspecies)

Mollugo flavescens Anderss. var. *floriana* Robins., Proc. Amer. Acad. 38: 143. 1902.

Annual or perennial; stems to 4 dm tall, glabrous or minutely glandular-puberulent; internodes (2.5)3.5–5(7.5) cm long; leaves linear or linear-oblanceolate, (8)12–32 mm long, 1–3 mm wide; pedicels glabrous or finely glandular-puberulent, (3)4–7(10) mm long, 3–5 from a node; sepals oblong-elliptical, (2)2.3–2.7 mm long, whitish with 3 green veins, glabrous; stamens (5)7–8(9); seed turgid, reniform, 0.4–0.6 mm in diameter, black or dark reddish brown, shiny, tuberculate, these arranged in rows dorsally; funiculus prominent or obscure.

Endemic. Three subspecies recognizable.

Mollugo floriana subsp. **floriana**

Leaves linear or linear-oblanceolate, 15–25(30) mm long, 1.5–2(3) mm wide; pedicels 3–7(8) mm long, glabrous; sepals (2)2.5–2.7 mm long; stamens (5)7–8; seeds 0.4–0.6 mm in diameter, the funiculus prominent.

Known only from Santa María and from Onslow, an adjacent islet.

Characteristically perennial with a woody base, but sometimes delicate; possibly flowering as a first-year, young plant.

Mollugo floriana subsp. **gypsophiloides** Howell, Proc. Calif. Acad. Sci. IV, 21: 21. 1933

Leaves linear-oblanceolate, (8)12–15(20) mm long, (1)1.3–2(3) mm wide; pedicels 6–8(10) mm long, finely glandular-puberulent; sepals 2.5–2.7 mm long; stamens about 8; seeds about 0.4 mm in diameter; funiculus prominent.

Known only from Isla Pinzón.

Intermediate in character between *M. floriana* subsp. *floriana* and *M. crockeri.* Also, in its small seeds and in its apparently annual habit it approaches *M. flavescens.*

Mollugo floriana subsp. **santacruziana** (Christoph.) Eliass., Madroño 20: 266. 1970

Mollugo snodgrassii Robins. var. *santacruziana* Christoph., Nyt Mag. Naturvid. 70: 75. 1932 (1931).

Leaves linear, (18)20–32 mm long, about 1.5 mm wide; pedicels 6–8 mm long, glabrous; sepals about 2 mm long; stamens about 9 (only 1 flower available in which stamens could be counted); seed about 0.5 mm in diameter; funiculus not particularly prominent.

Known only from the type specimen, collected on Isla Santa Cruz.

The type specimen (*Rorud 123* [O]) is obviously a perennial plant.

Mollugo snodgrassii Robins., Proc. Amer. Acad. 38: 144. 1902

Perennial; stems from strongly lignified base (often to 1.5 cm thick), to 4 dm tall, often broomlike, glabrous; internodes 4–9 cm long; leaves linear-oblanceolate, (10)15–25 mm long, 1–2 mm wide (the lowest in younger plants rarely to 7 mm wide); pedicels (5)15–30 mm long, filiform, glabrous, (1)3(4) from a single node; sepals normally 5, 3–3.5(3.8) mm long, elliptic, obtuse, whitish, usually with 3 green veins; stamens 7–10; seeds turgid-reniform, 0.5–0.6 mm in diameter, dark brown or black, surface shiny, tuberculate, tubercles arranged in dorsal rows; funiculus prominent.

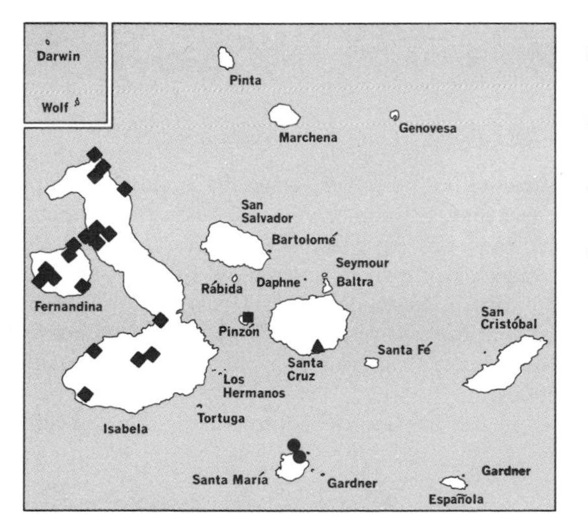

● *Mollugo floriana* subsp. *floriana*

■ *Mollugo floriana* subsp. *gypsophiloides*

▲ *Mollugo floriana* subsp. *santacruziana*

◆ *Mollugo snodgrassii*

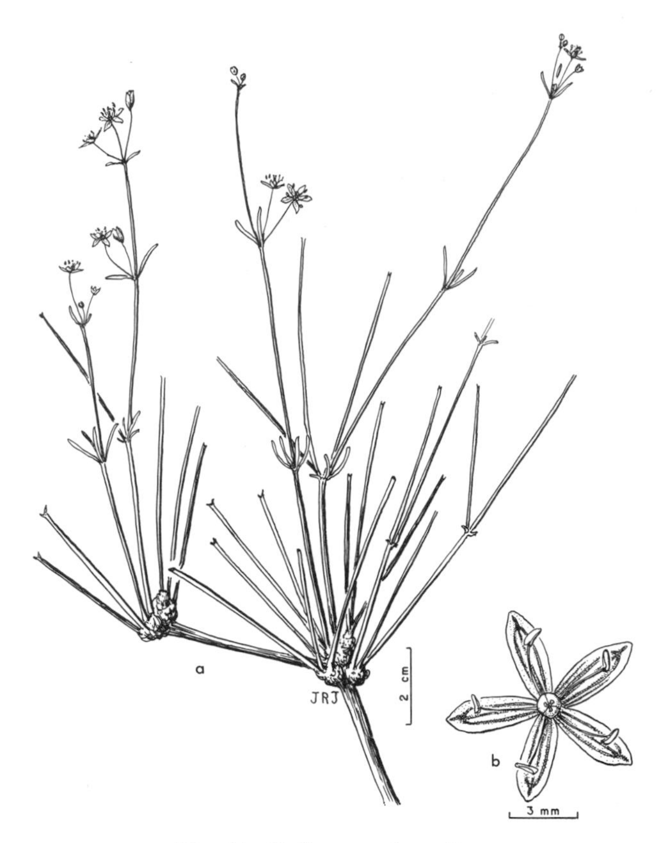

Fig. 43. *Mollugo snodgrassii.*

Endemic; common on relatively recent lava fields. Specimens examined: Fernandina, Isabela.

See Fig. 43.

NYCTAGINACEAE. FOUR-O'CLOCK FAMILY*

Trees, shrubs, or herbs. Leaves entire, without stipules, generally opposite, sometimes alternate. Flowers normally bisexual, actinomorphic, subtended by bracts. Bracts sometimes forming calyx-like involucre (in *Mirabilis*). Perianth uniseriate, segments 5, at least basally united, often forming an elongate tube. Base of perianth persisting around fruit. Fruit an anthocarp. Stamens 2–10(30), usually about 5, hypogynous, often unequal; filaments filiform, often connate at base. Anthers 2-celled, dehiscing longitudinally. Pistil 1; ovary superior, 1-loculed; carpel 1; ovule 1, basal, anatropous or campylotropous.

About 30 genera and 290 species in the tropics and subtropics of the Old and the New Worlds.

* Contributed by Uno Eliasson.

Flowers with a calyx-like involucre . . . . *Mirabilis*
Flowers with small bracts, not calyx-like:
  Plants herbs, sometimes somewhat lignescent at or near base:
    Anthocarp 5-sulcate or 5-angled in cross section; perianth about 2 mm long . . . . *Boerhaavia*
    Anthocarp 10-nerved, terete in cross section; perianth 8–10 mm long . . . *Commicarpus*
  Plants shrubs or trees, distinctly woody:
    Plant a shrub to 2–3 m high, with small, pyriform flowers; anthocarp about 1 mm long, viscid-puberulent . . . . *Cryptocarpus*
    Plant a tree to several meters high; anthocarp 10–15 mm long, 10-ribbed with stalked glands on ribs . . . . *Pisonia*

Boerhaavia L., Sp. Pl. 3. 1753; Gen. Pl. ed. 5, 4. 1754

Herbs, sometimes suffrutescent. Perianth funnelform or campanulate, about 2 mm long. Anthocarps obovoid or obpyramidal, usually 5-angled or 5-sulcate.

About 30 species in the warm regions of the world.

Anthocarp glabrous, truncate at apex; leaves finely red-dotted . . . . *B. erecta*
Anthocarp glandular-puberulent, obtuse, rounded, or acute at apex; leaves not red-dotted:
  Branches of inflorescence glandular-puberulent . . . . *B. caribaea*
  Branches of inflorescence glabrous . . . . *B. coccinea*

Boerhaavia caribaea Jacq., Obs. Bot. 4: 5. 1771

Boerhaavia hirsuta Willd., Phytogeogr. 1: 1. 1794.
Boerhaavia viscosa Lag. & Rodr., Anal. Cienc. Nat. 4: 256. 1801.
Boerhaavia diffusa sensu Anderss., Kongl. Svensk. Vet.-Akad. Handl. 1853: 171. 1855 et 64. 1861, non L. 1753.
Boerhaavia glandulosa Anderss., op. cit. 1853: 171. 1855.
Boerhaavia coccinea sensu Svens., Amer. Jour. Bot. 22: 230. 1935, non Mill. 1768.

Perennial; stems decumbent, viscid-puberulent to densely glandular-puberulent; leaves broadly ovate to suborbicular, mucronulate, 2–4 cm long, 1.5–3.5 cm wide, puberulent to densely villous; flowers in heads or congested cymes; inflorescence

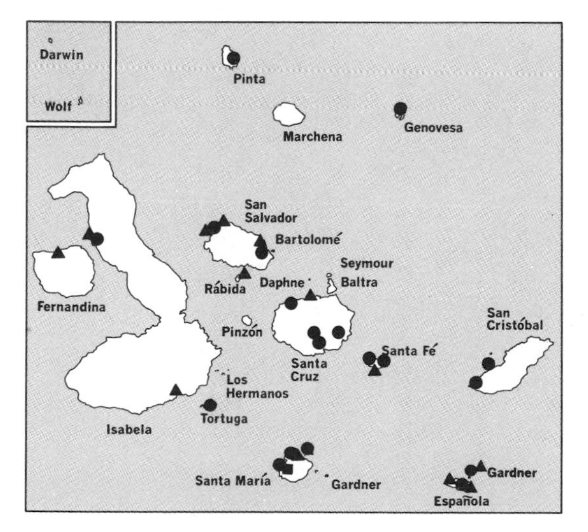

● *Boerhaavia caribaea*
■ *Boerhaavia coccinea*
▲ *Boerhaavia erecta*

branches finely to densely glandular-puberulent; pedicels to 1 mm long; perianth purplish red, about 1.5 mm long; stamens 1–3; anthocarp about 2.5 mm long, narrowly obovoid, 5-sulcate, glandular-puberulent.

Southern United States to Bolivia and Peru; common weed in many regions. Specimens examined: Champion, Española, Gardner (near Española), Genovesa, Isabela, Pinta, San Cristóbal, San Salvador, Santa Cruz, Santa Fé, Santa María, Tortuga.

Rather variable as regards amount and nature of pubescence.

Boerhaavia coccinea Mill., Gard. Dict. ed. 8. 1768

Stem decumbent or ascending, glabrous or nearly so; leaves broadly ovate, mucronulate, 2.5–6 cm long, 2–4.5 cm wide, glabrous; upper leaves puberulent below; inflorescence branches glabrous; flowers in headlike clusters arranged in cymes; perianth about 2 mm long; anthocarp about 4.5 mm long at maturity, narrowly obpyramidal, 5-ribbed, the ribs gland-bearing.

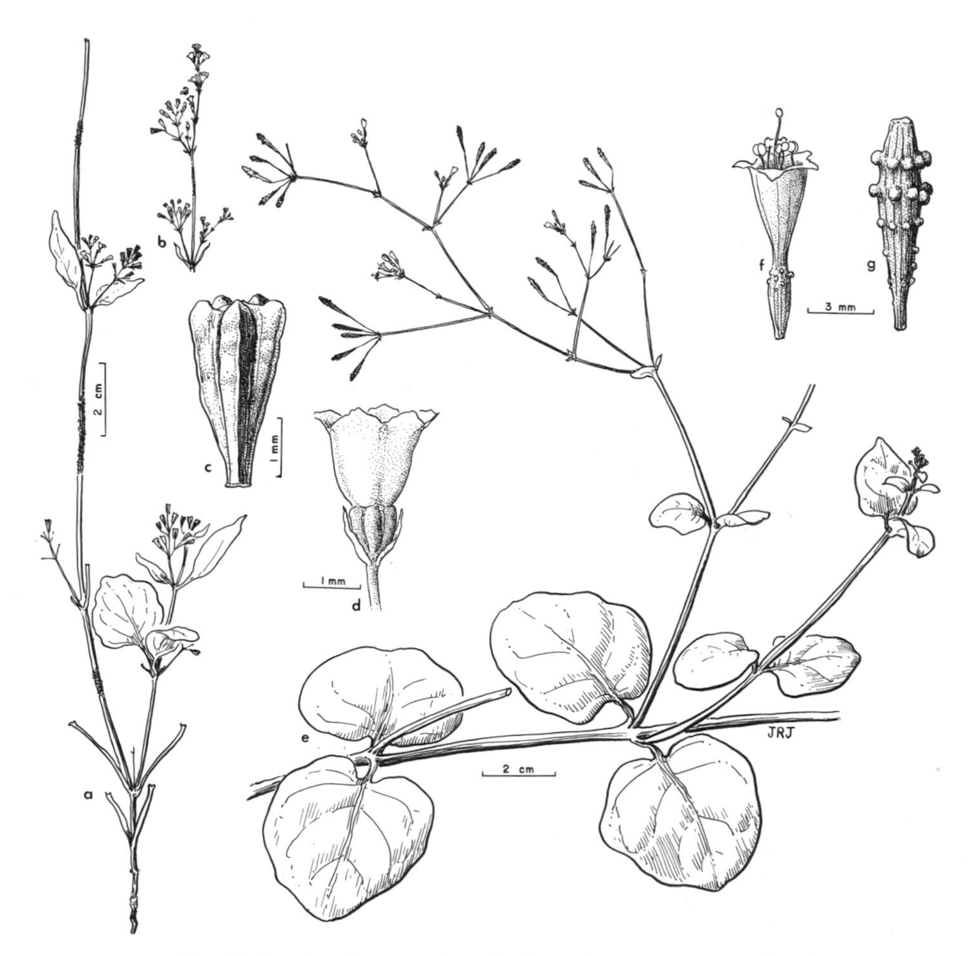

Fig. 44. *Boerhaavia erecta* (*a–d*). *Commicarpus tuberosus* (*e–g*).

About the same distribution as *B. caribaea*. Specimen examined: Santa María.
Only 1 collection has been seen from the Galápagos. Evidently the plant is much rarer there than *B. caribaea*.

Boerhaavia erecta L., Sp. Pl. 3. 1753

Boerhaavia virgata HBK., Nov. Gen. & Sp. Pl. 2: 215. 1817.
Boerhaavia discolor HBK., loc. cit.
Boerhaavia decumbens sensu Hook. f., Trans. Linn. Soc. Lond. 20: 193. 1847, non M. Vahl 1804.
Boerhaavia paniculata in litt., non Rich. 1792.

Annual herb; stems decumbent to erect, often greatly branched; branches usually puberulent on lower parts, glabrous on upper areas; middle of internodes usually with broad, brownish, viscous band on each; leaves variable in shape, narrowly linear to ovate or elliptic, 1–3 cm long, 0.3–1.5 cm wide, finely red-dotted on both surfaces; flowers in cymes; perianth 1–2 mm long, white or pink; stamens 2–3; anthocarp about 3 mm long, narrowly obpyramidal, 5-angled or 5-winged, truncate at apex, glabrous.

Southern United States to tropical South America; abundant in many neotropical regions. Specimens examined: Española, Fernandina, Gardner (near Española), Isabela, Rábida, San Salvador, Santa Cruz, Santa Fé, Santa María.

See Fig. 44.

Commicarpus Standl., Contr. U. S. Nat. Herb. 12: 373. 1909

Herbs, sometimes suffrutescent. Perianth short-funnelform, about 1 cm long. Anthocarp clavate, 10-nerved, terete or nearly so, gland-bearing near apex.

About 15 species in warm regions of the world.

Commicarpus tuberosus (Lam.) Standl., Contr. U. S. Nat. Herb. 18: 101. 1916

Boerhaavia tuberosa Lam., Encycl. Bot. Ill. Gen. 1: 10. 1791.
Boerhaavia excelsa Willd., Phytogeogr. 1: 1. 1794.
Boerhaavia litoralis HBK., Nov. Gen. & Sp. Pl. 2: 216. 1817.
Boerhaavia scandens in litt., non L. 1753.

Stems decumbent to erect, glabrous, richly branched; leaves broadly ovate to subcordate or suborbicular, 3–6 cm long, 2.5–4.5 cm wide; inflorescence of numerous, small, 4–7-flowered umbels, each umbel with 4–7 narrowly lanceolate bracts 2–3 mm long at its base, these caducous; pedicels 5–15 mm long, glabrous; perianth 8–10 mm long, pink or purplish red; stamens 2 or 3; anthocarp 6–10 mm long, clavate, obscurely 10-nerved to almost terete, with about 5 rounded, viscid glands near apex.

Ecuador and Peru. An abundant plant in the Galápagos, distributed widely in the archipelago. Specimens examined: Champion, Gardner (near Santa María), Isabela, Pinzón, San Cristóbal, San Salvador, Santa Cruz, Santa María.

See Fig. 44.

Cryptocarpus HBK., Nov. Gen. & Sp. Pl. 2: 187. 1817

Coastal bush with alternate leaves. Perianth with 4 or 5 lobes. Stamens 4 or 5. Stigma a bit eccentric, deeply divided into several segments. Anthocarp pyriform.

A single species along the coast of western South America.

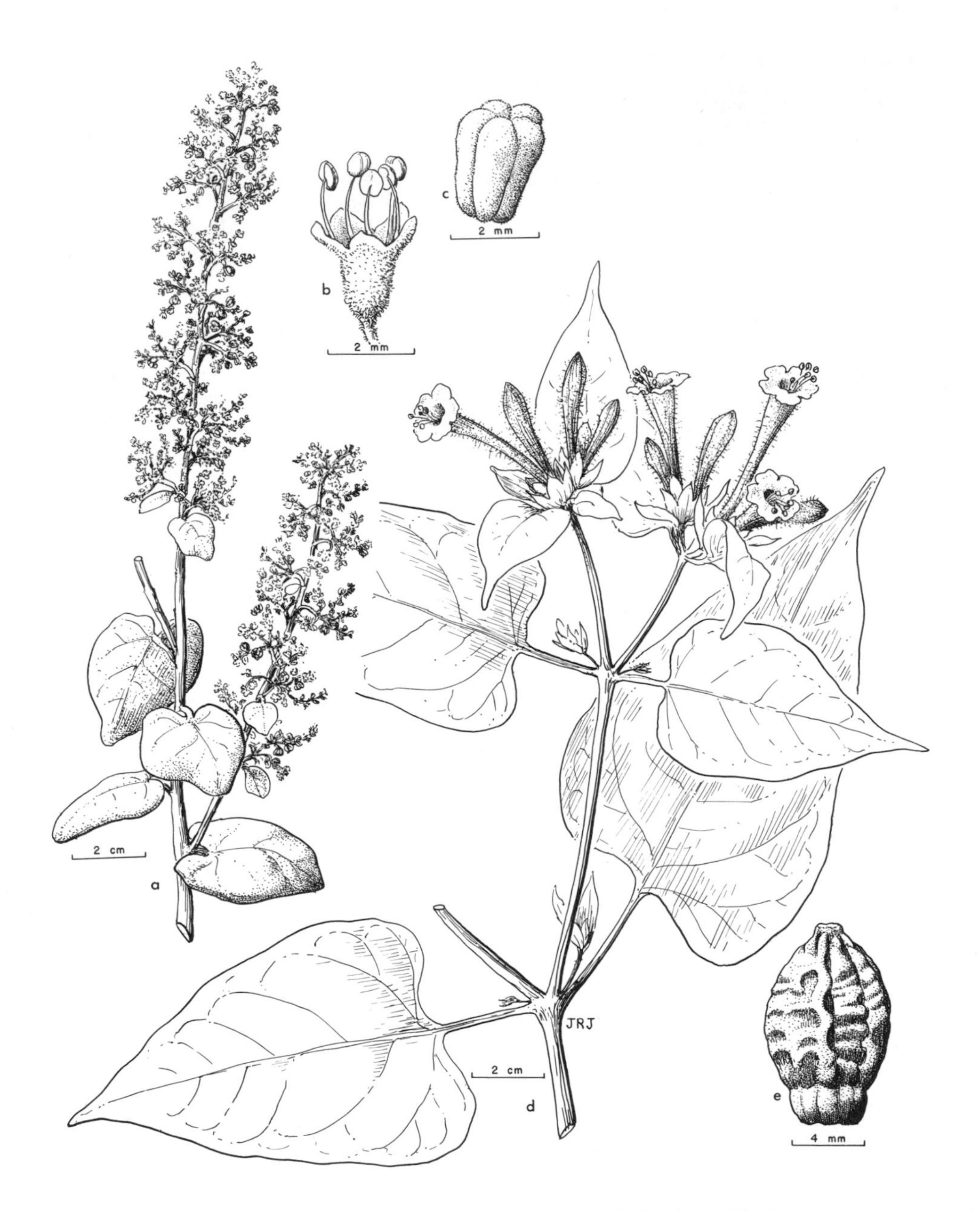

Fig. 45. *Cryptocarpus pyriformis* (*a–c*). *Mirabilis jalapa* (*d, e*).

Cryptocarpus pyriformis HBK., Nov. Gen. & Sp. Pl. 2: 188. 1817

Salpianthus pyriformis Standl., Publ. Field Mus. Nat. Hist. Bot. 11: 95. 1931.
Cryptocarpus cordifolius Moric., Pl. Nouv. d'Amérique 75. 1833.
Cryptocarpus pyriformis var. *cordifolius* Moq. in DC., Prodr. 13: 88. 1849.

Stems much-branched, generally scandent, to several meters long, viscid-puberulent; petioles about 1 cm long; leaves 3–5 cm long, 3–4.5 cm wide, truncate-cordate at base, obtusely rounded at apex, finely puberulent on both sides; flower clusters arranged in compound, often leafy panicles; flowers sessile or subsessile on pedicels about 0.5 mm long, these yellowish green, finely viscid-puberulent; flowers pyriform; perianth about 2 mm long, with (4)5 lobes, puberulent; anthocarp more or less pyriform, about 1.5 mm long.

Along the coast of Ecuador and Peru. Very abundant in the Galápagos. Specimens examined: Bartolomé, Champion, Española, Fernandina, Gardner (near Española), Genovesa, Isabela, Las Plazas, Marchena, Pinta, Pinzón, Rábida, San Cristóbal, San Salvador, Santa Cruz, Santa Fé, Santa María, Seymour.

See Fig. 45 and color plates 15–17, 21, 25, 26, 29, and 40.

Mirabilis L., Sp. Pl. 177. 1753; Gen. Pl. ed. 5, 82. 1754

Herbs, often suffrutescent. Leaves opposite. Flowers 1 to several, with 5-lobed, calyx-like involucre formed by bracts. Perianth with long tube and spreading, 5-lobed limb. Stamens usually 5, unequal, exserted. Stigma covered with stalked papillae. Anthocarp elliptic to obovoid, often obtusely angular.

About 50 species, all but 1 native to tropical America.

Mirabilis jalapa L., Sp. Pl. 177. 1753

Stem erect, 3–7 dm high, glabrous or nearly so; leaves ovate to subcordate, 4–8(10) cm long, 2–5 cm wide; petioles 2–4 cm long; segments of calyx-like involucre ovate to ovate-lanceolate, 7–10 mm long, 3–5 mm wide; perianth limb 2–2.5 cm thick, variable in color, red, purplish, white, or yellowish; perianth tube 3.5–5 cm long; anthocarp about 1.5 cm long.

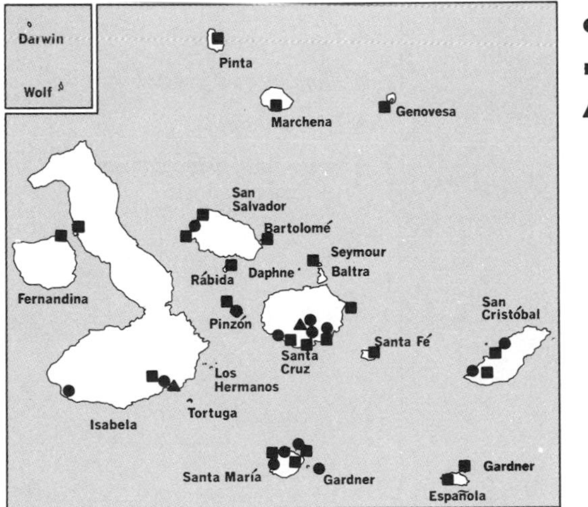

● *Commicarpus tuberosus*
■ *Cryptocarpus pyriformis*
▲ *Mirabilis jalapa*

A native of tropical America; introduced into the Galápagos Islands, and sometimes escaped from cultivation. Specimens examined: Isabela, Santa Cruz.
See Fig. 45.

Pisonia L., Sp. Pl. 1026. 1753; Gen. Pl. ed. 5, 451. 1754

Shrubs or trees. Leaves opposite or partly alternate. Flowers dioecious. Inflorescences of panicles arranged in umbel-like clusters. Perianth of staminate flowers funnelform to campanulate, that of pistillate flowers more or less tubular. Staminate flowers with 5–11 stamens, these of unequal length, at least some of them exserted. Pistillate flowers with 1 pistil and short, sterile stamens. Anthocarp oblong, ribbed, gland-bearing.

About 50 species, most of them native to tropical America, but some indigenous in Africa, Asia, and Australia, and 1 endemic in the Galápagos Islands.

Pisonia floribunda Hook. f., Trans. Linn. Soc. Lond. 20: 193. 1847
Guapira floribunda Lundell, Wrightia 4: 80. 1968.

Tree attaining height of 10–15 m; leaves elliptic to suborbicular, 3.5–8 cm long, 2–4.5(6) cm wide; glabrous or finely brown-puberulent; young leaves often brown-tomentose on lower surface; leaves dark, readily detached when dried; petioles 0.5–2 cm long, usually brown-puberulent; branches of inflorescences and basal parts of perianth brown-puberulent; staminate flowers with funnelform or campanulate perianths, 5–6 mm long, teeth about 2 mm long; stamens mostly 6, at least some of them exserted, surrounding rudimentary pistil 1.5–2 mm long; pistillate flowers ovoid-tubular, 3–4 mm long, brown-puberulent, perianth teeth about 1 mm long; stigma somewhat exserted, surrounded by 6–7 rudimentary stamens; ovary ovoid-oblong; anthocarp brown-puberulent in youth, glabrous at maturity, about 1 cm long, 4 mm broad, 10-ribbed, with stalked glands on ribs.

Endemic; occurs on all the larger islands at elevations above 50 m. Specimens examined: Isabela, Pinta, Pinzón, San Salvador, Santa Cruz, Santa María.
See Fig. 46 and color plates 44, 47, and 60.

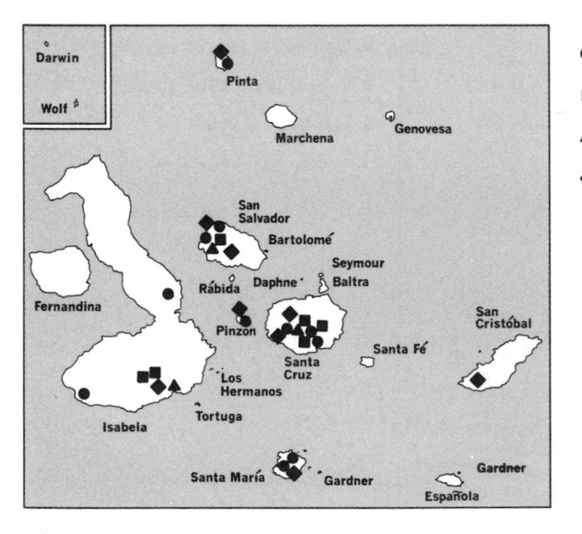

● *Pisonia floribunda*
■ *Phytolacca octandra*
▲ *Rivina humilis*
◆ *Peperomia galapagensis* var. *galapagensis*

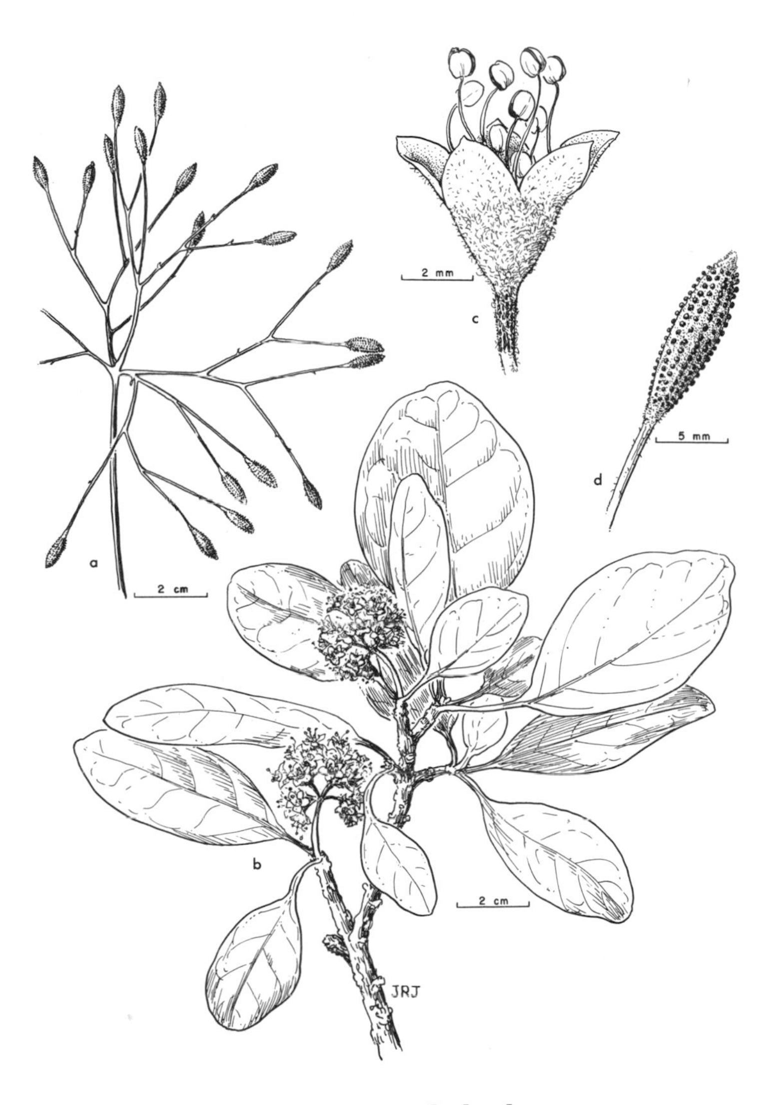

Fig. 46. *Pisonia floribunda.*

PHYTOLACCACEAE. POKEWEED FAMILY*

Herbs, shrubs, or trees, with alternate, entire leaves. Flowers usually bisexual, actinomorphic, in terminal or axillary racemes. Perianth with 4 or 5 segments. Petals in most genera usually lacking. Stamens same number as perianth-segments and alternate with them, or numerous. Ovary free, of 1 to several uniovulate carpels, these free or united into a several- to many-celled ovary. Stigmas usually same in number as carpels. Fruit of several uniovulate, free carpels, or these coalesced

* Contributed by Uno Eliasson.

into several-loculed fruit, each locule containing solitary seed. Seeds lenticular or reniform; embryo annular or semi-annular.

About 12 genera and 100 species, mostly in tropical and subtropical regions.

Perianth-segments 5; styles and pistils several. . . . . *Phytolacca*
Perianth-segments 4; pistil 1; style eccentric. . . . . *Rivina*

Phytolacca L., Sp. Pl. 441. 1753; Gen. Pl. ed. 5, 200. 1754

Herbs, shrubs, or trees. Flowers usually bisexual. Inflorescences racemose, usually opposite leaves. Perianth-segments 5. Stamens numerous (6–30, in the Galápagos

Fig. 47. *Phytolacca octandra.*

species 8). Ovary of 8–12 carpels. Styles same in number as carpels. Fruit berry-like, juicy. Seeds lenticular or reniform.

About 30 species in tropical, subtropical, and warm temperate regions.

Phytolacca octandra L., Sp. Pl. ed. 2, 631. 1762

Perennial herb; stems erect, often sulcate; leaves elliptical, pointed, 8–20 cm long, 4–9 cm wide; flowers bisexual, in lateral, spicate racemes; pedicels 0.5–2 mm long, in axil of narrowly lanceolate bract 2–5 mm long; perianth-segments oval, 3–4 mm long, greenish white to slightly violet; stamens 8; carpels about 8; fruit berry-like, dark purple to nearly black; seeds turgidly lenticular, about 2.5 mm in diameter, black, shiny.

From Mexico to Peru. Specimens examined: Isabela, San Salvador, Santa Cruz.

See Fig. 47.

Rivina L., Sp. Pl. 121. 1753; Gen. Pl. ed. 5, 57. 1754

Herbs, sometimes woody at base. Leaves ovate. Flowers bisexual. Perianth-segments 4, free. Stamens 4. Pistil 1, with distinctly eccentric style. Fruit a berry. Seed thickly lenticular, densely hairy.

Three species in tropical and subtropical America.

Rivina humilis L., Sp. Pl. 121. 1753

Stems much branched, glabrous, to 7 dm tall, erect or vinelike; leaves ovate, acuminate to cuspidate, 3–7 cm long, 1.5–4 cm wide; inflorescences of 10–20-flowered racemes; pedicels 3–5 mm long, each subtended by narrowly lanceolate, glandular bract 1.5–2 mm long; perianth-segments 4, pink, oblong to narrowly elliptic, 2–3 mm long, erect and partially enclosing fruit at maturity; berry globose, orange to red, 2–3 mm in diameter.

From Texas to Florida and southward to Chile and Argentina. Specimens examined: Isabela, San Salvador, Santa Cruz.

See Fig. 48.

PIPERACEAE. PEPPER FAMILY*

Erect or scandent herbs, or less commonly evergreen shrubs or trees, often with distinct, scattered vascular bundles as in monocotyledons. When herbaceous the stems often succulent and with swollen nodes. Leaves usually alternate, but sometimes opposite or whorled, petiolate, entire, palmately or pinnately nerved, often fleshy; stipules adnate to petioles or lacking. Flowers minute, bracteate, usually bisexual, in dense, more or less fleshy spikes, these solitary or in umbels. Perianth none. Stamens 1–10, hypogynous, with distinct but short filaments; anthers of 2 distinct or confluent chambers, dehiscing longitudinally. Pistil 1; ovary superior, 1-celled, 2–5-carpeled. Ovule solitary, basal, orthotropous. Style 1, or the 1–5 stigmas sessile, often brushlike and lateral in *Peperomia*. Fruit a small drupe. Seeds small; embryo and endosperm both minute.

* Contributed by Ira L. Wiggins.

A family of 10–12 genera and over 1,300 species, distributed throughout the tropics of the world.

Peperomia R. & P., Prodr. 8. 1794

Prostrate to erect, succulent, terrestrial, or epiphytic herbs, acaulescent or with distinct stems, and opposite, whorled, or alternate, simple leaves varying greatly in size and shape. Herbage glabrous or puberulent with microscopic hairs to hirsute, often pellucid-glandular. Petioles often clasping stem. Inflorescence of terminal, axillary or leaf-opposed, cylindric, clavate, or fusiform, pedunculate spikes, with flowers borne on surface or in pits in the rachis, each flower subtended by orbicular to ovate, peltate bract. Flowers perfect, sessile, consisting of 2 soon-deciduous stamens and pistil. Fruit 1-seeded, drupelike, with indurated endocarp,

Fig. 48. *Rivina humilis.*

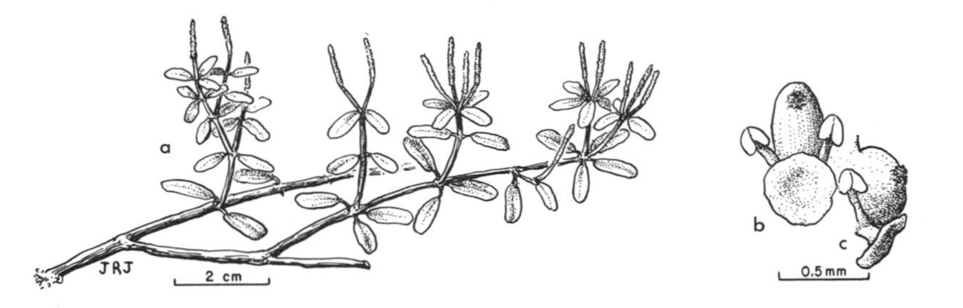

Fig. 49. *Peperomia galapagensis* var. *galapagensis*.

globose, ovoid, ellipsoid, subcylindric, obovoid, or turbinate, rounded but sometimes beaked apically, the surface smooth to verrucose and viscid.

Widely distributed in the tropics of the world.

Trelease and Yuncker (1950) divide the genera represented in northern South America into 8 groups and recognize 365 species in the area extending from Colombia and Venezuela into Brazil and Ecuador. All of the species known from the Galápagos Islands are included by them in the subgenus *Sphaerocarpidium*, and have globose-ovoid fruits with a slightly oblique, pointed apex and a subapical stigma.

Leaves predominantly 2 at a node, some alternate; peduncles 5–15 mm long; stems with upwardly curved, crisped hairs; petioles 5–15 mm long; blades broadly ovate to nearly orbicular, often shorter than petioles. . . . . *P. petiolata*
Leaves 3–9 (mostly 3–5) at a node, not alternate; peduncles rarely to 10 mm long, usually 3–5 mm long; stems puberulent, hirtellous, or glabrous, the hairs not curved upward; petioles 2 mm long or less; blades oblong to obovate, much longer than petioles:
  Leaf blades 3- to 5-nerved; spikes 2–15 cm long; stems puberulent to hirtellous but scarcely velvety:
    Petioles about 2 mm long; fruit ovoid, 0.5–0.6 mm long; stems to 15 cm long, puberulent; leaves 6–11 mm long. . . . . *P. obtusilimba*
    Petioles less than 1 mm long; fruit globose, 1 mm long; stems to 1 m long, densely hirtellous; leaves 5–30 mm long. . . . . *P. galioides*
  Leaf blades obscurely 1-nerved; spikes 8–15 mm long; stems densely velvety-puberulent:
    Leaves glabrous to sparingly puberulent. . . . . *P. galapagensis* var. *galapagensis*
    Leaves densely puberulent. . . . . *P. galapagensis* var. *ramulosa*

Peperomia galapagensis Hook. f. ex Miquel in Hook., Lond. Jour. Bot. 4: 426. 1845, var. **galapagensis**

Cespitose epiphytic and terrestrial plant with stems 2–3 mm thick when dry, to 15 cm tall, somewhat striate-ridged longitudinally, freely and divergently branched, tips of branches ascending, velvety-puberulent; internodes 1–3 cm long; leaves 3–4(7) in a verticil, oval-oblong to subspatulate, 2–3 mm wide, 5–8 mm long, rounded, obtuse, or retuse at apex, obtuse to somewhat acute at base, obscurely 1-nerved, glabrous or sparingly puberulent above when young, ciliate at apex, often drying coriaceous and involute; petioles 1–2 mm long, channeled along upper surface, puberulent; spikes many, terminal and in axils of upper leaves, 8–12(15) mm long; peduncles 3–4(8) mm long, puberulent; bracts round-peltate; fruit subglobose, 0.5–0.7 mm long, oblique at apex; stigma subapical.

Endemic; occurring in shaded, moist areas, mostly above 100 m. Specimens examined: Isabela, Pinta, Pinzón, San Cristóbal, San Salvador, Santa Cruz, Santa María.

See Fig. 49.

Peperomia galapagensis var. **ramulosa** (Anderss.) Yuncker, Bot. Publ. Bishop Mus. Occ. Papers 18: 230. 1946.

Peperomia ramulosa Anderss., Kongl. Svensk. Vet.-Akad. Handl. 1853: 158. 1855.
Peperomia snodgrassii DC. in Robins., Proc. Amer. Acad. 38: 131. 1902.

Differing from var. *galapagensis* mainly in having the leaves densely puberulent throughout.

Endemic. Specimens examined: Isabela, Pinta, Santa Cruz, Santa María.

Peperomia galioides HBK., Nov. Gen. & Sp. Pl. 1: 71, *pl. 17*. 1815

Piper galioides Poir. in Lam., Encycl. Méth. 4: Suppl. 470. 1816.
Peperomia suaveolens Hamilton, Prodr. Pl. Ind. Occ. 2: 1825.
?*Peperomia mollugo* Willd. in Miquel, Syst. Pip. 156. 1843, as synonym.
Peperomia flagelliformis Hook. f. ex Miquel in Hook., Lond. Jour. Bot. 4: 423. 1845.
Peperomia jamesonii Regel, Bull. Bot. Soc. Moscow 31: 544. 1858.
Peperomia agapatensis C. DC. in DC., Prodr. 16^1: 455. 1869.
Peperomia galioides var. *longifolia* C. DC. in DC., op. cit. 464.
Peperomia galioides var. *nigropunctulata* C. DC. in DC., loc. cit.
Peperomia galioides var. *aprica* Henschen, Nov. Act. Regiae Soc. Sc. Uppsal. ser. 3, 8: 37. 1873.
Peperomia galioides var. *umbrosa* Henschen, loc. cit.
Peperomia subcorymbosa Sodiro, Contr. al Conoc. Fl. Ecuator. Mongr. 1, ed. 2, 181. 1901.
Peperomia galioides var. *minutifolia* C. DC., Ann. Conserv. & Jard. Bot. Genève 21: 262. 1920.
Peperomia granata Trel., Rep. Spec. Nov. Fedde 23: 24. 1926.

Succulent, suberect, considerably branched terrestrial or epiphytic herb with stems 2–5(8) mm thick, to 1 m long, branching dichotomously, trichotomously, or in verticils, the branches slender and often wandlike, with variable internodes 1–10 cm long, moderately to densely hirtellous, the epidermis scurfy, exfoliating in sheets or scales; leaves (3)4–5(9) at a node, elliptic-oblong, subspatulate, or oblanceolate, 5–30 mm long, 2–5 mm wide, obtuse at apex, acute at base, obscurely palmately 3-nerved, the midrib often branching near tip, blades often puberulent along midrib and near base on undersurface, ciliolate near apex, otherwise glabrous, membranous and pellucid-dotted and yellow-glandular beneath when dry; petioles less than 1 mm long, hirtellous; spikes terminal and axillary, solitary or 3–6 in verticils, 1 mm across, 4–7(15) cm long, loosely flowered; peduncles 4–5(10) mm long, sparsely hirtellous to glabrate; bracts orbicular-peltate; fruit globose-ovoid, about 1 mm long, oblique at base and apex; stigma subapical.

Widely distributed in moist places in forests and jungles, West Indies, Mexico, and Central and South America; common in the Galápagos Islands. Specimens examined: Isabela, Pinta, Pinzón, San Cristóbal, San Salvador, Santa Cruz.

Trelease and Yuncker wrote as follows (p. 571): "An exceedingly variable species with the leaves varying from the minimum to the maximum size on the same plant in many specimens. Some specimens exhibit basal shoots with short internodes and numerous small leaves thus producing a bushy effect, while another part of the same specimen will show elongated internodes and longer leaves. Varieties based on the length of internodes and size of the leaves are not, therefore, believed to be tenable. The glandular dots on the leaves are commonly golden yellow, but occasionally may be nearly black. . . . The pubescence of the stem and the shape of the leaves within reasonable limits are fairly constant."

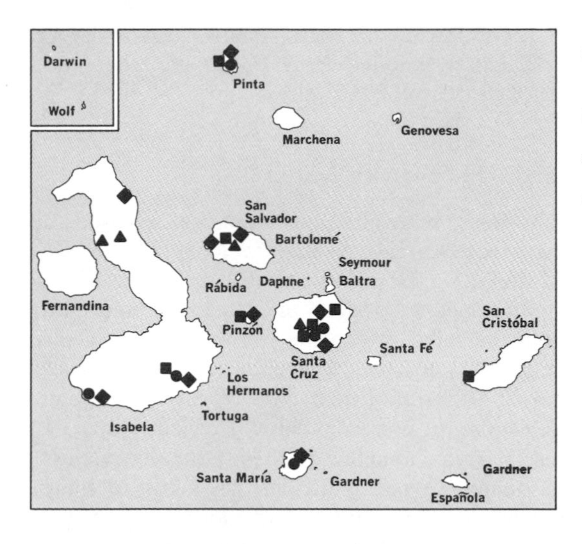

● *Peperomia galapagensis* var. *ramulosa*
■ *Peperomia galioides*
▲ *Peperomia obtusilimba*
◆ *Peperomia petiolata*

Peperomia obtusilimba C. DC. in Stewart, Proc. Calif. Acad. Sci. IV, 1: 49. 1911

Somewhat cespitose plant epiphytic on trees or on rocks, rarely to 15 cm tall, but freely and divaricately branched; stems 1.5–3 mm thick near base when dry; branches slender, moderately to densely puberulent, internodes 5–15 mm long; leaves 3–5 at a node, elliptic-oblong to narrowly obovate, 6–11 mm long, 3.5–5 mm wide, rounded or obtuse at apex, rounded to subacute at base, palmately 3- or faintly 5-nerved, the nerves more or less branching above, glabrous on both sides or sparingly puberulent near base on upper surface, drying thin and membranous; petioles about 2 mm long, puberulent; spikes several in axils of upper leaves and terminal, slender, 2–7 cm long; peduncles 3–5 mm long, sparingly puberulent to glabrate; bracts round-peltate, margins smooth or nearly so; fruit ovoid, 0.5–0.6 mm long, oblique at apex; stigma subapical.

Endemic; in shaded places in woods, from about 75 to 500 m altitude on several of the islands. Specimens examined: Isabela, San Salvador, Santa Cruz.

Peperomia petiolata Hook. f., Trans. Linn. Soc. Lond. 20: 181. 1847

Peperomia stewartii C. DC. in Stewart, Proc. Calif. Acad. Sci. IV, 1: 49. 1911.

Stems decumbent, creeping on soil, rocks, or trunks of trees, 1–2 mm thick when dry, crisp-pubescent with most hairs curving upward on young branches, glabrate in age, the stems branched freely with upright branches 10–15 cm tall; internodes 5–30 mm long; leaves opposite, or occasionally some alternate, elliptic to round-elliptic, 10–15(25) mm long, 5–15 mm wide, rounded, obtuse, or rarely slightly attenuate at apex, rounded to somewhat acute at base, sparsely crispate-puberulent to glabrate on upper surface near base, marginally ciliolate above middle, 3-nerved or large leaves obscurely 5-nerved, drying dark green above, yellowish and slightly dark-dotted beneath; petioles 5–10(15) mm long; spikes solitary or in groups of 3–5 in axils of upper leaves, 6–9(10) cm long; peduncles (5)8–10(15) mm long, glabrous or crisped-pubescent; bracts oval-orbicular, peltate, with irregular margins, bearing yellowish to blackish glandular dots; fruit subglobose, 0.7–0.8 mm long, oblique at apex; stigma subapical.

Endemic; usually in moderate to dense shade on rocks, on trunks and larger limbs of trees, or on the surface of the soil, in wooded areas. Specimens examined: Isabela, Pinta, Pinzón, San Salvador, Santa Cruz, Santa María.

POLYGONACEAE. Buckwheat Family*

Herbs, shrubs, or sometimes vines or trees, with alternate, opposite, whorled, or fasciculate leaves, sometimes forming rosette. Leaves minute to ample, usually simple, entire. Stipules connate and sheathing (forming ocreae), or absent. Flowers monoecious, dioecious, polygamodioecious, or perfect, on slender, jointed pedicels, variously arranged in clusters on a spike or raceme, rarely solitary. Calyx 2–6-cleft or -parted, lobes often petaloid, persistent, the outer usually narrower than the inner. Petals none. Stamens 2–9, inserted near base of calyx or hypanthium; anthers small, basifixed. Ovary superior, free from calyx, 1-celled, 1-ovulate; ovule orthotropous, pendulous or erect; style branches (2)3(4), usually distinct; stigmas capitate or peltate. Fruit an achene, ovoid, lenticular, trigonous, or sometimes 6-angled, in some species both the lenticular and the triquetrous achenes in same inflorescence, usually enclosed within the perianth envelope, the envelope persistent. Embryo straight or variously curved or folded. Endosperm copious, mealy.

Nearly 35 genera and over 800 species, widely distributed but with largest representation in the North Temperate Zone.

Polygonum L., Sp. Pl. 359. 1753; Gen. Pl. ed. 5, 170. 1754

Annual or perennial herbs from heavy rootstocks or fibrous roots. Leaves alternate, entire, bearing entire, ciliate, or lacerate connate sheathing stipules (ocreae). Inflorescences spicate or racemose, simple or somewhat branched, the flowers arranged in fascicles or rarely solitary on slender, jointed, often very short pedicels. Calyx 4–6-parted, greenish, rose, pink, purplish, or white, the lobes frequently petal-like, withering-persistent, usually erect in fruit. Stamens 3–9. Styles 2 and the achenes lenticular, or 3 and the achenes triquetrous. Achenes wholly enclosed in calyx or the tip slightly exserted, faces convex or concave. Embryo curved, lying in groove in endosperm at 1 angle of achene.

About 200 species widely distributed from the tropics into the temperate zones throughout the world.

Calyces conspicuously punctate-glandular, at least toward base; leaf blades uniformly and minutely punctate-glandular, the glands obvious but less than 0.1 mm in diameter .... *P. punctatum*
Calyces without punctate glands, or if present then microscopic and obscure; leaf blades glandless, or the glands 0.2–0.4 mm in diameter, irregularly scattered on undersurface:
  Leaf blades 2–8 cm wide, lance-ovate; achenes 2–4 mm long:
    Ocreae 2–4 cm long; achenes 2–2.5 mm long; leaf blades broadly lanceolate, acuminate, 15–30 cm long; petioles 1–3 cm long........ *P. acuminatum*
    Ocreae 1.5–2.5 cm long; achenes 2.5–3 mm long; leaf blades broadly ovate, not markedly acuminate, 6–15 cm long; petioles 5–10 cm long........ *P. galapagense*

---

* Contributed by Ira L. Wiggins.

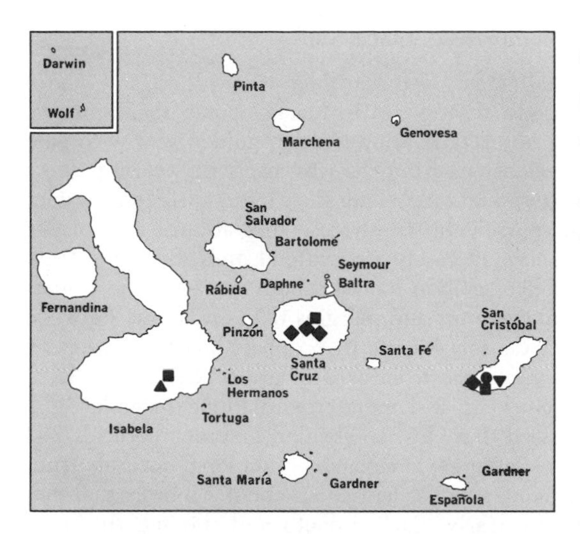

● *Polygonum acuminatum*

■ *Polygonum galapagense*

▲ *Polygonum hydropiperoides* var. *persicarioides*

◆ *Polygonum opelousanum*

▼ *Polygonum punctatum*

Leaf blades 5–20 mm wide, narrowly lanceolate or lance-elliptic; achenes 1.5–3 mm long:

Achenes elongate-ovoid, 2.5–3 mm long, usually wholly enclosed by calyx in fruit, both lenticular and triquetrous achenes usually in same inflorescence; leaf blades without conspicuous punctate glands on lower surface. . . . . *P. hydropiperoides* var. *persicarioides*

Achenes broadly ovoid, 1.2–2 mm long, apex usually slightly exserted beyond calyx, triquetrous, rarely if ever lenticular; leaf blades with irregularly scattered, grayish erumpent resinous dots 0.2–0.4 mm in diameter on undersurface. . . . . *P. opelousanum*

Polygonum acuminatum HBK., Nov. Gen. & Sp. Pl. 2: 178. 1817

Polygonum cuspidatum Willd. in Spreng., Syst. 2: 256. 1825.
Polygonum erectum Velloso, Fl. Flum. 4: *pl. 42.* 1827.
Polygonum acuminatum var. *humboldtii* Meisn. in Mart., Fl. Bras. 5: 14, *pl. 4, fig. 1.* 1855.

Stout perennial with strigose to glabrate stems 6–10 dm long; internodes rather short, stem thus often hidden by overlapping ocreae; ocreae tubular or at length split, 2–4 cm long, strigose and stoutish, with scattered, straight, ascending hairs on and between veins, truncate at apex, bearing bristle-like fimbriae 1–1.5 cm long at tip of each vein; petioles 1–3 cm long, sparsely strigose to glabrate; leaf blades lanceolate to oblong-lanceolate, 1–4 cm wide, 6–30 cm long, acuminate at apex, strigose on both surfaces or glabrate; inflorescence simple or paniculate, the racemes dense, 4–10 cm long; ocreolae oblique, more or less imbricate, about 3 mm long, bristly-fringed along upper margin, otherwise glabrous or sometimes sparingly strigose; pedicels slender, 3–4 mm long; calyx white, 3–3.2 mm long, (4)5-parted nearly to base, glabrous, not punctate-glandular, the lobes ovate or oblong, obtuse or rounded; stamens 5–9, usually 6; style branches 2, exserted; achenes lenticular, ovoid to oblong, 2–2.5 mm long, black, smooth and shining.

In marshy places, on wet soil, and at margins of pools and streams, West Indies, and Central and South America. Specimen examined: San Cristóbal.

Some of the reports of this species from the Galápagos Islands may be based on specimens of *P. galapagense*.

Polygonum galapagense Caruel, Rendic. Acad. Lincei 5: 624. 1889

Upright or decumbent, to 12 dm tall; stems 4–6 mm thick when dried, internodes 1–4.5 cm long, smooth and glabrous, drying nearly black; ocreae membranous, 15–25 mm long, the older usually ruptured, bearing slender, golden yellow to pale brown bristles 5–12 mm long as extensions from the ribs, each rib bearing a few bristles similar to but shorter than terminal ones along their outer surfaces; petioles somewhat stout, 5–10 mm long, sparsely bristly-strigose to glabrate; leaf blades lance-elliptic, acute at base, acute to moderately attenuate at apex, 8–14 cm long, 2–4.5 cm wide, beset with bristle-like, golden hairs on both surfaces, the pubescence denser and of stouter hairs along veins and margins than on lamina between veins, lower surface paler and more densely hirsute than upper; inflorescence racemose or less commonly paniculate-racemose from axils of upper leaves, simple or geminate, the flowering part 4–6 cm long, 2–4 mm across; peduncle slender, glabrous, 1–3 cm long; flowers appressed to rachis, on slender, glabrous pedicels 2–3 mm long; perianth-segments rounded at apex; ocreolae 3–4 mm long, obtusely truncate to obtuse, without bristles or hairs; achenes lenticular, about 3 mm long, 3 mm wide, 1.5 mm thick, dark brown to nearly black, smooth and shining, distinctly short-apiculate from base of the persistent style.

Endemic; in damp areas and around pools above the forest and shrubby zones. Specimens examined: Isabela, San Cristóbal, Santa Cruz.

Polygonum hydropiperoides Michx. var. **persicarioides** (HBK.) Stanford, Rhodora 28: 27. 1926

Polygonum persicarioides HBK., Nov. Gen. & Sp. Pl. 2: 143. 1817.
Persicaria persicarioides Small, Fl. SE. U. S. 378. 1903.

Decumbent to erect perennial 3–10 dm tall, with glabrous to strigillose simple or branched stems; ocreae cylindrical to narrowly funnelform, 1–2 cm long, glabrous to strigose, truncate or slightly oblique, bristly-ciliate along upper margin; petioles 1–2 cm long, or leaves sometimes subsessile; leaf blades narrowly lanceolate to oblong-lanceolate, acuminate or acute at both ends, 6–14 cm long, 5–15(20) mm wide, finely strigillose, inconspicuously and minutely punctate-glandular or glands obscure, more densely and coarsely strigose on margins and midribs than on surface of lamina between veins; inflorescences usually paniculate, the racemes pedunculate, narrow, loosely flowered, erect, 2–7 cm long; ocreolae funnelform, 2.5–3 mm long, bristles at apex stoutish, shorter than blades; pedicels 1–3 mm long; calyx 2–2.5 mm long, rose tinged with green, lobes oblong to narrowly ovate, obtuse; stamens usually 8; style branches 2 or 3; achenes lenticular or triquetrous in same inflorescence, 2.5–3 mm long, wholly enclosed within calyx at maturity, black, finely granular, shiny.

In shallow water of ponds or on wet soil, central United States south through Central America into South America. Specimen examined: Isabela.

Polygonum opelousanum Riddel ex Small, Bull. Torrey Bot. Club 19: 354. 1892

Polygonum hydropiperoides opelousanum W. Stone, Pl. So. New Jersey 422. 1912.
Persicaria opelousana Small, Fl. SE. U. S. 378. 1903.

Erect or ascending perennial herb from a decumbent or slightly creeping base, the stems 3–10 dm tall, slender, usually branching freely from base; internodes 3–6 cm long, glabrous; ocreae 10–15 mm long, scarious-membranous, sparsely strigose with stout, upwardly pointing, yellowish hairs, bristly-fringed with similar but longer hairs (to 5 mm long); petioles 4–5 mm long; leaf blades narrowly lanceolate, 3–12 cm long, 5–15 mm wide, glabrous or with appressed, short bristles along veins and margins, the lower surface paler than upper, conspicuously and irregularly dotted with grayish circular, resinous glands 0.2–0.4 mm in diameter; inflorescence simple, pedunculate racemes, or geminate, the slender peduncles glabrous, ascending, racemes ascending or slightly drooping, 1.5–4 cm long, fascicles few-flowered, rather closely arranged; ocreolae 1.5–2 mm long, glabrous, fringed with bristles 1–2 mm long along upper margin; pedicels 1–2.5 mm long; calyx 1.5–2 mm long, lobes pinkish, obtuse or rounded, not noticeably glandular; stamens 7–8; style 3-branched nearly to base; achenes triquetrous, 1.5–2 mm long, 1.2–2 mm wide, rounded at base, apiculate at apex, concave on sides, the tip usually slightly exserted from calyx at maturity, dark reddish brown, smooth and shining.

Around margins of streams and pools, and in wet, seepy soil, southeastern United States to Mexico and northern South America. Specimens examined: San Cristóbal, Santa Cruz.

See Fig. 50.

Fig. 50. *Polygonum opelousanum*.

Polygonum punctatum Ell., Bot. S. C. & Ga. 1: 455. 1817

Polygonum acre HBK., Nov. Gen. & Sp. Pl. 2: 143. 1817, non Lam. 1778.
Persicaria punctatum Small, Fl. SE. U. S. 379. 1903.

Nearly glabrous perennial (may flower first year), with simple to much-branched, slender, erect to ascending stems 3–15 dm tall, the base often rooting at nodes; ocreae cylindrical on younger parts, often ruptured on older stems, 1–1.5 cm long, truncate, glabrous or sparingly strigillose along the prominent veins, each rib prolonged into slender bristle 3–10 mm long; petioles 5–10 mm long, glabrous; leaf blades narrowly lanceolate to oblong-lanceolate, 3–15 cm long, 1–2(3) cm wide, moderately acuminate at both ends, glabrous over most of lamina but margins and midrib usually bearing stout, short, forward-pointing, subappressed, stiff hairs; inflorescence usually paniculate, short-pedunculate, racemes 3–8 cm long; ocreolae funnelform, 2–3(4) mm long, sparingly bristly-ciliate along the slightly flaring margin; pedicels slender, 3–4 mm long; calyx 2–2.5 mm long, greenish or lobes sometimes faintly roseate, conspicuously punctate-glandular, at least on basal part, often upward to lower part of ovate to oblong lobes; stamens 7–8; style branches 3, recurved and persistent in fruit; achenes triquetrous (very rarely lenticular), 2.2–2.5 mm long, black or very dark red-brown, smooth and shining.

In shallow water or on wet soil around margins of pools, along sluggish streams, and in seepy ground, British Columbia to the Atlantic Coast of North America, south through Mexico and Central America into South America. Specimens examined: San Cristóbal.

ULMACEAE. ELM FAMILY*

Trees or shrubs with watery juice and terete to slightly angled branchlets arising from lateral, scaly buds. Leaves alternate or rarely opposite, simple, serrate or entire, petiolate, the blade asymmetrical at base, deciduous. Stipules papery, caducous. Flowers perfect, polygamodioecious, or monoecious, in clusters or the pistillate solitary. Calyx 4–9-lobed or -parted. Petals none. Stamens 4–6, erect or with anthers turned downward toward base of (rudimentary) pistil in bud. Style sessile. Stigmas 2, more or less feathery. Ovary 1-celled, 1-ovuled. Fruit a papery samara, nut, or small drupe. Seed flat to angular-subglobose. Endosperm scanty or lacking.

About 15 genera and over 150 species, in the tropics, subtropics, and temperate zones of the world.

Trema Lour., Fl. Cochinch. 562. 1793

Trees or tall, unarmed shrubs with leaves often distichous, short-petiolate, and flowers in subsessile, axillary cymes. Flowers monoecious or polygamodioecious; staminate flowers with 4- or 5-parted calyces, the segments induplicate-valvate or slightly imbricate in bud; stamens 5 or rarely 4, their filaments short, erect or spreading at base, sharply curved upward and erect at tips, rudimentary ovary present or absent in flowers of same tree. Pistillate flowers with nearly flat segments, these somewhat imbricate or when abortive stamens present the lobes somewhat concave-induplicate. Ovary sessile; style central, divided to base, bearing moder-

* Contributed by Ira L. Wiggins.

ately fleshy, linear stigmatic branches; ovule 1, attached near apex of ovary, pendulous. Fruit a small drupe, ovoid to subglobose, the flesh juicy. Seed ovoid to globose, with thin testa. Embryo curved; endosperm scanty.

About 50 species, distributed in the moister tropics and subtropics of the Old and New Worlds.

Trema micrantha (L.) Blume, Ann. Mus. Bot. Lugd. Bat. 2: 58. 1853

Rhamnus micranthus L., Syst. Nat. ed. 10, 2: 937. 1759.
Celtis canescens HBK., Nov. Gen. & Sp. Pl. 2: 28. 1817.
Celtis schiedeana Schlecht., Linnaea 7: 140. 1832.
Sponia canescens Decne., Nouv. Ann. Mus. Paris 3: 498. 1834.
Sponia micrantha Decne., loc. cit.
Sponia schiedeana Planch., Ann. Sc. Nat. ser. iii, 10: 335. 1848.
Sponia crassifolia Liebm., Vidensk. Selsk. Skr. V, 2: 340. 1851.
Sponia grisea Liebm., loc. cit.

Large shrub or small tree to 10 m tall, the branches slender, ascending and outer ones more or less arcuate, young branches and twigs harshly spreading-pubescent; bark brown, narrowly fissured; leaves distributed along the terminal 3–5 dm of slender, gracefully arcuate twigs; petioles 5–12 mm long, densely spreading-pubescent; blades lanceolate to lance-ovate, to 15 cm long, 3–5 cm wide, asymmetrically rounded at base, attenuate at apex, finely serrate-dentate, scaberulous on both surfaces, triplinerved from base and pinnately veined throughout rest of blade; flowers borne in dense axillary cymes, the fruits persistent after leaves fall; flowers monoecious, the staminate less numerous than the pistillate; pedicels 1–2 mm long, scaberulous; staminate sepals usually 5, broadly ovate, 1.2–1.4 mm long, acute, pubescent along midvein, somewhat cucullate, pale greenish cream; stamens inserted beneath deeply fimbriate disk, filaments curved upward, anthers horizontally inflexed, rudimentary ovary flattened, about equaling stamens, lacking stigmas; pistillate flowers about 3 mm long, nearly as wide, sepals similar to those of staminate flowers, stigma branches broadly subulate, reddish, feathery, 1–1.4 mm long, spreading; drupes broadly ovoid to subglobose, 3–3.5 mm long, the flesh translucent cream to light red, juicy; seeds 2.5–2.8 mm long, coarsely and irregularly ridged, slightly compressed, pale brown.

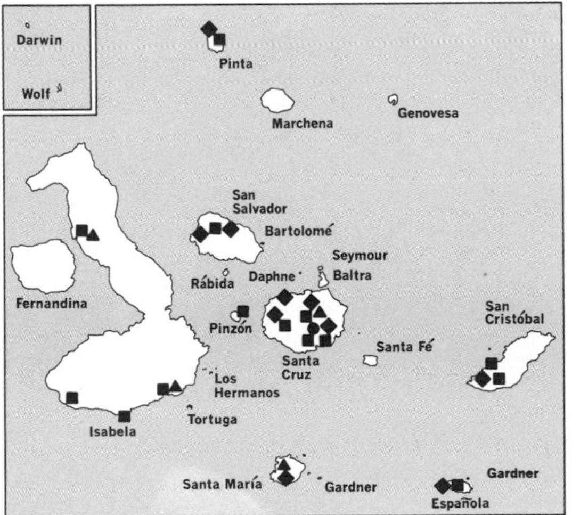

● *Trema micrantha*
■ *Fleurya aestuans*
▲ *Parietaria debilis*
◆ *Pilea baurii*

In forests and along roads, margins of fields, and waterways, from Florida, the West Indies, and from Sinaloa and Veracruz in Mexico through Central America into tropical and subtropical South America. Specimens examined: Santa Cruz, a new record for the Galápagos Islands.

A score or more colloquial names are applied to this species in various parts of its range. Only limited local use is made of the strong fibers that abound in its bark. The wood is soft and pale brown in color; it too is utilized infrequently. Some of the trees examined are fully 10 m tall, with trunks 15–20 cm in diameter. They were in full flower in early February in 1964. *Trema*, which is reported to be a common constituent of second-growth thickets in Guatemala, has not been reported from the Galápagos Islands previously. Since both the collections made are from along trails or roadsides, it is probable that it is a relatively recent introduction. It certainly is reproducing spontaneously at present.

See Fig. 51 and color plate 45.

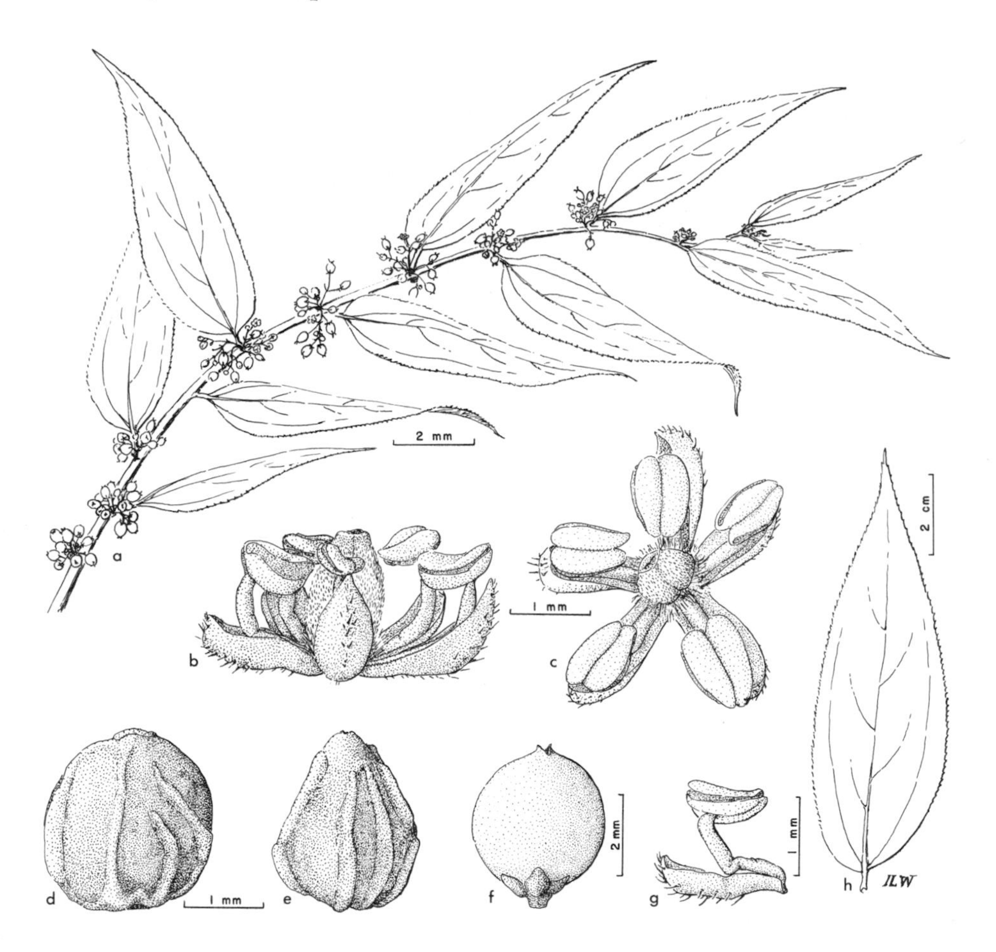

Fig. 51. *Trema micrantha*: *b, c*, flower; *d, e*, seeds with pulp removed; *f*, fruit; *g*, tepal and stamen.

URTICACEAE. NETTLE FAMILY*

Herbs, shrubs, or less commonly small trees, with alternate or opposite, stipulate leaves and frequently with stinging hairs, often with linear or elliptic cystoliths on 1 or both surfaces of leaves. Flowers unisexual or bisexual, monoecious or dioecious, axillary, cymose or paniculate, or in sessile or subsessile glomerules. Perianth simple, calycine, the segments valvate or imbricate, or in pistillate flowers sometimes flasklike or tubular with a small, toothed ostiole, or rarely totally lacking. Segments of staminate flowers generally 4, of the pistillate 2–4(5). Stamens as many as perianth segments and opposite them, the filaments sharply inflexed in bud, with reversed anthers, straightening explosively in anthesis, anthers then erect. Pistillate flowers with a superior, 1-celled ovary, 1-carpeled. Style undivided or none. Stigma penicillate-capitate or elongate. Ovule 1, attached at or near base, oblique, ascending, or erect, orthotropous. Fruit an achene, sometimes enclosed in perianth. Seeds erect, with or without endosperm.

About 42 genera and nearly 600 species, widely distributed in tropical, subtropical, and temperate regions.

Leaves opposite .... *Pilea*
Leaves alternate:
  Stinging hairs none; leaves without stipules .... *Parietaria*
  Stinging hairs present on stems or leaves or both; stipules present, intrapetiolar:
    Stigma on short style, becoming reflexed and hooked in fruit; plants annual herbs; fruit a dry achene, the perianth segments neither fleshy nor colored ..... *Fleurya*
    Stigma penicillate-capitate; shrubby or small trees; fruit (in ours) encased in fleshy orange or red perianth segments .... *Urera*

Fleurya Gaud. in Freyc., Bot. Voy. 497. 1826

Annual herbs with somewhat succulent, watery stems and alternate, petiolate, 3-nerved, dentate leaves beset with stinging hairs and linear cystoliths. Stipules connate. Flowers monoecious or dioecious, the glomerules unisexual or androgynous, paniculate or spicate in axils of leaves, usually shorter than leaves but sometimes exceeding them. Staminate flowers with 4–5 perianth segments, globose to slightly depressed in bud; stamens same in number as perianth lobes and opposite them. Pistillate flowers of 4 subequal to unequal, imbricate, more or less cucullate segments. Ovary straight at first, soon oblique, bearing linear to oblique-ovate stigma, this finally inflexed and forming small hook in fruit. Achene compressed, oblique, exserted from perianth.

About 8 species in the tropics of the world, but only 1 native in the Americas.

Fleurya aestuans (L.) Gaud. in Freyc., Bot. Voy. 497. 1826

Urtica aestuans L., Sp. Pl. ed. 2, 1397. 1763.
Fleurya glandulosa Wedd., Ann. Sci. Nat. III, 18: 205. 1852.
Fleurya aestuans var. *glandulosa* Wedd., Arch. Mus. Paris 9: 112. 1856–57.
Fleurya aestuans var. *racemosa* Wedd. in DC., Prodr. 16^1: 72. 1869.

Erect or somewhat decumbent herb to about 1 m tall, stems translucent-succulent, from densely hirsute to nearly glabrous, usually armed with few to many stinging

* Contributed by Ira L. Wiggins.

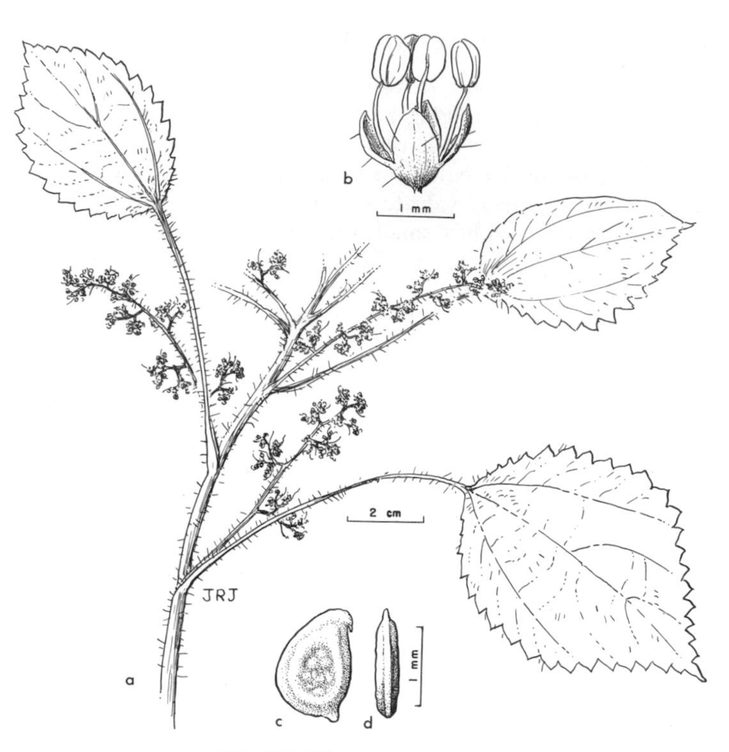

Fig. 52. *Fleurya aestuans.*

hairs and often very glandular; leaves long-petiolate, petiole 1–10 cm long, blades mostly 6–8 cm long (sometimes to 15 cm long, 12 cm wide), broadly ovate, acute or short-acuminate, obtuse, rounded or subcordate at base, coarsely dentate or crenate, thin, dark green above, paler beneath, sparsely hispid and with elliptic cystoliths on both surfaces, these more numerous above than beneath; inflorescences usually androgynous, paniculate, shorter than, equaling, or surpassing petioles; staminate flowers with 5 ovate perianth segments 3–4 mm long, each segment bearing straight, erect, rigid stinging hair about 2 mm long dorsally about halfway between base and tip; stamens 5, inflexed in bud, straight, erect or spreading after anthesis, anthers 1–1.3 mm long, filaments slender, about equaling perianth segments; pistillate flowers on slender pedicels 1–3 mm long, with 4 perianth segments, the adaxial and outer ones much smaller (0.2–0.3 mm long) than the lateral pair (0.6–0.9 mm long), adaxial one bearing stinging hair dorsally, other 3 unarmed; achene about 1.5–2 mm long at maturity, sharply deflexed-geniculate on pedicel, stinging hair on adaxial segment almost erect above base of achene; persistent stigma forming hook 0.2–0.3 mm long at apex of achene.

Tropical South America, the West Indies, and Africa; in crevices and potholes in the lava, along trails, under trees, and at margins of pools and drainage ways from near sea level to 615 m or more in the Galápagos Islands, usually in the shade but sometimes on nearly bare rock in virtually full sun. Known from all the larger islands and some of the smaller ones in the Galápagos archipelago. Specimens examined: Española, Isabela, Pinta, Pinzón, San Cristóbal, San Salvador, Santa Cruz.

The specimens from Isla Española (*Stewart 1329* [CAS]) were young seedlings 5–12 cm tall when collected, and still have the basal pair of leaves attached. These leaves are different from the later ones, being long-petioled and with blades only 3.5–5 mm wide, orbicular and emarginate, devoid of teeth, and very thin. Yet they have small glomerules of normal flowers in their axils, indicative of the adaptation to conditions marginal for growth, which permits them to produce at least a few flowers and viable fruit very early in the normal life span. Later, if growing conditions remain favorable, these same plants might become a meter high and produce thousands of seeds. Hendrickson noted on the sheet he collected at El Chato, Isla Santa Cruz (*H-17* [DS]) that this is 1 of 2 nettles growing around the small pools in the vicinity and avidly eaten by the Galápagos tortoises. He did not determine whether they preferred this or *Pilea baurii*, the other nettle.

See Fig. 52 and color plate 67.

Parietaria L., Sp. Pl. 1052. 1753

Annual or perennial herbs, often with rather weak, diffuse stems bearing simple, usually uncinate, prehensile, non-stinging hairs. Leaves alternate, petiolate, small, entire, usually 3-nerved, without stipules. Flowers polygamous, sessile, in dense axillary cymules or fascicles, subtended by bracts, the outer of these more or less connate in some species to form involucres containing 3 to several flowers each, in others the bracts free; perianth deeply 4-lobed, lobes valvate. Stamens 4. Ovary free within perianth, lenticular. Stigma linear, sessile or short-styled, penicillate and recurved. Ovule 1, erect from base of ovary chamber. Achene included in herbaceous perianth; pericarp thin and papery.

About 8 species in tropical and warm temperate regions of the world.

Parietaria debilis Forst., Fl. Ins. Austr. Prodr. 73. 1786

Parietaria floridana Nutt., Gen. N. Am. Pl. 1: 208. 1818.
Parietaria debilis var. *ceratocantha* Wedd., Arch. Mus. Paris 9: 515. 1856–57.

Plant diffuse and spreading to erect, usually much branched but rarely more than 3 dm tall, sparsely to densely soft-pilose throughout; petioles slender, 3–20 mm long; leaf blades suborbiculate, broadly ovate, or lance-ovate, 0.5–3 cm long, rounded to acute at base, obtuse to long-acuminate at apex, thin, entire, 3-nerved, dark green above, slightly paler beneath, densely white-puncticulate above, less conspicuously so beneath; flowers in crowded axillary cymules, bracts linear to narrowly lanceolate, 1–1.5 mm long, free; perianth segments of perfect flowers lanceolate, 1.5–2 mm long, margins irregularly ciliate-hispid; stamens slightly longer than perianth lobes at time of dehiscence of anthers; achenes terete, elliptic, pale brown and shining, about 1 mm long.

In shade under overhanging rocks, on cliffs, under shrubs and trees, and sometimes in cultivated areas as a minor weed; widely distributed in tropical and warm temperate areas of the world. Specimens examined: Isabela, Santa Cruz, Santa María.

See Fig. 53.

Fig. 53. *Parietaria debilis.*

Pilea Lindl., Coll. Bot. *pl. 4.* 1821

Annual or perennial herbs (rarely shrubby at base), diffuse and creeping to erect, slightly to moderately branched. Leaves opposite (members of a pair sometimes unequal), entire or toothed, strongly 3-nerved from base or almost nerveless; stipules connate into 1, intrapetiolar. Flowers monoecious or dioecious, in axillary cymes. Cymes solitary in axils, simple or laxly paniculate, sometimes densely capituliform, sessile or pedunculate. Bracts small and inconspicuous. Staminate flowers usually with 4-parted perianth, the segments green, regular, sometimes connate near base, subvalvate; ovary in staminate flowers rudimentary. Pistillate flowers with 2- or 3-divided perianth, 1 segment usually larger than the other 1 or 2, cucullate and partially enclosing achene; staminodes opposite segments, scale-like or resembling perianth segments, usually abruptly and tightly inflexed prior to maturity, then erect or spreading. Stigma sessile, minutely penicillate. Achene ovoid or roundish, lenticular-compressed, smooth or slightly roughened with microscopic papillae. Seed similar to achene in shape, with broad cotyledons and little or no endosperm.

A tropical and subtropical genus of about 200 species, most of them native in South America; not known in Australia.

Leaves coarsely serrate-dentate, 2.5–6 cm long; pistillate flowers often with only 2 perianth segments; staminodes similar to perianth segments, surpassing the smaller segment, sometimes nearly equaling the larger . . . . *P. baurii*
Leaves entire or nearly so, less than 15 mm long; pistillate flowers nearly or always with 3 perianth segments; staminodes scale-like:
Leaves obovate to oblanceolate, crowded; outer perianth segment of pistillate flower only slightly longer than lateral ones, shorter than achene; cystoliths on upper surface of leaf blades closely crowded . . . . *P. microphylla*
Leaves ovate-orbicular, 3–6 mm long, not markedly crowded; outer perianth segments of pistillate flowers much longer than lateral ones, partly enclosing achene; cystoliths less crowded . . . . *P. peploides*

Pilea baurii Robins., Proc. Amer. Acad. 24: 133. 1902

Annual or possibly short-lived perennial herb 1–6 dm tall, stems subsucculent, glabrous, 2–6 mm thick, translucently pale green, to deeply tinged with red; internodes 3–10 cm long; leaves bright green, thin, petioles about equaling or slightly

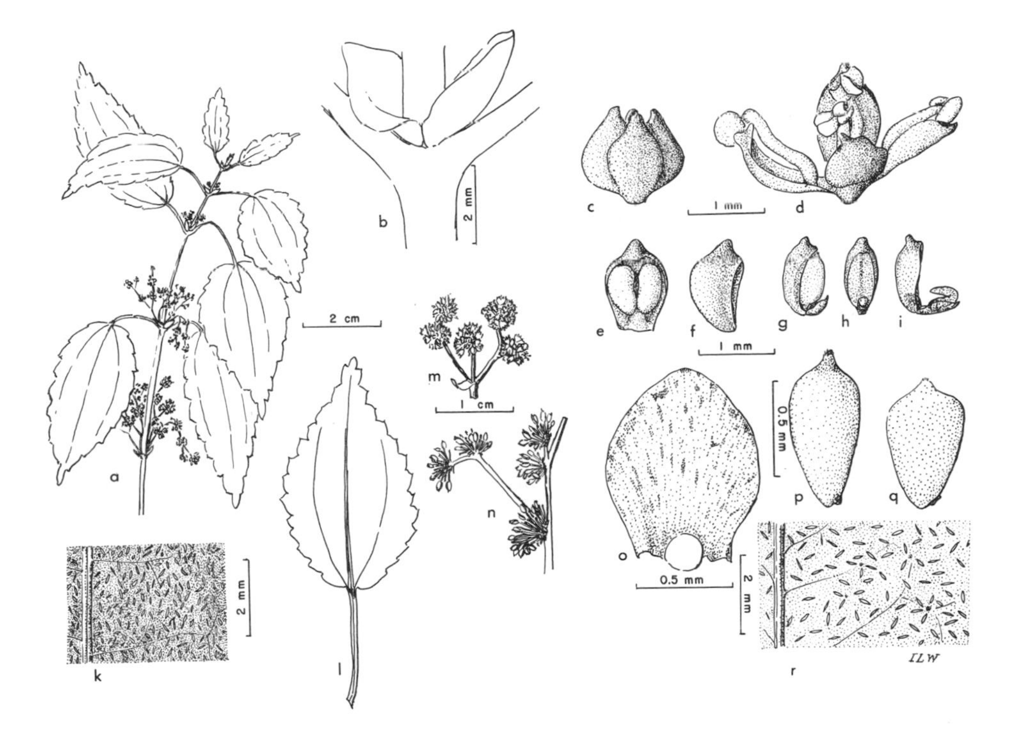

Fig. 54. *Pilea bauri*: *a*, tip of plant; *b*, node and bases of leaves; *c*, *d*, staminate flower before and immediately after anthesis; *e*, tepal and stamen before anthesis; *f*, tepal, side view; *g–i*, pistillate flowers; *k*, cystoliths, upper surface of leaf; *m*, *n*, staminate and pistillate inflorescences; *o*, stipule; *p*, *q*, utricles; *r*, cystoliths, undersurface of leaf.

longer than blades, these ovate, 2–10 cm long, 1.5–5 cm wide, rounded at base, rather bluntly serrate, usually somewhat attenuate at apex, bearing linear to very narrowly elliptic cystoliths on both surfaces, these more closely arranged above than below; stipules connate, broadly ovate, 1–3 mm long, nearly hyaline, dotted with flecks of red or brown; inflorescence short and compact to several times branched and moderately elongate, then nearly as long as subtending petiole, usually 2 panicles in axil of each leaf; bracts minute; staminate flowers usually near base of inflorescence or its branches, often occurring at base of cluster of pistillate flowers, their perianth segments 4, connate at extreme base, glabrous, 1–1.2 mm long, each segment bearing a minute, blunt, erect apiculation near tip; stamens 4, filaments sharply inflexed in bud, ovary in staminate flowers rudimentary or practically lacking; pistillate flowers crowded, on pedicels 1–3 mm long, articulate thereto, segments 2 (very rarely 3), outer one 1–1.4 mm long, cucullate, partly enclosing achene, apiculate at tip, brownish, the second segments usually minute, about 0.4 mm long, a staminode similar in texture to perianth segments in axil of each segment, inflexed in bud, straightening after anthesis, then about three-fourths as long as larger segment; ovary elliptic-lenticular; stigma sessile-capitate, microscopically penicillate; achene smooth, elliptic-obovate, compressed, 0.3–0.4 mm long at maturity, yellowish to pale brown, microscopically reticulate; cotyledons flat; endosperm little or none.

Endemic; in the shade of trees at lower elevations, in open grassland 300 to 840 m above sea level. Specimens examined: Española, Pinta, San Cristóbal, San Salvador, Santa Cruz, Santa María.

Locally called "ortega," according to an annotation of *J. T. Howell 9160* (CAS). Hendrickson noted that the Galápagos tortoise near El Chato, on Isla Santa Cruz, ate this plant and *Fleurya aestuans*. Both nettles grow in profusion in partial shade around the margins of shallow pools frequented by the tortoises.

See Fig. 54.

Pilea microphylla (L.) Liebm., Vidensk. Selsk. Skr. V, 2: 296. 1851

Parietaria microphylla L., Sp. Pl. ed. 2, 1492. 1763.
Pilea muscosa Lindl., Coll. Bot. *pl. 4.* 1821.
Pilea succulenta Hook. f., Trans. Linn. Soc. Lond. 20: 182. 1847.

Annual to short-lived perennial herb sometimes becoming woody at base; stems 0.5–5 dm tall, when woody grayish, glabrous, smooth; herbaceous stems succulent, 1–2 mm thick; leaves 1-nerved, entire, orbicular, obovate, elliptic, or oblong, 2–15 mm long, 1–5 mm wide, on short petioles 1–5 mm long or the smaller subsessile, those of a pair often very unequal in size, on other plants or on different branches both leaves of a pair equal, upper surface bearing crowded, elliptic cystoliths, lower surface finely reticulate-meshed; stipules narrowly ovate, brownish, about 1 mm long; flowers monoecious or dioecious; inflorescences of sessile glomerules or cymes on slender peduncles 1–6 mm long; staminate flowers 0.6–1 mm long, on pedicels about as long, perianth segments 4, broad, glabrous, bearing a short, blunt, dorsal apiculation near tip; pistillate flowers subsessile, 0.8–1 mm long, with 3 perianth segments, the outer one slightly longer than laterals, roseate to brownish, obtuse or rounded at apex, slightly cucullate, the dorsal apiculation very short, broad; staminodes nearly as long as perianth segments, nearly flat, narrower than segments, thin and nearly hyaline; achene elliptic, compressed, about 1 mm long, slightly longer than outer perianth segment.

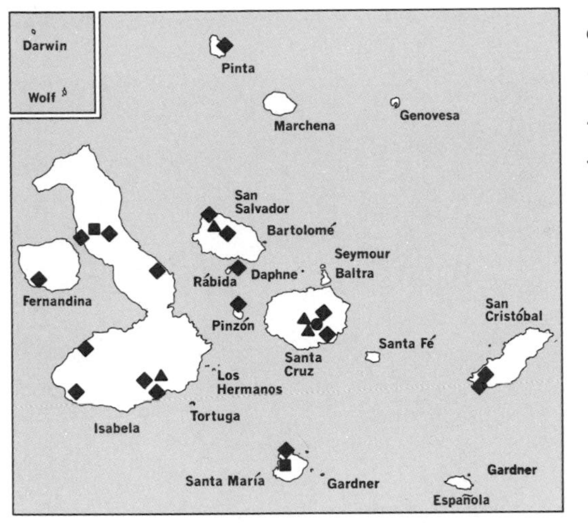

● *Pilea microphylla*
■ *Pilea peploides*
▲ *Urera caracasana*
◆ *Phoradendron henslovii*

On rocks, at bases of trees, under shrubs, and along trails and roadsides, the West Indies and the southern United States to Central America and tropical South America. Specimens examined: Santa Cruz.

Collected in the Galápagos Islands by Darwin in 1835, and apparently not taken again in the archipelago until 1964.

Pilea peploides (Gaud.) Hook. & Arn., Bot. Beechey 96. 1832
Dubreulia peploides Gaud. in Freyc., Bot. Voy. 495. 1826.

Slender, rather weak glabrous herb 5–15 cm tall, the stems erect, decumbent, or somewhat procumbent; stipules ovate, cuneate, 1–1.5 mm long; leaves broadly ovate to suborbicular, the lower 1–3 pairs entire, acute at apex, obtuse to truncate at base, 1–6 mm long, 1.2–6 mm wide, rather conspicuously beset with linear to elliptic or narrowly oblong cystoliths on both surfaces; petioles 1.5–5 times as long as blades, later leaves crenate-dentate, suborbicular-ovate, 4–12 mm long, 3.5–10 mm wide, mostly obtuse at base, obtuse to rounded at apex, bright green above, markedly paler and reddish beneath, cystoliths very pale on lower surface; petioles to 15 mm long, blades 3-nerved, nerves more obvious beneath than above; inflorescences of compact axillary fascicles 2–3 mm broad; staminate flowers 0.4–0.5 mm long, 4-lobed, perianth segments oblong, cucullate, with minute dorsal apiculation just below apex; pistillate flowers more numerous than staminate, on pedicels 1–1.5(2) mm long; lobes 3, the larger 0.6–0.8 mm long, the laterals about two-thirds as long, reddish, all beset with cystoliths; utricle about 0.6–0.8 mm long at maturity, smooth, tawny or pale cream, lenticular-ovate, stigma brownish, persistent.

In crevices and on moist rocks in forest, usually in dense shade, northern South America and the Hawaiian Islands. Specimens examined: Isabela, Santa María.

Urera Gaud. in Freyc., Bot. Voy. 496. 1826

Shrubs or small trees, usually with stinging hairs on younger stems, inflorescences, and leaves. Leaves alternate, entire, dentate or lobed, 3–5-nerved or pinnately nerved; stipules free or more or less connate. Flowers unisexual, dioecious or monoe-

Fig. 55. *Urera caracasana.*

cious, in dichotomously much-branched panicles or glomerules; bracts small or lacking. Staminate flowers with 4–5 ovate, slightly imbricate perianth segments; stamens 4–5, inserted at bases of perianth segments. Pistillate flowers with 4 subequal perianth segments, or outer ones smaller than inner, segments becoming fleshy in fruit; ovary straight or somewhat oblique; stigma subsessile, penicillate-capitate, persistent. Achene straight or oblique, lenticular-compressed to subglobose, often ventricose, surrounded by the fleshy, often brightly colored, juicy perianth.

About 15 species in tropical America, Africa, and Asia.

Urera caracasana (Jacq.) Griseb., Fl. Brit. W. Indies 154. 1859*

Urtica caracasana Jacq., Pl. Hort. Schoenbr. 3: 71, *pl. 386.* 1798.
Urera jacquinii Wedd., Ann. Asi. Nat. III, 18: 200. 1852.
Urera subpeltata Miquel in Mart., Fl. Bras. 4[1]: 189, *pl. 66.* 1853.
Urera acuminata Miquel, op. cit. 190, non Gaud.

Coarse shrub or small tree to 10 m tall, with thick, pale, smooth branches, all younger parts, leaves, and inflorescences with slender, straight, stiff, stinging hairs to 5–6 mm long; leaves broadly ovate to nearly orbicular, mostly 15–20 cm long, 8–15(30) cm broad, short-acuminate to obtuse at apex, rounded to cordate at base, crenate-dentate, dark green and more or less bullate above, cystoliths inconspicuous, lower surface paler than upper, usually moderately to densely velutinous-pilose, the cystoliths obscured by hairs; stipules connate part or all of their length, 4–5 mm long, pilose; flowers usually dioecious, in compact lax cymes in axils of leaves or at nodes where leaves have fallen; staminate flowers 1.2–1.5 mm high, 2–2.5 mm wide, the 4 perianth segments obovate, hyaline along margins, without dorsal appendages, finely and moderately hispidulous, bearing several orbicular-punctate cystoliths near margins and toward tips, tinged with dark green or black; stamens 2–3 mm long, erect to spreading; anthers 0.8–1 mm long; pistillate flowers on pedicels 0.4–1 mm long, perianth segments subequal, the 2 adjacent to lateral faces of ovary interior to, slightly longer than, and nearly twice as wide as the 2 exterior segments, interior ones becoming fleshy-juicy and about 3 mm long in fruit, outer pair increasing only slightly in length and only faintly fleshy in fruit; staminodes lacking; stigma about 0.3 mm broad, finely penicillate; fruit bright orange, 3–4 mm wide, 2.5–3 mm high; achene broadly ovate-lenticular, 1.2–1.5 mm long, 0.8–1 mm wide, surpassed and hidden by fleshy perianth.

In shady forest habitats, southern Mexico to Brazil and Peru; in the Galápagos mainly at 200–500 m. Specimens examined: Isabela, San Salvador, Santa Cruz.

See Fig. 55 and color plate 80.

VISCACEAE. MISTLETOE FAMILY†

Plants shrubby or herbaceous, aerial parasites on other seed plants, pubescent or glabrous, monoecious or dioecious. Stems evergreen, usually forked, brittle and

* The size, dentation, and shape of the leaf blades are all variable; thus several species and varieties unworthy of recognition have been described. For added synonymy and notes of interest, see Macbride (1937, 362), and Standley and Steyermark (1952, 426–27).

† Contributed by Delbert Wiens. Heretofore the American mistletoes have been placed in the single family Loranthaceae, which was divided into two subfamilies, the Loranthoideae and Viscoideae. These subfamilies have recently been formally raised to family status by Barlow (Proc. Linnaean Soc. New South Wales 89: 268–72. 1964).

much branched with generally swollen and articulated nodes. Leaves mostly opposite (rarely alternate), simple, entire, evergreen or sometimes reduced to scales. Flowers minute (about 2 mm long or less), monochlamydeous, unisexual, solitary or clustered at nodes, or in axillary spikes or cymes. Perianth segments 2–4, valvate. Staminate flowers with stamens opposite, adnate or free, and equal in number to tepals, sometimes with vestigial style. Pollen spherical. Pistillate flowers with simple style and terminal stigma. Ovary inferior, uniloculate and without ovules, the embryo sac originating from short placental column. Fruit baccate, with endosperm, a viscous outer layer, and persistent tepals.

About 7 genera, found on all continents but with the greatest development in tropical and subtropical regions.

Phoradendron Nutt., Jour. Acad. Nat. Sci. Phila. 1[2]: 185. 1848

Shrubs parasitic on other woody plants; leaves (and sometimes stems) chlorophyllous, opposite, evergreen, either well-developed or reduced to connate scales; flowers unisexual, the plants dioecious or monoecious, about 2 mm across, borne in

Fig. 56. *Phoradendron henslovii.*

spikelike or thin-cylindrical, axillary inflorescences from a few mm to a few cm long, the flowers sunken in rachis; perianth segments (2)3(4)-merous, distinct, deltoid, scale-like, persistent, seldom erect and never spreading; staminate flower bearing sessile biloculate anther at base of each perianth segment; pistillate flower manifestly epigynous, with 1 style; ovary 1-chambered; fruit a small drupe (but without an obviously indurated endocarp), 1-seeded with a glutinous, mucilaginous mesocarp, usually whitish.

A New World genus of wide distribution, but primarily tropical, with about 150 species.

Phoradendron henslovii (Hook. f.) Robins., Proc. Amer. Acad. 38: 133. 1902

Viscum henslovii Hook. f., Trans. Linn. Soc. Lond. 20: 216. 1845.
Viscum galapageium Hook. f., loc. cit.
Viscum florianum Anderss., Kongl. Svensk. Vet.-Akad. Handl. 1853: 92. 1855.
Phoradendron galapageium Robins., Proc. Amer. Acad. 38: 133. 1902.
Phoradendron florianum Robins., loc. cit.
Phoradendron uncinatum Robins., op. cit. 134.

Bushes to about 1 m tall, usually forked (occasionally whorled), glabrous, monoecious; internodes at maturity about 5–15 cm long, each bearing 1–3 cataphylls between the foliar nodes; leaves highly variable, but mostly lance-ovate to elliptical, extremes linear-lanceolate to orbicular, about 2–20 cm long, 1.5–10 cm wide, often slightly recurved and falcate, 4–6-veined, basinervous; inflorescences with up to 7 fertile segments, individual segments with flowers distributed in 2 laterally occurring series, each with central apical flower subtended by 3 ranks of flowers, distribution of sexes seemingly random; anthesis apparently continuous; fruit orbicular, 4–6 mm in diameter, translucent; pericarp watery, without the typical viscous layer; seed flattened. $n = 28$.

Endemic, but with a related species in the West Indies. Locally abundant throughout most of the archipelago where suitable hosts occur; from sea level to elevations of at least 750 m. Known hosts include: *Cordia revoluta, C. lutea, Croton scouleri, Lippia rosmarinifolia, Macraea laricifolia, Mollugo snodgrassii, Psidium galapageium, Sarcostemma angustissima, Scalesia pedunculata,* and *Zanthoxylum fagara*. Specimens examined: Fernandina, Isabela, Pinta, Pinzón, Rábida, San Cristóbal, San Salvador, Santa Cruz, Santa María.

The characteristics delimiting all previously described Galápagos mistletoes appear to vary randomly within and between populations and hosts, suggesting a single variable species. However, the plants occurring in the moist, higher elevations (especially on Santa Cruz) possess larger parts, particularly leaves, internodes, and fruit. The available material suggests the variation is continuous and perhaps clinal, but additional studies might support separation of these populations at the subspecific level. The single specimen from Isla Pinzón (*Stewart 1109* [CAS]) has distinctively smaller leaves, but additional collections are needed to determine its status.

See Fig. 56.

Dicotyledones
Series II: Gamopetalae

Dicotyledones
Series II: Gamopetalae

Dicotyledonous plants of herbaceous, shrubby, arborescent, or vinelike habit. Inner whorl of perianth segments (corolla) connate to one another, at least at the base, to form a continuous ring, cup, or tube, the lobes equal and similar or different in size and shape, the corollas thus regular or irregular. Other characters as in the subclass (Dicotyledones, p. 180).*

[In one family of the Polypetalae, the Malvaceae, plants appear gamopetalous, but the petals are free at the very base and fused only to form a ring at or near the apex of the ovary. For convenience, Malvaceae has been included in the key below, in each case with a page reference to its position in the text.]

KEY TO THE FAMILIES OF THE GAMOPETALAE †

Ovary superior, free from calyx tube except at very base:
  Corollas regular (actinomorphic), all lobes essentially same shape and size:
    Ovary or ovaries 1-celled:
      Fruit a fleshy berry; styles 5, each fimbriate at apex; leaves palmately lobed..... .... CARICACEAE, p. 299
      Fruit a dry capsule or utricle; style 1, simple or 2-lobed, or if 2 the stigmas not fimbriate; leaves entire, not lobed:
        Fruit a capsule, longitudinally valvate, many-seeded; corolla funnelform; calyx divided nearly to base, not stipitate-glandular; ovary appearing 2-celled owing to intrusion of placentae.............HYDROPHYLLACEAE, p. 398
        Fruit a utricle or tardily circumscissile, 1-seeded; corolla salverform, the tube slender; calyx tubular, stipitate-glandular; ovary obviously 1-celled....... .... PLUMBAGINACEAE, p. 417
    Ovary 2- to several-celled:
      Fruits indehiscent, berry-like, or consisting of mericarps or nutlets:
        Stamens numerous (often 50 or more), coalesced into a tube surrounding the style but not adnate to it; fruit consisting of mericarps, or a capsule....... ....MALVACEAE (Series III), p. 666
        Stamens 4, 5, or 10, same number or twice as many as corolla lobes, inserted on a disk or on corolla tube or throat:

* Synopsis of series and key to families by Ira L. Wiggins.

† A collection very doubtfully ascribed by Caruel to the Bignoniaceae is discussed briefly on p. 509; the family, however, is not included in the key.

Anthers opening through apical tubes, these equaling or slightly exceeding thecae; flowers usually nodding on sharply reflexed tips of erect pedicels .... ERICACEAE, p. 395
Anthers dehiscing longitudinally or by apical pores, but without apical tubes; flowers rarely nodding at tips of reflexed pedicels:
Fruits bony mericarps, usually inserted in 2 or more whorls, 1 above the other, around central axis.................... NOLANACEAE, p. 414
Fruits fleshy, berries or berry-like:
Flowers arranged in scorpioid spikes or racemes; corollas often with swelling or appendages partially constricting throat.................. .... BORAGINACEAE, p. 277
Flowers not in scorpioid spikes or racemes; corolla throat not constricted:
Stamens all inserted at same level, usually low in corolla throat; corolla funnelform, rotate, or campanulate, rarely salverform; fruit berries, each locule several-seeded........... SOLANACEAE, p. 454
Stamens inserted in corolla tube at different levels; corolla usually salverform, tube short to long, obvious; fruit consisting of drupelets .... VERBENACEAE, p. 482
Fruits dehiscent, consisting of capsules or follicles:
Fruits follicular, dehiscing along 1 side only; carpels 2, often distinct except at apex in flower, completely so in fruit; juice often milky:
Corolla cylindrical-campanulate, tubular-salverform, or funnelform; anthers free, without translators; corona and column none.. APOCYNACEAE, p. 269
Corolla rotate or shallowly saucer-shaped; corona appendages usually surrounding anthers; stamens coherent into a column, with translators between adjacent anthers (pollinia)........... ASCLEPIADACEAE, p. 271
Fruits capsular, dehiscing along 2 or more sutures; carpels varying in number, but adherent to one another most or all of their length; juice not milky:
Corolla scarious; corolla lobes 4; leaves mostly or entirely basal............. .... PLANTAGINACEAE, p. 415
Corolla not scarious; corolla lobes mostly 5; leaves mostly cauline, rarely basal:
Stamen filaments coalescent into a tube surrounding style; anthers many, borne on upper part of column on short filaments; petals adnate to column near its base, but free at insertion on receptacle or disk......... .... MALVACEAE (Series III), p. 666
Stamens free throughout or adnate to corolla tube, not coalesced into a tube; anthers usually 5; corolla lobes not adnate to a column, not separate at base:
Leaves opposite; capsule 1-seeded by abortion, laterally compressed; stamens 4, didynamous................... AVICENNIACEAE, p. 275
Leaves alternate; capsule 2- to several-seeded, not laterally compressed:
Corolla puberulent on outer surface; ovary 1-celled but falsely 2-celled owing to intrusion of placentae; seeds small (0.3–0.5 mm long), pitted or alveolate................. HYDROPHYLLACEAE, p. 398
Corolla usually glabrous without; ovary 2–5-celled; seeds mostly 3 mm long or more (sometimes 1–1.5 mm in *Cuscuta*), smooth or essentially so:
Capsule usually chartaceous, circumscissile, longitudinally or irregularly dehiscent; corolla usually contorted in bud; sepals free nearly full length............... CONVOLVULACEAE, p. 367
Capsule firmer, never circumscissile; corolla plicate or valvate in bud; sepals coalescent into a cup, only tips free.. SOLANACEAE, p. 454
Corollas irregular (zygomorphic), the lobes differing more or less in shape and/or size:
Fruits of 2 or 4 nutlets or sometimes drupaceous; stems often 4-angled:

Style 2-cleft or 2-lobed; ovary 4-lobed....................... LABIATAE, p. 401
Style entire; ovary not 4-lobed......................... VERBENACEAE, p. 482
Fruits capsular; stems mostly terete:
Plants aquatic or of very wet habitats; leaves much dissected into filiform segments, bearing insectivorous bladder-like traps; ovary 1-celled; endosperm none.... .... LENTIBULARIACEAE, p. 412
Plants terrestrial, or if semi-aquatic not insectivorous; ovary 2-celled; endosperm present or lacking:
Seeds borne on stout papillae or on hooked projections from placenta; stigmas entire, cupulate, or faintly 2-lobed; capsule often stout-stipitate; endosperm lacking .... ACANTHACEAE (below)
Seeds borne on slender funiculi, not on indurated papillae or hooked processes; stigmas usually distinctly 2-lobed; capsule sessile; endosperm present..... .... SCROPHULARIACEAE, p. 441
Ovary inferior, adnate to calyx most or all of its length:
Fruit a berry, capsule, mericarp, pepo, or nut, not an achene, not bearing pappus, wings, or awns; endosperm usually present (absent in Cucurbitaceae):
Flowers actinomorphic:
Stipules absent; flowers unisexual; plant climbing or scandent vine, often with tendrils; stamens coalesced in a column, anthers adnate to enlarged part of column; endosperm none......................... CUCURBITACEAE, p. 383
Stipules present; flowers bisexual; plant not a vine nor vinelike, without tendrils; stamens free and distinct; endosperm present........... RUBIACEAE, p. 419
Flowers zygomorphic:
Anthers connate into a tube (or partially so); calyx lobes obvious, usually exceeding calyx tube; fruit a capsule, dehiscing by apical pores................. .... CAMPANULACEAE, p. 296
Anthers free; calyx truncate at apex, lobes obsolete or very short; fruit a succulent drupe, indehiscent............................ GOODENIACEAE, p. 396
Fruit an achene, usually bearing awns, bristles, paleae, or silky hairs; endosperm lacking:
Inflorescence usually an involucrate head composed of few to many flowers; corollas not saccate at base of tube; stamens 5; anthers usually connate into a ring around the style, their filaments normally distinct............... COMPOSITAE, p. 300
Inflorescence (in ours) a dichasial cyme; corolla (in ours) slightly spurred or saccate at base; stamens 2; anthers free.................... VALERIANACEAE, p. 481

ACANTHACEAE. ACANTHUS FAMILY*

Herbs, shrubs, rarely small trees, with simple leaves, these usually entire, opposite, or sometimes, as in *Elytraria,* alternate or subopposite. Cystoliths commonly found as minute short lines on upper surface of leaves, on upper portions of stem, on branches of inflorescence, and on calyx. Flowers irregular to nearly regular, perfect. Calyx persistent, with 5 or occasionally fewer segments. Corolla gamopetalous, the limb 5-lobed or (1)2-lipped. Stamens 4, didynamous, or only 2, staminodes often present in flowers with 2 stamens, anther sacs 2 or 1, longitudinally dehiscent. Ovary 2-celled, ovules 2–10 in each locule. Style filiform, simple, stigmas 1 or 2. Fruit a capsule, 2-celled, 2-valved. Seeds usually flat, borne on retinaculae, these papilliform in a few genera but usually hook-shaped; testa smooth or roughened, often mucilaginous when moistened.

* Contributed by Dieter Wasshausen.

About 250 genera and over 3,000 species, widely distributed in tropical and subtropical regions.

Funicle papilliform or lacking . . . . *Elytraria*
Funicle hook-shaped (a retinaculum):
  Corollas contorted at aestivation; stamens 4:
    Flowers borne in dense terminal 4-sided spikes; bracts conspicuous, closely imbricate . . . . *Blechum*
    Flowers borne in axillary dichotomously branched panicles; bracts minute, inconspicuous . . . . *Ruellia*
  Corollas imbricate at aestivation; stamens 2:
    Stems 6-angled; flowers borne in axillary verticillasters; placentae separating from mature capsule valves . . . . *Dicliptera*
    Stems subterete or subquadrangular; flowers borne in terminal and axillary spikes; placentae remaining attached to capsule valves:
      Flowers borne in axillary dichotomously branched spikes; anther lobes superposed, obliquely affixed, lower lobe apiculate or calcarate . . . . *Justicia*
      Flowers borne in terminal or axillary, closely imbricate 4-angled spikes; anther lobes slightly, if at all, superposed, parallel, muticous . . . . *Tetramerium*

Blechum P. Br., Nat. Hist. Jam. 261. 1756

Perennial herbs; leaves petiolate, repand-crenate or entire. Flowers borne in dense, terminal spikes, the bracts imbricate. Calyx 5-parted, segments slightly unequal, linear-subulate. Corolla white or purplish, tube slender, limb nearly equally 5-lobed. Stamens 4, didynamous, anther sacs parallel. Ovules few in each cavity. Capsule broadly oblong, with short, narrowed base.

About 6 species, all native to tropical America.

Blechum brownei Juss. forma **puberulum** Leonard, Jour. Wash. Acad. Sci. 32: 184. 1942

Herb; stems erect or ascending, 20–70 cm high, branches slender, more or less puberulous; leaves ovate, 2–7 cm long, 1–5 cm wide, acute or obtuse at apex, nar-

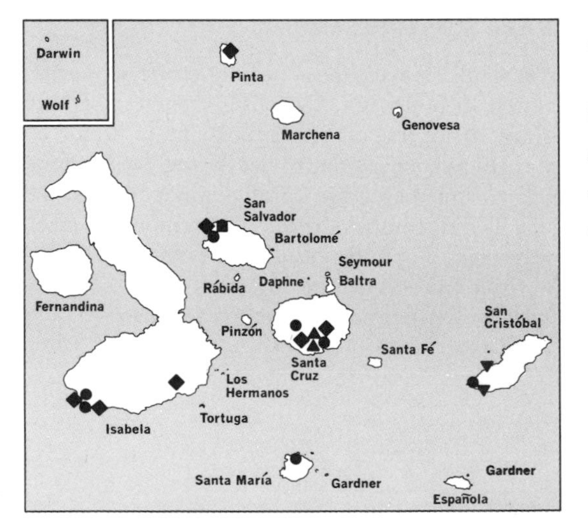

● *Blechum brownei* forma *puberulum*
■ *Dicliptera peruviana*
▲ *Elytraria imbricata*
◆ *Justicia galapagana*
▼ *Ruellia floribunda*

Fig. 57. *Blechum brownei.*

rowed at base, glabrous to sparingly pilose, thin, cystoliths rather conspicuous; spikes dense, 4-sided, 3–6 cm long; bracts ovate, pinnately nerved, 1–2.5 cm long, about 1 cm wide, subacute at apex, rounded at base, loosely strigose and ciliate, more or less softly and densely puberulous in addition to coarser hairs, softer hairs minute, curved, white; corolla white or purplish, 12–15 mm long, slightly exceeding subtending bracts, sparingly pubescent; capsule broadly oblong, about 6 mm long, puberulous, many-seeded; seeds flat, brown, about 1.5 mm in diameter, apparently glabrous when dry, if moistened the edge bearing a dense gelatinous band of minute, straight, white hairs.

Common in flat, open sunny places, from eastern and southern Mexico to northern South America. Specimens examined: Isabela, San Cristóbal, San Salvador, Santa Cruz, Santa María.

See Fig. 57.

Dicliptera Juss., Ann. Mus. Paris 9: 267. 1807, nom. conserv.

Herbs or shrubs, stems more or less hexagonal in cross section. Leaf blades entire or undulate, mostly ovate, petiolate. Flowers 1 to several, borne in often contracted cymes, these forming spikes of panicles subtended by an involucre of 2–4 pairs of conspicuous bracts. Calyx 5-parted, hyaline. Corolla narrow, slightly ampliate, limb 2-lipped. Stamens 2, anther sacs often unequal, longer one sometimes calcarate at base; staminodes none. Capsules ovate or suborbicular; placentae separating elastically from their walls and rupturing on dehiscence. Seeds 2 or 4.

About 300 species in the tropical and temperate regions of the world.

Dicliptera peruviana (Lam.) Juss., Ann. Mus. Paris 9: 268. 1807

Justicia peruviana Lam., Dict. 1: 633. 1783.
Dianthera mucronata R. & P., Fl. Peruv. 1: 11, *t. 16, fig. a.* 1798.

Herb; stems branching, erect or ascending, to 60 cm tall or more, subhexagonal, striate, hirtellous, the hairs retrorsely curved; leaf blades ovate, to 6 cm long, 2.5 cm wide, acute to mucronate, abruptly narrowed at base, entire, ciliate, the upper surface hirsute, venation obscure, the lower surface sparingly hirsute, confined mostly to costa and veins; petioles to 1.5 cm long, hirsute; flowers borne in axillary verticillasters, the uppermost confluent, forming narrow spikelike thyrse; peduncles to 3 mm long, hirsute; bracts subtending cymes narrowly lanceolate, about 5 mm long, 0.5 mm wide, slenderly acuminate, ciliate, involucre bracts 6, the larger and outer pair subtending the cymule obovate, the larger bract 13 mm long, 7 mm wide, the smaller 11 mm long, 6 mm wide, mucronate to cuspidate, narrowed gradually from middle to base, firm, veiny, margins ciliolate, both surfaces rather sparingly hirtellous, both green toward tip, whitish toward base; innermost pairs of bracts narrowly lanceolate, the larger pair 9 mm long, 1 mm wide about 3 mm above base, the smaller 5–6 mm long, about 0.75 mm wide, both pairs costate, ciliate, the tips hirtellous and greenish, whitish toward base; calyx deeply 5-parted, segments linear, 4.5 mm long, 0.5 mm wide at base, hirtellous; corolla purple, about 16 mm long, tube about 1.5 mm broad, glabrous to finely pubescent, upper lobe ovate, about 11 mm long, 5 mm wide, acute at tip, lower lip oblong, about 10 mm long, 2 mm wide, 3-lobed at apex, the lobes rounded, about 0.75 mm long and wide; stamens exserted about 5 mm beyond mouth of tube, filaments glabrous; capsule 5 mm long, about 2 mm broad, 0.75 mm thick, obtuse and apiculate at tip, pilose; retinacula

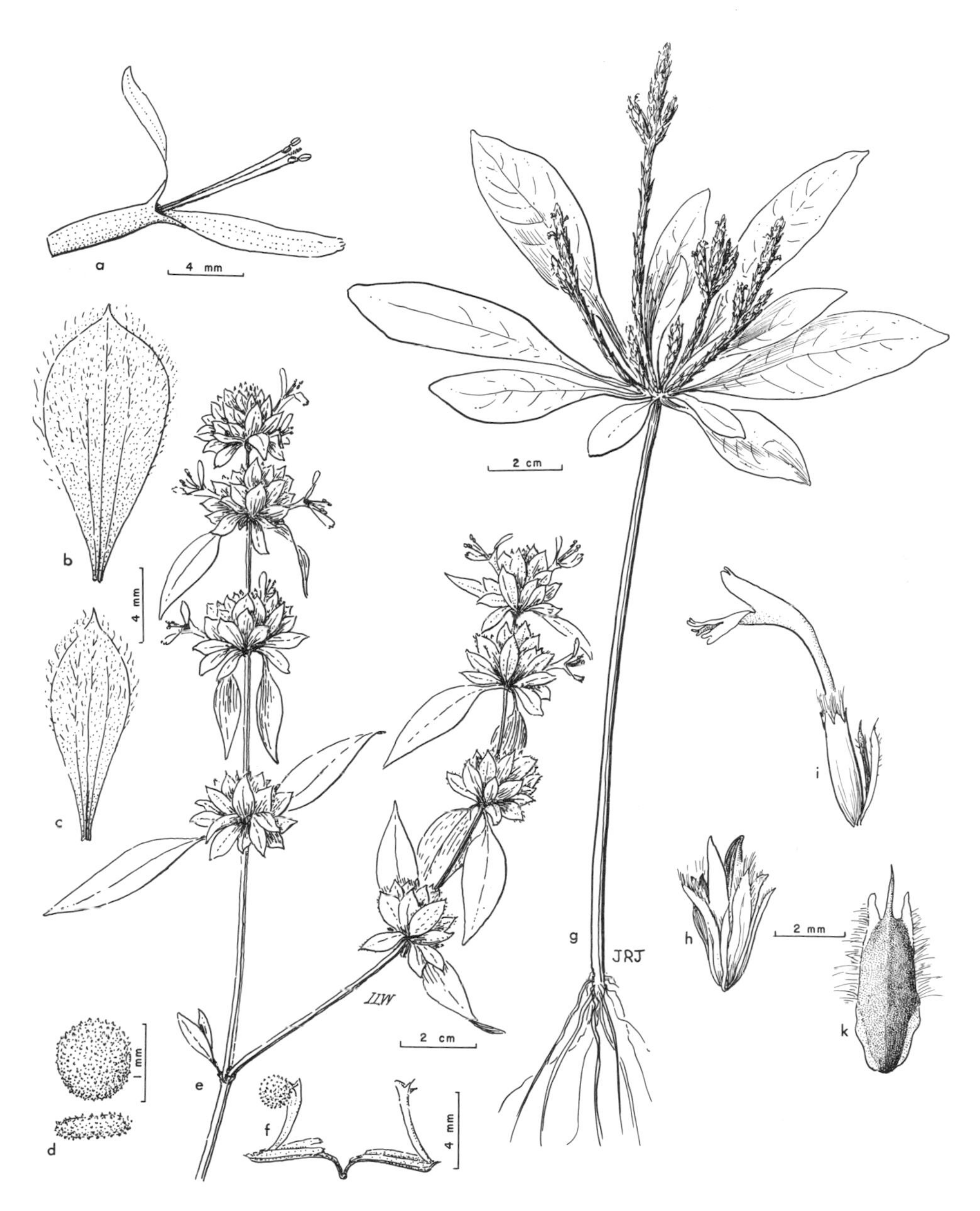

Fig. 58. *Dicliptera peruviana* (*a–f*): *b*, *c*, floral bracts.
Elytraria imbricata (*g–k*): *h*, capsule encased in calyx and bracts; *k*, primary bract.

triangular, acuminate, about 1 mm long; seeds suborbicular, flattened, 1–1.25 mm in diameter, 0.5 mm thick, brown, echinate.

In shaded habitats up to 3,000 m altitude, Peru and Ecuador. Specimens examined: San Salvador.

See Fig. 58.

Elytraria L. C. Rich. in Michx., Fl. Bor.-Am. 1: 8. 1803

Caulescent or acaulescent herbs. Leaves alternate or sometimes subopposite, basal or crowded at ends of branches. Flowers in dense peduncled spikes, both spikes and peduncles bearing imbricate coriaceous bracts. Calyx scarious, segments narrow, entire or toothed at apex. Corolla white or blue, tube slender, limb 2-lipped, lower lip 3-lobed. Stamens 2, barely exserted, anthers 2-celled, sacs equal, parallel, sometimes awn-tipped at base, staminodes usually wanting. Ovary 2-celled, ovules 6–10 in each locule. Capsules narrow, contracted at base, acute at apex.

A genus of 7 species, 6 native to temperate or tropical America, 1 found chiefly in Africa.

Elytraria imbricata (M. Vahl) Pers., Syn. 1: 23. 1805

Justicia imbricata M. Vahl, Ecol. Amer. 1: 1. 1796.
Verbena squamosa Jacq., Pl. Hort. Schoenbr. 1: 3, *pl.* 5. 1797.
Elytraria tridentata M. Vahl, Enum. Pl. 1: 107. 1804.
Elytraria frondosa HBK., Nov. Gen. & Sp. Pl. 2: 234. 1817.
Elytraria fasciculata HBK., op. cit. 235.
Elytraria ramosa HBK., loc. cit.
Elytraria scorpioides Roem. & Schult., Syst. Veg. Mant. 1: 128. 1822.
Elytraria apargiifolia Nees in DC., Prodr. 11: 65. 1847.
Elytraria squamosa Lindau, Anal. Inst. Fisico-Geogr. Costa Rica 8: 299. 1895.

Acaulescent, or if caulescent the leaves crowded at tips of glabrous or sparingly pilosulous stems to 60 cm long; leaf blades ovate to oblong or obovate, rarely linear-lanceolate, 1–18 cm long, 0.5–6 cm wide, blunt or subacute at apex, narrowed at base to a slender winged petiole, both surfaces appressed-pilose or glabrate, margins undulate; scapes numerous, axillary, 1–24 cm long, simple or branched (sometimes leafy at tip), covered by tightly appressed ovate to subulate scales; spikes 1 to several, to 6 cm long; bracts oblong or elliptic, 3–6 mm long, 1–2 mm wide, firm, awn-tipped and bearing a pair of triangular or rhombic hyaline teeth near apex; bractlets 3 mm long; calyx segments thin, the upper bidentate; corolla blue to lavender-purple, 5–8 mm long; capsule oblong, glabrous.

Usually growing on bushy slopes, in thickets, and in shrubby forests at low elevations from Arizona and Texas to northern and western South America. Specimens examined: Santa Cruz.

See Fig. 58.

Justicia Houst. ex L., Sp. Pl. 15. 1753; Gen. Pl. ed. 5, 10. 1754

Herbs or shrubs. Leaves opposite, petiolate, usually ovate to oblong, entire. Flowers spicate, paniculate or solitary. Bracts various, small, linear or subulate, distant to conspicuous and imbricate. Calyx segments usually narrow and nearly equal, 5 or (in some species) 4 in number. Corollas usually white, pink, or purple, sometimes with purple or white markings in throat, the tube usually narrow, short to long, limb 2-lipped, upper lip 2-lobed, lower 3-lobed. Stamens 2, often slightly exserted but usually not exceeding corolla lips; anther cells 2, more or less superposed, 1 or both cells apiculate or tailed, the connective narrow to broad, lobes parallel or obliquely affixed. Capsules clavate, 4-seeded.

A pantropical genus of about 300 species of herbs and shrubs.

Justicia galapagana Lindau in Robins., Proc. Amer. Acad. 38: 203. 1902

Branching herb to 1 m tall; stems ascending, more or less hexagonal or subterete, the hairs spreading-pilose to densely glandular-pilose; leaf blades lanceolate to ovate, to 5 cm long, 2.5 cm wide, acute to attenuate, rounded and abruptly nar-

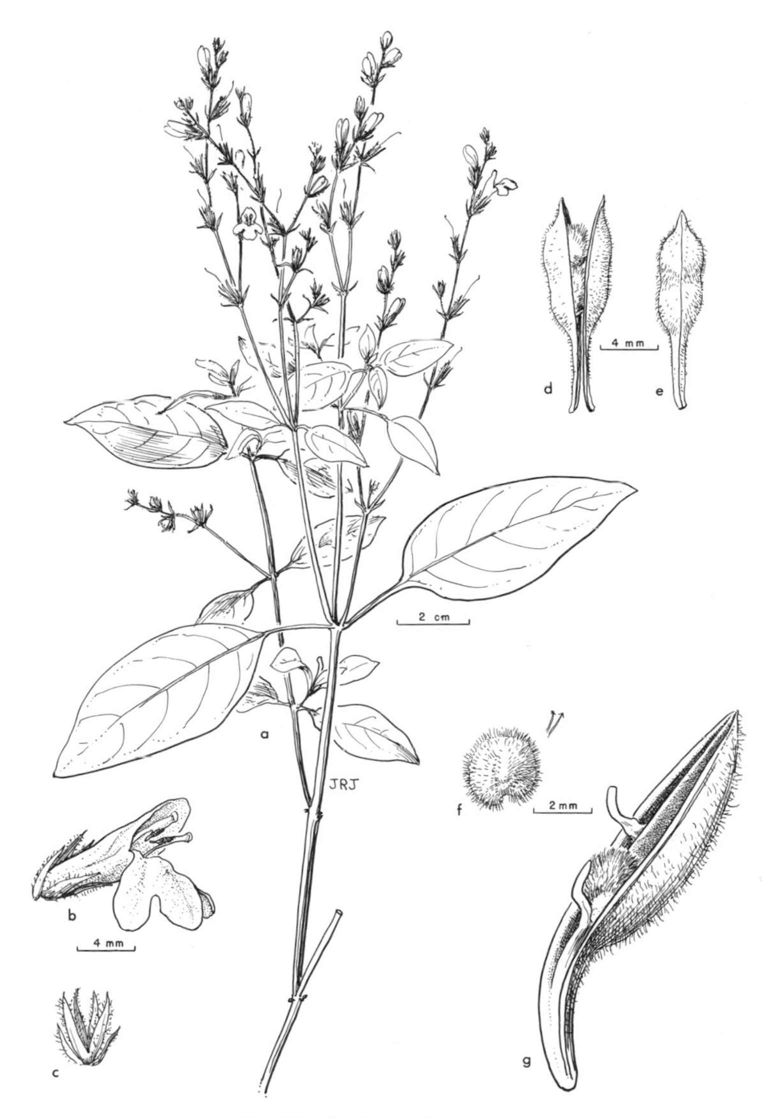

Fig. 59. *Justicia galapagana.*

rowed at base, moderately firm, entire, sparingly to densely pilose, the hairs confined chiefly to costa and lateral veins, cystoliths usually obscure; petioles to 10 mm long, pilose; inflorescences few-flowered, axillary, dichotomously branched, not exceeding terminal leaves, densely glandular-pilose; flowers alternate; pedicels almost none; bracts and bractlets subulate, 2 mm long, 0.5 mm wide at base, pubescence intermixed pilose and glandular-pilose; calyx deeply 4-parted, its lobes subulate, 4 mm long, 0.75 mm wide, pubescence as in bracts; corolla purple, to 14 mm long, densely pubescent externally, the tube 4–6 mm long, basally 2 mm broad, throat 3 mm broad, upper lip 4–6 mm long, basally 4 mm broad, apically somewhat dentate, folded longitudinally inside, lower lip 3-lobed, lobes 4 mm long, 2–3 mm across, rounded at tip; stamens barely exserted, filaments slender, flattened, 3–4 mm long, glabrous, the anther sacs obliquely attached to a flattened connective, upper lobe obtuse, 1 mm long, lower 1 mm long, apiculate at base; capsule 4-seeded, clavate, 12 mm long, 4 mm broad, pilose to glandular-pilose; retinacula 2 mm long, slightly curved; seeds suborbicular, brownish-tomentose, 2.5 mm in diameter.

Endemic; in moist and shady places. Specimens examined: Isabela, Pinta, San Salvador, Santa Cruz.

See Fig. 59.

Ruellia Plum. ex L., Sp. Pl. 634. 1753; Gen. Pl. ed. 5, 238. 1754

Perennial herbs or shrubs. Leaves petioled, entire, undulate, or rarely dentate. Flowers usually large and showy, solitary or clustered in axils or borne on terminal cymose panicles. Calyx usually 5-parted, segments often narrow. Corolla red, yellow, purple (usually mauve), or white, funnelform or salverform, sometimes saccate, the tube usually narrow below, upper portion more or less campanulate, the limb of 5 obtuse, spreading lobes. Stamens 4, didynamous, the anther sacs blunt at base. Stigma lobes unequal. Capsule oblong or clavate.

A large genus of upward of 200 species, the majority tropical or subtropical. The geographic center of distribution in the Western Hemisphere is somewhere in southern Mexico or Central America.

Ruellia floribunda Hook., Bot. Miscell. 2: 236. 1831

Dipteracanthus floribundus Nees in DC., Prodr. 11: 141. 1847.
Dipteracanthus viscidus Nees, loc. cit.
Aphragmia rotundifolia Nees in Benth., Bot. Sulph. 146. 1849.

Herbaceous or suffrutescent plant to 2 m high; stems erect, branching, nearly terete, tips glandular-pilose, lower portions becoming whitish and glabrate; leaf blades ovate to oblong-ovate, 2–6 cm long, 1–4 cm wide, entire, obtuse or slightly acute at apex, abruptly narrowed at base, glandular-pubescent; petioles slender, 0.5–2 cm long; inflorescences axillary, divaricate or ascending, dichotomously branched, glandular-pubescent; bracts leaflike, ovate, glandular-pubescent, those subtending flowers 2–3 mm long, 1.5–2 mm wide; calyx deeply 5-parted, segments unequal, odd segment to 1 cm long, 2.5 mm wide, remaining segments to 0.8 cm long, 1 mm wide, glandular-pubescent, linear-subulate; corolla magenta to purple, funnelform, finely pubescent, to 2 cm long, the lobes rounded, 4 mm long, 2.5 mm wide, limb 6–8 mm broad; stamens included; capsule clavate, to 1 cm long, 3 mm

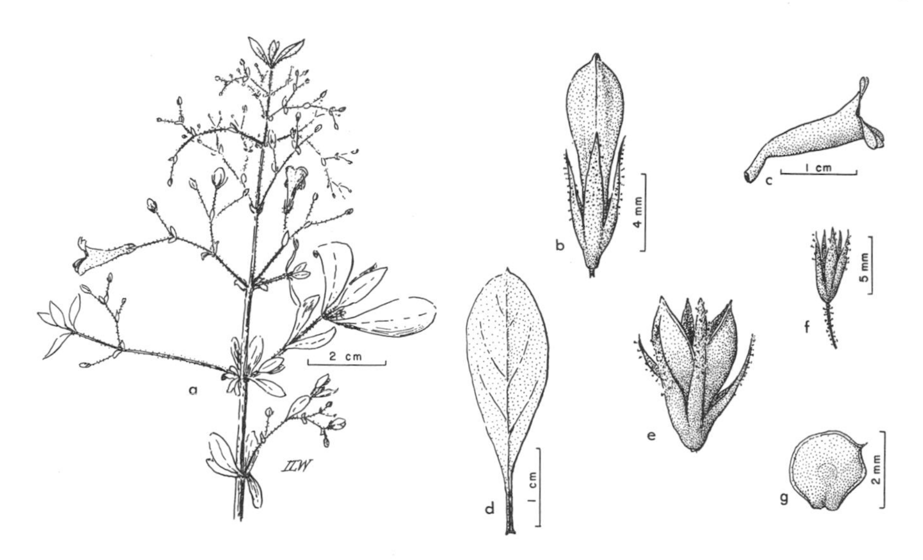

Fig. 60. *Ruellia floribunda.*

wide, glabrous, pointed, the tip shiny brown, 2–4-seeded; seeds flat, 2 mm long, 1.5 mm wide, mucilaginous-pubescent when moistened.

Occasional in shrubby areas and along watercourses of arid regions, Ecuador and Peru. Specimens examined: San Cristóbal.

See Fig. 60.

Tetramerium Nees in Benth., Bot. Sulph. 147. 1844

Fruticose or suffruticose, stems usually much branched, the pubescence often disposed in 2 opposite lateral lines. Leaves sessile or petioled, blades linear to ovate, entire, glabrous or pubescent. Flowers in terminal or axillary 4-angled spikes, the bracts conspicuous, usually closely imbricate, opposite, ciliate, cuspidate at tip. Calyx 4- or 5-parted, segments ciliate. Corolla infundibuliform to tubular, tube straight or slightly curved, upper lip entire or emarginate, lower lip 3-lobed, the lobes nearly equal. Stamens 2, included, anther sacs slightly converging toward apex, one somewhat longer than other. Capsule obovate, apiculate, contracted below into solid base, usually hispid and (2)4-seeded. Seeds flattened, tuberculate or muriculate.

About 25 species, from the southern United States to Colombia and Ecuador.

Tetramerium nervosum Nees in Benth., Bot. Sulph. 148, *pl. 48.* 1844

Blechum angustius Nees in DC., Prodr. 11: 467. 1847.

Branched, suffrutescent herb to 1 m high, the branches spreading, slender; stems subterete, at first evenly pubescent, later pubescent in lines or glabrous, the hairs white, spreading, to 1 mm long but usually shorter; leaf blades ovate, 1.5–6 cm long, 1–3 cm broad, acute to acuminate, subcordate to obtuse at base, the upper surface sparingly hirsute, costa and veins rather prominent and more or less puberulous with minute, appressed hairs, the lower surface sparingly hirsute to glabrate; petioles slender, 0.5–2.5 cm long, pilose to glabrous, the channel sometimes puberulous; inflorescence a closely imbricate spike, 1–4 cm long, rachis densely hirsute to glabrous; bracts round-ovate to ovate, 6–12 mm long, 5–9 mm broad, short-acuminate or cuspidate at apex, rounded or abruptly contracted below middle into subpetiolate base, sparingly hispidulous, small hairs intermixed with white spreading ones to 2 mm long, margins strongly hispid-ciliate, the hairs jointed, white, to 2.5 mm long, costa and 2 pairs of lateral veins prominent; bractlets linear-lanceolate, 2–3.2 mm long, acuminate, hispid-ciliate; calyx lobes 4, linear-lanceolate, 2–3 mm long, 0.5 mm wide at base, hispid; corolla white or purplish white, 10 mm long, essentially glabrous, the lips about 5.5 mm long, upper lip oblong-obovate and emarginate, lower 3-lobed, lobes oblong-obovate, 5–5.5 mm long, 1.8 mm wide at base, rounded at tip; capsule clavate, 5 mm long, 1.8 mm broad, minutely hirsute near apex, constricted base about 2 mm long; seeds 4, white to dark brown, flattened, 1.5 mm long, 1.2 mm broad, strongly papillose.

In dry or damp situations, thickets and open hillsides, low altitudes, Central and northern South America. Specimens examined: Isabela, San Salvador, Santa Cruz.

See Fig. 61.

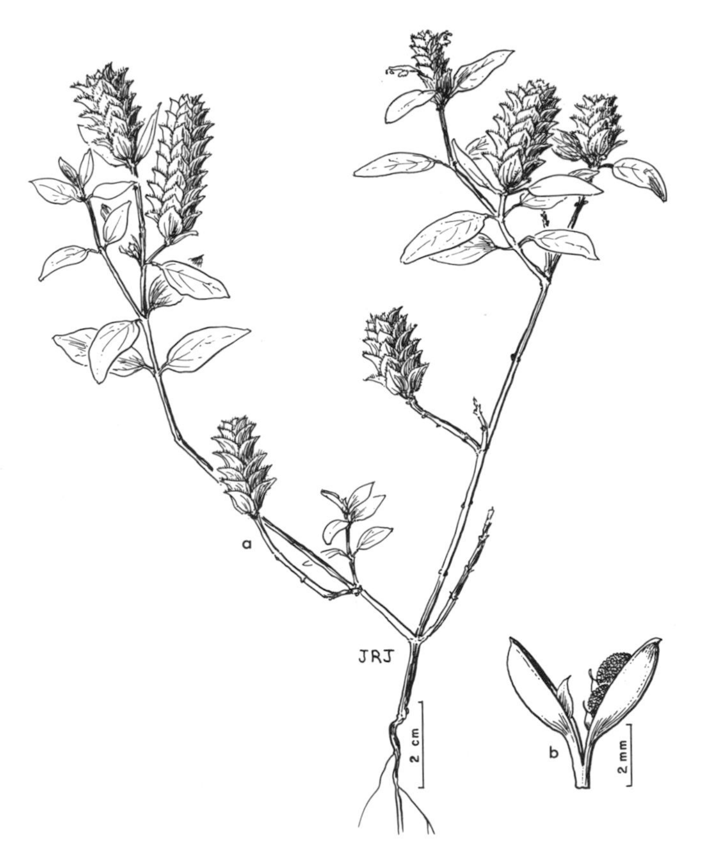

Fig. 61. *Tetramerium nervosum.*

APOCYNACEAE. DOGBANE FAMILY*

Perennial, herbaceous, suffrutescent, shrubby, or arborescent plants, often with milky or acrid juice. Leaves simple, entire, opposite or alternate, sessile or petiolate, exstipulate. Flowers perfect, regular. Calyx lobes 5, distinct to base or nearly so. Corolla cylindrical-campanulate, tubular-salverform, or funnelform; lobes 5, convolute in bud. Stamens 5, inserted separately on corolla; anthers often connivent and somewhat adherent to stigma, bearing appendages apically, basally, or in both positions or none. Pistils 2, separate; stigmas often united in flower; ovary superior or nearly so. Fruit a pair of erect or divergent follicles or a drupe. Seeds 1 to many, often comose; embryo straight.

About 300 genera and 1,300 species of nearly worldwide distribution; most abundant in the tropics.

Vallesia R. & P., Fl. Peruv. Prod. 28, *pl. 5.* 1794

Shrubs or small trees with alternate, short-petiolate, entire, persistent leaves and smooth, thin bark. Twigs slender. Flowers small, in cymes or on stoutish peduncles opposite leaves. Calyx deeply 5-parted into ovate lobes, eglandular. Corolla salverform; tube nearly cylindrical, enlarged into short throat just below the 5 lobes. Stamens inserted near apex of corolla tube, included; anthers cordate to lanceolate, not appendaged. Ovary 2-carpelled, 2–4-ovuled; style filiform; stigma cylindrical, puberulent. Fruit a 1- or 2-seeded drupe or berry. Seeds oblong, compressed; embryo curved.

A genus of 5 or 6 species in the tropical and subtropical regions of the New World.

Twigs, young leaves, and inflorescences glabrous. . . . . *V. glabra* var. *glabra*
Twigs, young leaves, and inflorescences finely puberulent. . . . . *V. glabra* var. *pubescens*

Vallesia glabra (Cav.) Link, Enum. Hort. Berol. 1: 207. 1821, var. **glabra**

Rauwolfia glabra Cav., Sc. 3: 50, *pl.* 297. 1794.
Vallesia cymbaefolia C. G. Ortega, Hort. Mat. 58. 1798.
Vallesia dichotoma R. & P., Fl. Peruv. 2: 26, *pl. 151.* 1799.
Rauwolfia oppositiflora Sesse & Moç., Pl. Nov. Hisp. 32. 1888.

Slenderly branched shrub or small tree to about 6 m tall; petioles 3–9 mm long; leaf blades narrowly lanceolate to oblong-lanceolate, 2.5–9 cm long, 6–25 mm wide, broadly cuneate to rounded at base, acute, slightly acuminate or sometimes obtuse at apex, slightly fleshy or leathery, with inconspicuous veins; cymes few-flowered, shorter than subtending leaves; pedicels 2–5 mm long; calyx lobes ovate, about 0.5 mm long; corolla 5–7 mm long, tube 4–5 mm long, greenish at base, white above, slightly constricted at orifice, lobes lance-ovate, 1–1.5 mm long, rotate-spreading; fruit translucent white, oblong to subglobose, 8–12 mm long, the flesh thin, watery, somewhat mucilaginous; seeds slightly compressed laterally or nearly terete, 2.5–3 mm broad, 6–10 mm long, weakly striate-grooved longitudinally.

In forests, on bushy ridges, along margins of lava flows, and at upper edges of beaches and margins of sand dunes, from sea level to about 325 m, Florida, the

* Contributed by Ira L. Wiggins.

West Indies, Mexico, and northern South America. Specimens examined: Española, Isabela, San Cristóbal, Santa Cruz.

The younger growth, including buds, twigs, leaves, and herbaceous parts of the inflorescences, is glabrous throughout; and the leaves are nearly always acute to acuminate at the apices, usually cuneate at the bases.

A note accompanying the Koford specimens (*Koford K-32* [DS]) stated "goats do not eat this." Our own observations on Isla Española and other islands in 1967 confirmed his field notes.

See Fig. 62.

Vallesia glabra var. **pubescens** (Anderss.) Wiggins, Madroño 20: 252. 1970

Vallesia pubescens Anderss., Kongl. Svensk. Vet.-Akad. Handl. 1853: 195. 1855.

Differing from var. *glabra* only in having a fine, closely arranged, erect indument of simple, nonglandular hairs on twigs, petioles, inflorescences, and undersurfaces of leaf blades, and a slightly sparser pubescence of the same kind on upper surfaces; the leaf blades seem slightly wider in proportion to length, and tend to be less acuminate or acute and more broadly rounded at the base in this variety than in the plants of var. *glabra*.

Known only from the Galápagos Islands, in habitats similar to those occupied by

Fig. 62. *Vallesia glabra* var. *glabra* (*a–d*). *V. glabra* var. *pubescens* (*e*).

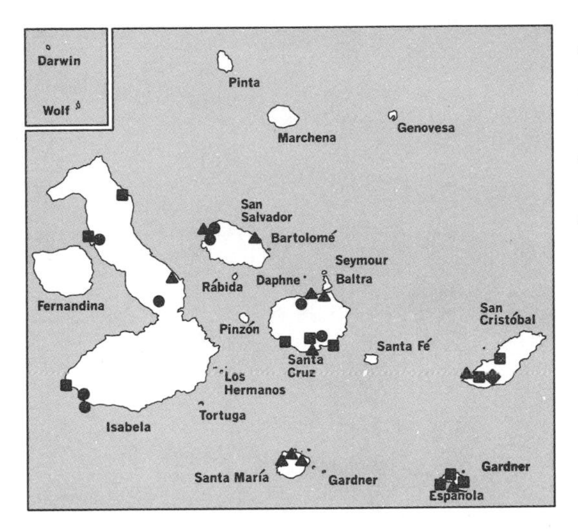

● *Tetramerium nervosum*
■ *Vallesia glabra* var. *glabra*
▲ *Vallesia glabra* var. *pubescens*
◆ *Asclepias curassavica*

var. *glabra*. Specimens examined: Española, Isabela, San Cristóbal, San Salvador, Santa Cruz, Santa María.

See Fig. 62.

ASCLEPIADACEAE. MILKWEED FAMILY*

Perennial, herbaceous, suffrutescent, or sometimes woody vines or upright to prostrate herbs; stems frequently twining or scandent. Leaves simple, mostly opposite or whorled, usually petiolate, entire. Flowers perfect, regular, 5-merous (except for 2 carpels), mostly in umbelliform cymes, extremely specialized in structure. Stamens and style coherent into a gynostegium, this adnate to base of corolla. Corona of separate or united, fleshy, laminate or hoodlike appendages usually present between corolla and anthers, adnate to column or corolla, or to both. Anthers winged and scariously tipped with a small, hyaline scale; pollen in narrowly pyriform masses (pollinia), usually 1 pollinium in each anther cell, the pollinia of adjacent anthers attached in pairs by transverse filaments or "translator arms" connected to small clip or "corpusculum," pendulous, horizontal, or erectly ascending from attachment of translators. Stigma conical or depressed, entire or shallowly lobed or angled. Ovary superior, of 2 pistils, these attached apically by common stigma but otherwise distinct. Fruit a pair of follicles, or 1 of them abortive. Seeds mostly ovoid, strongly flattened, concavo-convex, usually bearing terminal coma of silky hairs or fine bristles.

Credited variously with 75 to 320 genera, depending on the view of the taxonomist, and including probably about 1,800 species. Distributed worldwide from the temperate zones into the tropics; especially well represented in South America.

Stems erect or ascending, not twining; corona of 5 hoods, each with a hornlike crest within; corolla lobes bright red-orange (in ours) . . . . *Asclepias*

Stems twining or scandent; corona double, consisting of an outer ring adnate to corolla and 5 turgid vesicles attached to column within; corolla lobes greenish to purplish . . . . *Sarcostemma*

* Contributed by Ira L. Wiggins.

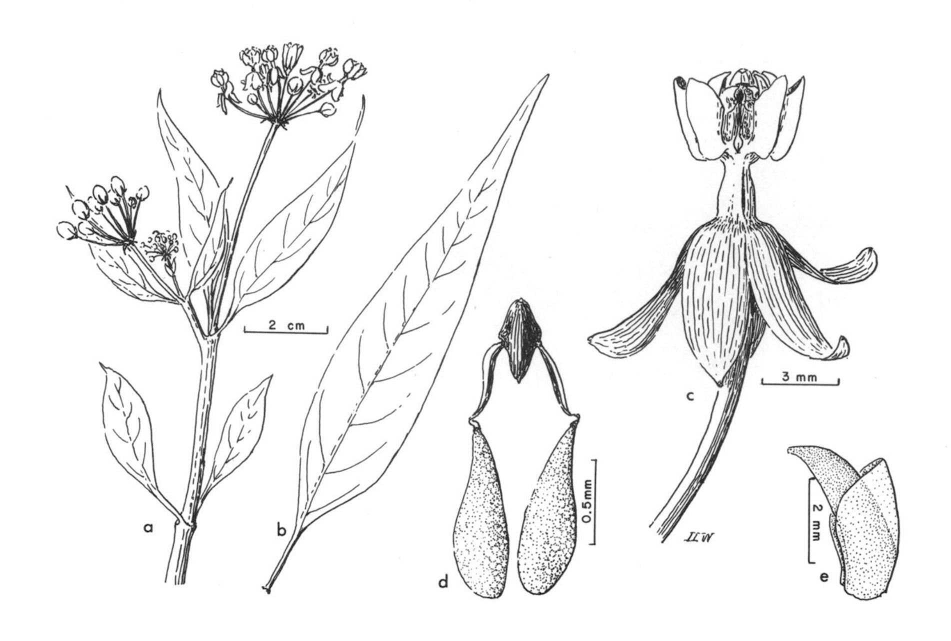

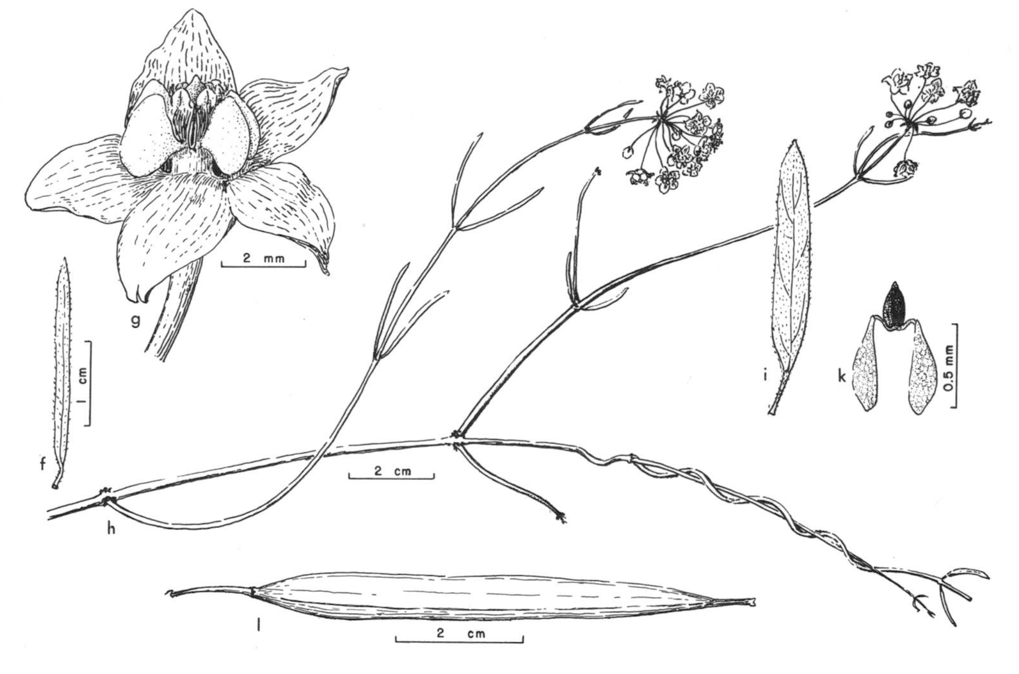

Fig. 63. *Asclepias curassavica* (*a–e*): *d*, pollinia removed from anther; *e*, horn and hood. *Sarcostemma angustissima* (*f–l*): *k*, pollinia removed from anthers; *l*, fruit before dehiscence.

Asclepias L., Sp. Pl. 214. 1753; Gen. Pl. ed. 5, 102. 1754

Perennial or annual herbaceous or suffrutescent plants with alternate, opposite, or whorled leaves, these entire or crispate-erose, glabrous to lanate. Flowers greenish, cream-colored, orange, red, or purple-tinged, pedicellate, borne in pedunculate, umbelliform, dichasial inflorescences. Calyx small, deeply 5-lobed, usually bearing 2 glands at each sinus. Corolla lobes 5, rotate, reflexed, or ascending in anthesis. Column often very short. Hoods mostly erect or ascending, involute-concave, shorter than, equaling, or surpassing anthers, often somewhat pitcher-shaped, pendulous, saccate at base, entire to 2–3-toothed at apex, with or without erect or ascending crests or horns, these, when present, included to exserted. Anther wings usually angulate at or below middle, corneous. Pollinia pendulous. Stigma flat-topped, 5-angled. Follicles 2, 1 often abortive, naked or echinate with soft, spreading processes. Seeds narrowly winged; coma silky, silvery white to sordid.

About 100 species, well represented in the temperate and tropical areas of the New World; present in the Old World also.

Asclepias curassavica L., Sp. Pl. 215. 1753

Asclepias bicolor Moench, Meth. 717. 1794.

Herbaceous to slightly woody at base; stems erect, to 2 m tall, at first sparsely strigose but soon glabrate or mostly so; leaves opposite, lance-linear to lance-elliptic, 10–15 cm long, 1–3 cm wide, acute, acuminate, or attenuate at each end, sparsely puberulent, soon glabrate, slightly paler beneath than above; peduncles terminal (appearing axillary in upper axils, but these branches actually terminal in a dichasial inflorescence), to 6 cm long; cymes few- to many-flowered; pedicels slender, 1–1.5 cm long; calyx lobes narrowly lanceolate, 3–3.5 mm long, puberulent, reddish; corolla lobes bright red-orange, lance-oblong, 7–8 mm long, nearly glabrous, reflexed, obtuse to acute at apex; corona and anthers on a slender column 2–2.5 mm high; hoods yellow, 3.5–4 mm high, saccate at base, distinctly short-stipitate, obliquely acute and slightly spreading at apex, 1–1.5 mm longer than anthers; horns yellowish, slender, exserted 2–2.5 mm and curving inward over column; follicles erect, fusiform, 6–10 cm long, 6–9 mm thick, purplish, often glaucescent, glabrous, or puberulent; seeds about 5 mm long, dark brown, with lighter-colored narrow wing; coma silky white, about 1.5 cm long.

Margins of fields and roads, and in other more or less disturbed areas, from central Baja California to Florida and south into tropical South America; warm regions of the Old World. Specimens examined: San Cristóbal.

Almost surely an introduced weed.

See Fig. 63.

Sarcostemma R. Br., Mem. Wern. Soc. 1: 50. 1809

Herbaceous or suffrutescent, scandent or spirally twining perennials with opposite, petiolate leaves. Flowers often fragrant, white or purplish, in pedunculate umbelliform cymes at tips of main branches and on short branches that appear lateral but are arrested main branches arranged dichasially. Calyx 5-parted, small. Corolla rotate to campanulate, 5-lobed; lobes often somewhat contorted (plane in ours). Corona consisting of both an outer, short ring adnate to corolla or gynostegium or

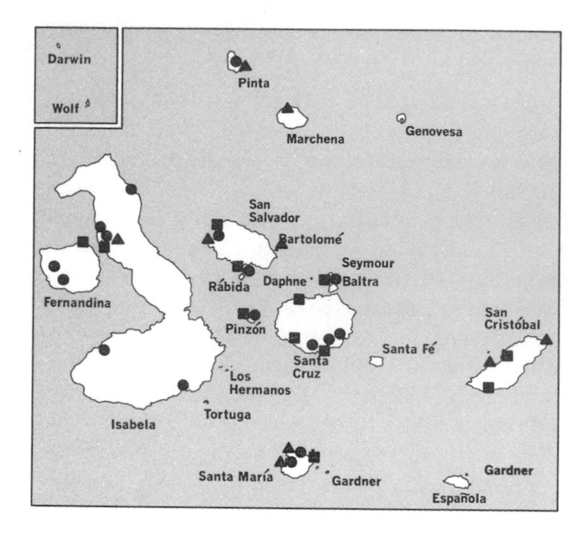

to both, and 5 turgid vesicles attached to stamen column. Pollinia pendulous. Follicles terete, smooth, slender (in ours), or turgid, few- to many-seeded. Seeds flattened, comose at one end.

About 25 species distributed from the southwestern United States to Argentina, and 4 or 5 additional species distributed from Africa to Australia.

Sarcostemma angustissima (Anderss.) R. W. Holm, Ann. Mo. Bot. Gard. 37: 545. 1950

Asclepias angustissima Anderss., Kongl. Svensk. Vet.-Akad. Handl. 1853: 196. 1855.
Funastrum angustissimum Fourn., Ann. Sci. Nat. Bot. VI, 19: 388. 1882.

Stems slender, scandent or weakly climbing, sometimes twining about one another, with many "lateral" branches to 5 m long, finely and sparsely puberulent when young, soon glabrate; internodes 5–8(15) cm long; leaves linear to narrowly oblong, rarely linear-oblanceolate, acute apically, cuneate at base, 1–6 cm long, 1.5–4 mm wide, narrowly revolute, without glands at base of blade, minutely and sparsely puberulent, soon glabrate on blades; petioles 1–5 mm long, more or less pubescent; inflorescences 5–25-flowered; peduncles obsolete or present and to 1 cm long, minutely spreading-pubescent; bracts linear, minute; pedicels slender, 1–1.5 cm long at anthesis; calyx lobes ovate, about 1.5 mm long, hispidulous without, glabrous within, ciliolate, acute, glandular at tip; corolla rotate-subcampanulate, purplish, turning black when dry, the lobes ovate, acute, 2.5–3 mm long, glabrous on both surfaces; gynostegium about 2 mm high, with very short column; anthers 1 mm long, with semiorbicular hyaline apical appendage; sacs of pollinia oblong, 0.5 mm long, corpusculum sagittate, purplish; corona ring very thin; vesicles of corona pale yellow to nearly white, narrowly adnate to ring, ovoid, slightly pointed at tip, 1.8–2 mm long; ovaries about 1 mm long, minutely puberulent; follicles narrowly fusiform, 8–12 cm long, 5–7 mm broad, minutely puberulent, greenish to purplish; seeds brown, flask-shaped, 6–7 mm long, 2.5–3 mm wide, concavo-convex, thin, the marginal wing 0.5–0.8 mm wide; coma silvery white, silky, about 2 cm long, of over 100 fine filaments.

Endemic; in openings among trees in *Opuntia-Bursera* stands of Arid Zone, less commonly in Transition Zone, on the larger islands of the archipelago. Specimens examined: Baltra, Fernandina, Isabela, Pinta, Pinzón, Rábida, San Salvador, Santa Cruz, Santa María.

Seen on Islas Champion and Española (in vegetative stages only) in 1967.

See Fig. 63 and color plate 88.

AVICENNIACEAE. BLACK-MANGROVE FAMILY*

Shrubs or trees of maritime regions with pneumatophores, mostly inhabiting saline or brackish mangrove lagoons. Branchlets and twigs commonly terete, prominently nodose, articulate. Leaves decussate-opposite, thick-textured, persistent, petiolate, exstipulate, entire. Inflorescences axillary or terminal, determinate and centrifugal (cymose), spicate or subcapitate, axillary ones mostly paired. Flowers small, sessile, perfect, hypogynous. Calyx of 5 nearly separate sepals, these ovate, imbricate, unchanged in fruit, subtended by pseudoinvolucre composed of a scale-like bractlet and 2 alternate scale-like prophylla slightly shorter than calyx, imbricate with each other and sepals. Corolla actinomorphic, gamopetalous at base, campanulate-rotate, 4-parted. Stamens 4, inserted in throat of corolla tube, equal or subdidynamous. Gynoecium composed of 2 carpels; ovary compound but with a free central, often more or less winged placenta; ovules 4, pendent, orthotropous, hanging from tip of central columella. Fruit a compressed, oblique capsule, dehiscent by 2 valves, by abortion regularly 1-seeded. Seeds without a testa; embryo viviparous; radicle hairy; cotyledons 2, folded lengthwise.

A single genus with 16 known living species and varieties, the chief constituent of most coastal lagoons, occurring in the maritime regions of the tropics and subtropics of the world. Four fossil species are known, 2 from the United States and 2 from Colombia; a modern species is also known in fossil form from Trinidad.

Avicennia L., Sp. Pl. 110. 1753; Gen. Pl. ed. 5, 49. 1754

Characters of the family. A single species in our area.

Avicennia germinans (L.) L., Sp. Pl. ed. 3, 891. 1764

Bontia germinans L., Syst. Nat., ed. 10, 2: 1122. 1759.
Avicennia nitida Jacq., Enum. Syst. Pl. Carib. 25. 1760.
Avicennia tomentosa Jacq., loc. cit.
Avicennia elliptica Holm in Thunb., Pl. Bras. Dec. 3: 37. 1821.
Avicennia floridana Raf., Atl. Jour. 1: 148. 1832.
Avicennia meyeri Miquel, Linnaea 18: 262. 1844.
Avicennia lamarckiana Presl, Bot. Bemerk. 99. 1844.
Avicennia oblongifolia Nutt. ex A. W. Chapm., Fl. South. U. S. 310. 1860.
Hilairanthus nitidus Van Tiegh., Jour. de Bot. 12: 357. 1898.
Hilairanthus tomentosus Van Tiegh., loc. cit.
Avicennia officinalis sensu Robins., Proc. Am. Acad. 38: 194. 1902, and sensu Stewart, Proc. Calif. Acad. Sci. IV, 1: 131. 1911, non L. 1753.

Erect or spreading low shrub to slender tree to 25 m tall; pneumatophores many, erect, pencil-thick, projecting 5–10 cm above water level, leafless; stilt-roots none; branches spreading, crooked; branchlets slender, brownish, more or less tetragonal, glabrous or minutely gray-pubescent, often shiny; lenticels scattered; nodes swollen, annulate; internodes 1.5–9 cm long; petioles slender, 2–27 mm long, flat-canalicu-

* Contributed by Harold N. Moldenke.

Fig. 64. *Avicennia germinans.*

late above, mealy or glabrous, slightly ampliate basally; leaf blades firmly chartaceous to coriaceous, gray-green to dark green and shiny above, mostly same color beneath but lamina often obscured by whitish furf, occasionally brunnescent or nigrescent in drying, lanceolate, lanceolate-oblong, elliptic, or obovate, 4.5–15 cm long, 1.8–4.4 cm wide, acute to obtuse at apex, acute or acuminate to cuneate at base, glabrous, densely impressed-punctulate above, uniformly and densely white- or gray-furfuraceous to pulverulent-tomentulose to glabrous and more or less punctate beneath; inflorescence axillary and terminal, spicate, 1.5–6.5 cm long, 1–1.5 cm wide at anthesis, axillary ones usually confined to 1 pair at base of terminal one, shorter than it or another pair in next lower axils, dense; flowers usually opposite, 1–15 pairs per spike, sessile, sometimes few and distant, 1–2 cm wide at anthesis, fragrant, highly nectariferous; bractlets and prophylla light green, ovate or oblong, sessile, closely appressed, obtuse or acute, appressed-cinereous; calyx lobes ovate, 3–5 mm long, 2–3 mm wide, densely appressed-pubescent outside; corolla campanulate, yellow or white, 12–20 mm long, about 10 mm wide, tube about equaling calyx, practically glabrous, lobes 4, spreading, unequal, 2–2.5 mm long, oblong or subquadrate, rounded, densely cinereous-pubescent without, velutinous-tomentose within; stamens slightly exserted from tube; pistil equaling stamens; stigma bilobed; fruiting calyx enlarged but not indurated, 5-parted almost to base; fruit yellowish, oblong or elliptic to obpyriform or ovate and asymmetric, 1.2–5 cm long, 7–13 mm wide, often turning plum color in full sun, apiculate, densely white- or gray-pulverulent, often also more or less white-strigose at apex and on apiculation, opening longitudinally.

Common in mangrove swamps and lagoons, from southern Florida, eastern Texas, Bermuda, and the Bahamas throughout the West Indies, and along both coasts of Mexico south to the coasts of Brazil and Peru. Specimens examined: Baltra, Fernandina, Isabela, Pinzón, Rábida, San Cristóbal, San Salvador, Santa Cruz, Santa María.

A very closely related species occurring on the coasts of West Africa is sometimes regarded as conspecific; it is very different from the Old World *A. officinalis* L., with which it has been confused. Several writers have reported sighting *A. germinans* at different localities in the Galápagos Islands without making collections to document their observations.

See Fig. 64 and color plate 26.

BORAGINACEAE. BORAGE FAMILY*

Herbs, shrubs, trees, or rarely lianas with hispid, puberulent, or sometimes glabrous herbage. Leaves usually alternate but lowermost sometimes opposite (in some annuals the earliest in rosettes from among which stems and branches arise), simple, entire but margins sometimes undulate, exstipulate, mostly bearing cystoliths on their surfaces. Inflorescence determinate, usually of 1 or more scorpioid spikelike cymes, sometimes glomerate, bracteate or not. Flowers usually bisexual, regular, hypogynous. Sepals (4)5, distinct or connate at base or nearly full length, usually imbricate, rarely valvate. Corolla sympetalous, 5-lobed, lobes imbricate to contorted in bud, rotate-salverform, funnelform, or campanulate, the tube often with folds or partially blocked by appendages at apex of throat. Stamens (4)5, inserted in corolla tube, equal or unequal, included or slightly exserted, alternate with corolla lobes; anthers 2-celled, basifixed or dorsifixed, introrse, dehiscing longitudinally. Nectary at base of stamens or corolla tube. Ovary superior, 2-carpelled but each carpel deeply bilobed, fruit thus appearing 4-carpelled; ovules 4, or fewer by abortion. Style 1, gynobasic or terminal; stigmas various. Fruit of 4 nutlets or a 1–4-seeded nut or drupe, often ornamented or sculptured, or smooth, glabrous and glossy or dull. Endosperm scanty or none.

About 100 genera and over 2,000 species, widely distributed from the tropics to the arctic zones of the world; particularly well represented in the arid parts of the New World.

Fruit dry; plants herbs or suffrutescent:
- Style deeply 2-cleft; stigmas faintly capitate, 1 on each branch of style; flowers never in scorpioid spikes or cymes, borne among the crowded leaves at nodes or on short spur branches; leaves minute, rarely over 6 mm long. . . . . *Coldenia*
- Style entire or shallowly bidentate at apex; stigmatic surface forming a ring below sterile apical portion of style; flowers in scorpioid cymes at tips of branches; leaves larger, often to 5 cm or more. . . . . *Heliotropium*

---

* Contributed by Ira L. Wiggins.

Fruit drupaceous; plants shrubs or small trees, occasionally suffrutescent:
Calyx tubular to narrowly campanulate, the teeth much shorter than the tube; style twice bifid into 4 clavate lobes; plant shrubby to a tree to 8 m tall........*Cordia*
Calyx 5-lobed nearly or completely to base; style once bifid into 2 lobes; plant a shrub to 3 m tall..*Tournefortia*

Coldenia L., Sp. Pl. 125. 1753; Gen. Pl. ed. 5, 61. 1754

Low herbs, suffrutescent plants, or small shrubs, often diffuse, decumbent or prostrate or less commonly ascending, the outer bark exfoliating in irregular strips from the slender internodes. Leaves small, alternate, often closely crowded on lateral twigs with extremely short internodes, thus appearing fasciculate, entire (in ours), or in a few species shallowly dentate or lobed, often crispate, densely pubescent. Calyx 4–5-parted, sepals linear to narrowly lanceolate or lanceolate-ovate. Corolla tubular-rotate to funnelform, tube short; limb 4–5-lobed, interior of throat naked (in ours) or bearing as many scales as corolla lobes. Stamens 4–5, inserted low in corolla tube. Ovary ovoid, shallowly 4-lobed, 4-celled or 2-celled but falsely 4-celled. Styles slender, deeply divided nearly or fully to base, attached to small conical projection of receptacle that rises just above apices of nutlets. Fruit drupaceous in some species, in ours consisting of 4 dry, hard, ovate nutlets, each deeply enfolded in lower part of subtending sepal, smooth or granular to tuberculate, grooved on ventral face.

About 12–15 species, generally distributed in the arid regions of the Western Hemisphere; 1 species in the Old World.*

Nutlets smooth and shining (sometimes sparingly tuberculate on sides near base); leaves and sepals bearing 1 to several coarse, stiff hairs near their apices in addition to finer, shorter hairs:
Pubescence on stems appressed-ascending or spreading; corollas 1.5–2.5 mm long, the tube campanulate-funnelform; stamens equaling or slightly exceeding corolla tube; plant prostrate-spreading ....*C. darwinii*
Pubescence on stems all very fine, retrorse-appressed; corollas 2.5–3 mm long, the tube cylindrical; stamens shorter than corolla tube; plant spreading-ascending, to 3–4 dm tall ....*C. nesiotica*
Nutlets granular-tuberculate throughout, the surface dull; leaves and sepals bearing only 1 type of hair, not markedly coarse:
Corolla 1–2 mm long, the tube campanulate-funnelform; style branches separate to very base; calyx about 1.5 mm long, fully as wide at anthesis; hairs on leaf blades short, stoutish, with pustulate-celled bases..............................*C. fusca*
Corolla 3–4 mm long, the tube cylindrical; style branches fused about 0.5–1 mm at base, free above; calyx 2–3 mm long, about half as wide as long at anthesis; hairs on leaves and sepals mostly about 1 mm long, slender, more crowded, not markedly pustulate at base ....*C. galapagoa*

Coldenia darwinii (Hook. f.) A. Gray, Proc. Amer. Acad. 5: 341. 1862

Galapagoa darwinii Hook. f., Trans. Linn. Soc. Lond. 20: 196. 1847.

Plant prostrate, forming low, often dense mats 4–10 dm in diameter, woody, dark brown to gray-black at base (outer layers lost by exfoliation); branches slender, numerous, much branching, cinereous, the hairs spreading to antrorsely appressed;

* For further information on this genus, see Johnston (1924) and Howell (1937).

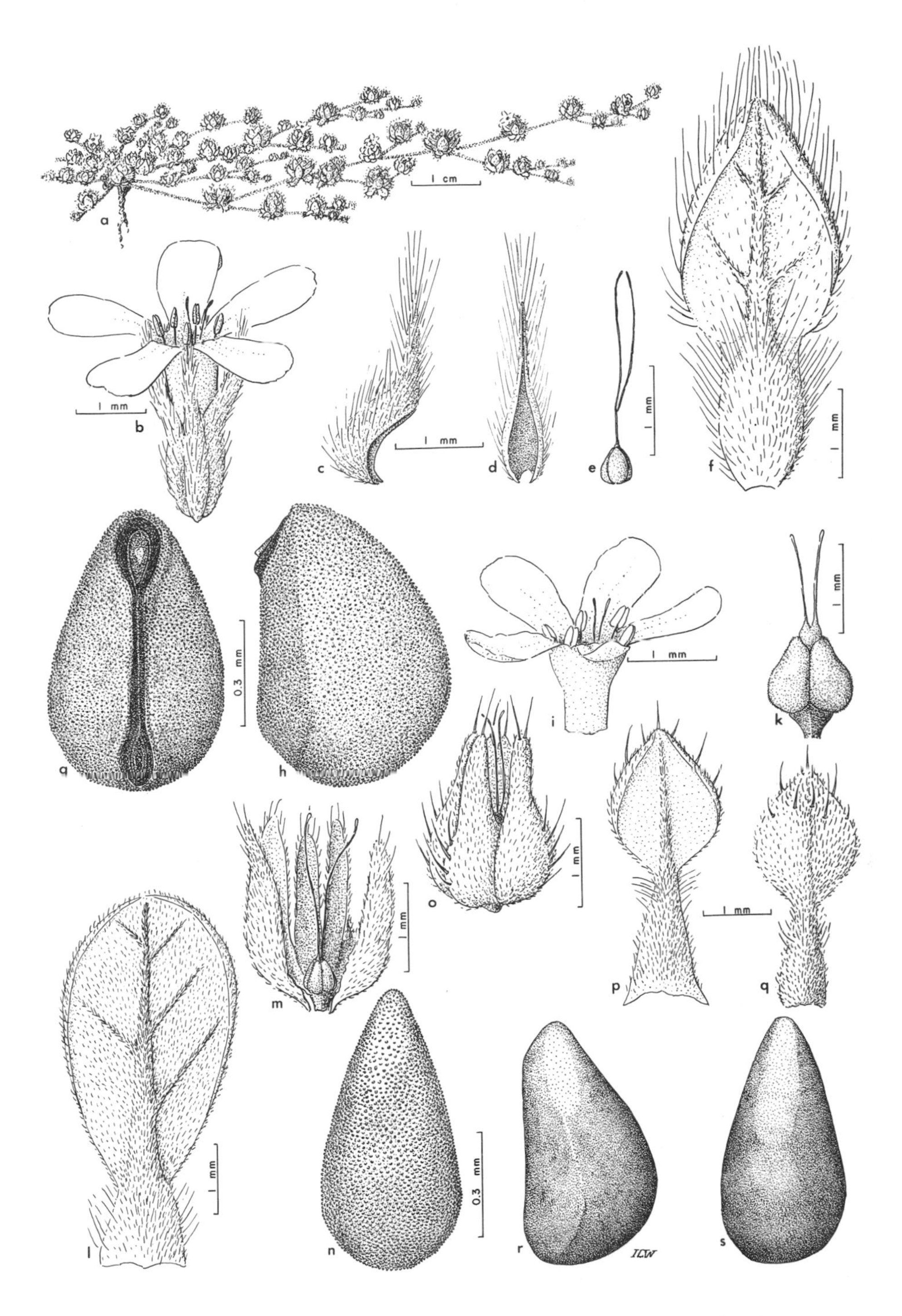

Fig. 65. *Coldenia galapagoa* (*a–h*). *C. fusca* (*i–n*). *C. darwinii* (*o–s*).

internodes 0.4–2 cm long; leaves crowded at tips of branches and short spur branches at most nodes, elliptic to ovate, the blades 2–3.5 mm long, 1–2 mm wide, acute at apex, rounded to broadly cuneate at base, margin tightly and narrowly revolute, midrib deeply impressed above, prominent beneath, lateral veins not evident, pubescence of 2 kinds, short, rather slender, antrorsely appressed ones and a few (usually 1–10) much stouter, longer (to 0.5 mm) pustulate-based ones near tip and pointing toward apex of blade; petiole nearly as long as blade, about half as wide at apex as at base; flowers borne among clusters of leaves; calyx lobes 1.5–2 mm long at anthesis, slightly unequal, with same kind of hairs as borne on leaves, the larger, setose ones along midrib of lobes on distal half to two-thirds, rarely present elsewhere on lobes, finer, antrorsely appressed hairs throughout rest of calyx exterior, the basal part of 4 of lobes enfolding a nutlet, enfolded margins thin, glabrous within, the distal one-half to two-thirds of lobe flat or only slightly enfolded, distinctly narrower than basal part, synsepalous in basal third at anthesis, later splitting apart; corolla sordid white, 1.5–2.5 mm long, tube campanulate-funnelform, lobes nearly orbicular, spreading-rotate, often bearing a few minute, fine hairs on exterior surfaces near tips; stamens about 1 mm long, inserted 0.2–0.4 mm above base of corolla tube, about equaling corolla tube or slightly exserted; style about 1 mm long, branches distinct to base; stigmas narrowly capitate; nutlets narrowly ovate as viewed dorsally, 0.7–0.8 mm long, smooth and shining, black, sometimes sparingly and minutely tuberculate on sides of basal third, smooth elsewhere, ventral groove very slightly if at all widened at either upper or lower end.

Endemic. Specimens examined: Bartolomé, Isabela, Marchena, Pinta, San Cristóbal, San Salvador, Santa María.

See Fig. 65.

Coldenia fusca (Hook. f.) A. Gray, Proc. Amer. Acad. 5: 341. 1862

Galapagoa fusca Hook. f., Trans. Linn. Soc. Lond. 20: 197. 1847.

Suffrutescent perennial with woody roots and basal parts, the stems spreading-prostrate, much branched, forming irregular mats 1–8 dm across; older branches usually glabrous owing to exfoliation, dark brown to nearly black; younger branches

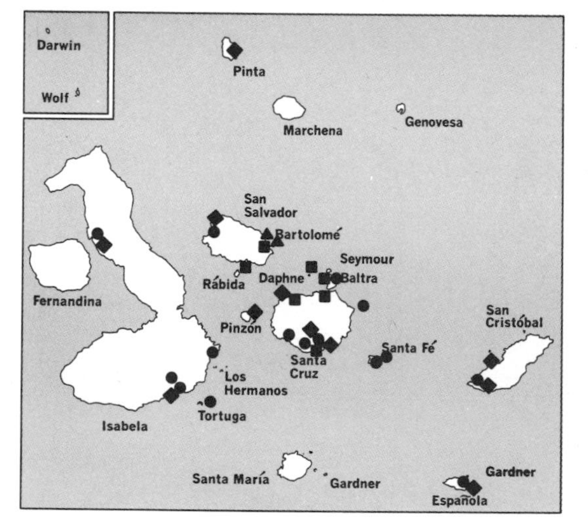

and twigs cinereous or fuscous, with subappressed-ascending or rarely spreading, fine hairs; leaves ovate, elliptic, obovate, or suborbicular, rounded or acute at apex, rounded at base, strigillose with more or less appressed hairs or subhirsute with erect, stoutish hairs from pustulate bases, usually subcinereous beneath, the margins narrowly but tightly revolute, midrib and lateral veins deeply impressed in upper surface, obvious and more densely pubescent than adjacent tissue beneath; calyx 1.5–2 mm long at anthesis, tube 0.4–0.6 mm deep, lobes erect, not noticeably inflexed, pustulate-hirsutulous with ascending, relatively fine hairs; corolla barely exceeding calyx lobes, campanulate-funnelform; stamens attached 0.2–0.3 mm above base of corolla tube, about 1 mm long, included; style 1–2 mm long, branches distinct to base or essentially so; nutlets dark brown to dull black, ovate in dorsal view, 0.7–0.8 mm long, acute at apex, rounded below, closely granular or finely tuberculate throughout, the ventral groove widened slightly at apex and base.

Endemic; in sandy soil near the coast, on most of the larger islands and on several smaller ones. Specimens examined: Baltra, Española, Isabela, Las Plazas, San Cristóbal, San Salvador, Santa Cruz, Santa Fé, Tortuga.

See Fig. 65.

Coldenia galapagoa Howell, Proc. Calif. Acad. Sci. IV, 22: 108. 1937

Low, spreading suffrutescent plant with semiprostrate to assurgent stems forming dense mats 1–6 dm across, 1–5 dm tall; woody basal branches with thin, brownish to black bark; younger branches numerous, villous with straight hairs to 1 mm long and somewhat retrorse (or spreading), finer, shorter ones; primary internodes to 3 cm long, subsequent ones much shorter, leafy-congested; leaves ovate, oblong-ovate, or oblong-obovate to oblong-lanceolate, 2–6 mm long, 1–4 mm wide, hirsute to hispid with hairs to 2 mm long; veins deeply impressed above, raised and obvious beneath; petioles densely hirsute-ciliolate; flowers nearly concealed by leaves; calyx 2–3 mm long, tube about 0.8 mm deep, lobes linear-oblong, hirsute with erect, white bristle-like hairs, the terminal third of each lobe narrowed by inflexion of margins in fruit; corolla sordid white, 3–4 mm long, tube subcylindrical, 2–2.5 mm long, lobes rounded-oblong, about 1 mm long; stamens inserted about 0.3 mm above base of corolla tube, 3–3.5 mm long, exserted about length of anthers; style 2.5–3 mm long, branches fused in basal 0.2–0.6 mm; nutlets broadly ovate in dorsal view, rounded at base, obtuse at apex, dull black, 0.7–0.8 mm long, about 0.5 mm wide, slightly deeper (from front to back) than wide, finely and uniformly tuberculate throughout, angled and ventrally grooved, the groove obviously widened toward apex, slightly less so at base.

Endemic; in sandy habitats on several islands in the archipelago. Specimens examined: Baltra, Daphne Major, Rábida, San Salvador, Santa Cruz.

See Fig. 65.

Coldenia nesiotica Howell, Proc. Calif. Acad. Sci. IV, 22: 237. 1941

Coldenia conspicua Howell, op. cit. 105. 1937, non Johnston 1935.

Small, suffrutescent plant 3–4 dm tall, with loosely spreading to suberect, cinereous branches, the hairs on upper ones crowded, retrorsely appressed or a few hairs spreading; secondary branchlets spurlike, bearing densely crowded clusters of small leaves; leaf blades ovate to ovate-lanceolate, 1.5–2.5 mm long, strigose-canescent;

petioles half as long as to equaling blades, hirsute-villous and more or less ciliate-margined, midveins prominent beneath, lateral veins not evident, margins revolute; flowers small, nearly hidden among leaves; calyx lobes oblong, faintly to distinctly unequal, 1.5–2 mm long; corolla clear white, 2.5–3 mm long, tube cylindrical, not plicate; stamens barely equaling tube or slightly shorter; style about equaling stamens, its branches distinct to base; nutlets partially enfolded by subtending calyx lobes, smooth, shining black, 0.75–0.85 mm long, distinctly deeper dorsoventrally than wide, rounded dorsally, angled ventrally, groove widened only slightly above, scarsely more so basally.

Endemic; on ashy talus slopes and windblown sand near beaches. Specimens examined: Bartolomé, San Salvador.

Cordia L., Sp. Pl. 190, 1753; Gen. Pl. ed. 5, 87. 1754

Shrubs or trees with leaves usually broad and ample but occasionally linear-revolute, alternate, petiolate. Flowers in spicate cymes, glomerules, or panicles, or in dense capitate spikes. Calyx 5-toothed or 5–10-lobed, mostly persistent, in some circumscissile in age. Corolla white, yellow, or rusty red, small to large and conspicuous, usually 5-lobed, salverform, funnelform, to subrotate, deciduous or in some species subpersistent and papery as corollas wither. Stamens as many as corolla lobes, opposite or alternate with lobes, included or slightly exserted, the filaments inserted in corolla tube, equal or slightly unequal. Ovary 4-celled, each cell 1-ovuled. Style terminal, 2-lobed or 2-parted and each branch bilobed or remaining entire. Stigmas (2)4, elliptic, short-clavate, or capitate. Fruit spherical or ovoid (broadly acorn-shaped), becoming a thin-fleshed drupe with bony pit and mucilaginous exocarp, or in some the walls becoming dry and papery, 1–4-celled. Cotyledons plicate. Endosperm lacking.

A large genus widely distributed in the tropical and warm temperate areas of the world; particularly well represented in the Americas, but the number of species still uncertain.

Leaf blades broadly ovate to subrotund, to 10 cm long, two-thirds or more as wide; corolla 3–4 cm in diameter, showy, pale yellow; calyx conspicuously costate, 10–15 mm long; plant a shrub or small tree. . . . . *C. lutea*

Leaf blades lanceolate or linear, smaller; corolla limb 3–7 mm in diameter, relatively inconspicuous; calyx not costate, 3–5 mm long; plant a slenderly branched shrub:

- Leaves linear, strongly revolute; corolla tubular, the tube 3–4 times as long as width of limb; stems and upper surface of leaves covered with stout, short, falcate, strongly appressed, roughened hairs . . . . *C. revoluta*
- Leaves lanceolate, margins not revolute; corolla funnelform, the tube less than twice as long as width of limb; upper surface of leaves with erect or ascending hairs, not appressed-pubescent:
  - Upper surface of leaves with both simple and stout, forked, or stellate and simple hairs with conspicuously pustulate bases; calyx lobes clothed with coarse simple and forked hairs much longer than those on lower part of cup. . . . *C. anderssonii*
  - Upper surface of leaves bearing only simple, erect or slightly curved, usually more slender hairs; calyx lobes with same-sized hairs as base of cup:
    - Stems and lower surface of leaves bearing only stiff, erect, simple hairs. . . . . *C. leucophlyctis*
    - Stems and lower surface of leaves bearing scattered, erect, simple hairs and a close cover of minute, appressed, stellate hairs. . . . . *C. scouleri*

Cordia anderssonii (Kuntze) Gürke in Engl. & Prantl, Nat. Pflanzenfam. IV, Abt. 3a: 83. 1893

Varronia canescens Anderss., Kongl. Svensk. Vet.-Akad. Handl. 1853: 203. 1855, non *Cordia canescens* HBK. 1818.
Lithocardium anderssonii Kuntze, Rev. Gen. 2: 976. 1891.

Erect, moderately branched shrub 1–3 m tall, often with several stems from base, with dark brown, thin, irregularly fissured bark on older parts; seasonal twigs and leaves harshly scabrous-puberulent with stiff, stoutish, erect or spreading, simple, forked hairs, and sometimes with a few stellate hairs; petioles relatively slender, 5–15 mm long; leaf blades lanceolate, somewhat acuminate at apex, broadly cuneate at base, 3–18 cm long, 1–4.5 cm wide, dark green above, paler green beneath, 7–11 veins on each side of midrib, the margins coarsely serrate-crenulate; peduncles slender, 1.5–10 cm long, inflorescence from subcapitate to short-spicate, about 1 cm broad, exclusive of corollas; calyces narrowly campanulate, 3–5 mm high, closely and finely puberulent on basal half to two-thirds with simple, forked and stellate, brownish hairs, with additional longer, coarser, mainly simple, spreading hairs on terminal part, the lobes deltoid, with attenuate tips erect in bud but geniculately and abruptly spreading in flower; corollas funnelform, 7–9 mm long, tube white or faintly tinged with lavender without, about 2 mm broad at throat, 5–6 mm long, limb white, rotate-spreading or often the lobes reflexed soon after anthesis, 5–8 mm across at anthesis, lobes 5(7), broadly rounded; stamens as many as corolla lobes, their filaments free only for terminal 1–1.5 mm, lower part completely fused with corolla tube, entire tube below level of free portions of filaments sparsely to densely clavate-pilose, hairs longest and most numerous directly over stamen traces about halfway between base of corolla tube and free portion of filaments, only 1 vein between each pair of stamen traces; style slender, 1–1.5 mm long to base of stigmatic branches, then bifid 0.6–1 mm, the lobes entire or again bifid halfway to bases, stigmatic area about 0.3–0.5 mm long; fruits globose to broadly ovoid, 6–10 mm long, nearly as wide, flesh thin, reddish, glabrous.

Endemic; in shrubby and forested areas from near sea level into lower part of *Scalesia* Zone on several of the larger islands. Specimens examined: Española, Isabela, Pinta, Pinzón, San Cristóbal, San Salvador, Santa Cruz.

The material from Isla Española (*Howell 8706* [CAS], *Koford K-24* [DS]) has particularly dense, fine-haired pubescence on the leaves, but many of these hairs on the upper surface of the blade are forked, and the flowers are within the range of variation among specimens from other islands.

Cordia leucophlyctis Hook. f., Trans. Linn. Soc. Lond. 20: 199. 1847

Varronia scaberrima Anderss., Kongl. Svensk. Vet.-Akad. Handl. 1853: 202. 1855, non *Cordia scaberrima* HBK. 1818.
Varronia leucophlyctis Hook. f. ex Anderss., op. cit. 203.
Lithocardium galapagosanum Kuntze, Rev. Gen. 2: 976. 1891.
Lithocardium leucophlyctis Kuntze, op. cit. 977.
Cordia galapagensis Gürke in Engl. & Prantl, Nat. Pflanzenfam. IV, Abt. 3a: 83. 1893.

Open shrub 1–2.5 m tall, with 1 to several erect ascending stems from root crown; bark dark brownish gray, with scattered, rusty tan, nearly circular lenticels and fine, inconspicuous, irregular fissures; younger twigs, petioles, and peduncles densely pilosulous with erect, stiff, simple hairs; petioles 4–8 mm long; leaf blades lanceolate, cuneate at base, acute to attenuate at apex, finely but obviously rugulose, dark green, scabrous-hirsute on upper surface with erect, simple, pustulate-

Fig. 66. *Cordia leucophlyctis* (*a–c*). *Heliotropium angiospermum* (*d–f*).

based hairs, and with many minute, globose, sessile, golden glands among hairs, lower surface markedly netted-veined, paler green, pubescent and glandular, the hairs mostly longer and finer and more densely crowded than above, margins irregularly serrate-dentate; inflorescences capitate to short-spicate, on terminal peduncles 1–5 cm long, flowers compactly crowded, the heads 1–2 cm long, 8–12 mm broad; calyces broadly cup-shaped to narrowly campanulate, about 6 mm long (including the slender lobes), 4 mm broad at anthesis, 4- or 5-lobed, each lobe deltoid, about 1 mm wide at base and markedly attenuate at apex, the narrow, linear part 1–2 mm long, erect in bud, spreading in flower, or in some plants the attenuation almost lacking, whole calyx evenly puberulent with stiff, spreading, simple hairs 0.2–1 mm long, their bases somewhat pustulate, the shafts minutely scaberulous-papillate almost to tip; corolla plicate in bud, funnelform, glabrous or sometimes a few minute hairs on outer surface of lobes, tube 7–8 mm long, limb about as wide as length of tube, lobes broad and short, usually 5, spreading-rotate to slightly reflexed, crinkly, white; stamens as many as corolla lobes and either opposite to or alternate with them, inserted about 4 mm above base of corolla tube, their vascular traces pilose from near base to free part of filaments, these about 1 mm long, glabrous; anthers ovoid, 0.8–1 mm long, versatile; ovary ovoid, glabrous, 1.5–2 mm long at anthesis; style linear-subulate, 1.4–1.8 mm long, the terminal 0.4–0.6 mm bifid into slenderly subulate, glabrous stigmatic lobes, or each lobe again bifid halfway to base; fruits broadly ovoid to subspherical, 4–6 mm long, scarlet when ripe.

Endemic; among lava blocks and flows and on rocky soil, from near sea level almost to summits, on several of the islands. Specimens examined: Española, Fernandina, Isabela, Santa Cruz, Santa Fé.

This species is quite similar in general appearance to *C. anderssonii*, but the differences in the types of hairs on the young twigs and surfaces of leaves seem to hold well. The hairs on leaves of *C. anderssonii* are distinctly more robust and with heavier, more pronounced pustulate bases than those on leaves of *C. leucophlyctis*. The single vascular trace between adjacent stamen traces in *C. leucophlyctis* and *C. anderssonii* further indicates a close relationship between the two, and critical field and herbarium studies need to be carried out to determine the real taxonomic status of the endemic species of the Galápagos Islands.

One of the plants examined—*Wiggins 18485* (DS, GB)—grew in dense shade and had leaves as much as 10 cm wide and 25 cm long! Its peduncles were to 15 cm long, and the pubescence throughout was sparse as compared with that of most of the Galápagos material. Because the inflorescences were in bud only, mature flowers were not available for comparison with those of smaller-leaved plants growing in the area.

See Fig. 66 and color plate 68.

Cordia lutea Lam., Ill. 1: 421. 1791

Cordia rotundifolia R. & P., Fl. Peruv. 2: 24, *t. 148, fig. a.* 1799.
Varronia flava Anderss., Kongl. Svensk. Vet.-Akad. Handl. 1853: 201. 1855.
Cordia marchionica Drake, Ill. Fl. Ins. Mar. Pacif. 240. 1886.
Lithocardium flavum Kuntze, Rev. Gen. 2: 977. 1891.
Lithocardium rotundifolium Kuntze, loc. cit.

Varying in habit from low shrub to round-crowned tree to 8 m tall, in larger specimens the trunk 2 dm or more thick, 1–3 m to lower branches, the bark light gray, smooth, exfoliating in irregular patches; twigs numerous, usually somewhat zigzag,

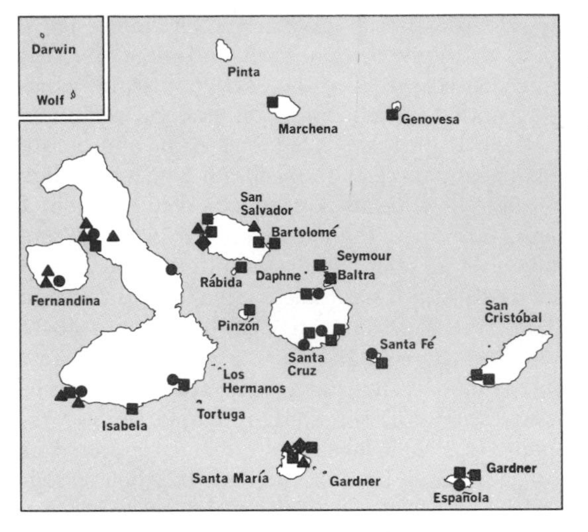

● *Cordia leucophlyctis*
■ *Cordia lutea*
▲ *Cordia revoluta*
◆ *Cordia scouleri*

closely puberulent with coarse, slightly curved, pale brownish simple hairs; petioles 5–25 mm long, densely pubescent with hairs like those on twigs; leaf blades ovate, obovate, or subrotund to broadly elliptic, mostly 4–10 cm long, 1.5–8 cm wide, dark green and scabrous with short, erect, bulbous-based hairs on upper surface, paler green and more densely pubescent with more slender hairs beneath, margins weakly and somewhat irregularly crenulate; flowers pale yellow, in several-flowered cymes extending beyond leaves at tips of twigs; pedicels stoutish, 1–12 mm long, densely pubescent with erect, crowded, yellowish hairs; calyces tubular at anthesis, 9–12 mm long, 3–4 mm wide, obviously costate with 15 well-marked, darker brown ribs and paler brown background, all densely hirsute with stellate, erect to subappressed hairs, the teeth deltoid-ovate, 2–3 mm long at anthesis, calyx soon becoming obconic as fruit develops; corolla tube about equaling calyx, then flaring into broad, funnelform throat about 1 cm long, 1.5 cm across at apex, limb rotate-spreading, 2–4 cm across, 5–8-lobed, becoming papery and subpersistent after anthesis but falling well before fruit matures; stamens as many as and opposite corolla lobes, filaments adnate to corolla tube 8–11 mm, remaining 2 cm free, sparsely hirsute from base of tube to free portion of filaments, then glabrous, 3 parallel veins in corolla tube and throat midway in each pair of stamen traces, the latter much stronger than intervening ones; anthers 3–4 mm long, linear, versatile, pale yellow; ovary smooth and glabrous; style slender, 10–12 mm long, apically divided into 2 branches 3–5 mm long, each branch again bilobed halfway to its base, these lobes linear-spatulate, stigmatic over much or all of their inner surfaces; fruit broadly ovoid, 8–12 mm wide, the fleshy, relatively thin exocarp drying firm, long-persistent, each bony cell 1-seeded. n = 36.*

Arid slopes and mesas along streams, western Ecuador, northwestern Peru, and the Marquesas Islands. Specimens examined: Baltra, Champion, Española, Gardner (near Española), Genovesa, Isabela, Marchena, Pinzón, Rábida, San Cristóbal, San Salvador, Santa Cruz, Santa Fé, Santa María, Seymour.

* Count made by D. Kyhos on buds from same plants as voucher *Wiggins 18325* (DS, GB), from Isla Santa Cruz.

Newly opened flowers are markedly fragrant. The label on the sheet collected by Meggs on Isla San Cristóbal (*Meggs s.n.* [CAS]) bears the annotation "Fls. yellow, crinkly, trumpet-shaped, with strong syringa-like scent which is wafted over the sea to approaching boats." The fragrance is particularly apparent immediately after a shower. Macbride reported that a decoction from this plant is used with good results for jaundice in Peru (1960, 582). Johnston (1952, 65–66) presents convincing arguments for believing that this species was introduced into the Marquesas by the French during the nineteenth century as an ornamental. The flowers are used there for making leis now, and the plants are well established in wild habitats.

See color plate 91.

Cordia revoluta Hook. f., Trans. Linn. Soc. Lond. 20: 199. 1847

Cordia linearis Hook. f., loc. cit., non DC. 1845.
Varronia linearis Hook. f. ex Anderss., Kongl. Svensk. Vet.-Akad. Handl. 1853: 204. 1855.
Varronia revoluta Hook. f. ex Anderss., loc. cit.
Lithocardium revolutum Kuntze, Rev. Gen. 2: 977. 1891.
Cordia hookeriana Gürke in Engl. & Prantl, Nat. Pflanzenfam. IV, Abt. 3a: 83. 1893.
Sebestena linearis von Friesen, Bull. Soc. Bot. Genève sér. 2, 24: 182. 1933.

Very slenderly branched shrub 2–4 m tall, with dark gray-brown, shallowly fissured bark marked by transversely elliptic lenticels; young twigs densely fine-pubescent with antrorsely appressed, simple hairs; leaves linear, strongly revolute, 4–7 cm long, 1.5–4 mm wide, dark green, closely appressed-pubescent, the midrib and lateral veins moderately impressed on upper surface, lower surface slightly paler and more densely appressed-pubescent, midrib markedly elevated; petioles 1–4 mm long, or the cuneate-based blade nearly sessile; inflorescence globose-capitate on a slender, appressed-pubescent, terminal peduncle 1–5 cm long; calyx broadly cup-shaped to nearly globose, 2.5–3.5 mm high, about 3 mm across, slightly contracted at apex of cup, densely puberulent with stoutish, antrorsely appressed, pale tan hairs, shallowly 5-toothed, the teeth broadly deltoid, 0.5–0.8 mm long, about as wide, erect or only slightly turned outward in flower; corolla white, tubular, 7–12 mm long, 2–2.5 mm broad, throat usually slightly narrowed, lobes very short, usually 0.5–0.8 mm long, ascending-spreading to rotate, usually strongly crinkled, tube sparingly pilose along stamen traces from base to free part of filaments, glabrous above this level; stamens 4–5, the slender filaments inserted slightly above middle of corolla tube, about 2 mm long, stamens wholly included; anthers slenderly ovoid, 0.8–1 mm long; style very slender, 8–10 mm long, glabrous, bifid to depth of 1–1.4 mm, each branch again bifid almost to its base and expanded into 2 elliptic, ascending-spreading stigmatic lobes 0.4–0.5 mm long; fruits reddish, ovoid-subglobose (acorn-shaped), 5–6 mm long and wide, acute at apex, rounded at base, minutely puberulent to glabrate and irregularly reticulate-pitted on exposed portion, the calyx widely campanulate in fruit.

Endemic; among shrubs and in forest as understory, from near sea level to 1,100 m or more on several of the larger islands. Specimens examined: Fernandina, Isabela, San Salvador, Santa María.

In contrast to the pattern among other species of endemic representatives of the genus in the Galápagos Islands, *C. revoluta* is morphologically uniform. The narrowly tubular flowers with well-included stamens, and the linear, revolute leaves, set the species well apart from the other small-flowered cordias occurring in the archipelago.

Cordia scouleri Hook. f., Trans. Linn. Soc. Lond. 20: 200. 1847

Varronia scouleri Hook. f. ex Anderss., Kongl. Svensk. Vet.-Akad. Handl. 1853: 204. 1855.
Lithocardium scouleri Kuntze, Rev. Gen. 2: 977. 1891.

Ascendingly branched shrub with dark gray (almost black) dullish branches sprinkled with circular or transversely lenticular lenticels; young twigs and undersurfaces of leaves densely appressed-pubescent with fine, several-rayed stellate hairs and a few erect, stiff, simple hairs; leaves narrowly ovate to ovate-lanceolate, 3.5–10 cm long, 1–2 cm wide, cuneate or rounded-cuneate at base, acute to attenuate at apex, dark green, minutely but obviously rugulose and hirsute-scabrous on upper surface, distinctly paler, finely reticulate-veined and more densely puberulent beneath, the midrib and veins elevated; margins shallowly undulate to short but rather coarsely serrate; petioles 5–10 mm long, pubescent like twigs; inflorescences globose-capitate, 7–10 mm across in bud, slightly elongated in fruit, on peduncles 3–5 cm long; calyx globose in bud, campanulate after anthesis, about 4 mm high exclusive of slender tips of lobes, 3–4 mm across, minutely puberulent over whole surface with mostly appressed or subappressed tan to light brown hairs, calyx lobes broadly deltoid, 1–1.2 mm long and wide, acute to blunt at apex; corollas white, tubular-funnelform, 6–8 mm high, gradually ampliate from near base to throat, this about 3–4 mm across, glabrous without, the lobes 5, very short, more or less crinkly-crispate, basal 1 mm of interior of tube glabrous, stamen traces pilose above glabrous ring up to attachment of filaments, these again glabrous; stamens inserted 2–3 mm below base of corolla limb, barely equaling funnelform part of throat, the veins between stamen traces usually starting as single trace, then branching to form 3 traces above basal 1–2 mm of corolla tube; style about equaling stamens, the bifurcation of branches delayed if present; fruits broadly ovoid, acute at apex, 3–5 mm high.

Endemic. Specimens examined: San Salvador, Santa María. Also reported from Isabela and Santa Cruz.

Heliotropium L., Sp. Pl. 130. 1753; Gen. Pl. ed. 5, 63. 1754

Herbaceous, suffrutescent, or shrubby plants with small to large leaves and glabrous to pubescent herbage. Inflorescences scorpioid spikes or racemes or flowers solitary along stems. Calyx 5-toothed or 5-lobed, persistent or deciduous, sometimes slightly accrescent in fruit. Corolla tubular with a rotate-spreading limb or narrowly funnelform, 5-lobed or 5-toothed, white, yellow, or blue. Stamens 5, inserted in corolla tube, included. Ovary 4-celled; style terminal; stigma sessile or on a distinct, slender style, the apex bearing a conical or cylindrical sterile appendage above the receptive stigmatic surface. Fruit dry, lobed or unlobed, breaking into 2 or 4 bony, 1- or 2-seeded nutlets.

Over 125 species, widely distributed in tropical and warm temperate regions; well represented in the Americas.*

Leaves succulent, glabrous and glaucous; plant a perennial herb, halophilous .... *H. curassavicum*

Leaves thin, not succulent, obviously pubescent, at least in youth; plant a suffrutescent, shrubby, or annual herb:

Plants weedy annual herbs; leaves usually 3–6 cm wide, often glabrate on upper sur-

* For further information on this genus, see Johnston (1928, 3–73).

face; nutlets pointed at apex, 2.5–3 mm high, strongly 2-costate on back; surface smooth and glabrous . . . . *H. indicum*

Plants suffrutescent, woody at base; leaves rarely over 1.5–2 cm wide, rarely glabrate; nutlets rounded at apex, if costate the ribs low, 3–5; surface closely appressed-puberulent or glabrous, conspicuously favose-reticulate:

Flowers 3–4 mm long; nutlets deeply favose, the ridges between pits thin, sharp; sepals ovate-acuminate, accrescent, to 5 mm long in fruit. . . . . *H. rufipilum* var. *anademum*

Flowers 1.5–2 mm long; nutlets smooth or weakly 3- to 5-costate, not reticulate-favose; sepals linear to lanceolate, not accrescent:

Pubescence on leaves closely appressed, antrorse, of tan to brownish hairs; sepals 2–2.5 mm long in fruit, exceeding nutlets; leaves 2–7 mm wide. . . . . *H. anderssonii*

Pubescence on leaves sparse, at least partially spreading, of white, slender hairs; sepals about 1 mm long in fruit, shorter than nutlets; leaves 1–2 cm wide. . . . . *H. angiospermum*

Heliotropium anderssonii Robins., Proc. Amer. Acad. 38: 192. 1902

Sarcanthus asperrimus Anderss., Kongl. Svensk. Vet.-Akad. Handl. 1853: 209. 1855.
Heliotropium asperrimum Anderss., Freg. Eugenes Resa, Bot. 86. 1861, non R. Br. 1810.

Small, subshrubby plant 1–4 dm tall, with woody roots and lower parts, the older branches brownish gray, antrorsely appressed-puberulent but eventually glabrate; leaves lance-ovate to linear-lanceolate, 1–2.5 cm long, 1.5–6 mm wide, acute to short-acuminate at apex, narrowly to broadly cuneate at base, entire, closely appressed-pubescent with stout, tan or brownish hairs closely crowded on both surfaces; petioles 1–2 mm long, pubescent like blade; inflorescence scorpioid, racemose or subspicate, usually 2-branched at end of flowering stems, to 10 cm long in fruit, the internodes between fruits 5–8 mm long at maturity; sepals ovate, 2–2.5 mm long, densely appressed-pubescent, slightly unequal, abaxial one longest and nearly twice as wide as others; corollas barely exceeding sepals, pale yellow, the outer surface densely and antrorsely appressed-pubescent with coarse, stiff hairs, tube 2–2.5 mm long, 0.4–0.6 mm across, lobes 0.8–1 mm long, acute, margins somewhat in-

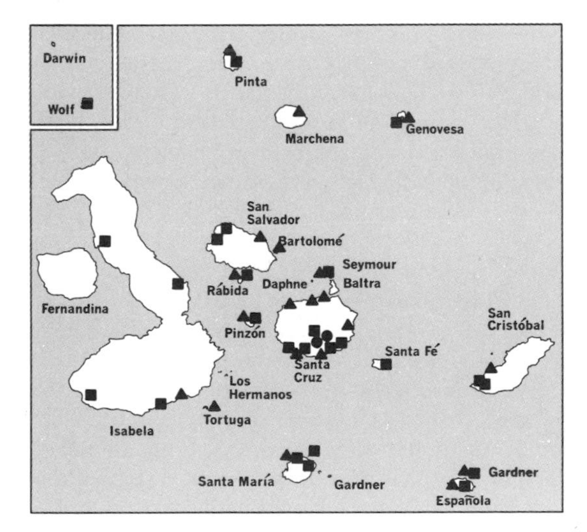

● *Heliotropium anderssonii*
■ *Heliotropium angiospermum*
▲ *Heliotropium curassavicum* var. *curassavicum*

flexed and undulate, midveins of lobes thickened, firm; stamens attached about at middle of corolla tube, filaments about 0.2 mm long; anthers narrowly lanceolate, 0.8 mm long, bearing a few minute hairs at apex; stigmatic column 0.8–1 mm high, a pronounced annulus 0.3–0.4 mm broad about one-third up from base, terminal part narrowly subulate, minutely papillate throughout, the papillae around apex nearly twice as long as those near base; fruits subglobose, flattened apically, 1–1.2 mm high, 2–3 mm broad, outer surfaces of nutlets smooth, neither sulcate nor costate, evenly rounded, closely appressed-strigose with whitish, stiff hairs conforming to curvature of nutlet; inner faces of nutlets smooth, glabrous, a conspicuous broadly elliptical pit in 1 or both faces, pale tan to nearly white, finely tuberculate pad on each face at base forming attachment of receptacle.

Endemic; arid slopes and mesa tops, in Arid Zone and lower parts of Transition Zone. Known only from Isla Santa Cruz.

Uncommon. All plants seen were in open areas, growing in potholes of clay soil or in swales where such soil has accumulated. The hairs on the leaves are strongly appressed and regularly arranged in parallel rows, and each hair is microscopically papillate-roughened. These characteristics, along with the pale yellow flowers and small, tawny, elliptical leaves, set this species apart from all other heliotropes in the Galápagos Islands.

Heliotropium angiospermum Murr., Prodr. Stirp. Götting. 217. 1770

Heliotropium parviflorum L., Mant. 2: 201. 1771.
Schobera angiospermum Murr. ex Scop., Intr. 158. 1777.
Heliotropium humile Lam., Tabl. Encycl. 1: 393. 1791.
Heliotropium synzystachyum R. & P., Fl. Peruv. 213, *t. 109a.* 1799.
Heliotropium latifolium Willd. ex Lehm., Nov. Act. Acad. Caes. Leop. Nat. Car. 9: 127. 1818.
Heliotropium scorpioides HBK., Nov. Gen. & Sp. Pl. 3: 89. 1818, non Willd. 1818.
Heliotropium patibilcense HBK., op. cit. 87.
Tournefortia synsystachya R. & S., Syst. 4: 539. 1819.
Synzistachium peruvianum Raf., Sylva Tellur. 89. 1838.
Heliotropium rugosum Mart. & Gal., Bull. Acad. Brux. 11^2: 336. 1844.
Heliotropium oblongifolium Mart. & Gal., loc. cit.
Heliophytum parviflorum DC., Prodr. 9: 553. 1845.
Heliotropium foetidum Salzm. ex DC., loc. cit.
Heliophytum portoricense Bello, Anal. Soc. Español Hist. Nat. 10: 297. 1881.

Low, suffrutescent plant 1–5 dm tall, base of stems and root woody, much of aerial parts herbaceous, the plants sometimes flowering first season; herbage moderately hirsute with simple, spreading to subappressed, whitish hairs, at least when young; leaves elliptical, ovate, or lanceolate, 1.5–10 cm long, mostly 1–2(3) cm wide, broadly cuneate at base, acute to slightly acuminate at apex, margin entire, blade dark green, smooth on upper surface, much paler beneath, bearing many globose, sessile, golden to pale brownish glands on both surfaces among hairs; petioles slender, 5–10 mm long, moderately hirsute; spikes scorpioid, tightly coiled at tip when in bud, to 2 dm long in fruit, without bracts; calyx 5-parted, sepals 1.5–2 mm long at anthesis, narrowly ovate to elliptical, acute at apex, sparsely hirsute along margins and midrib, glabrate elsewhere; corolla narrowly campanulate-rotate, white, tube and limb together about equaling calyx lobes, sparsely pilosulous along veins without, lobes broadly ovate, 0.8–1 mm long, somewhat crinkly, throat slightly constricted and densely pilose within; stamens inserted slightly above middle of corolla tube, filaments short, anthers ovoid, acute, more or less inflexed; style stout, short; stigmatic column hemispherical-ovoid, about 0.6 mm broad and high, faintly 4-lobed, minutely and sparingly penicillate toward apex and around lower margin; fruit 2-carpelled, 1.8–2 mm high, 2–2.8 mm wide, each carpel with rounded groove from base to apex on back, this flanked by 4 low costae, the whole outer surface densely covered with obtuse, whitish vesicles, each carpel biovulate. n = 13.*

In sandy or clay soils or on bare lava flows, Florida, the West Indies, and Central and South America to Argentina and Bolivia. Specimens examined: Champion, Española, Gardner (near Española), Genovesa, Isabela, Pinta, Pinzón, Rábida, San Cristóbal, San Salvador, Santa Cruz, Santa Fé, Santa María, Seymour, Wolf.

During January and February in 1964 and in the same months in 1967, this species was seen at many different localities and habitats. The inception of flowering varied greatly from one area to another. Some plants bore only buds; others already had mature fruits, with only a few flowers present. When dried material is boiled to facilitate dissection and examination, the wet herbage exudes a strong "mousy" odor.

See Fig. 66 and color plate 39.

Heliotropium curassavicum L., Sp. Pl. 130. 1753, var. **curassavicum**

Heliotropium glaucum Salisb., Prodr. 113. 1796.
Heliotropium glaucophyllum Moench, Suppl. Meth. 147. 1802.
Heliotropium chenopodioides Humb. & Bonpl. ex Willd., Enum. 175. 1809.
Heliotropium curassavicum var. *chenopodioides* Lehm., Asperif. 1: 34. 1818.
Heliotropium chilense Bertero, Mercurio Chileno 1828–29[14]: 647. 1829.
Heliotropium portulacoides DC. ex Bello, Anal. Soc. Español Hist. Nat. 10: 298. 1881.
Heliotropium curassavicum var. *parviflorum* Ball, Jour. Linn. Soc. 21: 227. 1884.
Heliotropium curassavicum var. *genuinum* Johnston, Contr. Gray Herb. 81: 14. 1928.

Much-branched decumbent perennial herb with glabrous, more or less glaucous, subfleshy herbage, forming dense mats to 1 m broad; leaves short-petiolate to subsessile, oblanceolate, narrowly elliptic, or spatulate, 1.5–6 cm long (or lowermost longer), 3–15 mm wide, rounded, acute, or short-acuminate at apex, gradually narrowing at cuneate base, petiole 1 cm long or less; inflorescence bifurcate, arising in upper axils or at tips of branchlets, scorpioid, without bracts, the peduncle to 5 cm long, flowering spikes to 10–15 cm long in fruit; sepals lanceolate-ovate, 1.4–1.8 mm long at anthesis, abaxial ones slightly larger than others, glabrous, glaucous; corolla tubular-funnelform, white with a yellowish or sometimes purplish throat, about 2 mm long, throat faintly constricted, glabrous, limb rotate-spreading, 1–2.5 mm broad; stamens inserted about 0.5 mm above base of corolla tube, filaments very short; anthers narrowly ovate, acute at apex, 1–1.2 mm long; style none; stigmatic column conical, about 0.6 mm high, 0.8 mm broad, acute or truncate at apex, faintly 4-lobed, sides minutely pilosulous; fruit depressed-globose, 1.5–2 mm high, 2–2.5 mm broad, nutlets ovoid, rounded and smooth or faintly rugulose on outer surfaces.

In wet or low places, especially in saline soils, southern California to Florida and the West Indies, and south to Patagonia. Specimens examined: Española, Gardner (near Española), Genovesa, Isabela, Marchena, Pinta, Pinzón, Rábida, San Cristóbal, San Salvador, Santa Cruz, Santa María, Seymour, Tortuga.

This species is common in the archipelago, and there has probably been a rapid expansion in its range within the area since the turn of the century. This is suported by the recent collections of the plant from a number of islands where Stewart failed to find it in 1905–6. The species does not give off the "mousy" odor when boiled to soften the flowers and other tissues for examination. Some plants were heavily infested by a white scale insect over much of the surface of the lower stems and branches.

* Count made by D. Kyhos on buds from same plant as herbarium specimens *Wiggins 18332* (DS, GB), from Isla Santa Cruz.

Heliotropium indicum L., Sp. Pl. 130. 1753

Heliotropium horminifolium Mill., Gard. Dict. ed. 8, no. 3. 1768.
Heliotropium cordifolium Moench, Meth. 415. 1794.
Heliotropium foetidum Salisb., Prodr. 112. 1796.
Tiaridium indicum Lehm., Asperif. 1: 14. 1818.
Eliopia riparia Raf., Sylva Tellur. 90. 1838.
Eliopia serrata Raf., loc. cit.
Heliophytum indicum DC., Prodr. 9: 556. 1845.

Coarse annual 1–10 dm tall, considerably branched above, herbage densely pilose on younger parts with tawny or whitish, slender, simple hairs 1–2 mm long, hairs spreading or slightly retrorse, leaves becoming sparsely pilose to nearly glabrous on upper surface in age; petioles 2.5–5 cm long, alate toward base of blades, at first pilose, later glabrate or nearly so; blades ovate to broadly lanceolate, 3–15 cm long, 2–10 cm wide, thin-herbaceous, acute or broadly acuminate at apex, acute, cuneate, rounded, or subcordate at base, dark green above, paler and prominently dark-veined beneath, the margins entire to undulate-crenate; spikes ebracteate, solitary instead of paired, scorpioid, 1–3 dm long at maturity; calyx lobes linear to linear-lanceolate, 2.5–3 mm long, hirsute along margins, otherwise nearly glabrous; corolla bluish or violet or sometimes white, tubular-funnelform, tube 3–5 mm long, antrorsely hispidulous without, limb 2–4 mm across, throat somewhat constricted, glabrous; anthers linear-subulate, 0.8–1 mm long, acute at apex, on very short, slender filament, inserted one-third up from base of corolla tube, glabrous; style about 0.5–0.6 mm long; stigma capitate, faintly 4-lobed, 0.2–0.3 mm high, 0.4 mm thick; fruit glabrous, prominently 4-lobed, lobes divergent, each nutlet eventually free, angled, 2.5–3.5 mm long, drawn out to acute beak about 0.5 mm long, 2- or 3-costate on back, black and smooth, 1-seeded.

In damp soil or in moderate shade and as a weed in clearings, the southern United States, the West Indies, and Mexico south to northern Argentina and Paraguay, and in the tropics of the Old World. Specimens examined: San Cristóbal, Santa Cruz, Santa María.

Apparently a relatively rare plant in the Galápagos Islands.

Heliotropium rufipilum (Benth.) Johnston var. **anademum** Johnston, Contr. Gray Herb. 81: 44. 1928

Slenderly branched suffrutescent plant or small shrub 0.5–2 m tall, the younger branches, leaves, and peduncles pilose with white to tawny, spreading, more or less tangled hairs 0.5–2 mm long, the larger trichomes conspicuously white-pustulate at base; petioles 2–4 mm long, densely pilose and with many spherical, sessile, yellowish or brownish glands among bases of hairs; leaf blades ovate, rounded at base, acute- to short-acuminate apically, 1.5–3.5 cm long, 5–15 mm wide, impressed-venose, rugulose and pilose above, veins elevated and surface densely pilosulous and glandular beneath; peduncles 1–3 cm long, bearing 2–3 scorpioid spikes, these 10–15 cm long in fruit; sepals broadly ovate, 3–3.5 mm long at anthesis, pilose with fine hairs and bearing a few coarser, larger trichomes along margins and on midribs, gland-dotted among hairs, accrescent and becoming 4–5 mm long and somewhat papery in fruit; corollas white, narrowly tubular-funnelform, about 4 mm long, limb 3.5–4.5 mm across, tube and lobes finely pilosulous without, throat plicate-constricted, glabrous within; anthers linear, 1.2–1.4 mm long, borne on minute filaments about 0.1 mm long, these inserted equally about 1–1.2 mm above base of corolla tube, this glabrous within; style stoutish, about 0.2 mm long; stigmatic col-

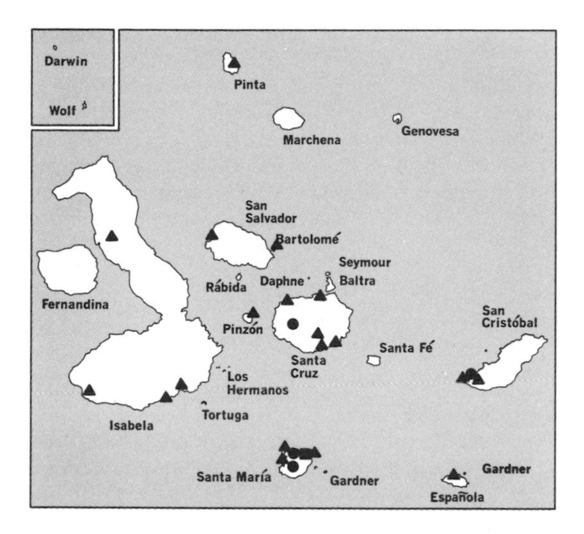

● *Heliotropium indicum*

■ *Heliotropium rufipilum* var. *anademum*

▲ *Tournefortia psilostachya*

umn 0.6–0.8 mm high, 0.4–0.5 mm thick at base, tapering slightly from base to apex, the latter 0.3–0.4 mm thick, faintly crateriform at apex; nutlets ovoid, dull greenish to dull black, 1.4–1.8 mm high, deeply and relatively coarsely favose-reticulate, the ridges thin, minutely papillate but devoid of glands.

In heavy clay soils, usually in openings in the forest cover, Ecuador, Peru, and Bolivia. Specimen examined: Santa María, a new record for the Galápagos Islands and, so far as is known, the only collection of this species from the islands. There are no traces of glandular structures on the nutlets, so it most certainly belongs with Johnston's var. *anademum*.

Macbride (1960, 567) states that he believes *H. submolle* Klotzsch and *H. urbanianum* Krause may prove to be within the limits of *H. rufipilum*. I am inclined to the same opinion, for the specimen examined (*Schimpff 200* [CAS]) has the rugulose, impressed-veined leaves that, according to Johnston's key, would lead to *H. urbanianum*. The characters used to separate these three "species" are not impressive.

Tournefortia L., Gen. Pl. ed. 5, 68. 1754

Shrubs or woody vines with ample leaves and flowers in scorpioid racemes or spikes borne in dichotomous panicles. Calyx usually deeply 5-lobed, persistent. Corolla small, white, greenish, or yellowish, with cylindrical tube and 5-lobed, spreading limb. Stamens usually 5, with short filaments inserted on corolla tube, included. Ovary 4-celled; style solitary, terminal. Stigma sessile or on short but distinct style, peltate or conical, receptive on sides, usually bifid. Fruit a drupe, lobed or unlobed, at maturity breaking into 2 or 4 bony nutlets, these 1- or 2-seeded, often with abortive cavities. Endosperm thin; cotyledons flat.

About 100 poorly defined species, widely distributed in the tropics of the Old and New Worlds; well represented in North and South America.

Plant more or less vinelike, stems usually more or less decumbent, with short, erect laterals 1–5 dm tall; corolla lobes subulate, several times as long as broad; anthers connivent and connate at apices . . . . *T. psilostachya*

Plant an erect shrub 1–3 m tall, stems not scandent-decumbent; corolla lobes short, rounded or truncate at apex, much broader than long; anthers erect, free, neither connivent nor connate:

Calyx cup-shaped, the lobes broadly deltoid-ovate, extending no more than halfway to base of calyx; stigma appendages triangular-subulate, equaling or exceeding style; pubescence short, of appressed white or whitish hairs . . . . *T. pubescens*

Calyx campanulate, the lobes distinct almost or fully to base of calyx; stigma appendages broadly deltoid-ovoid, shorter than style; pubescence of longer, rufous, often spreading hairs . . . . *T. rufo-sericea*

Tournefortia psilostachya HBK., Nov. Gen. & Sp. Pl. 3: 78. 1818

Tournefortia floribunda HBK., op. cit. 79.
Tournefortia difformis Anderss., Kongl. Svensk. Vet.-Akad. Handl. 1853: 206. 1855.
Tournefortia strigosa Anderss., op. cit. 207.
Tournefortia cirrhosa Vaupel, Bot. Jahrb. 54. Beih. 119: 3. 1916.
Tournefortia volubilis in litt., pro parte, non L.

Spreadingly branched scandent plant woody only at base, the main stems and branches slender, smooth, glabrous; bark grayish brown with pale, tawny elliptic lenticels 0.5–1 mm long, the long axes parallel with axis of stem; young branches often strigose; petioles slender, shallowly canaliculate, 3–10 mm long, puberulent to closely strigose; leaf blades thin, ovate-lanceolate to elliptic-lanceolate, 1–4 cm wide, rounded to cuneate basally, acute to acuminate at apex, entire, sparsely strigillose to densely strigose, the hairs often markedly white-pustulate at base; inflorescences in upper axils and terminal on many branchlets, usually 2–4-branched, branches scorpioid in youth, slender, strigillose, usually 2–6 cm long in fruit; flowers sessile, about 5–7 mm long at anthesis, calyx 5-parted to base, closely strigillose, lobes lance-subulate, 1.6–2 mm long, erect at anthesis, tips connivent after fall of corolla; corolla tube yellowish or greenish yellow, 4–5 mm long, flaring slightly just above calyx lobes, inwardly plicate below sinuses, strigillose without, lobes linear-subulate, 2–2.5 mm long, spreading at anthesis, reddish brown and glabrous within, slightly pale greenish brown and strigose without, margins of lobes inflexed, especially on broad basal part; anthers subsessile, inserted at base of flare in throat, subulate, about 1 mm long, connivent and connate at apices (sometimes connivent in pairs, with 1 free at tip), included; style relatively slender, glabrous, it plus stigma almost equalling and overarched by the connivent anthers; stigma annular, 0.8–1 mm thick, surmounted by bilobed appendage, it and stigma together appearing subconic, about 0.4–0.6 mm long; fruit 5–6 mm long, 2–4-lobed, or globose when only 1 nutlet matures, flesh thin, yellowish to orange-yellow, in ours always glabrous.

Mostly in shade of other shrubs and forest trees, sometimes in full sun, sea level to 400 m or more, northwestern South America. Specimens examined: Champion, Española, Isabela, Pinta, Pinzón, San Cristóbal, San Salvador, Santa Cruz, Santa María.

Tournefortia pubescens Hook. f., Trans. Linn. Soc. Lond. 20: 198. 1847

Tournefortia opaca Anderss., Kongl. Svensk. Vet.-Akad. Handl. 1853: 205. 1855.

Erect shrub 1–4 m tall, with pale cinnamon brown, shallowly fissured or smooth bark lacking noticeable lenticels, the younger branchlets glabrous or moderately to densely puberulent with white to rusty, often tangled hairs; petioles 0.5–2 cm long, mostly rufous-puberulent; leaf blades elliptical, ovate, or lanceolate-elliptical, 2.5–15 cm long, 1.5–8 cm wide, dark green and sparsely puberulent to glabrate on upper

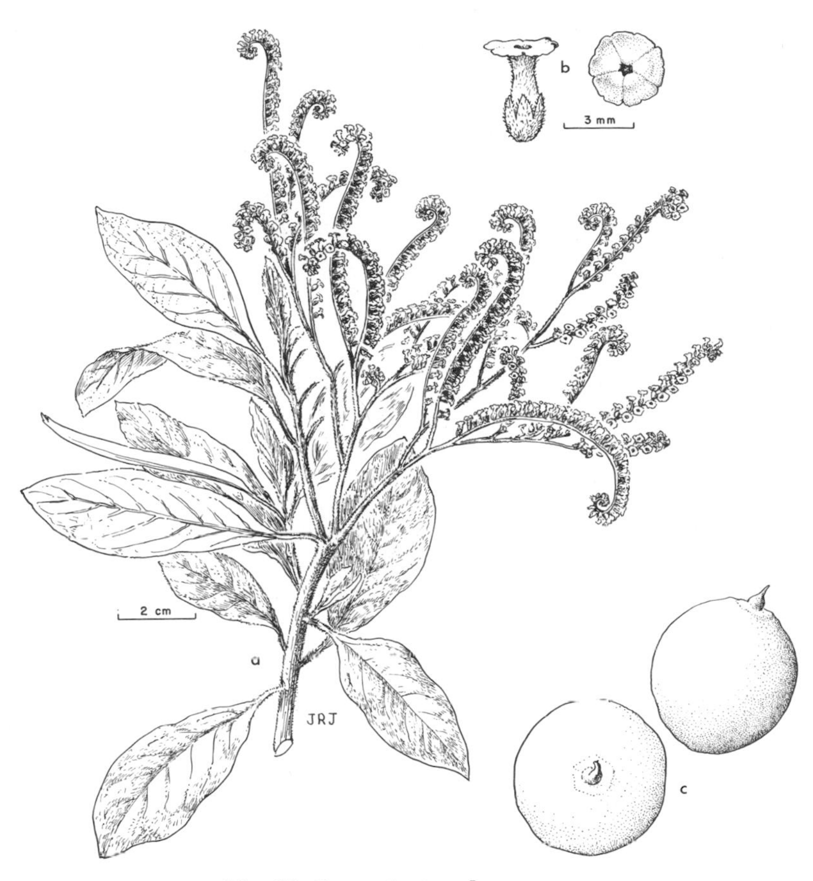

Fig. 67. *Tournefortia pubescens.*

surface, slightly paler and more densely puberulent to tomentulose beneath, rarely somewhat rugulose, margins entire, broadly cuneate to rounded at base, acute to short-acuminate at apex; inflorescences in upper axils and terminal on some branchlets, usually considerably branched, without bracts, the branches scorpioid, 1–6 cm long in fruit; flowers fragrant, sessile, 4–5 mm high at anthesis, corollas falling soon after opening; calyx broadly cup-shaped, 1.5–2 mm high, 5-lobed about to middle, lobes equal, broadly deltoid-ovate, acute to rounded, puberulent on outer surface; corolla white, often with yellow "eye," tube flaring at level of calyx lobes to form slightly wider throat, strigose-puberulent on outer surface, glabrous within, orifice slightly constricted, lobes spreading, broadly rounded to nearly truncate, limb about 4 mm across, puberulent on outer surface, sparingly so to glabrous within; anthers subsessile, 1 mm long, inserted in lower part of throat, tips thus nearly reaching orifice, all distinct, neither connivent nor connate; style stoutish, it and ovary together about 1–2 mm long, glabrous, stigmatic ring 0.6–0.8 mm across, ruminate-roughened, appendages conic, 1–1.2 rım long, greenish; fruit globose, 3–6 mm long, obscurely or not at all lobed, flesh thin, white, styles persisting for some time after maturity.

Endemic; in clay, cinders, sandy soil, or among lava blocks and flows, from near sea level to about 800 m or more. Specimens examined: Fernandina, Isabela, Pinzón, San Cristóbal, San Salvador, Santa Cruz, Santa María, Wolf.

Quite variable with respect to density of pubescence and size of leaves.

See Fig. 67 and color plates 45 and 46.

Tournefortia rufo-sericea Hook. f., Trans. Linn. Soc. Lond. 20: 197. 1847

Erect or ascending shrub 1–5 m tall, with reddish brown bark on older stems, younger ones spreading-pilose with reddish hairs to 2 mm long, these often densely tangled; petioles stoutish, 2–3 mm wide, canaliculate adaxially, 1–2.5 cm long, densely rufous-pilose; leaf blades broadly elliptic or sometimes ovate, 5–25 cm long, 2–13 cm wide, dark green, pubescent to glabrous, often rugulose above, velvety-rufous-pilose beneath, rounded or broadly cuneate at base, acute to short-acuminate at apex, entire; inflorescences terminal on branches and branchlets, compact and considerably branched, often 10–20 cm across in fruit, densely rufous-sericeous; spikes scorpioid; flowers 4–6 mm high, rather crowded; calyx 5-parted almost to base, the lobes 2.5–4 mm long, lanceolate, slightly unequal, densely rufous-sericeous on outer surface, acutish at apex; corollas white, fragrant, 4–6 mm long at anthesis, lobes about 3 mm long, glabrous within, sericeous on outer surface, broad and truncate to rounded, the expanded limb 3.5–4 mm across; anthers subsessile, included, 1–1.2 mm long, inserted 1.8–2.6 mm below orifice, erect, distinct, free; style very short, stout, it and ovary together 1.5–2.2 mm long; stigmatic ring about 1 mm across, roughened, the terminal appendages subconic, about 0.5 mm high; fruit globose, 5–6 mm in diameter, flesh white-translucent.

Endemic; mostly in shaded forest or among other shrubs, primarily above the Transition Zone. Specimens examined: Isabela, Pinta, San Cristóbal, San Salvador, Santa Cruz, Santa María.

In the habitats near sea level and in the Transition Zone, where *T. rufo-sericea* and *T. pubescens* grow together or contiguously, the former occupies the moister microhabitats, *T. pubescens* the drier ones.

CAMPANULACEAE. Bellflower Family*

Annual or perennial herbs or subshrubs, rarely arborescent, with watery or milky juice. Leaves alternate or rarely opposite or whorled, simple, exstipulate. Inflorescences basically cymose, but appearing racemose or thyrsiform, occasionally the flowers solitary in axils of leaves or involucrate heads, usually bracteate and often bracteolate. Flowers perfect, regular to slightly irregular; calyx lobes 3–10, usually 5, imbricate or valvate in bud. Corolla campanulate, tubular, or bilabiate, often split down 1 side in irregular flowers, the lobes or segments usually 5, mostly clearly gamopetalous, sometimes choripetalous and in a few genera lacking in cleistogamous flowers. Stamens as many as corolla lobes (or petals) and alternate with them, the filaments often broad at base, frequently connate into tube or domelike structure above disk and ovary, usually borne on base of corolla tube or hypogynous; anthers 2-celled, introrse, distinct or connate in ring or tube, dehiscing longitudinally along 1 side. Pistil 1; ovary inferior or half-inferior, 3–10-lobed, 2–5-carpelled, with axile placentation in most but with parietal placentation in some. Ovules numerous, anatropous. Style 1, slender, with 2–5 stigma lobes. Fruit a capsule de-

* Contributed by Ira L. Wiggins.

hiscing by apical slits, circumscissile, or by apical or basal pores (a berry in a few genera). Seeds small, numerous; embryo straight, endosperm fleshy and copious.

About 60 genera and 1,500 species, widely distributed, mostly in temperate and subtropical regions.

Lobelia L., Sp. Pl. 929. 1753; Gen. Pl. ed. 5, 401. 1754

Annual or perennial herbs or sometimes suffrutescent, with alternate leaves, the upper reduced to bracts in several genera. Flowers blue, red, yellowish, or white, appearing racemose. Calyx tube short, semiglobular, adnate wholly or in part to ovary, its lobes 5, equal, subequal, or slightly bilabiate. Corolla tube often split to base along 1 side; limb 5-lobed, lower 3 lobes forming lower lip, upper pair (1 on each side of split) erect or recurved, the flower often inverted to place the 3-lobed lower lip uppermost. Anthers united into a ring around style, 2 of them, or all 5, more or less hairy with straight, stiffish, spreading hairs. Ovary comparatively large, inferior, 2-loculed, many-ovuled. Fruit a 2-valved capsule dehiscing by 2 terminal pores. Seeds minute, numerous, ellipsoidal to subglobose, sometimes trigonous or angled, minutely striate, reticulate, tuberculate-scabrous or smooth and shining.

About 365 species widely distributed from cool temperate to tropical regions.

Lobelia xalapensis HBK., Nov. Gen. & Sp. Pl. 3: 315. 1818

Lobelia monticola HBK., op. cit. 316.
Lobelia mollis Graham, Edinb. New Phil. Jour. 185. 1829.
Rapuntium affine Presl, Prodr. Monogr. Lobel. 25. 1836.
Rapuntium monticolum Presl, loc. cit.
Rapuntium xalapense Presl, loc. cit.
Lobelia ocimodes Kunze, Linnaea 24: 178. 1851.
Lobelia cliffortiana var. *xalapensis* A. Gray, Synopt. Fl. N. Amer. 2[1]: 7. 1878.
Dortmannia cliffortiana Kuntze, Rev. Gen. Pl. 2: 380. 1891.
Dortmannia monticola Kuntze, op. cit. 972.
Dortmannia mollis Kuntze, loc. cit.
Dortmannia ocimodes Kuntze, op. cit. 973.
Dortmannia cliffortiana var. *xalapensis* Kuntze, op. cit. 3: 187. 1898.

Erect, moderately to profusely branched annual herb 1–5 dm tall, with puberulent to glabrescent branches and leaves; stems often reddish; leaf blades opaque, petio-

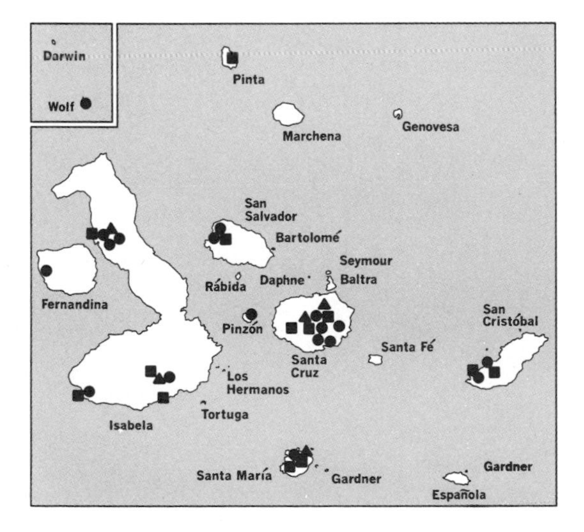

● *Tournefortia pubescens*
■ *Tournefortia rufo-sericea*
▲ *Lobelia xalapensis*

Fig. 68. *Lobelia xalapensis.*

late, ovate, 8–35 mm long, 5–20 mm wide, rounded to broadly cuneate at base, acute to short-attenuate at apex, sinuate-crenate; petioles slightly winged, gradually widening toward leaf blade, to 1.5 cm long; inflorescence a narrow, bracteate raceme, 1–3 dm long; bracts linear-subulate, 2–4 mm long; pedicels very slender, ascending, puberulent or rarely glabrescent, 5–9 mm long in flower; calyx tube campanulate to semiglobose, about 1 mm long, slightly wider at anthesis, sepals linear-subulate, 2–2.5 mm long, about 0.5 mm wide, purplish near tips or full length, glabrous in ours; corolla tube about equal to calyx lobes, persistent, parchment-like in fruit, the lobes about 0.5 mm long, pale lavender to white; 2 anthers hirsute along inner face; capsule ellipsoidal, 4–5 mm long, wholly inferior or slightly free at distal end; seeds ellipsoidal, slightly compressed, 0.4–0.5 mm long, reddish or reddish brown, smooth and shining. $n = 7$.*

In forests, along streams and trails, and rarely in open areas, central Mexico to Peru and Argentina. Specimens examined: Isabela, Santa Cruz, Santa María.

See Fig. 68.

CARICACEAE. PAPAYA FAMILY†

Trees, shrubs, or large perennial herbs, with milky latex in all parts; plants usually dioecious or monoecious. Leaves alternate, usually large and long-petiolate, palmately lobed or compound, rarely entire; stipules absent. Inflorescences in axils of upper nodes; staminate in extensive and highly compound thyrse; carpellate much less compound, with fewer and larger flowers. Flowers usually unisexual, 5-merous, hypogynous, sympetalous but usually imperfectly so in carpellate flowers. Calyx 5-lobed, small and inconspicuous, imbricate or nearly open in bud. Corolla slender and elongated or funnelform, more or less valvate or contorted in bud; staminate flowers with well-developed tube, limb 5-lobed; carpellate flowers short-tubed, more or less campanulate and imperfectly sympetalous. Stamens (5)10, in (1)2 series, inserted near mouth of corolla tube; filaments free or connate, reduced to staminodes or absent in carpellate flowers; anthers 2-loculed. Gynoecium 3–5-carpelled, syncarpous; ovary superior, usually 1-loculed and placentation parietal, rarely 3–5-loculed and placentation axile; ovules numerous, anatropous; style short or absent; stigmas 3–5, sessile or essentially so, reduced or absent in staminate flowers. Fruit a 1–5-loculed fleshy berry. Seeds numerous; endosperm fleshy; embryo straight.

A predominantly tropical American family of 5 genera and about 70 species, with 1 genus indigenous to tropical Africa.

Carica L., Sp. Pl. 1036. 1753; Gen. Pl. ed. 5, 458. 1754

Plants dioecious or monoecious; trunk soft-wooded, stout, usually simple and unbranched; leafy at apex. Leaves large, spreading, subpeltately palmate, sometimes digitately 3–9-foliolate, or rarely oblong. Inflorescences often long-pedunculate; bracts small or absent. Corolla slender and elongate in staminate flowers, lobes oblong to linear; campanulate in carpellate flowers, lobes oblong to linear. Stamens 10 in 2 series, free, connective occasionally produced at apex, 5 opposite calyx lobes

* Count made by D. Kyhos on buds from same plant as herbarium specimens *Wiggins 18783* (DS), from Isla Santa Cruz.

† Contributed by Duncan M. Porter.

sessile or subsessile, the 5 opposite corolla lobes on short filaments; staminodes absent or rarely present and minute in carpellate flowers. Ovary 1-loculed and usually septate basally, or 5-loculed with spurious septa, rudimentary in staminate flowers; ovules 2- or several-seriate on 5 parietal placentae; style very short or absent; stigmas 5, linear or variously divided. Berry small to very large. Seeds ovoid, arillate, the testa smooth, rugose, or echinate; embryo axile; cotyledons oblong, flat.

A tropical American genus of 57 species, occurring from Mexico to Argentina; cultivated in many tropical areas.

Carica papaya L., Sp. Pl. 1036. 1753*

Trees or shrubs 2–8 m high, usually glabrous or nearly so; trunk straight, unbranched, heavily ornamented with large, seemingly spirally arranged leaf scars; leaves in dense terminal crown, long-petiolate, to 6 dm or more broad, simple but usually 5–7-lobed, lobes pinnatifid and acute to obtuse; staminate inflorescences pendulous, a little shorter than subtending petioles, carpellate inflorescences much shorter and usually 1–3-flowered; staminate flowers with calyx lobes broadly deltoid, obtuse, 1–1.5 mm long; corolla creamy white, tube 1.5–2 cm long, about 2 mm broad, lobes lanceolate to oblong, 1–1.5 cm long, slightly spreading; anthers opposite corolla lobes about 2.5 mm long, nearly sessile and dehiscing basally into corolla tube, those opposite calyx lobes erect and wholly exserted, about 2 mm long, the connective 2-lobed, about as long as anther; rudimentary ovary slender, about 1 mm high; carpellate flowers with calyx lobes broadly deltoid, 5–10 mm long; corolla irregularly campanulate, creamy white, essentially polypetalous or petals very weakly connate, petals lanceolate, 5–7 cm long, 1.5–2 cm wide, irregularly reflexed apically; ovary ovoid, about 3 cm high including style; stigmas about 1 cm long, irregularly branched; berry extremely variable in size and shape, to 3 dm long and 2 dm wide, yellow or orange and more or less streaked with green at maturity, pericarp yellow or orange, sweet and juicy, developing at base of young growth and appearing cauliflorous.

Probably native to the eastern slopes of the Andes, but cultivated and widely adventive in lowland tropical areas. Andersson (1855) reported this species from Isla Santa María, and it has been observed along the trail between Academy Bay and Bella Vista on Isla Santa Cruz. Elsewhere in the Galápagos Islands it is likely to be found in the vicinity of habitations or around abandoned farm sites.

See Fig. 69.

COMPOSITAE (ASTERACEAE). Composite or Aster Family†

Herbs or less commonly shrubs or even trees, with opposite to alternate or whorled, simple to compound or dissected leaves without stipules. Flowers sessile in close head on common receptacle, flowering in centripetal sequence within head, sometimes individually subtended by a small bract (chaff), and nearly always collectively surrounded by an involucre of few to many bracts. Heads arranged in various sorts of secondary inflorescences nearly always showing some indication of cymose (centrifugal) sequence of development. Individual flowers epigynous, perfect

* For complete synonymy of this species, see Badillo (1967).

† Contributed by Arthur Cronquist.

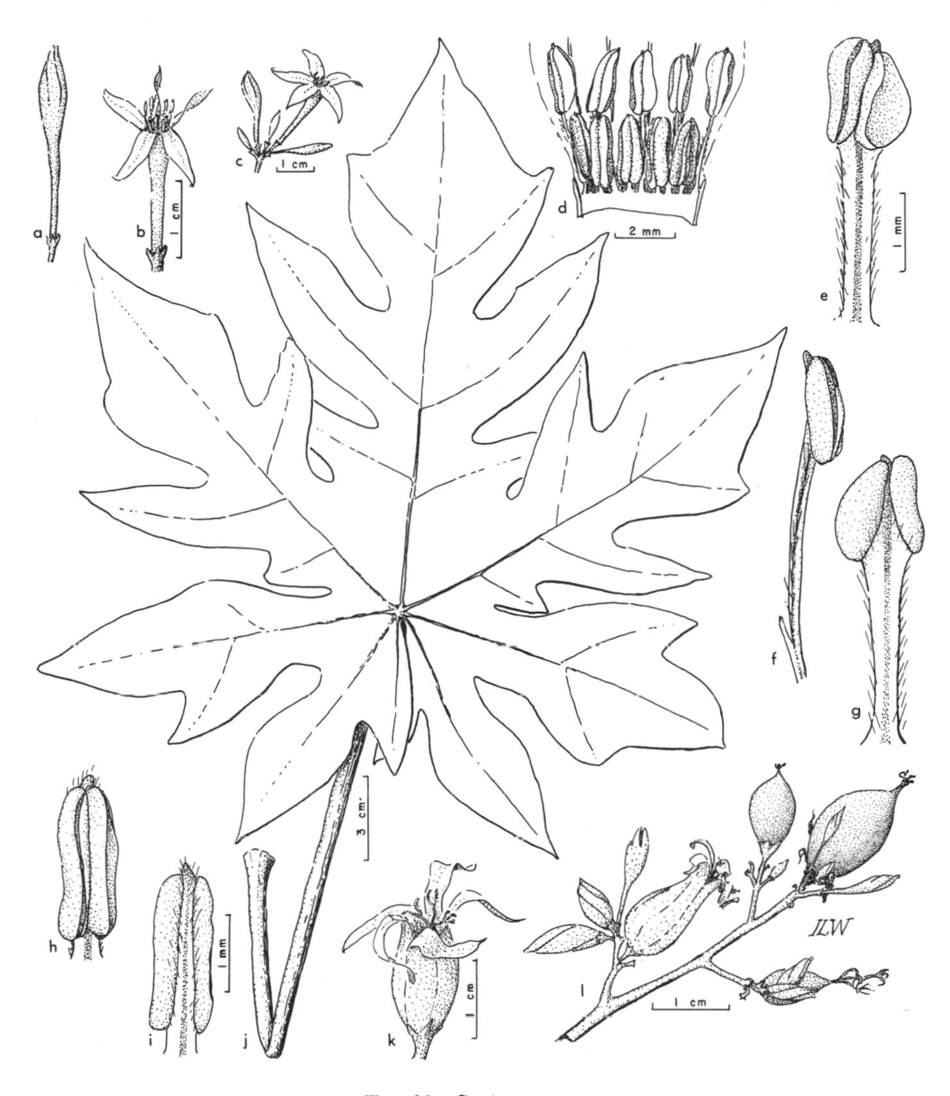

Fig. 69. *Carica papaya.*

or unisexual, gamopetalous, regular or irregular, commonly (4)5-merous, without definite calyx. Stamens as many as corolla lobes and alternate with them, epipetalous, with elongate anthers united into a tube, or anthers rarely free. Ovary inferior, of 2 carpels, 1-celled, normally with single erect anatropous ovule. Style usually 2-cleft. Fruit an achene, unappendaged, or more commonly crowned with a pappus of hairs or scales.

The involucral bracts are usually herbaceous or subherbaceous in texture, varying to scarious, hyaline, or cartilaginous; they may be few and in a single row, or numerous and imbricate, or modified into spines. The receptacle may be chaffy, with a bract subtending each flower, or may be covered with long stout bristles, or may be naked; when naked it may sometimes be minutely pitted, with slender chaffy partitions separating the pits (alveolate), and may even be shortly hairy. The number of flowers in each head may be as few as 1 (as in pistillate heads of species of *Ambrosia*) but is seldom less than 5, and sometimes runs to hundreds.

The flowers are of several general types. In one type they are perfect (or functionally staminate), the corolla tubular or trumpet-shaped, with typically 5 short terminal lobes. This is the *disk flower*; and a head composed wholly of disk flowers is *discoid*. In another type the flower is pistillate or neutral (without a style), and the corolla tubular only at the extreme base, above which it is flat, commonly bent to one side, and often exhibits traces of 2 or 3 of the lobes as small terminal teeth. The flattened part of a corolla of this type is a *ray* or *ligule*, and the flower bearing it is known as a *ray flower*. Except for a few dioecious groups, the head is never composed solely of flowers of this type; instead these pistillate or neutral ray flowers are found at the margin of the head, the center being occupied by perfect (or functionally staminate) flowers. Such a head, having both ray and disk flowers, is *radiate*. In some species the ray or ligule of the marginal, pistillate flowers fails to develop, so that the corolla is tubular. In addition to the absence of stamens, a corolla of this type may be distinguished from that of an ordinary disk flower by the absence of the regular terminal teeth, and often also by being much more slender. A head in which the pistillate flowers lack rays is *disciform*. Another type of flower, superficially resembling the ray flower of a radiate head, but differing in being perfect and in having 5 terminal teeth on the ligule, is characteristic of the tribe Cichorieae (elsewhere in the family it is found in fully developed form only in certain genera of the Mutisieae not represented in the Galápagos Islands); the heads of the Cichorieae, *ligulate heads*, consist solely of flowers of this type. Yet another type of flower, found chiefly in the Mutisieae (represented in the Islands by *Jungia*), is *bilabiate*, with the outer lip generally larger than the inner.

The pappus is highly diverse in structure. Located at the summit of the achene, it has commonly been regarded as a modified calyx (the interpretation here accepted), though it has also been thought to be merely an outgrowth from the achene. It may be composed of simple or plumose hairs in one or more series, or scales, or stout awns, or a mere projecting ring or crown, or combinations of these, or it may be lacking entirely.

The anthers are coherent by the lateral margins (rarely free); their bases vary from nearly truncate to slenderly caudate, in which case the slender tails are sterile, lacking pollen; usually there is also a short, sterile, terminal appendage. The anthers dehisce introrsely, and the pollen is pushed out through the anther tube by the growth of the style, which acts as a piston or plunger. The style branches commonly diverge above the anther tube, have various distinctive forms and textures, and tend to be stigmatic only on definite parts of their surface. The characteristic style branches of the various tribes are to be sought only in the fertile disk flowers; those of the rays are mostly very similar in all groups, and those of the sterile disk flowers are often reduced and undivided. The sterile disk flowers, when present, are said to be *functionally staminate*. Strictly staminate flowers, with no pistillate parts, are not found in the Compositae (except perhaps as an individual abnormality), since the style is necessary as a plunger to eject the pollen.

One of the largest families of flowering plants (in most parts of the temperate zones the largest), with certainly more than 15,000 species, cosmopolitan in distribution. The flower heads, often simulating a single flower and misinterpreted as such by the layman, vary from small to very large and are often brilliantly colored. Many species are cultivated for ornament.

KEY TO THE TRIBES OF THE COMPOSITAE*

(The characters used in these keys apply to the species in the Galápagos Islands.)

Heads ligulate, i.e., the flowers all perfect and with a distinct, minutely 5-toothed ligule; herbs with milky juice. . . . . .CICHORIEAE
Heads variously radiate (with both ray and disk flowers), disciform, or discoid (with only disk flowers), never truly ligulate; herbs, shrubs, or trees, the juice not milky:
  Heads with the flowers all bilabiate, the outer lip of the corolla longer than the inner, and, at least in the marginal flowers, liguloid; leaves alternate; anthers tailed; ours coarse herbs with uniseriate involucre, chaffy receptacle, and plumose pappus. . . . . .MUTISIEAE
  Heads with none of the flowers bilabiate, the corolla either tubular and regular or nearly so, or ligulate and without an inner lip; other characters diverse, but not combined as above:
    Plants either (1) with the receptacle chaffy, or (2) with conspicuous, embedded oil glands in leaves and bracts, or (3) with both opposite leaves and yellow flowers, or (4) with 2 or all 3 of these features. . . . . .HELIANTHEAE
    Plants with naked receptacle, without conspicuous, embedded oil glands in leaves and bracts (though sometimes glandular-hairy, or atomiferous-glandular, or finely glandular-punctate), and not at once with opposite leaves and yellow flowers:
      Heads radiate, or disciform (i.e., the marginal flowers pistillate, with tubular-filiform, eligulate corolla, the central flowers perfect or functionally staminate and with coarser, tubular corolla), or if truly discoid then unisexual; disk flowers often yellow:
        Anthers entire to merely sagittate at the base; plants not white-woolly; heads variously radiate to disciform. . . . . .ASTEREAE
        Anthers caudate at the base; plants white-woolly herbs; heads disciform. . . . . INULEAE
      Heads strictly discoid, the flowers all perfect, never yellow:
        Leaves opposite; style branches merely papillate (not hairy), often somewhat clavate . . . . .EUPATORIEAE
        Leaves alternate; style branches slender and gradually tapering to tip, minutely hispidulous externally . . . . .VERNONIEAE

KEY TO THE GENERA OF THE COMPOSITAE, BY TRIBE

ASTEREAE

Heads unisexual, the plants dioecious; pistillate flowers rayless; plant a shrub. . .*Baccharis*
Heads bisexual, obviously radiate to merely disciform; plant a shrub or herb:
  Plant a shrub; heads obviously radiate, the ligules mostly 2.5–5 mm long. . . . . *Darwiniothamnus*
  Plant an herb; heads inconspicuously radiate or disciform, the ligules less than 1 mm long . . . . *Conyza*

CICHORIEAE

A single genus . . . . *Sonchus*

---

* Three species of Compositae probably not occurring in the Galápagos Islands in recent times (*Tagetes erecta* L., *Hemizonia squalida* Hook., and *Haplopappus lanatus* Hook.) are discussed briefly on pp. 366–67; their genera, not otherwise represented in the archipelago, are excluded from these keys.

EUPATORIEAE

Anthers truncate at tip, without terminal appendages; pappus of 3–5 short, stout, blunt awns, these glandular above the middle .... *Adenostemma*
Anthers with a short, pointed, sterile appendage at tip; pappus otherwise:
 Pappus of 5 or fewer awn-tipped or blunt scales, or sometimes reduced to a mere crown, or obsolete; achenes as in *Eupatorium* .... *Ageratum*
 Pappus of more or less numerous capillary bristles:
  Achenes 5-angled .... *Eupatorium*
  Achenes 10-ribbed .... *Brickellia*

HELIANTHEAE

Involucre strongly flattened to triquetrous, consisting of 3–5 bracts, the lowermost flat and much the largest, dominating the involucre; opposite-leaved annual herbs with only 2–7 flowers per head .... *Elvira*
Involucre distinctly otherwise, cylindric to hemispheric or rotate, not dominated by any one bract; habit and flowers various:
 Heads unisexual, bearing either a pistillate flower without a corolla, or functionally staminate flowers with tubular corolla, both types produced on same plant. . *Ambrosia*
 Heads bisexual, with both ray and disk flowers, or with disk flowers perfect and fertile, or both; ray flowers pistillate or neutral, always provided with a corolla:
  Involucral bracts only (2)3 .... *Flaveria*
  Involucral bracts (4)5 or more:
   Receptacle naked; involucre and leaves dotted with conspicuous, embedded, translucent oil glands:
    Heads radiate (rays sometimes short and inconspicuous); disk yellow; involucre relatively small, less than 1 cm high .... *Pectis*
    Heads strictly discoid; disk tinged with purple; involucre large, 1.5 cm or more high .... *Porophyllum*
   Receptacle chaffy; involucre and leaves not dotted with oil glands, though leaves may be glandular-pubescent or finely glandular-punctate:
    Plants low, diffuse annuals with alternate, dissected leaves and small, yellow heads .... *Chrysanthellum*
    Plants otherwise, variously annual to perennial herbs or woody plants, the annuals all with principal leaves opposite:
     Disk flowers functionally staminate, only ray flowers fertile:
      Leaves entire or subentire to lyrately repand-dentate; rays minute; "fruits" (involucral bracts individually enclosing achenes) spiny along body as well as at top; plants annual, obviously herbaceous. . . *Acanthospermum*
      Leaves pinnately more or less cleft or dissected; rays well-developed, mostly 5–15 mm long; "fruits" unarmed or with single row of prominent spines around apical rim; plants shrubs or frutescent annuals. . . . . *Lecocarpus*
     Disk flowers perfect and ordinarily fertile:
      Receptacle high-conic or cylindric; annual herbs with few or no rays and numerous disk flowers:
       Receptacular bracts clasping the achenes, deciduous; leaves petiolate; disk corollas glabrous .... *Spilanthes*
       Receptacular bracts (except often some of the outermost) slender, not clasping the achenes, tending to persist after achenes fall; leaves sessile or nearly so; disk corollas spreading-hairy near base. . *Jaegeria*
      Receptacle flat or merely convex; heads and habit various:
       Achenes equably quadrangular or more or less compressed parallel to involucral bracts, at right angles to radii of head:

Achenes all wingless; heads pedunculate; involucre of 2 dissimilar series of bracts, the outer more or less herbaceous, the inner more membranous and often striate .... *Bidens*
Achenes of ray flowers with conspicuous, lacerate wings; heads sessile in glomerules; involucre imbricate, not of 2 conspicuously differentiated series .... *Synedrella*
Achenes, at least those of disk flowers, more or less compressed (or compressed-quadrangular) along radii of head:
Plants shrubs or trees; flowers white or yellow (disk sometimes purple):
Achenes wing-margined, those of marginal flowers triquetrous, with the broad back apposed to subtending bract; pappus a short crown, sometimes prolonged into 1 or 2 awns; leaves opposite, linear; heads yellow............................. *Macraea*
Achenes wingless, all compressed; pappus usually none; leaves and heads various:
Rays (when present) and disk flowers white or whitish; receptacular bracts trifid at tip; achenes glabrous........... *Scalesia*
Rays yellow; disk flowers purplish (in ours); receptacular bracts entire; achenes densely long-hairy, at least along margins distally .... *Encelia*
Plants herbs; flowers mostly white:
Heads few-flowered, the rays commonly 1–5, disk flowers commonly about 8; leaves distinctly petiolate, the petiole mostly (5)10–25 mm long .... *Blainvillea*
Heads many-flowered, the rays more than 10, disk flowers more than 15; leaves sessile or narrowed to a shortly petiolar base not over 5 mm long:
Heads sessile; receptacular bracts well-developed, embracing disk flowers .... *Enydra*
Heads slender-pedunculate; receptacular bracts slender and fragile, somewhat bristle-like, not embracing disk flowers. . *Eclipta*

INULEAE

A single genus .... *Gnaphalium*

MUTISIEAE

A single genus .... *Jungia*

VERNONIEAE

A single genus .... *Pseudelephantopus*

COMBINED KEY TO THE GENERA OF THE COMPOSITAE IN THE GALÁPAGOS ISLANDS

Plants shrubs or trees:
Leaves chiefly or entirely alternate:
Pappus none, or represented by pair of short callosities or slender awns; receptacle chaffy:
Rays (when present) and disk flowers white or whitish; achenes glabrous. . *Scalesia*
Rays yellow; disk flowers purplish; achenes densely long-hairy, at least along margins distally .... *Encelia*
Pappus of more or less numerous capillary bristles; receptacle naked:
Heads radiate, with short, white or faintly anthocyanic rays and yellow disk flowers .... *Darwiniothamnus*

Heads rayless, unisexual, the plants dioecious; pistillate flowers with tubular-filiform, eligulate corolla . . . . *Baccharis*
Leaves chiefly or entirely opposite:
Pappus of more or less numerous capillary bristles; receptacle naked. . . . . *Eupatorium*
Pappus of a short crown that may be produced into one or two awns, or of only one or two awns, or of only a pair of callosities, or obsolete; receptacle chaffy:
Heads white or whitish. . . . . *Scalesia*
Heads yellow:
Leaves narrowly linear and entire. . . . . *Macraea*
Leaves pinnately more or less cleft or dissected. . . . . *Lecocarpus*
Plants herbs:
Leaves chiefly or entirely alternate (sometimes most of them basal):
Pappus of plumose bristles; flowers all bilabiate, outer lip of corolla longer than inner and, at least in marginal flowers, liguloid. . . . . *Jungia*
Pappus various, but never of plumose bristles; flowers never bilabiate:
Heads ligulate, the flowers all perfect with a distinct, minutely 5-toothed ligule . . . . *Sonchus*
Heads radiate to disciform or discoid, the ligules (when present) 2- or 3-toothed or entire and borne only on the marginal, pistillate (or neutral) flowers (disk flowers deeply 5-lobed and subligulate in *Pseudelephantopus*):
Leaves 1–3 times pinnatifid or ternate-pinnatifid:
Heads radiate, with short yellow rays (0.5)2–3 mm long. . . . . *Chrysanthellum*
Heads unisexual, bearing either a pistillate flower without corolla, or functionally staminate flowers with tubular corolla, both types of heads borne on same plant . . . . *Ambrosia*
Leaves entire to merely toothed or wavy-margined:
Heads strictly discoid, the flowers all perfect and fertile:
Heads long-pedunculate, mostly terminating the branches. . . . . *Porophyllum*
Heads sessile in sessile glomerules or 1 to several, forming spikelike compound inflorescences . . . . *Pseudelephantopus*
Heads disciform to radiate, the central flowers perfect, marginal ones pistillate, with or without ligule:
Plants more or less tomentose, at least on lower surfaces of leaves; involucral bracts conspicuously scarious. . . . . *Gnaphalium*
Plants not at all tomentose, though sometimes otherwise hairy; involucral bracts not notably scarious. . . . . *Conyza*
Leaves chiefly or entirely opposite:
Involucre strongly flattened to triquetrous, consisting of 3–5 bracts, the lowermost flat and much the largest, dominating the involucre; flowers only 2–7 in a head. . . . . *Elvira*
Involucre distinctly otherwise, cylindric to hemispheric or rotate, not dominated by any one bract; flowers few to numerous:
Heads unisexual, bearing either pistillate flowers without corolla, or functionally staminate flowers with tubular corolla, both types borne on same plant. . . . . *Ambrosia*
Heads bisexual, with both ray and disk flowers, or with disk flowers perfect and fertile, or both; ray flowers (when present) pistillate or neutral, always provided with corolla:
Involucral bracts only (2)3. . . . . *Flaveria*
Involucral bracts (4)5 to many:
Receptacle naked:
Leaves and involucral bracts beset with conspicuous, scattered, embedded, translucent oil glands:

Involucre elongate, at least 1.5 cm long; heads anthocyanic. . *Porophyllum*
Involucre shorter, less than 1 cm long; heads yellow. . . . . *Pectis*
Leaves and involucral bracts without prominent oil glands, though sometimes glandular-hairy or minutely glandular-punctate:
Pappus of more or less numerous capillary bristles:
Achenes 5-angled . . . . *Eupatorium*
Achenes 10-ribbed . . . . *Brickellia*
Pappus otherwise:
Pappus of 3–5 short, stout, blunt awns, these glandular above middle; heads on slender peduncles mostly 1–2 cm long in a large, openly branched inflorescence . . . . *Adenostemma*
Pappus of 5 or fewer awn-tipped or blunt scales, or sometimes reduced to mere crown, or obsolete; heads short-pedunculate (peduncles less than 1 cm long) in compact, corymbiform inflorescences. . . . . *Ageratum*
Receptacle chaffy:
Disk flowers functionally staminate, only ray flowers fertile:
Leaves entire to lyrately repand-dentate; rays minute. . . *Acanthospermum*
Leaves pinnately more or less cleft or dissected; rays 5–10 mm long. . . . . *Lecocarpus*
Disk flowers perfect and ordinarily fertile:
Receptacle high-conic or cylindric:
Leaves evidently petiolate. . . . . *Spilanthes*
Leaves sessile or very nearly so . . . . *Jaegeria*
Receptacle flat or merely convex:
Leaves sessile or narrowed to shortly petiolar base not over 5 mm long; rays more than 10:
Heads sessile; receptacular bracts well-developed, embracing disk flowers . . . . *Enydra*
Heads slender-pedunculate; receptacular bracts slender and fragile, somewhat bristle-like, not embracing flowers. . . . . *Eclipta*
Leaves evidently slender-petiolate, the petiole mostly (5)10 mm long or more; rays less than 10:
Pappus awns retrorsely barbed near tip; leaves simple or more often compound or dissected . . . . *Bidens*
Pappus awns not retrorsely barbed; leaves simple:
Heads sessile in glomerules; achenes of rays with conspicuous, lacerate wings . . . . *Synedrella*
Heads (or many of them) slender-pedunculate; achenes not winged . . . . *Blainvillea*

Acanthospermum Schrank, Pl. Rar. Hort. Acad. Monac. 53. 1820

Branching, annual herbs; leaves rather small, opposite, entire or subentire to lyrately repand-dentate; heads solitary in axils and forks of stem, sessile or short-pedunculate, shortly and inconspicuously radiate. Involucre of 2 very dissimilar series, the outer of 4–6 relatively broad, foliaceous, loose or spreading bracts, the inner set indurate, individually enclosing the achenes, and evidently prickly or spiny. Receptacle small, convex, chaffy throughout, its bracts soft, loosely folded or convex and embracing disk flowers; rays mostly 5–10, minute, yellow or ochroleucous, pistillate and fertile, with thickened, epappose achene somewhat compressed parallel to radii of head. Disk flowers relatively few, mostly 5–15, yellow or yellowish, func-

Fig. 70. *Acanthospermum microcarpum.*

tionally staminate, with undivided style and abortive or infertile achene. Anthers entire or sagittate at base.

About 5 species, all weedy, native to tropical and subtropical America.

Acanthospermum microcarpum Robins., Proc. Amer. Acad. 38: 208. 1902

Acanthospermum donii Blake, Contr. U. S. Nat. Herb. 20: 388. 1921.
Acanthospermum simile Blake, loc. cit.*

Freely branching annual mostly 3–5(10) dm tall, the stem loosely villous or villous-hirsute, especially upward, with rather long, multicellular hairs; leaves sessile or

* These two names of Blake are based on material from the coast of Ecuador.

contracted to short, more or less petiolar base, blade elliptic or nearly rhombic to obovate or ovate, evidently to scarcely toothed, mostly 2.5–8 cm long, 1.2–4 cm wide, sparsely or moderately hirsute with more or less appressed to loosely spreading hairs shorter and firmer than those of stem; heads essentially sessile; outer involucral bracts 3–5 mm long, notably villous-ciliate on margins, less hairy or glabrous on back; "fruits" (mature inner involucral bracts) radially somewhat compressed, obtriangular in side view, resinous-glandular, and beset with stout, often apically finely hooked prickles, the body commonly 3.5–5 mm long exclusive of the 2 to several larger prickles about top.

A weed of coastal Ecuador and Peru; in the Galápagos Islands, established on Santa María.

Although the oldest name is based on a specimen from Santa María, it seems clear that the species originated on the mainland and subsequently (perhaps recently) was introduced into the Galápagos Islands. The islands have no native mammals suitable for distributing the prickly burs.

See Fig. 70.

Adenostemma Forst., Char. Gen. 89, *t. 45.* 1776

Annual to perennial, glandular-hairy to subglabrous herbs with simple, opposite, entire or toothed, commonly triplinerved leaves. Heads rather small, borne in openly branched inflorescence, discoid, the flowers all perfect and fertile, white. Involucre campanulate or hemispheric, its bracts narrow, green, and more or less herbaceous, in 1–3 series, often connate below; receptacle flat or nearly so, naked. Anthers truncate to obtuse at base, rounded-truncate at tip, without appendages. Style branches strongly clavate, minutely papillate, long-exserted, without conspicuous stigmatic lines. Achenes glandular or glandular-hairy, mostly 5-angled, crowned at angles with 3–5 short, stout, blunt awns, these glandular above middle.

About half a dozen species, one pantropical, the others tropical American.

Adenostemma lavenia (L.) Kuntze, Rev. Gen. 304. 1891

Verbesina lavenia L., Sp. Pl. 902. 1753.
Adenostemma viscosum Forst., Char. Gen. 90. 1776.
Verbesina brasiliana Pers., Syn. Pl. 2: 472. 1807.
Adenostemma brasilianum Cass., Dict. Sci. Nat. 25: 363. 1822.

Coarse, erect annual to about 1 m tall, the stem subglabrous to evidently hirsute-puberulent, at least above, with loose or spreading, multicellular hairs; leaves relatively large and soft, glabrous or inconspicuously hairy, largest ones (commonly borne near or above middle of plant) with ovate or elliptic to more often (at least on vigorous plants) deltoid to subcordate or subhastate blade 5–15 cm long, 3–15 cm wide, with evidently toothed to subentire margins, on a more or less evident, often broadly winged petiole; uppermost leaves sometimes alternate; inflorescence openly branched and nearly naked, with more or less numerous, rather small heads on slender peduncles mostly 1–2 cm long; involucre 3–5 mm high, its bracts herbaceous, not much if at all imbricate, rough-puberulent and often somewhat glandular (as also the peduncles), oblong or oblong-spatulate, broadly rounded distally, tending to be connate toward base; disk 5–8 mm wide; styles white like the hairy or glandular-hairy corolla; achenes short-clavate, densely stipitate-glandular or

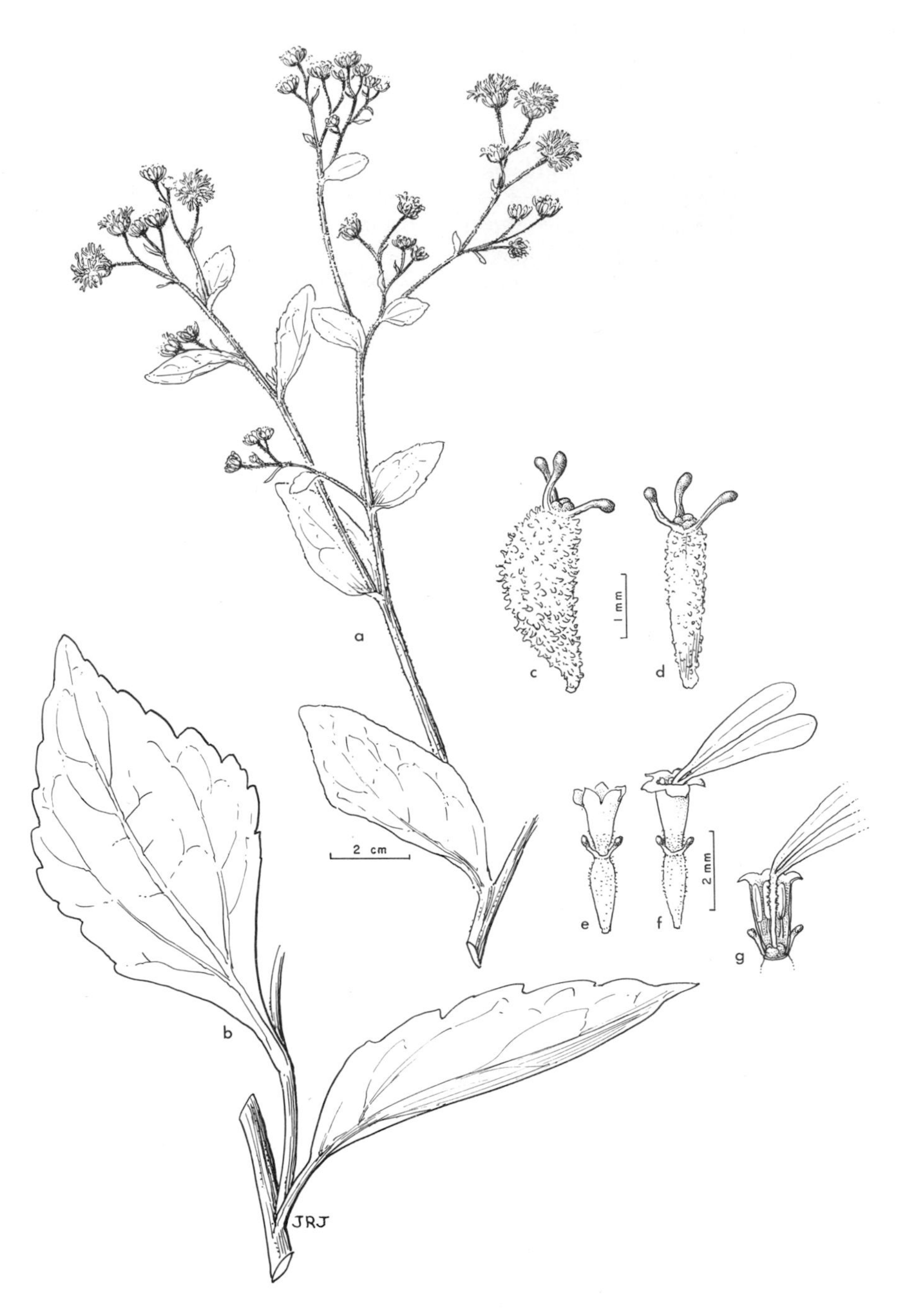

Fig. 71. *Adenostemma lavenia.*

glandular-warty, about 2.5 mm long; pappus members 0.5–1 mm long, obviously thickened above middle. n = 10.*

A pantropical weed, recently introduced into Isla Santa Cruz.

See Fig. 71.

Ageratum L., Sp. Pl. 839. 1753; Gen. Pl. ed. 5, 363. 1754

Annual to perennial herbs or shrubs. Leaves opposite (rarely alternate), sessile or petiolate, simple, toothed or seldom entire, usually more or less ovate, commonly glandular-punctate. Heads rather small, in a more or less corymbiform inflorescence, discoid, the flowers all tubular and perfect, blue to pink or white, never yellow. Involucral bracts 2- or 3-seriate, not notably imbricate, narrow, acute to attenuate, rather dry, very often 2-ribbed. Receptacle flat to conic, naked or chaffy. Anthers rounded at base, shortly appendiculate at tip. Style branches as in *Eupatorium*. Achenes 5-angled, more or less prismatic. Pappus typically of 5 well-developed, awn-tipped or blunt scales, or scales sometimes fewer or more or less reduced, or connate into entire or toothed crown, or pappus obsolete. n = 10, 20.

Some 30–40 species, chiefly of tropical and subtropical America, 1 species (*A. conyzoides*) a pantropical weed.†

Ageratum conyzoides L., Sp. Pl. 839. 1753

Ageratum latifolium Cav., Icon. 4: *t.* 357. 1797.‡
Coelestina microcarpa Benth., Vidensk. Medd. Kjob. 1852: 72. 1852.
Ageratum conyzoides var. *inaequipaleaceum* Hieron., Bot. Jahrb. 19: 44. 1894.
Alomia microcarpa Robins., Proc. Amer. Acad. 49: 452. 1913.
Ageratum latifolium var. *galapageium* Robins., op. cit. 466.

Simple to much-branched, malodorous annual weed, 1–10 dm tall, erect or short-creeping at base, fibrous-rooted or with short, deliquescent taproot, subglabrous to more often evidently hirsute, hirsute-strigose, or hirsute-puberulent; leaves opposite, evidently petiolate, the blade narrowly to more often broadly ovate or deltoid, or seldom subcordate, mostly 1–7(10) cm long, 0.7–6(7) cm wide, palmately or subpalmately 3(5)-nerved, crenate; heads several to numerous in compact, corymbiform inflorescences terminating stem and branches, individually short-pedunculate, rather small, the involucre campanulate to hemispheric, 2.5–4.5 mm high, the disk 2–5 mm wide; involucral bracts rather thin and dry, greenish or partly anthocyanic, not notably imbricate, acute to gradually long-acuminate; receptacle small, convex or low-conic, naked; flowers blue to lavender, pink, or white; achenes black or blackish, about 1.5 mm long, glabrous or nearly so; pappus diverse, typically of 5 awn-tipped scales 1.5–3 mm long, ranging through various degrees of equal or unequal reduction to set of awnless scales only 0.5 mm long, or still further reduced or even obsolete.

A pantropical weed of American origin; in America known from Mexico and southern Florida to Argentina. At present known only from 4 islands in the Galá-

* Count made by D. Kyhos on buds from same plant as herbarium specimens *Wiggins 18547* (*DS, NY*), from Isla Santa Cruz.

† *Ageratum* is very much like *Eupatorium*, except for the scaly (or obsolete) pappus. Most of the species commonly referred to *Alomia* are here considered to belong to *Ageratum*.

‡ Recent herbarium annotations by Miles F. Johnson indicate a new combination of the epithet *latifolium* in subspecific rank under *A. conyzoides*.

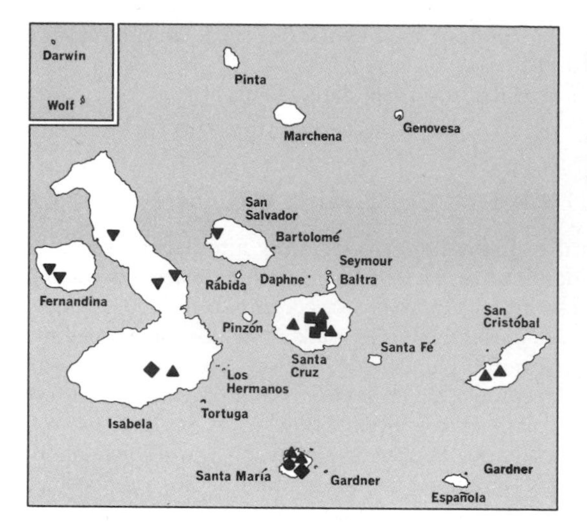

● *Acanthospermum microcarpum*
■ *Adenostemma lavenia*
▲ *Ageratum conyzoides*
◆ *Ambrosia artemisiifolia*
▼ *Baccharis gnidifolia*

pagos, but to be expected on some of the other islands as well. Specimens examined: Isabela, San Cristóbal, Santa Cruz, Santa María.

Typical *A. conyzoides* has 5 essentially equal pappus scales tapering gradually into an awn tip. This phase is scarcely represented in the Galápagos, though some achenes of *Bowman s.n.*, September 29, 1957, Isla Santa Cruz, are reasonably typical. The phase with the scales dissimilar and unequally reduced has been described as var. *inaequipaleaceum* Hieron. The phase with all of the scales short and awnless but still well developed has been called *A. latifolium*. The phase with the scales very much reduced but still recognizable has been called *A. latifolium* var. *galapageium*. The phase with the pappus nearly or quite obsolete has been called *Alomia microcarpa*. All of these phases, with intermediates, are represented on the Galápagos Islands, though not with equal frequency; nor are all represented on the same island.

See Fig. 72.

Ambrosia L., Sp. Pl. 987. 1753; Gen. Pl. ed. 5, 425. 1754

Coarse annual or perennial herbs or shrubs with opposite or alternate, mostly lobed or dissected leaves and numerous small, inconspicuous, unisexual, rayless heads, both sexes borne on same plant but not always in equal numbers. Sterile (staminate) heads in spiciform or racemiform, bractless inflorescence, the involucre subherbaceous, 5–12-lobed, the receptacle flat, chaffy, its bracts slender, filiform-setose. Filaments monadelphous. Anthers scarcely united. Style undivided. Fertile (pistillate) heads borne below sterile heads, in axils of leaves or bracts, the involucre closed, nutlike or burlike, with 1 to several series of tubercles or spines. Pistillate flowers 1 to several, without corolla, the style bifid. Pappus wanting.

About 40 species, mostly native to the New World.*

* The genus is now generally interpreted in the broad sense to include also *Franseria;* our single species would remain in *Ambrosia* even if the species of *Franseria* were excluded.

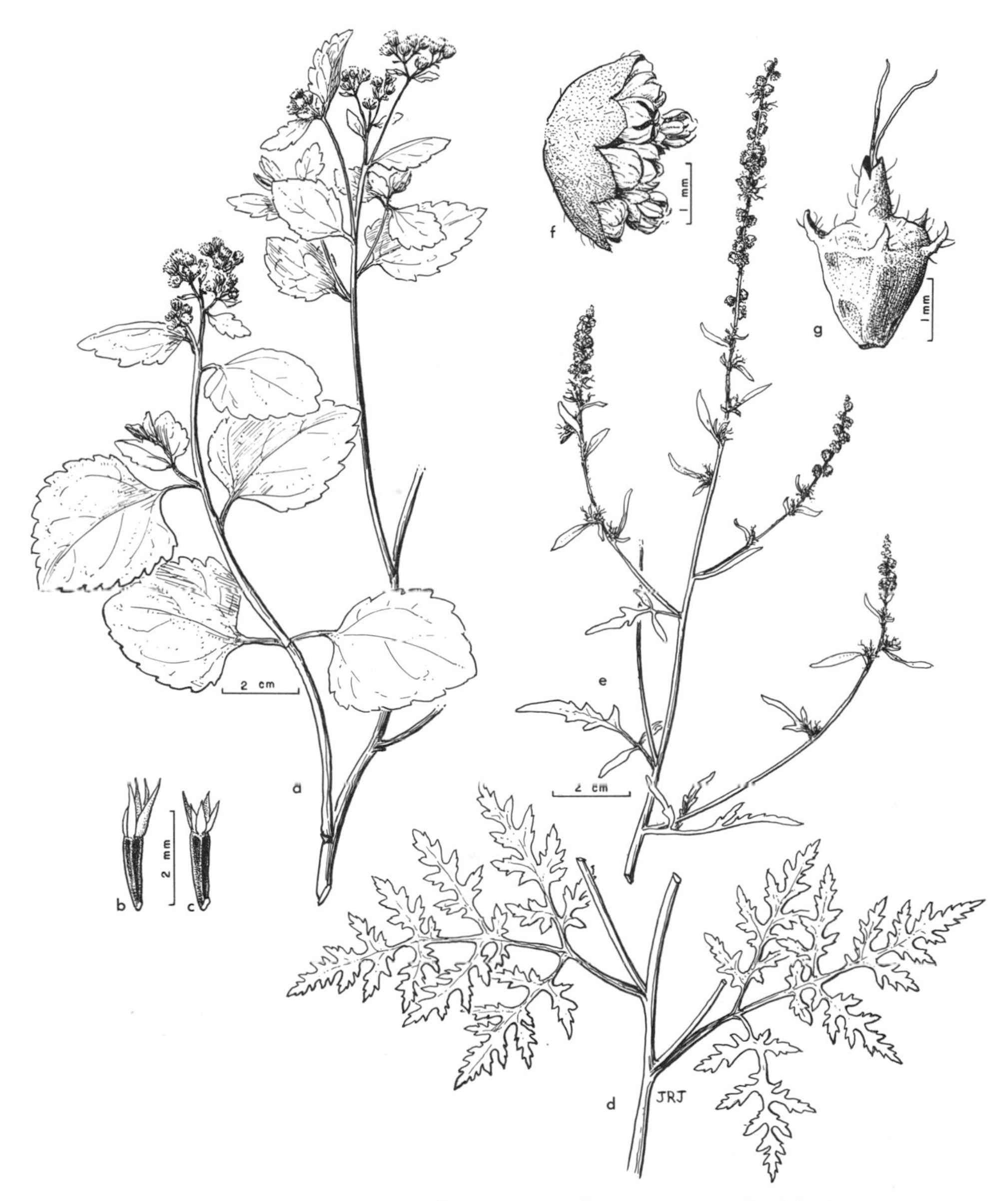

Fig. 72. *Ageratum conyzoides* (*a–c*). *Ambrosia artemisiifolia* (*d–g*).

Ambrosia artemisiifolia L., Sp. Pl. 988. 1753

Ambrosia elatior L., op. cit. 987.

Annual weed, branching at least above, variously hairy or in part subglabrous, 1–10 dm tall; leaves opposite below, then alternate above, pinnatifid or more commonly bipinnatifid, narrowly to broadly ovate or elliptic in outline, commonly 3–10 cm long, the middle and lower ones, at least, generally evidently petiolate; sterile heads short-pedunculate; fruiting involucre short-beaked, 3–5 mm long, with several short, sharp spines in whorl near or a bit above middle, producing single pistillate flower and achene.

A widespread weed of American origin. Specimens examined: Isabela, Santa María.

Perhaps a recent introduction.

See Fig. 72.

Baccharis L., Sp. Pl. 860. 1753

Dioecious shrubs (or seldom herbs) with alternate, simple leaves. Fertile heads with more or less numerous pistillate flowers having a tubular-filiform, eligulate corolla. Sterile heads with more or less numerous functionally staminate flowers, ovary abortive, style branches sometimes fused; anthers obtuse and entire or nearly so at base. Flowers white to yellowish or greenish. Involucral bracts subequal to strongly imbricate, chartaceous or subherbaceous. Receptacle flat or convex, naked. Pappus of numerous capillary bristles, those of sterile heads fewer and shorter than those of the fertile, which frequently elongate in fruit. Achenes generally somewhat compressed and ribbed.

A large genus of perhaps 300 species, all American, most of them in South America.

Leaves narrow, from linear-filiform to linear-elliptic, 0.5–7 mm wide, 5 to many times as long as broad .... *B. gnidiifolia*

Leaves broader, elliptic or elliptic-obovate, 2 or 3 times as long as wide, the better-developed 9–17 mm wide .... *B. steetzii*

Baccharis gnidiifolia HBK., Nov. Gen. & Sp. Pl. 4: 48 (folio ed.). 1818

Freely branching shrub to 2 m tall, minutely and inconspicuously granular-glandular, otherwise glabrous; leaves rather numerous but not densely crowded, narrow, linear-filiform to linear-elliptic, strictly entire, mostly 1.5–4 cm long, 0.5–7 mm wide, 5 to many times as long as wide, sessile or tapering to shortly petiolar base; heads mostly in small glomerules terminating many of the twigs or on short axillary shoots, subsessile or on peduncles to 2(7) mm long; pistillate involucre campanulate, 3–5 mm high, inconspicuously glandular-glutinous, its bracts imbricate in several series, chartaceous with a thicker and darker subterminal spot, acute or subacute, or inner ones blunter, these with hyaline-scarious, minutely erose or fimbriate distal margins; staminate involucre similar to pistillate or a bit smaller, sometimes with fewer series of bracts; pappus white or whitish, 2–3 mm long, not notably elongate in fruit; achenes glabrous or nearly so.

Ecuadorian Andes. Specimens examined: Fernandina, Isabela, San Salvador.

The Galápagos plants have often been confused with the widespread and highly variable mainland species *B. pingraea* DC., which, however, is a woody herb up to 1 m tall from running roots. The broader-leaved plants among our specimens are apparently not to be distinguished from *B. gnidiifolia* of the Ecuadorian Andes, as has been pointed out to me by Dr. José Cuatrecasas. The narrower-leaved specimens are not yet matched with mainland material, but I am unable to draw any taxonomic distinction.

Baccharis steetzii Anderss., Kongl. Svensk. Vet.-Akad. Handl. 1853: 177. 1855

Freely branching shrub 1–2.5 m tall, glabrous and somewhat glutinous; leaves numerous, short-petiolate, the blade elliptic or elliptic-obovate, subtriplinerved, strictly entire, 2–3 times as long as wide, acute or subacute and often mucronate, the better developed mostly 2–4 cm long, 9–18 mm wide; inflorescence of several or many small, corymbiform clusters terminating the twigs or on short axillary shoots, the heads on short peduncles seldom over 5 mm long; pistillate involucre campanulate, 2.5–4 mm high, inconspicuously glutinous, its bracts imbricate in several series, obtuse or subacute or the inner acute, chartaceous with thicker and darker central or subterminal region, at least inner ones with narrow, hyaline-scarious, sometimes erose margins; pappus whitish or discolored, 2–3 mm long, not notably elongate in fruit; achenes glabrous; staminate heads unknown.

Endemic. Specimens examined: San Cristóbal, Santa María.

Although very different in appearance because of its broader leaves, *B. steetzii* is in other respects rather similar to *B. gnidiifolia*. A specimen collected on Santa María by Edmonstone was identified by J. D. Hooker as *B. pilularis* DC. and so reported in Trans. Linn. Soc. Lond. 20: 261. 1847. *B. pilularis* is a Californian species, not otherwise known from the Galápagos. Although I have not seen the Edmonstone specimen, I suppose it to represent *B. steetzii* Anderss., which was described from Santa María in 1855.

See Fig. 73.

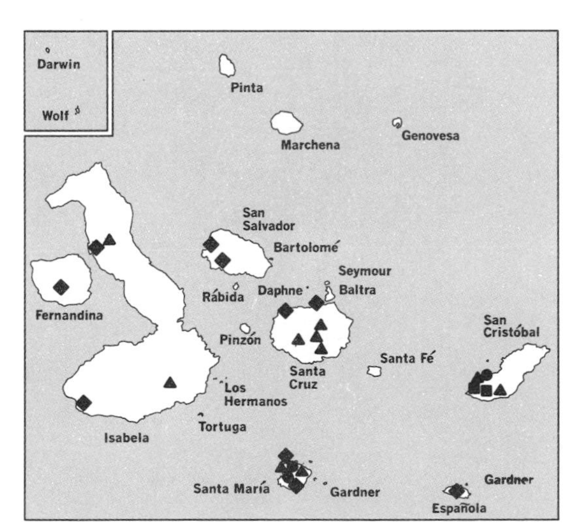

● *Baccharis steetzii*

■ *Bidens cynapiifolia*

▲ *Bidens pilosa*

◆ *Bidens riparia*

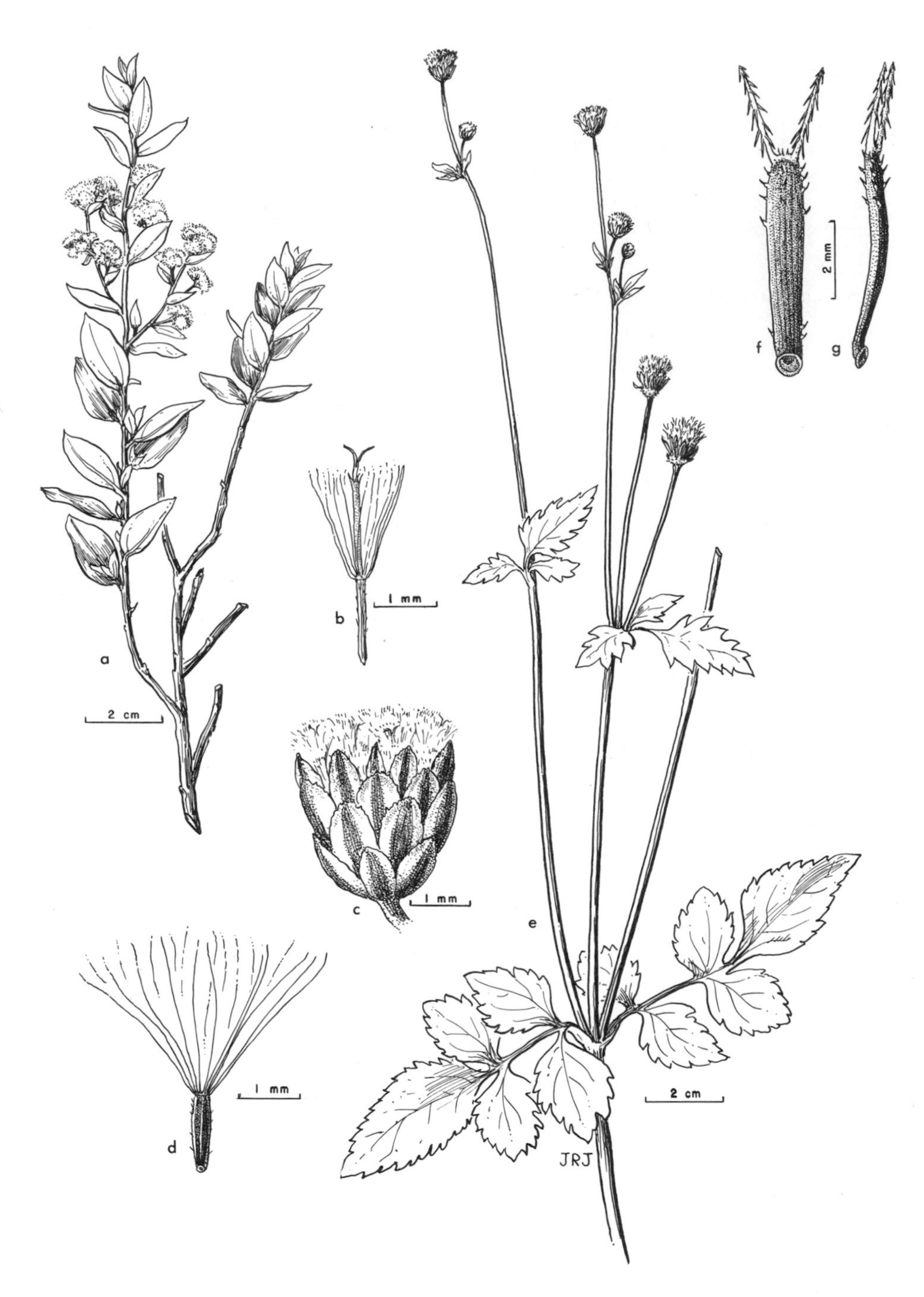

Fig. 73. *Baccharis steetzii* (*a–d*). *Bidens pilosa* (*e–g*).

Bidens L., Sp. Pl. 831. 1753; Gen. Pl. ed. 5, 362. 1754

Herbs or sometimes shrubs (ours all annual weeds) with opposite, simple to dissected leaves. Heads mostly terminal and on axillary peduncles, sometimes forming a more or less diffuse inflorescence, radiate, subradiate, disciform, or discoid, the rays (when present) white to yellow, neutral or only rarely pistillate. Involucre of 2 dissimilar series of bracts, the outer more or less herbaceous and sometimes very large, the inner membranous, often striate. Receptacle flat or a bit convex, chaffy, its bracts narrow, flat or nearly so. Disk flowers yellow or yellowish, perfect and fertile. Style branches flattened, with externally hairy, usually short appendages, without well-marked stigmatic lines. Anthers entire or minutely sagittate at base. Achenes tetragonal or more often flattened parallel to involucral bracts (at right angles to radii of head). Pappus of (1)2–4(8) awns or teeth, commonly (including all our species) retrorsely barbed.

Perhaps as many as 200 species, of wide geographic distribution.

Outer involucral bracts distinctly oblanceolate or even spatulate, commonly ciliate; achenes mostly somewhat flattened, glabrous or sparsely hispidulous distally, outer ones not notably more hairy than the inner; pappus of 2(–4) erect or ascending awns. . . . . *P. pilosa*

Outer involucral bracts linear or sometimes linear-oblanceolate, seldom at all ciliate; achenes slender, tetragonal, most of them glabrous or nearly so, usually some of outer ones conspicuously hispidulous; pappus of (3)4(5) awns:

- Pappus awns all erect or strongly ascending; outer achenes often arcuate, the fruiting head thus broad-topped. . . . . *B. cynapiifolia*
- Pappus awns, or some of them, spreading at right angles to body of achene, or deflexed, the other(s) erect or nearly so; outer achenes not markedly arcuate. . . . . . *B. riparia*

Bidens cynapiifolia HBK., Nov. Gen. & Sp. Pl. 4: 185 (folio ed.). 1818

Leafy-stemmed, usually freely branched annual mostly 3–10 dm tall, the stem glabrous or nearly so, leaves sparsely pubescent with flattened, white, multicellular hairs; leaves petiolate, highly variable, once or more often 2 or 3 times pinnate, with or without well-defined leaflets, the segments or leaflets with or more often without regularly serrate margins; heads long-pedunculate from upper axils and terminating the main stem and branches, inconspicuously radiate or discoid, the rays (when present) commonly 4–5, pale yellow or white, only about 2–4 mm long; disk yellow, 5–10 mm wide at anthesis; involucre with some long, multicellular, deflexed hairs at base (between base of involucral bracts and tip of peduncle); outer involucral bracts commonly about 8, linear or sometimes linear-oblanceolate, glabrous or nearly so, seldom inconspicuously and irregularly ciliate; achenes tetragonal, principal ones glabrous or nearly so, mostly 9–14 mm long and scarcely 1 mm wide, some of the outer commonly arcuate above at maturity, so that fruiting head is broad-topped; usually a few of outermost achenes notably shorter than others and copiously hispidulous throughout, sometimes sterile or imperfectly developed, or occasionally wanting; pappus of (3)4(5) retrorsely barbed, erect or ascending awns mostly 2–4 mm long.

A widespread weed in tropical America; in the Galápagos Islands known only from San Cristóbal.

Bidens pilosa L., Sp. Pl. 832. 1753

Bidens alausensis HBK., Nov. Gen. & Sp. Pl. 4: 184 (folio ed.). 1818.
Bidens chilensis DC., Prodr. 5: 603. 1836.
Bidens pilosa var. *alausensis* Sherff, Bot. Gaz. 81: 35. 1926.

Leafy-stemmed, sparingly to freely branched annual 1–10 dm tall, sparsely to moderately pubescent, especially on leaves, often with flattened, translucent, white hairs; leaves petiolate, highly variable, ranging from simple and merely serrate, with lance-ovate or ovate blade to 10 cm long and 3 cm wide, to often trifoliolate or once-pinnate with serrate leaflets or segments, to bi- or tripinnatifid, without well-defined leaflets; heads long-pedunculate from upper axils and terminating main stem and branches, ours discoid or inconspicuously radiate, the rays when present (in our forms) white or whitish; disk yellow, only about 5–7 mm wide at anthesis; outer involucral bracts commonly about 8, distinctly oblanceolate or spatulate, tending to be conspicuously white-ciliate and sometimes also hairy on back; achenes somewhat compressed, mostly linear-oblong, commonly 5–10 mm long, 1 mm wide, glabrous or somewhat hispidulous distally, outer ones sometimes evidently shorter but not otherwise much different from others; pappus of 2 (or on some achenes 3, rarely 4) retrorsely barbed, erect or closely ascending awns mostly 2–4 mm long.

A widespread weed in tropical and subtropical regions of the world, with many intergrading races that sometimes seem sharply marked in particular localities. Specimens examined: Isabela, San Cristóbal, Santa Cruz, Santa María.

Plants from the Galápagos Islands have mostly been referred by Sherff to the var. *alausensis* (HBK.) Sherff, said to occur throughout most of the Andean region of tropical South America, but the variety is scarcely distinctive.

See Fig. 73.

Bidens riparia HBK., Nov. Gen. & Sp. Pl. 4: 185 (folio ed.). 1818

Bidens refracta Brandegee, Zöe 1: 310. 1890.
Bidens riparia var. *refracta* O. E. Schulz, Urb. Symb. Antill. 7: 132. 1911.

Freely branched, sparsely pubescent annual 1–10 dm tall, the hairs commonly somewhat shorter than in our other 2 species; leaves petiolate, highly variable, ranging from trifoliolate or once-pinnate with distinct, serrate leaflets, to sometimes bi- or tripinnatifid, without well-defined leaflets; heads long-pedunculate from upper axils and terminating main stems and branches, discoid or inconspicuously radiate, the rays (when present) inconspicuous, yellow, only about 2–4 mm long; disk yellow, 5–8 mm wide at anthesis; involucre with some multicellular, deflexed hairs at base (between base of involucral bracts and tip of peduncle), but these shorter and less conspicuous than those of *B. cynapiifolia;* other involucral bracts commonly about 8, sometimes 13, linear or sometimes linear-oblanceolate, glabrous or nearly so, seldom inconspicuously and irregularly ciliate; achenes tetragonal, straight or nearly so, principal ones mostly 10–18 mm long, scarcely 1 mm wide, glabrous or somewhat hispidulous distally, usually a few of outermost notably shorter than others and copiously hispidulous throughout, sometimes sterile or imperfectly developed or occasionally wanting; pappus of (3)4(5) retrorsely barbed awns, at least 1 spreading at right angles to body of achene or deflexed, and at least 1 erect or nearly so.

A widespread weed in tropical and subtropical America. Specimens examined: Española, Fernandina, Isabela, San Salvador, Santa Cruz, Santa María.

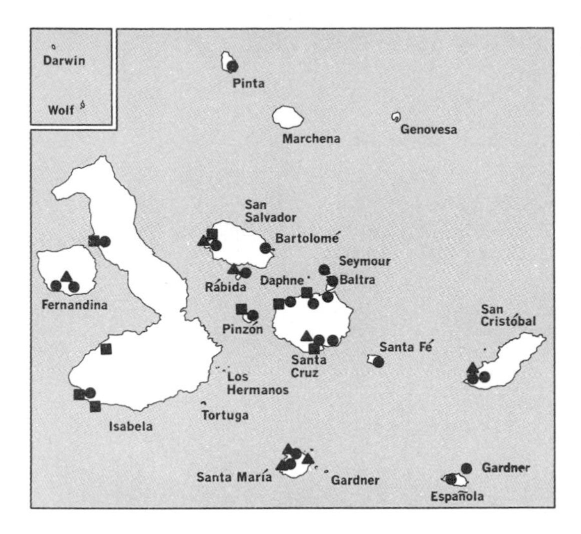

● *Blainvillea dichotoma*
■ *Brickellia diffusa*
▲ *Chrysanthellum pusillum*

Common on the Galápagos Islands. Most of our material has the leaves only once-pinnate or trifoliolate and thus represents what has been called var. *refracta* (Brandegee) O. E. Schulz.

Blainvillea Cass., Dict. Sci. Nat. 29: 493. 1823

Branching herbs with opposite (or the upper alternate), simple, petiolate leaves. Heads small, pedunculate or sessile at branch tips or in axils or in forks of stem, radiate or disciform, the pistillate flowers fertile, with short, tubular corolla that may or may not be tipped by short, white (or sometimes yellow) ray. Involucre of a few scarcely imbricate, more or less herbaceous or herbaceous-tipped bracts, inner ones passing into bracts of receptacle. Receptacle small, somewhat convex, chaffy, its bracts embracing disk flowers, shortly and irregularly lacerate-toothed at the broad tip. Disk flowers perfect and fertile, corolla white or somewhat yellowish. Style branches flattened, with submarginal stigmatic lines and shortly hairy terminal appendage. Anthers obtuse and entire at base. Achenes of rays triquetrous, those of disk somewhat compressed parallel to radii of head, both types with more or less truncate summit. Pappus of 2–5 awns or slender scales, sometimes connate at base, covering only small part of top of achene, well in from marginal angles.

A small genus, consisting chiefly or entirely of a pantropical weedy species of American origin.

Blainvillea dichotoma (Murr.) Stewart, Proc. Calif. Acad. Sci. IV, 2: 149. 1911*

Verbesina dichotoma Murr., Comm. Gött. 2: 15, *t. 4.* 1779.
Eclipta latifolia L. f., Suppl. 378. 1781.
Blainvillea rhomboidea Cass., Dict. Sci. Nat. 29: 493. 1823.
Blainvillea latifolia DC. in Wight, Contr. Bot. Ind. 17. 1834.

Freely branched annual weed (0.4) 1–10 dm tall, hirsute throughout; leaves petiolate, petiole mostly (0.5) 1–2.5 cm long, blade lanceolate or lance-elliptic to more

* The name *B. dichotoma* appears also in Hemsley, Biol. Centr. Am. Bot. 4: 112. 1886–88, attributed to Cassini but without description or reference. Presumably the present plant was intended, but the name is strictly a nomen nudum.

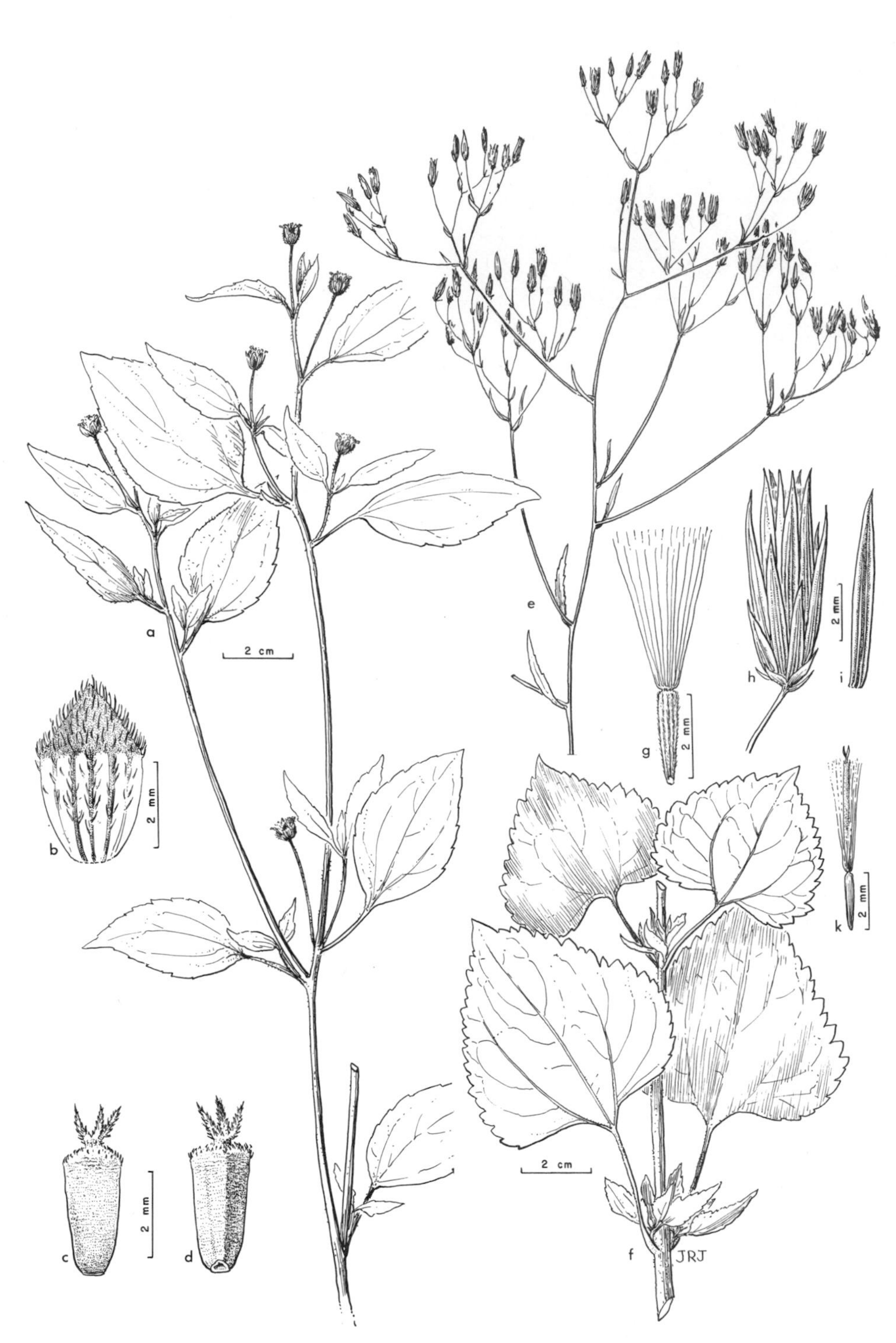

Fig. 74. *Blainvillea dichotoma* (*a–d*): *b*, involucral bract; *c*, *d*, achene, abaxial and adaxial views. *Brickellia diffusa* (*e–k*).

commonly ovate, often broadly so, mostly 1.5–8 cm long, 1–7 cm wide, acute or acuminate, evidently serrate or crenate-serrate to occasionally subentire; heads more or less numerous according to vigor and maturity of plant, on slender peduncles 0.5–4 cm long, relatively small and with not numerous flowers, mostly campanulate or turbinate-cylindric at anthesis, broader and more nearly hemispheric at maturity, the disk then 5–8 mm wide; involucre 3.5–6 mm high, its bracts green-tipped above the paler, striate, chartaceous base; ray flowers few, ligule white, inconspicuous, only about 1 mm long, or scarcely developed; disk flowers few, commonly about 8; achenes blackish, 2–3.5 mm long, minutely cross-rugulose, also irregularly tuberculate or roughened distally and hirtellous near summit with appressed or ascending hairs; pappus awns 0.5–1.5 mm long.

A weed in open or disturbed sites, from near sea level to about 675 m, pantropical. Specimens examined: Baltra, Española, Fernandina, Gardner (near Española), Isabela, Pinta, Pinzón, Rábida, San Cristóbal, San Salvador, Santa Cruz, Santa Fé, Santa María, Seymour.

See Fig. 74.

Brickellia Ell., Sk. Bot. S. C. 2: 290. 1824, nom. conserv.

Annual or more often perennial herbs, or shrubs, with simple, opposite or alternate leaves. Heads campanulate or cylindric, in various sorts of inflorescences, discoid, the flowers all tubular and perfect, white or creamy to pink-purple. Involucral bracts striate, imbricate in several series. Receptacle naked, mostly flat or nearly so. Anthers minutely rounded-sagittate at base, shortly appendiculate at tip. Style branches with short stigmatic lines and elongate, papillate appendages, filiform to more often somewhat clavate. Achenes 10-ribbed, prismatic. Pappus of 10–80 barbellate to nearly smooth or rarely subplumose capillary bristles.

Nearly 100 species, native to North and South America, chiefly in tropical and warm temperate regions.

Brickellia diffusa (M. Vahl) A. Gray, Pl. Wright. 1: 86. 1852

Eupatorium diffusum M. Vahl, Symb. 3: 94. 1794.
Bulbostylis diffusa DC., Prodr. 7: 268. 1838.
Coleosanthus diffusus Kuntze, Rev. Gen. 328. 1891.

Erect annual to 2 m tall, simple below and with main axis continuing nearly to tip, but more or less diffusely branched upward when well-developed; stem puberulent or woolly-puberulent, especially upward, often also somewhat viscid or glandular; basal leaves none; cauline leaves opposite, lower ones reduced and often soon deciduous, others well-developed, with slender petiole 1–8 cm long and a thin, deltoid-ovate to broadly subcordate blade mostly 2–8 cm long and wide, palmately trinerved, evidently crenate, often atomiferous-glandular, and otherwise sparsely hairy (especially along veins) or nearly glabrous; heads numerous to very numerous on slender peduncles mostly (2)5–15 mm long; heads mostly 8–13-flowered, cylindric, the involucre 6–9 mm high, its slender bracts strongly imbricate in several series, greenish with hyaline margins, commonly with 2 pale principal veins flanking midline, often with thinner vein delimiting each of the hyaline margins; corollas greenish white, very slender, scarcely expanded upward, about 4 mm long; style branches filiform; achenes about 2 mm long, blackish, antrorsely strigose or more loosely hirsute-strigose.

A common weed of much of tropical America. Specimens examined: Isabela, Pinzón, San Salvador, Santa Cruz.

Perhaps only recently introduced into the Galápagos Islands, where the earliest record dates from 1899.

See Fig. 74.

Chrysanthellum L. C. Rich. ex Pers., Syn. Pl. 2: 471. 1807

Slender annuals, diffusely branched when well-developed, with petiolate, alternate leaves (sometimes most of them clustered toward base) dissected into more or less numerous, small and narrow ultimate segments. Heads small, radiate, yellow, on long, slender peduncles in axils and terminating the branches. Involucre mostly campanulate to hemispheric, of several flat, equal or subequal, membranous bracts in 1 or 2 series, sometimes with 1 or more much shorter and narrower bracts outside others. Receptacle flat or nearly so, chaffy throughout, or naked toward middle, its bracts flat, slender, or the outermost broader and grading into bracts of involucre. Ray flowers pistillate and fertile, with short but usually evident, entire or bifid, 2-veined yellow ligule. Disk flowers perfect and fertile, or innermost sterile. Anthers obtuse and entire at base. Style branches flattened toward base and with short, introrsely marginal stigmatic lines, tipped with slender, elongate, hirsutulous appendage. Achenes thickened and only slightly compressed parallel to involucral bracts (at right angles to radii of head), or (especially the inner) more evidently compressed and even wing-margined; pappus none.

A small, sharply defined genus of about 4 or 5 species, widely distributed in tropical regions; probably of American origin.

Chrysanthellum pusillum Hook. f., Trans. Linn. Soc. Lond. 20: 214. 1847

Chrysanthellum erectum Anderss., Kongl. Svensk. Vet.-Akad. Handl. 1853: 188. 1855.
Chrysanthellum fagerlindii Eliass., Svensk. Bot. Tidskr. 61(1): 91. 1967.*

Plant glabrous or nearly so, branched at base and (when well-developed) freely branched upward, the stems sympodially branched, very slender, lax, often curved or prostrate at base, but not rooting, to about 2 dm tall; leaves soft, mostly crowded toward base or well distributed along stems, petiolate, blade pale beneath, to about 2 cm long and wide, pinnately or ternate-pinnately 2 or 3 times dissected into small, minutely apiculate, short and crowded to rather elongate and distant segments scarcely 1 mm wide, varying to merely trifid, with apically toothed segments to 3 mm wide; lower petioles elongate, generally surpassing blade, sometimes to 6 cm long, the upper progressively shorter; heads several to numerous, borne singly on elongate, slender, leaf-opposed peduncles mostly 1–8 cm long, and also terminating some of uppermost axillary branches; disk commonly 2–3 mm high, 2–5 mm wide; involucre mostly 2–3 mm high, with about (5)8–13 principal bracts, these membranous and more or less reddish brown, obscurely to evidently veined-striate, often with thinner and paler margins; rays about 13–21, ligule (0.5)2–3 mm long and less than 1 mm wide; achenes wingless, 1–1.5(2) mm long, the outer individually

* *Chrysanthellum fagerlindii* has been relegated to synonymy because it appears to be a minor form with very short rays and deeply cleft disk corollas, which probably will fail to persist on an island (San Salvador) with the typical *C. pusillum.* The latter will probably swamp the incipient species in a few generations.

Fig. 75. *Conyza bonariensis* (*a–c*). *Chrysanthellum pusillum* (*d, e*). *Darwiniothamnus tenuifolius* var. *tenuifolius* (*f–k*).

subtended by involucral bracts, thick, only slightly or scarcely compressed, obscurely to conspicuously tuberculate or roughened, the inner smoother and somewhat more compressed, or the innermost often not maturing.

Endemic; known from most of the larger islands. Specimens examined: Fernandina, Rábida, San Cristóbal, San Salvador, Santa Cruz, Santa María.

See Fig. 75.

Conyza Less., Syn. Gen. Comp. 203. 1832, nom. conserv.*

Annual or perennial herbs, often weedy, or seldom shrubs, with alternate, simple leaves and several to numerous, mostly rather small, disciform to minutely radiate heads. Involucral bracts more or less imbricate in several series, scarcely herbaceous. Receptacle flat or nearly so, naked. Pistillate flowers numerous to very numerous, in several series, with slender, tubular-filiform corolla, this produced in some species into short, narrow, inconspicuous, white or purplish ligule to 1 mm long and very slightly, if at all, exceeding pappus. Disk flowers few, much fewer than pistillate flowers, mostly yellow, perfect and fertile. Anthers entire at base. Style branches flattened, with ventromarginal stigmatic lines and short to somewhat elongate appendage. Achenes flattened, 1- or 2-nerved or nerveless. Pappus of capillary bristles, sometimes with short outer series. Embryo sac monosporic, with rather numerous (commonly 10–15) antipodal cells.

More than 50 species, chiefly tropical and subtropical, in the Old and New Worlds.

Conyza bonariensis (L.) Cronq., Bull. Torrey Bot. Club 70: 632. 1943

Erigeron bonariensis L., Sp. Pl. 863. 1753.
Erigeron linifolius Willd., Sp. Pl. 3: 1955. 1804.
Conyza floribunda HBK., Nov. Gen. & Sp. Pl. 4: 57. (folio ed.). 1818.
Erigeron floribundus Schultz-Bip., Bull. Soc. Bot. Fr. 12: 81. 1865.
Leptilon bonariense Small, Fl. SE. U.S. 1231, 1340. 1903.
Leptilon linifolium Small, loc. cit.
Erigeron bonariensis var. *leiothecus* Blake, Contr. Gray Herb. n.s. 52: 28. 1917.
Erigeron bonariensis var. *microcephalus* Cabrera, Rev. Mus. LaPlata Bot. n.s. 4: 88. 1941.
Conyza bonariensis var. *microcephala* Cabrera, Man. Fl. Buenos Aires 481. 1953.
Conyza bonariensis var. *leiotheca* Cuatr., Phytologia 9: 5. 1963.

Annual weed 1–10 dm tall, the herbage sparsely to copiously spreading-hirsute or partly strigose; leaves well-distributed along stem, relatively narrow, mostly linear to oblanceolate or linear-elliptic, sessile or tapering to shortly petiolar base, mostly 2–10 cm long overall and 1–10 mm wide, entire or some (especially the lower) rather coarsely few-toothed; heads more or less numerous in an inflorescence ranging from elongate and paniculiform to broadly open-corymbiform with lateral branches overtopping central axis, the disk in our forms well under 1 cm wide as pressed; involucre in ours 2.7–4.5 mm high, densely short-hirsute to glabrous; pistillate flowers more or less numerous, (45)50–100 or more, with an elongate, tubular-filiform corolla produced into very short or scarcely developed ligule to about 0.5 mm long, this generally surpassed by style and equaling or more often surpassed by pappus; disk flowers few, mostly 5–8, sometimes about 13 or even 21.

A weed of tropical American origin, now widespread in tropical and warm tem-

* All of the original Linnaean species of *Conyza* have been transferred to other genera, and the present usage, legitimized by the conserved list in the International Rules, dates from Lessing in 1832.

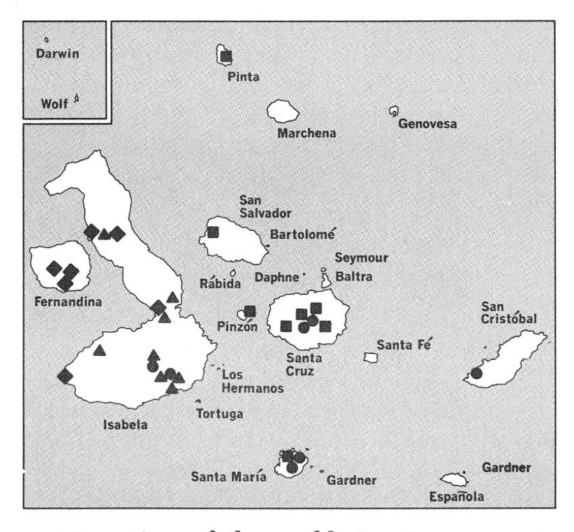

● *Conyza bonariensis*

■ *Darwiniothamnus tenuifolius* var. *tenuifolius*

▲ *Darwiniothamnus tenuifolius* var. *glabriusculus*

◆ *Darwiniothamnus tenuifolius* var. *glandulosus*

perate regions of the world. Specimens examined: Isabela, San Cristóbal, Santa Cruz, Santa María.

The Galápagos plants represent 2 very different phases that would surely be treated as different species if material from elsewhere in the world were not to be considered. The plants from Santa Cruz, which may be only recently introduced, have a glabrous or subglabrous involucre mostly 3.5–4.5 mm high. These can fairly easily be accommodated as a somewhat small-headed form of var. *leiotheca* (Blake) Cuatr., which more often has the involucre 4–6 mm high as in typical material of the species. These Santa Cruz specimens are somewhat like *C. canadensis* (L.) Cronq. in habit and have been confused with that chiefly more northern species. However, *C. canadensis* has fewer (commonly 25–40) pistillate flowers with a better-developed ligule commonly 0.5–1 mm long. Elucidation of the proper taxonomic status of *C. bonariensis* var. *leiotheca* (distinguished by its glabrous or subglabrous involucre, in contrast to the evidently hairy involucre of other varieties) awaits a comprehensive revision of the group. The plants from Isabela, San Cristóbal, and Santa María have strongly pubescent, notably small involucres only 2.7–3.5 mm high, resembling in this respect some plants from Bolivia. They might conceivably be accommodated under the var. *microcephala* (Cabrera) Cabrera, originally described from Buenos Aires, but the biological and taxonomic status of that variety is debatable.

See Fig. 75.

Darwiniothamnus Harling, Acta Horti Berg. 20(3): 108. 1962

Freely branching shrubs with alternate, narrow, entire, sessile or short-petiolate leaves clustered toward ends of branches. Heads narrowly campanulate to hemispheric, solitary or in leafy-bracteate corymbiform clusters terminating some branches, radiate, heterochromous; involucre of several series of narrow, chartaceous, strongly imbricate bracts; receptacle flat or slightly convex, naked, merely pitted, not alveolate; ray flowers pistillate and fertile, with short but evident, white or faintly anthocyanic ligule distinctly surpassing pappus, sometimes more numerous than disk flowers, these perfect and fertile, yellow. Anthers obtuse at base. Style branches flattened, with ventro-marginal stigmatic lines and short, broadly triangular or rounded, externally papillate-hairy appendage. Pappus of numerous capillary bristles in single series. Achenes somewhat dimorphic, those of rays compressed, inconspicuously 2(4)-nerved, those of disk flowers somewhat shorter and broader, evidently 3–5(6)-nerved; embryo sac monosporic, with few antipodal cells.

A single highly variable species, endemic to the Galápagos Islands, which in the past was referred to the genus *Erigeron*.*

Darwiniothamnus tenuifolius (Hook. f.) Harling, Acta Horti Berg. 20(3): 109. 1962

Erigeron tenuifolius Hook. f., Trans. Linn. Soc. Lond. 20: 207. 1847.

Bushy-branched shrub commonly (0.5) 1–3 (3.5) m tall, the leaves and twigs villous-puberulent to glandular or subglabrous; leaves aromatic, linear-filiform to oblanceolate, mostly 1.5–10 cm long, 0.5–8 mm wide, sessile or narrowed to shortly petiolar base seldom to 1 cm long; heads solitary or more often in open to compact, corymbiform, commonly leafy-bracteate inflorescences terminating some branches, on peduncles 0.3–5 cm long; involucre 4–6.5 mm high, glabrous to hairy or glandular, the bracts often purple-tipped; disk mostly 6–12 mm wide as pressed; ligules mostly 2.5–5 mm long and about 0.5 mm wide, often becoming circinately outrolled in drying.

Endemic; in a wide variety of habitats, from woodlands to open lava beds, from the coasts to the mountaintops; known from all of the larger islands except Española and San Cristóbal, as well as from some of the smaller islands.

The species consists of 3 confluent but geographically differentiated and morphologically distinctive varieties. Some of the narrower-leaved specimens of var. *glabriusculus* from Isabela would probably be referred to var. *tenuifolius* if they had been collected on more easterly islands, and likewise a specimen examined from San Salvador would have been called var. *glabriusculus* had it been taken on southern Isabela. Var. *glabriusculus* also intergrades with var. *glandulosus* in the vicinity of Tagus Cove on Isabela, and indeed on pure morphology all 3 varieties could be recognized from Tagus Cove. In spite of these difficulties, most of the specimens do sort out into 3 morphologic-geographic varieties.

See Fig. 75.

Leaves linear or linear-filiform, mostly 0.5–1.5(2) mm wide (or to 3.5 mm wide on vigorous vegetative, nonflowering shoots); pubescence, inflorescence, and head size as in var. *glabriusculus* . . . . var. *tenuifolius*

Leaves broader, linear to oblanceolate, mostly 1.5–8 mm wide:

Leaves puberulent or subglabrous but only inconspicuously or not at all glandular; heads relatively small, the disk less than 1 cm wide, forming corymbiform clusters at branch tips . . . . var. *glabriusculus*

Leaves evidently glandular-hairy or stipitate-glandular; heads larger, the disk mostly 1 cm or more wide as pressed, borne singly or 2–3 together at branch tips . . . . var. *glandulosus*

---

* *Darwiniothamnus* differs from *Erigeron* in being distinctly shrubby, and from the more typical species of that genus in having a strongly imbricate involucre, several-nerved achenes, more ray flowers than disk flowers, and a monosporic embryo sac. *Darwiniothamnus* is also related to *Conyza*, from which it differs in its well-developed ligules and few antipodal cells. The original material of the genus was collected by Darwin.

Darwiniothamnus tenuifolius (Hook. f.) Harling var. **tenuifolius**

Erigeron tenuifolius Hook. f., Trans. Linn. Soc. Lond. 20: 207. 1847.
Darwiniothamnus tenuifolius subsp. *santacruzensis* Harling, Acta Horti Berg. 20(3): 113. 1962.

Characters as in the key.

Specimens examined: Pinta, Pinzón, San Salvador, Santa Cruz, Santa María. Probably does not occur on Isabela or Fernandina.

Darwiniothamnus tenuifolius var. **glabriusculus** (Stewart) Cronq., Madroño 20: 255. 1970

Erigeron lancifolius Hook. f., Trans. Linn. Soc. Lond. 20: 208. 1847.
Erigeron lancifolius var. *glabriusculus* Stewart, Proc. Calif. Acad. Sci. IV, 1: 151. 1911.
Darwiniothamnus lancifolius Harling, Acta Horti Berg. 20(3): 115. 1962.
Erigeron tenuifolius subsp. *lancifolius* Solbrig, Contr. Gray Herb. 191: 43. 1962.

Characters as in the key.

Known only from Isla Isabela, from the southern part of the island north at least to Tagus Cove.

Darwiniothamnus tenuifolius var. **glandulosus** (Harling) Cronq., Madroño 20: 255. 1970

Darwiniothamnus lancifolius subsp. *glandulosus* Harling, Acta Horti Berg. 20(3): 117. 1962.

Characters as in the key.

Specimens examined: Fernandina, Isabela (chiefly north of Istmo Perry, notably about Tagus Cove).

See color plate 88.

Eclipta L., Mant. 2: 157, 286. 1771, nom. conserv.

Branching or diffuse, annual or perennial herbs with opposite, entire or toothed, simple leaves. Heads on axillary and terminal peduncles, radiate, the rays pistillate, fertile or sterile, white or rarely yellow, scarcely or barely exceeding disk; involucral bracts 1- or 2-seriate, herbaceous or herbaceous-tipped, subequal, or the inner narrower and shorter; receptacle flat or slightly convex, its bracts slender and fragile, somewhat bristle-like, the central sometimes wanting; disk flowers perfect and fertile, the corolla 4(5)-toothed. Style branches flattened, with short, obtuse, hairy appendages; anthers obtuse at base. Achenes thick, commonly tuberculate or transversely rugose, 3- or 4-angled, the marginal often triquetrous and with 1 surface apposed to an involucral bract, the others more commonly compressed-quadrangular, compressed along radii of head. Pappus none, or of 2 short awns, or more commonly a small, irregular, nearly obsolete crown well removed from margins of the truncate-topped achene.

About 4 species, mostly of tropical distribution.

Eclipta alba (L.) Hassk., Pl. Jav. Rarior. 528. 1848

Verbesina alba L., Sp. Pl. 902. 1753.
Verbesina prostrata L., loc. cit.
Eclipta erecta L., Mant. 286. 1771 (an illegitimate substitute for *Verbesina alba* L.).
Eclipta prostrata L., loc. cit.

Weak or spreading, strigose annual, the stem branching or diffuse, often rooting at nodes; leaves lanceolate, lance-elliptic, or lance-linear, seldom a little wider, nar-

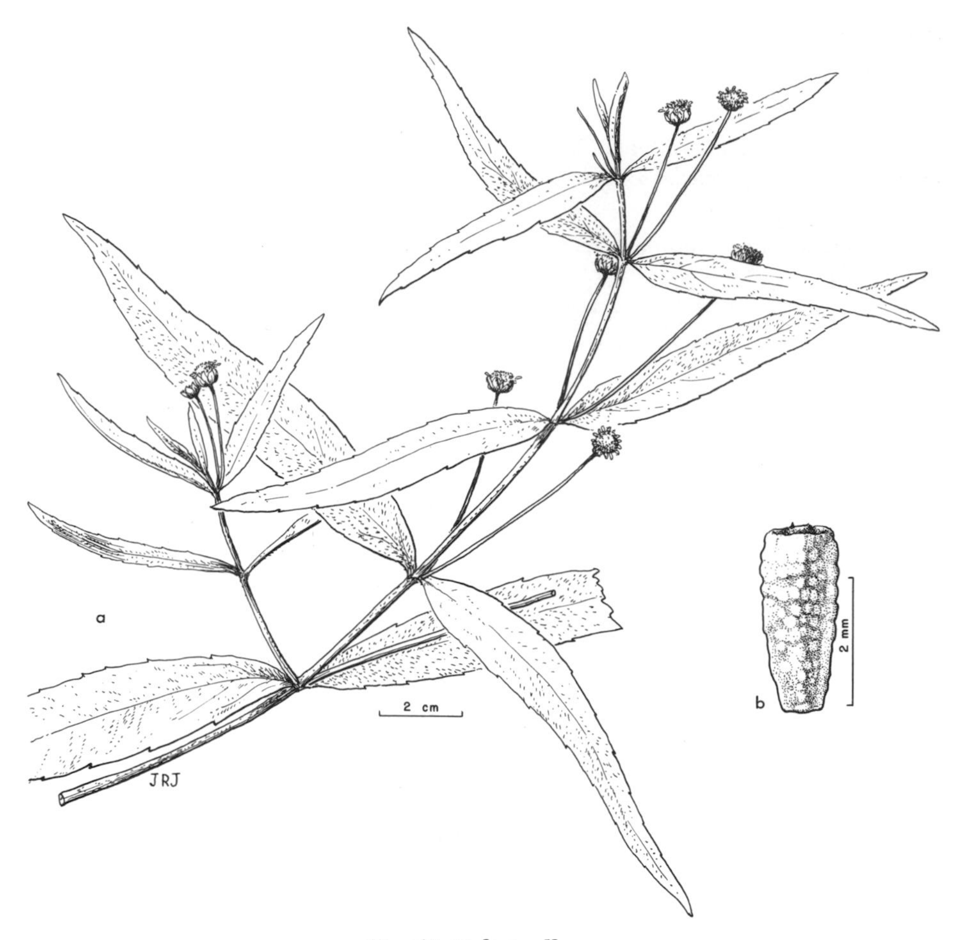

Fig. 76. *Eclipta alba.*

rowed to a sessile or shortly petiolar base, distinctly serrulate, commonly 2–10 cm long, 4–25(30) mm wide; heads mostly 1–3 in terminal or axillary clusters on the numerous branches, the slender peduncles mostly 1–5 cm long; rays white or whitish, rather numerous, slender, about 1 mm long; disk flowers numerous, white or whitish; achenes 2–2.5 mm long, minutely cross-rugulose to more often rugose or tuberculate; pappus nearly obsolete.

A weed in moist bottomlands and muddy places, native to the Americas, now found in tropical and warm temperate regions throughout the world. Specimens examined: Española, Isabela, San Cristóbal, Santa Cruz, Santa María.

See Fig. 76.

Elvira Cass., Dict. Sci. Nat. 30: 67. 1824

Branching, annual herbs; leaves opposite, entire to serrate or biserrate. Heads minutely radiate, pedunculate or sessile in upper axils and terminating stem and branches; involucre strongly flattened to tripterous, consisting of 3–5 bracts, the lowermost flat and much the largest, dominating involucre, the others more or less appressed to it or to each other and forming subsidiary wing, all joined at base and collectively embracing the achene(s). Pistillate flowers 1–3, fertile, with slender tube and very short, spreading, bifid or subentire, yellow to whitish or anthocyanic ligule; disk flowers 1–4, functionally staminate, with undivided style and 4- or 5-toothed, tubular corolla; receptacle small, without bracts; anthers entire at base. Achenes glabrous or distally hairy, obovate or obcuneate, flattened or flattened-triquetrous with 1 broad and 2 narrower sides, included within the persistent involucral bracts, the involucre eventually deciduous as a unit and often resembling small elm-fruit in gross aspect. Pappus none.

A genus of 3 species, 2 of them endemic to the Galápagos, the other widespread on the mainland of tropical America.

Pistillate flowers 3; staminate flowers 3–4. . . . . *E. inelegans*
Pistillate flower solitary; staminate flowers 1–2. . . . . *E. repens*

Elvira inelegans (Hook. f.) Robins., Proc. Amer. Acad. 38: 212. 1902

Desmocephalum inelegans Hook. f., Trans. Linn. Soc. Lond. 20: 209. 1847.

Erect annual weed about 3 dm tall, freely branching from base; leaves petiolate, blade ovate, more or less biserrate, coriaceous, scaberulous and shiny above, darker and hairier beneath, the principal ones about 4 cm long and 3 cm wide; heads numerous, closely crowded, forming depressed-spherical masses 1–1.5 cm thick in upper axils; principal involucral bract broadly ovate or rhombic, acuminate, about 4 mm long and nearly or fully as wide; pistillate flowers 3, with spreading-hairy tube and very short, bifid ligule; staminate flowers 3–4; achenes broadly obcuneate, compressed-trigonous, hairy toward tip.

Endemic; known only from the original collection by Darwin on Santa María.

Elvira repens (Hook. f.) Robins., Proc. Amer. Acad. 38: 212. 1902

Microcoecia repens Hook. f., Trans. Linn. Soc. Lond. 20: 209. 1847.
Elvira atripliciformis Howell, Proc. Calif. Acad. Sci. IV, 21: 333. 1935.

Freely branched annual, the stems rather coarsely white-strigose, 1–5 dm long, varying from merely decumbent and not rooting to trailing and rooting, with assurgent branches; leaves petiolate, petiole 1–5 mm long, blade ovate, inconspicuously serrate or subentire, 0.5–3(5) cm long, 0.4–2(2.5) cm wide, more or less tripli-nerved from near base, loosely and rather shortly hispid or scabrous-hispid above, the lower side similarly hairy to more often canescently hispid-strigose; heads relatively few, subsessile in axils; principal involucral bract obovate to truncate or reniform-obcordate, 2.5–4 mm long, 2.5–9 mm wide, strigose; pistillate flower solitary, with slender, nearly glabrous tube and very short, yellow to whitish or somewhat anthocyanic, minutely bidentate ligule; staminate flowers 1 or 2; achenes compressed-trigonous, 1.5–2 mm long, inconspicuously short-hairy distally.

Endemic. Specimens examined: Isabela, San Salvador, Santa Cruz.

The several collections are somewhat diverse among themselves in habit and measurements, but collectively they form a unit sharply set off from the mainland species, *E. biflora*, and apparently also from *E. inelegans*.

See Fig. 77.

Encelia Adans., Fam. Pl. 2: 128. 1763

Perennial herbs or less often shrubs, with alternate, simple, entire to deeply cleft leaves. Heads small to middle-sized, solitary or in corymbiform to paniculate inflorescences, radiate or less often discoid; involucre of 2 or 3 series of subequal to more or less imbricate bracts; receptacle convex, chaffy, its bracts scarious, soft, embracing achenes and falling away with them; rays, when present, neutral, yellow; disk flowers perfect and fertile, yellow or anthocyanic; style branches flattened, with introrsely marginal stigmatic lines and short, blunt, minutely papillate-hairy appendage; anthers minutely sagittate at base. Achenes strongly flattened along radii of head, but not winged, long-hairy at least on margins. Pappus none, or sometimes of 2 slender awns.

About 14 species, of the southwestern United States, Mexico, and Andean South America.

Encelia hispida Anderss., Kongl. Svensk. Vet.-Akad. Handl. 1853: 186. 1855

Freely branching shrub mostly 6–10 dm tall; leaves and young twigs more or less canescent, sericeous-strigose to strigose-hispidulous; leaves numerous and relatively small, the petiole slender, mostly 3–18 mm long, blades lance-ovate or narrowly ovate to more or less elliptic, mostly 1–5 cm long, 5–22 mm wide, entire or with a few remote, spreading teeth, rather firm, cuneate at base and more or less trinerved;

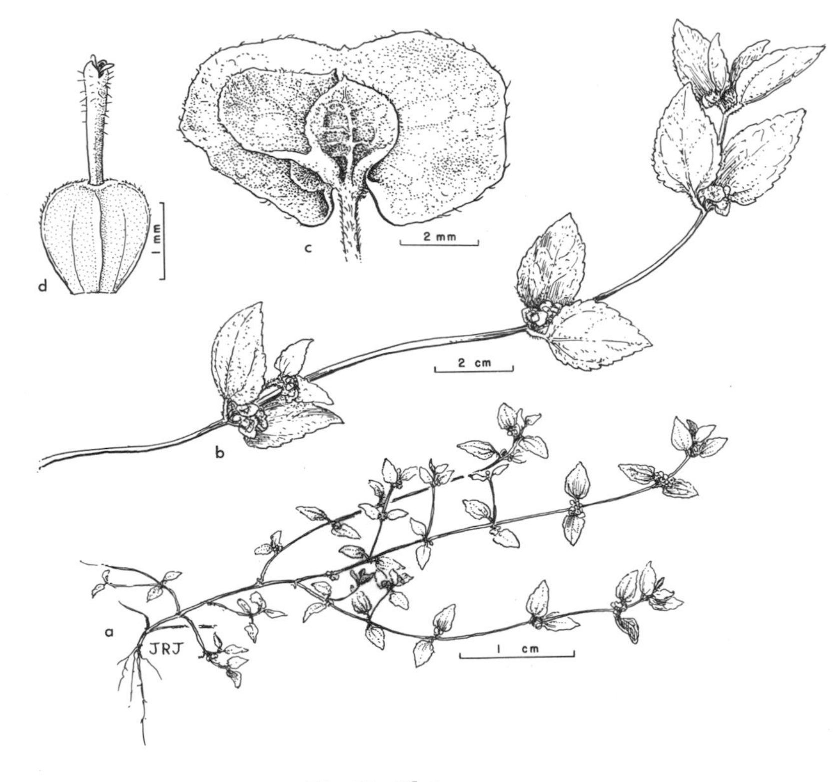

Fig. 77. *Elvira repens.*

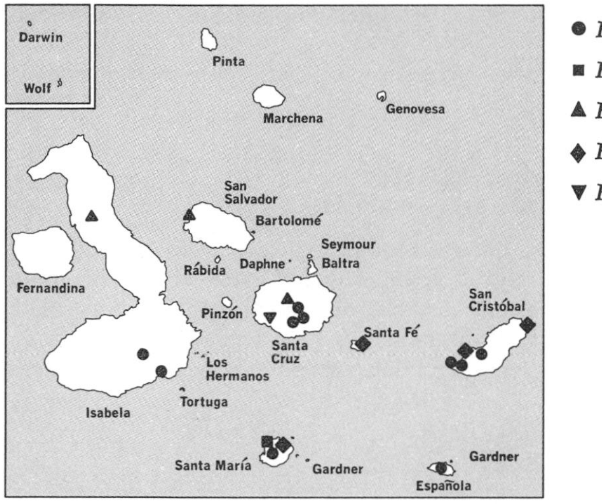

loose corymbiform inflorescences of 2–7 heads, terminating some branches, on slender, arcuate-ascending peduncles 0.5–10 cm long, more or less hemispheric, the disk 7–11 mm wide; involucre 4–6 mm high, canescent, its bracts firm, rather narrow, somewhat imbricate in 2–3 series, greenish under pubescence; rays about 8, yellow, short and broad, commonly 2–4 mm long and well over 1 mm wide; receptacular bracts about 5 mm long, distally villous; disk flowers (fide Blake) purple; achenes 4–5 mm long, densely pubescent along margins and to a lesser extent along midstrip on each surface, with straight, white, appressed-ascending hairs, the upper conspicuously surpassing body of achene (by more than 1 mm); pappus none.

Endemic. Specimens examined: San Cristóbal, Santa Fé, Santa María.

Encelia hispida is closely related to *E. canescens* Lam., of the Peruvian coast, and the 2 may eventually prove to be conspecific. However, *E. canescens* has larger heads, with the disk (1)1.5–2 cm wide and with distinctly longer receptacular bracts, which elongate to 8–10 mm as the achene nears maturity.

See Fig. 78.

Enydra Lour., Fl. Cochinch. 510. 1790*

Hollow-stemmed herbs of wet places, commonly creeping at least at base and rooting at nodes. Leaves opposite, simple, entire or merely toothed, sessile or obscurely short-petiolate. Heads sessile or subsessile in axils, minutely radiate; involucre of a few (commonly 4 or 5) broad, more or less herbaceous bracts, shorter than to somewhat surpassing disk, the 2 or 3 outermost commonly a bit larger than others; receptacle convex (in ours) to low-conic, chaffy throughout, its bracts well-developed, firm, enfolding achenes; pistillate flowers in several series, fertile, with tiny, 3- or 4-toothed, white or whitish ligule; disk flowers more or less numerous, white or whitish, perfect and fertile, or the innermost sometimes functionally staminate. Anthers basally obtuse and entire. Style branches flattened, with introrsely marginal stigmatic lines and blunt, scarcely appendiculate tip. Achenes subterete or somewhat

* The name is often spelled *Enhydra,* following DC., Prodr. 5: 636. 1836, but the original spelling is correct under the rules.

compressed (the inner along radii of head, the outer perpendicular to radii). Pappus none.

About half a dozen species, widespread in tropical and subtropical regions of the world.

Enydra maritima (HBK.) DC., Prodr. 5: 637. 1836

Meyera maritima HBK., Nov. Gen. & Sp. Pl. 4: 211 (folio ed.). 1818.

Stems sprawling, blackish red, soft, hollow, rooting at nodes, with more slender, assurgent branches; leaves fleshy, reddish green with red veins, the short, broad, membranous petioles 1–3 mm long and very shortly connate around stem, blade lance-oblong, rounded at base, acute at tip, mostly 2–6 cm long, 4–7 mm wide, inconspicuously few-toothed, atomiferous-glandular, especially beneath, and also ob-

Fig. 78. *Encelia hispida* (*a–e*): *c*, corolla of ray floret; *d*, receptacular bract. *Enydra maritima* (*f–o*): *g, i, m*, bracts encasing achenes; *k*, disk florets.

scurely pubescent beneath with multicellular hairs; heads borne singly in some axils, subglobose, 7–9 mm thick, involucral bracts striate toward the thin, membranous base, somewhat reticulate-veiny toward the more herbaceous tip; achenes black, about 2 mm long, somewhat striate, narrow, tapering to slender, shortly substipitate base.

Wet places along the Pacific Coast of tropical South America; in the Galápagos known only from Santa Cruz.

The name *E. maritima* is only tentatively applied to our plant, as here described, pending clarifications of the taxonomy of the genus. *Meyera maritima,* the basionym for *E. maritima,* was described as having a conic receptacle, whereas the receptacle in our plants is only slightly convex, but in other respects the original description fits our plants reasonably well.

See Fig. 78.

Eupatorium L., Sp. Pl. 836. 1753; Gen. Pl. ed. 5, 363. 1754

Mostly perennial herbs, vines, and shrubs, a few annual. Leaves mostly opposite, sometimes whorled or alternate, entire or toothed to occasionally dissected. Heads small to rather large in mostly corymbiform or sometimes paniculiform inflorescence, rarely solitary or forming true panicle, discoid, the flowers all tubular and perfect, pink or purple to blue, white, or somewhat greenish or yellowish green but never bright yellow; involucral bracts subequal to evidently imbricate, often striate, sometimes greenish, but tending to be dry rather than herbaceous in texture; receptacle flat to conic, naked or rarely chaffy. Anthers obtuse and entire at base to minutely sagittate, shortly appendiculate at tip. Style branches with short, inconspicuous stigmatic lines and slender, elongate, papillate, obtuse, often clavate appendage. Achenes mostly 5-angled, prismatic. Pappus of more or less numerous capillary bristles.

A vast genus, of certainly more than 500 species, nearly cosmopolitan, centering in the American tropics.

Involucral bracts strongly and obviously imbricate in several series, relatively broad and blunt, and carrying their width well toward tip, the inner ones generally well over 0.5 mm (often 1 mm) wide at a distance of 0.5 mm from tip . . . . . . *E. pycnocephalum*

Involucral bracts only inconspicuously imbricate, relatively narrow and tapering to slender tip, the inner ones seldom as much as 0.5 mm wide at a distance of 0.5 mm from tip . . . . *E. solidaginoides*

Eupatorium pycnocephalum Less., Linnaea 6: 404. 1831

Perennial, usually erect, herbaceous, and 3–15 dm tall, sometimes becoming woody toward base and more or less scrambling on other vegetation; stem and usually the leaves more or less viscid-villosulous or viscid-puberulent, especially upward, sometimes also atomiferous-glandular; leaves opposite, petiole mostly 1–5 cm long, blade mostly deltoid to broadly ovate or triangular-ovate, 2.5–8 cm long, 2–6 cm wide, evidently crenate, palmately or subpalmately 3–5-nerved; involucre 4–5 mm high, evidently puberulent or viscid-puberulent at least toward base (sometimes glabrous in mainland forms), its bracts obviously imbricate in several series, conspicuously ribbed-striate (often with pair of prominent ribs flanking midline), relatively broad and blunt, the inner generally well over 0.5 mm (often 1 mm) wide at 0.5 mm from the broadly rounded to merely obtuse tip, the reduced outer bracts narrower and obtuse to merely acute; flowers mostly 11–26 in each head, blue or lavender through pink to white; achenes black, glabrous or nearly so, about 1.5 mm long.

A widespread weed in tropical and subtropical America, extending from southwestern United States to at least Peru and Bolivia; in the Galápagos Islands known only from Isla Santa Cruz.

Doubtless introduced within historic times, the oldest known collection dating from 1905.

Eupatorium solidaginoides HBK., Nov. Gen. & Sp. Pl. 4: 99 (folio ed.). 1818

Eupatorium filicaule Schultz-Bip. ex A. Gray, Proc. Amer. Acad. 21: 384. 1886.
Ophryosorus solidaginoides Hieron., Bot. Jahrb. Syst. 29: 4. 1900.

Vegetatively much like *E. pycnocephalum*, perhaps oftener becoming subshrubby, and with the leaves sometimes cordate at base; inflorescence tending to be longer

Fig. 79. *Eupatorium solidaginoides.*

and more paniculate than in *E. pycnocephalum;* involucre 3–5 mm high, evidently puberulent or viscid-puberulent at least toward base, its bracts rather inconspicuously imbricate and inconspicuously striate, all relatively narrow and tapering gradually to the slender tip, the inner often less so than the outer, but still seldom as much as 0.5 mm wide at 0.5 mm from tip, the outer slender and very narrowly acute or acuminate; flowers mostly 9–16 in each head, pale greenish or yellowish green to white or seldom lilac or bluish, the corolla lobes glandular externally; achenes black, generally hispidulous at least on angles, 1.5–2.5 mm long.

In thickets, along banks of streams, at edges of forest, and at disturbed sites, Mexico, Central America, and western tropical South America, south at least as far as Bolivia; in the Galápagos Islands known only from Isla Isabela.

Doubtless introduced within historic times, the oldest known collection dating from 1898.

See Fig. 79.

Flaveria Juss., Gen. 186. 1789

Annual or perennial herbs with opposite, rather narrow leaves tapering to sessile or shortly petiolar base. Heads yellow, numerous, in most species more or less crowded and sometimes even forming secondary heads, radiate or some discoid; involucre of 2–8 rather thin, subequal bracts (or the outermost somewhat shorter than others), sometimes subtended by 1 or 2 smaller bracts at base; receptacle small, naked; pistillate flower solitary, with short to very short ligule, or wanting from some heads; disk flowers 1–15, perfect and fertile. Style branches flattened, with ventro-marginal stigmatic lines, truncate, minutely penicillate. Anthers entire at base. Achenes somewhat compressed along radii of head, glabrous, about 10-ribbed. Pappus none, or of a few small, irregular scales.

About a dozen species, occurring from the southern United States to South America, most abundantly in Mexico.

Flaveria bidentis (L.) Kuntze, Rev. Gen. 3: 148. 1898

Ethulia bidentis L., Mant. 110. 1767.
Eupatorium chilense Molina, Sagg. Chile 142. 1782.
Flaveria chilensis J. F. Gmel., Syst. Nat. ed. 13, 2: 1269. 1791.
Milleria contrayerba Cav., Icon. 1: 2, *t. 4.* 1791.
Flaveria contrayerba Pers., Syn. Pl. 2: 489. 1807.

Subglabrous, branching annual 1–10 dm tall or more, often subdichotomously branched above, with each main axis foreshortened and overtopped by uppermost pair of branches; leaves rather narrowly elliptic or lance-elliptic, evidently to obscurely serrate, acute, tapering to shortly petiolar or subsessile base, 3–12 cm long overall, 0.5–3 cm wide, trinerved from base; heads numerous, sessile, in several or many dense clusters whose compactly cymose-branched structure can be seen at 10×, each head generally subtended by 1 or 2 short bracts of the infloresence, which may simulate a short outer involucre; involucre 3–4 mm high, of (2)3 membranous, rather pale, striate bracts, the outer often shorter than other 2; each head usually with single ray flower and 1–8 disk flowers, or some heads without ray flower; ligule very short, 1 mm long or less, erect or spreading; achenes black, epappose, mostly 2–2.5 mm long, that of ray flower often a bit larger than those of disk.

A weed found in disturbed sites from the southern United States to Chile and

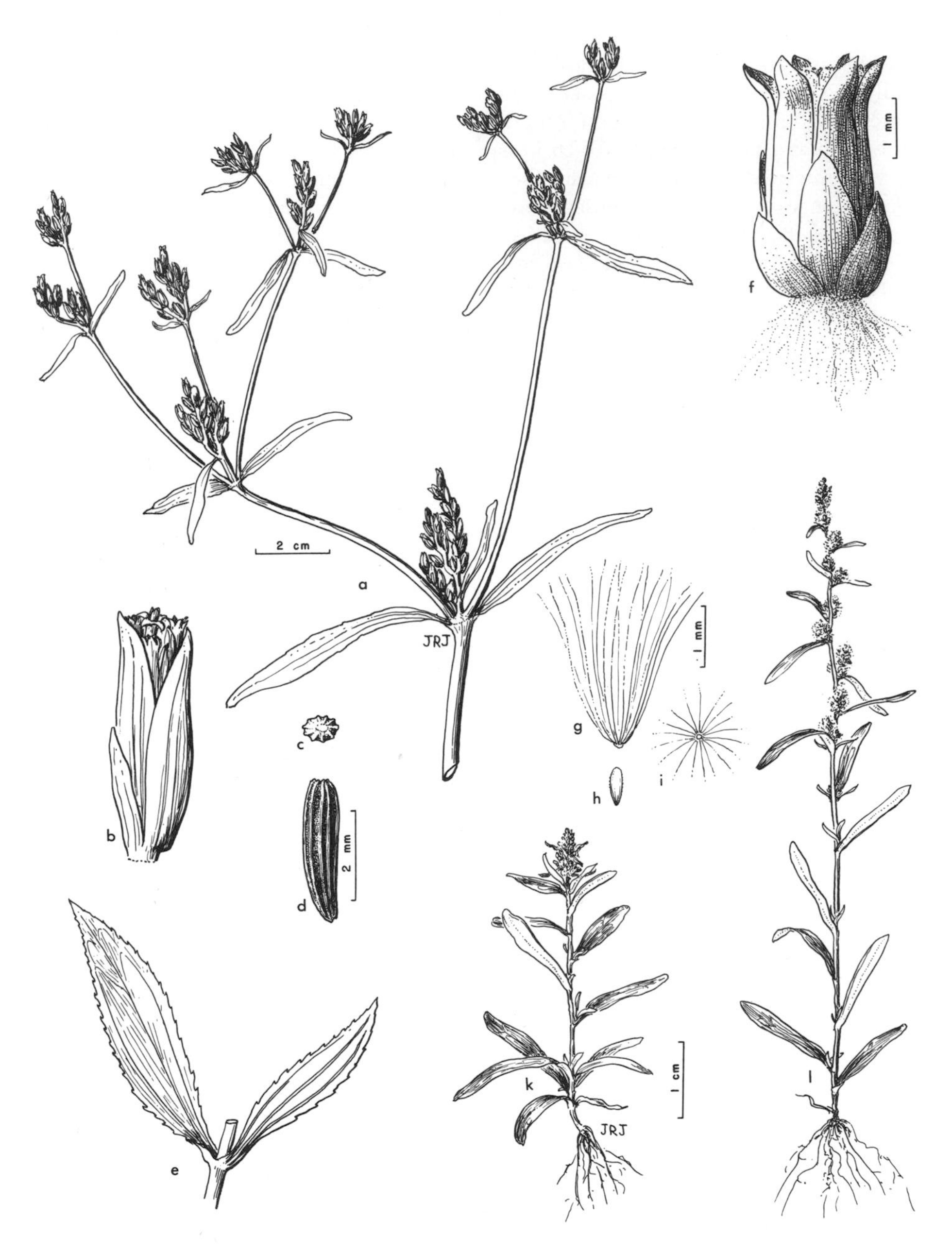

Fig. 80. *Flaveria bidentis* (*a–e*). *Gnaphalium purpureum* (*f–l*).

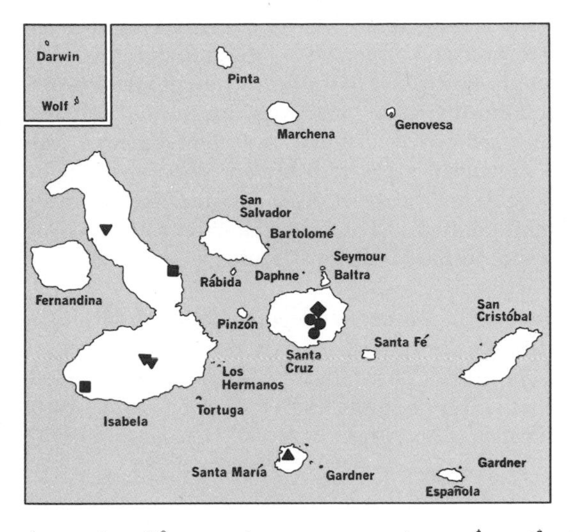

● *Eupatorium pycnocephalum*
■ *Eupatorium solidaginoides*
▲ *Flaveria bidentis*
◆ *Gnaphalium purpureum*
▼ *Gnaphalium vira-vira*

Argentina. I have not seen any specimens from the Galápagos Islands but Andersson reported it from Isla Santa María.

See Fig. 80.

Gnaphalium L., Sp. Pl. 850. 1753; Gen. Pl. ed. 5, 368. 1754

Annual to perennial, generally more or less white-woolly herbs with alternate, entire, commonly rather narrow leaves. Heads disciform, the numerous outer flowers pistillate, with slender, tubular-filiform corolla, the inner coarser and perfect. Involucres ovoid or campanulate, the bracts slightly to evidently imbricate, scarious at tip or nearly throughout; receptacle naked. Flowers yellow or reddish to whitish. Anthers caudate. Style branches of the perfect flowers flattened, truncate, without sharply differentiated stigmatic portion. Pappus of numerous capillary bristles, sometimes thickened at apex, sometimes more or less united at base; achenes small, nerveless.

Over 100 species, with worldwide distribution from tropical to frigid zones.*

Inflorescence narrow, dense, spiciform-thyrsoid; pappus bristles definitely united into a ring at base, falling as unit . . . . *G. purpureum*

Inflorescence of 1 or usually several cymosely arranged terminal glomerules; pappus bristles only lightly and imperfectly coherent at base, tending to fall singly or in small groups . . . . *G. vira-vira*

Gnaphalium purpureum L., Sp. Pl. 854. 1753

Gnaphalium americanum Mill., Gard. Dict. ed. 8. no. 17. 1768.
Gnaphalium spicatum Lam., Encycl. Méth. 2: 757. 1786.
Gamochaeta americana Wedd., Chloris Andina 1: 151. 1856.
Gnaphalium purpureum var. *spicatum* Baker in Mart., Fl. Bras. 6(3): 124. 1882.
Gamochaeta purpurea Cabrera, Bol. Soc. Arg. Bot. 9: 377. 1961.
Gamochaeta spicata Cabrera, op. cit. 380.

Annual or biennial, simple or branched, 1–4 dm tall, our forms with thinly tomentose or subglabrous stem and with leaves strongly discolored, white and pannose-

* The taxonomy of tropical American *Gnaphalium* is in such a state of confusion that little confidence can be placed in most identifications.

tomentose beneath, green and glabrous or subglabrous above; lowest leaves oblanceolate or spatulate, seldom to 10 cm long and 2 cm wide, those above generally gradually reduced, often not so strongly tapered at base; heads numerous in terminal, somewhat leafy-bracteate, spiciform-thyrsoid, sometimes interrupted inflorescence, or some clusters occasionally somewhat elongate and inflorescence thus branched; involucre 3–5 mm high, brownish or greenish brown, sometimes tinged with pink or purple, glabrous, or woolly only at base, its bracts imbricate, obtuse to acuminate; pappus bristles evidently united at base, deciduous in a ring.

A widespread native American weed; in the Galápagos Islands known only from Isla Santa Cruz.

Apparently only recently introduced, the earliest collection dating from 1963. Our plants belong to the phase of the species having strongly discolored leaves, to which the name *G. purpureum* var. *spicatum* (Lam.) Baker has often been applied. This phase does not occupy a distinctive range to the exclusion of other phases of the species, and its taxonomic significance is doubtful.

See Fig. 80.

Gnaphalium vira-vira Molina, Sagg. Chile 149, 354. 1782

Taprooted perennial, apparently rather short-lived, possibly sometimes blooming first or second year, often with stems of principal branches erect from decumbent or arcuate base; lowermost leaves oblanceolate, to 4.5 cm long and nearly 1 cm wide, often crowded, others usually progressively reduced; heads in 1 or usually several cymosely arranged tight glomerules, with somewhat tawny cast; involucre 5–6 mm high, loosely woolly at or toward base, otherwise glabrous and shining, its bracts hyaline, more or less sharply acute, in age finally radiately spreading; pappus bristles barbellate at the thickened base and tending to cohere by the intermeshed barbels, but falling separately or in small groups rather than in a definite ring.

Andean and coastal regions of Chile, Bolivia, and Peru; in the Galápagos Islands known only from Isla Isabela.

Our plants have usually been misidentified as *G. luteo-album* L., a Mediterranean annual weed with smaller, usually more distinctly yellowish involucres that are commonly 3–5 mm high and have broadly rounded or obtuse bracts. It seems reasonably likely that the Galápagos plants should properly be identified instead with an Andean species that has been going under the name *G. vira-vira* Molina. I have not seen the type, and I am not at all sure that the name is properly applied.

Jaegeria HBK., Nov. Gen. & Sp. Pl. 4: 218 (folio ed.). 1818

Annual or perennial herbs of shallow water or moist places, mostly decumbent and rooting at nodes. Leaves simple, opposite and often basally connate, sessile (in our species) or petiolate, entire or toothed. Heads rather small, slender-pedunculate from upper axils and terminating the stems, radiate or seldom discoid; involucre of 1 or 2 series of equal, herbaceous to somewhat coriaceous bracts, these individually subtending and usually clasping the ray achenes, typically with thin margins wrapped around edges of achene; ray flowers pistillate and fertile, of same number as phyllaries, with white, yellow, or anthocyanic ligule, or rarely wanting and phyllaries then subtending outermost disk flowers; receptacle high-conic, chaffy throughout, its bracts slender to broad, sometimes some of the outer broader than the inner and clasping achene; disk flowers perfect and fertile, yellow or yellow-green, the short, slender corolla-tube evidently spreading-hairy, in contrast to the expanded, glabrous limb. Anthers sagittate at base. Style branches flattened, with ventro-marginal stigmatic lines and very short, blunt, externally papillate appendage. Achenes black, glabrous, essentially epappose, those of rays broad-backed and somewhat flattened or trigonous, others scarcely to evidently compressed at right angles to radii of head.

Eight species, of moist places from Mexico to South America.

Involucral bracts thickened, wrapped around and fully enclosing the subtended achene, often some or even all of bracts much like outer receptacular bracts, these glabrous, strongly thickened, conspicuously grooved-striate, wrapped around and fully enclosing achene . . . . *J. crassa*

Involucral bracts relatively thin and with margins barely, if at all, wrapped around edges of the subtended achene, not meeting in center; outer receptacular bracts slightly to strongly differentiated from the middle and inner ones, often somewhat clasping their achenes, and occasionally approaching the structure and texture of those of *J. crassa*. . . . . *J. gracilis*

Jaegeria crassa Torres, Brittonia 20: 71. 1968

Resembling *J. gracilis* in general aspect, but larger and more robust, and often less hairy, the stem sometimes quite glabrous; larger leaves commonly 2.5–5 cm long, 13–20 mm wide; involucral bracts more or less thickened, wrapped around and fully enclosing the subtended achene, often some or all much like outer receptacular bracts, these glabrous, strongly thickened, conspicuously grooved-striate, wrapped around and fully enclosing achene. $2n = 36$.*

Endemic; in moist soil at disturbed sites. Specimens examined: Isabela, San Cristóbal, Santa Cruz.

Jaegeria gracilis Hook. f., Trans. Linn. Soc. Lond. 20: 213. 1847

Jaegeria prorepens Hook. f., op. cit. 214.

Slender annual, weakly erect to more often decumbent or trailing and rooting at lower nodes with upper internodes assurgent to erect, mostly 1–3(6) dm tall; herbage loosely hispid or hispid-hirsute throughout, or peduncles merely strigose; leaves small, sessile or nearly so, broad-based or more often tapering and sometimes shortly subpetiolate, elliptic to lanceolate or ovate, acute, mostly 1–3.5 cm long and 4–15 mm wide, entire or with a few short, divergent, callous teeth; peduncles very slender, mostly 1–8 cm long; heads small, radiate or discoid, disk 4–6 mm wide, involucre 2–3 mm high, spreading-hairy like the herbage, its bracts green, herbaceous, with margins barely, if at all, wrapped around subtended achene, not meeting in center; ray flowers (when present) usually as many as phyllaries and each provided with short and wide yellow ligule 1–3 mm long, the phyllaries subtending disk achenes when rays are absent; receptacular bracts chartaceous, tapering to sharp, often black or blackish purple tip, the middle and inner ones persistent, slender and narrower than subtended achene, a few of the outermost somewhat enlarged (surpassing involucral bracts) and expanded, tending to clasp subtended achene, occasionally approaching structure and texture of *J. crassa*; disk flowers numerous, the

* Torres, loc. cit.

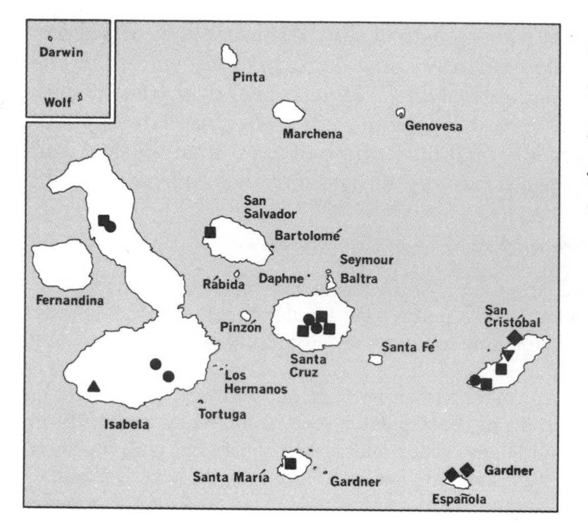

● *Jaegeria crassa*
■ *Jaegeria gracilis*
▲ *Jungia hirsuta*
◆ *Lecocarpus lecocarpoides*
▼ *Lecocarpus leptolobus*

yellow or greenish yellow corolla about 1 mm long, achenes about the same. 2n = 36.*

Endemic; in moist soil at disturbed sites. Specimens examined: Isabela, San Cristóbal, San Salvador, Santa Cruz, Santa María.

The species is closely allied to the widespread mainland species *J. hirta* (Lag.) Less., but the latter has relatively shorter receptacular bracts, which are not dark-tipped, and has the ray achenes more definitely enfolded by the involucral bracts.

See Fig. 81.

Jungia L. f., Suppl. 58. 1781†

Erect or subscandent herbs or subshrubs. Leaves alternate, commonly cordate-orbicular and angularly dentate or lobed. Heads small to middle-sized, more or less numerous in ample, paniculiform inflorescences; involucre of 1 or 2 series of equal or graduated bracts; receptacle chaffy throughout; flowers white to pink or yellow, all bilabiate, outer lip shortly 3-toothed and liguloid, inner more deeply cleft, with 2 slender segments, shorter (at least in marginal flowers) than outer. Anthers strongly caudate. Style branches flattened, truncate and penicillate. Achenes linear or oblong, subterete, 5-ribbed, beaked or with subapical constriction. Pappus a single series of generally plumose bristles.

Two or three dozen species, native to tropical America.

Jungia hirsuta Cuatr., Brittonia 8: 48. 1954

Coarse herb 10–15 dm tall, with stout, simple, loosely villous stem; leaves villous on both sides, more densely so beneath, lower ones on petioles 5–10 cm long, blade

* Torres, *Brittonia* 20: 68. 1968.

† *Jungia* L. f., is antedated by at least 2 earlier homonyms, *Jungia* Heister (1759) and *Jungia* Boehmer (1760). Continued use of the name for the present genus can be legitimized only by conservation. Since the entire genus is now under review by another botanist, no nomenclatural change is here proposed.

shallowly lobed, 10–15 cm wide, those above sessile, irregularly toothed, mostly 6–10 cm long, 4–7 cm wide; inflorescence to 5 dm long, glandular and villous; heads numerous, on densely glandular peduncles 1–2 cm long; involucre uniseriate, about 5 mm high, glandular; flowers about 21 in each head, white; corollas 4–6 mm long, tube equaling or a bit longer than the more or less liguloid outer lip; achenes glabrous or nearly so, with subapical constriction and expanded, pappiferous disk; pappus bristles 20–25, white.

A species of continental Ecuador, perhaps only recently introduced into the Galápagos Islands, where reported by Eliasson (1968) from Isla Isabela, a new record for the Galápagos Islands.

Fig. 81. *Lecocarpus lecocarpoides* (*a–d*): *c, d*, apically winged achene. *Jaegeria gracilis* (*e–h*): *e, g*, outer and inner bracts.

Lecocarpus Decne., Bot. Voy. Venus, Atlas, *t. 14.* 1846

Erect shrubs or at least sometimes frutescent annuals, with single stem at base. Leaves opposite, pinnately more or less cleft or dissected. Heads borne singly on peduncles arising terminally or from upper axils, radiate; involucre of 2 very dissimilar series, the outer of 4 or 5 broad, foliaceous, loose or spreading bracts, the inner indurated, individually enclosing ray achenes and usually either with 1 to several prominent spines around apical rim or with rim developed into broad, flaring collar, otherwise unarmed; receptacle flat or convex, chaffy throughout, its bracts loosely folded or convex and embracing disk flowers, tending to be laciniate or fimbriate distally; ray flowers well-developed but few, commonly 6–10, pistillate and fertile, with thickened, epappose achene and an evident yellow ligule mostly 5–15 mm long; disk flowers more or less numerous, yellow, functionally staminate, with undivided style and poorly developed, infertile achenes. Anthers entire or minutely sagittate at base.

A genus of 3 closely allied species, endemic to the Galápagos Islands. Two of the species have usually been referred to the related genus *Acanthospermum*, but the unity of the group is obvious and has often been remarked.

"Fruit" unarmed, at maturity developing a conspicuous, broad, saucer-like flange around top . . . . *L. pinnatifidus*
"Fruit" generally with 1 to several prominent spines around margin at top, but without a broad, expanded, saucer-like flange:
  Leaves deeply dissected into narrow, ultimate segments, the intact midstrip only 0.5–1.5 mm wide . . . . *L. leptolobus*
  Leaves less deeply dissected, the intact midstrip (1.5)2–5(10) mm wide . . . . *L. lecocarpoides*

Lecocarpus lecocarpoides (Robins. & Greenm.) Cronq. & Stuessy, Madroño 20: 256. 1970

Acanthospermum lecocarpoides Robins. & Greenm., Amer. Jour. Sci. III, 50: 141. 1895.
Acanthospermum brachyceratum Blake, Jour. Wash. Acad. Sci. 12: 203. 1922.

Plants 3–10 dm tall, freely branched upward, the herbage wholly spreading-hispidulous, sometimes also glandular, the lower, more woody part of stem often more or less glabrate; petioles mostly 0.5–2 cm long; leaf blades mostly 1.5–6 cm long, 1–3.5 cm wide, more or less elliptic in outline, pinnately cleft with fairly well-separated, again 1- or 2-toothed or -cleft segments, the undivided central midstrip mostly (1.5)2–5(12) mm wide; peduncles mostly 2–5 cm long; outer involucral bracts narrowly to broadly ovate, obtuse or acute, entire, 0.5–1 cm long; ligules 5–10 mm long, 2–3 mm wide; "fruits" 4–5 mm high and wide, glandular-hispidulous, generally with 1 to several variably developed spines around the prominent apical rim, these shorter to a bit longer than body of "fruit."

Endemic; chiefly on Española and associated islets. Specimens examined: Española, Gardner (near Española), San Cristóbal.

The discovery of *L. lecocarpoides* on San Cristóbal breaks down the geographic distinction between this species and *L. leptolobus*, which is apparently restricted to San Cristóbal. The possibility that the 2 species might better be treated as 1, with or without infraspecific segregation, presents itself but is not here resolved.

See Fig. 81.

Lecocarpus leptolobus (Blake) Cronq. & Stuessy, Madroño 20: 256. 1970

Acanthospermum leptolobum Blake, Jour. Wash. Acad. Sci. 12: 204. 1922.

Plants 3–12 dm tall, freely branched upward, the herbage wholly spreading-hispidulous, sometimes also glandular, the lower, more woody part of stem often more or less glabrate; petioles 0.5–1 cm long; leaf blades mostly 2.5–4 cm long, 1.3–4 cm wide, more or less elliptic or ovate in outline, rather openly bi- or tripinnatifid into small and rather narrow ultimate segments seldom over 1.5 mm wide, the undivided foliar midstrip only 0.5–1.5 mm wide; peduncles 0.8–2.5 cm long; outer involucral bracts narrowly to rather broadly ovate, acute or obtuse, commonly few-toothed, 6–10 mm long; ligules 8–15 mm long, 2.5–4.5 mm wide; "fruits" densely glandular and somewhat hispidulous, 2.5–3.5 mm high and nearly as wide, generally 1 to several variably developed spines around the prominent apical rim, these shorter to a bit longer than body of "fruit."

Endemic; known only from San Cristóbal.

Lecocarpus pinnatifidus Decne., Bot. Voy. Venus, Atlas, *t. 14.* 1846

Lecocarpus foliosus Decne., op. cit. *t. 20.* 1864.

Plants 2–20 dm tall, freely branched upward, the herbage wholly spreading-hispidulous, sometimes also glandular, the lower, more woody part of stem often more or less glabrate; petioles mostly 0.5–1.5 cm long; leaf blades mostly 3–7 cm long, 1.5–4 cm wide, elliptic or narrowly elliptic in outline, pinnatifid to tripinnatifid into mostly oblong or oblong-lanceolate, apically rounded ultimate segments, appearing more deeply dissected than in *L. lecocarpoides* but with larger, relatively broader, and more crowded ultimate segments than in *L. leptolobus*; peduncles 2–5 cm long; outer involucral bracts 8–13 mm long, ovate or broadly ovate, obtuse or acute, entire or obscurely toothed or undulate; ligules about 1 cm long, 3–5 mm wide; "fruits" glandular or glandular-puberulent but not hispidulous, unarmed, the body 2–4 mm high, provided with prominent collar when young, this much expanded at maturity into flaring, *Ipomoea*-like, chartaceous crown to 1 or even 1.5 cm wide, its margins commonly scalloped or wavy or toothed and sometimes with

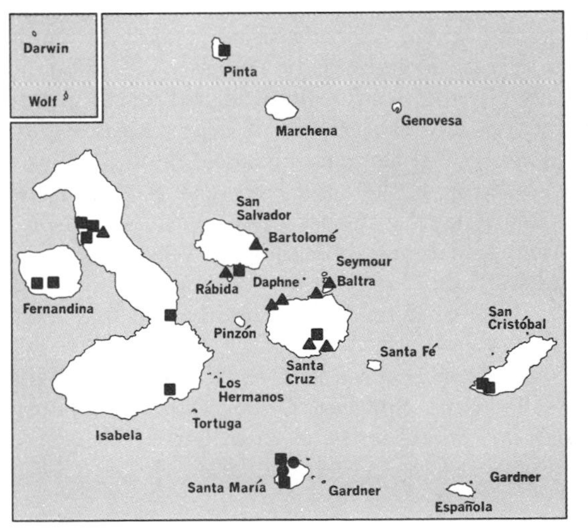

● *Lecocarpus pinnatifidus*
■ *Macraea laricifolia*
▲ *Pectis linifolia*

1 or more small, divergent processes, which may represent vestigial spines. n = 11, fide Eliasson (1970).

Endemic; known only from Santa María, where it is a pioneer on lava and cinders at low elevations.

See color plate 71.

Macraea Hook. f., Trans. Linn. Soc. Lond. 20: 209. 1847

Erect, much branched shrubs with slender, leafy branches. Leaves opposite, ericoid, narrowly linear, 1-nerved, with strongly revolute margins, often with axillary fascicles of smaller leaves. Heads on slender, naked peduncles terminating the branches, yellow, radiate to discoid or subradiate (i.e., with some marginal flowers intermediate in structure between typical ray flowers and typical disk flowers); involucre more or less hemispheric, with 2 or 3 series of firm, scarcely imbricate bracts, the outer equaling or slightly surpassing the inner; receptacle flat or polsterform, chaffy throughout, its bracts folded and embracing the florets or achenes, with a differentiated, more or less triangular, sometimes anthocyanic tip that tends to be inflexed and to cover young florets; rays pistillate (when normally developed) and fertile; disk florets perfect and fertile, corolla tube abruptly expanded into the swollen, tulip-shaped limb, lobes thickened and erect. Anthers sagittate at base, not caudate. Style branches flattened, with introrsely marginal stigmatic lines and short, more or less deltoid, hirsutulous appendage. Achenes with roughened surfaces and winged margins, the outermost triquetrous, with the broad back apposed to subtending bract, the others somewhat compressed parallel to radii of head, all tapering to substipitate base. Pappus a short but distinct, irregularly toothed or fimbriate crown, sometimes prolonged into 1 or 2 awns.

A single species, endemic to the Galápagos Islands.

Macraea laricifolia Hook. f., Trans. Linn. Soc. Lond. 20: 210. 1847

Trigonopterum ponteni Anderss., Kongl. Svensk. Vet.-Akad. Handl. 1853: 184. 1855.
Lipochaeta laricifolia A. Gray, Proc. Amer. Acad. 5: 131. 1861.

Shrub 1–2.5 m tall, profusely and slenderly branched from base, the young twigs densely white-strigose; leaves mucronate at tip, glossy and moderately to thinly strigose above, the lower side dull, strigose along midstrip, densely villosulous in the trough on each side between midstrip and margin, the principal leaves mostly 1.5–5 cm long, 1–2 mm wide, fascicled ones shorter but of nearly same width; heads on slender, strigose peduncles mostly 2–4 cm long; involucre mostly 3–4 mm high, shorter than disk, its bracts white-strigose, the outer lance-triangular to ovate or ovate-oblong, the inner mostly broader and a bit shorter, often transitional toward receptacular bracts in appearance and sometimes subtending flowers; rays when present mostly 5–13, light yellow, about 5 mm long; disk 6–12 mm wide; achenes 1.5–2 mm wide (including wings), those of outer flowers mostly 2–3 mm long, others mostly 2.5–4.5 mm long.

Endemic; in the less extreme habitats, from near sea level to 855 m, on all of the larger islands and many of the smaller ones. Specimens examined: Fernandina, Isabela, Pinta, Rábida, San Cristóbal, San Salvador, Santa Cruz, Santa María.

See Fig. 82.

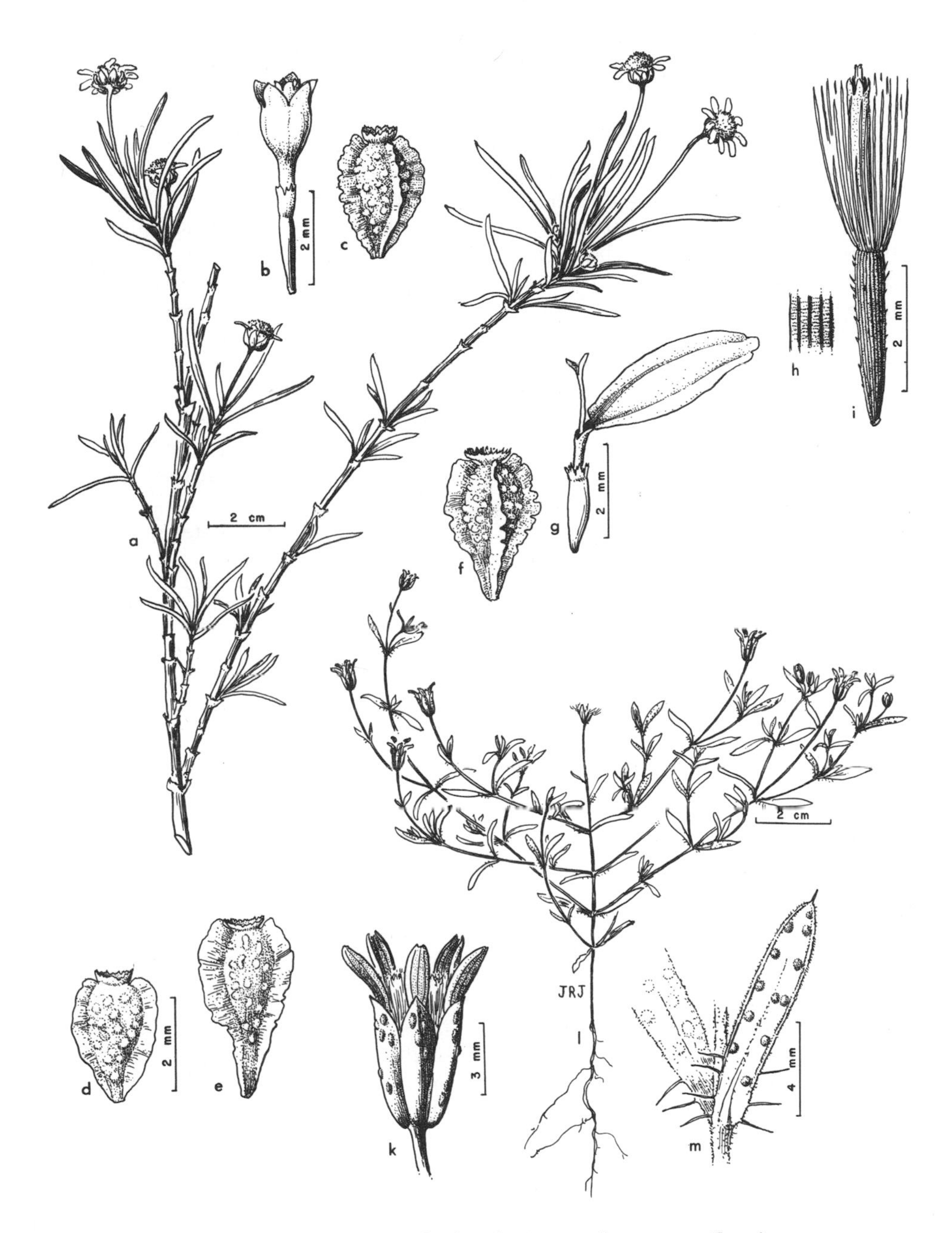

Fig. 82. *Macraea laricifolia* (*a–g*). *Pectis subsquarrosa* (*h–m*).

Pectis L., Syst. Nat. ed. 10, 1221. 1759

Aromatic, annual or perennial herbs, the leaves and usually the involucral bracts beset with conspicuous, scattered, embedded, translucent oil glands. Leaves opposite, simple, narrow, often linear, generally with some coarse, bristly, marginal cilia near base (or along whole margin), otherwise generally entire or occasionally serrate. Heads slender-pedunculate to sessile, commonly terminating the branches, radiate (sometimes inconspicuously so); involucre cylindric to campanulate or turbinate, of 3–12 separate bracts in single series, these rounded-carinate at least below, commonly gibbous at base, and individually subtending or partly clasping ray achenes; receptacle small, flat or a bit convex, naked; ray flowers pistillate and fertile, usually of same number as involucral bracts, with short to evident, yellow or partly anthocyanic ligule; disk flowers rather few, perfect and fertile, with yellow, regular or somewhat bilabiate corolla. Anthers entire at base. Style branches notably short, with ventro-marginal stigmatic lines, the tip blunt, scarcely appendiculate. Achenes narrow, terete or somewhat angular, inconspicuously striate-nerved, glabrous or hairy. Pappus various, of few to many awns, bristles, or short scales, or reduced to mere crown or obsolete.

More than 50 species, occurring from the southwestern United States to tropical South America, most abundant in Mexico.

Rays short and inconspicuous, the ligule mostly 0.5–1.5 mm long, equaling or generally shorter than its corolla tube; achenes generally all with 2–4 prominent, more or less divergent or divaricate, often curved pappus awns .... *P. linifolia*

Rays larger and more showy, the ligule mostly 2–5 mm long, longer than its corolla tube; achenes of the disk flowers commonly with pappus of more or less numerous bristles, or pappus sometimes much reduced or wanting, but never (in disk flowers) as in *P. linifolia*:

  Plant annual .... *P. subsquarrosa*

  Plant perennial .... *P. tenuifolia*

Pectis linifolia L., Syst. Nat. ed. 10, 1221. 1759

Verbesina linifolia L., op. cit. 1226.
Pectis punctata Jacq., Enum. Pl. Carib. 28. 1760.
Pectidium punctatum Less., Linnaea 6: 707. 1831.

Erect, freely branched, glabrous annual mostly 1.5–5(10) dm tall, sparsely leafy, leaves linear or nearly so, mostly 1–6 cm long, 1–4 mm wide, with few or no basal bristles; heads on slender peduncles mostly 0.5–3 cm long terminating stems, the peduncles naked or more often with a few small, inconspicuous, alternate bracts; involucre narrowly cylindric, mostly 4–6.5 mm high, composed of (4)5 linear or linear-oblong bracts, these anthocyanic distally or nearly throughout and often minutely ciliolate around the broadly rounded tip; ray flowers as many as involucral bracts, ligule inconspicuous, yellow (or anthocyanic externally), erect or spreading, mostly 0.5–1.5 mm long, equaling or generally shorter than its corolla tube; disk flowers few, commonly 2—5, yellow or anthocyanic; achenes black, hispidulous distally or throughout, the outer about as long as the subtending involucral bract, others often evidently shorter; pappus of all flowers of 2–4 prominent, more or less divergent or divaricate, often somewhat arcuate-curved, smooth awns (1)1.5–3(4) mm long, or the disk achenes sometimes without pappus.

A fairly common tropical weed of the Americas; in the Galápagos Islands found

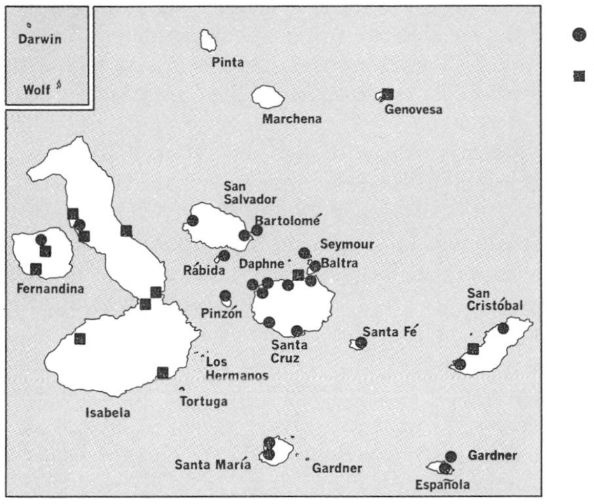

● *Pectis subsquarrosa*
■ *Pectis tenuifolia*

in both natural and disturbed habitats in open places on most of the larger islands and some of the smaller ones. Specimens examined: Baltra, Isabela, Rábida, San Salvador, Santa Cruz.

Pectis subsquarrosa (Hook. f.) Schultz-Bip. in Seem., Bot. Herald 309. 1856

Lorentea subsquarrosa Hook. f., Trans. Linn. Soc. Lond. 20: 206. 1847.
Lorentea gracilis Hook. f., loc. cit., non *Pectis gracilis* Baker 1884.
?*Lorentea linearis* Anderss., Kongl. Svensk. Vet.-Akad. Handl. 1853: 174. 1855.
Pectis anderssonii Robins., Proc. Amer. Acad. 38: 214. 1902, non LaLlave 1885.
Pectis hookeri Robins., loc. cit.
Pectis hookeri f. *stellulata* Howell, Proc. Calif. Acad. Sci. IV, 21: 335. 1935.
Pectis glabra Howell, op. cit. 333.
Pectis glabra f. *calvescens* Howell, op. cit. 334.
Pectis glabra f. *pubescens* Howell, op. cit. 335.
Pectis glabra f. *setulosa* Howell, loc. cit.

Taprooted, minutely hispidulous-puberulent to glabrous annual, 0.5–5 dm tall, freely (seldom only sparingly) branched from base or from as much as 6 cm above base, often forming rounded, bushy mass as wide as or wider than high; leaves well distributed along stem and branches, linear-oblong, mostly 0.4–2.5(5) cm long and (0.5)1–2(4) mm wide, with a few spreading setae along margins near base, mostly without axillary fascicles; peduncles slender, 0.5–5 cm long, naked or with a few small, inconspicuous, alternate bracts, or heads rarely (and perhaps abnormally) subsessile and closely subtended by somewhat reduced foliage leaves; involucre cylindric to narrowly campanulate, 4–6 mm high, composed of 5 linear-oblong or narrowly oblong-elliptic, more or less anthocyanic bracts, these sharply to slightly acute or even obtuse at tip; ray flowers as many as involucral bracts, with evident, spreading ligule 2–5 mm long, distinctly longer than its corolla tube, yellow above, and maroon beneath; disk flowers about 8 or about 13; achenes black, 2.5–3.5 mm long, minutely hispidulous or subglabrous, those of rays epappose or with pair (seldom several) of well-developed awn-scales about as long as pappus of disk flowers, or with reduced, irregular, chaffy pappus; pappus of disk flowers of numerous slender, basally flattened, sordid, barbellate bristles 2–3 mm long. n = 36.*

* Count by D. Kyhos, on *Wiggins 18709*, Santa Cruz.

Endemic; typically found in loose cinders, less often in crevices of raw lava, on all the larger islands and several of the smaller ones. Specimens examined: Baltra, Bartolomé, Eden, Española, Fernandina, Gardner (near Española), Isabela, Pinzón, Rábida, San Cristóbal, San Salvador, Santa Cruz, Santa Fé, Santa María, Seymour.

Hooker described *Lorentea subsquarrosa* as a perennial, but Howell has examined the type and reports (personal communication) that it lacks a root and is clearly the annual species he (Howell) described as *P. glabra* in 1935. The identity of *Lorentea linearis* Anderss. (the basionym for *P. anderssonii* Robins.) is uncertain, but it seems unlikely that there is another local species to be recognized here. Andersson described his plant as a perennial with a woody root, which would seem to make it the same as *P. tenuifolia*, but he also described the leaves as 3–4 mm wide, which is beyond the range or variation of *P. tenuifolia* and only just within that of *P. subsquarrosa*. A specimen at the Gray Herbarium that might be presumed to be an isotype consists of a few dwarf, relatively wide-leaved, clearly annual plants. Since Andersson also collected *P. tenuifolia* in the Galápagos Islands, one may wonder if the description included both elements.

See Fig. 82.

Pectis tenuifolia (DC.) Schultz-Bip. in Seem., Bot. Herald 309. 1856

Lorentea tenuifolia DC., Prodr. 5: 103. 1836.

Taprooted, copiously leafy perennial, freely branched and sprawling at base, with assurgent or erect stem-tips and branches, forming rather dense, rounded clump 1–3 dm wide and to 1–1.5 dm high, above which the subnaked peduncles arise; leaves glabrous, numerous, crowded, mostly linear-filiform, (1)1.5–4 cm long, 0.5–1(1.5) mm wide, prominently setose-bristly along margins below middle, sometimes with axillary fascicles of similar or somewhat smaller leaves; peduncles slender, 1.5–6 cm long, glabrous or minutely puberulent, with a few small, inconspicuous, alternate bracts; involucre cylindric or narrowly campanulate, 4–6.5 mm high, composed of 5 linear-oblong or narrowly oblong-elliptic to linear-spatulate, more or less anthocyanic bracts, these broadly rounded at tip; ray flowers as many as involucral bracts, with evident, spreading ligule 3–5 mm long, distinctly longer than its corolla tube, yellow above, maroon beneath; disk flowers several, about 8 or about 13, yellow; achenes black, 3–4 mm long, minutely hispidulous, those of rays epappose or with pair (seldom several) of well-developed awn-scales about as long as pappus of disk flowers; pappus of disk flowers of numerous slender, basally flattened, sordid, barbellate bristles 3–5 mm long.

Endemic; typically in crevices of raw lava, less commonly in cinders. Specimens examined: Fernandina, Isabela, Genovesa, San Cristóbal, Santa Cruz.

Porophyllum Adans., Fam. Pl. 2: 122. 1763

Odoriferous, annual or perennial herbs or low shrubs, glabrous and often conspicuously glaucous, the leaves (especially near margins) and involucral bracts beset with conspicuous, scattered, embedded, translucent oil glands. Leaves opposite or alternate (often both ways on same stem), simple, entire or toothed, varying from sessile and linear to petiolate and with broad blade. Heads solitary at ends of branches or sometimes on axillary peduncles, strictly discoid; involucre cylindric or narrowly campanulate, consisting of a few (mostly 5–9) bracts in single series, free or connate only at base, usually with 2 rows of oil glands on back; receptacle small, flat or a bit convex, naked; flowers all perfect and fertile, corolla yellow or purple, regular or sometimes more deeply cleft on the 2 sides of 1 lobe. Anthers entire at base. Style branches slender, elongate, hispidulous externally, with inconspicuous, ventro-marginal stigmatic lines. Achenes narrow, often tapering at 1 or both ends, multistriate, commonly hispidulous. Pappus of more or less numerous capillary or somewhat flattened bristles.

About 30 species, most abundant in Mexico and adjacent United States, but extending south to South America.

Porophyllum ruderale (Jacq.) Cass. var. **macrocephalum** (DC.) Cronq., Madroño 20: 255. 1970

Porophyllum macrocephalum DC., Prodr. 5: 648. 1836.
Porophyllum ruderale subsp. *macrocephalum* R. R. Johnston, Univ. Kansas Sci. Bull. 48: 233. 1969.

Erect, subsimple to freely branched, malodorous annual, 2–10 dm tall, conspicuously blue-glaucous throughout; blades thin, narrowly to more often broadly elliptic, obtuse or rounded to somewhat acute at both ends, broadly and shallowly undulate-toothed, commonly with prominent oil gland 1–1.5 mm long associated with each sinus, mostly 1.5–4 cm long, 0.5–2.5 cm wide on well-developed, slender petiole often nearly or fully as long; heads erect, terminating the branches, on clavate peduncles mostly 3–10 cm long; involucre cylindric at anthesis, its 5 or 6 bracts narrow and elongate, mostly 17–21 mm long, persistent and becoming reflexed in age; disk dark, only about 6–7 mm wide at anthesis; flowers more or less numerous, from about 21 to more than 65; corolla about 1 cm long, the very slender, pale tube many times longer than the very short, purple throat; corolla lobes about 1 mm long or a bit more, equaling or longer than throat, purple or greenish; achenes 9–13 mm long, with prominent, smooth, shining, pale, callous base; pappus sordid or brownish, nearly or fully 1 cm long.

A weed in disturbed soil, widespread in tropical America, extending north through Mexico into Arizona. Specimens examined: Española, Gardner (near Es-

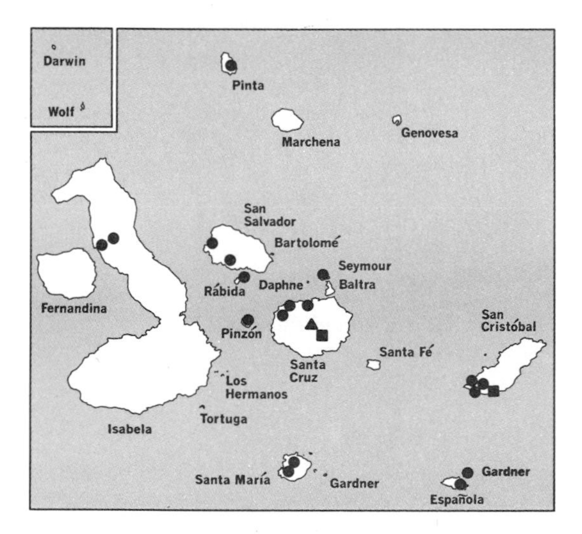

● *Porophyllum ruderale* var. *macrocephalum*

■ *Pseudelephantopus spicatus*

▲ *Pseudelephantopus spiralis*

Fig. 83. *Porophyllum ruderale* var. *macrocephalum* (*a–d*). *Pseudelephantopus spicatus* (*e–i*).

pañola), Isabela, Pinta, Pinzón, Rábida, San Cristóbal, San Salvador, Santa Cruz, Santa María, Seymour.

The Galápagos plants, as here described, represent the var. *macrocephalum* (DC.) Cronq., which otherwise occurs mainly from Arizona to Mexico and Central America, and south near the Pacific coast to Ecuador. The species may be a rather recent introduction to the Galápagos Islands, as was suggested by Robinson: "Now one of the most abundant and generally distributed plants of the islands, but apparently of recent introduction, as it was not noted by Darwin or any of the earlier collectors. Andersson in 1852 found it upon two islands, Baur in 1891 upon five, and Snodgrass and Heller in 1899 upon seven" (1902, 216). To this we may add that in 1932 Howell found it on 10 islands.

See Fig. 83.

Pseudelephantopus Rohr, Skr. Nat.-Selsk. Kjob. 2: 213. 1792

Perennial herbs. Leaves alternate, the lower larger, the others progressively reduced. Heads sessile, 1 to several in each glomerule, these in turn sessile on distal part of main stem and branches, forming leafy-bracteate to subnaked, spikelike compound infloresences; involucre of 4 pairs of decussate bracts, these pale and chartaceous or coriaceous toward base, greener distally, the 2 inner pairs about equal, the 2 outer successively shorter; receptacle flat or nearly so, naked; flowers 4 in each head, corolla blue or lavender to white, deeply 5-lobed, the adaxial cleft deeper than others, the more or less spreading corolla limb thus subligulate. Anthers sagittate at base. Style branches elongate, slender, gradually pointed, minutely hispidulous, without conspicuous stigmatic lines. Achenes 10-nerved, somewhat compressed. Pappus of several subequal to more often evidently unequal bristles or awns, these sometimes with a paleaceous base, some main ones with terminal part either doubly reverse-bent or loosely crisped or curled.

A genus apparently consisting of only 2 closely related but sharply distinct species, both native to tropical America.*

Principal pappus awns (normally 2) rigidly double-reversed above middle; achenes 5.5–8 mm long . . . . *P. spicatus*
Principal pappus awns (2 to several) with distal part soft and loosely curled or crisped; achenes 3–4 mm long . . . . *P. spiralis*

Pseudelephantopus spicatus (Juss.) C. F. Baker, Trans. Acad. Sci. St. Louis 12: 55. 1902

Elephantopus spicatus Juss. ex Aubl., Hist. Pl. Guian. 2: 808. 1775.
Distreptus spicatus Cass., Dict. Sci. Nat. 13: 367. 1819.
Matamoria spicata LaLlave & Lex., Nov. Veg. Descr. 1: 8. 1824.

Perennial herb with scattered solitary stems rising 2–10 dm, probably from a system of creeping roots, the herbage somewhat hirsute or coarsely strigose, especially up-

* The genus is named for its resemblance to the closely related genus *Elephantopus*, which differs in having a more ordinary pappus of scales or paleaceous bristles. In a broader view of genera, *Pseudelephantopus* might be submerged under *Elephantopus*. Although the name is customarily (as here) spelled *Pseudelephantopus*, it was originally spelled *Pseudo-Elephantopus* by Rohr, and a case might be made for reviving the original spelling. However, I interpret Article 73, Note 2 and Recommendation 73G(b) of the International Code of Botanical Nomenclature (1966), to at least permit the spelling I use.

ward, with straight, appressed or ascending hairs, or subglabrous; lower leaves oblanceolate to elliptic or narrowly obovate, tapering to a more or less evident, often winged petiole, commonly 5–25 cm long inclusive of petiole and 1.5–7 cm wide, petiole tending to be expanded and wrapped around stem at base; middle and upper leaves sessile and progressively reduced; stems loosely branched above, main axis and primary (or secondary) branches terminating in spiciform inflorescences 5–25 cm long, these narrowly leafy-bracteate toward base, nearly naked distally, the individual glomerules of heads readily distinguishable, or confluent distally; involucre glabrous or with only a few long hairs, mostly 9–14 mm high, its bracts firm, shortly awn-pointed, greenish distally between the raised midrib and pale, dry margins; corolla limb commonly 2–3 mm long; achenes 5.5–8 mm long, conspicuously pubescent with stiffly ascending straight hairs on the 10 prominent ribs, glandular in furrows; pappus of several paleaceous-based awns, 2 of these relatively large and stiff, elongate, abruptly and rigidly reflexed for about 1.2–2 mm at a point well above middle, then as abruptly reversed again for the terminal 1.5–2.5 mm, whole awn 5–7 mm long by straight measurement from base to tip; 2 or 3 other awns nearly as long as the main, but straight and more slender, the remaining shorter and distinctly unequal. $n = 14$.*

A common and widespread weed in tropical America, now widely introduced into the Old World tropics as well. Specimens examined: San Cristóbal and Santa Cruz.

Apparently only recently introduced, the earliest collection dating from 1932.

See Fig. 83.

Pseudelephantopus spiralis (Less.) Cronq., Madroño 20: 255. 1970

Distreptus spiralis Less., Linnaea 6: 690. 1831.
Spirochaeta funckii Turcz., Bull. Soc. Nat. Moscow 24: 167. 1851.
Chaetospira funckii Blake, Jour. Wash. Acad. Sci. 25: 311. 1935.
Pseudelephantopus funckii Philipson, Jour. Bot. Lond. 76: 301. 1938.

Similar to *P. spicatus* in general aspect, but spreading by conspicuous, long, slender stolons; herbage and inflorescences often more conspicuously pubescent, with more numerous, longer, looser hairs; inflorescence tending to be more compact, not so prominently leafy-bracteate below, and with glomerules more confluent; involucre softer, 7–8 mm high; achenes 3–4 mm long, evidently glandular, otherwise glabrous or minutely hispidulous on ribs; pappus of several slender awns, the larger (or all) loosely spiraled, crisped, or curled in distal half, 3–5 mm long by straight measurement from base to tip, all much softer than the 2 principal awns of *P. spicatus*.

A common weed in much of tropical South America, extending northward occasionally into the Lesser Antilles; in the Galápagos Islands known only from a single recent collection (1967) from Santa Cruz.

The name *P. crispus* (Cass.) Cabrera, Darwiniana 6: 371. 1944, based on *Distreptus crispus* Cass., Dict. Sci. Nat. 30: 601. 1830, might seem to apply to this species, but actually it does not. Cassini's description is clearly that of the present species, but his name was technically based on *Elephantopus nudiflorus* Willd., a quite different species now treated as a taxonomic synonym of *Orthopappus angustifolius* (Sw.) Gleason. The typification of *P. spiralis* presents a small problem, inasmuch as the original publication of *Distreptus spiralis* Less. gave the locality as Jamaica. The species is not otherwise known there, though it is found occasionally in the Lesser Antilles and is widespread on the South American mainland. However,

* Count by D. Kyhos, on *Wiggins 18627*, Santa Cruz.

the description is so clearly that of the present species that it seems necessary to take up Lessing's epithet.

Scalesia Arn. in Lindl., Nat. Syst. Bot. 443. 1836

Erect, usually single-stemmed and sparingly branched shrubs, or trees with trunk to about 3 dm thick, 0.3–20 m tall, with soft wood and resinous or gummy sap. Leaves alternate to opposite or subopposite, petiolate or sometimes merely tapering to sessile base, entire or toothed to irregularly cleft or pinnately dissected, tending to be clustered toward ends of branches, often with dead leaves of previous season hanging beneath. Heads 1 to several at branch tips, each on short or elongate axillary peduncle, in 2 species tending to form a corymbiform inflorescence, radiate in 1 species, in others discoid but sometimes some marginal flowers with ampliate, subradiate corolla and reduced or abortive stamens and style; involucre commonly more or less campanulate or urceolate, varying to subcylindric or nearly hemispheric, its bracts in 2–4 series, the outer commonly coriaceous-thickened and shorter than or merely equaling the inner, or sometimes the outer leafy-tipped and surpassing disk; receptacle flat or low-convex, its bracts trifid at tip, folded and embracing florets or achenes, sometimes anthocyanic; rays, when present, white, neutral, with sterile achenes; disk flowers white, perfect and fertile. Anthers sagittate at base, not caudate. Style branches somewhat flattened, with rather poorly marked, introrsely marginal stigmatic lines and deltoid to elongate-lanceolate, hirsutulous appendage. Achenes strongly compressed along radii of receptacle (perpendicular to involucral bracts), flat but not winged, scarcely nerved. Pappus none or sometimes represented by pair of callosities or short, slender awns.

A genus endemic to the Galápagos Islands, consisting, as here interpreted, of 11 species, 1 with 2 varieties. Some of the species are rather variable on individual islands, as well as from island to island, and the extreme form of 1 species on 1 island may approach the normal form of another species from another island. Hybridization has been suggested but remains to be demonstrated.

It may be suggested, on subjective, intuitive grounds, that the several species take their origins on different islands, as follows: S. *affinis*, north of the isthmus on Isabela; S. *aspera* and S. *crockeri*, Santa Cruz; S. *atractyloides* and S. *stewartii*, San Salvador; S. *cordata*, south of the isthmus on Isabela; S. *helleri*, Santa Fé; S. *incisa*, Pinta; S. *microcephala*, Fernandina; S. *pedunculata*, San Cristóbal; S. *villosa*, Santa María.

See color plates 8, 51, 55, and 59.

Leaves coarsely few-toothed to more often lobulate and toothed, or cleft, or laciniate, or pinnately dissected, usually not at all triplinerved, instead with lowest pair of lateral veins, like others, directed toward or into teeth or lobes:

Leaves bi- to quadripinnatifid into numerous small or narrow ultimate segments, primary divisions reaching more than two-thirds of way to midrib . . . . *S. helleri*

Leaves from coarsely few-toothed to more often lobulate or more or less cleft with segments again toothed or cleft, the divisions not extending more than two-thirds of way to midrib, leaving in any case a broad median undivided portion of blade . . . . S. *incisa*

Leaves entire or merely toothed (sometimes doubly toothed), but not at all laciniate or dissected, often (including all species with evidently toothed leaves) more or less distinctly triplinerved, with pair of prominent lateral veins ascending or arcuate-ascending from near base of blade and not associated with any teeth on margin:

Heads typically radiate, rarely discoid; leaves usually evidently toothed; young twigs conspicuously white-villous . . . . *S. affinis*
Heads typically discoid, occasionally some with 1 or 2 rays or with some subradiate marginal corollas; leaves and twigs various:
Outer phyllaries foliaceous-prolonged and exceeding disk; leaves not at all triplinerved, or tending only slightly thus:
Leaves neither evidently scabrous nor evidently glandular; involucre somewhat constricted near middle . . . . *S. atractyloides*
Leaves strongly scabrous above and densely glandular beneath; involucre not constricted . . . . *S. stewartii*
Outer phyllaries equaling or usually shorter than disk, not at all foliaceous-prolonged; leaves often triplinerved:
Leaves more or less strongly scabrous at least on upper surface, sometimes also glandular:
Leaves mostly alternate, scabrous but scarcely glandular, generally with all hairs under 1(1.5) mm long . . . . *S. aspera*
Leaves mostly opposite, evidently glandular as well as scabrous, often with some elongate hairs 2–5 mm long . . . . *S. crockeri*
Leaves rather softly hairy, not evidently scabrous:
Leaves elongate and narrow, blade mostly 4–8 times as long as wide, to 2(2.5) cm wide; many hairs of leaves and young twigs elongate, 2–5 mm long . . . . *S. villosa*
Leaves relatively or actually broader, larger ones commonly more than 2 cm wide and less than 4(5) times as long as wide; hairs all short, not more than 1 mm long:
Heads large and numerously flowered, usually or always with more than 25 and often more than 50 flowers; heads 1–3 at branch tips, not clustered in corymbiform inflorescence; leaf blades cuneate to sometimes broadly rounded or subtruncate at base, but never cordate, the lower surface villosulous or villous-puberulent, never sericeous-tomentose:
Heads relatively large, the disk mostly 2–3 cm wide . . . . *S. pedunculata* var. *pedunculata*
Heads smaller, the disk (0.7)1–2 cm wide . . . *S. pedunculata* var. *parviflora*
Heads smaller and few-flowered, with about 8 to about 21 flowers, tending to be grouped in a corymbiform cluster of 4–10:
Leaves cordate at base, loosely villous beneath, the epidermis readily visible at 10× magnification; trees, to 10 m tall . . . . *S. cordata*
Leaves more or less rounded to subtruncate at base, but not cordate, the lower surface sericeous-tomentose with appressed or subappressed hairs collectively hiding the epidermis, seldom merely densely villosulous; shrubs or small trees 0.5–3.5 m tall . . . . *S. microcephala*

Scalesia affinis Hook. f., Trans. Linn. Soc. Lond. 20: 212. 1847

Scalesia gummifera Hook. f., loc. cit.
Scalesia decurrens Anderss., Kongl. Svensk. Vet.-Akad. Handl. 1853: 182. 1855.
Scalesia decurrens f. *denudata* Anderss., loc. cit.
Scalesia narbonensis Robins., Proc. Amer. Acad. 38: 218. 1902.
Scalesia affinis subsp. *gummifera* Harling, Acta Horti Berg. 20(3): 71. 1962.
Scalesia affinis subsp. *brachyloba* Harling, op. cit. 72.

Shrub 0.4–3 m tall, or seldom a tree to 9 m; twigs of the season densely villous with soft, loosely spreading, white hairs commonly 0.5–2 mm long, often also finely glandular; leaves alternate or seldom opposite, blade serrate or biserrate to rarely subentire, lance-ovate or rather narrowly elliptic to broadly ovate or subrhombic, mostly 3–18 cm long, 1.2–12 cm wide, 1.5–3.5 times as long as wide, evidently triplinerved,

Fig. 84. *Scalesia affinis.*

the first pair of primary lateral veins prominent, strongly ascending or arcuate-ascending, base of blade broadly rounded to abruptly contracted or cuneate into the slender or evidently winged petiole, or blade sometimes tapering to an essentially sessile base; both surfaces of leaves densely white-villous when breaking from bud, at maturity glandular and loosely villous or villous-hirsute in varying proportions, often with the pustulate bases of hairs persistent after slender tip has fallen; heads 1–3 at branch tips, on peduncles mostly 1–7 cm long; involucre campanulate or hemispheric, sometimes constricted near middle, villous-hirsute and somewhat glandular, the outer bracts varying from wholly coriaceous and shorter than the inner to somewhat foliaceous-prolonged and equaling or slightly surpassing the inner and disk; rays few, about 8 or about 13, scarcely 1 cm long, or seldom apparently wanting; disk 1–2 cm wide, with numerous flowers; achenes 2.5–4 mm long. $n = 34$.*

Endemic; a pioneer on lava beds and cinder slopes and ridges, from sea level to about 600 m. Specimens examined: Fernandina, Isabela, Santa Cruz, Santa María.

See Fig. 84 and color plates 37, 40, and 76.

* Count by Uno Eliasson (1970, 152).

Scalesia aspera Anderss., Kongl. Svensk. Vet.-Akad. Handl. 1853: 180. 1855
Scalesia aspera var. *subintegerrima* Harling, Acta Horti Berg. 20(3): 74. 1962.

Shrub 0.5–3 m tall; twigs of the season rough-puberulent with coarse, incurved, more or less pustulate-based hairs, sometimes also with a few longer and straighter hairs to 1 mm long; leaves mostly alternate, blade evidently serrate to entire, narrowly to broadly ovate or elliptic, acute but not long-acuminate, commonly 3.5–10 cm long, 1.5–7 cm wide, mostly 1.5–2.5 times as long as wide, tending to be triplinerved, with 1 pair of primary lateral veins enlarged and strongly ascending or arcuate-ascending from near base; leaf surfaces scabrous with short, pustulate-based hairs, sometimes also with some longer, straighter, white hairs to 1 or even 1.5 mm long along midrib beneath; petiole distinct, with slender basal part 0.5–2 cm long, upper part flaring or more abruptly expanded into blade; heads solitary at branch tips, on short or elongate peduncle to 8 cm long, discoid, many-flowered, disk nearly or fully 2 cm wide; involucre imbricate, outer bracts distinctly shorter than inner, scabrous or scabrous-puberulent; achenes 4–5 mm long.

Endemic; near the shore but in less barren, more mesic sites than S. *affinis*, on the northern and western sides of Santa Cruz and adjacent offshore islets.

Scalesia atractyloides Arn. in Lindl., Nat. Syst. Bot. 443. 1836
Scalesia darwinii Hook. f., Trans. Linn. Soc. Lond. 20: 211. 1847.

Shrub or small tree mostly 1.5–3 m tall; twigs of the season more or less villous or villous-sericeous at least at first, with rather short to elongate hairs, these sometimes with swollen base that persists after slender tip has fallen; leaves alternate, blade slender and elongate, entire or nearly so, mostly 5–10 cm long, 4–12 mm wide, 8–12 times as long as wide, long-acuminate, gradually narrowed to the short (1 cm or less) petiole, pinnately veined, not at all triplinerved, villous to long-villous beneath, more shortly and not so softly pubescent above with hairs that often have a swollen base, this persisting after the long tip has fallen, the surface in any case soft or only slightly rough to touch, not notably scabrous; heads 1–3 at branch tips, on short peduncles mostly 0.5–1 cm long, discoid, disk mostly 7–15 mm wide, with 20–30 or

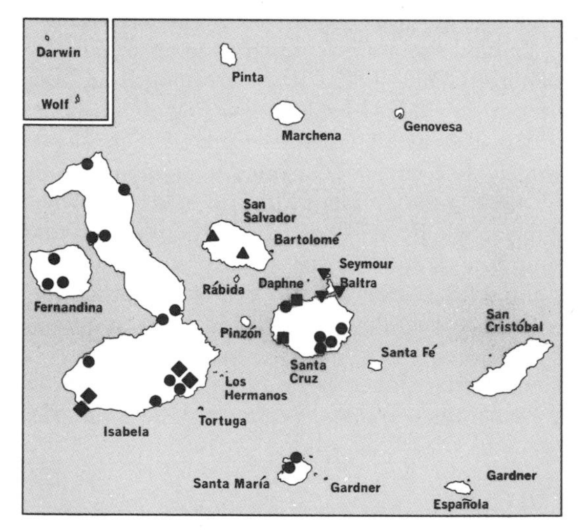

● *Scalesia affinis*
■ *Scalesia aspera*
▲ *Scalesia atractyloides*
◆ *Scalesia cordata*
▼ *Scalesia crockeri*

more flowers; involucre broad-based, tending to be somewhat constricted near middle, outer phyllaries, or some of them, with elongate, leafy tip more or less conspicuously surpassing disk; achenes 3–4 mm long.

Endemic; known only from San Salvador, where it is a pioneer on and around recent lava flows.

The specimens of *Stewart 670* (CAS, GH, NY, US) are similar in most respects but differ markedly in pubescence. The one at NY has the relatively short and appressed pubescence "typical" of S. *atractyloides*, whereas the one at CAS has the longer, looser, softer pubescence of the segregate S. *darwinii*.

Scalesia cordata Stewart, Proc. Calif. Acad. Sci. IV, 1: 156. 1911

Small tree to about 10 m tall; leaves crenate or subentire, ovate with distinctly cordate base, mostly 8–16 cm long (measured from insertion of petiole), 4–10 cm wide, pinnipalmately veined, with pair of prominent primary lateral veins arcuate-ascending from base, each of these veins with 1 or 2 prominent, divergent branches on basiscopic side near base; upper surface of leaf green and villous-puberulent or hirsute-puberulent with short hairs sometimes pustulate at base, the lower surface moderately to rather densely spreading-villous and sometimes also glandular, but pubescence not hiding surface when viewed at 10×; otherwise evidently much like S. *microcephala*, with similar inflorescence and small heads, but specimens in anthesis not yet known.

Endemic; known only from south of the isthmus on Isabela, where it is found in relatively mesic sites (and apparently not on raw lava) from about 50 m upward to 400 m or higher.

Scalesia crockeri Howell, Proc. Calif. Acad. Sci. IV, 22: 249. 1941

Low, rounded shrub, 0.6–1 m tall; twigs of the season densely glandular and often also with some rather coarse, spreading, white hairs 1–5 mm long; leaves mostly opposite, blade serrate or entire, narrowly to broadly ovate or elliptic, acute but not long-acuminate, commonly 2.5–7 cm long, 1–6.5 cm wide, with broadly winged, petiolar base or essentially sessile, tending to be triplinerved, with 1 pair of primary lateral veins enlarged and strongly ascending or arcuate-ascending from near base; leaf surfaces densely glandular and scabrous, more scabrous above, more glandular beneath; heads solitary at branch tips, on peduncles to about 4 cm long, discoid, many-flowered, disk 1.5–2 cm wide; involucral bracts all narrow, the outer equaling or somewhat shorter than inner, glandular and rough-hairy; achenes 4–5 mm long.

Endemic; on bluffs and in lava crevices near the shore. Specimens examined: Baltra, Santa Cruz, Seymour.

When the populations are better known it may become necessary to reduce S. *crockeri* to synonymy under S. *aspera*.

Scalesia helleri Robins., Proc. Amer. Acad. 38: 217. 1902

Scalesia helleri subsp. *santacruziana* Harling, Acta Horti Berg. 20(3): 90. 1962.

Shrub 0.4–3 m tall; twigs of the season sparsely to densely beset with short (under 0.5 mm) to elongate (to 5 mm), spreading white hairs, and commonly glandular; leaves alternate or opposite, slender-petiolate, blade more or less ovate or elliptic

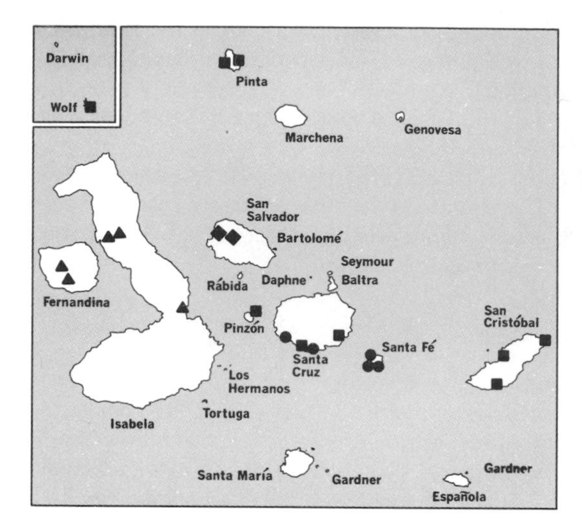

in outline, 2–10 cm long, 1.5–6 cm wide, bi- to quadripinnatifid into numerous small or narrow ultimate segments, the primary divisions reaching nearly or quite to midrib (at least more than two-thirds of way to midrib), glandular or glandular-scabrous on both sides, and often also with some more or less elongate, spreading, white hairs; heads 1–3 at branch tips, on slender peduncles 2–4 cm long, discoid (but sometimes some marginal flowers with ampliate corolla and reduced reproductive parts), disk mostly 1–1.7 cm wide; involucre shorter than or about equaling disk, its firm bracts imbricate, glandular and usually more or less hairy; achenes 2–3 mm long.

Endemic; a pioneer on cliffs and raw lava, especially near the shore. Specimens examined: Santa Cruz, Santa Fé.

[The contrast between Stewart's statement under his *No. 675* (CAS, DS, GH, NY, US)—"all over the island"—and the situation found in 1967 (61 years later) is indicative of the depredations to the vegetation of the island caused by the large herds of goats. In 1967 we found very few plants of *S. helleri* on Santa Fé, and the few we did see grew from crevices on the faces of vertical cliffs!—EDS.]

See color plate 42.

Scalesia incisa Hook. f., Trans. Linn. Soc. Lond. 20: 210. 1847

Scalesia divisa Anderss., Kongl. Svensk. Vet.-Akad. Handl. 1853: 179. 1855.
Scalesia baurii Robins. & Greenm., Amer. Jour. Sci. III, 50: 141. 1895.
Scalesia retroflexa Hemsley in Hook., Icon. Pl. 28: *pl. 2715.* 1901.
Scalesia baurii var. *glabrata* Robins., Proc. Amer. Acad. 38: 216. 1902.
Scalesia hopkinsii Robins., op. cit. 217.
Scalesia snodgrassii Robins., op. cit. 219.

Small to large or arborescent shrub 0.6–3.5 m tall; twigs of the season puberulent to loosely villous with hairs to about 1 or 1.5 mm long, sometimes also glandular; leaves alternate or opposite, with distinct, slender petiole, the blade ovate to cordate-ovate to lanceolate or lance-oblong in outline, mostly 2.5–13 cm long, 1–8 cm wide, sometimes decurrent onto upper part of petiole, coarsely few-toothed to more often lobulate or more or less cleft with segments again toothed or cleft, the divisions extending not more than two-thirds of way to midrib, leaving in any case a broad, undivided median portion of blade, the primary segments and their teeth variously rounded to acute or jagged; blade strictly pinnately veined with primary lateral veins widely spreading and directed toward or into lobes or teeth, varying to approaching the triplinerved condition of such species as S. *affinis* and S. *pedunculata*, puberulent with pustulate-based hairs on both sides, varying to more or less villous, especially beneath; heads 1–3 at branch tips, on prominent peduncles 1.5–8 cm long, discoid (sometimes with some marginal corollas subradiate and with reduced reproductive parts), disk commonly 1–1.5 cm wide, many-flowered; involucre equaling or shorter than disk, its bracts firm, imbricate, rough-puberulent; achenes 3–4 mm long.

Endemic; a pioneer of cliffs and raw lava, from sea level to at least 520 m, on most of the northern and eastern islands, but absent from Fernandina and Isabela. Specimens examined: Pinta, Pinzón, San Cristóbal, Santa Cruz, Wolf.

See color plates 54 and 57.

Scalesia microcephala Robins., Proc. Amer. Acad. 38: 218. 1902

Shrub or small tree, 0.5–3.5 m tall; twigs of the season puberulent or villous-puberulent; leaves mostly alternate, lanceolate or ovate, with broadly rounded or subtruncate (but not cordate) base, gradually tapering to slender tip, entire or nearly so, 2.5–12 cm long, 1–5 cm wide, 2–3 times as long as wide, tending to be somewhat triplinerved, 1 pair of primary lateral veins larger than others and strongly arcuate-ascending from near base of blade; upper surface of leaves green and persistently villous-puberulent, varying to sometimes merely inconspicuously puberulent, the lower surface commonly gray-white and thinly sericeous-tomentose or villous-tomentulose with the surface wholly hidden by pubescence, varying to densely villosulous with the surface visible among the hairs; heads mostly 4–10 at branch tips, forming somewhat corymbiform inflorescence, each head nonetheless axillary to its own leaf in the fashion of the genus, small and with relatively few (about 8–21) flowers, discoid or with single, inconspicuous ray; involucre narrow, cylindric to campanulate, 5–9 mm high, its bracts puberulent or villous-puberulent, not notably imbricate, a bit shorter than or equaling disk, this 4–8 mm wide; achenes 3–4 mm long.

Endemic; at middle and upper altitudes, from about 350 m upward, in less barren habitats than S. *affinis*. Specimens examined: Fernandina, Isabela.

Scalesia pedunculata Hook. f., Trans. Linn. Soc. Lond. 20: 211. 1847, var. **pedunculata**

Small to middle-sized tree, mostly 3–20 m tall, the trunk to about 3 dm thick; twigs of the season puberulent or villous-puberulent to rather shortly villous with appressed to spreading hairs less than 1 mm long; leaves alternate to opposite, ovate to lanceolate or lance-elliptic, gradually acuminate, usually cuneate to broadly rounded or even subtruncate at base and distinctly slender-petiolate, but sometimes merely tapering to an elongate, broadly winged, subpetiolar base, the blade 3–30 cm long, 1–12 cm wide, 1.5–5 times as long as wide, entire or toothed, more or less distinctly triplinerved, with pair of prominent primary lateral veins strongly arcuate-ascending from near base; upper surface of leaf puberulent or scabrous-puberulent and sometimes also somewhat glandular, the hairs with a persistent, thickened base and more slender, often deciduous tip, lower surface more villosulous-puberulent or villosulous, with somewhat longer and softer, often more numerous and more persistent hairs; heads 1–3 at branch tips, on conspicuous, often arching, slender peduncles mostly 4–15 cm long, discoid, typically larger and more numerously flowered than in S. *microcephala* or S. *cordata*, disk (0.7) 1–3 cm wide, flowers usually or always more than 25, often more than 50; involucre equaling or usually shorter than disk, its firm bracts villous-puberulent, imbricate or occasionally subequal.

Endemic; in relatively mesic or wet sites, not on raw lava, from about 60 m upward, often forming considerable forests, pure or mixed with other kinds of trees and shrubs; on Islas San Cristóbal, San Salvador, Santa Cruz, and Santa María.

Scalesia pedunculata var. **parviflora** Howell, Proc. Calif. Acad. Sci. IV, 22: 254. 1941

Scalesia ovata Anderss., Kongl. Svensk. Vet.-Akad. Handl. 1853: 181. 1855.
Scalesia pedunculata var. *svensoni* Howell, Proc. Calif. Acad. Sci. IV, 22: 254. 1941.
Scalesia pedunculata var. *indurata* Howell, op. cit. 255.
Scalesia pedunculata var. *pilosa* Howell, loc. cit.

Differing from var. *pedunculata* chiefly, or solely, in the smaller, fewer-flowered heads, and in its occurrence on islands other than San Salvador.

Specimens examined: San Cristóbal, Santa Cruz, Santa María.

See color plates 51–53 and 55.

Scalesia stewartii Riley, Kew Bull. 1925: 223. 1925

Scalesia stewartii var. *euryphylla* Harling, Acta Horti Berg. 20(3): 82. 1962.

Shrub or small tree mostly 1–3 m tall; twigs of the season densely glandular and sparsely to densely beset with long, ascending or spreading white hairs; leaves alternate, blade elliptic to narrowly lance-elliptic, 4–11 cm long, 0.5–3.5 cm wide, 2–10 times as long as wide, acute to long-acuminate at tip, cuneate or more often gradually tapering at base to the short (1 cm or less) petiole, strictly pinnately veined as in S. *atractyloides*, or with tendency toward triplinervation, densely scabrous and glandular, more scabrous above, more glandular beneath, often also with some scattered long hairs, these somewhat firmer and straighter than those of S. *atractyloides*; heads solitary at branch tips, on short peduncles to about 3 cm long, discoid, disk about 1.5 cm wide, with numerous (40 or more) flowers; involucre

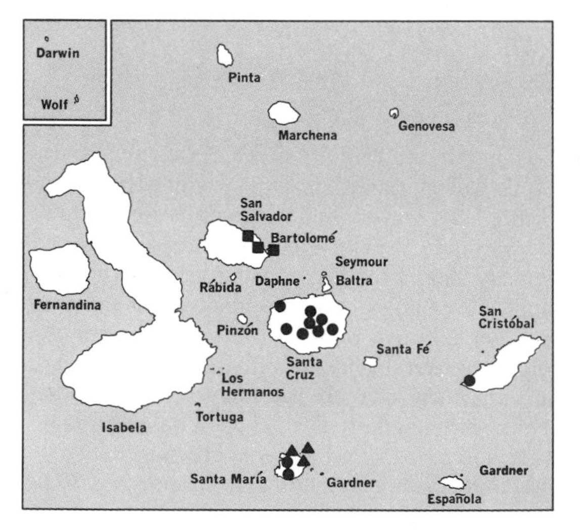

● *Scalesia pedunculata* var. *parviflora*
■ *Scalesia stewartii*
▲ *Scalesia villosa*

broad-based, pubescent like leaves, campanulate or hemispheric, not contracted above, the outer bracts, or some of them, with elongate, leafy tips conspicuously surpassing disk; achenes 3–4 mm long.

Endemic; in barren lava beds and cinders, from near the coast to above 200 m. Specimens examined: Bartolomé, San Salvador.

When the populations are better known it may become necessary to reduce S. *stewartii* to synonymy under S. *atractyloides*.

Scalesia villosa Stewart, Proc. Calif. Acad. Sci. IV, 1: 158. 1911

Scalesia villosa var. *championensis* Stewart, op. cit. 159.

Shrub 1–3 m tall; twigs of the season densely long-villous or villosulous-tomentose, with long, soft, flexuous white hairs mostly 2–5 mm long, these commonly with short, enlarged base persisting after slender terminal part has fallen; leaves mostly alternate, blade entire or somewhat serrulate, lanceolate or narrowly lance-elliptic, tapering gradually to both ends and without well-defined petiole, mostly 4–12 cm long, 0.5–2(2.5) cm wide, 4–8 times as long as wide, evidently triplinerved, loosely villous on both sides when young with long, soft, white hairs, some as much as 2–5 mm long, at maturity sometimes more thinly hairy and beset especially on upper surface with the persistent, pustulate bases of the otherwise deciduous hairs; heads 1–3 at branch tips, on peduncles to 6 cm long, discoid, relatively large and many-flowered, commonly rotund and 1.5–2.5 cm wide as pressed; involucre white-villous, its bracts not notably imbricate, about equaling outer flowers; achenes 2.5–3 mm long.

Endemic; a pioneer on barren cinder slopes and lava beds. Specimens examined: Champion, Santa María.

See color plate 71.

Sonchus L., Sp. Pl. 793. 1753; Gen. Pl. ed. 5, 347. 1754

Annual or perennial herbs with milky juice. Leaves alternate or all basal, entire to pinnatifid or dissected, mostly auriculate, often prickly-margined. Heads medium-sized to rather large, solitary or usually several or numerous in irregularly corymbose-paniculiform to subumbelliform inflorescence; involucre ovoid or campanulate, rarely narrower, its bracts generally imbricate in several series, occasionally merely calyculate, often basally thickened and indurate in age; flowers all ligulate and perfect, yellow; receptacle naked. Achenes flattened, usually 6–20-ribbed, merely narrowed at apex, beakless, often transversely rugulose, otherwise glabrous. Pappus of numerous white, capillary, often somewhat crisped bristles that tend to fall connected, commonly with some stouter outer bristles that fall separately; some of pappus bristles more than 4-celled in cross section at base.

About 70 species, native to Eurasia and Africa, several of them now cosmopolitan weeds.

Sonchus oleraceus L., Sp. Pl. 794. 1753

Annual with short taproot, commonly 1–10 dm tall, glabrous except sometimes for some spreading, gland-tipped hairs on involucre and peduncle; leaves (except the more or less ovate and petiolate lowermost) runcinate-pinnatifid to occasionally

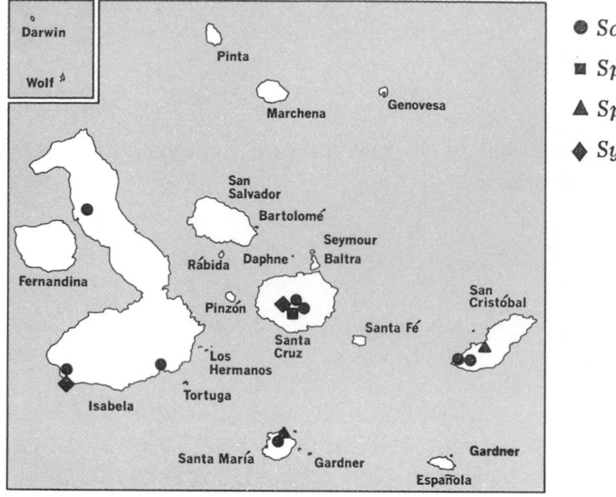

merely toothed, mostly 6–30 cm long and 1–15 cm wide, margins rather weakly or scarcely prickly, prominently auriculate, auricles with well-rounded margins but eventually sharply acute; leaves progressively less divided upward, and more or less reduced; heads several in corymbiform inflorescence, relatively small, usually only 1.5–2.5 cm wide in flower; flowers mostly (85)120–160(250) per head; fruiting involucre mostly 9–14 mm high; achenes 2.5–3 mm long, transversely rugulose and evidently to rather obscurely 3–5-ribbed on each face. n = 16, or apparently sometimes 8.*

A cosmopolitan weed of European origin. Specimens examined: Isabela, San Cristóbal, Santa Cruz, Santa María.

See Fig. 85.

Spilanthes Jacq., Enum. Pl. Carib. 8, 28. 1760

Annual or rarely perennial herbs with opposite, simple, toothed or entire leaves, these petiolate (in all our species) or sessile. Heads solitary to several or many, terminating the stem and pedunculate from upper axils, radiate or discoid, the rays (when present) pistillate, yellow or white; involucral bracts more or less herbaceous and subequal, in 1 or 2 series, individually subtending outermost flowers of head; receptacle elongate, high-conic to cylindric, its bracts more or less clasping achenes and eventually deciduous, often still holding achenes when shed; disk flowers perfect and fertile, glabrous, white to yellow. Anthers truncate at base. Style branches flattened, truncate, exappendiculate. Achenes of rays or of outermost disk flowers (i.e. those achenes subtended by involucral bracts) trigonous or flattened at right angles to radii of head, others radially compressed, often ciliate on margins, but not winged. Pappus of 1–3 slender, setiform awns, or absent.

Perhaps as many as 40 species, many of them weedy, widely distributed in the New and Old Worlds, chiefly in tropical and subtropical regions.

* Counts reported by Fedorov (1969) from non-Galápagos material.

Fig. 85. *Sonchus oleraceus.*

Leaves large, blade commonly 2.5–10 cm long; heads large, disk corollas mostly 1.5–2 mm long (dried), disk 7–15 mm high at maturity; achenes shortly ciliate-margined. . . . . *S. acmella*
Leaves smaller, blade only about 0.7–1.5 cm long; heads smaller, disk corollas only 1–1.3 mm long, disk only 5–6 mm high at maturity; achenes glabrous. . . . . *S. diffusa*

Spilanthes acmella (L.) Murr., Syst. Veg. ed. XIII, 610. 1774

Verbesina acmella L., Sp. Pl. 901. 1753.
Pyrethrum acmella Medic., Obs. Bot. 243. 1775.
Bidens acmella Lam., Encycl. Méth. Bot. 1: 415. 1783.
Bidens ocymifolia Lam., op. cit. 416.
Spilanthes ocymifolia A. H. Moore, Proc. Amer. Acad. 42: 531. 1907.

Simple to freely branched annual weed 1–8 dm tall, erect or more often sprawling at base and rooting at lower nodes; leaf blades ovate (sometimes broadly so) to less often elliptic or upper ones lanceolate, coarsely toothed to subentire, sparsely hairy or subglabrous, mostly 2.5–10 cm long, 1–5 cm wide, on well-developed petioles mostly 0.5–4 cm long; heads solitary to several or many on slender peduncles mostly 5–15 cm long; ray flowers typically (including our material) wanting, when present few and with short, white or less often yellow ligule only 1–3 mm long; involucre 3–5 mm high, its bracts lanceolate to narrowly oblong; disk flowers numerous, corolla mostly 1.5–2 mm long (dry), typically white or whitish, seldom yellow, with black anthers usually very shortly exserted; disk hemispheric to rounded-conic at maturity, 7–15 mm high, 6–9 mm wide; achenes 1.5–2 mm long, black, with shortly ciliate, often pale and thickened margins, the outermost (next to involucral bracts) triquetrous, others strongly compressed, but not winged; pappus of 2–3 short, slender awns, or none.

A pantropical weed in moist soil in disturbed sites; in the Galápagos Islands known only from Fernandina, Santa Cruz, and Santa María, but to be expected on the other islands as well. Specimen examined: Santa Cruz.

The New World plants are often distinguished from the Old World plants (typical *S. acmella*) as *S. ocymifolia* (Lam.) A. H. Moore, but I cannot consistently separate them.

Spilanthes diffusa Hook. f., Trans. Linn. Soc. Lond. 20: 214. 1847

Spilanthes diffusa var. *minor* Hook. f., loc. cit.

Diffuse and several-stemmed from base, or with weakly erect solitary stem to 2 dm tall, loosely hirsute or hispid-hirsute throughout, or (in San Cristóbal plant) only thinly and inconspicuously strigose; leaf blades deltoid or subcordate to lance-ovate, entire or inconspicuously few-toothed, mostly 0.7–1.5 cm long, 0.4–1.2 cm wide, on short petioles to 6 mm long; involucre 2.5–4 mm high, its bracts rather narrowly elliptic to lance-ovate, strigose (only thinly so or subglabrous in San Cristóbal plant); heads small, white, strictly discoid, the disk only about 5–6 mm high and wide at maturity; disk corollas 1–1.3 mm long (dry); achenes 1.4–1.5 mm long, black, smooth-margined, the outermost trigonous, others strongly compressed, but not winged; pappus none, or of a slender and inconspicuous awn.

Endemic; in moist soil in disturbed areas. Specimens examined: San Cristóbal, Santa María.

See Fig. 86.

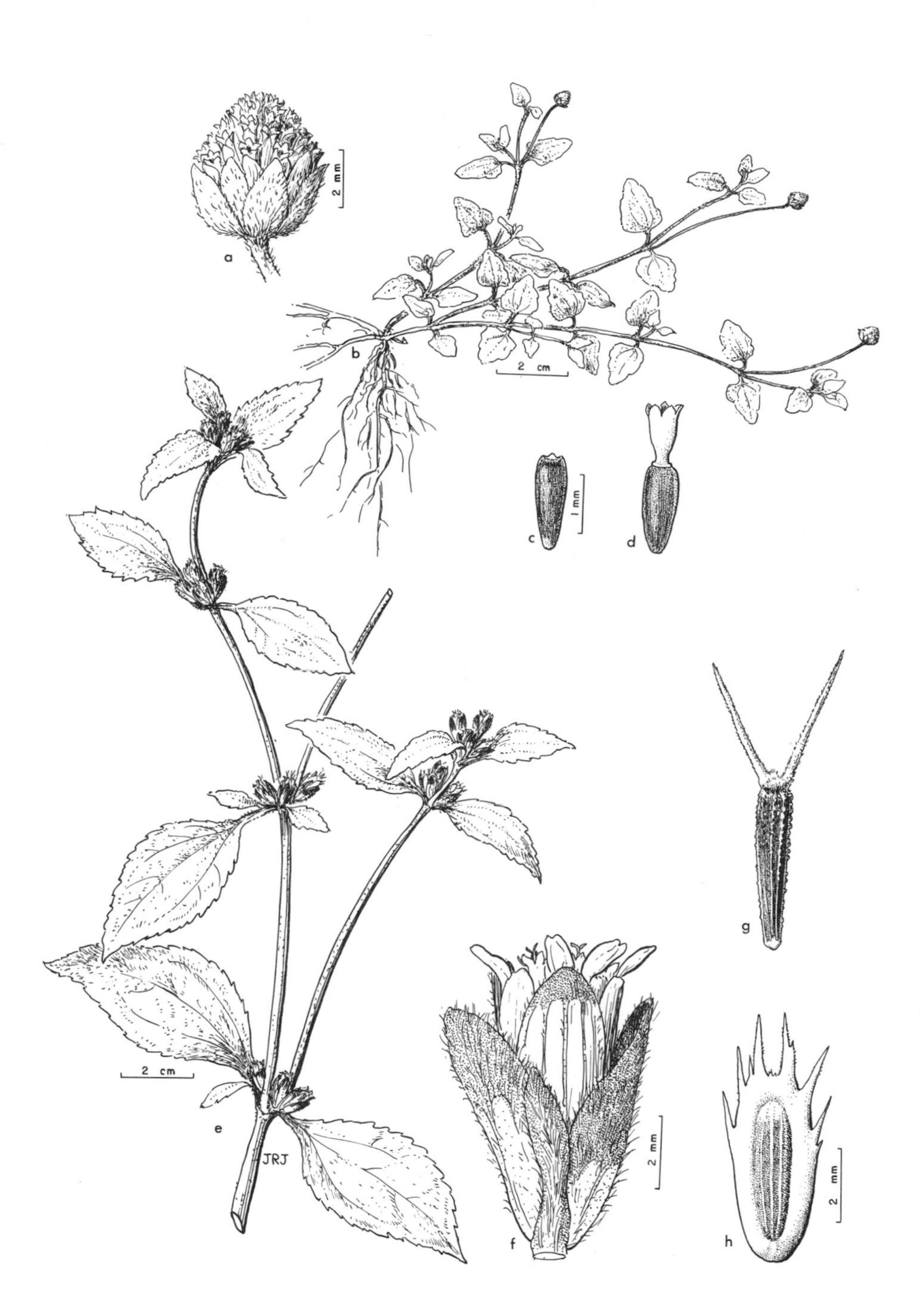

Fig. 86. *Spilanthes diffusa* (*a–d*). *Synedrella nodiflora* (*e–h*).

Synedrella Gaertn., Fruct. 2: 456, *t. 171*. 1791

Branching annual herbs with opposite, simple, petiolate, somewhat toothed leaves. Heads in axillary glomerules or pedunculate from axils and terminating the branches, radiate, yellow. Involucre ovoid to cylindric, its bracts few, not very unequal, individually subtending ray flowers, the outer more or less herbaceous or herbaceous-tipped, the inner more chartaceous, stramineous, and striate, resembling bracts of receptacle; ray flowers pistillate and fertile, with slender tube and short, broad, 2- or 3-toothed ligule; receptacle small, chaffy, its bracts stramineous, flat, with evident midvein but not folded, subtending but not embracing flowers; disk flowers perfect and fertile, with 4-toothed corolla. Anthers entire or minutely sagittate at base. Style branches flattened toward base, with short, introrsely marginal stigmatic lines and long, slender, hirsutulous appendage. Achenes dimorphic, those of rays broad, strongly flattened parallel to involucral bracts (perpendicular to radii of head), with conspicuous, lacerate wing-margins produced into pair of flattened pappus-awns; disk achenes narrower, but still flattened (or flattened-triquetrous), with 2 or 3 conspicuous pappus-awns.

A well-marked genus of 2 species, both native to tropical America.

Synedrella nodiflora (L.) Gaertn., Fruct. 2: 456. 1791

Verbesina nodiflora L., Cent. Pl. 1: 28. 1755.

Annual weed 1.5–10 dm tall, freely branched or subsimple, the herbage rather thinly and coarsely strigose; leaves narrowly to broadly ovate or less often elliptic, mostly 2–10 cm long, 0.8–5 cm wide, more or less toothed, on distinct petiole to about 2.5 cm long; heads mostly sessile or subsessile in small, axillary clusters, seldom some solitary; involucre subcylindric to campanulate, 7–9 mm high, 2 or 3 outer phyllaries more or less herbaceous or herbaceous-tipped, others passing into bracts of receptacle but tending to be broader and more striate; rays 2–9, with short ligule mostly 1–2 mm long; disk flowers 4–12, more numerous or occasionally fewer than rays, the innermost sometimes not subtended by bracts; achenes black, 4–5 mm long, those of rays smoothish, with pale, lacerate wing-margins, those of disk flowers narrower, not wing-margined, conspicuously muricate or tuberculate.

A widespread weed in tropical America; known from 2 recent collections in the Galápagos Islands. Specimens examined: Isabela, Santa Cruz.

See Fig. 86.

COMPOSITAE GENERA AND SPECIES EXCLUDED FROM THE FLORA

Tagetes L., Sp. Pl. 887. 1753; Gen. Pl. ed. 5, 378. 1754

Tagetes erecta L., Sp. Pl. 887. 1753

Tagetes major Gaertn., Fruct. 2: 434, *t. 122*. 1791.

The record of this species is based on a single collection on San Cristóbal by Chierchia in 1884. The collection was not widely distributed, and I have not seen a specimen. I have no particular reason to doubt the identification, as reported in a list published by Caruel (in Reconditi della Reale Accademia dei Lincei 5: 623. 1889), but I take the plants to represent a casual escape from cultivation. Apparently the species has not been re-collected in the Galápagos Islands.

Hemizonia DC., Prodr. 5: 692. 1836

Hemizonia squalida Hook. f., Trans. Linn. Soc. Lond. 20: 208. 1847

The name *Hemizonia squalida* was based on a Californian specimen of *H. angustifolia* DC., mistakenly thought to have been collected in the Galápagos Islands. The error resulted from an unfortunate confusion in labeling of specimens collected by naturalists of the French frigate *La Venus*, which under the command of Captain Du Petit-Thouars cruised around the world, stopping among other places at the Galápagos Islands and the coast of California (in 1837). A Californian sea gull was also mistakenly attributed to the Galápagos Islands in the materials from this voyage. A fuller account is given by John Thomas Howell in Leaflets of Western Botany 1: 189–91. 1935.

Haplopappus Cass., Dict. Sci. Nat. 56: 168. 1828

Haplopappus lanatus Hook. f., Trans. Linn. Soc. Lond. 20: 215. 1847
Aster lanatus Kuntze, Rev. Gen. Pl. 1: 318. 1891.

The most doubtful exclusion from the flora. Hooker based this name on an incomplete specimen also taken on the voyage of *La Venus*, purportedly in the Galápagos Islands. In his monograph of *Haplopappus* (Carn. Inst. Wash. Pub. 389. 1928), H. M. Hall excluded the species from *Haplopappus* and referred it instead to *Aster*, but he did not formally transfer it to *Aster*. From the description it would appear to be certainly unique in either of those genera. Inasmuch as nothing like it has since been found in the Galápagos Islands, the possibility presents itself that it too came from some other part of the world. Until I can examine the type, I prefer to leave the name in limbo.

CONVOLVULACEAE. MORNING-GLORY FAMILY*

Annual or perennial herbs or vines, less commonly shrubs or rarely trees (parasitic in *Cuscuta*). Sap often milky. Leaves alternate, exstipulate, simple and entire to lobed, incised, or parted in various patterns, reduced to functionless scales in *Cuscuta*. Flowers perfect, regular, solitary or in cymes. Bracts often present, minute to ample. Sepals 4 or 5, distinct or more or less united at base, often imbricate, persistent, equal or unequal, in 1 or 2 whorls. Corolla sympetalous, usually convolute in bud; limb entire, 5-angled, or shallowly 5-lobed, often involute after flowering. Stamens 5, alternate with corolla lobes; filaments equal or unequal, commonly dilated at base and pubescent, inserted in and adnate to corolla tube; anthers 2-celled, erect or incumbent, straight or spirally twisted. Ovary superior, 2–5-celled, 2–4 ovules per cell, seated in fleshy disk; placenta axile. Styles 2 or united into 1; stigmas capitate or sometimes 2-lobed, linear, oblong, or subglobose. Fruit a capsule, 2–4(5)-celled. Seeds comparatively large; coat coriaceous or membranous. Embryo curved; cotyledons foliaceous, usually folded. Endosperm mucilaginous.

About 50 genera and over 1,200 species, widely distributed in the warm temperate to tropical regions of the Old and New Worlds; particularly well represented in

* Contributed by Ira L. Wiggins.

tropical America and Asia. Conspicuous in most of the Galápagos Islands, from sea level upward through the *Miconia* Zone or higher.

Plants parasitic; stems devoid of leaves, yellow, orange, or reddish, about 1 mm thick, attached to host plants by knoblike haustoria; corolla tube with fringed appendages between bases of stamen filaments . . . . *Cuscuta*
Plants autophytic; herbaceous stems more robust, leaf-bearing, green in youth; appendages in corolla tubes lacking:
  Flowers small, usually less than 3.5 mm broad; corolla limb cleft halfway to throat; stems commonly rooting at nodes . . . . *Dichondra*
  Flowers larger (sometimes only 7 mm broad in *Evolvulus*), often to 5 cm broad or more; corolla limb only angulate or very shallowly lobed; stems rarely rooting at nodes:
    Styles 2, distinct nearly to base, each with 2 stigma lobes; corollas 3–10 mm long . . . . *Evolvulus*
    Style 1, or if shallowly bilobed, each lobe entire; corollas 1.5–15 cm long:
      Stigmas 2, ovate to spatulate, sessile or nearly so . . . . *Convolvulus*
      Stigmas entire, or of 2 globose, sessile lobes:
        Filaments markedly glandular at base; anthers spirally twisted at or soon after anthesis . . . . *Merremia*
        Filaments pilose or glabrous, not conspicuously glandular; anthers not spirally twisted:
          Capsule dehiscing regularly by 2–6 valves; dissepiment wingless; seeds more or less angular in cross section . . . . *Ipomoea*
          Capsule dehiscing irregularly to form 4 subterminal pores; dissepiments transversely winged; seeds not angular . . . . *Stictocardia*

Convolvulus L., Sp. Pl. 153. 1753; Gen. Pl. ed. 5, 76. 1754

Perennial, often rhizomatous plants with twining or creeping stems. Leaves alternate, petiolate, ovate, sagittate, or reniform, glabrous or pubescent, entire. Flowers axillary, solitary or clustered in axils, each peduncle usually bearing 2 bracts close to or distant from flower or flowers, often partially or wholly covering calyx. Sepals nearly equal, or outer pair longer than inner, ovate, ovate-lanceolate, or oblong, acute to rounded at apex, or sometimes apiculate, persistent and accrescent in fruit. Corolla campanulate to funnelform, plaited, usually 5-angled or shallowly 5-lobed, glabrous on outside, or mid-petal lines and distal borders of limb more or less pubescent, purple, white, or roseate. Stamens equal, included; anthers relatively large, sagittate at base; pollen grains spherical, 3-pored. Disk annular, more or less lobed. Ovary ovoid, glabrous, bilocular or unilocular toward apex, 4-ovuled. Style 1, glabrous; stigmas 2, oblong or ovoid. Capsule globose or subglobose, 4-valved, normally 4-seeded. Seeds dark brown to black, glabrous, smooth.

About 200 species, widely distributed in temperate and tropical regions in the Old and New Worlds.

Convolvulus soldanella L., Sp. Pl. 159. 1753

Calystegia soldanella R. Br., Prodr. 484. 1810.
Calystegia reniformis R. Br., loc. cit.

Short-creeping, much-branched, prostrate vine with rather fleshy, glabrous stems and leaves, from deep-seated rootstock; petioles 1–6 cm long; blades reniform, 8–25

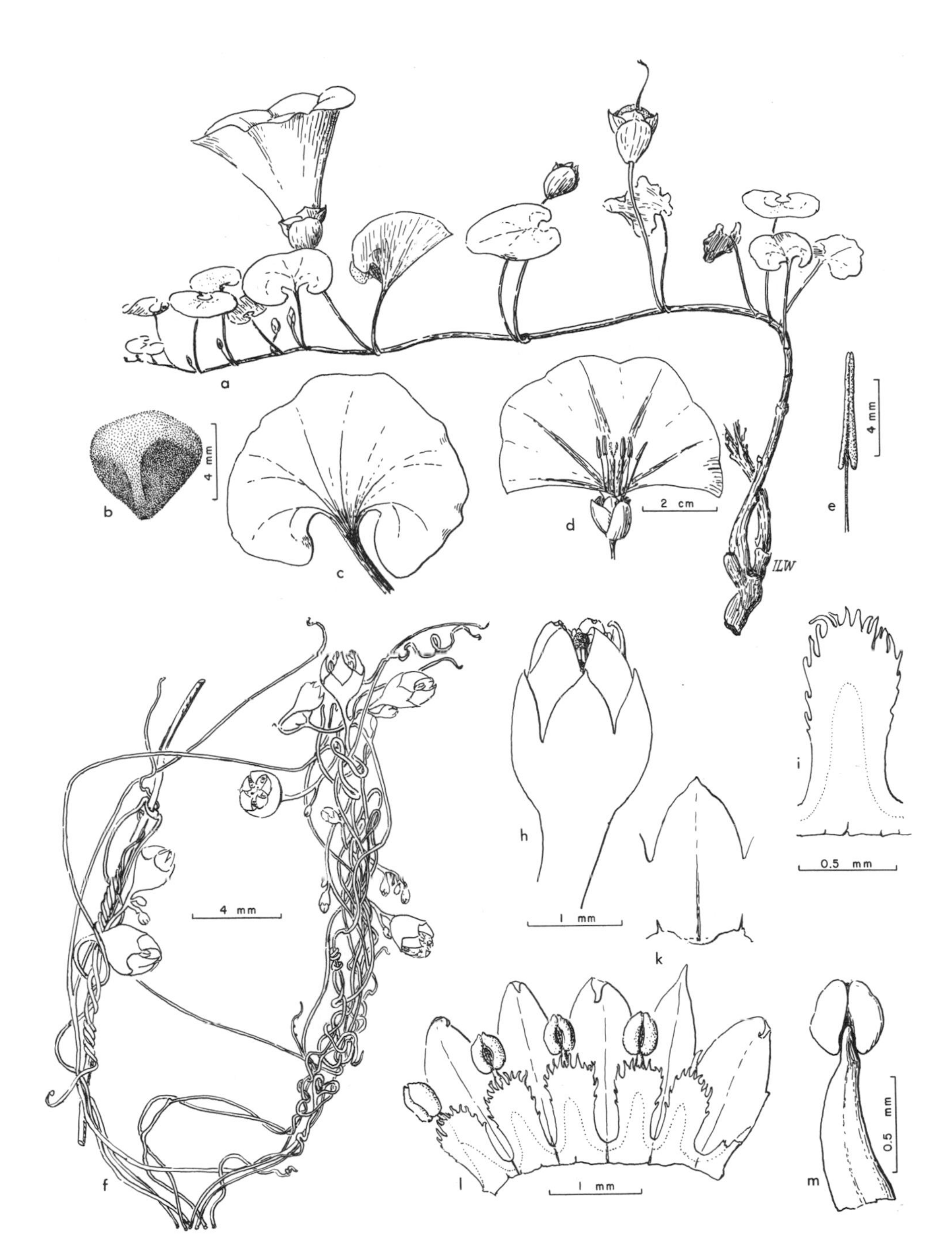

Fig. 87. *Convolvulus soldanella* (*a–e*). *Cuscuta acuta* (*f–m*): *i*, appendage: *k*, sepal.

mm long, 9–40 mm wide, entire, often emarginate, cordate at base, the margins sometimes undulate; bracts suborbicular, 8–12 mm long, obscurely cordate, equaling or slightly shorter than sepals; sepals broadly ovate, often mucronate, the outer 9–11 mm long, 7–8 mm wide, the inner slightly smaller; corolla rose-purple, short-funnelform, 4–6 cm long, shallowly 5-lobed, glabrous; stamens about 2 cm long, filaments glandular-hirsute at base; anthers 4–5 mm long; ovary ovoid, glabrous, incompletely bilocular; capsule subglobose, 1–1.5 cm in diameter, usually buried in sand, unilocular at maturity, 4-valved, enclosed by the accrescent bracts and sepals; seeds 5–6 mm long, smooth or minutely tuberculate.

Upper beaches and sand dunes, Washington to Baja California, Pacific coasts of Central and South America, Pacific islands, and maritime Europe from England to the Mediterranean. Specimen examined: designated only "Galápagos Islands," *Edmonstone s.n.* (GH).

See Fig. 87.

Cuscuta L., Sp. Pl. 124. 1753; Gen. Pl. ed. 5, 60. 1754

Herbaceous parasites with slender, often filiform, yellowish or orange-colored, leafless, twining stems nearly or quite devoid of chlorophyll, attaching themselves to stems and leaves of host plants by means of disklike haustoria and deeply penetrating, branching processes. Leaves reduced to functionless minute scales. Flowers small, borne in loose to dense, cymose clusters, sympetalous, 4- or 5-merous. Calyx campanulate to turbinate or rarely cylindrical, shallowly to deeply parted; lobes acuminate to orbicular, entire or denticulate. Corolla campanulate to cylindrical; lobes shorter than, equaling, or exceeding tube. Stamens inserted in throat of corolla, alternating with corolla lobes; filaments uniform in diameter, subulate. Fringed appendages same number as stamens, inserted below stamens in most species. Styles distinct (in ours) or united; stigmas capitate or elongate. Ovary 2-celled, each cell 2-ovuled. Fruit a capsule, circumscissile or rupturing irregularly. Seeds 1–4; embryo filiform or with knob at 1 end, without visible cotyledons.

Possibly 100 species, distributed from the temperate to the tropical zones in the New and Old Worlds.

Calyx equaling or longer than corolla tube; filaments slender, longer than anthers; corolla lobes longer than tube; styles usually divergent-spreading in fruit (usually erect in flower) . . . . *C. acuta*

Calyx shorter than corolla tube; filaments stoutish, about as long as anthers, broadly subulate; corolla lobes shorter than tube; styles erect or nearly so in fruit. . *C. gymnocarpa*

Cuscuta acuta Engelm., Trans. Acad. Sci. St. Louis 1: 497. 1859

Stems bright orange-yellow, slender, 0.2–0.8 mm thick, forming dense mats on herbs and surface of ground, the branches often twisting tightly around one another and on host plants; flowers in compact, more or less globose clusters of short cymes; inflorescences few to many; pedicels 2–3 mm long, slightly broader at apex than below; flowers 2–3 mm long at anthesis, pearly white; calyx campanulate, about equaling corolla tube, lobes deltoid-ovate, 1–1.5 mm long, acute, erect, closely appressed to corolla; corolla campanulate, lobes lanceolate, acute to attenuate-acuminate, usually erect or slightly spreading in fruit, tips nearly always abruptly

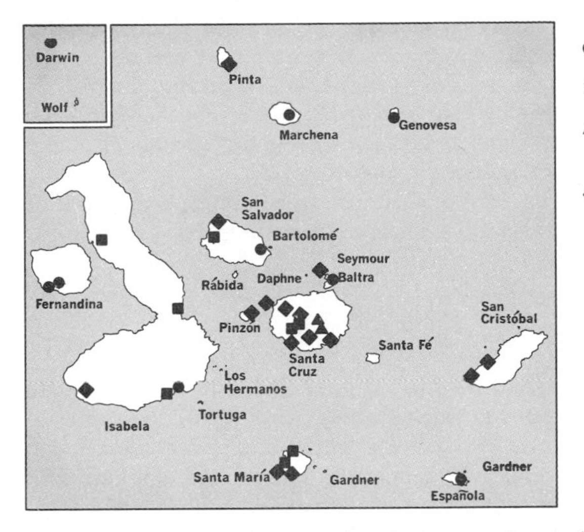

● *Cuscuta acuta*

■ *Cuscuta gymnocarpa*

▲ *Dichondra repens* var. *microcalyx*

◆ *Evolvulus glaber*

inflexed, about equaling or slightly longer than tube; stamens about two-thirds to three-fourths as long as corolla lobes, filaments narrowly subulate, longer than anthers; scales oblong, nearly twice as long as wide, thin, bluntly fringed on upper half, bridged below middle, reaching to or slightly above bases of filaments, adnate to corolla in lower half of scales; styles about equal to ovary, slightly subulate, slightly divergent at anthesis, strongly so in fruit; capsules globose to depressed-globose, 2.5–3 mm broad, with a relatively large intrastylar aperture, dehiscing irregularly near base but not circumscissile; seeds brownish, 1–1.2 mm long, ovoid, minutely papillate.

Possibly present on the coast of Peru, although long considered endemic in the Galápagos, where it is parasitic on many herbaceous and suffrutescent hosts at lower and middle altitudes on several of the islands. Specimens examined: Baltra, Darwin, Española, Fernandina, Genovesa, Isabela, Marchena, San Salvador.

See Fig. 87.

Cuscuta gymnocarpa Engelm., Trans. Acad. Sci. St. Louis 1: 496. 1859

Stems yellowish orange, 0.4–0.8 mm thick, twining and spreading over herbs and small subshrubs to 1 m high or more, often covering several square meters with close web; haustoria crowded, the disks often to 1.5 mm across; flowers rarely solitary, usually in globose clusters of short cymes, each flower about 2 mm long at anthesis, becoming 2.5–3 mm long in fruit; bracts lance-ovate to narrowly lanceolate, 2–3 mm long, 0.6–1.2 mm wide, membranous; pedicels slightly stouter than stems, 0.4–2 mm long, slightly dilated upward; calyx broadly campanulate, 1–1.4 mm long, almost equaling corolla tube, lobes broadly ovate, obtuse or abruptly acute; corolla campanulate, thin, lobes narrowly deltoid-ovate, acute, upright, tips often slightly inflexed, slightly shorter than tube; scales broadly ovate, nearly as wide as long, reaching bases of stamens, finely fringed with short, blunt processes from near base upward, bridged below middle; stamens about half as long as corolla lobes, inserted almost at margins of corolla tube below sinuses; anthers oval, about equaling the stout, broadly subulate filaments; styles distinct, erect both in flower and fruit (or rarely slightly divergent in fruit), equal to or slightly shorter than the globose ovary at anthesis; stigmas globose-capitate; fruit globose to depressed-globose, usually about 2 mm high, 2–3 mm thick, rupturing irregularly near base or along 1 side, subtended by the withered corolla; seeds 2–4 in each capsule, tawny to brownish, broadly ovoid to ellipsoidal, round-trigonous to planoconvex, 0.8–1.2 mm high, minutely papillate-roughened.

Endemic; parasitic on many herbaceous and suffrutescent plants, on several of the islands at low and middle altitudes. Specimens examined: Isabela, San Salvador, Santa Cruz, Santa María.

Dichondra Forst. & Forst., Char. Gen. Pl. 39. 1776

Creeping or scrambling perennial herbs with long-petioled, minutely stipulate leaves and slender rhizomes and stolons, freely rooting at nodes. Pubescence of medianly attached, more or less appressed hairs and some simple, spreading hairs. Leaf blades reniform to orbicular-reniform, rounded to emarginate, cordate with a narrow to broad sinus at base, entire. Flowers small, axillary, solitary (often appearing clustered owing to crowding of leaves), pedicellate, the pedicels straight at anthesis, often sharply recurved their full length or just beneath calyx soon after flowering. Calyx shallowly campanulate, sepals united only at base, more or less accrescent in fruit. Corolla valvate-induplicate in bud, deeply lobed more than half its length, shorter than to slightly exceeding sepals at anthesis, persistent but not accrescent. Stamens 5; filaments filiform to narrowly subulate, up to three-fourths as long as corolla lobes, inserted just below sinuses in corolla; anthers small, nearly orbicular. Ovary 2-carpeled, with 2 ovules in each carpel, erect, collateral, often 1 or more ovules abortive. Styles 2, distinct or united only at base, filiform, equal or subequal; stigmas capitate, papillate-roughened. Fruit a capsule, apically rounded (by abortion of 1 carpel), truncate, emarginate, or deeply bilobed. Seeds ovoid, erect, dark brown to nearly black.

A genus of 10 or 12 species, distributed from central-western California to Argentina and New Zealand, and along eastern coast of North America from Carolina south through the Caribbean and into South America; introduced in many parts of the world as a weed or as lawn cover.

Dichondra repens Forst. var. **microcalyx** Hallier f., Engl. Bot. Jahrb. 18: 84. 1894

Prostrate; stems 0.4–0.8 mm thick, internodes 1–6 cm long; petioles 3–7 cm long, densely appressed-pubescent in youth, at length glabrate; leaf blades 9–28 mm long, 12–33 mm wide, basal sinus 2–7 mm deep, densely appressed-pubescent beneath, subglabrous above; flowers usually 2 at a node; pedicels 5–10 mm long, abruptly reflexed just below calyx soon after anthesis; calyx 2–2.2 mm long at anthesis, 5-parted almost to base, ciliate, appressed-pubescent on outside; corolla about equaling sepals, lobes divided two-thirds to three-quarters of way to base, often undulate-crisped marginally, glabrous; stamens about half as long as corolla lobes, anthers ovoid-orbicular, about 0.4 mm across; styles united slightly at base, distinct and filiform above; fruit deeply bilobed or globose to obovoid when 1 carpel aborts, 2.6–3 mm long, sparsely spreading-pilosulous; seeds 1 in each functional carpel (2d ovule usually abortive), broadly ellipsoidal-globose, 1.8–2.2 mm high, about half as thick as wide, dark brown, smooth, dull.

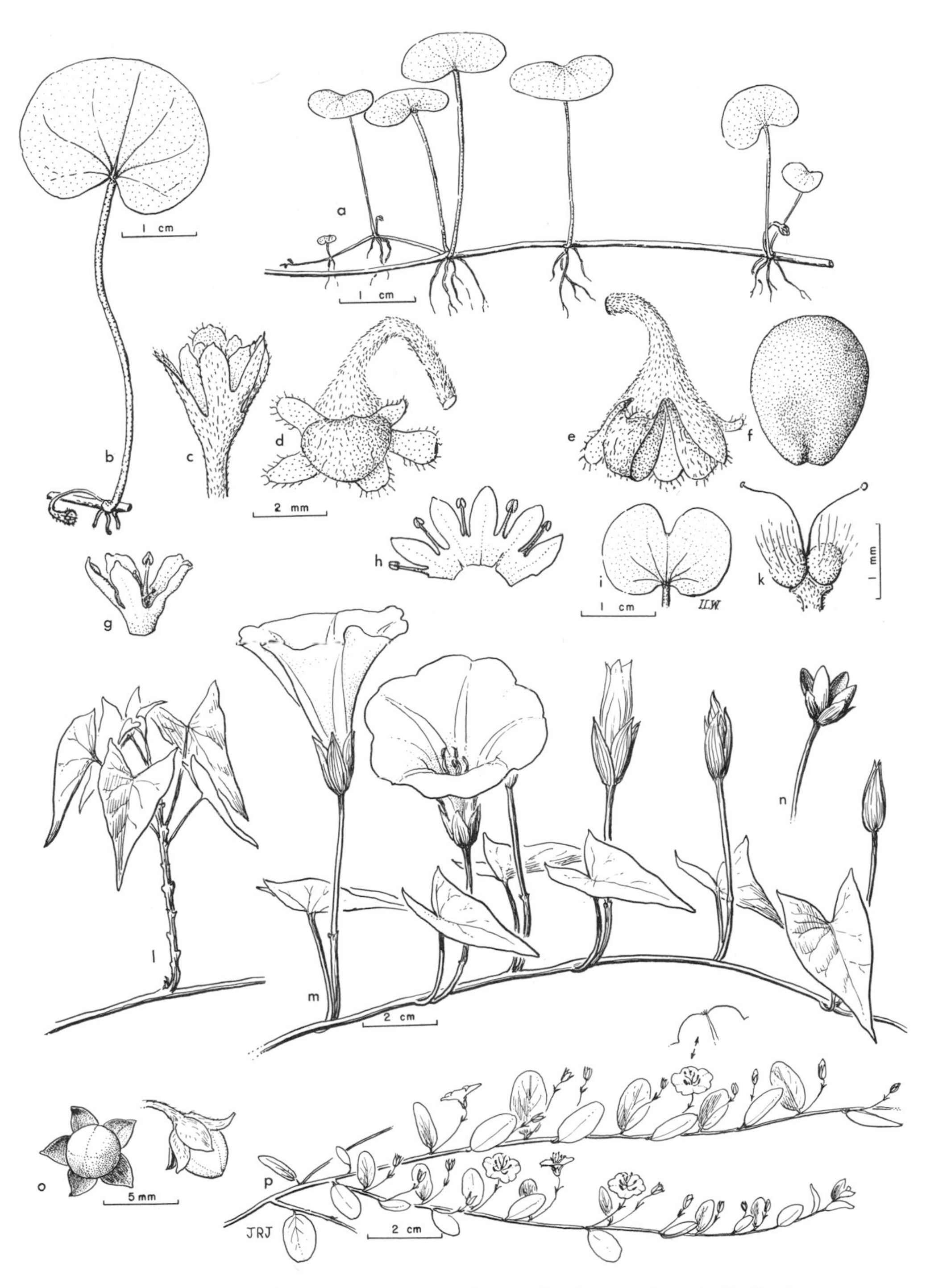

Fig. 88. *Dichondra repens* var. *microcalyx* (*a–k*). *Ipomoea linearifolia* (*l–n*). *Evolvulus glaber* (*o, p*).

In damp soil along rills and in meadows, around pounds, and along trails, widely distributed in northern South America from Brazil to Argentina and Colombia to Peru. Specimens examined: Santa Cruz.

The material examined (*Taylor TT-55* [CAS], *Wiggins 18781* [DS], *Itow 204* [DS]) is rather reluctantly placed in this variety, for it is less densely pubescent than several specimens annotated by Tharp and Johnston (1961), and its stamens are considerably longer than the sepals instead of equaling or slightly surpassing them. More adequate material is needed to determine the range of variation and relationship with other varieties.

See Fig. 88.

Evolvulus L., Sp. Pl. ed. 2, 391. 1762

Annual or perennial herbs or small shrubs. Stems prostrate to erect, not twining. Leaves simple, entire, petiolate or sessile. Flowers perfect, pedicellate to subsessile, axillary or on slender, 1- to few-flowered peduncles. Sepals 5, distinct, equal or subequal, obtuse to acuminate. Corolla small, rotate, funnelform, or salverform, blue, white, or yellow, limb obscurely 5-lobed or 5-angled. Stamens 5, inserted in tube and included or slightly exceeding it; anthers ovate to oblong. Ovary ovoid to globose, 2-celled, 4-ovuled. Styles 2, separate to base or nearly so, each bearing 2 stigma lobes; stigmas terete, elongate. Capsule ovoid to globose, 1–4-seeded. Seeds smooth or minutely roughened.

About 110 species in the warm temperate and tropical regions of the Old and New Worlds.

Stems prostrate; leaves broadly elliptic, at least 2 times as long as wide; flowers on filiform peduncles 5–12 mm long; corolla blue . . . . *E. glaber*

Stems upright or ascending; leaves spatulate, narrowly elliptic to lance-oblong, 3–5 times as long as wide; flowers on short peduncles 0.5–3 mm long, often subsessile; corolla white or slightly yellowish . . . . *E. simplex*

Evolvulus glaber Spreng., Syst. 1: 862. 1825

Evolvulus nummularius auct., non L.
Evolvulus hirsutus auct., non Lam.
Nama convolvuloides Willd. ex Schult., Syst. 6: 189. 1820.
Evolvulus mucronatus Sw. ex Wikstr., Kongl. Vetensk. Acad. Handl. 1827: 61. 1827.
Evolvulus glabriusculus Choisy, Mém. Soc. Phys. Genève 8: 78. 1837.
Evolvulus alsinoides var. *procumbens* f. 3 *obtusifolia glabrata* Schweinf., Beitr. Fl. Aeth. 94. 1867, pro parte.
Evolvulus linifolius var. *linearis* Meisn. in Mart., Fl. Bras. 7: 347. 1869.
Majera coerulea Karst. ex Peter in Engl. & Prantl, Nat. Pflanzenfam. IV, Abt. 3a: 19. 1897.
Evolvulus karstenii Peter, loc. cit.
Evolvulus campestris Brandegee, Univ. Calif. Pub. Bot. 6: 190. 1915.

Stems several to many, much-branched, 1–5 dm long or more, about 1 mm thick, prostrate, moderately strigose-pubescent with simple, tawny, antrorsely oriented hairs; stipules glandular-globose, about 0.5 mm thick, dark brown to nearly black; petioles slender, 1–3.5 mm long, strigose; blades broadly elliptic to nearly orbicular, 6–20 mm long, 4–15 mm wide, rarely slightly emarginate, margins entire, dark green and sparsely pubescent to glabrate above, paler and slightly more densely strigose beneath; peduncles 5–12 mm long, strigose; bracts lance-linear, 1.5–2.5 mm long, almost equaling pedicels; calyx 2.2–2.8 mm long, strigose, sepals acute; corolla narrowly funnelform-rotate, about 4 mm long, limb 7–10 mm across, pale blue; capsules broadly obovoid to subglobose, short-apiculate, about 3 mm long,

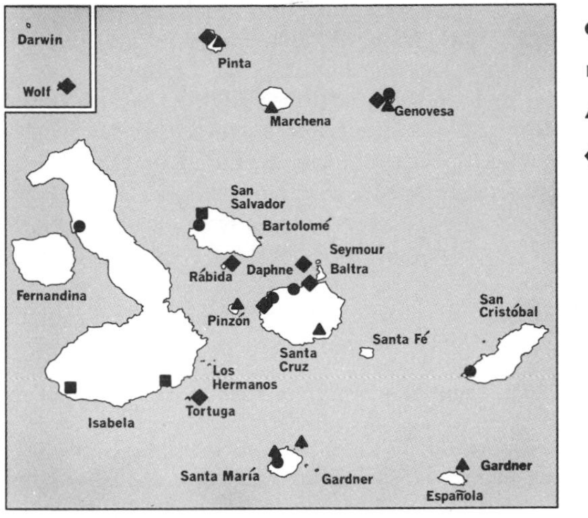

● *Evolvulus simplex*
■ *Ipomoea alba*
▲ *Ipomoea habeliana*
◆ *Ipomoea linearifolia*

2–2.5 mm across, glabrous, chartaceous; seeds dark brown, 1.8–2.2 mm long, about half as wide, minutely rounded-papillate. n = 13.*

In full sun or partial shade, often in heavy, packed soil, Florida and the West Indies, much of Mexico and Central America, and into South America; common at low and medium elevations in the Galápagos Islands. Specimens examined: Eden, Isabela, Pinta, Pinzón, San Cristóbal, San Salvador, Santa Cruz, Santa María, Seymour.

See Fig. 88.

Evolvulus simplex Anderss., Kongl. Svensk. Vet.-Akad. Handl. 1853: 211. 1855

Annual herb with simple, erect stems 1–4 dm tall or sometimes moderately branched from base or lower third of plant, copiously hirsute with appressed to spreading, tawny simple hairs throughout; leaves narrowly lanceolate to oblong-lanceolate, 10–22 mm long, 1.5–5 mm wide, acute at apex, acute or subacute at base, densely appressed-hirsute with short hairs mingled with longer ones on both faces, longer hairs dominant on margins and midrib beneath; petioles short, 1 mm long or less; flowers 1–3 on short axillary peduncles, usually appearing sessile; sepals narrowly lanceolate, 3–5 mm long, densely hirsute; corolla small, 3–4 mm long, lobes about 0.5 mm long, nearly erect, white; capsules membranous, globose, about 2 mm in diameter; seeds pale brown, about 1.5 mm long, minutely papillose, dull.

On open ground or bare lava, northern Peru at moderate altitudes. Specimens examined: Genovesa, Isabela, San Cristóbal, San Salvador, Santa Cruz, Santa María.

Ipomoea L., Sp. Pl. 159. 1753; Gen. Pl. ed. 5, 76. 1754

Annual or perennial, twining, trailing, or creeping vines, or rarely erect shrubs or trees with ample leaves and showy, axillary, solitary or cymose flowers on short to long peduncles, sometimes from fleshy tubers. Sepals 5, equal, subequal, or un-

* Count made by D. Kyhos.

equal, margins often thin, membranous or scarious. Corolla funnelform, campanulate, or rarely salverform; limb entire, 5-angled or 5-lobed, more or less plaited, particularly in bud, often tenaciously stuck together in withering. Stamens 5, equal or unequal, inserted on corolla tube, included or sometimes slightly exserted, filaments filiform or dilated at base, often hairy below; anthers ovate to linear. Ovary ovoid to globose, 2–4-celled, 4–6-ovuled. Styles filiform, included or slightly exserted; stigmas 1 or 2, capitate to globose. Capsule ovoid to globose, 2–4-valved, 2–6-seeded. Seeds usually more or less angled, glabrous, scantily puberulent, short-hirsute or silky-comose on angles, less commonly over most of seed. Cotyledons often lobed or divided.

About 400 species, widely distributed in the warm temperate and tropical regions of the Old and New Worlds.*

Corolla markedly salverform; tube slender, same diameter from base to apex; limb horizontally rotate-spreading; stamens exserted. . . . . *I. alba*

Corolla funnelform or campanulate; tube gradually ampliate upward; limb spreading-ascending, not horizontal; stamens included (slightly exserted in *I. habeliana* but leaves of this noncordate):

Plants annual (sometimes short-lived perennial, stems not woody):

Sepals 20–25 mm long, markedly hirsute at base; capsule depressed-globose, about 1.5 times as wide as long. . . . . *I. nil*

Sepals 6–9 mm long, glabrous or sparsely hirsute; capsule ovoid to globose, as long as wide or longer . . . . *I. triloba*

Plants perennial, stems more or less woody or subsucculent:

Stems creeping, succulent; plants of sandy beaches and dunes; leaf blades orbicular-cordate or obcordate, thick and subfleshy. . . . . *I. pes-caprae*

Stems clambering, woody at base; plants of rocky habitats; leaf blades sagittate-ovate or oblong:

Corolla white, 10–12 cm long, tube much longer than throat; leaves lance-ovate, rounded at base . . . . *I. habeliana*

Corolla roseate to pale purplish, 3–4 cm long, tube shorter than gradually ampliate throat; leaves sagittate-ovate, cordate at base. . . . . *I. linearifolia*

Ipomoea alba L., Sp. Pl. 161. 1753

Convolvulus aculeatus L., op. cit. 155.
Ipomoea bona-nox L., op. cit. ed. 2, 228. 1762.
Convolvulus bona-nox Spreng., Syst. Veg. 1: 600. 1825.
Convolvulus pulcherrimus Vell., Fl. Flum. 72. 1825.
Calonyction speciosum Choisy, Conv. Orient. 59. 1833.
Calonyction bona-nox Bojer, Hort. Maurit. 227. 1837.
Ipomoea aculeata Kuntze, Rev. Gen. Pl. 2: 442. 1891.
Calonyction pulcherrimum Parodi, Contr. Fl. Paraguay 24. 1892.
Convolvulus aculeatus var. *bona-nox* Kuntze, Rev. Gen. Pl. 3^2: 212. 1898.
Calonyction aculeatum House, Bull. Torrey Bot. Club 31: 590. 1904.

Robust perennial, creeping vine with glabrous or rarely sparsely hirsute herbage, stems and branches often markedly verrucose-tuberculate; petioles 3–18 cm long; leaf blades ovate to ovate-lanceolate, rounded-cordate, 4–20 cm long, 4–18 cm wide, auricles usually rounded but sometimes angulate to attenuate, apex usually acuminate; peduncles 3–25 cm long, rather stout; bracts more or less alternate; flowers 1 to several in each inflorescence, fragrant, nocturnal; pedicels 7–20 mm long, stout, becoming clavate in fruit; outer sepals ovate-lanceolate to oblong, 8–13

* Two species reported from the Galápagos but not collected recently are not included in the key. They are briefly discussed following the description of *I. triloba*.

mm long, 4–6 mm wide, with erect, fleshy, subapical appendages 5–12 mm long, inner sepals ovate to elliptic, 10–16 mm long, 7–10 mm wide, obtuse to acute, bearing a mucro 1–5 mm long; corolla salverform, 6–14 cm long, tube cylindrical, slender, 3–5 mm broad, limb white, rotate-spreading, 9–15 cm broad; stamens 8–13 cm long, exserted 1–2 cm; stigma equaling or slightly surpassing stamens, slightly bilobed; capsule broadly ovoid, 3–3.5 cm long, glabrous, 4-seeded; seeds 11–13 mm long, black, glabrous or pubescent along angles and around hilum.

Widely cultivated, of uncertain geographic origin; in tropical and warm temperate parts of South America and the West Indies, but rare in the Galápagos Islands. Specimens examined: Isabela, San Salvador.

Ipomoea habeliana Oliv. in Hook., Icon. Pl. *pl. 1099.* 1871

Widely creeping, subwoody perennial with stems 1–8 m long, upright leaves, and wholly glabrous herbage; stems often pendent from lips of cliffs and crevices; petioles slender, 2–4(5) cm long; leaf blades lanceolate to ovate-lanceolate, 6–15 cm long, 2–5 cm wide, acute, rounded, or obtuse at base, acute to slightly acuminate at apex, dark glossy green on upper surface, slightly paler and duller beneath, leaves mostly standing upright in 2 rows from creeping stems; veins prominulous beneath; peduncles shorter than petioles, 1–6 cm long, erect, 1- to several-flowered; bracts ovate, 1–3 mm long, caducous; pedicels stoutish, 5–25 mm long; sepals oblong-elliptic, unequal, the outer 12–18 mm long, acute to rounded, the inner successively longer, largest 20–22 mm long, thin but margins not scarious, becoming indurated and strongly reflexed in fruit; corollas white, opening late in evening or early part of night, closing by or before mid-morning of following day, narrowly funnelform to almost salverform, 10–12 cm long, the tube 5–6 mm broad at base, ampliate moderately to throat, this 1–1.5 cm broad, glabrous without, sparsely pilose within; limb 2.5–3.5 cm broad; stamens exserted 5–10 mm, filaments pilose at base; anthers 5–6 mm long; style bilobed, about equaling stamens, each lobe globose; capsules broadly ovoid, sharply apiculate, 12–16 mm thick, 20–25 mm long, glabrous, firm, often persisting into third year; seeds usually 4, obovoid, 8–10 mm long, 6–7 mm wide, rounded and closely fine-puberulent on back and inner face, the lateral margins closely set with silky, pale brown hairs 4–5 mm long, these mostly pointing backward, a small tuft of upright hairs at apex of each angle.

Endemic; a common vine on rocky areas, particularly at the lips of cliffs, in partial shade or sometimes in full sun. Specimens examined: Champion, Gardner (near Española), Genovesa, Marchena, Pinta, Pinzón, Santa Cruz, Santa María.

Ipomoea linearifolia Hook. f., Trans. Linn. Soc. Lond. 20: 204. 1847

Ipomoea kingbergii Anderss., Kongl. Svensk. Vet.-Akad. Handl. 1853: 212. 1855.

Perennial, mostly scrambling vine from woody rootstock and woody lower stems, branches of the year slender, herbaceous, glabrous, finely striate, slightly angulate; petioles 1–4.5(7) cm long, erect; blades ovate to broadly lance-sagittate, 2–7 cm long, 1–3.5 cm wide at broadest part, deeply cordate at base, basal lobes 0.5–2 cm long, acuminate to attenuate, apical lobe attenuate, whole blade set at angle of 45° to 60° from vertical, with apex lower than base, lamina thin, slightly paler beneath than above, sometimes all lobes greatly reduced and almost linear; peduncles surpassing leaves, 1–3-flowered, 1.5–10 cm long; bracts minute, 1–1.5 mm long, deltoid-ovate, caducous; pedicels 3–6 cm long; sepals ovate-lanceolate, slightly unequal, the outer 17–19 mm long, 3–5 mm wide, attenuate, strongly 4- or 5-veined, the inner 12–16 mm long, 7–9 mm wide, acute, short-apiculate, thin, paler than the outer; corollas roseate, broadly funnelform, 4.5–5 cm long, limb about as wide, glabrous, mid-petal line strongly marked, 4–5 mm wide near departure from calyx; capsules narrowly ovoid, 14–17 mm long, 6–8 mm thick, acute, firm, mostly 4-valved; seeds 2–4, 6–7 mm long, 3–4 mm wide, rounded on back, retrorsely appressed-strigose over all surfaces.

A vine of rocky habitats, often covering large areas with a dense mass of intertangled vines; according to Svenson, also in northwestern Peru. Specimens examined: Daphne Major, Genovesa, Pinta, Rábida, Santa Cruz, Tortuga, Wolf.

The species was extremely abundant on Isla Wolf in January 1967, nearly smothering stands of the low-growing *Opuntia*.

See Fig. 88.

Ipomoea nil (L.) Roth, Cat. Bot. 1: 36. 1797

Convolvulus nil L., Sp. Pl. ed. 2, 219. 1762.
Ipomoea scabra Forsk., Fl. Aegypt.-Arab. 44. 1775.
Ipomoea cuspidata R. & P., Fl. Per. & Chil. 2: 11. 1799.
Convolvulus tomentosus Vell., Fl. Flum. 74. 1825.
Pharbitis cuspidata G. Don, Gen. Hist. 4: 270. 1838.
Ipomoea longicuspis Meisn. in Mart., Fl. Bras. 7: 227. 1869.

Annual or sometimes short-lived perennial, with twining to decumbent-creeping, slender, somewhat angular stems 2–5 m long; herbage sparsely to moderately hirsute with tawny, simple hairs 1–4 mm long, hairs spreading to slightly retrorse; leaf blades suborbicular, entire to narrowly and shallowly to moderately 3-lobed, 3–18 cm long, 3–20 cm wide, cordate with narrow sinus, acute to short-acuminate at apex and tips of lateral lobes, auricles rounded and entire, dark green, sparsely subappressed-hirsute on upper surface, much paler and more densely hirsute beneath; inflorescence cymose on peduncles 1–15 cm long, retrorsely hirsute; bracts linear to lanceolate, 5–11 mm long; flowers 1 to several on each peduncle; sepals linear, 15–30 mm long, nearly equal, bases ovate, 2.5–3 mm wide, gradually nar-

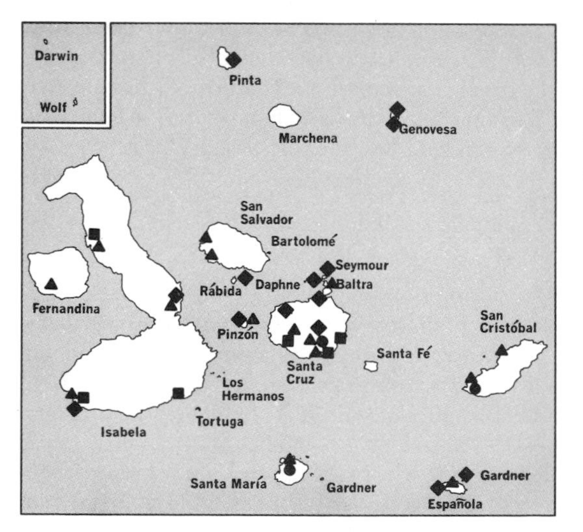

● *Ipomoea nil*
■ *Ipomoea pes-caprae*
▲ *Ipomoea triloba*
◆ *Merremia aegyptica*

rowing into linear terminal blade, densely hirsute; corolla funnelform, 2.5–6 cm long, pale blue to deep blue, throat and tube gradually ampliate, paler than limb, this 3–5 cm across, glabrous without; stamens subequal, the shorter 12–40 mm long, the longer 15–45 mm long, conspicuously glandular-hirsute toward base; style 15–30 mm long, slender; stigma shallowly bilobed, granular; capsules ovoid to depressed-subglobose, 7–8 mm high, 8–12 mm thick, 3–6-seeded; seeds 4.5–6 mm long, black or brown, sparsely to densely grayish-puberulent, or sometimes nearly glabrous.

A pantropical species, distributed from Mexico and the West Indies through Central America and South America to Bolivia, Paraguay, and Argentina; scantily represented in the Galápagos Islands. Specimens examined: San Cristóbal, Santa Cruz, Santa María.

See color plate 77.

Ipomoea pes-caprae (L.) R. Br. in Tuckey, Narr. Exped. Zoire 477. 1818

Convolvulus pes-caprae L., Sp. Pl. 159. 1753.
Convolvulus brasiliensis L., loc. cit.
Ipomoea biloba Forsk., Fl. Aegypt.-Arab. 44. 1775.
Ipomoea brasiliensis Sweet, Hort. Sup. Lond. 35. 1818.*

Creeping perennial with stout stems 1–10 m long, somewhat fleshy and with milky, astringent juice; branches subsucculent, often to 1 cm thick; herbage glabrous; petioles stout to slender, 1–7 cm long; blades orbicular-cordate, broadly ovate or obcordate, 5–10 cm long, 2.5–8 cm wide, leathery, obtuse, rounded or cordate, the veins conspicuous beneath; bracts ovate-lanceolate, 3–5 mm long, caducous; pedicels 1–3 cm long, slender in flower but becoming thickened in fruit; sepals unequal, the outermost broadly oblong, 4–5 mm long, the inner successively longer, 8–10 mm long, margins thin but not scarious, backs minutely and faintly granulate or smooth; corollas broadly funnelform, 2.5–3 cm long, purplish, limb about as wide as length of flower; capsules ovate, about 2 cm long, 12–15 mm wide, valves nearly woody; seeds angulate along inner edge, rounded on back, 8–10 mm high, densely pubescent with brownish spreading hairs 0.5–1 mm long.

Along beaches and on coastal sand dunes, Mexico and the West Indies into the warmer parts of South America, circumtropical; occasional in the Galápagos Islands. Specimens examined: Isabela, Santa Cruz.

Ipomoea triloba L., Sp. Pl. 161. 1753

Ipomoea galapagensis Anderss., Kongl. Svensk. Vet.-Akad. Handl. 1853: 213. 1855.†

Annual twining vine with slender procumbent to climbing stems, often to several meters tall; petioles slender, mostly shorter than blades; herbage sparsely hirsute to glabrous; blades ovate-cordate or hastately 3-lobed, 4–10 cm long, acuminate at apex, the sinus broad, shallow; peduncles usually longer than leaves, to 1 dm long, 1–5-flowered; bracts linear-lanceolate, 2–3 mm long, caducous; pedicels 6–12 mm long; sepals nearly equal, oblong, lanceolate, or sometimes the outer suborbicular, 6–9 mm long, apiculate, ciliate or sparsely hirsute on back; corolla 1.5–2 cm long,

* For additional synonymy, see van Ooststroom (1940, 532–39).

† The synonymy of this widespread species is in a state of confusion; study of the complex over a wide area is needed to clarify the taxonomy.

pink, lavender, purplish, or rarely white, the limb 10–15 mm broad; stamens and style included; capsules subglobose, 6–8 mm broad, sparsely pilose to glabrous; seeds dark brown, glabrous or sparsely puberulent along angles and around hilum.

In various habitats from open, sunny hillsides to relatively dense forest in partial sun or deep shade, southern Arizona and Florida through the West Indies, Mexico, and Central America into tropical and warm temperate parts of South America; very abundant at lower and middle altitudes in the Galápagos Islands. Specimens examined: Baltra, Española, Fernandina, Isabela, Pinzón, San Cristóbal, San Salvador, Santa Cruz, Santa María.

See color plate 44.

Ipomoea Species Excluded from the Flora

Ipomoea campanulata L., Sp. Pl. 160. 1753

This species was reported from Isla Isabela by Robinson (1902, 187), but the specimen that he cited (*Snodgrass & Heller 43* [GH]) is *Stictocardia tiliifolia* (Desr.) Hallier f.

Ipomoea tubiflora Hook. f., Trans. Linn. Soc. Lond. 20: 204. 1847

This species, described by Hooker on the basis of a collection taken by Darwin on Isla San Salvador, is uncertain. Nothing like it has been collected by any subsequent field workers, and the type has not been seen.

Merremia Dennst., Schlüssel Hort. Ind. Malobaricus 23. 1818

Twining vines or sometimes small, erect shrubs, with glabrous to densely pubescent herbage and entire, lobed, or divided (in ours) leaves. Flowers solitary or in few-flowered axillary clusters. Peduncles variable in length. Bracts linear to lanceolate. Sepals subequal or the outer longer than inner, glabrous, pubescent, or glandular-pubescent. Corolla campanulate to funnelform, fairly large. Stamens equal or essentially so; filaments usually glandular-hairy at base; anthers spirally twisted after dehiscence. Style filiform. Stigma globose or of 2 globose lobes, included. Ovary 2- or 3-celled, 4–6-ovuled. Capsule dehiscent by 4–6 longitudinal valves, rarely splitting irregularly. Seeds usually 4–6, rarely solitary, glabrous to densely pubescent.

Circumtropical, with about 30 species in Mexico, Central America, and northern South America, and additional species in the Old World.

Merremia aegyptica (L.) Urban, Symb. Ant. 4: 505. 1910

Ipomoea aegyptica L., Sp. Pl. 162. 1753.
Convolvulus pentaphyllus L., op. cit. ed. 2, 223. 1762, excluding var. *serpens* L.
Ipomoea pentaphylla Jacq., Coll. 2: 297. 1788.*

Prostrate scandent or climbing vine with slender, twining stems 1–4 m long; herbage sparsely to densely hirsute with simple, spreading to subappressed, silvery to tawny hairs 2–4 mm long, these especially conspicuous on sepals and pedicels; petioles

* For additional synonymy, see O'Donnell (1941, 489).

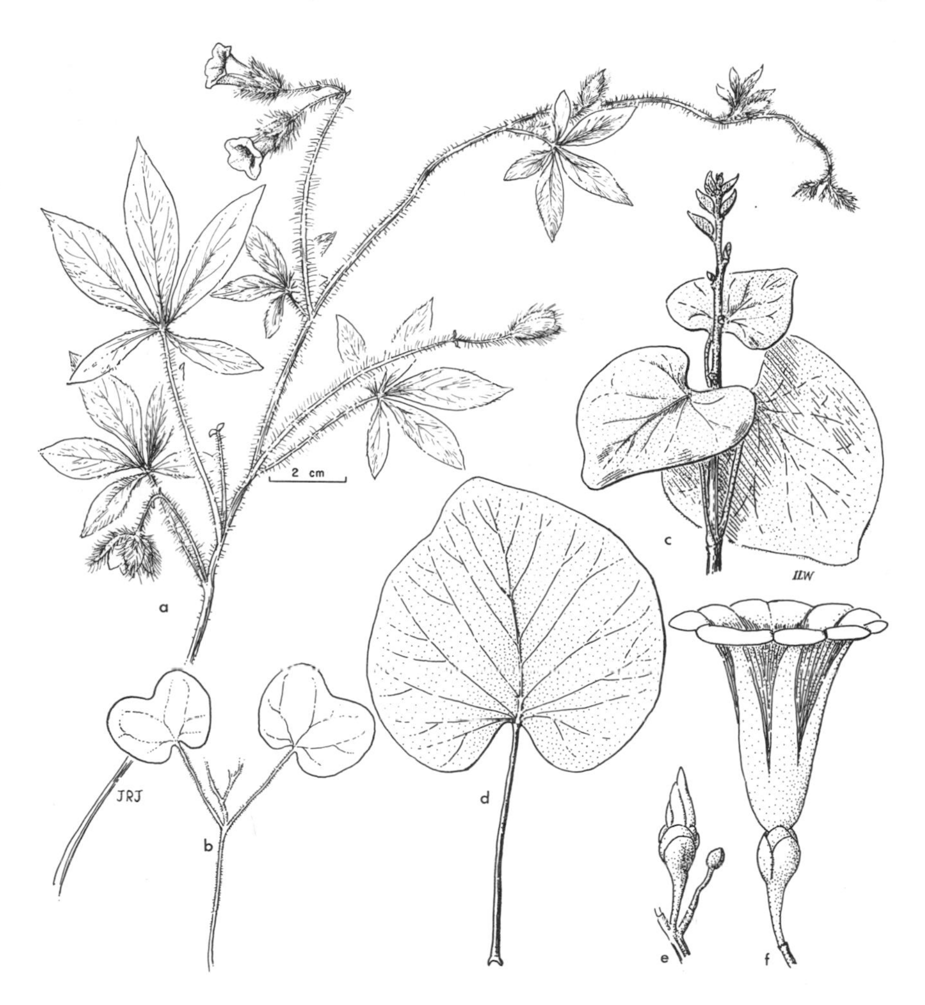

Fig. 89. *Merremia aegyptica* (*a, b*). *Stictocardia tiliifolia* (*c–f*).
(Figs. *e* and *f* redrawn from Svedelius.)

slender, 2.5–11 cm long, hirsute; leaves palmately 5-foliolate, leaflets elliptic to obovate, acuminate at apex, acute at base, 1.5–5 cm long, 2–25 mm wide, entire, unequal, on petiolules to 5 mm long or some nearly sessile, irregularly dotted beneath with globose glands 0.1–0.2 mm across, becoming black dots when dry; peduncles 1–10 cm long, 1- to several-flowered; bracts lanceolate, 2.5–5 mm long at anthesis, caducous; sepals oblong, acute, unequal, 3–6 mm long at anthesis, to 25 mm long in fruit, outer 3 densely hirsute, inner 2 glabrous; corolla campanulate-funnelform, 2–3 cm long, 2–2.5 cm across open limb, white, mid-petal lines distinct, glabrous; stamens 8–10 mm long, included, filaments glandular-pubescent at base; anthers spirally twisted after dehiscence; stigma of 2 globose lobes; capsule globose, 10–14 mm in diameter, glabrous, 4-locular, 4-seeded; seeds brownish, glabrous, dull, 4–5 mm high.

Common in open and in shrubby or forested areas, Mexico and the West Indies through Central America to Argentina, Bolivia, Peru, and Uruguay; very common on several islands in the Galápagos archipelago. Specimens examined: Daphne Major, Española, Gardner (near Española), Genovesa, Isabela, Pinta, Pinzón, Rábida, Santa Cruz, Seymour.

Probably native only in the New World in spite of the specific epithet. Seedlings too young to bear flowers were seen on several of the islands during January and February 1967. This plant flowers late in the evening or in the very early morning, and the corollas wilt and become hopelessly stuck in a tight ball soon after sunrise. It is usually found in open *Bursera* stands in the Arid Zone and in lower parts of the Transition Zone on most of the islands.

See Fig. 89.

Stictocardia Hallier f., Engl. Bot. Jahrb. 18: 159. 1894

Woody or herbaceous perennial vines with ovate to orbicular, mostly cordate, entire leaves, the lower surfaces of blades dotted with minute, partially sunken glands turning black with age. Flowers on axillary peduncles, in 1- to several-flowered cymes. Bracts small, caducous or deciduous. Sepals elliptic to orbicular, obtuse to emarginate, subcoriaceous, equal or the outer slightly shorter than inner. Corolla ample, regular, funnelform, the mid-petal stripes often moderately pilose and with glands similar to those on leaf blades. Stamens and style included; filaments inserted near base of corolla, slender. Style 1; stigma bilobed, each lobe globose-tipped. Ovary glabrous, 4-celled, 4-ovuled. Capsule globose, enclosed by accrescent sepals, dissepiments with 2 transverse wings at surface of fruit, the walls thin, separating irregularly from dissepiments and their wings to produce 4 openings. Seeds globose, pubescent.

A small genus of less than 10 species, circumtropical in distribution.

Stictocardia tiliifolia (Desr.) Hallier f., Engl. Bot. Jahrb. 18: 159. 1894

Convolvulus grandiflorus L. f., Suppl. 136. 1781.
Convolvulus tiliaefolius Desr. in Lam., Encycl. 3: 544. 1789.
Ipomoea grandiflora Lam., Tabl. Encycl. 1: 467. 1791.
Convolvulus gangetinus Roxb., Hort. Berg. 13. 1814, nomen nudum.
Ipomoea tiliaefolia R. & S., Syst. 4: 229. 1819.
Ipomoea benghalensis R. & S., loc. cit.
Convolvulus campanulatus Spreng., Syst. 1: 608. 1825.
Ipomoea pulchra Blume, Bijdr. 716. 1825.
Ipomoea gangetica Sweet, Hort. Brit. ed. 2, 288. 1830.
Convolvulus melanostictus Schlecht., Linnaea 6: 737. 1831.
Rivea tiliaefolia Choisy, Mém. Soc. Phys. Genève 6: 407. 1833.
Ipomoea melanosticta G. Don, Gen. Syst. 4: 271. 1838.
Argyreia tiliaefolia Wight, Icon. 4: 2. 1850.
Rivea campanulata House, Muhl. 5: 72. 1909.
Stictocardia campanulata Merrill, Phil. Jour. Sci. 9: 133. 1914.
Argyreia campanulata Alston in Trimen, Fl. Ceyl., Suppl. 201. 1931.

Climbing vine with rather robust, more or less woody stems and short-hirsute to subtomentose branches, leaves, and inflorescences; petioles 2–30 cm long, densely pubescent with short, multicellular, simple hairs; leaf blades orbicular-cordate to broadly ovate-cordate, 5–20 cm long and wide, densely to moderately short-pubescent with spreading, stout hairs, the sinus broad and shallow, midvein and 7–11 pairs of lateral veins conspicuous and more densely pubescent than adjacent blade beneath, margins entire, apex obtuse to rounded; peduncles 2–12 cm long, rather slender, 1- to few-flowered; pedicels 1–1.5 cm long, slightly wider above than at base; bracts orbicular to broadly obovate, 10–15 mm long, tomentulose; sepals

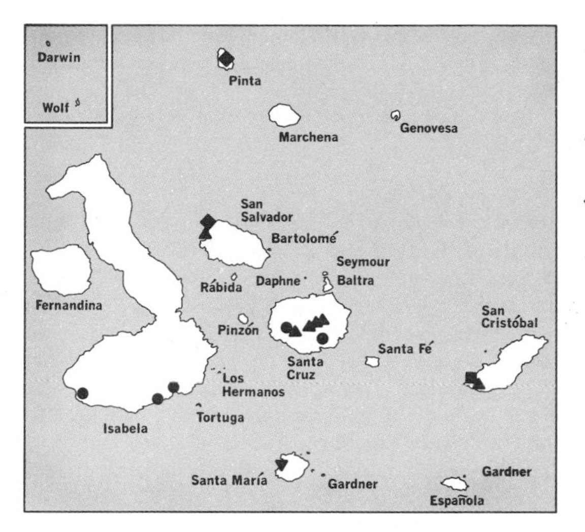

● *Stictocardia tiliifolia*
■ *Citrullus lanatus*
▲ *Elaterium carthagenense* var. *carthagenense*
◆ *Elaterium carthagenense* var. *cordatum*
▼ *Luffa astorii*

about 1.5–2 cm long, subequal, rounded at apex, accrescent in fruit; corolla narrowly funnelform, 8–15 cm long, limb about as broad, roseate to nearly white; capsule 2–2.5 cm broad; seeds brown, nearly globose, finely strigillose-pubescent.

In the tropics of the Old and New Worlds. Specimens examined: Isabela, Santa Cruz.

Probably introduced into the Galápagos Islands.

See Fig. 89.

CUCURBITACEAE. GOURD FAMILY*

Herbaceous or sometimes woody vines with coiled, simple or branched tendrils and alternate leaves. Stems and foliage usually more or less pubescent or scabrous. Leaves exstipulate, mostly petiolate, entire or commonly palmately lobed or divided. Flowers monoecious or dioecious, small to large. Calyx tube adnate to ovary, cup commonly 5-lobed or 5-toothed. Corolla campanulate to funnelform, sympetalous or of 5 distinct petals inserted alternately with sepals on rim of calyx tube, usually withering-persistent. Stamens (1,2)3(4,5), when 3, 2 bearing 2-celled anthers, the 3d with a 1-celled anther; filaments short, often united. Ovary 1–3-celled; style terminal, simple or lobate. Fruit a pepo, variable in form and size, indehiscent or in some dehiscent at apex or by subterminal pores. Seeds usually numerous, more or less compressed; endosperm lacking.

About 90 genera and 700 species, mainly in the tropics and subtropics of the Old and New Worlds but extending into the temperate regions.†

Ovary and fruit 3-celled or with 3 placentae, usually many-seeded; ovules horizontal, ascending or erect:
  Leaves cordate-ovate, neither lobed nor deeply parted; filaments of stamens connate into a column; fruit obviously beaked, this often nearly or fully as long as body of fruit; all anthers with 2 thecae each. . . . . *Elaterium*

---

* Contributed by Ira L. Wiggins.

† *Cucurbita pepo*, cultivated on two of the islands but clearly not persisting in the wild, is not included in the key, but is discussed briefly on p. 393.

Leaves distinctly 5–7-lobed or deeply parted; filaments of stamens free or nearly so, not connate into a column; ovary and fruit not markedly beaked; 1 anther with single locule, others with 2 locules each:

Fruit fibrous and dry at maturity, dehiscing by terminal, operculate pores; ovary and fruit elongate-oblong to cylindric.......................................*Luffa*

Fruit fleshy or pulpy at maturity, indehiscent or dehiscing by 3 irregular valves, not elongate-cylindric; not terminally operculate:

Pistillate flowers each with 3 setiform or ligulate rudimentary stamens; fruit indehiscent, globose to broadly oblong, the rind smooth, green or mottled........ .... *Citrullus*

Pistillate flowers without rudimentary stamens, or these, if present, represented by 3 glands at base of ovary; fruit dehiscent by 3 irregular, recurving valves, the rind warty or umbonate, bright orange-red....................*Momordica*

Ovary and fruit 1-celled, 1-ovuled; ovules pendulous from apex of ovary:

Pistillate flowers pedicellate, borne at base of long-pedunculate raceme, proximal to, but in same raceme with, staminate flowers; spines on fruit bent sharply upward, parallel with axis, retrorsely barbed; anthers straight, not flexuous; beak of fruit not oblique ....*Sicyocaulis*

Pistillate flowers sessile, borne in cluster at apex of short peduncle in same axil as, but distinct from, staminate raceme; spines or bristles on fruits radiate-spreading, not retrorsely barbed; anthers sigmoid- or conduplicate-flexuous; beak of fruit oblique, continuous with 1 side of fruit...................................*Sicyos*

Citrullus Schrad. ex Ecklon & Zeyher, Enum. Pl. Afr. Aust. 279. 1834

Annual or perennial, long-trailing vines, with more or less pubescent herbage and branched tendrils. Leaves erect or ascending from creeping stems, blades parallel to surface of soil; blades deeply pinnatifid, lobes further lobed or divided, more or less glaucous. Flowers monoecious, axillary, medium in size, pale yellow. Corolla rather deeply 5-lobed, limb spreading. Stigmas 3. Ovary with 3 placentae; seeds many. Seeds strongly flattened, white, mottled, or black.

An Old World genus of only 3 or 4 species, but with many variants and cultivars, widely distributed in the tropics and warm temperate regions of the world. Widely and extensively cultivated for the fruits (watermelon).

Citrullus lanatus (Thunb.) Mats. & Nakai, Cat. Sem. et Spor. Hort. Bot. Univ. Imp. Tokyo 30, no. 854. 1916

Cucurbita citrullus L., Sp. Pl. 1010. 1753.
Momordica lanata Thunb., Fl. Cap. 36. 1807.
Citrullus vulgaris Schrad. ex Ecklon & Zeyher, Enum. Pl. Afr. Aust. 279. 1834.
Citrullus edulis Spach, Hist. Nat. Veg. Phan. 6: 214. 1838.
Calocynthis citrullus Kuntze, Rev. Gen. Pl. 1: 256. 1891.

Characters of the genus. Blades often 6–15 cm long, ultimate segments broadened at apex, pale green and glaucous; fruit spherical to ellipsoid or oblong, externally green or mottled or striped with dark and light green, white within the outer, hard rind, the inner flesh a delicate pink or yellow, this filling whole melon; seeds in 2 broad bands along each of the 3 parietal placental limbs, varying considerably in size (3–12 mm or more), sordid white to black, or sometimes mottled, each surrounded by a delicate, transparent sac, slippery when moist.

Almost worldwide in the tropical and warm temperate regions; introduced into agricultural areas and escaped to a limited extent on Islas Isabela and Baltra. Specimen examined: San Cristóbal.

Fig. 90. *Citrullus lanatus.*

It is rare to find escaped plants producing fruit of significant value after the first 2 or 3 years.

See Fig. 90.

Elaterium Jacq., Stirp. Am. 241. 1763

Slender, scandent herbaceous vines with sparsely pubescent to glabrous herbage. Leaf blades cordate, entire, lobed, or parted. Tendrils simple or branched, in ours both kinds often occurring on one plant. Flowers rather small, monoecious, white, yellowish, or greenish, sometimes tinged with brown, the pedicels capillary or nearly so. Staminate flowers racemosely arranged or in fascicles at tips of slender peduncles; calyx tube slenderly cylindric; corolla rotate, with 5 subulate to filiform lobes parted nearly to base of disk. Stamens with filaments connate into slender column; anther cells linear, sigmoid-flexuous. Pistillate flowers solitary, axillary, on short, slender pedicels. Fruit obliquely ovoid, rostrate, somewhat gibbous, 1- to several-celled, at maturity rupturing elastically. Seeds with crenulate or rarely smooth margin.

Possibly a dozen species, ranging from Cuba and central Mexico to South America, at least as far south as Peru.

Leaves ovate-cordate, 2.5–6 cm long, 2–5 cm wide, nonlobed to obscurely 3-lobed at base . . . . *E. carthagenense* var. *carthagenense*
Leaves relatively wider, more deeply cordate, lateral lobes less evident near base . . . . *E. carthagenense* var. *cordatum*

Elaterium carthagenense Jacq., Enum. Pl. Carib. 31. 1760, var. **carthagenense**

Stems slender, to 3 m long or more, sparingly villous when very young, soon glabrate, angulate-striate; petioles slender, 1–5 cm long, markedly villous-ciliate along adaxial ridges, otherwise glabrous; leaf blades ovate-cordate, 2.5–6 cm long, 2–5 cm wide, acute at apex, nonlobed to obscurely 3-lobed near base, margins entire or faintly and irregularly denticulate, sparsely beset with pustulate-based, slender, simple hairs over upper surface, the hairs confined to veins beneath, these clothed with fine, white hairs often obscuring pustules of larger hairs; tendrils capillary, simple or once- or twice-branched near base, basal 1–2 cm of stalk straight, not coiled, distal to this, tendril branches tightly coiled in close spirals; staminate peduncles slender, 1.5–6 cm long, glabrous; staminate calyx tube 8–10 mm long, pale greenish yellow, glabrous, its teeth 0.2–0.3 mm long, situated at bases of sinuses between corolla lobes, glabrous; corolla lobes linear-subulate, 7–8 mm long, brownish green, ascending-spreading, minutely and sparsely granulose within; stamen column slender, 8–9 mm long, greenish; anther cells forming folds about 1.5 mm long and making collectively 5 or 6 loops at apex of column, pale yellow; pistillate flowers solitary, inserted in axils at bases of staminate peduncles, on thickish pedicels 4–8 mm long; fruit oblique, ovoid to subglobose, 10–15 mm long, gibbous toward one side at base, covered with softish bristles 1–2 mm long, rostrum none; seeds 5–6 mm long, distinctly tridentate at both ends, pale yellowish, considerably compressed.

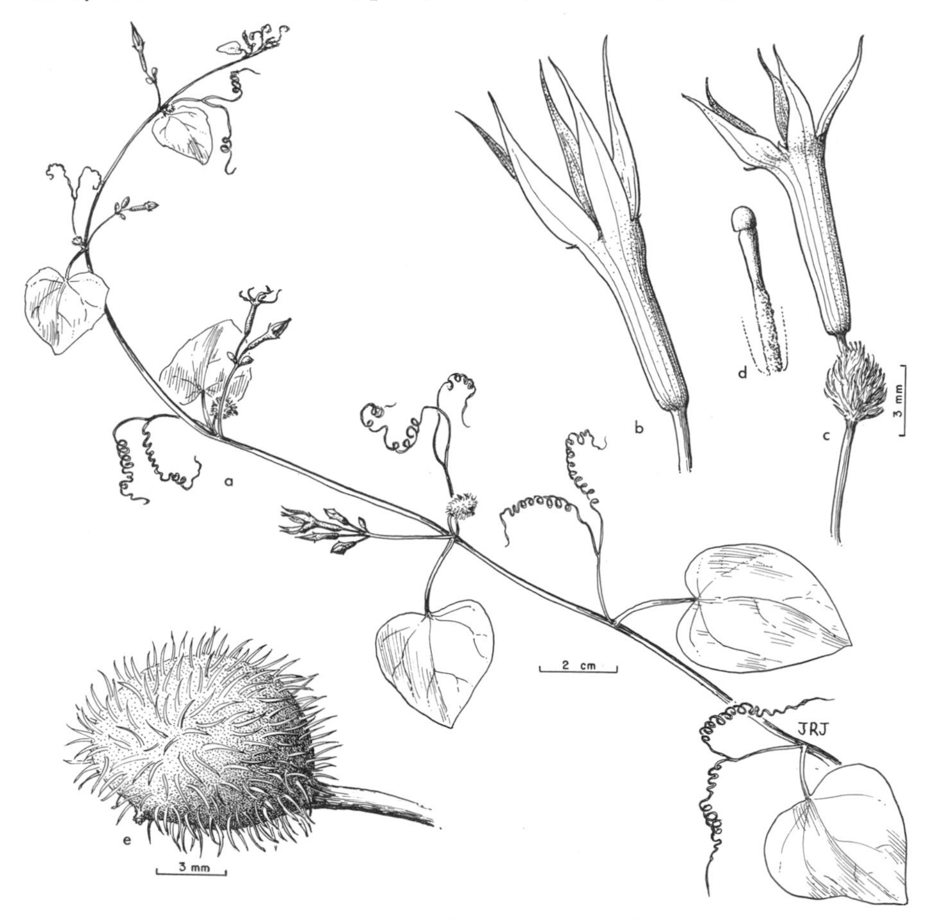

Fig. 91. *Elaterium carthagenense* var. *carthagenense*.

Throughout Ecuador and Peru at lower elevations. Specimens examined: San Cristóbal, San Salvador, Santa Cruz.

See Fig. 91.

Elaterium carthagenense var. **cordatum** (Hook. f.) Svens., Amer. Jour. Bot. 22: 256. 1935

Elaterium cordatum Hook. f., Trans. Linn. Soc. Lond. 20: 224. 1847.

Differing from var. *carthagenense* only in having relatively broader, more deeply cordate leaves with less evidence of lateral lobes near the base.

Known only from the Galápagos Islands. Specimens examined: Pinta, San Salvador.

According to the label on Snow's specimens (*164* [DS]), the plant is known locally as Espumilla. Ripe fruits dehisce explosively, casting seeds several meters.

Luffa Adans., Fam. Pl. 2: 138. 1763

Annual, scandent to moderately climbing vine with broad, glabrous to scabrous leaves and branched tendrils. Leaves 5–7-lobed or rarely subentire, petiolate, glandless. Flowers monoecious, staminate on axillary, long-pedunculate racemes; receptacle campanulate or turbinate; sepals 5, triangular to lanceolate; petals 5, distinct, spreading, obovate to obcordate, entire or erose-margined. Stamens 3 or rarely 4 or 5, inserted on receptacular tube, distinct; anthers linear, more or less sinuous, 1- or 2-loculed. Pistillate flowers solitary, receptacle tube extending well beyond ovary; staminodia 3–5, fleshy. Ovary ovoid to elongate, cylindrical, angulate or sulcate, containing 3 placentae. Style columnar. Stigma trilobate, branches more or less distorted and curled. Ovules many, horizontal. Fruit dry, fibrous within, ovoid, oblong, or cylindrical, terete or angulate to costate, trilocular, dehiscing by terminal operculum, this often beaked. Seeds ovoid to oblong, compressed.

A genus of 10 or 12 species, mostly native in the tropics and subtropics of the Old World, 1 or 2 species native in the tropics of the New World. Some species widely distributed and cultivated for commercial purposes.

Luffa astorii Svens., Amer. Jour. Bot. 22: 256. 1935

Scandent or climbing to height of 4 m or more, internodes angulate and more or less hispid to scabrous; petioles 4.5–8 cm long, 2–3 mm thick, finely hirsute with slender, white hairs; leaf blades orbicular to pentagonal, 8–5 cm long, fully as wide or slightly wider than long, cordate with narrow basal sinus 1–2.4 cm wide, angulate to shallowly 5-lobed, slightly scaberulous above, appressed-pubescent beneath, margins minutely denticulate, apex acute to obtuse, slightly apiculate; tendrils bifid, pedestal 2–5 cm long, the branches to 15 cm long, tightly coiled; staminate flowers yellow, on peduncles 8–18 cm long, 6–15-flowered near tips, pedicels 6–10 mm long, puberulent; receptacle broadly campanulate, 3–5 mm broad, puberulent; sepals ovate-lanceolate, 6–7 mm long, pubescent without; petals obovate, 18–20 mm long, 10–12 mm wide, irregularly erose at apex; stamens 3; anthers strongly sigmoid-sinuous, on filaments about 3 mm long, ciliate at base; pistillate flowers solitary in axils devoid of staminate peduncles, their pedicels 1–2 cm long, receptacle broadly campanulate, 3–4 mm wide, borne at apex of slender beak 0.5 mm broad, 10–12

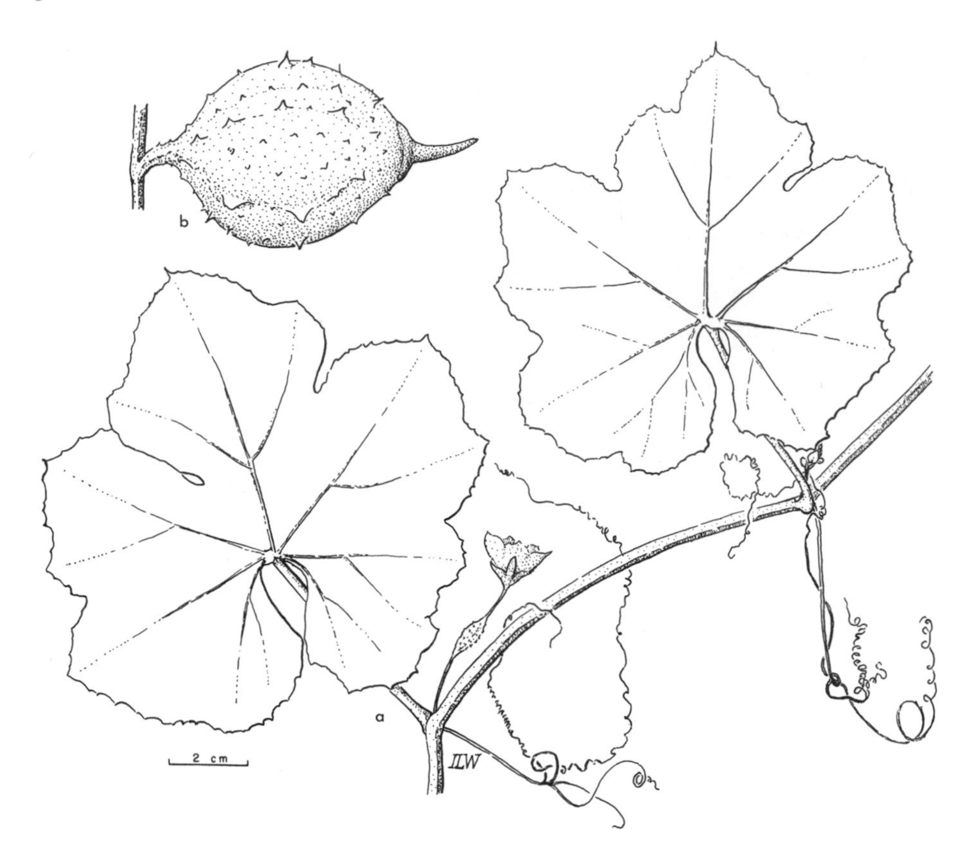

Fig. 92. *Luffa astorii.*

mm long, the sepals lance-ovate, 4–5 mm long, 1.5–2 mm wide, acuminate, the petals yellow, spreading, 12–15 mm long; ovary ellipsoidal, 12–15 mm long, 4–5 mm across at anthesis, distinctly low-costate, appressed-puberulent; stigma 3-lobed, branches irregularly curved and coralloid; fruit 5–8 cm long, 8–10-costate, ribs bearing conical spinelike processes 1–6 mm long, surfaces between costae grayish, smooth and glabrate; operculum conical, long-beaked, 1.5–2 cm broad at base, the beak 2–3 cm long; seeds gray, 6–8 mm long, 4–4.5 mm wide, smooth or irregularly roughened, dull and faintly mottled with slightly darker gray.

Growing among shrubs and trailing over lava, on low hills and plains near the sea, central Ecuador to Peru; in the Galápagos Islands known only from Santa María.

This plant is near *L. operculata*, and may not be distinct from it. The description of the pistillate flower was drawn from a specimen collected near Guayaquil, Ecuador (*Asplund 15346* [CAS]).

See Fig. 92.

Momordica L., Sp. Pl. 1009. 1753; Gen. Pl. ed. 5, 440. 1754

Slender, scandent herbaceous vines with simple or branched tendrils and sparsely scabrous herbage. Petioles erect, longer than the palmately lobed blades. Flowers yellow, dioecious or monoecious, the pistillate flowers solitary, the staminate often in few-flowered fascicles. Calyx deeply 5-lobed. Corolla subrotate, pale yellow, 5-lobed. Stamens 3, filaments short but distinct; anther cells flexuous. Fruit ovoid to cylindric, subfleshy, the 3 valves recurved in dehiscence. Seeds rather turgid, only moderately flattened, finely rugulose, each embedded in bright red, fleshy-juicy aril-like covering.

About 60 species, mostly in the Old World tropics.

Momordica charantia L., Sp. Pl. 1009. 1753

Momordica indica L. in Stickm., Herb. Amb. 24. 1754.
Momordica elegans Salisb., Prodr. 158. 1796.
Momordica operculata Vell., Fl. Flum. 10. 1827.*

Plant monoecious; stems scandent and climbing into shrubs and trees 2–5 m high, much branched, forming dense mats over rocks and shrubs; herbage sparsely scabrous-puberulent or younger parts more densely so; petioles villous to subglabrous, 3–6 cm long, erect; leaf blades reniform-suborbicular, 5–12 cm long and wide, 5–7-lobed to middle or deeper, lobes ovate to oblong, dentate to again more or less lobed, middle lobe larger than laterals, tips acute to obtuse or sometimes rounded, deep green above, paler beneath, puberulent to subglabrate; tendrils very slender, simple in ours, soon tightly coiled in close spirals; staminate flowers solitary, the pedicels and peduncles slender, 5–15 cm long, villous to subglabrous, bearing ovate to cordate bract 5–15 mm long near middle; sepals 5, ovate-lanceolate, 4–6 mm long, 2–3 mm wide, acute; corolla pale yellow, sometimes slightly irregular, the segments 15–20 mm long, 8–12 mm wide, obtuse to emarginate; pistillate peduncles 5–10 cm long, bracteate near base; ovary fusiform, rostrate, more or less mucronate; fruit bright orange-red, oblong or ovoid, 4–8 cm long, covered sparingly with acute to obtuse tubercles, 3-valved at apex at maturity, the valves curling back to expose the bright red, fleshy arils surrounding each seed; seeds 8–10 mm long, 4–6 mm wide, moderately compressed, slightly 3-toothed at apex and base, rugulose on surfaces.

In partial shade, in deep shade, or sometimes in full sun, in the tropics and subtropics of the Old and New Worlds; on several of the larger Galápagos Islands. Specimens examined: Isabela, Santa Cruz.

Common and abundant in the vicinity of Bahía de la Académia, Santa Cruz. The pulp around the seeds is eaten by children, but care is taken to avoid swallowing the seeds, which are reputed to be violently purgative and somewhat toxic. An extract of the roots in alcohol taken in small doses is claimed to be an aphrodisiac. Fatty oils, oleo-resins, and several other substances have been identified from the leaves, stems, roots, and fruits. Commonly used as a fence cover and as a massed ornamental plant.

See Fig. 93 and color plate 96.

Sicyocaulis Wiggins, Madroño 20: 251. 1970

Climbing and scandent herbaceous perennial vine. Leaves thin, shallowly 5-lobed or angular-pentagonal, petiole about equaling blade. Tendrils bifid; branches tightly

* For several additional synonyms, see A. Cogniaux and H. Harms, Das Pflanzenr. IV, 275. II: 24. 1924.

Fig. 93. *Momordica charantia.*

coiled. Flowers monoecious, minute, borne in long-pedunculate racemes, the pistillate flowers few (1–5) at base of raceme, the staminate more numerous and filling rest of inflorescence. Staminate flowers with broadly campanulate calyx, its sepals minute, narrowly triangular; corolla rotate-campanulate, 5-lobed, segments lance-ovate, filaments connate into slender column slightly shorter than corolla lobes; anthers 4, adnate to apex of column along their inner surfaces, tips free, locules slightly introrse-curved over tip of column but not flexuous nor curved sideways, splitting longitudinally. Pistillate flowers distant at base of raceme, pedicellate, calyx and corolla as in staminate flowers; style slender, two-thirds to three-fourths as long as corolla lobes; stigma oblong, weakly 3-lobed; staminodes none. Ovary 1-celled, slenderly ovoid, with a slender beak about equaling body, this bearing 4–8 low, narrow, irregularly crenate wings and several forward-pointing, retrorsely barbed spines near base, the wings reduced gradually toward beak. Ovule 1, pendulous from apex of locule. Fruit small, coriaceous, narrowly ovoid, slenderly beaked, slightly winged and bearing forward-pointing spines near base, surface between wings glabrate or sparsely puberulent. Seed somewhat flattened, tuberculate along edges and at both ends, testa firm to sub-bony.

One species, endemic; known only from the general area of Bella Vista, Isla Santa Cruz, where it grows in moderate to deep shade in *Scalesia* forest.

Sicyocaulis pentagonus Wiggins, Madroño 20: 252. 1970

Characters of the genus. Stems climbing to 5 m or more, the internodes to 2 dm long; unbranched basal part of tendril 1–15 cm long, bifid, the branches filiform,

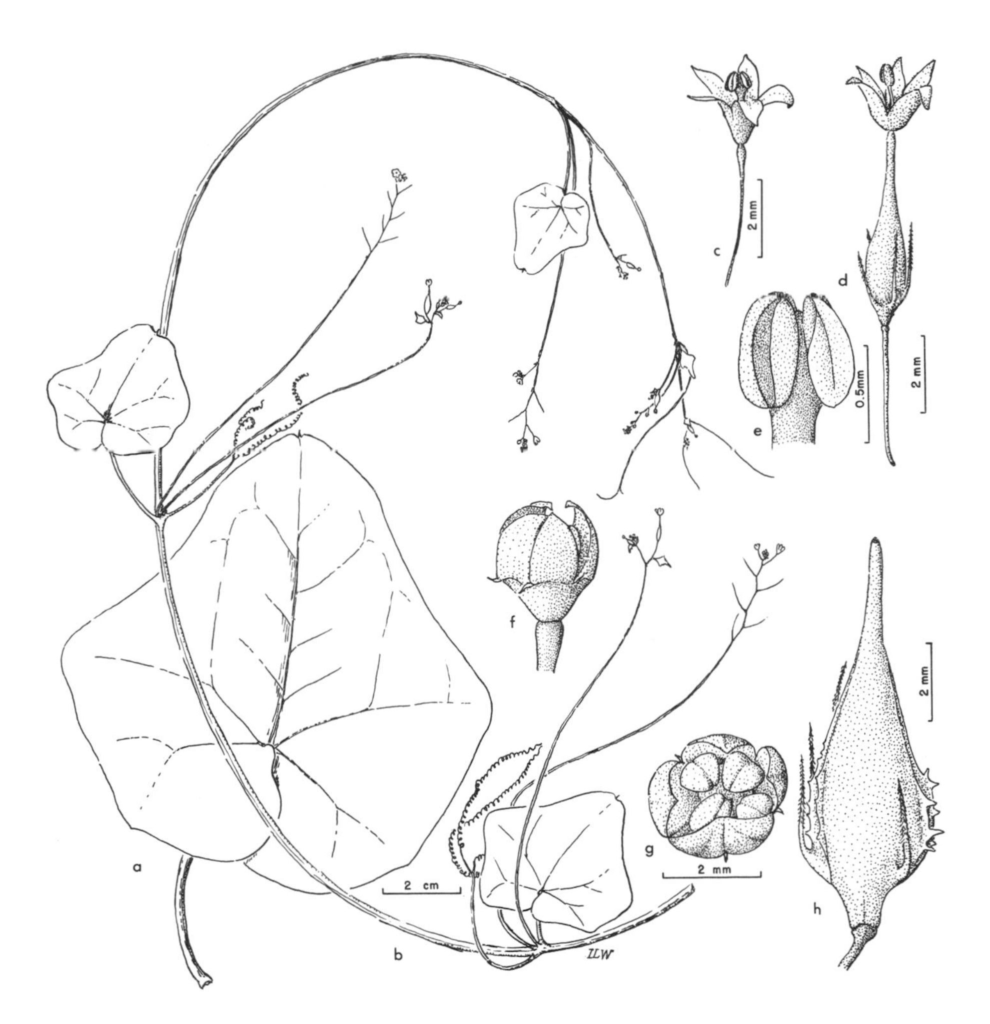

Fig. 94. *Sicyocaulis pentagonus*: *c*, staminate flower; *d*, pistillate flower; *e*, staminal column and anthers; *f*, staminate bud before anthesis; *g*, staminate bud, from above.

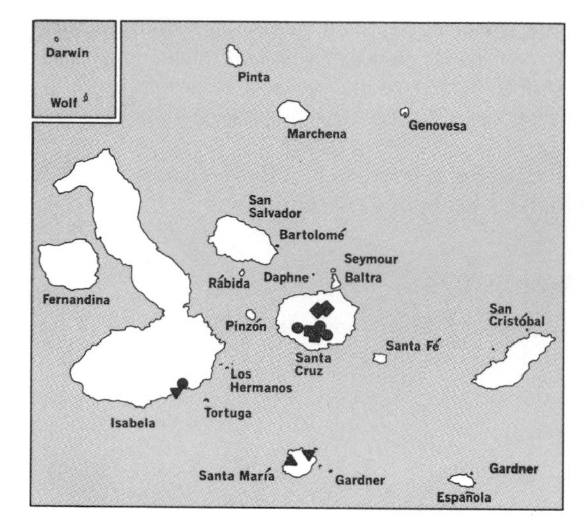

usually exceeding length of stalk; leaves with petioles 2–10 cm long, blades 2–12 cm long (measured from base of lower lobes to apex), sinus to 3 cm deep, adjacent lobes overlapping, thus closing sinus; rest of margin more angulate than lobed, upper surface dark green, sparsely scaberulous with short, pustulate-based, erect or curved hairs, the lower surface paler, more closely scaberulous; main veins 5, palmately arranged; peduncles axillary, 1–5 cm long at anthesis, elongating after anthesis, usually bearing 1–3(5) pistillate flowers at base of raceme, 5–15 or more staminate flowers above these; pedicels 3–6 mm long at anthesis, filiform, those of staminate flowers usually slightly clavate-swollen toward apex; calyx lobes about 0.3 mm long; calyx cup 0.3–0.5 mm deep; corolla lobes oblong to ovate-oblong, 0.6–1 mm long, acute to minutely apiculate, cream-colored, ascending-rotate, glabrous; stamen column 1–1.4 mm high, with subglobose terminal part about 0.6 mm broad; introrsely curved anthers about 0.6 mm long, apical 0.1–0.2 mm free above head of column; pistillate flowers with pedicels, calyx, and corolla like those of staminate, the ovary 4–5 mm long at anthesis, bearing forward-pointing spines as part of wings and along axis of ovary at about basal third to half; style 1–1.5 mm long, stigma about a third as long, faintly 3-lobed; base of fruit narrowed to short stipe, fertile part 10–15 mm long, maximally 4–5 mm thick (near base), beak 2.5–3 mm long; seed 8–11 mm long, closely enclosed in the fruit.

In shaded areas in *Scalesia* Zone near Bella Vista.

See Fig. 94.

Sicyos L., Sp. Pl. 1013. 1753; Gen. Pl. ed. 5, 443. 1754

Annual monoecious vines with angled or striate stems, mostly pubescent herbage, and capitate clusters of small fruits. Tendrils petiolate, with 3–5 branches. Leaves petiolate or uppermost sessile, subentire to deeply lobed, thin, angulate. Staminate flowers in racemes, panicles, or umbel-like clusters, inflorescences long-pedunculate. Calyx shallowly and broadly campanulate to rotate, with 5 remote, small teeth. Corolla salverform; limb 5-lobed, spreading. Stamens 3; connate filaments forming column; anthers connate or free. Pistillate flowers in sessile clusters or in short-pedunculate umbel-like heads, borne in same axils as staminate inflorescences. Style about equaling corolla tube; stigma broad, bilobed, flattened. Ovary 1-celled. Ovule solitary, pendulous. Fruit dry, ovoid, echinate with retrorsely barbellate prickles, often with pubescence among prickles. Seeds reticulate or pitted.

A genus of about 35 species, all in the New World, mostly distributed in the tropics but a few species found in temperate zones, to southern Canada and to Chile and Argentina.

Sicyos villosa Hook. f., Trans. Linn. Soc. Lond. 20: 223. 1847

Scrambling and climbing vine; stems to 8 mm thick, angulate and deeply striate, villous with whitish hairs 2–5 mm long, some gland-tipped; tendrils mostly 5-branched, on peduncles 8–10 cm long, branches tightly coiled, to 15 cm long; petioles 10–15 cm long, 3–4 mm across, villous; leaf blades broadly orbicular-ovate, obscurely 5–7-angulate, about 15–16 cm broad, sinus narrow, about 5–6 cm deep, 4–8 mm wide, margins of blade dentate with spreading teeth 1–1.5 mm long, 3–5 mm apart; upper surface bearing numerous white pustules and coarse, subarachnoid hairs along main veins, undersurface more densely villous; peduncles of staminate inflorescences 15–18 cm long to lowest branch, densely villous; staminate flowers 6–8 mm across, cream-colored, lobes deltoid-ovate, about 2 mm wide; filaments 2.5–3 mm long; anthers 3, coalesced into capitate structure 2–3 mm across; pistillate flowers unknown; fruiting clusters on peduncles 3–4 mm long; fruits closely clustered in headlike fructification, each fruit sessile, oblong-ovoid, 9–11 mm long, 5–6 mm wide, bearing apiculation about 1 mm long at apex, acute at base, moderately prickly with yellowish, stiff, acicular bristles 5–7 mm long, the intervening surface closely villous with arachnoid, gland-tipped white or tawny hairs.

Known only from the type collection, "Charles Island [Santa María], in great beds injurious to vegetation, Sept. 1835, C. Darwin."

The holotype, deposited in the Henslow Herbarium, Cambridge University, was loaned by the curator of the herbarium, Dr. S. M. Walters. Mature fruits were in a pocket on the type sheet, and a peduncle from which they had become detached was still an integral part of the specimen. No pistillate flowers had survived. A staminate inflorescence is mounted beside the main specimen and has the characteristic pubescence of the species.

See Fig. 95.

CUCURBITACEAE GENUS AND SPECIES EXCLUDED FROM THE FLORA

Cucurbita L., Sp. Pl. 1010. 1753; Gen. Pl. ed. 5, 441. 1754

Cucurbita pepo L., Sp. Pl. 1010. 1753

Cucurbita ovifera L., Mant. 1: 126. 1767.
Cucurbita esculenta S. F. Gray, Nat. Arr. Brit. Pl. 2: 552. 1821.
Cucurbita courgero Ser., Mém. Soc. Phys. Genève 3: 2, *pl. 1.* 1825.
Cucurbita elongata Bean ex Schrad., Linnaea 12: 407. 1838.

At least 20 additional specific epithets have been relegated to synonymy under this binomial by various authors (see Index Kewensis, 1895).

Given by Stewart as growing in gardens at Villamil, on Isla Isabela, and on Isla Santa María. Local inhabitants now declare that the vines do not persist in the wild, though they are still grown as vegetables.

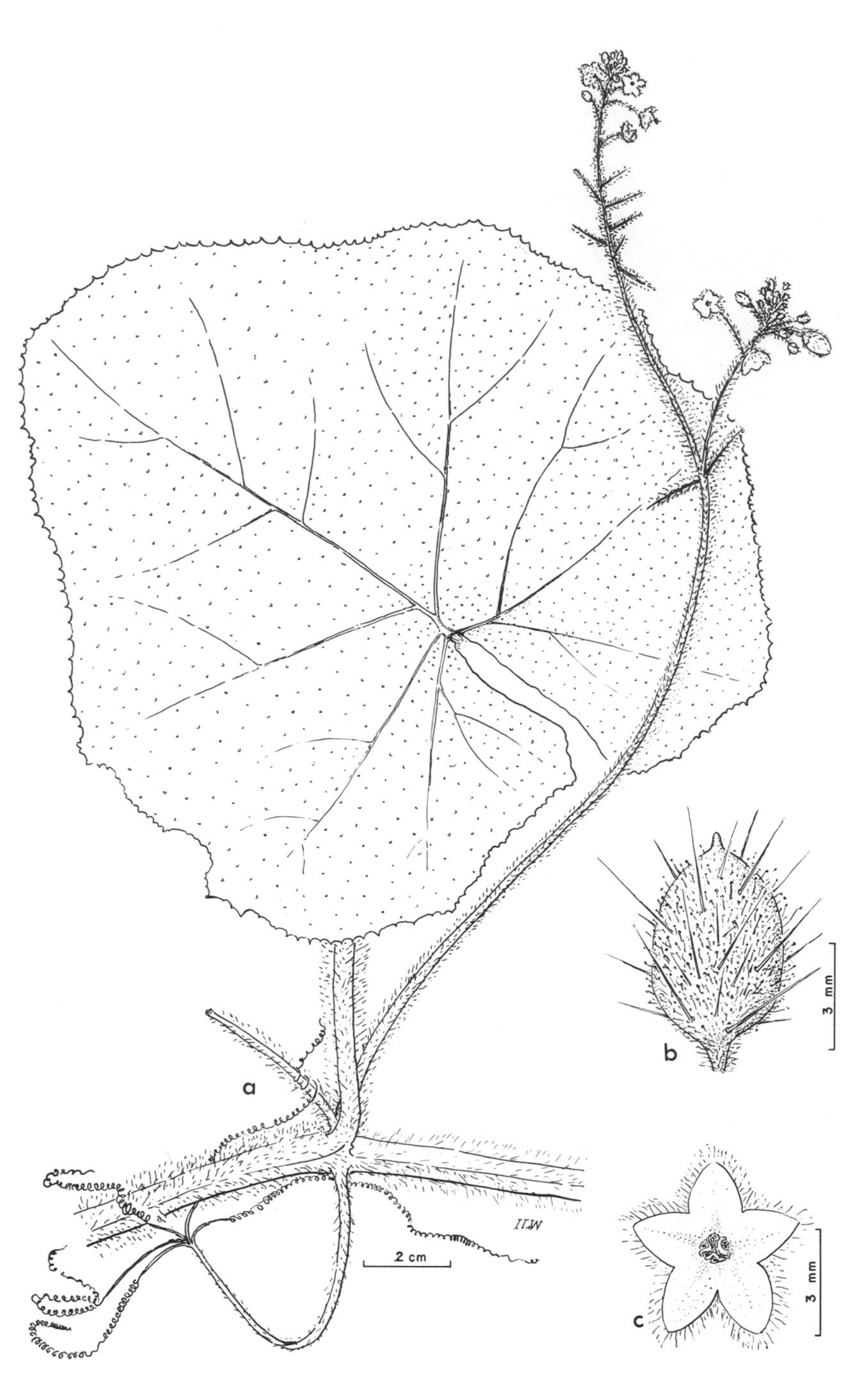

Fig. 95. *Sicyos villosa.*

ERICACEAE. Heath Family*

Perennial herbs, suffrutescent plants, shrubs, or trees, or rarely trailing or scandent vines with alternate, opposite, or whorled leaves, these simple, exstipulate, usually coriaceous and persistent. Flowers perfect, regular (slightly irregular in a few genera), borne solitary in axils of leaves or in terminal or axillary clusters, racemes, or panicles. Calyx of 4–7 lobes or sepals, these usually connate basally, but some distinct, mostly persistent, small. Corolla of 4–7 distinct or more commonly connate petals, thus campanulate, funnelform, or urceolate. Stamens usually twice as many as petals or corolla lobes, inserted at base of a disk (hypogynous or epigynous in some genera), distinct, the filaments often flattened or dilated, connate in some, straight or sigmoidly curved. Anthers 2-celled, often saccate basally, frequently appendaged dorsally or apically, each theca dehiscing introrsely by terminal pore or less commonly by longitudinal slit. Pistil 1; ovary superior or inferior, locules and carpels 4–10, usually 5, with axile placentation. Ovules often numerous in each locule. Fruit a capsule or berry. Seeds usually small; embryo straight; endosperm fleshy.

About 70 genera and approximately 1,900 species, widely distributed throughout the temperate regions of the world and at high altitudes in some tropical areas.

Pernettya Gaud., Ann. Sci. Nat. ser. I, 5: 102. 1825

Suffrutescent or shrubby plants with much-branched, procumbent, erect or ascending stems and glabrous to setose-pubescent herbage. Leaves small, closely spaced, alternate, coriaceous and entire to denticulate, persistent. Flowers axillary, on slender, recurved pedicels. Calyx of 5 distinct, persistent sepals. Corolla globose to ovoid-urceolate, white or pinkish, the orifice shallowly 5-dentate. Stamens 10, free from corolla, included, filaments subfleshy or dilated just above base; anthers bilocular, biaristate dorsally near apex. Ovary depressed-globose, 5-celled, subtended by hypogynous disk, this bearing 10 glands and more or less trilobate. Style short; stigma convex, shallowly 5-lobed. Fruit a berry with rather spongy flesh, white or pink, turning red to black in age. Seeds several in each locule, minute, ovoid to oblong.

A small genus whose relationship with *Gaultheria* is inadequately defined, making the distributional limits uncertain; presently considered to range from Mexico to southern South America.

Pernettya howellii Sleumer, Notizblatt 12: 649. 1935

Low, much-branched shrub 1–3 dm tall; younger twigs covered with appressed, brownish setae 1–1.5 mm long, and moderately pubescent with minute, white, erect hairs between bristles; leaves closely spaced, the internodes less than one-third as long as subtending leaves, petioles stoutish, 1–2 mm long, glabrous or nearly so; leaf blades broadly ovate to broadly elliptic, 5–9 mm long, 3–6 mm wide, markedly coriaceous, deep green above, paler beneath, glabrous except for a few hairs on midrib, the margin denticulate with 4–7 minute teeth on each side, faintly revolute, each tooth bearing a minute apical seta, the blade with apiculation 0.6–1.2 mm

* Contributed by Ira L. Wiggins.

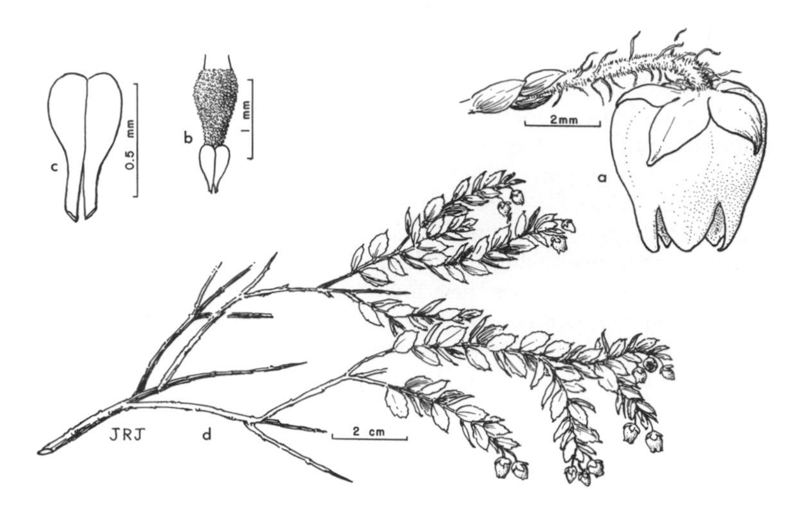

Fig. 96. *Pernettya howellii.*

long; pedicels 5–7 mm long, setose; sepals ovate, about 1.5 mm wide, 2–2.2 mm long, acuminate at apex, glabrous; corolla white, campanulate to urceolate, 4–5 mm long, nodding, the lobes deltoid, about 1.5 mm long, curved outward; stamens (8)10, about 1.6 mm long, filaments thin, smooth at base, swollen-fleshy from about 0.2 mm above insertion to base of anthers, surface of swollen portion minutely papillose-granular, anthers (including terminal tubes) 0.4–0.5 mm long, without appendages, tubes distinct about 0.2 mm, with terminal pore, orifice at right angles to axis or slightly oblique; styles erect, straight, 1–1.2 mm long, smooth and glabrous; stigma minutely 5-lobed; ovary ovoid at anthesis, about 2 mm broad, 1.5 mm long, greenish; berries globose, 4–5 mm across, white or more frequently delicate pink; seeds dark brown to nearly black.

Endemic; known only from ridges and bases of cliffs at elevations above 300 m on Isla Santa Cruz, but this attractive shrub should be sought in similar habitats on other islands.

See Fig. 96.

GOODENIACEAE. GOODENIA FAMILY*

Perennial herbs or small shrubs with watery juice and alternate or rarely opposite or basal, simple, exstipulate leaves. Inflorescence cymose or racemose, less commonly capitate, paniculate, or a single flower. Flowers perfect, usually more or less irregular. Calyx adnate to ovary, 5-lobed or consisting of narrow ring of tissue. Corolla tubular, 5-lobed, bilabiate or 1-lipped, valvate or induplicate in bud. Stamens 5, inserted on ovary alternate with corolla lobes, free or adnate to basal part of corolla tube, distinct at apex, or anthers connate into cylinder around style; anthers 2-celled, introrse, dehiscing longitudinally. Ovary inferior or sometimes partially superior, 1- or 2-celled, 2-carpeled. Ovules 1 to many, on basal or axile placenta. Style 1, slender; stigma simple or 2- or 3-branched, surrounded by shal-

* Contributed by Ira L. Wiggins.

low cup. Fruit a capsule, berry, drupe, or nut. Embryo straight. Endosperm fleshy.

About 13 genera and 300 species, mostly in Australia, but a few more widely distributed along and near the shore in warmer seas.

Scaevola L., Mant. 2: 145. 1771

Erect or scandent herbs or shrubs with alternate leaves. Corolla only slightly irregular (in ours), or more markedly so, lobes nearly equal or very dissimilar. Anthers free. Ovary 2-celled, 1 ovule in each cell. Style curved to sharply bent near apex; stigma entire or slightly bilobed, included in shallow cup. Fruit succulent, indehiscent, a drupe in ours, usually 1-seeded by abortion of 1 ovule.

About 80 species, mostly near the coast of all warm seas.

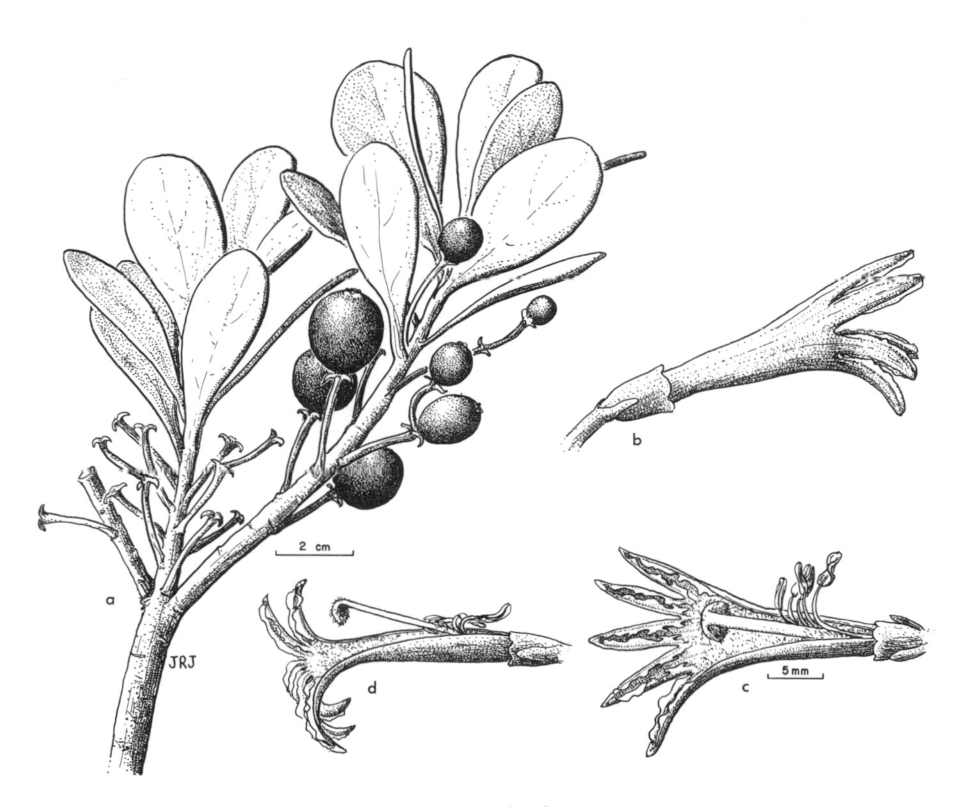

Fig. 97. *Scaevola plumieri.*

Scaevola plumieri (L.) M. Vahl, Symb. Bot. 2: 36. 1791

Lobelia plumieri L., Sp. Pl. 929. 1753.
Scaevola lobelia Murr., Syst. ed. 13, 178. 1774.*

Erect shrub 6–12 dm tall, with stoutish, terete, freely branching stems, glabrous except for tufts of slender white hairs in axils of leaves and above leaf scars; leaf

* Several additional names have been applied, but since the interspecific limitations are hazy they are not listed here.

blades fleshy, obovate-elliptic to oblong-obovate, obtuse to rounded at apex, 3–8 cm long, 1.5–3.5 cm wide, gradually narrowed to broad, winged petiole 5–12 mm long; leaf margins entire; cymes 1–3-flowered on peduncles 1–5 cm long; bracts linear-lanceolate, acute, 3–7 mm long; calyx truncate or very shallowly 5-toothed, the collar about 1 mm high; corolla white, 2–2.5 cm long, glabrous without, hirsute with much-branched hairs within, lobes oblong, short-acuminate, 10–14 mm long, 3–4 mm wide, the margins slightly undulate and lacerate-fringed near base; style about 15 mm long, hirsute on lower third and at tip, glabrous between; stigma cup 2–2.5 mm across, hirsute, turned sharply to 1 side; fruits broadly ellipsoidal to subglobose, 1–1.8 mm long, black or purplish, smooth and lustrous, flesh yellow or greenish inside skin.

In sandy soil and upper beaches, southern Baja California, western Mexico, Central America, the West Indies, and the tropics of Ceylon, Madagascar, Africa, and northern South America. Specimens examined: Isabela, Santa María.

See Fig. 97.

HYDROPHYLLACEAE. WATERLEAF FAMILY*

Annual, biennial, or perennial herbs or shrubs with alternate or opposite, simple, divided, or compound leaves, and cymose or solitary, axillary flowers. Flowers perfect, regular or essentially so. Calyx lobes usually 5, similar or unequal, with or without auricles, often accrescent in fruit. Corolla sympetalous, lobes usually 5, equal, longer or shorter than corolla tube, usually with pair of scales at base of each filament. Stamens mostly 5, hypogynous, exserted or included. Pistil 1, compound, of 2 united carpels; style deeply 2-parted to entire. Stigmas 2, capitate or rarely subulate. Fruit a loculicidal capsule dehiscent by 2 valves, or both loculicidally and septicidally dehiscent by 4 valves, or irregularly dehiscent. Capsule 1-celled or imperfectly 2-celled by intrusion of placentae. Ovules few to numerous. Seeds 1 to over 100. Endosperm present; cotyledons 2, entire.

About 20 genera and over 265 species, represented in all continental areas except Australia; abundant in western North America and extending southward to the Strait of Magellan.

Nama L., Syst. Nat. ed. 10, 950. 1759

Herbaceous or less commonly suffrutescent, erect, ascending or prostrate annuals or perennials with alternate entire leaves. Flowers solitary in axils of leaves or on short, lateral or terminal cymes. Calyx divided nearly or entirely to base; lobes 5, linear, lanceolate or spatulate. Corolla tubular to campanulate-funnelform, finely puberulent on outside; lobes rounded and usually spreading. Stamens unequal or subequal, usually included; filament bases somewhat dilated, adnate portion with free wing or margin or this lacking, glabrous with 1 exception. Styles 2, free to base or nearly so, or connate part or full length. Ovary superior, or partially inferior by adnation of basal part to calyx base, many-ovuled, 1-celled but falsely 2-celled by ingrowth of placentae, usually pubescent. Capsule ovoid to oblong, loculicidally dehiscent, sometimes septicidally dehiscent also. Seeds numerous, variously pitted,

* Contributed by Ira L. Wiggins.

reticulate or alveolate, sometimes angular, transversely corrugated or wrinkled in addition to pits.

About 40 species and a number of subspecific taxa, predominantly of the western United States and Mexico, with a few in South America and 1 in Hawaii.

Nama dichotomum (R. & P.) Choisy, Mém. Soc. Phys. Genève 6: 113. 1833

Hydrolea dichotoma R. & P., Fl. Peruv. 3: 22, *pl. 244, fig. b.* 1802.
Nama tetrandra Pavon ex Choisy, Mém. Soc. Phys. Genève 6: 113. 1833, in synonymy.
Nama stricta Philippii, Fl. Atac. 37. 1860.
Nama dichotomum var. *angustifolium* A. Gray, Proc. Amer. Acad. 8: 284. 1870.
Nama dichotomum β pauciflora Choisy ex A. Gray, loc. cit.
Marilaunidium strictum Kuntze, Rev. Gen. Pl. 2: 434. 1891.
Marilaunidium dichotomum Kuntze, loc. cit.
Conanthus angustifolius Heller, Bull. Torrey Bot. Club 24: 479. 1897.
Nama angustifolium A. Nels. in Coult. & Nels., New Man. Rocky Mt. Bot. 410. 1909.
Nama dichotomum subsp. *eu-dichotomum* Brand, Pflanzenr. 4^{251}: 151. 1913.
Nama dichotomum subsp. *eu-dichotomum* var. *stricta* Brand, loc. cit.
Nama dichotomum subsp. *angustifolium* Brand, loc. cit.
Marilaunidium tenue Woot. & Standl., Contr. U. S. Nat. Herb. 16: 162. 1913.

Erect or ascending, simple or branching annual 0.5–2.5 dm tall, with glandular-hirsute herbage; leaves few, on nodes 1–5 cm apart; leaves linear-oblanceolate to narrowly spatulate, 5–30 mm long, 1–3(4.5) mm wide, gradually narrowed to petiole, acute to obtuse at apex, antrorsely hispid with white hairs bent sharply just above pustulate base; flowers solitary or in pair in axils of upper branches; pedicels 1–3 mm long, hispid; calyx lobes linear to linear-spatulate, 3–4 mm long at anthesis, to 5–10 mm long in fruit; corolla tubular-campanulate, about 5 mm long, limb 5–6 mm broad, bluish; stamens inserted about 1 mm above corolla base, the adnate part with very narrow, delicate marginal wing whose free portions extend almost to base of corolla tube; styles 0.5–2 mm long; capsules 20–60-seeded, thin, splitting to base; seeds brown, about 0.5–0.6 mm long, squarish at ends, uniformly and minutely alveolate and more coarsely pitted over whole surface.

In sandy or loam soils and among rocks at middle to higher elevations, Colorado, New Mexico, and Arizona southward into central and southern Mexico, then from Ecuador to Chile. Specimen examined: Isabela.

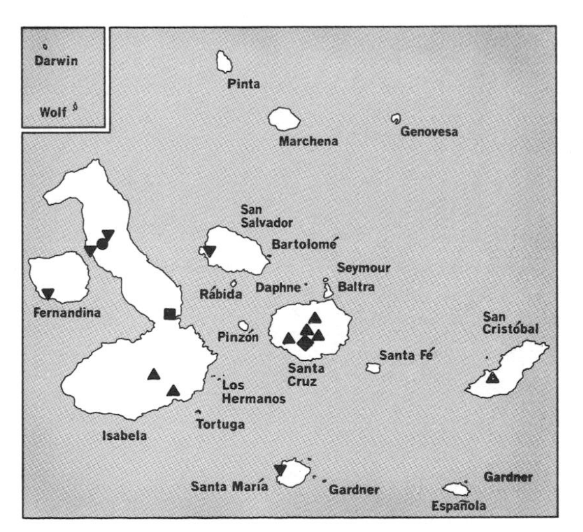

● *Nama dichotomum*
■ *Hyptis gymnocaulos*
▲ *Hyptis rhomboidea*
◆ *Hyptis sidaefolia*
▼ *Hyptis spicigera*

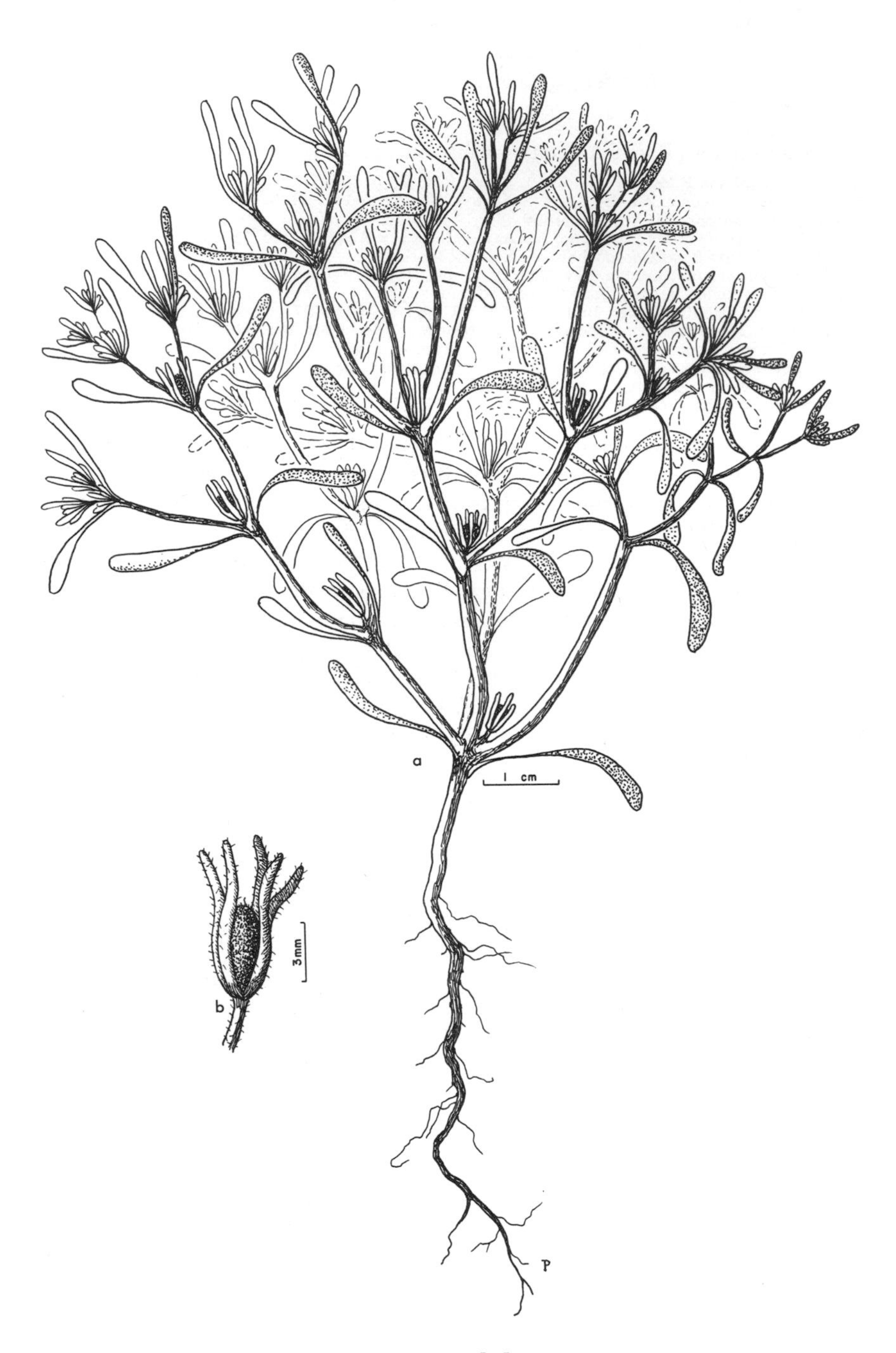

Fig. 98. *Nama dichotomum*.

Additional material is sorely needed for critical comparison of the Galápagos plants with those of the mainland and of North America.

See Fig. 98.

LABIATAE. MINT FAMILY*

Aromatic punctate-glandular herbs, shrubs, or rarely trees, with mostly quadrangular stems and opposite, simple exstipulate leaves. Flowers variously clustered, often in interrupted spicate, racemose, or cymose inflorescences, typically irregular and bilabiate, nearly regular in a few genera. Calyx regular or 2-lipped, usually 5-toothed or 5-lobed, persistent. Corolla usually bilabiate, upper lip entire, emarginate, or 2-lobed, lower 3-lobed. Stamens inserted in corolla tube alternate with lobes, usually 4 and didynamous, sometimes reduced to 2; staminodia present or absent; anthers 2-celled, introrse, confluently 1-celled, or sometimes 1-celled by abortion of 1 cell. Ovary 4-lobed or 4-parted, superior, each division 1-ovuled; ovules anatropous; style arising from central depression between ovary lobes; stigma 2-branched. Fruit of 4 nutlets. Seed usually erect; endosperm scanty or none; embryo small, usually straight.

About 200 genera and 3,200 species, cosmopolitan in distribution, with the greatest concentration in the Mediterranean area. Sometimes difficult to differentiate from genera in the Verbenaceae.

Calyx becoming inflated and with small orifice in fruit; ovary 4-lobed; style not basal to nutlets; nutlets almost completely united laterally. . . . . *Teucrium*

Calyx slightly enlarged but not inflated in fruit, the orifice about as wide as or wider than subtending tube; ovary 4-parted; style inserted at or near base of nutlets; nutlets attached basally, almost distinct laterally:

- Stamens declined along, or enveloped in, lower lip of corolla; calyx not obviously 2-lipped:
  - Calyx not accrescent in fruit, straight; upper lip of corolla emarginate; lower lip saccate; filaments usually pilose in whole or in part. . . . . *Hyptis*
  - Calyx moderately accrescent and curved in fruit; upper lip of corolla 2-lobed to base; lower lip of 2 narrow lobes and 1 broad lobe, not saccate; filaments glabrous (in ours) . . . . *Ocimum*
- Stamens ascending, not declined on or enveloped in lower lip of corolla; calyx distinctly 2-lipped:
  - Stamens 4; anther connectives neither elongate nor articulate with filaments; pollen sacs closely approximate, parallel; corolla slightly bilabiate, the upper lip not cucullate . . . . *Mentha*
  - Stamens 2; anther connectives elongate, articulate with filaments; pollen sacs distinctly separated; corolla strongly bilabiate, the upper lip arched and concave-cucullate . . . . *Salvia*

Hyptis Jacq., Collect. 1: 101. 1786

Annual or perennial herbs, suffrutescent subshrubs, or large shrubs with bilabiate flowers borne in headlike clusters in axils of reduced leaves or on axillary peduncles, often in verticils forming interrupted cymose or spicate infloresence. Calyx straight

* Contributed by Ira L. Wiggins.

or oblique, sessile or pedicellate; tube ovoid, campanulate or cylindric, 5-lobed or 5-toothed, often markedly ridged below each acute to awn-tipped tooth or lobe. Corolla 2-lipped, upper lip emarginate, erect or spreading, lower saccate, drooping or horizontal; tube cylindrical to infundibuliform. Stamens 4, inserted in corolla tube, declined and resting on lower lip; filaments mostly hairy, each anther-bearing; anthers confluently 2-celled. Style glabrous; stigma arms equal or unequal. Nutlets ovoid to subglobose, often slightly angled marginally, smooth to reticulate-roughened.

Over 400 species, widely distributed in the tropical to temperate regions of the world; particularly well represented in the New World.

Inflorescence of sessile, often crowded verticils, forming spike to 10 cm long; plant annual .... *H. spicigera*
Inflorescence of pedunculate heads, these axillary to foliage leaves or to leaflike bracts; plant perennial:
  Plant markedly suffrutescent to strongly shrubby; leaf blades mostly 5–7 mm long, finely rugulose .... *H. gymnocaulos*
  Plant herbaceous, not lignescent at base; leaf blades 2–12 cm long, not rugulose on upper surface:
    Leaf blades cuneate at base, sparsely pubescent to glabrate beneath; petioles winged marginally, gradually widening into blade; heads to 1.5 cm broad .... *H. rhomboidea*
    Leaf blades rounded to subcordate at base, finely velutinous-pubescent beneath; petioles not winged; heads 3–5 mm broad .... *H. sidaefolia*

Hyptis gymnocaulos Epling, Rep. Spec. Nov. Beih. 85: 246. 1936

Suffrutescent to definitely shrubby plant to about 1 m tall, with almost glabrous, divaricate, stiffly spreading to ascending, slender branches, internodes greatly exceeding subtending leaves; petioles short, 3–5 mm long; leaf blades broadly ovate, 5–7(12) mm long, obtuse at both ends, finely crenulate, minutely rugulose and dark green on upper surface, paler beneath, sparsely pubescent and with some pellucid short-stipitate glands on both surfaces; inflorescences laxly few-flowered, on filiform peduncles usually 2–3 cm long, these borne in axils of only slightly reduced leaves; flowering calyces 3–3.5 mm long, delicately hirsute, gland-dotted with short-stalked, capitate, golden glands, the teeth 0.6–0.8 mm long, tube becoming 3.5–4.5 mm long in fruit; corolla tube 4–5 mm long, throat and lobes 2.5–4 mm long; nutlets about 1.5 mm long.

Endemic. Specimens examined: Isabela.

This is a distinctive, slenderly branched shrub with remarkably slender peduncles.

Hyptis rhomboidea Mart. & Gal., Bull. Acad. Brux. 11^2: 188. 1844

Pycnanthemum decurrens Blanco, Fl. Filip. ed. 2, 33. 1846.
Mesosphaerum rhomboideum Kuntze, Rev. Gen. Pl. 2: 527. 1891.
Hyptis celebica Zoll. & Koorders, Medeleel. van's Lands Plants Buitenzorg 19: 561. 1898, nomen nudum.
Hyptis capitata var. *pilosa* Briq., Ann. Conserv. et Jard. Bot. Genève 2: 244. 1898.
Hyptis capitata var. *mexicana* Briq., loc. cit.
Hyptis decurrens Epling, Rep. Spec. Nov. 34: 120. 1933.

Perennial herb 2 dm to 2 m tall, the stems often slightly decumbent and rooting at nodes, angles moderately hirsute with several-celled, irregularly curved hairs, intermarginal surfaces glabrous, lower internodes shorter than subtending leaves, the

Fig. 99. *Hyptis rhomboidea.*

upper progressively longer, often exceeding leaves; petioles narrowly winged, to 4 cm long, wings gradually ampliate, merging into blades; leaf blades rhomboidally ovate to rhomboid-lanceolate, 3–12 cm long, 1–4 cm wide, acute to attenuate at apex, cuneate at base, coarsely and irregularly serrate, dark green and sparsely hispid above, paler, markedly pellucid-punctate and with more or less appressed, several-celled, coarse hairs along veins beneath, intervenous surfaces glabrous; inflorescences semiglobose, borne on axillary peduncles 1–5 cm long, peduncles hirsute; bracts lanceolate to oblanceolate, 6–12 mm long, acute to acuminate at apex, somewhat more densely hirsute than leaves, entire; flowers crowded in heads, sessile; calyx tube about 1.5–2 mm long, cylindro-campanulate, the lobes erect, linear-subulate, about 2 mm long, antrorsely short-ciliate, equal or nearly so; fruiting calyces coarsely reticulate-veined between ribs; corolla about 4 mm long, white, bearing hairs to 1 mm long on outer surfaces of upper part of tube and lobes, tube about 2 mm long, gradually ampliate upward; upper pair of stamens slightly shorter than lower; nutlets about 1 mm long.

In tropical parts of Mexico and Central America, from Jalisco and Vera Cruz to Nicaragua, and in the Galápagos Islands; also in the Philippines and eastern Taiwan to Borneo. Specimens examined: Isabela, San Cristóbal, Santa Cruz.

This species is closely related to *H. capitata* Jacq., but has somewhat narrower and smaller bracts below the heads of the flowers, and broader and shorter calyx teeth.

See Fig. 99.

Hyptis sidaefolia (L'Hér.) Briq., Ann. Conserv. et Jard. Bot. Genève 2: 204. 1898

Boystropogon sidaefolia L'Hér., Sert. Angl. 19. 1788.
Hyptis polyantha Poit., Ann. Mus. Paris 7: 470. 1806.
Mesosphaerum polyanthum Kuntze, Rev. Gen. Pl. 2: 526. 1891.

Erect perennial, aromatic herb to 1 m tall with moderately pubescent foliage and bearing numerous pellucid-punctate glands on upper stems, leaves, and calyces; petioles 1–2 cm long, pubescent; leaf blades broadly ovate, 4–6 cm long, 3–4 cm wide, slightly attenuate at apex, rounded at base, irregularly serrate, deep green and minutely hirsute on upper surface, silkily pubescent and usually glaucous beneath; peduncles 2–3 cm long, pubescent, axillary to the upper, reduced leaves or leaflike bracts, forming ample panicles, these more or less leafy; flowering calyces purplish, 2–2.5 mm long, velvety-puberulent, the teeth straight, 0.3–0.8 mm long; calyx tube to 3.5 mm long at maturity, orifice then bearded with fine, white hairs; corolla tube about 4 mm long, about twice as long as lobes; nutlets 1.5 mm long, dark brown.

In forests of moderate density and along margins of trails and fields, Colombia, Ecuador, and Peru. Specimen examined: Santa Cruz.

Hyptis spicigera Lam., Encycl. 3: 185. 1789

Nepeta americana Aubl., Hist. Pl. Guiane Fr. 2: 623. 1775.
Hyptis lophanta Mart. ex Benth., Lab. Gen. & Sp. 78. 1833.
Hyptis pohliana Jacq. ex Benth., op. cit. 141.
Hyptis subverticillata Anderss., Kongl. Svensk. Vet.-Akad. Handl. 1853. 197. 1855.
Hyptis gonocephala Wright ex Griseb., Cat. Pl. Carib. 212. 1866.
Mesosphaerum gonocephalum Kuntze, Rev. Gen. Pl. 2: 526. 1891.
Mesosphaerum lophantum Kuntze, loc. cit.
Mesosphaerum spicigerum Kuntze, op. cit. 527.
Mesosphaerum subverticillatum Kuntze, loc. cit.
Hyptis americana Urban, Rep. Spec. Nov. 15: 322. 1918, non Briq. 1897.

Annual herb in understory of forest, the stems slender, erect to ascending, often scrambling through shrubs to 1(3) m; herbage scaberulous with short, slightly retrorsely curved, stoutish hairs along angles of stems and petioles; petioles slender, 0.5–2.5 cm long; leaf blades broadly ovate, 2–8 cm long, about two-thirds as wide as long, deep green above, paler and with pellucid dots beneath, acute to acuminate at apex, broadly cuneate at base, irregularly and coarsely serrate, glabrous or nearly so on both surfaces, except for delicate, minute hairs along veins beneath; flowers rather closely crowded in verticils to form spikes to 1.5 cm broad and 10 cm long; bractlets linear to lance-linear, 3.5–5 mm long in flower, ciliate with white, slender hairs; calyx tube cylindrical to turbinate, 1.5–2 mm long, sparsely hispid, the teeth linear, erect, about as long as tube, fruiting tube to 5 mm long, with white-hirsute orifice; corolla white or pale lavender, 4–5.5 mm long, short-hispid on outer surfaces; nutlets dark brown, about 1.2–1.4 mm long.

Common in shaded forests of moist zones in tropics of the Old and New Worlds;

Mexico and Cuba to Brazil and Peru. Specimens examined: Fernandina, Isabela, San Salvador, Santa María.

Mentha L., Sp. Pl. 576. 1753; Gen. Pl. ed. 5, 250. 1754

Perennial herbs with erect or diffuse stems often rooting at lower nodes, leaves sessile or petiolate, punctate, the herbage pungently fragrant or odorous. Flowers small, in axillary whorls or in congested or interrupted terminal spikes. Calyx campanulate to cylindrical, 10-nerved, regular to slightly 2-lipped, 5-toothed. Corolla tube shorter than calyx, limb bilabiate; upper lip entire to emarginate; lower lip 3-lobed. Stamens 4, equal, erect, included or short-exserted; filaments glabrous; anther sacs oblong to linear, parallel. Ovary deeply 4-parted; style 2-cleft. Nutlets normally 4, ovoid, smooth.

A genus of 15 to 20 species, with a large number of named subspecies and varieties; chiefly native in the Old World but a few species in the New and some species widely distributed as escapes from cultivation or as weeds.

Mentha piperita L., Sp. Pl. 576. 1753

Perennial from slender, leafless stolons or sometimes the exposed parts leafy-bracted; stems slender, erect or decumbent, simple or branched, 3–8 dm tall, glabrous, often reddish or purplish; petioles slender, 3–5 mm long; leaf blades narrowly ovate to lanceolate, 1.5–4 cm long, 5–15 mm wide, acute to acuminate at apex, rounded to broadly cuneate at base, dark green above, paler and conspicuously punctate beneath; interrupted spikes with whorls 10 mm across, whole spike to 12 cm long; bracts narrowly lance-acuminate to linear; calyx tube narrowly campanulate, about 2.5–3 mm long, glabrous, closely gland-dotted, the teeth 1–1.5 mm long, straight, erect; corolla white to lavender or pink, upper lip shallowly emarginate, the lower with 3 equal lobes 1–1.4 mm long; stamens about equaling corolla lobes; styles slender, 1.5–3 mm longer than corolla lobes; nutlets smooth, dark brown, 1–1.3 mm long.

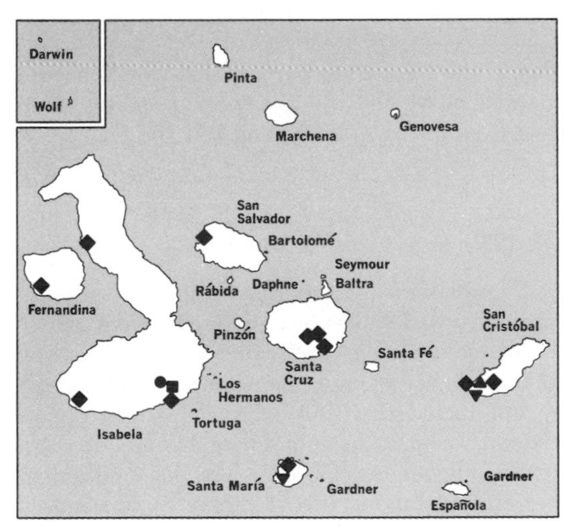

● *Mentha piperita*
■ *Ocimum micranthum*
▲ *Salvia insularum*
◆ *Salvia occidentalis*
▼ *Salvia prostrata*

Fig. 100. *Mentha piperita* (a–c). *Ocimum micranthum* (*d–g*).

Commonly escaped from cultivation in the temperate parts of the world, occurring along roadsides, streams, and margins of woods, and as a weed in poorly kept fields; in the Galápagos Islands known from a single locality on Isla Isabela.
See Fig. 100.

Ocimum L., Sp. Pl. 597. 1753; Gen. Pl. ed. 5, 259. 1754

Erect or ascendingly branched annual herbs with oval, petiolate, glabrate leaves, and flowers borne in verticils to form interrupted spikes or racemes. Calyces markedly bilabiate, upper lobe ovate to rounded, concave, decurrent to base of calyx; lateral teeth acuminate from deltoid base, lower pair similar to lateral but longer and more spinose. Corolla bilabiate, tube included within calyx; 2 lobes of upper lip about same size and shape as laterals, central lobe of lower lip longer than upper, somewhat cucullate. Stamens 4, didynamous, more or less declined along lower lip, slightly exserted, inserted at apex of tube or low in throat. Style slender, slightly shorter than stamens; stigma 2-lobed. Nutlets 4, smooth, ovoid to obovoid.

About 60 species, distributed throughout the tropical and subtropical parts of the Old and New Worlds.

Ocimum micranthum Willd., Enum. Hort. Berol. 630. 1809*

Annual herb 4–6 dm tall, with markedly tetragonal, reddish to purplish, glabrous or glabrate stems; internodes often 10–12 cm long; petioles slender, canaliculate, 2–3 cm long; leaf blades broadly elliptic to ovate, cuneate at base, acute, slightly acuminate, or obtuse at apex, 2–6 cm long, 1–3 cm wide, shallowly dentate, dark green above, much paler beneath, glabrous or sparsely puberulent along veins beneath, closely dotted with punctate glands; inflorescences interrupted racemes, the verticils usually containing 6 flowers, 3 in axil of each fugaceous bract; pedicels puberulent-glandular, about 1–1.5 mm long at anthesis, elongating to 4–5 mm in length and spreading-deflexed in fruit; calyces 2.5–3 mm long at anthesis, glandular-puberulent, the upper lobes oval, margins decurrent to base of tube and curved backward, to 8 mm long in fruit, arcuately curved upward, the 2 lateral lobes about 1 mm long, deltoid at base, acerose-spinose at tip, 2 lower lobes narrower and nearly twice as long as laterals; corollas pink to lavender, 3–4 mm long at anthesis, tube 0.6–0.7 mm long, throat ampliate, about 1.5–2 mm across, upper lip of nearly discrete, oblong lobes about 1.5 mm long, glabrous or essentially so, the central lobe of lower lip somewhat cucullate, slightly larger than laterals; stamens exserted 1–2 mm, declined along lower lip, filaments devoid of basal spurs; style sigmoid-curved in bud, slender, 2 mm long, nearly equaling stamens; stigmatic lobes linear, to 1.5 mm long; nutlets rich brown, smooth, obovoid, 1.5–1.8 mm long.

In waste places and in partially shaded areas, Florida, the West Indies, Mexico, and Central and South America. Specimen examined: Isabela.

Apparently cultivated locally throughout its range and used in the same manner as *O. basilicum*, with which it often is confused. It can be distinguished from *O. basilicum* by its smaller flowers (3–4 mm long in *O. micranthum*, 6–7 mm long in *O. basilicum*), by the absence of pilose spurs at the base of the posterior stamen filaments, and by the well-developed pedicels in contrast to sessile or subsessile flowers in *O. basilicum*.

See Fig. 100.

Salvia L., Sp. Pl. 23. 1753; Gen. Pl. ed. 5, 15. 1754

Annual herbs (rarely), suffrutescent perennials, or long-lived shrubs. Leaves sessile to long-petiolate, blades thin to subcoriaceous, often rugose or bullate, entire to pinnatifid or deeply lobed. Flowers sometimes in axils of leaves and bracts, more frequently in spiciform racemes or spikes, the number of flowers per verticil variable, each flower or verticil subtended by deciduous or caducous bract. Calyx bilobed or 2-lipped; upper lip frequently tridentate or trimucronate, 3–9(13)-veined, the teeth blunt, acute, or mucronate; lower lip usually 2-lobed or bidentate, lobes ovate to broadly lanceolate, acute to mucronate. Corolla blue, red, white, or with tints of blue, lavender, rose, or purple, the tube cylindric, straight or curved, often ven-

* A listing of synonyms is impractical owing to the lack of a modern monographic treatment of the genus.

tricose beneath lower lip, naked or papillate, rarely rugose transversely within, frequently somewhat invaginate near base, the upper corolla lip erect or nearly so, more or less galeate, the lower 3-lobed, its middle lobe largest, often emarginate. Stamens 2, ascending within the galeate upper lip, included or exserted, the connective between anther cells elongate, upper part dilated, thecae of anther distinctly separated by connective, lower theca often abortive. Style pilose to glabrous, branches equal or unequal. Gynobase incrassate, upper part dilated. Nutlets ovoid, hilum basal or sub-basal.

A large genus of nearly worldwide distribution. The subgenus *Calosphace* alone, which is confined to the New World, consists of 468 species, according to Epling; all of our representatives belong in this subgenus.

Stamens included within galeate apex of upper corolla lip, not exserted; calyx 2–3 mm long .... *S. occidentalis*
Stamens ascending within galeate upper lip, but exserted from it; calyx 4–5 mm long:
  Plants prostrate-creeping; corolla tube 2.5–3 mm long........ *S. prostrata*
  Plants erect, not prostrate nor creeping; corolla tube 3.5–6 mm long:
    Leaf blades densely and minutely white-tomentose beneath; upper calyx lip 2 mm long .... *S. insularum*
    Leaf blades sparsely cinereous-pubescent beneath; upper calyx lip barely 1 mm long .... *S. pseudoserotina*

Salvia insularum Epling, Rep. Spec. Nov. Beih. 85: 66. 1935

Ascendingly branched herb with hirsute pubescence on stems and leaves, pubescence especially apparent along angles of stems; petioles 1–2 cm long; leaf blades deltoid-ovate, rounded to obtuse at apex, subtruncate at base, crenate, 2–6 cm long, 1.5–4.5 cm wide, upper surface glabrous or nearly so, lower surface minutely and densely white-tomentose; flowers 2–25 in each verticil, the verticils 2–3 cm apart in lower part of inflorescence, forming dense spiciform raceme above; bracts ovate-acuminate, 2–4 mm long, caducous; pedicels 3–4 mm long, hirtellous; flowering calyces about 4 mm long, finely hirsute and with scattered sessile glands among the nonglandular hairs, upper lip subtruncate, short-mucronate, about 2 mm long; teeth of lower lip about 1.5 mm long, acuminate; corolla tube about 4.5 mm long, upper lip erect, 2 mm long, the lower suborbicular, 5–6 mm broad, blue or lavender; mature fruit unknown.

Endemic; known only from Isla San Cristóbal.

Salvia occidentalis Sw., Prodr. Veg. Ind. Occ. 14. 1788

Salvia procumbens R. & P., Fl. Peruv. 1: 27. 1798.
Salvia martinicensis Sesse & Moç., Fl. Mex. ed. 2, 9. 1894.

Diffuse annual or short-lived perennial herb with upright, moderately to profusely branched stems 1–3 dm tall, some branches sometimes procumbent in age; herbage hirtellous with crisped or curled, simple hairs, these densest along angles of stems and undersurfaces of leaves, or herbage at length glabrate, the gland-tipped hairs interspersed with nonglandular hairs in inflorescences; petioles slender, 3–10 mm long, wingless; blades ovate to rhombic, 2–3(6) cm long, 1–2.5(4) cm wide, acute at apex, narrowly cuneate at base, crenate-serrate from about middle to apex, sparsely appressed-hirsute and dark green above, paler and hirtellous beneath, both

Fig. 101. *Salvia occidentalis* (*a, b*). *Teucrium vesicarium* (*c–e*).

surfaces sometimes glabrate in age; flowers 2–6 in a verticil, these about 5–15 mm apart and in slender, interrupted racemes 10–30 cm long; bracts ovate to broadly ovate, 1.5–2.5 mm long, abruptly acuminate; calyx ovate-cylindric to campanulate, 2–3 mm long, closely glandular-pilose without, the upper lip obscurely trimucronulate, nearly truncate, the lower with 2 ovate-acuminate, sharp teeth 0.4–0.6 mm long at anthesis to 3.5 mm at maturity; pedicels slender, 1.5–2 mm long; corolla 5–6 mm long, pale blue, lavender, or nearly white, the tube about 2.5 mm long, concave-galeate, erect; lower lip 2.5 mm long, spreading, lateral lobes about half as long as middle lobe, margins of all lobes sparsely ciliate-pilose; nutlets ellipsoid, about 1 mm wide, 1.5–1.7 mm long, pale brown to tawny, glabrous, faintly granular, rarely more than 1 per flower maturing, others abortive.

In moist, shaded places, central Florida, most of Mexico, and Central America to Bolivia; common in the Galápagos Islands. Specimens examined: Fernandina, Isabela, San Cristóbal, San Salvador, Santa Cruz, Santa María.

See Fig. 101.

Salvia prostrata Hook. f., Trans. Linn. Soc. Lond. 20: 200. 1847

Small prostrate herb with slender, sparsely hirsute stems occasionally rooting at nodes; petioles 8–10 mm long, hirtellous; leaf blades deltoid-ovate, truncate at base, obtusely rounded at apex, 6–10 mm broad on lateral branches, to 4.5 cm broad on more vigorously growing stems, convex, crenate, sparsely hirtellous on both surfaces; bracts linear, 1.5–2.5 mm long; flowers solitary, in pairs, or in verticils of 5–6, these usually congested into short, hirsute racemes about 1.5–3 cm long toward tips of inflorescences; flowering calyces about 4 mm long, upper lip slightly larger than lower, both slightly flaring, moderately puberulent, especially along veins, gland-dotted between nonglandular hairs, to 5 mm long at maturity; corolla about 5–6 mm longer than flowering calyx; fruit unknown.

Endemic. Specimens examined: San Cristóbal, Santa María.

Salvia pseudoserotina Epling, Rep. Spec. Nov. Beih. 85: 67. 1935

Salvia floreana Howell, Proc. Calif. Acad. Sci. IV, 21: 332. 1935.

Perennial herb with crispate-pubescent, decumbent stems; petioles about 1 cm long, ascending to spreading, slender; leaf blades broadly ovate to suborbicular, obtuse to rounded at apex, rounded to subtruncate at base, serrulate and marginally somewhat undulate from near base (basal leaves often practically entire), dark green, rugulose, subglabrous but with subsessile golden glands on upper surface, 1.5–4 cm long, 1.5–3 cm wide, sparsely cinereous-tomentose beneath, veins rather prominent; flowers 2–4 per verticil, these in interrupted spiciform racemes 3–6 cm long; bracts lanceolate, 1–2 mm long, caducous; pedicels 1–2 mm long, ascending; flowering calyces 4–4.5 mm long, 13-veined, short-pubescent and with many stipitate glands on outer surface, especially on veins, tube campanulate, 2-lipped, upper lip rounded, about 1 mm long, lower of 2 equal ovate-attenuate lobes 1–1.5 mm long, margins of teeth sublanate with short, straight, white hairs near tips; corolla tube 3–3.5 mm long; upper corolla lip about 1.5 mm long, erect, lower corolla lip 2.5–3 mm long, ascending-spreading, pale lavender to nearly white; mature fruit unknown.

Endemic. Specimens examined: Santa María.

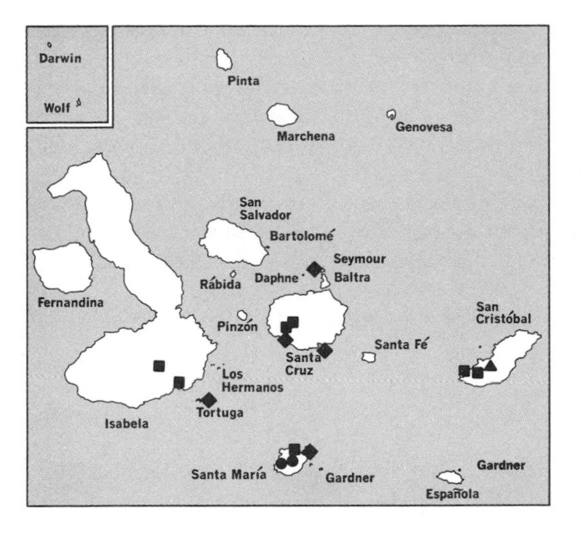

● *Salvia pseudoserotina*
■ *Teucrium vesicarium*
▲ *Utricularia foliosa*
◆ *Nolana galapagensis*

Teucrium L., Gen. Pl. ed. 5, 247. 1754

Perennial herbs from creeping rootstocks, suffrutescent plants, or rarely annual herbs. Leaves simple, lobed, or parted, petiolate or subsessile. Flowers solitary in axils of modified leaves or inflorescence or in bracteate, terminal spikes or racemes. Pedicels short and slender, or lacking. Calyx campanulate, 10-veined, 5-toothed, the teeth lance-deltoid, subequal, 2–3 times as long as tube, or unequal and shorter than tube, which then is cuplike or saccate. Corolla with short tube, bilabiate, the middle lobe of lower lip much larger than laterals; upper lip inconspicuous, 2-lobed, lobes about same size as lateral lobes of lower lip. Stamens 4, in pairs, anterior pair the longer, all inserted near middle or top of corolla tube, exserted from between 2 lobes of upper lip. Style 2-lobed, exserted with stamens. Nutlets oval, sculptured or smooth, pubescent or glabrous.

About 100 species, worldwide in distribution, in temperate and tropical zones at low or moderate altitudes.

Teucrium vesicarium Mill., Gard. Dict. ed. 8, under Teucrio no. 17. 1768

Teucrium inflatum Sav., Prodr. Veg. Ind. Occ. 88. 1797.
Teucrium palustre Kunth in HBK., Nov. Gen. & Sp. Pl. 2: 206. 1817, non Lam. 1778.
?*Teucrium hirtum* Willd. ex Spreng., Syst. 2: 710. 1825.
?*Teucrium carthaginense* Lange, Vidensk. Medd. Kjob. 1881: 97. 1881.
Teucrium picardae Krug & Urban in Urban, Symb. Antill. 1: 396. 1899.
Teucrium mollifolium Larranaga, Escrito D. A. Larranaga 1: 24. 1922.
Teucrium tenuipes Epling, Ann. Mo. Bot. Gard. 12: 119. 1925.
Teucrium vesicarium var. *palustre* Epling, Rep. Spec. Nov. Beih. 85: 3. 1935.

Perennial herb from creeping rootstock and erect stem 2–12 dm tall, branching almost entirely above, simple below, all herbage glandular-hirsute or the older stems glabrate, brownish; petioles slender, 0.5–3.5 cm long; leaf blades ovate, 4–12 cm long, 1–6 cm wide, acute at apex, rounded to subtruncate at base, crenate-serrate, upper surface dark green, hirtellous to glabrate, lower surface paler, with soft, curled hairs or less commonly glabrate; inflorescences terminal, spiciform-racemose, 2–15 cm long, 1.5–2 cm broad; bracts linear to lance-linear, 4–6 mm long, hirtellous; pedicels filiform, 1–3 mm long, or flowers subsessile; calyx 5–8 mm long at anthesis, pubescent with short, curled hairs, the teeth deltoid, 2–3 mm long, upper 3 more or less connate, obtuse, lower 2 acute to acuminate; corolla 9–13 mm long, finely glandular-hirsute without and on inner surface of lower lip, pale lilac to nearly white; stamens and style exserted 3–5 mm; fruiting calyx nearly globose, 4–5 mm in diameter; nutlets reddish brown, coarsely reticulate-foveolate, 2.5–3 mm long, glabrous.

In moist woods and shaded hillsides, around margins of ponds, and in swampy areas, central Mexico and the West Indies south to Argentina; occasional in the Galápagos Islands. Specimens examined: Isabela, San Cristóbal, Santa Cruz, Santa María.

Fruiting plants superficially resemble *Priva lappulacea*, their fruits being about the same shape and size.

See Fig. 101.

LENTIBULARIACEAE. Bladderwort Family*

Annual or perennial herbs, aquatic (then often without roots), or plants of wet habitats and usually bearing roots, nearly always insectivorous. Leaves in basal rosettes at bases of flowering stems, or alternate, nearly always dimorphic in submerged species, the submerged leaves finely divided and bearing insectivorous bladders of complex structure, aerial leaves appearing on aquatic representatives only in connection with flowering, then forming a floating rosette or reduced to small bracts along scape, or wholly absent, dimorphism in terrestrial species represented by tubular or pitcher-shaped, insectivorous leaves. Flowers borne on erect or ascending scape, or racemose, or solitary at tips of scapes, bracteate, pedicellate (pedicels of some bracteolate). Flowers perfect, irregular. Calyx 2–5-lobed or -parted, the segments spreading or imbricate. Corolla sympetalous, 5-lobed, imbricate in bud, bilabiate, lower lip saccate or spurred, often personate, with prominent palate. Stamens 2, attached to extreme base of corolla tube (2 staminodes also present in some); anthers 1-celled. Pistil 1; style 1 or obsolete; stigma sessile, 2-lobed. Ovary superior, 1-celled. Carpels 2; placenta central and free. Ovules many (2 in *Biovularia*). Fruit capsular, dehiscing by valves, or circumscissile, or splitting irregularly. Seeds minute, more or less peltate, often winged. Embryo poorly differentiated. Endosperm none.

A family of 5 genera and more than 300 species, widely distributed on all continents.

Utricularia L., Sp. Pl. 18. 1753; Gen. Pl. ed. 5, 11. 1754

Leaves alternate, submerged, 2–8 parted and finely divided, dichotomously or pinnately dissected, at least some divisions bearing insectivorous bladders, each bladder with pair of finely divided bristles at rim of mouth. Flowers in racemes or sometimes solitary at tips of scapes. Calyx 2-lobed, the lobes concave, persistent. Corolla 2-lipped, with prominent palate at base of lower lip, this often 2-lobed. Seeds more or less peltate, margins sometimes winged.

Over 200 species, widely distributed from the tropical to the arctic zones.

* Contributed by Ira L. Wiggins.

Utricularia foliosa L., Sp. Pl. 18. 1753

Aquatic herb with free-floating, much-branched stem, with leaves repeatedly divided into filiform segments, and bearing numerous ovoid bladders about 1.5 mm long; flowering peduncles 5–8 cm tall, these gradually narrowing toward base, bearing terminal 5–10-flowered raceme; bracts ovoid, obtuse, about 2 cm long, purplish; pedicels 5–8 mm long, reflexed or nodding in fruit; calyx 2-lipped, each lip about 3 mm long, 2.5–3 mm wide, striate, broadly rounded at apex, purplish; corolla yellow, 7–8 mm long, upper lip orbicular, 3–4 mm broad, the lower broadly deltoid, about 8 mm long, 5 mm wide, finely striate; spur 3.5–4 mm long, conical; capsule globose, 3–5 mm in diameter; seeds lenticular, about 2 mm in diameter, with thin, firm, narrow wing encircling body.

Widely distributed from the southeastern United States, the Antilles, and the West Indies to Central America, tropical South America, and tropical Africa. Specimens examined: San Cristóbal, the only island in the Galápagos where this species is known.

Lanier, who collected a specimen in 1932, was not a trained botanist; perhaps, had he been familiar with the unusual life history of the plant, he might have searched diligently enough to have secured flowers or fruits. All of his material, as well as that obtained by Dr. Colinvaux in the summer of 1966, consisted of sterile, submerged leaves with their attached, characteristic bladders. The plant was col-

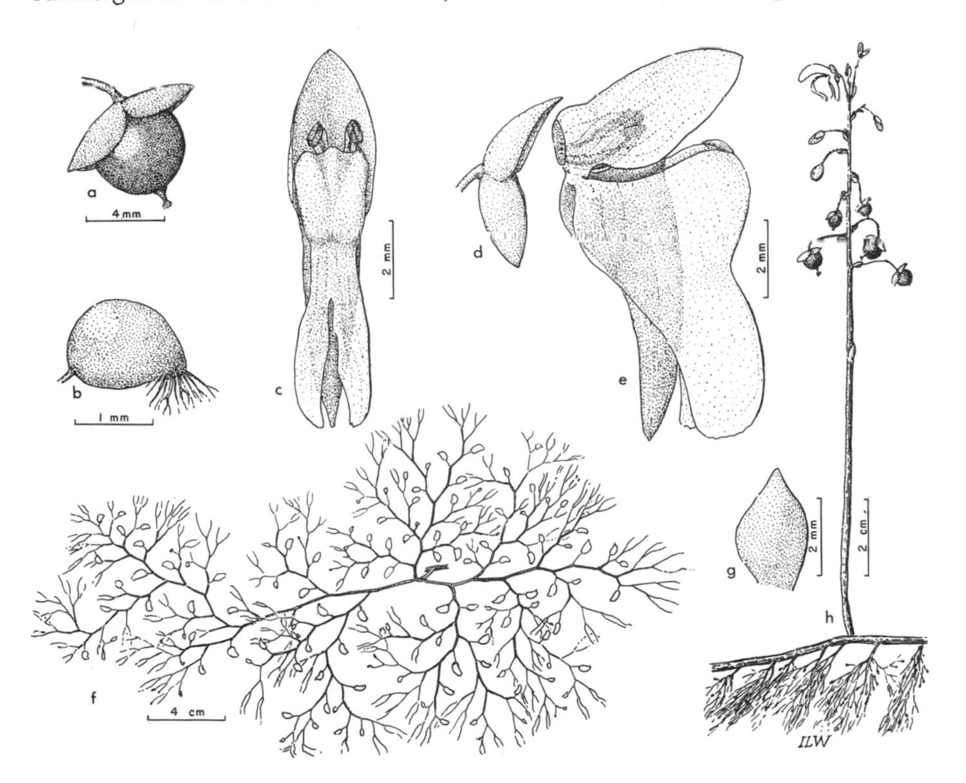

Fig. 102. *Utricularia foliosa*: *b*, float bladder from leaf; *c*, flower, face view; *d, e*, flower somewhat separated, side view; *g*, sepal.

lected at the same locality, in the same month (April) as Lanier's collection, and reported by Dr. and Mrs. Eliasson (Svensk. Bot. Tidskr. 62: 245–46. 1968), the first time the plant had been obtained in flower in the Galápagos Islands and the first published report of its presence there.

See Fig. 102.

NOLANACEAE. Nolana Family*

Herbs or shrubs, often with succulent or subfleshy leaves borne in crowded clusters on short spur branches. Leaves small and narrow, exstipulate. Flowers perfect, regular, solitary in leaf axils, but often appearing glomerate owing to crowding of leaves. Calyx 5-cleft or 5-lobed. Corolla sympetalous, plicate in bud, 5-lobed, hypogynous, campanulate to funnelform. Stamens 5, equal or slightly unequal, alternate with corolla lobes, inserted near base of corolla tube; anthers 2-celled, longitudinally dehiscent. Disk at base of ovary evident, often more or less lobed. Pistil solitary, superior, often radially or transversely lobed, usually 5-carpelled, each carpel unilocular or 2- or 3-locular by formation of transverse partitions. Style 1, terminal, capitate; stigma shallowly 2- to several-lobed. Fruit a 1- to several-seeded mericarp, with bony endocarp. Seeds small, with curved embryo and copious endosperm.

Two genera and 60–65 species, most abundantly represented in Chile and Peru.

Nolana L., Sp. Pl. ed. 2, 1: 202. 1762; Gen. Pl. ed. 6, 79. 1764

Plants often woody. Leaves entire, more or less succulent. Calyx symmetrical or nearly so, teeth broadest at base, or calyx asymmetrical and teeth broader above middle than at base. Corolla small and funnelform to ample (to 4 cm broad), obviously 5-lobed, or by secondary notching appearing 10-lobed, white to light blue. Fruit deeply lobed, of 3 to many, few- to many-seeded, distinct mericarps (nutlets) attached only to receptacle and free from one another or essentially so.

About 60 species, found on well-drained, sandy soils from sea level to elevations of 4,000 m.

Nolana galapagensis (Christoph.) Johnston, Contr. Gray Herb. 112: 32. 1936

Periloba galapagensis Christoph., Nyt Mag. Naturvid. 70: 89, *figs. F–H.* 1932.

Weak shrub with soft-ligneous stems 0.5–1.5 m tall, with numerous short spur branches, the younger growth closely and minutely puberulent with short, simple, erect hairs; leaves spatulate, entire, 5–35 mm long, 3–6 mm wide near apex, basal half of each leaf petiole-like, 1–1.5 mm wide, whole leaf succulent, puberulent; flowers axillary on spur branches or near apex of main branches; pedicels slender, closely glandular-puberulent, 1–6 mm long; calyx 2.5–3 mm wide, cup 2–3 mm deep, lobes ovate, about equaling cup, acute, glandular-puberulent; corollas tubular-funnelform, 12–20 mm long, 6–10 mm wide across limb, white, sparsely to densely glandular-perbulent without, the lobes deltoid, about 2 mm long and wide,

* Contributed by Ira L. Wiggins. For further information on the family, see Johnston (1936).

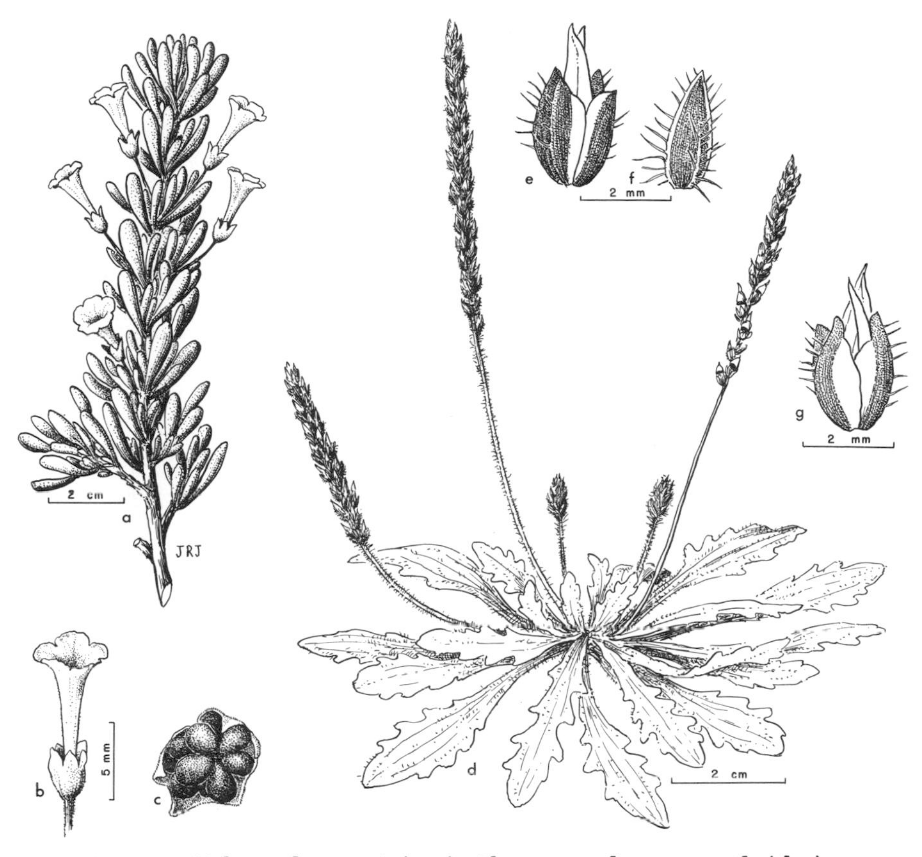

Fig. 103. *Nolana galapagensis* (*a–c*). *Plantago paralias* var. *pumila* (*d–g*).

glabrous within; stamens slightly unequal, 3 about equaling corolla tube, others slightly shorter; mericarps 15–25, in 2–3 series, 8–20 small ones 1–2 mm long below middle of gynobase, 3–5 larger ones 2–3 mm long above; seeds black, conspicuously favose-reticulate.

Endemic; near the sea on sandy or rocky areas and on islands of low relief. Specimens examined: Santa Cruz, Santa María, Seymour, Tortuga.

See Fig. 103.

PLANTAGINACEAE. Plantain Family*

Annual or perennial herbs or rarely suffrutescent subshrubs with basal, usually alternate leaves. Stipules none. Inflorescences scapose, spicate, or capitate, bracteate. Flowers regular, usually perfect, 4-merous. Calyx tubular and 4-toothed or deeply 4-divided, green or membranous. Corolla scarious, typically 4-lobed, lobes imbricate in bud, spreading or erect after anthesis. Stamens 4 (sometimes 1 or 2),

* Contributed by Ira L. Wiggins.

alternate with corolla lobes, inserted on corolla tube, exserted (sometimes shorter than corolla lobes); anthers 2-celled, versatile, comparatively large, dehiscent longitudinally. Pistil 1. Ovary superior, usually 2-lobed, 2-carpeled. Style 1, filiform, bifid. Ovules 1 to several in each cell. Fruit a circumscissile capsule or bony nutlet. Seeds with fleshy endosperm and small, straight embryo.

About 4 genera and several hundred species, *Plantago* being nearly cosmopolitan in all save arctic regions.

Plantago L., Sp. Pl. 112. 1753; Gen. Pl. ed. 5, 52. 1754

Characters of the family. Inflorescence spicate, slender. Calyx 4-parted to base or essentially so, anterior sepals slightly shorter than the posterior and asymmetrical, the wing on abaxial side much narrower than other. Seeds strongly mucilaginous when moistened.

Widely distributed from tropical to cool temperate zones.

Plant perennial; leaves broadly ovate, blades 5–25 cm long, 2–12 cm wide; seeds 6–18 in each capsule, minutely reticulate-favose, hilum face convex . . . . *P. major*

Plant annual; leaves elliptic to lanceolate, 2–10 cm long, 8–15 mm wide; seeds 2 in each capsule, minutely pebbled, not reticulate, hilum face plane . . . . *P. paralias* var. *pumila*

Plantago major L., Sp. Pl. 112. 1753

Plantago officinarum Crantz, Inst. Rei Herb. 2: 329. 1766.
Plantago latifolia Salisb., Prodr. Stirp. 46. 1796.

Perennial, sparsely puberulent to glabrous, acaulescent herb with deep, tough taproot and horizontally spreading leaves in basal rosette; petioles broad and thin, 2–25 cm long, channeled on upper side; leaf blades broadly ovate, mostly broadly cuneate to subtruncate at base, obtuse to rounded at apex, 5–25 cm long, 2–12 cm wide, mostly with 5 longitudinal veins, undulate or irregularly toothed; scapes erect to decumbent, 6–40 cm tall, sometimes with ovate to elliptic bracts 2–3.5 cm long at base of spikes; spikes dense, 5–6 mm broad and 5–35 cm long at anthesis, to 10 mm broad in fruit; bracts ovate to lance-ovate, 2–4 mm long; sepals broadly

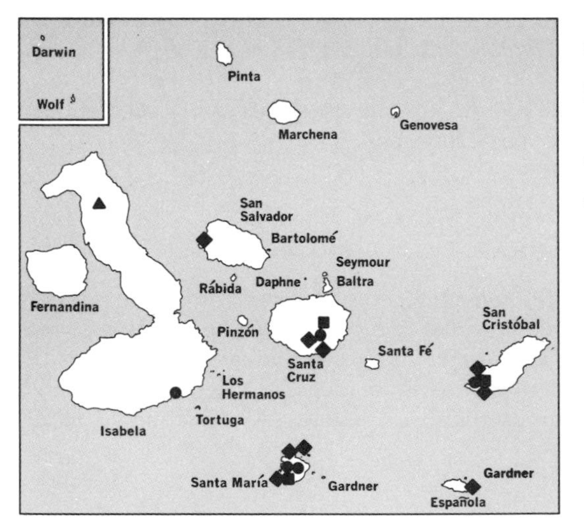

● *Plantago major*
■ *Plantago paralias* var. *pumila*
▲ *Plumbago coerulea*
◆ *Plumbago scandens*

ovate, 2–2.5 mm long; corolla lobes 1–1.5 mm long, reflexed, acute to slightly apiculate, margins involute; stamens well exserted; capsule broadly ovoid to subglobose, 2–2.5 mm broad, to 4 mm long, brownish, circumscissile near middle, 6–18-seeded; seeds red-brown, 1–1.2 mm long, minutely foveolate-reticulate, the inner (hilum) face distinctly convex.

A common and very widely distributed weed in fields, along roadsides and in disturbed soil, from the tropics to cool temperate regions; a native of Europe, occasional in the Galápagos Islands. Specimens examined: Isabela, San Cristóbal, Santa Cruz, Santa María.

Plantago paralias Decne. var. **pumila** (Hook. f.) Wiggins, Madroño 20: 252. 1970

Plantago tomentosa Lam. var. *pumila* Hook. f., Trans. Linn. Soc. Lond. 20: 194. 1847.

Small annual from slender taproot 3–10 cm long; leaves in basal rosette, these flat on ground to somewhat ascending, elliptic-lanceolate to oblanceolate, 2–12 cm long, 4–10 mm wide, acute to obtuse at apex, gradually narrowed toward base, entire or with 3–5 pairs of repand teeth 1–2 mm long, these rounded or blunt, nearly as wide as long, whole blade moderately hispid with coarse, spreading, white hairs, or pubescence mainly along margins, the rest subglabrate; inflorescences straight and erect to arcuate-ascending or sometimes scape decumbent, scapes hispid, 2–12 cm long; spikes slender, laxly flowered, 1.5–4 cm long, about 5 mm broad; bracts narrowly ovate, 0.6–1 mm wide, 2.5–2.8 mm long, carinate, marginal scarious or hyaline part 0.1–0.2 mm wide, ciliate, carina coarsely but sparsely hispid; anterior sepals ovate, 2–2.4 mm long, wider wing 0.4–0.5 mm wide, narrower almost lacking; posterior pair of sepals slightly longer than anterior, broadly and symmetrically ovate, acute at apex, rounded below; corolla 3–4 mm long, tube tawny, 1.5–1.8 mm long, lobes narrowly ovate, acute, erect with overlapping margins, the flower thus appearing closed even at anthesis; stamens 2–2.2 mm long, anthers 0.8–1 mm long; capsules narrowly ovoid, about 2.5 mm long, brownish, shiny, finely reticulate without, dehiscing slightly below middle; seeds reddish brown, broadly ellipsoidal, plano-convex, 1.8–2.2 mm long, 1–1.2 mm wide, markedly mucilaginous when moistened.

Endemic; in shrubby zones, mostly along trails and in openings in the cover; relatively uncommon in the Galápagos Islands. Specimens examined: San Cristóbal, Santa Cruz, Santa María.

See Fig. 103.

PLUMBAGINACEAE. PLUMBAGO FAMILY*

Perennial herbs or suffrutescent plants, or sometimes vinelike, with basal or alternate, simple, entire leaves. Flowers bracteate, regular, perfect, 5-merous, in panicles or spikelike racemes, in globose heads, or in racemes. Calyx tubular to funnelform, 5-lobed, often 5- or 10-ribbed. Corolla sympetalous, tubular or barely united at base. Stamens opposite corolla lobes and adnate to tube or base of claw or nearly free. Ovary superior, 1-celled, 1-ovuled, often 5-angled near apex; styles 5. Fruit usually included within withering calyx, indehiscent, circumscissile or valvate. Seed 1, pendulous.

* Contributed by Ira L. Wiggins.

Ten genera and about 300 species, especially well represented in the Mediterranean and central Asiatic regions; less common in the New World but present from Alaska to South America. A few species are sometimes cultivated as ornamentals.

Plumbago L., Sp. Pl. 151. 1753; Gen. Pl. ed. 5, 75. 1754

Perennial herbs or suffrutescent plants with slender, often scandent, branches and petiolate or auriculate-amplexicaule leaves. Flowers in paniculate spikes. Calyx tubular, stipitate-glandular, 5-lobed, plicate and with scarious or subscarious sinuses and margins near apex. Corolla salverform; tube slender; limb 5-lobed, spreading or rotate. Stamens inserted at base of corolla tube and free. Capsule membranous, circumscissile near base.

Corolla violet-blue, lobes acute; corolla tube 8–10 mm long; calyx glabrous on lower half; branchlets and rachises of inflorescences stipitate-glandular. . . . . *P. coerulea*
Corolla white or tinged with purple or lavender, lobes rounded and mucronulate; corolla tube 2–2.5 cm long; calyx stipitate-glandular full length; branchlets and rachises of inflorescences glabrous . . . . *P. scandens*

Plumbago coerulea HBK., Nov. Gen. & Sp. Pl. 2: 220. 1817

Stems procumbent with ascending branches, sometimes climbing into and over shrubs; lower parts glabrous; branchlets copiously stipitate-glandular, somewhat viscid; leaves sessile or subsessile, elliptic, acuminate, attenuate at base, 3–7 cm long, 1–3 cm wide, glabrous; flowers on bracteate spikes, lowermost sometimes short-pedicellate; rachis stipitate-glandular; each flower subtended by ovate, glandular bract 3–4 mm long and about 2 mm wide and by 2 lateral, lanceolate bractlets 1.5–2 mm long, 0.6–1 mm wide, glabrous or bearing a few minute stipitate glands; calyx tubular, 6–7 mm long, about 2 mm broad, deeply 5-lobed, the distal half stipitate-glandular, basal half glabrous, margins of lobes hyaline; corolla violet-blue, the tube 8–10 mm long, lobes acute; capsules thin, circumscissile near the base.

At scattered localities on the mainland of South America; in the Galápagos known only from the rim of Volcán Darwin, Isla Isabela.

Plumbago scandens L., Sp. Pl. ed. 2, 215. 1762

Decumbent to scandent, often straggly plant with slender, weak stems 1–6 m long; herbage glabrous except in the stipitate-glandular inflorescences, often deeply tinged with red when young; leaves elliptic to lance-elliptic, oblong or ovate, acute to acuminate, to 10 cm long, to 4 cm wide, acute basally, narrowed to short petiole 3–18 mm long; bracts lance-ovate, attenuate, 3–6 mm long; flowers sessile or subsessile, in elongated, many-flowered spikes paniculately arranged; calyx 8–12 mm long at anthesis, nerves of lobes closely beset with conspicuous, stipitate glands to 1.5 mm long; corolla white or faintly tinged with lilac or rose, 2–2.5 cm long, the tube very slender, lobes 6–10 mm long, rounded and minutely mucronulate; anthers exserted 1–2 mm, bluish; capsule ovoid, 5–7 mm long, membranous.

Generally distributed in tropical and subtropical areas from Arizona and Florida to northern South America; common in brushy and forested areas from sea level to

Fig. 104. *Plumbago scandens.*

altitudes of 615 m or more on many of the islands in the archipelago. Specimens examined: Champion, Española, San Cristóbal, San Salvador, Santa Cruz, Santa María.

See Fig. 104.

RUBIACEAE. Coffee Family*

Herbs, shrubs, or trees with stipules caducous, deciduous, or persistent, usually connate to form single structure between bases of adjacent leaves of pair or connate with petioles to form sheath at node. Leaves opposite or verticillate, petiolate or sessile, entire or rarely pinnatifid. Flowers minute to large and showy, perfect or rarely unisexual, usually regular. Calyx tube adnate to ovary (forming hypanthium), calyx lobes 4 or 5, varying from small teeth to well-developed lobes, these equal, subequal, or markedly unequal with 1 lobe enlarged and foliaceous, or in a few genera the calyx teeth or lobes obsolete. Corollas gamopetalous, variable in form. Stamens as many as corolla lobes and alternate with them, inserted in tube or throat of corolla; anthers 2-celled. Ovary inferior, 1- to 10-celled; style short to moder-

* Contributed by Ira L. Wiggins.

ately elongate, included or exserted; stigma capitate to bilobed. Fruit 1- to 10-celled, baccate, drupaceous, capsular, or of mericarps, dehiscing in various ways when capsular. Seeds 1 to many, small or large, sometimes winged, longitudinally ridged or finely sculptured, usually flat or concave on inner face.

Nearly 400 genera (many monotypic) and perhaps 5,000 species, generally throughout the tropics and well into the temperate or arctic regions.

Fruits dry (sometimes fleshy in *Galium*); seeds minutely foveolate or transversely rugulose-cerebriform:
  Stipules resembling leaves, and with them forming whorl at each node, not tubular, cupulate, nor fimbriate; sepals obsolete . . . . *Galium*
  Stipules forming a tubular to cupulate, fimbriate sheath, not leaflike; sepals present, often persisting in fruit; leaves opposite; fruits consisting of cocci:
    Cocci indehiscent, separating from one another and retaining their respective seeds when mature . . . . *Diodia*
    Cocci, or 1 of them, dehiscent and releasing their seeds at maturity:
      Cocci alike, both dehiscent and shedding their seeds; each coccus surmounted by 2 sepals . . . . *Borreria*
      Cocci different, 1 open on septal face and shedding its seed, the other remaining closed by persistence of septum; open coccus normally bearing 1 sepal, closed coccus surmounted by 3 sepals . . . . *Spermacoce*
Fruits somewhat fleshy or juicy, not dry; seeds smooth or longitudinally costate, not foveolate or transversely rugulose:
  Flesh of fruit white at maturity, lenticular, compressed contrary to septum; corolla narrowly funnelform . . . . *Chiococca*
  Flesh of fruit red to black at maturity, fruit globose to ellipsoidal, not compressed laterally; corolla broadly funnelform to broadly tubular-salverform:
    Corolla lobes contorted in bud, 1–2 cm long; carpels 2, the nutlets flattened on inner side, rounded and smooth on outer face . . . . *Coffea*
    Corolla lobes valvate in bud, not contorted, 3–6 mm long; carpels 2–5, longitudinally 2- to 5-sulcate on outer surface . . . . *Psychotria*

Borreria G. F. Mey., Prim. Fl. Esseq. 79, *t. 1.* 1818

Low annual or perennial herbs or suffrutescent plants with prostrate to erect stems and branches. Older stems of perennials usually terete, those of annuals and younger parts of suffrutescent plants tetragonal. Stipules connate with petioles into setiform sheath. Leaves opposite, often with smaller, crowded leaves in axils on short spurs of some to many primary leaves, all leaves plicate-nerved, the lateral veins rarely evident. Flowers small to minute, sessile in leaf axils or in dense terminal heads; sepals 2 or 4, equal, subequal, or unequal, often with minute interposed teeth. Corolla funnelform, 4-lobed, lobes slightly to strongly spreading, often inflexed at apex; throat usually more or less barbate with stoutish, moniliform, white hairs. Stamens 4, inserted in throat of corolla, alternate with corolla lobes; anthers ovoid to lance-ovoid, dorsifixed, included or partially exserted. Style slender, usually about equal to corolla tube; stigma slightly bilobed. Fruit coriaceous or membranous, 2-celled, cells septicidal. Seeds 1 in each cell, brown or black, oblong, rounded on outer face, broadly and shallowly excavate-sulcate on inner face, foveolate or transversely rugulose-sulcate.

About 100 species, widely distributed, known from the tropics and subtropics of the Old and New Worlds; at least 11 generic synonyms are given by some authors.

Plant herbaceous, mostly annual; leaves ample, ovate to broadly elliptic, to 3 cm wide, 2–7 cm long, lateral veins evident; seeds transversely rugulose-sulcate . . . . *B. laevis*
Plant perennial (sometimes flowering as seedlings); leaves small, coriaceous, rarely over 3 mm wide or 25 mm long, lateral veins not evident; seeds minutely foveolate, not transversely sulcate:
Leaves nearly as wide as long, 3–4 mm wide . . . . *B. rotundifolia*
Leaves usually much longer than wide, linear to narrowly ovate:
Body of fruit subspherical, about 1 mm long; seeds reddish brown, 0.8–1 mm long; stems and branches wholly prostrate; plant herbaceous or only faintly lignescent at base, not erect nor strongly ascending . . . . *B. perpusilla*
Body of fruit obovoid to oblong-obovoid, 2–2.4 mm long; seeds black, 1.6–2 mm long; branches erect or ascending, woody at least at base in mature plants, not prostrate:
Primary leaves mostly elliptic to ovate, wider at or near base or middle than above; axillary fascicles of leaves usually not conspicuous:
Ovary and mature fruit glabrous or essentially so; leaves usually paler beneath than above, glabrous to sparingly pubescent, larger ones 10–18 mm long; plants often flowering as seedlings . . . . *B. dispersa*
Ovary and mature fruits hispid with coarse, spreading to subappressed hairs; leaves rarely paler beneath than above, densely pubescent beneath, largest ones to 2.5 cm long; plants rarely flowering as seedlings . . . . . . *B. suberecta*
Primary leaves linear to narrowly linear-lanceolate, usually nearly same width their full length; fascicles of small leaves crowded in axils, conspicuous:
Leaves markedly mucronulate, 3–6 mm long, nearly glabrous; internodes usually less than 1 cm long; sepals strongly mucronulate . . . . *B. ericaefolia*
Leaves usually merely acute, to 25 mm long, rarely mucronulate, usually more or less pubescent, especially beneath; internodes 10–30 mm long; sepals acute, rarely mucronulate . . . . *B. linearifolia*

Borreria dispersa Hook. f., Trans. Linn. Soc. Lond. 20: 217. 1847

Borreria basalis Anderss., Kongl. Svensk. Vet.-Akad. Handl. 1853: 191. 1855.
Borreria ovalis Anderss., op. cit. 192.
Borreria baurii Robins. & Greenm., Amer. Jour. Sci. III, 50: 140. 1895.
Spermacoce baurii Robins. & Greenm., op. cit. 141, nomen nudum.
Borreria pacifica Robins. & Greenm., op. cit. 140.
Spermacoce pacifica Robins. & Greenm., op. cit. 141, nomen nudum.
Borreria ovalis f. *abingdonensis* Robins., Proc. Amer. Acad. 38: 205. 1902.

Small shrub or suffrutescent plant 1–6 dm tall, often flowering first season after germination, the seedlings moderately hirsute on herbage and inflorescences; branches lignescent in well-established plants, numerous, erect or ascending; internodes to 5 cm long; angles of tetragonal young branches scaberulous with short, stout, slightly curved to straight, white hairs; stipular sheaths 3–5 toothed; primary leaves ovate, elliptic, lanceolate, or rarely oblong-linear, often very diverse ones on 1 plant, 3–13 mm long, 1.5–5 mm wide, finely and sparsely scaberulous on upper surface and margins or sometimes glabrate, the undersurface paler to nearly glaucous and mostly glabrous, margins plane on leaves of seedlings, narrowly revolute on those of older branches, acute at apex or rarely slightly mucronulate, the midvein impressed above, slightly elevated beneath, lateral veins not evident; smaller, crowded leaves often in axillary fascicles on older plants; flowers axillary, usually 2–4 per axil, each subtended by linear bractlets about half as long as hypanthium, this 1.6–1.8 mm long at anthesis, sepals oblong to oblong-obovate, 1.2–1.6 mm long, short-mucronulate, sparsely hirsute and bearing several narrowly elliptic, white, semi-punctate glands on back, the margins usually inconspicuously ciliolate

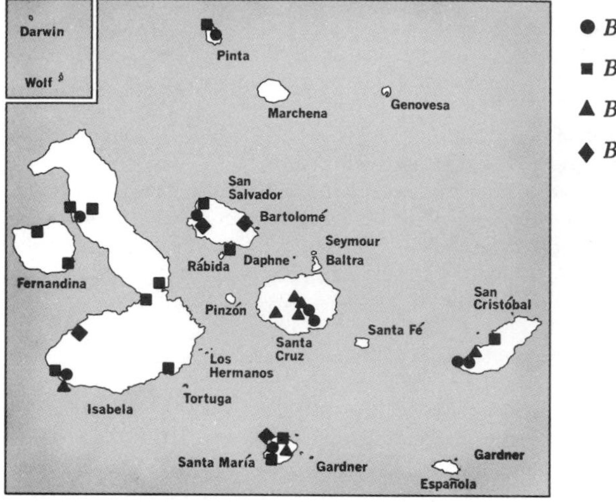

or scaberulous; corollas white, 1.5–2 mm long, tube about 1 mm long, glabrous, lobes broad and short, 0.4–0.6 mm long, acute and incurved at apex, bearing transverse band of moniliform hairs across their bases at apex of throat; stamens inserted 0.2–0.3 mm below sinuses of corolla, anthers exserted one-fourth to half their length; style 0.4–0.6 mm long; stigma bilobed; fruit 1.8–2.2 mm long, usually glabrate at maturity; seeds about 0.8 mm wide, 1.6–1.8 mm long, black, minutely foveolate in longitudinal lines.

Endemic; in sandy, rocky, or tufaceous soil, and on lava flows. Specimens examined: Isabela, Pinta, San Cristóbal, San Salvador, Santa Cruz, Santa María.

The Darwin collections examined (*Darwin s.n.* [CGE, isotype of *B. dispersa*]) are mounted on the same sheet in the Henslow Herbarium. Three branches are involved, 2 from Isla San Salvador and one from Santa María, the coloration of the leaves and the degree of scaberulous investiture of the stems being distinctive enough to make matching of the San Salvador parts quite easy. Hooker undoubtedly had both collections before him when he wrote the original description of *B. dispersa*, for he cited both collections, not indicating that either was more important in delimiting the new species. However, the material from San Salvador is somewhat better, in that it has both older parts and twigs of the year attached, and closely matches a large percentage of the specimens collected by more recent workers. Therefore, I selected the 2 parts of plants constituting the collection from "James Island" as the lectotype of *B. dispersa* Hook. f.

See Fig. 105 and color plate 37.

Borreria ericaefolia Hook. f., Trans. Linn. Soc. Lond. 20: 218. 1847

Borreria parvifolia Hook. f., loc. cit.
Borreria divaricata Hook. f., op. cit. 219.

Suffrutescent perennial 2–6(10) dm tall, with spreading-ascending branches, the bases of these terete and encased in reddish brown to gray, faintly fissured bark, the upper branches tetragonal, minutely puberulent to scaberulous, especially along angles; internodes rather short, mostly 5–10 mm long except on most rapidly growing twigs, minutely puberulent, scaberulous, or glabrate, numerous axillary twigs 2–12 mm long bearing closely crowded, smaller, fasciculate leaves and giving plant

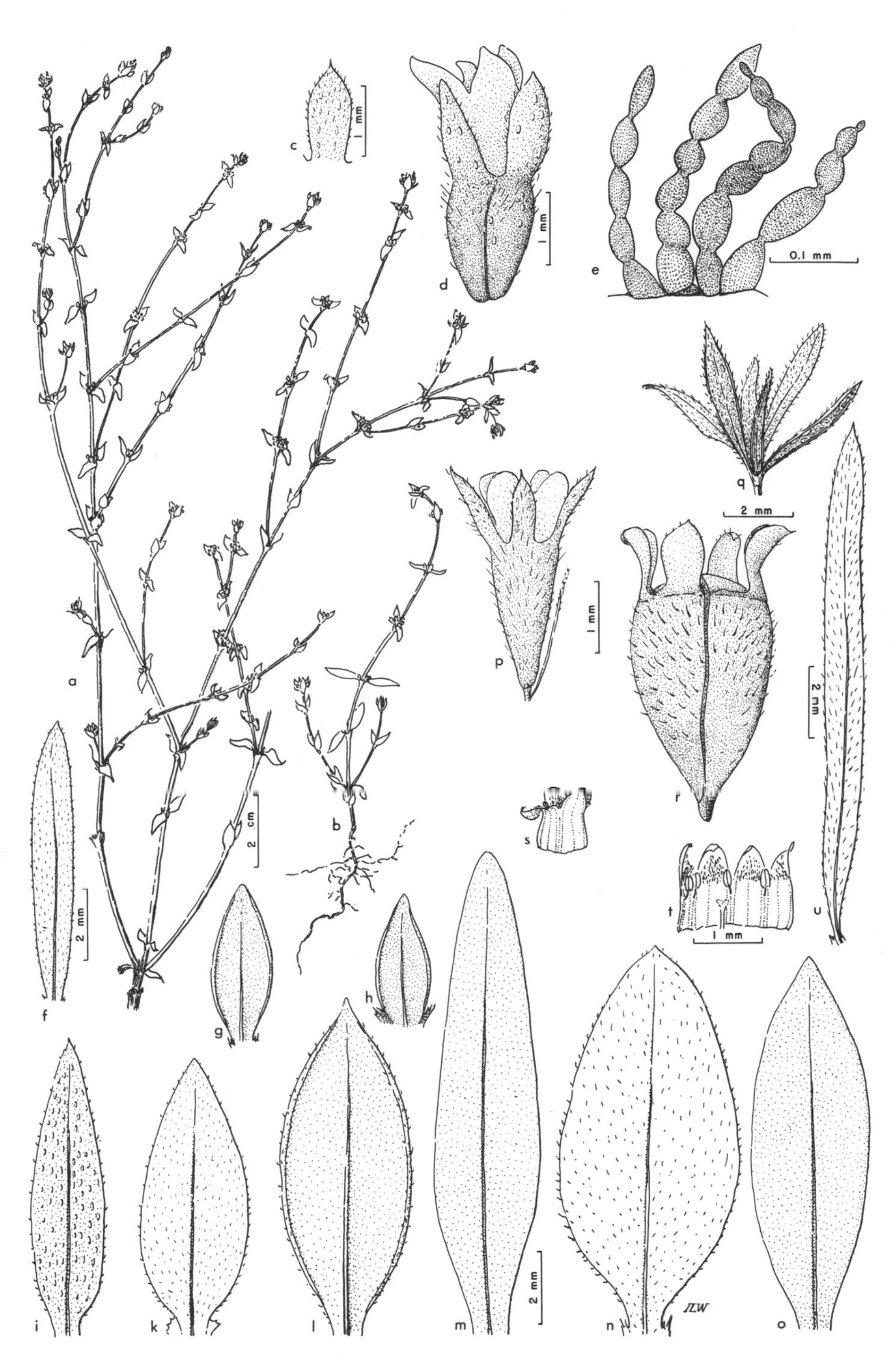

Fig. 105. *Borreria dispersa* (*a–o*): *e*, moniliform hairs from throat of corolla. *Borreria suberecta* (*p–u*).

a compact to crowded appearance; primary leaves linear to narrowly ovate, 3–6 mm long, 0.6–1.4(2) mm wide, mucronulate at apex, margins tightly revolute, blade bright green and shining above, usually glabrous, paler and glabrous or sparsely pubescent beneath; stipular sheaths very short, about 0.6 mm long, often twice as wide, bearing 3–7 slender, often ciliolate teeth to 1 mm long; flowers few, often only 2 or 3 at a node; hypanthium 1.2–1.4 mm long at anthesis, glabrous to sparingly hirsute, usually bearing a few narrowly elliptic, white, semipunctate glands, sepals ovate to oblong, about equalling hypanthium tube, mucronulate, glabrous or ciliolate; corolla white, 1.4–1.6 mm long, glabrous without, densely barbate with moniliform, coarse white hairs at orifice of throat, the lobes puberulent with stouter, erect hairs along midvein almost to apex; tips of anthers barely exserted; stigma about equaling throat; fruit obovoid, 1.8–2 mm long, punctate-glandular on sides, glabrous or glabrate; seeds 0.7–0.8 mm wide, 1.6–1.8 mm long, dull black, minutely foveolate.

Endemic; in pumice, among rocks, and in crevices of lava flows. Specimens examined: Fernandina, Isabela, Pinta, Rábida, San Cristóbal, San Salvador, Santa María.

See color plate 37.

Borreria laevis (Lam.) Griseb., Gött. Abh. 231. 1857

Spermacoce laevis Lam., Ill. Gen. 1: 273. 1791.
Spermacoce assurgens R. & P., Fl. Peruv. 1: 60, *pl.* 92, *fig. c.* 1798.

Herbaceous, annual or sometimes short-lived perennial with erect, ascending, or procumbent branches, 1–3(6) dm tall; stems and branches glabrous or retrorsely puberulent along angles, internodes 3–12 cm long; stipular sheaths 2–5 mm deep, the slender teeth 1–4 mm long, cup glabrous or more commonly finely puberulent, with connecting vascular bundle arching from each side to base of central tooth; petioles slender, 5–15 mm long, very narrowly winged, adaxial surface shallowly and broadly grooved, usually puberulent, abaxial side mostly glabrous; leaf blades broadly elliptic, ovate, elliptic-ovate, to lance-ovate, 2.5–7 cm long, 1–4 cm wide, abruptly decurrent-cuneate at base, acute to acuminate at apex, more or less pilose along veins, usually glabrous or glabrate between veins, dark green above, slightly paler beneath; lateral veins 5–7 pairs, alternate, departing from midrib at acute angle and running nearly parallel with margins toward apex, the tertiary veinlets anastomosing freely; flowers in dense axillary and terminal heads, each flower subtended by several linear, simple, ciliolate bracts, or some also by subcupulate, several-lobed larger bracts, all about half as long as to equaling hypanthium, this about 1.4–1.6 mm long at anthesis, usually puberulent near apex; sepals broadly deltoid, about 0.4 mm high, caducous or inconspicuous in fruit; corolla white, 2–2.6 mm long, lobes lance-ovate, slightly longer than tube, sparsely pilose on outer surface, moderately to densely barbate in throat and on inner face of each lobe; stamens inserted below sinuses of corolla, anthers about 2 mm long, exserted 2–3 mm; capsule about 2 mm long at maturity, usually puberulent near tip, sometimes glabrate; seeds plano-convex to concavo-convex, oblong, 1.8–2.2 mm long, 1–1.3 mm wide, reddish brown, irregularly rugulose-sulcate transversely and with some narrower connecting grooves running longitudinally, the inner face with pale tan or light brownish straplike structure running full length of median area.

Along trails, in clearings, and in uncrowded grassy areas throughout much of tropical America from sea level to 1,500 m or higher where sufficient moisture is available; a common weed in many areas. Specimens examined: Isabela, San Cristóbal, Santa Cruz, Santa María.

Few collections of this species were made in the Galápagos Islands prior to the turn of the century. Baur obtained one in 1891, Snodgrass and Heller got none in 1898–99, and Stewart collected only one number of the weed in 1905–6. Now it is fairly common around agricultural areas, and its weedy character suggests that it was introduced with crop seed or in feed for livestock by later settlers. Plants just coming into flower can be separated from *Diodia radula* only with difficulty, for the two are similar in general appearance.

See Fig. 106.

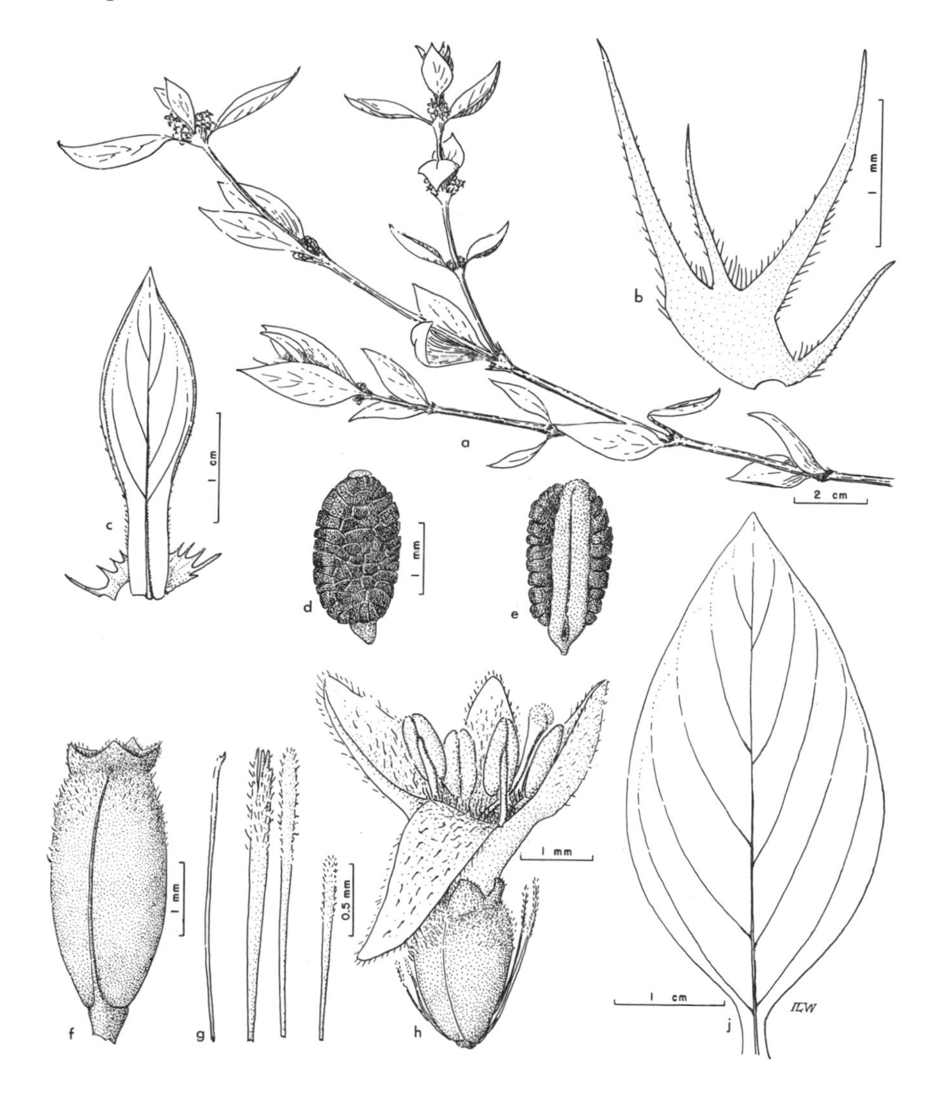

Fig. 106. *Borreria laevis*: *g*, bracts from base of flowers.

Borreria linearifolia Hook. f., Trans. Linn. Soc. Lond. 20: 217. 1847
Borreria falcifolia Hook. f., op. cit. 219.

Low shrub or suffrutescent perennial 2–5 dm tall, the lowermost branches often procumbent, the central erect or ascending, terete at base, with dark reddish brown, shallowly fissured bark, the upper tetragonal, 0.5–1.5 mm thick, glabrous or angles minutely scaberulous with short, whitish, uncinate hairs; internodes 5–25 mm long; primary leaves linear to linear-lanceolate, 5–20 mm long, 0.6–2.2 mm wide, sparingly scaberulous to glabrous above, paler and usually glabrous beneath, margins narrowly revolute to plane, axillary fascicles of leaves similar but smaller, rarely more than 3–5 mm long; flowers axillary, 1–4 per axil; hypanthium 1.8–2.2 mm long at anthesis, glabrous and usually shining, pale yellow-green, sepals narrowly obovate or oblong, 1.4–3 mm long, distinctly mucronulate, 0.6–1 mm wide, glabrous or margins sparingly ciliolate; corolla white, 2–2.6 mm long, lobes spreading, 1–1.4 mm long, inwardly uncinate at apex, each lobe bearing transverse band of coarse, blunt, moniliform hairs across its base at orifice of throat; stamens inserted short distance below sinuses of corolla lobes, the filaments very slender, about half as long as anthers, these narrowly ovate, 0.5–0.6 mm long, exserted about half their length; style short, stigma about level with bottom of anthers; body of fruit narrowly obovoid, 2–2.4 mm long, glabrous, bearing persistent sepals; seeds black, minutely foveolate in longitudinal lines, 1.6–2 mm long, 0.6–0.8 mm wide, dull to sublustrous.

Endemic; growing on tufaceous soil in openings in the vegetative cover, among rocks, or in crevices on lava flows. Specimens examined: Isabela, San Salvador, Santa María.

The specimens placed in this species differ from those of *B. suberecta* in having sepals that are significantly longer than the body of the fruit, narrower primary leaves, and markedly less puberulent leaves, hypanthium tubes, and sepals. Critical field studies may prove that the 2 populations should be merged into a single species. The type specimen has some leaves as much as 19.5 mm long, that are lance-elliptic and glabrous except for a few small hairs on the margins at maturity.

Borreria perpusilla Hook. f., Trans. Linn. Soc. Lond. 20: 218. 1847

Delicate, prostrate-creeping perennial herb with filiform, tetragonal branches 2–30 cm long, 0.2–0.4 mm thick, the internodes 1–1.5 cm long, narrowly sulcate on faces, microscopically and sparsely antrorse-scaberulous to spreading-scaberulous along angles, otherwise glabrous and shining; primary leaves oblong to oblong-ovate, 3–5 mm long, 1–1.6 mm wide, acute at apex, sparsely scaberulous marginally, dark green and lustrous above, slightly paler beneath; fasciculate leaves smaller, crowded; stipular sheaths 0.4–0.6 mm long, reddish, slightly puberulent, bearing 1–5 short, subulate teeth or fimbriae to 0.5 mm long; flowers few in axils of primary leaves, usually 2–4 per axil; hypanthium broadly obovoid, 0.8–1 mm long at anthesis, sparingly scaberulous near apex, usually with a few narrowly elliptic, whitish to reddish semipunctate glands on sides; sepals erect, ovate, 0.8–1.4 mm long, 0.6–0.8 mm wide, acute, minutely and sparsely ciliolate, persistent in fruit; corolla white, broadly tubular, tube about 0.8 mm long, 1 mm wide; lobes 4–5, broadly ovate, 0.4–0.5 mm long, spreading at anthesis, the acute tips inflexed, throat faintly constricted, bearing transverse line of moniliform, white hairs across base of each corolla lobe, the hairs erect, about 0.2 mm high; stamens 4 or sometimes 5, inserted below sinuses

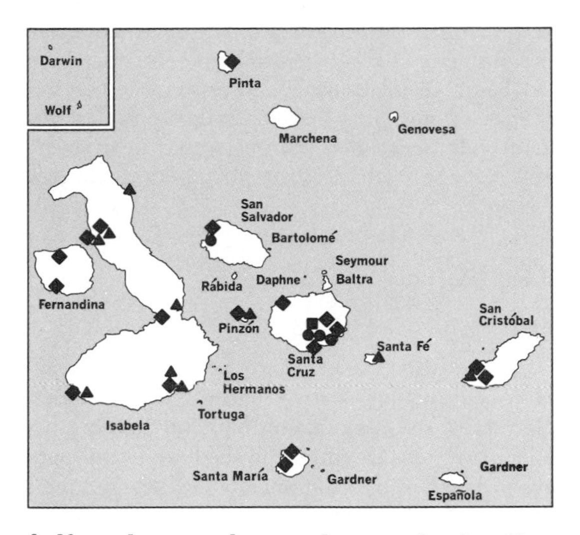

● *Borreria perpusilla*
■ *Borreria rotundifolia*
▲ *Borreria suberecta*
◆ *Chiococca alba*

halfway between base and apex of tube, filaments very short, anthers narrowly ovate, about 0.3–0.4 mm long, barely reaching base of sinuses, white or faintly bluish; dome of ovary dark purplish to nearly black at anthesis; style very short, 0.1–0.2 mm long, stigma shallowly bilobed; fruit subglobose to broadly obovoid, 0.8–1.2 mm long, sparsely puberulent to glabrate toward apex, some hairs stiff and nearly echinate, whitish, dome of capsule nearly white at maturity; seeds oblong, 0.8–1 mm long, 0.4–0.6 mm wide, about 0.3 mm thick, microscopically foveolate, dark reddish brown to black, deeply excavate-sulcate along inner face.

Endemic; in small pockets of soil and in crevices in lava, Arid and Transition zones, in partial shade or full sun. Specimens examined: San Salvador, Santa Cruz.

This tiny plant is delicate and easily overlooked. It grows in minute crevices and is closely applied to the bare lava; its color blends in with that of the dark lava. Though inconspicuous, it is very attractive when examined closely. The dark apex of the ovary is striking when the open flowers are viewed from above and the moniliform hairs in the throat glisten like tiny beads of crystal. Apparently the species is subject to considerable fluctuation in growth under slightly subnormal moisture supply, for although a careful search was made in 1967 at the place where it was found in 1964, not a plant could be found. Hooker's description was based on a collection from Isla San Salvador, but it has not been rediscovered on that island; it should be sought there during a year with normal or more than normal rainfall.

Darwin's collection (*s.n.* [CGE, type]) consisted of 5 depauperate plants, each of which had branches 2.5 cm long or less, these erect and bearing characteristic small fruits. Our material was all prostrate or decumbent, the larger branches up to 3 dm long.

Borreria rotundifolia Anderss., Kongl. Svensk. Vet.-Akad. Handl. 1861: 77. 1861

Prostrate subherbaceous plant with branches fanning out from filiform rootstock, the stems branching considerably near tips and very slender, acutely quadrangular, glabrous, lustrous; leaves 4 (at a node), 2 of them larger, sessile, broadly ovate to nearly orbicular, 3–5 mm long, about 4 mm wide, margins entire and at length sub-revolute, dull green, both surfaces and margins bearing simple, scattered hairs, the apex acute but not cuspidate; stipules obsolete, of short, setose hairs. (Ex char.)

"In grassy places on Indefatigable [Santa Cruz] Island": Andersson.

I have not seen the type of this species nor anything that fits Andersson's description. It conceivably could be a variant of *B. perpusilla,* but one would need to examine Andersson's specimens to be sure of the relationship to other representatives in the archipelago.

Borreria suberecta Hook. f., Trans. Linn. Soc. Lond. 20: 217. 1847

Borreria suberecta var. β *flaccida* Hook. f., loc. cit.
Borreria galapageia Robins. & Greenm., Amer. Jour. Sci. III, 50: 140. 1895.

Suffrutescent perennial 2–5 dm tall, with all branches strongly ascending to erect, the lower becoming terete, the upper quadrangular, scaberulous at least on angles, usually with some whitish to slightly rufous, stout, spreading hairs on surfaces between angles; internodes 5–30 mm long; stipules lacerate, the teeth 2–3 mm long; leaves usually bearing scattered, narrowly elliptic, semipunctate, white structures 0.1–0.2 mm long on both surfaces, the leaves on young, vigorously growing stems linear, 1–2.5 cm long, 1–2(4) mm wide, more or less pubescent, the margins often narrowly but tightly revolute, dark green above, paler and usually more densely pubescent beneath, mucronulate at apex; smaller, crowded leaves often in axillary fascicles; older stems with shorter internodes, leaves 3–6 mm long, 1–2.5 mm wide, usually more densely puberulent, margins tightly revolute; flowers axillary, usually 4–8 flowers per node, body of hypanthium 1.8–2 mm long at anthesis, hispid with spreading, whitish, rather stiff hairs, sepals oblong, mucronulate, 0.8–1.2 mm long, subequal or abaxial and adaxial sepals slightly longer than laterals, often with whitish "glands" similar to those on leaves on outer faces, hispid when young, later glabrate or nearly so; corolla white, 0.8–1.2 mm long, lobes spreading only slightly at anthesis, about 0.4–0.5 mm long and wide, laxly hirsute within, glabrous without; corolla tube about 0.6 mm long and wide, glabrous or slightly hairy at apex; stamens inserted 0.2–0.3 mm below sinuses between corolla lobes, filaments slender, glabrous; anthers ovate, about 0.3 mm long; stigma shallowly bilobed; fruit obovoid, 2–2.4 mm long, domed at apex, hirsute with stoutish, spreading-ascending hairs, sepals persistent, nearly glabrous in fruit; seeds concavo-convex, oblong, 1.6–2 mm long, about 0.8 mm wide, dark reddish brown to black, dull, minutely foveolate-sculptured in penicillate lines.

Endemic; on lava flows, in tufaceous soil, and on cinder ridges. Specimens examined: Isabela, Pinzón, San Cristóbal, Santa Fé.

As interpreted here, *B. suberecta* shows a wide range of variation in the width/length ratio of the leaves. However, a series of specimens collected by Howell shows that these variations are continuous, and that some characters formerly used to separate "species" are no more than the range extremes of a continuum from young, vigorously growing stems to the older, slow-growing ones with fascicles of leaves in their axils. Flower and fruit characters were uniform throughout the series.

See Fig. 105.

Chiococca P. Br., Nat. Hist. Jamaica 164. 1756

Shrubs or small trees with terete branches and opposite, petiolate, membranous to coriaceous, entire leaves. Stipules persistent. Flowers small, white, in axillary, sim-

Fig. 107. *Chiococca alba.*

ple, or paniculate clusters; hypanthium ovoid to turbinate, pubescent or glabrous, usually compressed laterally; calyx lobes 5, persistent. Corolla funnelform, lobes valvate. Stamens 5, included or exserted, inserted at base of corolla tube; filaments pubescent, coalesced at base; anthers basifixed. Ovary 2-celled, inferior; style filiform; stigma entire or bilobed; ovules solitary in each cell, compressed. Fruit a small coriaceous drupe, orbicular or nearly so, usually compressed at right angles to septum. Seeds laterally compressed. Endosperm fleshy. Embryo straight.

Some 10–12 species in the tropics and subtropics of the New World.

Chiococca alba (L.) Hitchc., Rep. Mo. Bot. Gard. 4: 94. 1893
Lonicera alba L., Sp. Pl. 175. 1753.*

Shrub or small tree 2–6 m tall, with spreading to scandent, slender, gray to brownish branches and green, glabrous or sparsely puberulent branchlets; stipules often broader than long, abruptly acuminate or mucronate, 1–2 mm long; petioles 2–12 mm long; blades variable, ovate, lanceolate, elliptic, or rounded-oval, 2–10 cm long, 1–5 cm wide, acute to short-acuminate at apex, obtuse to rounded and short-decurrent at base, coriaceous, glabrous, dark green and lustrous above, paler beneath, margins plane to slightly revolute; inflorescences racemose to paniculate, peduncle exceeding subtending petioles and sometimes surpassing leaf blade, glabrous, few- to many-flowered; calyx and hypanthium 2–2.5 mm high, glabrous or rarely obscurely puberulent; calyx lobes deltoid or deltoid-subulate, acute, 0.6–1 mm long; corolla 6–8 mm long, white, throat 2–3 mm wide, lobes triangular, one-third to half

* For additional synonyms, see Standley, N. Am. Fl. 32: 287. 1934.

as long as tube; stamens included; fruit white, orbicular-compressed, 4–6(8) mm long, glabrous, persistent calyx lobes erect; seeds 3–4 mm long, puncticulate.

Widely distributed in the tropics of the West Indies and South America, extending northward into Baja California, Sonora, and Florida; well represented in the forests and Transition Zone of the Galápagos Islands. Specimens examined: Española, Fernandina, Isabela, Marchena, Pinta, Pinzón, San Cristóbal, San Salvador, Santa Cruz, Santa María.

See Fig. 107.

Coffea L., Sp. Pl. 172. 1753; Gen. Pl. ed. 5, 80. 1754

Shrubs or small trees with glabrous twigs and leaves. Stipules deltoid-triangular, persistent. Leaves short-petiolate, usually opposite, entire, firm to coriaceous. Flowers in small clusters in axils of leaves, 4- or 5-parted; calyx very short or nearly obsolete, usually reduced to narrow ring. Corolla salverform or funnelform, lobes dextrorsely contorted in bud. Stamens inserted in corolla tube, exserted; anthers subsessile, dorsifixed. Fruit fleshy, indehiscent, 2-seeded. Seeds semi-ovoid, flattened, deeply furrowed on inner face.

Possibly 25 species, native in the tropics of the Old World, but 2 or 3 species widely cultivated and escaped commonly in the tropical regions of the Western Hemisphere.

Coffea arabica L., Sp. Pl. 172. 1753

Usually a shrub, but sometimes small tree with single trunk while young, in our area to 8 m tall, often producing additional branches from base and becoming shrubby in habit; stipules deltoid, 4–5 mm long, 3–4 mm wide, abruptly acuminate, firm to leathery in age, moderately persistent; petioles 1–2.5 cm long; leaf blades lance-elliptic, 8–20 cm long, 4–8 cm wide, broadly cuneate at base, cuspidate-acuminate at apex, submembranous, dark green and lustrous above, slightly paler beneath, wholly glabrous; flowers delicately fragrant, in 2–8 clusters per node, 1–4 flowers

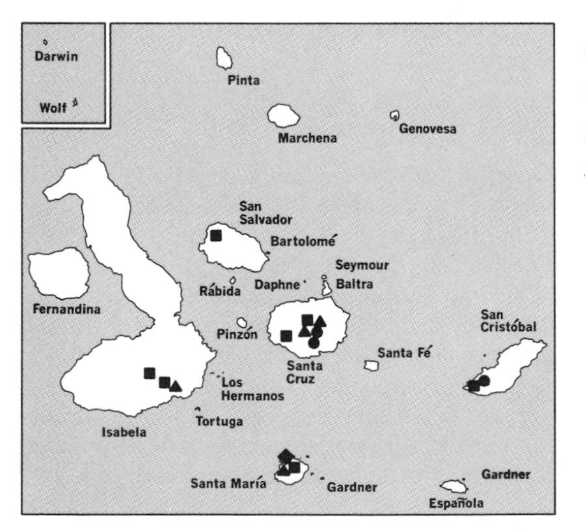

● *Coffea arabica*
■ *Diodia radula*
▲ *Galium galapagoense*
◆ *Psychotria angustata*

Fig. 108. *Coffea arabica.*

per cluster, short-pedicellate to subsessile; calyx narrow, usually no more than short ring at hypanthium apex; corolla white, 15–20 mm long, lobes narrowly spatulate to oblong, widely spreading; fruit ellipsoidal to subglobose, 10–16 mm long, smooth and glabrous, the flesh red to nearly black at maturity, thin; seeds 9–11 mm long, 7–8 mm wide, each encased in a firm-chartaceous, tan to nearly white coating, flattened and medianly grooved on inner face.

Frequently escaped in favorable habitats in tropical South America; cultivated in the wetter belts on some of the larger islands of the Galápagos, escaped and reproducing spontaneously in the forest, especially in areas abandoned and allowed to revert to jungle. Specimens examined: San Cristóbal, Santa Cruz.

See Fig. 108.

Diodia L., Sp. Pl. 104. 1753; Gen. Pl. ed. 5, 45. 1754

Annual or perennial herbs or sometimes suffrutescent at base, the stem often prostrate to decumbent. Stipules connate to base of petioles to form distinct sheath, this setiferous along upper margin. Leaves opposite, relatively small, often narrow, entire. Flowers white, minute or small, in axillary clusters or in spicate or cymose inflorescences. Sepals 2 or 4, equal or unequal, usually persistent. Corolla funnelform, 4-lobed, lobes valvate in bud. Stamens 4, inserted in throat or mouth of corolla, anthers dorsifixed. Fruit of 2 indehiscent carpels, these usually separating from the persistent column, membranous to firm or nearly woody. Seeds more or less flattened, usually more or less concavo-convex, broadly and shallowly sulcate on inner surface.

About 35 species, mainly occurring in tropical and subtropical areas of the New World; a few also in Africa.

Diodia radula (Roem. & Schult.) Cham. & Schlecht., Linnaea 3: 324. 1828

Spermacoce radula Willd. & Hoffm. in Roem. & Schult., Syst. Veg. 3: 531. 1818.
Diodia muriculata DC., Prodr. 4: 564. 1830.

Perennial, weedy herb with decumbent to ascending stems; branches to 6 dm long; base more or less lignescent; stems quadrangular, puberulent to glabrate, the internodes 3–10 cm long; stipules connate with bases of petioles forming sheaths 5–10 mm deep, these densely or moderately appressed-pubescent with reddish, coarse hairs, or the basal part glabrescent, the upper membranous and lacerate-ciliate; sheaths near apices of branches to 12 mm broad, many-flowered; leaves broadly elliptic to ovate-elliptic, cuneate to rounded at base, acute to short-acuminate at apex, to 1.5–2 cm wide, 2–4.5 cm long, dark green, pubescent on upper surface, much paler and more densely pubescent with longer, coarser hairs beneath; veins pinnately branched but ascending at a sharp angle, outermost pair not arising from midrib but running free to attachment of petiole, this to 12 mm long, or leaf sometimes nearly sessile; flowers many, densely crowded in axillary, subglobose clusters, subtended by numerous linear, short-ciliolate bracts about half as long as or equaling hypanthium; ovary obovoid, 1.5–2 mm long at anthesis, sparsely hirsute toward apex, glabrous below; sepals 4, deltoid to ovate-oblong, 0.4–0.8 mm long, ciliolate; corolla white, funnelform, 3.5–6 mm long, tube slender, 1–1.5 mm long, throat about as long as tube, flaring rapidly, lobes oblong, 2–3 mm long, subacute at apex, densely pilose with coarse, hyaline hairs over most or all of inner face, glabrous

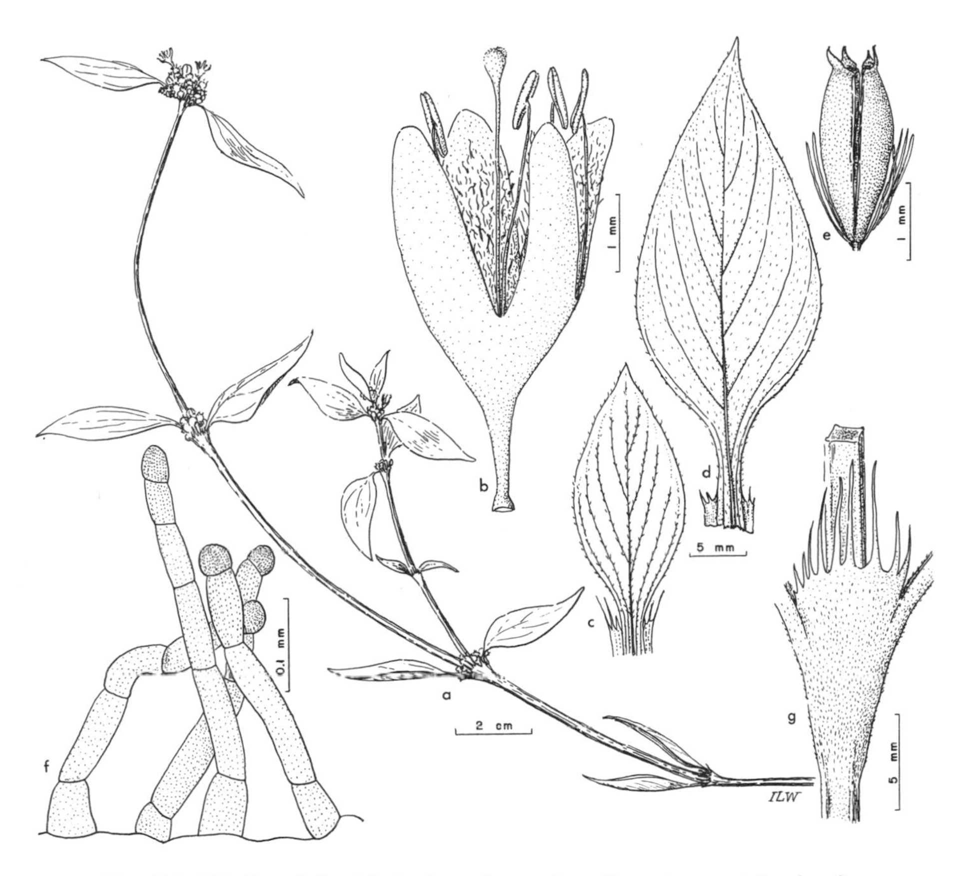

Fig. 109. *Diodia radula*: *f*, hairs from throat of corolla; *g*, interpetiolar sheath.

without; stamens inserted at bases of sinuses between corolla lobes, filaments very slender, equaling or slightly exceeding corolla lobes, anthers dorsifixed, linear, about 1 mm long; style filiform, equaling or slightly exceeding stamens, stigma capitate, faintly bilobed; fruit 2–3 mm long, sparsely pilosulous toward apex; seeds dark reddish brown to nearly black, ellipsoidal in outline, concavo-convex, broadly sulcate ventrally, about 1.4–1.8 mm long, minutely foveolate and somewhat rugose on back.

A weedy plant in openings in the forest and scrub, in moist areas, in Ecuador, Peru, and Brazil. Specimens examined: Isabela, San Cristóbal, San Salvador, Santa Cruz, Santa María.

The specimens here considered representative of this species have flowers slightly smaller than those of specimens from Brazil, but in all other characteristics they agree well with the description of *D. radula*. Because few fruits are present, none of them ripe, the seed has been described from immature examples.

The general appearance of this plant is similar to that of *Borreria laevis*, but the venation at the bases of the leaves separates them readily. In *B. laevis* the lowest lateral veins, like all of those above them, arise from the midrib, then curve upward near the margins of the blade. In *D. radula* the lowermost pair (or sometimes 2 pairs) of lateral veins is free from the midrib to the level of the attachment of leaf to stem, these veins running through the narrow wing flanking the midrib of the petiole. Furthermore, the herbage of *D. radula* is much more pubescent than that of *B. laevis*.

See Fig. 109.

Galium L., Sp. Pl. 105. 1753; Gen. Pl. ed. 5, 46. 1754

Annual or perennial plants with slender, tetragonal stems and branches, and opposite or whorled leaves. Flowers perfect, polygamo-monoecious or polygamo-dioecious, small, in axillary or terminal cymes or panicles, pedicels articulate with calyces. Calyx tube ovoid to globose, with minute teeth representing sepals, or these lacking. Corolla rotate to shallowly campanulate, usually 4-lobed, lobes involute and inflexed in bud, spreading in flower, often acuminate. Stamens (3)4, with short filaments and small anthers. Ovary inferior, 2-celled, 1 ovule in each cell. Styles 2, short; stigmas capitate. Fruit didymous, fleshy or dry, glabrous or more frequently hispid, separating into 2 indehiscent carpels. Seeds concavo-convex or spherical; endosperm horny; embryo curved; cotyledons foliaceous.

About 300 species, widely distributed in the Northern Hemisphere, less common in the Southern Hemisphere, extending from the tropics into the cool temperate regions.

Galium galapagoense Wiggins, Madroño 20: 250. 1970

Scandent perennial with slender, retrorsely aculeate-scabrous stems 1 m long or more, much branched and forming masses of tangled stems and branches over rocks and shrubs; primary stems 0.4–1 mm thick, internodes 3–7 cm long, their angles retrorsely scabrous or some hairs lax and spreading; "leaves" in 4's, or sometimes in pairs near tips of branches, elliptic, elliptic-ovate, or elliptic-obovate, 5–17 mm long, 2–9 mm wide, acute to rounded and abruptly apiculate at apex, narrowed toward the broadly cuneate base, petioles 1–2 mm long or blade subsessile, membranous-herbaceous, 3-nerved, moderately short-hispid to subglabrescent on both surfaces, the hairs along margins curved, subaculeate like those on stem angles, blades usually drying yellow-green, or on older parts turning dark to nearly black; inflorescences compound dichasia, the slender branches each terminated by sessile flower, with axillary branches 2–5 mm long originating in axils of subtending pair of leaves, axillary branches repeating the branching pattern 2–4 times, all parts of inflorescence moderately to densely pilose with spreading, lax, whitish hairs; flowers 2–2.4 mm high, corolla white to pale cream with faint greenish tinge, 2–2.5 mm across, lobes about 0.8 mm long, rotately spreading, turned upward abruptly at very tip of each short-apiculate lobe, finely and sparsely puberulent on outer surface; stamens inserted just below sinuses in throat of corolla, barely equaling corolla cup; ovary broadly obovoid, 0.6–0.8 mm high at anthesis, antrorsely uncinate-hispid in youth; style 1–1.4 mm long, bifid about one-third of distance to base; stigmas globose-capitate; fruit reddish to orange, thin-fleshed, slightly juicy, 1.5–2 mm in diameter, bases of hairs persisting as low-conical, minute bosses.

Endemic; in forest and shrubby zone above it. Specimens examined: Isabela, Santa Cruz, Santa María. Possibly occurs elsewhere in the archipelago.

See Fig. 110.

Fig. 110. *Galium galapagoense.*

Psychotria L., Syst. Nat. ed. 10, 929. 1759

Shrubs, small trees, or rarely suffrutescent plants, with opposite, coriaceous leaves and persistent or tardily dehiscent, commonly bilobed stipules, or in some species stipules entire and caducous. Inflorescences mostly terminal or in some axillary as well as terminal, variable in form, without an involucre or this inconspicuous and bracts distinct. Flowers small, inconspicuous, white, yellowish, or rarely roseate. Calyx lobes short or elongate, more or less united, or obsolete. Corolla short or elongate, tube straight, usually pilose-barbate in throat, lobes obovate or oblong-ovate. Stamens inserted in corolla tube, included or exserted. Ovary usually 2-celled but sometimes 5-celled. Fruit berry-like, containing 2–5 bony nutlets.

A large genus of questioned intergeneric limits and relationships, with over 100 species in South America; represented also in the tropics of Mexico, the Pacific islands, and the Old World.

Leaves, twigs, and inflorescences glabrous; corollas pink to rose; leaves 8–10 cm long, broadly lanceolate . . . . *P. angustata*
Leaves, twigs, and inflorescences more or less hirsute with reddish hairs, these particularly noticeable in angles of lateral veins on underside of blades; corollas white; leaves 8–20 cm long, elliptic-ovate to obovate . . . . *P. rufipes*

Psychotria angustata Anderss., Kongl. Svensk. Freg. Eugenies Resa Omk. Jorden 1861: 77. 1861

Divaricately branched shrub 2–4 m tall; nodes of twigs considerably swollen; twigs, leaves, and inflorescences glabrous, or a few setose-subulate scales in axils of some lateral veins, these appearing like hairs without a lens; stipules 5–6 mm long, about as wide, subulate to oval, castaneous, slightly subfleshy, truncate or rounded, often slightly lacerate at apex, caducous, the basal part indurated-persistent; petioles stoutish, 4–10 mm long; blades broadly lanceolate, 8–10 cm long, 3–4 cm wide, short-petiolate, entire, acute at apex, attenuate-cuneate at base, paler beneath than above; lateral veins arcuate, anastomosing freely, these and midrib subrugose-elevated on underside of blade; inflorescences axillary and terminal, peduncles to 5 cm long, repeatedly branched trichotomously (sometimes to 5 branches at a node); calyx urceolate, short, faintly 5-toothed or teeth obsolete, the upper margin ciliolate; corolla roseate, about 4 times as long as the calyx tube, limb 5-parted, lobes ovate-oblong, obtuse, revolute, throat densely white-barbate within; stamens inserted in corolla tube, filaments hirsute; anthers linear, yellowish; stigmas subfleshy, bilobed; fruit subglobose, glabrous, 4–6 mm in diameter, nutlets bony, subcoherent, obtusely 10-costate.

Endemic; found on Isla Santa María by both Darwin and Andersson but not obtained by subsequent collectors. Specimen examined: Santa María.

The syntype of *P. angustata*, a Darwin collection loaned to me through the kindness of Dr. S. M. Walters, curator of the herbarium at Cambridge University, consists of 3 terminal twigs, all bearing 3–7 leaves and very young inflorescences that had not expanded at the time of collection. One additional leaf, mounted with the undersurface exposed, and fragments of inflorescences (mostly unopened buds) in a pocket complete the material available for study. Examination of the undersurfaces of the leaves reveals that some of the angles between departing lateral veins and the midrib bear a few linear-subulate, reddish setae or scales 0.2–0.6 mm long set on the margins of a minute cupule adnate to the 2 veins. These setae are so insignificant in comparison with the dense coats of fine rufous hairs in the angles of the veins in *P. rufipes* that they are easily overlooked unless examined under a good lens or dissecting microscope.

Hooker (1847, 220) wrote: "The leaves and whole plant quite glabrous." This glabrous condition is the main character used by Andersson to separate *P. angustata* from *P. rufipes;* the roseate corollas and the smaller, uniformly broad-lanceolate leaves constitute the added distinguishing features for separating the 2 taxa. A careful search for glabrous specimens of *Psychotria* should be made on all islands that rise into the wet zone. Such search should be conducted with special zeal above an altitude of 250 m on Isla Santa María, where Darwin and Andersson obtained their specimens.

Psychotria rufipes Hook. f., Trans. Linn. Soc. Lond. 20: 220. 1847

Rounded, rather compact shrub 1–3 m tall; older bark gray, minutely and irregularly fissured, younger twigs with reddish brown, glabrous to sparsely hirsute, smooth bark with relatively few scattered, minute lenticels, or these sometimes not obvious; stipules connate between adjacent leaves, ovate, 7–10 mm long, obtuse, entire or slightly dentate, closely rufo-pilosulous, caducous, the scars left at nodes after dehiscence of stipules forming tan, transverse band 1–2 mm high; petioles short and rather stoutish, 7–20 mm long, rufo-pilosulous; blades elliptic, broadly elliptic-ovate, or broadly oblanceolate, cuneate at base, acute to short-acuminate at apex, 8–20 cm long, 4–7 cm wide, firmly coriaceous, dark green, white-veined, glabrous above, much paler and pilosulous beneath, the hairs more densely crowded in angles of lateral veins than on intervein surface of blade; veins in 10–13 pairs, arcuate-anastomosing at tips; inflorescences cymose, arranged in panicles, often appearing terminal but usually axillary to upper leaves, 5–10 cm long or more, peduncles mostly 2–3 cm long, glabrous to sparingly pilosulous; branches of inflorescences usually rufo-pilosulous; flowers sessile or subsessile, each subtended by

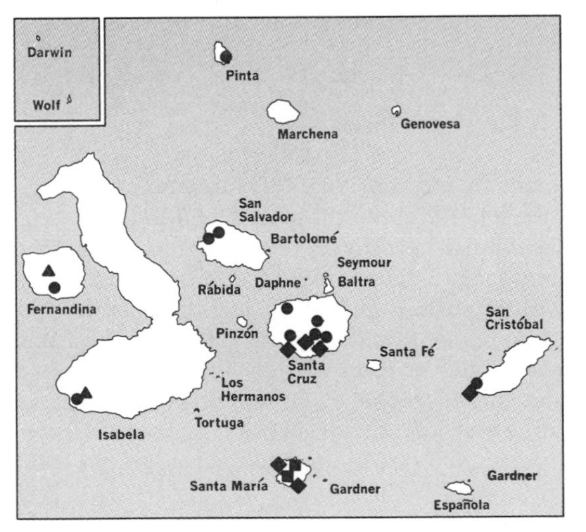

● *Psychotria rufipes*
■ *Spermacoce confusa*
▲ *Calceolaria meistantha*
◆ *Capraria biflora*

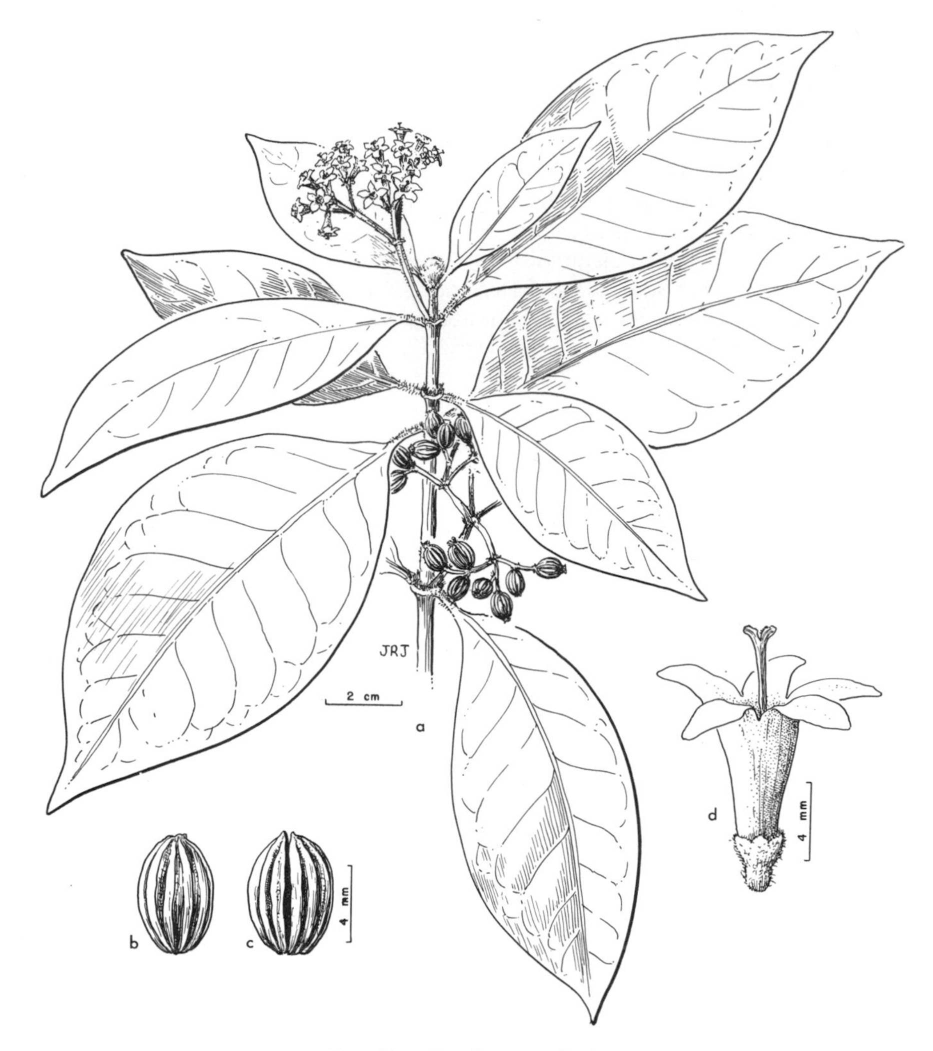

Fig. 111. *Psychotria rufipes.*

densely pilosulous, narrowly ovate bract about 1 mm long; calyx narrowly campanulate, 4–5 mm long, rufo-pilosulous, the lobes ovate to subspatulate, 0.8–1 mm long, acute or minutely apiculate; corolla tubular-salverform, tube 6–7 mm long, limb white, 8–10 mm broad, lobes 3–4 mm long, 2–2.5 mm wide, thickened and short-acuminate at apex, hirsutulous on outer surface and along margins, glabrous or nearly so within; throat densely white-barbate; anthers pale yellow, 1.2–1.4 mm long, subsessile, inserted near upper margins of throat, barely included; style 4–5 mm long, slender at base, clavate-broadened above, each lobe about 1 mm long, densely short-pilosulous on inner surface of lobes; fruit ellipsoid-globose, 7–9 mm long, 5–6 mm broad, the flesh red, juicy; 4–5 coarse, longitudinal ridges on outer face of each nutlet.

Endemic; mostly in moderate shade in *Scalesia* forest and intermingled with *Mi-*

conia shrubs in the brushy zone. Specimens examined: Fernandina, Isabela, Pinta, San Cristóbal, San Salvador, Santa Cruz.

See Fig. 111 and color plate 65.

Spermacoce L., Sp. Pl. 102. 1753; Gen. Pl. ed. 5, 55. 1754

Erect or spreading, weedy herbs, annual or perennial. Stipules connate with petioles to form tubular to urnlike, setiferous sheath. Leaves opposite, herbaceous, moderately ample, short-petiolate to sessile. Flowers small, densely crowded in axillary fascicles, 4-parted. Sepals small, partially short-connate at base, or 1 nearly free. Corolla white, funnelform, with valvate lobes, glabrous or sometimes villous within on at least part of corolla tube or lobes or both. Stamens 4, inserted at base of corolla tube, included to well-exserted; anthers versatile. Style slender, included or exserted; stigma capitate to slightly bilobed. Fruit dry, of 2 1-seeded carpels, these coherent at base, 1 carpel bearing 1 sepal and opening to release the included seed, the other bearing 3 sepals, indehiscent, retaining septum between 2 carpels and not shedding seed. Seeds elliptic to oblong, concavo-convex, with shallow, broad sulcus running longitudinally on ventral face, flanked by minute elliptic to short-linear, whitish scales, the convex surface finely foveolate or rugose.

A genus of 5 or 6 species in the tropics and subtropics of the New World.*

Spermacoce confusa Rendle, Jour. Bot. Brit. & For. 74: 12, *figs. d–f.* 1936

Spermacoce remota sensu Rendle, op. cit. 72: 329–33. 1934, non Lam.
Spermacoce tenuior auct., non L., pro parte.

Plant weedy, erect or less commonly prostrate-spreading, 1–6 dm high; stems and branches slender, quadrangular, glabrous or younger parts sometimes with minute scaberulous puberulence on angles; internodes mostly 3–7 cm long; stipular sheaths closely appressed to stem, tube 1–2 mm deep, teeth 5–10, 2–4 mm long; leaves ovate-lanceolate to linear-lanceolate, 2–5(7) cm long, to 15 mm wide, acuminate at apex, gradually cuneate at base and there narrowing to slender petiole 2–10 mm long, scaberulous on both surfaces and margins, or upper surface sometimes nearly glabrous, secondary veins evident or obscure, curving upward toward tip of leaf at a sharp angle to midrib; flowers densely crowded into axillary clusters, each glomerule several- to many-flowered; hypanthium 1.2–1.6 mm long at anthesis, sparsely beset with stout, short, upwardly curved setae; sepals narrowly ovate to ovate-lanceolate, unequal, the shortest 0.6–0.8 mm long, borne on the dehiscing carpel, longest 1.2–1.4 mm long, it and 2 intermediate sepals usually borne on indehiscent carpel; corolla white, tube about 1.4 mm long, lobes 0.5–0.6 mm long, erect or spreading, tube pilose within from level of anthers to top of throat or sometimes onto lower part of lobes, glabrous without; stamens inserted at bottom of corolla tube, the filiform filaments 0.2–0.3 mm long, glabrous, anthers broadly oval, about 0.5 mm long, completely included within corolla tube; style short, stout, barely equaling anthers or slightly surpassing them, stigma capitate or faintly bilobed; fruit subglobose to broadly obovoid, about 2–2.2 mm long, 2 carpels separating but septum remaining attached to larger of the 2, not shedding its seed, the other (smaller) bearing 1 sepal and shedding its seed; seeds dark reddish brown to nearly black,

* For further information on the genus, see Rendle (1934; 1936).

concavo-convex, 1.6–1.8 mm long, 0.8–1 mm wide, indistinctly and shallowly minute-foveolate.

In open places, along trails and roadsides, in fields and along fence rows, generally distributed in tropical America as a weed. Specimens examined: Santa María.

Because of a marked tendency for the fruits to shed the stoutish hairs that are on the upper part when young, the mature fruit may be almost or quite glabrate. The scaberulous leaves and the very short stamens and style appear identical with those structures in specimens collected in Brazil. There seems to be no reasonable doubt that our specimens should be determined as S. *confusa*.

See Fig. 112.

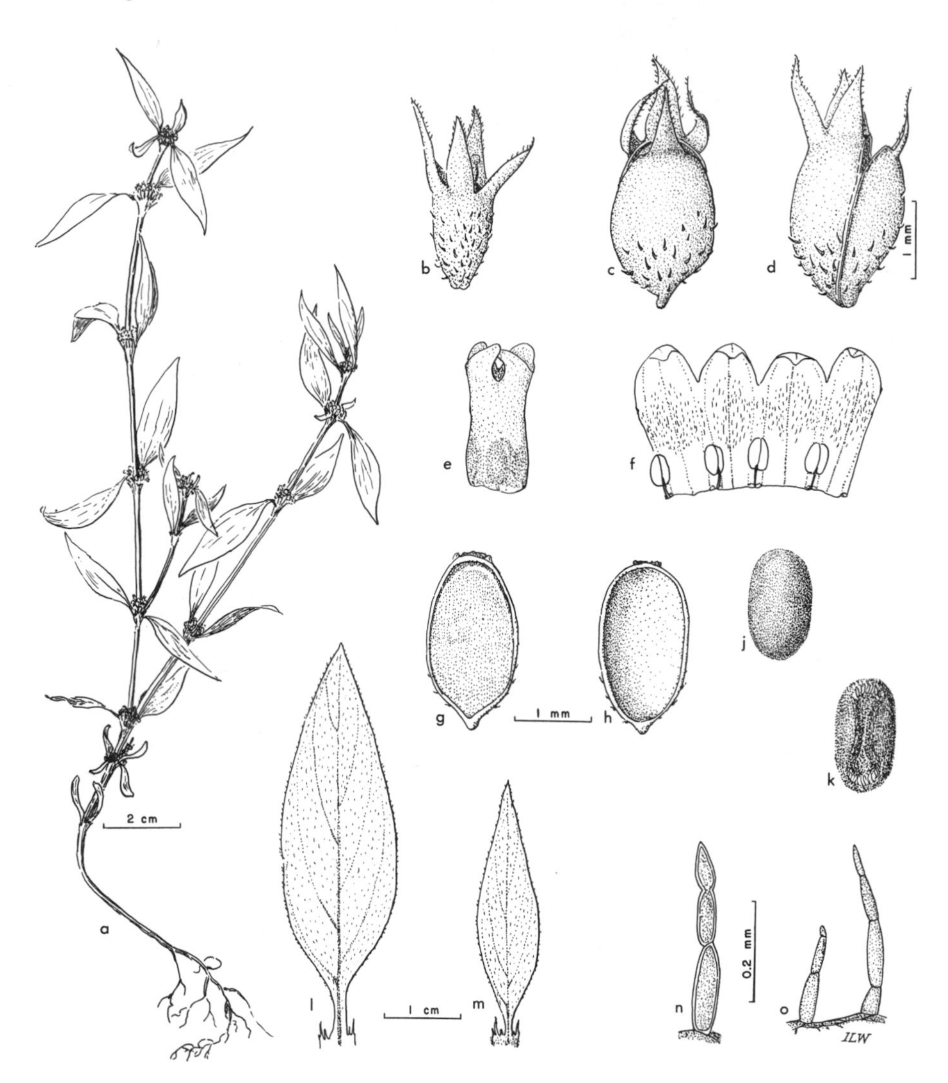

Fig. 112. *Spermacoce confusa*: *g, h,* carpels; *j, k,* exterior and interior faces of seed; *n, o,* moniliform hairs from inner surface of corolla lobes.

SCROPHULARIACEAE. FIGWORT FAMILY*

Annual or perennial herbs, or a few shrubby, partially parasitic in some genera. Leaves alternate, opposite, or rarely whorled, simple, entire to pinnately dissected, sessile or petiolate, the uppermost sometimes reduced to bracts or more or less colored. Flowers variously arranged, perfect, from nearly regular to strongly irregular. Calyx campanulate to tubular or of 4 or 5 practically distinct sepals, sometimes reduced to 1 or 2 more or less distinct leaflike or bractlike lobes, but when tubular or campanulate 4- or 5-toothed or -lobed. Corolla tubular, campanulate, or nearly rotate, or in some strongly galeate; throat often ventricose; base of tube spurred, saccate or gibbous in some genera. Stamens 2, 4, or 5, commonly 4 and didynamous, then a 5th stamen often represented by a scale, filament, or gland; anthers 1- or 2-celled, the cells variously arranged. Style 1; stigma entire, emarginate, or distinctly bilobed. Ovary superior, nearly or completely 2-celled; ovules many. Fruit a capsule, loculicidally or septicidally dehiscent or both, or opening by irregular, subterminal pores. Seeds numerous, small, smooth or more or less sculptured, roughened, or pitted, sometimes narrowly winged.

About 200 genera and over 3,000 species, widely distributed from the tropics to the cool temperate regions of the world.

Corolla saccate, gibbous at base, or slipper-shaped:
  Corolla yellow, lower lip slipper-shaped; annual; capsule septicidal, valves split about to middle; stamens 2 .... *Calceolaria*
  Corolla white, pinkish, or red, slightly saccate at base, lower lip not slipper-shaped; shrubby perennial; capsule dehiscing by irregular subapical pores; functional stamens 4 .... *Galvezia*
Corolla neither saccate nor gibbous nor slipper-shaped:
  Corolla markedly galeate, lower lip minute, much shorter than the erect upper lip; bracts of inflorescence leaflike and often more or less brightly colored......... *Castilleja*
  Corolla not galeate, lower lip nearly equaling or exceeding upper; bracts of inflorescence not leaflike nor brightly colored:
    Corolla lobes longer than tubes, nearly regular; stamens 4 or 5, not didynamous, usually slightly but distinctly exserted:
      Leaves alternate; capsule septicidal and loculicidally dehiscent nearly or quite to base; seeds obovoid; sepals 5.................................. *Capraria*
      Leaves opposite; capsule septicidal, the valves only shallowly split at apex; seeds ellipsoid-cylindric; sepals 4 .... *Scoparia*
    Corolla lobes shorter than tube; stamens 4, didynamous or reduced to 2 functional ones, not exserted:
      Anther cells separated by arms of connective, the arms at right angles to filament; corolla blue-violet; leaves in whorls at least at some nodes, the rest opposite .... *Stemodia*
      Anther cells confluent or closely appressed to each other, not separated; corolla yellow, cream, bluish, or white, often with spots or stripes of deeper color; leaves opposite:
        Functional stamens 2, anterior pair reduced to filaments only or completely lacking; corolla white or lavender; capsule septicidally dehiscent, septum persistent, valves not split; sepals all alike; bracteoles on pedicels none....... .... *Lindernia*
        Functional stamens 4; corolla yellow or cream; capsule septicidal and valves split

* Contributed by Ira L. Wiggins.

short way from apex; sepals dimorphic, outer 3 ovate, much broader than oblong or linear inner pair; a pair of bracteoles at base of each pedicel.... .... *Mecardonia*

Calceolaria L., Mant. 2: 143, 171. 1771

Aquatic or terrestrial annual herbs or suffrutescent plants with opposite, petiolate, sessile or amplexicaul, entire, dentate or dissected leaves, and glabrous to densely pubescent herbage, often glandular. Calyx 4-cleft or 4-parted, the lobes ascending or more commonly spreading. Corolla bilabiate from short tube, imbricate in bud; upper lip smaller than lower, cucullate, entire to bilobed; lower lip saccate-slipper-shaped, often conspicuously veined. Stamens 2; anthers varying widely, from thecae fully joined to widely separated by connective, the thecae then both functional or 1 abortive. Style short; stigma entire or faintly bilobed; ovary conical, 2-celled. Capsule septicidal, with valves splitting to halfway to base. Seeds numerous, minute, often finely tuberculate.

The largest genus in the family, with about 400 species distributed from southern Mexico to southern South America; most abundant in the Andes.

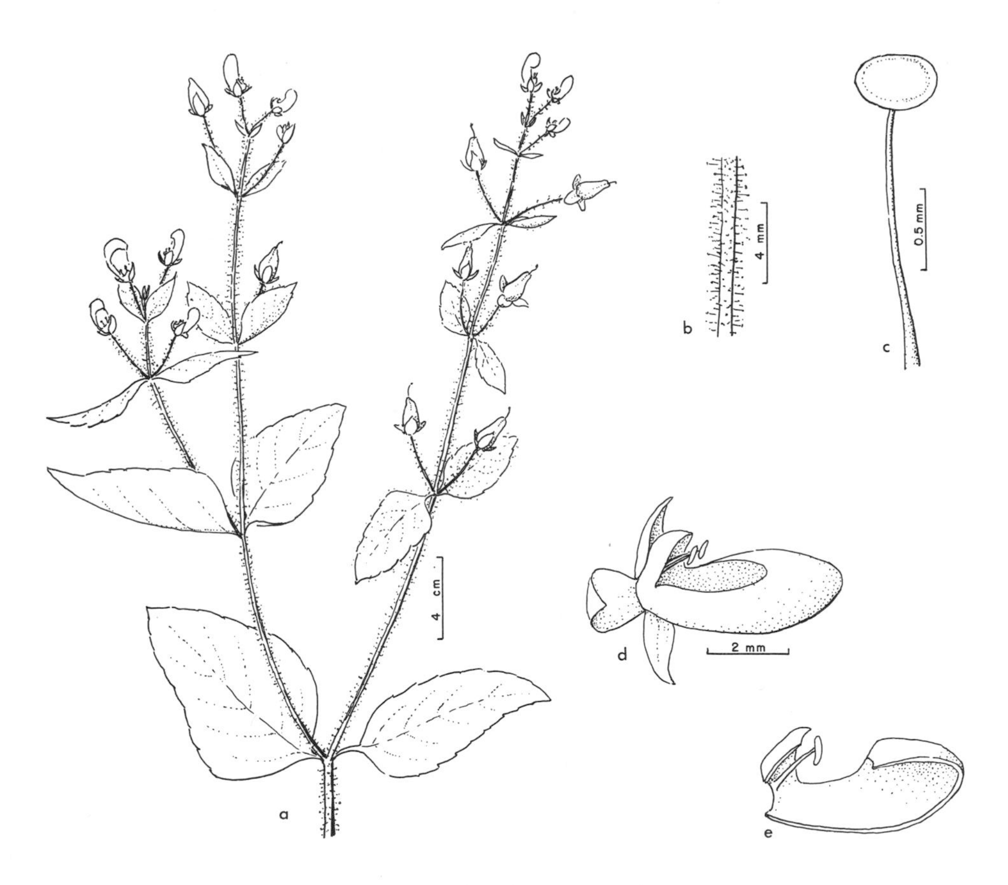

Fig. 113. *Calceolaria meistantha.*

Calceolaria meistantha Pennell, Proc. Acad. Nat. Sci. Phil. 103: 96, *f. 1.* 1951

Annual; stem laxly ascending, 3–15(30) cm tall, considerably branched above, densely villous-glandular on stems and pedicels, less densely so on leaf blades; petioles usually short, 2–10 mm long; blade ovate, 7–20(35) mm long, 4–16(20) mm wide, acute to obtuse, crenate to serrulate, dark green above, paler and slightly more densely villous beneath, the hairs less commonly glandular than on stems; pedicels filiform, 4–17 mm long, glandular-villous; sepals elliptic-ovate to suborbicular, 2–3.5 mm long; corolla yellow with 2 dark brown lateral spots within orifice, 5–6 mm long, upper lip ascending to arched-erect, lower 3–4 times as long as upper, broadly pouched, projecting forward and slightly curved upward distally; stamens 1.5 mm long, projecting forward and upward, filaments glabrous, anthers not lobed, thecae united to form broadly oval disk; style curved distally, 1–1.15 mm long; capsule ovoid, 5–6 mm long, narrowed to short, blunt beak.

In canyons and on moist banks and slopes, central Colombia south along the western flanks of the Andes into Peru, and recently reported from the Galápagos. Specimens recorded (but not seen by me): Fernandina, Isabela.

This exciting new record was reported by Eliasson (1968c, 244).

See Fig. 113.

Capraria L., Sp. Pl. 628. 1753; Gen. Pl. ed. 5, 276. 1754

Perennial herbs or low shrubs with alternate, exstipulate leaves and usually pubescent vegetative and floral elements. Leaves narrow, dentate, sometimes crowded on short spur branches. Flowers axillary, short- to long-pedicellate, borne singly or in few-flowered fascicles. Calyx of 5 distinct, narrow, acute, subequal sepals. Corolla white or faintly lined or tinged with lavender, green, or yellow, almost regular, narrowly to broadly campanulate, with 5 spreading lobes and short tube. Stamens 4 or 5, equaling throat or slightly exserted; anther sacs confluent and slightly divergent. Styles slender, usually slightly exceeding stamens; stigma bilobed, the 2 branches confluently in contact with each other soon after pollination. Capsules ovoid, firm, turgid, usually slightly longer than sepals, loculicidally and septicidally dehiscent. Seeds many, obovoid, pale reddish brown, finely reticulate.

A genus of 10 or 12 species distributed from Central America southward into central South America at low to moderate altitudes.

Leaves broadest above middle, obtuse at apex, strongly serrate-dentate on upper half, teeth to 1.5 mm long; herbage and inflorescences usually hirsute; pedicels mostly less than 5 mm long; style about 4 mm long, very slender........................*C. biflora*

Leaves broadest at or slightly below middle, acute-attenuate at apex, weakly and inconspicuously dentate-serrate; herbage and inflorescences glabrous; pedicels mostly 6–12 mm long; style 0.6–1 mm long, stouter...........................*C. peruviana*

Capraria biflora L., Sp. Pl. 628. 1753

Capraria semiserrata Willd., Sp. Pl. 3: 324. 1800.
Capraria hirsuta HBK., Nov. Gen. & Sp. Pl. 2: 355. 1818.
Capraria semiserrata var. *herteri* A. DC. ex Benth. in DC., Prodr. 10: 429. 1846.
Capraria biflora var. *pilosa* Griseb., Fl. W. Ind. 427. 1861.
Capraria biflora forma *hirta* Loes., Bull. Herb. Boiss. II, 3: 284. 1903.

Low, compact to straggly, ascendingly branched shrub 3–12 dm tall, with flowers usually borne in pairs in leaf axils; leaves oblanceolate, dentate-serrate, 1.5–6 cm

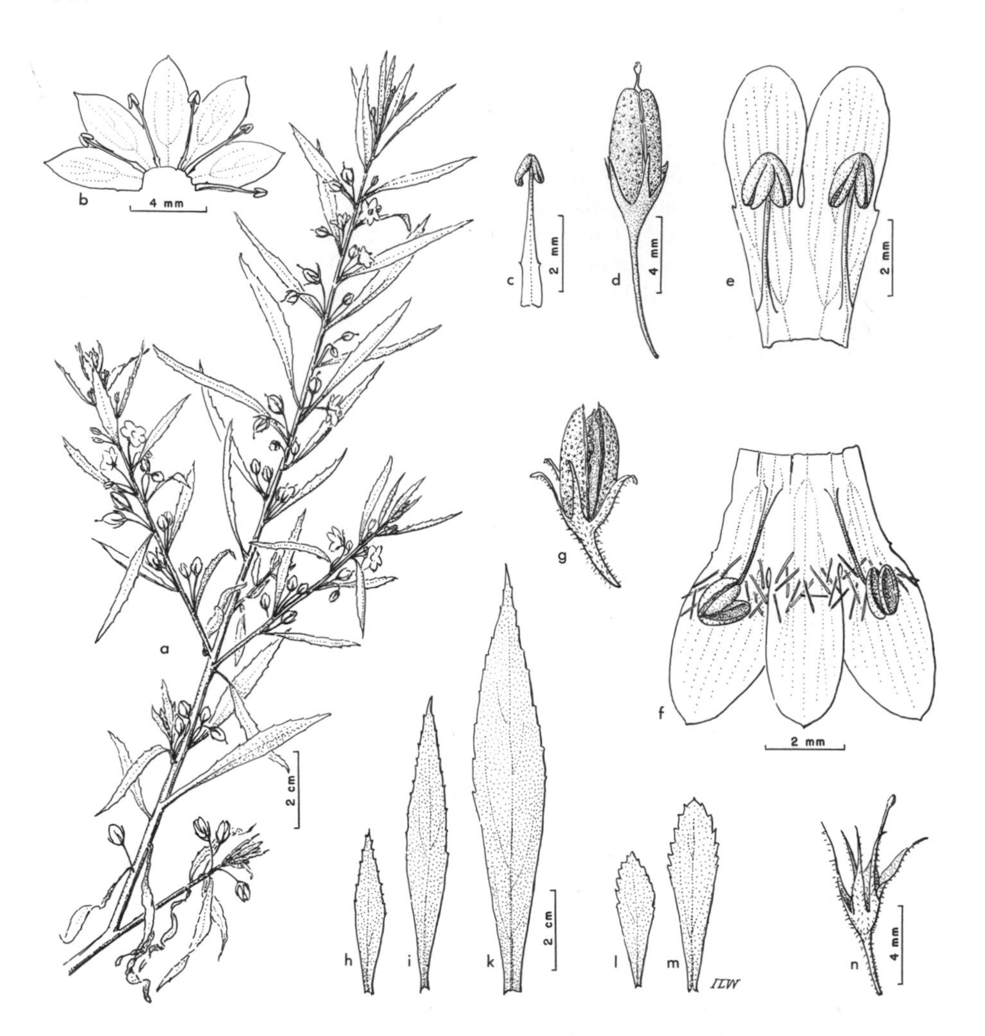

Fig. 114. *Capraria peruviana* (*a–k*). *C. biflora* (*l–n*).

long, 3–15 mm wide, sparsely to moderately hirsute, bright green above, paler beneath, closely set with sessile to subpunctate glands on both surfaces; pedicels slender, hirsute, 1–5 mm long; sepals lance-linear, 4–6 mm long, subequal, hirsute; corolla campanulate, white with lavender lines or sometimes pinkish, bilabiate, lobes spreading, limb 8–11 mm broad, tube 3.5–4.5 mm long, posterior lip of 2 connate lobes, anterior lip of 3 lobes, these about 2 mm wide, 4 mm long, rounded and slightly apiculate at apex, tips of all lobes rather tightly rolled backward soon after anthesis; throat with partial ring of stoutish, straight hairs 0.8–1.2 mm long below anterior lip; stamens 4, posterior pair exserted about length of their anthers, anterior pair about equaling corolla throat, all filaments glabrous; anthers bluish, 1.4–1.6 mm long; style about equaling posterior pair of stamens, stigma lobes about 0.5 mm long, spreading at anthesis, soon tightly adherent to each other; capsules 4–6 mm long, 3–4 mm wide, smooth, glabrous, brownish, closely set with punctate glands; seeds ellipsoidal, 0.4–0.5 mm long, pale tawny brown or reddish brown, finely reticulate.

Widely distributed in South America from Venezuela to Peru; common to scattered on several of the Galápagos Islands. Specimens examined: San Cristóbal, Santa Cruz, Santa María.

Because intergradation from plants that are subglabrate to those that are closely invested with spreading hairs is so gradual and complete, it seems futile to recognize var. *pilosa* Griseb.

See Fig. 114 and color plate 45.

Capraria peruviana Benth. in DC., Prodr. 10: 430. 1846

Xuaresia biflora R. & P., Syst. Veg. Fl. Peruv. 46. 1798.
?*Witheringia salicifolia* Hook., Bot. Misc. 2: 231. 1831.

Shrub 1–2 m tall with slender, ascending branches and glabrous herbage; leaves crowded, petioles 2–3 mm long; blades narrowly elliptic-lanceolate to linear-lanceolate, usually broadest about middle or slightly below, 3–11 cm long, 2–22 mm wide, dark green above, slightly paler beneath, glabrous, margins shallowly and irregularly serrate-denticulate from about one-third to halfway from base to tip upward, minutely glandular on both surfaces, acute-attenuate at tip; flowers 1–3 in axils of upper leaves; pedicels slender, 5–12 mm long, glabrous; sepals ovate-lanceolate, 3–4 mm long, 0.8–1.2 mm wide, about half as long as capsules, glabrous; corolla pale greenish yellow to greenish white, broadly campanulate, 5–6 mm long, 8–10 mm broad when fully open, the tube and throat together about 2 mm long, lobes 3–3.5 mm long, 2–2.3 mm wide at base, acute-apiculate at apex; stamens inserted at base of tube and adnate to it nearly length of tube, free part of filament 2.5–3 mm long; anthers about 0.8 mm long, grayish; stigma lobes 0.3 mm long; style stoutish, 0.5–0.8 mm long; capsules ovoid, slightly compressed contrary to septum, 5–7 mm long, 3–3.5 mm broad, blunt at apex; seeds many, obovoid, 0.3–0.4 mm long, pale reddish or tawny, minutely reticulate.

A shrub of the moister forest and brushy areas, Colombia to Peru. Specimens examined: Isabela, San Salvador, Santa Cruz, Santa María.

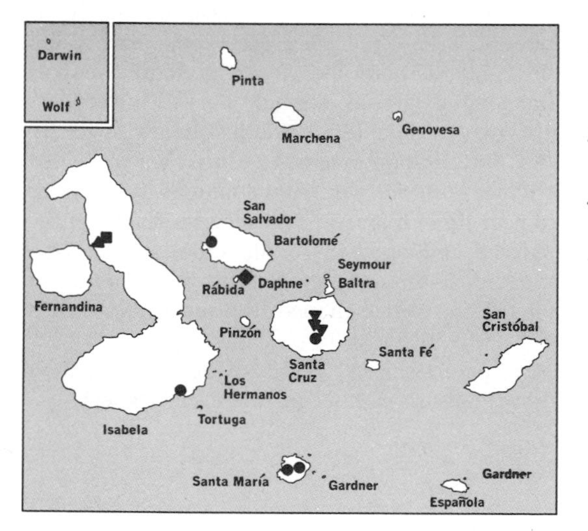

● *Capraria peruviana*
■ *Castilleja arvensis*
▲ *Galvezia leucantha* subsp. *leucantha*
◆ *Galvezia leucantha* subsp. *pubescens*
▼ *Lindernia anagallidea*

The leaves on one collection examined (*Schimpff* 99 [CAS, GH]) were nearly twice the size of those on most specimens, probably having come from a vigorous plant growing under especially favorable conditions.

See Fig. 114.

Castilleja Mutis ex L. f., Suppl. Syst. Veg. 47. 1781

Erect or ascending annual or perennial herbs or suffrutescent at base, with alternate leaves and spikelike racemes; hemiparasitic roots attached to other plants, at least among many perennial species. Bracts entire or variously lobed or divided, usually colored, at least at tip. Calyx of 2 or 4 lobes unequally separated anteriorly and posteriorly, the lateral sinuses shallower than other 2 or lacking. Corolla greenish to tinged with yellow or red along margins of galea, 2-lipped; upper lip (galea) united to tip into narrow, often hoodlike structure, this usually including or enfolding stamens and style; lower lip short, often rudimentary, usually of 3 short teeth. Stamens 4, didynamous; anther cells attached unequally at slightly different levels near tip of filament. Style slender; stigma bilobed to subcapitate. Capsule ovoid to cylindric-ovoid, turgid, glabrous, loculicidally dehiscent. Seeds many, small, cylindric to irregularly angled; testa loose, reticulate, thin.

About 200 species, all native to the New World except 1, which extends into Asia.

Castilleja arvensis Cham. & Schlecht., Linnaea 5: 103. 1830

Castilleja communis Benth. in DC., Prodr. 10: 529. 1846.

Upright annual herb 3–5 dm tall, hirsute with spreading, stiffish white hairs; stem erect and strict or ascendingly branched above middle; leaves linear to narrowly lanceolate, 3.5–5 cm long, 2–4.5 mm wide, acute at apex, sinuous-undulate along margins above middle, sessile or nearly so, scaberulous-puberulent with hairs most numerous along margins and on veins; inflorescences 10–20 cm long; bracts similar to leaves, 2–4 cm long, entire, tips of uppermost reddish, the calyx 12–13 mm long, divided into 2 equal, entire lobes about 4 mm wide; anterior sinus about 1 mm deeper than posterior, extending about one-third to halfway to base; corolla 10–12 mm long, included between calyx lobes, galea 5–6 mm long, erect, scarcely hooded, lower lip of 3 minute teeth, the laterals about 2 mm long, erect or slightly curved inward, central tooth less than half as long as laterals, strongly curved inward; corolla tube slender, with 2 oblong, shallow pits on anterior side just below lower lip, each pit about 1.5 mm long; stamens 4, anterior pair exserted 2–3 mm beyond galea, their anthers stuck to underside of stigma, posterior pair half as long as the anterior, included, their anther cells separated near tip of filament; style slender, exserted 2–3 mm beyond galea, often strongly curved or looped in corolla tube; stigma lobes about 1 mm long; capsules broadly ovoid, 5–6 mm long, dark brown, smooth and glabrous; seeds 0.8–1 mm long, cylindric to wedge-shaped, the testa pale brown, papery, finely reticulate.

In grassy areas and among openly spaced shrubs, central Mexico to northern South America; in the Galápagos known only from 1 record from Isla Isabela.

See Fig. 115.

Fig. 115. *Castilleja arvensis* (*a–e*): *c*, *e*, corolla, side and face; *d*, calyx, side. *Galvezia leucantha* subsp. *leucantha* (*f–r*): *i*, insertion of sterile stamen; *k*, corolla split down side; *l*, glandular hairs from stamen base; *n*, hairs from outside of corolla.

Galvezia Dombey in Juss., Gen. Pl. 119. 1789

Perennial herbs or shrubs with slender, elongate, glabrous to minutely glandular-pubescent branches and opposite (rarely alternate), or ternately whorled, entire leaves. Flowers axillary, the upper often subracemose by reduction of leaves. Pedicels slender, often sharply deflexed after anthesis. Calyx 5-parted into narrow, subequal sepals. Corolla tubular, bilabiate, somewhat saccate abaxially at base; throat open or partially closed by 2 palatine ridges just interior to sinuses; posterior lip erect or nearly so, shallowly bilobed; anterior lip 3-lobed, spreading. Functional stamens 4, more or less didynamous, anterior pair often equaling corolla tube, other pair smaller. Fifth stamen small and abortive. Anther cells confluent. Style slender; stigma weakly bilobed, lobes connivent. Capsule ovoid to subglobose, 2-celled, dehiscing by irregular subterminal pores. Seeds minute, truncate to acute at both ends, minutely ridged, or cristate-echinate to smooth, usually dull black.

About 6 species and several varieties in southern California, Baja California, and the western slopes of the Andes into Peru.

Foliage, young twigs, pedicels, and calyces glabrous or sparsely puberulent; corolla waxy white . . . . *G. leucantha* subsp. *leucantha*
Foliage, young twigs, pedicels, and calyces densely glandular-puberulent; interior of corolla sometimes reddish or pink . . . . *G. leucantha* subsp. *pubescens*

Galvezia leucantha Wiggins, Occ. Papers Calif. Acad. Sci. 65: 1, *figs. a–l.* 1968, subsp. **leucantha**

Intricately branched shrub to 1.5 m tall, with opposite, short-petiolate leaves, these bearing flowers singly in their axils; petioles 3–7 mm long; blades elliptic-lanceolate to ovate-lanceolate, 1.5–3.5 cm long, 5–15 mm wide, broadly cuneate to rounded at base, acute to obtuse at apex, entire, dark green above, paler beneath, glabrous or sparsely puberulent when young; pedicels slender, 5–12 mm long, glabrous, spreading-ascending, bent abruptly at apex, to 2.5 cm long in fruit; calyx with its shallow cup separated from sepals by sharp groove, sepals ovate, 3–4 mm long, adaxial one slightly longest, glabrous to sparsely puberulent; corolla waxy white, 10–12 mm long, outer surface puberulent with mostly nonglandular hairs, tube about 1.5–2 mm broad, slightly ampliate, the limb bilabiate, posterior lip 3–3.5 mm long, bilobed about halfway to base, spreading-ascending, moderately puberulent and sparsely glandular within, more densely so on outer surface, lower lip trilobate, 3.5–4 mm long, sinuses between lobes 2–2.5 mm deep, central lobe cucullate, the tissue immediately below sinuses folded into densely glandular-puberulent palatine ridges 1.5–2 mm long; posterior pair of stamens about 6 mm long, filaments glabrous, anterior pair 1.5–2 mm longer than the posterior, their filaments coarsely glandular-papillate 1.5–2 mm above geniculae, glabrous below and above this area; anthers about 1.5 mm long; fifth stamen nonfunctional, inserted 1.5–2 mm above base of tube, to 2 mm long; capsules subglobose, 3–4.5 mm long; seeds 0.4–0.5 mm long, broadly pointed to subtruncate at apex, more narrowly pointed at base, with 4–7 cristate-echinate ridges running longitudinally, dark brown to black at maturity.

Endemic. Specimens examined: Isabela, San Salvador.

See Fig. 115.

Galvezia leucantha subsp. **pubescens** Wiggins, Occ. Papers Calif. Acad. Sci. 65: 6, *f.m.* 1968

Similar to subsp. *leucantha* in all respects except that the foliage, young twigs, pedicels, and calyces are densely glandular-puberulent, and the inside of the corolla is sometimes reddish or pink.

Endemic; on lava ridges and cinder slopes, central Galápagos Islands. Specimens examined: Rábida.

Lindernia All., Misc. Taurin. 3: 178, *t. 5, f. 1.* 1766

Erect to diffusely branched or rarely decumbent annual herbs with tetragonal stems and opposite leaves. Flowers borne in axils of leaves, pedicellate. Calyx of (4)5 more or less distinct sepals. Corolla bluish to white with darker spots and lines, bilabiate, the posterior lip usually erect or slightly curved inward, smaller and shorter than anterior lip, this deeply 3-lobed, spreading. Functional stamens 2, their anthers connivent, frequently touching each other and stigma, locules of anthers confluent; anterior pair of stamens represented by filaments only. Style short, slender, usually sharply bent upward at tip; stigma 2-lipped. Ovary 2-celled. Capsule ovoid, septicidally dehiscent, the septum remaining attached at base and persistent after seeds fall. Seeds numerous, yellowish brown, longitudinally 4–6-ribbed and transversely lined with minute ridges.

About 70 species, mostly in the warmer parts of the Old World; some from northern United States into central South America.

Lindernia anagallidea (Michx.) Pennell, Acad. Nat. Sci. Phil. Monogr. 1: 152. 1946

?*Gratiola inaequalis* Walt., Fl. Carol. 61. 1788.
Gratiola anagallidea Michx., Fl. Bor.-Amer. 1: 6. 1803.
Lindernia dilatata Muhl. in Ell., Sketch Bot. So. Car. & Ga. 1: 16. 1816.
?*Ilysanthes geniculata* Raf., Antik. Bot. 45. 1840.
Ilysanthes anagallidea Raf., op. cit. 46.
Ilysanthes inaequalis Pennell, Torreya 19: 149. 1919.

Erect and strict to diffusely branched herb 0.5–3 dm tall, glabrous or nearly so; leaves broadly ovate, 3–15 mm long, 2–10 mm wide, sessile, rounded and subclasping at base, acute at apex, entire or with 1 or 2 small teeth above middle; pedicels slender, 5–10 mm long at anthesis, to 2.5 cm long at maturity; sepals 5, distinct almost to base, lance-oblong, 1.5–2.5 mm long, sparsely stipitate-glandular; corolla 3.5–8 mm long, pale lavender to nearly white, margins of lobes often bluish or purplish, upper lip about 1 mm long, exterior in bud, shallowly emarginate, the lower 3-lobed, lobes about 1.5 mm long, spreading, a small tuft of straight, thickish, yellowish hairs in each sinus between upper and lower lips; functional stamens 2, inserted above middle of corolla tube, anthers often in contact with each other and with stigma, filaments glabrous; style slender, 1.5–2 mm long, curved upward near tip; capsules 4–5 mm long, pale yellowish brown, septa persistent; seeds yellowish brown, truncate at base, obtuse at apex, about 0.3 mm long, 4–6-ridged longitudinally and transversely lined between ridges.

Widely distributed in moist soil from New Hampshire and Oregon southward through Mexico, Central America, and northern South America. Specimens examined: Santa Cruz, a new record for the Galápagos Islands.

See Fig. 116.

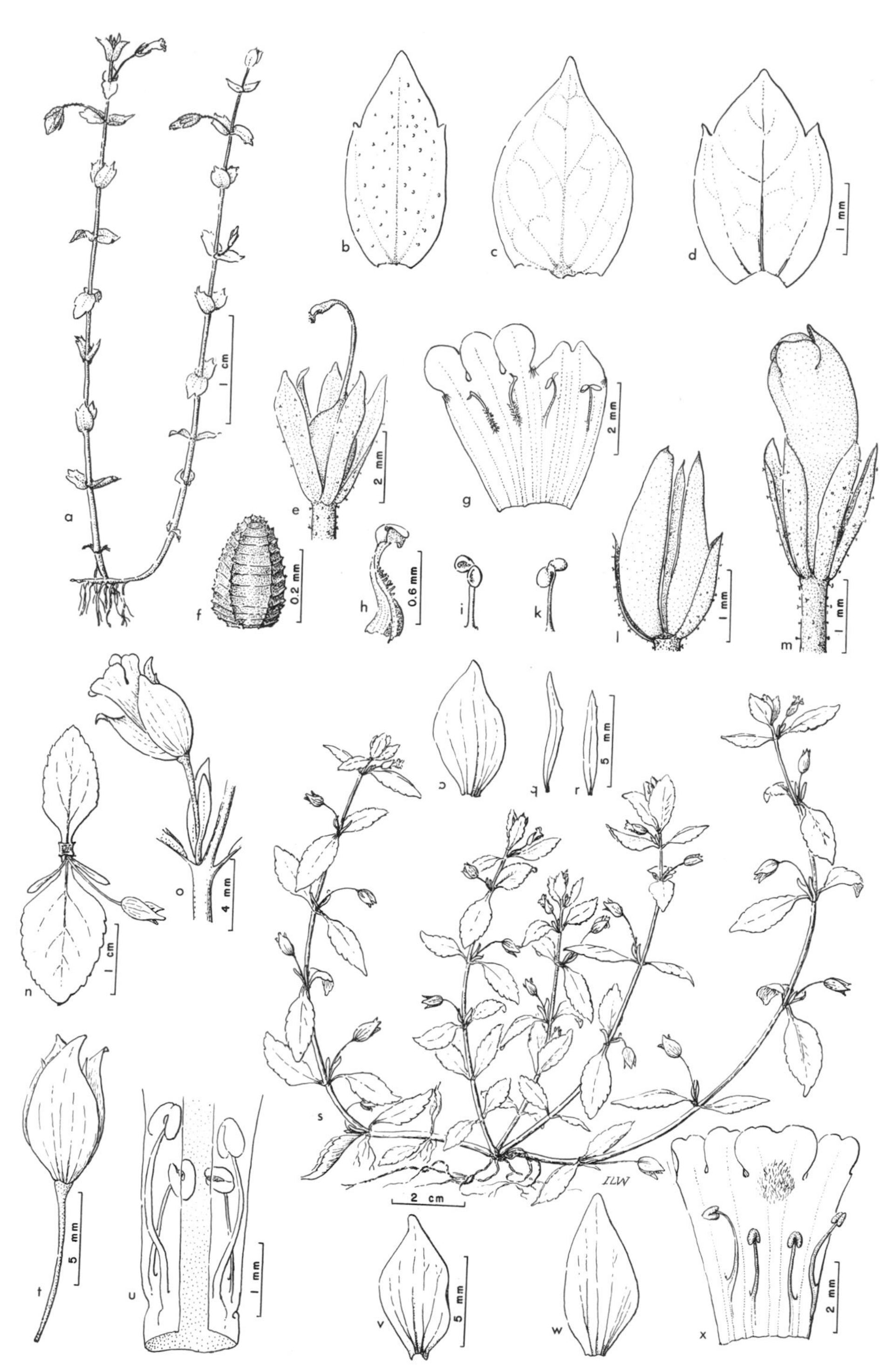

Fig. 116. *Lindernia anagallidea* (*a–m*): *g*, corolla, split along side; *h–k*, upper and lower stamens. *Mecardonia dianthera* (*n–x*): *u*, base of corolla tube and stamen insertion; *x*, corolla, split along side.

Mecardonia R. & P., Prodr. Fl. Peruv. et Chil. 95. 1794

Erect to diffuse, much-branched, glabrous herbs from perennial or annual roots. Stems slender, sometimes rooting at nodes, angled. Leaves opposite, short-petiolate or sessile, dentate, more or less punctate-glandular. Flowers axillary to upper leaves or bracts. Bractlets 2, at base of slender pedicel, often obscuring buds, exceeded by pedicels at anthesis. Sepals 5, unequal, outer 3 much wider. Corolla bilabiate, lobes shorter than tube, posterior lobes more or less united, basally pubescent within. Stamens 4, didynamous, posterior pair shorter than anterior. Style short, often abruptly turned upward; stigma lobes liplike. Capsule ovoid to subcylindric, acute, glabrous, septicidally dehiscent, valves again splitting short way at apex. Seeds minute, numerous, wingless, the testa finely reticulate.

Some 14 or 15 species, largely native to eastern temperate North America, the West Indies, and eastern South America, but some along the western parts of both continents.*

Mecardonia dianthera (Sw.) Pennell, Proc. Acad. Nat. Sci. Phil. 98: 87. 1946

Lindernia dianthera Sw., Prodr. Veg. Ind. Occ. 92. 1788.
Mecardonia ovata R. & P., Prodr. Fl. Peruv. et Chil. 95. 1794.
Herpestis chamaedryoides HBK., Nov. Gen. & Sp. Pl. 2: 369. 1818.

Erect to procumbent herb with slender stems 1–4 dm long; petioles 2–5 mm long, broadening gradually into blades, these broadly ovate, 6–22 mm long, 2–12 mm wide, crenate-dentate, the base broadly cuneate to rounded, apex acute to obtuse or emarginate; flowers borne singly in axils of all except lowermost leaves, each subtended by 2 bractlets at base of pedicels; bractlets leaflike, 3–7 mm long, entire or faintly denticulate; pedicels slender, 1–1.5 cm long at maturity; 3 outer sepals ovate, 4–5 mm long and 3 mm wide at anthesis, 5–9 mm long in fruit, rounded at base, acute at apex, entire; inner sepals subulate to narrowly lanceolate, 4.5–6 mm long, 0.4–0.6 mm wide; corolla bilabiate, 6–6.5 mm long, light yellow with brownish lines on lips, these moderately spreading; tube faintly ampliate, a patch of pale reddish brown hairs in throat just beneath posterior lip, rest of throat and tube glabrous; stamens 4, didynamous, posterior pair inserted slightly lower on tube and shorter than anterior pair; anthers 0.4–0.6 mm long; style about 1 mm long, curved upward sharply just beneath the liplike stigma lobes; capsule brownish, 4–5 mm long, cylindric, tapered very slightly upward, enfolded by the slightly accrescent sepals; seeds about 0.2 mm long, dark brown, the testa minutely favose-reticulate.

A weedy plant fairly common in tropical America. Specimens examined: Isabela, Santa Cruz, Santa María.

See Fig. 116.

Scoparia L., Sp. Pl. 116. 1753; Gen. Pl. ed. 5, 52. 1754

Erect, much-branched perennial herbs or in the tropics often suffrutescent, with opposite, petiolate leaves and axillary, small flowers. Leaf blades elliptic to ovate, serrate, minutely glandular-punctate. Flowers 1–3 in an axil, on slender pedicels. Sepals 4, equal. Corolla white or pinkish, tube short, densely hairy within, lobes 4, equal or nearly so, spreading to nearly rotate, much longer than tube, glabrous. Stamens 4,

* For further information on the genus, see Pennell (1946, 87).

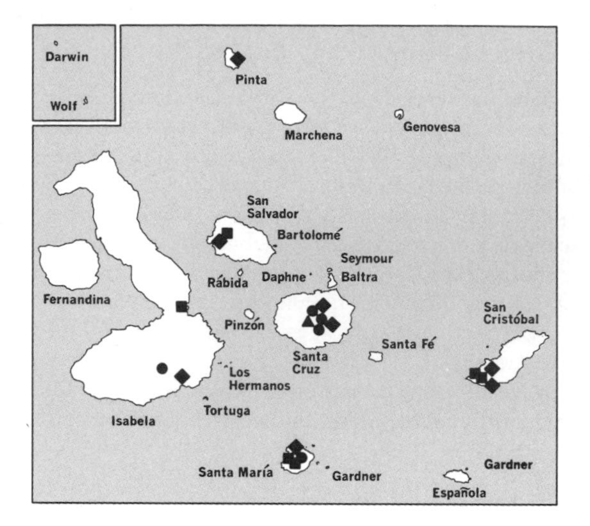

inserted at or near base of corolla tube. Ovary ovoid, acute at apex; style slender, about as long as ovary; stigma lobes flat and short, almost united. Capsule ovoid, acute, septicidally dehiscent and only slightly split at tips of valves. Seeds many, minute, ellipsoidal-cylindric, finely and irregularly tuberculate-reticulate in indefinite longitudinal rows or lines, often somewhat angulate.

About 20 species, mainly in tropical America but ranging as far north as Georgia; introduced into other parts of the world.

Scoparia dulcis L., Sp. Pl. 116. 1753

Gratiola micrantha Nutt., Amer. Jour. Sci. 5: 287. 1822.
Capraria dulcis Kuntze, Rev. Gen. Pl. 2: 459. 1891.
Scoparia grandifolia Nash, Bull. Torrey Bot. Club 23: 105. 1896.

Usually suffrutescent, to 1.5 m tall; younger stems pale green, glabrous, minutely punctate-glandular, obscurely tetragonal; petioles 3–8 mm long, merging into blade, this elliptic to lanceolate, 7–25 mm long, 3–12 mm wide, glabrous, dark green above, paler beneath, punctate-glandular on both surfaces, serrate; flowers 1–3 in an axil, appearing racemose by reduction of leaves upward; pedicels slender, 2–3 mm long at anthesis, to 10 mm long in fruit; sepals 4, ovate to oblong-elliptic, obtuse to rounded at apex, about 2 mm long, 3-veined, sparsely punctate-glandular; corolla rotate or nearly so, 3–4 mm broad, lobes about 2 mm long, white to pale pink, tube 0.5–0.8 mm long, this and apex of throat densely hirsute within; stamens 3–3.5 mm long, erect or slightly ascending-spreading, anthers 0.8–1.2 mm long, white; style 1–1.5 mm long; stigma lobes short, flattened, turned toward anthers by sharp bend at tip of style; capsule broadly ovoid, 3–3.5 mm long, septicidally dehiscent and valves split one-fourth to two-thirds the distance to base; seeds 0.3–0.4 mm long, light brown, usually minutely apiculate at funicular end, more or less angulate.

Widely distributed and often weedy in tropical America, mostly along trails, roads, and margins of fields. Specimens examined: Isabela, San Cristóbal, San Salvador, Santa María.

Said to have insecticidal properties, and sometimes steeped in bath water to kill body vermin.

See Fig. 117.

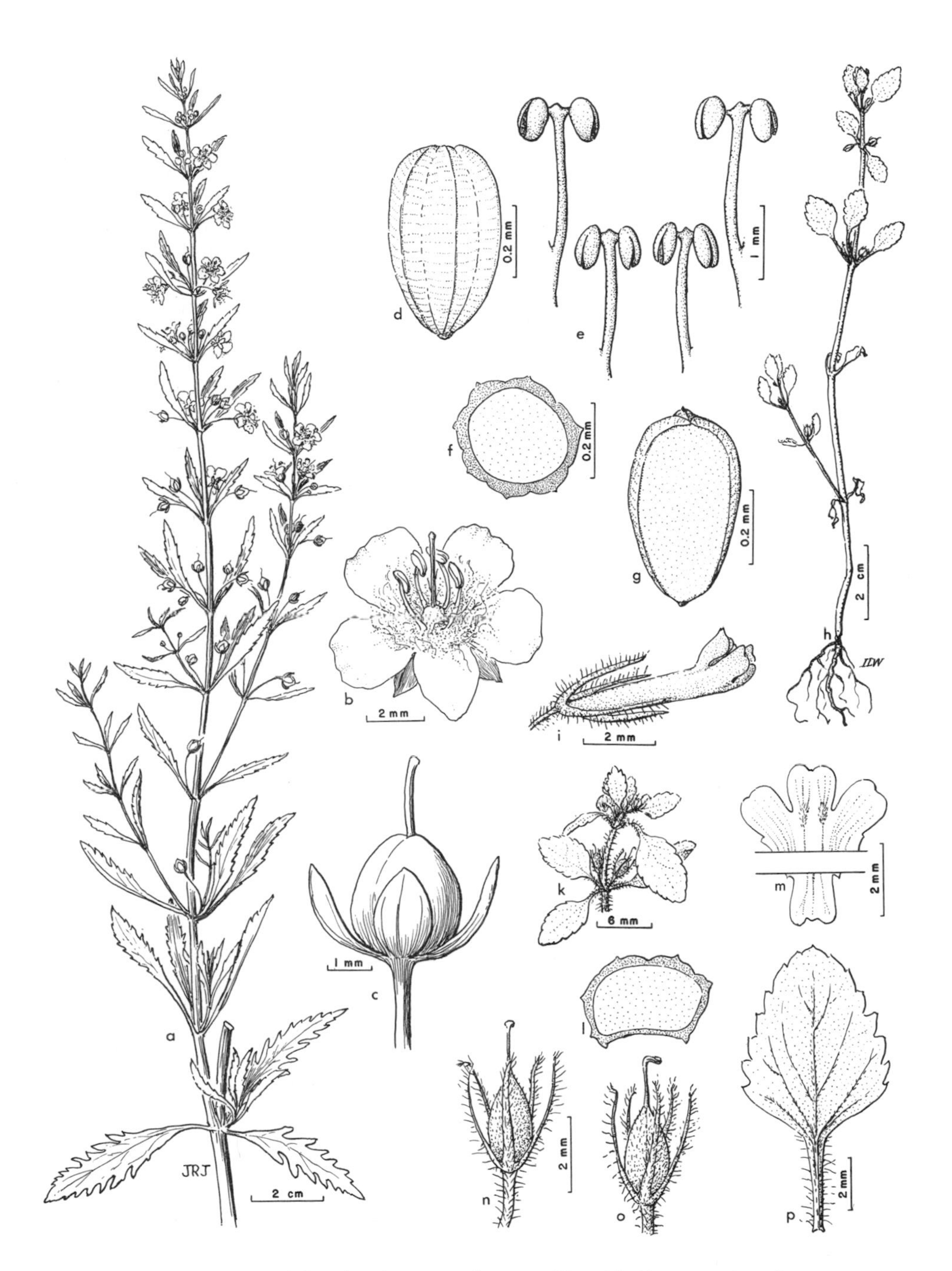

Fig. 117. *Scoparia dulcis* (*a–c*). *Stemodia verticillata* (*d–p*): *e*, spacing of stamens; *f*, *g*, seed, in transection and inner face; *l*, transection of seed crowded in capsule.

Stemodia L., Syst. Nat. ed. 10, 1118. 1759

Procumbent to erect annual or perennial herbs with opposite or verticillate leaves and small or medium-sized flowers in axils of leaves or of narrow bracts in loose racemes. Bracteoles 2, just below calyx or lacking. Sepals 5, distinct, linear-lanceolate, equal. Corolla usually blue-violet, tube cylindrical, limb 2-lipped, upper lip arched, its 2 lobes free one-third its length, lower lip somewhat deflexed-spreading, the 3 lobes distinct, hirsute-pubescent near bases. Stamens 4, didynamous, upper pair the shorter, anther cells separated by short arms of connective, these standing at right angles to filament. Stigmas 2, flattened, nearly equal. Capsule ovoid to cylindric-ovoid, septicidal, each valve splitting nearly to base. Seeds many, 1–3 times as long as broad, minutely and shallowly reticulate, white or faintly tinged with brown.

About 20 species, primarily of the American tropics but sparingly present in warm temperate regions.

Stemodia verticillata (Mill.) Sprague, Kew Bull. 1921: 211. 1921

Erinus verticillatus Mill., Gard. Dict. ed. 8, no. 5. 1768.
Stemodia parviflora Ait., Hort. Kew. ed. 2, 4: 52. 1812.
Stemodia arenaria HBK., Nov. Gen. & Sp. Pl. 2: 357, *pl. 175.* 1817.
Conobea pumila Spreng., Nov. Proc. 13. 1819.

Erect and strict to spreadingly branched annual herb 5–25 cm tall, with viscid-pubescent stems, petioles, calyces, and pedicels; leaves opposite or in verticils of 3, crowded near ends of branches; petioles 3–6 mm long, densely hirsute, with numerous sessile to subsessile glands among nonglandular hairs; leaf blades broadly ovate, 5–10 mm long, 3–8 mm wide, broadly cuneate to rounded at base, mostly obtuse at apex, crenate-serrate, sparsely pubescent adjacent to petioles and along major veins beneath, bearing small subpunctate glands; flowers axillary, 1–3 at a node, pedicels 0.5–1.5 mm long, glandular-hirsute; sepals distinct, 2.5–3.5 mm long, linear or linear-oblanceolate, long-hirsute, somewhat glandular; corolla 2-lipped, 4–6 mm long, blue-violet to purplish, upper lip emarginate, the lower with 3 distinct spreading lobes about 1–1.5 mm long; tube very short; throat gradually ampliate to about twice diameter of tube, bearing tuft of hairs just within orifice beneath upper lip, rest of throat and tube glabrous; stamens free from corolla tube about halfway up throat, the pair subtending upper lip slightly shorter than, and their anthers barely reaching level of, insertion of abaxial pair, the anther cells distinct, borne at ends of connective arms; capsules ovoid to subglobose, 2.5–3.5 mm long, spreading-hirsute with delicate hyaline hairs, septicidally dehiscent, both valves splitting nearly to base; seeds pale tan to nearly white, about 0.4 mm long, bearing 6–10 low, brownish, longitudinal ridges, obscurely striate transversely.

Widely distributed along trails and roads throughout much of tropical America. Specimen examined: Santa Cruz, a new record for the Galápagos Islands.

See Fig. 117.

SOLANACEAE. POTATO FAMILY*

Herbs or shrubs with mostly alternate leaves (upper leaves opposite in *Petunia*). Flowers perfect, regular, solitary, umbellate, cymose, or paniculate, axillary or termi-

* Exclusive of genera *Cacabus, Lycopersicon, Physalis,* and *Solanum,* contributed by Ira L. Wiggins.

nal. Calyx rotate, campanulate, or tubular, 5-toothed or 5-cleft (sometimes 2–4-toothed, as in some species of *Lycium*), usually persistent and in some inflated in fruit (*Physalis, Nicandra*). Corolla tubular, campanulate, funnelform, or rotate, 5-lobed; limb valvate or plicate in bud. Stamens 5, inserted in corolla tube and alternate with lobes; anthers ovoid to linear, dehiscing longtitudinally or by terminal pores or slits, often connivent and closely appressed to style. Ovary superior, 2–5-celled, several- to many-ovuled; style 1, slender or stout; stigma capitate, entire or shallowly 2-lobed. Fruit a berry or capsule. Seeds several to many, ovoid, subspherical, or strongly compressed. Embryo subperipheral, straight or strongly curved. Endosperm scanty.

About 85 genera, heavily concentrated in northern South America but extending into the temperate regions of the Old and New Worlds.

Corollas slightly irregular, the limb set slightly oblique to tube .... *Browallia*
Corollas regular, the limb at right angles to tube, not oblique:
  Corolla tubular, funnelform, or narrowly campanulate, not rotate:
    Plants shrubby, 1–3 m tall:
      Twigs not spinose; corollas 2.5–3 cm long, lobes much shorter than tube and throat combined; stamens included .... *Acnistus*
      Twigs spinose; corollas 5–8 mm long, the lobes equaling or exceeding the tube and throat; stamens exserted:
        Fruits drupe-like, with 4 2-seeded nutlets in each; corolla tube pubescent...... .... *Grabowskia*
        Fruits berry-like, several-seeded, without nutlet-like covering; corolla tube glabrous .... *Lycium*
    Plants herbaceous:
      Fruit a berry; calyx accrescent, nearly or totally investing the berry at maturity:
        Flowers truly axillary; sepals connate except at apex; corolla white or whitish; seeds wavy-lamellate .... *Cacabus*
        Flowers borne at one side of axils, not truly axillary; sepals distinct almost to base; corolla blue; seeds foveolate-punctate.................. *Nicandra*
      Fruit a capsule; calyx not accrescent nor investing major part of fruit at maturity:
        Capsules 3–5 cm long, armed with prickles or spines; calyx deciduous before maturity of fruit; corolla 6–8 cm long, broadly funnelform; seeds 3–5 mm long .... *Datura*
        Capsules 1–1.8 cm long, unarmed; calyx persistent; corolla 8–35 mm long, salverform or campanulate; seeds minute, less than 1 mm long........ *Nicotiana*
  Corolla broadly rotate-spreading, the tube very short:
    Calyx accrescent, bladdery-inflated, enclosing the berry through maturity.... *Physalis*
    Calyx not accrescent, never enclosing the berry:
      Flesh of fruits and seeds piquantly pungent (peppery); leaves usually geminate, one of each pair smaller than the other....................... *Capsicum*
      Flesh of fruits and the seeds not pungent; leaves rarely geminate or of different sizes at alternate nodes:
        Leaves compound; flowers in cymes, corymbs, or racemes; anthers with sterile tips, dehiscing longitudinally full length.................. *Lycopersicon*
        Leaves simple to unevenly pinnatifid; flowers in pedunculate inflorescences; anthers dehiscent by apical pores............................ *Solanum*

Acnistus Schott., Wien. Zeitschr. 4: 1180. 1829

Shrubs or trees with glabrous to slightly tomentose young growth and leaves, usually unarmed or rarely with solitary nodal spines. Leaves entire, thin and membranous to coriaceous. Flowers solitary or in fascicles, long-pedicellate. Calyx campanulate, truncate, shallowly 5-toothed or 5-lobed, slightly or not at all accrescent in fruit. Corolla tubular, subcampanulate or funnelform, 5(7)-lobed, more or less induplicate-valvate in bud, spreading in anthesis, the sinuses with introrse plications or naked. Stamens same in number as corolla lobes, inserted on and adnate to base of corolla tube, alternate with lobes; filaments slender, exserted or included; anthers ovate to oblong, basifixed, 2-celled, dehiscing longitudinally. Disk fleshy. Ovary 2-celled; style filiform; stigma bilobed, lobes short, flat, and spreading; ovules many. Berry globose, fleshy or pulpy, many-seeded. Seeds compressed, rugulose; embryo strongly curved, semiperipheral; cotyledons semiterete.

Some 40 to 50 species,* distributed from central Mexico and the West Indies to Peru and Argentina.

Acnistus ellipticus Hook. f. in Miers, Lond. Jour. Bot. 4: 343. 1845

Acnistus insularis Robins., Proc. Amer. Acad. 38: 198. 1902.

Erect to ascendingly branched shrub or small tree 2–5 m tall, with brown-barked, stout branches; leaves crowded near ends of main branches at tips of lateral twigs; youngest twigs sparsely hirsute with somewhat crispate hairs, soon glabrate; leaves variable in size; petioles slender, 0.5–4 cm long, slightly winged, channeled on upper side, crispate-pubescent to glabrate; blades thin, elliptic to obovate, 3–20 cm long, 1.5–8 cm wide, usually obtuse to rounded but sometimes acute or emarginate at apex, broadly cuneate to rounded at base, yellow-green, glabrous to sparsely crispate-hirsute above, glabrate to moderately crispate-tomentose beneath; flowers in fascicles of 2–8 in axils of leaves (or in axils of leaf scars on older wood) or rarely solitary, pendulous; pedicels 0.5–3.5 cm long; calyx broadly campanulate, 4–5 mm wide and about as long at anthesis, glabrous or sparingly crispate-hirsute without, 5-lobed to subtruncate, lobes rounded, thin, accrescent to 2 or 3 times their flowering size in fruit; corollas waxy white, drying brown, funnelform, 2.5–3 cm long, about 1 cm broad at throat, limb 2–2.5 cm broad, lobes 8–10 mm long, 6–7 mm wide, rounded, crispate-tomentose along margins and outside; sinuses unappendaged; stamens about equaling corolla throat, adnate to tube half their length; filaments slightly wider at base than above, subequal; anthers oblong-linear, 3–3.5 mm long; style 3–5 mm longer than stamens, glabrous; stigma bilobed; fruit globose, 10–12(25) mm in diameter, borne in the bowl-shaped, papery-thin calyx 8–10 mm deep, 2–2.5 cm across; berry black, the exocarp leathery; seeds angled, tuberculate-rugulose.

Endemic; in *Scalesia* forest and upper slopes of peaks at about 240 to 625 m altitude. Specimens examined: Isabela, Pinzón, San Cristóbal, San Salvador, Santa Cruz, Santa María.

See Fig. 118.

* The genus is badly in need of critical work. Sleumer (1950) reduced *Acnistus* to synonymy under *Dunalia* and transferred a number of species native in Argentina to the latter genus. He included *A. ellipticus* Hook. f. in Miers in his list of species "probably to be transferred into *Dunalia* or reduced to synonymy," but did not make the transfers or reductions. Until critical studies, including all genera within the Lyciinae are made, it seems hazardous to make uncertain transfers, and *Acnistus* is therefore retained as a recognizable genus. The Galápagos material has stamens with filaments adnate to the corolla tube fully half their length, and without a trace of basal or lateral appendages.

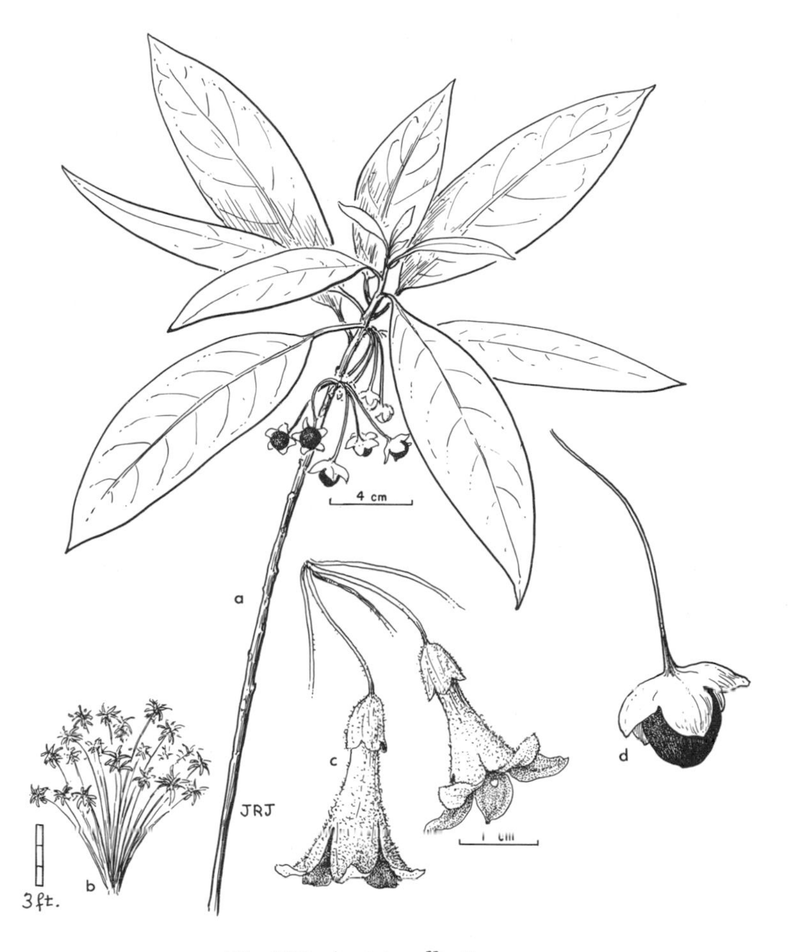

Fig. 118. *Acnistus ellipticus.*

Browallia L., Sp. Pl. 631. 1753; Gen. Pl. ed. 5, 278. 1754

Annual or perennial herbs, occasionally becoming woody near base, with thin, entire, petiolate, alternate to subopposite leaves and solitary to racemose, salverform flowers. Calyx narrowly to broadly campanulate, becoming broader in fruit, the lobes or teeth much shorter than to equaling tube. Corolla tube slender, slightly ampliate upward, sparingly hirsute, glandular, or glabrous; limb oblique, faintly bilabiate, blue with white and yellow spots near base of upper lip, or sometimes all white; aestivation imbricate-plicate, the adaxial lobes outermost in bud. Stamens didynamous, upper pair inserted lower in tube than abaxial pair, included, abaxial (lower) pair slightly exserted, with filaments dilated upward and densely ciliate along margins, incurved; pollen sacs of adaxial pair alike, both functional, abaxial pair of pollen sacs often dimorphic, the larger pollen-bearing, the smaller sterile, all anthers dehiscing longitudinally. Style slender, stigma bilobed, abruptly curved toward anthers of adaxial pair and in contact with dehiscing pollen sacs. Capsule ovoid, septicidally dehiscent, valves often again splitting to form 4 teeth, about equaling

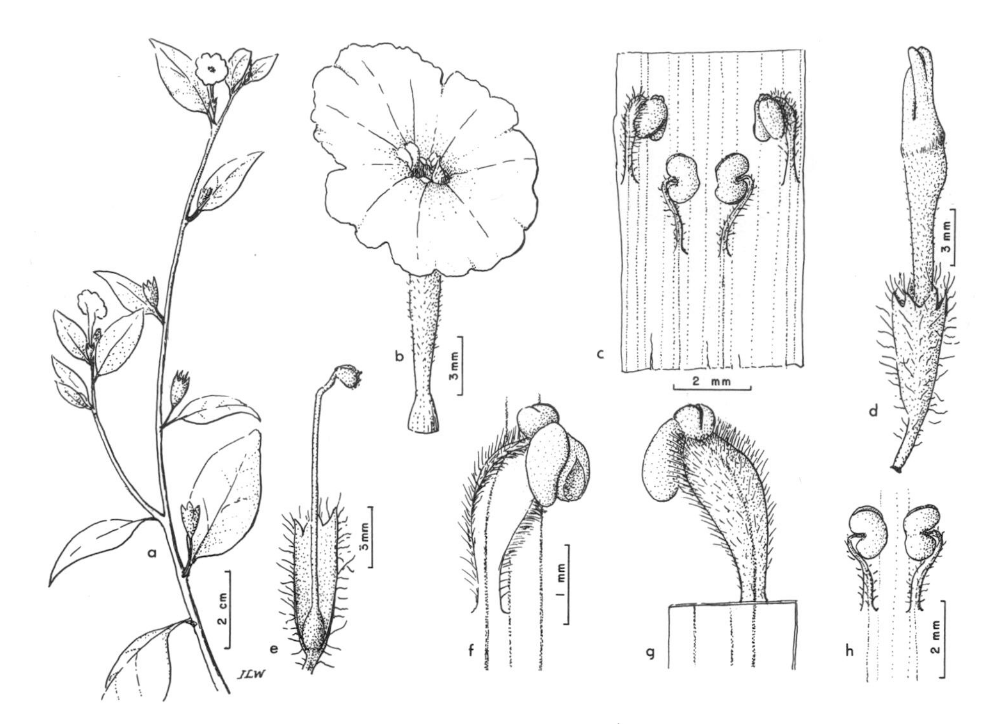

Fig. 119. *Browallia americana*: *f*, interior view of upper stamen; *g*, exterior view of upper stamen; *h*, pair of lower stamens.

calyx when mature. Ovules many, crowded; seeds dark brown to deep tan, more or less angled, rhomboidal, minutely foveolate, minute.

A small genus of 2 (some recognize 8–10) species, confined to the warm temperate and tropical regions of the New World.

Browallia americana L., Sp. Pl. 631. 1753

Browallia demissa L., Syst. Nat. ed. 10, 1118. 1758–59.*

Annual herb to 1 m tall, with erect, ascending, much-branched stems, sparingly hirsute to subglabrate leaves, and flowers borne solitary in axils of upper leaves; petioles 1–2 cm long, moderately to densely hirsute, finally glabrate or nearly so; blades thin, bright green, ovate to ovate-acuminate, 1.5–6 cm long, 1–3.5(4) cm wide, rounded to broadly cuneate at base, acute to short-acuminate at apex, sparsely to moderately hirsute along veins beneath, less so above, nearly glabrate in age; pedicels slender, 4–5 mm long at anthesis, 10–18 mm long and curving outward and upward in fruit, glabrate in age; calyx narrowly tubular-campanulate, 3.5–5 mm long, hirsute with lax, multicellular hairs to 1.5 mm long, the lobes 5, erect, 1–2 mm long, slightly unequal; corolla tubular-salverform, 10–13 mm long, hirsute without, the tube slightly ampliate into oblique throat 2–2.5 mm long, limb blue, 5-lobed, faintly 2-lipped, 10–12 mm broad, an arc of white overarching 2 small spots of yellow at

* At least a dozen binomials probably belong in the synonymy of this species, but the uncertainty surrounding the genus precludes listing them here.

base of upper lip, limb margins somewhat undulate; stamens 4, didynamous; ovary free, ovoid; style slender, glabrous, strongly curved at apex; capsule broadly ovoid, 6–8 mm long, sparsely puberulent, brownish, dehiscing into 4 teeth, the first split along septa; seeds numerous, rhomboidal, flattened, dark tan to pale brown, 0.4–0.6 mm long, minutely foveolate, dull. $n = 11$.*

Widely distributed in tropical America from Mexico to Peru; in the Galápagos Islands known only along trails and fence rows in the *Scalesia* Zone, Isla Santa Cruz, a new record for the archipelago.

See Fig. 119.

Cacabus Bernh., Linnaea 13: 360. 1839†

Annual herbs with alternate, petioled leaves; internodes sometimes condensed, the adjacent leaves then approximate. Flowers axillary, but in old plants a shortening of internodes and a reduction in leaf size in upper part of stem, or on short branches, produce aspect of a terminal, sometimes branched raceme; corolla subtubular to narrowly funnelform, shortly 5-lobed or 5-toothed, without the 5 hairy, glandular areas or 5 dark areas often associated with the corolla of *Physalis.* Stamens 5. Calyx 5-lobed or 5-toothed, in fruit 10-ribbed, enclosing and tightly investing the usually long-pediceled fruit.

Some 4 or 5 species, mostly of Ecuador and Peru.

Cacabus miersii (Hook. f.) Wettst. in Engl. & Prantl, Nat. Planzenfam. IV, Abt. 3b: 16. 1891

Dictyocalyx miersii Hook. f., Trans. Linn. Soc. Lond. 20: 202. 1847.
Thinogeton miersii Miers, Ann. & Mag. Nat. Hist. ser. 2, 4: 359. 1849.
Thinogeton hookeri Anderss., Kongl. Svensk. Vet.-Akad. Handl. 1853: 217. 1855.

Annual, reclining to procumbent, at least when well developed, branched, more or less vestite with septate hairs, usually spreading, often glandular; blades ovate to ovate-reniform, basally tapering to truncate to having a subreniform sinus, the margins coarsely and irregularly few-toothed to coarsely sinuate-dentate; principal blades 4–20 cm long, 4–18 cm wide; petioles 3–13 cm long; flowers in axils of alternate leaves; flowering calyx 5–12 mm long, divided above into 5 lanceolate to narrowly lanceolate lobes 3–5 mm long; pedicels 5–60 mm long; corolla white to whitish, immaculate, subtubular to narrowly funnelform, 3–7 cm long, expanded above into 5 teeth or lobes 2–10 mm long; limb 10–35 mm across expanded lobes; stamens 5, anthers oblong, 3–4 mm long, filaments filiform; fruiting calyx 10-ribbed, 15–25 mm long, tightly investing the elongate-ovate to obpyriform fruit; fruiting pedicel expanded below fruit, more or less curved-reflexed; seeds more or less irregularly wavy-lamellate.

Mainland of Peru and in the Galápagos. Along coastal sandy and rocky beaches, and around brackish pools several kilometers inland, sea level to 250 m. Specimens examined: Champion, Española, Fernandina, Gardner (near Española), Genovesa, Isabela, Pinta, Pinzón, Rábida, San Cristóbal, San Salvador, Santa Cruz, Santa María.

See Fig. 120.

* Count by D. Kyhos.
† Contributed by U. T. Waterfall.

Fig. 120. *Capsicum frutescens* (*a*, *b*). *Cacabus miersii* (*c*).

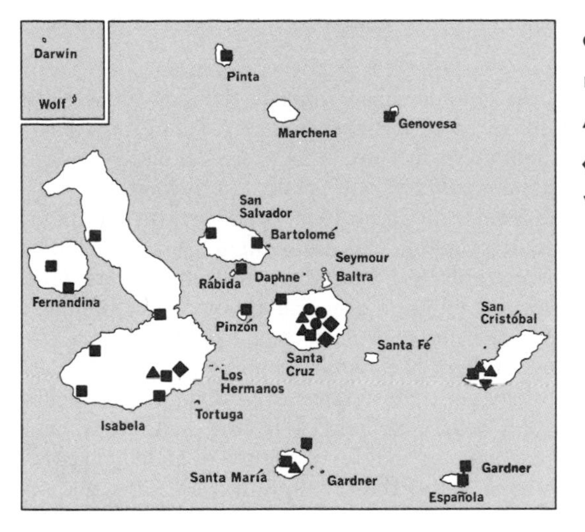

● *Browallia americana*
■ *Cacabus miersii*
▲ *Capsicum frutescens*
◆ *Capsicum galapagoense*
▼ *Capsicum pendulum*

Capsicum L., Sp. Pl. 188. 1753; Gen. Pl. ed. 5, 86. 1754

Herbs or frutescent or shrubby plants with profusely branched, procumbent, ascending, or erect habit. Leaves petiolate, entire or repand. Flowers usually solitary or in few-flowered clusters. Calyx short, broadly campanulate to saucer-shaped, truncate or shallowly 5-toothed. Corollas broadly campanulate to subrotate, 5- or 6-toothed or -lobed, limb plicate in bud, white, greenish, or yellowish, or less commonly purplish or blue; tube short. Stamens 5; filaments slender, inserted in corolla tube, glabrous or sparsely hirsute; anthers shorter than to slightly exceeding filaments, usually connivent, dehiscing longitudinally. Ovary 2(3)-celled; style slender, exceeding anthers; stigma slightly clavate or dilated apically. Fruit a fleshy berry, red, yellow, or sometimes white or brownish, globose to slenderly oblong, acute, truncate or rounded at apex, sometimes inflated. Seeds compressed, smooth to slightly reticulate-rugulose, mild to very pungent. Embryo curved to subspiral. Endosperm scanty, fleshy.

The number of species in this genus is uncertain, owing to the chaotic situation in the taxonomy of the group. Long-cultivated and many superfluous names have been applied. Believed to have originated in America, but introduced and cultivated in Europe very early.*

Flowering pedicels usually 2–4 at a node; corolla greenish white to greenish yellow; anthers blue-violet, equaling or longer than filaments . . . . *C. frutescens*

Flowering pedicels solitary; corolla white, with or without spots or streaks of tan or yellow in throat; anthers white or yellowish, mostly shorter than filaments:

- Leaves and stems densely tomentose with simple, stoutish, tawny hairs; leaf blades acute but not acuminate, mostly 1–3 cm long; pedicels not incrassate upward; corolla clear white, without streaks or spots; anthers white . . . . *C. galapagoense*
- Leaves and stems glabrous or nearly so; leaf blades acuminate, to 6 cm long; pedicels somewhat incrassate toward tip; corolla white with yellow or tan markings on throat; anthers yellow . . . . *C. pendulum*

* For further information on the genus, see Heiser and Smith (1953).

Capsicum frutescens L., Sp. Pl. 189. 1753*

Herbaceous to subshrubby plant 0.5–2 m tall, with glabrous or sparsely pubescent herbage; stems profusely branched, the younger striate-angled, the older terete; petioles 8–15(20) mm long; leaf blades ovate to elliptic-ovate, 2.5–6 cm long, 1.5–3 cm wide, broadly cuneate at base, acuminate at apex, dark green on upper surface, slightly paler beneath; flowers sometimes solitary, but usually 2 and sometimes to 6 at a node, the pedicels slender, 1–2.5 cm long at anthesis; calyx broadly campanulate, 2–2.5 mm long, nearly as wide, truncate or with faint suggestions of 5 obtuse teeth, the margin thin, nearly hyaline; corolla 4–5 mm long, 8–10 mm across, lobes ovate, separate almost to throat, greenish white to greenish yellow, waxy, immaculate, subrotate, the tips of lobes minutely puberulent or finely granular; filaments erect, about 1.5 mm long, glabrous; anthers blue-violet, sometimes with greenish tinge, 1.5–2 mm long; style 1–1.5 mm longer than stamens, stigma minute, truncate-capitate; fruit (in ours) oblong-ovoid, 8–15 mm long, 4–6 mm broad near base, bright red, shiny at maturity; seeds ovate, strongly compressed, pale yellowish, 3–3.5 mm long, 2–2.5 mm wide, about 1 mm thick, minutely reticulate, without ridge or thread around margin, shiny.

Currently cultivated in Mexico, Central America, and northern South America, the exact locality of origin uncertain; common in moist areas in the Galápagos Islands, mostly along trails and about habitations or their sites; flowering and fruiting throughout the year. Specimens examined: Isabela, San Cristóbal, Santa Cruz, Santa María.

See Fig. 120 and color plate 70.

Capsicum galapagoense Hunziker, VIIIe Cong. Int. Bot. Rapp. & Comm. Sect. 4(2): 73. 1958 (June)

Brachistus pubescens Stewart, Proc. Calif. Acad. Sci. IV, 1: 137. 1911.
Capsicum galapagense Heiser & Smith, Britt. 10: 200, *fig. 3.* 1958 (Oct.).

Much-branched shrub to 2 m tall, with densely pubescent herbage and twigs, the hairs tawny, simple, stout and spreading or slightly tangled; petioles stoutish, densely pubescent, 5–20 mm long; blades broadly elliptic-ovate, 10–35(40) mm long, 6–20 mm wide, slightly acuminate or merely acute at apex, rounded to broadly cuneate at base; flowers solitary at nodes; pedicels about equaling or shorter than subtending petioles; calyx campanulate, 2.3–2.6 mm long and wide, truncate or with short, rounded teeth about 0.5 mm high, 1–1.5 mm wide at base, densely lanate; corolla clear white, without streaks or spots, 5–6 mm long, 8–10 mm across, the lobes acute, ascending, distinct about halfway to base of tube, tips and upper margins minutely and densely short-puberulent to granular; anthers shorter than filaments, white; fruit globose or slightly flattened, 5–7 mm across, orange-red at maturity; seeds pale yellowish, shining.

Endemic; in shade of forest trees and shrubs, moist areas. Specimens examined: Isabela, Santa Cruz.

Stewart cited his *No. 3353* (CAS, type collection, GH, type) as collected at James Bay, Isla San Salvador, but no sheet under this number could be found either

* The present state of our knowledge of the taxonomy of the genus does not permit the citing of synonymy among the many names that have been proposed.

at the Gray Herbarium or at the California Academy of Sciences. Nor were the authors able to find this species at the northwestern end of the island, although a diligent search was made from sea level to about 520 m.

Capsicum pendulum Willd., Hort. Berol. 1: 242. 1809

Similar to *C. frutescens* in general appearance, perennial habit, and pungent fruits; differing in possessing pedicels somewhat incrassate toward tips, white corollas with yellow or tan spots traversed by veins at bases of corolla lobes, yellow anthers, and acute to obtuse calyx teeth 1–1.5 mm long.

Cultivated in western South America, where it may be native. Specimen examined: San Cristóbal.

Questionably reported from the Galápagos Islands on the basis of a single collection in which the anthers are yellow, the calyx margin is tipped by teeth 1–1.5 mm long, and there are faint suggestions of a spot at the base of each of the dried corolla lobes. The leaves on this specimen are close to those of *C. frutescens.*

Datura L., Sp. Pl. 179. 1753; Gen. Pl. ed. 5, 83. 1754

Coarse, erect or ascendingly branched, rank-smelling herbs, shrubs, or small trees with alternate, short-petiolate, entire, sinuate, or lobed leaves. Flowers large and showy, solitary in forks of branches, short-pedunculate, white, purple-tinged, yellow, or red, heavily scented. Calyx prismatic or cylindric, 5-toothed, usually circumscissile near base, lower part persistent below capsule as spreading or reflexed collar, upper part deciduous before fruit matures. Corolla funnelform to trumpet-shaped, convolute-plicate in bud, 5- or 10-angled or -lobed. Stamens included; filaments filiform or slender; anthers linear, dehiscent longitudinally. Stigma bilobed. Capsule ovoid to globose, smooth to markedly spiny, apically 4-valved or dehiscing irregularly, falsely 4-celled. Seeds many, subreniform to oblong, strongly compressed laterally. Embryo small, strongly curved.

About 25 species, found mostly in warmer parts of the world. Plants contain narcotic alkaloids, some of them poisonous. Some species are used in ceremonial rites by natives of the New World.

Corolla pendent, 15–20 cm long; plant large shrub or small tree; fruit without spines or prickles, indehiscent . . . . *D. arborea*
Corollas erect or ascending, 5–8 cm long; plant annual herb; fruit armed with prickles to 1.5 or 2 cm long, dehiscent by 4 apical valves:
  Corolla white; prickles of fruit much larger near apex than at base of fruit. . . . . *D. stramonium* var. *stramonium*
  Corolla tinged with purple in tube or purplish throughout; prickles subequal on whole fruit. . . . . *D. stramonium* var. *tatula*

Datura arborea L., Sp. Pl. 179. 1753

Brugmansia arborea Steud., Nom. Bot. ed. 2, 1: 230. 1840.

Shrub or small tree to 8–10 m tall, with ascending-spreading branches and softly pubescent twigs, leaves, pedicels, and calyces; petioles 2–8 cm long; blades ovate-lanceolate to oblong, 10–25 cm long, finely and closely pubescent on both surfaces, inequilaterally rounded at base, acute to acuminate at apex, acute to rounded at base; flowers pendent, heavily fragrant; peduncles 2.5–5 cm long; calyx tubular, 6–9 cm long, 2-cleft halfway to base, puberulent; corolla funnelform, to about 20 cm long, white, the limb 8–12 cm broad, with 5 narrowly lanceolate lobes 2.5–5 cm long; capsule ovoid to spindleform, unarmed and indehiscent, 4–8 cm long, mostly about half as wide as long.

A native of Peru; introduced into the Galápagos Islands, naturalized near Bella Vista, Isla Santa Cruz. Observed in 1964 and 1967, but no specimens collected.

See color plate 77.

Datura stramonium L., Sp. Pl. 179. 1753, var. **stramonium**

Annual, much-branched, robust herb 2–6 dm tall, with glabrous or very sparsely hirsute herbage; petioles to 8 cm long; blades ovate, angulate, to 20 cm long, 15 cm wide, or proportionately smaller, usually with 2 or 3 broad, irregularly margined lobes on each side, acute to slightly acuminate at apex, truncate, rounded or broadly cuneate at base, dark green, glabrous except for a few stoutish, curved trichomes along veins, these more numerous beneath than on upper surface; pedicels stoutish, erect or ascending, 1–2 cm long, glabrous or nearly so; calyx tubular, 3–4 cm long, about 1 cm broad, with 5 slightly unequal acute to subulate teeth 5–10 mm long; corolla white, narrowly funnelform, 8–10 cm long, limb 2–2.5 cm wide, bearing 5 delicate, subulate teeth 5–10 mm long; stamens about three-fourths as long as corolla tube; anthers white, 5–6 mm long; capsules broadly ovoid, 3–4 cm long, 2–3 cm wide, moderately to closely armed with stiff, subulate to narrowly triangular prickles 5–15 mm long, shorter ones at base of fruit, the longer near apex; seeds dark brown, 3–5 mm long, laterally flattened.

Widely distributed in the United States, where it is an introduced weed, and south to Peru; usually in disturbed soil along trails and roadsides or at margins of cultivated fields and in abandoned agricultural land; from Transition to *Miconia* zones in the Galápagos Islands. Specimens examined: San Cristóbal, Santa María.

Datura stramonium var. **tatula** (L.) Torr., Fl. N. Mid. U. S. 232. 1824

Datura tatula L., Sp. Pl. ed. 2, 1: 256. 1762.

Differing from var. *stramonium* in having the stems tinged with purple, the corolla throat and tube flushed with purple, and the spines more nearly uniform in size over whole capsule.

Occupying habitats similar to those in which var. *stramonium* flourishes and nearly as widely distributed. Specimens examined: Isabela, San Cristóbal, San Salvador, Santa María.

The density of pubescence and the spination on the fruits vary considerably from one specimen to another. The blue or purplish flush in the corolla is often lost or greatly faded if a specimen is dried slowly, thus adding to the uncertainties in distinguishing varietal positions of some herbarium specimens. *Snodgrass & Heller 124* (GH) from Iguana Cove, Isla Isabela, was annotated by A. S. Barclay as being "quite distinct. Possibly a new species or geographical var.?" Its leaves (badly wrinkled owing to withering before being put in press) are narrowly ovate to oblong. This, like many other taxa in the Galápagos, warrants careful study.

See Fig. 121.

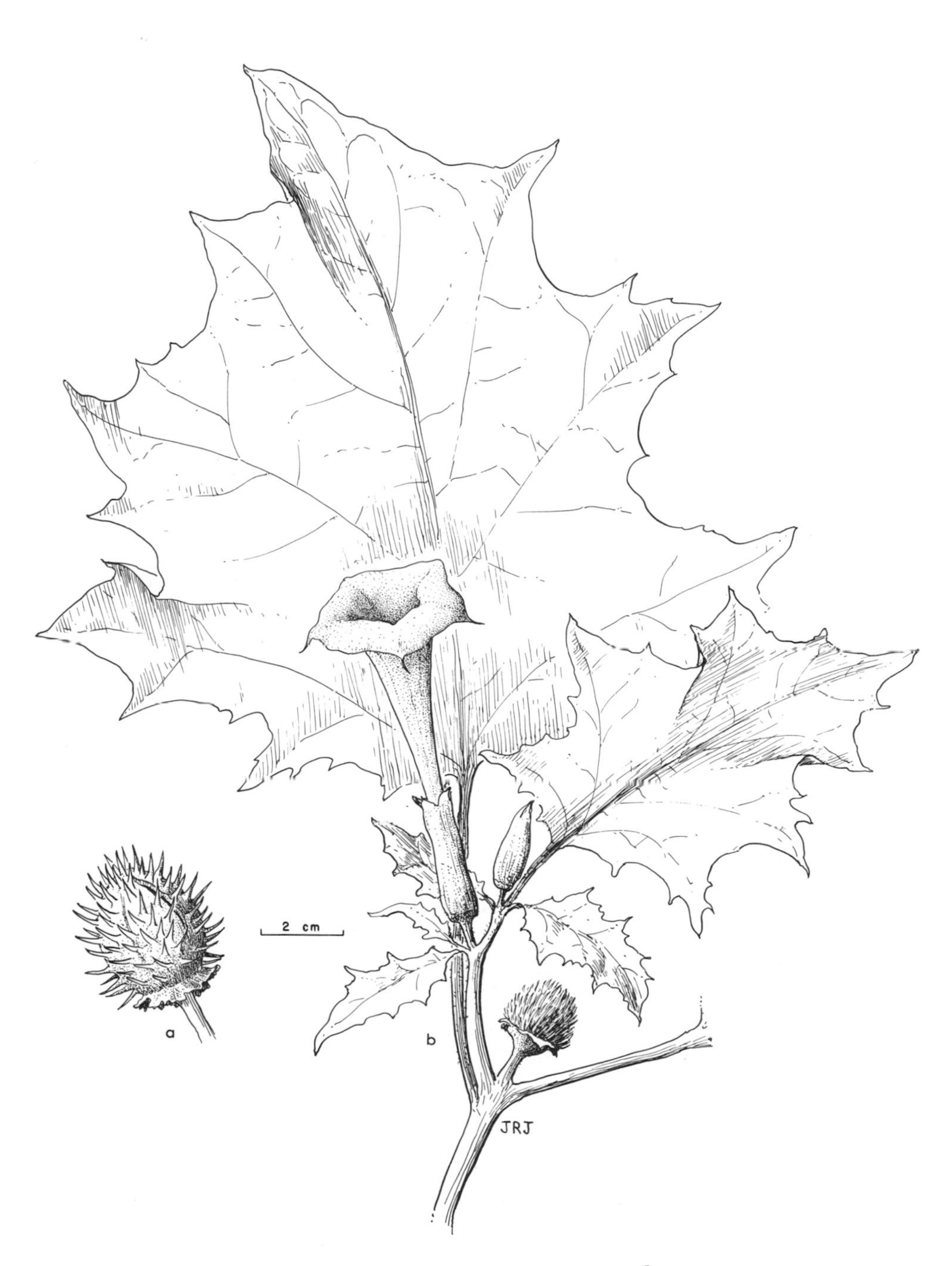

Fig. 121. *Datura stramonium* var. *tatula*.

Grabowskia Schlecht., Linnaea 7: 71. 1832

Shrubs with stout, rigid, axillary spines, subfleshy, firm, elliptic to rounded leaves, these often in fascicles, and with habit of *Lycium*. Flowers 1 to several in axillary clusters. Calyx shallowly campanulate, 5-toothed, on slender pedicels. Corolla funnelform, tube and throat about equaling ovate, white corolla lobes, these plicate-imbricate in bud. Stamens inserted on and adnate to corolla tube, exserted; anthers oval, not connivent. Style slender, exceeding stamens; stigma shallowly crateriform, slightly bilobed. Fruits 2-celled, bony, containing 4 nutlets, each 1- or 2-seeded.

Perhaps a dozen species, mostly in northern South America, sparingly represented in central Mexico.

Grabowskia boerhaaviaefolia (L. f.) Schlecht., Linnaea 7: 71. 1832

Lycium boerhaaviaefolium L. f., Suppl. 150. 1781.
Lycium heterophyllum Murr., Comm. Gött. 6: *pl. 2.* 1783.
Ehretica halimifolia L'Hér., Stirp. Nov. 45, *pl. 23.* 1785.

Stiffly branching erect shrub 1–2.5 m tall, with ascending to arching branches and smooth, glabrous, whitish to nearly silvery bark on young twigs, these turning reddish brown on older branches, often covered copiously with lichens; axillary twigs stiff, spinose, 2–3(5) cm long; leaves on new twigs with slender petioles 5–8 mm long, the blades broadly elliptic-ovate, 2–3 cm long, 8–22 mm wide, firm to subfleshy, glabrous, glaucous, indistinctly veined; leaves on spur branches in fascicles of 5–15 or more, similar to those on young growth except usually smaller and petioles shorter; flowers solitary or in few-flowered axillary clusters; pedicels slender, 10–15 mm long, sparsely and minutely tuberculate-scurfy or glabrous; calyx broadly bowl-shaped, 2–3 mm long, 3–4 mm wide, (4)5-toothed, teeth deltoid-subulate, 1.5–2 mm long; corolla white, usually faintly tinged with pale lavender or with purplish veins, tube 3–4 mm long, hirsute from base to lower part of throat within, throat equaling tube, about 3 mm long and same across mouth; lobes ovate, acute, 4–5 mm long, about 2 mm wide at base, spreading, often somewhat revolute; stamens exserted 4–5 mm, filaments adnate to corolla tube along basal 2–2.5 mm, moderately hirsute from base to apex of tube, then a ring of longer, more numerous

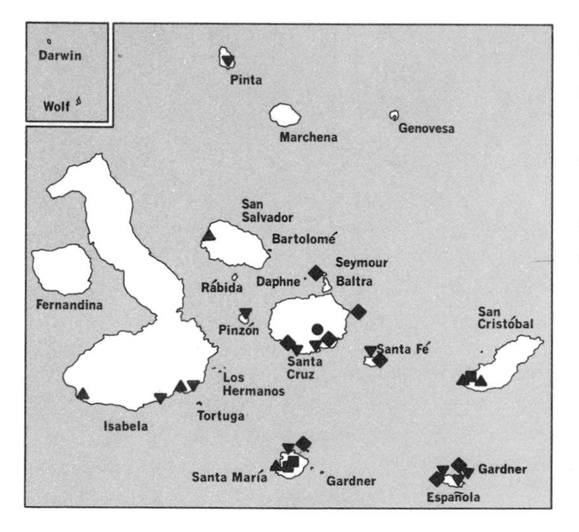

● *Datura arborea*
■ *Datura stramonium* var. *stramonium*
▲ *Datura stramonium* var. *tatula*
◆ *Grabowskia boerhaaviaefolia*
▼ *Lycium minimum*

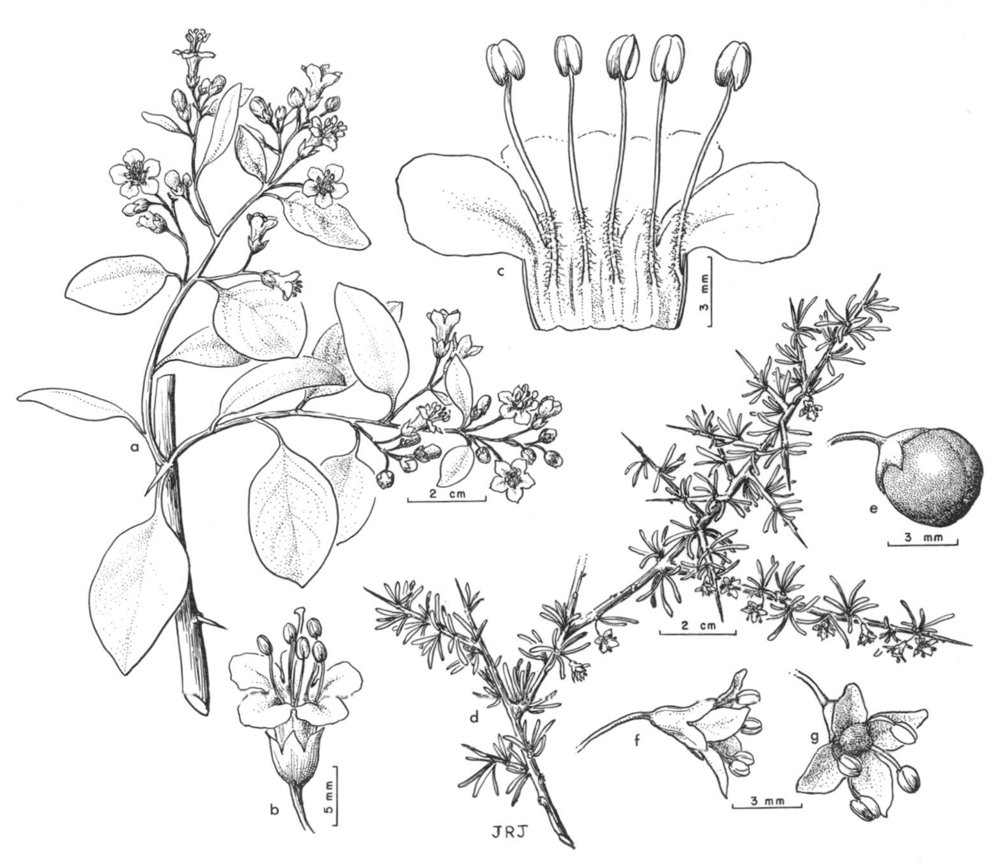

Fig. 122. *Grabowskia boerhaaviaefolia* (*a–c*). *Lycium minimum* (*d–g*).

tangled hairs around each filament; anthers broadly elliptic, about 2 mm long, dehiscing longitudinally; style 1–4 mm longer than stamens, glabrous; stigma crateriform or faintly bilobed; fruit ovoid, blue-black, with decided glaucous bloom, about 5 mm broad, 6–8 mm long. n = 12.*

At low elevations, often near the sea, coastal areas of Peru and Arid Zone of the Galápagos Islands. Specimens examined: Champion, Española, Gardner (near Española), Las Plazas, Santa Cruz, Santa Fé, Seymour.

The specific identity of this plant is somewhat uncertain. Macbride (1962, 7) described the fruit as "orange," but all material seen in the Galápagos had blue-black berries with a heavy glaucous bloom.

See Fig. 122 and color plate 93.

Lycium L., Sp. Pl. 191. 1753; Gen. Pl. ed. 5, 88. 1754

Shrubs or small trees with mostly rigid, spinose branches and alternate, fasciculate, glabrous to pubescent, more or less succulent leaves. Flowers solitary in axils or in groups of 2–6, short-pedicellate. Calyx tubular to campanulate, 4–6-toothed, slight-

* Count by D. Kyhos.

ly accrescent or ruptured by enlarging fruit. Corollas white, greenish, or tinged with purple, tubular to funnelform or narrowly campanulate, the lobes 4–5, rotate or recurved. Stamens 4–6, included or exserted; filaments unequal, adnate different distances to corolla tube, often pubescent near base of free part; anthers ovoid, versatile. Styles slender. Stigma shallowly bilobed. Ovary 2-carpelled, 2-celled, 2- to many-seeded, fleshy to dry and indurate. Seeds somewhat flattened, minutely pitted. Embryo coiled; cotyledons slender.

About 100 species, widely distributed in tropical to cool temperate regions of the Old and New Worlds.

Lycium minimum C. L. Hitchc., Ann. Mo. Bot. Gard. 19: 225, *pl. 14, figs. 23–25, pl. 21.* 1932

Stiffly branching, spinose shrub 1–3 m tall, with slender branches, glabrous twigs and foliage; leaves narrowly elliptic to nearly clavate, terete, 4–12 mm long, 0.25–2 mm broad, in fascicles of 2–5; pedicels slender, 3–5 mm long; calyx campanulate, tube 1.2–1.5 mm long, about as wide, glabrous, 4(5)-toothed, teeth deltoid, 1–1.5 mm long, somewhat spreading, glabrous basally, with short, spreading, closely crowded hairs at tips; corolla tube 2–3.5 mm long, about 1 mm broad at base, 2 mm broad at apex, lobes usually 4, spreading to recurved, 1.5–2 mm long, white with purplish veins, turning brown after anthesis; stamens exserted, about equaling corolla lobes, filaments adnate to corolla on basal 0.3–0.5 mm, densely soft-hirsute within corolla, glabrous on exserted part; anthers broadly elliptic, 0.8–1 mm long, white; berry obovoid, 5–6 mm long, red-orange, 2-seeded, each seed plano-convex, bony, 3–4 mm long. $n = 24$.*

Endemic; along the coast and inland in the Arid Zone on the larger islands and on many of the smaller ones. Specimens examined: Baltra, Española, Gardner (near Española), Isabela, Pinta, Pinzón, Santa Cruz, Santa Fé, Santa María.

Flowering and fruiting are irregular. A single shrub may bear numerous flowers (or fruits) while a score or more of plants in the same vicinity may be entirely devoid of either. Berries examined in the field showed each to contain 2 plano-convex seeds strikingly similar to those of *L. californicum*, which appears to be the closest relative to *L. minimum.*

See Fig. 122 and color plates 26, 33, and 94.

Lycopersicon Mill., Gard. Dict. abr. ed. 4. 1754†

Perennial herbs capable of producing seeds first year. Shoots dying back to crown annually or persistent for several years. Plants profusely branched. Leaves spirally arranged, pinnately segmented, or interrupted-pinnate with alternating large and small pairs of segments, or subdivided into higher order of pinnate segmentation. Leaf margins serrate, crenulate, or rarely entire. Inflorescence an extra-axillary raceme or furcate cyme, bracted or bractless at base of peduncle, at furcations, and at base of pedicels; pedicels articulate near middle or above. Flower regular or slightly zygomorphic. Calyx divided nearly to base into 5, usually lanceolate, seg-

* Count by D. Kyhos.

† Contributed by Charles M. Rick. Research supported in part by a grant from the National Science Foundation.

ments. Corolla yellow, gamopetalous, tube very short, limb shallowly or deeply divided into 5 lanceolate or triangular lobes, often reflexed. Stamens 5, subsessile or with short filaments; anther sacs elongate, terminating apically in ligulate sterile tips, dehiscent laterally by longitudinal sutures, the anthers connivent to form tube around pistil. Pistil of 2 united carpels; ovary globose; style elongate, variably exserted from stamen tube; stigma capitate, simple. Fruit a 2-loculed berry; placentae on central axis formed by locular partitions, many-seeded, fleshy, globose, red, yellow, greenish white, or green with purple or green markings. Number of floral parts increased in some cultivars of *L. esculentum*. Seeds obovate and flat to oblanceolate and thickened, each enclosed in mucilaginous sheath, glabrous to lanate (when cleaned and dried), the "hairs" being remains of lateral walls of outermost cell layer of testa. n = 12.

About 8 species in shady or exposed, well-drained situations from sea level to 2,000 m; native in Colombia, Ecuador, Peru, and Chile.

Lycopersicon cheesmanii Riley, Bull. Misc. Inform. Kew 1925: 227. 1925 (as *Lycopersicum*)

Lycopersicon peruvianum Mill. var. *parviflorum* Hook. f., Trans. Linn. Soc. Lond. 20: 202. 1847.
Lycopersicum peruvianum Mill. f. [1] Anderss., Kongl. Svensk. Vet.-Akad. Handl. 1853: 216. 1855.
Lycopersicon pimpinellifolium Mill., Gard. Dict. ed. 8. 1768 (sensu Muller, Misc. Pub. U.S.D.A. 382: 16. 1940, pro parte).

Profusely branched, prostrate-spreading or erect (when depauperate), occasionally climbing, short-lived perennial; all plant parts comparatively miniature for genus; stems slender, densely covered with 1-celled hairs, some capped with tetrads of glandular cells, sparsely or moderately pilose with multicellular hairs about 1 mm long; leaves oblong, 4–12 cm long, 3–8 cm wide, occasionally larger, crenately margined, light green or yellow-green; pseudostipules none; inflorescence of 1 raceme or 2 borne on a common bractless peduncle 1.5–3 cm long, 5–8-flowered, the pedicels 8–20 mm long at anthesis, usually articulated well above middle, pubescent like stems; calyx lobes 2–3 mm long at anthesis, glandular and pubescent like stems but less densely so; corolla 10–12 mm across, divided two-thirds the way to base into lanceolate, strongly reflexed lobes; staminal column 4–5 mm long, anthers sub-

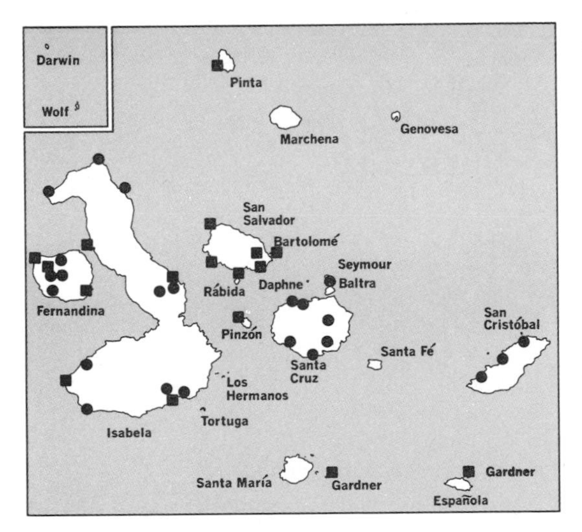

● *Lycopersicon cheesmanii*

■ *Lycopersicon cheesmanii* forma *minor*

sessile, tube 1.5 mm thick; style exserted about 0.5 mm from column; stigma capitate, scarcely exceeding diameter of style; ovary spherical, glabrous; fruit on pedicels to 2.5 cm long, spherical, 8–12 mm in diameter, greenish to golden yellow; pericarp thick and fruit firm at maturity; calyx in ripe fruit to 1 cm across, spreading or slightly accrescent; seeds minute (0.5 mm long or less), flat, oblanceolate, the strophiole acute, covered with false hairs, prone to dormancy and germinating only after digestion or erosion of testa.

Endemic; among lava flows, on well-drained tufaceous soil, and on cinder slopes, from lower to middle altitudes. Specimens examined: Baltra, Española, Fernandina, Isabela, San Cristóbal, Santa Cruz, Santa María.

Since our first investigations with preserved and living material in the field and in culture at Davis, Calif. (Rick, 1956), it has been apparent that the Galápagos tomatoes form a natural unit, concordant in the aforementioned syndrome of characters and genetic compatibilities, hence the basis for the broader concept of *L. cheesmanii* adopted here. Among their distinguishing features, the yellow to orange pigmentation (of which Muller and Luckwill were not aware) is unique among tomato species and serves as a dependable key character for the identification of *L. cheesmanii.*

All the many tested accessions of *L. cheesmanii* are freely intercompatible and suffer no barrier to artificial intercrossing with *L. esculentum* and *L. pimpinellifolium* (Sect. *Eulycopersicon,* according to Muller [1940]), and only a unilateral barrier with the other members of the "*esculentum* complex" (*L. minutum, L. hirsutum,* and *Solanum pennellii*), while they are separated by a complete barrier from *L. chilense* and *L. peruvianum* (Rick, 1963). Morphological comparisons reinforce the conclusion that *L. cheesmanii* shows its closest relationships with Sect. *Eulycopersicon.* Despite these compatibilities, the ecological limitations of *L. cheesmanii* differ radically from those of other species, and most individuals in segregating gen-

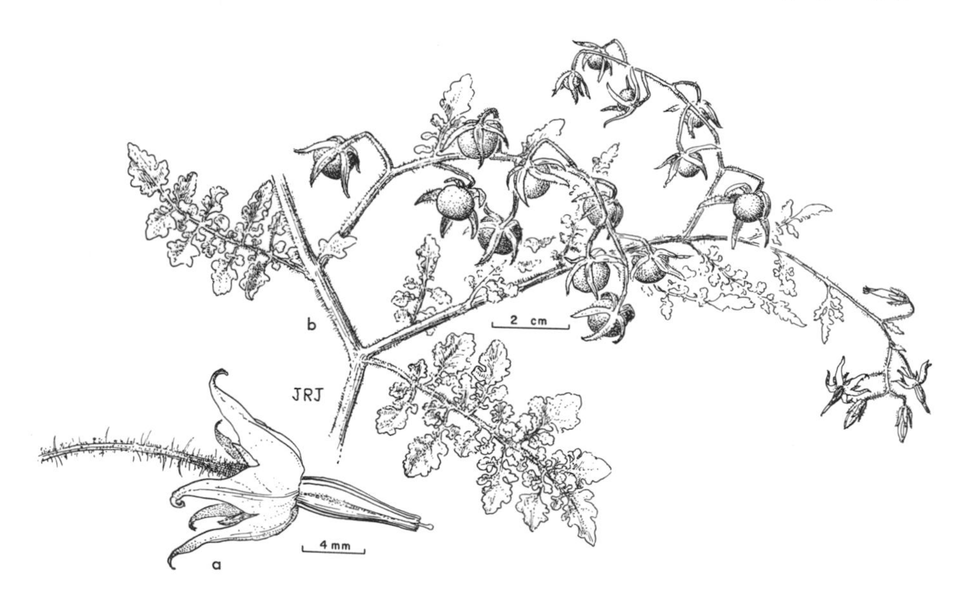

Fig. 123. *Lycopersicon cheesmanii* forma *minor.*

erations from crosses between *L. cheesmanii* and compatible species are unfruitful. These 2 factors cast doubt on the likelihood that *L. cheesmanii* and related species could exchange genes, even if they were not isolated geographically. Preferences of pollinating insects might constitute a third isolating factor. The recognition of a single Galápagean species thus appears justified on all tested biosystematic grounds.

Lycopersicon cheesmanii varies considerably within the described limits. Heritable variation is known for such characters as internode length, pubescence, leaf shape and color, presence of inflorescence bracts, articulation of the pedicel, fruit pigmentation, and pericarp thickness. As a consequence of its highly autogamous breeding system, its populations are exceedingly uniform, many behaving as if they were pure lines. At a higher level of organization, certain more or less discrete races can be recognized, none of which is restricted to a single island.

Some aspects of racial distribution are indicated below; for further details, see Rick (1963). In view of the morphological intergradations, the frequent sympatry, and the nature of the breeding systems, systematic designation of such subspecific differentiation is not warranted except for the following distinctive and widespread form.

Lycopersicon cheesmanii forma **minor** Muller, Misc. Pub. U.S.D.A. 382: 21. 1940

Lycopersicon esculentum Mill. var. *minor* Hook. f., Trans. Linn. Soc. Lond. 20: 202. 1847.
Lycopersicum peruvianum Mill. f. [2] Anderss., Kongl. Svensk. Vet.-Akad. Handl. 1853: 216. 1855.

Differs from the remainder of the assemblage in its highly divided (tri- or tetrapinnatifid) leaves; densely pubescent herbage with characteristic pungent odor; bifurcation of inflorescence subtended by pinnate foliar bract 2–5 cm long; calyx increasing in age to 15 mm, strongly accrescent with and exceeding the developing fruit; and ripe fruits deep orange-red, the pericarp thin. Its very short (2–3 cm) internodes also distinguish it from most other forms, though a few of the latter (e.g., UCD acc. LA746; *Cheesman 403*) exhibit the same character. This unique syndrome of characters is always found associated in members of this variety, though variation for such characters as amount of anthocyanic coloration of buds, internode length, and spreading of the leaves has been observed, and at least 1 instance of intergradation with the typical form, suggesting introgression, has been reported (Rick, 1963).

Habitat similar to that of var. *cheesmanii*. Specimens examined: Bartolomé, Fernandina, both Gardners, Isabela, Pinta, Pinzón, Rábida, San Salvador.

Forma *minor* is by far the commonest variant of the species, being represented on 6 of the major islands and at least 4 islets of the archipelago. It is the only form that has ever been found on San Salvador and its neighboring islands and islets; but despite extensive explorations, it has not been found on Isla Santa Cruz, which lies no more than 30 km away and offers similar habitats. It is distributed mostly in the northwestern part of the archipelago, and it generally appears at lower elevations, though not exclusively so. Although morphologically distinct from most other known biotypes and occupying a well-defined geographic distribution, this form does not merit specific rank any more than several other, less widely distributed races of the complex. All forms of the species are freely cross-compatible, and the aforementioned evidence (Rick, 1963) hints that introgression does occur, despite the highly autogamous nature of the breeding system.

See Fig. 123.

Nicandra Adans., Fam. Pl. 2: 219. 1763

Erect, much-branching herbs with alternate leaves and flowers borne singly at 1 side of upper axils. Leaf blades ample, to 15 cm long or more, angulate-sinuate marginally, decurrent onto petiole, cuneate at base, acute at apex. Herbage generally glabrous or nearly so. Calyx 5-parted almost to base, pentagonal, segments broadly ovate-cordate, conspicuously veined, enlarging and enclosing the berry at maturity. Corolla campanulate, limb shallowly 5-lobed to nearly entire, blue. Stamens 5, about equaling corolla throat; filaments somewhat dilated toward base, subconnivent at apex; anthers ovoid, dehiscing longitudinally. Style straight, almost equaling stamens; stigma subcapitate. Fruit a firm, dry berry, enclosed within calyx, 3–5-celled. Seeds many, compressed-reniform, foveolate-punctate. Embryo curved almost into a circle. Endosperm fleshy, ample.

A monotypic genus (possibly a second species exists), native in Peru and Bolivia but widely dispersed in tropical and subtropical areas in both the Old and New Worlds.

Nicandra physalodes (L.) Gaertn., Fruct. 2: 237, *pl. 131, fig. 2.* 1791

Atropa physalodes L., Sp. Pl. 181. 1753.
Nicandra minor Hort. ex Fisch., May & Avé-Lall., Ind. Sem. Hort. Bot. Petrop. 9: 81. 1843.
Physalodes peruviana Kuntze, Rev. Gen. Pl. 2: 452. 1891.
Boberella nicandra Krause in Sturm., Fl. Deutschl. ed. 2, 10: 61. 1903.

Fibrous-rooted herb to 1 m tall and wide; leaves to 15 cm long, 10 cm broad, margins shallowly angulate-sinuate, acute at apex, broadly cuneate at base; petioles slender, about 1 cm long at anthesis, to 3 cm long in fruit, more or less recurved; sepals about 1.5 cm long, 1 cm wide at anthesis, distinctly cordate, a small "tail" 1–2 mm long at each basal angle, the maturing sepals becoming chartaceous, with strongly netted veins, accrescent, 2–3.5 cm long in fruit, the tips connivent; corollas blue, 2–3 cm long, about as wide, limb moderately spreading; berry 10–12 mm in diameter, firm, not juicy, the wall rupturing irregularly above middle; seeds ovoid-reniform, compressed, brown, about 1.5 mm long, finely punctate-foveolate.

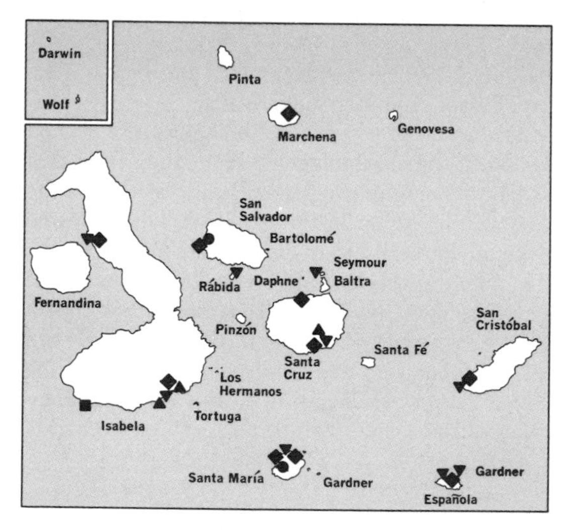

● *Nicandra physalodes*
■ *Nicotiana glutinosa*
▲ *Nicotiana tabacum*
◆ *Physalis angulata*
▼ *Physalis galapagoënsis*

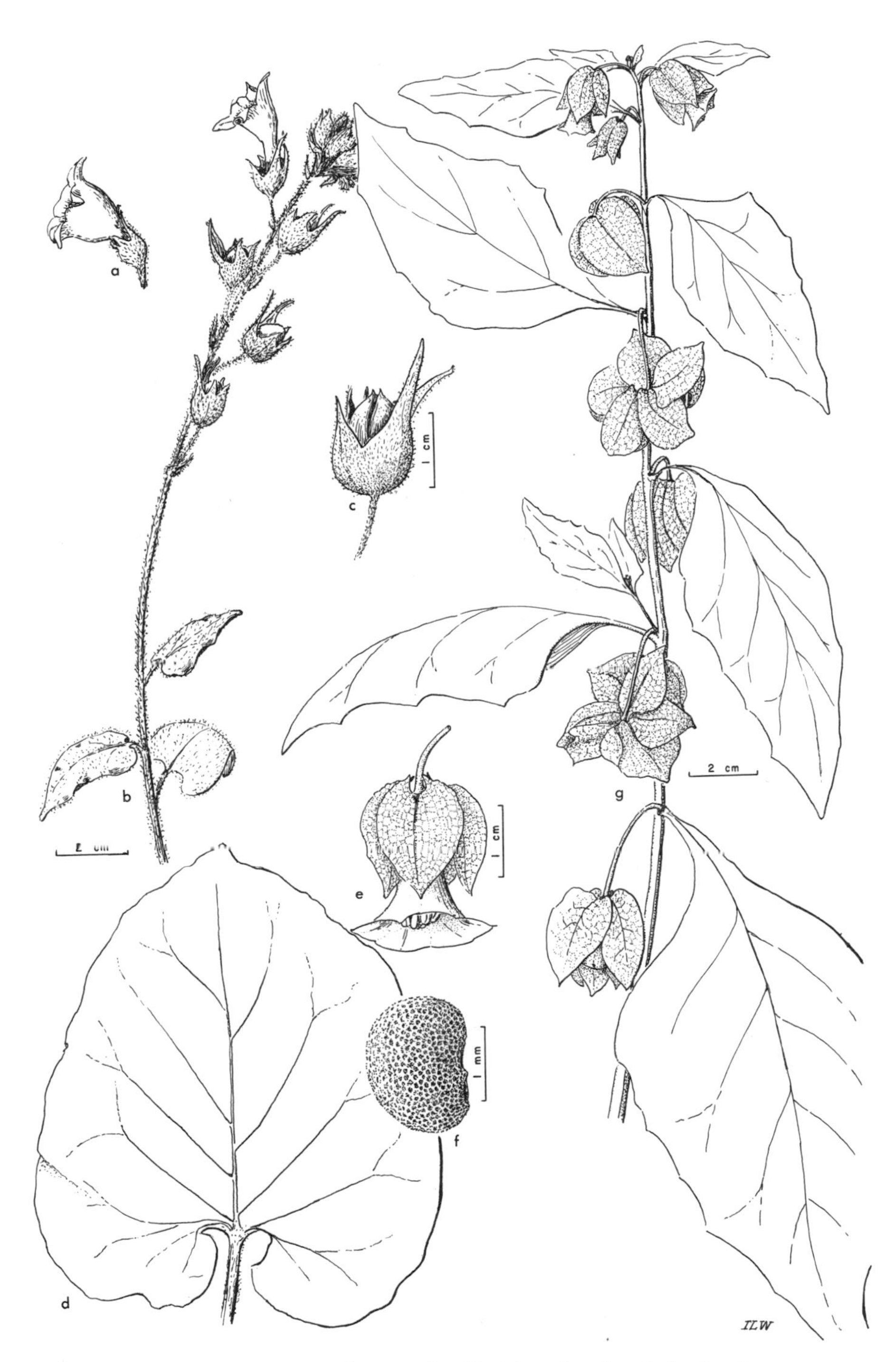

Fig. 124. *Nicotiana glutinosa* (*a–d*). *Nicandra physalodes* (*e–g*).

Near habitations or sites of former occupation. Specimens examined: San Salvador, Santa María.
See Fig. 124.

Nicotiana L., Sp. Pl. 180. 1753; Gen. Pl. ed. 5, 84. 1754

Annual or perennial herbs or shrubs, with heavy-scented, narcotic, poisonous, usually viscid-puberulent herbage. Leaves alternate, entire to repand, petiolate or sessile. Inflorescences few- to many-flowered, racemose to paniculate, terminal. Calyx tubular-campanulate or ovoid, 5-toothed or 5-cleft, persistent. Corolla tubular, salverform or campanulate, the limb shallowly 5-lobed, usually spreading, white, yellow, roseate, reddish, or purple. Stamens 5; filament slender, inserted near base of corolla tube or at base of throat, often more or less appressed to 1 side of corolla tube near apex, included or slightly exserted. Style slender; stigma capitate. Ovary 2-celled or sometimes 4-celled. Capsule ovoid, acute, 2- or 4-valved at summit. Seeds many, small, ovoid, reniform, or cylindric-oblong, minutely reticulate-punctate to ridged.

About 60 species native in the Americas, the South Pacific, and Australia; 1 or 2 species now widely cultivated from tropical to temperate zones worldwide.

Corolla throat broadly campanulate, as wide as long, curved somewhat to 1 side; lower leaves distinctly cordate; calyx lobes markedly unequal, the longer ones to 3 times as long as calyx tube . . . . *N. glutinosa*

Corolla throat tubular, longer than wide, straight, only slightly wider than tube; leaves decurrent, tapering or winged-petiolate at base; calyx lobes shorter than to barely equaling calyx tube . . . . *N. tabacum*

Nicotiana glutinosa L., Sp. Pl. 181. 1753

Tabacum viscidus Moench, Meth. Pl. Hort. Bot. Mart. 448. 1794.
Sairanthus glutinosa D. Don, Gen. Hist. 4: 467. 1837.

Coarse, viscid-pubescent annual to 2.5 m tall, with thickish, erect stems and narrowly divergent, stiff branches; blades on lower parts cordate-triangular to broadly cordate, 5–20 cm long, basal lobes turned upward, apex acute, often twisted, the petioles shorter than blades; leaves on upper branches less markedly cordate, often rounded or cuneate at base, 3–6 cm long; pedicels 5–12 mm long, stoutish; calyx campanulate, tube 5–7 mm long, lobes 1–3 times as long as tube and markedly unequal, linear to lance-linear above triangular base, usually recurved but sometimes erect and attenuate; corolla about 3–3.5 cm long, tube 3–5 mm long, throat white or greenish yellow, or on some plants reddish, limb 8–12 mm wide, densely viscid-pubescent without, lobes broadly ovate, the tips short-acuminate, usually recurved and channeled; stamens slightly exserted from throat but shorter than limb, filaments sparsely hirsute on lower half, oriented against adaxial side of corolla tube near tips; capsule broadly ovoid, 10–16 mm long, included by calyx lobes; seeds angular to broadly ovate, about 0.6 mm long, dark brown, surfaces reticulate-foveolate and with wavy ridges. $n = 12$.*

In semiarid regions, chiefly in the lowlands of central and northern Peru and southern Ecuador. Specimen examined: Isabela.

* Count by Goodspeed (1954, 369).

Goodspeed says of the collections known to him from the Galápagos Islands, "locality believed to be error of annotation." However, the arrangement of the flowers, the unequal, long calyx lobes, the capsules and seeds, and the shape and size of the leaves all match his description of the species. It is almost certain that specimens collected by Darwin, Edmonstone, and Andersson are from the Galápagos Islands.

See Fig. 124.

Nicotiana tabacum L., Sp. Pl. 180. 1753*

Stout, viscid annual or short-lived perennial herb 1–3 m tall, with thick, erect stems and large, decurrent leaves, these 1–6 dm long, ovate, elliptic, or lanceolate, tapering or winged-petiolate at base, acuminate at apex; flowers borne in compound branching panicles; pedicels 5–15 mm long at anthesis, to 25 cm long in fruit; calyx tubular-cylindric to cylindric-campanulate, 12–25 mm long, viscid-puberulent, teeth deltoid-acuminate, slightly unequal, shorter than to equaling calyx tube; corolla straight or essentially so, puberulent without, the tube 7–15 mm long, 2.5–3 mm across, throat 25–40 mm long, 3–5 mm across, pale greenish cream or upper part pink to red, limb 10–15 mm wide, white, pink, or red; stamens inserted on base of corolla throat, usually oriented along adaxial side of corolla throat, longer pairs slightly exserted, fifth stamen considerably shorter, included; capsule ovoid, orbicular, or narrowly ellipsoid, acute to rounded, 15–20 mm long; seeds nearly spherical to broadly ellipsoidal, 0.5 mm long, pale brown, sculptured with fluted ridges or these in labyrinthine patterns.

A widely distributed species as a cultivar and often escaping into areas along roadsides, fence rows, fallow fields, and other disturbed habitats; land of origin uncertain; an escape in the Galápagos. Specimens examined: Isabela, Santa Cruz.

Physalis L., Sp. Pl. 182. 1753; Gen. Pl. ed. 5, 85. 1754†

Annuals, though species from other areas may be perennial; leaves alternate, petiolate, the blades entire or toothed, often coarsely and irregularly so. Plants glabrous to varyingly vestite. Flowers pedicellate, usually singly in leaf axils; flowering calyx 5-toothed or 5-lobed; corolla campanulate to campanulate-rotate, limb sometimes reflexed, expanding fully only in bright light; 5 darker areas usually present near base of corolla limb, sometimes these not darkly contrasting, sometimes the corolla immaculate; varying amounts of hair usually present in tube near insertion of filaments, sometimes as 5 hairy areas, these sometimes more or less glandular; anthers 5, dehiscing longitudinally, oblong to ovate, yellow, blue, greenish blue, or violet, or so lined or tinged; fruiting calyx enlarged, usually inflated around the berry; fruiting pedicel usually filiform, not strongly thickened upward below calyx. Berry, which may have a thin, dry pericarp, usually sessile on calyx base, this often invaginated. Fruiting calyx 5-angled or 10-ribbed.

About 100 species, most abundant in the New World tropics and subtropics, extending into temperate areas; a few species in Europe and Asia, probably mostly introduced.

* For numerous synonyms, see Goodspeed (1954, 372).

† Contributed by U. T. Waterfall.

Fruiting calyx 10-angled or 10-ribbed, sometimes the 5 primary angles more prominent .... *P. angulata*
Fruiting calyx 5-angled:
  Corolla immaculate, or essentially so; fruiting calyces glabrous, 30–45 mm long, 37–40 mm wide .... *P. galapagoënsis*
  Corolla with 5 dark areas on limb near its base; fruiting calyces more or less pubescent, 18–30 mm long, 13–22 mm wide............................. *P. pubescens*

Physalis angulata L., Sp. Pl. 183. 1753

Physalis ramosissima Mill., Gard. Dict. ed. 8, no. 12. 1768.
Physalis capsicifolia Dunal in DC., Prodr. 13^1: 449. 1852.

Annual herb, 25–100 cm tall, branched, glabrous or with a few, short, antrorsely appressed hairs; blades ovate to ovate-lanceolate or oblong, coarsely and irregularly toothed to entire, principal ones 5–11 cm long, 3.5–8 cm wide, on petioles 4–8 cm long; flowering calyx usually 4–7 mm long, divided above into lobes 2–3 mm long, 2–4 mm wide at base; flowering pedicels 5–10 mm long; corolla usually immaculate, sometimes with indistinct spots, sometimes hairy on inside of tube, 6–12 mm long, 7–12 mm wide when fully open; anthers bluish or violet, 2–2.5 mm long, on slender filaments 3–5 mm long; fruiting calyces 10-ribbed or 10-angled, 20–35 mm long, 17–22 mm wide, on pedicels 8–15 mm long; berry 10–12 mm in diameter.

In shady places and open areas, eastern to western United States south to Peru; in the Galápagos along the shore and inland. Specimens examined: Española, Isabela, Marchena, San Cristóbal, San Salvador, Santa Cruz, Santa María.

Physalis galapagoënsis Waterfall, Rhodora 70: 408. 1968

Annual, 15–90 cm tall; branches, petioles, and pedicels more or less antrorsely hispidulous; blades ovate, unequally coarsely dentate, principal ones 3.5–8 cm long, 3–6 cm wide, with petioles 1.5–5 cm long; flowering calyces 2.5–5 mm long, 1.5–2 mm wide at bases of lobes; calyx lobes lanceolate to narrowly lanceolate-attenuate, (1)2–3 mm long; flowering pedicels 4–6 mm long; corollas yellow, immaculate, 4–6 mm long; anthers yellow, sometimes with violet margins, oblong to ovate-oblong, 1–1.5 mm long, filaments 1.5–3 mm long; fruiting calyx pentangular, glabrous, 40–45 mm long, 37–40 mm wide, divided above into lanceolate to narrowly lanceolate-attenuate lobes 4–8 mm long, these sometimes abruptly porrect; fruiting pedicels 15–30 mm long; fruit 12–15 mm wide.

Endemic; on beaches, along ridges, in shady places, and in *Bursera* forest. Specimens examined: Española, Gardner (near Española), Isabela, Rábida, San Cristóbal, Santa Cruz, Santa María, Seymour.

See Fig. 125 and color plate 45.

Physalis pubescens L., Sp. Pl. 183. 1753

Physalis villosa Mill., Gard. Dict. ed. 8, no. 13. 1768.
Physalis obscura Michx., Fl. Bor.-Amer. 1: 149. 1803.
Physalis hirsuta Dunal in DC., Prodr. 13^1: 445. 1852.

Annual, 8–90 cm tall, usually villous and somewhat viscid, varying to more or less glabrate; blades ovate, often acuminate, bases sometimes inequilateral, margins usually irregularly several-toothed, sometimes entire or with 1 to few teeth, sur-

Fig. 125. *Solanum nodiflorum* (*a, b*). *Physalis galapagoënsis* (*c, d*).

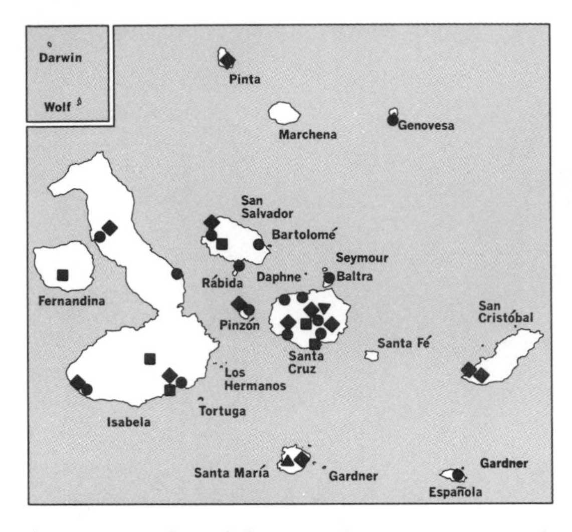

● *Physalis pubescens*
▲ *Solanum edmonstonei*
■ *Solanum erianthum*
◆ *Solanum nodiflorum*
▼ *Solanum quitoense*

faces varyingly soft-hairy with vestiture more abundant and appressed on veins of abaxial surface, sometimes nearly glabrous; principal blades usually 4–9 cm long, 2–4 cm wide, petioles 2–7 cm long; flowering calyx usually 4–10 mm long, 3–12 mm wide at base of lobes; upper 1–4 mm of calyx divided into ovate-deltoid to lanceolate lobes; flowering pedicels 3–6 mm long; corolla yellowish, dark-maculate, more or less matted-hairy in throat below maculations, 7–10(12) mm long, 10–15 mm wide; anthers bluish or violet, 1.5–3 mm long, on filaments 2–3 mm long; fruiting calyx 5-angled, usually prominently so, usually soft-hairy, 18–30 mm long and 13–22 mm wide, on pedicels 5–13 mm long; berry 10–18 mm in diameter, sessile, or subsessile on the invaginated calyx base.

In shady places, along paths, on sand dunes, and in wooded areas; widespread from northeastern United States to California, south into the South American subtropics. Specimens examined: Baltra, Española, Genovesa, Isabela, Pinzón, Rábida, San Salvador, Santa Cruz.

Solanum L., Sp. Pl. 184. 1753; Gen. Pl. ed. 5, 85. 1754*

Herbs, shrubs, or small trees, unarmed (in ours) or armed with prickles. Stems and foliage glabrous, pubescent, or tomentose, the hairs often variously stellate. Leaves alternate, simple, petiolate, entire, lobed, or pinnatifid. Flowers white to violet, in simple or compound cymes, these pseudoterminal or lateral, short- or long-pedunculate, sometimes umbelliform. Calyx campanulate, 5-lobed. Corolla rotate, tube very short, limb plicate in bud. Stamens 5, inserted on corolla tube near base; filaments short; anthers oblong or attenuate, connivent around style, dehiscent by pairs of large introrse, apical-lateral pores and longitudinal slits or only small terminal pores. Ovary usually 2-celled, stigma small, capitate or obscurely bilobed. Fruit a globose berry, fleshy or leathery. Seeds numerous, flattened, reticulate-punctate, mostly deltoid, reniform, or suborbicular.

A very large, mostly tropical and subtropical genus of more than 1,000 species, especially numerous in the New World.

* Contributed by Keith E. Roe.

Plant a shrub or small tree; stems and leaves pubescent to tomentose, the hairs mostly stellate; mature leaves mostly 15 cm long, 7 cm wide or larger:
Leaf margins entire; inflorescences usually long-pedunculate; anthers oblong; stem apices granular-stellate-tomentose . . . . S. *erianthum*
Leaf margins angular or sinuate; inflorescences very short-pedunculate; anthers attenuate; stems stellate-lanate . . . . S. *quitoense*
Plant an herb or subshrub; stems and leaves glabrous or sparsely pubescent, the hairs simple; mature leaves usually smaller than above:
Leaf margins unevenly sinuate-dentate or entire; inflorescences umbelliform; calyx not enlarging in fruit . . . . S. *nodiflorum*
Leaf margins pinnatifid; inflorescences distinctly cymose; calyx enlarging in fruit. . . . . S. *edmonstonei*

Solanum edmonstonei Hook. f., Trans. Linn. Soc. Lond. 20: 201. 1847

Perennial subshrub, straggling, about 0.3 m high, entire plant viscous-puberulent, stems terete; leaves generally ovate-oblong, 3–5 cm long, 1–2 cm wide, irregularly pinnatifid, the 4–9 lobes rounded or obtuse, bases attenuate; inflorescences cymose, rather long-pedunculate, about 5.5–7 cm long, lateral, 2–5-flowered; pedicels 1–1.8 cm long; flower buds obovoid; calyx lobed three-fourths way to base or more, enlarging in fruit, the linear to oblong lobes rounded or obtuse, about 6.3 mm long; corolla about 14 mm across, shallowly lobate; anthers oblong; ovary and style glabrous; fruit about 9 mm in diameter.

My description is based on the original description and a photograph of the type specimen. This species was collected by Edmonstone on Isla Santa María in 1846 and has not been taken since. Hooker relates it to the North American S. *triflorum* Nutt., but the inflorescences and leaves are quite different. *Solanum edmonstonei* has distinctly cymose inflorescences and leaves with rounded or obtuse lobes, whereas S. *triflorum* has umbelliform inflorescences and leaves with acute lobes. The supposedly endemic S. *edmonstonei* shows stronger affinities to some South American species of Sect. *Regmandra* and, indeed, may only represent an accidental introduction or waif.

Solanum erianthum D. Don, Fl. Nepal. 96. 1825

Solanum verbascifolium sensu auct., non L. 1753.
Solanum adulterinum Buch.-Ham. ex Wall., Cat. no. 2616h. 1831, nomen nudum.
Solanum verbascifolium var. *adulterinum* G. Don, Gen. Hist. iv: 415. 1838.
Solanum erianthum var. *adulterinum* Baker & Simmonds in R. O. Williams, Flora Trinidad & Tobago 2(4): 264. 1953.

Evergreen shrub or small tree 2–8 m high, dichotomously branched, with flat-topped, spreading crown; trunk 2–5 cm thick, wood soft, brittle, bark gray or brown, smooth-lenticellate; young branches often angled and grooved; younger parts, including stem apices and inflorescences, pale yellowish, granular-tomentose, at least some stellate hairs many-rayed; leaves entire, ovate or broadly elliptic, 12–37 cm long, 6–16 cm wide, apices acute or acuminate, bases rounded, obtuse, or acute, velutinous-tomentose above and below; petioles 1.5–3.5 cm long; inflorescences usually long-pedunculate, unbranched for 2–12 cm, pseudoterminal, becoming lateral with continued growth of stems, the several dichotomous branches terminating in many-flowered helicoid cymes; pedicels nodding in bud and flower, becoming erect in fruit, 3–6 mm long (11 mm in fruit); flower buds turbinoid, 6–9 mm long; calyx lobed halfway to base, lobes deltoid, acute, pubescent within;

corolla barely exserted from calyx at anthesis, white, 11–18(25) mm across; anthers oblong with large apical-lateral pores and longitudinal slits, 2.2–3 mm long; ovary pubescent, rays of hairs short, nonseptate, or ovary sometimes glabrous, the style glabrous; fruits 10–12 mm in diameter, yellow when ripe, usually pubescent; seeds 1.7–2 mm long, 1.4–1.6 mm wide, yellowish.

A common species in subtropical and tropical Mexico, Central America, and southeastern Asia; occasional to common in the Galápagos from sea level to 600 m in open habitats, forming part of undergrowth in *Scalesia* forest. Specimens examined: Fernandina, Isabela, San Salvador, Santa Cruz.

Solanum nodiflorum Jacq., Icon. Pl. Rar. 2: 288, *pl. 326*. 1788–89

Solanum chenopodioides Lam., Illustr. 2: 18. 1798, pro parte.
Solanum nigrum var. *minor* Hook. f., Trans. Linn. Soc. Lond. 20: 201. 1847.
Solanum nigrum var. *dillenii* A. Gray, Syn. Fl. N. Amer. 2(1): 228. 1878.
Solanum nigrum var. *nodiflorum* A. Gray, loc. cit.
Solanum nigrum sensu Jepson, Man. Fl. Pl. Calif. 892. 1925, non L.
Solanum photeinocarpum Nakamura & Odashima, Jour. Soc. Trop. Agr. Taiwan 8: 54, *fig. 2*. 1936.

Straggling perennial herb or weak subshrub to 6 dm high; stems terete or angled; younger parts, including stem apices and inflorescences, short-villous-pubescent, the hairs simple, subappressed, older parts often glabrescent; leaves entire or unevenly sinuate-dentate, generally ovate, rhomboid or lance-ovate, 2–14 cm long, 1–5 cm wide, apices acute or acuminate, bases attenuate, glabrous or sparsely pubescent above and below, the hairs simple; inflorescences lateral, umbelliform, peduncles 6–16 mm long; pedicels 3–5 mm long (to 9 mm in fruit), deflexed in fruit; flower buds oblong to subglobose, 1–3 mm long; calyx lobed one-third to halfway to base, lobes small, semiovate, rather thick, acute, glabrous or stipitate-glandular within, spreading or reflexed in fruit but not enlarging; corolla strongly exserted from calyx at anthesis, white to pale violet, 4–5 mm across; anthers oblong, with large, apical-lateral introrse pores and longitudinal slits, 1–1.8 mm long; ovary glabrous, style villous; fruits about 5 mm in diameter, purple to black when ripe; seeds 1.1–1.9 mm long, 1–1.4 mm wide, whitish to yellowish.

Occasional to common among rocks, along sand beaches and trails, in woodlands and fields; a ubiquitous weed, native in the Old World and widely distributed there and in the Americas. Specimens examined: Isabela, Pinta, Pinzón, San Cristóbal, San Salvador, Santa Cruz, Santa María.

See Fig. 125.

Solanum quitoense Lam., Illustr. 2: 16. 1797

Solanum angulatum R. & P., Fl. Peruv. 2: 36, *pl. 170, fig. a*. 1799.
Solanum quitense HBK., Nov. Gen. & Sp. Pl. 3: 25. 1818.

Robust shrub 1–2 m high, the young branches terete, stout, succulent, becoming soft-woody with age; stems, inflorescences, calyces and corollas densely stellate-pilose, rays per hair few; leaves angular or regularly sinuate-lobate, generally ovate, 27–34 cm long, 20–24 cm wide, petioles 5–8 cm long, apices acute, bases cordate, stellate-pubescent above and below; inflorescences cymose, lateral, very short-pedunculate, about 2.5 cm long, few-flowered; pedicels 3–10 mm long, densely purple-stellate-pilose; flower buds orbicular to obovoid, 12–14 mm long; calyx lobed half to two-thirds the way to base, the ovate-lanceolate lobes thick, acute, pubescent within only near tip; corolla barely exserted from calyx at anthesis, thick, white, about 2 cm across; anthers attenuate, with small apical pores, 7–8 mm long;

ovary tomentose, the rays of hairs long, septate; style glabrous or sparsely pubescent; fruit about 5 cm in diameter, orange-colored when ripe, pubescence deciduous; seeds about 4 mm in diameter.*

Occasional along trails in *Miconia* Zone (Isla Santa Cruz) and on southeast side of main crater (San Salvador), almost certainly introduced by settlers; Venezuela, Bolivia, and Ecuador, apparently known only in cultivation.

VALERIANACEAE. VALERIAN FAMILY†

Annual or perennial, strong-scented herbs, or rarely subshrubs, with opposite or basal, simple or pinnately divided, exstipulate leaves. Flowers small, perfect or unisexual, slightly irregular, in corymbose, paniculate, or capitate cymes, or rarely monochasial. Inflorescence bracteate and often bracteolate. Calyx reduced to inconspicuous ring or developing into awn or plumose pappus in fruit. Corolla usually tubular, 5-lobed, often spurred or saccate at base, the limb often only slightly irregular, usually spreading. Stamens 1–4, epipetalous, alternating with corolla lobes; anthers 2-celled, splitting longitudinally. Style 1, slender; stigma simple or 3-lobed. Pistil 1. Ovary inferior, 3-loculed but 2 locules usually suppressed. Ovule solitary in each cell, pendulous, anatropous. Fruit an achene, often winged, awned, or with a plumose pappus. Seeds and embryo straight. Endosperm none.

Ten genera and about 375 species, distributed widely in the North Temperate Zone, but a few species extending into the Andes of South America.

Astrephia Dufr., Hist. Nat. Valer. 50. 1811

Perennial herbs with erect, scandent, or lax stems, pinnatifid to pinnatisect leaves, and small flowers in capitate cymes at tips of axillary peduncles. Bracts ovate, small. Calyx a faint ring at apex of ovary. Corolla white to roseate, narrowly funnelform-tubular, only slightly gibbous at base, the limb 5-lobed, lobes spreading. Stamens 3, inserted slightly below alternate sinuses between corolla lobes. Style 3-branched, branches slender. Fruit ovoid, turgid, those of adjacent flowers often more or less coalesced.

Some 4 or 5 species, 3 or 4 of them native in South America, 1 native in China.

Astrephia chaerophylloides (Sm.) DC., Prodr. 4: 629. 1830

Valeriana chaerophylloides Sm., Pl. Icon. Ind., *pl.* 53. 1789–91.
Valeriana chaerophylla Pers., Syn. 1: 37. 1805.
Astrephia laciniata Dufr., Hist. Nat. Valer. 50. 1811.

Weak-stemmed herb scrambling through and over other plants to 1 or 2 m, internodes 5–20 cm long, glabrous, glossy, markedly striate; leaves bipinnatisect, 2–5 cm wide, to 10 cm long, primary segments ovate, to 3 cm long, again lobed nearly to rachilla, the lobes denticulate; petioles short, 3–10 mm long; inflorescences appearing umbelliform, but consisting of di- or trichotomously branched cymes, the ultimate branch of each usually bearing 3 flowers, main peduncles to 10–15 cm long in fruit, bracts lanceolate, to 1 cm long; calyx a faint rim, truncate or minutely 5-toothed; corolla white, about 2 mm long, tube and throat about equal, basal sac

* Diameter of fruit and seeds from R. E. Schultes and R. Romero-Castañeda (1962).
† Contributed by Ira L. Wiggins.

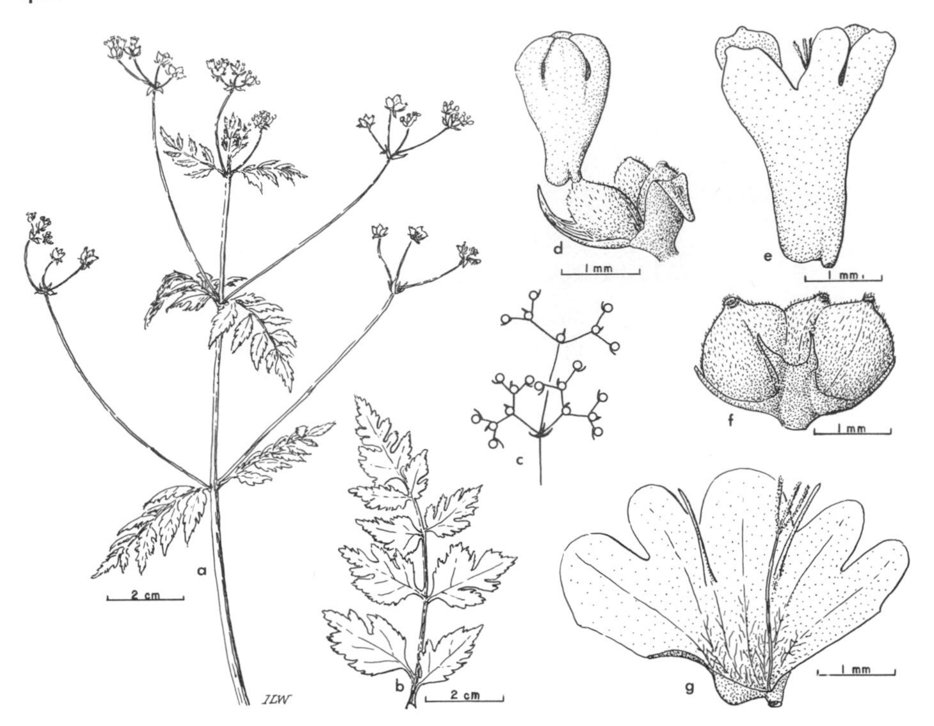

Fig. 126. *Astrephia chaerophylloides*: *c*, diagram of inflorescence.

small but about 3–4 times as wide as tube where attached to ovary, lobes narrowly ovate to oblong, the abaxial slightly narrower and longer than others; stamens inserted below 1 sinus, separating the lobe directly above the saccate pouch from others, then alternate with lobes; anthers 0.5–0.6 mm long; fruits usually in 3's, 3–4 mm long, puberulent, more or less coalesced in pairs or in 3's; embryo about 2 mm long, cotyledons strongly flattened.

In damp soil at margins of fields, along trails, and in woodland, tropics of Ecuador to Peru; in the Galápagos Islands known only from Isla Santa Cruz.

A new record for the Galápagos, though long known from Peru.

See Fig. 126.

VERBENACEAE. Vervain Family*

Herbs, shrubs, woody vines, or trees, with tetragonal twigs. Leaves decussate-opposite, whorled or alternate, deciduous, exstipulate, sessile or petiolate, simple or sometimes palmately compound or unifoliolate; blades entire to variously dentate, incised, or cleft. Inflorescences axillary or terminal, of cymes, racemes, spikes, panicles, thyrsi, heads, or false umbels, sometimes involucrate, axillary ones mostly solitary. Flowers perfect or imperfect, hypogynous, rarely polygamo-dioecious. Calyx gamosepalous, campanulate, tubular, or salverform, persistent, usually accrescent, mostly 4-lobed or 4-toothed. Corolla regular or irregular, gamopetalous,

* Contributed by Harold N. Moldenke.

funnelform or salverform, 4- or 5- to many-parted, often more or less 2-lipped. Stamens 4 and didynamous or reduced to 2, sometimes 4 or 5 and equal, inserted in corolla tube. Staminodes sometimes present. Gynoecium of 2 (rarely 4 or 5) united carpels, 1 sometimes aborted. Ovary mostly compound, sessile or subtended by a disk, mostly 4-lobed, at first 2- or 5-celled but usually becoming 4- or 10-celled through formation of false partitions; axile placenta in each cell bilobed, lobes each bearing 1 ovule, those cells subsequently not divided by cross-walls thus containing 2 ovules, those later divided by partitions containing 1 ovule in each cell; ovules anatropous and basal or hemianatropous and lateral. Fruit a dry schizocarp or drupe with a thin, dry, or fleshy exocarp and a more or less hard endocarp, 2–4-celled and indehiscent or dehiscent into 2(4–10) 1- or 2-celled pyrenes. Seeds testate. Radicle short, inferior. Cotyledons flat, more or less thickened, parallel.

An almost cosmopolitan family of 76 genera and about 3,377 specific and subspecific taxa, found in almost every type of habitat except on the Antarctic mainland and on the Arctic muskeg and tundra.

Inflorescence cymose, determinate and centrifugal:
  Flowers (especially the corolla) more or less zygomorphic; stamens 4, plainly didynamous . . . . . *Clerodendrum*
  Flowers essentially actinomorphic; stamens 5 or 6 . . . . . *Tectona*
Inflorescence spicate or racemiform, indeterminate and centripetal:
  Inflorescence racemiform and elongate; flowers pedicellate . . . . . *Duranta*
  Inflorescence spicate, often subcapitate during anthesis and elongating in fruit; flowers sessile or subsessile:
    Fruit composed of four 1-seeded schizocarps . . . . . *Verbena*
    Fruit composed of 2 pyrenes (or by abortion only 1):
      Pyrenes 2-celled and 2-seeded (or less by abortion), spiny (in ours); fruiting calyx markedly adhesive . . . . . *Priva*
      Pyrenes regularly 1-celled and 1-seeded, not spiny; fruiting calyx not adhesive:
        Perfect stamens 2, with a posterior pair of staminodes . . . . . *Stachytarpheta*
        Perfect stamens 4:
          Fruit drupaceous, usually with a fleshy and juicy exocarp . . . . . *Lantana*
          Fruit small, dry, with a hard and thin or papery exocarp:
            Plants mostly herbaceous, with trailing or ascending stems; stems rooting at the nodes, sometimes somewhat woody at extreme base; spikes elongating greatly in fruit; hairs malpighiaceous . . . . . *Phyla*
            Plants erect shrubs; stems not rooting at the nodes; spikes not elongating in fruit; hairs not malpighiaceous . . . . . *Lippia*

Clerodendrum Burm. ex L., Sp. Pl. 637. 1753; Gen. Pl. ed. 5, 285. 1754

Woody shrubs, trees, or sometimes vines, usually unarmed or rarely the petiole base spinescent, glabrous or pubescent throughout. Leaves simple, decussate-opposite or whorled, entire or variously dentate, exstipulate, deciduous. Inflorescence cymose, cymes mostly loose-flowered, pedunculate in upper leaf axils or paniculate at apex of branchlets or densely aggregate in terminal corymbs or heads. Flowers more or less zygomorphic, often large and showy, mostly white, blue, violet, or red. Calyx campanulate or rarely tubular, truncate, 5-toothed or 5-fid, often accrescent, subtending or enclosing fruit. Corolla hypocrateriform, gamopetalous, tube narrowly cylindric, straight or incurved, equal in diameter throughout or slightly ampliate at mouth, often elongate, more rarely slightly exceeding calyx, limb spreading or re-

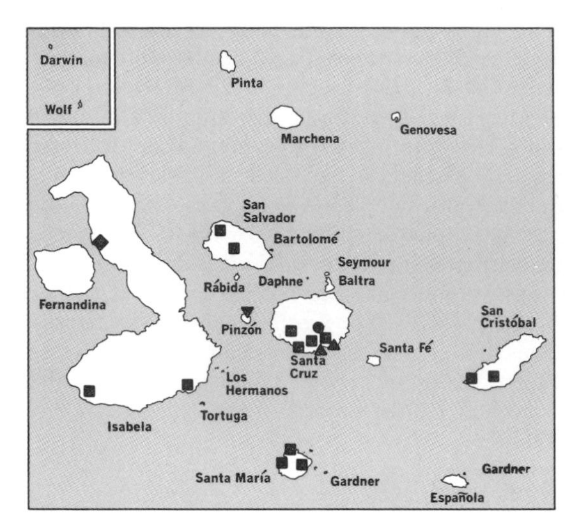

flexed, 5-parted, lobes subequal or posterior and exterior ones slightly shorter and the anterior larger, sometimes concave. Stamens 4, didynamous, inserted in corolla tube, long-exserted, involute in bud, perfect, alternate with corolla lobes; anthers ovate or oblong, 2-celled, dehiscing longitudinally. Pistil 1, compound, 2-carpelled; style terminal, elongate, shortly and acutely bifid at apex. Ovary imperfectly 4-celled, each cell 1-ovulate; ovules high-lateral, hemianatropous. Fruit drupaceous, globose or obovoid, often 4-sulcate and faintly 4-lobed, exocarp more or less fleshy, endocarp bony or crustaceous, smooth or variously rugose, separating into 4 pyrenes at maturity or the pyrenes sometimes cohering in pairs. Seeds oblong, without endosperm.

About 562 species and subspecific entities in forests and shrubby areas of tropical and subtropical regions; abundant in Asia and Africa, poorly represented in America except in cultivation and by naturalization, a few species in temperate zones.

Leaf blades densely short-tomentose beneath. . . . . *C. molle* var. *molle*
Leaf blades obscurely pulverulent or glabrescent beneath. . . . . . *C. molle* var. *glabrescens*

Clerodendrum molle HBK., Nov. Gen. & Sp. Pl. 2: 244. 1817, var. **molle**

Shrub to 5 m tall; branchlets slender, woody, subterete, gray or buff, conspicuously lenticellate, densely short-pubescent, nodes not annulate, prominently marked by persistent petiole bases; internodes 0.5–6 cm long; leaves decussate-opposite or ternate; petioles slender, 3–8 mm long, densely pubescent, lowest 1–2 mm persisting as corky protuberance after blades fall, not spine-tipped; blades chartaceous or membranous, dark green (nigrescent in drying) above, lighter beneath, elliptic, 1–6 cm long, 0.5–3.4 cm wide, acute to short-acuminate at apex, entire, acute at base; finely and more or less sparsely pubescent above, densely short-tomentose beneath; midrib slender, flat or subimpressed and usually more densely puberulent than lamina above, very slightly prominulent beneath; secondaries very slender, 3–6 per side, short, arcuate-ascending, not at all or but slightly prominulent beneath; vein and veinlet reticulation delicate, obscure or indiscernible above, faint beneath; inflorescence axillary; cymes ternate, solitary in each axil, divaricate, 5–7 cm long, 1–4 cm wide, loosely few-flowered or submultiflorous, bracteate, pubescent throughout; peduncles slender, 1–3.7 cm long, pubescent; foliaceous bracts usually none; bractlets and prophylla setaceous, about 1 mm long, densely pubescent; pedicels slender, 1–5 mm long, densely pubescent; flowers delicately fragrant; calyx campanulate, about 1 cm long, rather densely antrorsely pubescent on outer surface of tube, rim shortly 5-lobed in bud, buds growing rapidly to about 6 mm long during anthesis and calyx then deeply 5-parted, lobes elongate-lanceolate, attenuate, venose, subglabrescent; corolla hypocrateriform, 2.5–3 cm long, tube roseate or light lilac, puberulent on outside, limb spreading, its lobes white, punctate on back; filaments long-exserted, red-purple; anthers chocolate brown; fruits subglobose, less than 1 cm long and wide.

From Panama to Peru on sparsely covered hillsides and as undershrub in forests. Specimens examined: Isabela, San Cristóbal, San Salvador, Santa Cruz, Santa María.

See Fig. 127 and color plate 84.

Fig. 127. *Clerodendrum molle* var. *molle*.

Clerodendrum molle var. **glabrescens** Svens., Amer. Jour. Bot. 22: 251. 1935

Differs from var. *molle* in having its leaf blades merely pulverulent-puberulent or glabrate above and very lightly and obscurely puberulent and punctate or subglabrate beneath.

Endemic; known only from Isla Santa Cruz, at low altitudes in woodland or on open shrubby slopes.

Duranta L., Sp. Pl. 637. 1753; Gen. Pl. ed. 5, 284. 1754

Glabrous or pubescent, often spinose, shrubs or trees; leaves decussate-opposite or whorled, entire or dentate; inflorescence indeterminate, racemiform, elongate, mostly terminal, rarely axillary. Flowers small, pedicellate, each borne in axil of small bractlet. Calyx tubular or subcampanulate, truncate, 5-costate and subplicate, each rib terminating in short subulate tooth, the posterior tooth smallest. Corolla salverform, tube cylindric, straight or curved above, exserted from calyx, limb spreading, regular or oblique, 5-parted, mostly pubescent at mouth, lobes rounded, usually unequal, mostly pubescent on inner surface. Stamens 4, didynamous, included, inserted at or above middle of corolla tube; filaments very short; anthers sagittate, dorsifixed, erect. Ovary more or less completely 8-celled, composed of four 2-celled carpels, each cell 1-ovulate; style terminal, shorter than or equaling lower stamens; stigma obliquely subcapitate, very shortly and unequally 4-lobed. Fruiting calyx accrescent, flask-shaped, usually surpassing and closely appressed to, but not coalesced with, fruit, usually coarctate-rostrate at apex. Fruit drupaceous, mostly completely included by the mature calyx. Exocarp fleshy; endocarp hard; pyrenes 4, each 2-celled and 2-seeded.

A complex genus of about 48 species and varieties of tropical and subtropical America, from Florida, southern Texas, and Bermuda through the West Indies, Mexico, and Central America to Argentina; several species widely cultivated, one more or less naturalized in the warmer parts of the Old World.

Leaf blades densely tomentose beneath . . . . *D. dombeyana*
Leaf blades glabrous or subglabrate beneath:
  Venation usually conspicuously impressed above . . . . *D. mutisii*
  Venation usually not impressed above:
    Leaf blades mostly to 5 cm long when mature . . . . *D. repens* var. *repens*
    Leaf blades usually uniformly less than 2 cm long when mature . . . . *D. repens* var. *microphylla*

Duranta dombeyana Moldenke, Bull. Torrey Bot. Club 68: 500. 1941

Strict shrub 2–4 m tall; branches slender, obtusely tetragonal, glabrate, sometimes sparsely spinose; spines axillary, opposite, 5–7 mm long, stout at base; branchlets slender, densely fulvous-appressed-tomentose; main internodes 0.5–4 cm long; leaves decussate-opposite, often with additional pair or fascicle on short axillary twig; leaf scars corky, prominent, semicircular; petioles slender, 4–8 mm long, margined above, densely tomentose to glabrescent; blades chartaceous, uniformly dark green or sometimes brunnescent above when drying, sometimes paler beneath, elliptic, 1.7–9.3 cm long, 1.1–4 cm wide, obtuse, acute, or acuminate (occasionally emarginate) at apex, entire (rarely irregularly antrorse-serrate above widest part), rounded, acute or acuminate at base and usually prolonged into petiole, lightly short-pubescent or puberulent above, glabrescent in age, densely appressed-tomentose with fulvous or yellowish hairs beneath; midrib slender, flat or subimpressed above, prominent beneath; secondary veins 4–6 per side, ascending, flat or subimpressed above, prominulous beneath, joined in many loops near margins; veinlet reticulation sparse, the larger portions subimpressed above and subprominulous beneath, the rest obscure; inflorescences axillary and terminal, axillary 3–7 cm long, terminal to 20 cm long, spicate or subracemiform, secundly many-flowered; peduncles 4–13 mm long, these and rachis densely flavescent-tomentose; pedicels obsolete or to 2.5 mm long, densely flavescent-tomentose; bractlets and prophylla linear, 1–6 mm long, densely tomentose, caducous; calyx campanulate, 3–4 mm long, 2–3.5 mm wide, 5-costate, densely flavescent-tomentose outside, the rim truncate, obscurely and minutely 5-apiculate; corolla hypocrateriform, tube broadly cylindric, curved, 13–14 mm long, glabrous outside below bend, puberulent above, limb 8–11 mm across, densely puberulent on both surfaces; fruiting calyx obvolute and rostrate, completely enclosing fruit, firm, nigrescent, short-canescent (especially on apical beak), glabrate and shiny in age, eventually splitting; fruit subglobose, to 1 cm long and wide, glabrous, shiny, nigrescent in drying, umbonate-rostrate at apex.

A species of Ecuador and Peru; in the Galápagos known only on lava slopes at high altitudes on Isla Isabela.

Duranta mutisii L. f., Suppl. 291. 1781

Duranta obtusifolia HBK., Nov. Gen. & Sp. Pl. 2: 254. 1818.
Duranta repens var. *mutisii* Kuntze, Rev. Gen. Pl. 2: 507. 1891.
Duranta repens var. *obtusifolia* Kuntze, loc. cit.

Arching bush or somewhat scrambling shrub to 5 m tall or extensively branched tree to 10 m tall, sometimes almost a vine; bark gray, smooth or wrinkled; wood rosy purple; branches elongate, drooping or flexuous to decumbent, lax, spinose, gray-puberulent or appressed-pilosulous, often glabrescent in age; twigs slender, thorny on lower parts, densely strigose-pubescent with antrorse brownish hairs; spines large, stout, 2–4 per node, sometimes approximate, 5–10 mm long, recurved; leaves decussate-opposite or sometimes alternate; petioles 1–10 mm long, often weakly differentiated from the cuneate leaf base; blades dark green and glossy above, elliptic, 2.5–6.5 cm long, 1.3–4 cm wide, acute or acuminate at apex, acuminate or cuneate at base, entire or dentate, glabrous above, often more or less short-pubescent beneath when young, usually glabrous when mature; midrib and secondary veins deeply impressed above; inflorescences axillary and terminal, simple or branched, 5–15 cm long, the branches usually ascending, to 5 cm long; peduncles and rachises densely brown-pubescent or appressed-strigillose; flowers often few, about 15 mm long, slightly fragrant; calyx narrowly campanulate to broadly tubular, 6–10 mm long, 3–4 mm wide, antrorsely strigillose, rim 5-toothed, teeth apiculate; corolla hypocrateriform, blue or lavender to lilac, violet, or purple, occasionally white or white with purple tinge, pink with blue ribs, or pale blue with 2 darker stripes on lower segments, lobes often pale pink with deep purple tips, tube about 15 mm long, curved, densely canescent-puberulent outside above calyx, limb to 10 mm wide, densely canescent-puberulent on both surfaces; fruiting calyx rostrate, appressed-strigillose with antrorse hairs on upper part, glabrescent below; fruit yellow or orange, spherical, fleshy, 4-seeded, enclosed by the enlarged calyx.

Venezuela and Colombia to Peru; in the Galápagos known only among other shrubs at higher altitudes on Isla Pinzón.

Duranta repens L., Sp. Pl. 637. 1753, var. **repens**

Duranta erecta L., loc. cit.
Duranta ellisia Jacq., Enum. Pl. Carib. 26. 1760.
Duranta plumieri Jacq., Select. Am. 186, *pl. 176, fig. 76.* 1763.
Duranta racemosa Mill., Gard. Dict. ed. 8. 1768.
Duranta angustifolia Salisb., Prodr. 108. 1796.
Duranta latifolia Salisb., loc. cit.
Duranta xalapensis HBK., Nov. Gen. & Sp. Pl. 2: 255. 1818.

Extremely variable and polymorphic shrub or small tree to 6 m tall; branches slender, often drooping or trailing, unarmed or spiny; branchlets tetragonal; leaves numerous, ovate-elliptic, oval, or obovate, 1.5–5 cm long, 2–3.5 cm wide, obtuse, apiculate or acuminate at apex, entire or serrate above middle, cuneate into the short petiole, glabrate on both surfaces; racemes terminal and axillary, 5–15 cm long, loosely many-flowered, erect or usually recurved, often paniculate; bractlets minute, occasionally subfoliaceous; pedicels 1–5 mm long; calyx tubular, 3–4 mm long, angled, its teeth triangular at base, subulate at apex; corolla hypocrateriform, blue or lilac to lavender or sometimes purple, tube exceeding calyx, limb 7–9 mm wide; fruit yellow, globular, 7–11 mm in diameter, completely enclosed by the accrescent yellowish calyx, this produced into curved beak.

Found throughout tropical and subtropical America, south to Argentina; also introduced into the Old World tropics and widely cultivated. In our area known only from high altitudes on Isla Isabela, where it is represented by a form with entire leaves, perhaps best known as var. *integrifolia* (Tod.) Moldenke.

See Fig. 128.

Duranta repens var. **microphylla** (Desf.) Moldenke, Phytologia 1: 483. 1941

Duranta microphylla Willd., Enum. Hort. Berol. Suppl. 43. 1813, nomen nudum.
Duranta microphylla Desf., Cat. Hort. Paris, ed. 3, 392. 1829.

Differs from var. *repens* chiefly in its regularly smaller leaves, which are less than 2 cm long when mature.

Known from Panama and in cultivation; in the Galápagos, known only from high altitudes on Isla Isabela.

Lantana L., Sp. Pl. 626. 1753; Gen. Pl. ed. 5, 275. 1754

Erect herbs or shrubs, sometimes subscandent or scandent, usually more or less scabrous and hirtellous-pubescent or tomentose with simple hairs, sometimes spinose. Leaves opposite or ternate, dentate, often rugose, inflorescences dense cylindric spikes or contracted into compact heads, usually axillary, pedunculate. Flowers sessile, borne in axils of solitary bractlets, these oblong, lanceolate, or ovate, usually acute or acuminate at apex, sometimes obtuse, spreading, subimbricate or imbricate. Calyx small, membranous, truncate and entire or sinuate-dentate. Corolla hypocrateriform, red, yellow, blue, or white, often fading to various other shades, sometimes varicolored in same head, tube cylindric, slender, equal in diameter throughout or slightly ampliate above, limb spreading, regular or obscurely 2-lipped, 4- or 5-fid, lobes broadly obtuse or retuse at apex. Stamens 4, didynamous, inserted at about middle of corolla tube, included; anthers ovate, with parallel thecae.

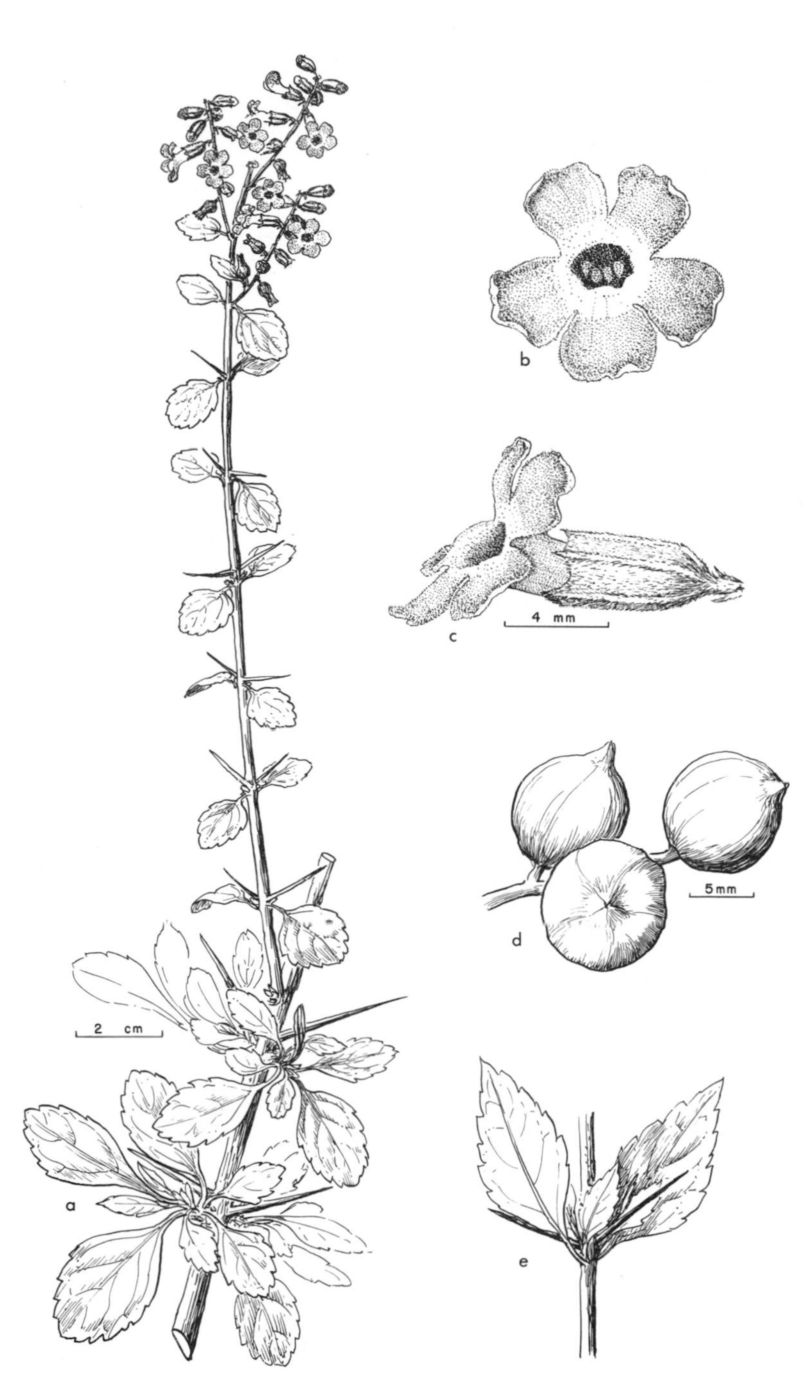

Fig. 128. *Duranta repens* var. *repens.*

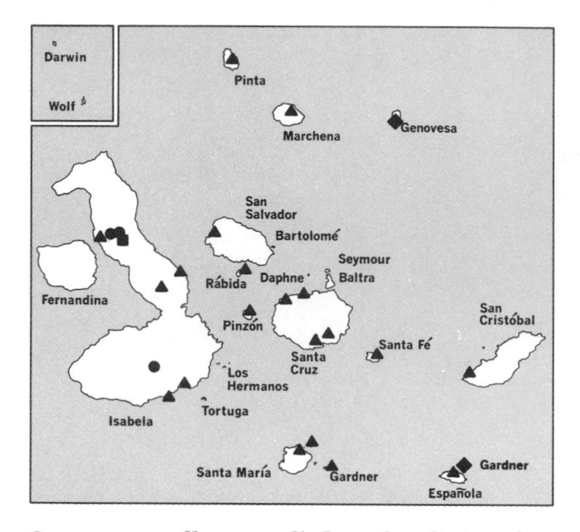

Ovary 1-carpellate, 2-celled, each cell 1-ovulate; style short; stigma rather thick, oblique or sublateral. Ovules basal and erect or attached laterally near base of each cell. Fruit drupaceous, exocarp usually more or less fleshy (rarely dry), endocarp hard, 2-celled or splitting into 2 1-celled pyrenes. Seeds without endosperm.

About 240 species and subspecific taxa, mostly natives of tropical and subtropical America; a few also in tropical Asia and Africa; several species widely cultivated, introduced and a pest in many parts of Australia and Oceania.

Leaf blades 1.5–5 cm long, to 2 cm wide, usually not persistently velutinous on both surfaces, mostly ovate-oblong or ovate-lanceolate to elliptic . . . . *L. peduncularis* var. *peduncularis*

Leaf blades to 7 cm long and 4 cm wide, persistently and very densely velutinous on both surfaces, distinctly ovate . . . . *L. peduncularis* var. *macrophylla*

Lantana peduncularis Anderss., Kongl. Svensk. Vet.-Akad. Handl. 1853: 200. 1855, var. **peduncularis**

Lantana recta sensu Hook. f., Trans. Linn. Soc. Lond. 20: 195. 1847, non Soland. in Ait. 1789.
Lantana canescens var. Hook. f., loc. cit.
Lantana odorata sensu Anderss., Galáp. Veg. 81. 1859, non L. 1767.
Lantana erecta sensu Robins., Proc. Amer. Acad. 38: 195. 1902.

Low canescent bush to 2 m tall; stems and branches slender, strict, tetragonal, subsulcate, glabrescent; twigs very slender, densely whitish-hispidulous with irregular hairs; principal internodes usually short, 1–6 cm long; leaves decussate-opposite; petioles very slender to filiform, 4–10 mm long, flattened above, densely white-hispidulous; blades thin, chartaceous, ovate or ovate-oblong to lanceolate or elliptic, 1.5–5 cm long, 6–20 mm wide, acute to obtuse or rounded at apex, rather uniformly serrate or serrulate to crenate along margins from apex to widest point or almost to base, with somewhat antrorse appressed or spreading rounded teeth, rounded or subcordate to acute or acuminate at base, often narrowed into petiole, densely white-pubescent on both surfaces when young, later less so or sparsely appressed-pilose above and densely tomentose beneath, occasionally velutinous; venation sometimes impressed above; inflorescences axillary, mostly solitary, usually surpassing subtending leaves, sometimes to 6 times as long, capitate; peduncles slender, 1.5–5 cm long, often tetragonal, densely or sparsely white-pubescent; heads many-flowered, 1–1.5 cm long, 1–2 cm wide, hemispheric, subrotund, or slightly elongate in fruit; bractlets laxly imbricate, involucrate, spreading, ovate-lanceolate to elliptic-lanceolate, the lowest to 10 mm long, 5 mm wide, abruptly or gradually acute, densely white-pilose, puberulent, or hispid on back, less so on inner face, the inner ones half as long as corolla; corolla hypocrateriform, white, 8–9 mm long, densely white-hispidulous on outside; style 2.2 mm long.

Endemic; widespread on at least 12 of the Galápagos Islands, on lava, talus slopes, and cinder fields, mostly at low altitudes. Specimens examined: Champion, Española, Gardner (near Española), Isabela, Marchena, Pinta, Pinzón, Rábida, San Cristóbal, San Salvador, Santa Cruz, Santa Fé, Santa María.

This is the plant referred to by Hooker (p. 195) as "*Lantana canescens* var. *ramis aphyllis apices versus foliosis, foliis 1–1½ unc. longis*" and by Svenson (p. 412) as "*L. pedunculata.*" Robinson (p. 195) suggested that this taxon may ulti-

Fig. 129. *Lantana peduncularis* var. *peduncularis* (*a, b*). *Lippia rosmarinifolia* var. *rosmarinifolia* (*c*).

mately prove to be conspecific with one of the continental species or may have to be segregated into several more or less distinct forms. Its closest affinity appears to be not with *L. lilacina* Desf. (= *L. fucata* Lindl.) or the *L. canescens* HBK. as suggested by him, but with *L. svensonii* Moldenke of coastal Ecuador and Peru.
See Fig. 129 and color plate 66.

Lantana peduncularis var. **macrophylla** Moldenke, Phytologia 14: 325. 1967

Differs from var. *peduncularis* in having leaf blades to 7 cm long and 4 cm wide when mature, distinctly ovate and permanently densely velutinous on both surfaces.
Endemic; dominant in scrub and open forest of Arid and Transition zones on Isla Genovesa and perhaps occasional elsewhere at low altitudes. Specimens examined: Gardner (near Española), Genovesa.

Lippia Houst. ex L., Sp. Pl. 633. 1753; Gen. Pl. ed. 5, 282. 1754

Erect shrubs, subshrubs, or rarely trees, with glabrous to variously pubescent herbage and twigs. Leaves opposite or ternate, rarely alternate or in 4's, decussate when opposite, simple, deciduous, entire, toothed, or lobed, exstipulate, petiolate or sessile, flat to rugose above, from thin-membranous to firmly coriaceous, mostly penninerved. Inflorescences spicate or capitate, solitary or fascicled in leaf axils or aggregate in terminal corymbs or panicles, the spikes mostly contracted into heads or cylindric, densely flowered, usually not elongate in fruit, conspicuously bracteate. Bractlets persistent, decussate or many-ranked, herbaceous, often folded, sometimes concave or flat, imbricate, sometimes forming an involucre, mostly rather large and ovate or lanceolate. Flowers sessile, small, often dimorphic, borne singly in axils of bractlets, often more or less 4-ranked; calyx small, membranous, gamosepalous, ovoid-campanulate or compressed and often 2-carinate or 2-alate, sometimes 2-lipped, the rim 2- or 4-fid or 4-dentate; corolla white or variously colored, hypocrateriform or infundibular, gamopetalous, zygomorphic, the tube cylindric, straight or curved, very slender, slightly exserted from calyx or elongate, the limb oblique, usually spreading, somewhat 2-lipped, 4-parted, the lobes broad, often retuse at apex, posterior lobe entire, emarginate to bifid about to middle, lateral lobes exterior, anterior lobe often larger. Stamens 4, didynamous, included to slightly exserted, sometimes absent in pistillate flowers. Pistil 1, often aborted or nonfunctional on staminate plants. Ovary globose, 2-celled, each cell 1-ovulate. Style single; stigma single, incrassate or capitate, oblique or recurved. Ovules basal and erect or affixed laterally near base. Fruit small, dry, ovoid, included in fruiting calyx and sometimes partially adnate to it, dividing into 2 pyrenes or nutlets at maturity, pericarp papery and hard, exocarp membranous and rarely distinct from pyrenes. Seeds without endosperm.
About 252 species and subspecific taxa, widely distributed in tropical and subtropical America; a few also in tropical parts of the Old World.

Leaves oblong-lanceolate, to 25 mm wide, short-petiolate, not revolute; peduncles shorter than mature leaves . . . . *L. salicifolia*
Leaves narrowly linear or narrowly elliptic, 4–7 mm wide, sessile, revolute; peduncles about twice as long as subtending leaves:
  Leaves pinnately lobed . . . . *L. rosmarinifolia* var. *stewartii*

Leaves entire or only obscurely toothed:
Leaves to about 4 mm wide, narrowly linear. . . . . *L. rosmarinifolia* var. *rosmarinifolia*
Leaves to about 7 mm wide, narrowly elliptic. . . . . *L. rosmarinifolia* var. *latifolia*

Lippia rosmarinifolia Anderss., Kongl. Svensk. Vet.-Akad. Handl. 1853: 198. 1855, var. **rosmarinifolia**

Shrub 2 m tall or more; branches strict, virgate, thick, terete, incrassate at nodes; bark dark gray; ultimate branchlets slightly tetragonal, brunnescent, strigose with short, deflexed-appressed hairs; leaves decussate-opposite, sessile, brunnescent in drying, narrowly linear, 1.2–2.7 cm long, about 4 mm wide, acute at apex, entire, conspicuously revolute, narrowed toward base, venation lineate-impressed above, upper surface much roughened with very short prickle-like hairs with tuberculate bases, short-setose-hispid beneath, midrib thick, prominent, lamina cinereous-tomentose; peduncles axillary, tetragonal, about twice as long as subtending leaves, spreading-erect, somewhat sulcate above, gray-strigose; heads subglobose to obovate, 20–30-flowered; bractlets imbricate, many-ranked, gray-hispid on outer surface, the lower ovate-cordate, acute, the upper spatulate, one-third to half as long as flowers, minutely apiculate; calyx tubular-campanulate, bilabiate, bicarinate, subcompressed, hispid-pilose, loosely enveloping mature fruit, the 2 entire segments surpassing fruit; corolla hypocrateriform, lilac, tube elongate, glabrous below, campanulate-dilated and gray-hispid above, lobes short, suberect, hispid on outer surface; fruit 1 mm long, osseous, globose-apiculate, greenish gray, glabrous, minutely punctulate; cocci 2, splitting apart when mature.

Endemic; occasional in cinder fields and on lava flows at low elevations, more common at higher altitudes on several of the islands. Specimens examined: Fernandina, Isabela, Pinta.

Robinson (p. 196) cited *Baur 181* and *182* from eastern Isla Isabela as probably representing this plant, although the branches were sterile. *Stewart 3308* (BISH, CAS, NY, US) has the leaves slightly toothed.

See Fig. 129.

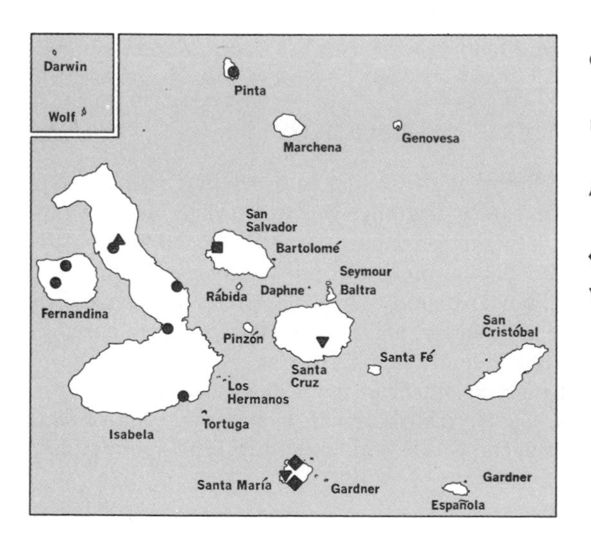

● *Lippia rosmarinifolia* var. *rosmarinifolia*

■ *Lippia rosmarinifolia* var. *latifolia*

▲ *Lippia rosmarinifolia* var. *stewartii*

◆ *Lippia salicifolia*

▼ *Phyla nodiflora* var. *reptans*

Lippia rosmarinifolia var. **latifolia** Moldenke, Phytologia 14: 217. 1967

Differs from var. *rosmarinifolia* in having its leaf blades narrow-elliptic and to 7 mm wide. At first glance this plant looks like *L. salicifolia*, but closer inspection reveals that its leaf blades are entire and revolute as in var. *rosmarinifolia*, but much wider and plainly elliptic in shape.

Endemic; known only from the original collection on Isla San Salvador (*Stewart 3309* [CAS]).

Lippia rosmarinifolia var. **stewartii** Moldenke, Phytologia 2: 415. 1948

Differs from var. *rosmarinifolia* in having pinnately lobed leaf blades, the lobes on the smaller leaves being toothlike, divergent, 1–3 per side.

Endemic; known only from high altitudes on Isla Isabela.

Lippia salicifolia Anderss., Kongl. Svensk. Vet.-Akad. Handl. 1853: 198. 1854

Shrub 2 m high or more; branches curving outward, terete or tetragonal, virgate, thickened at nodes; branchlets hirsute, appressed-strigose toward apex; leaves decussate-opposite; petioles very short, canaliculate above; blades oblong-lanceolate, 6.5–7.5 cm long, almost 2.5 cm wide, dark green, nigrescent in drying, rather obtuse to subrotund at apex, sharply but shallowly serrate or crenate-serrate, cuneate-narrowed and entire at base, not revolute, sparsely pilose with greatly appressed short hairs above, densely incanous-tomentose beneath, the midrib and secondary veins regularly impressed above, elevated and hispid beneath; peduncles axillary, spreading or somewhat erect, scarcely 5 cm long, incrassate, tetragonal, appressed-strigose; heads globose, becoming ovate-elongate after anthesis; bractlets imbricate in many ranks, ovate to broadly cuneate, apiculate, narrowed at base, hirtellous, half as long as corolla tube; calyx subglobose, slightly bicarinate-compressed when mature, cinereous-hirtellous below, glabrescent at base and on inner surface, completely and closely enveloping mature fruit but not surpassing it; corolla tube twice as long as calyx, ampliate above, smooth, limb deeply parted, lobes revolute; fruit globose, about 2 mm long, apiculate, glabrous, the cocci easily separable.

Endemic; known only from forested mountainsides on Isla Santa María.

Phyla Lour., Fl. Cochinch., 66. 1790

Perennial procumbent to creeping herbs, with trailing to ascending stems, sometimes somewhat woody at base or rarely shrubby; herbage subglabrate or appressed-strigose with more or less malpighiaceous hairs. Leaves decussate-opposite, variously dentate (except at base), flat or pinnately plicatulate above; inflorescences spicate, axillary. Spikes cylindric, densely many-flowered, usually greatly elongate in fruit, solitary, paired, or in 3's in leaf axils, never aggregate into corymbs or panicles. Flowers small, sessile, borne singly in axils of small cuneate-obovate or flabelliform bractlets, not at all 4-ranked; otherwise with characters of *Lippia*.

About 20 species and varieties, widely distributed in tropical and subtropical America; 1 or 2 in the warmer parts of the Old World, and a few in temperate parts of North and South America.

Leaf blades usually obovate, elliptic, elliptic-obovate, rhomboid, or spatulate, widest at or above middle . . . . *P. nodiflora* var. *reptans*
Leaf blades usually distinctly ovate, widest below middle . . . . *P. strigulosa*

Phyla nodiflora var. **reptans** (HBK.) Moldenke, Torreya 34: 9. 1934

Zapania reptans Spreng., Pl. Min. Cog. Pugill. 2: 70. 1813.
Verbena reptans Loisel. ex Spreng., loc. cit.
Lippia reptans HBK., Nov. Gen. & Sp. Pl. 2: 263. 1818.
Lippia nodiflora var. *reptans* Kuntze, Rev. Gen. Pl. 2: 508. 1891.

Herb; usually prostrate and creeping; branches sometimes ascending, 20–50 cm long, more or less densely appressed-strigillose; leaves decussate-opposite, not plicate in drying; petioles 5–10 mm long; blades obovate-cuneate, obovate, elliptic-obovate, elliptic, rhomboid, or spatulate, 2–6 cm long, 3–30 mm wide, widest at or above middle, obtuse to subacute at apex, cuneate and attenuate into petiole at base, coarsely and sharply dentate from about middle to apex, the teeth divergent to slightly recurved, more or less appressed-strigillose on both surfaces, green and not evidently canescent, the midrib and secondary veins usually prominent beneath; peduncles axillary, solitary, slender, mostly longer than subtending leaves; spikes ovoid, becoming subcylindric and to 2 cm long in fruit; bractlets closely imbricate, herbaceous, obovate-cuneate, short-acuminate, membranous and ciliate along margins, equaling corolla tube; calyx 2-toothed, 2-carinate, keels villosulous; corolla hypocrateriform, at first white, later fading to pink.

Widely distributed from southern Florida, Texas, Cuba, and Mexico through the West Indies and Central America to tropical and subtropical South America; introduced in parts of Europe, Africa, Asia, and Oceania. Specimens examined: Santa Cruz, Santa María.

Phyla strigulosa (Mart. & Gal.) Moldenke, Phytologia 2: 233. 1947

Lippa strigulosa Mart. & Gal., Bull. Acad. Roy. Brux. 11[2]: 319. 1844.
Lippia canescens sensu Robins., Proc. Amer. Acad. 38: 196. 1902, and sensu Stewart, Proc. Calif. Acad. Sci. IV, 1: 133. 1911, non HBK. 1818.
Phyla yucatana Moldenke, Phytologia 2: 140. 1948.

Procumbent herb, freely branching from base; branches slender, rooting at nodes, obtusely and irregularly tetragonal, often deeply and irregularly sulcate, often reddish or purplish toward base, canescent-strigillose with closely appressed antrorse hairs, the tips ascending to erect; secondary branches more slender, stramineous; nodes annulate; principal internodes 1–5 cm long; leaves decussate-opposite, numerous; petioles 1–5 mm long, mostly winged and merging into base of blades, obscurely canescent-strigillose or becoming glabrescent; blades ovate or ovate-elliptic, mostly conspicuously widest below middle, uniformly green on both surfaces, 1.5–4 cm long, 0.5–2 cm wide, rounded to acute at apex, acuminate and prolonged into petiole at base, regularly dentate with sharply acute or apiculate broadly triangular divergent teeth from apex to widest part, margins of teeth often more or less involute, both surfaces densely canescent-strigillose with short, closely appressed hairs; midrib slender, plane above, prominent beneath; secondary veins slender, 4–6 per side, plane above, prominent beneath, conspicuous to margins and ending in sinuses between teeth, often with 1 or 2 short branches issuing almost at tip and extending to apiculations of nearest teeth; tertiaries and veinlet reticulation not visible; inflorescences axillary, capitate; peduncles slender, 2.5–5.5 cm long, usually only 1 per node, deeply sulcate, densely canescent-strigillose to glabrescent; heads densely many-flowered, 4–8 mm long; bractlets ovate, about 3 mm long, 1.5 mm wide at base, sharply acute, strongly costate, densely canescent-strigillose; calyx minute; corolla white or sometimes tinged with lavender, throat yellow, about 3 mm long, limb about 1.5 mm broad.

A widespread species found from southern Texas and most of Mexico through Central America and from Cuba and Hispaniola through the West Indies to Ecuador, Peru, and Bolivia; variable. Specimens examined: Española, Pinzón, San Cristóbal, Santa Cruz.

See Fig. 130.

Fig. 130. *Phyla strigulosa.*

Priva Adans., Fam. Pl. 2: 505. 1763

Herbaceous caulescent annuals or perennials, mostly harshly pubescent throughout; stems, branchlets, and branches more or less tetragonal, often decumbent. Leaves decussate-opposite or subopposite, thin-membranous, often quite small, usually dentate, sessile or petiolate, deciduous, ovate. Inflorescence terminal or axillary, racemose to subspicate. Flowers hypogynous, small to medium-sized, each solitary in axil of small bractlet, arranged in spirally alternate or pseudosecund manner on elongate rachis, never whorled. Calyx tubular in anthesis, 5-ribbed, often slightly 5-plicate, slightly irregular or nearly regular, gamosepalous, terminating in 5 short equal or subequal teeth, persistent, accrescent and investing the fruit, usually contracted and more or less beaked at orifice at maturity. Corolla gamopetalous, more or less zygomorphic, hypocrateriform or infundibular, tube cylindric, straight or somewhat curved, more or less ampliate, not saccate, limb spreading, oblique, weakly to obviously bilabiate, abaxial lip 3-lobed with large central lobe, adaxial lip 2-lobed, the lobes usually small. Fertile stamens 4, didynamous, the upper (latero-anterior) pair more highly developed than, and inserted slightly above, lower (latero-posterior) pair, this pair inserted approximately at middle of tube, all included or equaling tube; filaments slender; anthers erect, ovate or oblong, 2-celled, dorsifixed at or below middle, introrse, the connective usually conspicuously thickened and sagittate, unappendaged, the thecae parallel or divergent at base, fifth (posterior) stamen reduced to minute staminode or absent. Pistil solitary, 2-carpelled, included; style terminal, usually equaling lower stamens, 2-lobed, anterior lobe longer and recurved or erect, stigmatiferous at tip, posterior lobe reduced, very minute, not stigmatiferous. Ovary 1, superior, 4-celled or (by abortion) 2-celled; ovules basal, erect, anatropous, solitary in each cell. Fruit a dry woody schizocarp, included within the often adhesive fruiting calyx, composed of 2 usually similar 2-celled or (by abortion) 1-celled cocci, these separating easily at maturity; pericarp hard, dorsal surface echinate, scrobiculate, or ridged, the commissural surface excavated, concave, or plane. Endosperm none.

A genus of 25 species and varieties in tropical and subtropical Asia, Africa, and America, the New World forms ranging from southern Florida and Texas through

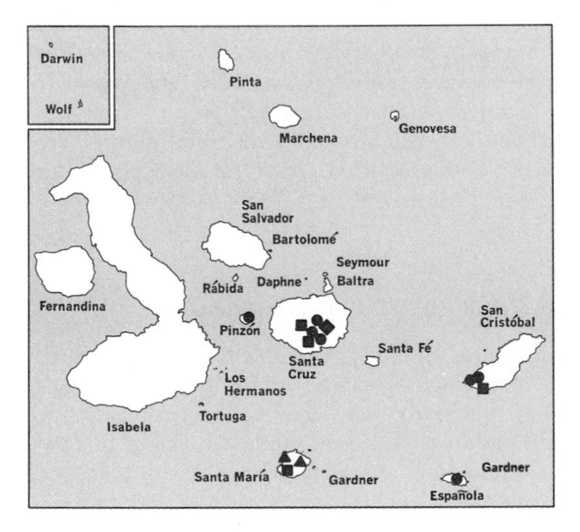

● *Phyla strigulosa*
■ *Priva lappulacea*
▲ *Stachytarpheta cayennensis*
◆ *Tectona grandis*

the West Indies, Mexico, and Central America to Brazil, Bolivia, Paraguay, and northernmost Argentina.

Priva lappulacea (L.) Pers., Syn. Pl. 2: 139. 1806

Verbena lappulacea L., Sp. Pl. 19. 1753.
Zapania lappulacea Lam., Tabl. Encycl. Méth. Bot. 1: 59. 1791.
Priva echinata Juss., Ann. Mus. Hist. Nat. Paris 7: 69. 1806.
Tamonea lappulacea Poir. in Lam., Encycl. Méth. Bot. 7: 568. 1806.
Blairia lappulacea Houst. ex Steud., Nom. Bot. ed. 2, 1: 208. 1840, in synonymy.
Priva lamiifolia Mart. & Gal., Bull. Acad. Roy. Brux. 11^2: 325. 1844.

Annual or perennial herb to 1 m tall; stems spreading and decumbent or procumbent; branches prostrate to ascending or erect, usually acutely tetragonal, often sulcate, pilose with curved or uncinate hairs, glabrescent below in age; larger nodes usually annulate with a band of longer hairs; petioles slender, 0.8–3 cm long, pilose; blades membranous, ovate, 1.4–14.5 cm long, 0.9–8.5 cm wide, acute or acuminate, uniformly and coarsely serrate, mostly subtruncate or subcordate (or acute when young) at base, pilose or strigose above, pilose beneath with scattered hairs and finer puberulence; inflorescences terminal on stems and branches, 4.5–21 cm long, 5–15 cm wide, many-flowered, the flowers loosely alternate on a puberulent-pilose rachis; peduncles slender, 0.8–5.8 cm long, more or less puberulent-pilose, leaves at base of peduncle often much reduced; bractlets usually surpassing pedicels during anthesis; calyx oblong-campanulate, 2–3.1 mm long, about 2.3 mm wide, densely short-tomentose with uncinate hairs about 0.3 mm long and interspersed straight hairs about 0.5 mm long, the calyx tube obscurely 5-ribbed only toward apex, its rim minutely 5-apiculate; corolla hypocrateriform, usually blue, pink, violet, or purple, occasionally lavender or white, with a few scattered hairs on both surfaces, tube broadly cylindric, straight, about 3.6 mm long adaxially, 3.3 mm long abaxially, 1.8 mm wide, the unequal lobes 1–1.8 mm long, all broadly elliptic-lingulate and rounded; upper pair of fertile stamens inserted about 1.5 mm, the lower pair about 1 mm, above base of corolla tube, included; ovary 4-celled, 4-ovulate; fruiting calyx broadly ovate, thin-membranous, conspicuously inflated, enclosing the fruit and short-beaked apically, 5–7 mm long, 3–4.5 mm wide, densely hispidulous with strongly clinging uncinate whitish hairs; schizocarp oblong, conspicuously quadrangular, glabrous, of 2 similar 2-celled, woody cocci, these about 3 mm long, 2 mm wide, the dorsal surface echinate with 2 parallel rows of short, straight spines 0.5–1 mm long, area between rows of spines obscurely scrobiculate-reticulate or transversely ridged, sides transversely and narrowly ridged, commissural faces plane or nearly so, not margined.

A nearly cosmopolitan weed from southern Florida and Texas through the West Indies and Mexico to Brazil, Bolivia, and Peru; naturalized in Java. Specimens examined: San Cristóbal, Santa Cruz, Santa María.

See Fig. 131.

Stachytarpheta M. Vahl, Enum. 1: 205. 1805, nom. conserv.

Herbs or low shrubs, glabrous throughout or villous with simple hairs; leaves decussate-opposite or alternate, dentate, often rugose. Inflorescences terminal, spicate, spikes abbreviated or elongate, loosely to densely many-flowered, occasionally few-flowered; flowers sessile or semi-immersed in rachis, each solitary in axil of bractlet, these small and narrow, appressed or spreading, or bractlets large, ovate to lanceo-

Fig. 131. *Priva lappulacea* (*a, b*). *Stachytarpheta cayennensis* (*c–f*).

late, imbricate. Calyx narrowly tubular, membranous or herbaceous, 5-costate, 5-dentate, sometimes variously split in fruit. Corolla hypocrateriform, white, blue, purple, or red, the tube cylindric, straight or incurved, slender throughout or ampliate above, limb spreading, 5-lobed, lobes broad, obtuse or retuse, equal or somewhat unequal. Stamens with 2 functional anthers, anterior, inserted above middle of corolla tube, included, filaments short; anthers unappendaged, thecae divergent and dehiscing in 1 continuous line; staminodes 2, posterior, small. Ovary 2-celled, each cell 1-ovulate; style elongate; stigma orbicular, subcapitate; ovules attached laterally near base of cells. Fruit oblong-linear, included within fruiting calyx, splitting at maturity into 2 hard, 1-seeded cocci. Seeds erect, without endosperm.

Some 139 species and subspecific taxa, widely distributed in tropical and subtropical America; several species found also in Africa, Asia, and Oceania, and others introduced there. Some species hybridize readily.

Stachytarpheta cayennensis (L. C. Rich.) M. Vahl, Enum. 1: 208 [as "*cajanensis*"]. 1804

Verbena cayennensis L. C. Rich., Act. Soc. Hist. Nat. Paris 1: 105. 1792.
Verbena dichotoma R. & P., Fl. Peruv. 1: 23, *pl. 34.* 1798.
Stachytarpheta dichotoma M. Vahl, Enum. 1: 207. 1804.
Stachytarpheta hirta HBK., Nov. Gen. & Sp. Pl. 2: 280. 1818.
Stachytarpheta veronicaefolia Cham., Linnaea 7: 246. 1832.
Lippia cylindrica Scheele, Linnaea 17: 351. 1843.
Valerianodes cayennense Kuntze, Rev. Gen. Pl. 2: 510. 1891.
Abena cayennensis Hitchc., Rep. Mo. Bot. Gard. 4: 117. 1893.
"*Bouchea* sp." Robins. & Greenm., Amer. Jour. Sci. III, 50: 147. 1895.

Shrub 1–2.5 m tall, much branched; stems and branches terete, from loosely pilose to lanuginous-pubescent with canescent hairs or glabrate; leaves decussate-opposite, blades membranous, ovate to elliptic, 3–7 cm long, 1.6–2.4 cm wide, obtuse to rounded (rarely subacute), regularly crenate-serrate, teeth subacuminate, blade narrowed or obtuse at base, more or less cuneate-decurrent into petiole, scabrous above with sparse strigillose hairs, not rugose, mostly glabrate beneath except appressed-strigillose along venation, often brunnescent in drying; spikes slender, flaccid, to 34 cm long, usually glabrous or subglabrate throughout; rachis scarcely or slightly incrassate, the furrows about as broad as rachis itself; flowers spreading during anthesis; bractlets narrowly linear or subulate, setaceous-acuminate or aristate, scarious along margins; calyx compressed, about 4 mm long, 4-costate, glabrate, rim 4-subulate-dentate, about equaling or surpassing subtending bractlets; corolla pale blue, hypocrateriform, tube about equaling calyx, limb small, about 5 mm wide; style included; fruiting calyx erect, half-immersed in furrows of rachis.

Widely distributed in tropical America, from Mexico and the Greater Antilles throughout Central America and the West Indies to Peru and Argentina; introduced into Hawaii and into parts of the Old World. Specimens examined: Santa María.

See Fig. 131.

Tectona L. f., Suppl. 20. 1781, nom. conserv.

Tall trees with soft bark and more or less tetragonal branches and branchlets. Leaves deciduous, petiolate or subsessile, mostly large and broad, entire or denticulate, rarely lobed, decussate-opposite or ternate. Inflorescences cymose, cymes numerous, many-flowered, borne in massive terminal panicles. Flowers regular; calyx gamosepalous, campanulate, shortly 5–7-lobed, persistent, greatly enlarged and often inflated in fruit, then including fruit and closed above it. Corolla gamopetalous, hypocrateriform, white or blue, tube cylindric, short, limb spreading or reflexed, 5–7-parted, lobes subequal, overlapping in bud. Stamens 5 or 6, equal, exserted; anthers ovate or elliptic-oblong, 2-celled, dorsifixed, thecae parallel, opening by longitudinal slits. Pistil single, elongate; style terminal, capillary; stigma very shallowly bifid, branches subequal. Ovary compound, 2-carpeled, completely 4-celled, each cell 1-ovulate; ovules lateral or high-lateral, hemianatropous. Fruit drupaceous, rounded or weakly 4-lobed, completely included within enlarging fruiting calyx, with thin, subcarneous exocarp and thick, bony, 4-celled endocarp, this with small central cavity between cells. Seeds without endosperm.

Only 5 species and subspecific taxa, native to southern and eastern Asia.

Tectona grandis L. f., Suppl. 151. 1781

Tectona theka Lour., Fl. Cochinch. 137. 1790.
Theka grandis Lam., Illustr. 2: 111. 1793.

Tree to 50 m tall; branches and branchlets stout, tetragonal, with large, quadrangular pith, densely furfuraceous-tomentose with cinereous or ochraceous tomentum; nodes distinctly annulate; leaves drooping, deciduous, decussate-opposite, short-petiolate or subsessile with clasping base; petioles stout, 1–5 cm long, margined to subalate, densely ochraceous-furfuraceous; blades firmly chartaceous to subcoriaceous, broadly elliptic, 11–85 cm long, 6–50 cm wide, acute or short-acuminate, entire or repand-denticulate, abruptly acute to acuminate basally, prolonged into the alate petiole or clasping base, densely squamose and rugose above, becoming glabrous and smooth, densely tomentose with ochraceous to reddish or brownish hairs and densely resinous-punctate beneath; midrib stout; secondary veins slender, 9–15 per side, ascending, arcuately joined at margins; inflorescences axillary and terminal, paniculate, the terminal panicles often 80 cm long and wide, densely furfuraceous and with cinereous or ochraceous tomentum; pedicels 1–4 mm long; bracts large and foliaceous, 2 subtending each pair of cymes; bractlets numerous, linear-lanceolate, to 15 mm long, 4 mm wide, sessile; calyx campanulate, 3–4.5 mm long, 3–3.5 mm wide, densely tomentose, 5–7-toothed, teeth ovate to ovate-oblong, 1.5–2.5 mm long, often reflexed, obtuse; corolla white, rosy on lobes, glabrous, tube broadly cylindric, 1.5–3 mm long, 1.5 mm wide, limb 5–7-parted, lobes obovate-elliptic, 2.5–3 mm long, erect or reflexed; stamens 5–6, inserted 1.3 mm below mouth of corolla tube, equal, slightly exserted; anthers yellow, ovate or oblong, about 1 mm long; style yellowish white, slender, 2.6–5.2 mm long, glabrate or pubescent; stigma bifid, branches 0.2–0.5 mm long; ovary ovate to conical, 1.5–2 mm long, densely pubescent, sordid white to rosy; fruiting calyx enlarged and inflated, to 2.5 cm long and wide, papyraceous, glabrate, mostly irregularly plaited or crumpled, bladder-like; fruit subglobose to tetragonally flattened, to 1.5 cm long and wide, densely tomentose with irregularly branched light brown hairs, umbilicate, 4-lobed at apex, 4-seeded (rarely 1- or 3-seeded by abortion), exocarp thin, somewhat fleshy, densely pubescent, endocarp thick, bony, corrugate; seeds oily.

Native and widely distributed in tropical and southern Asia and Malayan region; widely cultivated for its valuable timber. Specimen examined: Santa Cruz.

The tree examined (on Jacob Horneman's ranch) was round-crowned and about 40 m tall in February 1964. According to Mrs. Horneman, it was introduced as a seedling in 1938.

See Fig. 132.

Fig. 132. *Tectona grandis*.

Verbena L., Sp. Pl. 18. 1753; Gen. Pl. ed. 5, 12. 1754

Herbs, sometimes slightly woody at base; stems and branches procumbent, ascending, or erect, glabrous or variously pubescent. Leaves mostly decussate-opposite, rarely whorled, dentate (rarely entire), or variously lobed, incised, or pinnatifid. Inflorescences spicate, spikes terminal, usually densely many-flowered, often flat-topped and pseudo-umbellate or fasciculate-capitate, sometimes greatly elongate with densely crowded or scattered flowers, rarely also axillary, often greatly elongate only after anthesis. Flowers small or medium-sized, each solitary in axil of a usually narrow bractlet. Calyx gamosepalous, inferior, usually tubular, 5-angled, 5-ribbed, unequally 5-toothed, slightly or not at all changed in fruit. Corolla gamopetalous, salverform or funnelform, tube cylindric, straight or curved, often slightly ampliate or subinflated at apex, usually villous at insertions of stamens within and barbate at mouth, limb flat, spreading or oblique, weakly 2-lipped, lobes 5, usually elongate, more or less unequal, the 2 posterior outermost and anterior innermost in bud. Stamens 4, didynamous, inserted in upper half of corolla tube, included; anthers ovate, with 2 parallel or slightly divergent thecae, dehiscing by longitudinal slit, dorsal connective unappendaged and narrowly muticous or bearing glandular appendage. Style single, usually short and equaling stamens, somewhat dilated, shallowly bifid at apex, anterior lobe stigmatiferous and subpulvinate-papillose, posterior lobe narrower, smooth and horny, nonstigmatiferous. Ovary superior, compound, 2-carpellate, entire or 4-lobed at apex, situated on lobed annular disk, completely 4-celled even at anthesis, 1 ovule in each cell; ovules attached laterally at or near base, erect, anatropous. Fruiting calyx herbaceous, ventricose, eventually split, often connivent or contorted at apex, thin-textured. Fruit mostly enclosed by mature calyx, schizocarpous, pericarp hard and dry, readily separating at maturity into 4 1-seeded linear to linear-oblong crustaceous cocci (rarely elliptic-winged). Seeds 4, erect, oblong, triangular in cross section, outer face convex. Endosperm scanty or none.

About 380 species and subspecific taxa and named hybrids, native to temperate and tropical America; 2 species native to the Mediterranean region and the Near East and introduced elsewhere in the Old World; a few species widely cultivated; some American species now naturalized in parts of Europe, Asia, Africa, and Australia.

Leaves bipinnatifid . . . . *V. grisea*
Leaves not bipinnatifid:
  Leaves linear or nearly so, inconspicuous, the blades very narrow:
    Leaf blades deeply trifid, the segments 1–4.5 cm long, 1 mm wide. . . . *V. townsendii*
    Leaf blades not trifid:
      Leaf blades linear, the lower ones with a few short toothlike lobes; stems and branches glabrous and shiny; peduncles 6.5–7.5 cm long, glabrous and shiny . . . . *V. stewartii*
      Leaf blades linear or narrowly lanceolate, unlobed, entire; stems and branches sparsely pilose; peduncles 1–3 cm long, sparsely pilose. . . . . *V. galapagosensis*
  Leaves oblong, lanceolate, elliptic, or ovate, not linear, the blades broad:
    Fruiting calyx and bractlets glabrescent or lightly strigillose and ciliolate only, nigrescent in drying; peduncles and spikes extremely slender and lax; even uppermost leaves coarsely dentate. *V. glabrata* var. *tenuispicata*
    Fruiting calyx and bractlets rather densely and uniformly puberulent, not nigrescent; peduncles and spikes stout and firm, not lax; uppermost leaves usually entire:

Stems antrorsely scaberulous on angles when young; leaves lanceolate to oblong or oblanceolate, not brunnescent, to 1.5 cm wide, sessile or subsessile, sparsely strigillose on both surfaces. . . . . *V. litoralis*

Stems glabrous, or short-pilose with hairs readily visible only at 10×, or long-pilose with slender hairs; leaves elliptic, brunnescent in drying, to 2.7 cm wide, on petioles 1.5–2 cm long, variously pubescent on both surfaces:

Stems glabrous; leaf blades glabrous or with a few setulose hairs. . . . . *V. sedula* var. *sedula*

Stems long-pilose, or microscopically and more densely short-pilose; leaf blades more densely pubescent on both surfaces:

Lower surface of leaf blades markedly long-pilose. . . . . *V. sedula* var. *darwinii*

Lower surface of leaf blades densely short-pilose with hairs readily visible at 10×. . . . . *V. sedula* var. *fournieri*

Verbena galapagosensis Moldenke, Phytologia 2: 55. 1941*

Herb, nigrescent throughout in drying; stems and branches slender, acutely tetragonal, often sulcate and striate between angles, sparsely pilose with short, whitish, widely scattered antrorse hairs; nodes annulate; principal internodes 1–5.8 cm long; leaves decussate-opposite, sessile, subclasping at base, linear or very narrowly lanceolate, 5–15 cm long, rather abundantly pilose with appressed antrorse hairs on both surfaces; midrib and veinlet reticulation indiscernible; inflorescences terminal, spicate, 5–15 cm long, loosely many-flowered; peduncles slender, similar to branches in all respects, nigrescent in drying, acutely tetragonal, longitudinally striate, very sparsely scattered-pilose, 1–3 cm long; bractlets ovate-lanceolate, about 2.5 cm long, ciliolate-margined, acuminate at apex, usually glabrate or very obscurely pilosulous except for margins; mature flowers and fruits not known.

Endemic; known only (and common) at high elevations on Isla Isabela.

Verbena glabrata var. **tenuispicata** Moldenke, Phytologia 14: 283. 1967†

Spreading herb; stems often procumbent; branches spreading or erect, very slender, tetragonal, sulcate, with rounded angles, brunnescent on drying, glabrous or with few scattered erect hairs; branchlets abundant, spreading, similar to branches; leaves numerous, conspicuous, decussate-opposite, brunnescent or nigrescent throughout in drying, petiolate, those in inflorescence much reduced; petioles 1–5 mm long, winged and often difficult to differentiate from blades, scattered-pilose with a few stiff whitish hairs; blades narrowly elliptic, 2–4 cm long, 5–10 mm wide, acute or attenuate at apex, coarsely serrate with antrorsely divergent, narrowly elongate-triangular teeth, long-attenuate into winged petiole, scattered-pilose with whitish stiff hairs on both surfaces, most densely so on larger venation beneath; inflorescence terminal, paniculate, widely and loosely branched; peduncles indistinguishable from upper stems, 4–6 cm long; 1 pair of bracts beneath each pair of inflorescence branches, linear, 2–16 mm long, 0.5–1 mm wide, entire or wider ones appressed-serrulate, pilose on both surfaces; spikes short, slender, 4–12 cm long, about 5 mm wide at anthesis, flowers close and imbricate, more distant toward base of spike; bractlets

* Included in *V. litoralis* by Stewart (1911, 134).

† Included in *V. litoralis* by Stewart (1911, 134). Reported as *V. officinalis* L. by Andersson, Robinson, and Stewart.

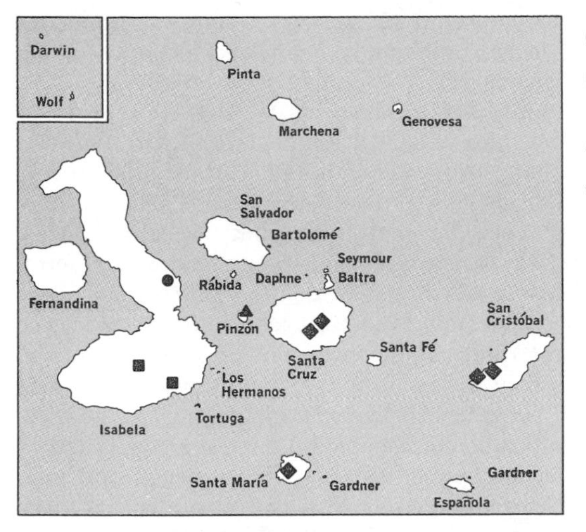

ovate-lanceolate, about 2 mm long, 1 mm wide, attenuate or acuminate at apex, glabrous or subglabrous except for the white-ciliolate margins; corolla hypocrateriform, very small, tube to 5 mm long, limb about 3 mm wide; fruit sparse, not contiguous.

Endemic; known only from the type collection, taken on Isla Isabela.

Verbena grisea Robins. & Greenm., Amer. Jour. Sci. III, 50: 142. 1895

Densely and somewhat sericeously gray-tomentose throughout, in parts brownish or ferruginous in dried state; stems tetragonal, somewhat furrowed; principal internodes considerably exceeding subtending leaves in most cases; leaves decussate-opposite, petiolate; blades ovate to deltoid, 2.5 cm long or more, dense pubescence obscuring venation on both surfaces, bipinnatifid, rachis and segments narrow, not 2 mm wide, frequently curled, obtuse at apex; inflorescences branched, flowers small, borne in slender, loose, elongate spikes, central spike floriferous almost to base, the other more or less pedunculate; bractlets small, subulate, one- to two-thirds length of calyx, pubescent; calyx cylindric, about 2 mm long, 5-ribbed, hirsute-pubescent on outer surface, rim shallowly and obtusely 5-toothed; corolla nearly twice as long as calyx, pubescent in throat, limb unequally 5-lobed, lobes rounded; anthers unappendaged.

Endemic; a rare plant at higher elevations on Isla Pinzón.

Robinson and Greenman's reference (p. 147) to this species on Isla Isabela was a typographical error.

Verbena litoralis HBK., Nov. Gen. & Sp. Pl. 2: 276, *pl. 137.* 1818

Verbena lanceolata Willd. ex Spreng. in L., Syst. Veg. ed. 16, 2: 748. 1825.
Verbena affinis Mart. & Gal., Bull. Acad. Brux. 11[2]: 322. 1844.
Verbena bonariensis β littoralis Hook. ex Walp., Repert. Bot. Syst. 4: 20. 1845.
Verbena littoralis var. ?*glabrior* Benth., Bot. Voy. Sulphur 153. 1846.
Verbena littoralis β leptostachya Schau. in A. DC., Prodr. 11: 542. 1847.
Verbena nudiflora Nutt. ex Turcz., Bull. Soc. Nat. Mosc. 36[2]: 195. 1863.
Verbena hanseni Greene, Pittonia 3: 308. 1898.

Rapidly growing, erect or suberect perennial herb, sometimes suffrutescent, 0.2–2 m tall, strict, shiny, sometimes large and robust, fastigiately few- to many-branched above, glabrous or sparingly strigillose; stems to 12 mm thick, tetragonal, internodes deeply sulcate above insertion of leaves and faintly many-sulcate between, antrorsely scaberulous on angles when young, green, slightly contracted at nodes; leaves decussate-opposite, lanceolate to oblong or oblanceolate, 3–10 cm long, 1–1.5 cm wide, acute at apex, tapering into short petiole or sessile, acuminate and somewhat clasping or subamplexicaul at base, antrorsely scabridulous throughout or scabrous and rugose above, upper two-thirds more or less sharply and coarsely serrate, commonly with 3–5 mucronate teeth per side, sparsely strigillose on both surfaces; spikes terminal, pedunculate, cymose or corymbose-paniculate, slender to filiform, pilosulous, at first congested and about 4 mm across, dense or interrupted, later loosely flowered and elongate, in fruit becoming 2–8 cm long, the fruits separated; bractlets ovate-lanceolate, marcescent, rigid, acuminate or subulate at apex, 1–3 mm long, usually subequal to or shorter than calyx, abruptly curved upward, glabrate; calyx often purple, 2–2.5 mm long, pubescent or finely strigillose outside, subtruncate, the 5 teeth unequal, minute, subulate; corolla about 3 mm long, blue, lilac, or violet to lavender, purple, mauve, or pink, puberulent on outer surface, tube somewhat longer than calyx, limb inconspicuous, 2.5–3 mm wide, puberulent on inner surface, lobes subequal, obtuse, glabrous; anthers greenish yellow; style about 1 mm long, green, glabrous; stigma 2-lobed, larger lobe stigmatiferous; ovary almost 1 mm long, green, glabrous; cocci trigonous, dark-stramineous, linear-oblong, to almost 2 mm long, glabrous, smoothish, striate, somewhat reticulate at apex, commissural face about as long as coccus, muricate-scabrous.

A polymorphic subtropical and tropical weed, widely distributed from the southern United States throughout Central America into South America; introduced in Oregon, California, South Africa, Hawaii, Australia, and parts of Oceania. Specimens examined: San Cristóbal, Santa Cruz, Santa María.

See Fig. 133.

Verbena sedula Moldenke, Phytologia 5: 229. 1955, var. **sedula**

Herb about 1 m tall; stems and branches tetragonal, sulcate and ribbed, brownish, glabrous; nodes annulate, often with white setulose hairs, principal internodes 3.5–

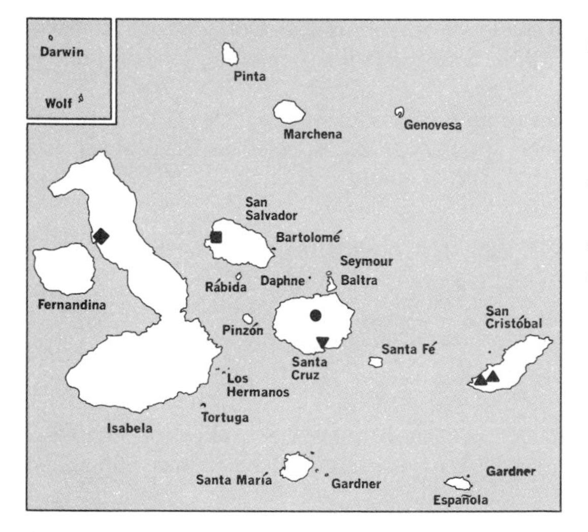

● *Verbena sedula* var. *sedula*
■ *Verbena sedula* var. *darwinii*
▲ *Verbena sedula* var. *fournieri*
◆ *Verbena stewartii*
▼ *Verbena townsendii*

Fig. 133. *Verbena litoralis.*

5.5 cm long; leaves decussate-opposite; petioles slender, 1.5–2 cm long on mature leaves, glabrous or with a few setulose hairs; leaf blades thin-chartaceous, elliptic, 4–8 cm long, 1–2.7 cm wide, acute, irregularly serrate from below middle to apex, glabrous or sparingly setulose-hairy; midrib slender, subimpressed above, prominulous beneath; veinlets often obscure; inflorescence terminal, openly paniculate, about 20 cm long, 15 cm wide, flowers borne in spicate fashion, spikes to 14 cm long, 5 mm wide, densely many-flowered, flowers contiguous and partly overlapping; peduncles slender, tetragonal, glabrous or sparsely setulose; bractlets lanceolate-ovate, about 2 mm long, acuminate, ciliate toward base; calyx tubular, about 3 mm long, sparsely pilosulous, rim 5-toothed; corolla white, hypocrateriform, about 4 mm long, limb about 3 mm wide.

Endemic; known only from the type collection, taken on Isla Santa Cruz.

Verbena sedula var. **darwinii** Moldenke, Phytologia 16: 341. 1968*

Differs from var. *sedula* in having its stems conspicuously and rather densely long-pilose with antrorsely substrigose or somewhat divergent hairs. The short branches, petioles, and lower leaf surfaces are also quite densely and conspicuously long-pilose and the upper leaf surfaces sparsely so.

Endemic; known only from the type collection, taken on Isla San Salvador.

Verbena sedula var. **fournieri** Moldenke, Phytologia 18: 211. 1969

Differs from var. *sedula* in having its petioles irregularly and rather sparsely short-pilose with hairs not plainly visible without a hand lens, and the lower leaf surfaces rather densely short-pilose. Differs from var. *darwinii* in having denser, shorter, finer hairs on all parts.

Endemic; known only from Isla San Cristóbal.

Verbena stewartii Moldenke, Phytologia 2: 56. 1941†

Herb, more or less nigrescent in drying; stems and branches slender, latter almost filiform, sharply tetragonal, glabrous and shiny throughout; nodes annulate; principal internodes elongate, 2–6 cm long; leaves sessile, upper linear, lower with 2 or 3 linear, widely divergent lobes, revolute, blunt-pointed, sparsely pilose with appressed, whitish, antrorse hairs on both surfaces; midrib prominent beneath; inflorescence terminal, spicate, few-flowered, dense toward apex and during anthesis, lower flowers often scattered after anthesis; peduncles slender, elongate, 6.5–7.5 cm long, glabrous and shiny; rachis filiform, glabrous and shiny or obscurely scattered-pulverulent; bractlets lanceolate, 1.5–2 mm long, acuminate, glabrous except along ciliolate margins; calyx tubular, about 2 mm long, minutely appressed-puberulent; corolla hypocrateriform, barely exceeding calyx, tube usually about 2 mm long, limb about 1.5 mm wide.

Endemic; known only from the type collection, taken from Isla Isabela.

* Reported by earlier workers as *Verbena polystachya* HBK., *V. caroliniana* Willd., *V. carolina* L., or *V. urticifolia* L.

† Stewart (1911, 134) included this in *V. litoralis* HBK.

Verbena townsendii Svens., Amer. Jour. Bot. 22: 253, *pl. 3, f. 1a–d.* 1935

Glabrous erect herb to 1 m tall; stems tetragonal, about 3 mm thick; principal internodes 5–7 cm long; branches rigid, suberect, paniculate; leaves decussate-opposite, deeply trifid; petioles similar to leaf segments, 1–2 cm long; blades with linear segments 1–4.5 cm long, about 1 mm wide, revolute; spikes elongate, to 20 cm long, borne at tips of branchlets, erect, filiform, lower flowers becoming remote, often separated 15 mm at maturity of fruit; bractlets ovate, about 2.6 mm long, acuminate, about two-thirds as long as calyx, ciliate-margined, otherwise glabrous; pedicels about 0.5 mm long; flowers very small; calyx to 2.7 mm long; corolla lilac or lavender, twice as long as calyx tube, bearded in throat, limb 5-parted, lobes 1.5 mm long, the lower 3 slightly emarginate; stamens 4, didynamous; anthers broadly elliptic, 0.3 mm long, equaling or exceeding filaments; filaments filiform, 0.1–0.3 mm long; style bilobed; fruiting calyx tetragonal, glabrous or rarely somewhat strigose, 4 of teeth incurved-plicate, slightly surpassing fruit, 5th (dorsal and innermost) shorter and straight; cocci 4, brown, 1.5 mm long, prominently venose on outer surface, white-roughened on commissural face.

Endemic; known only from the type collection, taken from Isla Santa Cruz.

Svenson cited *Stewart 3317, 3318,* and *3319* as belonging also in this species, but I am citing these as *V. glabrata* var. *tenuispicata* and *V. galapagosensis.*

BIGNONIACEAE. BIGNONIA FAMILY*

Caruel (1889, 622) reported "Bignoniacea? 12. Tecoma? – Chatham. – Sterile." However, this singularly attractive genus has not been reported again from the archipelago, nor has any other member of the family been found or seen there. Since the specimen was sterile, and since even the family was in doubt, it is impossible to determine the identity of the collection without examining it. It is very doubtful that it represents a member of the Bignoniaceae.

* Contributed by Duncan M. Porter.

Dicotyledones
Series III: Polypetalae

Dicotyledones
Series III: Polypetalae

Dicotyledonous plants with well-developed outer perianth whorl (calyx) (rarely reduced markedly) and inner perianth whorl (corolla) consisting of distinct, free segments (petals) usually white or of various colors or shades or tints, inserted on apex of receptacle, on a disk, or on the wall or rim of a hypanthium. All whorls of floral parts variable within rather broad limits. Otherwise with characters as in the subclass (Dicotyledones, p. 180).*

KEY TO THE FAMILIES OF THE POLYPETALAE

Ovary superior, free from calyx or adnate only at very base:
  Stamens and petals hypogynous, i.e., inserted on receptacle or on a narrow disk at base of pistil:
    Stamens more than 10:
      Ovary 1-celled:
        Leaves simple, entire or dissected, not compound:
          Petals undulate and crepe-like; leaves and stems armed with stiff prickles.... ....PAPAVERACEAE, p. 717
          Petals neither undulate nor crepe-like; leaves and stems unarmed (petiolar spines present in Subfam. Mimosoideae):
            Stamens spirally arranged; thecae of anthers overtopped by connective; endosperm large, ruminate.....................ANNONACEAE, p. 520
            Stamens in whorls or fascicles, not spirally arranged; connective shorter than thecae; endosperm smaller, not ruminate:
              Sepals 2; petals 4 or 5; sepals caducous..........PORTULACEAE, p. 728
              Sepals and petals same number, usually 5; sepals persistent:
                Gynoecium of several to many separate pistils, each 1-celled, 1- to several-ovuled; buds globose, to 10 cm broad, broadly rounded at apex; petals strongly crumpled in bud.....DILLENIACEAE, p. 565
                Gynoecium of a single pistil, the carpels fully united; buds less than 2 cm broad, acute; petals not markedly crumpled in bud:
                  Placentation parietal; ovary 1-celled (or falsely 2-celled by intrusion of placentae); fruit echinate; testa of seeds fleshy, bright red ....BIXACEAE, p. 523
                  Placentation axile; ovary 2–10-celled; fruit not echinate; testa of seeds not fleshy nor bright red............TILIACEAE, p. 758
        Leaves compound:
          Leaves digitately 3–5-foliolate; endosperm fleshy....CAPPARIDACEAE, p. 546

* Synopsis and key to families contributed by Ira L. Wiggins.

Leaves pinnately compound, leaflets often numerous; endosperm lacking or scanty and glassy (Subfam. Mimosoideae) . . . . LEGUMINOSAE, p. 599

Ovary 2- to several-celled:

Leaves pellucid-glandular or black-dotted; stipules lacking; stamens usually irregularly connate basally into partial rings or in 3–8 fascicles; endosperm none:

Fruit a capsule, dry; plant herbaceous or suffrutescent; stamens in fascicles, filaments slender, not fleshy . . . . HYPERICACEAE, p. 596

Fruit a hesperidium, juicy, or each seed in a fleshy aril; plant a large shrub or tree; stamens irregularly connate laterally at base to form segments of a circle, filaments more or less fleshy . . . . RUTACEAE, p. 741

Leaves not pellucid-glandular nor black-dotted; stipules usually present, often small; stamens free, or fused into a column, or connate at base, not in fascicles; endosperm usually present:

Stamens free from one another or faintly connate at base; endosperm scanty:

Leaves digitately compound; stipules caducous; seeds invested in long, silky hairs . . . . BOMBACACEAE, p. 524

Leaves simple; stipules more or less persistent, not caducous; seeds naked or short-puberulent . . . . TILIACEAE, p. 758

Stamens fused into a column or connate prominently at base; endosperm usually copious, often oily:

Ovary 3-carpellate; style branches again 2- or 3-branched; petioles and margins of leaves conspicuously cupulate- or capitate-glandular . . . . EUPHORBIACEAE, p. 568

Ovary mostly 5–30-carpellate; style branches simple; large cupulate or capitate glands lacking . . . . MALVACEAE, p. 666

Stamens 10 or fewer:

Placentation basal, i.e., ovules attached more or less centrally in bottom of gynoecial cavity (ovary 1-celled if more than 1 carpel):

Leaves opposite, mainly cauline (or some rosulate), not fleshy; petals markedly emarginate to deeply bifid; sepals 5, persistent . . . CARYOPHYLLACEAE, p. 548

Leaves alternate or rosulate, more or less fleshy; petals entire or shallowly emarginate; sepals 2, usually caducous . . . . PORTULACACEAE, p. 728

Placentation axile or parietal, not basal:

Placentation parietal, i.e., ovules borne along sutures between carpels surrounding gynoecial chamber(s):

Pistils 3–10(15), distinct, sessile or stipitate; fruit a capsule or follicle:

Plants creeping, ascending, or erect herbs (in ours), or shrubs, not twining vines; leaves markedly succulent; stigmas linear . . . . CRASSULACEAE, p. 559

Plants twining woody vines (in ours); anthers 4-celled (or falsely so); leaves not succulent; stigmas capitate or discoid . . . . MENISPERMACEAE, p. 700

Pistil 1, if more than 1-carpellate the carpels fused into 2-celled ovary; fruit a legume, silique, or silicle:

Flowers cruciferous, i.e., sepals 4, petals (2)4 (rarely absent); petals long-clawed; stamens 6 or sometimes 2; fruit a silique or silicle . . . . CRUCIFERAE, p. 560

Flowers not cruciferous; lower pair of petals often fused part of length, but distinct at base, claws short; stamens commonly diadelphous; fruit a legume . . . . LEGUMINOSAE, p. 599

Placentation axile, i.e., ovules attached to central axis of ovary (usually 2- to several-loculed):

Fruits indehiscent, or separating into indehiscent nutlets or mericarps:

Leaves simple (or appearing so), entire, lobed, or toothed:
Fruit drupaceous or berry-like, not capsular:
Leaves pellucidly punctate-glandular, aromatic; fruit a leathery-skinned berry (hesperidium) or brownish follicle. . . . . RUTACEAE, p. 741
Leaves not pellucidly punctate-glandular nor aromatic; fruit a drupe, the pulp fleshy, not juicy:
Branchlets unarmed; drupe 2–4 cm high; leaves ovate, to 10 cm long, plane; poisonous, corrosive latex in twigs and fruits (genus *Hippomane*) . . . . EUPHORBIACEAE, p. 568
Branchlets armed with axillary spines; drupes less than 16 mm high; leaves linear to elliptic, revolute; milky or corrosive latex none . . . . SIMAROUBACEAE, p. 754
Fruit capsular or a schizocarp, neither drupe-like nor berry-like:
Corolla irregular, lower petal fringed apically; fruit a winged capsule, this often indehiscent; stamens usually 8, connate into a sheath much of their length. . . . . POLYGALACEAE, p. 724
Corolla regular, the petals not fringed; fruit consisting of mericarps; stamens free, in 1–3 whorls of 5 each:
Mericarps smooth, 1- or 2-seeded; leaves simple, glabrous or nearly so; stipules globose-glandular. . . . . LINACEAE, p. 658
Mericarps rugose, tuberculate, or spiny, 1-seeded; leaves pinnately compound, pubescent; stipules foliaceous. . . . . ZYGOPHYLLACEAE, p. 772
Leaves compound, the leaflets distinct, few to numerous:
Stamens connate into a tube, only the anthers distinct; endosperm fleshy . . . . MELIACEAE, p. 697
Stamens free or connate only at base; endosperm scanty or none:
Drupes fleshy, juicy, not leathery nor valvate, to 5 cm long; leaflets 9–15, alternate . . . . ANACARDIACEAE, p. 518
Drupes with leathery, 2- or 3-valved exocarps, 12 mm high or less; leaflets 3–9, opposite. BURSERACEAE, p. 530
Fruits dehiscent, consisting of capsules or follicles:
Fruit of 1–5 coriaceous, thin-fleshy follicles, attached basally and spreading-ascending; follicles dehiscing along adaxial side. . . . . RUTACEAE, p. 741
Fruit capsular, dehiscing along 2 opposite sides or more, or apically:
Leaves digitately or pinnately compound:
Plants herbaceous (in ours), tendrils none; capsules oblong or cylindrical; apices of leaflets deeply notched; juice of herbage acid; ovary 5-carpellate . . . . OXALIDACEAE, p. 712
Plants vines or shrubs, tendrils often present; capsules inflated-globose; apices of leaflets acute to acuminate; juice not acid; ovary 3-carpellate (genus *Cardiospermum*). . . . . SAPINDACEAE, p. 747
Leaves simple, entire to lobed:
Capsules winged; corolla irregular, the lower petal fringed apically. . . . . POLYGALACEAE, p. 724
Capsules not winged; corolla regular:
Herbage (at least in youth) more or less stellate-pubescent:
Capsule firm, leathery or woody; ovules 1–2 per locule; stamens not as below. EUPHORBIACEAE, p. 568
Capsule thin-membranous, delicate; ovules 2 to several in each locule; stamens in 2 whorls, the inner fertile, the outer sterile . . . . STERCULIACEAE, p. 756
Herbage glabrous or bearing simple hairs, not stellate-pubescent:

Petals about 1 mm long, greenish; leaves leathery, distichous; seeds with a red or white aril........CELASTRACEAE, p. 551

Petals 2–10 mm long, yellow or yellowish white; leaves thin or moderately stiffened, not distichous; seeds naked or carunculate, not with a fleshy aril:

Sap not milky nor corrosive; leaf glands not present; ovary usually 5- or 10-celled..................LINACEAE, p. 658

Sap milky or colored, somewhat corrosive; marginal leaf glands often present; ovary usually (2)3-celled; seeds carunculate .... EUPHORBIACEAE, p. 568

Stamens and petals perigynous, i.e., borne on calyx cup or calyx tube or on a disk lining calyx cup, inserted distinctly above base of ovary but not adnate to it:

Ovary partially inferior, i.e., basal part adnate to calyx tube:

Sepals 2, often caducous; capsule circumscissile; embryo annular, nearly surrounding endosperm; plants herbaceous................PORTULACACEAE, p. 728

Sepals 3–14, basally connate, persistent; capsule (or berry) not circumscissile; embryo straight, surrounded by endosperm; large shrubs or small trees......... .... RHIZOPHORACEAE, p. 738

Ovary entirely superior, not adnate to calyx cup except at very base:

Corona of fimbriae present between perianth and androecium; pistil usually on raised androgynophore ........................PASSIFLORACEAE, p. 719

Corona none; androgynophore none:

Stamens as many as petals and opposite them; petals either hooded or caducous:

Petals hooded, each more or less enfolding its subtended stamen; inflorescences axillary or terminal..........................RHAMNACEAE, p. 733

Petals neither hooded nor enfolding stamens, apically connate, caducous; inflorescences cymose, each opposite a leaf............VITACEAE, p. 769

Stamens, when same in number as petals, alternate with them; petals neither hooded nor caducous:

Fruit a fleshy-pulped drupe; leaves compound, with 8–11 leaflets (genus *Sapindus*)..................................SAPINDACEAE, p. 747

Fruit a dry, membranous or leathery capsule or a 3-seeded schizocarp; leaves simple or of relatively few leaflets:

Ovary 1-celled; styles 3, fringed apically..........TURNERACEAE, p. 762

Ovary 2–6-celled (usually 3-celled); styles never fringed apically:

Flowers somewhat irregular (2 upper petals larger than lower 3 in Lythraceae, the reverse in Tropaeolaceae):

Petals markedly crumpled in bud; leaves opposite, their blades not peltate; dorsal sepal not produced into a spur; fruit a capsule; ovules several in each locule..............LYTHRACEAE, p. 663

Petals imbricate, not crumpled in bud; leaves alternate, their blades often peltate; dorsal sepal produced into a conspicuous spur; fruit a schizocarp of 3 indehiscent mericarps; ovule solitary in each locule .... TROPAEOLACEAE, p. 761

Flowers regular, all petals practically same size:

Capsule leathery; flowers greenish; leaves simple, thick-leathery, distichous .... CELASTRACEAE, p. 551

Capsule thin, nearly membranous, not leathery; leaves 2–5-foliolate, neither leathery nor distichous (genus *Cardiospermum*)........ .... SAPINDACEAE, p. 747

Ovary inferior, fully adnate to calyx tube:

Stamens numerous (15 or more):

Sepals 2; fruit a circumscissile capsule; seeds with central endosperm and annular embryo; stamens 6–20 . . . . PORTULACACEAE, p. 728

Sepals 4 to many; fruit a berry or longitudinally dehiscent capsule; endosperm scanty or none, never wholly central; stamens more than 20:

Ovary 1-celled:

Fruit a berry, usually armed with spines or glochids or both; sepals and petals indefinite, not strongly differentiated; stamens very numerous, not in clusters or tufts, inserted on whole perianth tube; stems usually armed with spines or glochids or both; plants shrubs or trees, succulent . . . . CACTACEAE, p. 533

Fruit a capsule; sepals 4 or 5; petals 4–10; stamens usually in clusters opposite petals; stems scabrous but not armed with spines or glochids; plants herbaceous (in ours), not succulent . . . . LOASACEAE, p. 660

Ovary 2- to several-celled:

Seeds usually germinating while fruit still attached to twig; petals thick, somewhat fleshy, hirsute marginally or within; stamens 4–20; leaves coriaceous; prop roots usually present . . . . RHIZOPHORACEAE, p. 738

Seeds not germinating while fruit attached to twig; petals not markedly fleshy, glabrous; stamens very numerous; leaves not coriaceous; prop roots none:

Hypanthium shallowly campanulate to saucer-shaped; ovary (1)2- or 3-celled; leaves more or less pellucid-glandular, often aromatic . . . . MYRTACEAE, p. 703

Hypanthium urceolate-tubular; ovary usually 8–12-celled; leaves not pellucid-glandular, not aromatic . . . . PUNICACEAE, p. 733

Stamens less than 15, usually 4–5:

Ovary 1-celled:

Fruit a leathery-coated drupe; ovules 2–6; inflorescence racemose, spicate, or flowers solitary . . . . COMBRETACEAE, p. 554

Fruit a dehiscent capsule; ovules numerous; inflorescence not racemose:

Capsule not circumscissile; ovary wholly inferior; leaves herbaceous, not succulent; sepals 4–5; endosperm none or scanty, not central (if present) . . . . LOASACEAE, p. 660

Capsule circumscissile; ovule only partly inferior; leaves more or less succulent; sepals 2; endosperm central . . . . PORTULACACEAE, p. 728

Ovary 2- to several-celled:

Petals hooded, more or less enfolding subtended stamens; fruit of 3 indehiscent, winged cocci (genus *Gouania*) . . . . RHAMNACEAE, p. 733

Petals not hooded, not investing stamens; fruit capsular or consisting of 2 mericarps, these usually wingless:

Plants aquatic; leaves pectinately divided into linear segments; stigma usually plumose . . . . HALORAGACEAE, p. 596

Plants terrestrial or, if growing in water, not submerged; leaves entire to divided, but segments not pectinate-linear; stigma not plumose:

Bases of leaves or petioles usually sheathing stems; fruit a schizocarp consisting of 2 mericarps; inflorescence an umbel . . . . UMBELLIFERAE, p. 763

Bases of leaves not sheathing stems; fruit a capsule (or berry); inflorescence not umbellate:

Filaments of stamens usually geniculate; anthers basifixed; fruit a juicy berry or a leathery, irregularly dehiscing capsule; leaves (in ours) with 3 main veins running length of leaf blade . . . . MELASTOMACEAE, p. 694

Filaments not geniculate; anthers versatile; fruit a dry capsule (rarely a berry), rupturing by regular sutures or valves; veins pinnate, only midvein running length of leaf blade . . . . ONAGRACEAE, p. 708

ANACARDIACEAE. Sumac Family*

Trees or shrubs, rarely subshrubs or vines; resin ducts present in bark and often in leaves, flowers, and fruits. Leaves alternate, very rarely opposite or verticillate, simple, 3-foliolate, or pinnate, persistent or deciduous; stipules none. Flowers small, usually 5-merous, regular, usually unisexual by abortion or both unisexual and bisexual, rarely all bisexual, hypogynous, rarely perigynous or epigynous, usually in axillary and/or terminal thyrses or panicles, or rarely in solitary or panicled spikelike racemes; perianth rarely simple or absent; bracts deciduous or persistent; bracteoles deciduous, persistent, or absent. Sepals usually connate basally, imbricate or valvate. Petals usually free, imbricate or valvate. Stamens 5–10 (or more), sometimes only 1 or 2 functional in staminate flowers; filaments free or basally connate; anthers 2-loculed, versatile, introrse, longitudinally dehiscent. Disk intrastaminal or extrastaminal, usually 5–10-lobed, occasionally produced into a gynophore, rarely absent. Gynoecium 3(5)-carpelled, or 1-carpelled by reduction, syncarpous or apocarpous, rudimentary or absent in staminate flowers; ovary 1(3–5)-loculed; ovules 1 per locule, apotropous, the funicle usually elongate, insertion basal, parietal, or apical; styles 1–3(5); stigmas 1–3(5). Fruit usually drupaceous, the mesocarp more or less resinous and sometimes waxy or oily, the endocarp crustaceous or bony. Seed 1; endosperm scant or absent; embryo straight or more or less curved.

About 70 genera and 600 species, mainly tropical and subtropical, but some genera extending into the temperate zones.

Leaves simple; stamens 1–5, 1 or 2 fertile; ovary 1-loculed; petals imbricate in bud.. *Mangifera*

Leaves pinnately compound, with 5–12 pairs of leaflets; stamens 9 or 10; ovary 4- or 5-loculed; petals valvate in bud.. *Spondias*

Mangifera L., Sp. Pl. 200. 1753; Gen. Pl. ed. 5, 93. 1754

Large trees with petiolate, entire, alternate leaves crowded near ends of branchlets; twigs slightly puberulent near tips, soon glabrate. Flowers small, numerous, polygamo-dioecious, bracteate, borne in much-branched terminal panicles. Sepals 4–5, imbricate in bud. Petals 4 or 5, imbricate, somewhat thickened basally and along midvein. Stamens 1, or 4 or 5, inserted along inner rim of disk, 1 or 2 fertile, the rest (if present) sterile; filaments slender, slightly dilated basally; anthers ovoid, dehiscing longitudinally around lateral margins. Ovary sessile, free, 1-loculed, compressed; style lateral to subterminal, curved; stigma simple, subulate; ovule solitary. Drupes ovoid to subreniform-oblong, the flesh containing masses of slender, tough fibers. Seed moderately flattened, enwrapped in fibers.

Some 6 to 10 species now accepted, but several hundred named varieties are in cultivation; native in tropical Asia, now pantropical in distribution.

Mangifera indica L., Sp. Pl. 200. 1753

Tree 20–45 m tall, with densely leafy spreading crown; leaf blades narrowly oblong-lanceolate, 10–20 cm long, 5–8 cm wide; panicles 20–35 cm long; flowers yellowish; sepals 2–2.5 mm long, pilose; petals 4–5 mm long, elliptic to oblong, the tips

* Contributed by Duncan M. Porter.

recurved; drupes inequilaterally oblong-ovoid to subreniform, 5–15 cm long, 6–8 cm wide, 2–6 cm thick, yellowish, often with a rosy flush, glabrous.

Mostly in cultivation, but escaping into tropical forests and around abandoned farm sites in practically all tropical and some subtropical regions. Widely cultivated for its delicious fruit, and becoming naturalized under favorable conditions. See Standley (1923, 660) for comments about the tree, its fruit, by-products, and uses.

This tree has been seen in cultivation in the village of Progreso on Isla San Cristóbal and on the Horneman ranch above Academy Bay, Isla Santa Cruz. It is probably under cultivation elsewhere in the Galápagos archipelago. The common name is "mango" in both English and Spanish.

Spondias L., Sp. Pl. 371. 1753; Gen. Pl. ed. 5, 174. 1754

Trees or shrubs; branches roughened by petiolar scars and covered with numerous lenticels, the apices densely foliose. Leaves alternate, odd-pinnate or rarely simple or bipinnate, deciduous; leaflets many, petiolulate or sessile, entire or crenate, all except the terminal more or less inequilateral and acuminate. Inflorescences terminal and compound or lateral and almost simple; flowers regular, 4- or 5-merous, polygamous, usually pedicellate. Calyx small, 4- or 5-lobed, lobes laciniate or dentate, triangular, acute, barely imbricate, deciduous. Petals 4 or 5, free, spreading, oblong-ovate, becoming reflexed, valvate. Stamens 8–10; filaments filiform-subulate and smooth or broadened and papillose, inserted below disk; anthers oblong-ovoid. Disk intrastaminal, moderately thick, annular, 8–10-lobed. Gynoecium 1- or 3–5-carpelled, syncarpous; ovary free, sessile, 1- or 3–5-loculed, ovoid, immersed in disk; ovules 1 per locule, pendulous; styles 1 or 3–5, connivent above; stigmas short, spatulate. Drupe elliptical, the mesocarp fleshy, the endocarp thick, bony, foraminate toward apex, usually 5-loculed and aborting to 1–3-loculed, the locules equal or unequal, erect or divergent. Seeds 1 per locule, oblong, filling locule; testa membranaceous; embryo straight; cotyledons elongate, plano-convex; radicle short, straight, superior.

Perhaps a dozen pantropical species, with centers of distribution in the Caribbean region and in southeast Asia.

Spondias purpurea L., Sp. Pl. ed. 2, 1: 613. 1762

Spondias cirouella Tuss., Fl. Antill. 3: 37. 1824.
Spondias crispula Beurl., Kongl. Svensk. Vet.-Akad. Handl. 1854: 119. 1856.
Warmingia pauciflora Engler in Mart., Fl. Bras. 12²: 281. 1874.
Spondias mexicana S. Wats., Proc. Amer. Acad. 22: 403. 1887.

Tree 5–8 m high; bark gray; branchlets reddish, striate, glabrous; leaves 8–19 cm long, 3.5–5.5 cm wide; leaflets 9–15, opposite or subopposite, all except terminal subsessile, 20–51 mm long, 10–26 mm wide, oblong to elliptic or obovate, the apex obtuse to retuse or occasionally acute, apiculate, the base narrowed, usually markedly inequilateral, membranaceous, crenulate-serrulate, glabrous except for usually pubescent veins and petiolules, the terminal leaflet obovate, decurrent to petiolule, 15–58 mm long, 8–18 mm wide, the petiolule 3–5 mm long, all deciduous; rachis slender, angled, slightly winged, pubescent, 9–13.5 cm long; flowers 5-merous, reddish, in few-flowered axillary panicles borne mostly on larger branches and often produced while tree is leafless; drupes borne laterally in small panicles on older

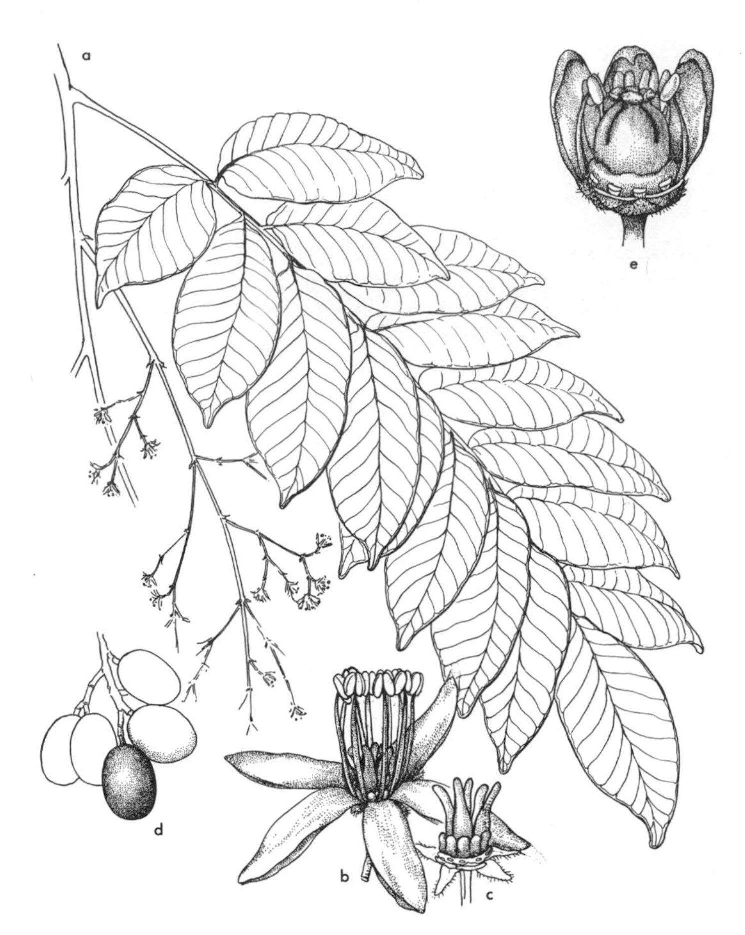

Fig. 134. *Spondias purpurea.*
(Reproduced by permission of Missouri Botanical Garden.)

branches, 2.5–5 cm long, 1.5–2.5 cm wide, the exocarp yellow with reddish blush, the mesocarp yellow and juicy, the endocarp 12–18 mm long.

A native of tropical America and naturalized throughout the tropics, where it is commonly cultivated for its edible fruit; in the Galápagos it is an escape from cultivation. Specimens examined: San Cristóbal, Santa Cruz. Reported also from Santa María by Stewart (1915).

See Fig. 134.

ANNONACEAE. Custard-Apple Family*

Trees, shrubs, or vines with aromatic foliage and wood. Leaves alternate, simple, entire, exstipulate, deciduous or persistent. Flowers normally perfect, regular, hypogynous; sepals 3 or rarely 2, small, obovate; petals commonly 6 in 2 series, free or bases connate, valvate or imbricate in bud, the inner whorl sometimes reduced or lacking. Stamens many, spirally arranged, with short, stout filaments and anthers

* Contributed by Ira L. Wiggins.

with 4 cells at maturity, the connective often expanded and variously shaped beyond anther cells. Carpels few or more commonly many, usually free; ovules 1 to several in each cell; fruiting carpels sessile or stipitate, free or united into a fleshy, multiple fruit. Seeds with or without aril. Endosperm copious and ruminate-wrinkled. Embryo minute.

About 75 to 80 genera and possibly 850 species in the tropics of the world, a few species ranging into the warm temperate zones.

Annona L., Sp. Pl. 536. 1753; Gen. Pl. ed. 5, 241. 1754

Trees or shrubs with simple or stellate hairs usually present on young twigs, buds, or leaves. Flowers solitary or in few-flowered inflorescences, these terminal, internodal, or opposite the leaves. Sepals 3, small, valvate. Petals 6, free or connate at base, in 2 series, the inner often reduced in size or none, concave at base or their full length. Stamens many, extrorse, the connective a deltoid-truncate disk above anthers. Carpels several to many, often connate. Ovules solitary, basal, erect. Fruit fleshy, of coalesced carpels.

About 100 species, native in the tropics to warm temperate regions of the New World.

Buds oblong, 5–8 mm broad, more or less triquetrous; fruit with rounded protuberances; petals linear, greenish within . . . . *A. cherimola*
Buds globose to broadly pyramidal, 2–2.5 cm broad; fruit with spinelike protuberances; petals orbicular-cordate, cucullate, yellowish or cream-colored . . . . *A. muricata*

Annona cherimola Mill., Gard. Dict. ed. 8, no. 5. 1768

Annona tripetala Ait., Hort. Kew. 2: 252. 1789.
Annona pubescens Salisb., Prodr. 380. 1796.

Rounded shrub to 5 m tall; young twigs persistently ferruginous-tomentose with very fine, short hairs; petioles stoutish, 6–10 mm long, densely tomentose; leaf blades mostly elliptic but occasionally lanceolate or broadly elliptic, cuneate to rounded at base, obtuse or acute at apex, 5–15 cm long, 3–10 cm wide, finely and

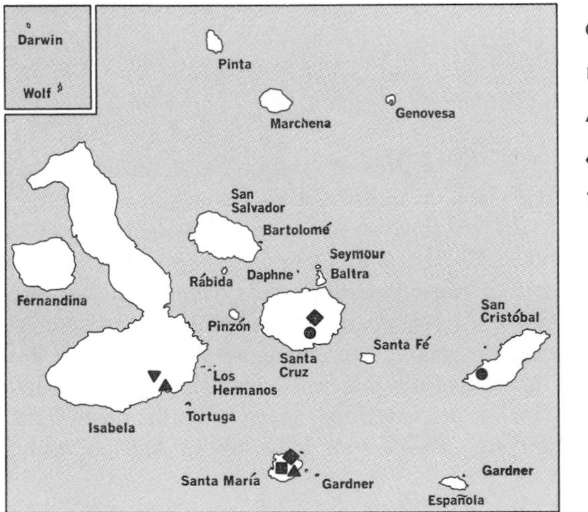

● *Spondias purpurea*
■ *Annona cherimola*
▲ *Annona muricata*
◆ *Bixa orellana*
▼ *Ceiba pentandra*

Fig. 135. *Annona cherimola.*

softly velutinous-pubescent with tangled, subappressed, simple hairs, these ferruginous on very young growth and along veins on undersurface, soon glabrate above, the blades much paler beneath than above; flowers opposite leaves, solitary or in pairs, the petioles ferruginous-tomentose, 8–12 mm long; buds 15–18 mm long, 5–8 mm broad at base, lanceolate-oblong, tomentose outside; sepals triangular, 2–4 mm long, tomentose; petals linear, 1.5–2.5 cm long at anthesis, ferruginous-tomentose on outer surface, greenish within; stamens many; fruit ovoid to globose, 5–10 cm in diameter, greenish to greenish-purple or nearly black, the surface with rounded protuberances and with U-shaped areoles, or sometimes nearly smooth; fleshy arils white, slightly acid, edible; seeds oblong, 10–12 mm long, about half as wide, smooth, black.

Cultivated and escaped locally, ranging from central Mexico, the West Indies, and Central America to Bolivia. Introduced in Asia, Africa, and the southern United States; probably native to mountains of Ecuador and Peru. Specimens examined: Santa María.

See Fig. 135.

Annona muricata L., Sp. Pl. 536. 1753

Annona bonplandiana HBK., Nov. Gen. & Sp. Pl. 5: 58. 1821.
Annona muricata var. *boriaquensis* Morales, Flor. Arb. Cub. 60. 1887.
Guanabamus muricatus Gomez de la Maza, Fl. Habanera 114. 1897.
Annona cearensis Barb. in Rodr., Pl. Nov. Cult. Jard. Bot. Rio de Janeiro 6: 3, *pl. 2, fig. a.* 1898.
Annona macrocarpa Werckle, Tropenpflanzen 428. 1903, non *A. macrocarpa* Barb., Rodr. 1898.

Small tree to 8 m tall, with ill-scented foliage and young branches at first ferruginous but soon glabrate; petioles stout, 4–5 mm long; leaf blades thin, obovate to oblong, 8–15 cm long, 3–6 cm wide, obtuse to abruptly acute at apex, acute at base, lustrous and glabrous above, sericeous beneath when very young, soon glabrate, with domatiae in axils of the nerves; flowers solitary, terminal or opposite leaves; pedicels stout, 1.5–2 cm long, sericeous; buds 2–2.5 cm thick just before anthesis, globose to broadly pyramidal, the outer petals rounded-ovate, 2.5–3.5 cm long, cream-color, abruptly acute at apex, cordate at base, thick, cucullate-concave, the inner petals similar but smaller; ovaries ferruginous-strigose; fruit ovoid to oblong-ovoid, 15–25 cm long, green, covered with curved, upwardly pointing, flexible, spinelike tubercles 4–6 mm long; seeds oblong, 12–15 mm long, smooth, black, enclosed in pulpy, white, acidulous arils.

Extensively cultivated in the tropical regions of the West Indies and of Central and South America; the native region is unknown. Specimens examined: Isabela, Santa María.

BIXACEAE. Anatto Family*

Shrubs or trees with red or yellow sap. Leaves alternate, simple, entire, membranaceous, long-petioled, palmately veined, pubescent with tufted and peltate trichomes; stipules paired, deciduous. Flowers regular, hypogynous, bisexual, large, in terminal panicles; pedicels with 5 apical glands at base of calyx. Sepals 5, free, imbricate, caducous. Petals 5, free, white to blue or red, imbricate, fugaceous. Stamens many, the filaments free, long, slender, inserted on thick receptacle; anthers 2-loculed, horseshoe-shaped, passing over filament apex and with both ends closely applied to filament, longitudinally dehiscent by 2 short pseudapical slits finally uniting into a single pore. Gynoecium 2-carpelled, syncarpous; ovary superior, 1-loculed, with 2 parietal placentae; ovules many, anatropous; style 1, terminal, slender, recurved in bud; stigma shortly 1–2-lobed. Fruit a densely echinate-setose or smooth capsule compressed contrary to placentae, dehiscing by 2 thick valves with placentae in middle. Seeds many, obovoid; testa fleshy, red or yellow, water-soluble; endosperm copious; embryo large, straight, axial; cotyledons broad, incurved apically.

A single genus and 2 or 3 species, native to tropical America; cultivated and escaped pantropically.

* Contributed by Duncan M. Porter.

Bixa L., Sp. Pl. 512. 1753; Gen. Pl. ed. 5, 228. 1754

Characters of the family.

Bixa orellana L., Sp. Pl. 512. 1753

Bixa americana Poir., Encycl. 6: 229. 1804.
Orellana americana Kuntze, Rev. Gen. Pl. 1: 44. 1891.
Orellana orellana Kuntze, op. cit. 3[2]: 9. 1898.*

Shrub or small tree to 3 m high, at least younger parts densely covered with both sessile and peltate minute rusty scales, at length glabrescent; bark dark brown, smooth to minutely fissured; branchlets ringed at nodes; leaves ovate to broadly ovate, the apex gradually long-acuminate, the base shallowly cordate, dark green above, paler beneath, 8–17.5 cm long, 5–12 cm wide, persistent; petioles slender, terete, thickened at base and apex, 15–67 mm long; panicles many-flowered; flowers about 5 cm broad; pedicels thick, terete, the apex thickened, 5–12 mm long; bracts caducous; sepals obovate, concave, obtuse, densely rusty-scaley, 9–10 mm long, 6–8 mm wide; petals white to pink or purplish, spreading, conspicuously veined, 20–28 mm long, 10–13 mm wide; ovary ovoid, densely clothed with short reddish bristles, 5 mm high; style thickened upward, 16 mm long; stigma terminal; capsule ovoid to globose, the apex usually acute or essentially so, reddish brown, densely clothed with long filiform, stiffish, smooth, reddish brown bristles and both sessile and peltate minute rusty scales, 2–4 cm long, 2–3.5 cm wide, dehiscing to base by 2 persistent valves, the endocarps becoming detached; seeds many, red, papillose, angular, chalazal end disciform, about 5 mm long.

Native to continental tropical America; in the Galápagos occurs as an escape from cultivation. Specimens examined: Santa Cruz, Santa María.

Cultivated throughout the tropics of the world for the red or yellow dye extracted from the pulpy covering of the seeds.

See Fig. 136.

BOMBACACEAE. Cotton-Tree Family†

Medium-sized to large trees with simple to digitately compound, alternate, deciduous leaves. Pubescence chiefly of stellate hairs and small, lepidote scales. Stipules caducous. Flowers bisexual, large and showy, solitary or in fascicles, often opposite a leaf, actinomorphic or rarely slightly zygomorphic, often appearing before leaves, usually bracteate; sepals 5, distinct or basally connate, valvate; petals 5 or sometimes absent, contorted in bud, oblong-linear to obovate, often more or less pubescent. Stamens 5 to many, distinct or monadelphous; anthers 1-celled (sometimes 2-celled), staminodes often present. Pistil 1; ovary superior, 2–5-loculed and 2–5-carpelled; placentation axile; ovules 2 to many in each locule. Fruit a loculici-

* The synonymy is incomplete. A number of names have been published in *Bixa,* but their correct application must await revision of the genus. Standley and Williams (1961, 65) wrote: "The numerous forms of this genus, at least those referred to *B. orellana,* are highly variable in the form of the fruit, as well as in the size of the flowers. A critical monograph of the variations, based upon adequate material (which apparently has never been available in any single herbarium) is greatly needed."

† Contributed by Ira L. Wiggins.

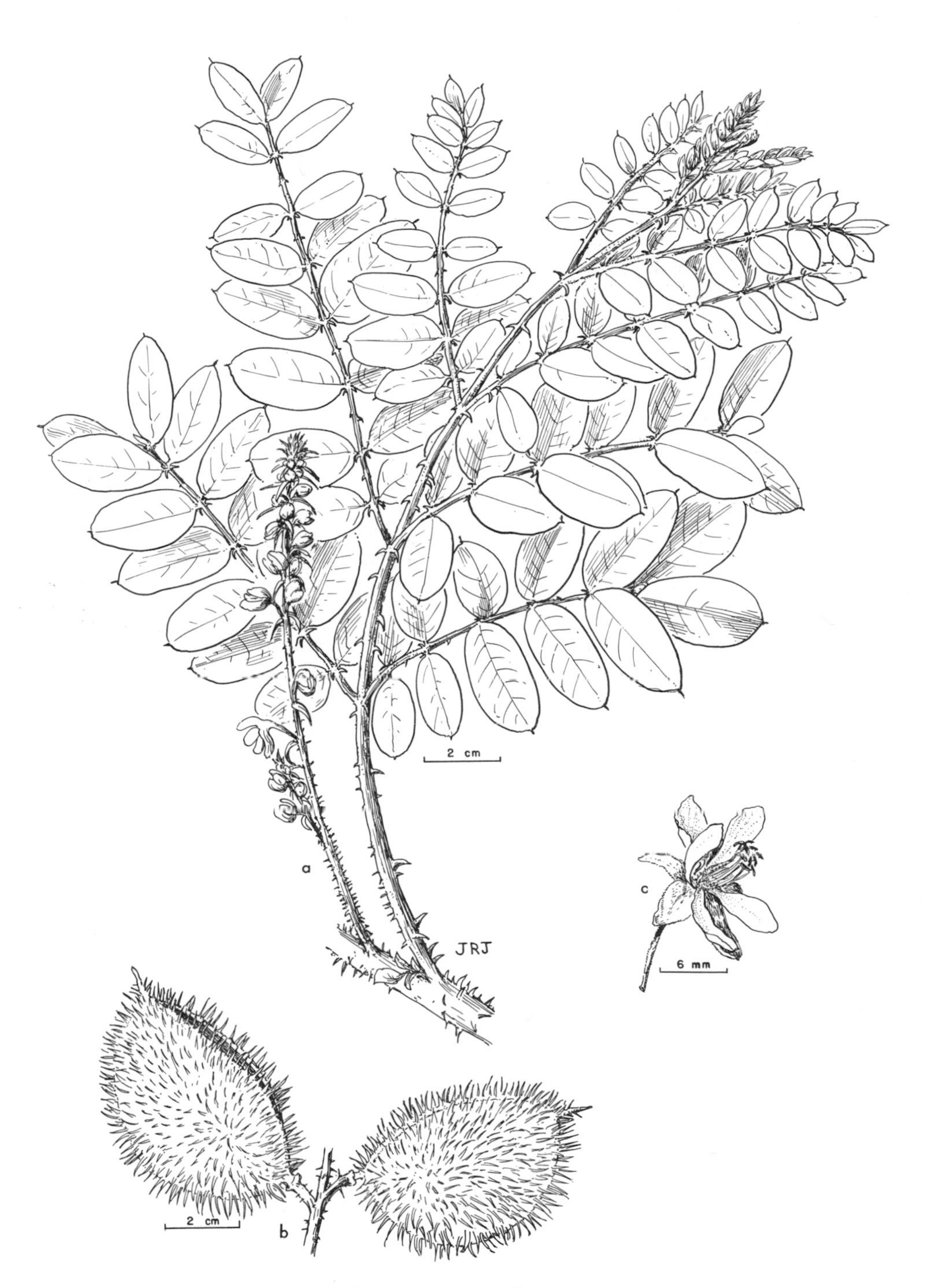

Fig. 136. *Bixa orellana*.

dal capsule (sometimes indehiscent), or berry-like. Seeds smooth, often embedded in woolly or pithy proliferations from pericarp, occasionally arillate; endosperm scanty or none.

About 22 genera and 140 species in the tropics and subtropics, mostly in the Americas.

Leaves digitately 2–7(9)-foliolate; calyx truncate or lobed, the lobes all alike, broadly rounded to truncate; trunk and twigs usually armed with corky, stout, conical spines or processes .... *Ceiba*
Leaves simple; calyx lobes dimorphic, 2 acute, 3 obtuse and much broader; trunk and twigs unarmed .... *Ochroma*

Ceiba Mill., Gard. Dict. ed. 4. 1754

Small to large trees with thick trunks, usually with thin buttresses at base; trunk, branches, and twigs often armed with stout, conical, transversely laminated spines or processes. Leaves digitately compound; leaflets 5–9, articulate, serrate or sometimes entire, long-petiolate. Flowers small to large and showy, solitary, borne at or near tips of branchlets. Calyx campanulate-tubular, truncate or 5-lobed. Petals 5, adnate to base of stamen column, fleshy, oblong to linear, velvety-tomentose without, deciduous. Staminal tube short, divided into 5 branches, sometimes 1 whorl of staminodes present, each staminal branch bearing several crowded anthers at or near tip. Ovary 5-celled, densely silky-comose with long, slender hairs within. Capsule woody, splitting into 5 valves. Seeds ovoid to subglobose, glabrous to minutely puberulent, 3–4 mm long or less, embedded in the silky fibers.

About 20 species in the tropics and subtropics of the world.

Ceiba pentandra (L.) Gaertn., Fruct. 2: 244, *pl. 133.* 1791

Bombax pentandra L., Sp. Pl. 511. 1753.
Ceiba caesaria Medic., Malvenfam. 16. 1787.
Eriodendron anfractuosum DC., Prodr. 1: 479. 1824.
Eriodendron occidentale G. Don, Gen. Hist. Pl. 1: 513. 1831.*

Medium to large tree, sometimes to 40 m tall in favorable sites, with massive trunk often supported by extensive basal buttresses; bark greenish to pale gray, smooth or with scattered, coarse, rather blunt to sharp conical spines 1–3 cm wide at base and about as high, the larger spines usually transversely laminated with alternating layers of corky and denser cells; branchlets usually aculeate with similar but smaller spines, these hard and sharp; leaflets mostly 5–7, oblanceolate to oblong-obovate, abruptly acuminate, 8–20 cm long, glabrous even in youth, usually finely serrate but sometimes entire; petioles about equaling leaflets; calyx 1–1.5 cm long, the lobes broad, only 1–2.5 mm long, 4–6 mm wide, the broadly campanulate cup about as wide as deep; petals obovate, 1.5–2 cm long, broadly short-clawed, white to pink, glabrous within, densely woolly with white to slightly yellowish, silky hairs without; fruit broadly elliptic, woody, 10–12 cm long, about one-third to half as wide as long, smooth and glabrous, greenish or turning brownish just before dehiscence, the valves curling more or less and falling at dehiscence; seeds dull brown, 3–4 mm long, embedded in the silky, silvery white to pale brownish hairs, the cot-

* For 25 additional synonyms, see Robyns (1964, 48–49).

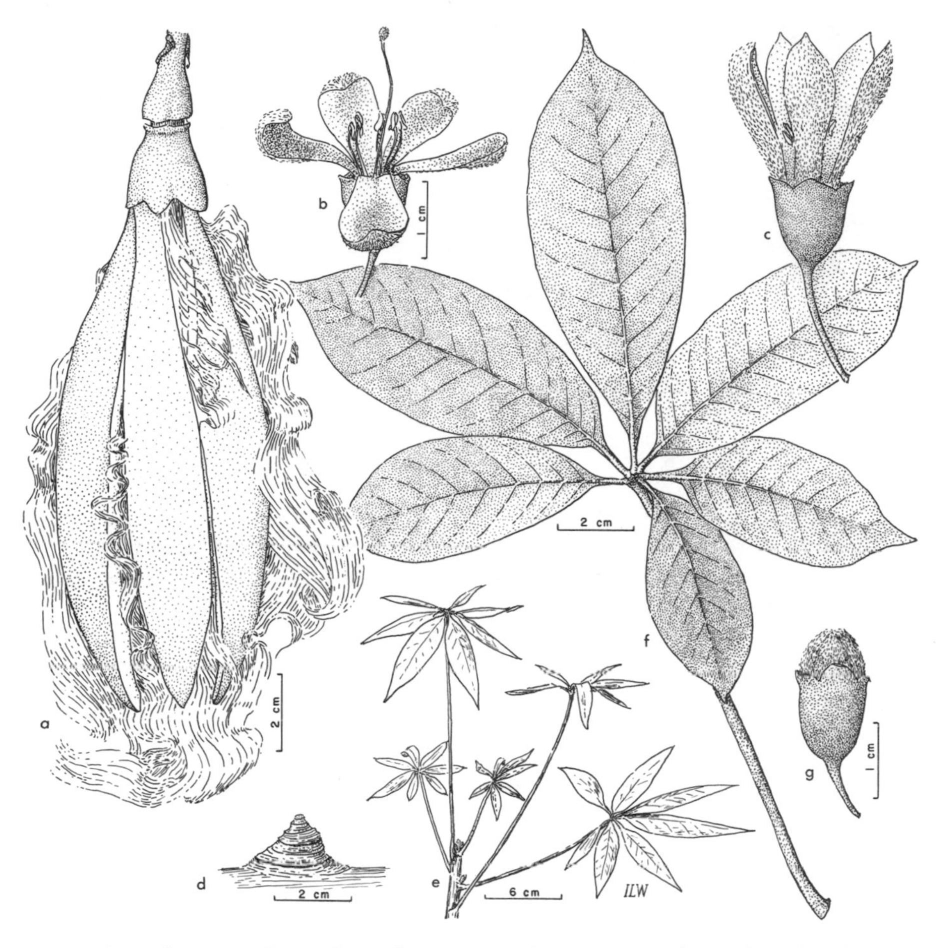

Fig. 137. *Ceiba pentandra*: *c*, flower beginning anthesis, outer surfaces of petals pubescent; *d*, corky excrescences on trunk or large branch; *g*, bud.

ton expanding into an ovoid to nearly spherical mass to 20–25 cm across before separating from axis and carrying the seeds considerable distances.

Widely distributed in tropical and subtropical America, Africa, and Asia; introduced into the Galápagos Islands and occurring sparingly near habitations. Specimen examined: Isabela.

See Fig. 137.

Ochroma Sw., Nov. Gen. Sp. Pl. Prodr. 97. 1788

Tall, straight-boled trees to 40 m high or more, with lower trunk free of branches, the crown depressed or flattened, occupying only small portion of total height; bark smooth, pale gray to pale reddish brown, usually mottled with white and sometimes white predominating; trunk, branches, and branchlets devoid of spines or processes. Leaves cordate-orbicular to angulately shallow-lobed or undulate, sinus

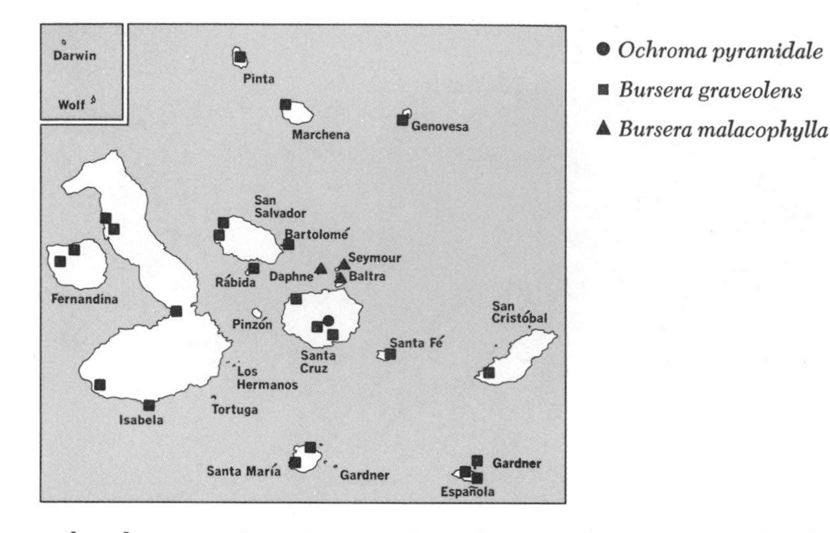

rather deep, margin entire, venation palmate and prominent, especially beneath, with obvious cross veinlets connecting main veins; petioles usually about equaling to slightly exceeding blades; stipules ovate to suborbicular, caducous. Flowers large, at least a few borne throughout year; calyx campanulate, with broad, unequal lobes, 2 of these acute, triangular-ovate, the other 3 obtuse to truncate and rather prominently carinate, about as wide as long, all minutely stellate-puberulent at anthesis; petals 5, adnate to base of staminal column, white, fleshy, 3–4 times as long as calyx lobes. Stamen tube with 5 broad, short lobes, these covered from about middle to tips with adnate, 1-celled, anfractuose, longitudinally dehiscent anthers. Ovary 5-celled; style sulcate; stigmas exserted slightly beyond stamen tube, spirally twisted. Fruit a ligneous capsule, narrowly elliptic to oblong; seeds many, obovoid, smooth or nearly so, dull, embedded in masses of silky, slender, white to pale brown "kapok" (very slender fibers).

A single, extremely variable species, widely distributed from central Mexico to Bolivia and introduced into other tropical regions.

Ochroma pyramidale (Cav. ex Lam.) Urban, Repert. Spec. Nov. Regni Veg. Beih. 5: 123. 1920

Bombax pyramidale Cav. ex Lam., Encycl. Méth. Bot. 2: 552. Apr. 1788.
Ochroma lagopus Sw., Nov. Gen. Sp. Pl. Prodr. 98. Jun.–Jul. 1788.
Ochroma peruviana Johnston, Contr. Gray Herb. 81: 95. 1928.
Ochroma pyramidale var. *concolor* R. E. Schultes, Bot. Mus. Leafl. Harv. Univ. 9: 177. 1941.
Ochroma tomentosa var. *ibarrensis* Benoist, Bull. Soc. Bot. Fr. 88: 439. 1941.
Ochroma lagopus var. *bicolor* Standl. & Steyerm., Field Mus. Nat. Hist., Bot. Ser. 23: 62. 1944.
Ochroma lagopus var. *occigranatensis* Cuatr., Phytologia 4: 480. 1954.
Ochroma pyramidale var. *bicolor* Brizicky, Trop. Woods 109: 63. 1958.*

Tall tree to as much as 30 m or more, the trunk 1–1.5 m thick, more or less buttressed at base; bark smooth, with varying amounts of whitish mottling; crown confined to uppermost part of tree, flattened; petioles 8–25 cm long, brownish, sulcate adaxially, finely stellate-puberulent but soon glabrate; blades broadly ovate to ovate-orbicular, 15–30 cm long, nearly as wide, rather thin, cordate, mostly 7-nerved, the nerves densely brown-puberulent beneath, sparingly puberulent to glabrate above,

* For additional synonyms, see Robyns (1964, 65–66).

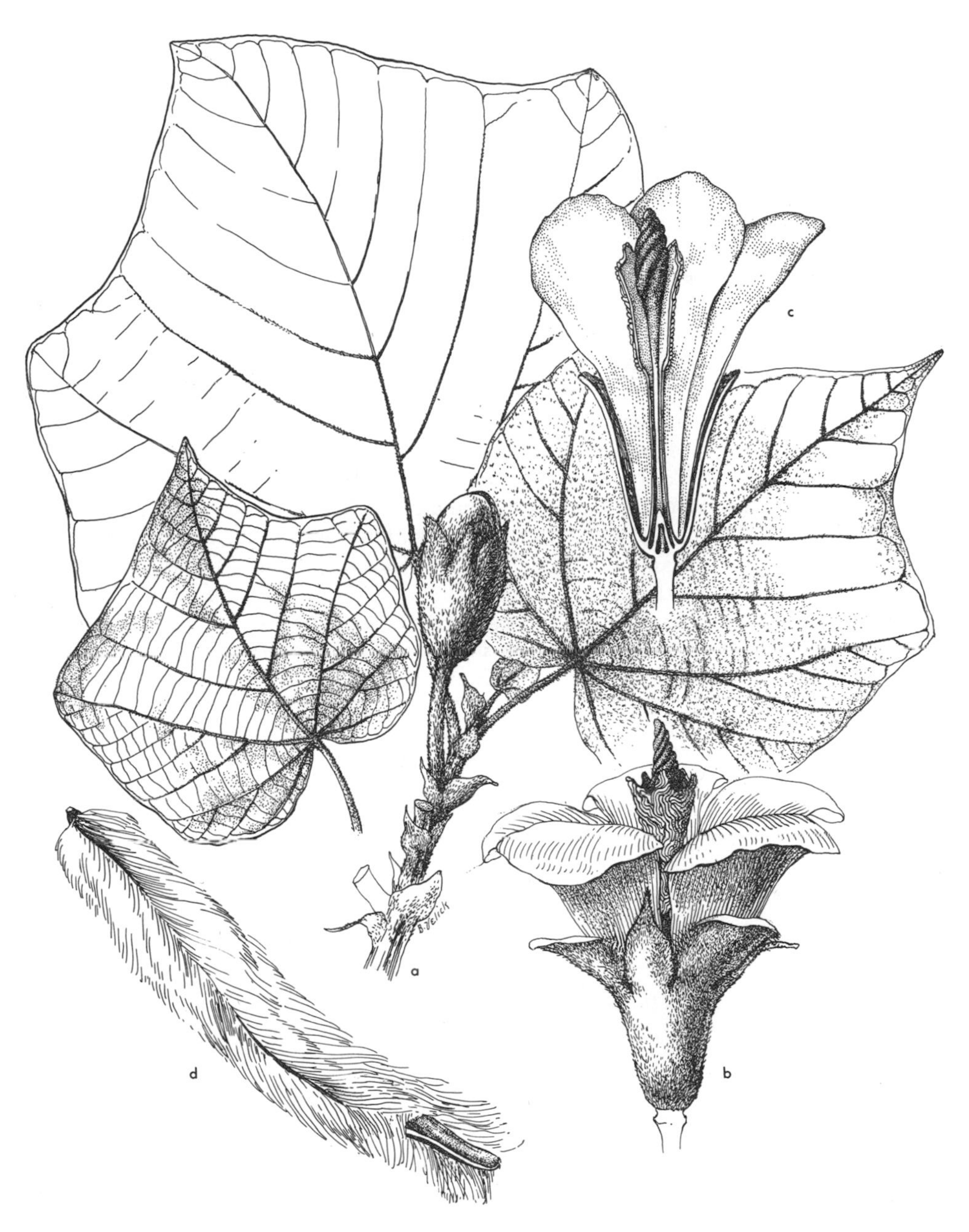

Fig. 138. *Ochroma pyramidale.*
(Reproduced by permission of Missouri Botanical Garden.)

the tissue dark green, nearly glabrate above, paler and moderately to densely pubescent beneath; flowers 10–15 cm long, calyx 6–8.5 cm long, lobes about 2 cm long, the broader, obtuse lobes to 3.5 cm wide, the acute about 2 cm wide at base, 1.5–2 cm long, less markedly carinate than the broad, the whole calyx finely and at first densely stellate-puberulent, finally subglabrate, the inner surface sparsely strigose with upwardly directed, closely appressed, yellowish hairs; petals white or whitish, the claws about 1 cm wide, blade to 5 cm wide, glabrous within, puberulent without, the staminal column slender, 7–8 cm long, stamen mass 3–4 cm long, 1.5–2.5 cm broad, spirally twisted-undulate; stigma twisted, exserted 1–3 cm beyond staminal column; capsule 15–23 cm long, closely packed with soft, light brown, silky fibers containing numerous dull brown, obovoid seeds 3–5 mm long.

Native in tropical Mexico and Central America south to Bolivia; introduced into the Galápagos Islands about 1940 and reproducing spontaneously. Specimens examined: Santa Cruz.

Jacob Horneman, a resident of Santa Cruz, states that the balsa tree was introduced in 1940 as an experimental planting. By January 1964 the trees were fully 30 m tall, and local Ecuadorians claimed that the trees were too large to produce high-quality balsa wood. According to Standley and Steyermark (1949, 396–99) the term "balsa" probably is Quechuan in derivation. The specific name applied by Swartz—*lagopus*—means "rabbit-foot," and in Oaxaca *pata de liebre* is applied to the balsa tree and has the same meaning. The furry masses of kapok right after the outer covering falls from the capsule do have a resemblance to the furry hind foot of a large rabbit. Pertinent literature includes Rowlee (1919), Tenny (1928), Pierce (1942a, 1942b), Standley and Steyermark (1949, 396–99), and Robyns (1964, 64–67).

See Fig. 138.

BURSERACEAE. TORCHWOOD FAMILY*

Aromatic shrubs or trees, with resin ducts in inner bark. Leaves alternate, usually simple-pinnate, deciduous or persistent. Stipules usually none. Flowers small, regular, hypogynous, 3–5-merous, usually unisexual by abortion; in axillary or terminal cymose panicles; plants mostly dioecious. Sepals 3–5, more or less connate, imbricate or valvate, persistent. Petals 3–5, or rarely absent, free or rarely connate basally, imbricate or valvate, persistent or deciduous. Stamens 6–10, in 1 or 2 whorls; filaments usually free, inserted below and sometimes adnate to disk; anthers 2-loculed, versatile, introrse, longitudinally dehiscent, usually sterile in carpellate flowers. Disk intrastaminal or rarely extrastaminal, nectariferous. Gynoecium 2–5-carpelled, syncarpous, usually rudimentary in staminate flowers; ovary 2–5-loculed; ovules 2 per locule, anatropous, placentation axile; style simple, short; stigma capitate or lobed, the lobes as many as locules. Fruit usually drupaceous, indehiscent or dehiscent by valves, the exocarp and mesocarp fleshy or dry, containing 1–5 pyrenes; usually 1-loculed by reduction or abortion. Seeds 1 per locule; endosperm none; embryo straight or curved, usually more or less deeply lobed; cotyledons contorted-plicate or flat; radicle superior.

A family of about 20 genera and 600 species, reaching its greatest development in tropical America, Malaysia, and northwestern Africa.

* Contributed by Duncan M. Porter.

Bursera Jacq. ex L., Sp. Pl. ed. 2, 1: 471. 1762;
Gen. Pl. ed. 6, 174. 1764, nom. conserv.

Bark smooth or rough, the older bark exfoliating in thin papery sheets or thick, platelike scales. Leaves odd-pinnate, bipinnate, or (1)3–9-foliolate, usually crowded at branchlet apices, deciduous. Leaflets opposite, entire to toothed, petiolulate to sessile, chartaceous to coriaceous. Flowers very small, 3- to 5-merous, unisexual and/or bisexual, in axillary raceme-like panicles, appearing just before or simultaneously with new leaves; plants dioecious or polygamodioecious. Sepals 3–5, minute, connate at least basally, imbricate. Petals 3–5, much longer than sepals, whitish to yellow, spreading and recurved, induplicate-valvate. Stamens 6–10; filaments subulate; anthers oblong, dorsifixed near base, smaller and sterile in carpellate flowers. Disk annual, 6–10-lobed, orange or red. Ovary 2- or 3-lobed, 2- or 3-loculed, ovoid, sessile, rudimentary in staminate flowers; ovules collateral; stigma 2- or 3-lobed. Fruit a drupe, subglobose to obliquely ellipsoid, indistinctly 2- or 3-angled, resinous, the exocarp and mesocarp leathery, separating at maturity into 3 valves; pyrenes usually 1.

A tropical American genus of perhaps 100 species.

Leaflets bright green, ovate-oblong or ovate-lanceolate, glabrous to puberulent, the margins crenate-serrate; inflorescences about as long as leaves; aril white, glabrous. . . . . *B. graveolens*

Leaflets grayish, ovate to rotundate (rarely ovate-lanceolate), densely tomentose or short-lanate, the margins coarsely crenate; inflorescences longer than leaves; aril whitish-pubescent . . . . *B. malacophylla*

Bursera graveolens (HBK.) Trian. & Planch., Ann. Sci. Nat. Bot. ser. 5, 14: 303. 1872

Elaphrium graveolens HBK., Nov. Gen. & Sp. Pl. 7: 24. 1825.
Elaphrium tacamaco Tul., Ann. Sci. Nat. Bot. ser. 3, 6: 368. 1846.
Spondias edmonstonei Hook. f., Trans. Linn. Soc. Lond. 20: 230. 1847.
Bursera tacamaco Trian. & Planch., Ann. Sci. Nat. Bot. ser. 5, 14: 303. 1872.
Amyris caranifera Willd. ex Engler in DC., Monogr. Phan. 4: 49. 1883, in synonymy.
Terebinthus graveolens Rose, Contr. U. S. Nat. Herb. 10: 119. 1906.

Tree or shrub 3–12 m high, usually with broadly spreading crown and short, thick trunk; branches ferruginous, glabrous or nearly so, terete, wrinkled, the leaf scars forming transverse striations; leaves densely clustered at branch tips, spreading, slightly to densely minutely orange-glandular-stipitate, 10–28 cm long, 5.5–13 cm wide when mature; petiole narrow, base dilated, glabrous or more or less puberulent, 1.5–6.5 cm long, 1–1.5 mm wide; rachis winged, glabrous to puberulent, the wing 2.5–4 mm wide, sometimes wider and adnate to base of subterminal pair of leaflets; leaflets (3)5–7(9), ovate-oblong or ovate-lanceolate, acute, the base usually cuneate, crenate-serrate, ciliate, sessile or short-petiolulate, opposite or subopposite, glabrous to puberulent, becoming more or less glabrate except for main veins on both surfaces, the lateral pairs inequilateral, unequal, 18–60 mm long, 13–38 mm wide, the terminal 35–75 mm long, 13–30 mm wide; inflorescences about as long as leaves, 10–13 cm long, slightly to densely stipitate-glandular with minute orange glands, glabrous or puberulent; bracts linear-lanceolate, ciliate, slightly to densely puberulent, 2 mm long or less; flowers 4-merous, functionally unisexual; sepals ovate or ovate-triangular, glabrous to puberulent, ciliate, glandular-stipitate, 1.5 mm long or less, persistent; petals oblong, thick, glabrous to minutely puberulent, glandular-stipitate, pale cream, drying orange, 4 mm long, 1–1.5

Fig. 139. *Bursera graveolens.*

mm wide, deciduous in pistillate flowers; stamens shorter than petals; filaments linear-subulate; anthers yellow; ovary ovoid, glabrous, about 1.5 mm high; style 1 mm long, persistent on fruit; stigma 2-lobed; drupe obovoid, acute at both ends, drying dark brown, wrinkled, minutely glandular, 2-valved, the endocarp fleshy, inner surface yellowish, 9–12 mm high, 7–9 mm wide; pyrenes black, with whitish, glabrous, fleshy aril.

Chiapas and Yucatan, Mexico, south to Peru. Specimens examined: Bartolomé, Española, Fernandina, Gardner (near Española), Genovesa, Isabela, Marchena, Pinta, Rábida, San Cristóbal, San Salvador, Santa Cruz, Santa Fé, Santa María.

This species has been reported also from Isla Darwin by Stewart (1911). Collections seen from the northern coast of Isla Santa Cruz (*Howell 9887* [CAS], *Wiggins & Porter 706* [CAS]) have leaves more pubescent than is usual for this species. This variation perhaps represents introgression from *B. malacophylla.* McVaugh and Rzedowski (1965) point out that *B. graveolens* does not differ greatly from the Mexican *B. penicillata* (DC.) Engler, and that the 2 may be conspecific. If this proves to be the case, *B. penicillata* has priority.

See Fig. 139 and color plates 1, 26, 29, 33, 39–41, 43, 44, and 46–50.

Bursera malacophylla Robins., Proc. Amer. Acad. 38: 160. 1902

Bursera graveolens f. *malacophylla* Macbr., Publ. Field Mus. Bot. 13: 712. 1949.

Shrub or small tree to 4 m high; branches ferruginous, tomentose toward apex, terete, wrinkled, the leaf scars forming transverse striations; leaves densely tomentose or short-lanate throughout, producing a grayish aspect, densely clustered at branch tips, spreading, densely and closely stipitate-glandular, the glands white or orange, 8–15 cm long, 5–11.5 cm wide; rachis winged, wing 1.5–3 mm wide, sometimes enlarged below subterminal leaflet pair and rotundate, auriculate, 5–6 mm wide or sometimes wider and adnate to bases of subterminal leaflet pair; leaflets 5–9, ovate to rotundate, rarely ovate-lanceolate, obtuse to acute, base cuneate to rounded, coarsely crenate, sessile or short-petiolulate, opposite or subopposite, the lateral pairs slightly inequilateral, unequal, 2–6 cm long, 1–4 cm wide, the terminal 3.5–7 cm long, 1–3 cm wide; inflorescences longer than leaves, 9–15 cm long, minutely orange-stipitate-glandular, puberulent; bracts linear-lanceolate, ciliate, puberulent, 2 mm long or less; flowers 4-merous; sepals ovate or ovate-triangular, puberulent, ciliate, glandular-stipitate, 1.5 mm long or less, persistent; petals oblong, thick, minutely puberulent, glandular-stipitate, drying orange, 4 mm long, 1–1.5 mm wide, deciduous in carpellate flowers; filaments linear-subulate; ovary ovoid, glandular-stipitate, about 1 mm in diameter; style less than 1 mm long, persistent; stigma 2-lobed; drupe obovoid, acute at both ends, drying brown or black, wrinkled, minutely glandular, 2-valved, the endocarp fleshy, inner surface yellowish, 8–13 mm high, 6–10 mm wide; pyrenes black, with a whitish-pubescent, fleshy, basal aril.

Known only from the Galápagos Islands. Specimens examined: Baltra, Daphne Major, Seymour.

Svenson (1946) and Macbride (1949) have considered *B. malacophylla* distinct only on the formal level, apparently because of its supposed presence in Peru. Collections from Ecuador and Peru that vary toward *B. malacophylla* in amount of pubescence differ from *B. graveolens* in leaflet shape and in pubescence type. They represent one extreme in the variation of the usually glabrous-leaved *B. graveolens*. Bullock (1936) has emphasized that the amount of pubescence is a poor taxonomic character in the genus, as it may be quite variable within a species, while at the same time recognizing the usefulness of pubescence type in specific delimitations. Further field studies are needed in the area on Isla Santa Cruz across the channel from Isla Baltra in order to determine whether these 2 species are sympatric. Perhaps hybridization occurs between them along the north coast of Isla Santa Cruz.

CACTACEAE. CACTUS FAMILY*

Fleshy, herbaceous, and woody plants with usually leafless simple or cespitose stems, these enlarged, succulent, flattened, cylindrical, or ribbed-cylindrical; leaves (when present), spines, trichomes, and in some groups glochids arising in areoles. Flowers usually bisexual, solitary, showy; perianth consisting of many somewhat coalescent segments intergrading from scale-like leaves to sepals to petals; ovary inferior, 1-loculed, with several parietal placentae; stamens numerous, arising from inner surface of hypanthium; anthers 2-celled, dehiscing longitudinally; pistil 1; carpels 4 to many; fruit a berry, either fleshy or dry at maturity, with spines, scales, hairs, or glochids, or naked. Seeds many, with straight or curved embryo, with or without endosperm and perisperm.

About 75 genera and 1,200–1,500 species, widely distributed in tropical, sub-

* Contributed by E. F. Anderson and D. L. Walkington.

tropical, and temperate regions of the New World; now found throughout the world.

Plants with flattened stem joints; glochids usually present; leaves present but early deciduous; flowers without long tube, yellow; fruits green or greenish yellow at maturity; seeds whitish, discoid, not punctate. . . . . *Opuntia*
Plants with cylindrical or ribbed-cylindrical stem joints; glochids absent; leaves absent; flowers with a long tube, usually not yellow; fruits usually reddish purple at maturity; seeds blackish, reddish, or brownish, reniform, roundish, or ellipsoidal, punctate:
Plants small, less than 1 m high, forming low, dense clusters; stem joints not obviously constricted; fruits spiny; central spines usually less than 3 cm long. . . *Brachycereus*
Plants large, more than 1 m high, columnar, stem usually solitary; stem joints usually obviously constricted; fruits with scales and hairs; central spines usually more than 3 cm long . . . . *Jasminocereus*

Brachycereus Britt. & Rose, The Cactaceae 2: 120. 1920

Plants dense, with numerous stems in low subcespitose clumps 40–60 cm high, 7.5–20 dm across; stems more or less erect, ribbed, cylindrical, yellow to brownish yellow, 10–50 cm long, 3–5 cm thick; ribs low, 16–22; areoles 2 mm across, typically 6–9 mm apart; spines unequal, yellowish, black at maturity, numerous (about 40 per areole), radiating in all directions but longer laterals usually pointing upward, stiff to somewhat flexible, 0.5–5 cm long; glochids absent; flowers 2–5.5 cm broad, 6–11 cm long; outer perianth parts with midribs brownish red and margins greenish to brownish red, the largest lanceolate, 10–18 mm long, 2–12 mm wide, acuminate, entire; inner perianth parts creamy white, the largest linear to lanceolate, 10–18 mm long, 2–8 mm wide, acuminate, entire; filaments creamy white, 1–2 mm long; anthers creamy white, about 1 mm long; style creamy white, 38–43 mm long, 1.5–2 mm in greatest diameter; stigmas 10–14, 7–10 mm long, papillate; hypanthium in anthesis with numerous spiny areoles containing trichomes and subtended by awl-shaped scales; fruit dark, fleshy at maturity, covered with yellow spines, 1.5–3.5 cm long, 1–1.4 cm broad, the floral parts persistent until maturity; seeds brownish black, rounded to ellipsoid, slightly longer than wide, slightly punctate, 1–1.5 mm long, 1 mm broad.

An endemic monotypic genus present on several islands, but exhibiting little morphological differentiation from population to population.

Brachycereus nesioticus (K. Schum.) Backbg. in Backbg. & Knuth, Kaktus-ABC 176. 1935

Cereus nesioticus K. Schum. in Robins., Proc. Amer. Acad. 38: 179. 1902.

Characters of the genus.

In xeric zones primarily on pahoehoe lava at sea level. Specimens examined: Bartolomé, Fernandina, Genovesa, Isabela, Pinta, San Salvador.

See Fig. 140.

Jasminocereus Britt. & Rose, The Cactaceae 2: 146. 1920

Plants erect, columnar, freely branching, usually arboreal, 3–7 m high; trunk usually single, well-formed; larger terminal joints green to greenish yellow, stout, cy-

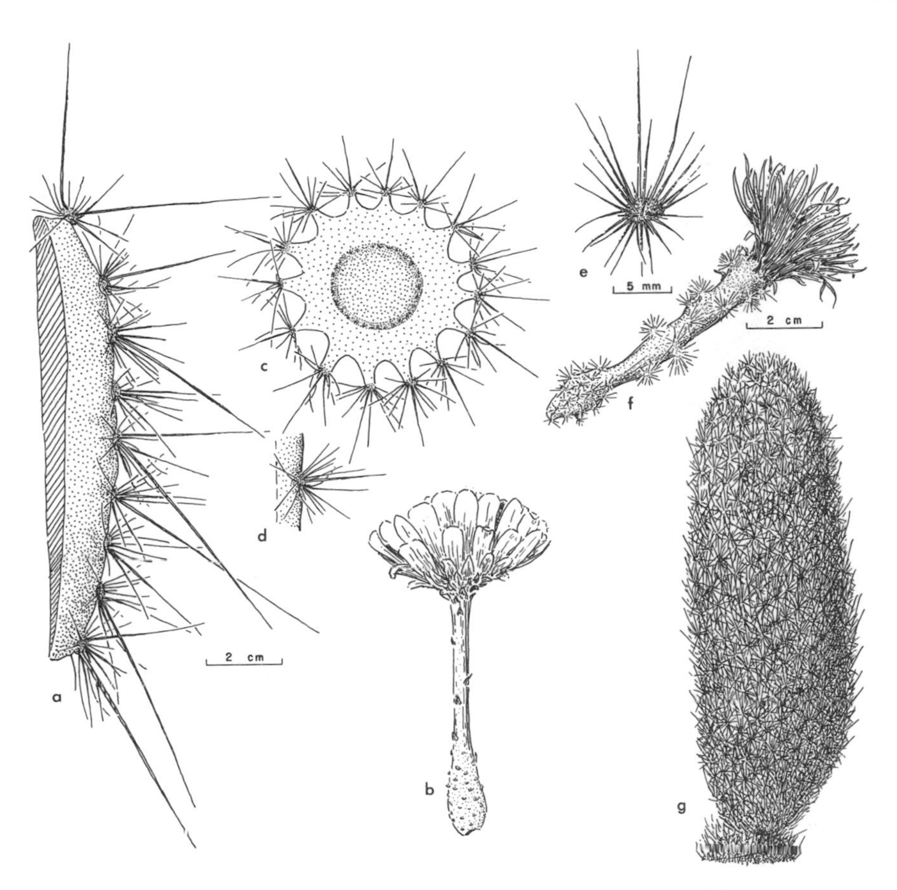

Fig. 140. *Jasminocereus thouarsii* var. *thouarsii* (*a, b*). *J. thouarsii* var. *delicatus* (*c, d*). *Brachycereus nesioticus* (*e–g*).

lindrical, rounded at ends, 12–30 cm long, 6–18 cm broad; ribs 11–22, triangular in cross section, 8–22 mm high; tubercles none; leaves none; areoles 3–6 mm across, typically 0.5–2.5 cm apart; spines along ribs, white with a tendency toward reddish brown or black, 8–36 per areole, usually with 2–4 centrals pointing upward and laterals radiating in all directions, twisted, 0.6–7.5 cm long, 0.25–0.5 mm thick basally, acicular, terete; glochids none; flowers 2–6 cm across, 5–8.8 cm long; outer perianth parts with midribs deep reddish, margins red to yellow to olive green, the largest oblong to cuneate, 5–20 mm long, 4–9 mm wide, truncate or mucronate, entire; inner perianth parts yellow to olive green, the largest oblong to obovate, 25–40 mm long, 5–10 mm wide, mucronate or rounded, entire; filaments whitish to yellow, 10–20 mm long; anthers whitish to yellow, 2 mm long; style whitish to yellow, 3.2–7 mm long, 0.5–1 mm in greatest diameter; stigmas 5–11, 8–15 mm long, reddish, papillate; hypanthium in anthesis with numerous awl-shaped scales and hairs; fruit greenish to reddish purple, somewhat fleshy but hard at maturity, with scales and hairs, globular to oblong, 15–57 mm long, 15–43 mm broad, the umbilicus small and inconspicuous; perianth parts often persistent and forming a flattened ring basally; seeds black to reddish, reniform, reticulate and

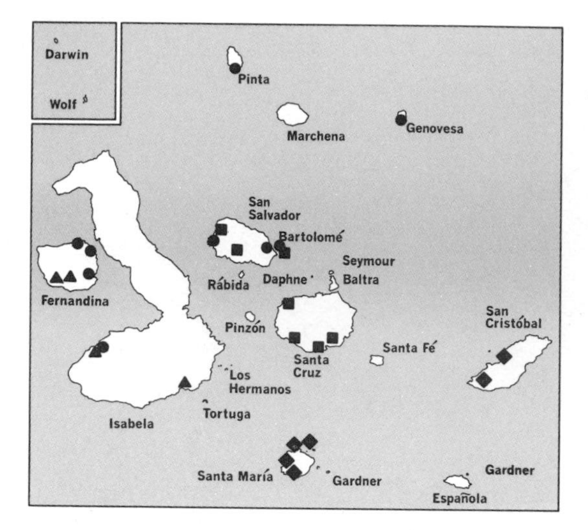

pitted, becoming tuberculate, 1–2 mm long, 1 mm wide, the hilum lateral and dark.

A monotypic endemic genus highly variable in morphological characters but exhibiting only early stages of taxonomic differentiation as a result of insular isolation.

Fruits to 7 cm long, tending to be elongate; spines to 35 in an areole; central spines usually lacking; flowers very waxy . . . . *J. thouarsii* var. *sclerocarpus*

Fruits 5 cm long or less, tending to be broadly ovoid to globular; spines to 22 in an areole; central spines usually present; flowers not markedly waxy (or unknown):

Flowers 5–7 cm long; fruits 3.5–4.2 cm in diameter; plant often 5 m high or more . . . . *J. thouarsii* var. *delicatus*

Flowers 6.5–8.8 cm long; fruits 1.8–3 cm in diameter; plant rarely exceeding 4 m in height . . . . *J. thouarsii* var. *thouarsii*

Jasminocereus thouarsii (Weber) Backbg., Die Cactaceae 2: 912. 1959, var. **thouarsii**

Cereus thouarsii Weber, Bull. Mus. d'Hist. Nat. 5: 312. 1899.
Cereus galapagensis Weber, loc. cit.*
Jasminocereus galapagensis Britt. & Rose, The Cactaceae 2: 146. 1920.
Brachycereus thouarsii Britt. & Rose, loc. cit.
Jasminocereus thouarsii var. *chathamensis* Dawson, Cact. & Succ. Jour. 34: 72. 1962.

Plant 3–4 m tall, ribs 14–22; areoles 3–5 mm across, typically 0.5–2 cm apart; spines 8–21 per areole, 0.6–6.6 cm long, centrals usually present; flowers 2–6 cm broad, 6.5–8.8 cm long, not markedly waxy; outer perianth parts with reddish yellow midribs and yellowish green margins, the largest obovate, 5–25 mm long, 6–9 mm wide, rounded or mucronate; hypanthium in anthesis with scales and small hairs; fruit olive green, with scales, nearly ovoid, 2.8–5 cm long, 1.8–3 cm across; seeds reddish becoming black, with small tuberculae or reticulate-pitted, 1 mm long, less than 1 mm wide.

*Although Britton and Rose stated that *C. thouarsii* Weber was the same as *C. nesioticus* K. Schum., herbarium and field studies indicate that the types of both of Weber's taxa came from the same population on Isla Santa María. His *C. thouarsii* apparently was based on a juvenile plant, his *C. galapagensis* on a mature one.

In Arid and lower Transition zones. Specimens examined: Champion, San Cristóbal, Santa María.

See Fig. 140.

Jasminocereus thouarsii var. **delicatus** (Dawson) Anderson & Walkington, Madroño 20: 256. 1970

Jasminocereus howellii Dawson, Cact. & Succ. Jour. 34: 71. 1962.
Jasminocereus howellii var. *delicatus* Dawson, op. cit. 72.

Plant often 5 m tall or more; ribs 12–19; areoles 3–6 mm broad, typically 0.7–2 cm apart; spines 10–22 per areole, 1–9 cm long, centrals usually present; flowers 2–4.8 cm broad, 5–7 cm long, waxiness not known; outer perianth parts with midribs reddish yellow and margins yellowish green, the largest elongate, 18–20 mm long, 4–5 mm wide, mucronate or truncate; hypanthium in anthesis with scales; fruit usually reddish, with a few scales, ovoid to globular, 1.5–4.4 cm long, 3.5–4.2 cm across; seeds blackish, minutely tuberculate, about 1 mm long, less than 1 mm wide.

In xeric habitats near the coast lines. Specimens examined: Bartolomé, San Salvador, Santa Cruz.

See Fig. 140 and color plates 25, 29, 33, and 40–42.

Jasminocereus thouarsii var. **sclerocarpus** (K. Schum.) Anderson & Walkington, Madroño 20: 256. 1970

Cereus sclerocarpus K. Schum. in Robins., Proc. Amer. Acad. 38: 179. 1902.
Jasminocereus sclerocarpus Backbg., Cactac., Jahrb. Deutsch. Kakt. Ges. 1943–44: 24. 1944.

Plant 3–6 m tall; ribs 11–16; areoles 3–6 mm across, typically 1–2.5 cm apart; spines 10–35 per areole, 0.8–6 cm long, centrals usually lacking; flowers 2.7–3.5 cm across, 6.5–9 cm long, very waxy; outer perianth parts with dark greenish red midribs, margins reddish green, largest ones spatulate, 10–20 mm long, 6–12 mm wide, mucronate; hypanthium in anthesis with a few scales; fruit green, suffused with red, ovoid, 4–7 cm long, 3–4.5 cm across, with scales; seeds reddish black or black, with small tuberculae, no pits and not reticulate, less than 1.5 mm long, less than 1 mm wide.

Widely distributed in Arid Zone, from near sea level to about 300 m. Specimens examined: Fernandina, Isabela.

See color plate 37.

Opuntia Mill., Gard. Dict. abr. ed. 4. 1754

Plants prostrate to arboreal, with branches pendulous to erect, 0.5–10 m high; trunk, when present, spiny or with large reddish plates; larger terminal joints greenish yellow to gray-green, orbicular to oblong; ribs absent; leaves green, subulate to conical, 2–11 mm long, early deciduous; trichomes usually present; spines highly variable, dimorphic in some species, usually present on all joints and more or less evenly distributed, yellow, becoming brownish or occasionally whitish with age, 1 to over 50 per areole, erect or spreading, stiff and pungent to soft and bristly, acicular to subulate, round to slightly flattened in cross section, not markedly barbed; glochids, when present, whitish to yellow-brown; flowers 2.5–10.5 cm across, 3–13 cm long; outer perianth parts with midribs green to yellow, margins greenish yellow to yellow, the largest obovate to obdeltoid, 11–48 mm long, 6–27 mm wide, mucronate, entire to occasionally ciliate; filaments yellow to white, anthers yellow to white; style yellow to white; stigmas 5–13; ovary at anthesis variable, with trichomes, leaves, spines, and glochids; fruits green becoming yellow-green and fleshy at maturity, with spines, glochids, and trichomes, spheroidal to turbinate, the umbilicus depressed, brownish or greenish, persistent until maturity, occasionally giving rise to new stem joints; seeds white to cream-colored with dark center, discoid.

About 300 species throughout most of the New World outside the arctic regions; rocky and sandy regions of volcanic origin, in xeric to mesic habitats from sea level to about 1,500 m on all islands and major islets of the archipelago.

A highly variable genus in the Galápagos, exhibiting rapid adaptive radiation that has resulted in marked morphological differences and establishment of several specific and infraspecific taxa within the archipelago.

Plants arborescent and/or arboreal (except *O. echios* var. *zacana,* which is shrubby to nearly prostrate), usually forming a distinct trunk; spines strongly dimorphic:
- Fruits usually more than 4 cm long when mature:
  - Glochids present; stem branches usually pendulous or drooping; seeds less than 5 mm long:
    - Fruits more than 7 cm long when mature; seeds 3–5 mm long; glochids few. . . . . *O. echios* var. *barringtonensis*
    - Fruits less than 7 cm long when mature; seeds 2–4 mm long; glochids abundant:
      - Branches strongly pendent, 1.5–5 m long; plants often becoming more than 10 m high. . . . . *O. echios* var. *gigantea*
      - Branches somewhat pendent but rarely more than 1.5 m long; plants rarely becoming more than 5 m high:
        - Fruits spheroidal; plants arboreal. . . . . *O. echios* var. *inermis*
        - Fruits turbinate; plants prostrate, shrubby, or arboreal:
          - Glochids absent or nearly so; seeds 2–3 mm long; spines variable; stem joints not tending to have distinct cuneate bases; plants usually arboreal but sometimes shrubby. . . . . *O. echios* var. *echios*
          - Glochids abundant; seeds 3–4 mm long; spines either very long or very short; stem joints tending to have distinct cuneate bases; plants shrubby to prostrate. . . . . *O. echios* var. *zacana*
  - Glochids few or absent; stem branches not pendent nor drooping; seeds more than 5 mm long:
    - Flowers more than 11 cm long; mature fruits usually more than 10 cm long; seeds 8–13 mm long; spines to 10 cm long. . . . . *O. megasperma* var. *megasperma*
    - Flowers less than 11 cm long; mature fruits usually less than 10 cm long; spines 5.5 cm long or less:
      - Fruits 4–6 cm long; flowers 8 cm long or less; spines golden brown, banded, becoming bone white with dark tip; stem joints dark green to grayish; areoles 1.5–2 mm across. . . . . *O. megasperma* var. *mesophytica*
      - Fruits 6–10(13) cm long; flowers more than 8 cm long; spines yellow becoming brownish white; stem joints green to yellow-green; areoles 2–3(6) mm across . . . . *O. megasperma* var. *orientalis*
- Fruits usually less than 4 cm long when mature:
  - Glochids absent or few; fruits without spines or with poorly developed spines:
    - Plants variable in habit; spines not strongly dimorphic or pungent; fruits 1.7–2.5 cm long, rarely bearing spines, profuse. . . . . *O. galapageia* var. *profusa*
    - Plants arborescent; spines strongly dimorphic and pungent; fruits 2–6 cm long, bearing some spines, not markedly profuse:

Trichomes many; leaves to 4 mm long . . . . *O. galapageia* var. *galapageia*
Trichomes few; leaves 4–9 mm long . . . . *O. galapageia* var. *macrocarpa*
Glochids present and abundant; fruits usually with spines:
Spines usually nearly equal in length, less than 5 cm long; flowers small, perianth parts 8–15 mm long; fruit round; areoles 1–1.8 cm apart . . . . *O. insularis*
Spines of varying lengths, usually more than 5 cm long; flowers large, perianth parts 2–2.5 cm long; fruit turbinate; areoles 1.5–2.7 cm apart . . . . *O. saxicola*
Plants shrubby or prostrate, rarely forming a distinct trunk; spines not strongly dimorphic:
Fruits less than 4 cm long, bearing well-formed spines; areoles 1–1.8 cm apart . . . . *O. insularis*
Fruits 4 cm or more in length, tending to have small or poorly developed spines; areoles 1.5–3(3.5) cm apart:
Seeds 4–6 mm long; stem joints not tending to have distinct cuneate bases; spines yellow becoming brown, 1–17 per areole, nearly erect, rigid; fruits spheroidal to oblong . . . . *O. helleri*
Seeds 3–4 mm long; stem joints tending to have distinctly cuneate bases; spines yellowish white becoming darker in age, 7–28 per areole, spreading, flexible; fruits turbinate . . . . *O. echios* var. *zacana*

Opuntia echios Howell, Proc. Calif. Acad. Sci. IV, 21: 49. 1933, var. **echios**

Opuntia myriacantha Weber in Bois, Dict. d'Hort. 894. 1894, non Link & Steud. 1841.
Opuntia galapageia var. *echios* Backbg., Die Cactaceae 1: 561. 1958.
Opuntia galapageia var. *myriacantha* Backbg., loc. cit.
Opuntia echios var. *prolifera* Dawson, Cact. & Succ. Jour. 34: 104. 1962.

Plant variable, mainly arboreal (some shrubby on Isla Daphne Major and in the mesic zone on Isla Santa Cruz), 1–3 m tall, the branches not strongly pendent; trunk sometimes absent, when present spiny or with reddish platelets, to 40 cm thick; larger terminal joints yellow-green to blue-green, orbicular, obovate or oblong, 25–45 cm long, 17–32 cm broad, 1–2.4 cm thick; leaves conical, 2–5 mm long; areoles round, 2–6 mm across, typically 1.3–3 cm apart; trichomes up to 3 mm long; spines strongly dimorphic, often sparse on younger stem joints, erect, stiff and pungent on young stems, to bristly and less pungent on mature plants, yellow becoming brown, 2–20 per areole or more, 1.2–12 cm long, 0.5–1.2 mm thick basally; glochids

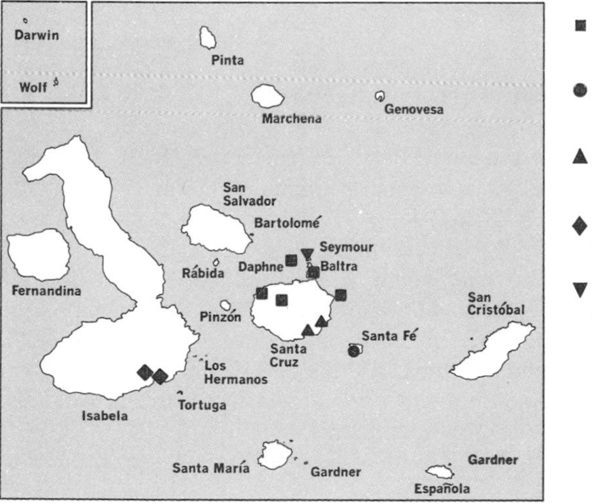

■ *Opuntia echios* var. *echios*
● *Opuntia echios* var. *barringtonensis*
▲ *Opuntia echios* var. *gigantea*
◆ *Opuntia echios* var. *inermis*
▼ *Opuntia echios* var. *zacana*

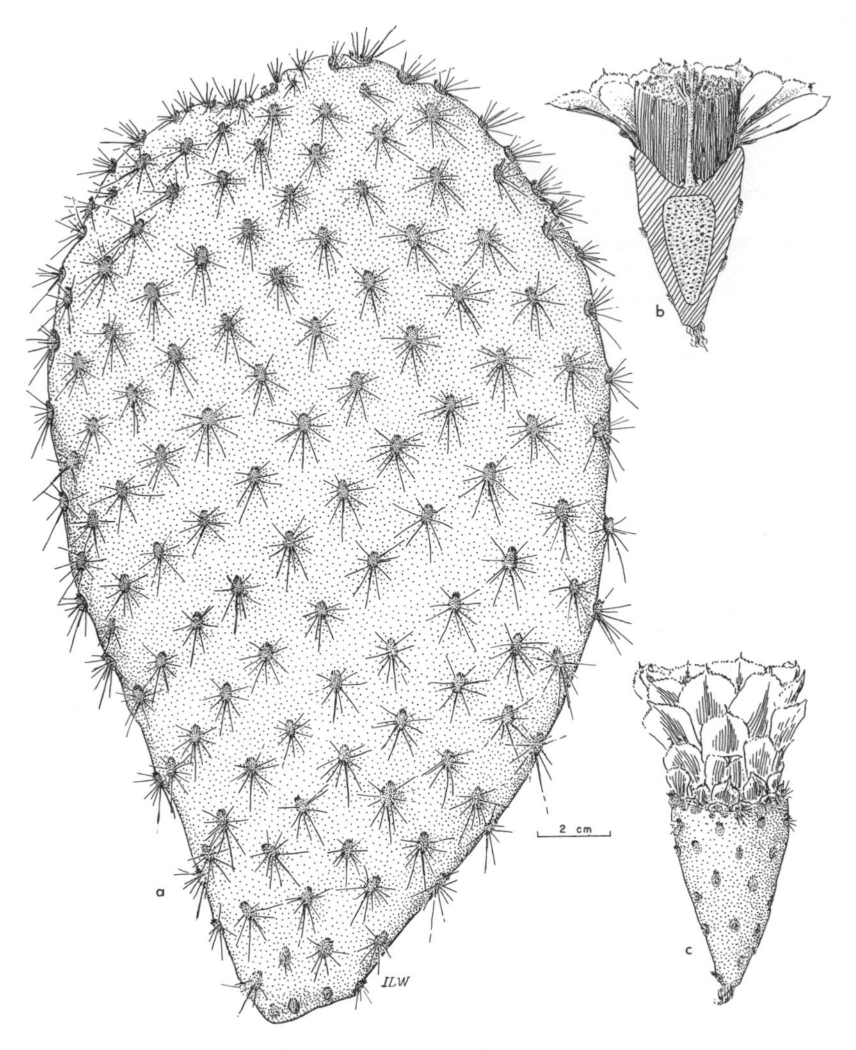

Fig. 141. *Opuntia echios* var. *echios*.

few or absent, yellow to yellowish green; flowers 5–7 cm across, 8–10 cm long; outer perianth parts with midribs yellow, green, or greenish red, the margins yellowish green, largest parts obovate, 14–48 mm long, 15–27 mm wide, mucronate to notched apically, entire; inner perianth parts yellow throughout, the largest obovate to oblong, 26–47 mm long, 12–31 mm wide, acuminate or rounded, entire; filaments yellowish white, 10–32 mm long; anthers yellowish white, 0.5–2.8 mm long; style whitish or reddish, 22–31 mm long, 3–6 mm in greatest diameter; stigmas 7–13, 4–9 mm long; fruits with spines and glochids, turbinate, 4–6 cm long, 3–3.5 cm broad, the umbilicus deeply impressed, green to brown; seeds whitish brown, discoid, 2–3 mm long, 1.5–2.5 mm wide, 0.5–1 mm thick.

Endemic; in thin soil or bare lava, Arid Zone into *Scalesia* Zone, from sea level to about 380 m or possibly higher. Specimens examined: Baltra, Daphne Major, Las Plazas, Santa Cruz.

See Fig. 141.

Opuntia echios var. **barringtonensis** Dawson, Cact. & Succ. Jour. 34: 104. 1962

Plant strongly arboreal, 3–6 m tall with somewhat pendent branches to 3 m long; trunk usually present, to 1.25 m thick, with spines at first, later covered with reddish platelets; larger terminal joints green or slightly yellow-green, 26–36 cm long, 19–24 cm wide, 1.3–1.9 cm thick; leaves 2–4.5 mm long; areoles 3–8 mm across, typically 1.6–3.3 cm apart; trichomes up to 5 mm long; spines yellowish white becoming brownish white or white, 2–20 or more per areole, 1.8–2.6 cm long, 0.5–1 mm thick basally; glochids rare or absent; flowers 6.8–9.1 cm across, 7.8–9.2 cm long; fruits with a few spines, glochids, and trichomes, ovate to turbinate, 7.2–11.7 cm long, 3.8–4.7 cm across; seeds 3–5 mm long, 2–3 mm wide, 1–2 mm thick. Characters otherwise as in var. *echios*.

Over most of Isla Santa Fé, and abundant in several areas.

See color plate 28.

Opuntia echios var. **gigantea** Howell, Proc. Calif. Acad. Sci. IV, 21: 51. 1933

Opuntia galapageia var. *gigantea* Backbg., Die Cactaceae 1: 562. 1958.

Plant strongly arboreal, the branches strongly pendent and from 1.5–5 m long, the tree 3–12 m tall; trunk well-developed, to 40 cm thick, with reddish, flaky platelets; larger terminal joints green to blue-green, 24–37 cm long, 15–26 cm broad, 1.2–1.9 cm thick; leaves 2–5 mm long; areoles 4–8 mm across, typically 1.3–2.8 cm apart; trichomes to 4 mm long; spines yellow becoming brownish white, 5–50 per areole, 1–7.5 cm long, 1–1.5 mm thick basally; glochids few, yellowish, 2–5 mm long; flowers 6–7.5 cm broad, 6–10 cm long; fruits with glochids, trichomes, and spines, spheroidal to oblong, rounded at base, 5.5–7.5 cm long, 3.1–3.8 cm across; seeds 2–3 mm long, 1.5–2 mm broad, 1 mm thick. Characters otherwise as in var. *echios*.

In thin to abundant soil, in sand near sea level, and in crevices in extensive lava flows, Arid and lower Transition zones, sea level to 150 m in the vicinity of Academy Bay, Isla Santa Cruz.

See color plates 25, 27, 29, 33, 38, 40, 42, and 46.

Opuntia echios var. **inermis** Dawson, Cact. & Succ. Jour. 34: 103. 1962

Plant arboreal, 2–6 m high, with somewhat pendent branches rarely more than 1.5 m long; trunk usually present, variable, to 40 cm thick, spiny at first, later with reddish platelets and blackish brown furrows; larger terminal joints green to yellow-green, 26–40 cm long, 14–27 cm wide, 1.5–2.1 cm thick; leaves 4–6 mm long; areoles 4–8 mm across, typically 1.7–2.8 cm apart; trichomes to 4 mm long; spines yellow becoming blackish white, 3–20 per areole or more, 5.5–6.2 cm long, 0.5–1 mm thick basally; glochids present, yellowish green; flowers 5.5–6.2 cm broad, 5.2–6.1 cm long; fruits with glochids, trichomes, and occasionally weakly developed spines, spheroidal, 4–5.2 cm long, 2.8–4 cm broad; seeds 2–3 mm long, 1.5–2 mm broad, 0.5–1 mm thick. Characters otherwise as in var. *echios*.

In Arid Zone, from sea level to about 100 m in the region around Villamil, Isla Isabela.

Opuntia echios var. **zacana** (Howell) Anderson & Walkington, Madroño, 20: 256. 1970

Opuntia zacana Howell, Proc. Calif. Acad. Sci. IV, 21: 48. 1933.
Opuntia galapageia var. *zacana* Backbg., Die Cactaceae 1: 562. 1958.

Plant shrubby to nearly prostrate, never becoming arboreal, 1–2 m high; trunk rarely present, but usually spiny if present, to 30 cm thick; larger terminal joints green or yellowish green, orbicular to elliptical, somewhat cuneate at base, 25–50 cm long, 20–30 cm wide, 2–3.5 cm thick; leaves 2–4 mm long; areoles 3–6 mm across, typically 1.5–3 cm apart; trichomes present, 2 mm long; spines nearly equal, evenly distributed, in some clones extremely long, yellow becoming brown, 1–17 per areole, nearly erect, rigid, 0.6–1.6 or to 3.7(5) cm long, 0.5–1 mm thick basally; glochids usually present, very evident on younger stem joints, yellow becoming brown before falling; flowers unknown; fruits turbinate, 4–8.5 cm long, 3–4.5 cm broad, bearing spines and glochids; seeds 3–4 mm long, 2–3 mm wide, 1–1.5 mm thick. Characters otherwise as in var. *echios*.

Known only in the Arid Zone, associated with *Bursera* in rocky soil and on lava beds, on Isla Seymour.

Opuntia galapageia Hensl., Mag. Zool. & Bot. 1: 467. 1837, var. **galapageia**

Plant arborescent and arboreal, with well-developed, rounded crowns, 2–5 m tall; trunk usually well-developed, bearing spines at first, later covered with reddish black platelets; larger terminal joints green to yellow-green, orbicular, oblong, or obovate, 22–38 cm long, 15–27 cm broad, 1–3.5 cm thick; leaves conical, large, 5–9 mm long; areoles 2–7 mm broad, typically 2.5–3.4 cm apart; trichomes few to many, to 4 mm long; spines highly dimorphic, few on young joints, becoming more numerous on older ones, pungent and long on young plants, bristly and not pungent on mature plants, yellow becoming reddish or brownish white, 5 to about 35 per

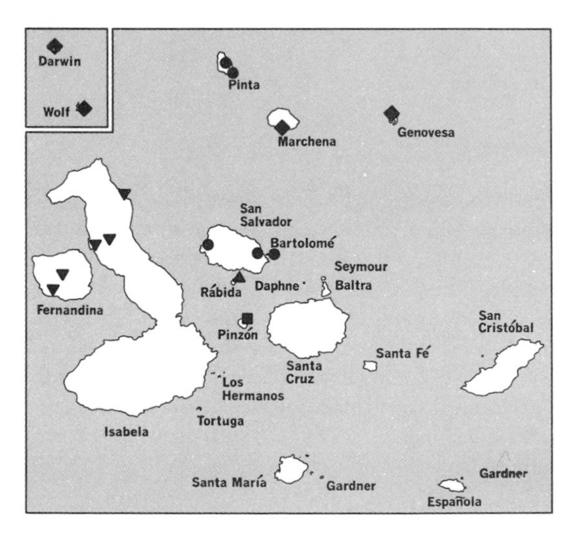

● *Opuntia galapageia* var. *galapageia*
■ *Opuntia galapageia* var. *macrocarpa*
▲ *Opuntia galapageia* var. *profusa*
◆ *Opuntia helleri*
▼ *Opuntia insularis*

areole, erect, 2.5–7.5 cm long, 0.5–1 mm thick basally; flowers 3.5–6 cm broad, 4–7 cm long; outer perianth parts with midribs yellow-green and margins greenish yellow, the largest parts obovate, 11–17 mm long, 6–13 mm wide, mucronate, entire or slightly ciliate; inner perianth parts yellow throughout, the largest obovate, 17–40 mm long, 9–16 mm wide, notched or rounded apically, entire or slightly ciliate; filaments yellowish white, 7–16 mm long; anthers same color, 0.7–1.2 mm long; style whitish, 17–20 mm long, 2–3 mm in greatest diameter; stigmas 7–10, 4–5 mm long; fruits green, with a few glochids, some spiny, globular to oblong, 2–6 cm long, 2–3.5 cm broad, the umbilicus fairly large but shallow, brownish; seeds brownish white with darker center, discoid, 2.5–3.3 mm long, 1.7–2.2 mm wide, 1–2 mm thick.

Endemic; in rocky areas and in crevices on lava flows, Arid Zone near sea level. Specimens examined: Bartolomé, Pinta, San Salvador.

See color plate 32.

Opuntia galapageia var. **macrocarpa** Dawson, Cact. & Succ. Jour. 37: 141. 1965

Plant arborescent and arboreal, with somewhat rounded crowns, 2–4 m tall; trunk usually well-developed, spiny at first, later having reddish, flaky platelets; larger terminal joints 22–30 cm long, 15–18 cm wide, 1.6–2.7 cm thick; leaves small, 2–4 mm long; areoles 4–6.5 mm across, typically 2.3–2.8 cm apart; spines highly dimorphic, pungent and long on young plants, bristly and not pungent on mature ones, yellow becoming whitish or brownish, 5 to about 25 per areole, 4.8–5.4 cm long; fruits green, with some trichomes, few or no glochids, some small bristly spines, more or less globular, 3.2–5 cm long, 2.8–4 cm broad, the umbilicus fairly large and shallow, brownish; seeds 3.5–5 mm long, 2.9–3.7 mm broad, 1.3–1.9 mm thick. Characters otherwise as in var. *galapageia*.

Endemic; found in shrubby and forested areas from about 175 to 300 m on Isla Pinzón.

Opuntia galapageia var. **profusa** Anderson & Walkington, Madroño 20: 256. 1970

Plant variable, prostrate, shrubby, arborescent, and arboreal, 1–3 m tall; trunk, when present, flaky and reddish; larger terminal joints 21–38 cm long, 18–26 cm wide, 1.8–2.7 cm thick; leaves small, up to 4 mm long; areoles 4–6 mm across, typically 2.2–3.3 cm apart; spines not strongly dimorphic, pungent only on new stem joints, bristly, yellow becoming brownish, 5 to about 22 per areole, 3–5.5 cm long; fruits greenish becoming yellow-green to brown, with glochids on some, spines absent, nearly globose, 1.7–2.5 cm long, 2.2–2.7 cm broad, with deep, small, greenish brown umbilicus, the fruits often profuse, up to 82 on a single joint; seeds 2–3 mm long, 1.5–2.5 mm wide, 1 mm thick. Characters otherwise as in var. *galapageia*.

Endemic; in Arid Zone from sea level to about 100 m on Isla Rábida.

Opuntia helleri K. Schum. in Robins., Proc. Amer. Acad. 38: 180. 1902

Opuntia galapageia var. *helleri* Backbg., Die Cactaceae 1: 562. 1958.

Plants prostrate and scrambling, often forming thickets, occasionally becoming shrubby and small-arborescent, 0.75–1.5(2) m high, the trunk usually absent but when rarely present becoming reddish brown; larger terminal joints yellowish green, orbicular, oblong, obovate, or ovate, 20–37 cm long, 10–22 cm wide, 1.3–2.6 cm thick; leaves conical with small terminal spine, 0.4–6 mm long; areoles ovate, 2.5–5 mm across, typically 1.5–2.6(3.5) cm apart; trichomes abundant, averaging 1.5–2 mm long; spines bristly, not pungent, evenly distributed, not dimorphic, yellowish white becoming darker with age, 7–28 per areole, spreading, flexible, varying in length, tending to be wavy, longer ones 1.2–4.1(5) cm long, less than 1 mm thick basally; glochids yellow, 2–6 mm long, sometimes absent; flowers 3–5.5 cm broad, 4–8 cm long; outer perianth parts with greenish yellow midribs, margins yellow, the largest obovate, 15–30 mm long, 10–15 mm wide, apically irregular but usually obtuse, entire; inner perianth parts yellow, the largest obovate, 25–35 mm long, 15–20 mm wide, obtuse to mucronate, entire; filaments yellow, 8–15 mm long; anthers yellow, 1–2 mm long; style yellow, 20–25 mm long, 2–3 mm in greatest diameter; stigmas 5–10, 5–10 mm long; fruits green at maturity, bearing trichomes, small spines, and glochids, spheroidal to oblong, 4–7 cm long, 2–4 cm broad, the umbilicus broadly depressed, brownish, often persistent for several years, tending to proliferate; seeds whitish yellow, discoid, 4–6 mm long, 3–5 mm wide, 1.5–2 mm thick.

Endemic; in Arid Zone near sea level. Specimens examined: Darwin, Genovesa, Marchena and Wolf.

See color plates 36 and 54.

Opuntia insularis Stewart, Proc. Calif. Acad. Sci. IV, 1: 113. 1911

Opuntia galapageia var. *insularis* Backbg., Die Cactaceae 1: 561. 1958.

Plant shrubby to arborescent, 1–2.5 m high; trunk poorly developed, spiny at first with reddish bark developing in older plants; larger terminal joints greenish to greenish yellow, primarily orbicular, to obovate and oblong, 10–52 cm long, 18–25 cm wide, 0.5–2.5 cm thick; leaves conical, 5–8 mm long; areoles 4–7 mm across, typically 1.4–1.8 cm apart; trichomes usually abundant, whitish; spines nearly equal and evenly distributed over stem joint, yellow becoming reddish or dark brown in age, 10 to about 50 per areole, most moderately pungent though some bristly, the longer 4–5 cm long (mostly 1–2 cm long), 0.5–1 mm thick basally; glochids yellow, 4–6 mm long; flowers poorly known; fruits greenish, with small spines, glochids, and trichomes, rounded, 2–4.2 cm long, 2–3 cm broad; seeds whitish brown, discoid, 2.5–3.5 mm long, 2–2.25 mm wide, 1–1.5 mm thick.

Endemic; in scoria and on ash fields, in the Arid, Transition, and *Scalesia* zones from sea level upward. Specimens examined: Fernandina, Isabela.

See color plates 31 and 50.

Opuntia megasperma Howell, Proc. Calif. Acad. Sci. IV, 21: 46. 1933, var. **megasperma**

Plant shrubby to arboreal, 2–6 m high; crowns rounded, branches proliferating freely; trunk spiny at first, later covered with platelets, to 1 m thick, reddish; larger terminal joints greenish yellow to blue-green, orbicular, ovate, oblong, 25–37 cm long, 15–25 cm wide, 1.8–3.4 cm thick; leaves conical, 4–6 mm long; areoles ovate, 2–6 mm across, typically 3–4.2 cm apart; trichomes few; spines highly dimorphic, stiff and pungent to bristly and lax, usually abundant, yellow becoming brownish or blackish white, 8–50 per areole, generally erect, 6–10 cm long, 1–2.2 mm thick

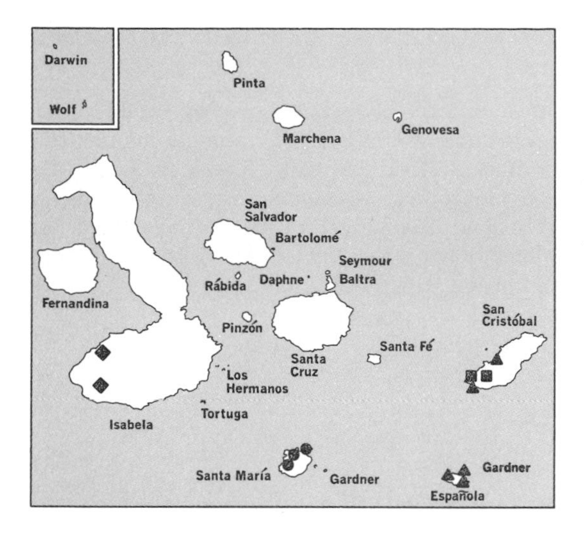

● *Opuntia megasperma* var. *megasperma*

■ *Opuntia megasperma* var. *mesophytica*

▲ *Opuntia megasperma* var. *orientalis*

◆ *Opuntia saxicola*

basally; glochids few or none; flowers 7–10.5 cm across, 11–13 cm long; outer perianth segments with greenish midribs and greenish yellow margins, the largest obovate, 18–30 mm long, 15–25 mm wide, mucronate to acuminate, entire; inner perianth segments yellow to reddish yellow in age, the largest obovate, 25–50 mm long, 20–33 mm wide, mucronate, entire; filaments whitish, 5–37 mm long; anthers light yellow, 1–2 mm long; style white, 25–37 mm long, 2–6 mm in greatest diameter; stigmas 5–9, 4–10 mm long; fruits green becoming yellowish green, bearing small bristly spines and trichomes, turbinate, 8–17 cm long, 3.5–5 cm across, the umbilicus large, depressed, brown; seeds brownish white, discoid to somewhat angular, 8–13 mm long, 6–10 mm wide, 4–6 mm thick.

Endemic; in scanty soil and among rocks and crevices in lava flows, Arid and lower Transition zones, sea level to about 50 m. Specimens examined: Champion, Santa María.

See color plates 34 and 35.

Opuntia megasperma var. **mesophytica** J. Lundh, Madroño 20: 254. 1970*

Plant arborescent to arboreal, the crown open with slender vertical branches, 2–6 m high; trunk to 40 cm thick, brownish; larger terminal joints dark green becoming gray-green, obovate to elongate, 12–20 cm long, 5.5–9.5 cm wide, 0.8–1.7 cm thick; leaves 3–7 mm long; areoles ovate, 1.5–2 mm across, typically 1.8–3.2 cm apart; spines often nearly absent, golden brown and banded, becoming bone white with dark tip, 1–17 per areole, 2.5–3.9 cm long, 0.5–1.5 mm thick basally; flowers about 6 cm across, 6–8 cm long; fruits 4–6 cm long, 2.7–3.6 cm across; seeds 7–10 mm long, 5–8 mm wide, 2–4 mm thick. Characters otherwise as in var. *megasperma*.

Endemic; in the upper Transition Zone and the *Scalesia* Zone, at about 250 m elevation on Isla San Cristóbal; rare.

* This taxon was recognized first as being distinct from var. *orientalis* by J. Lundh, who collected material and sent the proposed varietal epithet and descriptions to E. Yale Dawson and E. F. Anderson. The description was not published, but was utilized in part in drawing up our description, which here validates the Lundh variety.

Opuntia megasperma var. **orientalis** Howell, Proc. Calif. Acad. Sci. IV, 21: 48. 1933

Plant generally arboreal, with dense rounded crowns having heavy, proliferating branches, 2–5 m high; trunk to 60 cm thick, reddish; larger terminal joints green to yellow-green, orbicular, ovate, or oblong, 12–35 cm wide, 23–48 cm long, 1.2–2.7 cm thick; leaves 4–11 mm long; areoles round, 3–6 mm across, typically 1.8–4.5 cm apart; spines usually abundant, yellow, becoming brownish white, 20–40 per areole, 2.9–5.5 cm long, 0.5–2 mm thick; flowers 7–8.5 cm broad, 8.5–11 cm long; fruits 6–10(13) cm long, 3.5–5.5 cm broad; seeds 5–10 mm long, 4–9 mm broad, 2–4 mm thick. Characters otherwise as in var. *megasperma*.

Endemic; in Arid Zone and in lower Transition Zone, from sea level to about 75 m elevation. Specimens examined: Española, Gardner (near Española), San Cristóbal.

See color plate 30.

Opuntia saxicola Howell, Proc. Calif. Acad. Sci. IV, 21: 45. 1933

Opuntia galapageia var. *saxicola* Backbg., Die Cactaceae 1: 562. 1958.

Plant arborescent to shrubby, 1–2 m tall; larger terminal joints greenish yellow, round, oblong, and obovate, 17–29 cm long, 12–20 cm wide, 0.5–2.2 cm thick, the leaves conical, 3–4 mm long; aeroles 4–5 mm across, typically 1.5–2.7 cm apart; trichomes present, whitish to reddish; spines variable, mostly pungent, yellow becoming brown or reddish in age, 14–30 per areole, the longer 3–8 cm long, 0.5–1 mm thick basally, glochids present, yellow, 2–3 mm long; flowers 2.5–3.5 cm broad, 3–4 cm long, yellow; fruit green at maturity, with spines, glochids, and trichomes, globose, 2.5–3.9 cm long, 2.3–3.8 cm broad, the umbilicus depressed, brown; seeds whitish brown, discoid, 2–3 mm long, 1.5–2.5 mm wide, 1 mm thick.

Endemic; in dry habitats from sea level to about 1,500 m in the southwestern parts of Isla Isabela.

CAPPARIDACEAE. Caper Family*

Herbs, shrubs, or trees, usually with ill-scented foliage, without latex. Leaves alternate, simple or palmately compound, the stipules minute, glandular or spinose, or sometimes absent. Flowers perfect or sometimes monoecious, regular or more frequently zygomorphic, in racemes or in axils of upper leaves, bracteate. Sepals usually 4, occasionally 8 (or lacking in some), distinct or more or less connate. Petals 4 to many (rarely absent), equal or posterior pair larger, clawed or sessile, distinct. Stamens 4–30; anthers 2–4-celled, dehiscing longitudinally. Pistil 1, sessile or more commonly borne on a short to long gynophore. Ovary superior, unilocular, carpels 2–4; placentation parietal; style short to filiform-elongate; stigma capitate or bilobed. Ovules few to many on each placenta. Fruit a capsule dehiscing by valves, a berry, or an indehiscent nutlet. Seeds reniform. Embryo curved. Endosperm fleshy-oily.

About 46 genera and 700 species, widely distributed in the tropics and warm temperate areas of the world.

* Contributed by Ira L. Wiggins.

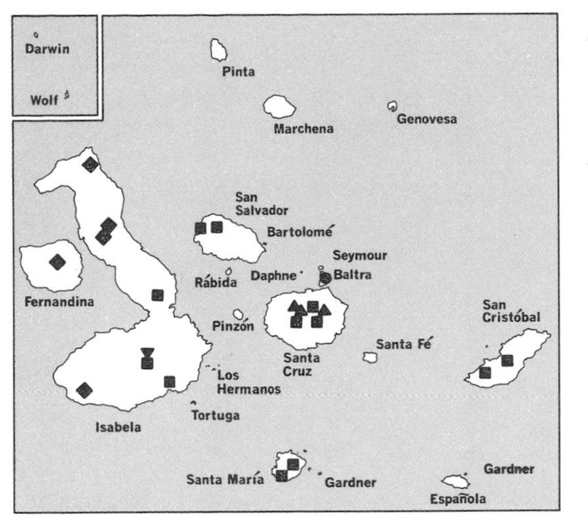

● *Cleome viscosa*
■ *Drymaria cordata*
▲ *Drymaria monticola*
◆ *Drymaria rotundifolia*
▼ *Stellaria media*

Cleome L., Sp. Pl. 671. 1753; Gen. Pl. ed. 5, 302. 1754

Annual or perennial herbs, or shrubs or trees, with simple or digitately 3–5-foliolate, compound leaves. Stipules none. Sepals 4, distinct or basally connate, deciduous or persistent. Petals 4, clawed, entire, overlapping in bud. Stamens 6–30, or sometimes only 4, included or distinctly exserted. Ovary usually stipitate (less commonly sessile as in ours), usually with basal gland (this lacking in ours). Fruit silique-like, linear to oblong, many-seeded. Seeds obovate-reniform, dull, usually more or less ridged or rugose.

About 200 species, native in the Old World but some widely distributed as pantropical weeds growing along roadsides, about margins of fields, in thickets, and in waste land.

Cleome viscosa L., Sp. Pl. 672. 1753*

Strict or considerably branched annual 2–16 dm tall, densely glandular-pubescent throughout, unarmed; petioles 1–6 cm long, exstipulate; leaflets 3–5, oblanceolate-elliptic, acute to obtuse at apex, cuneate to attenuate at base, the central leaflets 1–6 cm long, 0.5–3 cm wide, entire, densely glandular-ciliate, petiolules 1–3 mm long; flowers relatively few, solitary in axils of upper leaves; pedicels 5–23 mm long; sepals free, narrowly elliptic to oblanceolate-elliptic, acuminate, 5–10 mm long, 1–4 mm wide, deciduous; petals pale to deep yellow or orange-yellow, obovate to spatulate, rounded at apex, gradually attenuate to base, 6–14 mm long, 3–5 mm wide; disk obsolete; stamens 10–26, included, the filaments 5–8 mm long, 4–10 of the adaxial shorter than others and slightly clavate; anthers 2–3 mm long; ovary cylindric, densely glandular-pubescent, sessile; style about 1 mm long at anthesis, to 6 mm long in fruit; stigmas capitate; mature fruits linear-cylindric, attenuate toward apex, sessile, 2–4 mm broad, 3–10 cm long, erect or strongly ascending, glandular-pubescent, the valves longitudinally striate and dull without, lustrous within; seeds

* For synonymy, see Iltis, Brittonia 12: 281. 1960.

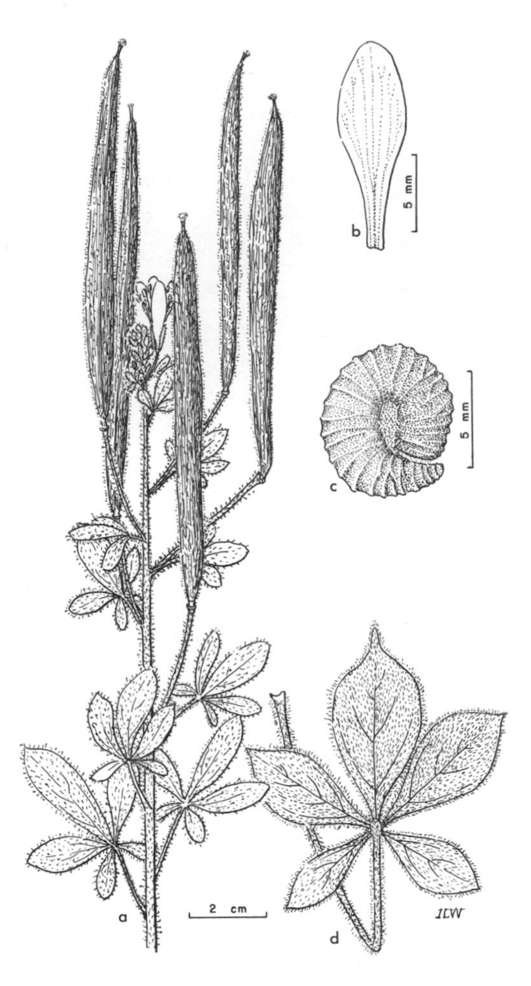

Fig. 142. *Cleome viscosa.*

obovoid-reniform, 1.2–1.8 mm long, dark reddish brown, strikingly rugulose transversely with about 30 rugae, these again minutely transversely rugulose.

A pantropical weed believed to be a native of tropical Asia but originally described from plants grown in the Botanical Gardens at Uppsala, Sweden; very common in the West Indies, fairly common in central Mexico and Central America south through Brazil, and sporadic in warm areas of other parts of South America. Specimen examined: Baltra; a new record for the family in the Galápagos Islands.

See Fig. 142.

CARYOPHYLLACEAE. Pink Family*

Annual or perennial herbs, sometimes suffrutescent. Leaves entire, with or without stipules. Flowers actinomorphic, usually bisexual, solitary or in cymes. Sepals free

* Contributed by Uno Eliasson.

or basally connate. Petals usually 4–5, free, often differentiated into claw and limb, sometimes lacking. Stamens up to 10, free, obdiplostemenous; anthers 2-celled, dehiscing longitudinally. Pistil 1; ovary superior, usually 1-loculed with free-central placentation, sometimes 3–5-loculed with axile placentation. Carpels 2–5; styles and stigmas usually same in number as carpels. Fruit usually a capsule opening by valves or apical teeth, sometimes an achene. Seeds 1 to many. Embryo usually amphitropous.

About 80 genera and 2,000 species, abundant in temperate and cool regions, and in mountainous areas in the tropics.

Stipules present; capsule dehiscing with 3 valves. . . . . *Drymaria*
Stipules none; capsule dehiscing with 6 teeth. . . . . *Stellaria*

Drymaria Willd. in Roem. & Schult., Syst. Veg. 5: 406. 1819

Annual herbs or suffrutescent perennials. Leaves entire, usually opposite, occasionally pseudoverticillate. Stipules usually present, entire to lacerate. Inflorescences terminal or axillary cymes, sometimes with racemose tendencies. Sepals 5, lanceolate to suborbicular. Petals 5, in a few cases 3 or lacking, deeply bifid. Stamens 2–5. Ovary 1-celled. Style usually simple at base, trifid above. Capsule dehiscing by 3 valves. Seeds 1 to many, amphitropous, characteristically sculptured.

About 50 species, predominantly subtropical in distribution. Most species are restricted to the New World.

Petals shorter than sepals; pedicels with girdle of dense, glandular pubescence. . *D. cordata*
Petals equal to or slightly exceeding sepals; pedicels glabrous:
  Flowers in axillary cymes; sepals about 5 mm long; leaves 10–25 mm long. . *D. monticola*
  Flowers in terminal cymes; sepals 2.5–3 mm long; leaves 3–7 mm long. . . *D. rotundifolia*

Drymaria cordata (L.) Willd. in Roem. & Schult., Syst. Veg. 5: 406. 1819

Holosteum cordatum L., Sp. Pl. 88. 1753.
Holosteum diandrum Sw., Prodr. 27. 1788.
Drymaria diandra MacFadyen, Fl. Jam. 1: 52. 1837.
Drymaria procumbens Rose, Contr. U. S. Nat. Herb. 1: 304. 1895.
Drymaria adenophora Urban, Fedde Repert. Spec. Nov. Regni Veg. 21: 213. 1925.
Stellaria adenophora León in León & Alain, Flora de Cuba 2: 154. 1950.

Annual; stems prostrate and spreading to erect, glabrous or slightly glandular; leaves reniform to suborbicular, 8–20 mm long, 10–20 mm wide; stipules lacerate; inflorescences of terminal or axillary dichasial cymes; pedicels 2–10 mm long, usually with girdle of dense, glandular pubescence; sepals narrowly elliptic or broadly lanceolate, 3-nerved, acute, 3–4 mm long, much exceeding petals; petals deeply bifid, 2–2.5 mm long; stamens usually 2–3; seeds dark reddish brown, about 1 mm in diameter, with low tubercles linearly arranged.

A more or less pantropical species occurring as a weed in many regions. Specimens examined: Isabela, San Cristóbal, San Salvador, Santa Cruz, Santa María.

See Fig. 143.

Drymaria monticola Howell, Proc. Calif. Acad. Sci. IV, 21: 329. 1935

Creeping, glabrous annual; leaves broadly ovate to subcordate, 10–25 mm long, 8–25 mm wide; stipules deeply lacerate; inflorescences usually axillary, dichásial cymes; pedicels to 5 mm long, glabrous; sepals elliptic-lanceolate, 3-nerved, acute, about 5 mm long (about as long as petals); petals (4)5, deeply bifid, about 5 mm long, the lobes about 1 mm wide, obtuse, often rounded at apex; stamens 4–5; seeds about 1 mm in diameter, tuberculate, dark reddish brown.

Endemic; known only from the highlands of Isla Santa Cruz.

Drymaria rotundifolia A. Gray, Bot. U.S. Expl. Exped. 123. 1854

Drymaria nitida Ball, Jour. Linn. Soc. Lond. 22: 31. 1887.

Annual; stems erect, to 20 cm high, dichotomously branched, glabrous; leaves broadly ovate to deltoid, acute, 3–7 mm long, 3–6 mm wide; stipules acicular; inflorescences of dense, terminal cymes; pedicels glabrous, about 1 mm long; sepals

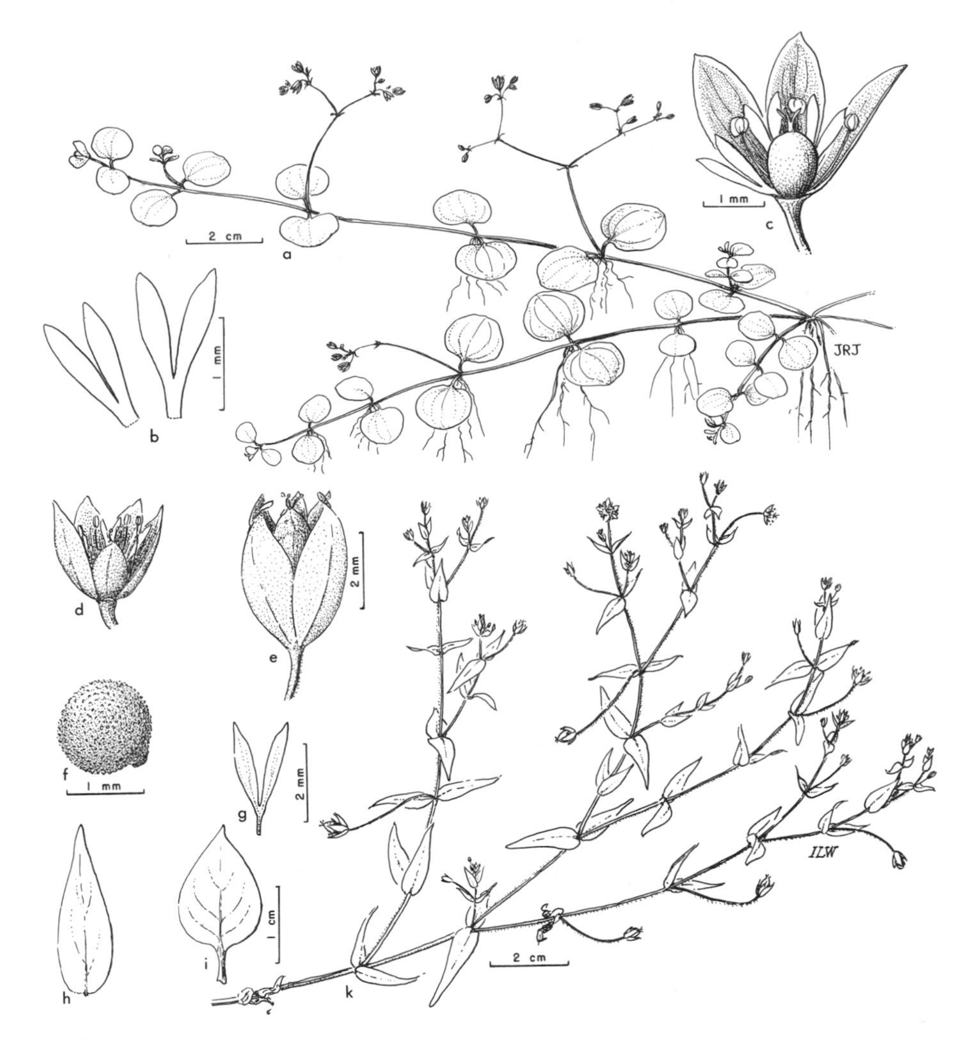

Fig. 143. *Drymaria cordata* (*a–c*). *Stellaria media* (*d–k*): *g*, petal; *h, i*, upper and lower leaves.

2.5–3 mm long, narrowly elliptic, obscurely 1–3-nerved, midnerve often dissipating near apex; petals 3–3.5 mm long, deeply bifid, the lobes obtuse at apex, about 0.5 mm wide; stamens 3–5; seeds about 0.8 mm in diameter, dark reddish brown or brown, with low tubercles arranged in rows.

Known from Ecuador, Peru, and the Galápagos Islands (where it grows at high elevations). Specimens examined: Fernandina, Isabela.

The Galápagos plant has been described as an endemic variety (var. *mattfeldiana* Howell, 1966, 351), characterized mainly by its lanceolate-elliptic sepals. It is, however, very closely related to the mainland varieties, *rotundifolia* and *nitida* (cf. Duke, 1961, 239).

Stellaria L., Sp. Pl. 521. 1753; Gen. Pl. ed. 5, 193. 1754

Annuals or perennials. Stipules absent. Inflorescences usually dichasial cymes, rarely the flowers solitary or 2 together. Sepals (4)5. Petals 5 or fewer, deeply bifid, sometimes absent. Stamens 10 or fewer. Styles 3, distinct to base. Capsule dehiscing with 6 teeth, the seeds few to many, amphitropous, sculptured.

A more or less cosmopolitan genus with about 120 species.

Stellaria media (L.) Vill., Hist. Pl. Dauph. 3: 615. 1789

Alsine media L., Sp. Pl. 272. 1753.

Annual to 30 cm tall; stems decumbent to ascending, villous on 1 side, (more or less profusely) branched; leaves opposite, ovate, glabrous, 5–30 mm long, 3–20 mm wide; upper leaves sessile, lower ones petiolate; petioles 5–10 mm long; pedicels (5)10–25 mm long; sepals 5, ovate-elliptic, green with white margins, about 5 mm long, 2 mm wide; petals 5, shorter than sepals, about 3 mm long, deeply bifid, each lobe 2-nerved; stamens about 5, 2–2.5 mm long; stigmas 3, to 1 mm long, curved outward; capsule many-seeded; seeds 1–1.2 mm in diameter, round-reniform, dark reddish brown, covered with low tubercles, these star-shaped at base.

A species of nearly cosmopolitan distribution; in the Galápagos known only from the single collection from Isabela, which was probably introduced recently.

The area of this collection is covered with grasslands where free-ranging cattle graze.

See Fig. 143.

CELASTRACEAE. STAFF-TREE FAMILY*

Trees or shrubs, rarely woody vines, usually glabrous. Leaves opposite or alternate, rarely minute or rudimentary, simple, membranaceous to coriaceous, petiolate, deciduous or persistent; stipules minute, usually caducous. Flowers small, regular, hypogynous, perigynous, or semiepigynous, bisexual by abortion and plants then monoecious or dioecious; pedicels rarely absent; inflorescences axillary and/or terminal, cymose and usually dichotomously branched, or racemose, or axillary and flowers solitary and/or fasciculate. Sepals 4–5, small, connate to nearly free, usually imbricate. Petals 4–5, free, inserted under disk, imbricate. Stamens 4–5(8–10), op-

* Contributed by Duncan M. Porter.

posite sepals, reduced or absent in carpellate flowers; filaments short, free, inserted on or under margin of disk; anthers basifixed or dorsifixed, introrse, longitudinally dehiscent. Disk intrastaminal, nectariferous, usually conspicuous, rarely absent, variously shaped. Gynoecium 2–5-carpelled, syncarpous, rudimentary in staminate flowers; ovary superior to semi-inferior, free or more or less immersed in and sometimes adnate to disk, 2–5-loculed; ovules (1)2(4–6 to many) per locule, anatropous, usually apotropous, pendulous from top or ascendent from base of placenta, placentation axile; style 1 or none; stigma 1 and more or less lobed, or 2–5. Fruit a loculicidal or rarely a septicidal capsule, or a drupe, berry, or samara, (1)2–5-loculed. Seeds 1–6 per locule, often arillate; endosperm present or rarely absent; embryo straight, axial; cotyledons large, foliaceous, often green; radicle short, superior or inferior.

A primarily pantropical family of about 50 genera and 800 species, extending into the temperate regions of the Old and New Worlds.

Maytenus Molina, Sagg. Stor. Nat. Chile 177. 1782; emend. Bosc., Nouv. Dict. Hist. Nat. 14: 211. 1803

Shrubs or small trees, usually glabrous, stems sometimes spinose. Leaves alternate, often distichous, entire or toothed, membranaceous to fleshy-coriaceous, short-petiolate, persistent; stipules minute and caducous, or absent. Flowers usually hypogynous, bisexual (or unisexual by abortion and plants monoecious or dioecious), in axillary dichasial cymes, short thyrses, racemes, or compact fascicles, rarely solitary; pedicels short. Sepals (4)5, connate basally, persistent. Petals (4)5, greenish white to red, spreading, longer than sepals. Stamens (4)5, shorter than petals, reduced and sterile in carpellate flowers; filaments subulate, inserted on or under margin of disk; anthers ovate-suborbicular, introrse, half as long as filaments. Disk conspicuous, flattish, (4)5-angled or -lobed in staminate flowers, annular and adnate to ovary in carpellate flowers. Gynoecium (2)3(5)-carpelled, rudimentary in staminate flowers; ovary superior, short-conical, slightly 3(5)-angled, completely or incompletely (2)3(5)-loculed, more or less immersed in and adnate to disk basally; ovules 1(2) per locule, ascending, placentation axile; style very short or absent;

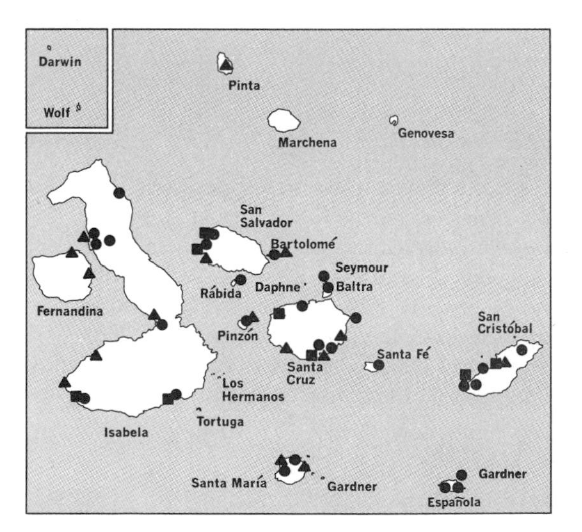

● *Maytenus octogona*
■ *Conocarpus erecta*
▲ *Laguncularia racemosa*

Fig. 144. *Maytenus octogona.*

stigma (2)3(5)-lobed. Capsule 1(3)-loculed, ellipsoid or obovoid, coriaceous, usually slightly angled, loculicidally 2–3(5)-valved. Seeds usually 3(1–6), ovoid-ellipsoid, ascending, covered entirely or partially by a fleshy red or white aril; testa crustaceous or coriaceous; endosperm fleshy or absent; radicle inferior.

A genus of perhaps 200 species or more, in the tropics and subtropics of the world.

Maytenus octogona (L'Hér.) DC., Prodr. 2: 9. 1825*

Celastrus octogonus L'Hér., Sert. Angl. 7. 1788.
Zizyphus peruviana Lam., Encycl. 3: 320. 1789.
Rhamnus peruvianus Lam., loc. cit., nomen nudum.
Senacia octogona (L'Hér.) Lam., Illustr. 2: 96. 1793.
Celastrus orbicularis Willd. ex Roem. & Schult., Syst. Veg. 5: 423. 1819.
Maytenus uliginosus HBK., Nov. Gen. & Sp. Pl. 7: 51. 1825.
Maytenus obovatus Hook. f., Trans. Linn. Soc. Lond. 20: 230. 1847.
Maytenus orbicularis (Roem. & Schult.) Loes., Bot. Jahrb. 50, Beibl. 111: 10. 1913.
Celastrus circumcissus Pavon ex Briq., Ann. Conserv. Jard. Bot. Genève 20: 354. 1919, in synonymy.

Shrub or tree, nearly prostrate to 8 m high; bark dark gray, fissured; branchlets tomentulose to glabrous; leaves elliptic or narrowly to broadly obovate, occasionally orbicular and rarely rhomboid, 20–53 mm long, 14–42 mm wide, dentate when young and becoming sinuate to entire, glabrous, subfleshy to coriaceous; petioles tomentulose to glabrous, 3–7 mm long; pedicels slender, tomentulose to glabrous, 2.5–5 mm long; flowers small, inconspicuous, green, in axillary panicles or solitary; calyx 5-lobed, lobes minute, rounded, ciliolate, tomentulose; petals 5, ovate to triangular, 1 mm long, persistent, margins scarious; stamens 5; anthers minute, yellow;

* Svenson (1935) was the first to determine that the Galápagos taxon is identical with the *Maytenus* on the adjacent mainland.

disk orbicular, turbinate and 5-lobed in staminate flowers; ovary 3-loculed, conical, the base spreading and immersed in disk, abortive in staminate flowers; style erect, short, thickened; stigma thickened, 3-lobed; capsule 3(4)-lobed, reddish, ovoid, obovoid, or globose, erect, glabrous, thick-walled, 9–12 mm high, 8–10 mm wide, dehiscing apically by 3(4) valves, these provided internally with prominent midribs projecting into lumen, equal, subtended by persistent recurved perianth; seeds 1(2–3), basifixed, erect, 5–7 mm long, covered by a pulpy, bright red aril.

Along the coastal parts of the mainland in Ecuador, Peru, and Chile; along beaches, around lagoons, and on arid hillsides, from sea level commonly to 300 m and occasionally to 1,200 m among lava flows and on cinder ridges in the Galápagos. Specimens examined: Baltra, Bartolomé, Española, Gardner (near Española), Isabela, Las Plazas, Pinzón, Rábida, San Cristóbal, San Salvador, Santa Cruz, Santa Fé, Santa María, Seymour; also reported from Fernandina.

See Fig. 144 and color plates 26, 29, 39, and 40.

COMBRETACEAE. COMBRETUM FAMILY*

Trees, shrubs, or lianas, spiny or unarmed. Leaves opposite, whorled, or alternate, crowded at tips of branches, simple, entire, the blades often with 2 basal glands and many small pits beneath, persistent or deciduous; stipules none. Flowers bisexual or both bisexual and staminate in same inflorescence, regular, epigynous, sessile, in axillary or extra-axillary panicles or globose heads, usually bracteate. Hypanthium glabrous, pubescent or scaly, lower part tubular or laterally flattened, more or less 2-winged, surrounding and adnate to ovary, the upper part deeply or shallowly cupuliform, produced beyond ovary; calyx lobes caducous or persistent, (4)5(8) or poorly developed and almost absent, valvate or imbricate, sometimes accrescent. Petals 4 or 5 or none, conspicuous or minute, free, imbricate or valvate, deciduous. Stamens usually twice as many as petals, exserted or included; filaments inserted inside upper part of hypanthium, inflexed in bud; anthers 2-loculed, orbicular or cordate, versatile or rarely adnate to filaments, dehiscing longitudinally. Intrastaminal disk usually surrounding base of style, glabrous or pubescent. Ovary inferior or half-inferior, 1-loculed; ovules 2(6) per locule, pendulous on slender funiculi from locule apex, anatropous, usually only 1 developing; style simple, usually free; stigma capitate or punctiform. Fruit a pseudocarp, variable in size and shape, fleshy or dry, usually indehiscent, often winged or ridged. Seed 1; endosperm scanty or none; testa membranaceous to coriaceous; embryo large, straight; cotyledons thick, fleshy, spirally convolute or plicate; radicle weakly developed, superior.

A pantropical family of 18 genera and about 500 species, its greatest development in tropical Africa.

Flowers aggregated into globose, conelike heads; stamens 5; fruit scale-like, less than 8 mm long; leaves pubescent .... *Conocarpus*
Flowers in racemes or spikes, not congested into heads; stamens 10; fruit drupaceous or winged, 1–6 cm long; leaves glabrous or essentially so:
  Petals 5, suborbicular; fruit 10–15 mm long, 10-ribbed; leaves opposite, 2–7 cm long .... *Laguncularia*
  Petals none; fruit compressed-ellipsoid, strongly ridged along lateral margins; leaves alternate, 10–30 cm long .... *Terminalia*

---

* Contributed by Duncan M. Porter.

Conocarpus L., Sp. Pl. 176. 1753; Gen. Pl. ed. 5, 81. 1754

Shrubs or small trees; stems erect to prostrate, glabrous to densely silvery-tomentose. Leaves alternate, with raised basal glands, acute at apex, acuminate or apiculate, rarely emarginate, usually with several small pits near vein angles beneath, persistent; petiole short or none. Flowers small, 5-merous, densely aggregated into globose sessile or pedunculate heads borne in small terminal panicles or racemes; bracts ovate, concave, acuminate. Hypanthium greenish white, the lower part laterally flattened and more or less 2-winged, the upper cupuliform, more or less persistent; calyx lobes 5, valvate. Petals none. Stamens 5(10), sometimes fewer by abortion, exserted anthers cordate, versatile. Intrastaminal disk fleshy, 5-lobed, surrounding base of style. Ovary inferior; ovules 2 per locule; style subulate; stigma punctiform. Pseudocarps small, coriaceous, obcordate, laterally compressed, 2-winged, scale-like, densely aggregated into a conelike subglobose head. Endosperm none; embryo completely filling seed-cavity; testa membranaceous; cotyledons convolute.

Two species, one widely distributed in tropical and subtropical America and tropical West Africa, the other in Somaliland and Arabia.

Conocarpus erecta L., Sp. Pl. 176. 1753

Conocarpus procumbens L., op. cit. 177.
Conocarpus supinus Crantz, Inst. Rei Herb. 1: 355. 1766.
Conocarpus acutifolia Humb. & Bonpl. ex Roem. & Schult., Syst. Veg. 5: 574. 1819.
Conocarpus erecta var. *arborea* DC., Prodr. 3: 16. 1828.
Conocarpus erecta var. *procumbens* DC., loc. cit.
Conocarpus erecta var. *sericea* Forst. ex DC., loc. cit.
Conocarpus pubescens K. Schum., Kongl. Dansk. Vid. Selsk. Naturvid. Math. Afh. 3: 135. 1828.
Conocarpus sericea G. Don, Gen. Syst. 2: 662. 1832.
Terminalia erecta Baill., Hist. Pl. 6: 266. 1877.

Shrub or tree 3–8 m tall, the bark thin, gray or brown, fissured, the crown rounded, spreading; branchlets slightly winged, brown or reddish, appressed-pubescent, younger parts tomentose; leaves oblanceolate or oval to narrowly elliptic, 2.5–9 cm long, 1–3 cm wide, entire to repand, subcoriaceous, rugose, more or less sericeous, especially on midrib, this prominent beneath, the base usually decurrent; petioles pubescent; flowers inconspicuous, about 1.5 mm across, aggregated into subglobose tomentose heads about 5 mm broad; peduncles tomentose, 2–10 mm long; bracts small, tomentose; hypanthium tubular, tomentose; calyx cupuliform, more or less pubescent abaxially, 2 mm long, calyx lobes triangular, about 1 mm long; filaments subulate; disk pilose; ovary tomentose, 1 mm high; style 2 mm long, exserted, persistent; pseudocarps densely imbricated into brownish subglobose head 8–16 mm broad, dry, brown, tomentose, recurved at apex, bearing persistent calyx, 3–3.5 mm long, separating at maturity; inflorescence to 7.5 cm long in fruit.

Widely distributed along the shores of tropical America from Florida and Mexico to Ecuador and Brazil; tropical West Africa. Specimens examined: Isabela, San Cristóbal, San Salvador, Santa Cruz.

Collections of this species from the Galápagos Islands vary in leaf pubescence toward what has been called *C. erecta* var. *sericea*. However, populations elsewhere are known in which this character is variable (Standley, 1924; Stern, 1958; Graham, 1964). Degrees of difference in pubescence do not seem adequate for recognizing varieties in this group.

See Fig. 145.

Fig. 145. *Conocarpus erecta* (*a*). *Laguncularia racemosa* (*b, c*).

Laguncularia Gaertn. f., Fruct. Sem. Pl. 3: 209. 1807

Shrubs or small trees, often with small, pencil-like pneumatophores. Leaves opposite, thick, coriaceous, oblong to obovate, glabrous, with numerous submarginal pits, the apex persistent and usually obtuse or retuse; petioles short, conspicuously 2-glandular at apex. Flowers 5-merous, fragrant, sessile, bisexual or a few staminate, in axillary or terminal elongated paniculate spikes; bracts deciduous. Hypanthium funnel-shaped, fleshy, scarcely produced beyond ovary, slightly ribbed, cinereous to glabrous, 2-bracteolate near apex, elongating as fruit develops, crowned by persistent calyx; calyx lobes 5, broadly triangular, imbricate. Petals 5, small, caducous. Stamens 10; filaments subulate, inserted in 2 whorls, scarcely exserted; anthers cordate. Disk 10-lobed, thick, flattened, surrounding base of style. Ovary inferior; ovules 2, collateral; style subulate; stigma capitate, 2-lobed. Pseudocarp almost sessile, ellipsoid to obovoid, coriaceous, laterally flattened, longitudinally ribbed or narrowly winged with 2 thick spongy lateral ribs, densely cinereous to glabrous, crowned by persistent calyx. Endosperm none; embryo filling seed cavity; testa membranaceous, reddish; cotyledons convolute, green.

One or two species found in the mangrove swamps of tropical and subtropical America and tropical West Africa.

Laguncularia racemosa (L.) Gaertn. f., Fruct. Sem. Pl. 3: 209. 1807

Conocarpus racemosa L., Syst. Nat. ed. 10, 2: 930. 1759.
Schousboa commutata Spreng., Syst. Veg. 2: 332. 1825.
Rhizaeris alba Raf., Sylva Tellur. 90. 1838.
Laguncularia obovata Miquel, Linnaea 18: 752. 1844.
Laguncularia racemosa f. *longifolia* Macbr., Field Mus. Nat. Hist. Publ. Bot. Ser. 8: 125. 1930.

Low spreading shrub or small tree 2–10 m high, the crown rounded; branchlets terete, reddish brown becoming brown with age, the nodes slightly swollen; leaves oblong to oval, the apex rounded to retuse, the base rounded to cuneate, entire or slightly undulate, decussate, spreading; midrib prominent, 55–92 mm long, 21–47 mm wide; petioles reddish, stout, glabrous, 6–13 mm long; flowers whitish, campanulate; spikes densely appressed-tomentose, 7–8 cm long; pedicels tomentose; bracts ovate, concave, tomentose, 2 mm long, about 1 mm wide; hypanthium whitish, tomentulose, 3–4 mm long; calyx lobes whitish, tomentose, 2 mm long, persistent; petals white-tomentulose, equaling calyx lobes; filaments inserted on hypanthium above disk, 1 mm long; anthers apiculate, versatile; disk fleshy; style 1 mm long, slenderly conical, persistent; stigma minutely bilobed; pseudocarp sessile to short-pedicellate, flask-shaped, cinereous, more or less ribbed, gray-green, 15–21 mm long, 7–9 mm wide.

Along the shores of tropical America from Mexico and Florida to Brazil and Peru; tropical West Africa. Specimens examined: Bartolomé, Fernandina, Isabela, Pinta, Pinzón, San Cristóbal, San Salvador, Santa Cruz, Santa María.

Laguncularia glabrifolia Presl, Rel. Haenk. 2: 22. 1831, described from specimens collected at Guayaquil, Ecuador, is doubtfully distinct from *L. racemosa*.

See Fig. 145 and color plates 8–10 and 14.

Terminalia L., Mant. Pl. 1: 21. 1771

Trees; crown dense, rounded in mature trees. Leaves alternate, crowded at ends of branches, deciduous, petiolate. Flowers perfect or polygamous, small, green. Calyx tube of perfect flowers narrowed just above ovary, limb campanulate, 5-dentate. Petals none. Stamens 10, exserted. Fruit ovoid to ellipsoid, compressed or winged, 1-seeded, drupaceous or thin and leathery with a narrow to broad wing; flesh of mesocarp intermeshed with tough fibers. Seed woody or bony, often enwrapped with the persistent fibers from mesocarp.

About 200 species in the tropics of the Old and New Worlds, with largest representation in the Indo-Malaysian and African areas.

Terminalia catappa L., Mant. Pl. 2: 219. 1771

Tree to 25 m tall with conspicuously whorled branches; leaves broadly obovate, 10–30 cm long, 6–15 cm wide, dark green, shining, apiculate to emarginate at apex, strongly 8–12-veined on each side of midrib; petioles stout, 8–22 mm long, minutely brown-puberulent, at least in youth; spikes 1 to several at tips of branchlets, 10–25 cm long, androgynous, minutely bracteate, the rachis puberulent; perfect flowers only at base of inflorescence, their calyx tubes distinctly narrowed just above ovary, puberulent, the limb campanulate, 4–5 mm wide, deltoid teeth 1–1.5 mm long and wide, strigulose without; calyx tube of staminate flowers elongated at anthesis; stamens 1.5–1.8 mm long, erect; anthers 0.5–0.6 mm wide, slightly shorter than wide; staminate flowers globose in bud, about 1.5 mm wide in bud, broadly campanulate,

2 mm wide in anthesis, on slender pedicels 1.5–2.5 mm long; fruit greenish, often with rosy flush when ripe, ellipsoid, moderately compressed, 2–6 cm long, 2–3 cm wide, 1–2 cm thick, narrowly winged or markedly 2-edged along lateral margins, acute to apiculate apically, the mesocarp firm to leathery, fibrous, astringent; endocarp stony, surrounded by mesh of fibers from mesocarp; seed edible.

Along streams, on moderate slopes, and in lowlands near the coast, native of Indo-Malaysian and Polynesian areas and widely planted in nearly all tropical regions for shade and for its edible seeds. In the Galápagos cultivated at Bella Vista, Isla Santa Cruz. It can be expected elsewhere in the Galápagos Islands, particularly close to the coast near habitations.

See Fig. 146.

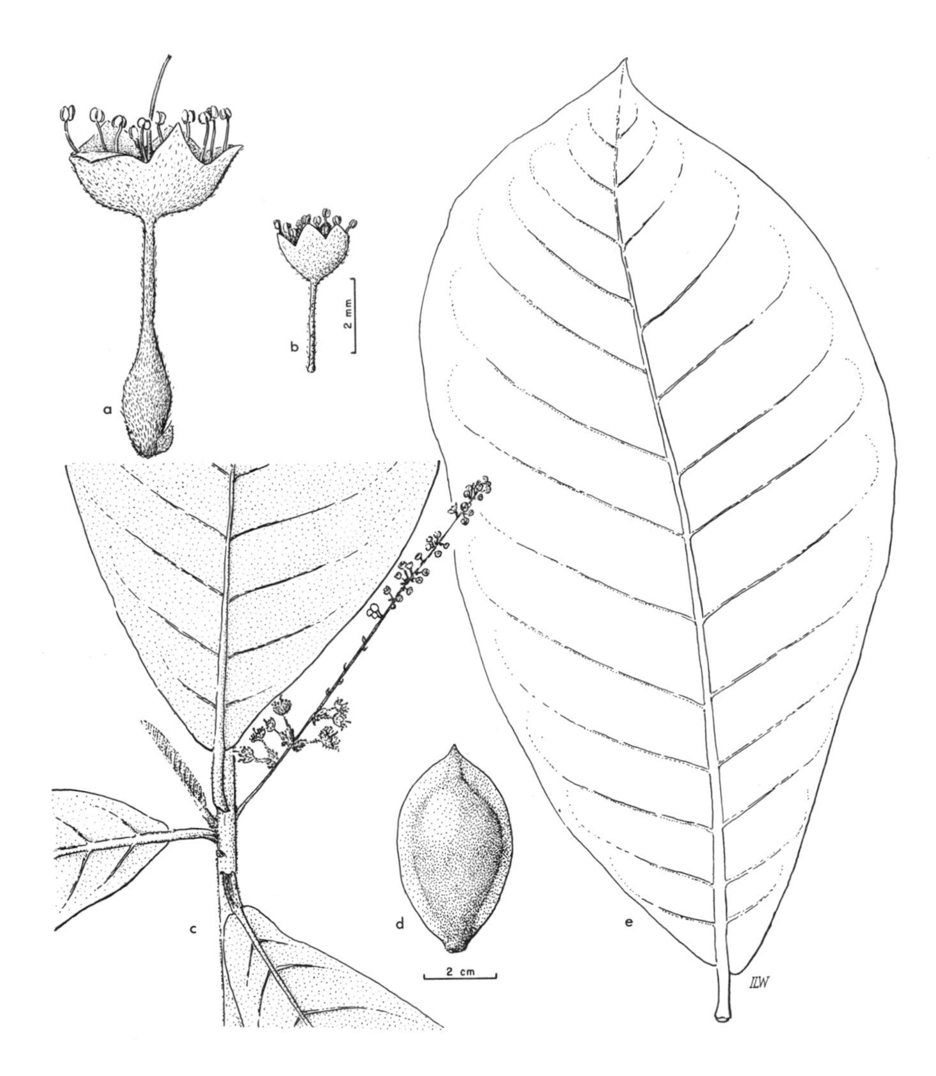

Fig. 146. *Terminalia catappa.*

CRASSULACEAE. ORPINE FAMILY*

Annual to perennial succulent herbs or subshrubs. Leaves alternate, opposite, or whorled, exstipulate, simple or pinnately compound. Flowers mostly perfect, mostly regular, subperigynous, 3–35-merous. Sepals 3–35, distinct or united, persistent. Petals as many, distinct or united, mostly persistent. Stamens as many or twice as many as petals, free or inserted on corolla, the anthers 2-celled, dehiscing longitudinally. Pistils mostly as many as and opposite petals, mostly distinct or nearly so, each 1-celled, with mostly several to many ovules on 2 adaxial placentae. Nectar glands usually present, 1 at base of each pistil. Fruit follicular, dehiscing ventrally. Seeds small; embryo straight.

About 36 genera and 1,200 species, especially widespread in temperate and subtropical regions of the Northern Hemisphere and Africa.

Kalanchoe Adans., Fam. Pl. 2: 248. 1763

Perennial or monocarpic succulent herbs or subshrubs. Leaves mostly opposite or whorled or some near apex alternate, mostly petiolate, simple or pinnate, the base with several vascular bundles, often broad, the margins laciniate to entire (in Sect. *Bryophyllum* often bulbiferous). Inflorescence mostly a terminal thyrse or cyme, with biparous or uniparous branches. Flowers 4-merous. Calyx tubular at base and sometimes inflated, or sepals free. Corolla tubular, commonly constricted above ovaries, the tube and throat together mostly exceeding lobes, these convolute in bud. Stamens 8, inserted in 2 series on tube or throat, rarely 4 sterile. Nectar glands semiorbicular to linear. Pistils slender, multiovulate. x = 17, 18, 20.†

A tropical and subtropical genus of perhaps 125 species, native mostly in Africa and especially Madagascar.

Kalanchoe pinnata (Lam.) Pers., Syn. Pl. 1: 446. 1805

Cotyledon pinnata Lam., Encycl. 2: 141. 1786.
Crassuvia floripendia Comm. ex Lam., loc. cit.
Bryophyllum calycinum Salisb., Parad. Londin. 1: *pl.* 3. 1805.
Bryophyllum pinnatum Oken, Allg. Naturg. 3: 1966. 1841.‡

Perennial, glabrous except for corolla; stems mostly erect, simple or offsetting at base, terete or subquadrangular, 0.5–2 m high, to 2 cm thick, becoming woody below, often almost bare at anthesis; leaves opposite, the lower simple, the middle mostly pinnate with 3–5 leaflets, the petioles 2–10 cm long, amplexicaul, the blades or leaflets elliptic, obtuse, 5–20 cm long, 2–12 cm wide, the margins crenate, bulbiferous; inflorescence a paniculate cyme 1–8 dm long; flowers pendent on pedicels 10–25 mm long; calyx cylindric, inflated, papery, pale green to light yellow, often marked with red, 2.5–4.5 cm long, the lobes deltoid, 6–12 mm long; corolla 3–6 cm long, sparsely glandular-pubescent, exserted part maroon, the tube subglobular, about 1 cm long, with 4 prominent and 8 lower longitudinal ridges, the throat subquadrangular, about 3 mm wide below and 10 mm above, lobes triangular-lanceolate, acuminate, 1–2 cm long; filaments greenish, 2.5–4 cm long; nectar glands

* Contributed by Reid Moran.
† Counts by Baldwin (1938); Uhl (1948).
‡ For additional synonymy, see Hamlet, Bull. Herb. Boiss. ser. 2, 8: 21–22. 1908.

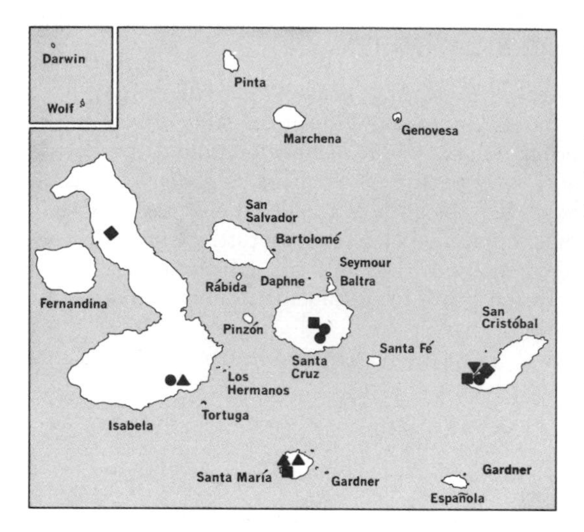

oblong, 2–2.5 mm long; pistils green, 3–4 cm long, styles 2–3.5 times exceeding ovaries, the ovules about 500; seeds oblong, about 0.8 mm long, obscurely striate longitudinally. n = 20.*

A weedy plant in open rocky places; pantropical, probably originally from Africa or Madagascar. Specimens examined: Isabela, San Cristóbal, Santa Cruz.

Reported under the name *Crassuvia floripendia* Comm. by Stewart as occurring at "Wreck Bay, around habitations," in 1905–6, but apparently he collected no specimens (cf. Stewart 1911, 68).

See Fig. 147.

CRUCIFERAE. Mustard Family†

Herbs, infrequently suffrutescent, with watery juice that is mostly pungent. Leaves alternate (rarely opposite), entire to lobed or pinnately divided and without stipules. Flowers bisexual, usually tetradynamous, mostly ebracteate in terminal racemes, infrequently solitary and pedunculate. Sepals 4, deciduous, usually oblong, erect and appressed to corolla or spreading at anthesis; petals 4 (rarely absent), hypogynous, entire or emarginate, rarely lobed or fimbriate, yellow, white, or lavender. Stamens 6 (rarely fewer or more) in 2 whorls, the outer single stamens 2, the inner paired stamens 4. Ovary superior with 2 locules (rarely 1). Fruit a dry, usually dehiscent silique with a wide range of shapes from narrowly linear to depressed-globose. Seeds without endosperm. Cotyledons with radicle usually folded retrorsely along their margins (accumbent) or along back of 1 cotyledon (incumbent). Embryo rarely straight.

About 375 genera and over 3,000 species, nearly cosmopolitan, but mostly in the temperate and cold parts of the world.‡

* Count by Baldwin (1938).

† Contributed by Reed C. Rollins.

‡ Only species introduced from elsewhere or escaped from cultivation occur in the Galápagos Islands. Only one species in each of the genera in the key has been found growing without cultivation.

Siliques linear with a gradually tapering beak:
  Flowers yellow; siliques dehiscent by valves . . . . *Brassica*
  Flowers lavender to whitish; siliques indehiscent . . . . *Raphanus*
Siliques orbicular or didymous:
  Siliques didymous; valves rugose, not strongly flattened; leaves pinnate to bipinnately lobed . . . . *Coronopus*
  Siliques orbicular; valves plane, strongly flattened at right angles to septum; leaves entire to dentate . . . . *Lepidium*

Brassica L., Sp. Pl. 666. 1753; Gen. Pl. ed. 5, 299. 1754

Annual, biennial, or sometimes perennial herbs with erect or ascendingly branched stems and alternate leaves, the basal pinnatifid or pinnately divided, the upper toothed or entire and frequently sessile. Inflorescences racemose, more or less elon-

Fig. 147. *Kalanchoe pinnata.*

gate, especially in fruit. Sepals slightly dimorphic, the lateral slightly gibbous at base. Petals yellow, large, the claw abruptly expanding into a broad blade. Nectary glands 4, papilla-like, green, episepalous. Silique terete or slightly angled, terminating in stoutish, subulate beak; valves 1–3-nerved. Stigmas truncate or bilobed. Seeds in 1 row in each cell, globose. Cotyledons conduplicate, incumbent.

Widely distributed in tropical to temperate zones.

Brassica campestris L., Sp. Pl. 666. 1753

Annual or biennial; stems erect, 3–10 dm tall, glabrous and somewhat glaucous, or very slightly hirsute below, especially when young; basal leaves lyrate-pinnatifid to dentate, long-petioled, often with a large terminal lobe; upper leaves reduced, lanceolate to linear, mostly entire or sparingly dentate, auriculate and usually clasp-

Fig. 148. *Brassica campestris.*

ing stem; sepals erect; petals yellow, 6–10 mm long; pedicels divaricately ascending, straight, 1–3 cm long; siliques erect to somewhat divaricate, 4–7 cm long, about 3 mm thick, often constricted between seeds, the beak 1–2 cm long; seeds spheroidal to slightly compressed laterally, reddish gray, 1.5–2 mm in diameter; cotyledons conduplicate.

In cleared fields, on waste ground, and around habitations from tropics to cool temperate zones in the Old and New Worlds. Specimens examined: San Cristóbal, Santa Cruz, Santa María.

A second species, commonly referred to as *Brassica* [*B. kaber* (DC.) Wheeler], was reported by Andersson under the name *Sinapis arvensis* L. We have not seen specimens of this species from the Galápagos Islands, and it seems possible that the material is referable to *B. campestris*. There is no reason why a second weedy species should not be on the islands, but it should have been re-collected by now if it is actually growing there.

See Fig. 148.

Coronopus Zinn, Cat. Pl. Gött. 325. 1757

Prostrate, heavy-scented annuals, with pinnatifid leaves and short racemes of minute greenish white flowers. Sepals oval, spreading. Stamens 6 or sometimes only 4, less commonly 2. Silicle small, more or less didymous, flattened at right angles to the narrow partition, the surface tuberculate or strongly wrinkled; valves falling as closed or nearly closed nutlets at maturity. Cotyledons incumbent.

A small genus of wide distribution, mostly as an introduced weed.

Coronopus didymus (L.) J. E. Smith, Fl. Brit. 2: 691. 1800

Lepidium didymum L., Mont. Pl. 92. 1767.
Senebiera pinnatifida DC., Mem. Soc. Nat. Hist. Paris 7: 144, *t*. 9. 1799.
Senebiera didyma Pers., Syn. Pl. 2: 185. 1807.

Prostrate herb with numerous branches, sparsely pubescent with simple trichomes, the outer branches 1–4 dm long; leaves monomorphic, bipinnately dissected, the basal oblong in outline and with margined petioles, 3–10 cm long, the cauline similar but with shorter petioles; flowering racemes axillary in leafy bracts; petals white, minute, about 1 mm long; stamens 2; silicles didymous, 2–2.5 mm broad, finely wrinkled; valves indehiscent; styles wanting; seeds cochleate, 1–1.4 mm long; cotyledons incumbent and transversely folded.

Weed of disturbed soils and open places; introduced originally from Europe, probably first to the mainland of South America where it is widespread and abundant. Specimens examined: Isabela, Santa María.

See Fig. 149.

Lepidium L., Sp. Pl. 643. 1753; Gen. Pl. ed. 5, 291. 1754

Erect, decumbent, or diffuse annuals, biennials, or perennials with pinnatifid or lobed lower leaves and reduced, often entire, upper leaves. Flowers small, white, yellowish, or greenish, arranged in terminal, compact to loose racemes, these elongate in fruit. Sepals obtuse or rounded at apex, often thin and subscarious at margins, caducous. Petals small, sometimes wanting. Stamens 6, 4, or 2. Styles slender, half as long as ovary or shorter, or stigma sessile. Fruit a silicle, orbicular, oblong, or obovate, flattened contrary to partition, more or less emarginate and often winged at apex, the valves separating from septum at dehiscence. Seeds 1 in each cell, somewhat flattened; cotyledons incumbent or less commonly oblique or accumbent.

A genus with representatives in both the Old and the New Worlds, ranging from cool temperate to tropical regions.

Lepidium virginicum L., Sp. Pl. 645. 1753

Erect to slightly decumbent annual or biennial, pubescent with simple trichomes particularly on stems and pedicels; stems usually single from base, branched above, up to 6 dm tall; basal leaves usually pinnatifid, 5–15 cm long, 2–5 cm wide, lower cauline leaves similar to basal, the cauline reduced upward, the upper linear to linear-lanceolate, entire or nearly so; sepals scarious-margined, usually caducous;

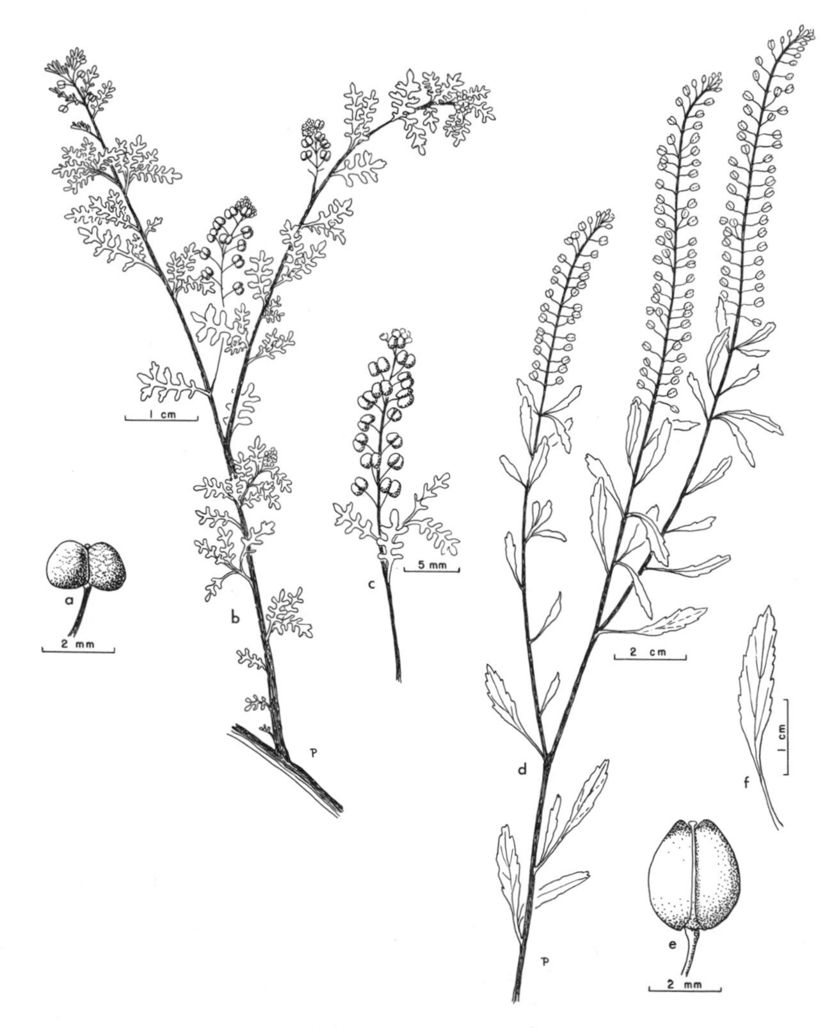

Fig. 149. *Coronopus didymus* (*a–c*). *Lepidium virginicum* (*d–f*).

petals white, obovate-spatulate, 2–3 times longer than sepals; stamens usually 2, rarely 4 or 6; fruiting racemes dense, usually elongated; pedicels slender, straight, divaricately ascending to nearly horizontal, pubescent on upper surface; silicles nearly orbicular to broadly ovate-elliptical, 2.5–4 mm long, glabrous, slightly winged distally, the notch shallow; styles nearly obsolete; seeds narrowly obovate to oval, flattened, narrowly winged; cotyledons accumbent.

Beside roads and trails, and in disturbed areas; native to eastern North America; in South America known from Guiana, Venezuela, and Brazil. Specimens examined: Isabela, San Cristóbal.

Hitchcock (1945) raised the possibility that the specimens of *Lepidium* from the Galápagos represented an undescribed species or were conspecific with *L. horstii* from the island of San Ambrosio. Nevertheless, he placed them in *L. virginicum*, and this appears to be the appropriate determination of the material.

See Fig. 149.

Raphanus L., Sp. Pl. 669. 1753; Gen. Pl. ed. 5, 300. 1754

Erect or much-branched annual or biennial herbs with lyrate leaves and showy flowers. Pods elongated, linear, fleshy or corky, constricted or continuous and spongy between seeds, indehiscent, tapering above into a persistent slender style. Seeds subglobose; cotyledons conduplicate.

A genus of 3 or 4 species, native in Europe and temperate Asia.

Raphanus sativus L., Sp. Pl. 669. 1753

Annual herb, often with fleshy root; stems stout, erect, much-branched above, 3–6 dm tall; lower leaves lyrate-pinnatifid, petiolate, sparsely hirsute to glabrous, the lobes dentate or denticulate, the leaves monomorphic, reduced upward, racemes elongate; sepals erect, the lateral pair slightly saccate, sparsely hirsute to glabrous; petals purplish to nearly white, with dark purple veins, 1.5–2 cm long, unguiculate, obovate with a long narrowed claw; pedicels straight, divaricately ascending, 1–2 cm long; siliques indehiscent, cylindrical to conical, 6–10 mm in maximum diameter, tapering from above base to a narrow beak, spongy within, lower member abortive; seeds subglobose, somewhat ridged, 3–3.5 mm in diameter; cotyledons conduplicate.

In waste places and disturbed ground; a native of Europe, a weed in both Old and New Worlds; escaped from cultivation. Specimens examined: San Cristóbal.

See Fig. 150.

DILLENIACEAE. DILLENIA FAMILY*

Scandent or upright shrubs and tall trees with simple, alternate, exstipulate leaves. Flowers paniculate, terminal, or axillary, perfect or polygamous. Sepals 3–5, distinct or nearly so, imbricate in bud, persistent. Petals usually 5, crumpled in bud, ample, broad-clawed. Stamens numerous, hypogynous in several ranks, distinct or basally fasciculate; anthers crowded in bud, the connective apiculate; thecae dehiscing longitudinally. Gynoecium of several to many 1-celled carpels attached to

* Contributed by Ira L. Wiggins.

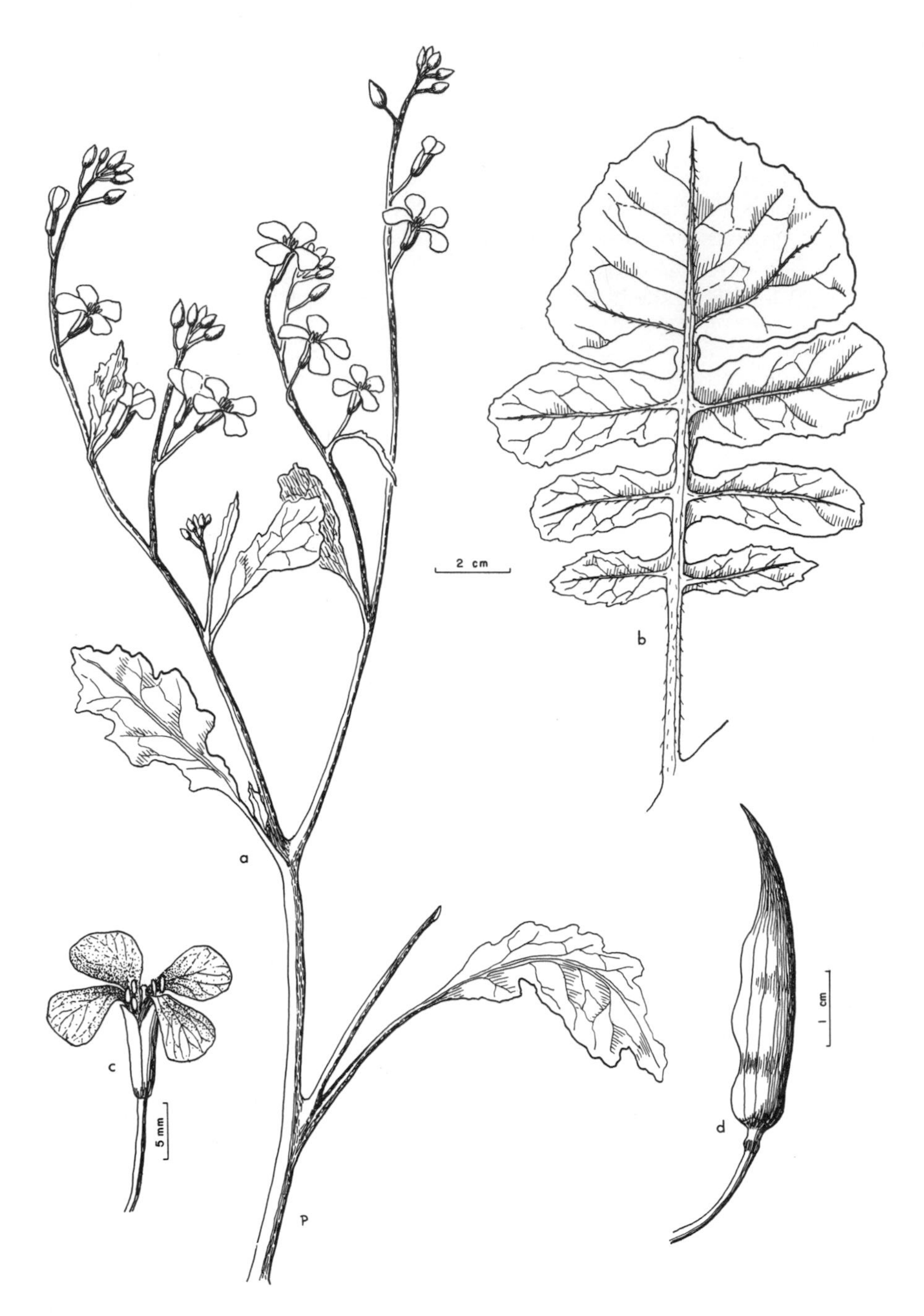

Fig. 150. *Raphanus sativus*.

central column, free laterally, many-ovuled; stigmas as many as carpels, elongate, spreading. Fruit a follicle or berry-like, the cavity full of hairs attached to back of seeds at maturity. Endosperm copious, fleshy; embryo minute.

Some 11 genera and 275 species, mostly Australasian and in tropical America.

Dillenia L., Sp. Pl. 535. 1753; Gen. Pl. ed. 5, 239. 1754

Trees or rarely shrubs. Leaves penninerved, the lateral veins close and parallel. Buds globose or compressed-subglobose, very large. Sepals 5 or sometimes 10–15, hard and leathery, very thick at base. Petals 5, white or yellow. Anthers linear, basifixed. Carpels 5–20. Seeds usually with aril around hilum, the back long-villous.

Fifteen to 20 species from the Philippines to Australia and through South Asia.

Dillenia indica L., Sp. Pl. 535. 1753

Dillenia speciosa Thunb., Trans. Linn. Soc. Lond. 1: 200. 1791.

Tree to 15 m tall or more, with dense, rounded, spreading crown and sturdy trunk 2–4 dm thick; leaves obovate, 1–3 dm long, serrate, a lateral vein running to each

Fig. 151. *Dillenia indica.*

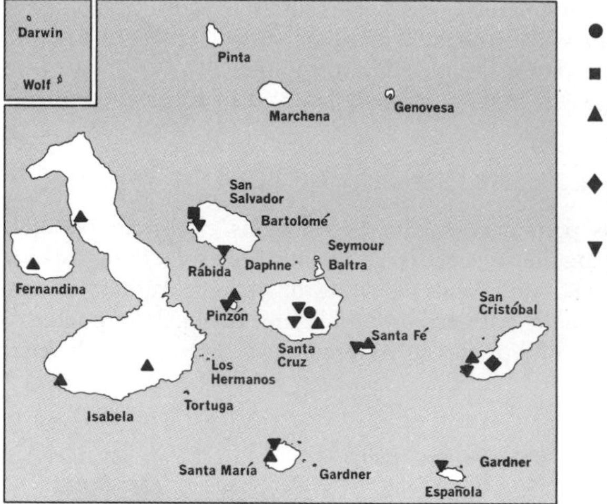

● *Dillenia indica*
■ *Acalypha flaccida*
▲ *Acalypha parvula* var. *parvula*
◆ *Acalypha parvula* var. *chathamensis*
▼ *Acalypha parvula* var. *reniformis*

tooth; buds 12–16 cm thick, three-fourths as long, sepals tightly imbricated, rounded, thick at base; petals broadly obovate, about twice as long as sepals, broad-clawed; carpels usually 20.

The East Indies and Malaysia; introduced on Isla Santa Cruz in the Galápagos.

Locally called Leña de Indigo. Introduced by Jacob Horneman, of Academy Bay, in the mid-1930's and doing well in 1964. According to his wife, the fruit can be used to make jelly. The odor of the buds just before opening is rather unpleasant.

See Fig. 151.

EUPHORBIACEAE. Spurge Family*

Trees, shrubs, herbs, or vines, the stems sometimes succulent or with milky sap; leaves alternate (rarely opposite or whorled), petiolate, usually stipulate. Flowers unisexual inflorescences monoecious or dioecious, cymose, racemose, spicate, or pseudanthial. Perianth usually actinomorphic; calyx mostly of 3–6 imbricate or valvate lobes or segments; corolla of 3–6 petals, often reduced or absent; glandular disk usually present and sometimes conspicuous. Stamens mostly 3–20 (rarely reduced to 1 or up to 400), filaments free or united, pollen grains mostly 3- or 4-colporate, sometimes porate or inaperturate. Gynoecium of (1)3 or 4 (10) united carpels; ovary superior, with axile placentation; ovules 1 or 2 per locule, mostly anatropous, obturate. Fruits usually capsular and elastically dehiscent, or sometimes drupaceous; seeds usually with endosperm, the embryo large and typically with broad cotyledons.

A diversified family of more than 300 genera and about 7,000 species, largely tropical in distribution.

Ovules paired in each locule of ovary; sap never milky; leaves entire, without glands; petals absent; stamens 3 . . . . *Phyllanthus*
Ovules solitary in each locule; sap milky or colored or else foliar glands present or styles dissected:

* All except *Chamaesyce* and *Euphorbia* contributed by G. L. Webster.

Flowers not in pseudanthia; leaves alternate:
Petals present in staminate flowers, the calyx not petaloid:
Anthers erect (not deflexed) in bud; trichomes unbranched. . . . . *Jatropha*
Anthers deflexed in bud; trichomes (at least in part) stellate. . . . . *Croton*
Petals absent:
Stamens fasciculate and branched, the androecium with hundreds of anthers; leaves peltate, palmately lobed, with glandular petiole. *Ricinus*
Stamens unbranched, the androecium with 10 anthers or less; leaves not peltate nor with glandular petiole:
Calyx petaloid; inflorescences obviously cymose; stamens 10. . . . . *Manihot*
Calyx not petaloid; inflorescences not cymose; stamens fewer:
Floral bracts without conspicuous basal glands; bracts of pistillate flowers much larger than staminate; styles dissected; fruits capsular, 3-locular. . *Acalypha*
Floral bracts with conspicuous basal glands; bracts of pistillate flowers not conspicuously larger than staminate; styles unbranched; fruits drupaceous, 6–9-locular . . . . *Hippomane*
Flowers in pseudanthia; leaves opposite; stems with milky latex:
Leaves whorled, reduced to scales or caducous; cyathial glands 5. . . . . *Euphorbia*
Leaves opposite, expanded, persistent; cyathial glands 4. . . . . *Chamaesyce*

Acalypha L., Sp. Pl. 1003. 1753; Gen. Pl. ed. 5, 436. 1754

Herbs, shrubs, or trees; sap watery, stems and foliage glabrous or with simple hairs or glands. Leaves alternate, unlobed, entire or toothed, petiolate, pinnately or palmately veined, stipulate. Inflorescences spicate, terminal or axillary; spikes unisexual or bisexual; staminate flowers subtended by minute bracts, pistillate flowers (in most species) subtended by large foliaceous bracts. Staminate flowers with calyx splitting into 4 segments; petals and disk absent; stamens 4–8, filaments free or basally connate, anther-sacs pendent, unilocular, vermiform; rudimentary ovary absent. Pistillate flowers subsessile; sepals 3–5, imbricate; petals and disk absent; ovary (1–2)3-carpeled, the styles usually lacerate (rarely bifid or entire), ovules solitary in each locule. Fruits capsular, sometimes surrounded by the accrescent pistillate bract; seeds smooth to pitted or tuberculate, usually caruncolate, the endosperm copious.

About 400 species of tropical and warm temperate regions, the majority American. The Galápagos species, all endemic, form a difficult and imperfectly understood complex. Much careful field work is still needed. No specimens have yet been collected from Isla Genovesa, where some species are likely to occur, though the genus may be truly absent from the isolated islets Darwin and Wolf.

Stalked glandular trichomes absent (rarely sparsely present on pistillate bracts); young leaves and stem tips densely villous or hirsute (usually with hairs over 0.5 mm long); peduncles of staminate spikes 1–8 mm long:
Pistillate bracts not over 2 mm long, sparsely hirsute but not hispidulous; spikes all axillary; leaf blades with 14–18 teeth per side. . . . . *A. flaccida*
Pistillate bracts well over 2 mm long, hispidulous and often hirsute as well:
Spikes all axillary, with solitary pistillate bracts; leaves with 8–15 teeth per side, distinctly bullate-rugose above . . . . *A. velutina*
Terminal spike of several pistillate bracts present; leaves with 10–30 teeth per side, slightly (if at all) rugose above:
Staminate spikes 30–60 mm long; stipules 1.5–2 mm long; styles 3–3.5 mm long. . . . . *A. sericea* var. *baurii*

Staminate spikes mostly much less than 30 mm long; stipules 0.5–1.5 mm long; styles 2–2.5 mm long:
Leaves with 17–26 teeth per side; seeds 1.2–1.3 mm long ....*A. sericea* var. *indefessus*
Leaves with 8–15 teeth per side; seeds 1–1.2 mm long.....*A. sericea* var. *sericea*
Stalked glandular trichomes present at least on pistillate bracts:
Staminate calyces and bracts smooth and glabrous; young stems densely villous; leaf blades mostly 3–7 cm long; stipules 1.7–3.5 mm long; spikes all axillary; seeds 1.3–1.5 mm long..*A. wigginsii*
Staminate calyces and bracts hispidulous (sometimes sparsely so); young stems not densely villous; leaf blades mostly 0.5–4.5 cm long; stipules 0.4–1 mm long; spikes terminal or axillary; seeds 0.7–1.6 mm long:
Seeds large, 1.2–1.6 mm long; stems usually densely glandular, rarely hirsute; spikes terminal and/or axillary; pistillate bracts subsessile or pedunculate; stems erect or spreading, usually densely glandular and not hirsute. .*A. parvula* var. *strobilifera*
Seeds small, 0.7–1.2 mm long; stems and bracts often hirsute:
Spikes usually terminal as well as axillary, pistillate bracts subsessile; staminate spikes short (0.5–2 mm long); stems often prostrate, usually not densely glandular but often hirsute; leaves typically cordate or subcordate, 1 cm long or less ....*A. parvula* var. *reniformis*
Spikes usually axillary, pistillate bracts often distinctly pedunculate; staminate spikes longer (mostly 2–10 mm or more), exserted far beyond pistillate bract; stems erect to prostrate, often densely glandular and/or hirsute; leaves rounded or truncate to cordate at base:
Stems erect, sparsely branched; larger leaves mostly 2.5–4.5 cm long; staminate spikes 6–15 mm long.....................*A. parvula* var. *chathamensis*
Stems erect or prostrate, sometimes diffusely branched; larger leaves mostly 0.5–2 cm long; staminate spikes mostly 2–10 mm long....*A. parvula* var. *parvula*

Acalypha flaccida Hook. f., Trans. Linn. Soc. Lond. 20: 186. 1847

Acalypha parvula η flaccida Muell.-Arg., Linnaea 34: 48. 1865.

Erect annual herb; stems recurved-puberulent and hirsute, densely hispid-tomentose at young tips, not glandular; leaves ovate, subacute, rounded to subcordate at base, about 2–4 cm long, 1.5–3 cm wide, with 14–18 bluntish teeth on a side, strigose-hirsute on both faces; petioles 10–20 mm long; stipules lanceolate, 0.5–0.8 mm long; spikes axillary, androgynous, with single basal pistillate bract; staminate spikes 2.5–5 mm long, with peduncles about 2 mm long; bracts subtending staminate flowers 0.4–5 mm long, ciliate; staminate calyces sparsely hispidulous; pistillate bracts 1.3–1.8 mm long, subsessile, cut about halfway into 3–6 lobes, sparsely hirsute but otherwise glabrous (neither hispidulous nor glandular); ovary hispid; styles laciniate, 2–2.5 mm long, hispid; seeds about 1.2 mm long.

Known only from the type collection, made by Darwin on Isla San Salvador.

Acalypha parvula Hook. f., Trans. Linn. Soc. Lond. 20: 185. 1847, var. **parvula**

Acalypha cordifolia Hook. f., op. cit. 186.
Acalypha diffusa Anderss., Kongl. Svensk. Vet.-Akad. Handl. 1853: 240. 1855.
Acalypha spicata Anderss., op. cit. 239.
Acalypha parvula β cordifolia Muell.-Arg., Linnaea 34: 47. 1865.
Acalypha parvula δ procumbens Muell.-Arg., op. cit. 48.
Acalypha albemarlensis Robins., Proc. Amer. Acad. 38: 163. 1902.

Annual; stems erect or spreading, puberulent with recurved hairs, usually distinctly glandular, sometimes hirsute; leaves ovate to elliptic, rather thin, often obtuse or blunt at tip, mostly truncate to shallowly cordate at base, 0.5–2.5(3) cm long, with 7–15(22) teeth per side; petioles 5–25 mm long; stipules ovate to lanceolate, 0.4–1 mm long; spikes usually axillary, solitary, androgynous, often pedunculate, with 1–5 pistillate bracts at base; staminate spikes 0.5–11 mm long, the peduncles 2–20(35) mm long; staminate bracts 0.3–0.6 mm long, usually hispidulous; staminate flowers hispidulous; pistillate bracts 1–3 per axillary spike, (2)3–5 mm long, hispidulous and glandular, often hirsute, the peduncles (0.5)2–10 mm long; pistillate flowers 2–3 per bract, ovary hispidulous and often hirsute as well; styles laciniate (rarely bifid or entire), 1.5–3(5) mm long, smooth or hispidulous; seeds finely pitted, 0.9–1.1(1.2) mm long.

Endemic; sea level to over 1,000 m, in various vegetation types. Specimens examined: Fernandina, Isabela, Pinzón, San Cristóbal, Santa Cruz, Santa Fé, Santa María.

Plants from Santa María are somewhat divergent and may prove not to belong here. *A. parvula* is the most widespread and undoubtedly the commonest Galápagos species of *Acalypha*, occurring from Fernandina to Española and San Cristóbal;

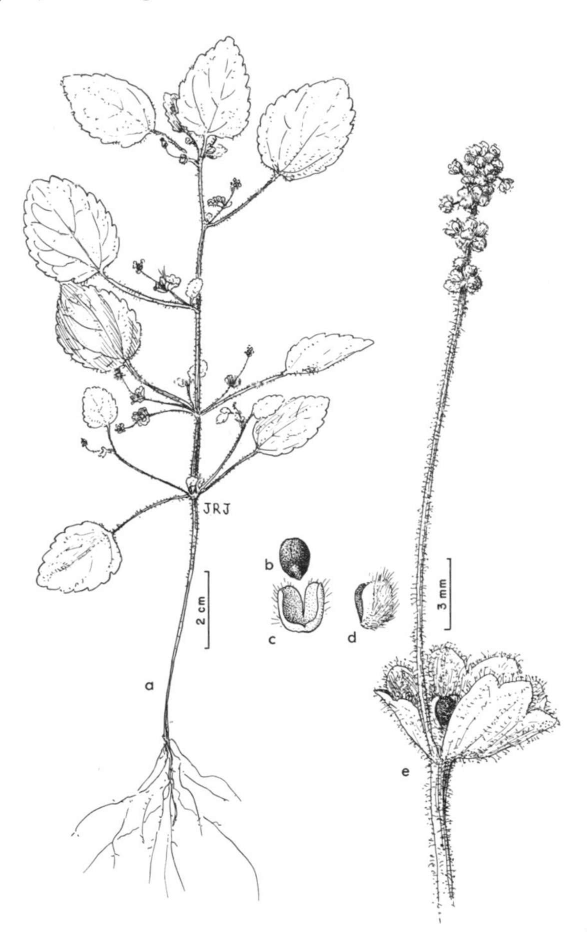

Fig. 152. *Acalypha parvula* var. *parvula*.

absent, however, from the more isolated northern islands (Genovesa, Marchena, Pinta, Wolf, and Darwin).

See Fig. 152.

Acalypha parvula var. **chathamensis** (Robins.) Webster, Madroño 20: 263. 1970

Acalypha chathamensis Robins., Proc. Amer. Acad. 38: 163. 1902.

Stems erect, slender, sparsely branched, 3–5 dm high, glandular, not hirsute; leaves thin, ovate, 2.5–4.5 cm long, with 8–15 teeth per side, acute at tip, cuneate to rounded at base, the petioles 15–35 mm long; staminate spikes (3)6–15 mm long, the peduncles 6–18 mm long; pistillate bracts 1–3 per axillary spike, (3)4–6 mm long, hirsute; peduncles (1)2–10 mm long; seeds 1–1.1 mm long. Characters otherwise as in var. *parvula*.

Endemic to Isla San Cristóbal, exact habitat unknown.

Acalypha parvula var. **reniformis** (Hook. f.) Muell.-Arg., Linnaea 34: 48. 1865

Acalypha reniformis Hook. f., Trans. Linn. Soc. Lond. 20: 186. 1847.
Acalypha adamsii Robins., Proc. Amer. Acad. 38: 161. 1902.

Stems usually spreading or prostrate, sometimes clustered from a perennial base, usually eglandular or sparsely glandular, sometimes hirsute; leaves mostly reniform to suborbicular, obtuse or rounded to acute at tip, subcordate to deeply cordate at base, mostly 0.5–1.5(2.5) cm long, with 6–12 teeth per side; petioles mostly 2–10(15) mm long; terminal pistillate spikes usually present; staminate spikes 0.5–4 mm long, the peduncles 0.5–7 mm long; pistillate bracts usually 2 to several per axil (less commonly solitary), 1.5–4 mm long, usually lobed halfway or more, often hirsute, subsessile or with peduncles to 2 (rarely to 7) mm long; seeds 0.7–1.1(1.2) mm long. Characters otherwise as in var. *parvula*.

Mostly at low altitudes (200 m and below) in xeric vegetation types. Specimens examined: Española, Pinzón, Rábida, San Cristóbal, San Salvador, Santa Cruz, Santa Fé, Santa María.

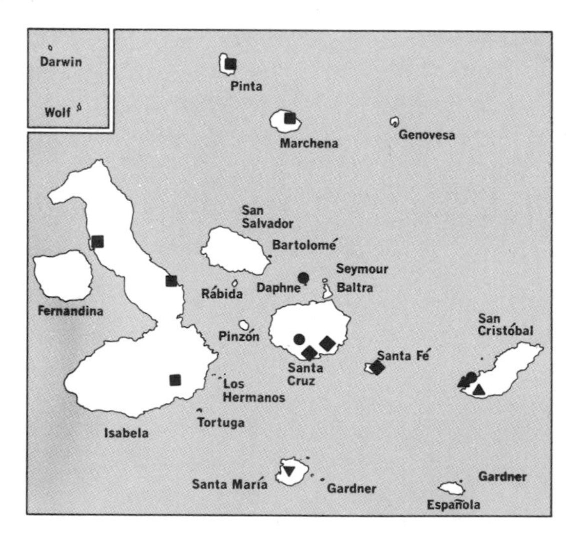

● *Acalypha parvula* var. *strobilifera*
■ *Acalypha sericea* var. *sericea*
▲ *Acalypha sericea* var. *baurii*
◆ *Acalypha sericea* var. *indefessus*
▼ *Acalypha velutina*

Acalypha parvula var. **strobilifera** (Hook. f.) Muell.-Arg., Linnaea 34: 47. 1865

Acalypha strobilifera Hook. f., Trans. Linn. Soc. Lond. 20: 187. 1847.

Stems erect, usually distinctly glandular but not hirsute; leaves ovate or elliptic, thin, often blunt at tip, obtuse to cordate at base, 0.7–2 cm long, with 7–15(18) teeth per side; petioles mostly 5–20 mm long; stipules 0.4–0.8 mm long; terminal pistillate spikes often present; staminate spikes (1)3–10(15) mm long, the peduncles (1)5–20(30) mm long; pistillate bracts 1 to several per axillary spike, most 4–6(7) mm long, often shallowly lobed, sometimes hirsute, the peduncles (0.5)3–18 mm long; seeds (1.2)1.3–1.6 mm long. Characters otherwise as in var. *parvula.*

At lower altitudes, often near sea level. Specimens examined: Daphne Major, San Cristóbal, Santa Cruz.

Acalypha sericea Anderss., Kongl. Svensk. Vet.-Akad. Handl. 1853: 238. 1855, var. **sericea**

Acalypha parvula γ pubescens a. *sericea* Muell.-Arg., Linnaea 34: 47. 1865.

Herb or subshrub to 1 or 2 m high; stems loosely recurved-puberulent to somewhat hirsute, densely villous-hirsute at young tip, not glandular; leaves mostly ovate, acute to short-acuminate, cuneate to rounded at base, mostly 1–2(3) cm long, with 8–15 bluntish teeth per side, strigose-hirsute above, villous or densely strigose beneath; petioles mostly 10–30 mm long; stipules lanceolate, 0.7–1.4 mm long; terminal pistillate spikes normally present; staminate spikes 3–15(30) mm long, the peduncles 3–10 mm long; bracts subtending staminate flowers 0.4–0.5 mm long, ciliate; staminate flowers sparsely hispidulous (sometimes nearly glabrous); pistillate bracts cut about halfway into 4–9 oblong lobes, hirsute, sometimes glandular, 3–5(6) mm long; pistillate flowers 1–3 per bract, the ovary hispidulous; styles laciniate, hispidulous, about 2–2.5 mm long; seeds 1–1.1 mm long.

Known from Isla Isabela only on the basis of Andersson's collection; also recorded from Marchena and Pinta.

Acalypha sericea var. **baurii** (Robins. & Greenm.) Webster, Madroño 20: 261. 1970

Acalypha baurii Robins. & Greenm., Amer. Jour. Sci. 50: 144. 1895.

Subshrub or shrub; leaves mostly 4–6 cm long, with 17–31 teeth per side; petioles 25–30 mm long; stipules 1.5–2 mm long; staminate spikes 30–60 mm long, the peduncles 2–8 mm long; pistillate bracts eglandular, about 5–6 mm long; styles 3–3.5 mm long; seeds about 1.2 mm long.

Known only from Baur's type collection on the southwest end of San Cristóbal.

Acalypha sericea var. **indefessus** Webster, Madroño 20: 261. 1970

Shrub; leaves mostly 2.5–5.5 cm long, 1.3–3.3 cm broad, with (14)17–26 teeth per side; petioles (7)10–30 mm long; stipules 0.6–1.8 mm long; staminate spikes 5–15 (22) mm long, the peduncles 1–5 mm long; pistillate bracts eglandular, 3.2–4.8 mm long; styles 1.8–2.5 mm long; seeds 1.2–1.3 mm long.

Endemic to Isla Santa Cruz, in the Transition Zone and lower *Scalesia* Zone, between 100 and 200 m.

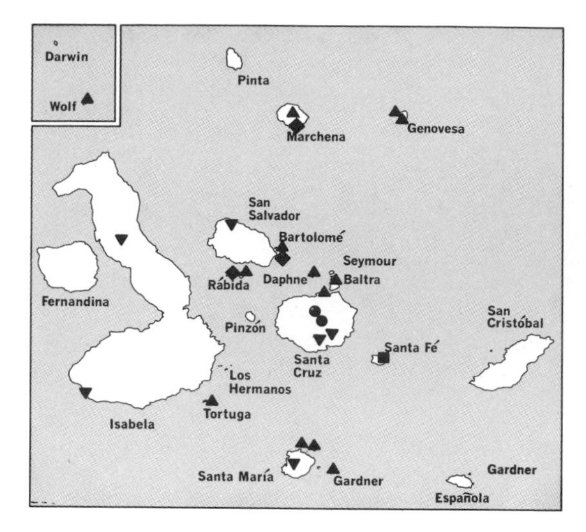

Acalypha velutina Hook. f., Trans. Linn. Soc. Lond. 20: 186. 1847

Acalypha velutina β minor Hook. f., op. cit. 187.
Acalypha parvula γ pubescens b. *velutina* Muell.-Arg., Linnaea 34: 48. 1865.

Herb with erect branches about 2–4 dm high; stems densely villous-hirsute or recurved-puberulent; leaves thick, rugose, ovate or elliptic, acute or obtuse, cuneate to rounded at base, 0.7–2 cm long, 0.5–1 cm broad, with 8–11 bluntish teeth per side, bullate and hispidulous or hirsute on both sides, veins prominently raised; petioles 2–10 mm long; stipules lanceolate, 1–1.8 mm long; spikes axillary, androgynous, each with single basal bract; staminate spikes 1.5–5 mm long, peduncles 2.5–5 mm long; staminate bracts 0.6–0.9 mm long, ciliate; staminate flowers hispidulous; pistillate bracts 2.4–3.2 mm long, bluntly 4–8-lobed, hispidulous and sparsely to distinctly hirsute at base; pistillate flowers 2 or 3 per bract, ovary hispidulous; styles laciniate, 2.5–3.5 mm long, copiously hispidulous; seeds 1–1.1 mm long.

Known only from Isla Santa María.

Acalypha wigginsii Webster, Madroño 20: 261. 1970

Erect herb about 2–4 dm high; stems villous with both long-pointed hairs and long-stalked glandular hairs; leaves ovate, pointed at tip, cordate at base, 2–7 cm long, 2–4.5 cm wide, 13–23 bluntish teeth per side, strigose-hirsute and glandular on both sides; petioles 10–50 mm long; stipules lanceolate-acuminate, sparsely hirsute, 1.7–3.5 mm long; spikes axillary, androgynous, each with 2–3 basal bracts; staminate spikes 1.5–9 mm long, peduncles 1–7 mm long; staminate bracts 0.4–0.6 mm long, glabrous; staminate flowers glabrous; pistillate bracts 4–5 mm long, cut about halfway or more into 6–10 pointed lobes, glandular and hispidulous, basally sparsely hirsute; pistillate flowers usually 2 per bract, ovary sparsely hirsute; styles laciniate, smooth or sparsely hispidulous, 2–3.5 mm long; seeds 1.3–1.5 mm long.

Known only from the central elevated region of Isla Santa Cruz, in the moist region above 350 m.

Chamaesyce S. F. Gray, Nat. Arr. Britt. Pl. 2: 260. 1821*

Herbs, subshrubs, or rarely shrubs, with milky latex in all parts; monoecious; main axis aborting above first pair of true leaves, secondary axes few to many, rarely rooting at nodes. Leaves opposite, simple, petiolate with interpetiolar stipules distinct or joined; blades expanded, with chlorophyll-bearing cells mainly in a sheath around veins and colorless areas between, the base inequilateral. Inflorescence a cyathium, terminal but appearing axillary, solitary at nodes or clustered in cymules, cup-shaped, more or less actinomorphic, the 5 lobes free only at their tips, the 4 glands (sometimes a vestigial fifth) at mouth of cup and alternating with lobes, the glands with petaloid appendages (these rarely obsolete). Staminate flowers few to many in 5 monochasia, naked, monandrous, each monochasium usually subtended by a bracteole fused for all or part of its length to cyathium and variously fringed or cut. Pistillate flowers terminal, solitary, naked or rarely with a vestigial calyx; ovary 3-celled, each cell with single ovule; styles 3, free or basally connate, partly bifid. Fruit a capsule, usually fully exserted from cyathium at maturity, columella persistent. Seeds ovoid, angled or terete, the coat smooth or variously sculptured, ecarunculate; endosperm copious; cotyledons fleshy, broader than radicle.

About 250 species, the vast majority of them in the New World.

Capsule glabrous:
  Leaves linear, clustered on short spurs; mature stems leafless on primary axes......... ....*C. viminea*
  Leaves linear-lanceolate or broader, mostly on laterals when stems mature, laterals not spurlike; mature stems bearing some leaves, at least near tips:
    Leaf blades broadly ovoid to orbicular, sometimes lobed; seeds plump, obscurely angled, faces smooth:
      Base of leaves deeply cordate, enveloping stem; seeds about 1.2 mm long........ .... *C. amplexicaulis*
      Base of leaves obtuse to cordate, not enveloping stem; seeds about 0.8–1 mm long .... *C. nummularia* var. *glabra*
    Leaf blades elliptic or narrower; seeds strongly angled, faces flat, variously sculptured or smooth:
      Leaves linear-lanceolate; seed faces irregular with broken, more or less transverse ridges ....*C. punctulata*
      Leaves elliptic; seed faces smooth or with unbroken transverse ridges:
        Stipules about 1 mm long, bifid or with lacerate margin; cyathial gland appendage merely a light rim to gland; seed more than 1.2 mm long, faces with broad transverse ridges ....*C. recurva*
        Stipules about 0.5 mm long, lacerate to base; cyathial gland appendage prominent, to twice the width of gland; seed less than 1 mm long, faces smooth .... *C. bindloensis*
Capsule pubescent:
  Cyathia rather few, not grouped in dense glomerules:
    Leaves narrowly elliptic to linear, 12–25 mm long; plant tomentose throughout; seed faces with fine, reticulate markings...........................*C. galapageia*
    Leaves ovoid to suborbicular or, if elliptic, less than 10 mm long; plant hirsute or strigose throughout; seed faces essentially smooth:
      Seeds plump, obscurely angled; leaf blades broadly ovoid to suborbicular with cordate base; plant hirsute throughout.........*C. nummularia* var. *nummularia*

* Contributed by Derek Burch.

Seeds strongly angled, the faces flat; leaf blades elliptic with base oblique; plant strigose throughout .... *C. abdita*
Cyathia many, collected into dense glomerules:
Glomerules terminal on main axes (often in pairs) or on leafy laterals.............
.... *C. ophthalmica*
Glomerules axillary on short leafless peduncles or terminal................ *C. hirta*

Chamaesyce abdita Burch, Ann. Mo. Bot. Gard. 56: 177. 1969

Herb, somewhat woody at base, apparently perennial from the rootstock; stems numerous, to 1 mm thick, branching freely from base, nodes somewhat enlarged, stems prostrate, to about 1 dm long, sparsely strigose; leaves membranous; stipules joined or distinct, entire, to about 0.3 mm long; blade elliptic, 1.5–3.5 mm long, 1–1.5 mm wide, base oblique, margin entire, apex obtuse, sparsely strigose, particularly on lower surface; cyathia solitary; glands small, somewhat stipitate, deep purple, appendages obsolete; male flowers 3–7 per cyathium; styles joined at base, bifid and somewhat clavate for one-third their length; capsules broadly ovoid, strigose, to 1.2 mm long, 1.6 mm wide; seed narrowly ovoid, 4-angled, to 0.7 mm long, 0.5 mm wide, the faces flat, smooth.

Endemic. Specimens examined: Santa Fé.

This tiny species is known only from one collection. It is possible that it has been overlooked by other collectors because of its small size, and it is to be hoped that further specimens will become available in order that its relationship within the genus may be clarified.

Chamaesyce amplexicaulis (Hook. f.) Burch, Ann. Mo. Bot. Gard. 56: 175. 1969

Euphorbia amplexicaulis Hook. f., Trans. Linn. Soc. Lond. 20: 183. 1847.

Subshrub; stems to 2 cm thick, swollen at nodes and pseudoarticulate, erect, to 6 dm tall, branching through their length and with new shoots arising from old wood, glabrous or with occasional short hairs; leaves membranous to coriaceous; stipules joined, lacerate to base, to 1 mm long; blades suborbicular, 3–11 mm long, 4–11 mm wide, size gradually reduced on side shoots, base broadly and deeply cordate, wrapping around stem, margin entire, apex rounded to emarginate, glabrous or rarely short-pubescent on lower surface; cyathia solitary, produced freely near tips of laterals; glands elliptic, black, appendages to 3 times as wide as glands, yellowish green; male flowers 8–14 per cyathium; styles joined at base, bifid for one-third their length; capsule broadly ovoid, to 1.8 mm long, 2.3 mm wide; seed ovoid, to 1.2 mm long, 1 mm wide, scarcely angled, the seed faces plumply convex, smooth.

Endemic; locally common in the northern and central islands. Specimens examined: Baltra, Bartolomé, Enderby, Daphne Major, Gardner, Genovesa, Marchena, Onslow, Rábida, San Cristóbal, Santa Cruz, Tortuga, Wolf.

In his original publication Hooker noted a resemblance between this species and one from the Bahamas, referring probably to *C. buxifolia* (Lam.) Small, a common species from sandy beaches throughout the Caribbean. The two differ markedly on closer examination, however, *C. buxifolia* having less strongly clasping leaf bases, larger stipules that are at most fimbriate, brown cyathial glands, and larger capsules and seed.

Chamaesyce bindloensis (Stewart) Burch, Ann. Mo. Bot. Gard. 56: 176. 1969

Euphorbia articulata Anderss. var. *bindloensis* Stewart, Proc. Calif. Acad. Sci. IV, 1: 91. 1911.

Subshrub; stems few, to 5 mm thick; slightly swollen at nodes and pseudoarticulate, ascending to 5 dm, glabrous or new shoots from old wood sparsely short-pubescent; leaves membranous; stipules joined, lacerate to base, 0.3–0.5 mm long; blades broadly lanceolate, 3–6 mm long, 2.5–3 mm wide, smaller on laterals, base cordate and somewhat enfolding stem, margin entire, apex rounded to obtuse; cyathia solitary; glands elliptic, purple-black, appendages to twice width of glands, rarely obsolete; male flowers 4–8 per cyathium; styles joined at base, bifid for half their length; capsule bluntly angled, to 1 mm long, 1.3 mm wide; seed ovoid, unequally 4-angled, about 0.7 mm long, 0.4 mm wide, the faces flat or slightly convex, smooth.

Endemic; growing as scattered individuals on cinder slopes and among blocks of lava on some of the northern islands. Specimens examined: Bartolomé, Marchena, Rábida.

Stewart described this taxon as a variety of *Euphorbia articulata* Anderss., citing a total of three collections from Pinta and Marchena. The specimen from Pinta has not been seen, but the other sheets agree closely in vegetative characters with two collections by Wiggins and Porter from Isla Bartolomé. This material is fertile and differs from other taxa in having rather inconspicuous stipules, prominent appendages to the cyathial glands, and small smooth-faced seed. The somewhat clasping leaf base is more like that in *C. amplexicaulis* than that in *C. punctulata*, but the leaf shape is quite different in other respects. As described here, the taxon is restricted to Marchena and Bartolomé, though it is possible that it occurs also on Pinta.

Chamaesyce galapageia (Robins. & Greenm.) Burch, Ann. Mo. Bot. Gard. 56: 177. 1969

Euphorbia galapageia Robins. & Greenm., Amer. Jour. Sci. III, 50: 144. 1895.
Euphorbia stevensii Stewart, Proc. Calif. Acad. Sci. IV, 1: 92. 1911.

Herb, somewhat woody at base; stems to 2 mm thick, somewhat swollen at nodes and pseudoarticulate, branching throughout their length, ascending to 4 dm, tomentose, usually only on 1 side of stem; leaves membranous; stipules joined, lacerate or 2–3-parted, to 1.5 mm long; blades linear to narrowly elliptic, 12–25 mm long, 1.5–5 mm wide, base deeply cordate, margin strongly to obscurely serrate in lower leaves, often entire at tips of shoots, apex acute, both surfaces glabrous to sparsely hairy; cyathia in small clusters on somewhat condensed laterals or solitary by abortion of all but 1 on the branch, tomentose, the glands small, elliptical, somewhat stipitate, brown to purple (at least in drying); appendages obsolete; male flowers 4–8 per cyathium; styles joined at base, bifid half their length; capsule ovoid, tomentose, 1.4 mm long, 1.4 mm wide; seed ovoid, to 1 mm long, 0.7 mm wide, roundly angled, the faces convex, obscurely reticulate.

Endemic; in arid habitats in full sun or partial shade. Specimens examined: Isabela, San Salvador, Santa Cruz, Santa María.

This endemic species has obvious similarities with the group that includes *C. hyssopifolia* (L.) Small and *C. lasiocarpa* (Kl.) Arthur of the American tropics and *C. nutans* (Lag.) Small from more temperate areas. These all carry cyathia in more or less dense clusters made up of dichotomously branching leafless laterals, and the character of the Galápagos species represents a reduction from this. None of the other species in the group consistently shows a comparable reduction; moreover, the combination of seed, cyathial, and pubescence characters in the Galápagos plants also differs from that in any of the other species.

Chamaesyce hirta (L.) Millsp., Publ. Field Mus. Nat. Hist., Bot. Ser. 2: 303. 1909

Euphorbia hirta L., Sp. Pl. 454. 1753.*

Herb, annual; stems few, to 2 mm thick, decumbent to 3 dm long, sparingly branched, tomentose and usually with abundant brown or purple multicellular hairs; leaves membranous, stipules distinct or joined at base, somewhat lacerate, to 1 mm long; blades ovate to lanceolate, often rhombiform, 1–35 mm long, 5–15 mm wide, base rounded to cuneate, margins serrate, apex obtuse to acute; cyathia in dense terminal and axillary glomerules formed of condensed leafless dichasia, strigose; glands minute, purple, the appendages obsolete to 3 times as wide as glands; male flowers 6–12 per cyathium; styles joined at base, bifid for half their length; capsule strigose, ovoid, to 1 mm long, 1 mm wide; seeds cuneiform, 0.8 mm long, 0.4 mm wide, angles subequal, faces concave, obscurely transversely ridged or wrinkled.

A pantropical weed. Specimen examined: Santa María.

It is curious that this species, which is probably the most common of all species of *Chamaesyce*, should have been collected only once in the Galápagos archipelago, particularly since the closely related *C. ophthalmica* is apparently well established on several of the islands. Where these two grow together in other parts of the New World, it is *C. hirta* that is usually the more aggressive and successful.

Chamaesyce nummularia (Hook. f.) Burch, Ann. Mo. Bot. Gard. 56: 177. 1969, var. **nummularia**

Euphorbia nummularia Hook. f., Trans. Linn. Soc. Lond. 20: 183. 1847.

Subshrub; stems several, to 4 mm thick, enlarged at nodes and pseudoarticulate, prostrate to decumbent (or young shoots ascending), to 5 dm long, densely hirsute; leaves membranous to coriaceous; stipules joined, lacerate to base, to 1 mm long; blades broadly ovoid to suborbicular, 3–8 mm long, 2–6 mm wide, base broadly cordate, margin entire, apex rounded, sparsely to densely hirsute on each surface; cyathia solitary; glands broadly elliptical, dark brown, the appendages a slight rim or somewhat wider; male flowers 10–18 per cyathium; styles joined at base, bifid one-third to half their length, slightly flattened at tips; capsule broadly ovoid, strongly angled, hirsute, to 1.6 mm long, 2 mm wide; seed ovoid, obscurely and unequally angled, to 1 mm long, 0.8 mm wide, the faces plumply convex, smooth.

Endemic and known from only a few of the islands. Specimens examined: Bartolomé, San Cristóbal, Wolf.

The appearance of this endemic species is quite variable, even among the few specimens examined. The collection from Isla Bartolomé (*Wiggins & Porter 307* [CAS, MO]) has a dark cast to the rather small leaves and a blackish stem; it does not, however, differ in any characters that would allow a consistent separation from the other collections.

* For additional synonymy, see Wheeler, Rhodora 43: 170. 1941.

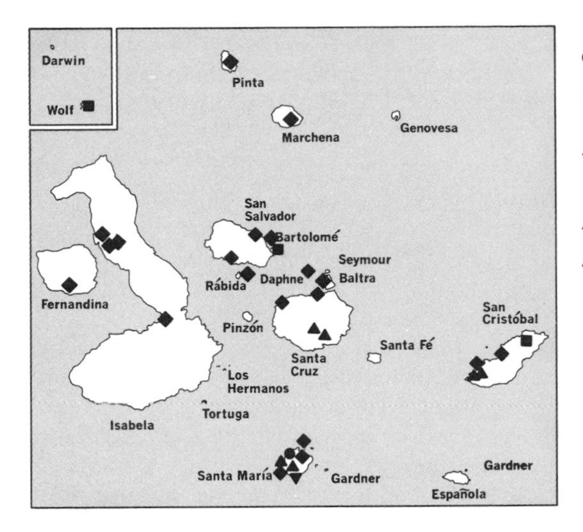

Chamaesyce nummularia var. **glabra** (Robins. & Greenm.) Burch, Madroño 20: 253. 1970

Euphorbia nummularia var. *glabra* Robins. & Greenm., Amer. Jour. Sci. III, 50: 144. 1895.

Differs from the typical variety only in being completely glabrous.

The leaves are frequently 2–3-lobed with a deep sinus between the unequal lobes. This leaf shape has been seen occasionally in the typical variety but is not known elsewhere in the genus.

This variety is known only from Isla Santa María. It was re-collected by Eliasson in 1966 and 1967 and is apparently well established at altitudes of 5–15 m at Las Cuevas and at Black Beach.

Chamaesyce ophthalmica (Pers.) Burch, Ann. Mo. Bot. Gard. 53: 98. 1966

Euphorbia ophthalmica Pers., Syn. Pl. 2: 13. 1807.
Euphorbia hirta var. *procumbens* N. E. Brown in Thistleton-Dyer, Fl. Trop. Afr. 6(1): 497. 1911.*

Herb, annual; stems to several, to 1 mm thick, prostrate or decumbent to 15 cm long, branching throughout their length, densely tomentose at tips, usually with long multicellular hairs; leaves membranous; stipules distinct or joined at base, somewhat lacerate to 1 mm long; blades elliptic or markedly rhombic, 8–20 mm long, 3–12 mm wide, base rounded to cuneate, margin serrate, sometimes obscurely so, apex obtuse to acute, cyathia in dense terminal glomerules, often 2 at apex of each stem or on short leafy laterals; glands minute, purple, the appendages obsolete; male flowers 2–9 per cyathium; style joined at base, bifid for half its length; capsule ovoid, strigose, to 1 mm long, 1 mm wide; seed cuneiform, 0.8 mm long, 0.4 mm wide, strongly 4-angled, angles subequal, the faces concave, obscurely transversely ridged or wrinkled.

A common weed throughout much of the American tropics, from the southern United States to southern Brazil. Specimens examined: San Cristóbal, Santa Cruz, Santa María.

* For additional synonymy, see Wheeler, Rhodora 43: 172. 1941.

There is considerable variation through the range of *C. ophthalmica* in the Caribbean and northern South American areas, but all the specimens seen from the Galápagos Islands are a close match for the large, soft-leaved form that is commonest in South America.

Chamaesyce punctulata (Anderss.) Burch, Ann. Mo. Bot. Gard. 56: 175. 1969

Euphorbia punctulata Anderss., Kongl. Svensk. Vet.-Akad. Handl. 1853: 235. 1855.
Euphorbia articulata Anderss., op. cit. 236, non Aubl. 1775.
Euphorbia diffusa Hook. f., Trans. Linn. Soc. Lond. 20: 184. 1847, non Jacq. 1781, nec Dufour. 1860.
Euphorbia anderssonii Millsp., Publ. Field Mus. Nat. Hist. Bot. Ser. 2: 63. 1900.

Subshrub; stems several, to 6 mm thick, swollen at nodes and pseudoarticulate, prostrate to decumbent or ascending to 1 m, glabrous; leaves membranous to coriaceous; stipules joined, trifid or lacerate, about 0.4 mm long; blades linear-lanceolate or somewhat broader, 4–12 mm long, 0.5–1.5 mm wide (to 24 mm long, 3 mm wide in seedlings), base obliquely sagittate, margin entire, apex acute to rounded; cyathia solitary; glands elliptic, purple-black, appendages a narrow white rim or obsolete; male flowers 4–7 per cyathium; styles joined at base, bifid at tip; capsule strongly angled, to 1.2 mm long, 1.2 mm wide, seed cuneiform, sharply 4-angled, to 0.8 mm long, 0.4 mm wide, the faces flat with more or less transverse, broken ridges.

Fig. 153. *Chamaesyce punctulata* (*a–e*). *C. viminea* (*f–r*): *g–j*, branches bearing small leaves on spur branches; *k*, *l*, twigs with mostly primary leaves; *n–r*, variation in leaves from different parts of plants. (Figs. *f–r* reproduced by permission of American Journal of Botany and Brooklyn Botanic Garden.)

Endemic and rather widely distributed through the islands. Specimens examined: Baltra, Bartolomé, Eden, Daphne Major, Fernandina, Isabela, Marchena, Pinta, Rábida, San Cristóbal, San Salvador, Santa Cruz, Santa María.

See Fig. 153.

Chamaesyce recurva (Hook. f.) Burch, Ann. Mo. Bot. Gard. 56: 175. 1969

Euphorbia recurva Hook. f., Trans. Linn. Soc. Lond. 20: 182. 1847.
Euphorbia apiculata Anderss., Kongl. Svensk. Vet.-Akad. Handl. 1853: 234. 1855.
Euphorbia flabellaris Anderss. ex Boiss. in DC., Prodr. 15^2: 17. 1862.
Euphorbia nesiotica Robins., Proc. Amer. Acad. 38: 167. 1902.

Herb; woody at base; stems several, to 4 mm thick, erect to ascending, the young shoots rarely decumbent, to 8 dm long, branching throughout length of branches, glabrous; leaves membranous to coriaceous; stipules joined, some more or less bifid, most lacerate, to 1 mm long; blades subelliptic, 6–14 mm long, 2–5 mm wide, leaves on main axes sometimes much larger, to 20 mm long, 12 mm wide, the leaves on main shoots usually markedly different in size from those on lateral branches, the base cordate, margin entire or rarely coarsely serrate, apex rounded to obtuse; cyathia solitary; glands elliptic, deep purple to black, the appendages a white rim or rarely obsolete; male flowers 5–12 per cyathium; styles joined at base, bifid and obscurely clavate in upper third; capsule broadly ovoid to 1.5 mm long, 2 mm wide; seed ovoid, 4-angled, to 1.4 mm long, 1 mm wide, the faces convex with more or less obvious transverse ridges.

Endemic; in the lower, arid parts of a number of the islands. Specimens examined: Baltra, Eden, Española, Gardner (near Española), Genovesa, Pinta, Pinzón, Rábida, San Cristóbal, San Salvador, Santa Cruz, Santa Fé, Seymour.

This is by far the most variable of the endemic species of *Chamaesyce*. The characters such as stipule and seed shape, which have been found to be conservative in other studies of the genus in the New World, are rather constant through all the collections, but habit, overall appearance and leaf structural characters show great variation. The taxa that are here placed in synonymy were distinguished primarily on the basis of these variable characters. The extremes are quite well marked, but there is no obvious correlation with geography and there are enough intermediates to suggest that they are better considered parts of a single variable species.

Chamaesyce viminea (Hook. f.) Burch, Ann. Mo. Bot. Gard. 56: 174. 1969

Euphorbia viminea Hook. f., Trans. Linn. Soc. Lond. 20: 184. 1847.
Euphorbia viminea f. *albemarlensis* Robins. & Greenm., Amer. Jour. Sci. III, 50: 138. 1895.
Euphorbia viminea f. *jacobensis* Robins. & Greenm., loc. cit.
Euphorbia viminea f. *castellana* Robins. & Greenm., loc. cit.
Euphorbia viminea f. *chathamensis* Robins. & Greenm., loc. cit.
Euphorbia viminea f. *carolensis* Robins. & Greenm., op. cit. 139.
Euphorbia viminea f. *barringtonensis* Robins. & Greenm., loc. cit.
Euphorbia viminea f. *jervensis* Robins. & Greenm., loc. cit.
Euphorbia viminea f. *abingdonensis* Robins. & Greenm., loc. cit.

Subshrub; stems several, to 1 cm thick or rarely more, erect or spreading to 1.5 m high, branching throughout their length and with new shoots arising from old wood, glabrous throughout or with occasional short hairs; young shoots leafy, older branches pseudoarticulate, leafless but with short spurs bearing clusters of leaves; leaves coriaceous; stipules joined, prominent on young shoots, triangular or somewhat acuminate, to 1 mm long, light colored, darker and forming a collar around stem on leafy spurs and older branches; blades linear, 10–42 mm long, 1–1.5 mm wide on young stems, much shorter on spurs (sometimes only 1.5 mm long), base

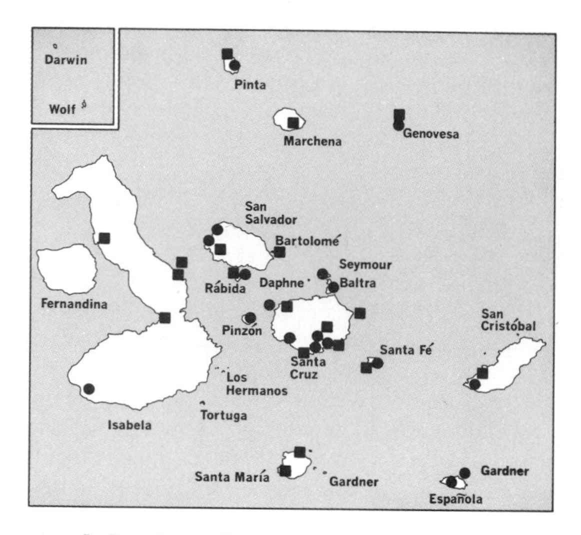

rounded, subequal, margin entire, often recurved on drying, apex obtuse to truncate or emarginate and minutely mucronate; cyathia solitary on young stems or somewhat crowded on spurs, glands orbicular, prominent, purple-black, the appendages obsolete; male flowers 4–16 per cyathium, sometimes only poorly developed; female flowers vestigial in some cyathia (usually those with many well-formed male flowers), styles joined at base, bifid in upper half; mature capsule and seed not seen.

Endemic and widely distributed through the archipelago. Specimens examined: Bartolomé, Genovesa, Isabela, Las Plazas, Marchena, Pinta, Rábida, San Cristóbal, San Salvador, Santa Cruz, Santa Fé, Santa María.

Several forms and one variety proposed by Robinson and Greenman are not distinguishable when all the collections now available are studied. The separation was made on leaf size, shape, and arrangement, but these characters are so variable even within one collection that no consistent patterns of variation can be found. There seems to be a tendency for separation of the sexes in the cyathia of this species: those with a normal-looking female flower usually have only tiny male flowers, whereas others are filled with normal male flowers and have only traces of a female flower. At least one collection (*Fournier 140* [DS, MO]) includes both "male" and "female" cyathia, and there are others in which the cyathia appear to contain well-developed flowers of each sex. It is curious that with such a relatively large number of collections available, no mature fruits have been seen; it is tempting to relate this to the separation of the sexes. One species in the Bahamas, *C. vaginulata* (Griseb.) Millsp., is similar in appearance to this endemic plant.

See Fig. 153 and color plates 36 and 39.

Croton L., Sp. Pl. 1004. 1753; Gen. Pl. ed. 5, 436. 1754

Trees, shrubs, or herbs; stems usually with colored sap, the indument (at least in part) stellate or lepidote. Leaves alternate, pinnately or palmately veined or lobed, entire to serrate, petiolate (the petiole often glandular at top), stipulate. Inflorescenses racemose to spicate (sometimes paniculate or subcapitate), monoecious or dioecious, the staminate flowers distal in bisexual inflorescences and with 4–6 imbricate or valvate sepals; petals usually 5, imbricate (rarely absent); disk entire or dissected; stamens mostly 8–50, free, the filaments inflexed in bud; rudimentary ovary absent. Pistillate flowers sessile or pedicellate, with mostly 5–7 imbricate or valvate sepals; petals 5, often reduced or absent; disk usually entire; ovary of 3 (rarely 2) carpels, the styles once to several times bifid; ovules solitary in each locule. Fruits capsular; seeds carunculate, endosperm copious.

A very diversified genus of nearly 1,000 species, conspicuous in most tropical and many subtropical areas, most abundant in tropical America.

Croton scouleri Hook. f., Trans. Linn. Soc. Lond. 20: 188. 1847.

Slender arborescent shrub or small tree 2–6 m high with smooth, pale bark, the leaves clustered near tips of branches; stipules obsolete or less than 1 mm long; petioles mostly 4–10(20) mm long, sparsely to densely stellate-canescent; blades linear to elliptic or ovate, 2–25 cm long, 0.1–8.5 cm broad, sparsely to densely stellate above, sparsely to densely stellate or tomentose beneath; basal foliar glands orbicular to elliptic, usually sessile, 0.4–1.2 mm broad, often obscured by pubescence; racemes terminal, erect or ascending, unisexual (plants dioecious), rachis angulate, stellate-puberulent; staminate flowers 1–3 in axil of each bract; bracts ovate-acuminate, 1.5–2 mm long, stellate-puberulent; pedicels slender, about 2–6 mm long, sparsely to densely stellate; calyx about 4–5 mm broad, the sepals spreading, 1.3–2.5(3.5) mm long, sparsely to densely stellate-puberulent without; petals linear to elliptic or obovate, 1.8–3.6 mm long, 0.5–1.4 mm broad, acute to obtuse at apex, glabrous except for basal tuft of long hairs; disk segments 5, obovate; receptacle villous with hairs 0.4–0.7 mm long; stamens 14–22, filaments mostly 2–3 mm long, glabrous, anthers ellipsoidal, 0.7–1.3 mm long; pistillate flowers subsessile or with pedicels up to 1 mm long; sepals oblong to lanceolate, 1.4–2(4) mm long, sparsely to densely stellate-pubescent outside, glabrous within; petals absent or rarely up to 3.7 mm long; disk usually 5-lobed; ovary densely stellate-tomentose; styles 3, distinct, slender, 2–3(4) mm long, bilobed half to three-fourths their length, usually sparsely stellate proximally; capsules subglobose, 5–7 mm in diameter, slightly 3-lobed, sparsely to rather densely stellate-canescent, the persistent columella trifid apically, 2.8–3.6 mm long; seeds oblong-ellipsoid to obovate-ellipsoid, somewhat compressed, olivaceous to plumbeous brown, 2.6–5 mm long, faintly to distinctly punctate; caruncle pale, yellowish, 0.8–1.2 mm broad.

Endemic; a highly variable species, probably present on all of the Islands and one of the dominant woody plants on many of them. Although combined by Svenson (1946, 459) with *Croton rivinifolius* HBK., that species has different seeds (distinctly grooved or ridged instead of pitted), and it seems preferable to maintain *C. scouleri* as a species endemic to the Galápagos.

Variation patterns in *C. scouleri* are nearly intractable to rational taxonomic treatment. Despite the striking extremes, such as the linear-leaved "*macraei*" forms on the one hand and the broad-leaved "*grandifolius*" forms on the other, variation appears to be clinal, nearly continuous, and with little geographic correlation of characters. The taxa here recognized at the varietal rank are delimited arbitrarily, and it is impossible to provide a key that will lead to easy determination of collections; this is especially true of staminate specimens, which of course lack the critical character of seed size.

See color plates 26, 29, 33, 39, 40, and 50.

Indument of mature leaves sparsely stellate to stellate-tomentose beneath but not distinctly flocculent (under 50× magnification); plants mostly of lower elevations:
Seeds mostly 2.6–3.7 mm long; leaves ovate or elliptic to linear, most often pointed, often densely tomentose beneath. . . . . var. *scouleri*
Seeds mostly 3.9–4.5 mm long; leaves ovate to suborbicular, obtuse to emarginate, sometimes sparsely pubescent. . . . . var. *darwinii*
Indument of mature leaves more or less loosely flocculent beneath (at 50× magnification), many of the stellate trichomes distinctly stalked; leaves ovate to lanceolate; plants mostly of higher elevations:
Leaves mostly 2–5 cm long, the stellate trichomes on the upper surface mostly 0.4 mm broad or less; pistillate flowers apetalous; seeds 2.8–3 mm long. . . . . var. *brevifolius*
Leaves mostly 6–15 cm long, the stellate trichomes on the upper surface 0.2–0.9 mm broad; pistillate flowers sometimes petaliferous; seeds 3.7–4.7 mm long. . . . . var. *grandifolius*

Croton scouleri var. scouleri

Croton macraei Hook. f., Trans. Linn. Soc. Lond. 20: 188. 1847.
Croton albescens Anderss., Kongl. Svensk. Vet.-Akad. Handl. 1853: 242. 1855.
Croton incanus Anderss., loc. cit., non *C. incanus* Kunth.
Croton incanus var. microphyllus Anderss., loc. cit.
Croton scouleri var. *albescens* Muell.-Arg. in DC., Prodr. 15²: 605. 1866.
Croton scouleri var. *macraei* Muell.-Arg., loc. cit.
Croton scouleri f. *incanus* Muell.-Arg., loc. cit.
Croton scouleri f. *microphyllus* Muell.-Arg., loc. cit.
Croton scouleri var. *castellanus* Svens., Amer. Jour. Bot. 22: 239. 1935.
Croton rivinifolius var. *scouleri* Svens., Amer. Jour. Bot. 33: 459. 1946.
Croton rivinifolius var. *macraei* Svens., loc. cit.
Croton rivinifolius var. *albescens* Svens., loc. cit.

Leaf blades narrowly oblong or elliptic to lanceolate, about 2–12 cm long, 8–23 mm broad, entire or denticulate, obtuse to acuminate at tip, sparsely to moderately stellate-puberulent above with depressed trichomes 0.2–0.4 mm across, paler beneath and more densely stellate-puberulent with mostly depressed trichomes; petioles (2)4–12 mm long; staminate flowers with pedicels 1.8–4.5 mm long; sepals usually densely stellate, 1.3–2.1 mm long; petals linear to narrowly elliptic, 1.8–2.3 mm long, 0.5–1.2 mm broad; stamens 15–22; anthers 0.7–0.9 mm long; pistillate flowers with rather densely stellate sepals 1.4–2 mm long; petals absent; styles 2–2.8 mm long; seeds 2.6–3.6 mm long (rarely to 4.1 mm).

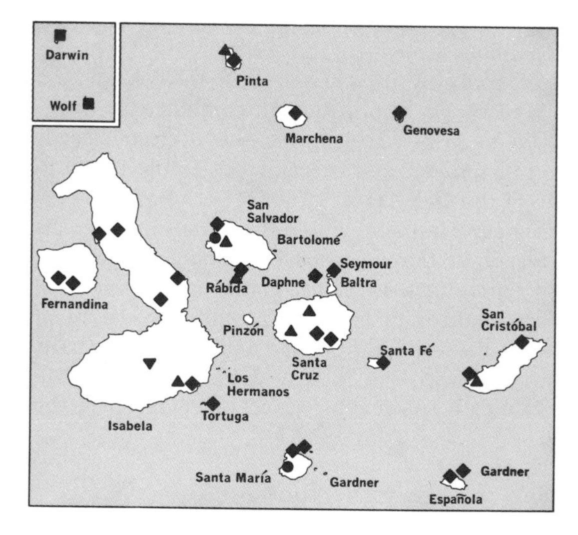

◆ *Croton scouleri* var. *scouleri*
● *Croton scouleri* var. *brevifolius*
■ *Croton scouleri* var. *darwinii*
▲ *Croton scouleri* var. *grandifolius*
▼ *Euphorbia equisetiformis*

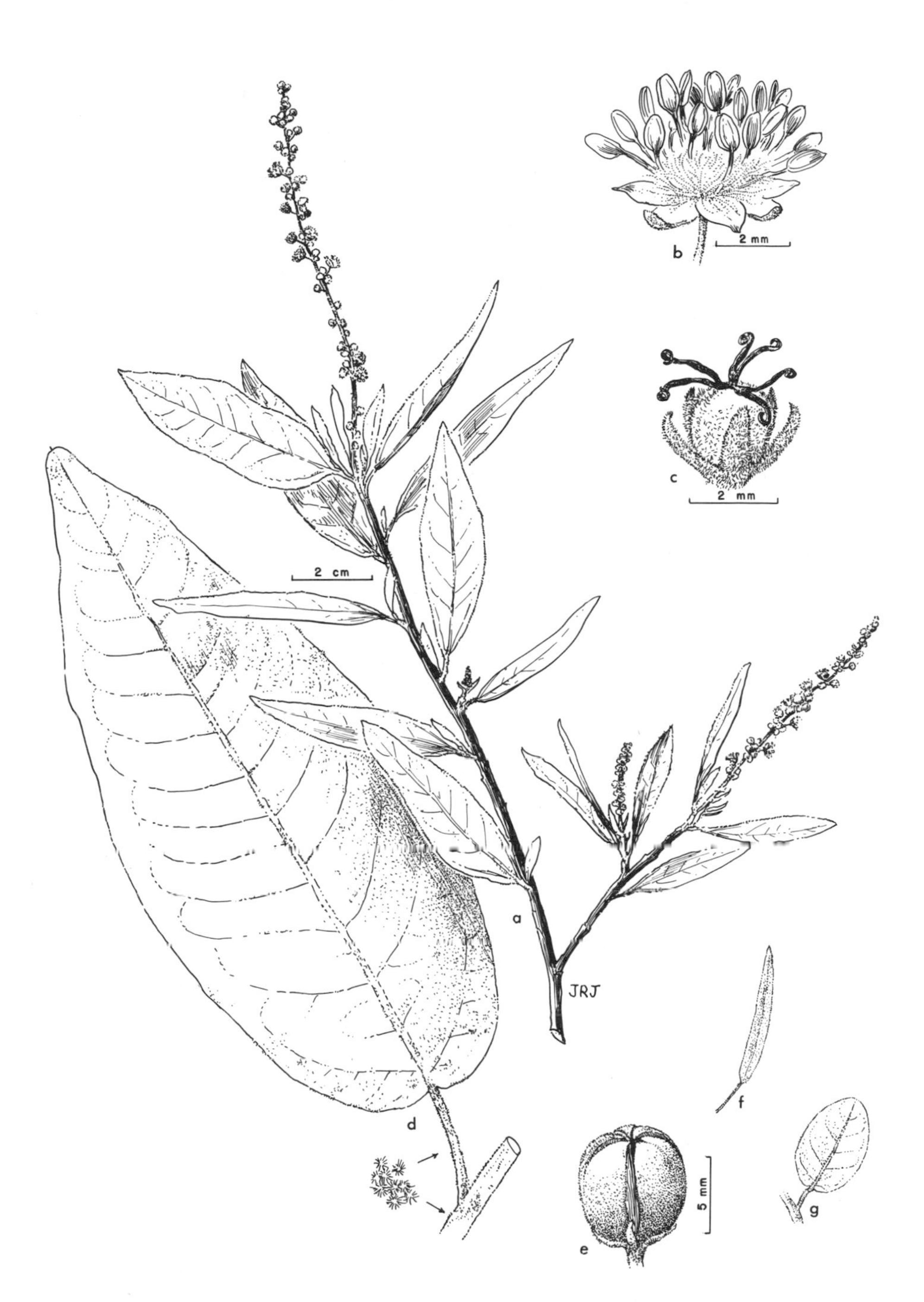

Fig. 154. *Croton scouleri* var. *scouleri* (*a–c, f*). *C. scouleri* var. *grandifolius* (*d, e*). *C. scouleri* forma *macraei* (*f*). *C. scouleri* var. *brevifolius* (*g*).

Widespread on most of the islands (except Darwin and Wolf), occurring mainly at lower altitudes; on some islands it is replaced above 200 m by var. *brevifolius* or var. *grandifolius*. Specimens examined: Champion, Daphne Major, Española, Fernandina, Gardner (near Española), Genovesa, Isabela, Marchena, Pinta, Rábida, San Cristóbal, San Salvador, Santa Cruz, Santa Fé, Santa María, Seymour, Tortuga.

As here interpreted, var. *scouleri* is extremely variable in leaf shape; some plants on Isabela, San Salvador, and Santa Cruz have extremely narrow leaf blades. These have been distinguished as var. *macraei* by most previous workers, but the available collections appear to present a continuum, and the narrow-leaved plants can only be distinguished at the rank of forma [f. *macraei* (Hook. f.) Webster, Madroño 20: 258. 1970]. Plants occurring on some of the smaller, drier islands (Española, Gardner (near Española), Genovesa, Santa Fé, Seymour) tend to be distinctive in having broader leaves, succulent stems, and sometimes larger flowers; but here again no real taxonomic discontinuity is evident. A peculiar plant from Pinta, with subsimple trichomes [f. *glabriusculus* (Stewart) Webster, Madroño 20: 258. 1970], merits further study but does not appear to represent a distinct variety. Pending further investigation, therefore, it seems best to refer all of the lowland forms in the main part of the Galápagos Islands to the single typical variety.

See Fig. 154 and color plates 32, 49, and 75.

Croton scouleri var. **brevifolius** (Anderss.) Muell.-Arg. in DC., Prodr. 15^2: 605. 1866

Croton brevifolius Anderss., Kongl. Svensk. Vet.-Akad. Handl. 1853: 105. 1855.

Twigs not succulent, scarcely if at all sulcate in drying; leaves with petioles mostly 3–12(18) mm long; blades ovate or elliptic, mostly 2–5 cm long, 1–2.7 cm broad, about 1–3 times as long as broad, above sparsely flocculent-stellate with trichomes 0.15–0.3 mm across, paler and densely flocculent-tomentose with distinctly stalked trichomes beneath; sepals and petals about 2–2.5 mm long, stamens 14–17, anthers 0.5–0.9 mm long; pistillate flowers apetalous, sepals 1.7–2 mm long; seeds 2.8–3 mm long.

Upper slopes, sometimes in *Scalesia* forest, to 310 m and above. Restricted to Santa María.

Specimens from San Cristóbal and Santa Cruz that have been referred to var. *brevifolius* probably represent plants from the transition between var. *scouleri* and var. *grandifolius*. Specimens from other islands that were cited by Stewart as var. *brevifolius* belong to var. *darwinii* or var. *scouleri*.

See Fig. 154.

Croton scouleri var. **darwinii** Webster, Madroño 20: 259. 1970

Twigs sometimes succulent and then inflated-sulcate in drying; leaves with petioles mostly 5–20 mm long; blades broadly elliptic or ovate to suborbicular, acute to rounded or emarginate at tip, 3–7(9) cm long, 1.5–6(8) cm broad, about 1–2(2.5) times as long as broad, above sparsely stellate with trichomes 0.2–0.5 mm across, beneath more densely stellate, sometimes flocculent-tomentose or glabrescent; staminate flowers with pedicels 1.5–3.5 mm long, sepals 1.7–2.2 mm long, petals

1.8–2.4 mm long, stamens 16–20, anthers 0.6–0.8 mm long; pistillate flowers ape talous, sepals 1.8–2 mm long; seeds 3.9–4.5 mm long.

Dominating the vegetation on Islas Darwin and Wolf.

Specimens of var. *darwinii* have previously been treated under the name var. *brevifolius* (in a misapplied sense) because of their superficial resemblance to plants from other small dry Galápagos islands. One specimen, *Pool 290* (BKL) from Isla Daphne Major, has even larger seeds (5.1–5.5 mm) than the Wolf and Darwin plants; it may prove to belong to var. *darwinii*, or on further investigation may represent yet a different xerophytic variety.

Croton scouleri var. **grandifolius** Muell.-Arg. in DC., Prodr. 15^2: 605. 1866

Twigs apparently not succulent, somewhat sulcate in drying; leaves with petioles 10–33 mm long; blades ovate, mostly 6–15 cm long, 3–8 cm broad, about 1.5–3.5 times as long as broad, above rather sparsely stellate with spreading trichomes about 0.5–0.8 mm across, beneath densely flocculent-tomentose with stalked stellate trichomes; staminate flowers with pedicels about 3–4 mm long, the sepals densely stellate, about 2 mm long; petals 2.3–3 mm long; stamens 15–18, anthers 0.8–1 mm long; pistillate flowers with sepals sparsely to densely stellate, 1.8–3.5 mm long; petals sometimes developed; seeds 3.7–4.7 mm long.

Occurring at higher altitudes, often in the *Scalesia* forest, on the larger islands of the archipelago. Specimens examined: Isabela, Pinta, Rábida, San Cristóbal, San Salvador, Santa Cruz.

See Fig. 154 and color plate 57.

Euphorbia L., Sp. Pl. 450. 1753; Gen. Pl. ed. 5, 208. 1754*

Herbs, shrubs, or trees, sometimes succulent, milky latex in all parts; monoecious or rarely dioecious. Leaves opposite, whorled, or alternate, often serially so on same plant, simple, sometimes caducous, particularly in succulent forms; usually petiolate except in succulents; stipules present or absent, sometimes glandular. Inflorescence a cyathium, the 5 cuplike lobes alternating at their tips with 4 or 5 glands, rarely fewer, these with or without appendages. Staminate flowers in 4 or 5 monochasia, the subtending bracteoles partly fused to involucre or reduced or absent, naked, monandrous. Pistillate flowers terminal, solitary; perianth of 3–6 united sepals or absent; ovary 3-celled, each with a single ovule, the styles 3, free or joined basally, usually partly bifid. Fruit capsular (rarely drupaceous); seed more or less ovoid, angled or terete, the surface smooth or variously sculptured, with or without caruncle; endosperm copious; cotyledons fleshy, broader than radicle.

About 1,500 species in the broad sense used by Linnaeus, in the tropical and warm temperate regions of the Old and New Worlds.

The genus shows great diversity in morphological and anatomical characters, and several suggestions have been made to split it into elements more nearly equivalent to genera as defined by workers in other families. One group, segregated as the genus *Chamaesyce* S. F. Gray, appears to be natural and relatively well defined. This is accepted as distinct here, while the other elements included by Linnaeus are retained.

* Contributed by Derek Burch.

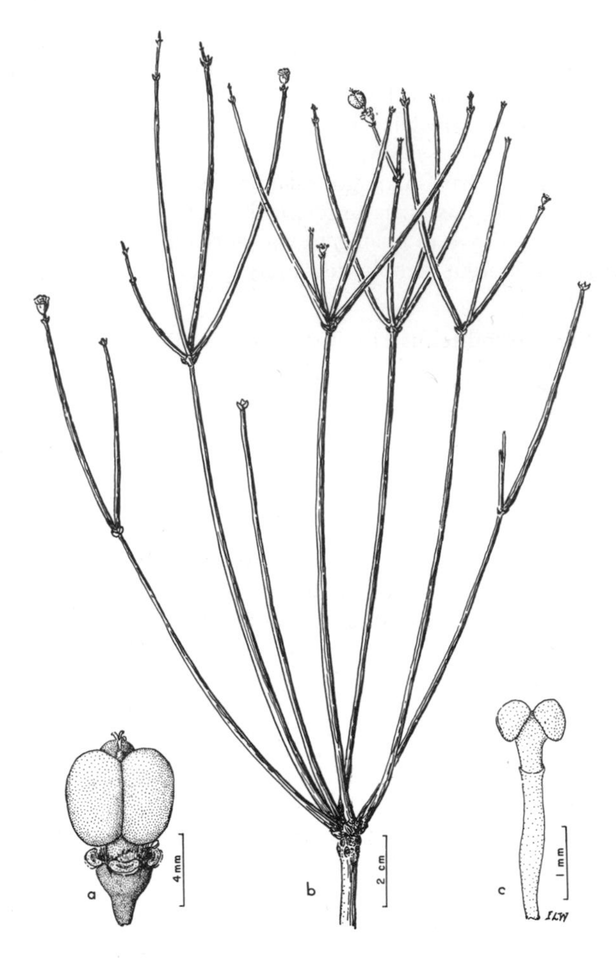

Fig. 155. *Euphorbia equisetiformis.*

Euphorbia equisetiformis Stewart, Proc. Calif. Acad. Sci. IV, 1: 91. 1911

Shrub; stems several, with clusters of branches at upper nodes, terete, somewhat articulate at nodes, ascending to 1.3 m, glabrous; leaves scalelike (or if blades expanded then caducous), whorled or opposite, exstipulate; cyathia solitary at tips of branches, apparently subtended by 2 bracts but these scalelike or caducous; glands 5, elliptical, dark, the appendages at most a narrow rim to gland; male flowers about 15 per cyathium; ovary subspherical, the styles connate at least halfway from base, bifid in upper third; mature capsule and seed not seen.

Known only from the type collection, taken from Isla Isabela.

This species does not seem closely related to any Central or South American taxon. The character of the mature seed is not known nor is it clear from the specimen whether the scales at the nodes are the remnants of caducous leaves or are themselves the leaves. The articulate stems and whorled scales (?leaves) suggest Boissier's Sect. *Alectoroctonum*, but additional material is needed to confirm or disprove this.

See Fig. 155.

Hippomane L., Sp. Pl. 1191. 1753; Gen. Pl. ed. 5, 499. 1754

Trees or shrubs with milky latex, monoecious. Leaves alternate, simple, petiolate, petiole glandular at apex; stipules caducous. Flowers in terminal spicate inflorescences; bracts glandular. Staminate flowers with 2- or 3-lobed calyx, lobes imbricate; petals and disk absent; stamens 2, filaments connate, anthers extrorse; rudimentary ovary absent. Pistillate flowers sessile, with 3-parted calyx; petals and disk absent; ovary of 6–9 carpels, styles simple; ovules solitary in each locule. Fruit drupaceous, reddish or yellowish; seeds compressed, ecarunculate, endosperm copious.

A genus of 2 or 3 species, one widespread in the Caribbean area, the others endemic to Hispaniola.

Hippomane mancinella L., Sp. Pl. 1191. 1753

Tree to 10 m high with rounded crown; twigs and foliage glabrous; leaves ovate to elliptic, 3–7 cm long, 2–4.5 cm broad, short-acuminate, crenulate; petioles slender, 1.5–4 cm long; stipules triangular, papery, 3–4 mm long, soon deciduous and represented by scars; spicate inflorescences 3–10 cm long, rachis thickened, with 1 or 2 solitary pistillate flowers at base and dense glomerules of staminate flowers above; staminate flowers with calyx about 1 mm long, lobes broadly ovate; stamens 2, filaments connate into a tube 1–1.5 mm high, anthers 0.5–0.6 mm long; pistillate flowers with a calyx about 3 mm long, lobes ovate, short-acuminate; styles entire, basally thickened, about 2 mm long, tips involute; fruits apple-like, about 3 cm thick (when fresh), with poisonous latex; seeds dark, about 6 mm long.

A widespread Caribbean species from Florida and the Bahamas south to the shores of Panama, Colombia, and Venezuela; not recorded from the South Ameri-

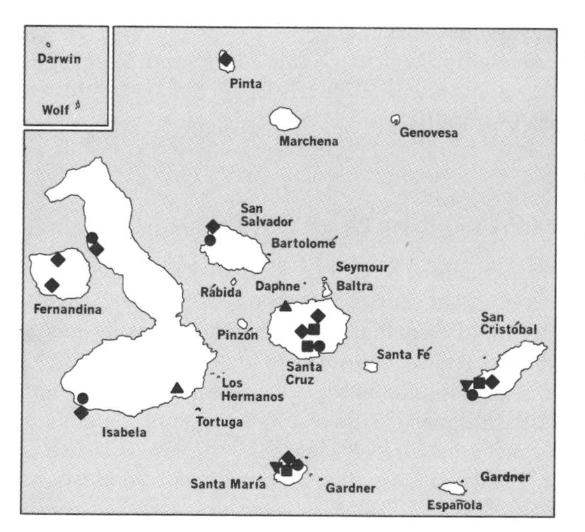

● *Hippomane mancinella*

■ *Jatropha curcas*

▲ *Manihot esculenta*

◆ *Phyllanthus caroliniensis* subsp. *caroliniensis*

▼ *Ricinus communis*

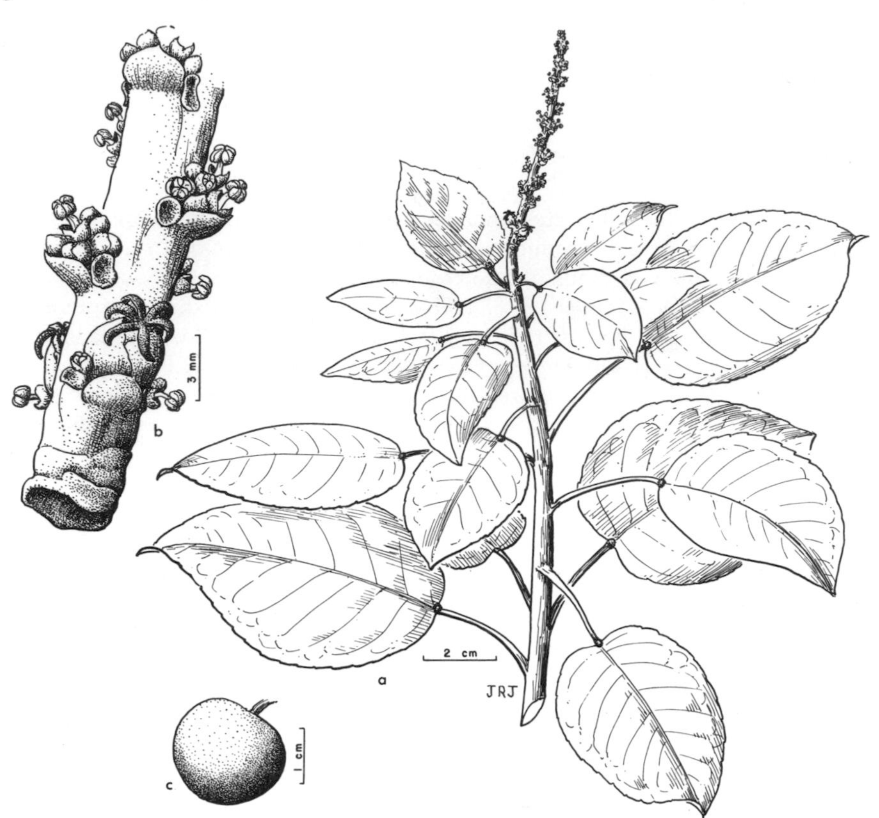

Fig. 156. *Hippomane mancinella.*

can mainland coast opposite the Galápagos. Specimens examined: Isabela, San Cristóbal, San Salvador, Santa Cruz, Santa María.

Abundant around the shores of Academy Bay, Isla Santa Cruz, and seen along margins of lagoon at James Bay, Isla San Salvador. Rarely collected owing to the poisonous latex, which causes severe dermatitis.

See Fig. 156 and color plate 73.

Jatropha L., Sp. Pl. 1006. 1753; Gen. Pl. ed. 5, 437. 1754

Trees, shrubs, or perennial herbs, monoecious or dioecious; stems with colored latex. Leaves alternate, simple to lobed or divided, entire to serrate, stipulate (stipules sometimes glandular). Flowers in terminal or axillary cymes; bracts usually inconspicuous. Staminate flowers with 5 sepals and 5 imbricate petals; disk entire or dissected; stamens mostly 8–12, filaments connate, anthers usually in 2 whorls; rudimentary ovary small or absent. Pistillate flowers with 5 imbricate sepals; petals 5, imbricate; disk entire or dissected; ovary usually of 3 carpels, the styles entire to bifid; ovules solitary in each locule. Fruits capsular; seeds carunculate, endosperm copious.

A pantropical genus of about 150 species, almost equally divided between the New and Old Worlds.

Jatropha curcas L., Sp. Pl. 1006. 1753

Shrub or small tree about 1–5 m high, the branches and mature foliage glabrous; leaves ovate, unlobed or 3–7-lobed, mostly 10–25 cm long, 9–15 cm wide, palmatinerved, with petioles (3)9–15 cm long; stipules almost obsolete (not evident on the mature leaves); cymes bisexual or sometimes unisexual, on peduncles 4–10 cm long; staminate flowers with puberulent pedicels 1–5 mm long; sepals 5, elliptic, entire, scarcely imbricate, 2.8–3.5 mm long; petals obovate, greenish or yellowish white, hirsute, 5–6 mm long; disk segments 5, massive; stamens usually 10, the filaments 2.2–3.7 mm long, connate below middle, the anthers elliptic, 1–1.6 mm long, rudimentary ovary absent; pistillate flowers with pedicels becoming 10–13 mm long; calyx lobes 5, oblong to lanceolate, entire, becoming 7–9 mm long; petals

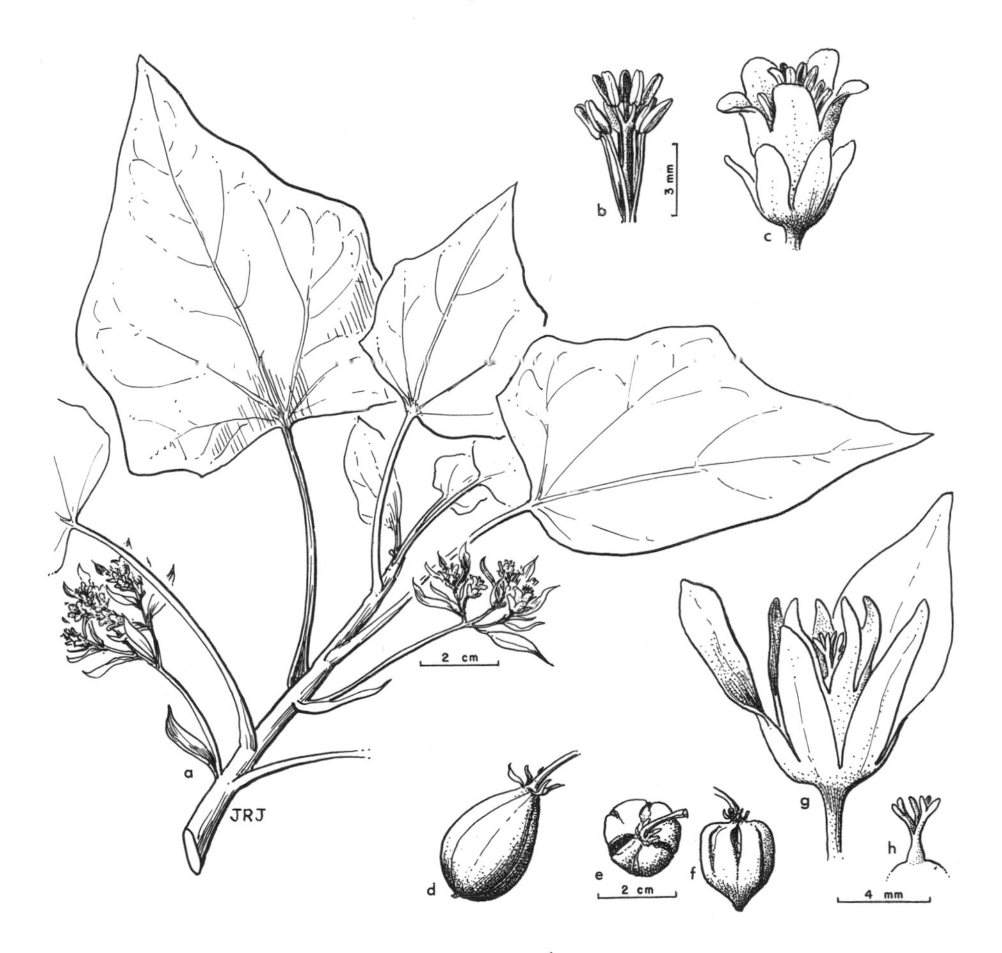

Fig. 157. *Jatropha curcas.*

as in staminate flowers; ovary glabrous, styles slender, 1.5 mm long, connate below, stigmas dilated; capsules ellipsoidal, 3 cm long, tardily dehiscent; seeds blackish, incrustate, 15–22 mm long, beaked, caruncle appressed and deeply lobed.

Native to Mexico and Guatemala, but commonly planted throughout the tropics and often escaping. Specimens examined: San Cristóbal, Santa Cruz, Santa María.

See Fig. 157.

Manihot Mill., Gard. Dict. ed. 4. 1754

Trees, shrubs, or herbs, monoecious; stems with milky latex; roots often tuberous. Leaves alternate, mostly palmately lobed or divided, lobes usually entire, stipulate. Flowers in terminal panicles (these sometimes pseudoaxillary). Staminate flowers gamosepalous, 5-lobed, lobes imbricate; petals absent; disk intrastaminal, entire or lobed; stamens 10, biseriate, filaments free, anthers introrse; rudimentary ovary small or absent. Pistillate flowers long-pedicellate; sepals barely connate; petals absent; disk massive, pulviniform, sometimes with staminodia; ovary of 3 carpels, styles with dilated or lacerate stigmas, ovules solitary in each locule. Fruits capsular; seeds carunculate, the cotyledons broad and thin; endosperm copious.

More than 100 American species, largely concentrated in Brazil; several, especially the common manioc, are cultivated in the tropics and occasionally escape.

Manihot esculenta Crantz, Inst. 1: 167. 1767

Herb or shrub 1–3 m high, the stems and foliage glabrous; leaves mostly deeply 3–5-lobed, lobes acuminate, glaucous, the middle lobe 9–17 cm long; petioles 5–17 cm long; stipules lanceolate, 7–8 mm long, deciduous; panicles 5–10 cm long at anthesis, flowers on spreading or reflexed pedicels; bracts greatly reduced; staminate flowers with pedicels 4–6 mm long, calyx campanulate, yellowish green, 3–4.5 mm long, glabrous outside, puberulent within; disk 10-lobed, 2 mm broad; stamens 10, filaments unequal, 2 mm long or less; anthers 1.8–2 mm long; rudimentary ovary absent; pistillate flowers with pedicels becoming 15–20 mm long; sepals 5, oblong-lanceolate, acute, 7.5–10 mm long; disk massive, 3–3.5 mm broad; ovary with 6 sharp ribs or wings, the styles 1.8–2 mm long, dilated and divided into several captitate tips; capsules 1.5 cm long, rugose, 6-winged; seeds compressed, bluntly beaked, dark-flecked, 7.1–9.3 mm long; caruncle deltoid, entire.

Apparently native to South America, but now widely grown throughout the tropics. Specimens examined: Isabela, Santa Cruz.

See Fig. 158.

Phyllanthus L., Sp. Pl. 981. 1753; Gen. Pl. ed. 5, 422. 1754

Trees, shrubs, or herbs, monoecious or dioecious; branches persistent in the local taxon). Leaves alternate, entire, stipulate. Flowers mostly solitary and axillary or in axillary, reduced cymes; bracts inconspicuous. Staminate flowers with 4–6-lobed calyx, lobes imbricate; petals absent; disk usually present; stamens mostly 3–6, free or connate; rudimentary ovary absent. Pistillate flowers pedicellate or subsessile; calyx 4–6-lobed, lobes imbricate; petals absent; disk entire to segmented (rarely absent); carpels usually 3 with bifid styles; ovules 2 in each locule. Fruits mostly

Fig. 158. *Manihot esculenta.*

capsular; seeds paired, ecarunculate, with dry seed-coats, the endosperm copious.
A primarily tropical genus of about 750 species, especially well developed in the Old World and in insular regions.

Phyllanthus caroliniensis Walt., Fl. Carol. 228. 1788, subsp. **caroliniensis**

Herb, annual or short-lived perennial, mostly 1–3 dm high, monoecious; stems unbranched to pinnately branched, the branches terete or slightly angled, smooth and glabrous; leaves elliptic to obovate, thin, entire, 3–15 mm long, 2–8 mm broad, with petioles 0.5–1 mm long; stipules auriculate, pointed, 0.5–1.2(1.8) mm long; flowers in axillary cymules of 1 or 2 male and 1 or 2 female flowers; staminate flowers with articulate pedicels 0.5–1 mm long; sepals 6, broadly elliptic, 0.5–0.8 mm long; disk segments 6, elliptic, 0.2–0.25 mm long; stamens 3, filaments free, 0.2–0.3 mm long; anthers dehiscing transversely, 0.25–0.3 mm broad; pistillate flowers with geniculate pedicels 1 mm long or less; sepals 6, narrowly obovate to linear-spatulate, 0.7–1.4 mm long, sometimes reddish-stained; ovary smooth, styles bifid, less than 0.5 mm long; capsules 1.5–1.9 mm broad; seeds dull to dark grayish brown, finely verruculose, 0.6–1 mm long.

A widespread polymorphic American species, extending from temperate latitudes in the eastern United States (Pennsylvania) to Argentina and across the American tropics; in open at margins of forest, among trees and shrubs, or on grassy hillsides. Specimens examined: Fernandina, Isabela, Pinta, San Cristóbal, San Salvador, Santa Cruz, Santa María.

See Fig. 159.

Ricinus L., Sp. Pl. 1007. 1753; Gen. Pl. ed. 5, 437. 1754

Shrubs or trees (behaving as herbs in temperate regions); sap watery, stems and foliage glabrous. Leaves alternate, palmately 7–11-lobed, serrate; petioles long, peltately inserted, glanduliferous at apex; stipules fused into caducous sheath. Inflorescences terminal (sometimes appearing pseudoaxillary), paniculate, monecious; lowermost nodes with cymules of staminate flowers, upper nodes bisexual or pistillate; bracts papery, glandular at base. Staminate flowers with calyx splitting into 3–5 segments; petals and disk absent; stamens many (up to 1,000), filaments irregularly branched and partially connate basally into fascicles; rudimentary ovary absent. Pistillate flowers pedicellate, with calyx similar to that of staminate flowers; petals and disk absent; ovary of 3 muricate carpels, styles bifid, ovules solitary in each locule. Fruits capsular, usually echinate; seeds mottled, carunculate, cotyledons broad and thin, endosperm copious.

A single variable species, Old World in origin but now widely cultivated in tropical and warm temperate regions and often escaping.

Ricinus communis L., Sp. Pl. 1007. 1753

Shrub or tree, usually 2–5 m high, twigs usually glaucous; leaves mostly 7–9-lobed, 1–10 dm across, palmatinerved but lobes pinnately veined, glabrous; petioles peltately inserted, 1–2 dm long or more, with 1 or 2 glands at apex; stipules 1–1.5 cm long, caducous; flowers in terminal panicles; bracts papery, with dark basal glands; staminate flowers with pedicels 5–15 mm long; calyx segments acute, 4–9 mm

Fig. 159. *Ricinus communis* (*a–d*): *b*, pistillate involucre and flower; *c*, staminate inflorescence. *Phyllanthus caroliniensis* subsp. *caroliniensis* (*e–g*).

long; androecium usually 5–7 mm across; pistillate flowers with pedicels becoming 1.5–3 cm long, the calyx segments mostly 5, acute, 3–4 mm long; ovary coarsely muricate, styles bifid, densely papillose, 3–4 mm long; capsules 12–21 mm across; seeds ellipsoidal, diversely mottled, 9–22 mm long, 4.5–9 mm broad.

Sparingly introduced in the Galápagos, but probably present on other islands in addition to those listed here. Specimens examined: San Cristóbal, Santa María.

See Fig. 159.

HALORAGACEAE. Water Milfoil Family*

Aquatic or terrestrial herbs or plants rarely suffrutescent. Leaves alternate, opposite, or whorled, extremely variable in size, those on submerged branches pectinately pinnatifid into slender, often linear segments; stipules none (or in the genus *Gunnera* forming an ochreate sheath or reduced to scales). Flowers usually unisexual and plants mostly monoecious, sometimes polygamous, flowers small to minute, solitary, in axillary clusters, in corymbs or panicles. Perianth biseriate, uniseriate, or lacking, when present adnate to ovary; calyx 2–4-lobed or lacking; petals 2–4, larger than calyx lobes, imbricate or valvate in bud, or lacking. Stamens 4 and alternate with petals, or 8 and outer whorl opposite petals; anthers basifixed, 2-celled, dehiscing longitudinally. Pistil 1; ovary inferior, 4-carpeled, 1–4-locular, each locule with a solitary, pendulous, anatropous ovule. Styles as many as locules; stigmas often plumose. Fruit a small drupe, nut, or mericarp, sometimes winged. Embryo straight; endosperm present.

Eight genera and about 100 species widely distributed in all continents except the arctic and antarctic regions.

Myriophyllum L., Sp. Pl. 992. 1753; Gen. Pl. ed. 5, 429. 1754

Leaves entire, dentate-serrate or pectinately pinnatifid, the submerged leaves with linear to filiform divisions. Stipules none. Flowers mostly solitary in axils of leaves. Petals 2–4, sessile, cucullate. Stamens 4 or 8. Ovary 4-locular. Fruit of 4 small mericarps.

About 30 to 35 species ranging from the tropics into boreal zones in the Old and New Worlds; well represented in Australia.

Wolf (1879, 284) reported an unidentified species of *Myriophyllum* from "a brook near the hacienda," on Isla Santa María, but made no collection. This genus has not been found subsequently by any botanist, either on Isla Santa María or elsewhere in the archipelago. Stewart (1911, 119–20) wrote, "At the times we visited this island there were no brooks except in the immediate vicinity of two small springs." The report of *Myriophyllum* in the Galápagos Islands remains suspect. No illustration of this genus is supplied.

HYPERICACEAE. Hypericum Family*

Herbs, shrubs, trees, or rarely lianas with resinous sap. Leaves exstipulate, opposite, whorled, or rarely alternate, sessile to petiolate, blades simple, usually with pel-

* Haloragaceae and Hypericaceae contributed by Ira L. Wiggins.

lucid or black-glandular dots. Flowers perfect, regular, hypogynous, in terminal cymes. Sepals 5 or sometimes 4, imbricate, green, persistent, often somewhat connate basally. Petals (4)5, sessile or clawed, with nectar pit or groove on claw, imbricate in bud, often oblique and more or less contorted. Stamens few to many, free (in ours) or united basally into 3–5 fascicles; anthers small, 2-celled, dehiscing longitudinally. Pistil 1, ovary superior, 1- or 3–5-celled, of 1–7 carpels; placentation usually axile. Styles as many as carpels, linear, distinct or connate basally. Ovules numerous, anatropous. Fruit a capsule or berry, capsules usually septicidally dehiscent. Seeds many, small. Embryo straight or curved; endosperm none.

A family of 8–10 genera and about 350 species, widely distributed in temperate to tropical regions; often a troublesome weed.

Hypericum L., Sp. Pl. 783. 1753; Gen. Pl. ed. 5, 341. 1754

Herbs (ours) or shrubs with opposite, sessile, more or less pellucid-punctate leaves and yellow flowers borne in cymes. Sepals 5, equal or nearly so. Petals 5, convolute in bud, mostly oblique, deciduous or marcescent. Stamens 5 to many, united in clusters or less commonly distinct. Ovary superior, 1-celled with 3 carpels or 3–5-celled and with axile placentation; styles 3–6. Capsule 1–5(6)-celled. Seeds minute.

About 300 species, widely distributed from the tropics to cool temperate regions, but predominantly subtropical.

Hypericum uliginosum Kunth var. **pratense** (Cham. & Schlecht.) Keller, Bull. Herb. Boiss. ser. 2, 8: 190. 1908

Hypericum pratense Cham. & Schlecht., Linnaea 5: 218. 1830.
Hypericum uliginosum var. *nigropunctatum* Keller, Bull. Herb. Boiss. ser. 2, 6: 264. 1898.

Perennial herb 1–5 dm tall, slightly woody at base, moderately to profusely branched from about middle upward; upper branches and twigs terete, with 4 low subalate ridges from initiation of a branch to next bifurcation; leaves lanceolate to linear-lanceolate, or lowest ovate-lanceolate, 8–25 mm long, 2–4(6) mm wide,

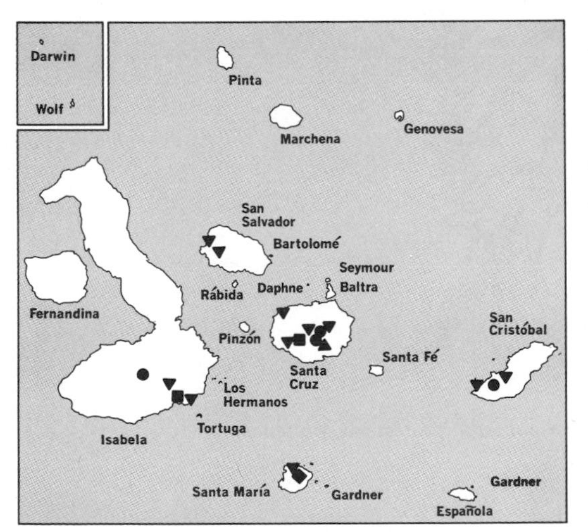

● *Hypericum uliginosum* var. *pratense*
■ *Caesalpinia bonduc*
▲ *Cassia bicapsularis*
◆ *Cassia hirsuta*
▼ *Cassia occidentalis*

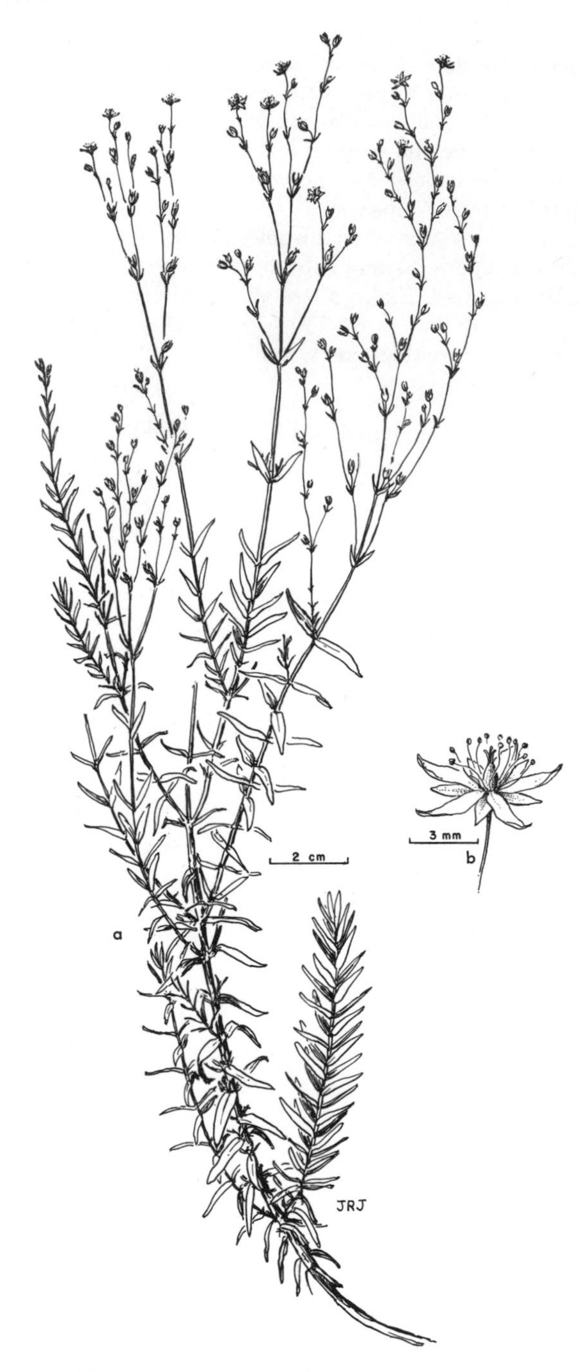

Fig. 160. *Hypericum uliginosum* var. *pratense*.

entire, sessile, acute to short-mucronate at apex, liberally sprinkled with pale yellow pellucid dots about 0.05 mm across on both surfaces, these turning black in age; cymes 5–10 cm long; bracts lance-ovate, 4–5 mm long; pedicels, 2.5–4 mm long, weakly 4-alate like stems; sepals ovate-lanceolate, 2.4–2.6 mm long, 0.8–1 mm wide, 5-nerved, pellucid-punctate; petals oblanceolate, 3–3.5 mm long, slightly asymmetrical; stamens 9–10, distinct, about equaling sepals; anthers minute, about 0.2 mm long, 0.3 mm wide; styles 3, distinct to base, 0.6–0.8 mm long; stigmas capitate, about 0.2 mm broad; capsules narrowly ovoid, about equaling sepals, a tuft of 10–15 stiff, black hairs usually at base of styles, these and style persistent; seeds oblong-ellipsoidal, about 0.3 mm broad, 0.4–0.5 mm long, reddish brown, faintly striate-ribbed and transversely lined.

From Mexico into Central America and northern South America; in damp soil, among rocks, and in tufaceous material from *Scalesia* Zone upward into Fern-Sedge Zone in the Galápagos Islands. Specimens examined: Isabela, San Cristóbal, Santa Cruz.

See Fig. 160.

LEGUMINOSAE. BEAN FAMILY*

Trees, shrubs, or herbs, the stems sometimes twining; leaves commonly alternate, rarely opposite, compound, sometimes simple, usually stipulate, sometimes stipellate. Flowers commonly bisexual, 5-merous, solitary or in compound inflorescences, axillary or terminal. Sepals free or united, 4 or 5, rarely 1, then spathaceous. Petals free or united, 5, sometimes fewer or lacking. Stamens commonly 5 to many, sometimes reduced to only 1 fertile member, the sterile members present or sometimes reduced to staminodes, the filaments free or united, the anthers 2-celled, dehiscing lengthwise or by terminal pores. Pistil usually 1, the ovary superior, 1-locular, 1- to many-ovulate; fruit a legume, dehiscent or indehiscent, 1- to many-seeded, 2-valved, but sometimes modified as a drupe, samara, follicle, or loment. Seed commonly with a coriaceous testa, reniform, lenticular, or spherical, sometimes alate or arillate, the hilum orbicular to linear, sometimes circumcinct, endosperm little or none.

A family of about 600 genera and nearly 18,000 species of worldwide distribution.

KEY TO THE SUBFAMILIES OF THE LEGUMINOSAE†

Flowers actinomorphic, radiate and regular; corolla and calyx valvate in bud; stamens 4 to numerous; leaves commonly bipinnate, sometimes pinnate or reduced to phyllodes . . . . MIMOSOIDEAE, p. 642

Flowers generally zygomorphic, sometimes subactinomorphic; corolla and calyx imbricate in bud or sometimes valvate:

- Uppermost (adaxial) petal enveloped by other petals in bud; stamens (fertile) 1 to numerous, commonly 10 or fewer; leaves usually pinnate, sometimes bipinnate, rarely simple . . . . CAESALPINIOIDEAE, p. 600
- Uppermost (adaxial) petal exterior in bud, enveloping other petals; stamens 10, rarely fewer; leaves simple or pinnate, never bipinnate . . . . FABOIDEAE, p. 608

---

* Family synopsis and key to subfamilies contributed by Velva E. Rudd.

† The subfamilies are arranged alphabetically as intact units, with the genera in each alphabetically in order to facilitate identifying the portions of the whole family treated by the two authors involved.

Subfamily CAESALPINIOIDEAE*

Typically shrubs or trees, sometimes vines or herbs, with impari- or paripinnate, or simple leaves, usually stipulate and alternate. Flowers mostly bisexual, irregular to nearly regular. Hypanthium usually present and sometimes well developed. Sepals usually 5, free or united, the petals occasionally lacking or reduced to 1, usually 5, adaxial one overlapped in bud by lateral ones. Stamens usually 5 or 10 but sometimes fewer or indefinite, the filaments distinct or variously united, the anthers opening by slits or pores. Ovary often stipitate.

About 125 genera and 2,500 species, chiefly in tropical regions. The Galápagos representatives of this subfamily are mostly pantropical weeds, with the exception of *Cassia picta* G. Don, which is indigenous to Ecuador and Peru as well as to the Galápagos Islands. They are frequently pioneers on old lava flows and in disturbed areas, such as old fields and roadsides, and occur from sea level to several hundred meters elevation.

Leaves pinnate; stamens usually at least dimorphic, the anthers opening by terminal pores .... *Cassia*
Leaves bipinnate; stamens usually uniform in size and form, the anthers opening by longitudinal slits:
  Pinnae more than 2, the rachis of each rounded and not spine-tipped; leaflets not regularly deciduous .... *Caesalpinia*
  Pinnae 2, borne on short rachis in form of a spine, the axis of pinnae ribbon-like and spine-tipped; leaflets deciduous .... *Parkinsonia*

Caesalpinia L., Sp. Pl. 380. 1753; Gen. Pl. ed. 5, 178. 1754

Trees or shrubs, sometimes clambering, unarmed or with scattered prickles. Leaves bipinnate, the leaflets many and small or few and large. Flowers yellow to red, in loose, axillary or terminal, simple or compound racemes; bracts usually small and caducous, bracteoles lacking. Receptacle short, bearing 5 imbricate sepals, the abaxial larger and concave. Petals 5, rounded or oblong, imbricate, equal or the adaxial smaller. Stamens 10, free, base of filaments pubescent or glandular. Ovary sessile or short-stalked, the stigma clavate, punctiform, truncate, or concave. Pods indehiscent or tardily dehiscent, ovate, oblong, lanceolate, or crescent-shaped, laterally compressed, the valves leathery. Seeds transverse, ovate to globose.

A tropical and subtropical genus of about 45 species.

Caesalpinia bonduc (L.) Roxb., Fl. Ind. 2: 362. 1832

Guilandina Bonduc L., Sp. Pl. 1: 381. 1753.
Caesalpinia criste L., op. cit. 380, pro parte.

Low, prickly shrub 0.75–2 m tall, the branchlets softly puberulous, with scattered prickles, the stipules large, conspicuous, simple or lobed, foliose; petiole 7–11 cm long, prickly and/or with short recurved spines or puberulous, the rachis 2 or 3 times as long as petiole and similarly invested with prickles or spines, sparsely to densely puberulous, and with paired recurved spines often present at each pair of leaflets and between each pair; pinnae 5–7-jugate, the leaflets 6–8-jugate, oval to

* Contributed by Richard S. Cowan.

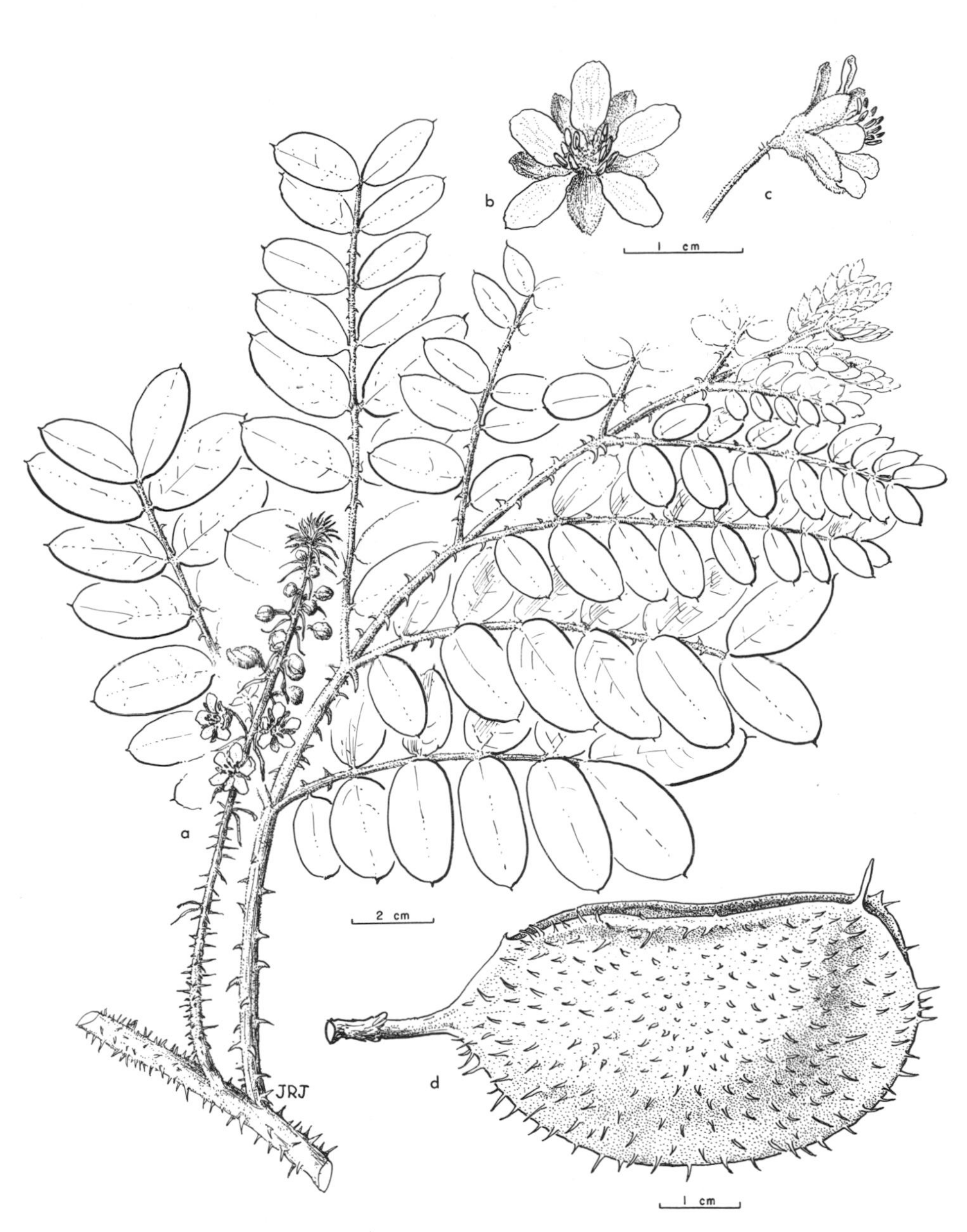

Fig. 161. *Caesalpinia bonduc.*

oblong or elliptic, obtuse at both ends or apex acute, sparingly flexuose-pilose on both sides; inflorescences racemose, axillary near branch tips, 15–30 cm long, puberulous and sometimes prickly, the bracts conspicuous, about 6–10 mm long, linear-lanceolate, puberulous on both surfaces; pedicels 4–7 mm long, puberulous, the receptacle cupular, about 2 mm deep, petals yellow, narrowly oblanceolate; stamens about 5 mm long, the filaments tomentose; ovary densely pubescent, short-stalked, the style very short, stigma truncate; pods oval-oblong to oblong, 5–6.5 cm long, 3.5–4.5 cm wide, flattened laterally, prickly, prickles and valve surfaces densely pubescent; seeds almost globose, stony, gray.

A cosmopolitan beach shrub in sandy places in subtropical and tropical regions; distribution is by seed-flotation in the sea; apparently not common in the Galápagos Islands. Specimens examined: Isabela, Santa Cruz.

See Fig. 161.

Cassia L., Sp. Pl. 376. 1753; Gen. Pl. ed. 5, 178. 1754

Trees, shrubs, lianas, or herbs. Leaves paripinnate, the leaflets of each pair usually opposite, rarely lacking, the axis bearing 1 or more obvious glands on petiole or rachis. Flowers usually more or less irregular, borne singly or in axillary or terminal racemes or panicles, bracteate and often bracteolate. Sepals 5, usually imbricate, often somewhat unequal. Petals 5, free, usually more or less unequal, yellow, red, pink, or white. Stamens 5–10, often unequal, staminodia sometimes present; anthers basifixed, opening by 2 apical or basal pores; filaments free, sometimes stout and rigid. Ovary sessile or stipitate, often curved, the stigma generally punctiform. Pods flat to cylindrical, often divided by cross-walls into chambers, few- to many-seeded.

About 500 species in tropical to warm temperate regions around the world.*

Leaflets obtuse to broadly rounded apically, sometimes mucronulate:
- Leaves glabrous; flowers about 3 cm across:
  - Leaflets usually 3–4-jugate, more or less obovate; larger anthers not aristate basally; pods turgid, cylindrical . . . . . *C. bicapsularis*
  - Leaflets usually 5–7-jugate, about oblong; larger anthers aristate basally; pods flat, oblong . . . . . *C. picta*
- Leaves more or less pubescent; flowers smaller:
  - Plant with short, white, appressed hairs . . . . . *C. tora*
  - Plant with red-brown, spreading pubescence on most parts . . . . . *C. uniflora*

Leaflets sharply acute:
- Plant densely hirsute . . . . . *C. hirsuta*
- Plant glabrous or nearly so . . . . . *C. occidentalis*

Cassia bicapsularis L., Sp. Pl. 376. 1753

Cassia sennoides Jacq., Coll. 1: 74. 1786.
Cassia coluteoides Collad., Hist. Cass. 102. 1816.
Senna bicapsularis Roxb., Fl. Ind. 2: 342. 1832.
Adipera biscapsularis Britt. & Rose, Sci. Surv. Pto. Rico 5: 370. 1924.

Erect or scrambling, glabrous shrub 2–6 m tall, the stipules linear-setaceous, caducous; leaves glabrous or with a few hairs at base of leaflet costa beneath, the rachis

* The genus has been subdivided in a number of ways, even into a series of genera by N. L. Britton and his associates; it is treated here in the aggregate sense.

with clavate gland between lowermost pair of leaflets, the leaflets 3–4(5)-jugate, obovate-oblong to obovate, base inequilateral, apex rounded-obtuse, 1.5–3.5 cm long, 12–20 mm wide; inflorescences axillary and racemose or terminal and paniculate-racemose, often longer than leaves, bracts linear, caducous, bracteoles lacking, pedicels 5–10 mm long; sepals about 8 mm long, petals about 10–25 mm long, yellow; perfect stamens 7, 2 with elongate filaments and large, oblong anthers, smaller stamens subsessile, the anthers rostrate apically but not aristate at base; ovary linear, curved, glabrous, stipitate; pods septate, turgid, cylindric, nearly straight, 4–15 cm long, 1–1.5 cm broad; seeds many in 1-seeded loculi, transverse, apparently surrounded by fleshy pulp.

Widely distributed weedy shrub in lowland tropical America; known from the Galápagos Islands only from a single collection taken on Isla Santa Cruz.

See color plate 86.

Cassia hirsuta L., Sp. Pl. 378. 1753

Ditremexa hirsuta Britt. & Rose, Sci. Surv. Pto. Rico 5: 372. 1924.

Hirsute herb with fetid odor, sometimes becoming woody near base, up to 1 m tall or more, the stipules persistent, linear, about 1 cm long; leaf axis with sessile or stalked, peltate or sometimes cylindrical gland near base of petiole, the leaflets (3)4(5)-jugate, hirsute on both sides, elliptic, acute at apex, membranous-chartaceous, about 4–9 cm long, 3 cm wide; inflorescences axillary or terminal, few-flowered racemes; bracts persistent, linear; pedicels 10–16 mm long, hirsute; sepals hirsute on inner and outer surfaces, 4–10 mm long, oval, the petals pale yellow or infrequently orange, glabrous, 10–14 mm long, obovate or oblanceolate; fertile stamens 6, the 2 larger with elongate filaments and arcuate-oblong, rostrate anthers, anthers of the 4 smaller oblong, shortly rostrate; stigma clavate, style hirsute, arcuate; ovary sessile, densely hirsute, more or less arcuate-linear; pods mostly 2 per peduncle, straight or more often arcuate, linear, 10–20 cm long, 4–6 mm wide, divided internally into 1-seeded loculi; seeds many, flat, broadly ovate, with a black line extending most of length of hilum, the lateral surfaces each with sharply delineated, oval area of paler color and apparently different structure.

Widely distributed weed in the world's tropical regions, principally in disturbed areas; apparently not common in the Galápagos but may be expected on all the islands. Specimens examined: Santa María.

Cassia occidentalis L., Sp. Pl. 377. 1753

Cassia falcata L., loc. cit.
Cassia frutescens Mill., Gard. Dict. ed. 8, no. 2. 1768.
Cassia longisiliqua L. f., Suppl. 230. 1781.
Cassia caroliniana Walt., Fl. Car. 135. 1788.
Cassia linearis Michx., Fl. Bor. Amer. 1: 261. 1803.
Cassia ciliata Raf., Fl. Ludov. 100. 1817.
Senna occidentalis Link, Handb. 2: 140. 1831.
Ditremexa occidentalis Britt. & Rose, Sci. Surv. Pto. Rico 5: 372. 1924.

Low, herbaceous to shrubby herb, to 2 m tall, the branches becoming woody toward base, the stipules caducous, broadly ovate, 4–6 mm long; basal part of petiole bearing small, columnar gland, the leaflets (4)5(6)-jugate, glabrous except for ciliolate margins, elliptic, the apex acute, sometimes mucronulate; inflorescences shorter than leaves, axillary or terminal, few-flowered, racemose, bracts deciduous, larger

Fig. 162. *Cassia occidentalis.*

than stipules, ovate, glabrous, bracteoles deciduous to persistent, 7–15 mm long, lanceolate to elliptic, glabrous, thin-chartaceous except for thickened, indurate margins; pedicels sparingly puberulous, longer than flower; sepals elliptic, about 6–7 mm long, 3–4 mm wide, glabrous; petals yellow to orange, glabrous, elliptic, distinctly clawed, about 12–15 mm long, 4–6 mm wide; fertile stamens 6, the 2 larger with longer filaments and arcuate-oblong, rostrate anthers, the 4 smaller with oblong, straight, slightly rostrate anthers; stigma clavate, style glabrous, sharply curved apically; ovary strigulose, arcuate-linear, subsessile; pods 8–10 cm long, 8–10 mm wide, linear, only slightly arcuate, minutely and sparingly strigulose, many-seeded, divided internally by papery cross-walls into 1-seeded loculi; seeds grayish, olive drab, broadly ovate in outline, somewhat flattened laterally, lateral surfaces with sharply defined oval-elliptic area of different structure.

Pantropical weed in waste places, in meadows, along roadsides, and in crevices in lava flows, from near sea level to 250 m. Specimens examined: Isabela, San Cristóbal, San Salvador, Santa Cruz, Santa María.

See Fig. 162 and color plate 50.

Cassia picta G. Don, Gen. Syst. 2: 444, 552. 1832

Cassia applanata Anderss., Kongl. Svensk. Vet.-Akad. Handl. 1853: 254. 1855.

Low, glabrous shrub 1–2 m tall, the stipules deciduous, linear-lanceolate; leaf rachis with small, slender gland between lowest pair of leaflets and usually aristate at apex, the leaflets 5–7-jugate, oval to oblong, the median 5–8 cm long, 2.5–4 cm wide, the basal smaller, the terminal longer and somewhat obovate-oblong, the apex rounded-obtuse and mucronulate; inflorescences axillary and racemose or a terminal panicle of racemes, about as long as leaves, the bracts arcuate, lanceolate, striate, deciduous; bracteoles lacking; pedicels 10–16 mm long, slender, glabrous; sepals glabrous, 10–12 mm long, oval, concave, deciduous, the clawed petals glabrous, bright yellow, drying with black venation, 15–22 mm long, 5–18 mm wide, oblanceolate to obovate; fertile stamens 6, the 2 larger with longer filaments and arcuate, narrowly oblong, basally appendaged anthers, the 4 smaller with oblong, basally appendaged and apically rostrate anthers; gynoecium glabrous, stigma clavate, arcuate; ovary arcuate-linear, borne on stipe about 2 mm long; pods flat, oblong, almost straight, 7–11 cm long, 1.5–2 cm wide, valves papery, the seed-producing area forming prominent, longitudinal ridge; seeds many, each enclosed in distinct loculus, not surrounded by fleshy pulp.

Frequent to occasional shrub in lowlands of coastal Ecuador and Peru; in the Galápagos on bare lava flows, in clearings, and in fence rows from 6 to 600 m elevation. Specimens examined: Fernandina, Isabela, San Cristóbal, San Salvador, Santa Cruz, Santa María.

Apparently an aggressive pioneer, able to accommodate to a wide range of altitude and ecological conditions.

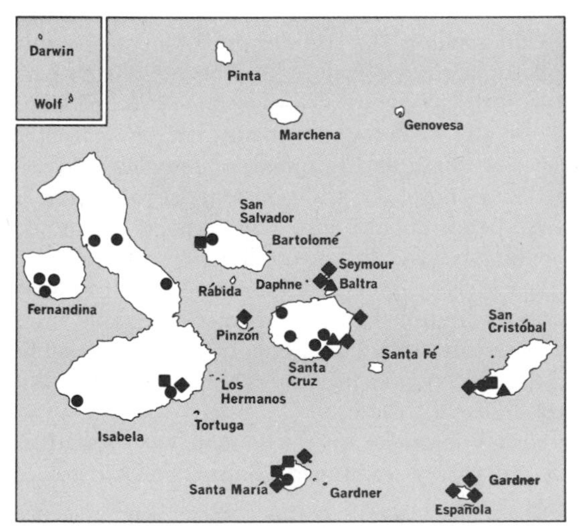

● *Cassia picta*
■ *Cassia tora*
▲ *Cassia uniflora*
◆ *Parkinsonia aculeata*

Cassia tora L., Sp. Pl. 376. 1753

Cassia obtusifolia L., op. cit. 377.
Cassia pentagonia Mill., Gard. Dict. ed. 8, no. 18. 1768.
Cassia humilis Collad., Hist. Cass. 96. 1816.
Diallobus falcatus Raf., Sylva Tellur. 128. 1838.
Diallobus uniflorus Raf., loc. cit.
Emelista tora Britt. & Rose, Sci. Surv. Pto. Rico 5: 371. 1924.

Herb to 1 m tall, the basal parts of main stem becoming woody; stipules persistent to deciduous, arcuate-linear, 8–15 mm long; leaf rachis with slender conical gland between lowermost pair of leaflets, leaflets 2- or 3-jugate, the median 2.5–5.5 cm long, 1.5–3.5 cm wide, oval to obovate, the apex rounded-obtuse and often mucronulate, the blades ciliolate, glabrous or appressed-puberulous on undersurfaces; inflorescences almost sessile, few-flowered (often only 2), racemose, or flowers solitary; bracts narrowly lanceolate, 2–4 mm long, ciliolate; pedicels 12–18 mm long, slender, glabrous; sepals glabrous except for ciliolate margins, 6–8 mm long, oval; petals bright yellow, glabrous, 9–12 mm long; fertile stamens 7; anthers oblong, erostrate apically and nonaristate basally, the 3 larger with longer filaments and somewhat longer anthers; stigma truncate; ovary arcuate-linear, appressed-puberulous, sessile; pods arcuate, 10–20 cm long, 3–5 mm wide, sparingly appressed-puberulous or glabrous, more or less quadrate, divided internally by papery septa into 1-seeded loculi, the margins incrassate; seeds many, quadrate, with sharply defined band on each side, the band of a paler color and apparently of a different structure.

Widely distributed woody herb in tropical regions of the world; in the Galápagos Islands it ranges from near the shore to middle elevations (about 100 m). Specimens examined: Isabela, San Cristóbal, San Salvador, Santa María.

Cassia uniflora Mill., Gard. Dict. ed. 8, no. 5. 1768

Cassia ornithopoides Lam., Encyl. 1: 644. 1785.
Cassia sericea Sw., Fl. Ind. Occ. 724. 1788.
Cassia sensitiva Jacq., Coll. 2: 362. 1788.
Cassia ciliata Hoffm. in DC., Prodr. 2: 493. 1825.
Cassia monantha DC., op. cit. 506.
Sericeocassia uniflora Britt., No. Amer. Fl. 23: 246. 1930.

Herb to about 5 dm tall, most parts rufous-pilose, the stipules persistent, setaceous, 4–12 mm long; leaf axis sparingly pilose, bearing cornute gland between each pair of leaflets; leaflets 2–4-jugate, oval, elliptic, or obovate, the median 1–3 cm long, 1–2 cm wide, apex rounded-obtuse, usually mucronulate, blades ciliate, sparingly appressed-pilose beneath; inflorescences axillary and terminal, of few-flowered racemes, much shorter than leaves; pedicels pilose, about 3 mm long; sepals about 3 mm long, sparingly pilose and ciliolate; petals about 5 mm long, obovate, glabrous, yellow; fertile stamens 7, larger 3 with oblong, rostrate anthers, smaller 4 oval-oblong; stigma funnel-shaped; style sparingly strigose and about one-fourth as long as the arcuate, pilose, sessile ovary; pods linear, slightly curved, 60–70 mm long, 4 mm wide, subcylindric, margins incrassate, divided internally by papery septa into 1-seeded loculi; seeds subquadrate, the lateral surfaces ornamented by sharply delineated, paler tissue of apparently different structure.

Dry savannas and llanos of northern Venezuela; an uncommon herb found in arid areas in the Galápagos Islands. Specimens examined: Baltra, San Cristóbal, Santa Cruz.

Parkinsonia L., Sp. Pl. 375. 1753; Gen. Pl. ed. 5, 177. 1754

Trees or shrubs with spinose stipules. Leaves bipinnate, the 2 or 4 pinnae borne in pairs on a short spinose leaf-rachis, the axis of pinnae ribbon-like, the leaflets few to many, small. Flowers in loose axillary racemes; bracts small, deciduous; bracteoles lacking. Hypanthium short; sepals 5, slightly imbricate. Petals 5, yellow, the adaxial wider than others. Stamens 10, filaments free, densely pubescent basally. Ovary short-stipitate from base of hypanthium; style slender; stigma puberulous. Pods indehiscent or weakly dehiscent, more or less moniliform, the wall leathery. Seeds 1 to 10 per pod, brown.

About 4 species in the Old and New World tropics and subtropics.

Fig. 163. *Parkinsonia aculeata.*

Parkinsonia aculeata L., Sp. Pl. 375. 1753

Tree or shrub 2–10 m tall, branching near ground or with trunk 7–35 cm thick, the bark thin, yellow-green or blue-green, becoming brown on large trunks; stipules spinose, persistent, to 25 mm long; leaves glabrous to sparingly strigulose, the rachis spine 10–25 mm long; pinnae 2 or 4, 17–45 cm long, axes ribbon-like, green, about 5 mm wide; leaflets 20–30-jugate, deciduous, oblong to narrowly elliptic, short-petiolulate, 3–10 mm long, 1–3 mm wide, obtuse apically, attenuate at base; flowers in loose axillary racemes near apex of branchlets, 6–20 cm long, the axes sparsely strigulose; bracts soon caducous; pedicels slender, 10–15 mm long; hypanthium cupular, about 1 mm deep; sepals 5, reflexed, 6–8 mm long, about 2.5 mm wide, acute, glabrous; petals 5, rounded, glabrous except at base, bright yellow, 10–14 mm long and wide, the adaxial broader, red-spotted; stamens 10, filaments free, densely hairy on basal half, 8–10 mm long; anthers glabrous, oval, 1.5 mm long; ovary linear, strigulose with glabrous style half as long; pods 5–20 cm long, glabrous but strongly micropunctate with white, regularly spaced dots; seeds oblong, to about 10 mm long, 5 mm wide, shining, brown.

Widely distributed in subtropical America, naturally or by cultivation, from Florida and Georgia to Texas and California south to the West Indies and Argentina; also naturalized in the Old World tropics; in the Galápagos Islands occasional to common from sea level to about 60 m, in open flats just above high tide, in open scrub forest, among lava blocks, and on old lava flows; reported from many of the islands and probably present on all of those supporting vegetation. Specimens examined: Baltra, Champion, Española, Gardner (near Española), Isabela, Pinzón, San Cristóbal, Santa Cruz, Santa María, Seymour.

See Fig. 163 and color plate 81.

Subfamily FABOIDEAE*

Trees, shrubs, or herbs, the stems sometimes twining, sometimes armed but more often unarmed. Leaves usually compound, 1- to many-foliolate, pinnate or digitate, rarely simple, usually stipulate, sometimes also stipellate. Flowers commonly bisexual, usually zygomorphic, papilionaceous, 5-merous, usually in racemose inflorescences, sometimes solitary. Sepals usually united, at least in part. Petals free except keel petals usually united in part, the standard (vexillum) exterior in bud; stamens commonly 10, sometimes fewer, the filaments free or united, diadelphous 5:5 or 9:1 or monadelphous, the anthers dorsifixed or basifixed, usually dehiscing lengthwise, sometimes apically, commonly eglandular. Fruit 2-valved, dehiscent or indehiscent, stipitate or sessile, sometimes modified as a drupe, samara, follicle, or loment. Seeds 1 to many, hilum apical or lateral.

Over 400 genera and 13,000 species occupying diverse habitats and with a worldwide distribution.†

Leaves 1-, 2-, or 3-foliolate:
  Stems predominantly erect, sometimes decumbent:

---

* Contributed by Velva E. Rudd.

† Two species of *Astragalus*, erroneously reported from the Galápagos, are not included in the key; they are briefly discussed following the description of *Zornia piurensis*.

Plants arborescent; flowers red . . . . *Erythrina*
Plants herbaceous or suffrutescent; flowers yellow, white, pink, or bluish to purple, not red:
Fruit inflated; leaves 3-foliolate:
Leaves digitately 3-foliolate; flowers yellow . . . . *Crotalaria*
Leaves pinnately 3-foliolate; flowers white, pink, or bluish to purple . . . . *Glycine*
Fruit compressed, jointed; leaves 1- to 3-foliolate:
Leaves 2-foliolate; flowers yellow; bracts foliaceous, peltate . . . . *Zornia*
Leaves 1- or 3-foliolate; flowers yellow or white to purplish:
Stipules adnate to petiole; stipels absent; flowers yellow; fruit 1- or 2-jointed with an uncinate beak . . . . *Stylosanthes*
Stipules free from petiole; stipels present; flowers usually white or pink to purplish, or sometimes greenish or yellowish; fruit about 4–9-jointed, without an uncinate beak . . . . *Desmodium*
Stems scandent or prostrate:
Flowers about 5 cm long, orange or scarlet, usually turning blackish on drying; fruit with stinging hairs . . . . *Mucuna*
Flowers about 2 cm long or less, yellow, white, pink, or purplish; fruit without stinging hairs:
Keel petals and style elongate, spirally coiled . . . . *Phaseolus*
Keel petals and style somewhat curved but not spirally coiled:
Calyx lobes conspicuously unequal in size, the vexillar lobes much longer than others . . . . *Canavalia*
Calyx lobes similar in size and shape:
Leaflets with resinous dots on lower surface . . . . *Rhynchosia*
Leaflets without glandular dots:
Flowers about 10 mm long or less . . . . *Galactia*
Flowers 15–20 mm long:
Petals yellow; fruit smooth on both margins . . . . *Vigna*
Petals white or pink to purplish; fruit with lower margin serrulate . . *Lablab*
Leaves 5- to many-foliolate (occasional 3-foliolate leaf may be present):
Plants woody, trees or shrubs:
Fruit drupaceous . . . . *Geoffroea*
Fruit alate . . . . *Piscidia*
Plants herbaceous or suffrutescent:
Stems and leaflets glabrous, glandular-punctate; fruit small, enclosed by calyx . . *Dalea*
Stems and leaflets pubescent, eglandular; fruit 2–5 cm long, extending far beyond calyx . . . . *Tephrosia*

Canavalia DC., Prodr. 2: 403. 1825, nom. conserv.

Slender, herbaceous to stout woody vines, unarmed. Leaves alternate, pinnately 3-foliolate. Stipules small, caducous, stipels minute, linear, caducous. Flowers in axillary thyrses or racemes. Calyx bilabiate, tubular at base, the vexillar lip with 2 lobes, the carinal with 3 small teeth. Petals purple to reddish, blue, or white, glabrous. Stamens 10, usually monadelphous, sometimes with vexillar stamen more or less free; anthers uniform, dorsifixed. Ovary many-ovulate, stipitate; style glabrous; stigma capitate. Legume 2-valved, dehiscent, somewhat compressed, usually longitudinally ribbed, many-seeded. Seeds elliptic, lenticular, or subreniform, brown to white; hilum lateral, elliptic to linear.

A pantropical and warm temperate genus of about 50 species.

Fig. 164. *Canavalia maritima.*

Canavalia maritima (Aubl.) Thouars, Jour. Bot. Desv. 1: 80. 1813

Dolichos maritimus Aubl., Pl. Guian. 765. 1775.
Canavalia obtusifolia (Lam.) DC., Prodr. 2: 404. 1825.*

Prostrate, herbaceous vine, the stems sparsely pubescent or glabrous; stipules peltate, about 5 mm long, lanceolate above point of attachment, rounded below, caducous; leaflets coriaceous, orbicular to obovate or elliptic, 4–12 cm long, 3.5–9 cm wide, obtuse to emarginate, cuneate or acute at base, moderately appressed-pubescent, glabrescent on both surfaces, the secondary veins evident; inflorescences racemose, few- to many-flowered, the nodes enlarged; flowers about 22–30 mm long; bracts minute, deltoid, caducous; bracteoles ovate, about 1.5 mm long; calyx 10–12 mm long, moderately appressed-pubescent, glabrescent, the vexillar lobes shorter than tube, the carinal teeth about 2 mm long with lowermost tooth slightly longer than laterals; petals purplish to blue or pink; fruit pubescent with subappressed

* For a much more extensive synonymy, see Sauer (1964, 163–65).

hairs, glabrescent, 6–15 cm long, 2.5 cm wide, about 1 cm thick, with stipe about 1.5 cm long and longitudinal rib about 3 mm from ventral suture; seeds brown, mottled, elliptic, about 15–18 mm long, 12–13 mm wide, 9–10 mm thick, the hilum about 7 mm long.

A widespread species found among rocks or in good soil of forests and scrub, in shade or sun, on tropical and subtropical seacoasts of the New and Old Worlds. Specimens examined: Marchena. (Usually on beaches only, but Stewart (1911, 70) says, "on the shore and in the interior of the island.")

See Fig. 164.

Crotalaria L., Sp. Pl. 714. 1753; Gen. Pl. ed. 5, 320. 1754

Herbs or shrubs, unarmed. Leaves alternate, digitately 3-foliolate, 1-foliolate, or rarely simple or 5- or 7-foliolate; stipules present or absent, often decurrent; stipels absent. Flowers commonly yellow, sometimes reddish- or brownish-tinged, rarely bluish, solitary or in racemes, terminal or leaf-opposed. Bracts small or foliaceous, sometimes lacking; bracteoles usually present on pedicel. Calyx 5-lobed, often deeply so. Stamens 10, monadelphous, filaments alternately subequal, anthers alternately small, dorsifixed, or long, basifixed. Ovary 1- to many-ovulate, usually sessile, style usually bearded longitudinally; stigma capitate. Legume 2-valved, inflated, dehiscent. Seeds essentially reniform or cordate with hilum lateral.

Some 300 to 350 species in the tropical and warm temperate regions of the world.

Leaflets predominantly obovate to oblong, less than 1 cm broad, rarely to about 2 cm broad; flowers 6–12 mm long, the calyx 4–5 mm long; fruit 1–2 cm long, 0.4–1 cm broad . . . . *C. pumila*

Leaflets predominantly elliptic or orbicular, more than 1 cm broad; flowers about 15–18 mm long, the calyx 12–14 mm long; fruit 2–5 cm long, 1–1.5 cm broad:

Leaflets generally pubescent, fruits often densely hirsute . . . . *C. incana* var. *incana*

Leaflets and fruit generally glabrous . . . . *C. incana* var. *nicaraguensis*

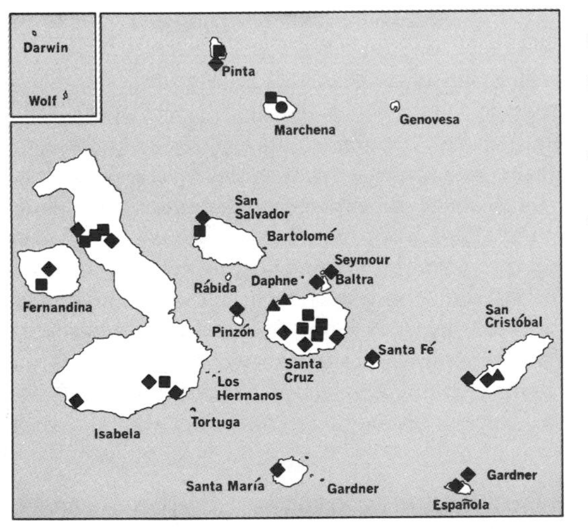

● *Canavalia maritima*

■ *Crotalaria incana* var. *incana*

▲ *Crotalaria incana* var. *nicaraguensis*

◆ *Crotalaria pumila*

Crotalaria incana L., Sp. Pl. 716. 1753, var. **incana**

Crotalaria setifera DC., Prodr. 2: 131. 1825.*

Herb, often suffrutescent, to about 2 m tall; young stems commonly pubescent with spreading hairs, sometimes glabrous; stipules setaceous, about 5–10 mm long; leaves 3-foliolate, the petiole 1–6 cm long, the leaflets elliptic or orbicular to obovate, rounded at apex, about 1–5 cm long and wide, sometimes darkening on drying, the surfaces pubescent with lax hairs or glabrous, the secondary veins relatively inconspicuous; inflorescence racemose; flowers about 15–18 mm long, calyx 12–14 mm long, pubescent or glabrous, lobes about 10 mm long; petals yellow or greenish yellow with the standard sometimes reddish or brownish; fruit about 2–5 cm long, 1–1.5 cm broad, 0.6–1 cm thick, densely hirsute to glabrous; seeds obliquely cordate, mostly yellowish to brown, often mottled, 2.5–4 mm long, 2–3 mm wide, 1.25–1.5 mm thick.

Tropical and subtropical regions of the Old and New Worlds; native in New World, introduced into Old World. Specimens examined: Fernandina, Isabela, Marchena, Pinta, San Salvador, Santa Cruz.

Specimens of the typical variety are generally pubescent; the fruits, especially, are often densely hirsute. There is a considerable range of variation in size of leaflets as well as in degree of pubescence. A few collections from Isla Isabela have glabrate fruits and subglabrous vegetative parts, showing transition to var. *nicaraguensis*.

See Fig. 165.

Crotalaria incana L. var. **nicaraguensis** Senn, Rhodora 41: 353. 1939

Crotalaria glabrescens Anderss., Kongl. Svensk. Vet.-Akad. Handl. 1853: 248. 1855, non Benth. 1843.

The type of var. *nicaraguensis* is completely glabrous, the extreme in a species that ranges from densely pubescent to glabrous. The specimens examined exhibit few if any hairs on the fruit or vegetative parts.

Known from Nicaragua and the Galápagos Islands; a wayside weed and in open forest. Specimens examined: Santa Cruz, San Cristóbal.

Crotalaria pumila C. G. Ortega, Hort. Bot. Matrit. 2: 23. 1797, non Schrank 1820, nec Raf. 1836, nec Blanco 1845, nec Hochst. & Steud. 1871

Crotalaria lupulina HBK., Nov. Gen. & Sp. Pl. 6: 402, *pl. 590*. 1824; DC., Prodr. 2: 133. 1825.
Crotalaria puberula Hook. f., Trans. Linn. Soc. Lond. 20: 225. 1847.†

Annual or perennial herb to about 2 m tall; young stems appressed-pubescent, sometimes glabrescent; stipules setaceous, about 1–3 mm long; leaves 3-foliolate, the petiole about 5–15 mm long, the leaflets obovate or oblanceolate to oblong, rounded at apex, about 0.5–3.5 cm long, 0.2–1.5(2) cm wide, glabrous above, moderately appressed-pubescent below, the secondary veins inconspicuous; inflorescences racemose; bracts and bracteoles setaceous; flowers about 6–12 mm long, calyx 4–5 mm long, appressed-puberulent, lobes 2–3 mm long; petals yellow with the standard sometimes reddish-tipped, somewhat pubescent on outer face, the keel rostrate, bent at right angle; fruit 1–2 cm long, 0.4–0.6 cm thick, 0.4–1 cm wide, puberulous; seeds subcordate, yellowish brown, 2–2.5 mm long, about as wide, 1.25–1.5 mm thick.

* For additional synonymy, see Senn (1939, 350; 1943, 455).

† For additional synonymy, see Senn (1939, 358).

Fig. 165. *Crotalaria incana* var. *incana* (*a–c*). *Dalea tenuicaulis* (*d–g*).

Widespread in tropical and subtropical America; a weedy plant in various habitats, possibly introduced into the Old World tropics. Specimens examined: Baltra, Fernandina, Gardner (near Española), Isabela, Pinta, Pinzón, San Cristóbal, San Salvador, Santa Cruz, Santa Fé, Santa María, Seymour.

Dalea L., Opera varia (Soulsby no. 9) 244. 1758, nom. conserv.*

Herbs or shrubs, unarmed, usually glandular; leaves alternate, imparipinnate, 3 to many, rarely 1-foliolate; stipules small, setiform or sometimes glandular; stipels

* There are differing opinions concerning the correct authority for this genus and the admissibility of the work cited above. See Barneby (1965, 160–64); Dandy (1967, 15); Stafleu (1967, 286).

present or absent, usually glandular; flowers commonly purple or bluish to white, sometimes yellow, in terminal or leaf-opposed spikes or racemes. Bracts foliaceous or setaceous, often caducous. Calyx campanulate with 5 subequal teeth and 10 ribs. Wing and keel petals often inserted on staminal tube. Stamens 10 or sometimes reduced to 5–9, monadelphous, the anthers small, uniform, dorsifixed, sometimes gland-tipped. Ovary 2- or 3-ovulate, essentially sessile, style glabrous, stigma capitate. Legume usually included in calyx, small, indehiscent, 1- or 2-seeded. Seeds subreniform.

Some 150–200 species, known from temperate and tropical America and apparently introduced into the Philippine Islands.

Dalea tenuicaulis Hook. f., Trans. Linn. Soc. Lond. 20: 226. 1847

Dalea parvifolia Hook. f., op. cit. 225.
Dalea tenuicaulis var. *goniocymbe* Riley, Kew Bull. 1925: 221. 1925.

Diffuse shrub to about 2 m tall; stems glabrous, glandular-punctate; stipules deltoid, 1–2 mm long; stipels minute, glandular; leaves 7–17-foliolate, petioles 2–4 mm long; leaflets obovate, obtuse, cuneate at base, 1–4(8) mm long, 1–5 mm wide, glabrous on both surfaces, glandular-punctate, the secondary veins not evident; inflorescences brevispicate, about 1–2 cm long; bracts 3–4 mm long, ovate, villous or glabrous, glandular-punctate; flowers 5–7 mm long; calyx villous, 3–5 mm long, the tube about 2 mm long, attenuate; corolla 5–7 mm long, reddish purple; stamens gland-tipped; fruit membranaceous, included in calyx, 1-seeded (or 2-seeded, fide Hooker, under *D. parvifolia*).

Endemic; on cinder slopes, in crevices in lava flows, and in open cover of Arid and Transition zones. Specimens examined: Isabela, San Cristóbal, San Salvador, Santa Cruz, Santa María.

Apparently only 1 species of *Dalea* occurs in the Galápagos Islands. I am following the advice of Rupert C. Barneby, who is engaged in a monographic study of the genus. He believes that the types of *D. tenuicaulis* and *D. parvifolia* represent seasonal aspects of a single species, and that the type of var. *goniocymbe* "is certainly nothing out of the ordinary, the shorter stamen tube being observed from a bud; the flower is not small as claimed" (personal correspondence). There are some

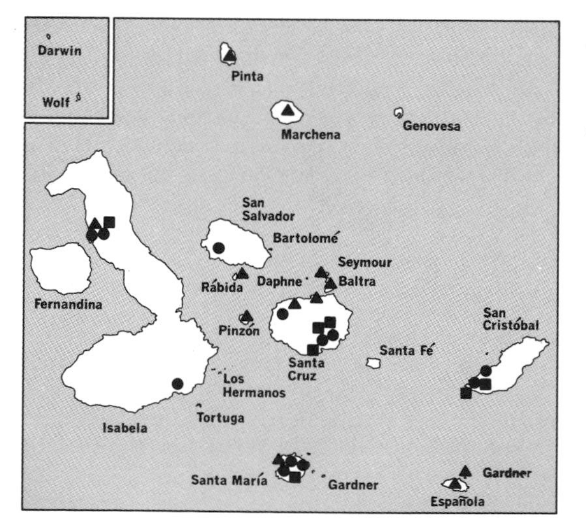

● *Dalea tenuicaulis*
■ *Desmodium canum*
▲ *Desmodium glabrum*

differences in size of leaflets, as indicated in the above description; the larger leaflets are usually on sterile or nearly sterile specimens.

See Fig. 165.

Desmodium Desv., Jour. Bot. ser. 2(1): 122, *t.* 5, 1813, nom. conserv.*

Herbs or shrubs, erect or prostrate, unarmed. Leaves alternate, pinnately 3-foliolate, sometimes unifoliolate, rarely 5-foliolate; stipules and stipels present, persistent or caducous. Flowers white, blue, or pinkish to purplish; inflorescences racemose, sometimes paniculate, axillary and terminal; primary bracts striate, subulate to ovate; secondary bracts usually subulate, occurring with a bract at base of a pedicel; calyx 5-lobed, bilabiate, tube relatively short; petals glabrous or nearly so. Stamens diadelphous, 9:1, vexillar filament partially fused with others, or sometimes monadelphous, the tube usually persistent at base of fruit, the anthers uniform, dorsifixed. Ovary sessile or stipitate, 2- to several-ovulate; style incurved, glabrous, stigma terminal, minute, capitate. Fruit a 1- to many-seeded loment, sessile or stipitate, usually indehiscent, compressed. Seeds orbicular to reniform, compressed, the hilum lateral.

Somc 200–300 species in tropical and warm temperate regions of the world.

Flowers about 2.5–4 mm long; fruit somewhat twisted:
  Fruit with terminal article about 7 mm long and 6 mm wide, the lower articles small, aborted, sometimes lacking; flowers 3.5–4 mm long; leaves 3-foliolate, leaflets moderately to densely pubescent . . . . *D. glabrum*
  Fruit with all articles essentially similar, about 2.5–5 mm long and 1.5–3 mm wide; flowers 2.5–3 mm long; leaves 3-foliolate or sometimes unifoliolate, glabrous to moderately pubescent . . . . *D. procumbens*
Flowers about 5–10 mm long; fruit not twisted:
  Upper suture of fruit straight or nearly so; articles of fruit semiorbicular; flowers about 5 mm long; bracts lanceolate or subulate . . . . *D. canum*
  Upper suture of fruit somewhat zigzag; articles of fruit rhombic or subrhombic; flowers about 6–10 mm long; bracts ovate, acuminate . . . . *D. limense*

Desmodium canum (J. F. Gmel.) Schinz & Thell., Mem. Soc. Neuchat. Sci. Nat. 5: 371. 1914

Hedysarum racemosum Aubl., Pl. Guian. 2: 774. 1775, non Thunb. 1784, nec *D. racemosum* (Thunb.) DC. 1825.
Hedysarum frutescens sensu Jacq., Hort. Vindob. 3: 47, *pl.* 89. 1776, non L. 1753.
Hedysarum incanum Sw., Prodr. Vcg. Ind. Occ. 107. 1788, non Thunb. 1784.
Hedysarum supinum Sw., op. cit. 196. 1788, non Vill. 1779.
Hedysarum canum J. F. Gmel., Syst. Veg. 2: 1124. 1791.
Desmodium incanum DC., Prodr. 2: 322. 1825.
Desmodium supinum DC., loc. cit.
Meibomia cana Blake, Bot. Gaz. 78: 276. 1924.
Desmodium frutescens Schindl., Repert. Spec. Nov. Regni Veg. 21: 9. 1925.†

Perennial herb, sometimes suffrutescent, to about 1 m high; stems erect or sometimes decumbent, puberulent, often glabrescent; stipules deltoid, attenuate, striate, sometimes connate, about 5–10 mm long; stipels subulate, 2–5 mm long; leaves 3-foliolate, the petiole 4–25 mm long; leaflets commonly elliptic, sometimes ovate, about 2–11 cm long, 1–4 cm wide, apex obtuse to acute, base obtuse, upper sur-

* Dr. Bernice G. Schubert has very kindly provided determinations for the numerous collections examined.—V. E. R.

† For additional synonymy, see Macbride (1943, 423f).

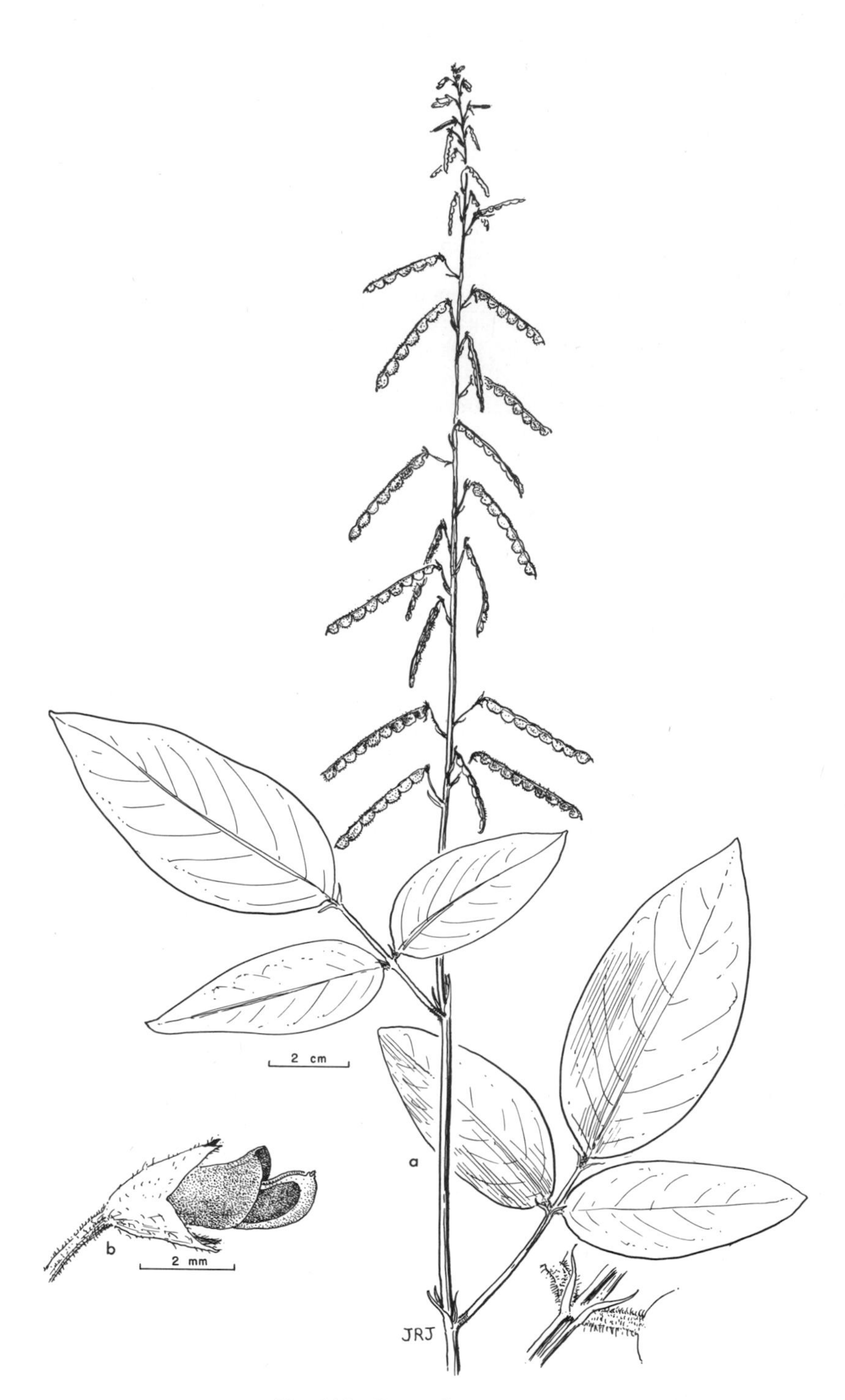

Fig. 166. *Desmodium canum.*

face glabrous or subglabrous, lower surface lighter in color, puberulent, the secondary veins moderately pubescent; inflorescences racemose, terminal or axillary; primary bracts persistent, lanceolate or subulate, about 2–4.5 mm long; secondary bracts subulate, about 1–1.5 mm long; flowers about 5 mm long, calyx 3–4 mm long, deeply lobed, puberulent, sometimes viscid, petals purplish, often fading to bluish pink; fruit brevistipitate, stipe shorter than calyx, pods 3- to 9-articulate, upper suture straight or concave, articles semiorbicular, 3.5–4 mm long, 2.5 mm wide, crisp-pubescent; seeds light brown, 2–2.5 mm long and 1–1.5 mm broad.

A common species in sunny or shaded habitats, often in poor soil, tropical and subtropical America; also introduced into the Old World. Specimens examined: Isabela, San Cristóbal, Santa Cruz, Santa María.

In general the collections from the Galápagos Islands have leaflets larger than average and fruits with more numerous articles and a straighter upper suture. The specimens from Isla Isabela were originally reported as *Rhynchosia*, but though fragmentary they appear to be of *D. canum*.

See Fig. 166.

Desmodium glabrum (Mill.) DC., Prodr. 2: 338. 1825

Hedysarum glabrum Mill., Gard. Dict. ed. 8, *Hedysarum* no. 12. 1768.
Hedysarum molle M. Vahl, Symb. Bot. 2: 83. 1791.
Hedysarum terminale Rich., Act. Soc. Hist. Nat. Paris 1: 112. 1792.
Desmodium molle DC., Prodr. 2: 332. 1825.
Desmodium terminale DC., op. cit. 327.
Meibomia glabra Kuntze, Rev. Gen. 1: 198. 1891.
Meibomia mollis Kuntze, loc. cit.
Meibomia terminalis Kuntze, loc. cit.
Desmodium campestre Brandegee, Univ. Calif. Pub. Bot. 6: 53. 1914.

Perennial herb, erect, to about 1 m tall; stems striate, pubescent; stipules deltoid-subulate, 3–5.5 mm long; stipels subulate, about 1–3.5 mm long; leaves 3-foliolate, the petiole about 5–45 mm long, leaflets ovate or lanceolate, sometimes rhombic, about 2–9 cm long, 2–5 cm broad, acute, mucronate at apex, rounded at base, moderately pubescent with subappressed, sometimes glandular hairs above, lower surface densely pilose, secondary veins conspicuous; inflorescences racemose, sometimes paniculate, terminal or axillary; primary bracts subulate, 2.5–5.5 mm long; secondary bracts subulate, about 1.5 mm long, caducous; flowers about 3.5–4 mm long, calyx 2.5–3 mm long, deeply lobate, puberulent, sometimes viscid; petals purplish or pinkish (greenish yellow when immature); fruit essentially sessile, 1–6-articulate, twisted, the terminal article orbicular to reniform, 7–9 mm long, 5–6 mm wide, puberulent, reticulate, lower articles sterile, about 2 mm long and wide; seeds light brown, about 3 mm long, 2 mm broad.

A species of tropical America, in shrubby cover in various soil types, from sea level to 1,000 m, southern Mexico southward to Peru and in the West Indies. Specimens examined: Baltra, Española, Gardner (near Española), Isabela, Marchena, Pinta, Pinzón, Rábida, Santa Cruz, Santa María, Seymour.

Desmodium limense Hook., Bot. Misc. 2: 2, 5. 1831

Desmodium filiforme Hook. f., Trans. Linn. Soc. Lond. 20: 227. 1847, non Zoll. & Mor. 1847.
Desmodium mandonii Britt., Bull. Torrey Bot. Club 16: 261. 1889.
Meibomia limensis Kuntze, Rev. Gen. Pl. 1: 198. 1891.
Desmodium galapagense Robins., Proc. Amer. Acad. 38: 150. 1902, nom. nov. for *D. filiforme* Hook. f.

Perennial herb, sometimes suffrutescent, erect or procumbent, to about 60 cm tall; stems striate, crisp-pubescent to subglabrous; stipules subulate, about 2–6.5 mm

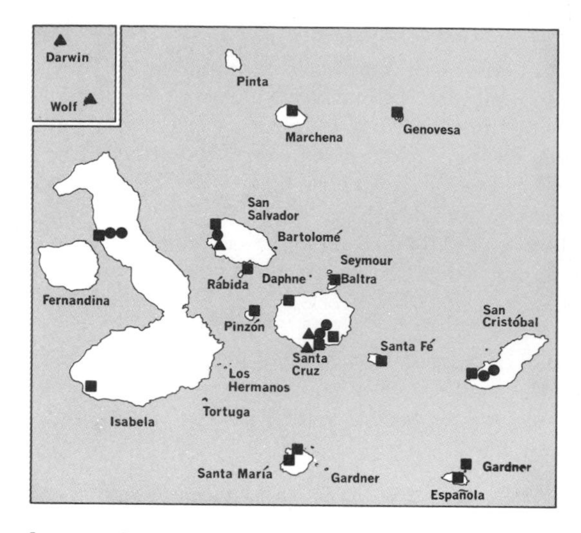

long, caducous; stipels linear, 2–3.5 mm long; leaves 3-foliolate, petioles 1.5–7 cm long, leaflets ovate, sometimes rhombic, about 2–10 cm long, 1–6 cm broad, acute or subacute, base rounded, upper surface moderately pubescent with appressed or subappressed hairs, glabrescent, rarely paler in color along midvein, lower surface densely appressed-pubescent, secondary veins evident; inflorescences racemose, terminal or axillary; primary bracts ovate, striate, acuminate, 5–15 mm long; secondary bracts apparently lacking; flowers about 6–10 mm long, calyx about 4 mm long, deeply lobate, puberulent or subglabrous, petals white to purplish pink or blue; fruit short-stipitate, stipe about as long as calyx, sometimes appearing longer by abortion of lower articles, 4–8-articulate, upper suture usually zigzag, sometimes almost straight, articles rhombic or subrhombic, uncinate-pubescent, 4–5 mm long, 2–3 mm broad; seeds light brown, about 3 mm long, 1.5 mm broad.

Among shrubs and forest or on barren slopes, from sea level to 400 m or more in Ecuador, Peru, Chile, and Bolivia. Specimens examined: Isabela, San Cristóbal, San Salvador, Santa Cruz.

Desmodium limense is very closely related to *D. intortum* (Mill.) Urban and to *D. uncinatum* (Jacq.) DC., and specimens from the Galápagos Islands were previously referred to those species. The type of *D. filiforme* Hook. f. was collected by Darwin on San Salvador, but apparently it has not been collected there since.

Desmodium procumbens (Mill.) Hitchc., Rep. Mo. Bot. Gard. 4: 76. 1893

Hedysarum procumbens Mill., Gard. Dict. ed. 8, *Hedysarum* no. 10. 1768.
Hedysarum spirale Sw., Prodr. Veg. Ind. Occ. 107. 1788.
Desmodium spirale DC., Prodr. 2: 332. 1825.
Desmodium tenuiculum DC., op. cit. 333.

Herb, apparently annual, to about 1 m high, commonly about 20–50 cm tall; stems usually procumbent or decumbent, sometimes erect, uncinulate-puberulent, glabrescent, subangulate; stipules ovate, acuminate, striate, ciliate, about 1–4.6 mm long; stipels subulate, ciliate, about 0.5–2.5 mm long; leaves 3-foliolate or some 1-foliolate, the petiole about 5–25 mm long, leaflets predominantly ovate to rhombic, sometimes elliptic, 0.5–5.5 cm long, 0.3–3 cm broad, apex acute or obtuse, base acute, obtuse, or truncate, surfaces glabrous to moderately pubescent with subappressed or uncinate hairs, venation usually conspicuous; inflorescences racemose-paniculate, terminal or axillary; bracts linear, about 1–3 mm long; flowers 2.5–3 mm long, calyx about 2 mm long, deeply lobate, strigillose, petals pink to white; fruit short-stipitate, about equal to or slightly longer than calyx, 3–6-articulate, somewhat twisted, upper suture zigzag, articles rhombic, about 2.5–5 mm long, 1.5–3 mm wide, uncinate-pubescent; seeds light brown, about 2.5 mm long, 1.5 mm wide.

Known from Mexico, Central America, the West Indies, and northern South America in diverse habitats; apparently introduced into tropical Africa and the Philippine Islands. Specimens examined: Baltra, Española, Gardner (near Española), Genovesa, Isabela, Marchena, Pinzón, Rábida, San Cristóbal, San Salvador, Santa Cruz, Santa Fé, Santa María.

The majority of the specimens examined appear to be typical *D. procumbens*, but a few tend toward var. *exiguum* (A. Gray) Schubert and var. *transversum* (Robins. & Greenm.) Schubert. These 2 varieties are characterized by numerous unifoliolate leaves with leaflets differing somewhat in size and shape from those of the other, 3-foliate leaves on the same plant. The complexities of the species and its varieties have been discussed by Schubert (1940, 3–16).

Erythrina L., Sp. Pl. 706. 1753; Gen. Pl. ed. 5, 316. 1754

Trees, shrubs, or perennial herbs, often armed, sometimes leafless at time of flowering. Leaves alternate, pinnately 3-foliolate; stipules linear or lanceolate, caducous; stipels commonly glandlike, ovate to oblong. Flowers usually red to orange, relatively large and showy, single or in axillary or terminal racemes. Bracts and bracteoles present. Calyx 5-dentate, campanulate or spathaceous. Corolla with vexillum much longer than wings and keel. Stamens 10, monadelphous or the vexillar stamen free, filmaments usually alternately unequal in length, anthers dorsifixed. Ovary usually many-ovulate, stipitate, style glabrous, stigma small, capitate. Legume 2-valved, dehiscent, many-seeded. Seeds reniform-ellipsoid, variously colored, often red or brownish, hilum lateral.

Some 150–200 species in the tropics and subtropics of the Old and New Worlds.

Erythrina velutina Willd., Ges. Nat. Freunde Berline Neue Schr. 3: 426. 1801

Chirocalyx velutinus Walp., Flora 36: 148. 1853.
Corallodendron velutinum Kuntze, Rev. Gen. Pl. 173. 1891.
Erythrina splendida Diels, Bibl. Bot. 116: 96. 1937.*

Tree to about 12 m tall, usually armed with stout spines to 1.5 cm long, leafless at anthesis; young stems densely tomentose with stellate hairs, glabrescent; stipules linear, about 1 mm long, caducous; petioles 4–27 cm long, densely stellate-pubescent; leaflets mostly deltoid-rhombic, 4–16 cm long, 4–19 cm broad, obtuse to emarginate at apex, truncate, subcordate, or obtuse at base, surfaces densely stellate-pubescent, upper surface glabrate, secondary veins evident; stipels elliptic to

* There is also a forma *aurantiaca* (Ridley) Krukoff (1939, 329) from Brazil that is similar to the typical form except for the seeds, which are black with a red band around the hilum.

oblong, glandlike, persistent, 1–2 mm long; bracts ovate, 2.3–2.9 mm long, 1.5–1.8 mm wide; bracteoles linear-lanceolate, 0.7–1.2 mm long, 0.4 mm wide; flowers about 4–6 cm long in axillary racemes, calyx spathaceous, stellate-tomentose, sometimes glabrescent, 2–3 cm long, petals orange-red, sparsely stellate-pubescent, subglabrous with the vexillum 3–3.5 cm long, wings and keel shorter; stamens monadelphous; fruit stellate-tomentose but usually glabrous at maturity, 1- to 4-seeded, rarely 5- or 6-seeded, 7.5–14 cm long, 1.2–1.7 cm broad, somewhat constricted between seeds; seeds 11–17 mm long, 6–11 mm broad, red with a black line about 3 mm long extending from hilum toward chalazal end.

Known from the West Indies and northern South America on cinder slopes, lava flows, and rocky hillsides; beyond the natural range as an ornamental. Specimens examined: Darwin, San Salvador, Santa Cruz, Wolf.

See Fig. 167.

Fig. 167. *Erythrina velutina.*

Galactia P. Br., Nat. Hist. Jam. 298. 1756

Herbs, climbing, prostrate, or suberect, unarmed, the stems sometimes lignescent. Leaves alternate, pinnately 3-foliolate, rarely 1-, 5-, or 7-foliolate. Stipules linear to lanceolate; stipels linear. Flowers axillary racemes or fascicles, sometimes only 1- or 2-flowered. Calyx 5-lobed, vexillar lobes usually connate. Petals red or purple to white, the vexillum glabrous on outer face. Stamens 10, diadelphous 9:1, vexillar stamen free, anthers uniformly dorsifixed. Ovary many-ovulate, essentially sessile, style glabrous, stigma terminal, minute, subcapitate. Legume 2-valved, dehiscent, compressed, many-seeded. Seeds elliptic-subreniform, brownish to black, often mottled, the hilum lateral, sometimes apical.

About 50 species distributed in the tropics and warm temperate regions of the world.

Leaflets pubescent below with short, appressed hairs; calyx sparsely to moderately appressed-pubescent; fruit 3–4 cm long, 4–6 mm wide, sparsely to densely pubescent with appressed to subappressed hairs . . . . *G. tenuiflora*
Leaflets densely to moderately tomentulose below; calyx tomentulose; fruit tomentulose, 3–7 cm long, 5–9 mm wide . . . . *G. striata*

Galactia striata (Jacq.) Urban, Symb. Antill. 2: 320. 1900

Glycine striata Jacq., Hort. Vindob. 1: 32, *pl. 76.* 1770.
Galactia velutina Benth. in Tayl., Ann. Nat. Hist. 3: 437. 1839, non *Collaea velutina* Benth. 1837.
Odonia tomentosa Bertol., Lucubr. 35. 1822.
Phaseolus tomentosus Anderss., Kongl. Svensk. Vet.-Akad. Handl. 1853: 250. 1854 (preprint). 1855.
Galactia jussiaeana var. *volubilis* Benth. in Mart., Fl. Bras. 15^1: 143. 1859, non *G. volubilis* Britt. 1894.
Galactia tomentosa Urban, Symb. Antill. 1: 472. 1900, non Spreng. 1826.
Galactia striata (Jacq.) Urban var. *tomentosa* Urban, op. cit. 321.
Odonia refusa Rose, Contr. U.S. Nat. Herb. 10: 102. 1907.

Herbaceous vine, sometimes lignescent, the stems pubescent with hairs more or less spreading to somewhat retrorse; stipules linear to subulate, 1–4 mm long; leaflets ovate to elliptic or suborbicular, about 1.5–6.5 cm long, 1–4 cm wide, apex obtuse or sometimes acute, base rounded, upper surface puberulent, glabrescent, lower surface densely to moderately tomentulose, secondary veins evident; inflorescences racemose, often short, few- to many-flowered; bracts and bracteoles deltoid to linear, about 1–3 mm long; calyx tomentulose, 5–9 mm long, deeply lobed; petals pink or lavender to white; fruit essentially sessile, 3–7 cm long, 5–9 mm wide, tomentulose, sometimes glabrescent, usually about 1- to 10-seeded; seeds 3–4 mm long, 2–3 mm wide, 1–1.5 mm thick, dark brown and black mottled.

The West Indies and northern South America, and possibly also in other parts of South America and Mexico. Specimens examined: Champion, Fernandina, Isabela, Pinta, Pinzón, Rábida, San Cristóbal, San Salvador, Santa Cruz, Santa María, Seymour.

Only partial synonymy for this species has been given above. Both the nomenclature and the taxonomy of the genus *Galactia* are unbelievably confused, chiefly because the taxa have been studied piecemeal and not considered on a worldwide basis. After examining types of the taxa listed above, I believe them to be essentially the same. Fawcett and Rendle earlier, in their treatment of *Galactia* for the Flora of Jamaica (4: 57. 1920), pointed out that Urban's "var. *tomentosa* agrees with Jacquin's type of the species." Specimens from Africa and Asia that have been

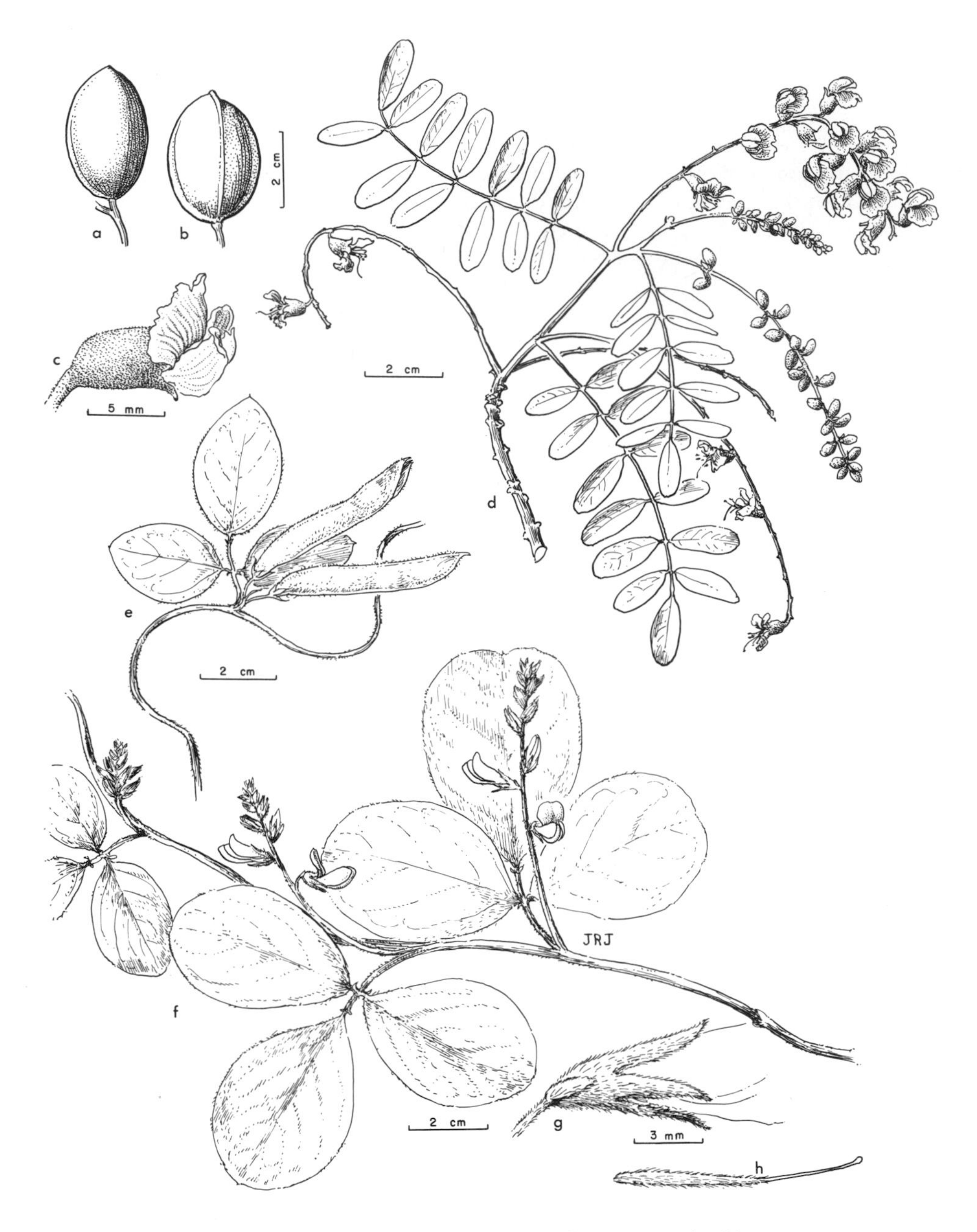

Fig. 168. *Geoffroea spinosa* (*a–d*). *Galactia striata* (*e–h*).

determined as *G. tenuiflora* var. *villosa* (Wight & Arn.) Benth. also appear to be referable to *G. striata.*
See Fig. 168.

Galactia tenuiflora (Klein ex Willd.) Wight & Arn., Prodr. 206. 1834
Glycine tenuiflora Klein ex Willd., Sp. Pl. 3: 1059. 1800.
Glycine dubia DC., Prodr. 2: 238. 1825.

Herbaceous vine; stems retrorsely strigillose, glabrescent; stipules linear-lanceolate, to about 2 mm long; leaves 3-foliolate, petiole 1.5–3 cm long; stipels linear, minute; leaflets ovate to elliptic, about 1–4 cm long, 0.5–3 cm wide, apex acute to emarginate, base rounded to subcordate, upper surface pubescent along midvein, otherwise sparsely and minutely appressed-pubescent to glabrous, lower surface sparsely to densely pubescent with appressed hairs, secondary veins usually inconspicuous; inflorescences racemose, few-flowered; flowers about 8–13 mm long; bracts and bracteoles linear, 1–2 mm long; calyx about 5 mm long, deeply lobed, sparsely appressed-pubescent; petals pink or bluish to lavender; fruit essentially sessile, 3–4 cm long, 4–6 mm wide, usually about 5- to 8-seeded, appressed-pubescent or hairs subappressed; seeds brown, mottled, about 3 mm long, 2 mm wide, 1 mm thick.

A weedy species, ranging through several zones, from eastern Asia, Australia, and the Pacific islands to the West Indies, western Mexico, and South America. Specimens examined: Española, Gardner (near Española), Isabela, San Cristóbal, Santa Cruz, Santa María.

Geoffroea Jacq., Enum. Pl. Carib. 7, 28. 1760;
Select. Stirp. Amer. Hist. 207, *tab. 180, f. 62.* 1763

Trees or shrubs, sometimes spiny. Leaves alternate, imparipinnate, usually 5- to many-foliolate; leaflets alternate to subopposite; stipules caducous; stipels lacking. Flowers yellowish to orange, in axillary racemes. Bracts small, caducous; bracteoles

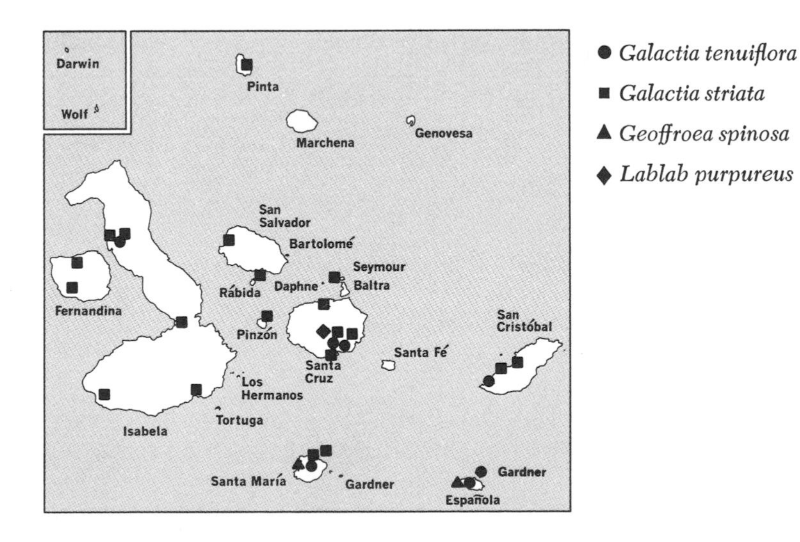

apparently absent. Calyx campanulate with 5 subequal lobes. Petals 5, free. Stamens 10 with the filaments variously connate, free almost to base, or with the vexillar stamen free and others connate, anthers dorsifixed. Ovary sessile, 2- to 5-ovulate, style glabrous, stigma small, terminal. Legume indehiscent, drupaceous, woody, 1-seeded. Seeds subreniform with lateral hilum or lenticular with subapical hilum.

Two species, both native to South America.

Geoffroea spinosa Jacq., Enum. Pl. Carib. 28. 1760; Select. Stirp. Amer. Hist. 207, *tab. 180, f. 62.* 1763

Robinia striata Willd., Sp. Pl. 3: 1132. 1803.
Geoffroea superba Humb. & Bonpl., Pl. Equinox. 2: 69, *t. 100.* 1809.
Geoffroea bredemeyeri HBK., Nov. Gen. & Sp. Pl. 6: 379. 1823.
Geoffroea striata Morong in Morong & Britt., Ann. N. Y. Acad. Sci. 7: 87. 1892.

Tree or shrub to about 11 m tall; young stems tomentulose, glabrescent; stipules subdeltoid to ovate, about 2–5 mm long; leaves about 9–17-foliolate, oblong to obovate, 1.5–4.5 cm long, 0.5–2 cm wide, obtuse to emarginate, surfaces puberulent, glabrescent, secondary veins numerous, parallel, moderately conspicuous; bracts ovate, about 3 mm long; flowers 12–14 mm long; calyx about 5–7 mm long, tomentulose, lobes 2–3 mm long; fruit tomentulose, 3–5 cm long, 2–3.5 cm wide, about 1.5–2.5 cm thick; seeds light brown, lenticular, hilum apical.

From Colombia and Venezuela to Argentina, as a forest tree and often planted as an ornamental. Specimens examined: Española, Santa María.

See Fig. 168.

Glycine L., Sp. Pl. 753. 1753; Gen. Pl. ed. 5, 334. 1754

Herbs, unarmed. Leaves alternate, pinnately or rarely digitately 3-foliolate; stipules and stipels small, deltoid. Flowers white, pink, blue, or purplish, usually in axillary racemes, sometimes solitary or in terminal panicles; bracts and bracteoles present, usually lanceolate to ovate; calyx 5-dentate, vexillar teeth somewhat connate. Stamens 10, uniform, monadelphous or belatedly diadelphous with vexillar filament separating from others, anthers small, uniform, dorsifixed. Ovary subsessile, few- to many-ovulate, style glabrous, slightly curved, stigma capitate, sometimes penicillate. Legume 2-valved, dehiscent, compressed or inflated. Seeds subreniform, oblong or subspherical with small, lateral hilum and small, scalelike, papyraceous caruncle.

A genus of 10 species native to the Old World tropics, some of them introduced into the New World.

Glycine max (L.) Merr., Interpr. Rumph. Herb. Amboin. 274. 1917*

Phaseolus max L., Sp. Pl. 725. 1753.
Dolichos soja L., op. cit. 727, non *Glycine soja* Sieb. & Zucc. 1843.
Soja hispida Moench, Meth. 153. 1794.
Glycine hispida Maxim, Acad. Pétersb. Bull. 18: 398. 1873.
Soja max Piper, Amer. Soc. Agron. Jour. 6: 84. 1914.

Annual, bushy herb to about 2 m high, a few shoots sometimes vinelike; young stems somewhat angled, pubescent with spreading hairs; stipules ovate, acute, 3–7

* A more complete description and additional synonyms are to be found in Hermann (1962).

Fig. 169. *Glycine max.*

mm long, striate; stipels lanceolate to setaceous, 1–3.5 mm long; leaves pinnately 3-foliolate, petiole (2)5–20 cm long; leaflets broadly ovate to suborbicular or elliptic-lanceolate, usually acute at apex, sometimes obtuse, 3–14 cm long, 2.5–10 cm wide, surfaces moderately pubescent with lax hairs, often glabrescent, secondary veins evident; inflorescences racemose, axillary, few-flowered; bracts lanceolate, striate, 4.5–5.5 mm long; bracteoles lanceolate, 2–3 mm long, caducous; flowers usually 6–7 mm long, calyx 4–7 mm long, villous or subappressed-pubescent, teeth about 2–3 mm long, attenuate, petals white, pink, or bluish to purple; fruit oblong, subfalcate, 2.5–7.5 cm long, 8–15 mm wide, coarsely hirsute or setose; seeds ovoid to subspherical, greenish tan or olive to reddish black, 6–11 mm long, 5–8 mm in diameter.

A cultivated species (soybean) native to Asia, introduced into other areas and occasionally found as an escape.

I have seen no specimens of *Glycine* from the Galápagos Islands, but Christopherson (1932, 78) cited as *Glycine soja* (L.) Sieb. & Zucc.: "Indefatigable: In the moist zone, Academy Bay, December 1926, no. 5. Not before reported as growing spontaneously on the Galápagos Islands."

See Fig. 169.

Lablab Adans., Fam. Pl. 2: 325. 1763

Herbs, usually climbing, unarmed. Leaves alternate, pinnately 3-foliolate. Stipules and stipels deltoid, striate. Flowers in long-pedunculate, axillary, nodose racemes. Calyx campanulate, 5-lobed, vexillar lobes somewhat adnate. Petals purple or pink to white, vexillum glabrous on outer face, keel rostrate. Stamens 10, diadelphous 9:1 with vexillar filament free, anthers uniform, dorsifixed. Ovary sessile, few- tò many-ovulate; style bent at a right angle, longitudinally bearded toward apex; stigma glabrous, terminal, subcapitate. Fruit 2-valved, dehiscent, oblong, falcate, usually few-seeded. Seeds dark brown to black, subellipsoid, hilum linear.

About 2 or 3 tropical and subtropical species.

Lablab purpureus (L.) Sweet, Hort. Brit. 481. 1827

Dolichos lablab L., Sp. Pl. 725. 1753.
Dolichos purpureus L., Sp. Pl. ed. 2, 1021. 1762.
Lablab niger Medik., Vorl. Churpf. Phys. Ges. 2: 354. 1787.
Lablab vulgaris Savi, Diss. 19. 1821.
Vigna aristata Piper, Contr. U. S. Nat. Herb. 22: 665. 1926.*

Climbing herb, stems puberulent, glabrescent; stipules deltoid, about 5 mm long; stipels narrowly deltoid, about 3–5 mm long; leaflets ovate to rhombic, acuminate, 4–12 cm long, 3–12 cm broad, cuneate to subtruncate at base, the surfaces glabrous or sparsely pubescent, secondary veins evident; inflorescences usually many-flowered; bracts and bracteoles ovate, about 2–5 mm long; flowers 1.5–2 cm long; calyx 5–7 mm long, glabrous or sparsely pubescent, teeth 1–2 mm long; petals purple or pink to white; fruit 4–10 cm long, 2–2.5 cm broad, glabrous, minutely serrate along margins, commonly 1- to 4-seeded; seeds black or dark brown, about 13 mm long, 8 mm broad, and 7 mm thick, the hilum raised, white, apical to midlateral, about 12 mm long.

Widespread in the tropics and introduced into other areas as an ornamental. Specimen examined: Santa Cruz.

See Fig. 170.

Mucuna Adans., Fam. Pl. 2: 325. 1763, nom. conserv.

Woody or herbaceous vines, shrubs, or erect herbs, unarmed. Leaves alternate, pinnately 3-foliolate, stipulate and often stipellate. Flowers usually large and showy, red, purple, or greenish yellow, borne in axillary, racemose or subcymose inflorescences. Bracts small and caducous. Calyx campanulate, 5-dentate but with the 2 vexillar teeth connate. Corolla with vexillum shorter than wings and keel, the vexillum usually shortest petal. Stamens diadelphous 9:1, vexillar filament free, anthers alternately long and basifixed or shorter and dorsifixed, usually barbate. Ovary sessile, villous, few-ovulate, style pubescent below, glabrous toward apex, stigma terminal, capitate, penicillate. Legume relatively large, oblong or linear, 2-valved, dehiscent, often covered with stinging hairs. Seeds lenticular or oblong, hilum elliptic or linear.

About 100 species in the tropics and subtropics of the Old and New Worlds.

* In addition to the names listed above, there are several putative synonyms reported from various parts of the world.

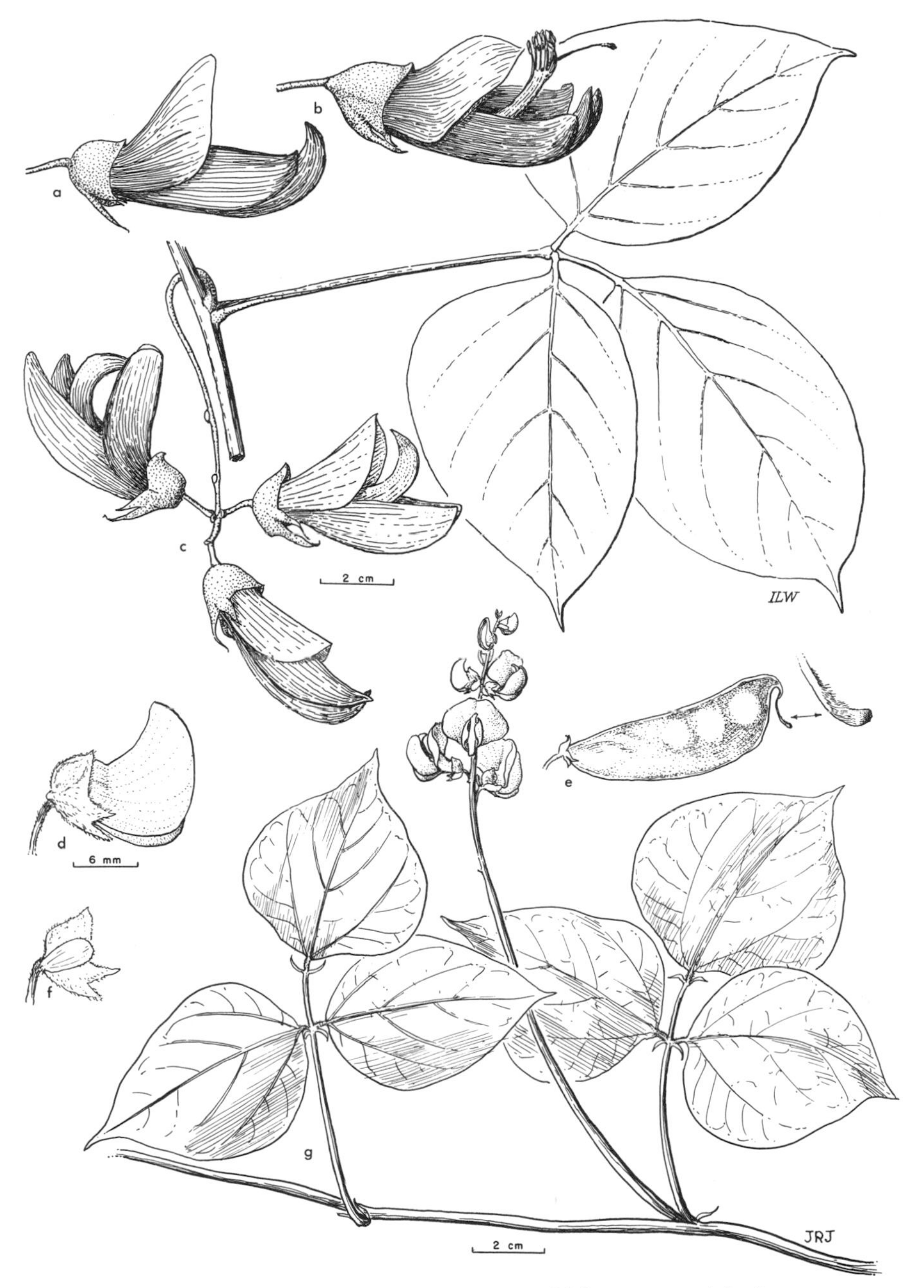

Fig. 170. *Mucuna rostrata* (*a–c*). *Lablab purpureus* (*d–g*).

Mucuna rostrata Benth. in Mart., Fl. Bras. 15^{1}: 171, *pl. 47.* 1859

High-climbing vine, herbaceous or somewhat woody; young stems pubescent, glabrescent; stipules lanceolate, about 5 mm long; stipels not seen; leaves with petioles about 5–15 cm long; leaflets ovate, rhombic, elliptic, about 9–16 cm long, 5–10 cm wide, abruptly acuminate, rounded at base, upper surface with some pubescence along midvein, otherwise glabrous, lower surface sparsely to moderately strigillose, secondary veins evident; inflorescences racemose, many-flowered, with enlarged nodes; bracts minute, caducous; flowers 6–8 cm long, calyx sericeous and sometimes also setose, 2–2.5 cm long, teeth 5–10 mm long, petals glabrous, orange-red, sometimes spotted, usually darkening on drying, the standard 4.5–6 cm long, wings 6–7 cm long, keel rostrate, 6.5–8 cm long; fruit oblong, 8–15 cm long, 3.5–4 cm wide, 1.5 cm thick, transversely rugose, densely pubescent with stinging hairs; seeds lenticular, about 2 cm long, 1.8 cm wide, 1.3 cm thick, grayish brown with a darker, circumcinct hilum.

Among forest trees and shrubs at moderate altitudes, Guatemala and British Honduras southward to Peru, Bolivia, and Brazil. Specimen examined: Santa Cruz.

A single specimen, consisting of flowers only, is known from the Galápagos Islands.

See Fig. 170.

Phaseolus L., Sp. Pl. 723. 1753; Gen. Pl. ed. 5, 323. 1754

Herbaceous vines, rarely somewhat woody at base, unarmed. Leaves alternate, pinnately 3-foliolate, rarely 1-foliolate. Stipules and stipels present, usually striate. Flowers in racemose, usually long-pedunculate, nodose inflorescences. Calyx campanulate, 5-toothed or apparently 4-toothed owing to fusion of vexillar pair of teeth. Petals white or yellowish to red or purple, vexillum usually glabrous on outer face, keel spirally contorted. Stamens 10, diadelphous 9:1 with vexillar stamen free, filament often appendaged or thickened above base; anthers uniform, dorsifixed. Ovary subsessile, many-ovulate, style spirally contorted, thickened within keel, longitudi-

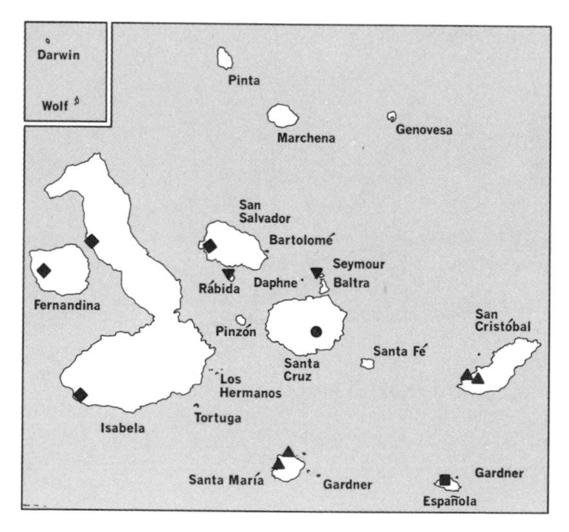

● *Mucuna rostrata*
■ *Phaseolus adenanthus*
▼ *Phaseolus atropurpureus*
▲ *Phaseolus lathyroides*
◆ *Phaseolus mollis*

nally bearded toward apex; stigma lateral or oblique. Fruit 2-valved, dehiscent, linear or oblong, sometimes falcate, compressed or subterete, few- to many-seeded. Seeds reniform to ellipsoid, hilum elliptic or linear, lateral.

About 150 species in the tropics and warm temperate regions of the world.

Legumes linear, 2–3 mm broad; flowers 15–20 mm long; keel 1-spiralled:
 Plant sprawling or climbing, generally moderately to densely pubescent; fruit tomentulose or subsericeous, about 3 mm broad; petals dark purple. . . . *P. atropurpureus*
 Plant erect, generally moderately to sparsely pubescent; fruit moderately appressed-pubescent, 2–3 mm broad; petals red to purple. . . . . *P. lathyroides*
Legumes oblong, 6–10 mm broad; flowers less than 15 mm long, or much larger and 20–30 mm long; keel 2-spiralled:
 Flowers purplish, 10–12 mm long; fruit about 2.5–3 cm long, 6–7 mm broad. . . *P. mollis*
 Flowers pink or lavender to white, 20–30 mm long; fruit about 7–12 cm long, 8–10 mm broad . . . . *P. adenanthus*

Phaseolus adenanthus G. F. W. Mey., Prim. Fl. Esseq. 239. 1818

Phaseolus truxillensis HBK., Nov. Gen. & Sp. Pl. 6: 451. 1824.

Herbaceous, perennial, climbing vine; stems puberulent with spreading hairs, glabrescent; stipules ovate, striate, about 3 mm long; stipels deltoid-ovate, 1 mm long; leaves 3-foliolate, petioles about 4–9 cm long; leaflets ovate, acute, base rounded to subtruncate, surfaces moderately pubescent with appressed hairs, glabrescent, secondary veins evident; inflorescences elongate, about 5–15 cm long, relatively few-flowered, solitary or in pairs; bracts and bracteoles ovate, striate, about 2–5 mm long, caducous; flowers 20–30 mm long; calyx glabrous or subglabrous, about 5 mm long, carinal tooth broad, emarginate, about 1.5 mm long, other teeth attenuate, about 1.5–2 mm long; petals pink or lavender to white, vexillum glabrous, keel and style 2-spiralled; fruit oblong, about 7–12 cm long, 8–10 mm wide, many-seeded, pubescent, glabrate; seeds subreniform, brown, about 5 mm long, 3.5 mm wide, 2 mm thick, hilum lighter in color, lateral.

A common species in forests and shrubby cover to open hillsides, from southern Mexico and the West Indies to Peru and Argentina. Specimen examined: Española.

Phaseolus atropurpureus DC., Prodr. 2: 395. 1825

Phaseolus vestitus Hook., Bot. Misc. 2: 213. 1831.
Phaseolus atropurpureus var. *vestitus* (Hook.) Hassler, Candollea 1: 457. 1923.

Sprawling or climbing vine, stems pubescent with spreading hairs; stipules densely pubescent, striate, attenuate, about 2–3 mm long; stipels linear, 1–1.5 mm long; leaves 3-foliolate, petioles about 2–5 cm long; leaflets ovate to rhombic or suborbicular, sometimes with a lateral lobe, obtuse to acute, 2.5–5 cm long, 1–3 cm wide, surfaces moderately to densely pubescent with soft, subappressed hairs, secondary veins evident, sometimes conspicuous; inflorescences elongate, about 15–20 cm long, relatively few-flowered, flowers solitary or in pairs; bracts and bracteoles linear, about 1–2 mm long, caducous; flowers, exclusive of coiled keel and style, about 15–20 mm long; calyx tomentulose, 4–8 mm long, teeth attenuate, 1–2 mm long; petals dark purple, vexillum glabrous, keel and style 1-spiralled; fruit linear, essentially straight, tomentulose or subsericeous, sometimes glabrescent, many-seeded, 5–9 cm long, about 3–4 mm wide; seeds elliptic-subreniform, light to dark brown

Fig. 171. *Phaseolus mollis.*

with darker spots, 2–4 mm long, 1.5–2 mm broad, the hilum 1 mm long or shorter.

A common species in various habitats from southwestern United States to Peru and Bolivia. Specimens examined: Rábida, Seymour.

Phaseolus lathyroides L., Sp. Pl. ed. 2, 1018. 1763

Phaseolus semierectus L., Mant. Pl. 1: 100. 1767.
Phaseolus lathyroides var. *semierectus* Hassler, Candollea 1: 447. 1923.
Macroptilium lathyroides Urban, Symb. Antill. 9: 457. 1928.

Herbaceous annual, stems erect, puberulent, glabrescent; stipules striate, lanceolate, about 5–10 mm long; leaves 3-foliolate, petioles 1–3 cm long; stipels linear-lanceolate, about 1 mm long; leaflets oblong to ovate, 1.5–7 cm long, 1–3 cm wide, apex obtuse to acute, base rounded, puberulent to glabrous on both surfaces, secondary veins evident but inconspicuous; inflorescences elongate, 7–30 cm long, usually many-flowered, flowers in pairs; bracts and bracteoles subulate, 2–6 mm long, caducous; flowers about 15–20 mm long exclusive of coiled keel and style; calyx moderately appressed-pubescent, 4–6 mm long with 5 deltoid teeth about 1 mm long; petals blood red to dark purple, vexillum glabrous, keel and style 1-spiralled; fruit linear, straight, many-seeded, moderately pubescent with appressed hairs, about 6–10 cm long, 2–3 mm broad; seeds ellipsoid, dark brown and black mottled, about 3 mm long, 1.5 mm broad, hilum white, about 1 mm long.

A widespread species in forests and open scrub of the tropics and subtropics. Specimens examined: San Cristóbal, Santa María.

Phaseolus mollis Hook. f., Trans. Linn. Soc. Lond. 20: 228. 1847

Sprawling vine, stems pubescent with spreading hairs, glabrescent; stipules lanceolate, about 2–3 mm long; stipels about 1 mm long; leaves 3-foliolate, petioles about 2–6 cm long; leaflets ovate, obtuse to short-acuminate, about 3–5.5 cm long, 1.5–4.5 cm wide, base rounded to subtruncate, surfaces puberulent, secondary veins evident; inflorescences elongate, to about 30 cm long, many-flowered, flowers solitary or in pairs; bracts and bracteoles ovate, about 1–1.5 mm long, caducous; flowers exclusive of coiled keel and style about 10–12 mm long; calyx tomentulose, about 3 mm long, with connate vexillar teeth, truncate, other teeth deltoid, sometimes obtuse, about 1 mm long; petals purplish, vexillum glabrous, keel and style 2-spiralled; fruit oblong, compressed, somewhat falcate, tomentulose, about 2–4-seeded, 2.5–3 cm long, 6–7 mm wide, mature seeds not seen.

Endemic, or possibly a variant of the widespread species *P. lunatus* L.; in forest, thickets, and in shade of cliffs. Specimens examined: Fernandina, Isabela, San Salvador.

The type of the species was collected by Darwin on San Salvador.

See Fig. 171.

Piscidia L., Syst. Nat. ed. 2, 1155, 1376. 1759, nom. conserv.*

Trees or shrubs, unarmed. Leaves alternate, imparipinnate, 5–19-foliolate; leaflets opposite; stipules (or bud scales?) broadly ovate-deltoid or semiorbicular, early

* This is one of the genera included in the so-called fish-poison plants, known in Spanish as *barbasco*.

caducous; stipels absent. Flowers white to pink or lavender in short axillary or lateral panicles. Bracts apparently lacking; bracteoles ovate to oblong, paired, at base of calyx. Calyx campanulate with 5 short, subequal teeth, vexillar pair often connate. Corolla with vexillum usually pubescent on outer face but glabrous in 1 species, wing petals commonly a little longer than vexillum, adherent to keel. Stamens 10, vexillar filament free at base, united with others above to form a tube; anthers dorsifixed. Ovary sessile, many-ovulate, style glabrous above; stigma penicillate. Fruit indehiscent, 1–8-seeded, compressed, with 4 longitudinal wings. Seeds reniform, compressed, hilum lateral, elliptic.

About 10 species occurring in the tropics and subtropics of the New World from the Caribbean area southward to Peru.

Piscidia carthagenensis Jacq., Enum. Pl. Carib. 27. 1760

Lonchocarpus guaricensis Pittier, Trab. Mus. Com. Venez. 4: 231. 1928; Arbol y Arbust. Legum. 3: 301. 1928.
Piscidia guaricensis Pittier, Mesa Guanipa 49. 1942.

Tree or shrub to about 15 m tall; young stems puberulent, glabrescent; stipules broadly ovate, pubescent, 3–4 mm long; leaves 7–13-foliolate, petioles about 2–6 cm long, leaflets commonly obovate, sometimes elliptic or ovate, rounded to subacute, about 4–20 cm long, 3–9 cm wide, glabrous above, sparsely to densely puberulent below with lax hairs, secondary veins evident; bracteoles ovate, 1.5 mm long, caducous; flowers 13–18 mm long, calyx 5–7 mm long, cinereo-sericeous, lobes obtusely deltoid, about 1 mm long; petals white to pink or lavender, somewhat pubescent; fruit 5–11 cm long, 3–4 cm wide, pubescent, glabrescent, body 4–5 mm wide, wings 13–18 mm wide, stipe 5–8 mm long or sometimes appearing longer owing to abortion of lower ovules; seeds reddish brown, about 5 mm long, 3 mm wide.

In Transition and moist zones, Colombia, Venezuela, and Ecuador. Specimens examined: San Cristóbal, Santa Cruz.

See Fig. 172 and color plates 44 and 45.

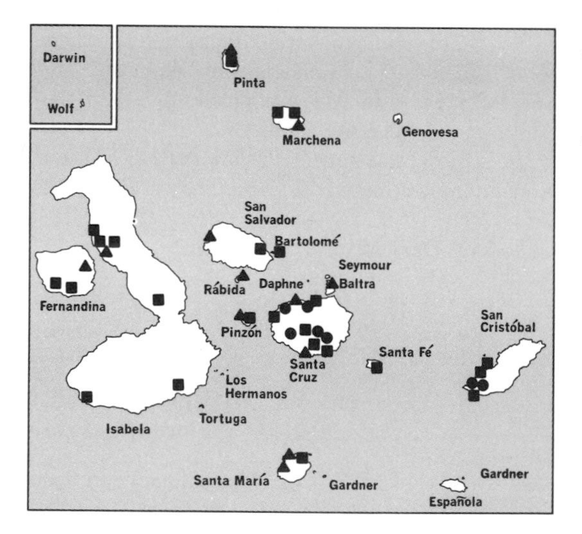

● *Piscidia carthagenensis*
■ *Rhynchosia minima*
▲ *Stylosanthes sympodialis*

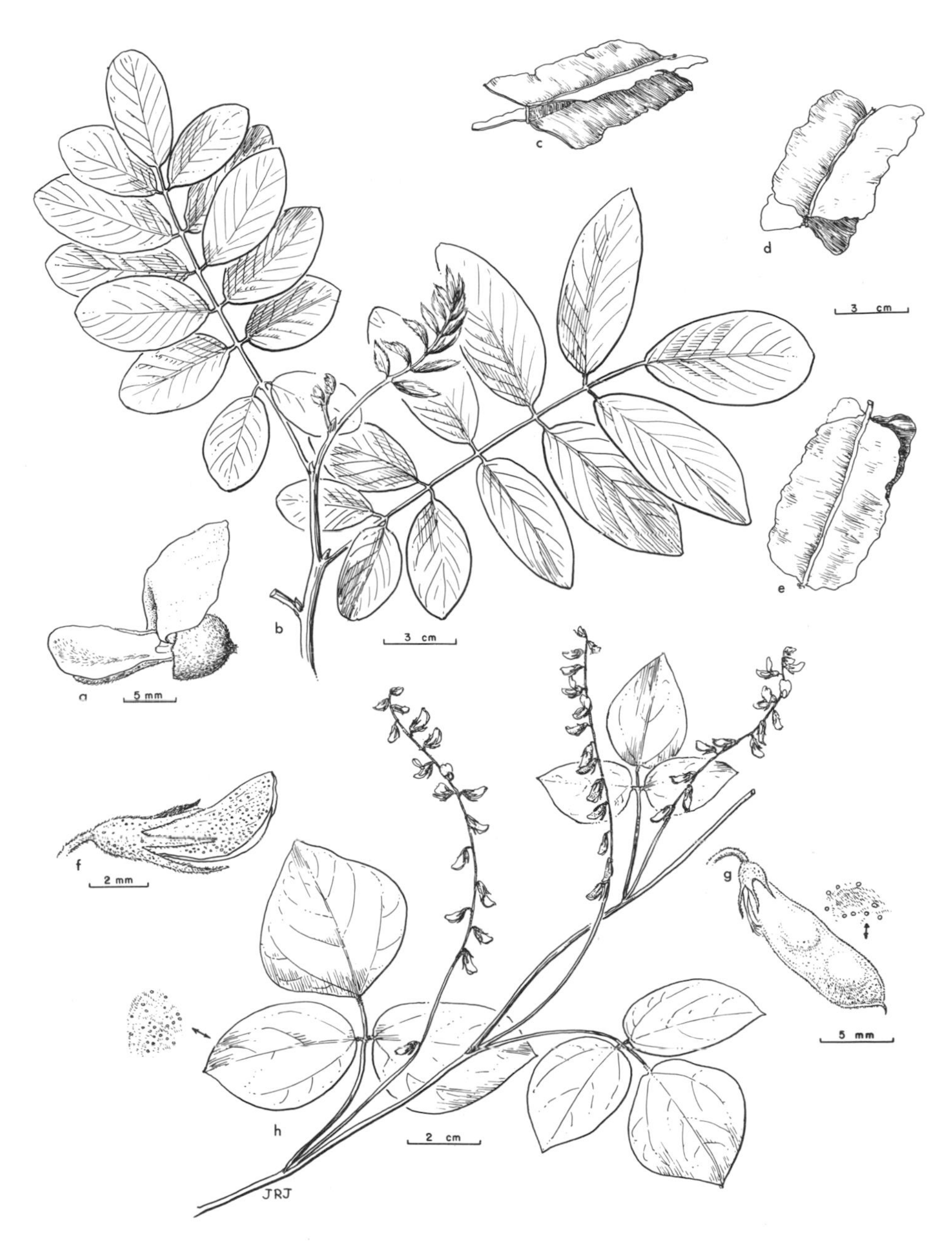

Fig. 172. *Piscidia carthagenensis* (*a–e*): *c–e*, segments of winged pod. *Rhynchosia minima* (*f–h*).

Rhynchosia Lour., Fl. Cochinch. 460. 1790, nom. conserv.

Herbs, unarmed, sometimes suffrutescent, usually climbing or prostrate, rarely erect. Leaves pinnately 3-foliolate or sometimes 1-foliolate; leaflets and other parts glandular-dotted. Stipules ovate to lanceolate. Stipels present or absent. Flowers in axillary racemes, rarely solitary. Bracts small, caducous; bracteoles lacking. Calyx 5-lobed, sometimes apparently 4-lobed by fusion of vexillar lobes. Petals yellow, sometimes purplish markings present, vexillum usually pubescent and glandular. Stamens 10, diadelphous 9:1, anthers uniform, dorsifixed. Ovary 1- to 3-ovulate, commonly 2-ovulate, subsessile, style glabrous, stigma minute, capitate. Legume 2-valved, dehiscent, compressed, 1- or 2-seeded. Seeds subglobose, elliptic, or reniform, brown, black, or red, hilum small, lateral.

A tropical and subtropical genus of about 150 species.

Rhynchosia minima (L.) DC., Prodr. 2: 385. 1825

Dolichos minimus L., Sp. Pl. 726. 1753.
Dolicholus minimus Medik., in Vorl. Churpf. Phys. Ges. 2: 354. 1787.
Rhynchosia punctata DC., loc. cit.
Rhynchosia aureo-guttata Anderss., Kongl. Svensk. Vet.-Akad. Handl. 1853: 252. 1855.
Rhynchosia exigua Anderss., loc. cit.

Perennial vine or rarely the plant erect, stems herbaceous, lignescent, puberulent to glabrous; leaves 3-foliolate; stipules linear-lanceolate, 1–3 mm long; stipels linear, 1 mm long or less, caducous; leaflets ovate to rhombic, about 4–80 mm long, 3–70 mm wide, apex acute or obtuse, acute to truncate at base, surfaces densely to sparsely pubescent, often glabrescent, secondary veins evident; inflorescences with the axes short or long, few- to many-flowered; bracts lanceolate, about 1–2 mm long, caducous; flowers about 5–7 mm long; calyx puberulent, usually glandular, about 4 mm long with 5 attenuate lobes or teeth, these longer than tube; petals yellow, sometimes with darker markings; fruit sessile, 1- or 2-seeded, puberulent, glabrescent, about 1.5 cm long, 3–5 mm wide; seeds reniform, dark gray to black, often mottled, about 3 mm long, 2 mm wide, 1 mm thick.

A common, widespread, pantropical and subtropical weed occupying diverse habitats. Specimens examined: Bartholomé, Fernandina, Isabela, Marchena, Pinta, Pinzón, San Cristóbal, San Salvador, Santa Cruz, Santa Fé, Santa María.

Because this species is so widely distributed and variable in leaflet size and degree of pubescence, apparently reflecting edaphic variability, many names have been published. The fact that the flowers and fruits are remarkably stable in their characters has been ignored. Only partial synonymy has been listed above. Bentham (in Martius 1859, 204f) and van Steenis (1961, 439), for example, cite additional synonyms; a revision of the genus as a whole will undoubtedly reveal a number of others.

See Fig. 172.

Stylosanthes Sw., Prodr. Veg. Ind. Occ. 108. 1788;
Fl. Ind. Occ. 3: 1280, *t. 25.* 1806

Herbs, sometimes suffrutescent, unarmed. Leaves alternate, pinnately 3-foliolate; stipules present, adnate to petiole; stipels absent. Flowers yellow or orange, sometimes with red or purplish stripes, in terminal or axillary spikes or capitula, usually few-flowered. Bracts lanceolate to ovate, foliaceous; bracteoles 1 or 2, small, scarious. Calyx with slender, elongate tube below, upper portion broader, 5-lobed, usually with carinal lobe separate, others connate at least at base. Corolla and stamens inserted at apex of calyx tube. Stamens 10, monadelphous, anthers alternately longer, sub-basifixed, and shorter, dorsifixed. Ovary subsessile, 2- or 3-ovulate, style glabrous or puberulent, recurved or revolute, lower portion persistent, stigma minute, terminal. Fruit a 1- or 2-seeded loment, sessile, indehiscent, hooked at apex. Seeds ovate or lenticular, compressed, hilum lateral or subapical.

About 25–30 species, occurring in tropical and warm temperate America, Africa, and southeast Asia.

Stylosanthes sympodialis Taub., Verh. Bot. Ver. Brandenb. 32: 19. 1890

Stylosanthes psammophila Harms, Repert. Spec. Nov. Regni Veg. 19: 69. 1924.

Herb, apparently annual, stems to about 45 cm long, puberulent, glabrescent, sometimes glandular-setose; stipules connate to petiole, bidentate, about 7–10 mm long, teeth about 2–3 mm long, glabrous to pubescent, sometimes glandular-setose; leaves with petiole extending about 10 mm beyond stipule; leaflets lanceolate-oblong, acute and apiculate at apex, cuneate at base, 10–20(40) mm long, 2–5.5 mm broad, upper surface puberulent to glabrous, lower surface pubescent with subappressed or lax hairs, sometimes glabrescent; inflorescences brevispicate, few-flowered; outer bracts 3-foliolate, inner 1-foliolate, outer bracteole (1)2.5–3 mm long, somewhat pubescent toward apex, inner bracteoles (2)2.5 mm long; axis rudiment present, plumose; flowers yellow; calyx about 5–7 mm long, tube 3–5 mm long, lobes 2 mm long, ciliate, otherwise glabrous; petals glabrous, 4–4.5 mm long; style pubescent; fruit villous to glabrous, reticulate, 2-jointed, about 10 mm long, 2 mm wide, sometimes shorter by abortion of lower ovule, lower article usually 3–4.5 mm long, upper 5–6.5 mm long including an uncinate beak 2–2.5 mm long; seeds brown, 2–2.5 mm long, 1 mm wide.

In arid, rocky situations, Ecuador and Peru. Specimens examined: Baltra, Fernandina, Isabela, Marchena, Pinta, Pinzón, Rábida, San Salvador, Santa Cruz, Santa María.

According to R. H. Mohlenbrock (1957, 317), the collections of *Stylosanthes* from the Galápagos Islands are all referable to *S. sympodialis*. In previous treatments such material was cited as *S. scabra* Vog., a more suffrutescent and viscid species, known only from the mainland, especially Brazil. *S. psammophila*, a variant with longer leaflets, is also to be found on the mainland, in Ecuador and Peru.

See Fig. 173.

Tephrosia Pers., Syn. 2: 328. 1807, nom. conserv.

Herbs or shrubs, unarmed. Leaves alternate, imparipinnate, 5 to many, rarely 1- or 3-foliolate; stipules setaceous or broader; stipels absent. Flowers purple, red, or white, in terminal, axillary, or leaf-opposed racemes. Bracts setaceous or broader; bracteoles absent. Calyx campanulate with 5 subequal lobes or teeth, or vexillar teeth connate. Stamens 10, diadelphous 9:1, or submonadelphous with vexillar filament free at base but united above; anthers uniform, dorsifixed. Ovary sessile, usually many-ovulate; style incurved or inflexed, glabrous or bearded; stigma com-

Fig. 173. *Stylosanthes sympodialis* (*a–c*). *Tephrosia decumbens* (*d, e*).

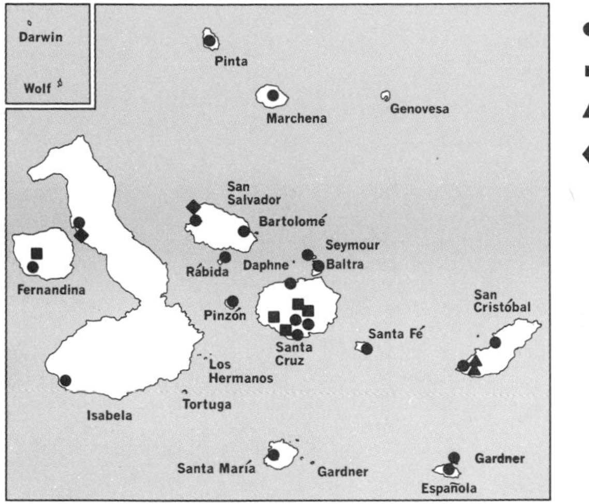

● *Tephrosia decumbens*

■ *Vigna luteola*

▲ *Zornia curvata*

◆ *Zornia piurensis*

monly penicillate. Legume linear, 2-valved, dehiscent, usually many-seeded. Seeds reniform or oblong with hilum lateral.

About 400 species in the tropics and subtropics of the world.

Tephrosia decumbens Benth. in Benth. & Oerst., Vidensk. Medd. Kjob. 1853: 8. 1854

Tephrosia cinerea var. *villosior* Benth. in Mart., Fl. Bras. 15^{1}: 48. 1859.

Decumbent herb, apparently annual, stems about 50 cm long, cinereous-pubescent with spreading or subappressed hairs; stipules setaceous, 2–5 mm long; leaves (3)5–9(11)-foliolate, petiole about 8–15 mm long, leaflets obovate-oblong or suborbicular, obtuse to retuse at apex, base cuneate, 5–20(30) mm long, 3–10(15) mm wide, densely to moderately subappressed-pubescent above, sometimes glabrescent, subsericeous below, secondary veins inconspicuous or conspicuously darkened (when dry); inflorescences opposite leaves, few-flowered; bracts setaceous; flowers 7–8 mm long; calyx pubescent, 4–5 mm long, teeth subequal, 2–4 mm long, attenuate; petals pink or blue to purplish, banner pubescent on outer face; style glabrous, stigma penicillate; fruit pubescent with lax or spreading hairs, 2–5 cm long, 3–4 mm wide, margins slightly thickened; seeds 3–4.5 mm long, 2–2.5 mm broad, light brown, somewhat sculptured, hilum small.

On cinder slopes, lava flows, and open forest, Arid and Transition zones in the Galápagos; known also from Central America. Specimens examined: Baltra, Española, Fernandina, Gardner (near Española), Isabela, Marchena, Pinta, Pinzón, Rábida, San Cristóbal, San Salvador, Santa Cruz, Santa Fé, Santa María, Seymour.

This taxon belongs to the large and widely distributed weedy complex that includes *T. cinerea* (L.) Pers., *T. littoralis* (Jacq.) Pers., *T. purpurea* (L.) Pers., and *T. senna* HBK. Until such time as the entire complex can be unraveled, it seems convenient to cite the collections examined as a species, *T. decumbens,* rather than to obscure them as a variety of *T. cinerea.* I am indebted to Dr. C. E. Wood, Jr., for alerting me to the affinity of the Galápagos specimens with those of Central America. In previous treatments of the Galápagos flora, collections of *Tephrosia* have been identified as *T. cinerea* or *T. littoralis.*

See Fig. 173.

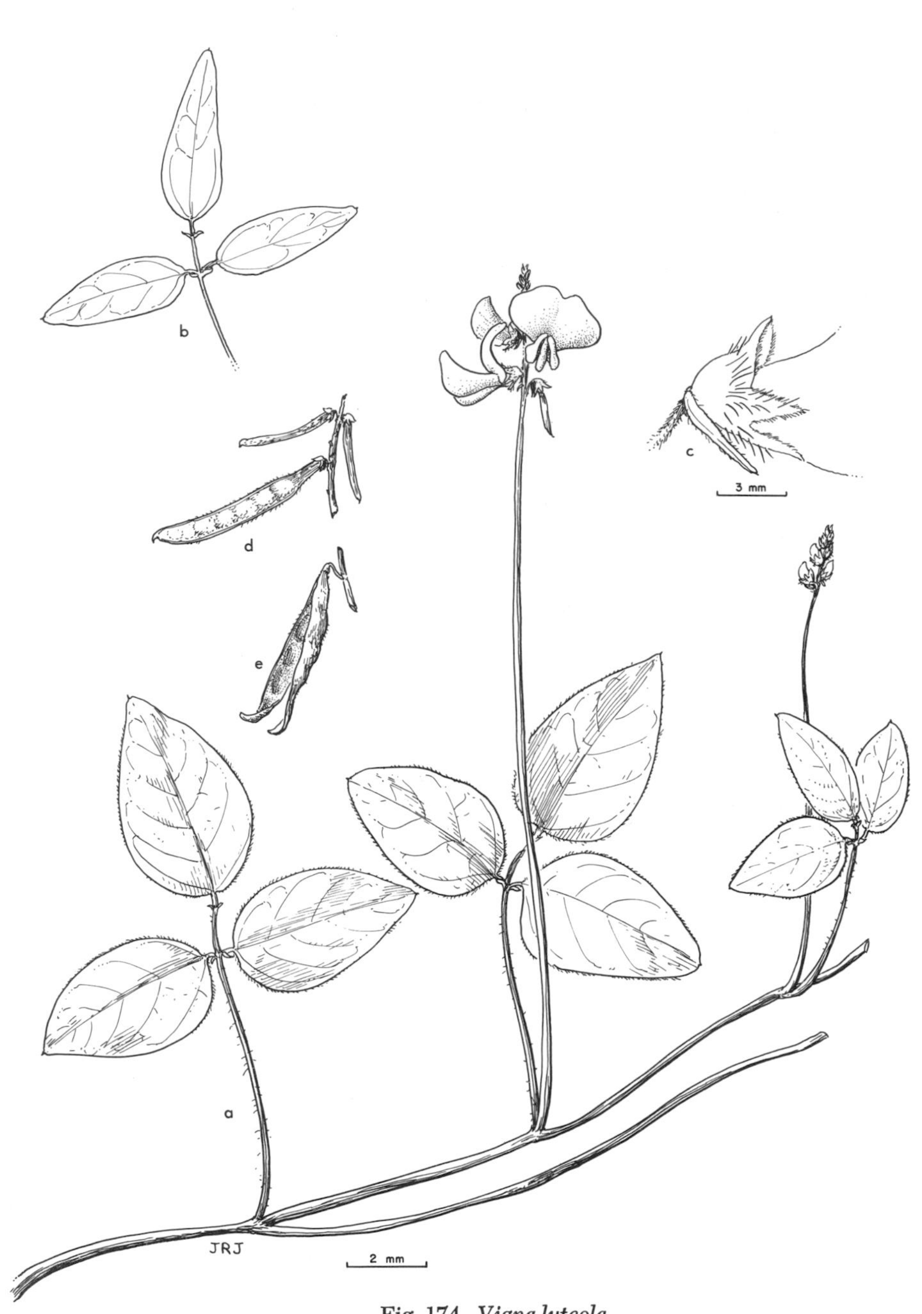

Fig. 174. *Vigna luteola.*

Vigna Savi, Mém. Phaséol. 3: 7. 1824

Herbs, climbing, prostrate, or erect, unarmed. Leaves alternate, pinnately 3-foliolate. Stipules linear or lanceolate; stipels lanceolate to ovate. Flowers in axillary, usually long-pedunculate, nodose racemes. Bracts and bracteoles minute, caducous. Calyx campanulate, 5-toothed or apparently 4-toothed by fusion of the vexillar pair of teeth. Petals yellow to pink or purplish, vexillum glabrous on outer surface, keel somewhat rostrate. Stamens 10, diadelphous 9:1 with the vexillar filament free; anthers uniform, dorsifixed. Ovary sessile, many-ovulate; style filiform, somewhat bent, longitudinally bearded toward apex; stigma lateral or oblique. Fruit 2-valved, dehiscent, linear or oblong, subterete, few- to many-seeded. Seeds reniform to ellipsoid, hilum small, elliptic, lateral.

A genus of about 100 species, occurring in the tropics and warm temperate regions of the world.

Vigna luteola (Jacq.) Benth. in Mart., Fl. Bras. 15[1]: 194, *pl. 50.* 1859

Dolichos luteolus Jacq., Hort. Vindob. 1: 399, *pl. 90.* 1770.
Dolichos repens L., Syst. ed. 10, 1163. 1759.
Vigna glabra Savi, Mém. Phaséol. 3: 8. 1824.
Vigna villosa Savi, op. cit. 9.
Vigna brachystachya Benth., Bot. Voy. Sulph. 86. 1844.
Vigna repens Kuntze, Rev. Gen. Pl. 1: 212. 1891, non Baker 1876.*

Prostrate or climbing herb, the stems glabrous or pubescent with retrorse, somewhat spreading hairs; stipules lanceolate, about 5 mm long; stipels lanceolate, about 1 mm long; leaflets ovate to lanceolate, about 2–9 cm long, 1–5 cm wide, obtuse to acute, base rounded, cuneate or subtruncate, surfaces glabrous to moderately pubescent with appressed hairs, secondary veins evident but not conspicuous; inflorescences long-pedunculate, relatively few-flowered; bracts and bracteoles deltoid, about 1–1.5 mm long, caducous; flowers 15–20 mm long; calyx 4-dentate, glabrous or sparsely pubescent, 4–5 mm long, 2 vexillar teeth fused appearing as 1, obtuse, other 3 teeth deltoid, about 2 mm long; petals yellow; fruit glabrous or pilose, about 4–7 cm long, 4–6 mm wide; seeds black, ellipsoid, 4–5 mm long, 3 mm broad, hilum about 1.5 mm long.

Worldwide in the tropics and subtropics. Specimens examined: Fernandina, Santa Cruz.

Because it is almost hopeless to sort out the synonymy of the infraspecific taxa, I have treated all of the material as 1 variable species. The specimen from Isla Fernandina has narrow leaflets, a variant that is not uncommon in other parts of the range. The other specimens examined have the more common ovate leaflets, with some variability in the degree of pubescence.

See Fig. 174 and color plate 52.

Zornia J. F. Gmel., Syst. 1076, 1096. 1792

Herbs, sometimes suffrutescent, unarmed. Leaves alternate, digitately 2- or 4-foliolate, leaflets of upper and lower leaves sometimes of different size and shape; stipules usually peltate; stipels absent. Flowers yellow to orange, rarely white, solitary or in spikes, axillary or terminal. Bracts minute, ovate, apiculate, hidden by stipules;

* Another probable synonym is *V. owahuensis* Vog. (Linnaea 10: 585. 1836).

bracteoles relatively large, peltate, usually glandular-punctate. Calyx 5-lobed, bilabiate, usually hyaline. Stamens monadelphous, lower portion of filament tube persistent; anthers alternately smaller, dorsifixed, and longer, sub-basifixed. Ovary essentially sessile, 2- to several-ovulate; style somewhat curved, base sometimes persistent, glabrous toward apex; stigma small, subcapitate, terminal. Fruit a 2–15-seeded loment, sessile, indehiscent. Seeds orbicular to subreniform, compressed, hilum lateral, subapical.

About 75 species in tropical and warm temperate areas of America, Africa, and Asia.

Inflorescence with bracteoles obliquely ovate, ciliate, about 8–12 mm long, 3–6 mm wide, overlapping by about 2–3 mm; fruit 3–5-jointed, usually included within bracteoles; articles about 2–2.3 mm long, 2 mm wide. . . . . *Z. piurensis*

Inflorescence with bracteoles obliquely ovate to subfalcate, eciliate, about 6–10 mm long, 2–3 mm wide, separated, usually not overlapping; fruit 5–7-jointed, exserted beyond bracteoles; articles 1.5–2 mm long, 1–1.5 mm wide. . . . . *Z. curvata*

Zornia curvata Mohlenbrock, Webbia 16: 132. 1961

Herb, apparently annual, to about 40 cm tall, the stems glabrous; stipules striate, about 5 mm long, 1 mm wide, attenuate above point of insertion, acute below, sometimes glandular-punctate; leaves 2-foliolate, with petiole about 10–15 mm

Fig. 175. *Zornia curvata.*

long; leaflets of lower leaves suborbicular to elliptic or ovate, about 5–22 mm long, 4–10 mm wide, obtuse to acute at apex and base, leaflets of upper leaves lanceolate, 10–22 mm long, 1–6 mm wide, acute at apex, rounded or acute at base, glabrous, glandular-punctate, especially along margin, secondary veins inconspicuous; inflorescences spicate, about 6–10 cm long, 3–8-flowered; bracts minute; bracteoles not overlapping, obliquely ovate, sometimes subfalcate, 6–10 mm long, 2–3 mm wide, 5-costate, glabrous, eciliate, sparingly punctate, acute at apex, lobe below point of attachment subacute, about 1 mm long; calyx about 2.5 mm long, hyaline, ciliate; petals yellow, glabrous, 6–8 mm long; fruit 5–7-jointed, exserted beyond bracteoles, articles 1.5–2 mm long, 1–1.5 mm wide, reticulate, puberulent and echinate with retrorsely barbed spines about 0.5 mm long; seeds light brown, suborbicular, about 1 mm broad, 0.5 mm thick.

A species of the Caribbean area and, possibly, southeast Asia. Specimens examined: San Cristóbal.

In referring the specimens examined to Z. *curvata*, I am following Mohlenbrock's citation of *Schimpff 174* (BM, CAS, S) under that species. I have reservations, however, because the holotype of Z. *curvata* differs considerably from many of the paratypes. The specimens from Isla San Cristóbal appear to me to be more similar to specimens from the West Indies cited by Mohlenbrock as Z. *microphylla* Desv., but that name may be a later synonym of Z. *reticulata* Smith. They also resemble material of Z. *cantoniensis* Mohlenbrock. Without examining types of the earlier species, I cannot give a more definite opinion. The description given here is based on the 2 collections examined, not on Z. *curvata* sensu lato.

See Fig. 175.

Zornia piurensis Mohlenbrock, Webbia 16: 120. 1961

Annual to about 50 cm tall, the stems puberulent with subappressed hairs, glabrescent; stipules striate, about 7–10 mm long, 1–2 mm wide, attenuate above point of insertion, rounded to acute below, sometimes glandular-punctate; leaves 2-foliolate with petiole about 10–15 mm long; leaflets of lower leaves ovate-elliptic, 10–25 mm long, 4–10 mm wide, obtuse to acute at apex and base, leaflets of upper leaves ovate to lanceolate, 10–20 mm long, 2–7 mm wide, glabrous or subglabrous above, moderately appressed-pubescent, glabrescent, glandular-punctate below, secondary veins inconspicuous; inflorescences spicate, to about 12 cm long, 5–15-flowered; bracts minute; bracteoles overlapping by 2–3 mm, obliquely ovate, 8–12 mm long, 3–6 mm wide, 5-costate, acute at apex, ciliate, otherwise puberulent to glabrous, sometimes glandular-punctate, portion below point of attachment subacute, about 1–1.5 mm long; flowers yellow; calyx about 3 mm long, hyaline, ciliate; petals glabrous, about 8 mm long; fruit 3–5-jointed, included within, or the terminal 1 or 2 articles exserted from, the bracteoles, the articles 2–2.3 mm long, about 2 mm wide, reticulate, puberulent and echinate with retrorsely barbed spines about 1 mm long; seeds light brown, sometimes spotted, about 1.2 mm long, 1 mm broad, hilum lateral, subapical.

In the Galápagos found in tufaceous soil, among rocks, and on lava; known also from coastal Ecuador and Peru. Specimens examined: Isabela, San Salvador.

According to the original description, the loments of this species are not reticulate. However, mature fruits on some of the specimens examined clearly show a reticulate surface. The bracteoles of specimens from the Galápagos Islands are slightly smaller than those of the mainland collections. In previous papers, all collections of *Zornia* from the Galápagos Islands have been identified as *Z. diphylla* (L.) Pers., but that species, according to Dandy and Milne-Redhead (1963, 73f), is known only from Ceylon and India.

LEGUMINOSAE (SUBFAM. FABOIDEAE) SPECIES EXCLUDED FROM THE FLORA

Astragalus L., Sp. Pl. 755. 1753; Gen. Pl. ed. 5, 335. 1754

Astragalus brevidentatus C. H. Wright, Kew Bull. 1906: 200. 1906

This species was cited by Wright as "Galápagos Islands. Without collector's name; received [by Hooker] from M. Decaisne, 1844." According to I. M. Johnston (1938, 95f), "This specimen, apparently a duplicate of the collection by Du Petit Thouars mentioned by Hooker when he described *Phaca edmonstonei*, though attributed to the Galápagos Islands, was probably collected at Monterey, California. The specimen seems to be characteristic of *A. menziesii* A. Gray of the California coast. It is not a South American species and was not collected on the Galápagos Islands." According to R. C. Barneby (1964, 803), *A. brevidentatus* and *A. menziesii* fall into synonymy under *A. nuttallii* (Torr. & Gray) Howell var. *nuttallii.*

Astragalus edmonstonei (Hook. f.) Robins., Proc. Amer. Acad. 38: 148. 1902
Phaca edmonstonei Hook. f., Trans. Linn. Soc. Lond. 20: 227. 1847.

Hooker cited two collections: "*Hab.* Galapagos, *Adm. Du Petit Thouars.* Charles Island, *T. Edmonstone Esq.*" According to Johnston, the Edmonstone collection was made in Chile and that of Du Petit Thouars probably in California. Johnston, referring to the Edmonstone collection as the type of this species, wrote: "Edmonstone, who was botanist on the exploring ship *Herald*, was killed while in Ecuador. There is other evidence that parts of his collection became confused after his death. There can be little doubt that he collected material of the *Astragalus* near Valparaiso where he is known to have botanized in Nov. 1846. The species is not a plant of the Galápagos Islands."

Subfamily MIMOSOIDEAE*

Trees, shrubs, or rarely herbs, sometimes scandent, often armed. Leaves commonly bipinnate but sometimes pinnate; stipules often spinose; stipels present or absent. Flowers mostly bisexual and regular, (4)5(6)-merous, usually in compound globose or spicate inflorescences. Sepals free or united. Petals usually united, at least below, sometimes free. Stamens 4 to numerous, filaments free or united, anthers small, dehiscing lengthwise, sometimes bearing apical gland. Fruit dehiscent or indehiscent, stipitate or sessile, sometimes breaking into 1-seeded articles. Seeds with apical hilum.

About 50–55 genera and 2,500 species, occurring chiefly in the tropics and subtropics, especially of the Southern Hemisphere. Apparently none of the species is truly endemic to the Galápagos Islands.

* Contributed by Velva E. Rudd.

Leaves simple-pinnate; leaflets large, commonly 4 cm long, 2 cm wide or more; flowers with filaments united, at least at base, forming a tube . . . . *Inga*
Leaves bipinnate; leaflets mostly small, 2 cm long, 1 cm wide or less (in *Mimosa albida* larger); flowers with filaments free to base:
Stamens numerous, more than twice as many as sepals or petals . . . . *Acacia*
Stamens as many or twice as many as sepals or petals (usually 10 or less):
Inflorescences long-spicate . . . . *Prosopis*
Inflorescences globose or short-spicate:
Plants spiny; flowers white or red; fruit hispid or prickly, usually breaking into articles . . . . *Mimosa*
Plants unarmed; flowers white or yellow; fruit glabrous, not breaking into articles:
Flowers white; anthers eglandular; fruit elongate, 3–4 mm wide . . . . *Desmanthus*
Flowers yellow; anthers glandular; fruit obliquely oblong, 7–11 mm wide . . . . *Neptunia*

Acacia Mill., Gard. Dict. abr. ed. 4. 1754

Trees, shrubs, or suffrutescent herbs, armed or unarmed. Leaves alternate, bipinnate, or sometimes modified to phyllodes, the pinnae opposite or subopposite with 2 to many leaflets. Leaflets opposite or nearly so; petiole and rachis essentially terete, commonly bearing 1 or more glands; stipules small or sometimes large and spinescent. Inflorescences capitate or spicate, terminal or axillary. Bracts mostly small and caducous. Flowers white or yellow, commonly 5-merous, calyx tubular, corolla tubular or with petals almost separate; stamens numerous, filaments separate to base, anthers small, usually eglandular. Legume compressed or terete, dehiscent or indehiscent, elliptic to linear, valves chartaceous to woody. Seeds sublenticular.

A pantropical genus of about 900 species, some ranging to warm temperate areas.

Involucel borne at or near middle of peduncle; legume commonly compressed, gray-tomentose, moniliform . . . . *A. nilotica*
Involucel borne at apex of peduncle, at immediate base of capitulum; legume compressed to subterete, glabrous to moderately pubescent, straight-margined or submoniliform:
Leaflets 3–15 mm long, 1–4 mm wide, predominantly more than 5 mm long, secondary veinlets evident; inflorescence (capitulum) about 10 mm broad at anthesis . . . . *A. insulae-iacobi*
Leaflets 1–5 mm long, 0.2–1 mm wide, secondary veins not evident; inflorescence (capitulum) 6–8 mm broad at anthesis:
Capitulum 6–7 mm broad at anthesis; leaflets 0.5–1 mm long, 0.2–0.4 mm wide, predominantly less than 1 mm long, ciliate, midvein scarcely evident . . . *A. rorudiana*
Capitulum about 8 mm broad at anthesis; leaflets 1–3 mm long, 0.5–1 mm wide, predominantly more than 1 mm long, sometimes ciliate, midvein evident though not conspicuous . . . . *A. macracantha*

Acacia insulae-iacobi Riley, Kew Bull. 1925: 220. 1925

Tree to about 7 m tall; young branches terete, glabrous, sometimes with prominent, whitish lenticels; stipular thorns straight, terete to about 2 cm long and 1.5 mm thick at base; leaves with 1–5 pairs of pinnae, each pinna with about 5–10 pairs of leaflets, axes glabrous or sparsely pubescent, bearing sessile, cupulate glands about 0.5–2 mm broad on petiole and rachis; leaflets oblong, 3–15 mm long, 1–4 mm wide, sometimes ciliate, otherwise glabrous, the apex obtuse or acute, base obliquely rounded, midvein slightly excentric, secondary veins evident; inflorescences axillary, globose, the capitulum about 10 mm broad, peduncle bearing an involucel at apex, just below capitulum; bracteoles spatulate, ciliate, 2–3 mm long; flowers orange-yellow, sessile, calyx glabrous, 2–2.5 mm long, about 1 mm broad, corolla glabrous, about 3 mm long, stamens exserted, to about 5 mm long; fruit linear, slightly compressed, submoniliform, 3–15-seeded, glabrous or somewhat pubescent, 10–17 cm long and 1–1.3 cm broad.

In cracks in lava, upper beaches, and rocky soil, from sea level to about 100 m; apparently also known from mainland Ecuador, Bolivia, and Argentina. Specimens examined: Baltra, San Cristóbal, San Salvador, Santa Cruz.

After comparing specimens of *A. insulae-iacobi,* including the type, with material of *A. albicorticata* Burkhart (1947, 504), I believe that only 1 species is involved, and that the latter name should be placed in synonymy. Burkhart points out that in the original description of *A. insulae-iacobi* the involucel is said to be 3–5 mm below the capitulum, but I can find no involucel occurring other than at the immediate base of the capitulum. The position of the seeds in the pod varies from transverse to oblique to longitudinal, depending on the width, i.e. lateral expansion, of the pod. In the data with the collections from the Galápagos Islands, no mention seems to have been made of the white bark so characteristic of *A. albicorticata.* This may be because the trees are relatively small and must endure unfavorable ecological conditions. It is sometimes difficult to distinguish sterile specimens of *A. insulae-iacobi* from those of *Prosopis* spp. (see pp. 656–58). Some of the leaflets and the fruits are so similar as to suggest possible introgression. Many of the collections examined had been identified previously as *A. farnesiana, A. tortuosa, "A tortuosa β glabrior,"* A. sp., or *Prosopis* spp.

Acacia macracantha Humb. & Bonpl. ex Willd., Sp. Pl. 4: 1080. 1806

Mimosa lutea Mill., Gard. Dict. ed. 8, no. 17. 1768.
Acacia flexuosa Humb. & Bonpl. ex Willd., Sp. Pl. 4: 1082. 1806.
Acacia macracanthoides Bert. ex DC., Prodr. 2: 463. 1825.
Acacia subinermis Bert. ex DC., loc. cit.
Acacia lutea Hitchc., Rep. Mo. Bot. Gard. 4: 83. 1893, non Leavenw. 1824.
Poponax macracantha Killip, Carib. For. 9: 248. 1948.*

Shrub or small tree to about 8 m tall; young branches puberulent to glabrous, slightly angular; stipular thorns straight, terete, or angular, sometimes compressed, puberulent, glabrate, to about 4 cm long and 2 mm thick at base; leaves with (3)5–35 pairs of pinnae, each pinna with about (10)20–35 pairs of leaflets; petiole 3–7 mm long, bearing raised gland about 0.5–1 mm across; leaflets slightly fleshy, 1–3 mm long, 0.5–1 mm wide, sometimes ciliate, otherwise glabrous, the apex obtuse, base obliquely rounded, midvein subcentral, secondary veins not evident; inflorescences axillary, globose, the capitulum about 8 mm broad, peduncles puberulent, bearing the involucel at apex immediately below capitulum; bracteoles spatulate, ciliate, about 2 mm long; flowers yellow to orange, sessile, calyx ciliate, glabrous at

* Only partial synonymy has been included. Bentham (1875, 500) lists about a dozen synonyms, reflecting the variability and wide distribution of *A. macracantha.* In addition, most of the taxa have been treated as species of the segregate genus *Poponax* (Britt. & Rose 1928, 89f; Britt. & Killip 1936, 139). Svenson (1946, 446) erroneously included in synonymy *A. cochliacantha* Humb. & Bonpl. ex Willd., a distinct species from Mexico (Rudd 1966, 257–62).

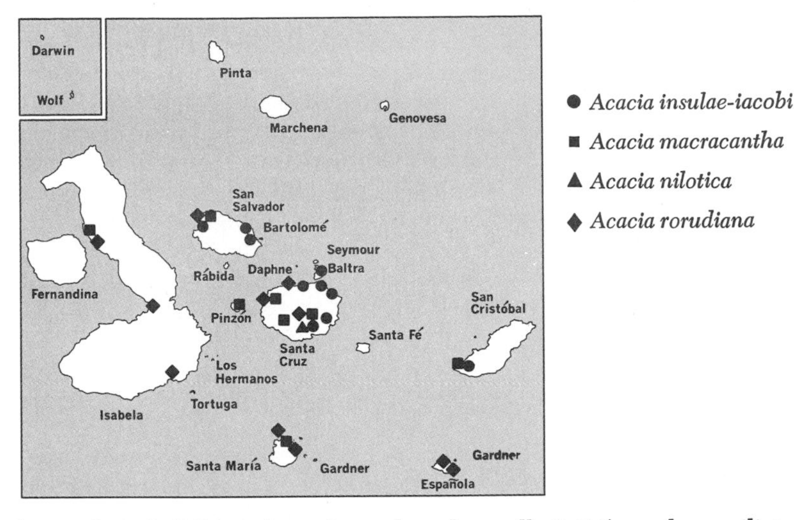

base, about 1–1.5 mm long, 1 mm broad; corolla 2–2.5 mm long, ciliate, otherwise glabrous, the petals united below, separate above; stamens exserted, about 3 mm long; fruit essentially sessile, stipe about 2 mm long, surface reticulate, puberulent, glabrate, 10–15 mm wide, 8–14 cm long, compressed or subterete.

Widespread in the Caribbean area, including southern Florida, Mexico, Venezuela, and Colombia, southward to Peru and Bolivia. In the Galápagos found mostly in the Arid and Transition zones. Specimens examined: Isabela, Pinzón, San Cristóbal, San Salvador, Santa Cruz, Santa María.

My description is based chiefly on the specimens examined from the Galápagos Islands. The type, collected at Guayaquil by Humboldt and Bonpland (Field Mus. Neg. 1278 ex B), is characterized by flat, sword-shaped thorns about 6.5 cm long and 5 mm broad at the base. A collection from Durán, Ecuador (*Rose & Rose 23584*) has thorns about 7 cm long by 1 cm broad at the base and fruit 11.5 cm long and 2.5 cm wide. A couple of specimens from Peru bear spines 8 cm long! A sterile collection of Andersson, "hab. in insulis Chatham & Charles," may be of *A. macracantha.*

Acacia nilotica (L.) DeLisle, Fl. Aegypt. 31. 1813

Mimosa nilotica L., Sp. Pl. 521. 1753.
Mimosa scorpioides L., loc. cit.
Acacia arabica Willd., Sp. Pl. 4: 1085. 1806.
Acacia vera Willd., loc. cit.
Acacia scorpioides Wight in Safford, Contr. U. S. Nat. Herb. 9: 173. 1905.*

Tree to about 8 mm tall; young branches gray-tomentose, glabrescent, essentially terete or sometimes slightly ridged; stipular thorns essentially straight, terete to about 6 cm long and 2 mm thick at base; leaves with about 3–16 pairs of pinnae,

* Only a partial synonymy is included. Rafinesque placed *A. nilotica, A. arabica,* and *A. vera* in his genus *Gumifera*. Various varietal combinations have been made, for example *A. arabica* var. *nilotica* Benth. and *A. scorpioides* var. *nilotica* Chevalier. Several other African species have been cited as synonyms by Brenan (1959, 109–11) and others. For sizeable list of synonymy, see Hill (1940, 97–99).

each pinna with about 10–30 pairs of leaflets, petiole 1–2 cm long, sometimes bearing sessile, cup-shaped gland about 1.5–2 mm broad; leaflets oblong, 3–5 mm long, 1 mm wide, sometimes ciliate, otherwise glabrous, the apex obtuse or subacute, base obliquely truncate, midvein slightly excentric, secondary veins inconspicuous; inflorescences axillary, globose, capitulum about 6–15 mm broad, peduncle bearing an involucel near middle; bracteoles spatulate, ciliate, about 1 mm long; flowers yellow, sessile, calyx glabrous, sometimes pubescent, 1 mm long and about 1 mm broad, corolla glabrous with petals united, 3–4 mm long; stamens exserted, 6–7 mm long; fruit linear, compressed, frequently moniliform, tardily dehiscent, white-tomentose, sometimes subglabrous, about 10–25 cm long, 1.5 cm wide.

Native to northwestern Africa; introduced into the American tropics, especially the West Indies. Specimens examined: Santa Cruz.

Acacia rorudiana Christoph., Nyt Mag. Naturv. 70: 76. 1932

Shrub or small tree to about 8 m tall; young branches puberulent, somewhat angular, tortuose; stipular thorns straight, essentially terete, puberulent, glabrescent, to about 4 cm long and 2 mm thick at base; leaves with about 5–30 pairs of pinnae, each pinna with about 14–24 pairs of leaflets, petiole about 4 mm long, bearing raised gland about 0.5–1 mm across; leaflets fleshy, oblong, 0.5–1 mm long, 0.2–0.4 mm wide, ciliate, otherwise glabrous or pubescent, apex obtuse or subacute, base obliquely rounded, midvein subcentral, inconspicuous, secondary veins not evident; inflorescences axillary, globose, capitulum 6–7 mm broad; peduncles pubescent, bearing an involucel at apex immediately at base of capitulum; bracteoles spatulate, ciliate, about 1 mm long; flowers yellow to orange, sessile; calyx ciliate, glabrous at base, about 1 mm long, 1 mm broad, corolla ciliate, otherwise glabrous, 1–2 mm long (2–2.5 mm fide Christopherson), the petals united below, separate above; stamens exserted to about 4–5 mm; fruit linear-oblong, 7–16 cm long, 10–17 mm wide, essentially sessile, the stipe about 2 mm long or less, surface somewhat reticulate, glabrous, puberulent, or finely brown-glandular-velutinous.

Possibly also on the mainland, in Chile; among rocks or cinders in the Galápagos. Specimens examined: Española, Isabela, San Salvador, Santa Cruz, Santa María.

Svenson (1946, 446f) may well be correct in treating *A. rorudiana* as synonymous with *A. macracantha*. The latter, sensu latior, is extremely variable and is difficult to interpret satisfactorily. However, as indicated in the key, I am segregating *A. rorudiana*, though with some reservations. My circumscription is somewhat more restricted than Christopherson's, which included several collections that I refer to *A. macracantha*. Robinson (1902, 147) recognized 2 of the collections I examined—*Snodgrass & Heller 161* and *283*—as distinct from *A. macracantha*, but cautiously cited them as "*A.* sp. affin. *A. macracantha*, H. & B." with the comment, "The species [*A. macracantha*] . . . has been so widely drawn by Bentham . . . and contains such an aggregation of old species, that it is impossible to characterize with confidence any new species of this affinity until the whole group can be worked over again." Comprehensive population studies are long overdue. There appears to be some intergradation between the 2 species in localities where both occur, but from other parts of the range of *A. macracantha* I have not found specimens that suggest *A. rorudiana* except for a few collections from Chile, some previously identified as *A. macracantha*, others as *A. caven* (Molina) Molina.

See Fig. 176 and color plates 46, 74, and 86.

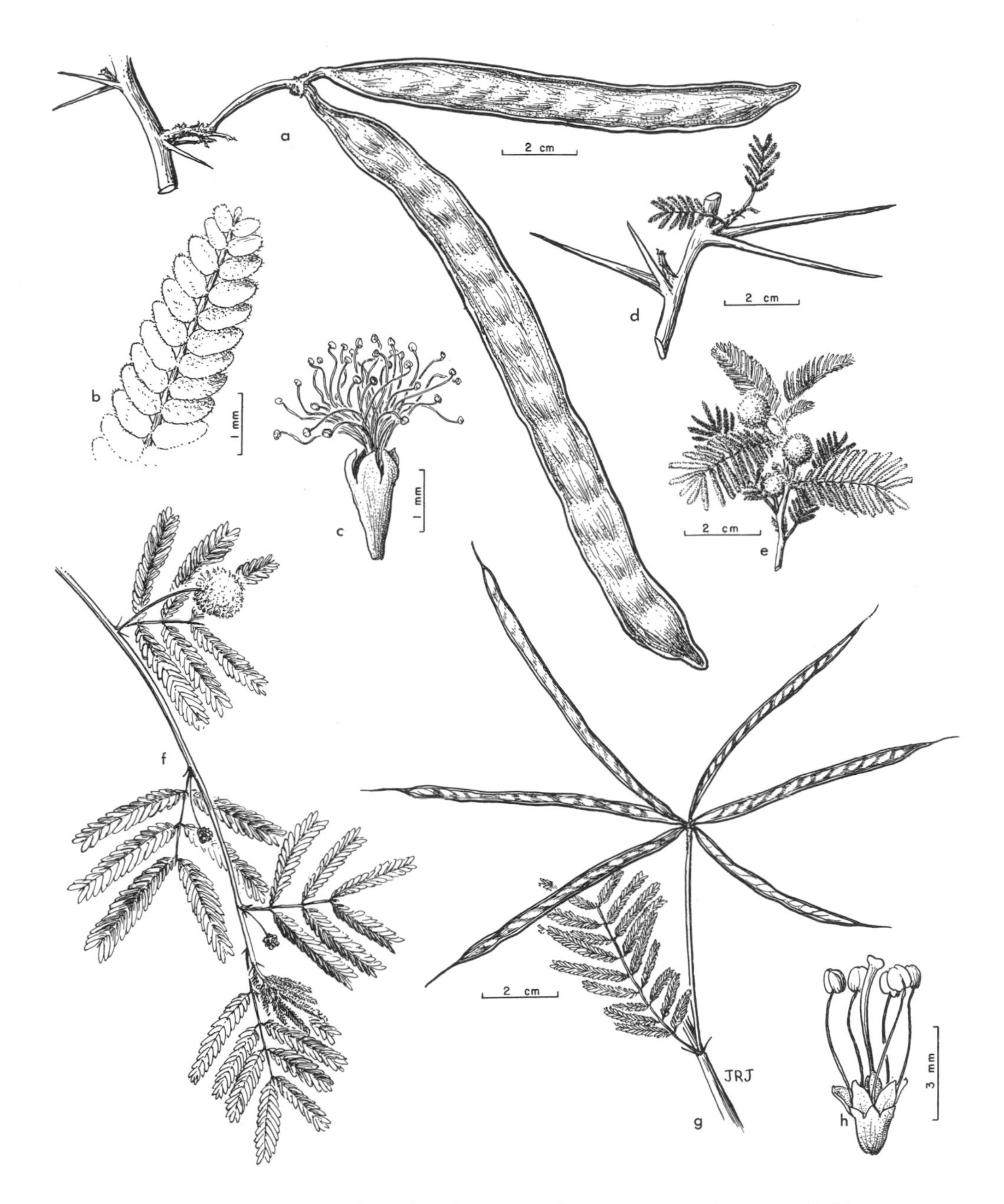

Fig. 176. *Acacia rorudiana* (*a–e*). *Desmanthus virgatus* var. *depressus* (*f–h*).

Desmanthus Willd., Sp. Pl. 4: 1044. 1806, nom. conserv.

Unarmed shrubs or herbs to about 4 m tall. Leaves alternate, bipinnate, the pinnae opposite, with many leaflets. Leaflets opposite; petiole and rachis essentially terete, canaliculate, commonly with gland at insertion of lower pinnae; stipules setaceous, persistent. Inflorescences capitate, terminal or axillary. Bracts setaceous. Flowers whitish or greenish, 5-merous, calyx campanulate, 5-dentate, petals essentially separate; stamens 5 or 10, filaments separate to base, anthers small, eglandular. Stigma truncate-concave. Legumes compressed, dehiscent, linear, straight or slightly falcate; valves chartaceous or slightly woody. Seeds sublenticular with apical hilum.

About 20 species, native to the New World tropics and warm temperate regions; introduced into the Old World.

Desmanthus virgatus (L.) Willd. var. **depressus** (Humb. & Bonpl. ex Willd.) B. L. Turner, Field & Labr. 18: 61. 1950

Desmanthus depressus Humb. & Bonpl. ex Willd., Sp. Pl. 4: 1046. 1806.
Mimosa depressa Poir. in Lam., Encycl. Suppl. 1: 1810.
Acuan depressum Kuntze, Rev. Gen. Pl. 1: 158. 1891, as *Acuania*.

Shrub or suffrutescent herb, erect or decumbent, to about 2 m tall; young stems essentially glabrous, somewhat angled; stipules setiform, to about 5 mm long, scarious-margined at base; leaves with 1–7, commonly 2–4, pairs of pinnae, each pinna with about 10–20 pairs of leaflets, axes glabrous or puberulent, bearing suborbicular, cupulate gland about 0.5–1 mm long at insertion of lowest pair of pinnae; leaflets oblong, about 2–8 mm long, 1–2 mm wide, sometimes ciliate, otherwise glabrous, the apex obtuse or acute, midvein slightly excentric, secondary veins not evident; inflorescences axillary, pedunculate, capitate with a few erect flowers; bracts setiform; bracteoles lanceolate, about 2 mm long; flowers greenish-whitish, sessile, calyx glabrous or nearly so, 1.5–3 mm long, about 1 mm broad, petals glabrous, obovate, 2.5–4 mm long, stamens 10, exserted to about 6–7 mm; fruit linear, glabrous, about 1.5–5.5 cm long, 2–3 mm wide, 10–25-seeded; seeds brown, lenticular, sometimes obliquely compressed, about 3 mm long, 2 mm wide, 1 mm thick.

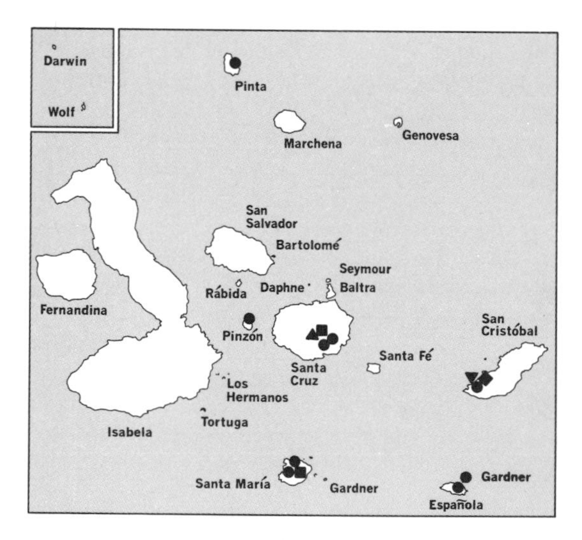

● *Desmanthus virgatus* var. *depressus*
■ *Inga edulis*
▲ *Inga schimpffii*
◆ *Mimosa acantholoba*
▼ *Mimosa albida* var. *aequatoriana*

A widespread and variable weed in diverse habitats of the American tropics and warm temperate regions; introduced into the Old World. Specimens examined: Española, Gardner (near Española), Pinta, Pinzón, San Cristóbal, Santa Cruz, Santa María.

Probably all of the plants from the Galápagos Islands can be referred to var. *depressus*, which is distinguished from the typical variety by smaller plants with less erect habit, leaves with fewer pinnae and smaller glands, and, generally, smaller fruits. Some of the specimens are from erect plants, but in other characters tend toward the "*depressus*" end of the spectrum. There is no sharp demarcation of morphological characters or geographical separation.

See Fig. 176.

Inga Mill., Gard. Dict. abr. ed. 4. 1754

Trees, unarmed. Leaves alternate, paripinnate, 2- to many-foliolate; leaflets opposite; petiole and rachis terete or winged, commonly with cup-shaped glands at insertion of leaflets; stipules commonly small and caducous. Inflorescences axillary or terminal, fasciculate or paniculate, capitate or spicate. Bracts small or large, caducous or persistent. Flowers white or yellowish, 5(6)-merous, calyx and corolla tubular or funnelform; stamens numerous, filaments united, at least at base, forming a tube, the anthers small, eglandular. Stigma truncate or lobed. Legume compressed, 4-angled, or subterete, sometimes striate, usually elongate, irregularly dehiscent, margins sometimes thickened. Seeds oblong, covered with a loose pulp.

A tropical and subtropical New World genus of about 300 species.

Young branches tomentose; leaflets subcoriaceous, pubescent, with petioles 1–2 mm long; fruit subterete, deeply sulcate, subsericeous, about 1 cm broad. . . . . *I. edulis*

Young branches sericeous; leaflets coriaceous, glabrous, essentially sessile; fruit somewhat compressed, glabrous, 4–4.5 cm broad. . . . . *I. schimpffii*

Inga edulis Mart., Flora 20: Beibl. 113. 1837

Mimosa ynga Vell., Fl. Flum. Ic. 11: *pl.* 3. 1835, non *M. inga* L., 1753.
Inga ynga J. W. Moore, Bernice P. Bishop Mus. Occ. Paper 10, 19: 6. 1934.

Tree to about 15 m tall; young branches tomentose, glabrescent, terete or somewhat angular; leaves 8–12-foliolate, axis about 8–20 cm long, tomentose, petiole 2–6 cm long, rachis alate, to about 2 cm broad, with sessile glands about 2 mm broad; leaflets with petiolules 1–2 mm long, blades elliptic, oblong, to lanceolate, 4–18 cm long, 2–8 cm broad, apex acute or obtuse, base rounded, upper surface densely pubescent along midvein, otherwise sparsely scabrous-pubescent, lower surface moderately pubescent with somewhat crinkled-crispate hairs, secondary veins prominent; inflorescence axillary, paniculate, axes fulvo-tomentose to ferrugino-tomentose; bracts and bracteoles tomentose, caducous, bracts lanceolate, bracteoles ovate; flowers sessile, 20–50 mm long, calyx tomentulose, 7–9 mm long, about 3 mm broad, corolla 14–20 mm long, subsericeous, stamens exserted, 20–50 mm long; fruit subterete, deeply sulcate, fulvo-sericeous, about 30 cm long (to 120 cm fide León).

Widespread in southern Central America and northern South America, largely because of its introduction for use as shade for coffee and cacao trees. Known as "guaica de machete." Specimens examined: Santa Cruz, Santa María.

See Fig. 177.

Fig. 177. *Inga edulis.*

Inga schimpffii Harms, Repert. Spec. Nov. Regni Veg. 43: 112. 1938

Inga spectabilis (Vahl) Willd. var. *schimpffii* (Harms) Little, Phytologia 19: 268. 1970.

Tree to about 10 m tall; young branches sericeous, glabrate, sharply angular; leaves 6–8-foliolate, axis about 11–25 cm long, glabrous, petiole about 1–7 cm long, rachis alate, to about 1 cm broad; glands sessile, about 1.5–2 mm broad; leaflets essentially sessile, blades coriaceous, oblong to obovate-oblong, 7–25 cm long, 3.5–12 cm broad, apex obtuse to acute, base rounded or subcordate, surfaces glabrous, secondary veins prominent; inflorescences axillary, paniculate, the axes sericeous with fulvous hairs, glabrate; bracts and bracteoles sericeous, caducous, the bracts lanceolate, bracteoles ovate; flowers sessile, about 30 mm long, calyx subsericeous, 6–8 mm long, 2–2.5 mm broad, corolla sericeous, 15–17 mm long, stamens about 30 mm long; fruit somewhat compressed, without strong angles or margins, glabrous, 30–50 cm long, 4–5.5 cm wide, 1–3 cm thick.

Otherwise known only from central Ecuador; locally called "guava de machete," or "guavo machete." Specimens examined: Santa Cruz.

The description of the legume, otherwise unknown, is based on "old fruit from the ground," *Little 6495* (US) from Pichilinque, Los Rios, Ecuador. He states that this is a "common shade tree in cacao plantation."

Mimosa L., Sp. Pl. 516. 1753; Gen. Pl. ed. 5, 233. 1754

Trees, shrubs, or herbs, sometimes scandent, commonly armed. Leaves alternate, bipinnate, pinnae 2 to many, opposite or subopposite; petiole and rachis essentially terete, canaliculate, sometimes bearing glands; stipules setaceous to foliaceous, sometimes small and inconspicuous, often caducous. Inflorescences spicate or globose, axillary, pedunculate. Bracts mostly small and caducous. Flowers white or red to purplish, 3–6-merous, calyx usually minute, tubular or sometimes paleaceous, corolla more or less tubular, petals sometimes separate. Stamens equal to or sometimes double the number of petals, filaments separate to base, anthers small, eglandular. Legume compressed, oblong to linear, dehiscent with the valves chartaceous to coriaceous, usually articulating and separating from the persistent margin. Seeds sublenticular with apical hilum.

A pantropical genus of about 600 species, with some ranging to temperate areas.

Leaflets large, about 2–4 cm long, 1–2.5 cm wide; pinnae 1 pair; stamens 4; fruit 4–5 mm wide . . . . *M. albida* var. *aequatoriana*

Leaflets smaller, mostly less than 1 cm long and 2 mm wide; pinnae 2 to many pairs; stamens 8–10; fruit about 5–15 mm wide:

Fruit essentially glabrous, not breaking into articles, the margins aculeate; leaflets oblong, 3–6 mm long, about 1–1.5 mm wide, the costa slightly eccentric, secondary veins not evident . . . . *M. acantholoba*

Fruit hispid, breaking into articles; leaflets linear-oblong, 2–8(13) mm long and 0.5–1(2) mm wide, multicostate, the costae essentially parallel. . . . . *M. pigra*

Mimosa acantholoba (Humb. & Bonpl. ex Willd.) Poir. in Lam., Encycl. Suppl. 1: 83. 1810

Acacia acantholoba Humb. & Bonpl. ex Willd., Sp. Pl. 4: 1089. 1806.

Shrub or small tree to about 4 m tall; young branches puberulent, glabrescent, usually armed with scattered, recurved thorns about 7 mm long or less; stipules setaceous, caducous; leaves with about 4–15 pairs of pinnae, each pinna with 5–30 pairs of leaflets, axes puberulent, eglandular; leaflets oblong, 3–6 mm long, about 1–1.5 mm wide, puberulent or glabrous, obtuse or acute, the costa slightly excentric, secondary veins not evident; inflorescences globose, borne in racemes in axils of upper branches; bracts deltoid, about 1 mm long; bracteoles spatulate, ciliate, 1 mm long or less; capitula 12–18 mm across; flowers white to pinkish, calyx glabrous or lightly puberulent, about 1 mm long, 1 mm broad, corolla 3–4 mm long with ciliate lobes, otherwise glabrous, stamens 8–10, exserted to about 7–9 mm; fruit essentially glabrous, elliptic-oblong with an aculeate margin, apex and base acute, 4–6 cm long, including stipe about 0.5–1 cm long, dehiscent by separation of margin, the valves not articulating.

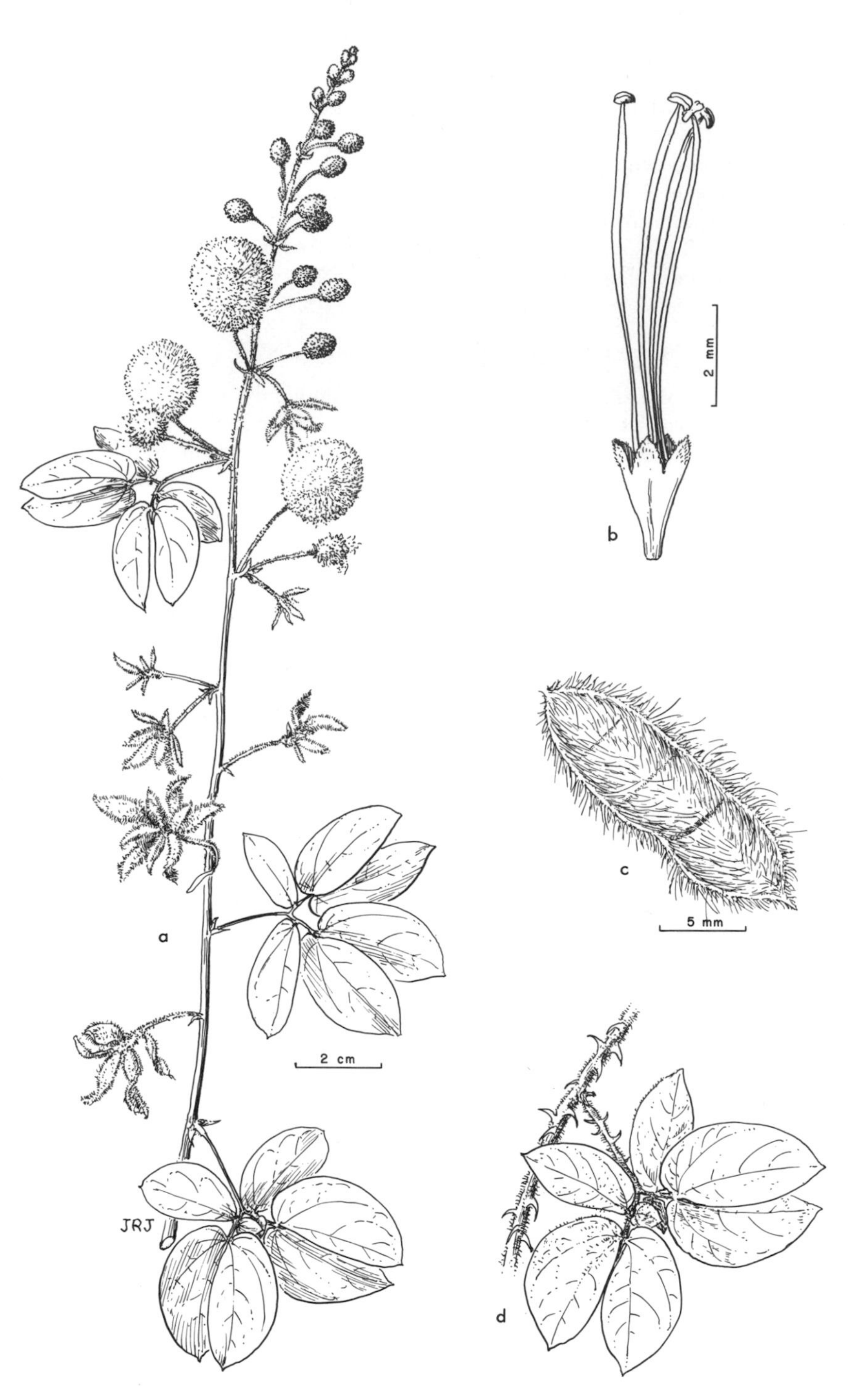

Fig. 178. *Mimosa albida* var. *aequatoriana*.

Known from the mainland, western Ecuador and Peru. Specimens examined: San Cristóbal.

Mimosa albida var. **aequatoriana** Rudd, Phytologia 16: 440. 1968

Shrub to about 3 m tall; young branches essentially terete, canotomentulose to hirsute, sometimes armed with scattered, recurved thorns about 3–5 mm long; stipules deltoid, about 2–4 mm long and 1 mm broad at base; leaves with 1 pair of pinnae, each pinna with 2 pairs of leaflets, the petiole and rachis eglandular, puberulent and setose with yellowish hairs; leaflets obliquely ovate or subelliptic, puberulent, glabrescent above, the lower surface puberulent and strigose, terminal pair and outermost of lower pair about 2–4 cm long, 1.5–2.5 cm wide, the innermost, lower leaflet much smaller, about 1 cm long, 0.5 cm broad, apex acute, base obliquely rounded, costa excentric, secondary veins evident; inflorescences globose, about 10–13 mm across, borne in axillary or terminal racemes; bracts deltoid like stipules, bracteoles linear, about 1–1.5 mm long, minutely pectinate; flowers pinkish, sessile, calyx glabrous, laciniate, about 0.2 mm long and broad; corolla about 2 mm long, 0.5 mm broad, pubescent on lobes, otherwise glabrous; stamens 4, exserted about 5 mm; fruit sessile, oblong, 4–6-seeded, 1.5–2.5 cm long, 4–5 mm wide, densely hirsute with setae 1–2 mm long, the valves breaking into articles.

This variety is elsewhere known only from Ecuador, though some tendency toward it is found in other areas; *M. albida* sensu latior ranges from Mexico southward to Venezuela and Bolivia. Specimens examined: San Cristóbal.

The hirsute pods distinguish var. *aequatoriana* from typical *M. albida*, which has strigose and often slightly larger fruits.

See Fig. 178.

Mimosa pigra L., Cent. Pl. 1: 13. 1755

Mimosa asperata L., Syst. Nat. ed. 10, 1312. 1759.
Mimosa asperata [var.] β *pigra* Willd., Sp. Pl. 4: 1035. 1806.
Mimosa polyacantha Willd., op. cit. 1034.
Mimosa hispida Willd., op. cit. 1037.
Mimosa pellita Humb. & Bonpl. ex Willd., op. cit. 1038.

Shrub about 0.5–3(4.5) m tall, sometimes scandent; young branches setose-hispid and armed with recurved thorns to 5 mm long; stipules ovate-lanceolate, 3–8 mm long; leaves with about 6–16 pairs of pinnae, each pinna with 20–30 pairs of leaflets, the axes hispid, eglandular; leaflets linear, 2–8(13) mm long, about 0.5–1(2) mm wide, acute, margins ciliate, both surfaces appressed-pubescent, venation multicostate (midrib and 2–5 secondary veins), essentially parallel; inflorescences globose or subglobose, axillary, pedunculate, peduncles about 5 cm long; capitula about 1 cm across; bracts lanceolate, about 3 mm long; bracteoles pectinate, 2–3 mm long; flowers pinkish; calyx laciniate, about 1 mm long and broad; petals 2.5–3 mm long, pubescent on lobes, otherwise glabrous; stamens 8, exserted to about 5 mm; fruit hispid, oblong, about 3–8 cm long, (0.8)1–1.5(1.8) cm wide, essentially sessile, the stipe 3–4 mm long, valves separating from persistent margins and also breaking into articles.

A widespread, weedy plant, native to the New World from Texas to Argentina; introduced into Africa and Australia. Specimens examined: San Cristóbal.

There is considerable variation in the fruit of this species, sensu latior, as indicated by the partial synonymy listed above and the range of measurements given

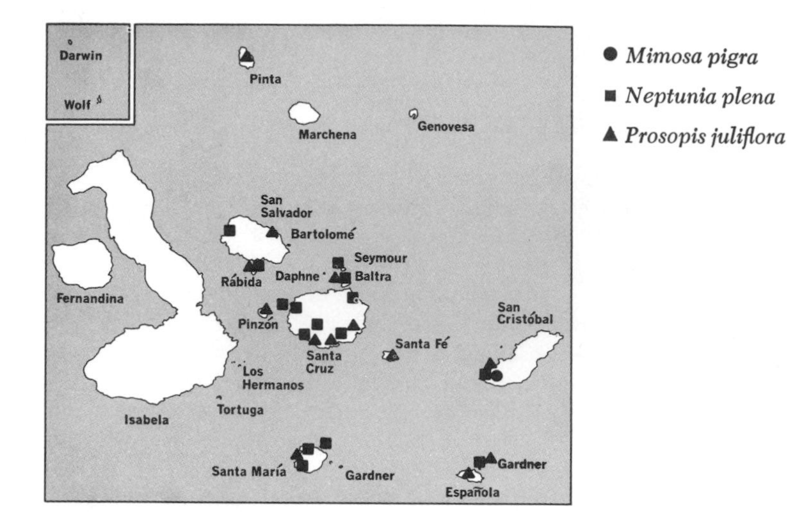

in the description. Commonly, the fruits are hispid, with setae 2–3 mm long, but along the Texas-Mexican border, the fruits of var. *berlandieri* (A. Gray in Torr.) B. L. Turner are sparsely hispid, with the setae shorter and somewhat appressed. Similar fruits have been collected in scattered localities in South America. Hassler (1910, 14) published a number of new varieties and forms based on material from Paraguay. The specimens from the Galápagos Islands I examined are sterile, the fruits unknown.

Neptunia Lour., Fl. Cochinch. 2: 641, 653. 1790

Herbs, unarmed, terrestrial or aquatic, erect, prostrate, or floating. Leaves alternate, bipinnate, pinnae 2 to many, opposite, with many leaflets. Leaflets opposite; petiole and rachis essentially terete, canaliculate, sometimes with small gland at or below insertion of lowest pair of pinnae; stipules lanceolate, persistent or deciduous. Inflorescences capitate, pedunculate, axillary. Bracts ovate, cordate. Flowers yellow or greenish, dimorphic with upper flowers perfect, lower staminate; calyx campanulate, 5-dentate; petals 5, free or somewhat united; stamens 5 or 10, filaments separate to base, sometimes sterile and petaloid, anthers about 1 mm long, usually bearing terminal, stipitate, often caducous gland. Stigma truncate. Legume compressed, dehiscent, oblong to suborbicular, the valves chartaceous. Seeds sublenticular, hilum terminal.

A pantropical and warm temperate genus of about 11 species.

Neptunia plena (L.) Benth., Hook. Jour. Bot. 4: 355. 1842

Mimosa plena L., Sp. Pl. 519. 1753.
Desmanthus plenus Willd., Sp. Pl. 4: 1045. 1806.
Neptunia surinamensis Steud., Flora 26: 759. 1843.*

Suffrutescent herb, erect or prostrate, to about 2 m tall; young stems glabrous, terete; stipules lanceolate, 4–12 mm long, the base cordate, about 1–3 mm broad;

* For additional synonymy, see Amshoff (1939, 267) and Windler (1966, 398).

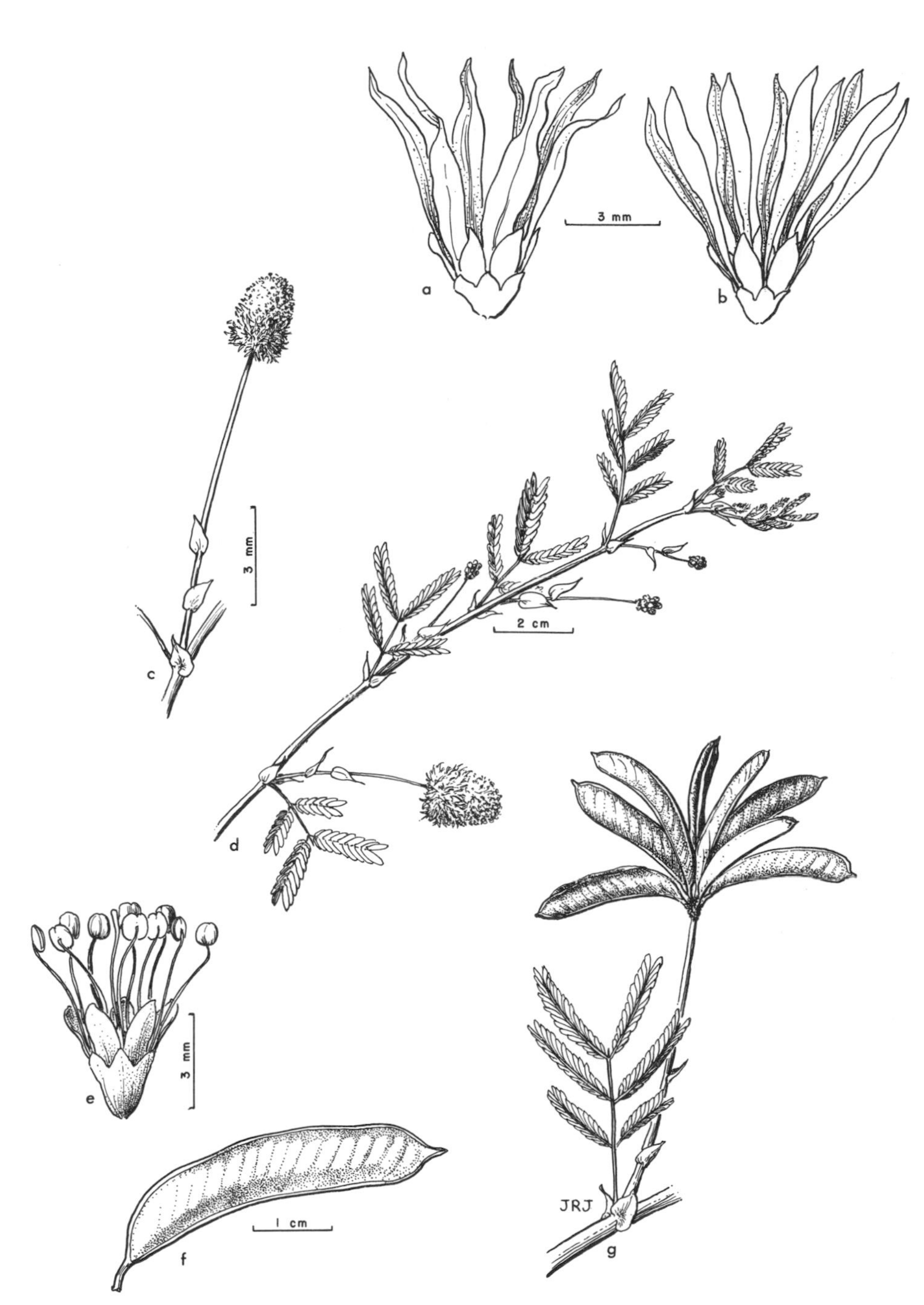

Fig. 179. *Neptunia plena.*

leaves with 2–5 pairs of pinnae, each pinna with 9–38 pairs of leaflets, the axes glabrous or nearly so, often bearing orbicular, cupulate gland just below insertion of lowest pair of pinnae; leaflets oblong, about 3–18 mm long, 1–3.5 mm wide, obtuse, sometimes mucronate, the midvein slightly excentric; inflorescences axillary, pedunculate, capitate or nearly so; bracts ovate, cordate; bracteoles lanceolate; flowers yellow, sessile, the upper perfect, the lower staminate or infertile with 10 petaloid staminodia 7–16 mm long; calyx glabrous, about 1–2 mm long, 1 mm broad; petals glabrous, spatulate, separate, about 3 mm long; stamens 10, exserted about 6.5–9 mm; fruit glabrous, 1.5–5.5 cm long, including stipe 4–6 mm long, 0.7–1.1 cm wide; seeds approximately at right angles to long axis of pod.

Widely distributed in America in wet areas from southern Texas to Brazil and Peru; introduced into the Old World. Specimens examined: Baltra, Champion, Eden, Gardner (near Española), Rábida, San Cristóbal, San Salvador, Santa Cruz, Santa María, Seymour.

See Fig. 179.

Prosopis L., Syst. Nat. ed. 12, 2: 282, 293. 1767; Mant. Pl. 10, 68. 1767

Trees or shrubs, armed or unarmed. Leaves alternate, bipinnate, rarely absent or reduced, the pinnae opposite, often with sessile gland between pinnae of each pair. Leaflets opposite, rarely alternate, few to many; stipules spinescent. Inflorescences spicate or capitate. Bracts small and caducous. Flowers usually greenish yellow, 5-merous, calyx tubular with 5 teeth, petals free or united at base; stamens 10, filaments free; anthers usually with apical gland. Legume indehiscent, straight, curved, or spirally coiled, terete or compressed, many-seeded. Seeds sublenticular, compressed, with apical hilum.

A tropical and subtropical genus of about 40 species, found in the Old and New Worlds.

Prosopis juliflora (Sw.) DC., Prodr. 2: 447. 1825

Mimosa juliflora Sw., Prodr. 85. 1788.
Mimosa piliflora Sw. (as typographical error), Fl. Ind. Occ. 986. 1800.
Acacia juliflora Willd., Sp. Pl. 4: 1076. 1806.
Neltuma juliflora Raf., Sylva Tellur. 119. 1838.

Tree or shrub, erect or prostrate, to about 10 m tall; young branches glabrous, finely striate, terete or slightly angled; stipules spinescent, about 1.5 cm long, 2 mm thick at base; leaves with 1 or 2(3) pairs of pinnae, each pinna with about 6–20 pairs of leaflets, the petiole and rachis glabrous or puberulent, often with raised, barrel-shaped gland at insertion of each pair of pinnae; leaflets coriaceous, oblong, about 5–24 mm long, 1.5–6 mm wide, obtuse, sometimes breviapiculate, the midvein slightly excentric, secondary veins evident, surface glabrous or somewhat puberulent; inflorescences axillary, spicate, many-flowered, about 5–12 cm long, 1–2 cm broad; bracts and bracteoles small, caducous; flowers yellowish, sessile or brevipedicellate, calyx glabrous or puberulent, about 1 mm long, 1.5 mm broad, teeth deltoid, 0.5 mm long or less; petals 3–4 mm long, glabrous without, pubescent within; stamens exserted 5–8 mm, anthers oblong, about 1 mm long; ovary pubescent; fruit glabrous, straight or curved, 9–25 cm long, 8–15 mm wide, 6–10 mm thick, sometimes slightly constricted between seeds, the joints shorter than broad, apex obtuse or apiculate, apiculation 5–15 mm long.

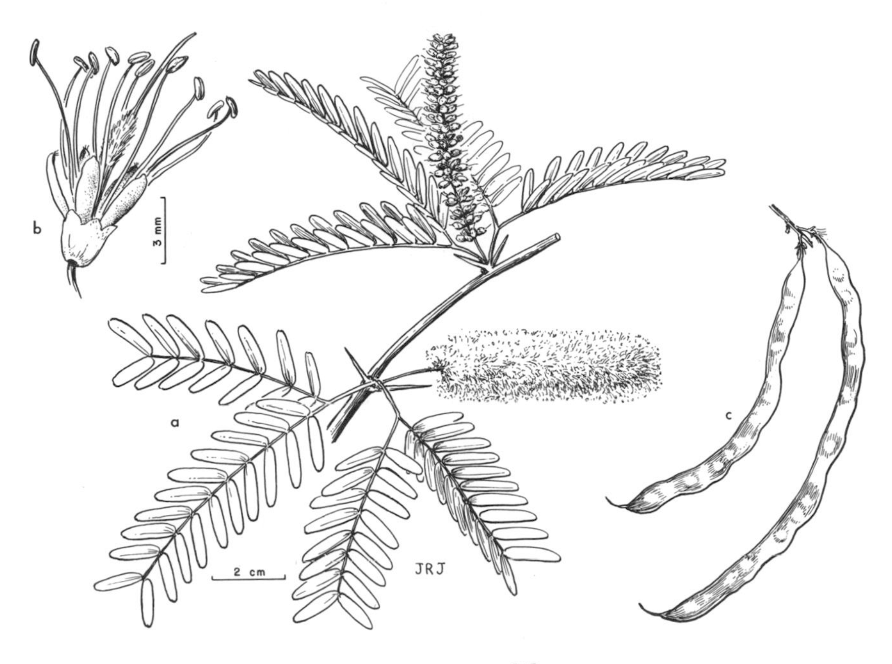

Fig. 180. *Prosopis juliflora.*

In arid and semiarid areas, Mexico, West Indies, and northern South America; introduced in other parts of the world. Specimens examined: Baltra, Champion, Española, Gardner (near Española), Pinta, Pinzón, Rábida, San Cristóbal, San Salvador, Santa Cruz, Santa Fé, Santa María.

It is difficult to separate sterile specimens of *Prosopis* spp. from *Acacia insulae-iacobi.* Even in fruit they are somewhat similar, but with remains of the axis of the inflorescence one can readily see if it was capitate, as in *A. insulae-iacobi,* or spicate, as in the local species of *Prosopis.* There are slight differences in the leaflet venation, which can be recognized when compared with known material. In the case of *A. insulae-iacobi,* a petiolar gland is usually present somewhat below the insertion of the lower pair of pinnae, but in *P. juliflora* the lowest gland is at or immediately below the insertion of the lower pair of pinnae. The young twigs provide another useful character; those of *A. insulae-iacobi* bear numerous, fairly conspicuous lenticels, those of *Prosopis* do not.

A more difficult problem is to identify the species of *Prosopis* that occur on the Galápagos Islands. The name *P. juliflora* is used above, but I do not believe the population is "pure." Some specimens resemble those from western Mexico and Central America, and others are similar to those from the Caribbean area, including Jamaica, the type locality. A few specimens show characters that suggest *P. pallida* (Willd.) HBK. from the mainland of Ecuador, Peru, and Colombia. These are: puberulent leaves with 3 pairs of pinnae, instead of glabrous with 1 or 2 pairs; inflorescences longer and wider than "normal" *P. juliflora*; and fruit with curved apiculum about 5–15 mm long. In *P. juliflora,* such an apiculation is shorter or absent; in *P. pallida,* it is usually about 10–20 mm long. Probably, there have been numerous separate introductions of *Prosopis,* each from a different source.

Unfortunately, I have not seen the Humboldt and Bonpland specimen from Peru that typifies *P. pallida*, but am basing my interpretation on the original description and on herbarium specimens so identified. Two other species based on collections of Humboldt and Bonpland, *P. inermis* HBK. and *P. horrida* Kunth, may possibly be synonymous with *P. juliflora* or *P. pallida*. As has been pointed out by Svenson (1946, 451–53), neither of the names used by earlier authors, *P. dulcis* Kunth or *P. chilensis* (Molina) Stuntz, are applicable to Galápagean taxa. Additional discussion and synonymy may be found in papers by Burkhart (1940, 57–128) and M. C. Johnston (1962, 72–90). My description is based only on the specimens examined, and so does not necessarily present a purview of *P. juliflora*, sensu lato.

See Fig. 180 and color plate 85.

LINACEAE. FLAX FAMILY*

Herbs, shrubs, trees, or vines. Leaves alternate, opposite, or whorled, simple; stipules present or absent, sometimes glandular. Inflorescences terminal or axillary cymes, the flowers rarely solitary. Flowers bisexual, regular, hypogynous or rarely perigynous, usually heterostylous. Sepals (4)5, free or partly connate basally, imbricate. Petals (4)5, free or rarely connate basally, often clawed, claw naked or crested, convolute, fugacious. Extra- or intrastaminal disk or glands sometimes present. Stamens same in number to 2 or 3 times as many as petals, the whorl opposite petals sometimes represented by small staminodes or absent; filaments more or less connate basally, sometimes glandular basally, inserted on disk; anthers 2-loculed, versatile, introrse, dehiscing longitudinally. Gynoecium (2)5-carpelled, syncarpous; ovary superior, (2)5-loculed, locules often subdivided by formation of incomplete or sometimes complete false septa from dorsum of each; ovules 2 per locule, anatropous, placentation axile; styles (2)5, filiform, free or variously connate; stigmas simple, capitate or clavate. Fruit a capsule septicidally dehiscent into (4)5 2-seeded mericarps or septicidally and loculicidally (between false septa) into 4–10 1-seeded mericarps, sometimes a drupe. Seeds 1–2 per locule, flattened to rounded; endosperm copious or scanty or absent; embryo straight or slightly curved; cotyledons flat.

About 12 genera and 300 species, found throughout the temperate and tropical regions of the Old and New Worlds.

Linum L., Sp. Pl. 277. 1753; Gen. Pl. ed. 5, 135. 1754

Annual or perennial, glabrous or pubescent herbs, sometimes basally woody or rarely shrubby. Leaves alternate, or occasionally opposite, or whorled or opposite on lower part of stem and alternate above, sessile or subsessile, narrow, entire to glandular-dentate. Inflorescence a terminal monochasial or dichasial cyme, the flowers rarely solitary. Flowers 5-merous, homostylous or heterostylous. Sepals 5, free, entire to glandular-dentate, persistent. Petals 5, free or rarely connate basally, blue, red, pink, yellow, purplish, or white, narrowly clawed. Stamens 5, usually alternate with minute staminodes; filaments connate basally to form short tube; extrastaminal glands 5, adnate to staminal tube opposite petals. Ovary 2–5-loculed, be-

* Contributed by Duncan M. Porter.

coming more or less completely 4–10-loculed by intrusion of a false septum in each locule; styles 2–5, free or connate. Capsule dehiscing into 4–10 mericarps or 5 2-seeded mericarps. Seeds often mucilaginous when wet; endosperm scanty or absent; embryo straight.

Some 150 to 200 species, almost worldwide in distribution, but most abundant in the southwestern United States and Mexico and in the Mediterranean region.

Stipules absent; sepals lacking basal glands. . . . . *L. cratericola*
Stipules glandular, conspicuous; sepals biglandular basally. . . . . *L. harlingii*

Linum cratericola Eliass., Bot. Not. 121: 634. 1968

Suffrutescent, 4–5 dm high, glabrous; stems and branches numerous, erect and ascending; branches striate; leaves alternate, narrowly oblong-lanceolate, subacute apically, narrowed basally, 1-nerved, numerous, covering stem, 7.5–11 mm long, about 1 mm wide; stipules absent; flowers solitary at tips of branchlets; sepals subovate to subrhomboid, acuminate, obscurely 3–5-nerved, midnerve prominent basally, the sepals 3–3.5 mm long, about 2 mm wide; petals yellow, broadly obovate, about 8 mm long, 5 mm wide; filaments dilated basally, alternating with toothlike staminodes; styles 5, free; capsule globose, light brown, about 2.5 mm broad; seeds subelliptic, flat, brown, shiny, about 1.5 mm long, 1 mm wide.

Endemic; known only from Isla Santa María.

Eliasson lists two collections from small craters northeast of Floreana Peak. I have seen neither of his specimens and have adapted my description from his type description.

Linum harlingii Eliass., Bot. Not. 121: 636. 1968

Suffrutescent perennial with glabrous ascending stems; branches striate, glabrous; branchlets slightly winged; leaves chiefly on new branches, the upper alternate, the lower opposite, linear-lanceolate, erect, entire, the margin involute apically, 3–7 mm long, to 1 mm wide, adaxial surface puberulent basally, 1-nerved, nerve in-

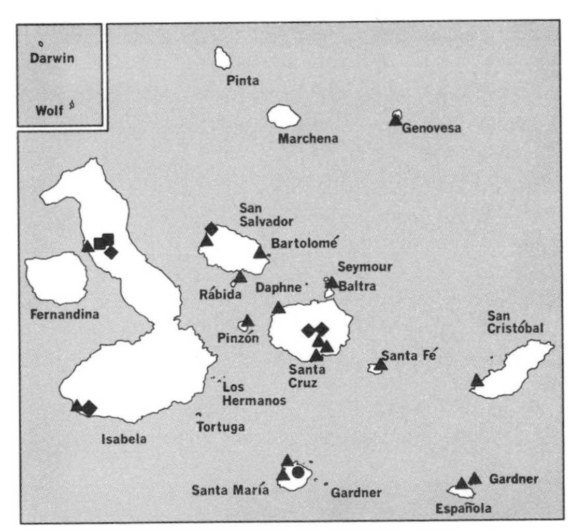

● *Linum cratericola*
■ *Linum harlingii*
▲ *Mentzelia aspera*
◆ *Sclerothrix fasciculata*

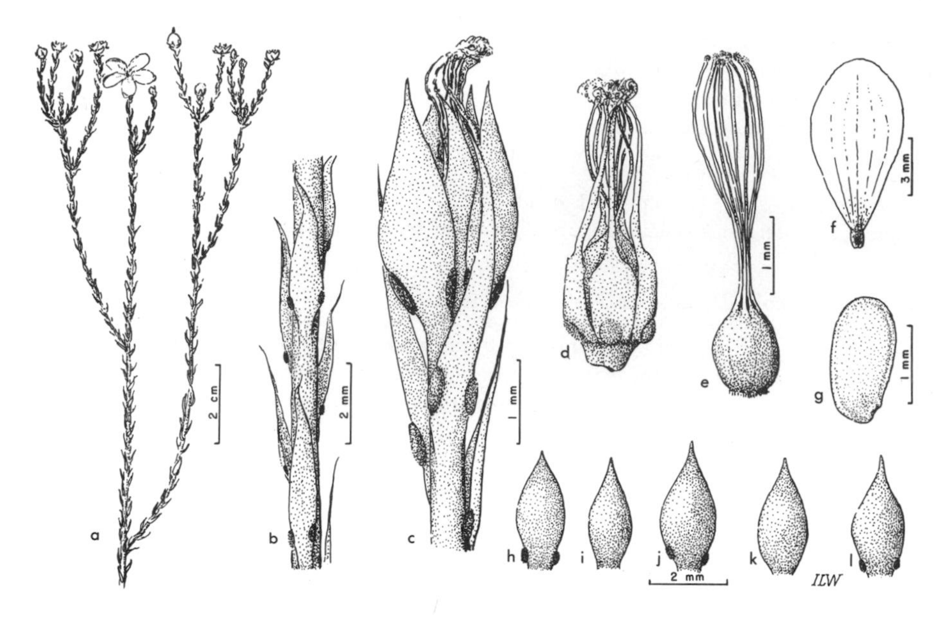

Fig. 181. ***Linum harlingii.***

conspicuous adaxially, dark but not prominent abaxially; stipules glandular, globose, reddish or dark, less than 1 mm thick; pedicels erect, slightly winged, about 1 mm long; sepals subequal, 3–3.5 mm long, about 1.5 mm wide, acuminate and revolute apically, biglandular basally, 1-nerved, the exterior slightly longer and ovate to obovate, the interior slightly wider and obovate, glabrous, the nerve abaxially elevated above and thickened below into a callus; petals yellow, free, obovate-cuneate, apex obtuse and rounded to mucronate, slightly crested below adaxially, 7–9 mm long, 3–4 mm wide; filaments connate into a tube less than 1 mm high, subulate, dilated basally, glabrous, about 3 mm long, two-thirds as long as styles; anthers globose, less than 1 mm broad; staminodes absent; ovary 5-loculed, ovoid, about 1.5 mm broad; styles 5, slightly connate basally, glabrous, 3–4 mm long; stigmas capitate; capsule globose, a little longer than surrounding sepals, dehiscing into 10 1-seeded mericarps; seeds more or less elliptic, flat, rugulose, 1.5 mm long.

Endemic; known only from Isla Isabela.

This species was referred by Robinson (1902) and by Stewart (1911) to *L. oligophyllum* Willd. ex Schult., presumably a Peruvian species, though its type locality is unknown.

See Fig. 181.

LOASACEAE. STICKLEAF FAMILY*

Annual or perennial herbs (to woody) with barbed (sometimes stinging) hairs. Leaves alternate or opposite, entire or lobed, petiolate. Flowers regular, bisexual;

* Contributed by H. J. Thompson.

sepals persistent or not; corolla apopetalous to sympetalous, sometimes with petaloid staminodes alternate with petals; stamens 5 to many, sometimes fascicled; filaments narrow to petaloid; style 1, persistent or not, the stigmas usually connate. Ovary inferior or semi-inferior, 1-locular with 1–5 placentae; ovules 1 to many. Fruit dry, dehiscent (or indehiscent and 1-seeded). Seeds variable; embryo straight or sometimes hooked.

About 14 genera and 200 species in temperate and tropical regions of the New World and 1 genus and 2 species in the Old World.

Leaves alternate . . . . *Mentzelia*
Leaves opposite . . . . *Sclerothrix*

Mentzelia L., Sp. Pl. 516. 1753; Gen. Pl. ed. 5, 233. 1754

Adhesive, usually brittle annual (or perennial) herbs; pubescent throughout with barbed but not stinging hairs. Leaves alternate, lobed or entire, petiolate or sessile, usually reduced upward. Flowers solitary and axillary (or in terminal, bracteate, more or less cymose inflorescences); perianth 5-merous, usually 2-seriate; stamens few to many, filaments elongate, filiform or dilated, staminodes present (or absent). Style filiform; stigma represented by 3 furrows or tuft of hairs. Ovary inferior; ovules few to many. Capsules variable in size and shape, sessile to short-pedicellate. Seeds variable.

About 60 species in temperate and tropical America, mostly in the southwestern United States and Mexico.

Mentzelia aspera L., Sp. Pl. 516. 1753

Mentzelia aspera var. *canescens* Anderss., Kongl. Svensk. Vet.-Akad. Handl. 1853: 222. 1855.
Mentzelia aspera var. *virescens* Anderss., loc. cit.
Mentzelia aspera var. *lobata* Anderss., loc. cit.

Annual herb with more or less sprawling stems to about 4 dm tall, these straw-colored, pubescent with large glochidiate hairs and many pointed hairs; leaves ovate to trilobed, margins irregularly dentate or entire, blades pubescent with long glochidiate and pointed hairs, more dense on lower surface; petioles 1–4 cm long; flowers produced February–June; calyx deciduous, lobes lanceolate, 3–5 mm long; petals yellow, broadly ovate, tip rounded to somewhat acute, 3–6 mm long, 4–4.5 mm wide; stamens about 20, filaments all narrow, 3–4 mm long; style 3–4 mm long; capsules cylindrical with narrowed base, sessile, 5–11 mm long; seeds flattened, pyriform, about 4 mm long, the surface irregularly striate.

In soil and rock crevices, in the open or among trees and shrubs, Mexico, West Indies, and South America; mainly in the Arid and Transition zones and generally distributed in the Galápagos Islands. Specimens examined: Baltra, Española, Gardner (near Española), Genovesa, Isabela, Pinzón, Rábida, San Cristóbal, San Salvador, Santa Cruz, Santa Fé, Santa María.

The Galápagos specimens of *M. aspera* differ consistently from those of the mainland. The Galápagos plants, from which my description is drawn, have shorter capsules and smaller flowers in which all filaments are filiform, whereas plants of the continent have longer capsules and larger flowers in which the 5 outer filaments are dilated. This difference in flower size probably indicates that the Galápagos plants are more readily self-pollinating than the mainland form. Andersson used

Fig. 182. *Sclerothrix fasciculata* (*a–c*). *Mentzelia aspera* (*d–f*).

vegetative differences and subtle and minor differences in length of calyx lobes in relation to corolla size to differentiate the varieties *canescens, lobata,* and *virescens.* Studies of other *Mentzelia* species have shown such characters to be either environmental modifications or minor variations within local populations.

See Fig. 182 and color plate 45.

Sclerothrix Presl, Symb. Bot. 2(6): 4, *t.* 53. 1833

Annual herbs with weak ascending-scandent stems to 2 m tall; stems straw-colored, glabrous in lower portions, retrorsely pubescent in the upper, and younger portions with pointed hairs (petioles pubescent but not retrorsely so). Leaves opposite, ovate, entire, the margins finely toothed; petioles to 1 cm long; blades pubescent with pointed hairs, somewhat more densely pubescent on lower than upper surfaces. Flowers minute, 4-merous, produced February–June. Calyx lobes triangular, 1–1.5 mm long, eventually deciduous; petals white, broadly ovate, cucullate, densely pubescent on outer surface, 1–1.8 mm long. Stamens 4–14, usually 1–4 opposite each petal; filaments narrow, slightly expanded at base and adnate to petals, 1–1.5 mm long; staminodes present, varying in width and number, alternate with petals, pubescent. Ovary semi-inferior with pubescent conical projection above calyx tube; style 0.5–1.2 mm long; capsules stipitate, short-cylindrical, 4–6 mm long, about 1 mm wide, twisted and splitting longitudinally when mature by 4 valves. Seeds 8–16 per capsule, usually half the seeds small (about 0.5 mm long) and half larger (about 1 mm long), oblong, brown, the surface with raised striae.

A monotypic genus occurring from Mexico to Peru, east to Brazil.

Sclerothrix fasciculata Presl, Symb. Bot. 2(6): 4, *t.* 53. 1833

Characters of the genus.

Scrambling over shrubs and rocks in shady areas in the *Scalesia* Zone, on several of the larger islands in the archipelago. Specimens examined: Isabela, San Salvador, Santa Cruz.

See Fig. 182.

LYTHRACEAE. LOOSESTRIFE FAMILY*

Herbs, shrubs, or rarely small trees; stems quadrangular or terete. Leaves opposite, rarely alternate or whorled, simple, entire; stipules none. Flowers bisexual, regular or irregular, rarely cleistogamous, 3–6-merous, sometimes di- or trimorphic in style or stamen length, perigynous, axillary or in terminal racemes or spikes; pedicels usually with 2 opposite bracteoles. Hypanthium campanulate to tubular, often conspicuously nerved, persistent; calyx lobes 4–6, mostly alternating with 3–5 triangular appendages. Petals 0–6, free, inserted on inner surface of hypanthium between calyx lobes, crumpled, deciduous or rarely persistent. Stamens as many as, twice as many as, or more numerous than petals, included to exserted; filaments inserted on inner surface of hypanthium below petals, often alternately unequal; anthers 2-loculed, versatile or rarely basifixed, longitudinally dehiscent. Disk hypogynous,

* Contributed by Duncan M. Porter.

often surrounding base of ovary, but in irregular flowers on adaxial side only. Gynoecium syncarpous; ovary superior, free, 2- to 6-loculed, the septa complete or reduced; ovules 2 to many per placenta, anatropous and epitropous, the placentation axile or rarely parietal; style filiform; stigma capitate, rarely 2-lobed. Fruit a membranaceous or coriaceous capsule enclosed by persistent hypanthium, septicidally or loculicidally dehiscent or indehiscent and splitting irregularly. Seeds 3 to many, pyramidal, ovoid, or discoid, occasionally slightly winged; endosperm little or none; embryo straight; cotyledons flat or rarely convolute, often auriculate-cordate; radicle short.

About 24 genera and 450 species, in temperate and tropical areas throughout the world, most abundant in the American tropics and subtropics.

Cuphea P. Br., Nat. Hist. Jamaica 216. 1756

Herbaceous or woody annuals or short-lived perennials, usually with viscid glandular trichomes. Leaves decussate, verticellate, or rarely alternate, ovate to lanceolate, elliptic, or linear, membranaceous to coriaceous, sessile to petiolate. Flowers irregular, 6-merous, homomorphic, 1–3 at a node, sessile or pedicellate; when alternate on stem, 1 at each node interaxillary, others if present axillary; when opposite or verticellate on stem, all interaxillary and often internodal. Hypanthium tubular, greenish or purple, base gibbous or distinctly spurred, spur curving downward toward or upward away from pedicel, tube distinctly 12-nerved; calyx lobes 6, triangular, the apex acute to acuminate, rarely apiculate, equal or adaxial lobe larger; appendages shorter to longer than calyx lobes. Petals 0–6, pale to deep purple, equal or adaxial 2 or 4 largest, caducous or nearly persistent. Stamens 5–12, included to exserted, 2 adaxial inserted lower in hypanthium than others; filaments alternately unequal, often densely covered with trichomes. Disk on adaxial side only, free from ovary. Ovary incompletely 2-loculed, appearing 1-loculed by reduction of septum, abaxial locule smaller and sometimes sterile; style included or exserted; stigma capitate, rarely 2-lobed. Capsule membranaceous, loculicidally dehiscent along adaxial wall, this wall of hypanthium also splitting and placenta projecting out of capsule and hypanthium. Seeds 3 to many, orbicular to ovoid, usually flattened, surface pebbled, often also narrowly winged; cotyledons flattened, cordate to auriculate.

About 250 species in New World tropical and subtropical regions. "The entire genus is in need of revision, with emphasis on a more natural arrangement of the species" (Graham 1964, 247).

Cuphea carthagenensis (Jacq.) Macbr., Field Mus. Nat. Hist. Bot. Ser. 8: 124. 1930

Lythrum carthagenense Jacq., Stirp. Amer. Hist. 148. 1763.
Balsamona pinto Vand., Fasc. Pl. 15. 1771.
Cuphea balsamona Cham. & Schlecht., Linnaea 2: 363. 1827.
Cuphea divaricata Pohl ex Koehne in Mart., Fl. Bras. 13²: 255. 1877.
Cuphea peplidioides Mart. ex Koehne, loc. cit.
Parsonia pinto Heller, Minn. Bot. Stud. 1: 862. 1897.
Cuphea pinto Koehne, Pflanzenr. IV, 216: 123. 1903.
Parsonia balsamona Standl. in Standl. & Calderon, Lista Prelim. Pl. El Salvador 159. 1925.

Annual herb 15–50 cm high, basally subligneous; stems terete, more or less reddish, white-tomentose and sparsely hispid with larger more or less glandular hairs; leaves opposite, obovate to oblong, apex obtuse, apiculate, base cuneate, decurrent

Fig. 183. *Cuphea carthagenensis* (*a–c*). *Abutilon depauperatum* (*d, e*).

into petiole, green to occasionally reddish, more or less scabrous, some glandular hairs beneath, margins ciliate, 2–6 cm long, 12–26 mm wide, lowest leaves largest; petioles 1.5–2 mm long, or absent; flowers small, in few-flowered terminal clusters, some axillary; pedicels about 1 mm long, 2-bracteolate; hypanthium tubular, pale green, sometimes becoming reddish, hispidulous, some hairs glandular, pubescence usually restricted to nerves, tube 4–7 mm long, flask-shaped and apically constricted in fruit; calyx lobes ovate, apiculate, alternating with 5 small triangular pubescent, sometimes hispid, exterior appendages; petals 6, pale pink, drying violet, equal, 1.5 mm long; stamens 11, included; filaments short, filiform, curved, inserted in 2 unequal whorls at constriction of hypanthium; anthers versatile, very small; disk small, suborbicular, horizontal or ascending, not reflexed; ovary ovoid, membranaceous, glabrous, 2.5 mm high; ovules 6; style curved, glabrous, included, about 1 mm long; stigma capitate; capsule ovoid, 1-loculed, 4 mm high, 2.5 mm wide; seeds 4, subcordate, brownish red, finely pitted, narrowly marginate, about 2 mm in diameter.

A native of South America, but widespread as a weed from Texas to North Carolina and south through Mexico, the West Indies, and Central America to Argentina and Paraguay; introduced into the Hawaiian Islands, Fiji, and the Philippines, and perhaps into the Galápagos Islands. Specimens examined: San Cristóbal, Santa Cruz.

This species was first reported from the Galápagos Islands as *C. patula* St. Hil. (Caruel, 1889; Stewart, 1911), but its correct identity was ascertained by Bacigalupi (1931).

See Fig. 183.

MALVACEAE. Mallow Family*

Annual or perennial herbs, shrubs, or trees, usually fibrous-stemmed, mucilaginous and pubescent, the hairs variously stellate and/or simple or glandular, sometimes aculeate. Leaves alternate, with few exceptions petiolate and stipulate, the blades mostly simple but sometimes palmately compound, unlobed or palmately lobed to dissected, divisions often pinnately lobate, major nerves usually palmate. Flowers bisexual or rarely unisexual, actinomorphic or essentially so, axillary and/or terminal, solitary or in cymes or spicate to paniculate or capitulate inflorescences. Calyx gamosepalous, valvate, mostly 5-lobed and persistent in fruit, usually nectariferous at base within, often subtended by an epicalyx of (1)3 to many free or laterally connate bracts. Petals mostly 5, convolute, distinct but adnate at base to staminal column. Stamens monadelphous, column antheriferous from near base or only apically, apex 5-toothed or terminated by filaments, anthers 5 to many, dorsifixed, reniform, 1-celled, longitudinally dehiscent, the pollen usually large, echinate. Ovary superior, (1)3- to many-carpeled and -loculed, carpels disposed in a single whorl but sometimes appearing as several superposed whorls, the placentation axile, ovules 1 to several in each locule, amphi- to anatropous, ascending to pendulous. Styles usually joined at base, with branches as many or twice as many as carpels and stigmas capitate or decurrent, sometimes connate to apex and terminated by a lobed or parted stigma. Fruit a dry, rarely fleshy, loculicidally dehiscent capsule or a schizocarp separating into indehiscent or loculicidally dehiscent mericarps, rarely

* Contributed by David M. Bates.

woody and completely indehiscent. Seeds 1 to several in each cell or mericarp, subreniform to subglobose, embryo usually curved, endosperm usually present, cotyledons mostly foliaceous and plicate.

About 92 genera and 1,500 species, distributed throughout the world but found principally in warm temperate and tropical regions.

Petals more than 3 cm long; fruit a more or less woody, loculicidally dehiscent capsule more than 1.5 cm long:
Style unbranched, the apex with a ribbed or lobed stigma; plants black-punctate; involucral bracts 3, foliaceous, lacerate-margined....................... *Gossypium*
Style 5-branched, the branches ultimately spreading; plants not black-punctate:
Calyx spathaceous, splitting on 1 side at anthesis, minutely 5-toothed apically, adnate to corolla and falling away with it after anthesis............... *Abelmoschus*
Calyx not spathaceous, 5-lobed, not adnate to corolla, persistent after flowering.. *Hibiscus*
Petals less than 2 cm long; fruit a schizocarp or if appearing capsular not woody and less than 1 cm long:
Style branches 10, twice as many as mericarps; flowers in axillary glomerules subtended by conduplicate, veiny, foliaceous bracts.......................... *Malachra*
Style branches 5 to many, same in number as mericarps; inflorescences ebracteate or bracts not foliaceous:
Involucral bracts absent (also see *Urocarpidium*); mericarps uniovulate with a pendulous ovule or bi- or triovulate:
Mericarps inflated, 5–20 mm long, rounded-reniform, muticous, laterally the walls often silvery .... *Herissantia*
Mericarps not inflated, less than 5 mm long or, if more than 5 mm, armed apically with an erect or deflexed spine:
Mericarps 5.8–9 mm long, triovulate............................ *Abutilon*
Mericarps less than 5 mm long, uniovulate:
Fruit capsular, the loculicidally dehiscent valves of adjacent mericarps connate .... *Bastardia*
Fruit a schizocarp, the mericarps separating septicidally from each other:
Mericarps beaked apically, pericarp unilamellar, lateral walls persistent, striate-reticulate .... *Sida*
Mericarps beaked dorsally, pericarp bilamellar with dorsally reticulate endocarp, lateral walls evanescent.......................... *Anoda*
Involucral bracts 2 or 3 (in *Urocarpidium* caducous); mericarps uniovulate with an ascending ovule:
Stigmas introrsely decurrent on filiform style branches; mericarps unarmed and subtended by the scarious, accrescent calyx...................... *Malva*
Stigmas terminal on style branches; mericarps armed or enclosed by calyx:
Petals yellow, 4–9 mm long; mericarps reddish brown, armed dorsally and sometimes ventrally or, if unarmed, flowers in dense, bracteate spikes; leaves unlobed .... *Malvastrum*
Petals roseate, about 3.5 mm long; mericarps blackish, reticulate but unarmed; leaves usually shallowly 3-lobed....................... *Urocarpidium*

Abelmoschus Medic., Kunstl. Geschl. Malven-Fam. 45. 1787

Annual or perennial herbs or subshrubs, often hispid. Leaves long-petioled, stipulate, the blades palmately angulate, lobed, or parted, the divisions usually toothed or lobate, base often hastate. Flowers axillary, solitary or in racemes by reduction of upper leaves, pedicels mostly short, not articulate. Involucral bracts 4–16, distinct or slightly connate at base, persistent or caducous. Calyx spathaceous, 5-toothed at apex, splitting on 1 side at anthesis, adnate to base of corolla and deciduous with it. Corolla mostly yellow, dark purple at center, sometimes white or pinkish, funnelform, the petals 5, obovate, short-clawed. Staminal column shorter than petals, antheriferous from near base to below 5-toothed apex. Ovary of 5 cells in single whorl, ovules many, style shortly 5-branched, stigmas terminal, expanded. Fruit a loculicidally dehiscent capsule, usually elongate, acuminate, each cell many-seeded. Seeds reniform, glabrous or pubescent.

About 16 species in southern and southeast Asia and northern Australia; 2 species, *A. esculentus* (L.) Moench and *A. moschatus* Medic., and less commonly a third, *A. manihot* (L.) Medic., are cultivated and sometimes found as escapes in the New World tropics.

Abelmoschus manihot (L.) Medic., Kunstl. Geschl. Malven-Fam. 46. 1787

Hibiscus manihot L., Sp. Pl. 696. 1753.

Erect, perennial herb 1–3 m or more tall, usually hispid with patent, simple or few-branched hairs to 4 mm long, sometimes glabrous or stellulate-pubescent; leaves variable, petioles longer than blades, stipules lanceolate, to 15 mm long, blades to 45 cm long, deeply and palmately 3-, 5-, or 7-lobed or -parted, divisions linear to narrowly oblong, usually coarsely serrate and pinnilobate, sometimes nearly entire, surfaces glabrous to moderately pubescent; flowers solitary in axils or in racemes toward ends of branches, pedicels accrescent, in fruit rather stout; involucral bracts 4–6(8), distinct, ovate to oblong, 1–3 cm long, 0.5–1 cm broad; calyx 2–3 cm long, velutinous; corolla white or yellow with small purple center, petals broadly obovate, 4–8 cm long; staminal column to 3 cm long, slightly exceeded by the spatulate-discoid, purple stigmas; capsule erect, ovate-elliptic, apiculate, 3–6 cm long, to 2.5 cm in diameter, deeply 5-angled, usually densely hispid, valves shiny white within; seeds many in each cell, ovoid-reniform, 3–4 mm long, dark brown or black, striate, pubescent.

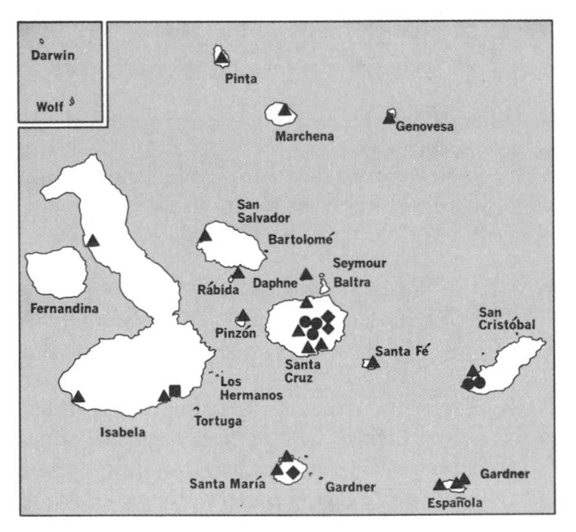

● *Cuphea carthagenensis*
■ *Abelmoschus manihot*
▲ *Abutilon depauperatum*
◆ *Anoda acerifolia*

Widely distributed in southern and southeast Asia and the Pacific islands, where it is cultivated as a vegetable; sometimes grown as an ornamental and occasionally is an escape in the New World. Reported by Stewart from Isla Isabela, near habitations at Villamil.

Abutilon Mill., Gard. Dict. abr. ed. 4. 1754

Perennial herbs or shrubs, less often annual herbs or small trees, usually pubescent. Leaves stipulate, mostly petiolate but those above sometimes sessile or nearly so, blades simple, often cordate-ovate, unlobed or palmately lobed. Flowers axillary, solitary or in cymes, often racemose to paniculate by reduction of leaves, pedicels usually articulated. Involucral bracts wanting. Calyx campanulate to cupuliform, sometimes reflexed, 5-lobed. Petals 5, white, yellow, orange, or reddish, forming a tubular-campanulate to broadly campanulate corolla or patent to reflexed. Staminal column usually shorter than petals, often ventricose at base, the filaments borne apically and terminating column. Ovary of 5 to many mericarps in a single whorl, each enclosing 2–9 ovules in a single vertical file, style branches same in number as mericarps, stigmas terminal, often abruptly expanded. Fruit a schizocarp but sometimes appearing capsular by lateral coherence of mericarps, usually more or less cylindrical, mericarps sometimes inflated, mostly ovate-oblong to oblong in lateral view, apex muticous or awned, endoglossum rarely present, lateral walls smooth or sometimes with obscure reticulum basally, dehiscence various, valves of each mericarp separating completely from each other or remaining connate at base dorsally and/or ventrally. Seeds 1 or more in each mericarp, reniform, glabrous or pubescent.

Perhaps 150 species, mostly in tropical and subtropical regions of the world but some entering temperate regions.

Abutilon depauperatum (Hook. f.) Anderss. ex Robins., Proc. Amer. Acad. 38: 173. 1902

Sida depauperata Hook. f., Trans. Linn. Soc. Lond. 20: 232. 1847.
Abutilon anderssonianum Garcke in Anderss., Kongl. Svensk. Vet.-Akad. Handl. 1853: 230. 1855.
Pseudoabutilon depauperatum Kearney, Madroño 11: 287. 1952.

Suffrutescent or shrubby, mostly less than 1 m but up to 3 m tall with strictly ascending branches, the stems, petioles, and peduncles copiously to densely but closely stellate-pubescent, the hairs often somewhat rufous distally and on calyx; leaf blades to 10(20) cm long and broad, suborbicular to deltoid- or lance-ovate, acute to short-acuminate, nearly entire or crenulate, crenate, or serrate, often irregularly so, cordate, sometimes shallowly but acutely 3-lobed, thin or thickish, concolorous or somewhat discolorous, stellate-pubescent above and more densely so below, veins rather prominent below, petioles as long as or shorter than blades, stipules subulate to filiform, 4–7 mm long, caducous; flowers axillary, solitary or in 2–6-flowered subumbellate clusters, peduncles and pedicels to 5 cm long, filiform, the pedicels articulate above middle; calyx campanulate, densely stellate-pubescent, 4–6 mm long in flower, to 9 mm long in fruit, lobed to or below middle, lobes ovate to triangular, acute or apiculate; petals pale to deep yellow, fading white, spreading, cuneate-obovate, mostly 7–10 mm long, base indistinctly clawed, ciliate, without auricles; staminal column 3–6 mm long, slender above the ventricose base, stellate-pubescent, the filaments terminal, slender, radiate, the anthers about 30–40; mericarps 6–8(10), each triovulate, style branches slender, stigmas capitate, somewhat discoid; fruit cylindrical, surpassing calyx, densely stellate-pubescent, the mericarps chartaceous, ovate-oblong to oblong in lateral view, 6–9 mm long including an erect to somewhat deflexed terminal awn 1–2.5 mm long, lateral walls smooth and mericarp without endoglossum, or lower lateral wall very faintly striate-reticulate and dorsal wall with internal endoglossum extending halfway to three-quarters across the locule above lowest ovule, dehiscence varying from complete, with the 2 valves of each mericarp separating from each other, to partial, with the valves remaining joined the lower third ventrally and/or dorsally; seeds 1–3, triangular-reniform, to 1.7 mm long, openly but conspicuously papillate.

Endemic. Specimens examined: Daphne, Española, Gardner (near Española), Genovesa, Isabela, Marchena, Pinta, Pinzón, Rábida, San Cristóbal, San Salvador, Santa Cruz, Santa Fé, Santa María.

Abutilon depauperatum is a polymorphic species closely related to if not conspecific with the more widely distributed, tropical American *A. umbellatum* (L.) Sweet. By various criteria previous authors have generally recognized 2 taxa in what herein is treated as a single species. On the basis of differences in the mericarps, 2 groups may be recognized. In one of these the mericarps have a small endoglossum above the lowermost ovule and the lateral walls are minutely striate basally. In the other fruiting collections, the mericarps have neither an endoglossum nor lateral wall markings. The presence or absence of these mericarp characters may or may not be constant to species, but in the absence of other distinguishing characters, the mericarp differences should probably not be regarded as definitive. In all other respects the collections of both fruiting types encompass the same range of character variation.

See Fig. 183 and color plate 44.

Anoda Cav., Diss. 1: 38. 1785

Annual herbs or suffrutescent, glabrate to densely pubescent, sometimes hispid plants. Leaves petiolate, stipulate, the blades often highly variable even on same plant, lance-ovate to deltoid, unlobed or palmately lobed, the margins obscurely to coarsely and often irregularly crenate to dentate, the base cordate to cuneate but typically hastate. Flowers solitary in axils or in racemes or panicles by reduction of leaves, pedicels usually elongate, often markedly accrescent, articulate at or above middle. Involucral bracts wanting. Calyx cupuliform, sometimes prominently nerved, mostly 5-lobed to middle or beyond. Petals 5, white, yellowish, blue, or violet, equal to or greatly exceeding calyx, mostly patent, obovate, bearded marginally at base. Staminal column included, filaments apical and terminating column, sometimes appearing in fascicles of 5. Ovary of 5–20 mericarps in single whorl, each enclosing a single, more or less pendulous ovule, style branches same in number as mericarps; stigmas terminal, often conspicuously dilated. Fruit a discoid or depressed-globose, mostly pubescent schizocarp, mericarps unarmed or dorsally bearing single, median, patent or deflexed spine, pericarp unilamellar or more typically bilamellar, endocarp plane or fenestrate, separating and partially or completely surrounding seed and sometimes adhering to it, lateral walls often thin, disintegrating or adnate to seed to form single circular chamber in schizocarp. Seeds angulate-reniform.

About 15 species from the southern United States and the West Indies to central South America; 2 species now naturalized in the Old World.

Anoda acerifolia DC., Prodr. 1: 459. 1824

Sida acerifolia Zucc. in Roem., Coll. 148. 1806, non Medic. 1787, nec Lag. 1816.

Erect or decumbent, branching annual to about 7 dm tall, the stems, petioles, and peduncles moderately simple-setose and usually stellulate as well, eventually glabrate; leaves petiolate, blades polymorphic, mostly broadly deltoid to narrowly triangular, unlobed or hastately lobed to palmately 5-lobed, apex acute, margins entire or sometimes coarsely serrate toward base, this broadly obtuse, truncate, or subcordate, the surfaces above and below openly and inconspicuously appressed simple-setose, more densely so at margins, stipules filiform to lanceolate, to 6 mm long; flowers solitary in axils on pedicels about 3–8 cm long, these to 15 cm long in fruit; calyx campanulate, sparsely hispid, to (5)7–9 mm long, ovate- to triangular-lobed, in fruit rotate and subtending the mericarps, to 20 mm broad; petals white (in ours), lavender, or purplish, 12–15 mm long, obovate, obliquely rounded to subemarginate apically, ciliate at basal margins; staminal column included, about 4 mm long, filaments terminal, to 2 mm long; schizocarp discoid, yellowish-hispid above, 11–15 mm broad, including spines, the mericarps 9–15, dorsally gibbous or with horizontal spine 1–1.5(2.5) mm long, lateral walls evanescent, in part adhering to seed coat, dorsal and apical wall bilamellar, inner portion forming white to black, coarsely fenestrate dorsal hood over the free, brownish, muricate seed.

Widely distributed in Central America, northern South America, and the West Indies, and naturalized locally in parts of the Old World. Specimens examined: Santa Cruz, Santa María.

See Fig. 184.

Bastardia HBK., Nov. Gen. & Sp. Pl. 5: 254. 1822

Herbs or shrubs, variously pubescent, the hairs stellate, simple, and glandular. Leaves petiolate, stipulate, the blades simple, cordate-ovate, unlobed or shallowly 3-lobate. Flowers axillary, solitary or in loose cymes, often exceeded by subtending leaves. Involucral bracts wanting. Calyx cupuliform, 5-lobed, persistent. Petals 5, white to yellow or yellow-orange, patent, obovate, broadly unguiculate, not auriculate. Staminal column included, ventricose basally, anthers few on terminal filaments. Ovary of 5–8 mericarps in single whorl, each enclosing a single, pendulous ovule; style branches as many as mericarps, greatly exceeding anthers; stigmas capitate, discoid. Fruit capsular, depressed-globose, 5–8-angled, mericarps obovoid, muticous or apically awned, loculicidal dehiscence complete, the valves of adjacent mericarps remaining laterally coherent and falling away from columella together. Seed trigonous.

Six species in the West Indies and Central and South America.

Bastardia viscosa (L.) HBK., Nov. Gen. & Sp. Pl. 5: 256. 1822

Sida viscosa L., Syst. Nat. ed. 10, 1145. 1759.

Subshrub 3–10 dm tall, the stems, petioles and pedicels usually densely glandular and moderately to densely pubescent with patent, simple hairs to 3 mm long and

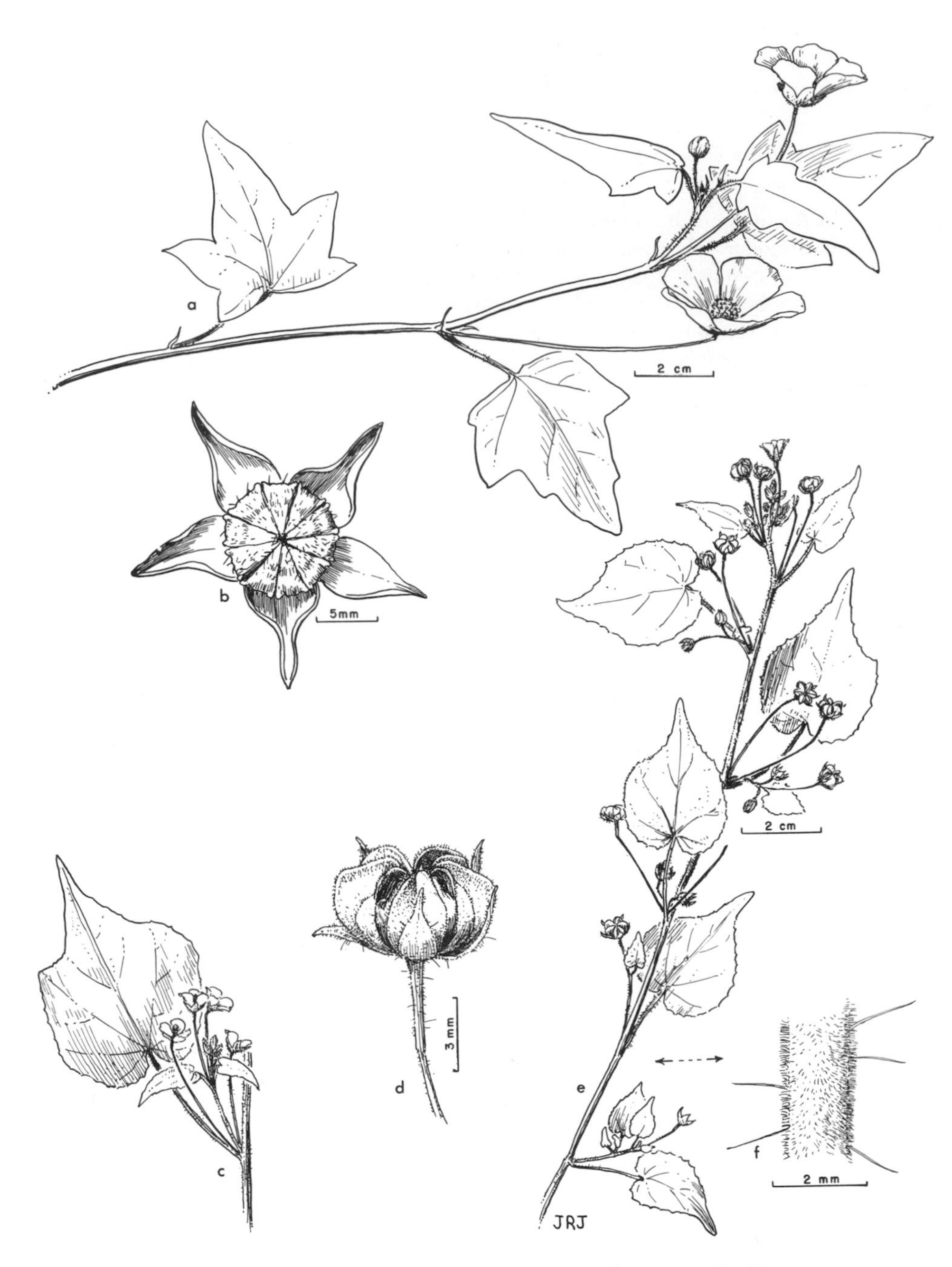

Fig. 184. *Anoda acerifolia* (*a, b*). *Bastardia viscosa* (*c–f*).

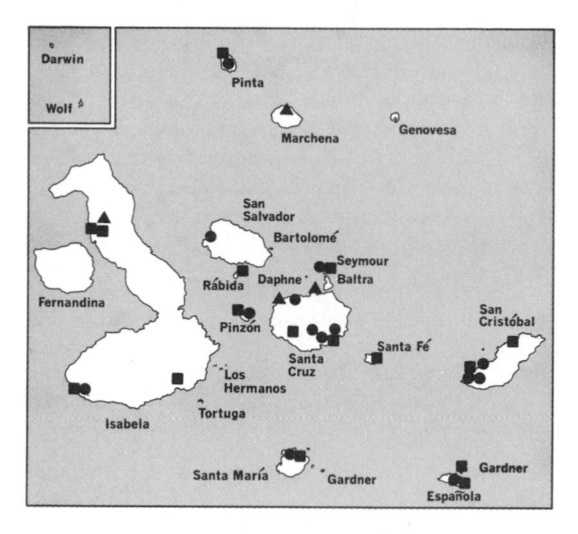

● *Bastardia viscosa*
■ *Gossypium barbadense* var. *darwinii*
▲ *Gossypium klotzschianum*

sparse stellulate hairs; petioles mostly half to three-fourths as long as blades, stipules filiform, caducous, the blades thin, green or yellowish green, to 10 cm long, but mostly smaller, cordate-ovate, acute to acuminate, entire to serrulate, unlobed, sparsely to copiously stellate above and below; flowers axillary, solitary or up to 3 per node, usually on elongate, apically articulate pedicels to 3 cm long, sometimes appearing subsessile; calyx campanulate, copiously pubescent, 4–5 mm long, lobes oblong-ovate, short-acuminate, 2.6–3 mm long; petals pale to deep yellow or yellow-orange, 5–6 mm long, obovate, the claw slightly puberulent; staminal column to 2.5 mm long, ventricose basally, glabrous or puberulent between petals, filaments terminal, radiate, 1–1.5 mm long, anthers about 20; fruit depressed-globose, (5)6–7-angled, copiously stellate, the mericarps (5)6–7, asymmetrically ovoid, unarmed, chartaceous, 3.3–3.8 mm long, 3–3.5 mm broad, lateral walls of adjacent mericarps remaining coherent, the valves falling away together; seed more or less rounded-trigonous, about 2 mm long and broad, brownish, silky-pubescent.

Widely distributed from the West Indies and southern Mexico south to northern Argentina and Peru; reported from several of the Galápagos Islands at elevations up to 310 m in rather bushy, mostly open habitats. Specimens examined: Española, Isabela, Pinta, Pinzón, San Cristóbal, San Salvador, Santa Cruz, Santa María, Seymour.

See Fig. 184.

Gossypium L., Sp. Pl. 693. 1753; Gen. Pl. ed. 5, 309. 1754

Annual or perennial herbs, shrubs, or small trees, glabrous to densely pubescent, black-punctate with oil glands throughout. Leaves petiolate, stipulate, the blades palmately lobed or parted or unlobed, entire, the primary nerve(s) beneath usually with nectary near base. Flowers solitary, axillary, often on sympodial branchlets, pedicels not articulate. Involucral bracts 3, distinct or slightly connate, usually foliaceous, entire or lacerate, persistent or sometimes caducous. Calyx relatively small, cupular, truncate, undulate, or 5-lobed. Petals 5, obovate, white to yellow or purple, often with purplish spot toward base. Staminal column elongate but included in corolla, antheriferous from near base to below the 5-lobed apex. Ovary 3–5-celled, each cell 2– to many-ovuled, style unbranched, stigma clavate, sulcate, rarely somewhat divided at apex. Fruit a globose to ovoid or fusiform, mostly apiculate, loculicidally dehiscent capsule, the walls somewhat coriaceous to woody. Seeds ovoid, often angulate, usually covered with short, fine tomentum (fuzz) and/or long, woolly tomentum (lint or floss), or sometimes nearly glabrous.

About 32 species, mostly in rather xeric habitats in the tropics and subtropics of the world; the cultivated species, the cottons of commerce, are occasional escapes.

Leaves lobed to middle or beyond, or sometimes distal ones unlobed or very shallowly lobed, lobes narrowly triangular to broadly ovate; seeds with long, whitish lint and short brownish fuzz . . . . *G. barbadense* var. *darwinii*

Leaves unlobed or rarely with a pair of angulate teeth above middle; seeds with a fine, inconspicuous fuzz . . . . *G. klotzschianum*

Gossypium barbadense var. **darwinii** (Watt) J. B. Hutch. in Hutchinson, Silow, and Stephens, The Evolution of Gossypium 51. 1947

Gossypium darwinii Watt, The Wild and Cultivated Cotton Plants of the World 68. 1907.

Few- to many-branched subshrub or shrub 1–3 m tall, copiously black-punctate throughout, new growth glabrous to densely whitish or yellowish stellate-pubescent; leaf petioles usually shorter than blade, stipules linear-triangular to falcate, usually caducous, the blades 3- or 5-lobed to beyond middle, rarely ovate, unlobed, up to 15 cm long, 20 cm broad, flowering leaves usually much smaller, lobes narrowly triangular-ovate to broadly ovate, acute or acuminate, usually narrowed slightly at base, often undulate in sinuses, margins entire, surfaces smooth; involucral bracts distinct, broadly cordate-ovate, 2.5–5 cm long, including the lacerate margins, teeth triangular-acuminate, to 1.8 cm long; flowers axillary, solitary or on short leafy branches, pedicels stout, about 2 cm long; calyx cupuliform, to 10 mm long, apex nearly truncate to crenate; corolla campanulate-rotate; petals yellow, often tinged reddish, with purple basal spot, 4–8 cm long; staminal column included, 2.5–4 cm long; capsule 1.8–2.8 cm long, ovoid to oblong-ovoid, suborbicular to triangular in cross section, apiculate to acuminate at apex, surface mostly roughened with oil glands sunken in pits, cells 3, sometimes 4, valves not reflexed, sutures glabrous; seeds free, usually 5–8 per cell, with brownish to nearly white lint surpassing dense coat of brownish fuzz.

Endemic; on rocky slopes and cinder fields, and in cracks in lava. Specimens examined: Española, Gardner (near Española), Isabela, Pinta, Pinzón, Rábida, San Cristóbal, Santa Cruz, Santa Fé, Santa María, Seymour.

Gossypium barbadense is the more common of the 2 species in the Galápagos. Even within the archipelago it is polymorphic, including both narrowly and broadly lobed leaf forms as well as densely pubescent to nearly glabrous forms. It apparently differs from the typical variety of the mainland in the smaller size of the capsule and in the character of the lint and fuzz. In var. *barbadense* the capsules are more than 3 cm long, the lint is long, and fuzz may or may not be present. These differences may not hold, however, for Hutchinson (1947, 52) states that collections similar to var. *darwinii* have been made in Peru. Hutchinson states also that the typical variety has been collected in the Galápagos Islands, but all the fruiting material among the collections examined falls within the range of var. *darwinii*.

See Fig. 185.

Fig. 185. *Gossypium barbadense* var. *darwinii* (*a–d*). *Herissantia crispa* (*e–i*).

Gossypium klotzschianum Anderss., Kongl. Svensk. Vet.-Akad. Handl. 1853: 228. 1855

Shrub to 4 m tall, the stems, leaves and involucral bracts velvety-stellate-tomentose, becoming glabrous in age, the black punctations rather obscure; leaves short-petiolate, stipules linear, caducous, the blades cordate-ovate, up to 10 cm long, acuminate, unlobed or with 1–2 shallow, acute lobes in upper half; flowers axillary, solitary or several on leafy, jointed branchlets; involucral bracts distinct, to 3 cm long, cordate-ovate, lacerate with 10–15 narrowly triangular-acuminate teeth up to 10 mm long; calyx about 10 mm long, truncate to undulate apically; petals yellow, striate-purplish at base, to 5 cm long, rather strongly plicate; capsule ovoid-fusiform, up to 2.5 cm long, apex apiculate, cells 3 or 4, apparently sometimes with fine line of hairs along sutures; seeds blackish with inconspicuous coat of fine fuzz.

Endemic; in cracks in lava, on cinder fields, and in rocky soil into Transition Zone. Specimens examined: Isabela, Marchena, Santa Cruz.

Notwithstanding the large number of collections that have been identified as this species, only 5 actually fit its circumscription. Of these only 2—*Snodgrass & Heller 656* (DS, GH) and *Wiggins & Porter 253* (CAS)—are fruiting collections. Without fruit this species may be mistaken for *G. barbadense* var. *darwinii*, in which the inflorescence leaves may also be rather tomentose and unlobed or nearly so. In general, however, the leaf differences seem to be definitive. *G. klotzschianum* also appears to be less conspicuously black-punctate and to have smaller floral parts than *G. barbadense* var. *darwinii*.

Herissantia Medic., Vorles. Churpf. Phys.-Oekon. Ges. 4: 244. 1788

Subshrubs, pubescent, sometimes hispid. Leaves stipulate, petiolate or distally sessile, the blades simple, cordate-ovate, unlobed, palmately nerved. Flowers axillary, solitary on elongate, articulate, geniculate pedicels. Involucral bracts wanting. Calyx broadly campanulate, 5-lobed, lobes persistent and reflexed in fruit. Petals 5, white or yellowish, obovate, slightly exceeding calyx, claw margins ciliolate but not auriculate. Staminal column included, filaments apical and terminal, anthers about 20–30.

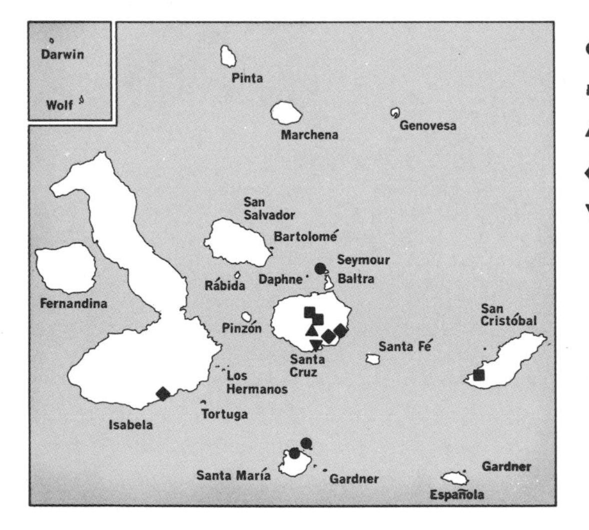

● *Herissantia crispa*
■ *Hibiscus diversifolius*
▲ *Hibiscus rosa-sinensis*
◆ *Hibiscus tiliaceus*
▼ *Malachra alceifolia*

Ovary of 8–14 mericarps in single whorl, each enclosing 1–3 vertically superposed ovules, style branches as many as mericarps; stigmas terminal. Fruit an inflated, depressed-globose, setose-pubescent schizocarp, mericarps scarious, reniform, muticous, without endoglossum, lateral walls smooth but beyond the contact surface with superficial, anastomosing venation pattern, dehiscence variable, each mericarp splitting vertically into 2 free valves or valves remaining joined basally, valves of adjacent mericarps sometimes coherent. Seeds 1–3, ovoid-reniform, black, glabrous or densely pubescent.

One polymorphic species native to warm temperate and tropical regions of the New World but now naturalized throughout the world in similar regions.

Herissantia crispa (L.) Brizicky, Jour. Arnold Arb. 49: 279. 1968

Sida crispa L., Sp. Pl. 685. 1753.
Abutilon crispum Medic., Kunstl. Geschl. Malven-Fam. 29. 1787.
Gayoides crispum Small, Fl. SE. U. S. 764. 1903.
Pseudobastardia crispa Hassler, Bull. Soc. Bot. Genève sér. 2, 1: 211. 1909.
Bogenhardia crispa Kearney, Leafl. West. Bot. 7: 120. 1954.

Diffuse subshrub 0.5–1.5 m tall, with very slender, sometimes twining branches, the branches, petioles, and peduncles usually pilose with spreading, simple hairs 0.5–2 mm long overlying sparse to dense, stellulate and glandular indumentum, sometimes the simple hairs wanting; leaves long-petioled below, above often sessile with the deeply cordate base amplexicaul, blades narrowly to broadly ovate, long-acuminate, crenulate to crenate or serrate, up to 10 cm long or more, upper surface plane, stellate-velutinous, greenish, lower surface more densely pubescent, whitish, the nerves prominent, reticulate; stipules 2.5 mm long; flowers solitary in axils, sometimes appearing more by congestion of the nodes; pedicels filiform, mostly 1–4 cm long, articulate and geniculate above middle; calyx 4–6 mm long, densely pubescent, reflexed in fruit, tube slightly 5-angled, lobes ovate, short- to long-acuminate, 2–4 mm long; petals about 6–9 mm long, 4.5–7 mm broad, claw margins ciliolate, dorsal surface and apical margin often glandular-papillate; staminal column to 3 mm long, filaments radiate, to 2 mm long; mericarps 5–20 mm long, 3.5–14 mm broad, lateral walls smooth, silvery on contact surface, superficially reticulate-veined on free surface.

Distribution as for the genus. In scanty to deep soil in partial shade, from sea level to Transition Zone. Specimens examined: Champion, Santa María, Seymour. Also reported from Daphne Major by Stewart.

See Fig. 185.

Hibiscus L., Sp. Pl. 693. 1753; Gen. Pl. ed. 5, 310. 1754

Herbs, subshrubs, shrubs, or trees, usually pubescent, sometimes spiny. Leaves petiolate, stipulate, simple and unlobed or palmately lobed or divided, occasionally palmately compound. Flowers axillary, solitary or cymosely clustered, or in racemes, panicles, or corymbs by reduction of upper leaves, pedicels usually articulate. Involucral bracts usually present, 4–20, distinct or basally connate or adnate to calyx. Calyx campanulate to tubular-campanulate, 5-lobed or -parted, usually accrescent, sometimes inflated, the nerves obscure or prominent, sometimes glandular. Corolla mostly rotate-campanulate, white, yellow, or pinkish to purplish or red, often with maroon or more intensely colored basal spot, petals typically 5, obovate, glabrous or pubescent. Staminal column included or exserted, apex usually 5-toothed, antheriferous from near base or only apically. Ovary 5-celled or rarely appearing 10-celled by vertical partitions, each cell with 3 or more ovules; style branched, branches as many as cells and terminated by expanded stigmas, or sometimes style unbranched and capped by 5-lobed or -branched stigma. Fruit a loculicidally dehiscent capsule. Seeds angular- or rounded-reniform, glabrous or pubescent.

About 250 species, in tropical to warm-temperate regions of the world.

Involucral bracts laterally connate, usually more than half their length, to form a 7–12-toothed epicalyx .... *H. tiliaceus*
Involucral bracts distinct:
Corolla 4–6 cm long, yellowish or purplish with maroon spot at base; staminal column included; branches and calyx usually spiny...................... *H. diversifolius*
Corolla 5–9 cm long, red; staminal column exserted; branches and calyx unarmed...... .... *H. rosa-sinensis*

Hibiscus diversifolius Jacq., Coll. Bot. 2: 307. 1789, Icon. Pl. Rar. 3: *t. 551.* 1792

Suffrutescent to arborescent shrub 1–10 m tall, the young stems generally stout-spiny, openly to densely stellate and sometimes simple-pubescent; leaves to 15 cm long, mostly orbicular in outline, palmately angled to parted, below inflorescences sometimes narrowly ovate and unlobed, margins irregularly crenate to crenate-dentate, base cordate to truncate, surfaces pubescent with nerves below often spinulose or setose; flowers solitary in upper leaf axils or congested in naked spikelike racemes, pedicels less than 2 cm long; involucral bracts mostly less than 10, 7–12 mm long, linear, simple; calyx usually 11–18 mm long in flower to 30 mm in fruit, densely yellowish-setose, often stellulate-pubescent as well, lobed to about middle, nerves prominent, midvein to each lobe usually with elongate gland; petals 4–6 cm long, nearly white, yellow or reddish purple, maroon-spotted basally; staminal column included, to 3 cm long; capsules to about 2 cm long, ovoid, beaked, setose; seeds to 4 mm long, minutely scaly.

African and Asian tropics and subtropics and reportedly introduced into similar regions in the New World. Specimens examined: San Cristóbal, Santa Cruz.

Hibiscus rosa-sinensis L., Sp. Pl. 694. 1753

Glabrate shrub 1–4(6) m tall; leaves mostly shiny, 4–15 cm long, 2–12 cm wide, ovate to broadly ovate, acute, usually serrate to deeply serrate-dentate from base or middle to apex, base cordate to truncate; flowers solitary in upper axils, patent-erect to subpendulous, pedicels 1.5–10 cm long; involucral bracts mostly 6–9, 5–18 mm long, linear-lanceolate, usually much shorter than calyx; calyx 17–30 mm long, tubular-campanulate, with narrowly deltoid-acute lobes in upper half or less; petals mostly 5–9 cm long, reddish or deep red with blackish red eye at base (in cultivars the corollas vary from single to double and from white to various shadings of red, yellow, and orange, and petals differ in conformation though basically obovate in outline); staminal column 5–10 cm long, exserted, the filaments to 2.5 cm long, spreading, arising over upper third or quarter of column; capsules about 2 cm long, ovoid, apiculate, glabrous.

Though generally considered to be native to continental tropical Asia, the species is unknown in the wild, and its origin and area of nativity are conjectural. It has been reported as an escape on Isla Santa Cruz.

Hibiscus tiliaceus L., Sp. Pl. 694. 1753

Evergreen shrub or small tree 2–6(15) m tall, the young stems, petioles, and pedicels glabrous or closely stellulate-pubescent; leaves petiolate, chartaceous to coriaceous, green and nearly glabrous above, whitish-stellate-tomentose below, 8–20(30) cm long, ovate to orbicular, unlobed, cordate, abruptly acuminate apically, margins entire or crenulate, median nerve or the 3–5 major nerves below with elongate, bordered gland near base; stipules foliaceous, oblong, 1–6 cm long, clasping stems in pairs, early deciduous to leave a circular scar; flowers solitary or in few-flowered, naked cymes often clustered toward ends of branches; pedicels stout, not articulate, 1–2 cm long, somewhat accrescent; involucral bracts connate basally to form 7–12-toothed cup mostly shorter than calyx, teeth deltoid to ovate, acute, usually shorter than tube; calyx 15–30 mm long, narrowly triangular-lobed to middle or beyond; corolla at anthesis yellow or rarely nearly white, with or without deep brownish or reddish basal spot, fading through day to orange-yellow then deep red, often drying greenish, petals 4–8.5 cm long, obovate, somewhat fleshy, stellate-pubescent dorsally; staminal column included, antheriferous from near base, exceeded by the 5-branched style; stigmas terminal, discoid; capsule 1.5–3 cm long, ovoid, short-beaked, sericeous-pubescent, 5-valved and 5-carpelled, each carpel

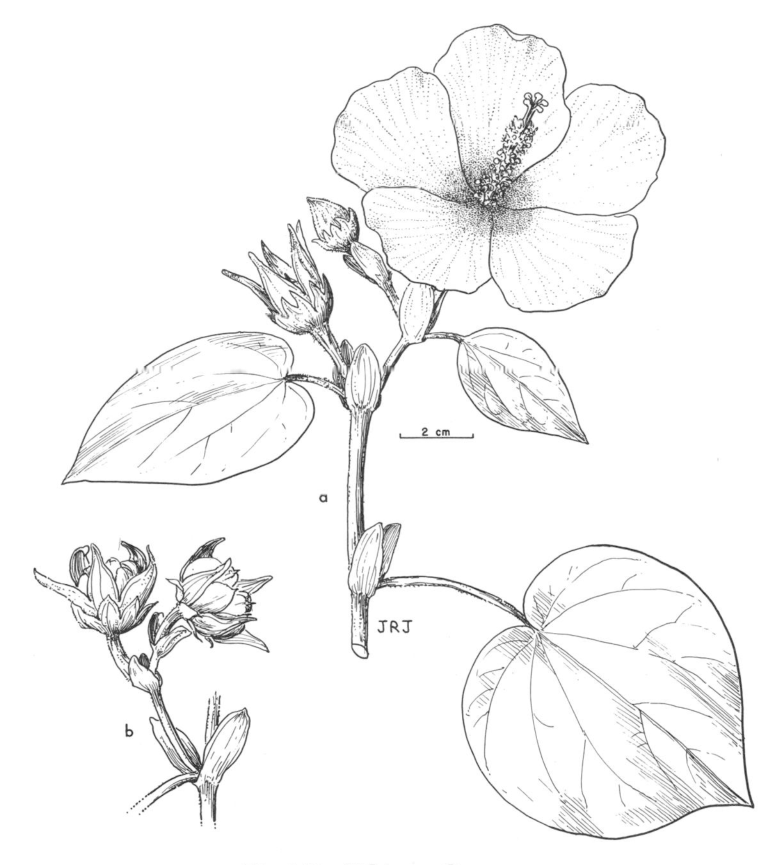

Fig. 186. *Hibiscus tiliaceus.*

2-celled by longitudinal septum; seeds to 7 in each cell, ovoid-reniform, minutely stellulate-tufted.

Native throughout the Old and New World tropics, mostly at low elevations in riparian or littoral habitats. Specimens examined: Isabela, Santa Cruz; also reported to have been collected on Santa María by Edmonstone.

See Fig. 186.

Malachra L., Syst. Nat. ed. 12, 2: 458. 1767; Mant. Pl. 13. 1767

Annual or perennial herbs or shrubs, often with hispid indumentum. Leaves petiolate, stipulate, the blades unlobed to angulate but most often palmately 3- or 5-lobed or -parted. Flowers in dense, axillary and terminal, subsessile or capitulate clusters, these subtended by foliaceous, plane, or conduplicate, mostly cordate-ovate to cordate-triangular, often scarious bracts. Involucral bracts absent or 9–12, distinct, filiform to subulate. Calyx campanulate to narrow-campanulate, 5-lobed, nerves often conspicuously colored. Petals 5, reddish, yellow, or white, obovate, unguiculate, rounded or subbilobate at apex. Staminal column included, 5-dentate at apex, anthers rather sparse on short filaments on upper half or so of column. Ovary of 5 mericarps, each with single ascending ovule, style branches 10, stigmas capitate, expanded. Schizocarp globose, mericarps falling from columella, indehiscent, trigonous-obovoid, rounded apically, more acute at base, dorsally convex, reticulate-veined, glabrous or pubescent. Seeds obovoid-reniform, glabrous.

About 8 poorly understood species, probably native to the New World tropics but 2 or 3 now distributed in the Old World.

Stems and bracts copiously hispid with simple or bifurcate hairs 4–5 mm long, sometimes also sparsely stellulate-pubescent . . . . *M. alceifolia*
Stems and bracts densely velvety-tomentose with stellate hairs, simple-hispid or stellate hairs to 2 mm long usually intermixed . . . . *M. capitata*

Malachra alceifolia Jacq., Coll. Bot. 2: 350. 1789; Icon. Pl. Rar. 3: *t. 549.* 1792

Stiff-branched, erect herb or subshrub 0.5–2.5 m tall; stems, branches, petioles, and bracts usually rather copiously hispid-pubescent with simple to bifurcate, usually pustular-based hairs to 4 or 5 mm long and copious to sparse stellulate indumentum about nodes; petioles often as long as or longer than blade; stipules filiform to subulate, simple or apically bifurcate or divided to base, to about 2 cm long, blades nearly orbicular to ovate in outline, to 15 cm long and broad but mostly smaller, unlobed or angulate or shallowly 3-, 5-, or 7-lobed, the angles of lobes acute to obtuse, margins serrate, base broadly subcuneate to subcordate, surfaces glabrous to minutely and inconspicuously stellulate, sometimes with hispid hairs on nerves; flowers in axillary, mostly subsessile or short-pedunculate heads, bracts short-stipitate, subtended by paired, filiform bracteoles, conduplicate, ovate to broadly triangular, 1–2.5 cm long, subcordate to deeply cordate, acute to subacuminate at apex, nearly entire to irregularly dentate-lobate, whitish and green-veiny with long-hispid hairs on veins toward center, in fruit brownish, scarious; involucral bracts wanting; calyx tubular-campanulate, 4–8 mm long, accrescent, whitish with reddish brown veins, lobes narrowly triangular, acute, long-aristate; petals pale to deep yellow, 10–20 mm long, glabrous or puberulent dorsally; staminal column nearly equaling petals; mericarps 3–3.5 mm long, 2–2.5 mm broad, muticous, papery, reddish veined, glabrous or puberulent; seed ovoid-cuneate, about 2.5 mm long, black, glabrous or with tuft of whitish hairs about hilum.

Native to the New World tropics and subtropics, but now naturalized in similar Old World regions. Specimen examined: Santa Cruz.

A variable taxon that differs consistently from *M. capitata* only in the character of the pubescence and probably is only a form of that species.

See Fig. 187.

Malachra capitata (L.) L., Syst. Nat. ed. 12, 2: 458. 1767

Sida capitata L., Sp. Pl. 685. 1753.

Mostly erect, coarse, annual or perennial herb 1–2 m tall, throughout densely whitish- or yellowish-tomentose with stellate hairs and usually also moderately to copiously hispid with simple or stellate hairs to 2 mm long; leaves long-petioled; stipules

Fig. 187. *Malachra alceifolia*.

lanceolate, 5–15 mm long; blades orbicular to ovate, 2–10 cm long, palmately sinuate to 3-, 5-, or 7-lobed, lobes mostly obtuse, crenate to serrate, the base obtuse or truncate; flowers in axillary, pedunculate, bracteate heads, bracts 1–2 cm long, stipitate and subtended by paired, filiform bracteoles, conduplicate, suborbicular to ovate, obtuse or acute, entire or once or twice dentate, obtuse to cordate at base, prominently veined and whitish basocentrally; involucral bracts wanting; calyx tubular-campanulate, 4–8 mm long, 5-lobed to below middle, lobes ovate-lanceolate, white with brownish or reddish nerves; petals yellow, obovate, 10–15 mm long, slightly exceeding staminal column; mericarps 3–3.5 mm long, muticous, reddish veined, puberulent; seed obovoid-cuneate, about 2.5 mm long, black, whitish-pubescent about hilum.

Native to the New World tropics but now introduced in similar Old World regions. Reported by Hooker to have been collected by Darwin on Isla San Salvador, but the specimen has not been seen.

Malva L., Sp. Pl. 687. 1753; Gen. Pl. ed. 5, 308. 1754

Annual, biennial, or perennial herbs, sometimes suffruticose, usually pubescent. Leaves stipulate, mostly long-petioled, the blades simple and unlobed to palmate-lobed or -dissected, occasionally palmately compound, mostly ovate to suborbicular in outline, sometimes more or less dimorphic with unlobed or scarcely lobed rosette leaves and dissected cauline leaves. Flowers axillary, solitary or in cymose glomerules usually exceeded by subtending leaves, less often in terminal racemes. Involucral bracts 3, distinct. Calyx campanulate to rotate, 5-lobed, persistent and sometimes accrescent. Petals 5, white, mauve, or violet-blue to reddish purple, obovate, more or less bilobed, sometimes clawed, glabrous or ciliate at base. Staminal column included, filaments apical and terminal. Ovary of 9–15 mericarps in single whorl, each with single ascending ovule; style branches same in number as mericarps, filiform, the stigmas introrsely decurrent. Fruit enclosed or subtended by calyx, a discoid, glabrous or pubescent schizocarp, mericarps indehiscent or dehiscent only ventro-apically, more or less depressed-reniform laterally, muticous or strongly reticulate, lateral wall acute or rounded. Seeds ovoid-reniform.

About 30 species of the Old World temperate and subtropical regions, with several species now naturalized in the New World.

Malva parviflora L., Demonstr. Pl. Hort. Ups. 18. 1753

Prostrate or ascending, annual herb 2–8 dm tall, stellate-pubescent at first, then glabrous; petioles to 3 times as long as blade; stipules to 6 mm long, triangular-ovate, ciliate; blades to about 8 cm long and broad, more or less orbicular, shallowly 3-, 5-, or 7-lobed, margins serrate, cordate basally, surfaces sparsely and minutely puberulent; flowers axillary, solitary or few in rather dense cymes greatly exceeded by subtending leaves; involucral bracts 3, linear, 1–4 mm long; calyx in flower 3–5 mm long, lobes ovate-acuminate, greenish, about equal to the nearly white tube, in fruit rotate, to 15 mm broad, scarious, veiny, apiculate; petals pinkish to purplish, scarcely surpassing calyx, claws glabrous; staminal column included, anthers few, terminal; mericarps 10–11, indehiscent, pale brown, glabrous or puberulent, depressed-reniform in lateral view, 2.3 mm high, to 3 mm broad, lateral walls transversely striate, edge acute and often raised or slightly winged, dorsal and dorso-

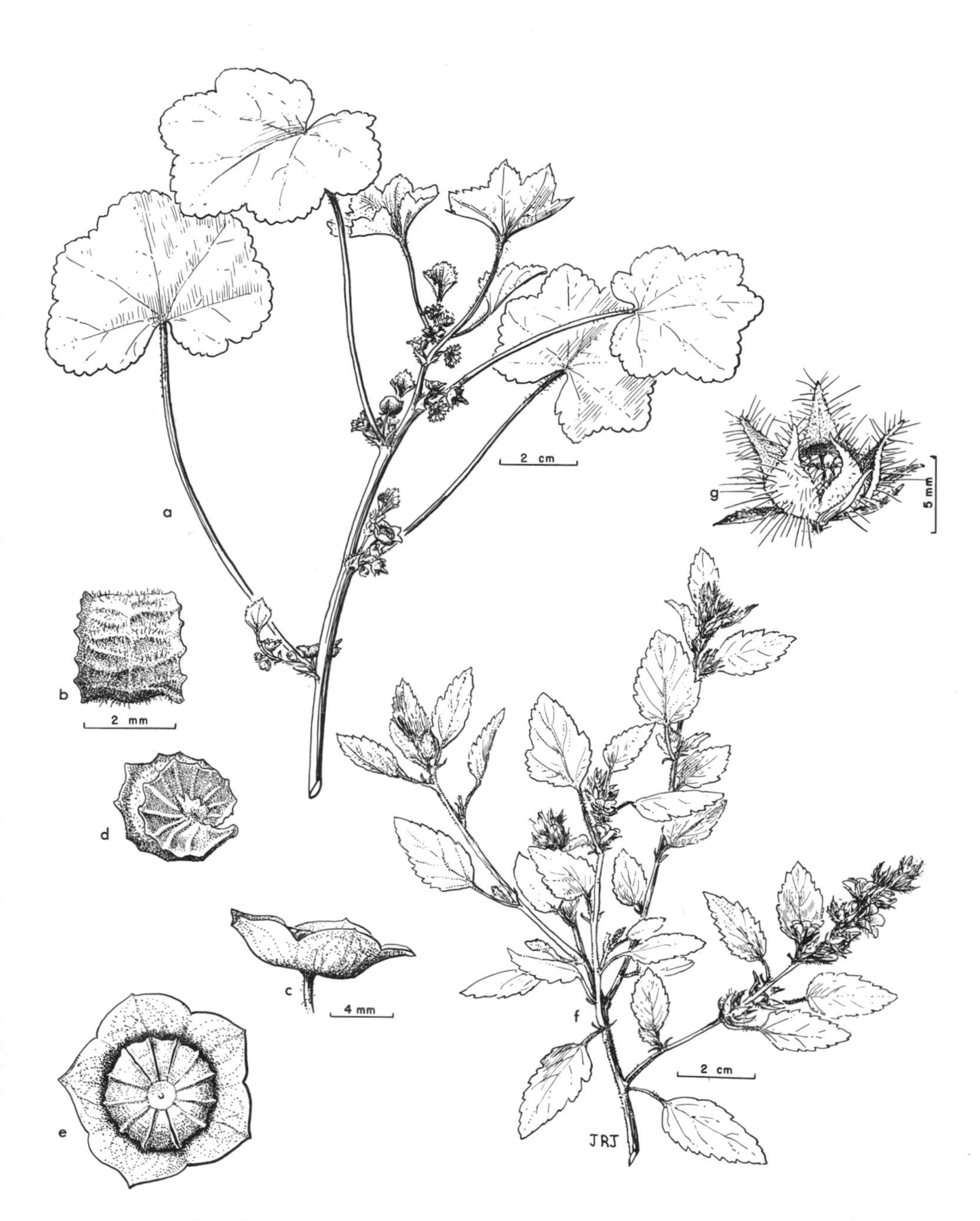

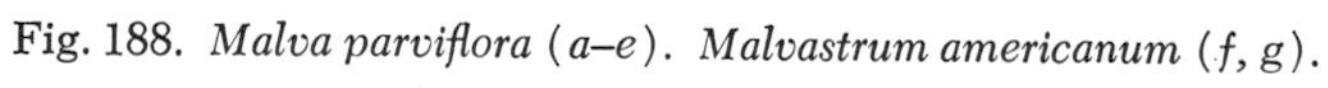

Fig. 188. *Malva parviflora* (*a–e*). *Malvastrum americanum* (*f, g*).

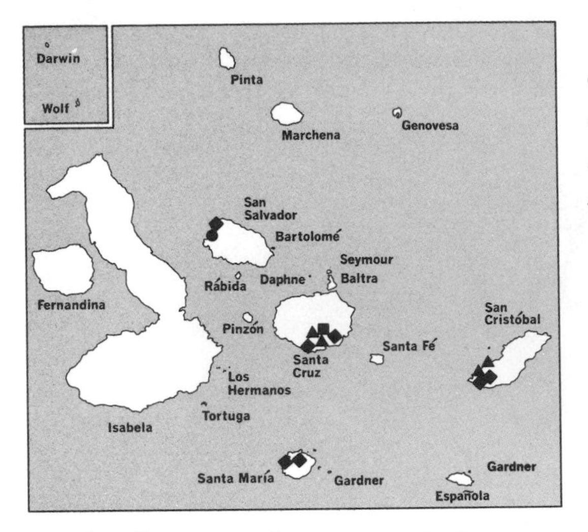

apical wall transversely rugose, endoglossum wanting; seeds reniform, about 1.8 mm long, reddish brown, glabrous.

Mediterranean region to western Asia; introduced into many areas of the world. Specimen examined: Santa Cruz.

See Fig. 188.

Malvastrum A. Gray, Mem. Amer. Acad. n.s. 4: 21. 1849

Annual or perennial subshrubs variously pubescent with simple and stellate hairs. Leaves short-petioled, linear-lanceolate to deltoid-ovate, unlobed or obscurely 3-lobed, apex acute to obtuse, margins crenate to serrate, base subcordate to cuneate, surfaces glabrate or pubescent, nerves sometimes impressed above; stipules subulate to filiform, often caducous or drying early. Flowers axillary and/or terminal, solitary, in cymose clusters, or in dense spikes, sometimes conspicuously bracteate. Involucral bracts 3, distinct or basally adnate to calyx tube, filiform to oblanceolate or elliptic-ovate or sometimes stalked with abruptly expanded deltoid-ovate blade. Calyx campanulate or subrotate, 5-lobed, slightly accrescent, often hispid-pubescent, usually drying brown and scarious. Corolla campanulate, petals 5, mostly scarcely exceeding calyx, yellow or yellow-orange, obovate, the apices obliquely subemarginate, base not auriculate, glabrous or with a few hairs at petal insertions. Staminal column included, yellowish, glabrous or puberulent between petals, filaments more or less terminal, anthers 5 to about 50. Ovary of 5–16 1-ovuled mericarps in a single whorl, ovule ascending, style branches as many as carpels, exserted and recurved within anthers before or at anthesis, stigmas capitate. Fruit a schizocarp, discoid, reddish or yellowish brown, sometimes glabrous but usually pubescent, often hispidly so, mericarps indehiscent, partially dehiscent along midvein basally, or completely dehiscent into free valves, angular-orbicular to transverse-elliptic in lateral view, dorso-lateral edge rounded or acute, sometimes raised in slight wing or prolonged dorso-apically into ascending to deflexed, laterally compressed knobs or spines, lateral walls often fused ventro-apically, sometimes forming indehiscent or partially dehiscent, erect knob or spine, smooth or transversely ridged below, the dorsal wall often canaliculate, an endoglossum present or absent. Seeds reniform, black or reddish brown to reddish black, glabrous.

About 12 species native to the warm temperate and tropical regions of the New World, with 2 of them naturalized in similar regions of the Old World.

Flowers in dense, elongate, axillary and terminal, bracteate spikes; mericarps muticous .... *M. americanum*
Flowers solitary or in few-flowered axillary glomerules; mericarps awned:
 Pubescence strigose, the hairs mostly 4-armed, the arms in opposing pairs; calyx clasping fruit; mericarps with erect ventro-apical awn mostly 1–1.5 mm long and 2 shorter dorso-lateral awns .... *M. coromandelianum*
 Pubescence appearing velvety, the hairs mostly with 6 or more radiate arms; calyx subtending fruit; mericarps with 2 prominent dorsal spines, ventro-apically acute or apiculate, sometimes with a spine to 0.5 mm long................ *M. scoparium*

Malvastrum americanum (L.) Torrey, Rep. U. S. Mex. Bound. Surv. 2: 38. 1859

Malva americana L., Sp. Pl. 687. 1753.
Malva spicata L., Syst. Nat. ed. 10, 1146. 1759.
Malvastrum spicatum A. Gray, Mem. Amer. Acad. n.s. 4: 21. 1849.

Perennial herb or subshrub to 3 m tall, copiously to densely pubescent, hairs principally stellate, 6-armed; petioles mostly more than one-third the length of blade; stipules subulate, 4–6 mm long; blades ovate, 2–7 cm long, unlobed or obscurely 3-lobed, apex acute, base truncate, margins serrate to dentate, surfaces stellate-puberulent above and below; flowers sometimes solitary but usually in dense, terminal or axillary, bracteate spikes 1–10 cm long, bracts usually bifurcate, ovate, to 10 mm long, each subtending a single flower; involucral bracts distinct or adnate at base to calyx tube, linear-lanceolate or narrow-lance-ovate, acute or long-acuminate, 4.3–4.8 mm long in flower, to 9 mm in fruit, like calyx with simple and bifurcate-hispid hairs 2 mm long overlying a stellulate indumentum; calyx open-campanulate, 5–5.5 mm long, lobes ovate, gradually acuminate, to 3.1 mm long, 1.9–2.3 mm broad, in fruit scarious, veiny, reddish, to 11 mm long with lobes to 7.5 mm long, 6 mm broad; petals yellow, asymmetrically obovate, to 6 mm long, 4 mm broad, claw ciliate; staminal column included, to 3 mm long, stellate-pubescent; schizocarp discoid, 4.5–5.2 mm broad, reddish brown, apically hispid-pubescent, mericarps 11–14, depressed, 1.2–2 mm high, 1.7–2.5 mm broad, ventro-apical edge often projecting conspicuously beyond endoglossum and over columella, lateral walls obscurely to prominently transversely ridged beyond contact surface, edge acute, eventually partially dehiscent around basal and dorsal wall with lateral walls turning upward; seeds 1.2–1.5 mm long.

A weedy species native to the American tropics and subtropics but now pantropical in distribution. Specimens examined: San Cristóbal, Santa Cruz.

See Fig. 188.

Malvastrum coromandelianum (L.) Garcke, Bonplandia 5: 295. 1857

Malva coromandeliana L., Sp. Pl. 687. 1753.
Malva tricuspidata R. Br. in Ait., Hort. Kew. ed. 2, 4: 210. 1812, nom. illegit.
Malvastrum carpinifolium sensu A. Gray, Mem. Amer. Acad. n.s. 4: 22. 1849.
Malvastrum tricuspidatum A. Gray, Pl. Wright. 1: 16. 1852, nom. illegit.

Annual or short-lived suffruticose perennial to 1(1.5) m tall, copiously pubescent with simple and 4-armed, mostly appressed, strigose hairs, the arms generally in opposing pairs; leaves to 8 cm long, lanceolate to oblong or oblong-ovate, sometimes nearly orbicular, unlobed, acute or obtuse at apex, cuneate to obtuse at base, margins serrate to dentate, surface strigose-pubescent above and below; stipules linear to lanceolate, to 10 mm long; flowers axillary, solitary or in few-flowered glomerules, pedicels mostly less than 5 mm long; involucral bracts distinct, shorter than calyx, linear to lanceolate, 3–7 mm long; calyx campanulate, 5-ribbed, strigose, 5–9 mm long, lobes deltoid to ovate, acuminate, to 5 mm long; corolla rotate-campanulate, petals yellow, obovate, to 9 mm long, ciliate at base; staminal column included, yellow, glabrous, to 4 mm long; schizocarp clasped by calyx, 5–7.3 mm broad, reddish brown, patent-hispid apically, the mericarps 10–14, indehiscent, transversely oblong-elliptic laterally, about 2 mm high, to 3 mm broad, ventro-apically with slender spine 0.5–1.8 mm long, dorsally or dorso-apically each of the acute lateral edges with single spine to 0.7 mm long, lateral walls transversely ridged; seeds to 1.5–2 mm long.

Of New World origin, this species is now widely distributed in tropical and subtropical regions throughout the world and sometimes enters into warm temperate associations. Specimens examined: San Cristóbal, San Salvador, Santa Cruz, Santa María.

Malvastrum scoparium (L'Hér.) A. Gray, Bot. U. S. Expl. Exped. 147. 1854

Malva scoparia L'Hér., Stirp. Nov. 53, *t. 27.* 1786, non Jacq. 1787.
Malva scabra Cav., Diss. 5: 281, *t. 138, f. 1.* 1788.
Sida depressa Benth., Bot. Sulph. 69. 1844.
Malvastrum scabrum A. Gray, Bot. U. S. Expl. Exped. 147. 1854.
Malvastrum dimorphum Howell, Proc. Calif. Acad. Sci. IV, 21: 331. 1935.
Malvastrum depressum Svens., Amer. Jour. Bot. 33: 465. 1946.

Annual or perennial herb with herbaceous stems to 2 m tall, densely but closely (often yellowish) pubescent throughout, the hairs chiefly stellate but with few simple hairs in marginal positions, larger hairs underlain by minute stellulate indumentum; petioles mostly much shorter than blades; stipules filiform, reddish, to 5 mm long, caducous; blades to 7 cm long, ovate to broadly ovate, unlobed or obscurely 3-lobed, coarsely dentate to crenate-dentate or serrate, apex acute, base broadly

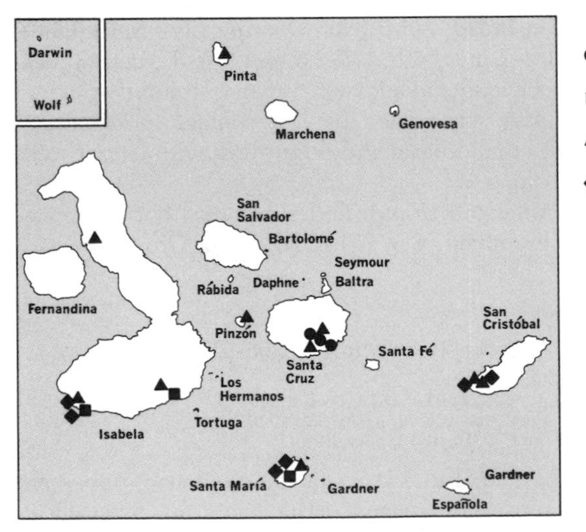

● *Malvastrum scoparium*
■ *Sida acuta*
▲ *Sida hederifolia*
◆ *Sida paniculata*

cuneate, surfaces often rather velvety-pubescent; flowers solitary on pedicels to 2 cm long in axils of primary leaves or more commonly in few-flowered, glomerate cymes in axils of secondary leaves; involucral bracts distinct, filiform, 3–5 mm long; calyx campanulate, in flower 6–8 mm long, lobes ovate to cordate-ovate, 3–5 mm long, 2–4 mm broad, in fruit accrescent, rotate, to 20 mm across, often reddish-punctate at bases of hairs; petals yellow, scarcely exceeding calyx, obovate, claw ciliate; staminal column included, stellulate; schizocarp stellate-discoid, to 9 mm across, subtended by calyx, mericarps 10–13, to 2 mm high, 4 mm broad, apically and dorsally densely stellate-pubescent or sometimes simple-pubescent, lateral edge acute, dorsally bearing 2 horizontal or deflexed pointed awns to 1 mm long, lateral walls more or less transversely ridged, ventral apex with inconspicuous bifurcate awn, inside with small endoglossum dividing locule into an indehiscent, 1-seeded lower cell and a barren, partially dehiscent, reduced, upper cell; seeds about 1.2 mm long.

Northwestern South America. Specimens examined: Santa Cruz.

The collections from Isla Santa Cruz have been segregated as *M. dimorphum* Howell, but they fall within the variation pattern of the mainland species.

Sida L., Sp. Pl. 683. 1753; Gen. Pl. ed. 5, 306. 1754

Annual or perennial herbs or subshrubs, usually pubescent. Leaves stipulate, mostly petiolate, the blades simple, linear to broadly ovate, unlobed or rarely palmately lobed, palmately or pinnately nerved. Flowers axillary, solitary or clustered or by reduction of leaves in racemose, paniculate, or capitulate inflorescences. Involucral bracts wanting or rarely 1–3, distinct, usually caducous. Calyx campanulate, often conspicuously angulate and nerved, 5-lobed, sometimes markedly accrescent. Corolla campanulate to reflexed, petals 5, white to yellow, orange, or purplish red. Staminal column usually included, filaments borne apically and terminating column. Ovary of 5 to many mericarps in single whorl, each enclosing single pendulous ovule; style branches same in number as mericarps, stigmas terminal, capitate. Fruit a schizocarp, depressed-globose to ovoid-conical, the mericarps indehiscent or dehiscent apically or ventrally, usually more or less trigonous, apex muticous or awned, lateral walls persistent, smooth or striate to fenestrate, dorsal wall and dorso-lateral edge often ridged. Seeds trigonous to reniform, glabrous or pubescent.

About 150 species, principally of the tropical and subtropical regions of the world.

Leaves palmately lobed .... *S. rupo*
Leaves unlobed:
  Mericarps 6–14:
    Mericarps 6–7; leaves mostly linear to linear-lanceolate .... *S. salviifolia*
    Mericarps usually more than 7; leaves generally oblong-ovate to rhombic:
      Leaves distichous; stipules mostly 8–15 mm long; pedicels usually less than 1 cm long .... *S. acuta*
      Leaves not distichous; stipules 3–8 mm long; pedicels mostly 1–4 cm long .... *S. rhombifolia*
  Mericarps 5:
    Leaves linear to oblong or sometimes narrowly ovate; stems usually with small, blackish, infrapetiolar tubercles .... *S. spinosa*
    Leaves ovate to suborbicular; stems without infrapetiolar tubercles:

Plant prostrate herb with slender wiry stems sometimes rooting at nodes; flowers solitary in axils; corolla campanulate, yellow-orange . . . . *S. hederifolia*

Plant coarse, erect herb, densely fulvous-pubescent throughout; flowers mostly in terminal panicles; corolla reflexed, maroon to purple . . . . *S. paniculata*

Sida acuta Burm. f., Fl. Ind. 147. 1768

Sida carpinifolia L. f., Suppl. Pl. 307. 1781.
Sida acuta var. *carpinifolia* K. Schum. in Mart., Fl. Bras. 12³: 326. 1891.
Sida acuta var. *hispida* K. Schum., op. cit. 327.

Much-branched, erect or sometimes spreading, herb or subshrub 0.5–1.5 m tall, glabrescent to moderately pubescent, the hairs stellulate or sometimes with few to many, simple, patent hairs to 2 mm long intermixed; leaves distichous; blades linear-lanceolate to oblong-ovate, elliptic or subrhombic, 3–7(10) cm long, greatly exceeding petioles, the apex narrowly to broadly acute, base cuneate to obtuse, often asymmetrically so, margins serrate to dentate, surfaces quite plane, minutely stellate-puberulent, nerves below sometimes with simple hairs; stipules usually slightly dimorphic, filiform to oblong-ovate, mostly 8–15 mm long; flowers axillary, usually exceeded by leaves, solitary or in loose clusters of up to about 8; pedicels less than 15 mm long, articulate near middle; calyx angled-campanulate, often somewhat plicate in bud, slightly accrescent and usually turning blackish in fruit, 5–7(9) mm long, lobes broadly triangular to oblong, acute to abruptly cuspidate, 3–5.5 mm long; petals white to yellow, 6–10 mm long, obovate, the apex obliquely truncate to subretuse, claw glabrous or slightly ciliate, not auriculate; staminal column included, to 3 mm long, glabrous or stellulate, the filaments terminal, to 2 mm long, anthers about 50; schizocarp depressed-conical, about 6 mm broad, apically somewhat fluted and sometimes puberulent, the mericarps 7–12, chartaceous, indehiscent, laterally compressed, more or less trigonous, about 2.5–3.5 mm high, lower dorsal wall and lateral walls rugose-reticulate, upper dorsal wall bearing 2 longitudinal ridges terminating at apex in erect beaks 0.5–1.3 mm long; seeds trigonous, about 2 mm long, pilose about raphe, otherwise glabrous.

A polymorphic, pantropical weed. Specimens examined: Isabela (an aberrant form with more or less spiral branching and somewhat reduced stipules), Santa María.

The Galápagos specimens have long-simple hairs on the stems and lower leaf surface and thus fall within the circumscription of Schumann's var. *hirsuta*. However, in such a variable species little seems to be gained by naming sporadic variants that occur through the range of the species.

Sida hederifolia Cav., Diss. 1: 8, *t.* 9, *f.* 3. 1785

Sida veronicifolia var. *hederifolia* K. Schum. in Mart., Fl. Bras. 12³: 320. 1891.

Small prostrate herb with slender, wiry stems very rarely rooting at nodes, stems minutely or rather coarsely pubescent with mostly few-armed or simple hairs, often glabrous; petioles usually half or more the length of blade; stipules filiform, to 5(7) mm long, caducous; blade ovate to transverse-ovate, mostly less than 25 mm but to 7 cm long, acute apically, margins crenate to crenate-dentate, base cordate, surfaces plane, copiously appressed-hispidulous-pubescent or sometimes soft-pubescent, the hairs simple or few-rayed; flowers axillary, solitary, on filiform, articulate pedicels 1.5–3 cm long; calyx somewhat rhomboidal in bud, campanulate at anthesis, 3.2–5 mm long, 10-ribbed, ribs and lobe margins pubescent, otherwise mostly glabrous, the lobes deltoid to ovate, 1.5–3 mm long; corolla yellow-orange, somewhat reddish toward base, exceeding calyx; petals 4.5–6 mm long, obovate, glabrous, the apex slightly oblique, rounded to irregularly lobate; staminal column included, to about 2.5 mm long, slender, puberulent, the filaments terminal, anthers about 10; mericarps 5, deltoid in cross section, ovoid-trigonous, 1.8–2 mm long, the walls membranous, reticulate, more or less coherent to seed coat, more chartaceous at apex, dehiscent and antrorsely pubescent, each valve terminated by divergent, slender awn 0.75–2 mm long; seeds trigonous, to 2 mm long, blackish, glabrous.

One of a complex of pantropical species or forms. Specimens examined: Isabela, Pinta, Pinzón, San Cristóbal, Santa Cruz, Santa María.

The Galápagos Islands collections, with the exception of *Wiggins & Porter 396* (BH, CAS), are relatively uniform. They all have awned mericarps, and because of this fall within the circumscription of *S. hederifolia*. The *Wiggins & Porter* collection differs from the others in its abundant, simple, patent stem hairs to 4 mm long and its larger or longer leaves, pedicels, calyces, and petals. The ovate-acuminate leaves are up to 7 cm long, the pedicels are thicker and up to 4 cm long, the calyces are up to 7.5 mm long with ovate-acuminate lobes, and the petals reach 8 mm long. The immature mericarps are 5 and awned. It is included here only because it is more like *S. hederifolia* than any other South or North American species of the genus.

Sida paniculata L., Syst. Nat. ed. 10, 1145. 1759

Erect, much-branched subshrub to 1.2 m tall, the new growth usually copiously to densely fulvous-stellate-pubescent, eventually glabrescent; lower leaves mostly suborbicular to broadly ovate, cordate, acute to abruptly acuminate apically with crenate to serrate margins, to 15 cm long on petioles two-thirds as long as blades, progressively smaller toward inflorescences, mostly about 6 cm long or less, ovate to lance-ovate with cordate to obtuse base; stipules subulate to filiform, 3–10 mm long; flowers on filiform pedicels to 4 cm long, articulated above middle, at first solitary in axils but at length forming large, many-branched and -flowered panicles; calyx campanulate, not conspicuously nerved, stellate-pubescent especially along margins of lobes, 2.5–3.5 mm long, the lobes deltoid, acute, to 2 mm long; petals purple to maroon, reflexed, 4–5.5 mm long, obovate, obliquely rounded, somewhat irregularly notched apically, glabrous basally, without auricles; staminal column slender, exserted, 4–5 mm long, pale purplish, minutely stellulate, the filaments terminal, spreading, 1.5–4 mm long, anthers yellow, about 20; mericarps 5, indehiscent, ovoid-trigonous, 2.8–3.7 mm long, dorsal wall chartaceous, muricate, puberulent on upper half, acute or with 2 beaks to 0.75 mm long at apex, lateral walls membranous, striate-reticulate; seeds ovoid-trigonous, about 1.7 mm long, sparsely puberulent.

West Indies and southern Texas south through Central and South America to northern Argentina; tropical Africa. Specimens examined: Isabela, San Cristóbal, Santa María.

A sterile specimen (*Snodgrass & Heller 860*) was erroneously reported by Robinson (1902) as *S. cordifolia* L., a species not known to occur in the Galápagos Islands.

See Fig. 189.

Fig. 189. *Sida paniculata* (*a–c*). *S. spinosa* (*d–f*).

Sida rhombifolia L., Sp. Pl. 684. 1753

Usually profusely branched, minutely stellate-puberulent, erect subshrub to about 1.5 m tall, sometimes annual or stems with simple patent hairs; leaf blades mostly 2–6(10) cm long, greatly exceeding petioles, more or less rhombic to elliptic, oblong-ovate or oblong-obovate, acute or sometimes obtuse or even emarginate apically, the base usually cuneate to narrowly obtuse, margins serrulate to serrate above base, upper surface plane, greenish, minutely stellate, lower surface whitish, more densely pubescent; stipules filiform to linear, 3–8 mm long; flowers axillary, solitary or few at a node, sometimes appearing corymbose terminally; pedicels usually elongate, filiform, to 4 cm long but mostly less than 2 cm long, articulated above middle; calyx campanulate, 10-ribbed, angulate and yellowish at base, minutely stellate-puberulent, 5–7(9) mm long, the lobes ovate to deltoid, acute to apiculate, 2.5–5 mm long; petals white to yellow or orange-yellow, sometimes purple-spotted at base, slightly exceeding to twice as long as calyx, 7–10(15) mm long, asymmetrically obovate, obliquely emarginate, claw not auriculate, ciliolate at base; staminal column included, 3–5 mm long, glabrous or puberulent, the filaments terminal, to 1.5 mm long, anthers about 25–40; schizocarp more or less depressed-ovate, 5–6 mm across, apex apiculate or beaked, glabrate or stellulate, the mericarps 7–14, chartaceous, dorso-apically dehiscent, laterally compressed, trigonous, about 2.5–3 mm high, apex with 2 erect cusps or awns 0.2–2.5 mm long, dorsal walls more or less rugose-reticulate; seeds trigonous, about 2 mm long, white-pubescent about raphe.

A polymorphic species even within the Galápagos Islands; distributed throughout the tropics and subtropics and collected principally along trails and in open habitats. Specimens examined: San Cristóbal, Santa Cruz, Santa María.

Sida rupo Ulbr., Bot. Jahrb. Beibl. 54 (117): 75. 1916

Erect annual herb to 6 dm tall, branching and densely pilose with simple hairs on the upper half, unbranched and glabrous below; leaf blades to about 5 cm long, deeply 5- or 7-lobed, lobes subovate to subelliptical, acuminate, irregularly serrate, pubescent on both surfaces but more densely so below; petioles to 5 cm long; flowers in loose panicles on pedicels to 5 mm long; calyx about 5 mm long, divided to about the middle in 5 triangular-acute lobes; petals violet, scarcely exceeding the calyx, obovate, glabrous; staminal column included, 3–4 mm long, simple-pubescent toward base, anthers 10; mericarps 5–7, triangular ovate, to 6 mm long, lateral walls smooth and glabrous below, becoming wrinkled apically and passing into 2 short retrorsely armed apical spines.

Previously known only from Peru, but Eliasson (1970) has reported it from Cerro Azul on Isabela, on the southwestern part of the rim of the caldera.

Sida salviifolia Presl, Rel. Haenk. 2: 110. 1835

Sida salviifolia var. *submutica* Howell, Leafl. West. Bot. 6: 169. 1952.

Erect herb 1–8 dm tall, copiously but rather obscurely pubescent, the hairs chiefly stellate but sometimes glandular and/or patent simple hairs intermixed; petioles slender, mostly one-third to half as long as blade, stipules filiform, to 5 mm long, caducous; blades linear to linear-lanceolate, rarely lance-ovate, mostly 1.5–3(6) cm long, apex acute, margins entire to denticulate or serrate, obtuse basally, sur-

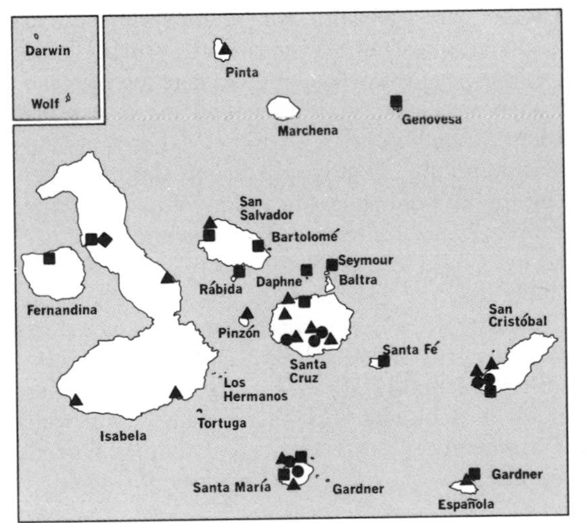

● *Sida rhombifolia*
■ *Sida salviifolia*
▲ *Sida spinosa*
◆ *Urocarpidium insulare*

faces stellate-pubescent, slightly discolorous with more greenish upper surface; flowers axillary, solitary or appearing as few-flowered clusters, usually corymbose toward ends of branches, usually shorter than subtending leaves; pedicels slender, to 10(15) mm long, articulate near apex; calyx narrow-campanulate, moderately 5-ribbed, stellate-pubescent, 4–6.5 mm long, slightly accrescent, lobes deltoid to narrow-triangular, acute to acuminate, 3.5–5 mm long; petals apparently white, brownish toward base, obtriangular, 5–6 mm long, apex obliquely emarginate, ciliolate, base without auricles, glabrous; staminal column included, to 2.5 mm long; anthers about 20, clustered terminally on short filaments; mericarps 6 or 7, trigonous, 2–3.5 mm long, lower half enclosing seed, dorsally and laterally reticulate with dorso-lateral edge often strongly tuberculate, upper half pubescent with walls smooth, loculicidally dehiscent and eventually divergent, each terminated by retrorsely pubescent beak 0.1–2.5 mm long; seeds rotund-trigonous, to 1.5 mm long, glabrous or pubescent about raphe.

West Indies and Mexico south to northern South America. Specimens examined: Daphne Major, Gardner (near Española), Fernandina, Genovesa, Isabela, Rábida, San Cristóbal, San Salvador, Santa Cruz, Santa Fé, Santa María, Seymour.

Often confused in the Galápagos with S. *spinosa*. The Galápagos collections have mericarps with awns less than 0.75 mm long in contrast to the continental and West Indian collections, which typically but not invariably have mericarp awns exceeding 1 mm. Since some mainland collections are indistinguishable from those on the islands, the material from the Galápagos need not be considered distinct as var. *submutica*.

Sida spinosa L., Sp. Pl. 683. 1753

Sida angustifolia Lam., Encycl. Méth. Bot. 1: 4. 1783, non Mill. 1768.
Sida tenuicaulis Hook. f., Trans. Linn. Soc. Lond. 20: 232. 1847.
Sida spinosa var. *angustifolia* Griseb., Fl. Brit. W. Ind. Isl. 74. 1859.

Erect annual or perennial herb, mostly 0.3–0.8(1.5) m tall, copiously but minutely stellate-puberulent, eventually glabrescent; petioles mostly less than 15 mm long, exceeded by blades and usually subtended by 2 small tubercles to 0.75 mm long below the filiform to narrow-subulate stipules; blades linear to oblong or ovate, mostly 1–4 cm long, apex obtuse to acute, base obtuse to truncate or rarely subcordate, upper surface dark green, stellulate-pubescent, lower surface whitish, more densely pubescent; flowers axillary, solitary or in few-flowered clusters, often somewhat corymbose toward ends of branches, exceeding or exceeded by subtending leaves; pedicels slender, 2–2.5 mm long, articulate above middle; calyx campanulate, 10-ribbed, stellate-tomentose, 3–6 mm long, slightly accrescent, the lobes deltoid, acute or acuminate, 2–3 mm long; petals pale yellow to yellow-orange, broadly obtriangular, 5–6 mm long, oblique apically, asymmetrically emarginate to retuse, ciliate, the base not distinctly clawed but broad and ciliolate, abruptly narrowed at point of insertion; staminal column included, to 3 mm long, yellow, minutely papillate; anthers about 20–24, terminally clustered on very short filaments; mericarps 5, elongate-trigonous, to 4 mm long, lower half or so enclosing seed, dorsally reticulate, scarious, eventually disintegrating, laterally striate-reticulate, chartaceous, upper half stellate-pubescent, loculicidally dehiscent with the smooth lateral walls diverging, each terminated by erect, antrorsely pubescent awn to 1 mm long; seeds rotund-trigonous, about 1.5 mm long, glabrous or with a few hairs about raphe.

A polymorphic species widely distributed in tropical and temperate regions of the world; found mostly at lower elevations in the Galápagos Islands. Specimens examined: Española, Isabela, Pinta, Pinzón, San Cristóbal, San Salvador, Santa Cruz, Santa María.

See Fig. 189 and color plate 45.

Urocarpidium Ulbr., Bot. Jahrb. Beibl. 54 (117): 63. 1916

Annual or sometimes perennial herbs, mostly rather diffusely branched and moderately pubescent. Leaves stipulate, usually rather long-petiolate, the blades simple, shallowly to deeply palmately lobed or parted, mostly ovate to suborbicular in outline. Flowers axillary, sometimes solitary but usually subsessile, in many-flowered, bracteate, helicoid cymes often exceeding subtending leaves. Involucral bracts (2)3, distinct, filiform to lanceolate, often caducous. Calyx campanulate to subrotate, 5-lobed, slightly accrescent, the lobes usually green above a whitish tube. Corolla campanulate to subrotate, petals 5, white to mauve or rose-magenta, equaling to twice as long as calyx, obovate, not auriculate basally, glabrous or slightly bearded. Staminal column included, filaments essentially terminal; anthers 5–15, rarely more. Ovary of 8–14 mericarps in single whorl, each with single ascending ovule; style branches as many as carpels, stigmas capitate. Fruit a discoid schizocarp usually clasped by calyx, pale brown to black, glabrous or apically puberulent, the mericarps indehiscent or slightly dehiscent ventrally, transversely reniform to asymmetrically ovate-reniform laterally, muticous or ventro-apically with laterally compressed, acute knob or elongate, recurved awn; lateral wall smooth to reticulate or ridged, edge rounded to subspinose, endoglossum present or absent. Seeds reniform, laterally compressed with acute dorsal edge, reddish black, glabrous.

Thirteen species occurring principally in drier, mountainous regions of southern Mexico, Guatemala, and western South America; one is found only in the Galápagos Islands.

Urocarpidium insulare (Kearney) Krapov., Darwiniana 10: 631. 1954

Malvastrum insulare Kearney, Leafl. West. Bot. 6: 167. 1952.

Erect, simple or sparingly branched, slender-stemmed, annual herb to 12 cm tall, the herbage moderately pubescent throughout, hairs chiefly stellate but some simple hairs intermixed, especially in marginal positions; petioles very slender, to 2 cm long; stipules subulate, to 3 mm long; blades thin, ovate to deltoid-ovate, to 12 mm long and broad, unlobed or shallowly 3-lobed, the apex obtuse to acute, base subcordate to truncate, margins rather deeply and irregularly crenate to dentate; flowers exceeded by leaves, solitary in axils on filiform pedicels to 10 mm long; involucral bracts caducous, apparently 2 or 3, filiform, about two-thirds as long as calyx; calyx campanulate, thin, veiny, 2.5–3 mm long, slightly accrescent, (4)5-lobed to below middle, lobes to 2 mm long, deltoid, acute; petals 3–5, roseate or lavender, scarcely exceeding calyx, obovate, rounded at apex, glabrous, not auricled; staminal column included, to 1 mm long; anthers 5 or 6, on erect, terminal filaments to 1 mm long; mericarps 7–10, indehiscent, black, stellate-pubescent then glabrous, transverse-elliptic-reniform laterally, to 1.3 mm high, about as broad, the lateral walls smooth or slightly reticulate but conspicuously ridged or muricate

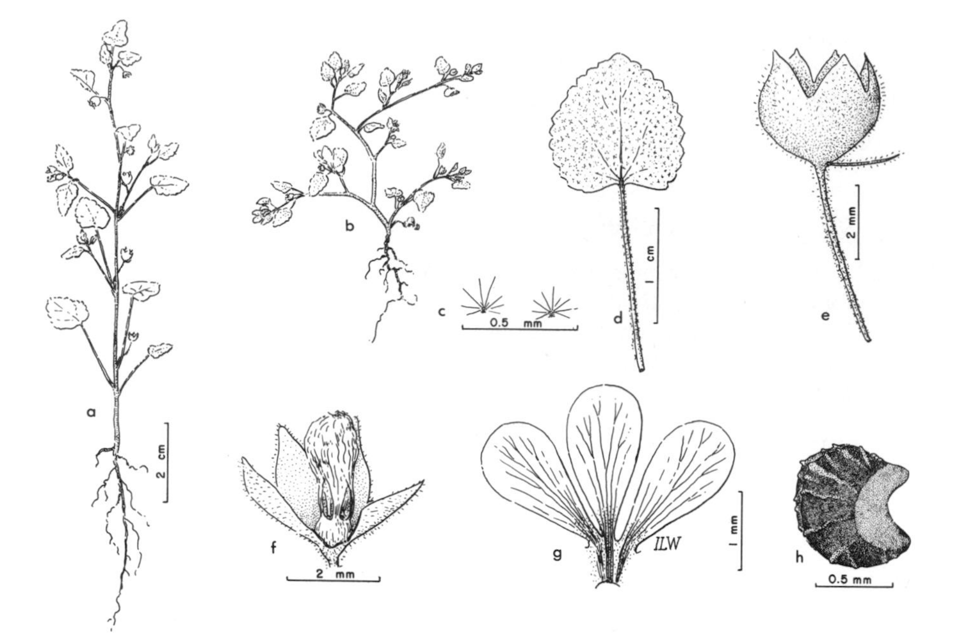

Fig. 190. *Urocarpidium insulare*: *c*, stellate hairs from stem and petiole; *e*, calyx with single bractlet; *f*, flower, corolla withered; *h*, mericarp, side view.

toward edges; chamber with obscure, ventro-apical endoglossum; seeds to 1 mm long.

Endemic; Isabela at Tagus Cove and reported by Eliasson (1970) from the crater rims of Cerro Azul and Volcán Wolf at elevations of 1,500 to 1,600 m.

See Fig. 190.

MELASTOMACEAE. Melastome Family*

Annual or perennial herbs, shrubs, or trees, rarely scandent or epiphytic. Leaves opposite, decussate, simple, often unequal, occasionally apparently alternate by abortion of 1 of a pair, longitudinally (1)3- to several-veined, veins arising at or near base of blade, extending in regular curves to apex. Inflorescences cymose, axillary or terminal, or flowers solitary. Flowers usually 4- to 6-merous, bisexual, usually slightly irregular, perigynous to epigynous. Hypanthium well developed, tubular, campanulate, or subglobose, bearing sepals, petals, and stamens apically. Sepals free or connate into a short tube or deciduous calyptra, rarely foliaceous, usually open in bud. Petals free, convolute. Stamens same in number as, twice as many as, or more numerous than petals, commonly dimorphic; anthers 2- to 4-loculed, ovoid to subulate, basifixed, usually dehiscing by 1 terminal pore, rarely by 2 or 4 pores or by short subterminal slits, inflexed in bud; connectives simple to appendaged. Gynoecium syncarpous; ovary superior or partly to wholly inferior, (1)2–10-loculed; ovules usually many, placentation axile or rarely basal; style 1, more or less elongate;

* Contributed by Duncan M. Porter.

stigma punctiform to capitate or peltate, simple or rarely lobed. Fruit a loculicidally to irregularly dehiscent capsule enclosed by persistent hypanthium, or a fleshy to leathery berry surmounted by persistent calyx. Seeds small, ovoid to linear, often coiled.

A pantropical family of about 200 genera and 4,500 species, very few extending into temperate areas.

Miconia R. & P., Prodr. 60. 1794

Shrubs or trees. Leaves petiolate or sessile, entire to toothed, membranous to coriaceous, pubescent to glabrous, those of a pair sometimes unequal. Inflorescences terminal or rarely axillary, usually paniculate. Flowers 4- to 7-merous, sessile or short-pedicelled, usually small. Hypanthium subglobose to tubular. Sepals variable in size and shape, usually with short, more or less adnate, exterior teeth, usually connate into a short tube, open in bud. Petals mostly small, often retuse or inequilateral. Stamens iso- or dimorphic, all perfect; filaments slender or flattened; anthers 2–4-loculed, variously shaped, dehiscing by 1, 2, or 4 terminal pores or by 1 or 2 longitudinal slits; connective simple to lobed or basally appendaged. Ovary partly to wholly inferior, 2- to several-loculed; ovules usually many; style straight or curved, usually elongate; stigma variably shaped. Fruit a many-seeded berry.

The largest genus in the family, with about 900 species; Mexico, the West Indies, and tropical America south to Argentina.

Miconia robinsoniana Cogn. in Robins., Proc. Amer. Acad. 38: 183. 1902

Shrub 2–5 m high, glabrous; branchlets acutely 4-angled, almost winged; leaves opposite and equal, oblanceolate to elliptic or oblong, obtusely acuminate, the base cuneate to slightly cordate, entire, coriaceous, somewhat rigid, 3-veined or obscurely 5-veined, the transverse veins numerous and slender, main veins prominent beneath, 10–29 cm long, 3.5–8 cm wide; petioles slightly twisted, acutely 4-angled, 11–55 mm long; inflorescences terminal, pyramidal, paniculate, divaricately 3- or

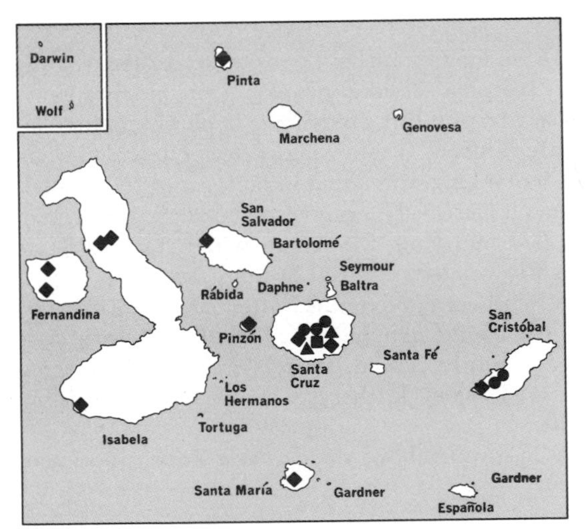

● *Miconia robinsoniana*
■ *Melia azedarach*
▲ *Cissampelos galapagensis*
◆ *Cissampelos pareira*

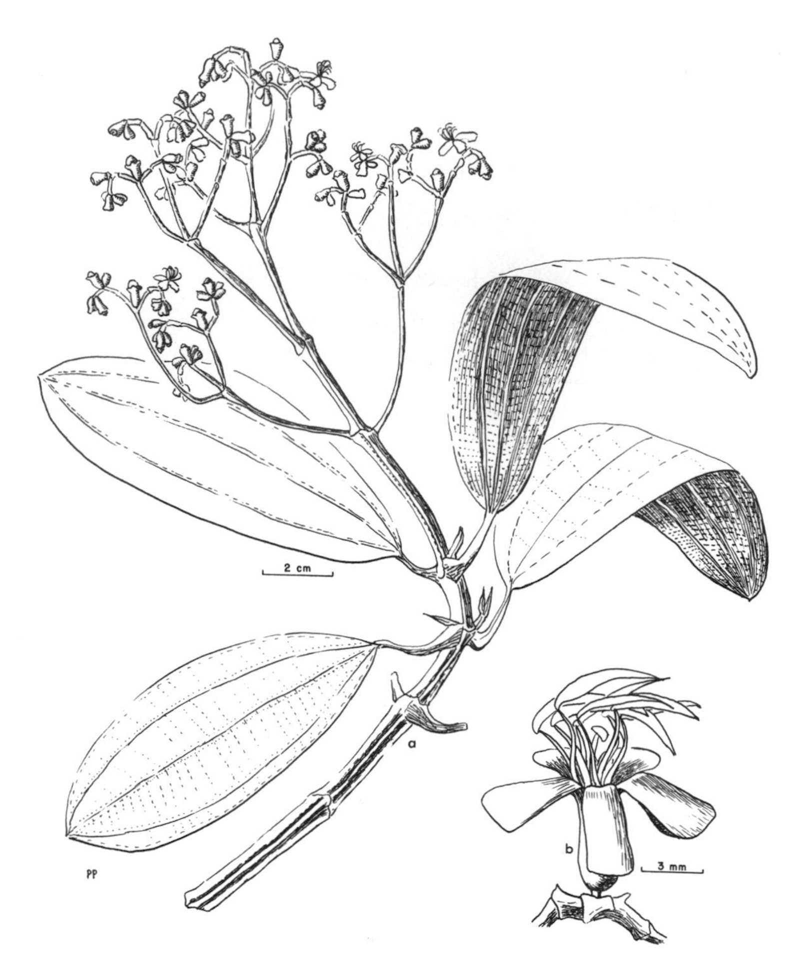

Fig. 191. *Miconia robinsoniana.*

4-branched, many-flowered, 10.5–18 cm long; branches opposite, spreading, slightly flattened, articulate, each branchlet scorpioid in floral development; bracts spreading, triangular, 1 mm long; flowers sessile, purplish, epigynous; hypanthium tubular, smooth, 3–4 mm long; calyx shortly 5-lobed, 3 mm long, exterior teeth present, lobes broadly rounded, ciliate, 1–1.5 mm long, spreading in flower and fruit; petals 4 or 5, narrowly obovate, obtuse, sinuate, inserted on ring of tissue at apex of hypanthium, adnate basally to stamens, 6–7 mm long, 2.5–3 mm wide; stamens 8–10, isomorphic; filaments glabrous, slightly winged, inserted in single series on ring at apex of hypanthium, 3–4 mm long; anthers curved, apex attenuate, dehiscing by 1 terminal pore, 4–4.5 mm long; connective simple; ovary wholly inferior; style subulate, ridged, glabrous, the apex slightly curved, straight below, 5 mm long; stigma capitate, 1 mm across; berry red, ovoid, leathery, surmounted by persistent calyx, 5–6 mm high, 4–4.5 mm wide.

Endemic; forming dense stands of shrubs above the *Scalesia* Zone. Specimens examined: San Cristóbal, Santa Cruz.

Several collections of *Miconia* from Panama have been referred to *M. robinsoniana* (cf. Gleason, 1958), but these appear to represent a separate, though perhaps closely related, species.

See Fig. 191 and color plates 59–62.

MELIACEAE. Chinaberry Family*

Trees or shrubs. Leaves alternate or rarely opposite, usually pinnate or palmate, occasionally simple, persistent or deciduous; leaflets opposite or alternate, usually entire, sometimes with pellucid dots or lines; stipules none. Flowers regular, hypogynous, (4)5(6)-merous, bisexual or rarely unisexual and plants polygamo-dioecious, in axillary or terminal, usually paniculate, inflorescences. Sepals connate basally or sometimes free, imbricate or valvate. Petals free or rarely slightly connate or adnate to lower part of staminal tube or disk; imbricate or valvate. Stamens same number as, twice as many as, or more numerous than petals; filaments usually connate partly or wholly into a tube, rarely free; anthers bilocular, exserted or included, introrse, longitudinally dehiscent. Intrastaminal disk usually present, annular or cupular, free or adnate to staminal tube or ovary. Gynoecium 2–6-carpelled, syncarpous; ovary 2–12-loculed; ovules (1)2 to many per locule, collateral or superposed, anatropous, pendulous; placentation axile; style 1, elongate or absent; stigma simple or grooved, discoid or capitate. Fruit usually a septicidal or loculicidal capsule, sometimes drupaceous or baccate. Seeds 1 to many per locule, sometimes winged; endosperm fleshy or absent; embryo straight or transverse; cotyledons fleshy or foliaceous; radicle superior or lateral.

About 45 genera and over 1,000 species in the tropics and subtropics.

Filaments of stamens free; anthers 5(6); petals erect in anthesis . . . . *Cedrela*
Filaments of stamens united into a tube, this exceeding ovary; anthers 10(12); petals spreading in anthesis:
  Fruit a drupe; leaves bipinnate; stamen tube cylindric . . . . *Melia*
  Fruit a septicidal capsule; leaves pinnate; stamen tube urceolate . . . . *Swietenia*

Cedrela P. Br., Nat. Hist. Jamaica 158. 1756

Tall trees with soft, light-colored wood. Leaves equally pinnate with many pairs of leaflets, these opposite or subopposite. Calyx small, 4- or 5-lobed. Petals 4 or 5, erect, imbricate, keeled to middle along inner side of midrib, keel adherent to disk. Disk columnar, 4–6-lobed. Stamens 4–6, filaments free; anthers versatile. Ovary 5-loculed, sessile on top of disk; style filiform; stigma discoid, often 4- or 5-lobed. Ovules 8–12 in each locule. Fruit a coriaceous, 5-valved capsule, septicidal almost to base. Seeds winged. Embryo straight; endosperm fleshy.

Some 6 or 7 species of the American tropics, distributed from Mexico to Argentina.

An unidentified tree that may be a species of *Cedrela*, having reddish, fragrant wood and called "cedro" is reported to have been brought down from the highlands of Santa Cruz and used in boat-building. Its identity is somewhat uncertain and no herbarium specimens are known.

* Contributed by Duncan M. Porter.

Melia L., Sp. Pl. 384. 1753; Gen. Pl. ed. 5, 182. 1754

Trees. Leaves alternate, pinnate or bipinnate; leaflets often numerous, dentate or serrate, sometimes entire. Flowers purple or rarely white, showy, bisexual, in axillary panicles. Sepals 5 or 6, free. Petals 5 or 6, free, contorted, spreading, imbricate. Stamens 10–12; filaments connate to apex into a cylindrical staminal tube, this dilated above, 10–12-dentate, each tooth cleft; anthers included, inserted within tube near apex. Disk annular. Ovary 3–6-loculed; ovules 2 per locule, superposed; style columnar, nearly as long as staminal tube; stigma 5- or 6-lobed. Fruit a drupe, endocarp 1–6-loculed. Seeds usually 1 per locule; testa crustaceous; endosperm fleshy; cotyledons foliaceous; radicle superior, terete.

About 10 species, native to the tropics and subtropics of the Old World.

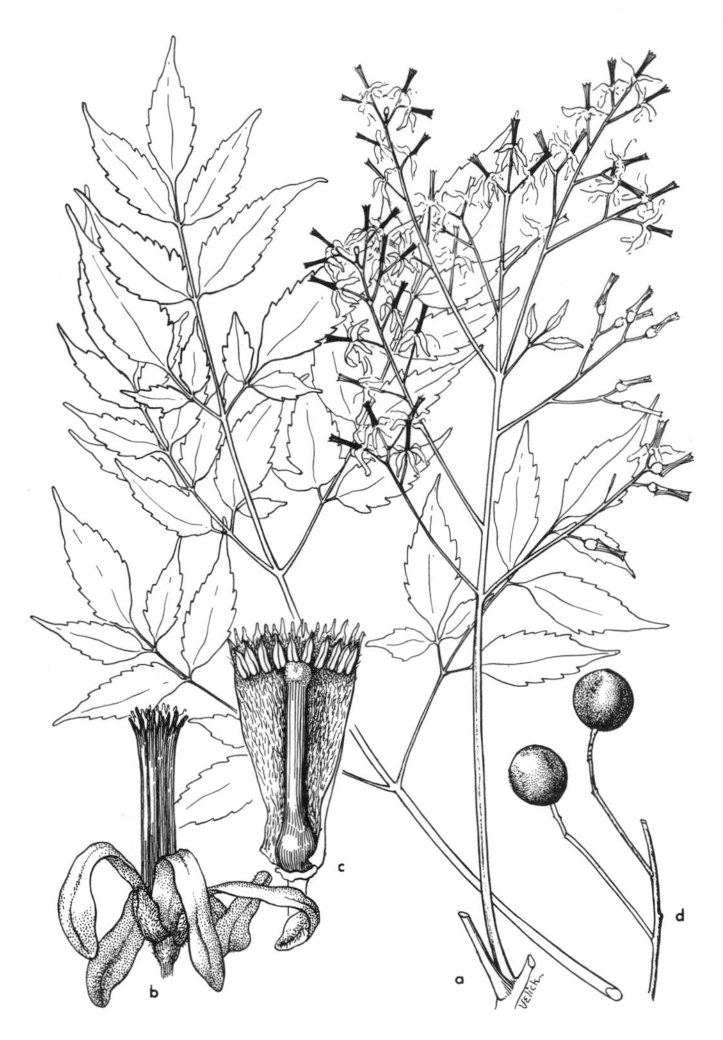

Fig. 192. *Melia azederach.*
(Reproduced by permission of Missouri Botanical Garden.)

Melia azedarach L., Sp. Pl. 384. 1753*

Tree usually 10 m high or less, the branches brittle; younger parts at first puberulent, becoming more or less glabrate; leaves bipinnate, more or less puberulent, 18–23 cm long, to 12 cm wide, deciduous; leaflets numerous, opposite to subopposite, petioled or subsessile, lanceolate to elliptical or ovate, acuminate at apex, inequilateral basally, 3–6.5 cm long, 12–21 mm wide, margin incised-serrate, ciliate; panicles open, borne near ends of branches, about equaling leaves; pedicels slender, more or less puberulent; flowers whitish; sepals lanceolate to ovate, acute, pubescent, 2–3 mm long, 1–2 mm wide; petals oblanceolate or narrowly oblong, spreading and recurved, puberulent to glabrate, 8–10 mm long, 1.5–3 mm wide; margins papillose; staminal tube as long as petals; ovary globose, about 1 mm long; style dilated above, slightly shorter than staminal tube; stigma capitate; drupe globose, smooth, yellow, 1.5–2 cm in diameter, endocarp bony and furrowed.

A native of southern Asia, cultivated as an ornamental and naturalized throughout the tropics and subtropics; the Galápagos representatives are escapes or grow on abandoned farm sites. Specimen examined: Santa Cruz.

See Fig. 192.

Swietenia Jacq., Enum. Pl. Carib. 4, 20. 1760

Tropical evergreen trees, with heavy wood. Leaves alternate, equally pinnate; leaflets opposite or subopposite, inequilateral basally. Flowers small, perfect, in axillary or subterminal panicles. Calyx (4)5-lobed. Petals (4)5, free, contorted, spreading. Stamen tube urn-shaped, 10-toothed. Anthers 10, inserted on outside below sinuses of stamen tube. Disk saucer-shaped, fluted exteriorly, crenate. Ovary (4)5-loculed, sessile in disk; style columnar; stigma discoid. Ovules several to many in each locule. Fruit septicidally (4)5-valved from base, the valves in two layers, separating from a (4)5-winged axis, this thickened at apex. Seeds in 2 rows in each locule, 10–14 per cavity, samaroid, the wing chartaceous, above the seed. Endosperm fleshy, thin; embryo transverse.

About 4–6 species in the tropics and subtropics from southern Florida through the West Indies, Mexico, and Central America into Brazil and Peru.

Swietenia macrophylla King in Hook., Icon. Pl. *pl. 1550.* 1886

Large tree; leaves 15–35 cm long; leaflets in 6–12 pairs, lanceolate, 7–15 cm long, 2.5–7 cm wide, inequilateral at base, acuminate at apex, petioluled, entire, coriaceous, glabrous; panicles 9–20 cm long; bracts and bracteoles orbicular or nearly so, deciduous; calyx cup 2–2.5 mm wide, lobes rounded; petals obovate to spatulate, 5–6 mm long, 2–3 mm wide; capsule ovoid, 12–15 cm long, 7–7.5 cm wide, acute to acuminate apically; seeds 7.5–8.5 cm long, the wing about twice as long as body, reddish brown.

Moist forests; southern Mexico to Peru and Brazil.

A tree believed to be this species has been seen near Bella Vista, Isla Santa Cruz. It was reported to have been introduced into the Galápagos Islands during the past

* A number of names have been published in *Melia*, the majority of them probably referring to *M. azedarach*. However, a listing of the synonymy is precluded by the lack of a modern revision of the genus.

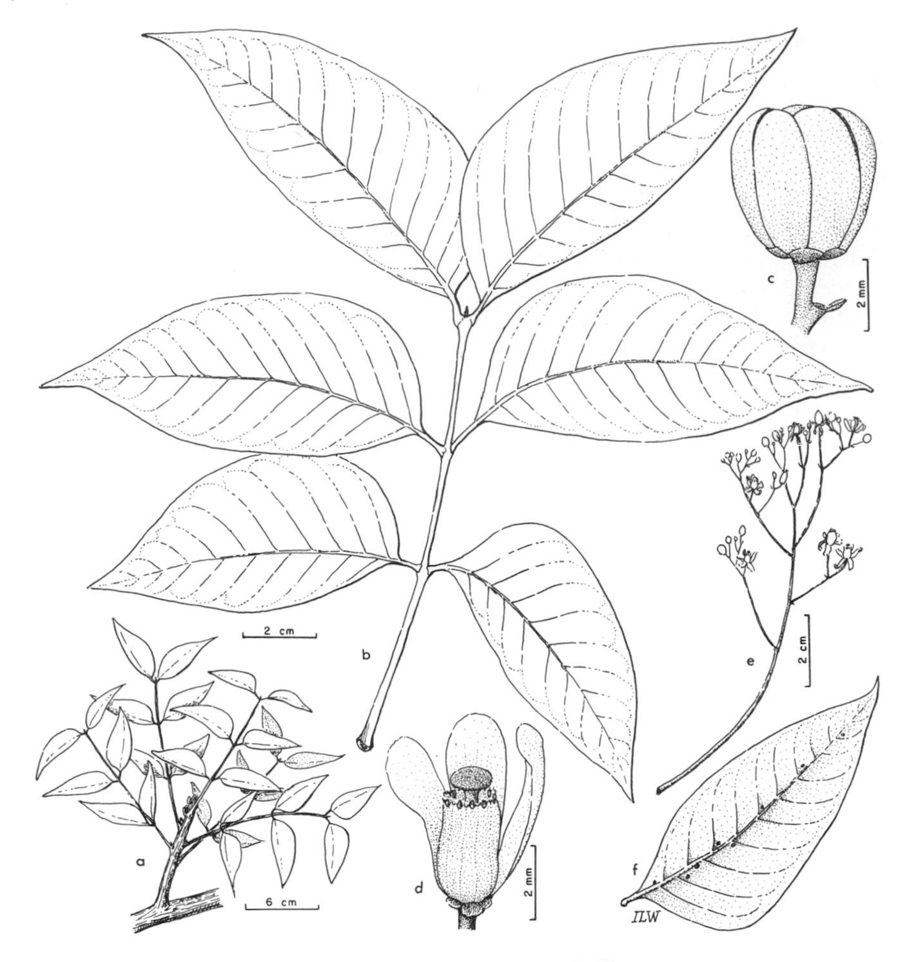

Fig. 193. *Swietenia macrophylla.*

two or three decades. It may occur also on San Cristóbal and Isabela, but has not been reported as an established escape.

See Fig. 193.

MENISPERMACEAE. Moonseed Family*

Twining woody, dioecious vines or sometimes shrubs or small trees. Leaves alternate, persistent or deciduous, simple or in some 3-foliolate, entire to palmately lobed, the petiole exstipulate. Flowers minute, unisexual, actinomorphic, greenish; sepals and petals 2–6, distinct or connate. Staminate flowers usually with 3 or 6 stamens (4 in ours) opposite petals when of same number; anthers 4-celled, dehiscing longitudinally. Pistillate flowers with (1) 3–6 distinct, sessile or stipitate pistils; staminodia present or lacking. Ovary superior, 1-celled, 2-ovulate, 1 ovule abort-

* Contributed by Ira L. Wiggins.

ing; ovule anatropous. Carpel 1, with a short style, or stigma sessile; stigma capitate to discoid or divided into 2–3 lobes. Fruit a drupe or achene. Seed usually curved, endosperm fleshy or lacking.

About 70 genera and 400 species, primarily in the tropics and subtropics of the Old and New Worlds but extending into warm temperate zones in some areas.

Cissampelos L., Sp. Pl. 1031. 1753; Gen. Pl. ed. 5, 455. 1754

Vines; woody at base. Leaf blades more or less peltate, glabrous to pubescent, ovate, cordate, or suborbicular; petioles slender. Staminate inflorescences paniculate-corymbose, cymose, or paniculate, in axils of normal leaves or of reduced, bractlike leaves. Pistillate inflorescences fewer-flowered than staminate, in simple cymes or fascicles in axils of more or less imbricate bracts. Staminate flowers with 4 sepals, often pubescent dorsally, the sepals obovate, rotate-spreading; petals 2–4, free, spreading, or cupuliform and connate below; stamens connate into a short column; anthers attached horizontally around rim of column. Pistillate flowers usually with single obovate sepal and 1–3 petals, these free or connate at base; carpel 1, more or less pubescent or glabrous. Drupe often pilose, exocarp thin-fleshy; endocarp crustaceous-bony, often bearing 1 or 2(3) low, irregular ridges around dorsal margin, transversely verrucose or ridged.

About 25 species, native in North and South America, Africa, Asia, and Australia.

Leaves and branches of season glabrous; fruits glabrous; leaves distinctly glaucous beneath . . . . *C. galapagensis*

Leaves and branches of season densely velvety-puberulent to villous; fruits pilose; leaves not glaucous beneath . . . . *C. pareira*

Cissampelos galapagensis Stewart, Proc. Calif. Acad. Sci. IV, 1: 66. 1911

Slender scandent vine with stems and branches 1–6 m long or more, glabrous throughout; petioles slender, 1.5–11 cm long, 0.4–0.8 mm in diameter; leaf blades broadly ovate to suborbicular, (2)4–7 cm long and wide, a slender mucro 1–2.5 mm long at tip of midrib, mostly 5-veined, dark green above, paler and glaucous beneath; staminate inflorescences cymose, in axils of normal leaves, diffuse, 1–4 cm long; peduncles filiform, 8–15 mm long; pedicels filiform, 0.5–4 mm long; staminate flowers 2.4–2.8 mm broad; sepals 4, spreading, obovate, 1.4–1.6 mm long, 0.6–0.9 mm wide, midrib prominent; corolla patelliform, 1.2–1.4 mm across, margin nearly entire; staminal column 0.4–0.5 mm wide at flattened apex; anthers 4, arranged as in *C. pareira;* pistillate flowers in few- to several-flowered fascicles in axils of bractlike reduced, imbricate leaves and in axils of normal foliage leaves, their structure similar to that in *C. pareira* but slightly smaller and wholly glabrous; fruit glabrous, 4–5 mm long, 3.5–4 mm wide, about 1.5–2 mm thick, with 3 rows of low, rounded teeth or bosses around outer rim, 12 or 13 teeth in each row, inconspicuous transverse ridges connecting bosses of 3 rows.

Endemic; in shaded and open forest, growing over lava, shrubs and small trees from near the sea in the Arid Zone upward to about 100 m or slightly more. Specimens examined: Santa Cruz; also reported by Christopherson to have been collected on San Salvador by Rorud.

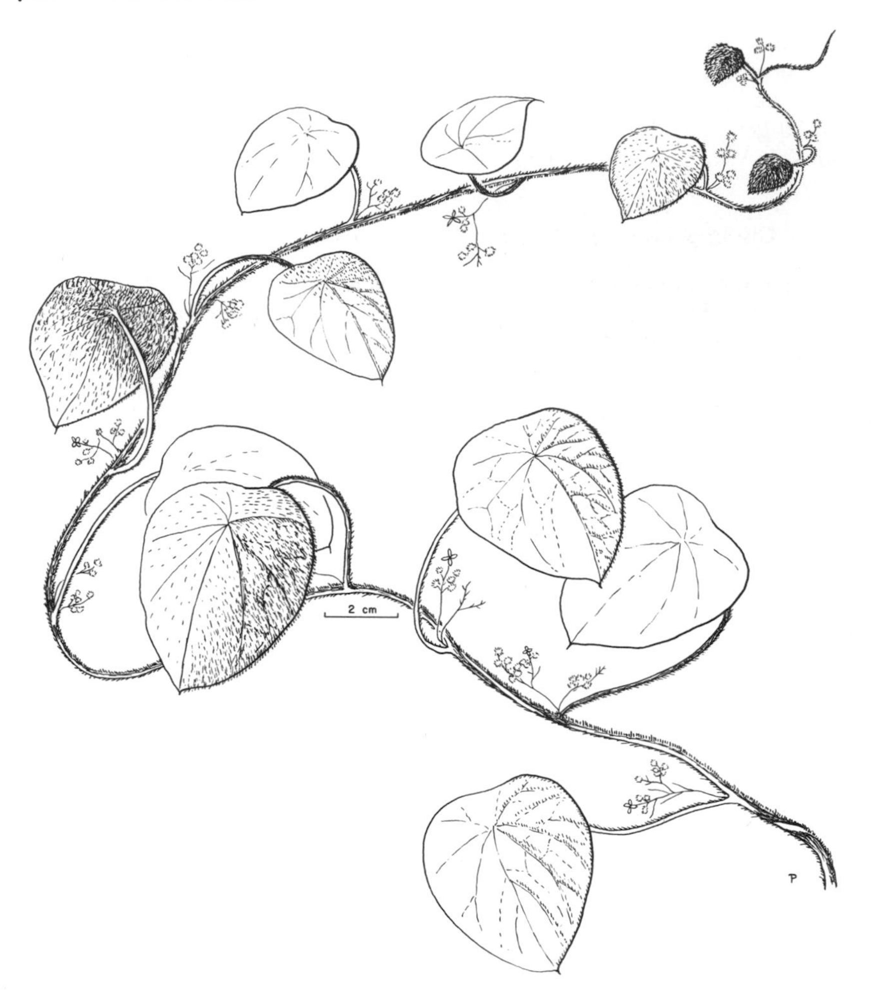

Fig. 194. *Cissampelos pareira.*

Cissampelos pareira L., Sp. Pl. 1031. 1753

Cissampelos cocculus Poir. in Lam., Encycl. 5: 9. 1804.*

Slender vine climbing to 5 m or more; lateral shoots numerous, often no more than 1 dm long; petioles slender, 1.5–8(12) cm long, usually densely soft-villous with slightly tawny spreading hairs 1–1.5 mm long; leaf blades broadly ovate to suborbicular, 3–11 cm long, 2.5–10 cm wide, palmately 5–7-veined, velvety-puberulent on both surfaces, usually slightly peltate, insertion of petiole 5–15 mm from margin of blade; staminate inflorescences divaricately cymose to subcorymbose, 1–2 cm long, each branch bearing 2–15 flowers, peduncles filiform, 3–8 mm long, villous; pedicels filiform, 1–2.5 mm long, villous; flowers 2–2.6 mm broad, glabrous within,

* For further synonymy, see Rhodes (1962, 170f).

appressed-villous without; corolla patelliform; sepals 4, 1–1.2 mm long, 0.6–0.9 mm wide, margins thin, somewhat erose; corolla 1–1.4 mm across, greenish; stamens 4, coalescent into low, broad column 0.3–0.4 mm wide at truncate apex; anthers about 0.4 mm long, inserted at right angles to axis of column; pistillate flowers borne in fascicles in axils of distinct to imbricate bracts on small axillary branches and toward tips of branches; bracts 2–6 mm long; sepal solitary, obovate, green, 2–2.4 mm long, 1–1.3 mm wide, appressed-villous on back; petals usually 2, obovate, free or connate at base or to half their length, appearing then as single, deeply emarginate petal 0.9–1.2 mm long, 0.6–0.8 mm wide; pistil strongly curved, villous; style stout, about 0.4 mm long, glabrous; stigma 2- or 3-lobed; fruits curved, U-shaped, with stigma almost against apex of pedicel, body 4–5 mm long, 3.5–4.2 mm wide, about 1.5–2 mm thick when ripe, with low keel around dorsal margin, rugose or verrucose with 10–12 transverse ridges, exocarp slightly fleshy, salmon to scarlet.

In shade and partially open forest and scrub throughout the tropical regions of the world; mostly above 75 m in the Galápagos Islands. Specimens examined: Fernandina, Isabela, Pinta, Pinzón, San Cristóbal, San Salvador, Santa Cruz, Santa María.

See Fig. 194.

MYRTACEAE. MYRTLE FAMILY*

Shrubs or trees, rarely subherbaceous. Leaves usually opposite, simple, entire, or rarely crenate, punctate with resinous or pellucid glands, venation usually pinnate, midvein usually elevated and prominent on lower surface, often with continuous submarginal vein; stipules none. Flowers bisexual, or rarely unisexual by abortion, regular or essentially so, epigynous, borne in axillary or rarely terminal branches, solitary or in racemose, glomerular, or umbelliform, dichasial, or paniculate inflorescences. Hypanthium adnate to ovary its entire length, or prolonged beyond ovary so floral parts appear to arise from distal margin of a short tube surrounding ovary summit. Calyx lobes usually 4 or 5, free and imbricate, or calyx calyptrate and circumscissile or rupturing irregularly at anthesis. Petals usually 4 or 5, sometimes reduced in number or size, rarely absent. Stamens usually many and inflexed in bud, inserted about margin of disk in 1 to many series; filaments usually free and filiform; anthers 2-loculed, usually short, versatile or basifixed, longitudinally dehiscent. Disk usually thickened, calycine. Gynoecium syncarpous; ovary inferior, 2- to many-loculed, placentae affixed to axis or parietal and coalesced into a central axis; ovules 2 or more per placenta, anatropous; style 1, elongate; stigma small, capitate or peltate. Fruit a capsule, berry, or drupe.

A primarily tropical family of perhaps 80 genera and 3,000 species, with centers of distribution in Australia and the Americas.

Calyx limb not closed in bud; sepals free, imbricate; calyx tube globose or turbinate; ovary 2-loculed . . . . *Eugenia*
Calyx limb closed in bud; sepals more or less united in bud, valvate; calyx tube campanulate or urceolate; ovary 4- or 5-loculed . . . . *Psidium*

* Contributed by Duncan M. Porter.

Eugenia L., Sp. Pl. 470. 1753; Gen. Pl. ed. 5, 211. 1754

Shrubs or trees with dense, rounded crown. Inflorescences racemose, cymose, corymbose, or fasciculate, or rarely flowers solitary in axils. Calyx tube subglobose to turbinate, usually 4-lobed, the lobes suborbicular, imbricate. Petals 4, spreading. Stamens many, in several whorls, free. Ovary 2-loculed. Style slender. Ovules few to many in each locule. Fruit drupaceous or berry-like, with 1 to few seeds. Embryo slightly fleshy.

About 600 to 800 species in tropical and subtropical regions of the world.

Racemes terminal; leaves 12–20 cm long, 3–5 cm wide; flowers white or greenish white; fruits depressed-globose, broader than long . . . . *E. jambos*
Racemes axillary; leaves to 35 cm long, 18 cm wide; flowers red or purplish; fruits oblong or obovoid, longer than broad . . . . *E. malaccensis*

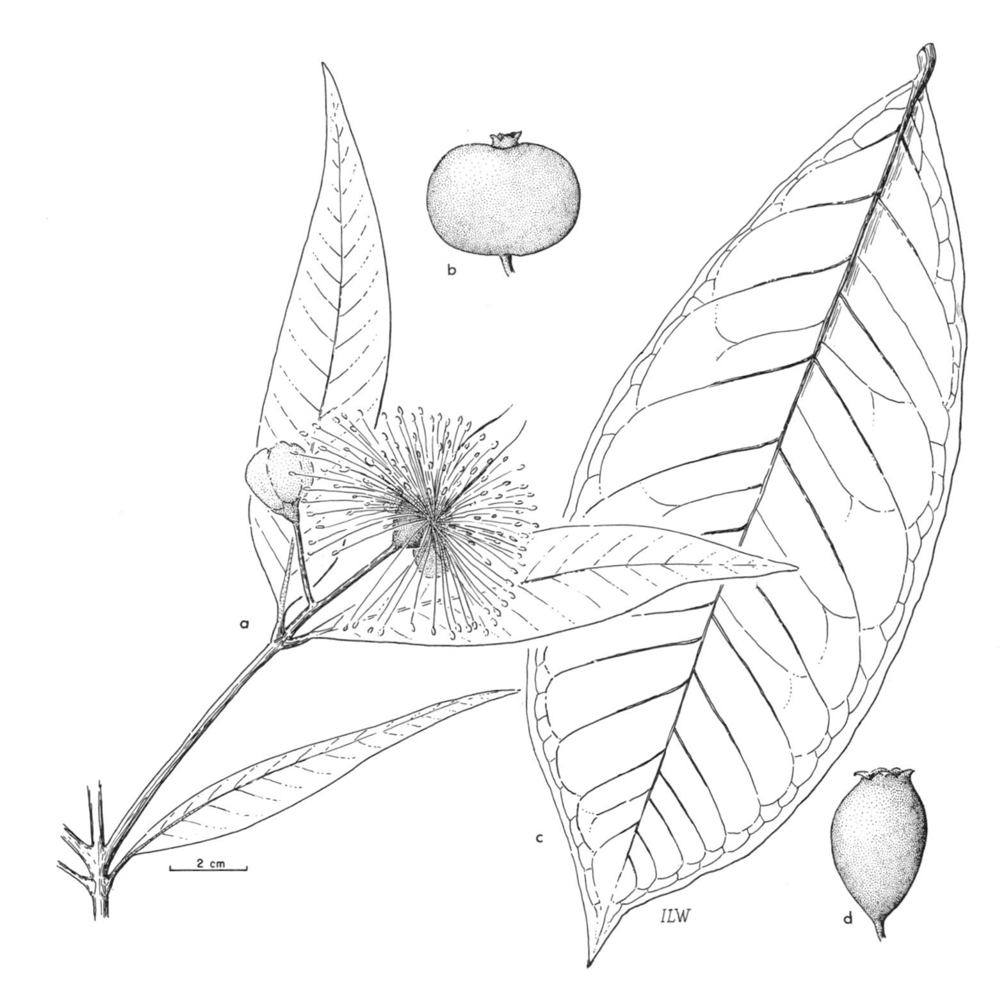

Fig. 195. *Eugenia jambos* (*a, b*). *E. malacensis* (*c, d*).

Eugenia jambos L., Sp. Pl. 470. 1753

Jambos vulgaris DC., Prodr. 3: 286. 1823.
Jambosa jambos Millsp., Publ. Field Mus. Nat. Hist. Bot. 2: 80. 1900.
Syzygium jambos Alston in Trimen, Fl. Ceylon (Suppl.) 6: 115. 1931.

Tree to 15 m tall; leaves lanceolate-elliptic, 12–20 cm long, 3–5 cm wide, narrowly acuminate at apex; racemes terminal, about 2.5 cm long, bearing 2–4 decussate pairs of flowers, these 7–8 cm across when stamens are spread; petals spreading, white or greenish white; fruit depressed-globose, 3–4 cm long, 5–6 cm broad, yellowish or faintly flushed with rose, fragrant, edible, 1-seeded.

Native of the Indo-Malaysian region; cultivated pantropically and naturalized in many areas. Known in the Galápagos only from cultivated trees at Wreck Bay and near Progreso, San Cristóbal, and from Bella Vista, Santa Cruz.

See Fig. 195.

Eugenia malaccensis L., Sp. Pl. 470. 1753

Eugenia purpurea Roxb., Fl. Ind. 2: 483. 1824.
Syzygium malaccense Merr. & Perry, Jour. Arnold Arb. 19: 215. 1938.

Tree similar in general appearance to *E. jambos*, but leaves to 35 cm long, abruptly acuminate; racemes axillary, about 1.5 cm long; flowers nearly sessile on the peduncle, in 2–5 pairs, red or purplish, the stamens much longer than petals; fruit obovoid to oblong, 4–7.5 cm long, 2.5–6 cm wide, reddish, 1-seeded, very fragrant.

Native of Malaysian region, cultivated widely for shade and fruit, occasionally becoming naturalized. Collected from cultivated trees at Bella Vista, Santa Cruz. Naturalized less commonly than *E. jambos*.

See Fig. 195.

Psidium L., Sp. Pl. 470. 1753; Gen. Pl. ed. 5, 211. 1754

Shrubs or small trees. Leaves opposite or subopposite, petiolate, glabrous to pubescent, usually persistent; continuous submarginal vein lacking. Flowers axillary, in a 3(4–7)-flowered dichasium or solitary. Hypanthium prolonged above summit of ovary. Calyx appearing tubular basally, 4- or 5-lobed or -toothed, completely closed, or with an apical pore in bud, at anthesis usually splitting longitudinally to summit of ovary into 2–5 somewhat irregular, usually widely spreading and often deciduous lobes. Petals free, white, often showy. Stamens many; filaments inserted on inner wall of hypanthium. Ovary (2)3–4(7)-loculed; sporophores parietal, directed toward center of ovary and there reflexed and somewhat produced into locules, mostly connate in pairs and appearing to arise from a central axis; placentae usually appearing bilamellate, but ovules often in single cluster in 2-loculed ovaries. Fruit a berry, pericarp often fleshy. Seeds many, hypocrateriform or reniform; testa bony or hard; embryo hooked or curved.

A genus of 100 or more species, indigenous to the tropics and subtropics of the New World. *Psidium* is badly in need of revision, using modern taxonomic techniques.

Flowers about 2.5 cm broad, in 1–3-flowered dichasia; leaves ovate to ovate-lanceolate, slightly inequilateral, 5–14 cm long, 2–6 cm wide, persistent . . . . *P. guajava*
Flowers about 1–1.5 cm broad, solitary; leaves elliptic to ovate or occasionally suborbicular, equilateral, 18–55 mm long, 9–26 mm wide, deciduous:

Fig. 196. *Psidium galapageium* var. *galapageium.*

Buds not lobed, closed at anthesis except for a terminal pore, distally glabrous; calyx opening by a circumscissile calyptra . . . . *P. galapageium* var. *galapageium*
Buds 5-lobed apically, open at anthesis, tomentose; calyx splitting irregularly into 4 unequal lobes . . . . *P. galapageium* var. *howellii*

Psidium galapageium Hook. f., Trans. Linn. Soc. Lond. 20: 224. 1847, var. **galapageium***

Shrub or small tree to 8 m tall; trunk to 1 m thick; bark smooth, pinkish gray; branches divaricate; branchlets terete, gray, punctate-glandular, tomentose to lanate with reddish to white or yellowish trichomes, becoming glabrate, bark becoming stringy, ultimate branchlets and leaves sometimes covered with scurfy reddish bloom; leaves opposite, elliptic to ovate, equilateral, apex acute to acuminate, base narrowly cuneate, decurrent on petiole, drying flat and usually black above, paler or sometimes reddish brown beneath, subcoriaceous, shortly appressed-pubescent to glabrate except for sparingly pubescent, reddish, more or less prominent midvein, both surfaces punctate-glandular, paler and more prominently so beneath, margins slightly revolute, blade 21–54 mm long, 9–26 mm wide, deciduous; peti-

* For a discussion of the endemic *P. galapageium* and its relationships, see Porter (1969b).

oles tomentose, flattened, slightly twisted, 1–3 mm long; peduncles solitary, opposite, slightly curved, spreading, 1-flowered, tomentose, 6–10 mm long; buds pyriform, punctate-glandular, contracted basally into a conic, more or less pubescent pseudostalk 1–1.5 mm long, distal portion subglobose and glabrous, apex obtuse, at anthesis closed except for a minute apical pore bearing a few projecting trichomes; flowers on branches of recent growth, about 1–1.5 cm broad; bracteoles 2, tomentose, 3–3.5 mm long, caducous; hypanthium slightly constricted above ovary, tomentose below constriction, punctate-glandular, about 2 mm long; calyx calyptrate, circumscissile about 1 mm above summit of ovary, calyptra glabrous, more or less persistent along 1 side, pubescent and punctate-glandular within, calyx further splitting longitudinally into several segments following anthesis, forming persistent ring about 1 mm high on summit of fruit; petals 5, obovate, concave, 8–9 mm long, punctate-glandular, caducous, filaments filiform, white, spreading, to 4 mm long, inserted on tomentose, punctate-glandular inner wall of hypanthium; anthers versatile, ovoid; ovary 3-loculed; style equaling stamens, lower three-fourths pilose, base persistent; stigma capitate; berry globose to subglobose, roughened with punctate glands, glabrous, yellow at maturity, drying black to reddish brown, 6–13 mm in diameter, pericarp about 1 mm thick; seeds several per locule, angular, dark, about 5 mm long; testa bony.

Endemic; a common member of the more mesic forests in the islands. Specimens examined: Fernandina, Isabela, Pinta, San Salvador, Santa Cruz.

See Fig. 196 and color plates 44, 45, and 47.

Psidium galapageium var. **howellii** Porter, Ann. Mo. Bot. Gard. 55: 370. 1969

Shrub or small tree 2–6 m high; leaves elliptic to ovate, occasionally suborbicular, apex acute to acuminate or rarely obtuse or retuse, base obtuse to narrowly cuneate, usually decurrent on petiole, shortly appressed-pubescent, especially along midvein, sometimes becoming glabrate except for midvein, blade 18–55 mm long, 10–21 mm wide; petioles 2–3 mm long; buds tomentose, more thickly so below, distal portion ovoid, apically 5-lobed, acute, open at maturity; flowers about 1 cm broad; hypanthium tomentose; calyx open in bud, 5-lobed, lobes free in mature

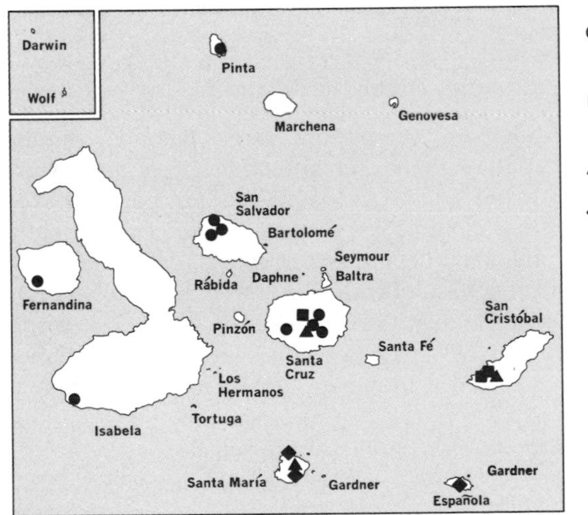

● *Psidium galapageium* var. *galapageium*

■ *Psidium galapageium* var. *howellii*

▲ *Psidium guajava*

◆ *Ludwigia erecta*

bud, tomentose within, at anthesis splitting irregularly between 4 lobes into 4 persistent segments 2.5–3 mm long, 1 segment larger and terminated by 2 lobes; petals 4–5.5 mm long, 3 mm wide; filaments white, greenish basally, 4–5 mm long; style pilose basally; berry drying reddish brown, 8–11 mm in diameter; seeds reddish brown. Characters otherwise as in var. *galapageium*.

Endemic; in mesic forests. Specimens examined: San Cristóbal, Santa Cruz.

Psidium guajava L., Sp. Pl. 470. 1753*

Small tree to 8 m high; branchlets 4-angled to slightly 4-winged, inconspicuously punctate-glandular, tomentose with white or gray trichomes; leaves opposite, ovate to ovate-lanceolate, slightly inequilateral, apex obtuse to apiculate, usually abruptly rounded or occasionally narrowed to obtusely pointed or rounded tip, base abruptly rounded or obtuse, decurrent on petiole, inconspicuously punctate-glandular and thinly pubescent to glabrate above, thickly and more conspicuously punctate-glandular and tomentose beneath, especially on veins, main veins reddish and prominent beneath, impressed above, blades 5–14 cm long, 2–6 cm wide, persistent; petioles tomentose, 4–5 mm long; buds tomentose, punctate-glandular, constricted under calyx, distal portion enlarged and ovoid to ellipsoid, apex acute, completely closed before anthesis, 7–10 mm long; flowers on branches of recent growth, about 2.5 cm broad, in 1–3-flowered dichasia (usually 1-flowered), terminal flower sessile, laterals on tomentose pedicels 10–12 mm long; peduncles tomentose, 10–20 mm long; bracteoles 2, subulate, tomentose, 2–2.5 mm long; hypanthium tomentose, glandular-punctate, slightly constricted above ovary, 4–6 mm long; calyx appressed-pubescent, basally punctate-glandular, splitting irregularly at anthesis into 4 or 5 lobes about 1 cm long, whitish and sericeous within, persistent on apex of fruit; petals 5, thin, delicate, broadly oval to elliptic, concave, 12–19 mm long, 6–10 mm wide, caducous; filaments white, spreading, longest as long as style; anthers versatile; style glabrous, 8–11 mm long; stigma capitate, slightly 2-lobed; berry globose, roughened with punctate glands, glabrous, pale yellow, about 2 cm in diameter at maturity.

A native of tropical America and widely cultivated and well established as an escape throughout the tropics; here, common in the more mesic forests of the larger islands. Specimens examined: San Cristóbal, Santa Cruz, Santa María.

ONAGRACEAE. Evening-Primrose Family†

Herbs, shrubs, or rarely trees with alternate or opposite, rarely whorled, simple, entire, lobed, or pinnatifid leaves; stipules minute or absent. Flowers actinomorphic or zygomorphic, perfect or rarely unisexual, (2)4–5(6)-merous, borne in axils of usually reduced foliage leaves or in a more or less distinct inflorescence, usually a corymb or raceme. Hypanthium prolonged beyond ovary or not, deciduous after anthesis or persistent. Sepals free or rarely fused. Petals free (rarely fused to sepals), sometimes absent. Stamens usually twice as many as sepals and in 2 sets, 1 opposite sepals and 1 opposite petals, or as many as sepals, sometimes reduced in number to 2; anthers versatile but sometimes attached near base, dehiscing longitudinally.

* Citation of full synonymy for this species is impossible at present.

† Contributed by Peter H. Raven.

Ovary inferior. Style 1; stigma variously lobed, discoid, capitate, or elongate. Fruit a loculicidally or more rarely an irregularly dehiscent capsule, sometimes indehiscent, or a berry, 1- to many-seeded. Seeds small, anatropous, lacking endosperm.

About 20 genera and 650 species, of worldwide distribution but best represented in western North America.

Ludwigia L., Sp. Pl. 118. 1753; Gen. Pl. 55. 1754

Slender herbs, erect or creeping and rooting at nodes, to large shrubs. Underwater parts often swollen and spongy or bearing inflated white spongy pneumatophores. Leaves alternate or opposite, mostly entire. Stipules absent or reduced, then deltoid. Flowers borne singly, clustered, or arranged in an inflorescence. Hypanthium not prolonged beyond ovary. Sepals 3–7, persistent after anthesis. Petals as many as sepals or absent, caducous, yellow or white, with contorted aestivation. Stamens as many, or twice as many, as sepals, or flowers rarely with an intermediate number of stamens; anthers usually versatile but sometimes apparently basifixed by reduction. Pollen shed in tetrads or singly. Disk (summit of ovary) flat to conical, often with depressed nectaries surrounding bases of the epipetalous stamens. Stigma hemispherical or capitate, upper half to two-thirds receptive, often lobed, the number of lobes corresponding to number of locules. Bracteoles lacking or conspicuous, usually 2, at or near base of ovary. Ovary with number of locules equaling number of sepals, rarely more; placentation axile; ovules pluriseriate or uniseriate in each locule, in 1 species uniseriate below, pluriseriate above; if uniseriate, the seeds sometimes embedded in powdery or woody endocarp. Dehiscence of capsule irregular, by a terminal pore or by flaps separating from the valvelike top. Seeds rounded or elongate, raphe usually easily visible and in some sections equal or nearly equal in size to body of seed.

About 75 species, mostly of tropical regions and best represented in the New World.

Sepals 4; petals 3.5–5 mm long; seeds free . . . . *L. erecta*
Sepals 5; petals 5–17 mm long; seeds embedded in persistent endocarp:
  Plant floating; seeds firmly embedded in woody coherent endocarp, pendulous, appearing as bumps in capsule wall about 1.5 mm apart; pollen grains falling singly . . . . *L. peploides* subsp. *peploides*
  Plant erect; seeds loosely embedded in horseshoe-shaped pieces of endocarp, appearing as bumps in capsule wall about 0.5 mm apart; pollen grains falling in tetrads . . . . *L. leptocarpa*

Ludwigia erecta (L.) Hara, Jour. Jap. Bot. 28: 292. 1953

Jussiaea erecta L., Sp. Pl. 1: 388. 1753.
Jussieua onagra Mill., Gard. Dict. ed. 8, no. 4. 1768.
Jussieua acuminata Sw., Fl. Ind. Occ. 2: 745. 1800.
Jussieua ramosa Jacq. f. ex Rchb., Icon. Bot. Exot. 54, *t.* 75. 1827.
Jussiaea erecta var. *sebana* DC., Prodr. 3: 55. 1828.
Jussiaea erecta var. *plumeriana* DC., loc. cit.
Jussiaea altissima Perrottet ex DC., loc. cit.
Isnardia discolor Klotzsch in Peters, Reise Mossamb. Bot. 70. 1861.
Jussiaea acuminata var. *longifolia* Griseb., Cat. Pl. Cubens. 107. 1866.
Jussiaea acuminata var. *latifolia* Griseb., loc. cit.
Jussiaea plumeriana Bello, An. Soc. Esp. Hist. Nat. 10: 267. 1881.

Subglabrous erect herb from 3 cm to more than 3 m tall, sometimes more or less woody at base, freely branched, the stems sharply angled from the decurrent leaf bases; leaves lanceolate to elliptical, rarely ovate, 2–13 cm long, 0.2–4.5 cm wide, narrowly cuneate at base, apex acuminate to acute, rarely obtuse, main veins 16–27 on each side of midrib, submarginal vein fairly prominent, petioles 2–15 mm long; flowers solitary in upper axils; sepals 4, lance-acuminate, 2–6 mm long, 1–1.5 mm wide; petals yellow, obovate, 3.5–5 mm long, 2–2.5 mm wide; stamens 8, subequal; filaments about 1.5 mm long; anthers about 0.6 mm long, shedding pollen directly on stigma at anthesis; pollen shed in tetrads; disk not elevated, with a sunken white-hairy nectary around base of each epipetalous stamen; style 0.5–1 mm long; stigma globose, 1–1.1 mm thick, its upper two-thirds receptive; bracteoles about 0.5 mm long; capsule glabrous, rarely puberulent, 1–1.9 cm long, 2–2.5 mm thick, pale brown with 4 prominent dark brown ribs, sharply 4-angled with 4 nearly flat walls, irregularly and readily loculicidal, subsessile or on a pedicel to 2 mm long; seeds pluriseriate in each locule of capsule, free, pale brown, minutely cellular-pitted, elongate-obovoid, 0.3–0.4(0.5) mm long, 0.2–0.3 mm thick; raphe about one-fifth diameter of body. $2n = 16$ (not determined in the Galápagos material).

Common from central Mexico and Florida to Paraguay and central Brazil, mostly within the tropics; introduced widely in tropical Africa and locally in India. Specimens examined: Española, Santa María.

Ludwigia leptocarpa (Nutt.) Hara, Jour. Jap. Bot. 28: 292. 1953

Jussiaea leptocarpa Nutt., Gen. N. Am. Pl. 1: 279. 1818.
Jussiaea pilosa Kunth, Nov. Gen. & Sp. Pl. 6: 101, *pl.* 532. 1823.
Jussiaea velutina G. Don, Gen. Syst. 2: 695. 1832.
Jussiaea surinamensis Miquel, Linnaea 18: 370. 1844.
Jussiaea schottii Mich., Flora 57: 302. 1874.
Jussiaea pilosa var. *robustior* J. Donn. Sm., Bot. Gaz. 16: 6. 1891.
Jussiaea variabilis var. *pilosa* (Kunth) Kuntze, Rev. Gen. Pl. 1: 251. 1891.
Jussieua pilosa var. *pterocarpa* Hassler, Rep. Spec. Nov. Fedde 12: 274. 1913.
Jussiaea leptocarpa var. *genuina* Munz, Darwiniana 4: 255. 1942.
Jussiaea seminuda H. Perr., Not. Syst. ed. Humb. 13: 146. 1947.
Ludwigia leptocarpa var. *meyeriana* Alain, Bull. Torrey Bot. Club 90: 191. 1963.

Erect hairy to subglabrous plant to 1 m tall, well branched, with erect floating pneumatophores arising from roots under water; leaves broadly lanceolate, 3.5–18 cm long, 1–4 cm wide, narrowly cuneate at base, apex acuminate, main veins on

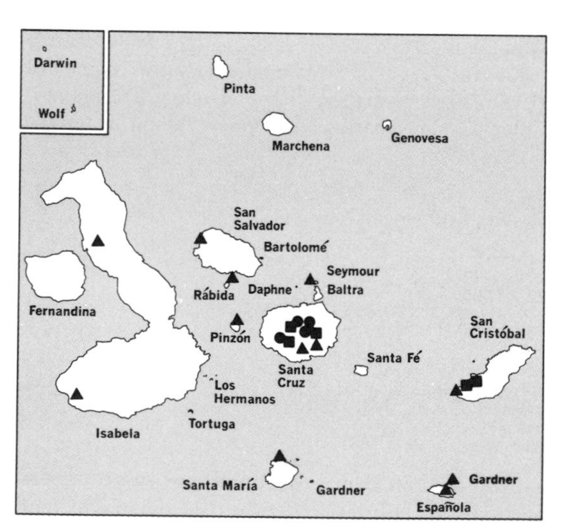

● *Ludwigia leptocarpa*
■ *Ludwigia peploides* subsp. *peploides*
▲ *Oxalis cornellii*

each side of midrib 11–20, submarginal vein inconspicuous, petiole 0.2–3.5 cm long; sepals (4)5(7), deltoid-acuminate, 5.5–11 mm long, 1.5–3 mm wide, with narrow wing running down from each sinus between sepals to apical portion of ovary; petals orange-yellow, obovate, 5–11 mm long, 4–8 mm wide; stamens twice as many as sepals, filaments 2–4 mm long, epipetalous ones shorter, anthers 1.2–1.6 mm long, extrorse, thus not shedding pollen directly on stigma; pollen shed in tetrads; disk slightly elevated, the base of each epipetalous stamen surrounded by a depressed nectary densely covered with matted white hairs; style 3–4.5 mm long, glabrous, stigma globose, 2–2.5 mm across, about 1 mm high, upper two-thirds receptive; bracteoles at base of ovary absent or rarely present, narrowly deltoid; capsule relatively thin-walled, long-hairy, 1.5–5 cm long, 2.5–4 mm thick, terete, dull light brown, with prominent ribs over locules and less prominent ribs over septa, marked on outside by bumps about 0.5 mm apart, corresponding to position of seeds, slowly and irregularly loculicidal; pedicels 2–20 mm long; seeds uniseriate in each locule of capsule, horizontal, shiny pale brown, finely pitted, obovoid, 1–1.2 mm long; raphe much narrower than body of seed; each seed loosely embedded in easily detached horseshoe-shaped segment of firm pale-brown endocarp, about 1–1.5 mm thick and 1 mm high.

Common from the southeastern United States and the West Indies to Peru and Argentina; also widespread in tropical Africa. Specimens examined: Santa Cruz.

Many of the plants that I have seen from the Galápagos are subglabrous and thus correspond to var. *meyeriana*, a taxon that I do not recognize formally; others, e.g. *Bowman*, Sept. 29, 1957, are pilose. Both forms occur widely on the mainland.

Ludwigia peploides (HBK.) Raven, Reinwardtia 6: 393. 1964, subsp. **peploides**

Jussiaea peploides HBK., Nov. Gen. & Sp. Pl. 6: 97. 1823.
Jussiaea polygonoides HBK., loc. cit.
Jussiaea patibilcensis HBK., loc. cit.
Jussiaea swartziana DC., Prodr. 3: 54. 1828.
Jussiaea ramulosa DC., loc. cit.
Jussiaea fluitans G. Don, Gen. Syst. 2: 692. 1832.
Jussiaea repens var. *peploides* HBK. in Griseb., Cat. Pl. Cubens. 107. 1866.
Jussiaea repens var. *ramulosa* (DC.) Griseb., loc. cit.
Jussiaea repens var. *minor* Mich. in Mart., Fl. Bras. 13^2: 166. 1875.
Jussiaea repens var. *californica* S. Wats., Bot. Calif. 1: 217. 1876.
Ludwigia diffusa var. *californica* (S. Wats.) Greene, Fl. Francisc. 1: 227. 1891.
Jussiaea diffusa var. *californica* (S. Wats.) Greene ex Jepson, Erythea 1: 244. 1893.
Ludwigia ramulosa (DC.) Gómez, An. Hist. Nat. Madrid 23: 66. 1894.
Jussiaea californica (S. Wats.) Jepson, Fl. W. Mid. Calif. 326. 1901.
Jussiaea gomezii Goyena, Fl. Nicaragüense 1: 406. 1909.
Jussieua repens subsp. *hirsuta* var. *ramulosa* (DC.) Hassler, Rep. Spec. Nov. Regni Veg. 12: 275. 1913.
Jussieua repens subsp. *hirsuta* var. *typica* Hassler, loc. cit.
Ludwigia adscendens var. *peploides* (HBK.) Hara, Jour. Jap. Bot. 28: 291. 1953.

Plant glabrous or minutely pubescent, sometimes long-hairy on creeping terrestrial branches; leaves oblong to oblong-spatulate, 1–4(6) cm long, 0.5–2 cm wide, stipules not conspicuous; flowers borne singly in upper leaf axils; sepals 4–7 mm long; petals 7–14 mm long, 4–10 mm wide; stamens 10, epipetalous ones slightly shorter, filaments bright yellow, 2.5–5 mm long, anthers pale yellow, 1–1.8 mm long, extrorse but often twisting and shedding pollen directly on stigma; pollen grains shed singly; disk slightly elevated, with a depressed white-hairy nectary surrounding base of each epipetalous stamen; style yellow, 2.5–5 mm long, densely long-hairy in lower half or higher, stigma lemon yellow, depressed-globose, 1.2–2 mm across, about 1 mm deep, deeply 5-lobed, usually surrounded by or elevated slightly above anthers at anthesis, upper two-thirds receptive; bracteoles at base

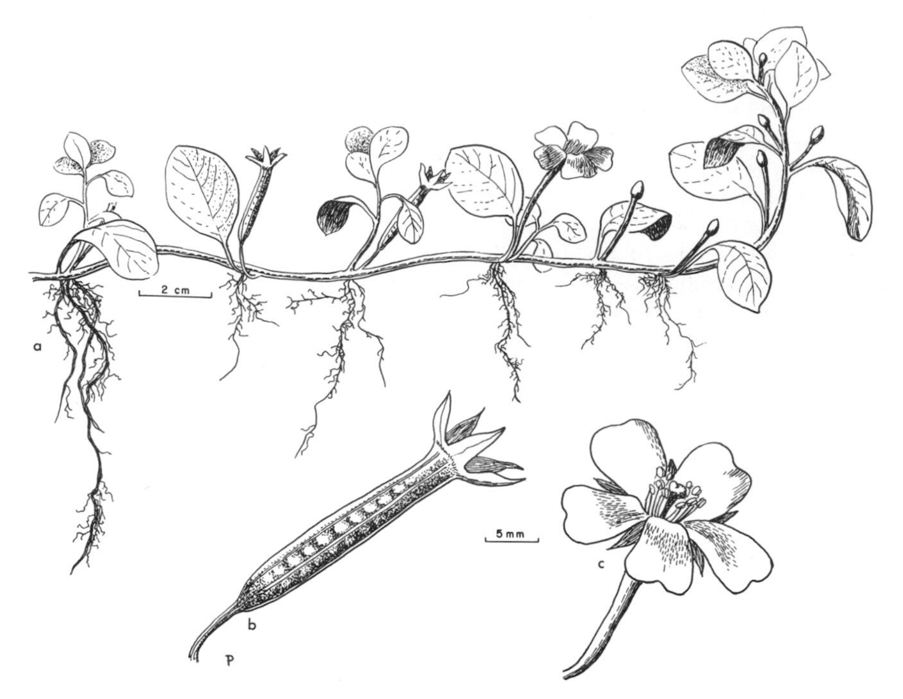

Fig. 197. *Ludwigia peploides* subsp. *peploides*.

of ovary; capsule glabrous or villous, 1–2.5 cm long, 3–4 mm thick, light brown, with 10 conspicuous darker brown ribs, terete, the seeds evident between ribs as bumps about 1.5 mm apart; capsules thick-walled, very tardily and irregularly dehiscent; fruiting pedicels 1–3 cm long; seeds uniseriate in each locule of capsule, pale brown, 1.1–1.3 mm long, more or less vertical, firmly embedded in coherent cubes of woody endocarp 1.2–1.5 mm high, 1–1.2 mm thick, this firmly fused to capsule wall.

Common, often in standing water, from the central United States throughout eastern and central South America to Argentina; introduced on the Pacific islands of Tahiti, Moorea, Rapa, and Rarotonga; also occurs in temperate Australia and temperate Asia, and on the North Island of New Zealand. Specimens examined: San Cristóbal, Santa Cruz.

See Fig. 197.

OXALIDACEAE. Woodsorrel Family*

Annual or perennial caulescent or acaulescent herbs, shrubs, or rarely trees, usually with watery sap containing oxalic acid. Leaves pinnately or digitately compound, usually 3-foliolate; leaflets opposite or alternate, ovate or obcordate, entire or apex decidedly notched; stipules free, adnate to petiole, or absent. Peduncles axillary, basal, cauline, or terminal, 1- to several-flowered, cymose, umbellate, or paniculate.

* Contributed by Duncan M. Porter.

Flowers 5-merous, bisexual, regular or nearly so, hypogynous. Sepals 5, free or shortly connate basally, equal or unequal, herbaceous or rarely petaloid, usually imbricate. Petals 5, short-clawed, free, connate basally or loosely connivent, imbricate, valvate or contorted. Disk and nectariferous glands absent. Stamens 10(15), usually in 2 unequal whorls of 5 each, outer whorl opposite petals, inner opposite sepals, 5 sometimes reduced to staminodes; filaments connate basally to differing degrees; anthers 2-loculed, versatile, introrse, longitudinally dehiscent. Gynoecium 5-carpeled, syncarpous; ovary superior, 5-lobed, 5-loculed; ovules (1)several to many per locule, anatropous, placentation axile; styles 5, free, connate basally, or coherent, persistent; stigmas terminal and capitate or sometimes slightly divided. Fruit a more or less 5-lobed, 5-loculed, small globose to columnar capsule, rarely a large, oblong berry. Seeds small, (1)several to many per locule; endosperm fleshy, copious or scanty, rarely absent; embryo straight.

A family of 7 genera and about 1,000 species, found throughout the world in the tropics and warm temperate zones.

Oxalis L., Sp. Pl. 433. 1753; Gen. Pl. ed. 5, 198. 1754

Annual or perennial herbs, rarely low shrubs; caulescent with elongate, erect, or repent stems, or acaulescent; stems arising from taproots, creeping rhizomes, or bulbs. Leaves alternate, cauline, basal, or clustered at branch tips, pinnately or digitately (1)3- to many-foliolate, strongly nyctitropic; petioles usually elongate, rarely phylloideous; leaflets mostly obcordate and notched at apex; stipules present or absent. Peduncles cymosely or umbellately 1- to several-flowered, usually about as long as leaves. Flowers small, homostylous to strongly heterostylous, sometimes minute, apetalous, and cleistogamous. Sepals often unequal, imbricate, persistent. Petals white, yellow, rose, or purple, free or shortly connate basally, contorted, deciduous, usually fugaceous. Stamens 10, in 2 unequal whorls of 5 each, outer whorl shorter and opposite petals; filaments usually connate basally into a short tube, the longer sometimes appendaged. Styles filiform or subulate. Fruit a globular to columnar capsule, dehiscing longitudinally, seeds ejected explosively, persistent valves remaining attached to axis. Seeds 1 to many per locule, pendulous, testa variously wrinkled, grooved, pitted, or striate; endosperm copious, only rarely absent.

About 850 species, wide-ranging in tropical and temperate regions but most numerous in South America and South Africa.*

Plant acaulescent, stems arising from bulb; flowers pink. . . . . *O. corymbosa*
Plant caulescent, bulbs lacking; flowers yellow:
  Leaves pinnately 3-foliolate; petiolule of middle leaflet 2–6 mm long; stems pubescent with glandular and acicular trichomes, these more prevalent below. . . . *O. cornellii*
  Leaves digitately 3-foliolate; petiolule none, all leaflets sessile; stems appressed-pubescent to glabrous, glandular trichomes absent:
    Leaves and peduncles borne terminally on short, stout, simple or branching stem; petioles minutely glandular-stipitate . . . . *O. megalorrhiza*
    Leaves and peduncles borne along the several, slender, long, creeping stems that root at the nodes; petioles villous . . . . *O. corniculata*

* The genus is in need of revision. See Standley & Steyermark (1946, 378) and Young (1958, 53) for pertinent remarks regarding the taxonomy of *Oxalis*.

Fig. 198. *Oxalis cornellii* (*a–c*). *O. corymbosa* (*d–g*).

Oxalis cornellii Anderss., Kongl. Svensk. Vet.-Akad. Handl. 1853: 246. 1855

Oxalis agassizii Rose, Contr. U. S. Nat. Herb. 1: 136. 1892.*

Herbaceous annual, becoming subligneous below and perhaps perennial; stems simple or branched, ascending or erect, green to reddish or dark brown, pubescent with acicular and glandular trichomes more prevalent below, 3.5–50 cm tall; roots slender; leaflets 3, appressed-pubescent, becoming glabrate and papillose, ciliate, surface finely reticulate, apex slightly retuse, base cuneate, margins more or less undulate, terminal leaflets broadly obovate, sometimes wider than long, 5–23 mm long, 6–24 mm wide, lateral leaflets narrowly to broadly obovate, longer than wide, 5–17 mm long, 3.5–11 mm wide; rachis 2–6 mm long, glabrous; petioles filiform, strict, more or less pilose and sometimes also glandular-pilose, 1–3.5 cm long; stipules small, reduced and adnate to petiole; peduncles in upper axils, ascending, filiform, sparingly pilose, apex more or less bifurcate, usually exceeding subtending leaves, 2.5–4 cm long; bracts lanceolate, pubescent, to about 3 mm long; pedicels glabrous, 1–3 mm long; flowers campanulate, 6–8 per inflorescence; sepals lanceolate, glabrous, equal, 4–6 mm long, 1–2 mm wide, margins membranous; petals deep yellow, orange along veins within, drying yellow or purplish, 5.5–7 mm long; longer filaments exceeding styles, pubescent, shorter filaments equaling styles, glabrous; ovary ovoid, glabrous, about 1 mm in diameter; styles pubescent, 1–2 mm long; stigmas terminal; capsules oblong, pendulous, glabrous, 7–10 mm long, 4 mm thick; seeds reddish, elliptical, transversely rugose, about 1 mm long.

Peru, Brazil, and Ecuador; apparently indigenous to the Galápagos Islands. Specimens examined: Española, Gardner (near Española), Isabela, Pinzón, Rábida, San Cristóbal, San Salvador, Santa Cruz, Santa María, Seymour; also reported from Santa Fé by Robinson.

This species is encountered mainly in cracks and potholes on bare lava between shrubs at lower elevations. It is easily recognized by its erect habit, pinnate, petiolulate leaves, and pendulous, glabrous fruits.

See Fig. 198.

Oxalis corniculata L., Sp. Pl. 435. 1753†

Herbaceous annual; stems creeping or decumbent, slender, rooting at nodes and becoming perennial, green to reddish, appressed-pubescent, trichomes nonglandular; root slender; leaflets 3, equal, sessile, obcordate, wider than long, 5–11 mm long, 8.5–18 mm wide, pubescent to nearly glabrate, ciliate, green to purple, apical notch prominent; petioles filiform, strict, villous, 3–5.5 cm long; stipules small, adnate to petiole; peduncles axillary, ascending, filiform, pubescent, shorter than to equaling or rarely exceeding subtending leaves, 1–6 cm long; bracts linear-lanceolate, pubescent, about 2 mm long; pedicels pubescent, reflexed in fruit, 1–1.5 cm long; flowers (1)2 to several per inflorescence; sepals lanceolate, equal, pubescent,

* For discussions of further synonymy, see Svenson (1946b, 455) and Macbride (1949, 560). Several specimens that have been seen were annotated as *O. dombeyi* St. Hil., and this may prove to be the correct name for this species. The original description of *O. dombeyi* has not been seen. Both Andersson (1855; 1861) and Svenson (1935) reported this species from the Galápagos Islands as *O. barrelieri* L.

† For a discussion of the taxonomy, synonymy, and relationships of this species, see Eiten (1963).

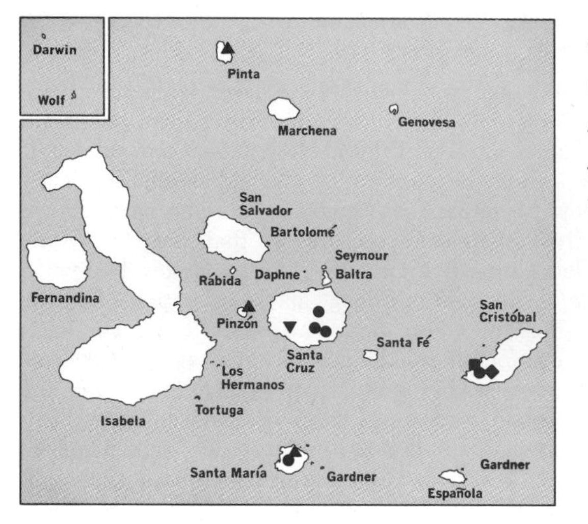

● *Oxalis corniculata*
■ *Oxalis corymbosa*
▲ *Oxalis megalorrhiza*
◆ *Argemone mexicana*
▼ *Passiflora colinvauxii*

margins reddish or scarious, ciliate, 3–4 mm long, about 1 mm wide; petals yellow, 6 mm long; longer stamens equaling styles, shorter stamens two-thirds as long, filaments glabrous; ovary cylindrical, pubescent, 3 mm long; styles pubescent, 1 mm long; capsules slender, cinereous, erect, 10–16 mm long, 2–3 mm wide; seeds brown, elliptical, transversely rugose, about 1 mm long.

A cosmopolitan weedy plant, apparently of Australasian origin. Specimens examined: San Cristóbal, Santa Cruz, Santa María.

Stewart (1911, 80) records *O. corniculata* from both San Cristóbal and Santa María, and also from Isabela. Svenson (1946b, 455) merely reports it as "not uncommon on the Galápagos Islands." His comments indicated its presence on San Cristóbal and Santa María, but he does not mention Santa Cruz. Apparently specimens collected on Santa Cruz by Wiggins and by Colinvaux are a new insular record for the species.

Oxalis corymbosa DC., Prodr. 1: 696. 1824

Oxalis martiana Zucc., Denkschr. K. Bayer. Akad. Wiss. München 9: 144. 1825.
Oxalis urbica St. Hil., Fl. Bras. Merid. 1: 126. 1825.
Oxalis bipunctata Graham, Edinb. New Phil. Jour. 16: 176. 1827.
Oxalis multibulbosa Turcz., Bull. Soc. Nat. Moscow 36: 595. 1863.
Acetosella martiana Kuntze, Rev. Gen. Pl. 1: 90. 1891.
Oxalis caripensis Hieron., Bot. Jahrb. 20: Beibl. 49: 32. 1895.
Ionoxalis martiana Small, Fl. SE. U. S. 665. 1903.

Acaulescent herb 15–30 cm high; bulb ovoid, 2–2.5 cm high, 1.5–2 cm broad, covered with brownish 3-ribbed, ovate scales, maturing into large masses of brown, ovoid bulbils 3–6 mm in diameter and sessile in axils of old scales; leaves 3-foliolate; leaflets rotund-obcordate, sessile, thin, often drooping, 14–30 cm long, 20–45 mm wide, apical sinus rather shallow, margin lightly but often acutely incised, ciliate, more or less appressed-pubescent, minute reddish tubercles scattered over surface, especially prominent beneath; petioles slender, flexuous, ascending, more or less villous, pubescence prominent at insertion of leaflets, 10–25 cm long; stipules small, membranous, adnate to petiole; peduncles few, slender, flexuose, erect, more or less villous, surpassing leaves, 15–30 cm long; bracts several, ovate-lanceolate, membranous, pubescent, 2.5–3 mm long; pedicels filiform, nodding, more or less spreading, appressed-pubescent, 1–3.5 cm long; inflorescence cymose, 5–12-flowered; flowers conic, deep; sepals equal, lanceolate, sparsely pubescent, with 2 reddish apical tubercles, 5 mm long, 2 mm wide; petals pink, paler basally, 15 mm long; filaments unappendaged, shorter 5 glabrous, spreading slightly, longer 5 pubescent, erect, 7 mm long; anthers cordate, about 1 mm in diameter; ovary oblong, pubescent, 3 mm high; styles free, pubescent, 2 mm high; stigmas capitate, lobed.

A native of Brazil, Bolivia, Paraguay, and Argentina, but extensively naturalized in the tropics and subtropics of the world. Specimen examined: San Cristóbal.

Spreads by the bulbils and sometimes becomes a troublesome weed. The colony from which our specimen was taken probably represents a recent escape from cultivation. It is a new record for the Galápagos.

See Fig. 198.

Oxalis megalorrhiza Jacq., Oxalis Monogr. 33. 1794

Oxalis bicolor Savigny, Encycl. Méth. Bot. 4: 687. 1797.
Oxalis rubrocincta Lindl., Bot. Reg. 28: 64. 1842.
Acetosella megalorrhiza Kuntze, Rev. Gen. Pl. 1: 92. 1891.
Acetosella rubrocincta Kuntze, loc. cit.
Otoxalis rubrocincta Small, N. Amer. Fl. 25: 27. 1907.

Perennial; stems and rhizomes stout, dark, fleshy, elongating upward with growth, simple or sparingly forked, leafy and floriferous only at apex, scaly apically with persistent bases of fallen petioles, to 1.5 cm thick; leaves numerous; leaflets 3, equal, subsessile, obcordate, emarginate, sparingly appressed-pubescent beneath, surface finely reticulate, midvein prominent, 10–12 mm long, 9–12 mm wide; petioles filiform, strict, minutely glandular-stipitate, dehiscing near base, to 7 cm long; peduncles ascending, filiform, strict, glabrous, to 7 cm long; flowers 1–3 per inflorescence; bracts ovate-lanceolate, ciliate, about 1 mm long; sepals unequal, in 2 series, 2 outer broadly ovate, apex obtuse, base nearly hastate, 4–5 mm long, 3–4 mm wide, 3 inner narrowly oblong, 4–5 mm long, to 2 mm wide, nearly hidden by outer pair in bud, all more or less connate basally; petals 6 mm long; filaments linear-subulate, glabrous; ovary oblong-cylindrical, glabrous; styles filiform, connate basally, slightly pubescent, 2 mm long.

Indigenous to Peru, Bolivia, Chile, and the Galápagos Islands. Specimens examined: Santa María; also reported from Pinzón by Robinson & Greenman and Stewart, and from Pinta by Stewart.

The species has long been identified as *O. carnosa* Molina, but Dandy and Young (1959) have shown the latter to be a superfluous substitute for *O. magellanica* Forst. f.

PAPAVERACEAE. Poppy Family*

Herbs or shrubs with watery, milky, or yellow sap and alternate or rarely opposite, exstipulate, entire or dissected, decompound leaves. Flowers perfect, regular, showy. Sepals 2(4), distinct or united into a cuplike calyptra. Petals 4 or 6 or sometimes lacking, deciduous, often more or less crinkled along margins. Stamens mostly numerous, distinct, hypogynous; anthers dehiscent by longitudinal slits. Pistil of 2 to several carpels united into a 1- or several-celled ovary, or sometimes these be-

* Contributed by Ira L. Wiggins.

coming distinct in fruit; stigma entire or divided, on a short style or sessile. Ovules mostly numerous on parietal placentae. Fruit a capsule, usually opening by terminal or subterminal pores. Seeds small, numerous, with a minute embryo and copious oily or mealy endosperm.

A family of 28 genera and about 250 species, most common in subtropical and temperate regions of North America and east Asia; sparingly present elsewhere.

Argemone L., Sp. Pl. 508. 1753; Gen. Pl. ed. 5, 225. 1754

Annual or biennial prickly herbs (1 species in Mexico a shrub), with acrid, white or yellow sap, and alternate, sinuate to pinnatifid, spinose (rarely spineless) leaves. Flowers erect in bud, showy. Sepals 2 or 3, each bearing a spine-tipped appendage at or near summit. Petals 4 or 6, white, yellow, or orange. Stamens many. Stigmas sessile or nearly so, more or less radiate; ovary with 4–6 slender placentae. Capsules oblong or fusiform, opening apically by 4–6 valves. Seeds numerous, globose, favose-reticulate.

Fig. 199. *Argemone mexicana.*

A genus of 23 species and about a dozen subspecific entities, distributed from the central, eastern, and western United States through the West Indies, Mexico, and Central America into northern South America (see Gerald B. Ownbey, 1958).

Argemone mexicana L., Sp. Pl. 108. 1753

Ectrus trivialis Lour., Fl. Cochinch. 1: 344. 1790.
Argemone spinosa Moench, Meth. Pl. 227. 1794.
Argemone versicolor Salisb., Prodr. Stirp. 376. 1796.
Argemone vulgaris Spach, Hist. Nat. Veg. 7: 26. 1839.
Argemone mucronata Dum. Cours. ex Steud., Nomencl. Bot. ed. 2, 1: 128. 1840.
Argemone mexicana a lutea Kuntze, Rev. Gen. Pl. 1: 12. 1891.
Argemone mexicana a lutea var. *parviflora* Kuntze, loc. cit.
Argemone mexicana var. *ochroleuca* Britt., Manual 439. 1901, non *A. ochroleuca* Sweet 1829.
Ectrus mexicanus Nieuwl., Am. Midl. Nat. 3: 350. 1914.

Erect annual herb 2–10 dm tall, with glabrous, glaucous, bristly or spinose stems, peduncles, calyces, and leaves; leaves sinuate-pinnatifid, sessile and more or less clasping, 6–20 cm long, to 8 cm wide, sparingly spinose on margins and on the whitish veins, spines 2–8 mm long; buds ovoid, 10–18 mm long, 6–8 mm broad; sepals sparsely bristly, horns slightly divergent and ascending, 5–8 mm long, spine-tipped but devoid of lateral bristles; petals in ours 4, occasionally white (in ours), but usually yellow or orange, broadly obovate, 2–3 cm long; style about 2 mm long; stigma lobes suberect, 1–1.5 mm long; capsule ovoid-fusiform to oblong, 2.5–4 cm long, moderately bristly with coarse, yellow, spreading bristles 4–10 mm long; seeds dark brown to dull black, finely favose-reticulate, 1.2–1.5 mm in diameter.

Common as an introduced weed in the tropics and subtropics around the world, probably a native of the West Indies; in the Galápagos known only from Isla San Cristóbal.

Thanks are extended to Dr. Wallace R. Ernst for supplying the determination of our collection, based on a specimen sent to the U. S. National Herbarium (*Wiggins & Porter 425*). This is the first record of the genus from the Galápagos Islands.

See Fig. 199.

PASSIFLORACEAE. PASSION-FLOWER FAMILY*

Herbs, shrubs, or commonly vines with axillary tendrils. Leaves alternate, usually simple but sometimes compound, stipulate, petiolate. Tendrils opposite leaves or axillary, or terminal on 1–3-flowered peduncles. Flowers perfect, rarely monoecious or dioecious, regular, axillary, solitary, in pairs or in fascicles, bracteate. Sepals 4 or 5, distinct or connate basally, imbricate, often petaloid or fleshy, persistent. Petals 4 or 5 or lacking, distinct or connate basally, often smaller than sepals. A corona of numerous filaments in a whorl between corolla and androecium. Stamens 5 or more, opposite petals, borne on base of corona or on a disk near apex of gynophore (then the 2 constituting an androgynophore), distinct or basally connate. Anthers 2-celled, versatile, dehiscing longitudinally. Pistil 1, borne on tip of androgynophore or gynophore. Ovary superior, 3–5-carpellate, 1-celled, with parietal, more or less intruding placentae. Style branches as many as carpels, distinct or connate; stigmas capitate or discoid. Ovules many on each placenta, anatropous. Fruit a berry or loculicidal capsule. Seeds smooth or rugose, usually embedded in pulpy or juicy aril. Embryo straight. Endosperm fleshy.

* Contributed by Ira L. Wiggins.

A family of 11 genera and about 600 species, found mainly in the tropics but extending into the central United States and south into New Zealand, Madagascar, and temperate Africa and Asia.

Passiflora L., Sp. Pl. 955. 1753; Gen. Pl. ed. 5, 410. 1754

Herbaceous or woody scandent or climbing vines or rarely erect shrubs, herbs, or small trees. Leaves lobed or not, lobes or lamina usually entire, sometimes serrate. Petioles often glanduliferous. Stipules small, setose to leaflike. Inflorescence axillary, simple or occasionally bearing 2 or more flowers; bracts small, usually more or less separated, in some species large and involucre-like just below flower. Flowers often highly colored. Calyx tube patelliform, campanulate, funnelform, or cylindrical. Sepals 5, fleshy or membranous, sometimes lacking. Petals 5 (or lacking), thin, about as long as sepals or slightly shorter, waxy white to deeply colored. Corona of 1 to several series of distinct or connate filaments. Operculum variously lacerate, plicate, or entire, borne below corona, or sometimes lacking. Limen borne near base of gynophore, cupuliform or annular, or lacking. Stamens 5, filaments monadelphous in a tube closely surrounding gynophore, distinct above, free parts of filaments spreading. Ovary borne on the elongated gynophore. Styles 3; stigmas capitate. Fruit baccate, indehiscent, globose, ovoid, or fusiform, filled with mucilaginous pulp. Seeds more or less compressed, reticulate, puncticulate, or transversely rugulose, each enclosed in pulpy aril, this sometimes brightly colored.

A diverse genus of about 400 species, chiefly of tropical and subtropical America but some species well into temperate regions.

Bracts of peduncles bi- to quadripinnatifid or pinnatisect, with gland-tipped segments; stipules semiannular, deeply cleft, divisions gland-tipped; herbage malodorous; leaf blades pubescent, margins glandular-ciliate . . . . *P. foetida* var. *galapagensis*

Bracts of peduncles setaceous to linear-lanceolate, entire, eglandular; stipules like bracts, entire, eglandular; herbage not malodorous; leaf blades glabrous or essentially so:

 Leaf blades distinctly 3–5-lobed, 5–7 cm wide, central lobe equaling or surpassing laterals; petioles bearing 1–2 circular, short-stipitate glands at or above middle; ocellae on blade none . . . . *P. suberosa*

 Leaf blades divergently 2-lobed, 10–16 cm from tip to tip of lobes; petioles without glands; ocellae obvious at base of blade and outward three-fourths way to tips of lobes . . . . *P. colinvauxii*

Passiflora colinvauxii Wiggins, Madroño 20: 251. 1970

Slender vine with tightly coiled tendrils to 15 cm long after coiling; stems 1–1.5 mm thick, markedly grooved, sparsely hirsutulous, soon glabrate; internodes 5–6 cm long; stipules subulate, 2–3 mm long, usually strongly falcate, yellowish, persistent; petioles 2.5–6 cm long, minutely and sparsely puberulent, eglandular; leaf blades lunate, broadly rounded at base, truncate across distal median part (or nearly so), 2 lobes ovate to ovate-lanceolate, blade 3–5 cm from base to upper edge along midrib, 7–16 cm tip to tip of spreading lobes, these 1.5–4 cm wide at their bases, obtuse to acute at apex, midveins of lobes 5–11 cm long; lamina bright green above, pale and subglaucous beneath, minutely and sparsely puberulent on both surfaces, a few short hairs gland-tipped; 3–6 circular, yellowish or brownish ocellae 0.4–0.8

mm broad between bases of main veins, several similar ones scattered toward distal parts of lobes; flowers solitary or in pairs on peduncles 1.5–2 cm long at anthesis, these glandless; bracts subulate, 0.5–1.5 mm long, near base of peduncle, deciduous; pedicels 2.5–4 mm long, articulation with peduncle obvious even in bud; flower 2–2.5 cm broad, calyx patelliform, 6–7 mm across, 2–3 mm deep, pale green, sepals broadly oblong, 6–7 mm long, 3–4 mm wide, thin, margins whitish; petals white, spreading, about three-fourths as long as sepals, 2–2.5 mm wide; corona 2-ranked, outer series of about 30 linguiform to narrowly linear-oblanceolate filaments, these 5–6 mm long, purplish, ascending and slightly spreading, inner rank slenderly filiform, erect, about 2 mm long, more numerous than outer filaments, white, forming dense fringe inside calyx cup; operculum thin, markedly plicate, curved inward, margin faintly denticulate; androgynophore 6–7 mm tall, glabrous, whitish, distinctly 5-ridged; free stamen filaments 3–3.5 mm long, strongly flattened, spreading-

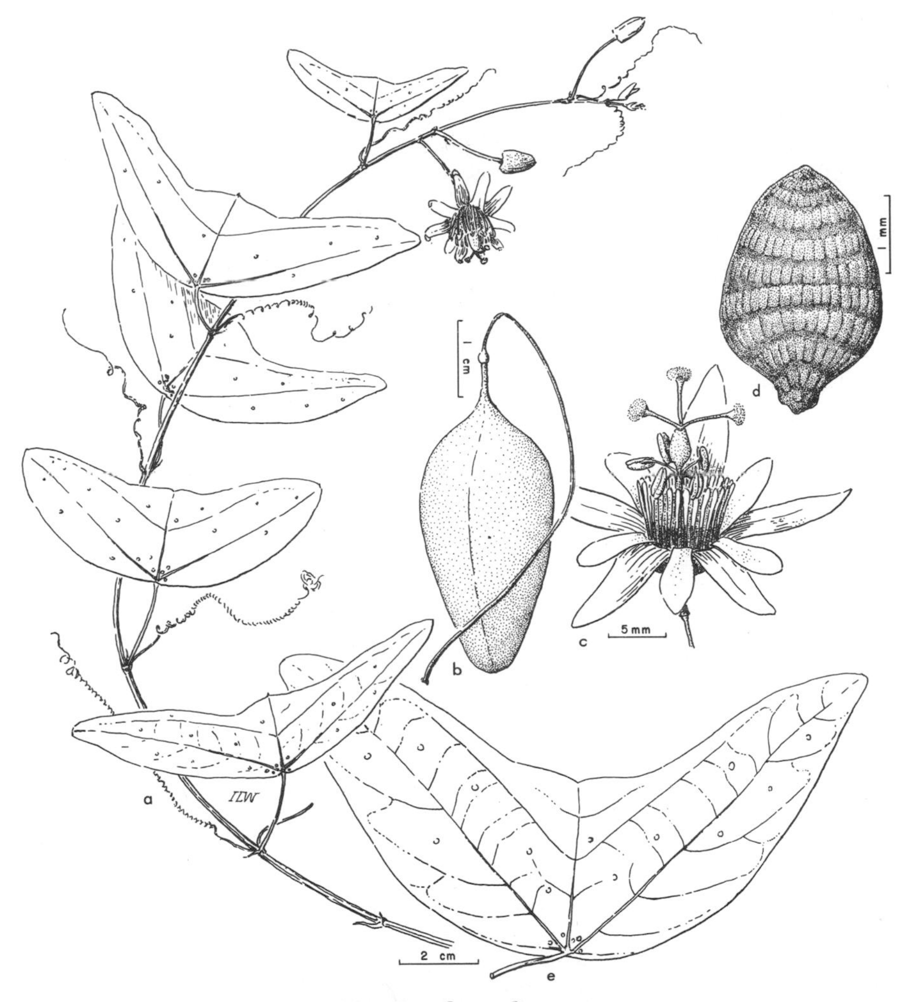

Fig. 200. *Passiflora colinvauxii.*

recurved, anthers greenish, about 4 mm long, 2 mm wide at anthesis; ovary ovoid, 6–7 mm long, 4.5–5 mm across at anthesis, glabrous, pale green; styles slender, 3.5–4 mm long, pale cream or whitish, slightly ascending-spreading; stigmas capitate-globose, 1–1.4 mm across, granular on receptive surface; fruiting pedicels and peduncles very slender, to 6 cm long or more, the fruit pendant, its body narrowly elliptic-ovate, 1–1.5 cm broad, 2.5–4 cm long, the base drawn out into a slender stipe 2–4 mm long, green, surface glabrous; seeds lenticular-ovoid, laterally compressed, about 2 mm long, slightly narrower, about 1 mm thick or slightly less, the flattened surfaces transversely corrugated and the ridges again transversely minutely striate, grayish, enclosed in mucilaginous aril, this nearly colorless.

Known only from open region west of Bella Vista, Isla Santa Cruz, scandent over lava and bushes, and occasionally climbing into trees to 5 m or more.

See Fig. 200 and color plate 82.

Passiflora foetida var. **galapagensis** Killip, Field Mus. Nat. Hist. Publ. Bot. 19: 505. 1938

Vine climbing over trees and shrubs or scrambling over rocks, to 6 m or more; herbage oily-viscid and malodorous; slender stems softly rufo-hirsutulous with spreading hairs 0.3–1 mm long; stipules semiannular, 2–3 mm high, 4–7 mm wide (around stem), margin closely set with filaments 1–3.5 mm long, these tipped with ovoid to clavate glands 0.4–0.8 mm long; petioles 1.5–2.5 cm long, hirsute, bearing gland-tipped filaments like those on stipules, cupuliform glands lacking; leaf blades ovate, 6–9 cm long, 4–7 cm wide, distinctly but shallowly 3-lobed, subcordate, margins irregularly and remotely denticulate-serrate, lobes acute to short-acuminate, main veins 3, middle lobe pinnately veined, both surfaces soft-hirsute with reddish or brownish, simple, spreading to subappressed hairs, margins bearing gland-tipped filaments; tendrils 10–20 cm long, sparsely hirsutulous; peduncles solitary, 2–2.5 cm long; bracts 2–3.5 cm long, bi- or tripinnatisect, glandular-ciliate, glabrous, borne immediately below calyx; flowers 4–5 cm across; sepals oblong, 1.5–2 cm long, 4–5 mm wide, petaloid, nearly white with greenish veins, margins membranous, tinged with purple, tipped by single glandular filament; petals white,

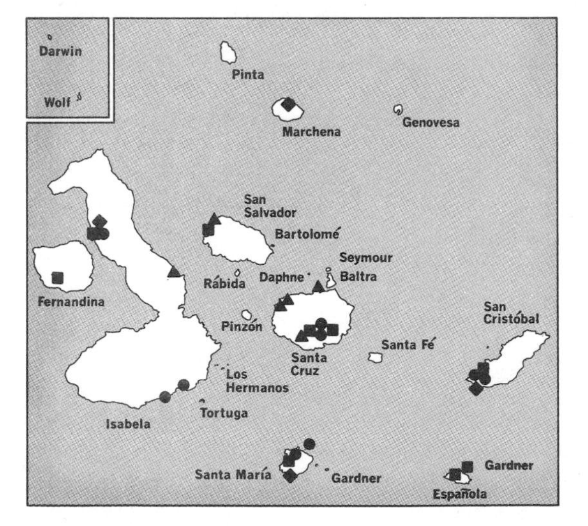

● *Passiflora foetida* var. *galapagensis*

■ *Passiflora suberosa*

▲ *Polygala anderssonii*

◆ *Polygala galapageia* var. *galapageia*

Fig. 201. *Passiflora foetida* var. *galapagensis*.

equaling sepals; filaments of corona purplish at base, white distally; gynophore 1–1.2 cm long; ovary glabrous; fruit obovoid to subglobose, 2.5–3 cm long, yellow, glabrous; seeds coarsely reticulate, 3–4 mm long, moderately compressed laterally, each enclosed in grayish aril.

Endemic; scrambling over rocks and climbing into trees and shrubs, from near sea level upward into the *Scalesia* and *Miconia* zones. Specimens examined: Champion, Isabela, San Cristóbal, Santa Cruz, Santa María.

See Fig. 201 and color plate 63.

Passiflora suberosa L., Sp. Pl. 958. 1753

Passiflora leariloba Hook. f., Trans. Linn. Soc. Lond. 20: 222. 1847.
Passiflora tridactylites Hook. f., loc. cit.
Passiflora puberula Hook. f., op. cit. 223.*

Young stems and leaves glabrous to sparingly puberulent, older stems becoming corky and irregularly fissured and winged; stipules linear-lanceolate to subulate,

* For further synonymy, see Field Mus. Nat. Hist. Publ. Bot. 19: 88–91. 1938.

1.5–2.5 cm long, yellowish; petioles slender, 5–15(40) mm long, bearing (1)2 circular, subsessile to distinctly stipitate glands 0.3–0.8 mm across at or above middle; leaf blades extremely variable, entire, 3-lobed or 3-parted, in ours rarely without marginal denticulations, 3-lobed ones rounded to subtruncate at base, 3–7 cm long, 2.5–8 cm wide (between tips of lateral lobes), central lobe broadly lance-ovate, 1–3 cm wide at base, lateral lobes diverging at 45° to 60° angle, ovate, 8–20 mm wide at base of sinuses between lateral lobes and central one, lamina with 3 primary veins, 1 for each lobe, central lobe with pinnately arranged secondary veins, lamina thin, submembranous, dark green above, slightly paler beneath, without ocellae; leaves of tripartite form with linear lobes 5–10 cm long, 2–5 mm wide; flowers solitary or in pairs in axils of leaves; peduncles 1.5–2.5 cm long; pedicels 5–18 mm long; bracts setaceous, minute, 1–2.5 mm long, caducous; calyx 1–3 cm broad, sepals ovate-lanceolate, 8–10 mm long, greenish yellow, obtuse, withering and tip then acute; petals none; corona filaments in 2 series, filiform, 5–8 mm long, outer series white with yellow tips and purplish bases, recurved, those of inner series capitellate; operculum membranous, plicate, minutely fimbriate, white, margin incurved; limen annular, inconspicuous; gynophore 7–8 mm long; stamens about 3 mm long, filaments subulate; fruit oblong-ovoid, dark purple to black, to 1.5 cm long, glabrous, glaucous when young; seeds obovoid, 3–4 mm long, 2–2.2 mm wide, flattened, slightly curved, abruptly short-acuminate at apex, tapered to base, coarsely reticulate, the ridges low, broad, distinctly rugulose transversely.

Florida to Texas southward into the tropics of Mexico and Central and South America, introduced into the Old World; among shrubs and rocks, mostly in shady areas, from near sea level to about 615 m in the Galápagos Islands. Specimens examined: Española, Fernandina, Gardner (near Española), Isabela, San Cristóbal, San Salvador, Santa Cruz, Santa María.

POLYGALACEAE. MILKWORT FAMILY*

Herbs, shrubs, or trees, sometimes scandent. Leaves alternate, rarely opposite or whorled, simple, entire, short-petiolate, often with pellucid glands, these sometimes also in flowers and fruits; stipules none or represented by small glands. Flowers bisexual, more or less irregular, pseudo-papillionaceous, hypogynous, subtended by a bract or 2 bracteoles, in racemes or spikes, or rarely in panicles or solitary; pedicels often jointed. Sepals (4)5, free or lower 2 connate, unequal, 1 posterior, 2 anterior, 2 lateral and interior and usually much larger, petaloid, winglike, imbricate. Petals 3(5), lower one boat-shaped, forming a keel, often with an apical beak or fimbriate crest, upper 2 usually ligulate or ovate, usually adnate basally to staminal sheath and often also connate to keel, lateral 2 short, free, exterior, rarely present. Stamens (4, 5)8; filaments connate most of their length into a sheath split along upper side and usually adnate basally to keel or upper petals or both, rarely free; anthers usually confluently 1-loculed, erect, basifixed, dehiscing by an apical or introrse-apical pore. Intrastaminal disk present, reduced to a gland at base of ovary, or absent. Gynoecium 1–2(3–5)-carpelled, syncarpous; ovary superior, 1-loculed with parietal placentation and many pendulous ovules, or 2- to 5-loculed with axile placentation and 1–2 ovules per locule; style simple; stigma bilobed, often tufted.

* Contributed by John Thomas Howell and Duncan M. Porter.

Fruit a capsule, samara, or drupe, dehiscent loculicidally or indehiscent. Seeds usually 1–2 per locule, usually pubescent and arillate; endosperm usually present; embryo straight, axial.

A family of 18 genera and more than 1,000 species, widely distributed throughout the Tropical and Temperate zones.

Polygala L., Sp. Pl. 701. 1753; Gen. Pl. ed. 5, 315. 1754

Herbs, shrubs, or trees, the branches sometimes spinose. Leaves alternate, sometimes opposite or whorled; stipular glands rarely present. Flowers in terminal or axillary or rarely extra-axillary racemes or spikes, sometimes contracted into heads, rarely paniculate, often showy, subtended by a small caducous or persistent bract; bracteoles 2, caducous or persistent. Sepals 5, outer 3 herbaceous or lower 2 rarely petaloid, free, or lower 2 connate, inner 2 petaloid or rarely herbaceous, usually much larger, all deciduous or persistent. Petals 3(5), upper 2 ligulate to ovate, sometimes galeate, adnate to staminal sheath or connate to keel at least basally, keel clawed, usually crested or beaked apically, sometimes 3-lobed or unappendaged, lateral 2 rarely present, always minute. Stamens 8; filaments connate nearly to apex into a sheath split along upper side, adnate basally to keel and upper petals. Disk present or absent. Gynoecium 2-carpeled; ovary 2-loculed; ovules 1 per locule, pendulous, anatropous; style usually slender and elongate, incurved and dilated apically, entire to 2- to 4-lobed. Fruit a small, loculicidal or indehiscent capsule, equally or unequally 2-loculed, 1 locule rarely more or less completely abortive, often margined or winged, compressed contrary to partition, usually membranaceous-herbaceous, rarely subcoriaceous. Seeds 1 per locule, globose to fusiform or conic, usually pubescent and arillate; testa crustaceous; endosperm present or absent.

A genus of 500 species or more, widely distributed in both tropical and warm temperate regions.

The 5 taxa of *Polygala* present in the Galápagos Islands are endemic and are quite similar morphologically. They appear to be most closely related to *P. paludosa* St. Hil. or *P. paniculata* L., two widespread species of the New World tropics, and they perhaps represent a radiation from a single past introduction from the mainland of South America (Howell & Porter, 1968).

Stems puberulent; wings 5–7-nerved; aril one-third as long as seed........*P. anderssonii*
Stems glabrous; wings 3–5-nerved; aril about one-fifth as long as seed:
  Apical crest of keel inconspicuous, not showy or petaloid, divisions simple or bilobed, final lobes 6–8; wings 3.5–5 mm long, 3-nerved:
    Leaves linear to oblanceolate-elliptic, acute to cuspidate; racemes becoming elongate, acute, 1.5–12 cm long; wings 3.5–4(5) mm long...*P. galapageia* var. *galapageia*
    Leaves oblanceolate to linear-lanceolate, obtuse to acute; racemes more congested, oblong or rounded, obtuse to subacute, 1–4 cm long; wings 4–5 mm long...... .... *P. galapageia* var. *insularis*
  Apical crest of keel conspicuous, showy and petaloid, divisions frequently 3-lobed, final lobes 12–16; wings 4–6.5 mm long, 5-nerved:
    Leaves broadly spatulate to rotundate, 6–13 mm long, 3–8 mm wide............. .... *P. sancti-georgii* var. *sancti-georgii*
    Leaves narrowly to broadly oblanceolate, 8–25 mm long, 1–4 mm wide........... .... *P. sancti-georgii* var. *oblanceolata*

Polygala anderssonii Robins., Proc. Amer. Acad. 38: 160. 1902

Polygala puberula Anderss., Kongl. Svensk. Vet.-Akad. Handl. 1853: 232. 1855, non A. Gray 1852.

Suffruticose plant to 1 m high, rarely branched basally; stems spreading, puberulent, becoming glabrate below, yellowish or rarely reddish; leaves alternate, linear-lanceolate to oblanceolate, acuminate, puberulent to subglabrate, coriaceous, 4–17 mm long, 1–1.5 mm wide; petioles less than 1 mm long; racemes becoming elongate, 3–9 cm long, about 1 cm wide, axis puberulent; flowers white to pale purple; pedicels 1 mm long, glabrous; bracteoles caducous; outer sepals ovate, acute, glabrous, green, margins white, more or less equal, 1.5–2 mm long, 1 mm wide; wings petaloid, obovate-elliptic, obtuse, glabrous, 5- to 7-nerved, 3.5–4.5 mm long, about 2 mm wide, longer than capsule; corolla about 3 mm long; apical crest of keel 12- to 14-lobed, less than 1 mm long; capsule oblong, glabrous, equally 2-loculed, loculicidally dehiscent, 3.5–4 mm long, about 2 mm wide; seeds 1 per locule, obconical, apex obtuse and more or less apiculate, base attenuate, dark, short-lanate, trichomes more or less appressed apically, 2.5–3 mm long; aril elliptical, bilobed, obtuse, white, a third as long as seed.

Mostly in cinders or light ashy soil at lower elevations. Specimens examined: Isabela, San Salvador, Santa Cruz.

Polygala galapageia Hook. f., Trans. Linn. Soc. Lond. 20: 233. 1847, var. **galapageia**

Suffrutescent perennial to about 1 m high, herbage more or less glaucous; stems spreading, glabrous, reddish below, yellowish above; leaves alternate, numerous, linear to oblanceolate-elliptic, acute to cuspidate, glabrous, coriaceous, 5–15 mm long, 1–1.5 mm wide; petioles less than 1 mm long; racemes becoming elongate, acute, 1.5–12 cm long, about 1 cm wide; flowers white or whitish; pedicels less than 1 mm long, glabrous; bracteoles caducous; outer sepals ovate, acute, glabrous, green, margins white, more or less equal, about 1.5 mm long; wings petaloid, obovate-elliptic, attenuate, glabrous, 3-nerved, 3.5–4(5) mm long, 2 mm wide, more or less equaling capsule; corolla 2–3 mm long; apical crest of keel inconspicuous, 4-parted and each part simple or bilobed, less than 1 mm long; capsule oblong, glabrous, equally 2-loculed, loculicidally dehiscent, 3–3.5 mm long, 1.5 mm wide; seeds 1 per locule, obconical, apex obtuse and more or less apiculate, base attenuate, dark, short-lanate, trichomes more or less appressed apically, 2–2.5 mm long; aril narrowly elliptical, 2-lobed, white, about a fifth as long as seed.

In sandy or tufaceous soil at low altitudes, often near the shore. Specimens examined: Isabela, Marchena, San Cristóbal, Santa María.

Polygala galapageia var. **insularis** (A. W. Bennett) Robins., Proc. Amer. Acad. 38: 161. 1902

Polygala obovata Hook. f., Trans. Linn. Soc. Lond. 20: 233. 1847, non St. Hil. 1835.
Polygala obovata f. *latifolia* Anderss., Kongl. Svensk. Vet.-Akad. Handl. 1853: 231. 1855.
Polygala obovata f. *angustifolia* Anderss., op. cit. 232.
Polygala chathamensis Anderss., loc. cit.
Polygala insularis A. W. Bennett, Jour. Bot. 17: 204. 1879.
Polygala obovata var. *angustifolia* Riley, Kew Bull. 1925: 219. 1925, pro syn.
Polygala obovata var. *latifolia* Riley, loc. cit., pro syn.

Differs from var. *galapageia* in having stems 30–45 cm high, reddish; leaves oblanceolate to linear-lanceolate, obtuse to acute, more or less apiculate, 6–10 mm long, 1–2 mm wide; racemes oblong or rounded, obtuse to subacute, 1–4 cm long, about 1 cm wide; wings 4–5 mm long.

At low elevations on lava beds and in rocky soil. Specimens examined: Marchena, Pinta, San Cristóbal, Santa Cruz, Santa María.

Polygala sancti-georgii Riley, Kew Bull. 1925: 218. 1925, var. **sancti-georgii**

Herbaceous annual 15–50 cm high, herbage glaucous; stems spreading, glabrous, reddish; leaves alternate, usually numerous and crowded, broadly spatulate to rotundate or rarely narrowly spatulate, obtuse, cuspidate, glabrous, coriaceous, blade decurrent into petiole, 6–13 mm long, 3–8 mm wide; petioles less than 1 mm long; racemes 1.5–3.5(6) cm long, 1.5 cm wide; flowers purplish to white; pedicels 1 mm long, glabrous; bracteoles caducous; outer sepals ovate, acute, glabrous, green, margins white, more or less equal, 1.5–2 mm long, 1 mm wide; wings petaloid, ovate-elliptic, more or less obtuse, glabrous, 5-nerved, 6–6.5 mm long, 2–3 mm wide, longer than capsule; corolla about 3 mm long; apical crest of keel conspicu-

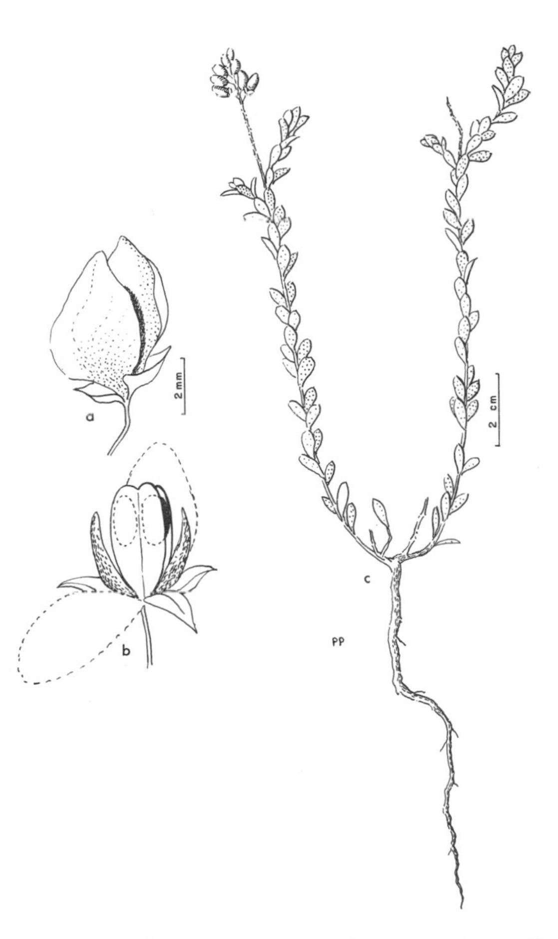

Fig. 202. *Polygala sancti-georgii* var. *sancti-georgii.*

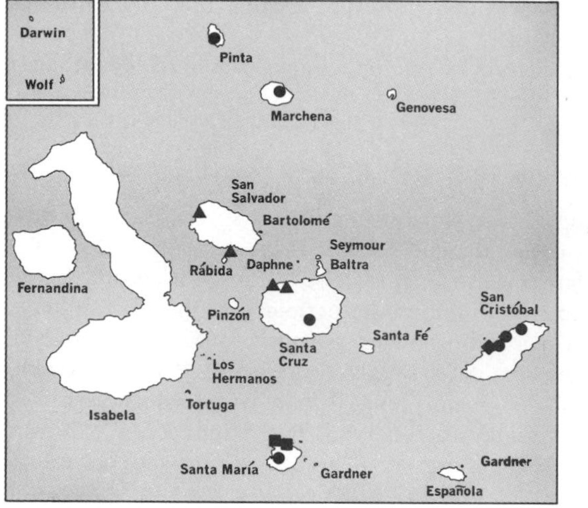

ous, petaloid, 14- to 16-lobed, about 1 mm long; capsule oblong, glabrous, equally 2-loculed, longitudinally dehiscent, 3.5–4 mm long, 2 mm wide; seeds 1 per locule, obconical, apex obtuse and more or less apiculate, base attenuate, dark, short-lanate, trichomes more or less appressed apically, 2.5–3 mm long; aril narrowly elliptical, 2-lobed, white, about a fifth as long as seed.

Apparently restricted to a rather limited area in the northeastern part of Isla Santa María.

See Fig. 202.

Polygala sancti-georgii var. **oblanceolata** Howell, Leafl. West. Bot. 10: 351. 1966

Differs from var. *sancti-georgii* in having stems 25–60 cm high; leaves narrowly to broadly oblanceolate, obtuse to acute, 8–25 mm long, 1–4 mm wide; racemes 2–7(12) cm long, 1–1.5 cm wide; flowers light to deep lavender; wings oblanceolate-elliptic, 4–6 mm long, 2–2.5 mm wide; corolla 3.5 mm long; apical crest of keel 12–14-lobed, about 1.5 mm long, lobes broad and pink; seeds 2.5 mm long.

In sandy or ashy soil near the coast. Specimens examined: Rábida, San Salvador, Santa Cruz.

PORTULACACEAE. Purslane Family*

Annual or perennial herbs, sometimes suffrutescent, often somewhat succulent. Leaves entire. Flowers bisexual, solitary or in cymose or racemose inflorescences. Involucral bracts generally 2, rarely several (sometimes interpreted as sepals). Perianth segments 4–6, rarely 2 or 3, free or united near base. Stamens 3 to numerous; anthers 2-celled, introrse. Pistil 1; carpels 2–3; ovary unilocular with 1 to numerous amphitropous ovules. Fruit a capsule. Embryo curved.

About 19 genera and 500 species in tropical and temperate regions.

* Contributed by Uno Eliasson.

Ovary half-inferior; capsule dehiscing with a transverse operculum . . . . *Portulaca*
Ovary superior; capsule dehiscing with 3 valves:
Involucral bracts persistent; seeds without a strophiole; leaves (in Galápagos species) linear . . . . *Calandrinia*
Involucral bracts fugacious; seeds with a strophiole; leaves broad . . . . *Talinum*

Calandrinia HBK., Nov. Gen. & Sp. Pl. 6: 77. 1823, nom. conserv.

Herbs, sometimes suffrutescent. Leaves of various shapes, rounded to linear. Flowers axillary or in cymes, racemes, or panicles. Involucral bracts 2, persistent. Tepals generally 5. Stamens 5 to many. Carpels 3. Ovary subglobose. Stigmas 3, sometimes united below and forming a distinct style. Capsule dehiscing into 3 valves. Seeds without a strophiole.

About 150 species in western America and Australia.

Calandrinia galapagosa St. John, Amer. Jour. Bot. 24: 95. 1937

Roots several, more or less horizontal; stems succulent, to 60 cm long, 2–3 cm thick at base; bark smooth and somewhat shining; leaves alternate, linear, succulent, 3–7 cm long, about 3 mm wide (only 1 mm wide when dried); inflorescence several-flowered cymes; peduncles 10–16 mm long; bracts triangular, about 1 mm long; pedicels to 14 mm long; involucral bracts subovate to subelliptic, 3.5–4.5 mm long, 2.5–3 mm wide; tepals obovate, about 10 mm long, pinkish white; stamens 12–15; style 4–6 mm long; capsule globose, about 4 mm in diameter, yellowish; seeds 1–1.3 mm in diameter, dark reddish brown, finely elongate-reticulate.

Endemic; in lava fissures and among lava slabs; known only from Sappho Cove on Isla San Cristóbal.

See Fig. 203.

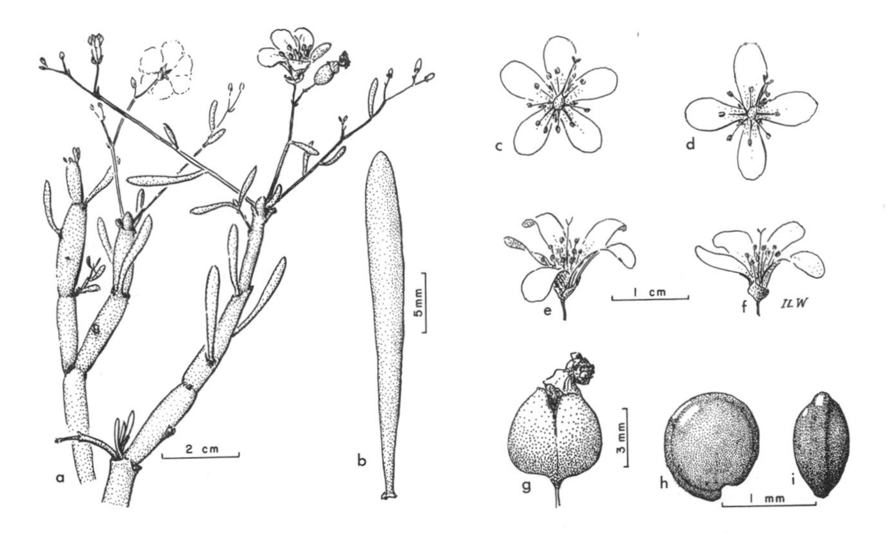

Fig. 203. *Calandrinia galapagosa.*

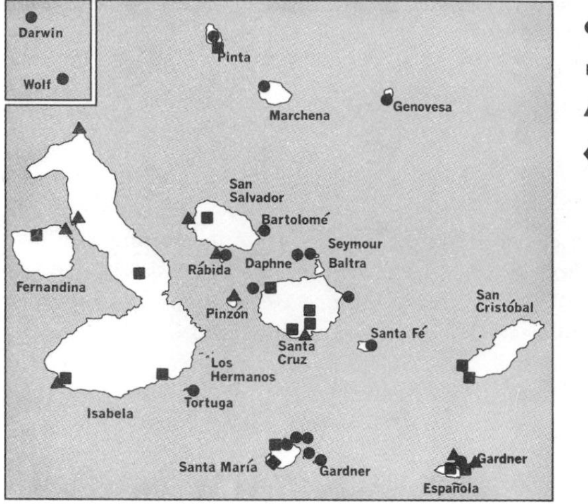

Portulaca L., Sp. Pl. 445. 1753; Gen. Pl. ed. 5, 204. 1754

Annual or perennial herbs; glabrous or fleshy. Leaves alternate or opposite; stipules setaceous. Flowers sessile, solitary or in few-flowered clusters, axillary or terminal. Tepals usually 5, rarely 4 or 6. Stamens 5 to numerous. Ovary half-inferior. Capsule dehiscing with a transverse lid. Seeds numerous, round-reniform; distinctly papillate.

More than 100 species in tropical and subtropical regions.

Tepals about 15 mm long; leaves fugacious; stems about 1 cm thick, very fleshy. . . . . *P. howellii*
Tepals 3–10 mm long; leaves persistent; stems 2–4 mm thick:
Seeds black, shining; operculum of capsule thin and translucent. . . . . *P. oleracea*
Seeds gray, lustrous; operculum thick and hard, not translucent. . . . . *P. umbraticola*

Portulaca howellii (Legr.) Eliass., Svensk Bot. Tidskr. 60: 428. 1966
Portulaca lutea var. *howellii* Legr., Comun. Bot. Mus. Mont. 2: 11. 1953.

Stems thick, fleshy, often 1 cm thick, decumbent to ascending; leaves cuneate to obovate, to 3 cm long, 1.5 cm wide; flowers large (to 4 cm broad when fully open); tepals 15 mm long or more, yellow; capsule 7–8 mm long, 6–7 mm broad; operculum firm, not translucent, with pronounced apical tumefaction; seeds about 1 mm in diameter, black, with star-shaped papillae arranged in rows.

Endemic; distributed over much of the archipelago; now apparently entirely restricted to small rocky islets, possibly as a result of heavy browsing by introduced animals, particularly the goats.* Specimens examined: Bartolomé, Caldwell, Cham-

* Isla Champion, which lies about 1 km off the northeasterly shore of Isla Santa María, has not been subjected to ravages by goats or other feral stock, hence the plants grow on moderate slopes instead of on the faces and rims of cliffs. The plants of *P. howellii* were much larger than those seen at any other locality, one being fully 1 m across and about 4 dm tall, covered with a profusion of large, yellow flowers.—I.L.W.

Fig. 204. *Portulaca howellii* (*a–c*). *Talinum paniculatum* (*d–f*).

pion, Daphne Major, Darwin, Eden, Enderby, Gardner (near Española), Gardner (near Santa María), Genovesa, Las Plazas, Marchena (by senior author), Pinta, Rábida, Santa Fé, Santa María, Seymour, Tortuga, Wolf.

The species bears flowers and leaves during only a few weeks. It passes the rest of the year as naked, gray, smooth stems.

See Fig. 204 and color plate 79.

Portulaca oleracea L., Sp. Pl. 445. 1753

Stems glabrous, often profusely branched, prostrate or decumbent; leaves cuneate to obovate, fleshy and glabrous, 1–2.5 cm long, 0.5–1.5 cm wide; flowers axillary; tepals 3–8 mm long, yellow; capsule circumscissile, enclosed by the 2 partially united pseudosepals (involucral bracts); upper part of pseudosepals tightly enclosing operculum and falling away together with operculum at maturity; operculum thin and translucent, usually with an apical tumefaction containing 1 or 2 seeds; seeds about 0.6 mm in diameter, black, shining, surface with rows of small papillae.

Nearly cosmopolitan in distribution, though a native of the Old World; well distributed in several of the Galápagos Islands, especially along trails and near habitations. Specimens examined: Española, Fernandina, Gardner (near Española) (by senior author), Isabela, Pinta, San Cristóbal, San Salvador, Santa Cruz, Santa María.

See color plate 78.

Portulaca umbraticola HBK., Nov. Gen. & Sp. Pl. 6: 72. 1823

Stems glabrous, prostrate to erect; leaves lanceolate, elliptic to obovate, 1–2 cm long, 0.4–1 cm wide; flowers terminal or axillary, sessile or nearly so; tepals about 10 mm long, purplish to yellowish; capsule circumscissile, enclosed by the 2 pseudosepals (involucral bracts); upper part of pseudosepals not tightly enclosing operculum, falling early; operculum thick and hard, not translucent, usually with minute apical tumefaction devoid of seeds; seeds about 0.8 mm in diameter, gray, lustrous, with large, star-shaped papillae.

The southern United States to Argentina; on lava slopes in the Galápagos. Specimens examined: Española, Gardner (near Española), Fernandina, Isabela, Pinzón (by senior author), Rábida, San Salvador, Santa Cruz.

Talinum Adans., Fam. Pl. 2: 245. 1763

Herbs, sometimes suffrutescent. Leaves flat, entire, without stipules. Inflorescence racemose or paniculate. Involucral bracts fugacious. Tepals 5, free or united at base. Stamens 10–30. Ovary free. Style trifid. Capsule dehiscing with 3 valves. Seeds with strophioles.

About 50 species in tropical and subtropical regions.

Talinum paniculatum (Jacq.) Gaertn., Fruct. 2: 219. 1791

Portulaca paniculata Jacq., Enum. Pl. Carib. 22. 1760.

Stems erect, somewhat shrubby, 3–8 dm high; leaves obovate, tapering toward base, 5–10 cm long, 3–4 cm wide; petioles 0.5–1 cm long; inflorescence a terminal panicle whose branches are cymes; involucral bracts round-triangular, membranous, about 2 mm long, deciduous; tepals rose or yellow, ephemeral, 2.5–4 mm long; stamens 15–20; capsule globose, about 3 mm wide, pale green; seeds numerous, 0.8–1 mm in diameter, dark reddish brown to black, shining, the surface elongate-reticulate; funiculus prominent.

A neotropical species; in the Galápagos Islands hitherto known from a single collection, and possibly a recent introduction. Specimen examined: Santa María.

See Fig. 204.

PUNICACEAE. POMEGRANATE FAMILY*

Shrubs or small trees, somewhat spiny, with opposite, exstipulate leaves, and twigs caducously 4-angled. Flowers perfect and regular, terminal and solitary or in cymes, perigynous-epigynous, the hypanthium short-tubular or urceolate, adnate to ovary. Sepals 5–8, valvate, persistent. Petals 5–7, imbricate and crumpled in bud, inserted on rim of hypanthium. Stamens many in several whorls on upper half of hypanthium tube, their filaments distinct. Pistil 1. Ovary inferior, 3–9-loculed. Ovules many. Fruit a leathery berry. Seeds with fleshy or juicy testa. Embryo straight; endosperm none.

A single genus of 2 species, native to semitropical Asia but widely naturalized and cultivated in all tropics and many subtropical regions.

Punica L., Sp. Pl. 472. 1753; Gen. Pl. ed. 5, 212. 1754

Characters of the family.

Punica granatum L., Sp. Pl. 472. 1753

Spiny shrub or small tree to 6 m high, usually branched from the base, with opposite or fasciculate, entire, elliptic to lanceolate leaves 2–6 cm long; large red flowers, thick, fleshy, persistent calyx lobes; lanceolate petals inserted in throat of hypanthium; and a red, globose, thick-skinned berry 5–10 cm in diameter crowned by the persistent calyx.

This plant has been collected on Isla Santa María, and observed on San Cristóbal and Santa Cruz. All shrubs of this species in the Galápagos were introduced, and are always near dwellings. None has been seen as an escape away from habitations.

See Fig. 205.

RHAMNACEAE. BUCKTHORN FAMILY†

Shrubs, small trees, or rarely vines, often with thorny branches, and simple, alternate or less commonly opposite leaves. Stipules small and deciduous, or corky and persistent. Flowers small, regular, perfect or polygamous, usually in axillary or terminal panicles or cymes, less commonly solitary or in pairs. Calyx cup-shaped, campanulate, or short-tubular, 4- or 5-toothed. Petals 4 or 5, inserted near or on rim

* Contributed by Duncan M. Porter.

† Contributed by Ira L. Wiggins.

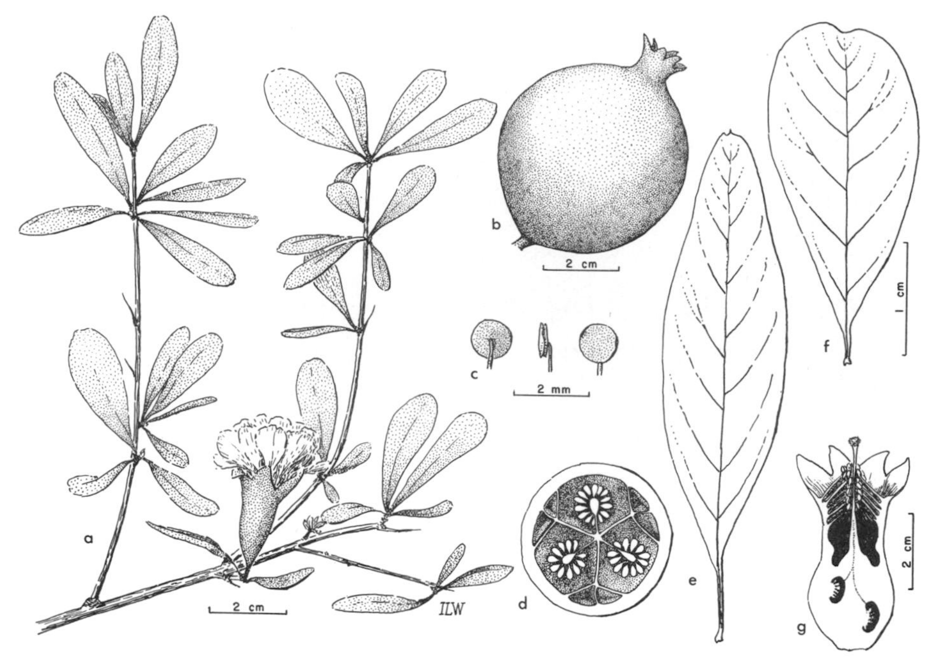

Fig. 205. *Punica granatum.*

of calyx, plane or cucullate, or lacking. Stamens 4 or 5, opposite petals; anthers versatile, short. Disk thin to fleshy, margin often undulate or lobed. Ovary sessile, 2–5-celled, free from disk or partially immersed in it. Ovules solitary in each locule, anatropous. Fruit a drupe, berry, or capsule, 2–5-celled. Seeds plano-convex to trigonous in cross section; endosperm fleshy or sometimes lacking; embryo large; cotyledons flat.

About 45 genera and 550 species, almost cosmopolitan in distribution, particularly well represented in the Western Hemisphere.

Plants with tendrils at bases of inflorescences, spineless; young growth densely brown-tomentose; inflorescences spicate-racemose; fruits dry, cocci narrowly winged at angles . . . . *Gouania*

Plants without tendrils, spiny-branched; young growth glabrous or essentially so; inflorescences not spicate, flowers borne in small fascicles; fruit a fleshy berry . . . . *Scutia*

Gouania Jacq., Sel. Stirp. Amer. Hist. 263. 1763

Usually scandent to climbing shrubs with tendrils borne on the peduncles. Leaves alternate, entire or dentate. Stipules deciduous or rarely persistent. Inflorescences terminal or axillary, of racemes or spikes. Flowers polygamodioecious; calyx tube short-obconic, adnate to ovary; lobes 5; petals 5, cucullate; disk 5-angled or 5-horned; stamens 5, more or less enfolded by petals. Ovary superior but appearing inferior owing to adnation to disk, 3–5-celled; styles 3-lobed or -parted. Fruit coriaceous, crowned by the persistent calyx, narrowly to strongly 3-winged, the 3 cocci indehiscent but separating from central axis and from each other.

About 40 species, distributed in tropical and subtropical areas of the Old and New Worlds. The genus is badly in need of monographic study.

Gouania polygama (Jacq.) Urban, Symb. Antill. 4: 378. 1910

Rhamnus polygama Jacq., Enum. Pl. Carib. 17. 1760.
Gouania tomentosa Jacq., Sel. Stirp. Amer. Hist. 263. 1763.

Mostly vinelike with slender branches, often climbing to 5 m or more; twigs and leaves densely (or leaves rarely sparsely) brown-tomentose with stiff, spreading, simple to few-rayed hairs; stipules lance-triangular, attenuate, 3–6 mm long, ciliolate, caducous, hirsute on outer surface, subglabrate within; petioles somewhat stoutish, 5–25 mm long, tomentose or hirtellous; leaf blades ovate to broadly elliptic, obtuse, rounded or truncate at base, short-acuminate apically, 5–12 cm long, 2–6 cm wide, dark green and sparsely to moderately pubescent above, brownish and more densely tomentose beneath, margins shallowly and coarsely crenate-serrate, tip of each tooth cartilaginous; veins 4–7 on each side, arcuately curved upward, prominulous; inflorescences terminal on branchlets and axillary to upper leaves, simple or paniculately branched, the branches slender, 6–15 cm long, spicate-racemose, flowers in fascicles along axis; bracts lance-linear, 2–4 mm long, mostly deciduous; calyx lobes triangular, about 1 mm long and wide, tomentose without, glabrous within; petals 1–1.4 mm long, cucullate, each partially enfolding a stamen; stamens inserted near margin of calyx tube, filaments dilated basally; disk 5-lobed, conspicuous, almost equaling calyx tube, its margin often 2- or 3-toothed at tip of each lobe; fruits about 5–8 mm long, 6–12 mm broad, each coccus bearing a thin wing 1–1.5 mm wide along its outer angle, outer surface inconspicuously reticulate-veined and sparsely puberulent to glabrate on the rounded back, pale brown; seeds obovoid-compressed, about 4 mm long, 3 mm wide, brown and shining, slightly rounded abaxially, low-carinate adaxially.

In shrubby or openly forested areas, central Mexico and the West Indies to northern South America. Reported from the Galápagos Islands by Christophersen on the basis of 2 collections by Rorud (*Nos. 126, 127*), obtained near Bella Vista, Isla Santa Cruz. According to Christophersen (1932, 82), Borghild Rorud saw the plant

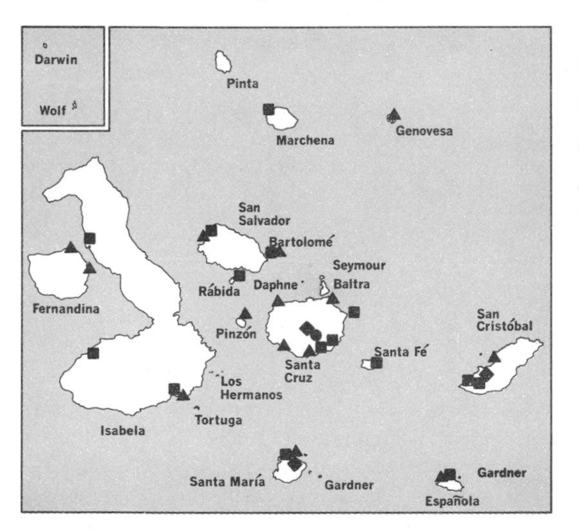

● *Gouania polygama*
■ *Scutia pauciflora*
▲ *Rhizophora mangle*
◆ *Citrus limetta*

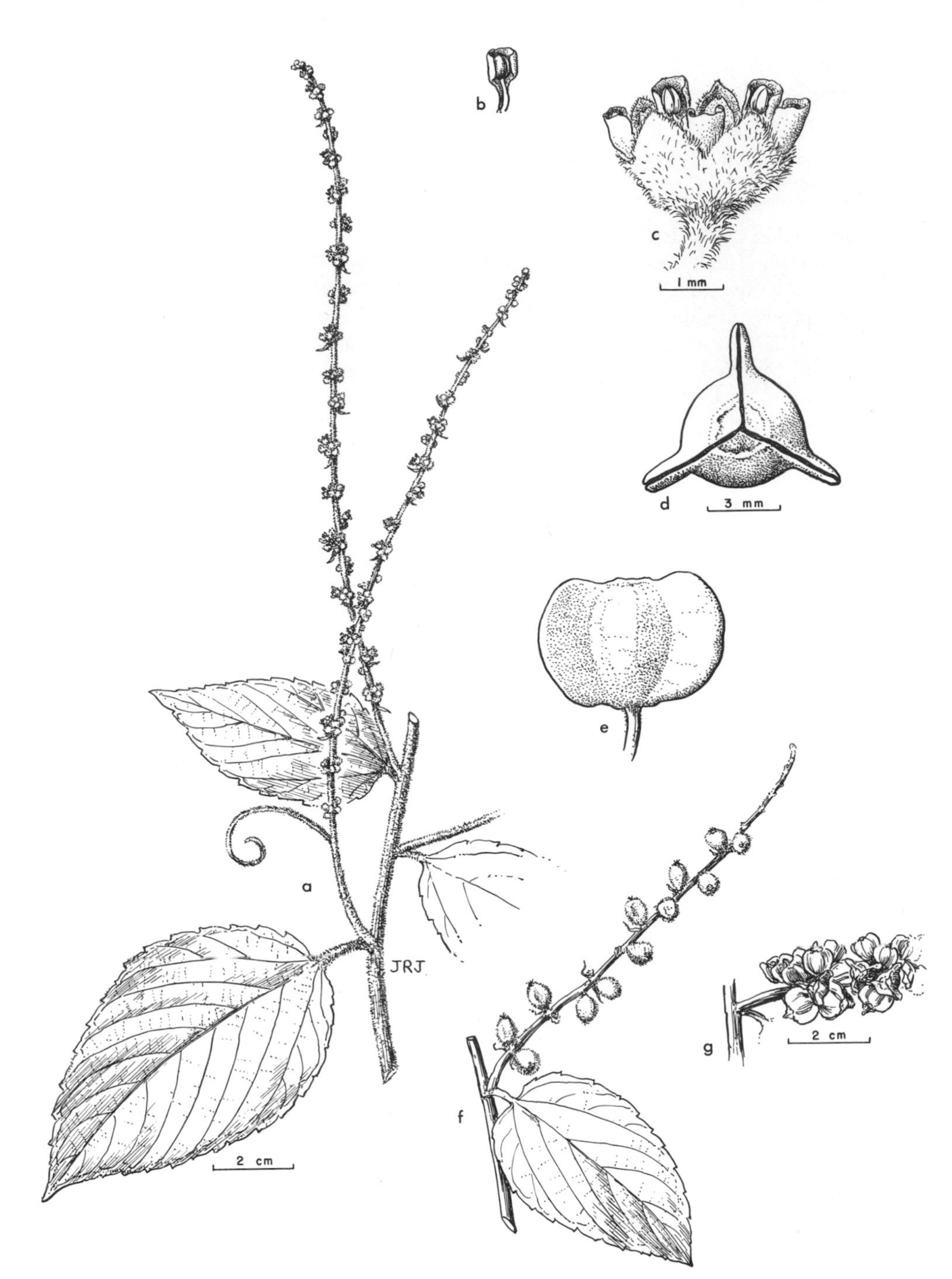

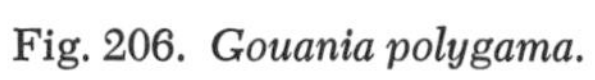
Fig. 206. *Gouania polygama.*

on Isla Santa María also, but apparently did not collect it at that locality. My description is drawn from Mexican and West Indian material.

See Fig. 206.

Scutia Comm. ex Brongn., Ann. Sci. Nat. Bot. 10: 362. 1827

Spiny shrubs with more or less angled branches, coriaceous, alternate (or rarely opposite) leaves, and small, 5-merous flowers in axillary fascicles or small umbels. Calyx tube hemispheric to turbinate, its lobes acute, thickened apically. Petals erect, plane or cucullate, equaling or slightly exceeding stamens. Disk basal, its margins free, undulate and lobed. Ovary free or slightly immersed at base, 2–4-celled. Fruit globose to obovoid, slightly fleshy, with 2–4 pyrenes, basally enclosed in the persistent calyx tube.

About a dozen species in the Southern Hemisphere.

Scutia pauciflora (Hook. f.) Weberb., Field Mus. Nat. Hist. Publ. Bot. 8: 84. 1930

Discaria pauciflora Hook. f., Trans. Linn. Soc. Lond. 20: 229. 1847.
Scypharia pauciflora Miers, Contr. Bot. 1: 301, *pl. 42.* 1861 (as *parviflora*).
Discaria parviflora Hook. f. ex Miers, loc. cit.

Low, rounded shrub 0.5–2.5 m tall, with green oppositely or suboppositely spined, intricately branched stems and twigs; leaves usually alternate but rarely opposite, caducous, 6–45 mm long, 3–18 mm wide, elliptic to oblong, broadly lanceolate or lance-elliptic, obtuse to rounded at base, acute to apiculate at apex, entire, minutely and sparsely puberulent beneath when young, soon glabrous on both sides, coriaceous, often more or less deflexed; petioles 1–2 mm long, adaxially canaliculate; stipules ovate, attenuate, 1–1.5 mm long, dark brown; spines terete, opposite or subopposite, 1–2.5 mm thick, 1–6 cm long, smooth or slightly striate, green except

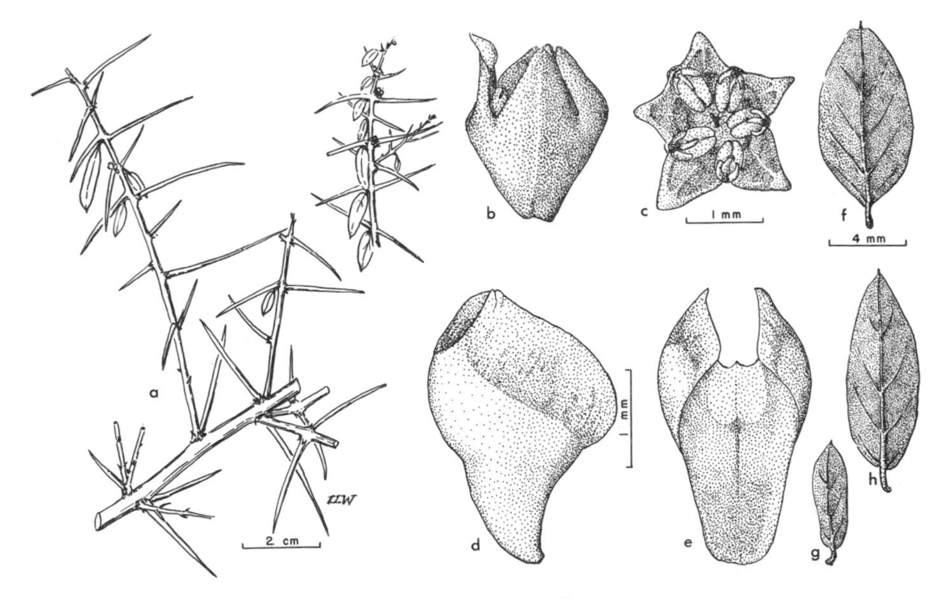

Fig. 207. *Scutia pauciflora.*

at dark brown tip, those opposite a leaf bearing 3 small bractlets similar to stipules near base; flowers sessile, in small fascicles of 2–4, or sometimes solitary, the fascicles in axils of lateral bracts near base of a spine, not in axils of foliage leaves, 2–2.4 mm across, 2–2.6 mm long at anthesis, pale greenish yellow with brown markings within and at tips of calyx lobes; calyx broadly campanulate, lobes deltoid-ovate, thickened at tips, 0.8–1 mm long, nearly as wide; petals obcordate-cucullate, 0.6–0.7 mm long, bifid about one-third of way to base, erect or slightly inflexed, closely enfolding stamens after anthesis, yellowish with brownish streaks and spots on terminal two-thirds of blade; stamens inflexed to horizontal before anthesis, later erect; anthers broadly elliptic, about 0.4 mm long, filaments rather thick; disk broadly 5-lobed, sinuses subtending petals and stamens, the margin thin, narrow, free part about 0.1–0.3 mm wide; fruit broadly obovate, about 5 mm long, usually bilocular, sometimes 3-locular, 2- or 3-seeded; seeds plano-convex or trigonous in cross section, about 1 mm thick.

On mainland coast northwest of Guayaquil; common and abundant near shore and less abundant to 400 m, often forming impenetrable thickets, on most of the islands. Specimens examined: Bartolomé, Española, Isabela, Las Plazas, Marchena, Rábida, San Cristóbal, San Salvador, Santa Cruz, Santa Fe, Santa María.

Svenson (1935, 242) reduced this species to synonymy under S. *spicata* (Willd.) Weberb., but I cannot agree with him that the material from the arid coast of Ecuador and from the Galápagos Islands is "identical" with the Peruvian S. *spicata*. The petals of S. *pauciflora* are more deeply notched and their blades more strongly cucullate than those of S. *spicata*, and the flowers of S. *spicata* are borne in small fascicles at several nodes on slender lateral twigs that are axillary to the lateral bracts on the main spine, whereas those of S. *pauciflora* are borne directly in the axils of the lateral bracts on the main spine at the base of the spine. Weberbauer's tentative assignment of S. *pauciflora* to Peru as well as to Ecuador has not been substantiated.

The bright red fruits, though very sour, are eaten by the children living around Academy Bay; they are taken also by at least 2 types of finches and by the native flycatcher, near the Darwin Research Station. Land iguanas on South Islet at Las Plazas were seen among the sturdier branches as much as 1.5 m above the ground, eating both the fruit and the tender tips of twigs, together with the attached young leaves.

See Fig. 207 and color plates 28 and 32.

RHIZOPHORACEAE. Mangrove Family*

Trees or shrubs, often with conspicuous prop- or stilt-roots; branches often swollen at nodes. Leaves opposite, decussate, simple, persistent, with glandlike corky protuberances sometimes present as small black spots beneath; stipules large, conspicuous, foliose, caducous, leaving an annular scar. Flowers bisexual or rarely unisexual or polygamous and plants monoecious, usually basally bracteolate, perigynous to epigynous; in axillary, simple or branched, usually cymose lax or condensed cluster-like, racemose, or fascicled inflorescences, rarely solitary. Hypanthium more or less adnate to and produced beyond ovary, rarely free; calyx lobes 3–6, valvate, persistent. Petals same in number as and alternate with calyx lobes, usually fleshy,

* Contributed by Duncan M. Porter.

entire, bifid, lacerate, or fringed apically, usually clawed, sometimes embracing 1–2 stamens, caducous or rarely persistent. Stamens usually twice as many as petals, in 1(2) whorl(s); filaments free, basally connate, or adnate to petals or hypanthium; anthers 4(many)-loculed, versatile, introrse, dehiscing longitudinally or by a large ventral valve. Disk fleshy, annular, crenate, flat or lobed, rarely absent. Gynoecium 2–5-carpeled, syncarpous; ovary inferior or rarely semi-inferior or superior, (1)2–12-loculed; ovules (1)2(many) per locule, pendulous, anatropous, the placentation axile or on a columella-like axis running from base to apex of 1-loculed ovaries; style 1(several), persistent; stigmas simple or more or less lobed. Fruit a coriaceous berry or drupe, or rarely a capsule, crowned by the persistent calyx. Seeds 1(many), pendulous, sometimes arillate, often viviparous; endosperm fleshy or small and soon obsolete; embryo straight or curved.

A family of 16 genera and about 120 species in the tropics and subtropics, occurring mainly in the Old World.

Rhizophora L., Sp. Pl. 443. 1753; Gen. Pl. ed. 5, 212. 1754

Trees or shrubs, with conspicuous bowed prop-roots at base and adventitious roots from lower branches; branching opposite, rarely alternate. Leaves entire, coriaceous, petioled, elliptic or oblong to obovate or lanceolate, punctate beneath; stipules convolute, surrounding unexpanded leaves, caducous. Flowers 4-merous, bisexual, regular, pedicellate or sessile, borne among or below leaves of the year in dichotomously branched 2–16-flowered cymes, rarely solitary; peduncles and pedicels subtended by connate, 2–4-lobed bracteoles. Calyx lobes 4(5), more or less free, coriaceous, prominently ribbed on inner surface. Petals 4, free, lanceolate, coriaceous, white or yellowish, equal to or shorter than calyx lobes, in bud folded lengthwise into depressions between ribs of adjacent calyx lobes, deciduous, margins densely villous to glabrous. Stamens 8(12), in 1 whorl; filaments short or absent; anthers elongate-deltoid, apiculate, areolate, thin membrane covering areolae dehiscing apically and along 2 longest sides. Gynoecium 2-carpeled; ovary semi-inferior, upper portion solid, lower portion 2(4)-loculed; ovules 2 per locule, axile, surrounded by spongy tissue; style filiform and apically forked or much reduced; stigma punctate. Fruit a conical, coriaceous berry. Seeds 1(2), placentation apparently parietal by displacement of placenta and abortive ovules by developing seed; testa thick, fleshy; endosperm abundant; embryo straight, green, viviparous; hypocotyl pendulous, clavate, penetrating apex of fruit and falling away; cotyledons thick, connate into a green tube, protruding from old fruit and withering with it after seed has fallen.

A genus of 6 or possibly more species, widely distributed along tropical coastal swamps and tidal streams.

Rhizophora mangle L., Sp. Pl. 443. 1753

Rhizophora americana Nutt., N. Amer. Sylv. 1: 95. 1842.
Rhizophora mangle var. *samoensis* Hochr., Candollea 2: 447. 1925.
Rhizophora samoensis Salvoza, Nat. Appl. Sci. Bull. Univ. Philippines 5: 220. 1936.

Tree or shrub 3–7 m tall; bark gray, twigs smooth, reddish; leaves obovate to elliptic, coriaceous and fleshy, usually slightly revolute, dark green above, paler beneath, glabrous, crowded at branch apices, 6.5–12.5 cm long, 3.5–6.5 cm wide, narrowed apically, more or less narrowed basally and decurrent into petiole, apex recurved and appearing obtuse, midrib prominent, veins obsolete; petioles slightly flattened,

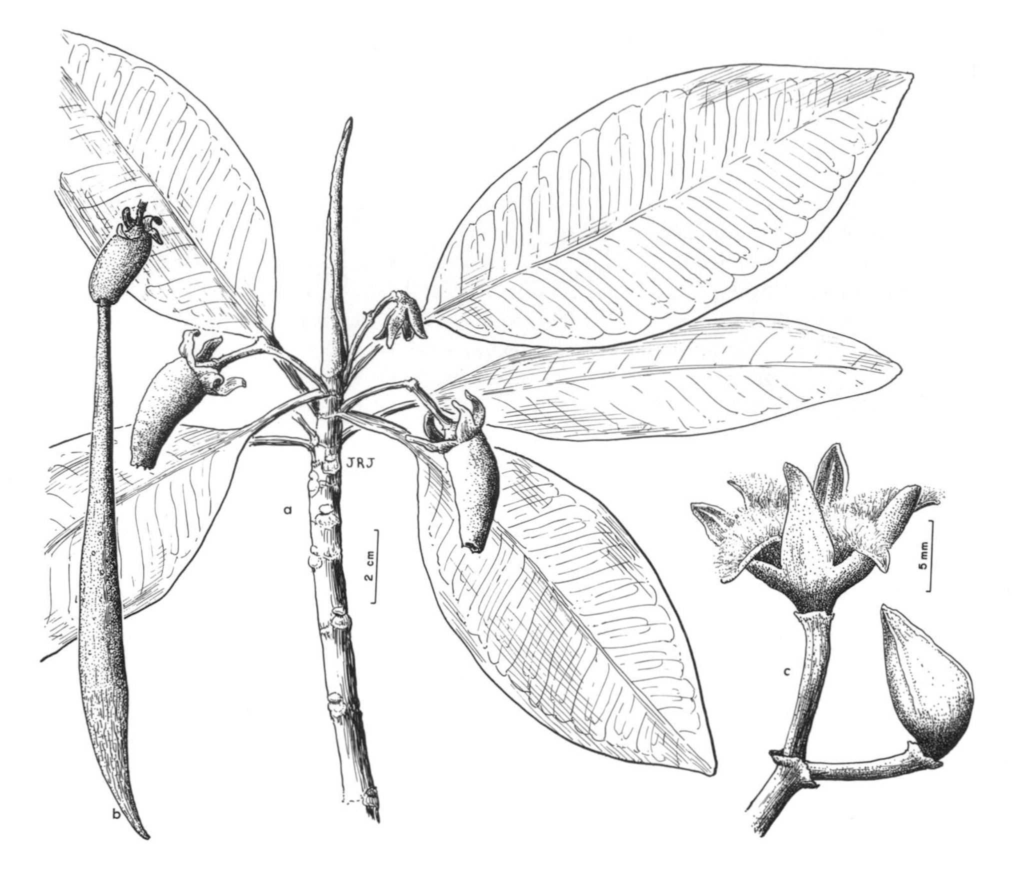

Fig. 208. *Rhizophora mangle.*

8–15 mm long, yellowish green; stipules conspicuous in bud, narrow, pointed, 2–5.5 cm long; cymes 2-flowered, borne among leaves of year; buds blunt, slightly 4-angled, slightly curved; peduncles 18–22 mm long; pedicels 8–10 mm long, recurved and flowers nodding; bracteoles 2, opposite, apical, persistent; hypanthium campanulate, 3.5–4 mm high; calyx lobes 4, lanceolate, obtuse, fleshy, thick, yellowish green, 1 slightly longer and hooded, 8–9 mm long, 3.5–4 mm wide, persistent, spreading or reflexed in fruit; petals white, turning brown, 7–8 mm long, about 2.5 mm wide, margins villous and revolute; stamens 8, 4 opposite petals, 4 opposite calyx lobes, anthers sessile, apiculate, about 6 mm long, deciduous; ovary conical at apex; style forked, persistent; berry narrowly conical, slightly curved, 1.5–2.5 cm long, the protruding radicle and hypocotyl becoming 15–25 cm long before falling.

Widely distributed along tropical shores from West Africa through the Caribbean region, from southern Florida to Brazil, and along the west coast of the Americas from Baja California to Ecuador, thence into Oceania and New Caledonia; probably commoner in the Galápagos than our specimens would indicate. Specimens examined: Bartolomé, Española, Fernandina, Genovesa, Isabela, Pinzón, San Cristóbal, Santa Cruz, Santa María. Reported by Stewart from San Salvador.

Populations of *R. mangle* within the Pacific Basin have been recognized as a separate species (*R. samoensis*) by Salvoza (1936) and Gregory (1958). Hou (1960, 627), while acknowledging "a slight indication of racial differentiation," concludes that "the Pacific specimens reckoned to *R. samoensis* have generally a shorter style and ovary apex than those (cf. *R. mangle*) in the Caribbean, but there is a range of intermediates defeating the use of this character even for varietal distinction."

See Fig. 208 and color plates 1, 8–11, 14, 21–25, and 29.

RUTACEAE. Citrus Family*

Armed or unarmed trees, shrubs, or perennial herbs, sometimes scandent or xeromorphic; stems, leaves, flowers, and fruits usually glandular-punctate, glands containing aromatic volatile oils. Leaves alternate or rarely opposite, compound, 1–3-foliolate, or simple, persistent or deciduous; petioles sometimes winged; stipules absent. Flowers bisexual and/or unisexual, regular or rarely irregular, hyogynous, usually 3–5-merous, solitary and axillary or in various axillary or terminal, often cymose inflorescences; plants dioecious, monoecious, or polygamous. Sepals almost always present, free or connate, usually imbricate. Petals free, rarely connate or absent, imbricate or valvate. Stamens as many as petals and in 1 whorl, or twice as many to numerous and in 2 whorls, outer whorl often shorter than inner or occasionally reduced to staminodes or transformed into ovaries; filaments free or connate, often conspicuously dilated or rarely appendaged basally; anthers 2(4)-loculed, introrse, dehiscent longitudinally, often glandular-tipped, versatile. Intrastaminal disk annular, cupular, or cushion-like, nectariferous, rarely obscure or absent. Gynoecium of (1)2–5(several) free or partially to completely connate carpels, occasionally rudimentary or absent; ovary sessile to stipitate, (1)2–5(several)-loculed; ovules 1–2 or more per locule, usually anatropous and epitropous, rarely apotropous, placentation axile or rarely parietal; styles basal, more or less lateral, or terminal, free, connivent, or connate, persistent or deciduous; stigmas free or connate, simple or lobed. Fruit of (1)2–5(several) follicles, or drupes, or a berry, capsule, samara, or schizocarp. Seeds 1 or 2 to several per locule, sessile or funiculate; endosperm present or absent; embryo relatively large, straight or curved; cotyledons planoconvex, sometimes convolute or rarely plicate; radicle superior.

Perhaps 150 genera and 1,500 species, widely distributed in the tropical and temperate regions of the world but most abundant in tropical America, South Africa, and Australia.

Leaves 1-foliolate, petioles articulated with leaflets; flowers white, 1 cm broad or larger; fruit a green, yellow, or orange hesperidium several cm in diameter. . . . . *Citrus*

Leaves odd-pinnate, leaflets 5–11, petioles not articulated; flowers dark, less than 5 mm broad; fruit a brownish follicle 3.5–5 mm in diameter. . . . . *Zanthoxylum*

Citrus L., Sp. Pl. 782. 1753; Gen. Pl. ed. 5, 341. 1754

Aromatic glabrous shrubs or small trees, with solitary axillary sharp, brown-tipped green thorns; young twigs angled, soon terete, green, older branches often thorn-

* Contributed by Duncan M. Porter.

less. Leaves alternate, 1-foliolate, persistent; petioles usually more or less winged and articulated with leaflets; leaflets subcoriaceous, usually thin, glandular-punctate, entire to serrate. Flowers usually relatively large, 2–5 cm broad, bisexual or staminate by more or less complete abortion of gynoecium, regular, axillary, solitary, paired, or in short corymbose cymes, often fragrant; plants dioecious or polygamous. Calyx shallowly cupular, 4- or 5-lobed, persistent. Petals (4)5(8), free, white, pink, or purplish pink, slightly fleshy, more or less oblong, strongly glandular-punctate, imbricate. Stamens 20–60, usually 4 times as many as petals, but sometimes 6–10 times as many, polydelphous or free; filaments linear-lanceolate, subulate upward, white, usually variously connate; anthers oblong or somewhat sagittate. Intrastaminal disk annular to cushion-like, short, supporting gynoecium. Gynoecium syncarpous; ovary sessile, subglobose and sharply distinct from the much narrower style, or truncate, or fusiform, or subcylindrical and merging gradually into a style nearly as thick as upper part of ovary, 8–18-loculed; ovules 4–8 (or more) per locule, in 2 collateral rows; style abruptly expanded into stigma, deciduous; stigma capitate, subglobose or oblate-spheroid, sometimes slightly lobed. Fruit a hesperidium, ellipsoidal and apex often mammillate, or pyriform to globular and apex sometimes depressed, pericarp differentiated into a leathery exocarp yellowish green to red-orange at maturity and dotted with numerous oil glands, a thick spongy white mesocarp, and a membranous endocarp filled with stalked fusiform pulp-vesicles containing a watery, acid to sweet, juice, the thin membranous radial locule walls often loosely coherent and easily separated from one another as well as from the spongy white fruit axis. Seeds ellipsoidal to obovoid, plump or flattened, more or less angular, apex sometimes beaked, usually several per locule at inner angle; testa coriaceous; endosperm absent; embryos 1 to many, white or greenish; cotyledons fleshy, plano-convex, often unequal.

Plants of southern and southeastern Asia and Malaysia and widely cultivated, often escaping in the warm regions of the world.*

Citrus limetta Risso, Ann. Mus. Hist. Nat. Paris 20: 195. 1813

Citrus medica subsp. *limonium* var. *limetta* Engler, Nat. Pflanzenfam. III, Abt. 4: 200. 1896.
Citrus limonia var. *limetta* Engler, op. cit. ed. 2, 19a: 338. 1931.
Citrus limettioides Tanaka, Jour. Indian Bot. Soc. 16: 236. 1937.

Small tree to 8 m high; branches irregular, conspicuously glandular-punctate, thorns stiff and numerous, becoming 1.5–7.5 cm long; bark brownish gray, relatively smooth; petioles narrowly but distinctly winged, spatulate, 8–29 mm long, wing 2–8 mm wide; leaflets obovate or ovate to elliptic, crenulate, 5.5–17 cm long, 2.8–8 cm wide, apex more or less obtusely acuminate, base rounded; flowers white in bud; calyx with 5 triangular, acuminate lobes about 4 mm long; petals white, oblong-lanceolate, 16–17 mm long, 6–7 mm wide; stamens 20–25; filaments about 10 mm long, connate in variable groups; anthers sub-basifixed; disk prominent, annular;

* Swingle (1943) ascribes 16 species to *Citrus*; Tanaka (1954) lists 145. Specific distinctions are complicated not only by hybridization within the genus and with closely related genera but also by such phenomena as nucellar polyembryony, rejuvenation by nucellar progeny of more or less senescent varieties long propagated asexually, and the spontaneous production of autotetraploids (cf. Swingle, 1943, 388–92). Failure to take into account such processes has led Tanaka to recognize an overabundance of species in *Citrus*.

Fig. 209. *Citrus limetta.*

ovary globose, apex depressed, glandular-punctate, 3 mm in diameter; style stout, 7–8 mm long, sharply distinct from ovary, soon deciduous; stigma globose, depressed apically; hesperidium globose, about 5 cm in diameter, skin light yellow at maturity, rind white, about 5 mm thick, pulp abundant, greenish, juice abundant, insipid, not acid.

Escaped from cultivation and widely naturalized in the tropics of the Old and New Worlds; very abundant in some areas in the Galápagos Islands. Specimens examined: San Cristóbal, Santa Cruz, Santa María.

The taxonomic disposition of the sweet lime is debatable. Tanaka (1937) applied the name *C. limettioides* to this taxon, reserving *C. limetta* for the sweet lemon. Other citrologists believe that sweet limes may be of hybrid origin, notably Webber (1943, 619–20), who wrote: "The best suggestion that can now be made concerning their relationship is that they are possibly hybrids of a type of lime like the Mexican [*C. aurantium* var.] with a type of sweet lemon like the Dorshape [*C. limonia* var.] or a sweet citron like the Corsican [*C. medica* var.]."

According to Koford (1966), the grapefruit (*C. paradisi* MacFadyen), lemon (*C. limonia* Osbeck), and sweet orange (*C. sinensis* [L.] Osbeck) are cultivated in the archipelago for their edible fruits. Only a specimen of the sweet orange has been seen. Other *Citrus* species that are found commonly in Latin America and may be expected as cultivated plants in the Galápagos Islands are the true lime (*C. aurantifolia* [Christm.] Swingle) and sour orange (*C. aurantium* L.). But the sweet lime appears to be the only *Citrus* that has become truly naturalized in the Galápagos Islands, being extremely common in some of the more mesic areas of Islas San Cristóbal and Santa María. Lemons, limes, and oranges have been reported as adventitious, but such reports seem to be based on *C. limetta* or to refer to cultivated specimens persisting where habitations have been abandoned. When the fruits of the sweet lime are green, they are easily confused with oranges or true limes. When ripe they look like lemons. However, when tasted their identity is evident. Although Latin Americans are fond of the fruits, "Just why it should be so esteemed is hard to determine, for to the northern palate it seems the most insipid and least appetizing of the citrus fruits" (Standley & Steyermark 1946, 406).

See Fig. 209.

Zanthoxylum L., Sp. Pl. 270. 1753; Gen. Pl. ed. 5, 130. 1754

Shrubs or trees, rarely scandent; often armed with pseudostipular prickles or corky excrescences; bark aromatic. Leaves alternate, compound, persistent or deciduous; petiole and rachis terete or angled, often winged, unarmed or prickly; leaflets opposite or alternate, entire or crenulate, frequently inequilateral. Flowers small, unisexual and/or bisexual, regular, white to greenish yellow; inflorescences axillary spikes or cymose fascicles, or terminal, sometimes corymbiform panicles; plants dioecious, monoecious, or polygamous. Sepals 3–10, or apparently absent, free to more or less connate, persistent or deciduous. Petals 3–8, free, imbricate or valvate. Stamens 3–8, absent in carpellate flowers, reduced to staminodes or transformed into ovaries; filaments free, unappendaged; anthers elliptic to ovate. Intrastaminal disk small or obscure. Gynoecium of 1(5) sessile to stipitate, free or connate carpels, rudimentary or absent in staminate flowers; ovary 1(5)-loculed; ovules 2 per locule, collateral, pendulous; styles sublateral, connate to divergent; stigmas capitate, free or connate. Fruit of 1–5 sessile to stipitate, coriaceous or fleshy, 2-valved, 1- or 2-seeded, free or basally connate, glandular-punctate follicles. Seeds obovoid, subglobose, or lenticular, often remaining attached to placenta at maturity; testa crustaceous, black, brownish, or reddish, shiny; endosperm fleshy; embryo axial, straight or curved; cotyledons flat, almost circular; radicle short.

About 200 species, primarily pantropical but extending into the North Temperate Zone in eastern Asia and North America.*

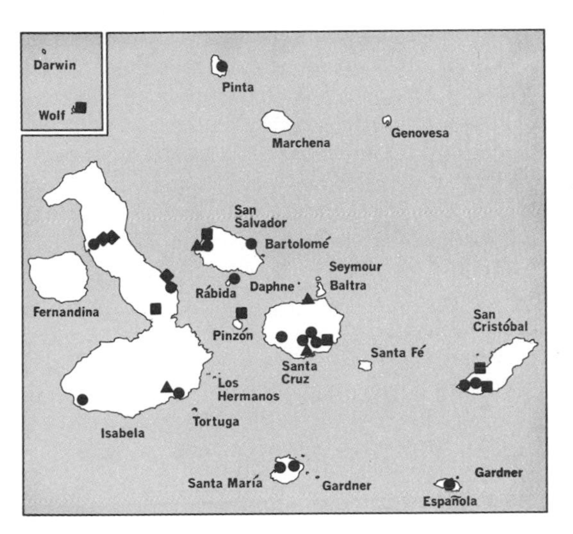

● *Zanthoxylum fagara*

■ *Cardiospermum corindum*

▲ *Cardiospermum galapageium*

◆ *Dodonaea viscosa* var. *galapagensis*

Zanthoxylum fagara (L.) Sarg., Gard. & For. 3: 186. 1890

Schinus fagara L., Sp. Pl. 389. 1753.
Fagara pterota L., Syst. Nat. ed. 10, 2: 897. 1759.
Pterota fagara Crantz, Inst. Rei Herb. 2: 417. 1766.
Fagara lentiscifolia Humb. & Bonpl. ex Willd., Enum. Pl. 165. 1809.
Zanthoxylum pterota HBK., Nov. Gen. & Sp. Pl. 6: 3. 1823.
Zanthoxylum lentiscifolium Anderss., Kongl. Svensk. Vet.-Akad. Handl. 1853: 244. 1855, non Champ. ex Benth. 1851.
Fagaras fagara Kuntze, Rev. Gen. Pl. 3^{2}: 34. 1898.
Fagara fagara Small, Fl. SE. U. S. 675. 1903.

Shrub or small tree to 10 m high; bark light gray; branches terete, glabrate, armed with paired subulate glabrous, dark, sharply hooked, deciduous, pseudostipular spines 2–6 mm long; twigs slender, commonly zigzag, light gray through reddish brown to dark brown, more or less puberulent, more or less angled, sometimes ridged between nodes; leaves odd-pinnate, 3.5–11 cm long, 2–5 cm wide, persistent; petiole winged, grooved above, more or less puberulent, 14–20 mm long, about 2 mm wide; rachis winged, grooved above, puberulent, 2–4.5 mm wide; leaflets 5–11, opposite or subopposite, obovate to broadly elliptic, inequilateral, acute, rounded

* The genus as here circumscribed is regarded by some as two separate genera, *Zanthoxylum* and *Fagara* L. (nom. conserv.), with the species below (as *Fagara pterota* L.) as the type of the latter. However, morphological evidence (cf. Brizicky, 1962b) indicates the recognition of a single genus. In this case, *Fagara* is treated as a subgenus (subgen. *Fagara* [L.] Triana & Planch.) or as a section (Sect. *Fagara* [L.] G. Don) of *Zanthoxylum*.

or retuse at apex, more or less cuneate or rarely rounded at base, crenulate, coriaceous, shiny above, paler beneath, darkly glandular-punctate, sessile or short-petiolulate, 9–29 mm long, 6–19 mm wide, lowest pair smallest; inflorescence a short axillary spike about 5 mm long, glabrous; flowers subsessile, unisexual, plants dioecious; staminate flowers with 4 broadly obovate imbricate sepals about 1 mm long, 4 obovate petals, and 4 stamens with subulate, glabrous filaments less than 1 mm long; anthers ovate; carpellate flowers similar to staminate, except bearing 2 glabrous, bean-shaped, basally connate carpels about 1 mm long, these sessile on an annular disk; follicles short-stipitate, obovoid, 3.5–5 mm in diameter, coriaceous, slightly rugose, darkly glandular-punctate, brownish, inner surface wrinkled, the persistent style forming a beak about 1 mm long; seeds solitary, globose, black, 4 mm in diameter, remaining attached by funiculus after valves dehisce.

Florida and Texas south through Mexico and the Caribbean region to Peru. Specimens examined: Española, Isabela, Pinta, Rábida, San Cristóbal, San Salvador, Santa Cruz, Santa María. Also reported from Fernandina (Robinson), Daphne Major (Stewart), and Gardner (near Española) (Robinson & Greenman).

See Fig. 210 and color plates 56 and 83.

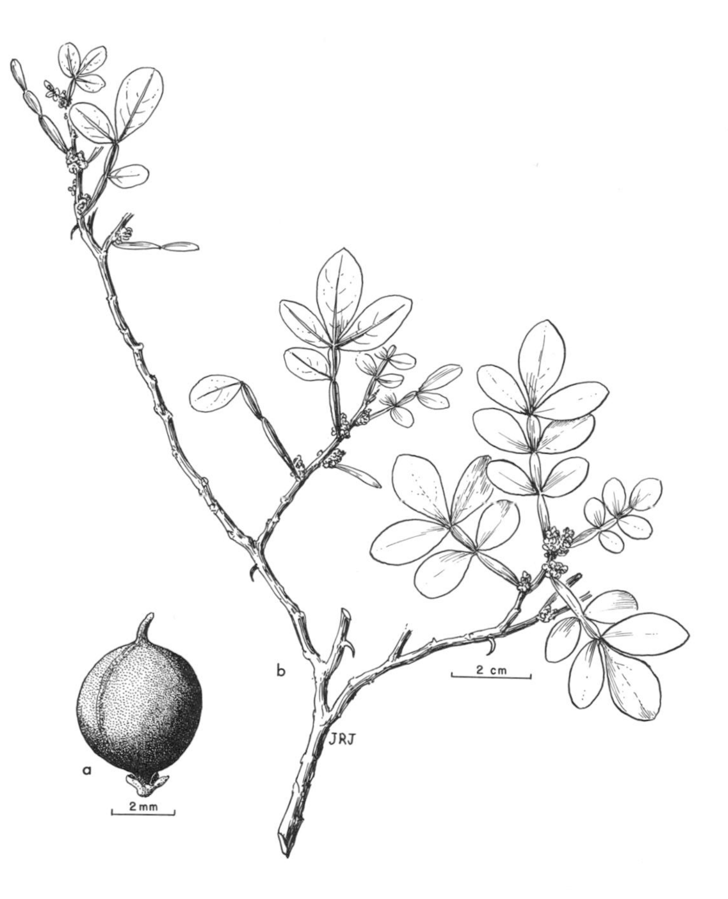

Fig. 210. *Zanthoxylum fagara.*

SAPINDACEAE. SOAPBERRY FAMILY*

Trees, shrubs, woody to rarely herbaceous climbing vines, or very rarely erect herbs. Leaves alternate, compound to decompound, often even-pinnate, rarely simple, persistent or deciduous; stipules rarely present. Flowers regular or rarely irregular, small, hypogynous, usually unisexual by abortion, occasionally also bisexual (plants polygamo-monoecious or polygamo-dioecious), in terminal and/or axillary thryses or simple or compound racemes, rarely solitary. Sepals 4 or 5, free or more or less connate, usually unequal, imbricate or valvate, deciduous or persistent. Petals 4 or 5, rarely absent, free, clawed, often with petaloid appendages on interior surface just above claw, imbricate. Disk usually extrastaminal, nectariferous, continuous and regular or unilateral and oblique. Stamens (6)7–8(12), short and sterile or absent in carpellate flowers; filaments free or basally connate, inserted interior to or on disk; anthers versatile, introrse or rarely extrorse, longitudinally dehiscent. Gynoecium syncarpous, (2)3(6)-carpeled, rudimentary in staminate flowers; ovary superior, locules same in number as carpels; ovules usually 1–2 per locule, campylotropous to anatropous, apotropous, or rarely epitropous, placentation axile; style short or elongate, sometimes more or less cleft; stigma simple or lobed, or stigmas 3. Fruit a drupe, berry, or capsule (sometimes winged), or a schizocarp splitting at maturity into drupelike, nutlike, or samaroid mericarps; seeds 1 per locule, often arillate; testa bony, crustaceous, or coriaceous, rarely fleshy; endosperm usually absent or scanty; embryo usually curved.

About 150 genera and 2,000 species, primarily in the New World; pantropical in distribution but a few species extending into warm temperate regions.

Plants suffrutescent scrambling vines climbing by tendrils.............. *Cardiospermum*
Plants erect trees or shrubs, tendrils absent:
  Leaves simple; fruit a 2- or 3-winged capsule.......................... *Dodonaea*
  Leaves compound; fruit a schizocarp of 1 or 2(3) wingless, obovoid drupelike mericarps .... *Sapindus*

Cardiospermum L., Sp. Pl. 366. 1753; Gen. Pl. ed. 5, 171. 1754

Herbaceous or somewhat shrubby vines, climbing by tendrils. Leaves alternate, biternate, sometimes also 3-foliolate with 3-lobed leaflets, or sub-bipinnate, petioled; leaflets usually coarsely toothed; stipules minute. Flowers small, irregular, unisexual by abortion, sometimes also bisexual, in axillary corymb-like reduced thyrses; peduncles bearing 2 opposite subapical tendrils. Sepals 4(5), 2 exterior about half as long as interior, imbricate. Petals 4, short-clawed, appendaged; appendages of 2 upper petals equilateral, hood-crested, petaloid, bearing a tongue-like downward-directed appendage below apex; appendages of 2 lower petals inequilateral, with a dorsal winglike crest. Disk extrastaminal, unilateral, with a gland opposite each upper petal. Stamens 8, deflexed, shorter and nonfunctional in carpellate flowers; filaments unequal, slightly connate basally, inserted on disk. Gynoecium 3-carpellate; ovary 3-loculed; ovules 1 per locule, anatropous, apotropous, ascendent; style short; stigmas 3, filiform, spreading. Fruit a membranous or subchartaceous inflated capsule, subglobose, obovoid, or turbinate, 3-angled and slightly 3-lobed,

* Contributed by Duncan M. Porter.

septicidally or loculicidally dehiscent. Seeds 1 per locule, subglobose, black, with an inconspicuous aril abscissing and leaving a whitish, reniform to semicircular pseudohilum near micropyle; testa thick, crustaceous; endosperm scanty; embryo curved; cotyledons fleshy, unequal; radicle short.

A pantropical but mainly New World genus of about 12 species; several species have been introduced and have become naturalized in warm temperate regions.

Ultimate divisions of leaves ovate to broadly elliptic, serrate, dentate, or lobed.......... .... *C. corindum*
Ultimate divisions of leaves narrowly to broadly lanceolate, entire or rarely bilobed basally .... *C. galapageium*

Cardiospermum corindum L., Sp. Pl. ed. 2, 1: 526. 1762

Cardiospermum corindum f. *villosum* Radlk. in Mart., Fl. Bras. 13^3: 447. 1897.*

Suffruticose vine scrambling over rocks and shrubs to height of 3–4 m; stems slender, ridged, puberulent or villosulous to hispidulous, the amount of pubescence variable; leaves biternate, dark green above, paler beneath, 3.5–14 cm long, 3–11 cm wide; petioles ribbed, puberulent; leaflets 3, each divided into 3 parts, ultimate divisions ovate to broadly elliptic, variously serrate, dentate, or lobed, acute to abruptly acuminate or rounded, cuneate to rounded basally, sparsely to densely puberulent or hispidulous, ciliate, lateral parts of each division smaller than terminal, terminal leaflets to 11.5 cm long, 9.5 cm wide; petiolules slightly winged, puberulent; peduncles spreading, puberulent, usually longer than leaves, to 10 cm long, bearing 2 strongly coiled subapical tendrils; flowers 5–11 mm broad; sepals 4, persistent, outer pair ovate, punctate-glandular, sparsely pubescent, 2 mm long, inner pair obovate, 3.5 mm long; petals white, 4.5 mm long, appendages white with yellow apices, about two-thirds as long as petals, upper petaloid, lower club-shaped and slightly flattened, free; glands on upper side of disk rounded, about 1 mm high; stamens unequal, nonfunctional in carpellate flowers; filaments pubescent, about 2 mm long; anthers sagittate, versatile; ovary pubescent, rudimentary in staminate flowers; style glabrous, exserted, persistent; capsule obovoid to globose, membranaceous, pubescent to glabrate, sharply 3-angled, light brown at maturity, septicidally dehiscent, 2.5–3.5 cm high, 1.5–3 cm wide; seeds glabrous, globose, 3 mm in diameter, pseudohilum subcordate, 2 mm high, 2.5 mm wide.

Pantropical in the Old and New Worlds; indigenous in the Galápagos Islands. Specimens examined: Isabela, Pinzón, San Cristóbal, San Salvador, Santa Cruz, Wolf.

See Fig. 211.

Cardiospermum galapageium Robins. & Greenm., Proc. Amer. Acad. 32: 38. 1896

Suffruticose vine, the stems slender, ridged, densely to sparsely puberulent, scrambling over rocks and shrubs to height of about 3 m; leaves biternate, dark green above, paler beneath, 2.5–9 cm long, 1.5–8.5 cm wide; petioles ribbed, puberulent; leaflets 3, each in 3 parts, ultimate divisions narrowly to broadly lanceolate,

* Clarification of the synonymy of this rather variable species, in which Radlkofer (1932) described 14 forms, awaits a modern revision of the genus, which may also answer the question of whether *C. corindum* is specifically distinct from *C. halicacabum* L.

Fig. 211. *Dodonaea viscosa* var. *galapagensis* (*a–c*). *Cardiospermum corindum* (*d*).

entire or rarely bilobed basally, acuminate, base attenuate to rounded, sparsely to densely puberulent, ciliate, the lateral parts of each division smaller than the terminal, terminal leaflet to 7 cm long, 4 cm wide; petiolules slightly winged, puberulent; peduncles bearing 2 strongly coiled subapical tendrils, spreading, puberulent, usually longer than leaves, to 8 cm long; sepals 4, persistent, outer pair ovate, punctate-glandular, ciliate, 2 mm long, inner pair obovate, 3–4 mm long; petals white, punctate-glandular, 5 mm long, appendages white with yellow tips, about two-thirds as long as petals, upper petaloid, lower club-shaped and slightly flattened, free or connate; glands on upper side of disk rounded, 1 mm high; stamens unequal, nonfunctional in carpellate flowers, filaments pubescent, slightly winged, 2.5–3 mm long; anthers sagittate, versatile; ovary obovoid, sharply 3-angled, densely pubescent, 5 mm high, rudimentary in staminate flowers; style glabrous, exserted, persistent, 1.5 mm long; stigma 1 mm long; capsule obovoid to ovoid, membranaceous, pubescent to glabrate, 3-angled, light brown at maturity, septicidally dehiscent, 2 cm high, 1 cm wide; seeds glabrous, 4 mm in diameter, pseudohilum subcordate, 2 mm in diameter.

Endemic; in forested areas from near sea level to about 400 m. Specimens examined: Isabela, San Salvador, Santa Cruz.

The specimen represented by *Howell 9658* (CAS) from James Bay, San Salvador, is morphologically intermediate between *C. galapageium* and *C. corindum*, and may represent a hybrid between the two. *Howell 9084* (CAS, GH), *Snow 497* (DS), and *Stewart 1941* (CAS) all vary toward *C. corindum* in leaf shape. This variation perhaps represents introgression of genes from the latter into populations of *C. galapageium*.

See color plate 90.

Dodonaea Mill., Gard. Dict. abr. ed. 4, 1: 1754

Shrubs or small trees, sometimes low and ericoid; usually resinous-viscous. Leaves alternate or rarely subopposite, simple or pinnate, entire or repand to more or less toothed, usually appearing varnished, more or less covered with resiniferous glands, petioled to sessile, persistent; stipules absent. Flowers regular, bisexual or unisexual by abortion, solitary or in terminal or axillary thyrses or panicles; plants monoecious, dioecious, or polygamodioecious; pedicels subtended by small bracts. Sepals (3)4–5(7), more or less connate basally, valvate or narrowly imbricate, usually deciduous. Petals absent. Stamens 6–10(12) in perfect and staminate flowers, sterile or absent in carpellate flowers; filaments free, very short; anthers oblong to linear-oblong, quadrangular, apiculate, much longer than filaments. Disk intrastaminal, prominent in perfect and carpellate flowers, rudimentary in staminate flowers. Gynoecium (2)3(6)-carpellate, rudimentary in staminate flowers; ovary (2)3(6)-lobed and -loculed, slightly raised on disk, mostly closely covered with resiniferous glands; ovules 2 per locule, superposed, campylotropous; style filiform, 2–3 times as long as ovary or longer, apex sometimes 2–4-lobed, deciduous; stigmas (2)3(4), small. Capsule chartaceous, membranaceous, or coriaceous, (2)3(6)-lobed, -loculed, and -winged, septicidally or septifragally dehiscent, wings papery. Seeds 1 or 2 per locule, lenticular to obovoid or subglobose; testa crustaceous; endosperm absent; embryo coiled; cotyledons linear; radicle elongate.

About 60 species, mostly in Australia but one pantropical and subtropical and several others described from Java, Hawaii, and Madagascar; a large number of varieties and forms have been described.

Leaves broadly elliptic to broadly oblong, tomentulose, the base cuneate; blade 29–58 mm long, 14–32 mm wide. . . . . *D. viscosa* var. *galapagensis*
Leaves oblanceolate or elliptic, glabrous or base of midrib tomentulose, the base attenuate to petiole; blade 34–55 mm long, 8–16 mm wide. . . . . *D. viscosa* var. *spatulata*

Dodonaea viscosa var. **galapagensis** (Sherff) Porter, Occ. Papers Calif. Acad. Sci. 81: 3. 1970

Dodonaea eriocarpa var. *obtusior* f. *galapagensis* Sherff, Amer. Jour. Bot. 32: 207. 1945.

Shrub or small tree 1–2 m high, dotted with globular orange sessile glands, these especially common on branchlets, petioles, and leaves; branches slender, dark-ferruginous, becoming gray, bark with numerous longitudinal cracks, becoming stringy; branchlets angled, tomentulose; leaves simple, more or less viscid, dark green or brownish above and paler beneath in dried specimens, tomentulose, broadly elliptic to broadly oblong, entire, subapically obtuse and apically acuminate, base cuneate, punctate-glandular, margins revolute, blade 29–58 mm long, 14–32 mm wide; flowers bisexual, 4–5 mm broad; inflorescences terminal, tomentose throughout; pedicels 3–4 mm long in flower, longer in fruit; sepals 4, broadly ovate, obtuse or acute, tomentose, concave, about 3 mm long, 2 mm wide, a bit shorter than stamens; stamens 8 or 9, inserted at edge of disk; anthers oblong, 2.5 mm long, apex pubescent; ovary ovoid, 3-lobed, pubescent, yellow-brown, 1.5–2.5 mm high; style dark brown, furrowed, not twisted, glabrous, 2–3 mm long, 3-lobed; capsule 2- or 3-winged, stramineous, punctate-glandular, pubescent, wings papery, conspicuously veined, continuous from rounded apex to truncate base, 12–16 mm high, 18 mm wide including wings; seeds dark gray, dull, lenticular, 3 mm high, 2.5 mm wide, 1 mm thick.

Endemic; growing among rocks and on lava flows, mostly above 900 m. Specimens examined: Isabela.

Reported from the Galápagos Islands by Robinson (1902) and Stewart (1911) as *D. viscosa* var. *spatulata.*

See Fig. 211.

Dodonaea viscosa var. **spatulata** (Smith in Rees) Benth., Fl. Austral. 1: 476. 1863

Dodonaea spatulata Smith in Rees, Cyclop. 12. 1809.
Dodonaea eriocarpa var. *vaccinioides* Sherff, Amer. Jour. Bot. 32: 210. 1945.
Dodonaea viscosa var. *arborescens* f. *spatulata* Sherff, op. cit. 214.

Shrub to 1.6 m tall, dotted with globular orange sessile glands, these especially numerous on branchlets, petioles, and leaves; branches slender, dark-ferruginous, becoming gray, bark with numerous longitudinal cracks, stringy; branchlets more or less tomentulose; leaves simple, viscid, green and shiny above, paler and slightly duller beneath, glabrous or midrib tomentulose at base, oblanceolate or elliptic, entire or repand, apex acute to short-acuminate or obtuse, base attenuate to petiole,

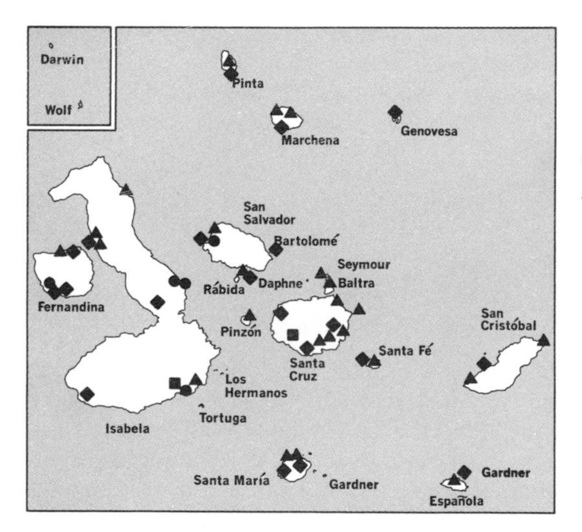

● *Dodonaea viscosa* var. *spatulata*
■ *Sapindus saponaria*
▲ *Castela galapageia*
◆ *Waltheria ovata*

punctate-glandular, margins revolute, blade 34–55 mm long, 8–16 mm wide; flowers bisexual, 4–5 mm broad; inflorescences terminal or axillary, glabrous except for pedicels, these glandular, tomentulose, 3–6 mm long in flower, 6–8 mm long in fruit; sepals 3 or 4, narrowly to broadly oblong, obtuse, apex and margins puberulent, concave, 2.5–3 mm long, 1–1.5 mm wide, a bit longer than stamens, becoming reflexed, more or less persistent; stamens 6–8, twice as many as sepals, inserted at edge of disk, becoming reflexed; anthers oblong, yellow or greenish, 2 mm long, apex pubescent; disk annular; ovary 3-lobed, pubescent, yellow-brown, 1–2 mm high; style dark brown, twisted, glabrous, 3–5 mm long; capsule (2)3-winged, stramineous to purplish, punctate-glandular, pubescence masked by viscid exudate, the wings papery, conspicuously veined, continuous from rounded apex to truncate base, 8–13 mm high, 10–18 mm wide including wings.

Nearly pantropical in distribution; apparently indigenous to the Galápagos Islands. Specimens examined: Fernandina, Isabela, San Salvador.

This variety was reported from the archipelago by Robinson (1902) and Stewart (1911) simply as *D. viscosa*.

Sapindus L., Sp. Pl. 367. 1753; Gen. Pl. ed. 5, 171. 1764

Trees or shrubs. Leaves alternate, even- or odd-pinnate (rarely simple), petiolate, persistent or deciduous; leaflets 4–18, entire; rachis winged or wingless; stipules absent. Flowers regular, unisexual by abortion and plants monoecious or dioecious, borne in terminal thyrses; bracts and bracteoles minute, deciduous. Sepals 4 or 5, connate basally, unequal, 2 outer smaller than 2 inner, imbricate, deciduous. Petals 4 or 5, equal, exceeding sepals, with bifid scale or 2 scales above claw on anterior surface, or scales absent. Disk annular, fleshy, lobed. Stamens 8(10), exserted in staminate flowers, short and anthers rudimentary in carpellate flowers; filaments inserted within disk. Gynoecium (2)3(4)-carpeled; ovary sessile, (2)3(4)-lobed and -loculed; ovules 1 per locule, ascending, campylotropous, apotropous; style short, columnar, (2)3(4)-furrowed; stigma small, 3-lobed. Fruit a schizocarp of 1–3 drupelike, subglobose or obovoid, yellow to black, 1-loculed mericarps and rudiments of aborted locules, pericarp fleshy, resinous, crustaceous when dried. Seeds 1 per locule, globose or obovoid, not arillate; testa bony, smooth, black or dark brown; endosperm absent; embryo curved; cotyledons fleshy, superposed, the dorsal incurved and almost enclosing the ventral; radicle short, inferior.

About 13 species, occurring in the tropics and subtropics of Asia, Oceania, and the New World.

Sapindus saponaria L., Sp. Pl. 367. 1753

Sapindus forsythii DC., Prodr. 1: 607. 1824.
Sapindus peruvianus Walp., Nov. Act. Acad. Caes. Leop. 19. Suppl. 1: 312. 1843.
Sapindus peruvianus var. *meyenianus* Walp., loc. cit.
Sapindus peruvianus var. *dombeyanus* Walp., loc. cit.
Sapindus saponaria f. *genuinus* Radlk. in Mart., Fl. Bras. 13^3: 517. 1900.

Tree; branchlets slender, striate, yellow-green to light gray, dotted with minute brownish lenticels, tomentulose when young; leaves odd- or even-pinnate, bright green, soft and fleshy, 12–30 cm long; leaflets 8–11 (terminal sometimes present only as a mucro), subopposite, subsessile, obovate to elliptic or oblanceolate, apex acute to acuminate, base inequilateral, margin slightly wavy, blade tomentulose basally and on midvein above, 4–14 cm long, 15–29 mm wide; petiole and rachis

Fig. 212. *Sapindus saponaria.*

winged, tomentulose above, rachis interrupted, wing 3–7 mm wide, inequilateral; flowers whitish, 4–5 mm across, mostly staminate; inflorescences about as long as leaves; sepals 4 or 5, petaloid, spreading; petals 4 or 5, pubescent, unappendaged; schizocarps 1–2.5 cm in diameter.

The West Indies, Florida, and Mexico to northern Argentina; indigenous to the Galápagos archipelago. Specimens examined: Isabela, Santa Cruz.

Two Isabela collections (*Baur 57* [GH] and *Stewart 1947* [CAS, GH]) have been cited by Radlkofer (1932) as S. *saponaria* f. *inaequalis* (DC.) Radlk. However, the specimens examined appear to be the typical form, though the wing of the petiole and rachis is narrower than that usually found on mainland collections. According to Stewart (1911; 1915), S. *saponaria* is the largest forest tree in the islands, forming dense stands on Isabela near Villamil. It occurs also in Africa and on some Pacific islands, probably introduced through cultivation and naturalization.

See Fig. 212.

SIMAROUBACEAE. SIMAROUBA FAMILY*

Trees, shrubs, or rarely subshrubs; wood often bitter. Leaves usually alternate, pinnately compound, simple, or rarely rudimentary, never glandular-punctate, usually entire; stipules usually none. Flowers small, (3)5(8)-merous, regular, hypogynous, unisexual, bisexual, or both, in terminal or axillary, many- to few-flowered cymose panicles or racemes or clustered, rarely solitary; plants usually dioecious. Sepals free or connate basally, imbricate or valvate. Petals as many as sepals, rarely absent, free, imbricate or valvate. Stamens as many or twice as many as petals, rarely more numerous, rudimentary or absent in carpellate flowers; filaments free, sometimes the base adaxially appendaged, usually inserted at base of disk; anthers 2(4)-loculed, usually versatile, introrse. Intrastaminal disk usually present, annular, cupular, cushion-like, or columnar, usually lobed or crenate, rarely obscure or absent. Gynoecium 1–8-carpeled, apocarpous or syncarpous, sessile, or gynophore rarely present, usually inserted on, or base encircled by, disk; ovary 1–8-loculed, rudimentary or absent in staminate flowers; ovules 1–2(many) per locule, collateral, superposed, pendulous from top or ascending from base of locule, anatropous or rarely orthotropous to campylotropous; styles basal, lateral, or terminal, free or partly to completely connate; stigmas free to connate, terminal or introrse. Fruit apocarpous and of 2–8 1-carpeled and 1-loculed drupes, berries, or samaras, or a syncarpous 2–4-carpeled and 1–4-loculed berry, samaroid capsule, drupe, or schizocarp. Seeds 1(many) per locule; endosperm scanty or none; embryo large, straight or rarely curved; cotyledons narrow, mostly fleshy, plano-convex or flat; radicle usually very short.

About 30 genera and 200 species, pantropical in distribution; a few species in warm temperate and at least 1 commonly adventive in cool temperate areas.

Castela Turp., Ann. Mus. Hist. Nat. Paris 7: 78. 1806, nom. conserv.

Shrubs or small trees; branches with or without small axillary spines. Leaves alternate, simple, small, mostly less than 5 cm long; petioles short; stipules absent. Flowers reddish, 4(8)-merous, axillary, 1 to several in a cluster, unisexual; plants dioe-

* Contributed by Duncan M. Porter.

cious. Sepals 4(8), connate basally, imbricate, more or less persistent. Petals 4(8), free, larger than sepals, imbricate, deciduous. Stamens mostly 2 or 3 times as many as petals, included, inserted around a glandular disk; anthers reduced and nonfunctional in carpellate flowers. Gynoecium of 4(8) carpels, free basally, in staminate flowers absent, rudimentary, or very rare; ovules 1 per locule; styles 4(8), terminal, connate basally, recurved and spreading, adaxial surface stigmatic. Fruit of several small, free to more or less basally connate, 1-loculed drupes; endocarp crustaceous, mesocarp fleshy. Seeds 1 per locule.

About 14 species, found from southern Texas, Arizona, and Baja California to Argentina, most of the species occurring in the Caribbean region.

Castela galapageia Hook. f., Trans. Linn. Soc. Lond. 20: 229. 1847

Castela galapageia f. *albemarlensis* Robins., Proc. Amer. Acad. 38: 158. 1902.
Castela galapageia f. *bindloensis* Robins., loc. cit.
Castela galapageia f. *carolensis* Robins., loc. cit.
Castela galapageia f. *duncanensis* Robins., op. cit. 159.
Castela galapageia f. *jacobensis* Robins., loc. cit.
Castela galapageia f. *jervensis* Robins., loc. cit.
Castelaria galapageia Moldenke, Phytologia 1: 8. 1933.
Castela erecta subsp. *galapageia* Cronq., Brittonia 5: 469. 1945.

Densely branched, spiny, prostrate to erect shrub 0.5–5 m high; bark gray to light brown or reddish; young twigs conspicuously white-tomentose; younger branches armed with small sharp tomentose axillary spines 1.5–7 mm long, or rarely unarmed, lateral branches often with spinelike tips; leaves dense to sparse, in clusters

Fig. 213. *Castela galapageia.*

of 1–5, 8–40 mm long, 2–21 mm wide, linear-lanceolate, lanceolate, oblanceolate, oblong, elliptic, or spatulate, sometimes inequilateral, mucronate, the apex obtuse or rounded, rarely retuse, base acute to obtuse, margins revolute, entire to wavy or sometimes furnished with 1 to several small lateral teeth, yellowish green to dark green, shiny, glabrous or slightly pubescent above, densely white-tomentose beneath, the network of veins beneath raised and conspicuous on older leaves, usually obscured by tomentum on younger, becoming less noticeably tomentose with age, the youngest red; petioles tomentose, 1–4 mm long; peduncles tomentose, to 2 mm long in flower, 2–3 mm in fruit; flowers in small axillary clusters, rarely solitary, red or orange in bud, red without and yellow within at maturity, about 4 mm broad; sepals 4, 1–1.5 mm long, obtusely triangular, pubescent, ciliate; petals 4, 3–5 mm long, 2–3 mm wide, boat-shaped, more or less pubescent and 1-nerved dorsally; stamens 8; filaments 1–2 mm long, subulate, thickened and pilose below; anthers 1.5 mm long, sagittate; disk annular; carpels 4, bean-shaped, red, glabrous; style branches basally connate, stiffly spreading, 1 mm long, red, deciduous; drupes lenticular, smooth, fleshy, green, turning bright red at maturity, 1–3 together, 8–12 mm high, 6–8 mm wide, 3–5 mm thick, drying dark and wrinkled.

Endemic; on bare lava, in small patches of soil, and on cinder accumulations in arid areas. Specimens examined: Baltra, Española, Fernandina, Isabela, Las Plazas, Marchena, Pinta, Pinzón, Rábida, San Cristóbal, San Salvador, Santa Cruz, Santa Fé, Santa María, Seymour.

Students of the Galápagos flora, from Hooker (1847b) to the present, have indicated the close relationship of the endemic species *C. galapageia* to *C. erecta* Turp. and to *C. tortuosa* Liebm. (= *C. erecta* subsp. *texana* [Torr. & Gray] Cronq.), found in the West Indies, in adjacent northern South America, and from Texas into Mexico. These two taxa are too similar to be regarded as anything other than subspecies. The young leaves of *C. galapageia* are similar to those of the 2 subspecies of *C. erecta,* but the larger, mature leaves are quite distinct. It should be noted that the flowers in most species of *Castela* are very similar, and that these species have been delimited by vegetative characters and geographical distribution. For a discussion of these 3 closely related taxa, see Cronquist (1945). Stewart (1915b) showed that Robinson's (1902) reliance on such variable characters as leaf shape and arming of the stem to delimit forms in *C. galapageia* is of little taxonomic value. Such characters vary infraspecifically throughout the genus.

See Fig. 213 and color plates 29, 40, and 42.

STERCULIACEAE. STERCULIA FAMILY*

Shrubs, trees, vines, or rarely herbs, with alternate, simple, entire or palmately lobed or compound leaves and stellate pubescence. Stipules caducous. Flowers mostly perfect, but monoecious in some genera, regular or slightly irregular, in various types of inflorescences or solitary in axils. Sepals 3–5, basally connate, valvate in bud. Petals same in number as sepals, small and reduced, or lacking, hypogynous, sometimes adnate to base of androecial column, contorted in bud. Stamens in 2 whorls, distinct, or connate into a tube, inner whorl fertile, outer 5 stamens reduced to sterile staminodia; anthers 2-celled, dehiscing longitudinally, sometimes connate apically. Pistil 1; ovary superior, 1–5-celled, the placentae axile; ovules 2 to several

* Contributed by Ira L. Wiggins.

in each locule, ascending or horizontal, anatropous. Styles as many as carpels, distinct or connate. Fruit membranous, leathery, fleshy, or woody, dehiscent or indehiscent, the carpels sometimes separating into cocci. Seeds 1 to several per locule. Embryo straight or curved. Endosperm copious, somewhat oily.

About 50 genera and 750 species, distributed in the tropical and subtropical parts of the world.

Waltheria L., Sp. Pl. 673. 1753; Gen. Pl. ed. 5, 304. 1754

Herbs or shrubs with pubescence mainly or wholly of stellate hairs. Leaves petiolate, serrate, with small linear or narrowly lanceolate stipules. Flowers small, in axillary glomerules or cymose or capitate in terminal racemes or panicles. Calyx narrowly campanulate, 5-toothed. Petals 5, spatulate to narrowly obovate, persistent. Stamens 5, connate into a delicately hyaline tube at base, free above, opposite petals; staminodia none; anther cells parallel. Ovary sessile, 1-carpelled, 1-celled, 2-ovulate but usually only 1-seeded. Style slightly excentric, clavate or fimbriate at tip. Capsule dorsally bivalvate; walls thin and delicate. Seeds oblong to subglobose.

About 30 species, mostly in the New World tropics and subtropics.

Waltheria ovata Cav., Diss. 6: 317, *t. 171, f. 1.* 1788

Waltheria reticulata Hook. f., Trans. Linn. Soc. Lond. 20: 231. 1847.
Waltheria sericea Turcz., Bull. Soc. Nat. Moscow 31: 214. 1888.
Waltheria reticulata f. *acamata* Robins., Proc. Amer. Acad. 38: 176. 1902.
Waltheria reticulata f. *anderssonii* Robins., loc. cit.
Waltheria reticulata f. *intermedia* Robins., op. cit. 177.
Waltheria ovata f. *acamata* Svens., Amer. Jour. Bot. 22: 244. 1935.
Waltheria ovata f. *intermedia* Svens., loc. cit.
Waltheria ovata f. *reticulata* Svens., loc. cit.

Slender, amply branching shrub 0.5–2 m tall, the bark reddish brown to dark gray on older stems, younger twigs densely and minutely stellate-pubescent; stipules linear, narrowly subulate or linear-lanceolate, 2–8 mm long, ascending and often falcately curved, minutely stellate-puberulent; petioles stoutish, 4–10 mm long,

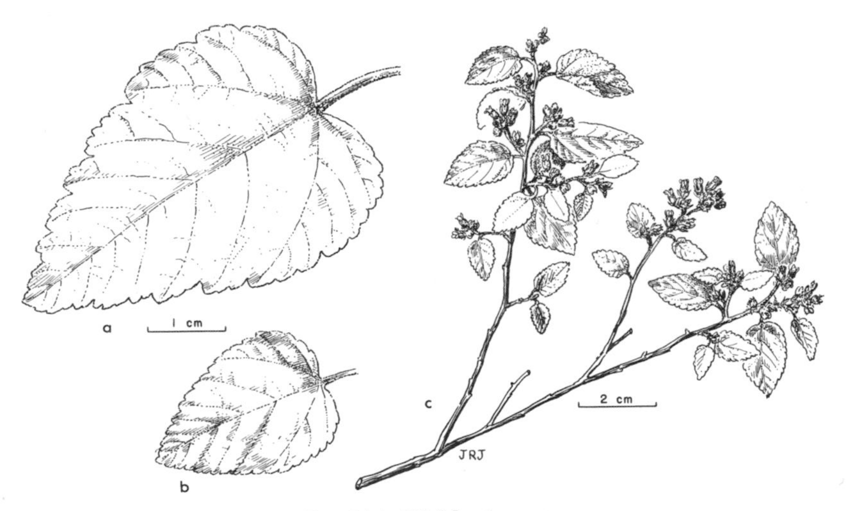

Fig. 214. *Waltheria ovata.*

finely stellate-canescent; blades broadly ovate, sometimes very shallowly trilobate, 1–6 cm long, 0.5–4 cm wide, rounded to cordate at base, acute to obtuse at apex, shallowly serrate, with 4–8 pairs of pinnately arranged veins prominent beneath, canescent with minute, closely crowded stellate-puberulent hairs on both surfaces, slightly paler beneath than above; glomerules sessile or in 2–5 separate aggregates along short, lateral branches and at tips of main branches, closely crowded in each glomerule; bracts simple, oblong, or sometimes 2- or 3-lobed, 4–6 mm long, 1–2.5 mm wide, densely stellate-puberulent; calyx narrowly campanulate, densely stellate-puberulent, 5–6 mm long, the 5 teeth deltoid-ovate, 2.5–3 mm long, valvate in bud; petals yellow, spatulate to narrowly obovate, 6–7 mm long, 1.8–2.5 mm wide at distal end, glabrous throughout, with minute marginal teeth or cilia; stamens about equaling petals, connate into a tube 1.5–2.2 mm long at base, interstaminal membrane hyaline, delicate; ovary and style pilose with simple, straight hairs, these ascending on ovary, shorter and spreading on style; stigma 1.2–2 mm long, fimbriate-feathery; capsule thin-walled, delicate, 2–3 mm long.

On rocky slopes and tops of ridges, from near sea level to 300 m or more in the Galápagos Islands and to 900 m in arid Peru. Specimens examined: Bartolomé, Fernandina, Gardner (near Española), Genovesa, Isabela, Marchena, Pinta, Rábida, San Cristóbal, San Salvador, Santa Cruz, Santa Fé, Santa María.

See Fig. 214.

TILIACEAE. Linden Family*

Trees or shrubs (rarely herbs) with mostly alternate, simple petiolate leaves and simple or, more commonly, stellately branched trichomes. Stipules geminate or none. Leaf blades often oblique at base, pinnately nerved. Flowers small or large, cymose, regular and usually perfect. Sepals 3–5, valvate or partially imbricate. Petals 3–5 or none, distinct when present, occasionally sepaloid, contorted, imbricate or valvate. Stamens 10 to many, hypogynous, free or connate at base into fascicles of 5–10; anthers 2-celled, dehiscing longitudinally or through apical pores. Pistil 1. Ovary superior, sessile, 2–10-celled; style simple but divided at apex, or stigmas rarely sessile. Ovules 1 to several in each locule; placentation axile. Fruit fleshy or dry, dehiscent or indehiscent, sometimes with transverse dissepiments. Seeds 1 to many in each cell, smooth and glabrous or sometimes hairy. Embryo straight or slightly curved; endosperm copious or scanty.

Some 41 genera and about 400 species, distributed mainly in the tropics of the Old and New Worlds, a few genera extending into the temperate regions.

Plants (in ours) herbaceous; fruits linear to oblong, unarmed capsules; pubescence (in ours) of simple hairs; petals naked basally . . . . *Corchorus*

Plants (in ours) shrubs or small trees; fruits globose or subglobose, cocci or indehiscent capsules, invested with stiff bristles or spines; pubescence of stellate trichomes; petals glandular-thickened or foveolate at base . . . . *Triumfetta*

Corchorus L., Sp. Pl. 529. 1753; Gen. Pl. ed. 5, 234. 1754

Herbs or sometimes small shrubs with simple, serrate or crenate, alternate leaves. Flowers small, yellow, on short, 1- to several-flowered peduncles opposite leaves or

* Contributed by Ira L. Wiggins.

axillary. Sepals (4)5. Petals as many as sepals, naked at base. Stamens usually many, sometimes only 10, free, inserted on flat torus. Ovary 2–5-celled, each cell with several to many ovules, these separated by transverse dissepiments. Style short, stigmatic at apex. Fruit a linear to oblong, silique-like capsule, cells dehiscing longitudinally. Seeds pendulous or horizontal, usually numerous. Embryo curved; cotyledons foliaceous.

About 30 widely distributed species.

Corchorus orinocensis HBK., Nov. Gen. & Sp. Pl. 5: 337. 1821

Corchorus hirtus var. *orinocensis* K. Schum. in Mart., Fl. Bras. 12[3]: 127. 1886.

Erect or spreadingly branched annual 1–5 dm tall, the stems sparsely to moderately pubescent with subappressed hairs or glabrate; stipules linear-subulate, 3–6 mm long, sparsely hirsute; petioles 4–10 mm long, hirsute to glabrate; leaf blades ovate to lance-oblong, 2.5–5 cm long, 1–2.5 cm wide, rounded and often slightly asymmetrical at base, acute to short-acuminate at apex, regularly crenate along margins, sparsely to moderately hirsute with subappressed hairs antrorsely oriented, the pubescence most pronounced along main veins beneath; flowers solitary, short-pedunculate, 6–7 mm long; sepals oblong-lanceolate, 5–6 mm long, more or less pilose on back, petals about equaling sepals, yellow; capsule linear, 2–5 cm long, 2.5–3 mm broad, acuminate into a beak 4–6 mm long, pilose to glabrate, hairs more or less appressed; seeds terete in cross section, truncate at both ends, upper end with a short, erect lip on 1 side, dull black, finely punctate-reticulate, 1.2–1.4 mm long, nearly as wide.

Along margins of pools and streams, on moist or seepy ground, and in disturbed soil, Texas, Mexico, and the West Indies south into South America; occasional in the Galápagos Islands. Specimens examined: Española, Gardner (near Española), Isabela, Pinzón, Santa Cruz, Santa María.

See Fig. 215.

Triumfetta L., Sp. Pl. 444. 1753; Gen. Pl. ed. 5, 203. 1754

Shrubs or less commonly herbs with stellate pubescent branches, leaves, and inflorescences. Leaves simple, often 2–5-lobed. Flowers axillary or opposite leaves, cymose or solitary. Sepals 5, distinct, often dorsally short-appendaged near apex. Petals 5 or rarely lacking. Stamens 10 to many. Ovary 2–5-celled, superior, each cell 2-ovuled; style slender; stigma 2–5-toothed. Fruit a subglobose, setose or echinate, small capsule, indehiscent or breaking into cocci. Seeds 1–2 in each cell, separated by dissepiments when 2 are present. Embryo straight.

About 50 to 70 species in the tropical and subtropical regions of the world.

Triumfetta semitriloba Jacq., Sel. Stirp. Amer. Hist. 147. 1763*

Slender shrub 0.5–3 m tall, with the young stems, branches, and inflorescences finely stellate-pubescent; leaves ovate to lance-ovate, larger shallowly 3-lobed, lateral lobes short and ascending, middle lobe (or whole leaf) acuminate, rounded or broadly obtuse and sometimes slightly asymmetrical at base, 3–10 cm long, 1–5 cm wide, irregularly serrate, 3-veined, dark green and sparsely pubescent above,

* For an extensive list of synonymy, see Lay (1950).

Fig. 215. *Triumfetta semitriloba* (*a, b*). *Corchorus orinocensis* (*c–f*).

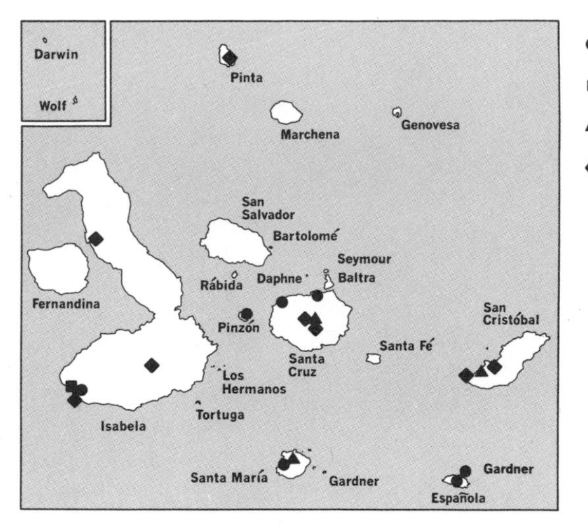

● *Corchorus orinocensis*
■ *Triumfetta semitriloba*
▲ *Apium laciniatum*
◆ *Apium leptophyllum*

paler and more densely pubescent beneath; petioles 1–5 cm long; inflorescences of few-flowered fascicles opposite leaves, pedicels 3–6 mm long; peduncles of few-flowered fascicles, to 1 cm long; sepals minutely stellate-puberulent, 5–7 mm long, each bearing a filiform subapical appendage 1.5–2 mm long; petals yellow, nearly equaling sepals; stamens about 20, 6–8 mm long; styles 2–3 mm longer than stamens; fruit (including spines) 6–9 mm in diameter, body brown, usually glabrate at maturity, the spines brown, slender, 2–3 mm long, rigid, sparsely puberulent most of their length, abruptly short-hooked at tip.

Baja California, central Mexico, and the West Indies to northern South America; in the archipelago, occasional among rocks and in open areas, from sea level into the *Scalesia* Zone. Specimens examined: Isabela.

There is no good explanation for the absence of this species from collections made since the turn of the century in areas where our specimens were taken.

See Fig. 215.

TROPAEOLACEAE. NASTURTIUM FAMILY*

Moderately succulent herbs with acrid, watery juice, mostly prostrate, twining or clambering stems, and alternate, mostly simple (sometimes digitately divided or compound), exstipulate, peltate leaves. Flowers perfect, zygomorphic, spurred, borne solitary and axillary or rarely in umbels. Calyx bilabiate, of 5 distinct sepals, the dorsal one produced into slender spur. Petals 5, clawed, imbricate, the upper 2 smaller than the lower 3 and inserted in opening of calyx spur. Stamens 8, in 2 whorls, distinct, declinate, unequal; anthers dehiscing longitudinally. Pistil 1, of 3 carpels, 3-locular, with axile placentation. Style 1, apical; stigmas 3, linear. Fruit a schizocarp of 3 1-seeded mericarps. Embryo straight; endosperm usually none.

A single genus and about 50 species in mountainous regions from Mexico to Chile and Argentina; *Tropaeolum majus* is widely cultivated and frequently naturalized.

* Contributed by Duncan M. Porter.

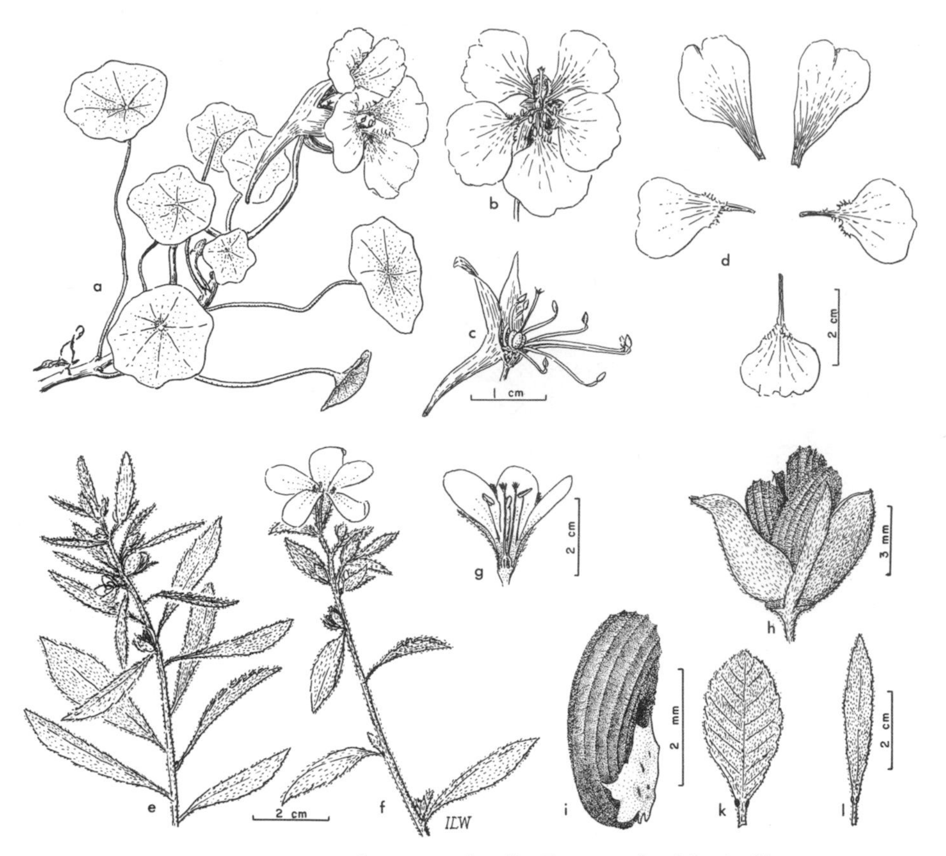

Fig. 216. *Tropaeolum majus* (*a–d*). *Turnera ulmifolia* (*e–l*).

Tropaeolum L., Sp. Pl. 345. 1753; Gen. Pl. ed. 5, 162. 1754

Characters of the family.

Tropaeolum majus L., Sp. Pl. 345. 1753

Scandent, slightly succulent herb with orbicular, long-petiolate, entire or shallowly lobed leaves; yellow, orange, or dark red flowers 3–7 cm across; fruit often rugose or fluted, about 1–2 cm broad, the mericarps separating readily at maturity.

This Andean native was seen in cultivation near Bella Vista, Isla Santa Cruz. It may be expected as an escape near dwellings in the moist regions of the Islands. See Fig. 216.

TURNERACEAE. Turnera Family*

Herbs or occasionally shrubs with entire to lobed, alternate, exstipulate leaves. Flowers perfect, regular, yellowish. Calyx tubular or campanulate, 5-toothed. Petals

* Contributed by Duncan M. Porter.

5, inserted on calyx tube, distinct, contorted in bud. Stamens 5, inserted at base of corolla tube; anthers 2-loculed, dehiscing longitudinally. Ovary superior, 1-loculed, with 3 parietal placentae. Style branches 3; stigmas fringed apically. Capsule loculicidally 3-valved, a placenta in middle of each valve. Seeds pitted, arillate. Embryo straight; endosperm horny or fleshy.

Six genera and about 110 species, native to the tropical and subtropical regions of North and South America.

Turnera L., Sp. Pl. 271. 1753; Gen. Pl. ed. 5, 131. 1754

Leaves serrate, often bearing 2 glands at base. Flowers axillary and usually solitary. Capsule thin-walled. Aril submembranous. Characters otherwise as in the family.

About 60 species, native from southernmost United States into central South America.

Turnera ulmifolia L., Sp. Pl. 271. 1753

Turnera angustifolia Mill., Gard. Dict. ed. 8. *Turnera* no. 2. 1768.
Turnera trioniflora Sims, Curtis's Bot. Mag. *pl. 2106*. 1820.
Turnera mollis HBK., Nov. Gen. & Sp. Pl. 6: 126. 1823.
Turnera caerulea DC., Prodr. 3: 346. 1828.
Turnera velutina Presl, Rel. Haenk. 2: 44. 1836.
Turnera alba Liebm., Ann. Sci. Nat. III, 9: 318. 1848.

Reported by Hooker (1847b) from Isla Santa María on the basis of an Edmonstone collection. The specimen cannot be found in the Gray Herbarium, and the species has not been collected in the Galápagos by recent workers.

See Fig. 216.

UMBELLIFERAE. CARROT FAMILY*

Herbs, or less commonly shrubs or small trees, with mostly alternate or basal, usually compound or conspicuously lobed, incised, or divided simple leaves with sheathing petioles or rarely with stipules. Flowers regular, bisporangiate, or occasionally monoecious or dioecious, in simple or compound, lax or capitate involucrate or naked umbels, these variously arranged. Perianth double, calyx tube wholly adnate to ovary, limb divided into 5 teeth, these frequently small or obsolete, corolla of 5 petals usually with inflexed tip. Stamens 5, opposite calyx teeth, inserted on an epigynous disk. Ovary inferior, bilocular, with 1 anatropous ovule in each locule; styles 2, usually swollen at base to form a stylopodium. Fruit a schizocarp, the 2 mericarps united by their commissural face, usually separating at maturity but remaining suspended from apex of an entire or divided carpophore, the fruit terete or variously compressed, pericarp frequently ribbed or winged, traversed lengthwise by 10 or more vittae or oil ducts. Seed erect; endosperm cartilaginous.

A cosmopolitan family of some 480 genera and 2,600 species, especially abundant in temperate areas of the world.

Plants low, creeping perennials with dentate, crenate, or entire simple leaves; fruits flattened laterally:
Leaves nonpeltate; involucre of a pair of conspicuous, scarious bracts; mericarps with 7–9 anastomosing ribs .... *Centella*

---

* Contributed by Mildred E. Mathias and Lincoln Constance.

Leaves peltate; involucre of numerous small bracts, or inconspicuous; mericarps 3-ribbed .... *Hydrocotyle*
Plants erect or ascending annuals or biennials with pinnate to pinnately decompound leaves; fruits nearly terete:
Plants annual; leaves decompound with linear to filiform divisions; involucel lacking; flowers white or slightly greenish.......... *Apium*
Plants biennial; basal leaves with large leaflets; involucel present; flowers yellow.......... *Petroselinum*

Apium L., Sp. Pl. 264. 1753; Gen. Pl. ed. 5, 128. 1754

Herbs, erect, ascending, or some prostrate, branching, annual, biennial, or perennial from taproots or rootstocks. Leaves petiolate, pinnate to decompound, the divisions large and orbicular to small and filiform; petioles sheathing. Flowers perfect, or some staminate, in lax, pedunculate or sessile, simple or compound umbels; involucre and involucel usually lacking. Calyx teeth minute or obsolete. Petals white or slightly greenish, ovate to suborbicular, plane or with a somewhat inflexed apex. Fruit oval to orbicular or broader than long, slightly compressed laterally, ribs filiform, subequal, prominent. Oil tubes solitary in the intervals, 2 on the commissure. Woody endocarp lacking.

About 30 species, occurring principally in temperate Eurasia and the Southern Hemisphere.

Leaflets linear, not many times longer than broad; fertile flowers fewer than 10 in each umbellet.......... *A. laciniatum*
Leaflets filiform; fertile flowers more than 10 in each umbellet.......... *A. leptophyllum*

Apium laciniatum (DC.) Urban in Mart., Fl. Bras. 11[1]: 343. 1879

Helosciadium laciniatum DC., Mém. Soc. Phys. Genève 4: 495. 1829.
Helosciadium gracile Clos ex Gay, Fl. Chil. 3: 124. 1847.
Helosciadium deserticola Philippi, Fl. Atacam. 26. 1860.
Apium biternatum Reiche, Anal. Univ. Chile 104: 836. 1899.

Annual herb, branching above, 1.5–2 dm tall; leaves oblong-ovate, 0.5–3 cm long, 0.5–3 cm broad, ternate-pinnately compound, the divisions linear, to 4 mm long, about 1 mm broad, petioles to 3.5 cm long, sheath white scarious-margined; umbels compound, sessile, umbellets usually fewer than 10-flowered; involucre and involucel lacking; rays to 1 cm long; fruit ovoid, 1.5–2 mm long, 1–2 mm broad, occasionally rugulose or hispidulous.

An inhabitant of coastal lomas and similar habitats, from southern Ecuador to central Chile; it apparently occurs naturally in the archipelago. Specimens examined: San Cristóbal, Santa Cruz, Santa María.

Apium leptophyllum (Pers.) F. Muell. in Benth. & Muell., Fl. Austral. 3: 372. 1886

Pimpinella leptophylla Pers., Syn. Pl. 1: 324. 1805.
Ciclospermum ammi Lag., Amen. Nat. 2: 101. 1821, non *Sison ammi* L. 1753.
Helosciadium leptophyllum DC., Mém. Soc. Phys. Genève 4: 493. 1829.
Apium ammi Urban in Mart., Fl. Bras. 11[1]: 341. 1879, non *Sison ammi* L. 1753.

Annual herb, diffusely branching, 0.5–6 dm tall; leaves oblong-ovate, 3.5–10 cm long, 3.5–8 cm wide, pinnately or ternate-pinnately decompound, divisions linear to filiform, 1.5–7 mm long, 0.5–1 mm wide, petioles 2.5–11 cm long, sheath white scarious-margined; umbels compound or some simple, sessile or pedunculate, um-

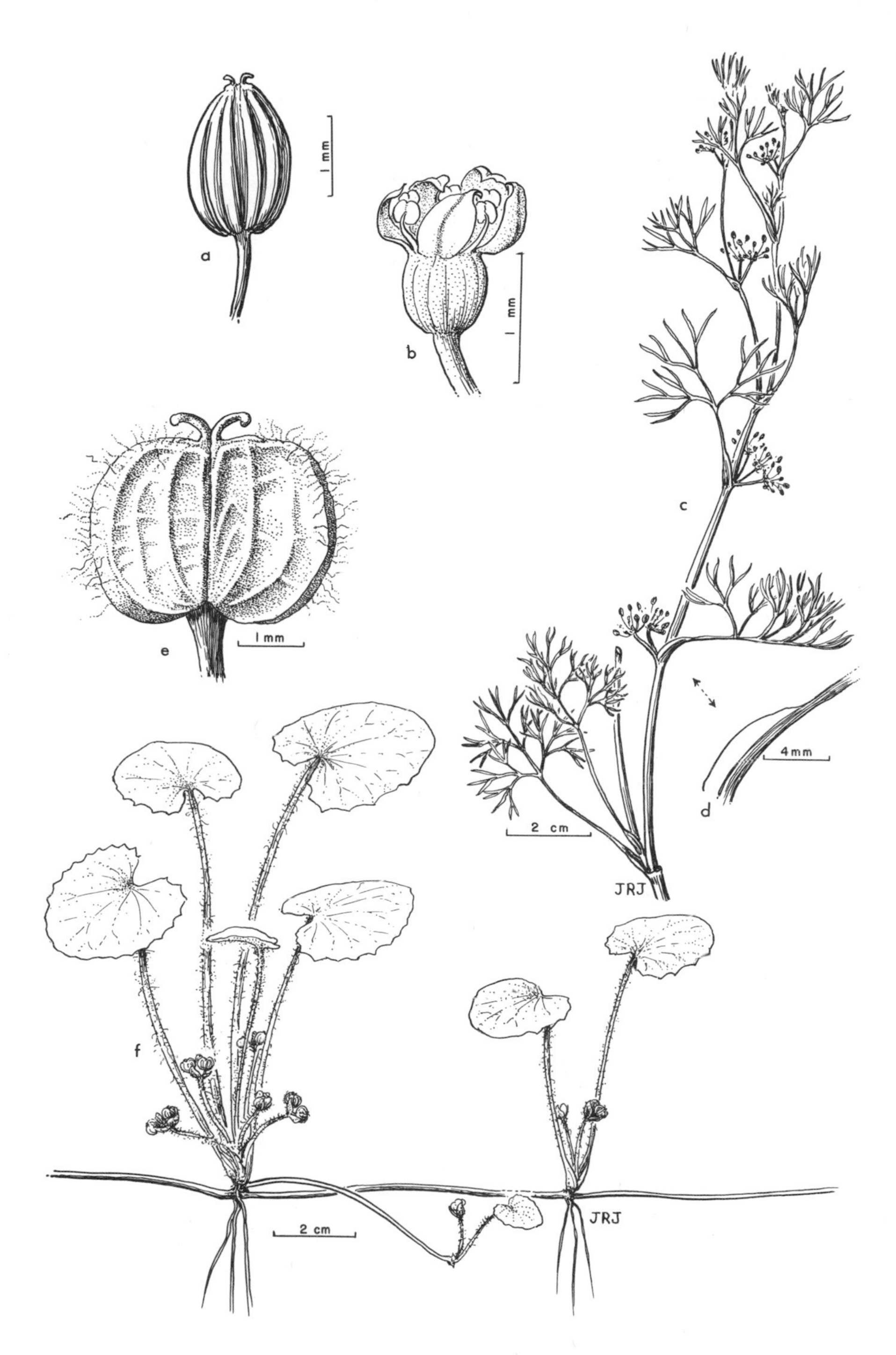

Fig. 217. *Apium leptophyllum* (*a–d*). *Centella asiatica* (*e, f*).

bellets 15–20-flowered, involucre and involucel lacking; rays to 2 cm long; fruit ovoid, 1.2–3 mm long, 1.5–2 mm broad.

A circumtropical weed, occurring throughout the warmer parts of the New World. Specimens examined: Isabela, Pinta, San Cristóbal, Santa Cruz.

See Fig. 217.

Centella L., Gen. Pl. ed. 6, 485. 1764

Perennial herbs, creeping and rooting at nodes. Leaves clustered at nodes, entire to crenate-dentate; petioles slender, sheathing at base. Flowers perfect, in simple, few-flowered, involucrate, axillary umbels. Calyx teeth obsolete. Petals white or purplish, orbicular, slightly incurved, acute or obtuse. Fruit orbicular to broader than long, strongly flattened laterally, both primary and secondary ribs developed and anastomosing. Oil-bearing cells in pericarp, a woody endocarp surrounding seed cavity.

A largely South African genus of perhaps 20 species, with a single species complex (*C. asiatica*) widespread in the warmer parts of the world.

Centella asiatica (L.) Urban in Mart., Fl. Bras. 11[1]: 287. 1879

Hydrocotyle asiatica L., Sp. Pl. 234. 1753.

Glabrous to villous; rhizome 1 to several dm long; leaves clustered at nodes, ovate-cordate to oval, 0.5–10 cm long, 1.5–9 cm broad, obtuse, cordate to truncate at base, entire to crenate or repand-dentate; petioles erect or ascending, 0.3–5 cm long, glabrous-sheathing; peduncles 3–10 cm long, usually shorter than leaves; umbels 2–5-flowered; involucral bracts ovate to suborbicular, membranaceous; pedicels 0.5–4 mm long; the fruit 2–4 mm long and 3–5 mm broad, subcordate at the base.

In moist, protected places, southeastern United States and Mexico to the West Indies, Venezuela, and Colombia; eastern South America to Argentina; and southern Chile; known in the Galápagos Islands from several islands at elevations of 350 to 800 m, but not reported from the Ecuadorian or Peruvian coast. Specimens ex-

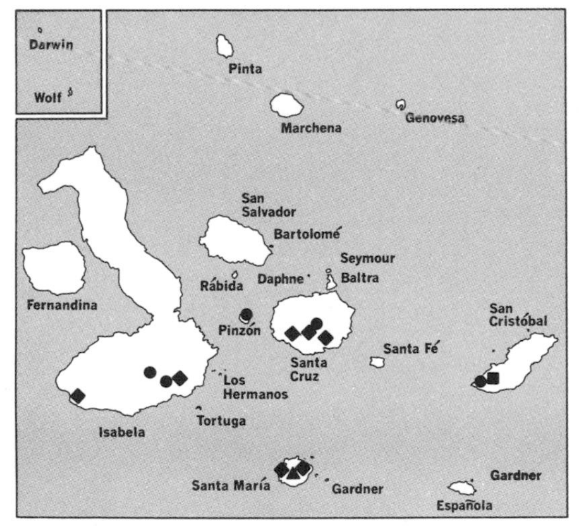

● *Centella asiatica*
■ *Hydrocotyle galapagensis*
▲ *Petroselinum crispum*
◆ *Cissus sicyoides*

amined: Isabela, San Cristóbal, Santa Cruz; also reported from Pinzón by Stewart. See Fig. 217.

Hydrocotyle [Tourn.] L., Sp. Pl. 234. 1753; Gen. Pl. ed. 5, 109. 1754

Perennial herbs, creeping and rooting at nodes. Leaves alternate, peltate or non-peltate, entire or variously lobed or parted; petioles slender, with pair of small stipules at base. Flowers usually perfect, rarely staminate, in simple involucrate umbels, these sometimes proliferating or spicate. Calyx teeth minute or obsolete. Petals white or greenish to purplish, ovate, plane, without inflexed tip. Fruit orbicular to broader than long, strongly flattened laterally, ribs prominent to obscure. Oil-bearing cells (rather than oil tubes) conspicuous to obsolete in pericarp, groups of sclerenchyma (rather than woody endocarp) surrounding seed cavity.

A largely Southern Hemisphere genus of about 75 species, especially well represented in South America but extending into North America and Eurasia.

Hydrocotyle galapagensis Robins., Proc. Amer. Acad. 38: 184. 1902

Glabrous creeping perennial, rooting at nodes; leaves peltate, orbicular, 12- or 13-palmately nerved, nerves and veinlets translucent, margins doubly dentate; peduncles erect, umbels to 16-flowered, non-proliferating; involucral bracts ovate, much shorter than pedicels; calyx teeth obsolete; petals white, ovate, obtuse; fruit broader than long, rounded or subcordate at base.

Endemic; known only from the upper part of Isla San Cristóbal.

See Fig. 218.

Petroselinum Hoffm., Gen. Umbell. 25: 78. 1814

Biennial herbs, erect, branching, from taproot. Leaves petiolate, pinnate to decompound, divisions large and ovate to linear; petioles sheathing. Flowers perfect, in lax, pedunculate, compound umbels; involucre usually lacking and involucel present. Calyx teeth obsolete. Petals yellow or greenish yellow, obovate, with narrower inflexed apex. Fruit ovoid to oblong, slightly compressed laterally, ribs filiform, subequal, prominent. Oil tubes solitary in the intervals, 2 on the commissure. Woody endocarp lacking.

A Eurasian genus of several species, one of which—the cultivated parsley—has become naturalized throughout the world.

Petroselinum crispum (Mill.) A. W. Hill, Hand-list Herb. Pl. Kew. ed. 3, 122. 1925

Apium petroselinum L., Sp. Pl. 264. 1753.
Apium crispum Mill., Gard. Dict. ed. 8, *Apium* no. 2. 1768.
Petroselinum hortense Hoffm., Gen. Umbell. 25: 78. 1814, nomen nudum.
Petroselinum sativum Hoffm., op. cit. 177, nomen nudum.
Petroselinum vulgare Lag., Amen. Nat. 103. 1821.

Biennial herb 3–13 dm tall; leaves ovate, ultimate divisions ovate to linear but relatively large, 2–5 cm long, 1–4 cm broad, distinct, petiolulate, dentate to lobed, the petioles 1–2 dm long; umbels compound, pedunculate, 10–20-rayed; involucre inconspicuous or lacking; involucel of 4–6 linear, entire bractlets shorter than flowers; rays subequal to unequal, 1–5 cm long; fruit ovoid-oblong, 2–4 mm long, 1–3 mm broad.

Fig. 218. *Petroselinum crispum* (*a–c*). *Hydrocotyle galapagensis* (*d–f*).

Central and northern Europe; widely adventive in the Western Hemisphere. Reported from Santa María; also seen by the senior author in a garden on Isla Santa Cruz in 1964, but not collected.

See Fig. 218.

VITACEAE. GRAPE FAMILY*

Climbing vines, rarely shrubs or trees, with sympodial branching in which tendrils representing the main axis are subordinated to the more vigorous growth of axillary branches in opposing leaf axils, the tendrils apparently lateral, and opposite the leaves; nodes often swollen and prominent. Leaves alternate, simple, palmately or pinnately compound, frequently with pellucid punctate dots on lamina; stipules petiolar or none. Flowers minute, perfect or unisexual and the plants monoecious, regular, borne in cymose or paniculate inflorescences arising opposite a leaf. Sepals 4 or 5, distinct or connate at base, subimbricate. Petals 4 or 5, small or obsolete, distinct or connate at tips but separating from one another at base, whole corolla then soon deciduous as cap. Disk evident to prominent, annular or lobed. Stamens as many as petals and opposite them, perigynous from base of disk; anthers 2-celled, distinct or connate, longitudinally dehiscent. Pistil 1; ovary superior, 2–6-loculed; placentation axile; ovules 1–2 on each placenta, anatropous. Style 1, short; stigma capitate or discoid. Fruit a berry. Seeds cartilaginous to bony. Embryo straight; endosperm copious, mealy.

A family of 11 or 12 genera and about 600 species ranging from the tropics into the temperate zones of the world.

Petals 4, free at their tips, spreading; disk 4-lobed; inflorescence umbellate-cymose; bark with many obvious lenticels . . . . *Cissus*

Petals 5, connate at their tips to form deciduous cap; disk 5-lobed or of 5 glands adnate to base of ovary; inflorescence paniculate; lenticels few or none, bark shreddy . . . . . *Vitis*

Cissus L., Sp. Pl. 117. 1753; Gen. Pl. ed. 5, 53. 1754

Vines herbaceous to woody, stems subfleshy, usually from large, fleshy tubers and with tightly coiling tendrils. Leaves simple or 3-foliolate, thin to succulent. Flowers 4-merous, usually perfect. Calyx short, 4-toothed or sometimes entire. Petals free at tips and spreading at anthesis, caducous. Ovary 2-celled, the base partially immersed in the 4-lobed disk. Berries 1–4-seeded, the flesh dry, inedible.

About 300 species in tropical and subtropical areas, primarily in the Southern Hemisphere but some species occurring throughout Mexico and barely entering Arizona.

Cissus sicyoides L., Syst. Nat. ed. 10, 2: 897. 1759

Cissus elliptica Schlecht. & Cham., Linnaea 5: 221. 1830.
Vitis sicyoides Morales in Poey, Repert. 1: 206. 1866.

Scrambling over rocks and climbing into shrubs and trees, sometimes to considerable heights; leaves simple, ovate, 4–15 cm long, 3–10 cm wide, obtuse to subcordate at base, acute to acuminate at apex, sharply and finely but remotely serrate,

* Contributed by Ira L. Wiggins.

Fig. 219. *Cissus sicyoides.*

conspicuously veined and paler beneath than above, glabrous to densely puberulent on lower surface or both surfaces, deep green above, many pellucid dots irregularly spaced on both surfaces, petioles 5–12 mm long; inflorescences opposite leaves, decompound in subglobose, open cymes 2–5 cm long and wide; bracts subulate, 1–2 mm long, ciliolate; pedicels 1–4 mm long, glabrous; calyx disk shallowly lobed; petals about 2 mm long, caducous; fruits ovoid to globose, 5–8 mm in diameter, black, usually 1-seeded.

Widely distributed in tropical America northward to southern Sonora, Mexico; in shady areas or in small openings in the forest or brush, from near sea level to a bit over 300 m in the Galápagos. Specimens examined: Isabela, Santa Cruz, Santa María; reported (but not collected) by Heller from Marchena and Fernandina.

A smut attacks this plant and often causes gross deformation of the parasitized parts. On an infected branch the leaf blades are greatly reduced, leaving little more than the midrib and the blade tip; the latter is slightly expanded into a tiny flabelliform structure, or may give rise to a fascicle of minute projections that at first glance look like small green flowers. This deformation has been seen on specimens collected on Isabela, Santa María, and Santa Cruz.

See Fig. 219.

Vitis L., Sp. Pl. 202. 1753; Gen. Pl. ed. 5, 95. 1754

Climbing or scrambling woody vines, usually with tendrils opposite some of the leaves, and small, dioecious, polygamo-dioecious, or rarely perfect flowers. Leaves simple, dentate and usually palmately lobed. Stipules caducous. Flowers in panicles or racemes. Calyx minute, limb entire. Petals 4 or 5, hypogynous or perigynous, deciduous in a coherent cap without expanding. Stamens opposite petals. Ovary 2(4)-celled, each cell 2-ovulate. Fruit a globose or ovoid berry, pulpy and juicy.

About 50–60 species in temperate to subtropical regions of the Northern Hemisphere, with greatest representation in North America.

Vitis vinifera L., Sp. Pl. 202. 1753

A vine but trained into a low bush in cultivation; leaves suborbicular with deep basal notch, irregularly and coarsely dentate, tomentose to glabrate beneath, 5–25

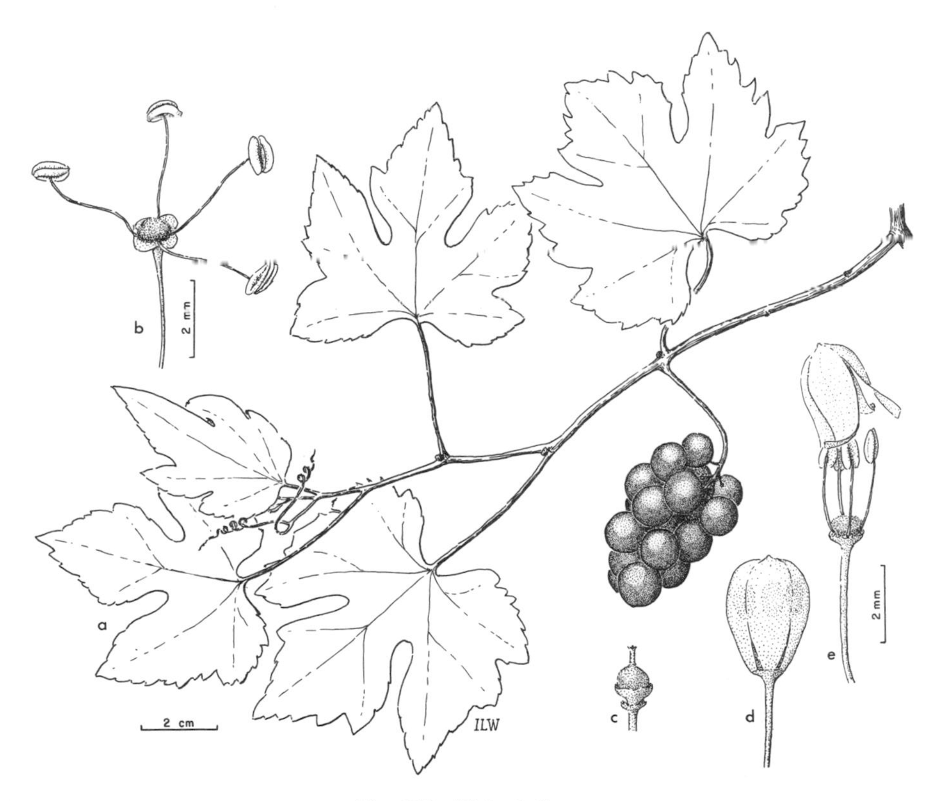

Fig. 220. *Vitis vinifera.*

cm long and wide; inflorescences often large (to 25 cm long); berries usually oval to oblong, 1–2.5 cm long, white, pale yellow, reddish to deep purple or blackish.

Believed native to Caspian and Caucasus regions and to western India; cultivated extensively in both the Old and the New World.

Reported from Santa María by Chierchia, according to both Robinson (1902) and Stewart (1911). Collections of *Vitis* from the Galápagos have not been found in herbaria and it seems probable that Chierchia's report was based on vines persisting after a homesite had been abandoned. No naturalized grapevines are known now on any islands in the Galápagos.

See Fig. 220.

ZYGOPHYLLACEAE. Caltrop Family*

Shrubs, annual to perennial often suffrutescent herbs, or occasionally trees; branches usually divaricate, with angled or swollen nodes, growth sympodial. Leaves opposite or occasionally alternate, usually even-pinnate, occasionally simple or 2-foliolate, rarely 3–7-foliolate, often fleshy to coriaceous, petiolate to subsessile, persistent; leaflets usually entire, inequilateral, petiolulate to subsessile; stipules paired, free, foliaceous, fleshy or spinescent, persistent or rarely deciduous. Flowers (4)5(6)-merous, bisexual, hypogynous, regular or slightly irregular; peduncles terminal or pseudoaxillary, solitary or few to many; sepals (4)5(6), free or slightly connate basally, imbricate, persistent or deciduous; petals (4)5(6), free or rarely connate basally, often clawed, sometimes twisted, imbricate, or convolute, deciduous, rarely marcescent; extrastaminal and/or intrastaminal disk usually present; stamens in (1)2(3) whorls of 5 each, outer whorl usually opposite petals, often alternately unequal in length or sterile; filaments free, subulate to filiform or rarely winged, frequently glandular or appendaged basally, occasionally outer whorl basally adnate to petals, inserted on or below disk; anthers 2-loculed, sub-basifixed to versatile, introrse, longitudinally dehiscent; gynoecium (2)5-carpeled, syncarpous; ovary superior, (2)5(12)-lobed and -loculed, sessile or rarely on a short gynophore; ovules (1)2 to many per locule, pendulous or ascending, anatropous, placentation axile or rarely basal; style terminal, usually simple; stigma minutely and obscurely lobed to distinctly ridged. Fruit a (2)5-lobed loculicidal or septicidal capsule or a schizocarp splitting lengthwise into 5 or 10(12) hard, tubercled to spiny or winged mericarps, rarely a drupe or drupaceous berry. Seeds 1(many) per locule; endosperm present or lacking.

About 27 genera and 250 species, widely distributed, mainly in the warmer, drier regions of the world.

Fruits tuberculate, at maturity separating into 10 mericarps, beak persisting; intrastaminal glands lacking . . . . *Kallstroemia*
Fruits usually spiny, at maturity separating into 5 mericarps, beak falling with mericarps; intrastaminal glands present . . . . *Tribulus*

Kallstroemia Scop., Introd. 212. 1777

Annual or perennial herbs; stems herbaceous to suffrutescent, diffusely branched, prostrate, decumbent, or ascending, terete, somewhat succulent, becoming striate

* Contributed by Duncan M. Porter.

on drying, densely pubescent to glabrate, spreading from central taproot to 1 m or more long. Leaves opposite, elliptic to broadly obovate, abruptly even-pinnate, 1 of each pair alternately smaller or sometimes abortive; leaflets 2–10 pairs, opposite, entire, subsessile, elliptic to broadly oblong or obovate, somewhat unequal, those on 1 side of rachis slightly smaller, lowest pair markedly unequal, terminal pair pointed forward and falcate, pubescent to glabrate; stipules foliaceous. Flowers solitary, pseudoaxillary, regular; peduncles emerging from axils of alternately smaller leaves; sepals 5(6), free, pubescent, persistent or rarely caducous; petals 5(6), white to orange, base same color or green to red, free, spreading hemispherically, fugaceous, usually marcescent, convolute; disk fleshy, annular, obscurely 10(12)-lobed; stamens 10(12), 5(6) opposite petals exterior, somewhat longer, adnate basally to petals, 5(6) opposite sepals subtended to exterior by small bilobed gland; filaments filiform, subulate, or rarely basally winged, unappendaged, inserted on disk; anthers globose or ovoid to linear-oblong or rarely linear, those opposite sepals rarely abortive; ovary sessile, 10(12)-lobed and -loculed, globose or ovoid or occasionally conical or pyramidal, glabrous to pubescent; ovules 1 per locule, pendulous, sometimes 1 or more aborting, placentation axile; style simple, cylindrical to broadly conical, more or less 10(12)-ridged, persisting to form beak on the fruit; stigma capitate, oblong, or clavate, 10(12)-ridged or -lobed, papillose or rarely coarsely canescent, terminal or rarely extending to base of style. Fruit 10(12)-lobed, ovoid, occasionally conical, rarely pyramidal, glabrous or pubescent, at maturity dividing septicidally and separating from a persistent styliferous axis into 10(12) or occasionally fewer mericarps; mericarps 1-loculed, 1-seeded, obliquely triangular, wedge-shaped, more or less tuberculate or rugose abaxially; seeds oblong-ovoid, obliquely pendulous; testa membranaceous; embryo straight; endosperm absent.

The largest New World genus in the family, with 17 species occurring in warm arid and tropical areas from the southern United States to central Argentina; also introduced into West Africa and India.

Kallstroemia adscendens (Anderss.) Robins., Proc. Amer. Acad. 38: 156. 1902

Tribulus adscendens Anderss., Kongl. Svensk. Vet.-Akad. Handl. 1853: 245. 1855.
Tribulus maximus var. *adscendens* Anderss., op. cit. 1857: 107. 1861.

Annual herb; stems prostrate to decumbent, sericeous and hirsute with apically directed white trichomes; stipules 3–4 mm long, about 1 mm wide; leaves elliptic, to 3.5 cm long, to 2 cm wide; leaflets 2–3 pairs, oblong to subfalcate, 9–18 mm long, 3.5–7 mm wide, penultimate pair largest, appressed-hirsute, veins and margins sericeous to almost glabrate; peduncles exceeding subtending leaves in fruit, bent sharply basally, straight above, slightly thickened distally, 5–21 mm long; flowers 5-merous, less than 1 cm broad; sepals 5, subulate, 3–3.5 mm long, about 1 mm wide, slightly shorter than petals, hirsute and strigose, exceeding style in flower but not extending beyond mature mericarps, spreading from base of mature fruit, margins sharply involute, persistent; petals 5, yellow or orange-yellow, obovate, about 4 mm long, 2–3 mm wide, marcescent; stamens 10, equaling style; anthers globose, less than 1 mm across, yellow; ovary ovoid, about 1.5 mm in diameter, strigose; style about 1 mm long, stout, conical, strigose; stigma clavate, 10-ridged, less than 1 mm long, papillose; fruit 10-lobed, ovoid, 3–4 mm in diameter, strigose; beak about 2 mm long, about half as long as body, conical, strigose; mericarps about 3 mm high, 1 mm wide, abaxially cross-ridged and tubercled, sides pitted, adaxial edge angled.

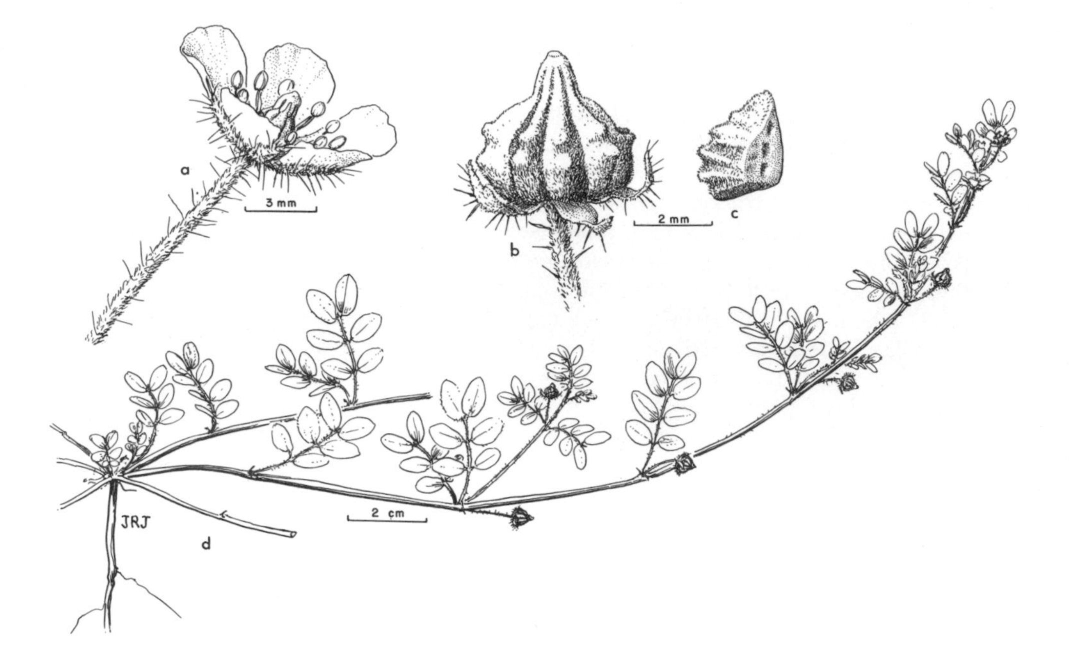

Fig. 221. *Kallstroemia adscendens.*

Endemic; a weed of the arid regions and extremely common in some localities. Specimens examined: Champion, Española, Gardner (near Española), Pinzón, San Cristóbal, Santa Fé, Santa María.

See Fig. 221.

Tribulus L., Sp. Pl. 386.1753; Gen. Pl. ed. 5, 183. 1754

Annual or perennial herbs, rarely shrubby; stems herbaceous to suffrutescent, diffusely branched, prostrate to decumbent or ascending, terete, slightly succulent, becoming striate in drying, densely pubescent to glabrate, spreading from central taproot, to 3 m long. Leaves opposite, even-pinnate, 1 of each pair alternately smaller or sometimes abortive; leaflets 3–10 pairs, opposite, entire, sessile to shortly petiolulate, oblong to ovate or elliptic, terminal pair pointed forward, pubescent; stipules foliaceous. Flowers solitary, pseudoaxillary; peduncles emerging from axils of alternately smaller leaves; sepals 5, free, pubescent, caducous; petals 5, bright yellow or rarely white, base darker, free, spreading hemispherically, deciduous, imbricate; disk fleshy, annular, 10-lobed; stamens 10, 5 opposite petals exterior, somewhat larger, adnate to petals basally, 5 opposite sepals subtended both exteriorly and interiorly by nectariferous glands, intrastaminal glands free or connate to form an urceolate ring around base of ovary; filaments filiform or subulate, unappendaged, inserted on disk; anthers cordate to sagittate; ovary sessile, 5-lobed and -loculed, ovoid or globose, densely pubescent; ovules 3–5 per locule, pendulous, superposed in 2 vertical rows on placenta; placentation axile; style simple, stout, cylindrical, 5-ridged, deciduous; stigma terminal, pyramidal or globose, 5-lobed, papillose. Fruit 5-angled, horizontally depressed, pubescent, at maturity dividing septicidally into 5 or rarely fewer mericarps and leaving no central axis; mericarps broadly triangular, each divided internally by oblique transverse septa into 2–5 1-seeded compartments, spiny or winged or rarely only tuberculate abaxially; seeds oblong-ovoid, obliquely pendulous, horizontally arranged one above the other; testa membranaceous; embryo straight; endosperm absent.

Perhaps several dozen species, of the Old World deserts; three species occur as introduced weeds in the New World.

Flowers 15–25 mm broad; sepals lanceolate, densely strigose and silky-pubescent, 5–9 mm long, 1.5–3.5 mm wide; buds acute, 3–7 mm long. *T. cistoides*
Flowers 5–10 mm broad; sepals ovate-lanceolate, hirsute, 3.5–4 mm long, 1.5–2 mm wide; buds obtuse, 2–3 mm long. . . . . *T. terrestris*

Tribulus cistoides L., Sp. Pl. 387. 1753

Tribulus terrestris var. *moluccensis* Blume, Bijdr. Fl. Nederl. Ind. 5: 243. 1825.
Tribulus moluccanus Decne., Nouv. Ann. Mus. Hist. Nat. Paris 3: 446. 1834.
Kallstroemia cistoides Endl., Ann. Naturg. Mus. Wien 1: 184. 1836.
Tribulus sericeus Anderss., Kongl. Svensk. Vet.-Akad. Handl. 1853: 245. 1855.
Tribulus sericeus var. *erectus* Anderss., loc. cit.
Tribulus sericeus var. *humifusus* Anderss., loc. cit.
Tribulus macranthus Hassk., Flora 48: 403. 1865.
Tribulus terrestris var. *cistoides* Oliv., Fl. Trop. Afr. 1: 284. 1868.
Tribulus alacranensis Millsp., Field Columb. Mus. Publ. Bot. Ser. 2: 54. 1900.
Tribulus cistoides var. *anacanthus* Robins., Proc. Amer. Acad. 38: 157. 1902.
Tribulus cistoides var. *galapagensis* Svens., Amer. Jour. Bot. 22: 235. 1935.
Tribulus cistoides var. *galapagensis* f. *anacanthus* Svens., op. cit. 236.
Tribulus terrestris var. *sericeus* Anderss. ex Svens., op. cit. 33: 457. 1946, nom. illeg.

Prostrate to suberect perennial from woody rootstock; herbage often silvery gray, glabrate in age; stems densely sericeous, more or less hirsute, especially at nodes, to 75 cm long; leaves 2.5–8.5 cm long, 10–26 mm wide; leaflets 6–8(10) pairs, short-petiolulate, obliquely oblong to elliptic, inequilateral, acute or obtuse, middle pairs largest, densely appressed silky-pubescent in youth, whitish beneath, largest 6–21 mm long, 2.5–9 mm wide; stipules subulate to falcate, pubescent, 3–9 mm long, 1–4 mm wide; buds ovoid, acute, 3–7 mm long; flowers 1.5–2.5 cm broad;

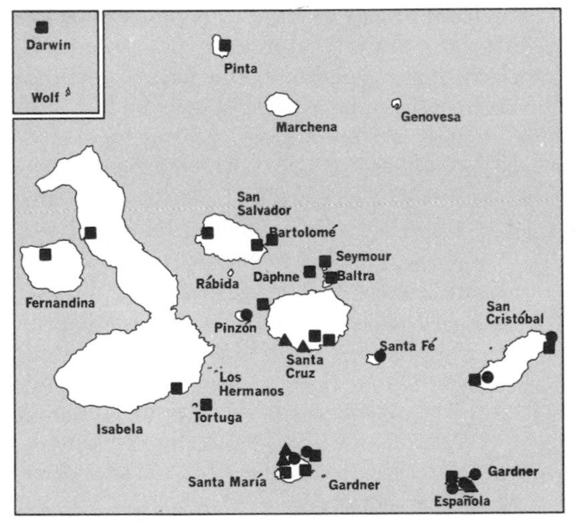

● *Kallstroemia adscendens*
■ *Tribulus cistoides*
▲ *Tribulus terrestris*

peduncles longer than shorter pair of leaves, hirsute and strigose, bent and slightly thickened distally in fruit, (6)19–35 mm long in flower, 11–34 mm long in fruit; sepals lanceolate, acute, densely strigose and silky-pubescent, margins scarious, ciliate, 5–9 mm long, 1.5–3.5 mm wide; petals bright yellow, obovate-cuneate, (5.5)7–17 mm long, (3)5–11 mm wide; outer nectaries bilobed, green; filaments subulate, 2.5–5 mm long; anthers oblong-cordate to narrowly sagittate, versatile, 1–3 mm long; inner whorl of nectaries broadly triangular, yellow, basally connate into 5-lobed urceolate ring surrounding base of ovary, to about 1 mm high; ovary 1.5–3 mm high, hirsute with stiff, upward-spreading bulbous-based trichomes; style 1–2 mm long; stigma globose to pyramidal, slightly asymmetrical, 1 mm long or 1 mm across; fruit 8–10 mm in diameter exclusive of spines, about 7 mm high; mericarps dorsally crested and tuberculate, bearing 2 conical spreading dorsal spines 5–7 mm long and 2 smaller spines near base directed downward or latter absent, rarely spines absent and mericarps tuberculate, or spine 1, body hispid and densely sericeous to strigose or almost glabrate, green to gray, sometimes 1 or more abortive.

A native of the Old World, now a pantropical weed; especially common in the Caribbean region and in tropical Mexico and usually occurring in maritime habitats. (It apparently is easily distributed by adhering to the feet of sea birds.) Specimens examined: Baltra, Bartolomé, Champion, Daphne Major, Darwin, Eden, Española, Fernandina, Isabela, Pinta, San Cristóbal, San Salvador, Santa Cruz, Santa María, Seymour, Tortuga.

The species here shown varies considerably in flower size, in size and amount of pubescence on the herbage, and in spininess of the mericarps. However, the range of morphological variation seen falls within that shown by *T. cistoides* in Mexico. Our present knowledge of the species does not warrant recognition of subspecific taxa.

See Fig. 222 and color plate 87.

Tribulus terrestris L., Sp. Pl. 387. 1753*

Prostrate annual; herbage whitish-pubescent, especially on young shoots, glabrate in age; stems more or less hirsute and sericeous, to 48 cm long; leaves 2–3 cm long, about 1 cm wide; leaflets 4–8 pairs, subsessile, ovate to elliptic, inequilateral, unequal, acute or obtuse, the middle pairs largest, densely appressed silky-pubescent, giving younger parts silvery appearance, becoming glabrate, the largest 4–6 mm long, 2–3 mm wide; stipules subulate to falcate, pubescent, 2–4 mm long, 1 mm wide; buds ovoid, obtuse, 2–3 mm long; flowers 5–10 mm broad; peduncles shorter than to exceeding shortest pair of leaves, hirsute and strigose, bent but little thickened distally in fruit, 5 mm long in flower, 6–10 mm long in fruit; sepals ovate-lanceolate, acute, hirsute, margins scarious, minutely ciliate, 3.5–4 mm long, 1.5–2 mm wide; petals yellow, oblong, 2.5–5 mm long, 1–3 mm wide; outer nectaries more or less bilobed, yellowish; filaments subulate, 2–3 mm long; anthers cordate, versatile, 1 mm long; inner whorl of nectaries yellow, triangular, free or basally connate into 5-lobed urceolate ring surrounding base of ovary, this hirsute with stiff upward-spreading bulbous-based trichomes, 1.5 mm high; style 1–1.5 mm long; stigma globose, slightly asymmetrical, about 1 mm across; fruit 12–13 mm across

* Though it is not feasible at present to supply a list of synonyms, most of the names published in the genus probably are referable to this species.

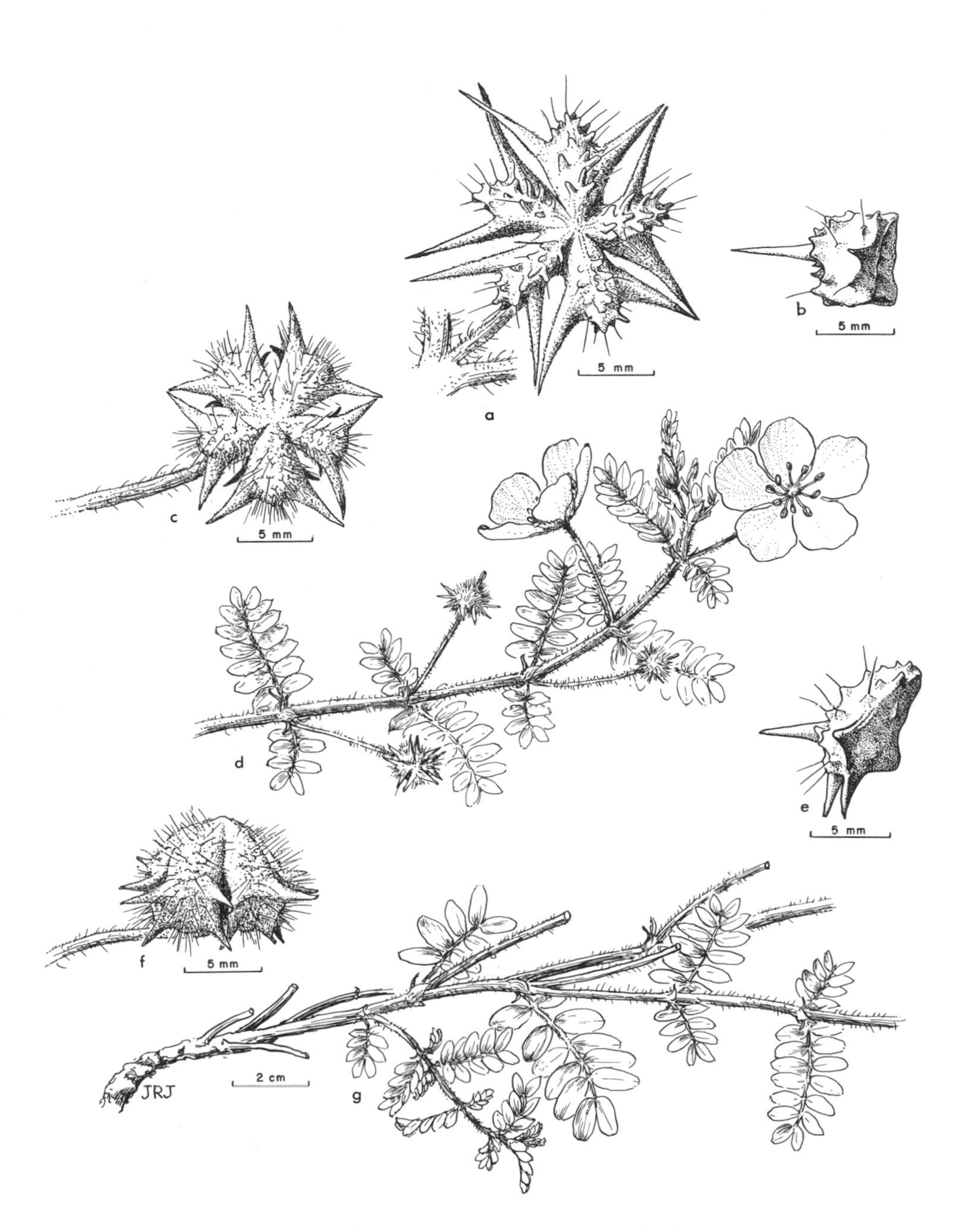

Fig. 222. *Tribulus terrestris* (*a, b*). *T. cistoides* (*c–g*).

exclusive of spines, 4–12 mm high; mericarps dorsally crested and tuberculate, bearing 2 conical spreading dorsal spines 3–4 mm long and 2 smaller ones near base directed downward or the latter absent, rarely spines absent and mericarps with short spinules, hispid and densely sericeous to strigose or almost glabrate, gray to green, sometimes 1 or more abortive.

A native of the Mediterranean region, now a widespread weed in the warm temperate parts of the world; collected only rarely in the New World tropics and hardly ever at low elevations. Specimens examined: Española, Santa Cruz, Santa María.

See Fig. 222.

Monocotyledones

Monocotyledones

Stems composed of parenchymatous ground tissue with vascular bundles scattered irregularly throughout the mass. Cambium none, or present in limited areas and forming new "closed bundles" in some arborescent species; no definite differentiation into pith, wood, and bark. Leaves usually parallel-veined, mostly alternate and entire, or margins dentate, bases often sheathing stem with little or no distinction into blade and petiole. Flowers mostly 3-merous, or some whorls suppressed. Embryo with a single cotyledon and first leaves of seedling alternate.*

Key to the Families of the Monocotyledones

Plants strictly aquatic, floating or submerged in water:
  Plants minute, thallus-like, free-floating, rarely rooted in mud; fruits extremely small, less than 0.2 mm long; each plant (in ours) producing a solitary, unbranched, pendent root . . . . Lemnaceae, p. 896
  Plants immersed in water, rooted to bottom; fruits 2–4 mm long; each plant with several to many roots:
    Stigmas slender, subulate; leaves spiny-toothed, whorled or opposite . . . . Najadaceae, p. 899
    Stigmas sessile or essentially so, capitate or peltate; leaves entire, crenulate, or microscopically serrulate, alternate or opposite, but not whorled:
      Drupelets sessile; peduncles not coiled; stamens 4 . . . . . Potamogetonaceae, p. 923
      Drupelets stipitate; peduncles coiled after anthesis; stamens only 2 . . . . Ruppiaceae, p. 926
Plants terrestrial, or if growing in water, only partially submerged, not floating on surface, always rooted in soil or epiphytic on other plants:
  Inflorescence a spadix (minute flowers borne on a central column, crowded closely together); spadix subtended and/or ensheathed by a white or colored spathe; bisexual flowers at base of spadix, unisexual (staminate) flowers above and usually more numerous . . . . Araceae, p. 785
  Inflorescence not a spadix; spathe, if present, either smaller or enfolding a single flower, not a whole inflorescence:
    Perianth reduced or lacking, segments not petaloid, the perianth parts consisting of bristles or scales, these in a single whorl:
      Stems usually solid, often angular; leaf sheaths cylindric, not split down 1 side (sometimes tardily ruptured in age); each flower subtended by 1 chaffy bract; leaves mostly 3-ranked . . . . Cyperaceae, p. 794
      Stems usually hollow in internodes, terete; leaf sheaths split down 1 side; each

* Synopsis and key to families contributed by Ira L. Wiggins.

flower subtended by 2 chaffy bracts, the lemma and the palea (latter sometimes lacking); leaves mostly 2-ranked. . . . . GRAMINEAE, p. 823

Perianth of 2 distinct whorls, the inner one (and in some families also the outer) petaloid:

Herbage bearing water-absorbing scales, at least when young; plants (in ours) usually epiphytic, rarely on rocks or cliffs. . . . . BROMELIACEAE, p. 787

Herbage not bearing water-absorbing scales; plants mostly terrestrial (some genera in Orchidaceae epiphytic):

Petioles ensheathing nearly entire stem and young, unexpanded leaves, forming false trunk to 8 m tall; stamens 6, 5 antheriferous, the 6th reduced to a staminode; fruit (in ours) a pulpy, edible berry. . . . . MUSACEAE, p. 898

Petioles not ensheathing stem and young leaves, not forming false trunk; stamens usually 3 or 6, not 5 functional and 1 staminode; fruit not an edible berry:

Outer perianth segments herbaceous, much smaller than those of inner whorl; inner whorl of perianth segments deliquescent; flowers borne in boat-shaped, tightly folded spathe; stamens hypogynous; ovary superior; endosperm mealy. . . . . COMMELINACEAE, p. 791

Outer perianth segments not markedly herbaceous (somewhat so in some Orchidaceae), usually not much smaller than those of inner whorl; inner whorl of perianth segments not deliquescent (or if so, outer one also); flowers not in spathe; stamens epipetalous; ovary inferior; endosperm fleshy or horny, or lacking:

Filaments of stamens neither petaloid nor united with style and stigma; stigmas 3, or if 1, then 3-lobed; flowers regular:

Perianth segments subfleshy, not at all deliquescent; leaves stiff, rigid, ensiform, armed with hard, curved, marginal teeth and a terminal spine . . . . AGAVACEAE, below

Perianth segments thin, often deliquescent (both inner and outer whorls) after anthesis; leaves not ensiform, often equitant, unarmed:

Plants rhizomatous; stamens 6. . . . . HYPOXIDACEAE, p. 892

Plants not rhizomatous (in ours); stamens 3. . . . . IRIDACEAE, p. 895

Filaments of stamens petaloid, or connate with style and stigma to form column (gynandrium); flowers markedly irregular:

Stamens 6 (or 4 by reduction or abortion), style petaloid, winged, stigmas linear along margin of winged style; seeds subglobose, 2–4 mm in diameter; endosperm copious, bony; pollen grains not agglutinated into pollinia . . . . CANNACEAE, p. 788

Stamens 1 or 2, connate with style and stigma; seeds minute, very light in weight, less than 0.1 mm long; endosperm lacking; pollen grains often agglutinated into masses (pollinia). . . . . ORCHIDACEAE, p. 902

AGAVACEAE. AGAVE FAMILY*

Perennial plants from rootstocks or rhizomes. Leaves usually crowded on or at base of stem, narrow, often markedly fibrous, thick or fleshy, the margin often armed with horny, persistent spines or teeth. Inflorescences terminal on a scape in racemes, panicles, or large thyrse, the branches subtended by bracts. Flowers perfect, polygamous, or dioecious, regular or slightly irregular. Perianth tube short or long; lobes or segments equal or subequal. Stamens 6, inserted at base of lobes or on tube; filaments free, filiform or thickened toward base; anthers introrse, linear, usually

* Contributed by Ira L. Wiggins.

dorsifixed, dehiscing by longitudinal slits. Ovary superior or inferior, 3-locular, with axile placentae. Style slender. Ovules 1 to many in each locule, when more than 1 superposed in 2 rows, anatropous. Fruit a loculicidal capsule or a berry. Seeds 1 to many, usually compressed. Embryo small, straight. Endosperm fleshy, surrounding embryo.

About 20 genera and possibly 1,000 species, widely distributed on all continents from tropical to temperate zones.

Furcraea Vent., Bull. Soc. Philom. 1: 65. 1793

Large, scapose plants from an erect, often subterranean caudex. Leaves long and narrow, densely crowded at apex of caudex, margins spinose-dentate to subentire, apex subspinose, attenuate, base broad, usually arranged in compact rosette with outer leaves often drooping or prostrate, others stiffly spreading-ascending to erect, imprint of marginal teeth of overlapping leaves often readily visible on most blades. Panicles large, pyramidal, terminal, with many flowers solitary or fasciculate on relatively short, stoutish pedicels, the flowers often replaced by bulblets. Perianth white or greenish white, rotate to broadly campanulate, segments distinct almost to base, oval-oblong, equal. Stamens inserted on bases of perianth segments and shorter than segments, filaments dilated and more or less fleshy below, subulate at apex; anthers linear-oblong, dorsifixed, often shallowly sagittate-lobed at base. Ovary 3-celled, inferior, oblong, short-rostrate at apex; style columnar but swollen toward base; stigma small, capitate to faintly 3-lobed or rim shallowly lacerate. Ovules many, biseriate in each cell. Capsule ovoid to oblong, loculicidally 3-valved. Seeds many, strongly flattened.

Some 15–20 species in tropical and subtropical Mexico, Central America, and western South America.

Furcraea cubensis (Jacq.) Vent., Bull. Soc. Philom. 1: 66. 1793

Agave cubensis Jacq., Sel. Stirp. Amer. Hist. 100, *pl. 175, fig. 28.* 1763.
Agave odorata Pers., Syn. Pl. 1: 380. 1805.
Agave bulbifera Salm-Dyck, Hort. 8, 303. 1834.

Acaulescent or the stem below ground; leaves 25–40 in compact to open rosette to 1 m across and nearly as high, individual leaves lanceolate, 4–9 dm long, 3–6 cm broad at base, 8–15(20) cm wide at or near middle, bright green, rigid, slightly concave and smooth adaxially, more or less scabrous abaxially, bearing a subspinose, rather blunt, horny, brownish tip 2–3 cm long, this usually more or less grooved on adaxial side, margins of leaves armed with antrorsely hooked brown teeth or prickles 2–8 mm long; flowering scape erect, including panicle often to 5–7 m tall, the basal stalk 8–10 cm thick; panicle to 2 m tall, 1–1.5 mm broad at base, lower branches spreading, others gradually more ascending to nearly erect, 1–7 dm long, bearing numerous flowers (or bulblets) on slender, slightly drooping pedicels 3–10 mm long; perianth broadly campanulate to subspreading, oblong segments 2–2.5 cm long, about 8–9 mm wide; stamens about half as long as perianth segments, filaments 4–6 mm wide below, subulate above; style slightly longer than stamens; stigma minute, irregularly fimbriate or lacerate around rim; flowers often replaced by bulblets 1–3 cm high on many plants, sometimes almost entire panicle bearing only bulblets and few or no flowers.

Fig. 223. *Furcraea cubensis* (*a–d*). *Colocasia esculenta* (*e*).

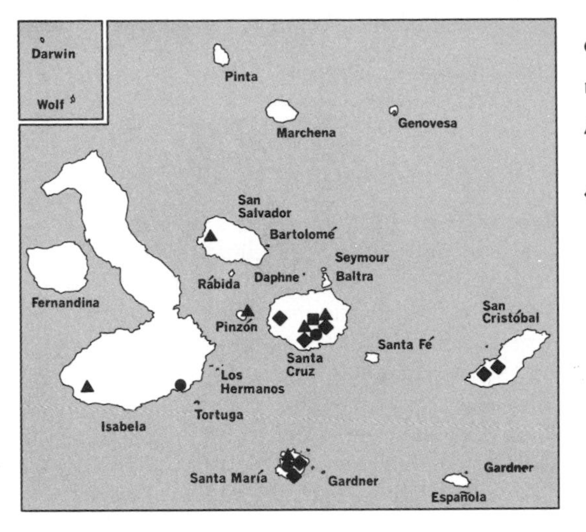

● *Furcraea cubensis*

■ *Colocasia esculenta*

▲ *Tillandsia insularis* var. *insularis*

◆ *Tillandsia insularis* var. *latilamina*

The West Indies to northwestern South America; introduced into the Galápagos Islands; in openings in forest, along trails, and around abandoned ranch sites, or occasionally among shrubs on ridges, often cultivated as an ornamental or to form hedges around fields and pastures. Specimens examined: Isabela, Santa Cruz, Santa María.

Stewart reported *F. cubensis* to be extensively present from 120 to over 360 m altitude on the northwest side of Santa Cruz, in hedges and around plantations on San Cristóbal, and in similar situations near habitations on Isabela in the vicinity of Villamil, but apparently he made no collections of the plant on any of these three islands. Extensive thickets are often virtually impenetrable and thoroughly exclude most other plants in such a colony. The local inhabitants occasionally use the leaves to make light cordage and fairly good rope.

See Fig. 223.

ARACEAE. ARUM FAMILY*

Terrestrial, epiphytic, aquatic, or rarely floating herbs with watery, milky, or acrid sap; acaulescent or with erect or scandent stems; tuberous or rhizomatous. Leaves simple or compound, basal and solitary or clustered, or cauline and alternate; petioles sheathing basally. Flowers bisexual, or unisexual and plants monoecious or dioecious, small; inflorescence a spadix; an herbaceous spathe subtending or enveloping spadix or absent, persistent or deciduous in fruit. Perianth present and of 2–6 free or connate segments, or absent. Stamens 1–6, free or connate into a synandrium. Gynoecium 2- to several-carpeled, syncarpous; ovary 1- to many-loculed, superior or partially to completely inferior; ovules 1 to many per locule. Fruit usually a berry. Seeds 1 to many; endosperm present or absent; embryo large.

Over 100 genera and about 1,800 species, primarily tropical and subtropical in distribution.

* Contributed by Duncan M. Porter.

Leaf blades ovate-cordate, peltate; spathe tardily deciduous, yellowish; ovules orthotropous; style short but present. . . . . *Colocasia*
Leaf blades sagittate, never peltate; spathe persistent, purplish near base; ovules anatropous; style none, the stigma sessile on top of ovary. . . . . *Xanthosoma*

Colocasia Schott in Schott & Endl., Melet. Bot. 18. 1832

Perennial terrestrial herbs, acaulescent or with erect stem, roots tuberous. Leaves often large, simple, peltate, ovate-cordate; petioles long. Peduncles stout, shorter than petioles; spathe basally convolute, tubular, thick, and persistent, blade ovate-lanceolate, constricted below, enclosing spadix, deciduous; flowers unisexual and plants monoecious; spadix thick, carpellate flowers basal, staminate flowers above, fertile portions separated by a zone of sterile staminate flowers, spadix apex naked or covered by sterile flowers. Perianth absent. Staminate flowers of 3–6 stamens connate into a peltate synandrium; anthers elongate, adnate laterally or partially free and pendent, dehiscing by apical slit. Carpellate flowers of a 1-loculed ovary with 2–4 parietal placentae; ovules many, orthotropous; style short. Seeds several per berry; endosperm abundant.

About 7 species, native to tropical Asia but cultivated and escaped throughout the tropics and subtropics of the world.

Colocasia esculenta (L.) Schott in Schott & Endl., Melet. Bot. 18. 1832

Arum esculentum L., Sp. Pl. 965. 1753.
Arum colocasia L., loc. cit.
Colocasia antiquorum Schott in Schott & Endl., loc. cit.
Colocasia antiquorum var. *esculenta* Schott, Syn. Aroid. 42. 1856.

Fleshy, usually acaulescent herb to about 1 m high; leaves ovate, base cordate, anterior lobe broadly ovate and abruptly short-acuminate, posterior lobes rounded, to 50 cm long or more; petioles terete, purplish, basal half sheathing, inserted well within margin of basal notch of blade, to 1 m long; peduncles terete, much shorter than petioles; spathe yellowish, basal portion green, blade lanceolate, elongate, inclined backward at anthesis and exposing the erect, cream-colored portion of spadix; spadix much shorter than spathe; berries numerous.

Taro is cultivated throughout the tropics and subtropics for its starch-yielding roots and for its leaves, which are eaten as a vegetable. It escapes readily from cultivation and becomes naturalized, as it has done on Isla Santa Cruz. Numerous varieties and cultivars have been described, but in the absence of a modern treatment, it is impossible to assign the specimen examined to one of them.

See Fig. 223.

Xanthosoma Schott, emend. Engler in Mart., Fl. Bras. 3^2: 188. 1878

A genus superficially similar to *Colocasia* but differing by characters set forth in the key to the genera. The spathe in *Xanthosoma* is less markedly constricted near its middle than in *Colocasia*; each ovary is 3- or 4-loculed in *Xanthosoma,* unilocular in *Colocasia;* the stigma in *Xanthosoma* is seated on a broad, flat disk that equals or exceeds the diameter of the ovary, whereas no such disk is present in *Colocasia.*

About 40 species in the well-watered parts of the American tropics from central Mexico and the West Indies to southern Brazil and tropical eastern Peru.

Xanthosoma violaceum Schott, Österr. Bot. Wochenbl. 3: 370. 1853
Xanthosoma ianthinum C. Koch, Ind. Sem. Hort. Berol. 2. 1854.

The basal, carpellate portion of the spadix is narrower and much shorter than the part bearing the fertile stamens; petioles 3–7 dm long, blades oblong-sagittate, 2–5 dm long; pistillate part of spadix 3–4 cm long, staminate part to 15 cm long.

Native to the West Indian region, but widely cultivated throughout much of tropical America for the starchy, edible roots. Collected from cultivated plants at Fortuna, Isla Santa Cruz (*Howell 9276* [CAS]). This species, like *Colocasia esculenta,* is likely to occur as an escape in the rain forest areas (*Scalesia* Zone or its equivalent) in the Galápagos Islands.

BROMELIACEAE. PINEAPPLE FAMILY*

Herbs or shrubs, largely epiphytic; roots usually present but often serving merely as holdfasts in the epiphytic species. Leaves usually rosulate, always simple, entire or spinose-serrate, bearing peltate scales (lepidote) at least when young, these serving to absorb moisture. Inflorescence simple or compound. Perianth heterochlamydeous. Stamens 6, in 2 series. Styles 3-parted. Ovary superior or inferior, 3-celled. Placentae axile. Fruit capsular or baccate. Seeds plumose, winged or naked. Embryo small. Endosperm mealy, usually copious.

Easily separable into 3 distinct subfamilies, totaling nearly 2,000 species, most of them restricted to the New World and occurring from the southern United States to northern Chile and Argentina, including Cocos Island.

Tillandsia L., Sp. Pl. 286. 1753; Gen. Pl. ed. 5, 138. 1754

Leaves rosulate or fasciculate, always entire, lingulate to narrowly triangular or linear; scape usually distinct. Inflorescence usually of distichous-flowered spikes or sometimes reduced to a single polystichous-flowered spike. Flowers perfect; sepals symmetric or asymmetric, free or joined; petals free, without nectar scales. Ovary superior, glabrous, ovules usually numerous, caudate. Capsule septicidal, seeds erect with a plumose appendage.

Over 300 species, 1 endemic to the Galápagos Islands, the others occurring from the southeastern United States to northern Argentina and Chile.†

Inflorescence glabrous; leaf sheaths not partly purple; blades 2.4–4 cm wide, spikes spreading to recurved . . . . *T. insularis* var. *insularis*
Inflorescence lepidote; leaf sheaths partly purple, blades 4.3–7 cm wide; spikes ascending . . . . *T. insularis* var. *latilamina*

Tillandsia insularis Mez in DC., Monogr. Phan. 9: 756. 1896, var. **insularis**

Plant 20–150 cm tall; leaves 13–45 cm long, blades 2.5–4 cm wide, lingulate, coriaceous, the margins with or without some purple or red, moderately lepidote, usually apiculate at apex, sheath 2–6 cm long, inconspicuous, blending into and usually

* Contributed by Amy Jean Gilmartin.
† Locally, the Galápagos plant is called "Guicundo," a name commonly applied to many epiphytic species of *Tillandsia* on the mainland of Ecuador.

concolorous with blade; scape extremely variable in length, from shorter than the leaf rosette to almost 75 cm long exclusive of inflorescence, erect; scape bracts imbricate or not, not foliaceous, approximately same length through the scape, about 3–15 cm broad, some depauperate specimens with inflorescences as small as 4 cm long, 2 cm wide, usually tripinnate, lax, the main rachis exposed, erect; primary bracts 1–6 cm long, always exceeding branch stipe, erect-spreading, apex acute, lepidote to glabrous; branches 3–15 cm long, 2–3 cm wide, at intervals of 0.5–4 cm, branch stipe 0.5–5 cm long, with several sterile bracts; spikes 2–9 cm long, 0.8–1.1 cm wide, often with several sterile apical bracts, spreading to recurved at and following anthesis, rachis geniculate or undulate, having 4–14 flowers per spike; floral bracts 4–7 mm long, about 5 mm wide, not imbricate, sometimes becoming secund, subcarinate, nerved, ovate, glabrous without, lepidote within, papery, apex acute, shorter than sepals; sepals 4–7 mm long, 3 mm wide, all 3 carinate, free, glabrous without and glabrous or lepidote within, subcoriaceous, ovate to elliptic, subobtuse to acute; petals scarcely exserted beyond calyx, white; capsule to 2.3 cm long; seeds red or black.

Endemic; epiphytic and terrestrial where moss and leafy liverworts are on the ground, at about 200 m altitude or slightly higher, frequent in the *Scalesia* Zone on lianas and *Psidium* trees. Specimens examined: Isabela, Pinzón, San Salvador, Santa Cruz; also known from San Cristóbal.

See color plate 65.

Tillandsia insularis var. **latilamina** Gilmartin, Phytologia 16: 163. 1968

Leaf blades 4.3–7 cm wide, some purple evident at least on sheath or on blade or on both, outer surface of floral bracts lepidote, spikes ascending and floral bracts often slightly exceeding sepals. Characters otherwise as in var. *insularis*.

Endemic; epiphytic and terrestrial, growing from about 170 m to nearly 400 m. Specimens examined: San Cristóbal, Santa Cruz, Santa María.

See Fig. 224.

CANNACEAE. CANNA FAMILY*

Erect perennial herbs, arising from slender to stout, sympodially branched, sometimes tuberiferous, segmented rhizomes, each segment ending in a stem; stems usually stout, usually triangular, glabrous to woolly-pubescent, nodes slightly swollen. Leaves alternate, spirally arranged, simple, large, lanceolate to ovate or elliptic, slightly coriaceous, pinnately veined, glabrous above, glabrous to woolly-pubescent beneath, margin membranous, apex obtuse to cirrhous, base sessile to short-petiolate, clasping stem in a long, eligulate sheath. Inflorescences terminal, racemose or paniculate, sometimes spicate, usually branched, with primary and secondary bracts; flowers scattered to crowded along a flexuous rachis. Flowers irregular, more or less showy, epigynous, bisexual, sessile to short-pedicellate, single or paired, rarely 3, bracteolate; sepals 3, small, erect, green to purplish, free to slightly connate basally, oblong to broadly lanceolate, obtuse to acute, more or less equal, more or less imbricate, persisting at apex of fruit; petals 3, similar to, but longer than, and alternating with sepals, erect to spreading or reflexed, connate basally into a tube or rarely

* Contributed by Duncan M. Porter.

Fig. 224. *Tillandsia insularis* var. *latilamina*.

free, narrowly to broadly lanceolate, obtuse to acute, 1 usually smaller; androecium modified into 0–4 petaloid staminodes and 1 petaloid stamen, all connate basally into a tube usually longer than, and adnate basally to, corolla tube, the free portion of various shapes and sizes; anterior staminode generally smaller and usually forming a more or less revolute lip; anther 1-loculed, inserted on margin of stamen slightly below apex; gynoecium 3-carpeled, syncarpous; ovary inferior, 3-loculed, ovoid to oblong or ellipsoid, warty, tuberculate, or papillose; ovules numerous in each locule, in 2 vertical rows on axile placentae; style petaloid, usually equaling stamen and staminodes, usually adnate to stamen; stigma 1, truncate or slightly oblique near apex. Fruit a capsule, 3-loculed, subglobose to ellipsoid, warty or spiny, spines or papillae falling prior to dehiscence, leaving a membranous covering; dehiscence irregular. Seeds several per locule, round, smooth, dark brown to spotted, very hard; endosperm thick, white; embryo straight.

A single genus with about 50 species.

Canna L., Sp. Pl. 1. 1753; Gen. Pl. ed. 5, 1. 1754

Characters of the family.

Canna is native to the tropics and subtropics of the New World, with centers of distribution in South America and the Caribbean region; species are now naturalized throughout the world in both tropical and temperate regions. Long cultivation by man and the resulting artificial and natural hybridization has led to a taxonomic and nomenclatural mare's nest in the genus. Such problems are discussed by Woodson (1945).

Herbage woolly-pubescent, green; flowers in pairs; staminodes 3; lip erect. . . . *C. lambertii*
Herbage glabrous, purplish; flowers solitary; staminodes 2; lip recurved. . . . . *C. lutea*

Canna lambertii Lindl., Bot. Reg. *pl. 470.* 1820

Canna poeppigii Bouche, Linnaea 12: 143. 1838.

Stems green, woolly-pubescent, to about 2 m high; leaves elliptic to ovate, narrowly acuminate apically, basally inequilateral and decurrent into a winged petiole, green, paler beneath, sparsely woolly-pubescent along upper midrib, pubescence readily deciduous, the blades 36–50 cm long, 12–19 cm wide, both surfaces minutely punctate and with a series of indentations in rows parallel to and between veins, sheath to 12 cm long, woolly-pubescent on both surfaces; panicles erect, elongate, sometimes slightly glaucous, to 30 cm long; flowers in pairs; pedicels 2 mm long in flower, 4 mm long in fruit; floral bracts oblanceolate, revolute, 6.5 cm long, caducous; bracteoles ovate, 14 mm long, 5 mm wide, deciduous; sepals lanceolate, equal, 11 mm long, 4.5 mm wide; petals lanceolate, revolute, acute apically, erect, equal, 4.5 cm long, 8 mm wide, slightly connate basally, the innermost adnate basally to staminodial tube; staminodes 3, narrowly spatulate, erect, subequal, to 5 cm long, 5 mm wide, lower half connate into a tube; lip linear, revolute, erect, 5 cm long, lower half adnate to staminodial tube; stamen spatulate, 5 cm long, 6 mm wide, acute apically, lower half adnate to staminodial tube and style; anther 7 mm long; style narrowly spatulate, 4.5 cm long, 4 mm wide, erect, lower half adnate to staminodial tube and stamen; stigma more or less apical, 4 mm long; ovary ellipsoid, papillose, 5–6 mm high; capsule ellipsoid, papillose, 2.5–4 cm high, papillae deciduous and seeds remaining in an irregularly dehiscent basket-like network of veins; seeds subglobose,

glabrous, dark brown, minutely pitted, 5–7 mm in diameter, micropylar end black.
From the West Indies to Brazil; probably indigenous in the Galápagos Islands. Specimens examined: San Salvador, Santa Cruz.

Canna lutea Mill., Gard. Dict. ed. 8, no. 4. 1768

Canna aurantiaca Roscoe, Monandr. Pl. *pl. 21.* 1828.
Canna maculata Link, Handb. 1: 227. 1829.
Canna commutata Bouche, Linnaea 8: 147. 1833.
Canna floribunda Bouche, op. cit. 18: 489. 1844.
Canna densiflora Bouche, loc. cit.
Canna lutea var. *maculata* Regel, Ind. Sem. Hort. Petrop. 87. 1866.
Canna lutea var. *aurantiaca* Regel, loc. cit.

Stems purplish, glabrous; leaves lanceolate, equilateral, acuminate apically, cuneate at base, purplish, lighter beneath than above, minutely punctate on both surfaces, sessile or sometimes decurrent into a winged petiole, 20–25 cm long, 9–13.5 cm wide; sheath to 12 cm long; racemes simple or branched, about 15 cm long, erect, elongate, slightly glaucous; primary bracts to 9 cm long, 4.5 cm wide, leaflike, ovate, acuminate, spreading apically, sheathing basally; secondary bracts lanceolate, 6 cm long, deciduous, sheathing, obtuse apically, the margins membranaceous; flowers solitary; pedicels 2–4 mm long in flower, 4–10 mm in fruit; floral bracts persistent, 6–11 mm long, 4–7 mm wide, obovate, truncate apically, more or less membranaceous; bracteoles lanceolate, 6 mm long, about 2 mm wide; sepals subulate, subequal, 6–9 mm long, 2 mm wide; petals subequal, 3.5–4 cm long, lanceolate, acute apically, erect, connate basally into a tube about as long as longest sepals, free part involute basally; staminodes 2, unequal, to 5 cm long, 6 mm wide, spatulate, narrowed below, notched apically, slightly spreading; lip linear, 4.5 cm long, revolute, recurved, notched apically; stamen lanceolate, erect, about 5 cm long; anther 7 mm long; style linear, 4.5 cm long, erect, basal half adnate to stamen; stigma more or less apical, 1.5 mm long; ovary 4–5 mm high, ellipsoid, papillose; capsule ellipsoid, spiny, about 2 cm high, 1.5 cm wide.

From Mexico and the Caribbean region to southern Brazil; also widely planted as an ornamental. Specimen examined: Santa Cruz.

This species has been introduced into the Galápagos Islands; a note on the label of the specimen examined reads: "A pest in fields, spreads badly." Two collections of *Canna* listed by Stewart (1911) as from Isabela and Santa Cruz have not been seen; their identity is uncertain.

See Fig. 225.

COMMELINACEAE. SPIDERWORT FAMILY*

Annual or perennial, usually succulent herbs with a repent, procumbent, or erect habit, the nodose stems often rooting at nodes. Leaves alternate; petioles dilated into a sheathing base; blades flat or trough-like, entire, parallel-veined. Flowers perfect, usually small, umbellate, cymose (in ours), racemose, or capitate, the bracts small or large, often spathe-like; perianth 2-seriate, the outer of 3 green, distinct, herbaceous sepals, the inner a blue, pink, or white ephemeral and deliquescent corolla of 3 equal or slightly unequal, free, or basally connate petals. Stamens 6, sometimes fewer by abortion, the abortive usually persistent as staminodes, hypogynous; anthers 2-celled. Ovary superior, sessile, 2- or 3-celled; placentation axile;

* Contributed by Ira L. Wiggins.

Fig. 225. *Commelina diffusa* (*a*). *Canna lutea* (*b–d*).

ovules few, orthotropous. Stigma entire to faintly lobed. Fruit a capsule or crustaceous and indehiscent. Seeds with a linear or punctate funicular scar, their coats usually reticulate, muricate, or ridged; embryo situated beneath a disklike callosity; endosperm mealy, copious.

About 25 genera, mainly of the tropics and subtropics, distributed worldwide.

Commelina L., Sp. Pl. 40. 1753; Gen. Pl. ed. 5, 25. 1754

Perennial herbs with simple or branched, glabrous or pubescent stems and fibrous or fusiform-fleshy roots. Leaves ovate to lanceolate, petioles sheathing stems, somewhat succulent. Peduncles solitary or aggregate, subtended by spathiform bracts, crowded at ends of branches or scattered in axils of leaves, often bifid to give rise to racemose branches, one producing 1–3 flowers, the other often with a somewhat secund arrangement of 2–12 flowers, those of 1 raceme often sterile, and 1 or more flowers of other racemes fertile. Sepals 3, the outer oblong-elliptic, cucullate, the inner 2 obovate to falcate, all green or in some species petaloid, persistent. Petals 3, the outer short-clawed, ovate, often smaller than the inner 2 or sometimes absent, inner 2 petals ovate to cordate, long-clawed, usually blue but sometimes white. Fertile stamens 3, anthers oblong, filaments glabrous; sterile stamens usually 2 or 3, anthers cruciate. Ovary glabrous, sessile, 2- or 3-celled, 2 ventral cells usually 1- or 2-ovuled, dorsal cell 1-ovuled or abortive. Capsule chartaceous, longitudinally 2- or 3-valved. Seeds small, ellipsoid to pyramidal, coat smooth or roughened.

About 90 species, widely distributed throughout the world.

Commelina diffusa Burm. f., Fl. Ind. 18, *pl.* 7, *fig.* 2. 1768

Commelina nodiflora sensu Burm. f., op. cit. 17, and auct., non L.
Commelina longicaulis Jacq., Coll. Bot. 3: 234. 1789.
Commelina pacifica M. Vahl, Enum. 2: 168. 1806.
Commelina cespitosa Roxb., Fl. Ind. 1: 178. 1820.
Commelina ochreata Schau., Nova Acta Acad. Leop.-Carol. Nat. Cur. 19: Suppl. 1: 447. 1843.

Stems prostrate, ascending, scandent or erect, moderately to much branched, the branches to 8 dm long, usually longer and more slender in deep shade than in open

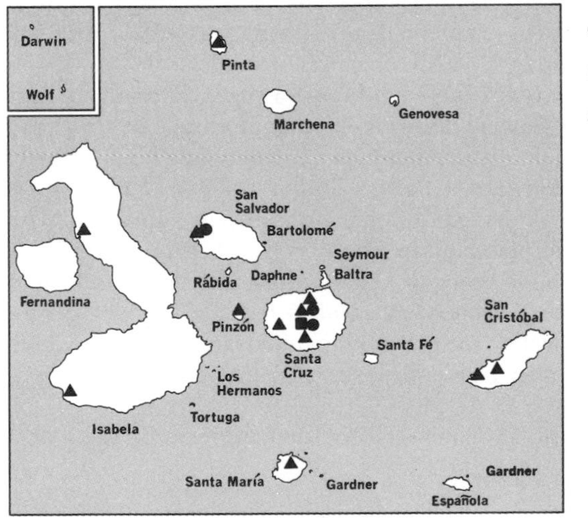

● *Canna lambertii*
■ *Canna lutea*
▲ *Commelina diffusa*

habitats, glabrous except along margins of spathes and edges of basal sheaths of leaves, rooting freely at nodes; internodes 2–8 cm long; leaves ovate, 5–20 mm wide, 1.5–5 cm long, acute to acuminate at apex, rounded at base; sheaths thin, scarious, 2.5–5 mm broad, 5–15 mm long, glabrous except for white-ciliate margins; spathes ovate to lanceolate-ovate, acute, rarely acuminate, 5–7 mm deep, 10–15 mm long, green, sometimes tinged with purple, glabrous but with white marginal cilia on lower third to two-thirds; peduncles 1–5 cm long, or sometimes nearly obsolete; lower raceme of inflorescence 1–3-flowered, the lowermost usually fertile, the others abortive, upper peduncle longer than lower, producing several flowers, these usually all abortive; sepals 3–4 mm long, green, marginally scarious; petals blue or sometimes white, the 2 upper 4–5 mm long, the lower about two-thirds as long; capsule 4–5 mm long, broadly ellipsoid, on sharply reflexed pedicel 1–2 mm long; seeds usually 5 per capsule, black, faintly reticulate, about 3 mm long.

Florida and Mexico south to Peru; terrestrial or occasionally on large limbs of trees with *Tillandsia* and cryptogamic epiphytes, often in deep shade of trees and large shrubs, on shaded cliffs, and under overhanging ledges; in some areas a common weed around cultivated fields. Specimens examined: Isabela, Pinta, Pinzón, San Cristóbal, San Salvador, Santa Cruz, Santa María.

Said to be eaten by cattle and grazed by the tortoises on Isla Santa Cruz. White-flowered forms are not uncommon.

See Fig. 225.

CYPERACEAE. SEDGE FAMILY*

Grasslike or rushlike annual or perennial herbaceous plants, sometimes with knotty ligneous rhizomes or stolons. Stems (culms) scapelike or nodose and leaved, trigonous, generally solid. Leaves radical and/or cauline, with linear or lance-elliptic blade and sheathing base, occasionally reduced to bladeless sheaths; ligules usually weakly developed, but ventral side of leaf sheaths sometimes projected beyond sheath orifice into a tongue-like appendage (contra-ligule). Inflorescences borne at apex of culms and often also at upper nodes of culms, corymbose, paniculate, spicate, racemose, or congested in a head, subtended by leafy bracts, with 1 to many compact units of flower-bearing glumes (spikelets or cymelets). Ultimate units of inflorescence a spikelet or (in Mapanioideae) a cymelet, with a prophyll at base; spikelets with glumes bearing an axillary flower; cymelets (in Mapanioideae) consisting of empty glumes imbricated or 2-ranked on a determinate axis terminated by a single pistillate flower. Flowers hermaphroditic or unisexual, rarely dioecious, without a perianth or with perianth of 3 to many bristles or of scaly segments. Stamens 1–3 (rarely more) to a flower; anther basifixed, opening by longitudinal slit, connective often projected beyond anther apex into a subulate appendage. Pistil usually 2- or 3-carpeled; style elongate or abbreviated, apex divided into 2 or 3 (rarely to 8) stigmatic branches; ovary unilocular with single anatropous ovule, maturing into achene. Achenes lenticular or trigonous, rarely globular, with sclerenchymatous achene walls, free or (in Mapanioideae) adnate in whole or in part to surrounding utricular structure and developing into compound fructification.

More than 3,800 species in about 48 genera, distributed throughout the world

* Contributed by Tetsuo Koyama.

under all ecological conditions, with highest generic concentration in tropical South America. Three subfamilies—Mapanioideae, Cyperoideae, and Caricoideae—are recognized, the Caricoideae not represented in the Galápagos Islands.

Achenes terminal at apex of cymelet axis, globular or very obtusely trigonous, bony, exposing upper part over subtending glumes; flowers unisexual; all leaves of 2 parts, a closed sheath and a grasslike blade, the blades mostly near or above middle of culms (Subfam. Mapanioideae) . . . . *Scleria*
Achenes borne laterally at axil of glumes of spikelets, 3-sided or lenticular, not bony, completely hidden by subtending glumes; fruit-producing flowers hermaphroditic; all or most blade-bearing leaves in basal tufts, or culms leafless (Subfam. Cyperoideae):
  Spikelet-bearing glumes of various shapes and sizes, the lower glumes empty and smaller than the flower-bearing middle glumes, only 1 to few glumes of a spikelet flower-bearing . . . . *Rhynchospora*
  Spikelet-bearing glumes similar in shape and size, usually all component glumes of a spikelet flower-bearing:
    Glumes 2-ranked . . . . *Cyperus*
    Glumes imbricated:
      Inner scale (small and hyaline) present between achene and rachilla of spikelets; style base neither thickened nor jointed at base . . . . *Hemicarpha*
      Inner scale lacking; style base thickened, persistent at apex of achene, or jointed at base and falling from mature achene:
        Style base jointed between its base and apex of achene, falling from mature achene . . . . *Fimbristylis*
        Style base persistent at apex of achene as verruciform or conical appendage:
          Inflorescence an open umbel with few to many spikelets; leaves with elongated blade; style base verruciform . . . . *Bulbostylis*
          Inflorescence a single terminal spikelet; leaves reduced to bladeless sheaths at base of naked culms; style base conical . . . . *Eleocharis*

Bulbostylis Kunth, Enum. Pl. 2: 205. 1837

Small or medium-sized annual or perennial sedges, rarely with ligneous caudex. Culms tufted, bearing leaves only at base. Leaves few to a culm; blades filiform or capillary, convolute, long-sheathing at base; ligule absent. Inflorescence an umbel-like corymb, a head, or rarely a single terminal spikelet. Spikelets consisting of many glumes imbricated or rarely subdistichously disposed on a continuous simple axis. Each glume bearing axillary hermaphroditic flower. Flowers with pistil and 1 or 3 stamens; hypogynous bristles absent. Achenes trigonous or lenticular; style bases persisting as tubercle-shaped or conical appendage at apex of achene; stigmas 2 or 3.

About 90 species in tropical and warm regions of the world; more abundant in African and South American savanna areas than in tropical Asia.

Bulbostylis hirtella (Schrad.) Nees ex Urban, Symb. Antill. 2: 166. 1900

Isolepis hirtella Schrad. in Schult., Mant. Pl. 2: 70. 1824.
Fimbristylis capillaris sensu Robins., Proc. Amer. Acad. 38: 129. 1902, and sensu Stewart, Proc. Calif. Acad. Sci. IV, 1: 44. 1911, non A. Gray.
Stenophyllus hirtellus H. Pfeiffer, Bot. Archiv. 6: 190. 1924.

Densely tufted annual with fibrous roots; culms filiform, 5–25 cm tall, 0.2–0.3 mm thick, costate, hirtellous, few-leaved at base; leaves one-third to half as long as culms, blades capillary, hirtellous, involute, sheaths pale brownish, membranous,

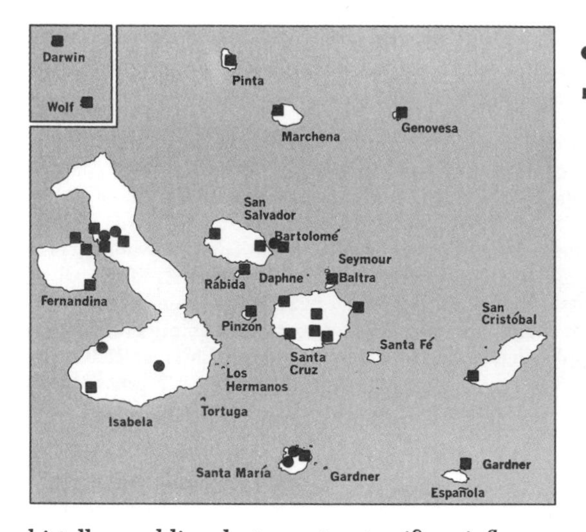

hirtellous, obliquely truncate at orifice; inflorescence umbelliform, 1–2.5 cm long, loosely bearing 3–8 spikelets; involucral bracts 1 or 2 setaceous, one-fourth to half as long as inflorescence; rays 1–3, 0.5–2 cm long, terminated by spikelet or by secondary umbel of 2 or 3 spikelets; spikelets ovoid or ovoid-ellipsoid, 3–5 mm long, 1.2–1.5 mm wide, fuscous; glumes somewhat loosely imbricated; ovate, strongly boat-shaped, 1.4–1.6 mm long, herbaceous, wholly hirtellous, dark reddish brown on both sides, the keel pale green, 3-veined, ending in a straight or recurved mucro at apex of glume; achene broadly obovate, 3-sided, 0.6 mm long, pale yellowish, finely papillose throughout; style base small, tubercle-shaped, fulvous, persistent on achene; style trifid at apex.

Widespread in tropical America; common in dry grasslands. Specimens examined: Bartolomé, Isabela, Santa María.

See Fig. 228.

Cyperus L., Sp. Pl. 74. 1753; Gen. Pl. ed. 5, 26. 1754

Annual or perennial sedges, sometimes with rhizomes or stolons. Leaves radical and subradical, with linear blades, occasionally reduced to bladeless sheaths; sheaths closed, without ligule. Culms simple, without nodes, leaves only at base, terminated by single inflorescence. Inflorescence a simple or compound umbel-like corymb or congested into a head, subtended by involucral bracts. Rays when present arising from a tubular prophyll. Spikelets few- to many-flowered, rachilla simple, or rarely jointed at base of spikelet or at base of each glume. Glumes 2-ranked (rarely slightly imbricated), base often decurrent along angles of rachilla, rachilla thus winged between glumes; all glumes bearing an axillary hermaphroditic flower (but in some only the middle glumes floriferous). Flowers of pistil and 1–3 stamens; hypogynous bristles none. Achenes trigonous or lenticular, in latter case either bilaterally or dorsiventrally compressed; style tri- or bifid, base not thickened.

A cosmopolitan genus, about 630 species occurring under all habitat conditions.

Spikelets not jointed at base, rachilla remaining on axis of spike; glumes falling individually from rachilla of spikelet:

Stigmas 3; achenes 3-sided:
  Spikelets spicately disposed on the elongated axis of spike; leaves and culms not septate-nodose:
    Plant annual, with fibrous roots only; spikelets greenish or yellow-green, strongly compressed; rachilla with acute angles but not winged; axis of spike relatively short, the upper spikelets thus becoming radiate.............*C. compressus*
    Plant perennial, with decumbent rhizome or stolons; spikelets sanguineous or straw-brown, only slightly compressed; rachilla conspicuously winged with decurrent bases of glumes; axis of spike markedly elongate, all spikelets spaced:
      Plant a robust perennial with short, decumbent, ligneous rhizome without a tuber; inflorescence large, decompound, 10–20 cm long...............*C. distans*
      Plant a medium-sized or small perennial with elongate slender stolons terminated by a tuber; inflorescence small, simple or in part compound, less than 10 cm long:
        Glumes 7-nerved, yellow-brown or reddish brown, veins relatively evenly spaced on both sides of glumes..............................*C. esculentus*
        Glumes 3- or 5-veined, deeply sanguineous, veins close together on obtuse keel, the sides of glumes thus sparsely veined.............*C. rotundus*
  Spikelets capitate or digitate at apex of rays, axis of spike hardly elongate; leaves and culms more or less septate-nodose:
    Glumes broadly ovate, 2–2.5 mm long, deeply red-brown or sanguineous, distinctly 2-nerved along both sides of keel, this scabrous toward apex; achenes blackish at maturity.............................*C. elegans* subsp. *rubiginosus*
    Glumes ovate or ovate-elliptic, 1.2–1.5 mm long, cinereous-green and ferruginous- or fulvous-tinged, the sides nerveless, minutely cellulate, keel smooth; achenes grayish brown at maturity..................*C. virens* subsp. *drummondii*
Stigmas 2; achenes lenticular:
  Achenes dorsiventrally compressed; culms leafless, clothed at base with bladeless sheaths only; lowest bract culmlike, erect, continuing culm.......*C. laevigatus*
  Achenes bilaterally compressed; culms leafy at base, leaves with blades; bracts leaflike, spreading:
    Spikelets linear or linear-lanceolate, 1.2–1.8 mm wide; glumes light yellowish brown or pale brown; inflorescence with many spikelets.......................
    ....*C. polystachyos* subsp. *holosericeus*
    Spikelets ovate-elliptic, 2–2.5 mm wide; glumes dark chestnut brown or pale and variegated with dark chestnut brown; inflorescence with few to several spikelets.....................................*C. rivularis* subsp. *lagunetto*
Spikelets jointed at base immediately above prophyll, the spikelets thus falling intact from axis of spike; glumes often semipersistent on spikelet:
  Achenes lenticular; stigmas 2; inflorescence a head....................*C. brevifolius*
  Achenes trigonous; stigmas 3; inflorescence umbelliform:
    Apex of glumes obtuse or mucronulate, not recurved; spikes cylindrical to oblong with elongate axis; spikelets slightly compressed:
      Spikelets linear, 10–15 mm long, somewhat loosely 9–14-flowered..*C. grandifolius*
      Spikelets elliptic, 2.5–6 mm long, densely 2–4-flowered:
        Leaves 2–6 mm wide, deep green, not septate-nodose; spikelets 2.5–4 mm long
        .... *C. anderssonii*
        Leaves 8–12 mm wide, cinerous-green, septate-nodose; spikelets 4–6 mm long
        .... *C. ligularis*
    Apex of glumes with recurved awn or cusp; spikes broadly ovoid to nearly globose with abbreviated axis; spikelets markedly compressed:
      Plant small annual; base of culms not thickened; glumes 1–1.5 mm long, reddish brown, apex with a recurved awn..........................*C. aristatus*
      Plant medium-sized perennial; base of culms later becoming thickened with a

cormlike enlargement; glumes 2–2.5 mm long, straw-colored or yellow-brownish, apex recurved-cuspidate .... *C. confertus*

Cyperus anderssonii Boeck., Linnaea 36: 388. 1870

Mariscus brachystachys Hook. f., Trans. Linn. Soc. Lond. 20: 179. 1847.
Mariscus mutisii sensu Hook. f., op. cit. 178. 1847, non HBK.
Cyperus brachystachys Anderss., Svensk. Freg. Eugen. Res. Bot. 2: 53, *t. 12, f. 2.* 1861, non Presl 1820, nec Nees 1869.
Cyperus mutisii Anderss., op. cit. 53. 1861, excluding basionym.

Perennial with solitary culms or these loosely tufted in small clumps, (15)30–70 cm tall, smooth throughout, the base thickened and slightly corm-shaped, clothed with purple-brown leaf sheaths and their fibrous remnants; leaves radical, a few to a culm; blades linear, shorter than to surpassing culm, 2–6 mm wide, occasionally minutely spinulate-scaberulous on margins and midrib; sheaths 2–15 cm long, the lower purple-brown, the upper reddish brown, orifice hyaline, ventrally obliquely truncate; inflorescence an umbel-like corymb, occasionally congested into a lobed head; involucral bracts 3–7, leaflike, much longer than umbel rays, the longer to 30 cm long and 6 mm wide, spreading; umbel rays, when developed, to 9, slender, smooth, 1.5–9 cm long, terminated by spike or pyramidal cluster of 3 or 4 sessile spikes; prophylls tubular, to 1 cm long, tinged with reddish purple; spikes oblong-cylindrical, crowded with spikelets, apex obtuse, middle spike 1–2 cm long, 7 mm thick, lateral ones 4–7 mm long; spikelets divergent or obliquely patent, jointed at base, oblong-elliptic, 3 mm long, obscurely subcompressed-tetragonous, generally 2(3)-flowered; rachilla broadly winged with pair of oblong hyaline decurrent bases of glumes; glumes erect, ovate, 1.5 mm long, herbaceous, apex subacute or subobtuse, yellowish brown or reddish brown on both sides, pale toward base, more or less 9-nerved including keel, upper margins white-hyaline; achenes 1–1.25 mm long, elliptical or oblong-elliptical, 3-sided, puncticulate, dark brown at maturity; style trifid.

Endemic; in wet grasslands and open sandy seashore from about sea level to 430 m; widely spread over all the islands, often maintaining a pure colony in dry lava crevices. Specimens examined: Baltra, Bartolomé, Darwin, Fernandina, Gardner (near Española), Genovesa, Isabela, Las Plazas, Marchena, Pinta, Pinzón, Rábida, San Cristóbal, San Salvador, Santa Cruz, Santa María, Wolf.

See Fig. 226.

Cyperus aristatus Rottb., Descr. Icon. 23, *t. 6, f. 1.* 1773

Cyperus inflexus Muhl., Descr. Goam. Pl. Calamar. 16. 1817.
Cyperus aristatus Rottb. var. *inflexus* Boeck., Linnaea 35: 500. 1868.

More or less tufted small annual, with fibrous roots; culms (2)5–20 cm tall, slender, smooth, few-leaved at base; leaves generally half to three-fourths as long as culms, narrowly linear, 1–2 mm wide, soft, smooth; basal sheaths purplish brown, to 15 mm long, only the outer short-bladed; inflorescence umbelliform with 1–3(5) rays or congested in head; involucral bracts 2–3(5), leaflike, patent, the longer to 7 cm long, 2 mm wide, much exceeding inflorescence; rays, when developed, slender, patent, 1–4 cm long; prophylls 2–3 mm long, tubular, red-brown, obliquely truncate at the bimucronate orifice; spikes ovoid or broadly ovoid, 6–15 mm long and as wide, densely bearing (3)7–16 spikelets; spikelets spreading, jointed at base, linear-oblong or broadly linear, 4–15 mm long, 1–1.5 mm wide, compressed, 8–30-

Fig. 226. *Cyperus anderssonii*: *b*, ray of inflorescence; *c*, *e*, prophylls; *f*, bractlet; *g*, *h*, glumes and achene.

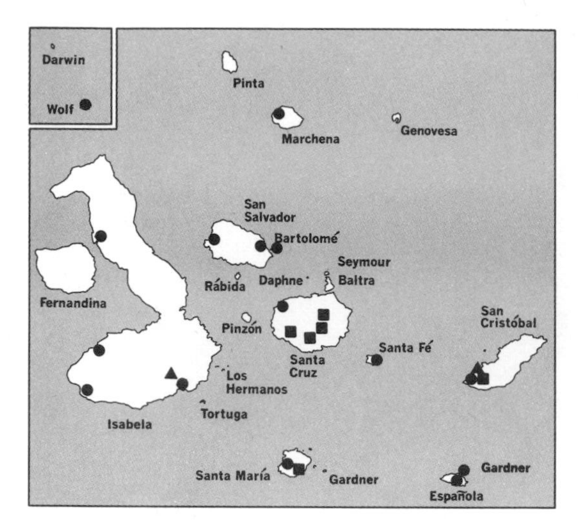

● *Cyperus aristatus*
■ *Cyperus brevifolius*
▲ *Cyperus compressus*

flowered; glumes ovate or ovate-elliptic, 1.5 mm long, thickly membranous, reddish brown or straw-colored, 7-nerved including keel, gradually narrowed to strongly recurved, acute apex, keel projected beyond glume apex into recurved awn 0.4–0.5 mm long; achenes narrowly obovate, 3-sided, 0.6–0.7 mm long, blackish brown at maturity, minutely puncticulate, abruptly contracted at the apiculate apex; style trifid.

Widely distributed in the tropical regions of the world; widespread in the Galápagos Islands in wet grasslands, open margins of wet places, and wet sand. Specimens examined: Bartolomé, Española, Gardner (near Española), Isabela, Marchena, San Cristóbal, San Salvador, Santa Cruz, Santa Fé, Santa María, Wolf.

Cyperus brevifolius (Rottb.) Hassk., Cat. Hort. Bogor. 24. 1844

Kyllinga brevifolia Rottb., Descr. Icon. 13, *t. 4, f. 3.* 1773.

Perennial with long creeping slender rhizomes; culms remotely to closely arranged in single row along rhizome, 7–30 cm tall, soft, slender, few-leaved at base; leaves radical; blades shorter than or occasionally equaling culms, narrowly linear, 2–3 mm wide, soft, herbaceous, scabrid on margins and on abaxial midvein; sheaths membranous, brownish or purplish brown, lower ones almost bladeless; inflorescence a single, terminal, globose head (rarely 2 or 3 heads); involucral bracts 3, leaflike, very unequal in length; head globose or broadly ovoid-globose, 5–10 mm long and as wide, pale green and often becoming straw-colored at maturity, crowded with spikelets; spikelets lance-oblong, compressed, 3–3.5 mm long, jointed at base, 4- or 5-squamose, 1-flowered; glumes ovate-elliptic, folded with an acute keel, membranous, pale green, sometimes with resinous spots, 7-nerved including midvein, cuspidate at apex, keel sparsely spinulose toward apex, projecting beyond glume apex into straight or slightly recurved, short cusp; achene 1.5 mm long, obovate, laterally lenticular, brownish, puncticulate; style 2-cleft.

Widely distributed as a weed in tropical and subtropical regions of the world, with a number of variants; in wet soil of both open and shady habitats, and in wet sand near the shore. Specimens examined: San Cristóbal, Santa Cruz, Santa María.

Cyperus compressus L., Sp. Pl. 46. 1753

Tufted annual with fibrous roots only; culms patent, 8–35 cm tall, relatively stout, 3-sided, smooth; leaves few to a culm, basal; blades linear, flat, 1–3 mm wide, shorter than culms, light green; sheaths membranous, pale brownish, striate; inflorescence umbelliform or congested, 2–10 cm long and as wide; involucral bracts 2–4, leaflike, unequal in length, the longest 2 or 3 times as long as inflorescence; rays when present 2–5, patent, 0.8–5 cm long, slightly compressed; spike bearing 3–10 spikelets on an abbreviated axis, oval or somewhat flabelliform, 3 cm long; spikelets linear-oblong, 10–25 mm long, 2.5–3 mm wide, 15–40-flowered, compressed, greenish and turning straw-colored at full maturity; rachilla not winged; glumes ovate or broadly so, 3–3.5 mm long, herbaceous or thinly coriaceous, strongly folded, with acute keel, 3-nerved on each side, keel green, finely many-veined, apex acute with straight mucro about 0.8 mm long; achenes 1–1.25 mm long, broadly obovate, 3-sided, dark brown, shiny, minutely puncticulate; style elongate, trifid at apex.

A widespread weed in open grassland and farmland, and along wet sandy riverbanks and coastal shores, in the tropical, subtropical, and warm regions of the world. Specimens examined: Isabela, San Cristóbal.

Cyperus confertus Sw., Prodr. Veg. Ind. Occ. 20. 1788

Cyperus dissitiflorus Anderss., Kongl. Svensk. Vet.-Akad. Handl. 1853: 153. 1855, non Nees ex Torr. 1836.
Cyperus biuncialis Anderss., op. cit. 156.
Cyperus confertus var. *biuncialis* Boeck., Linnaea 35: 515, 1868.
Mariscus confertus Clarke in Urban, Symb. Antill. 2[1]: 50. 1900.
Cyperus confertus f. *gracillimus* Kükenth. in Engler, Pflanzenr. IV[20], Heft 101: 498, 1936 (Ex typo!).
Cyperus confertus f. *biuncialis* Kükenth., loc. cit. 498. 1936.

Perennial, variable in size; culms tufted or solitary, (2)6–35 cm tall, (0.2)1–2 mm thick, trigonous, smooth, leafy at base, this later forming small cormlike enlargement; leaves few to a culm, all basal or rarely to one-third the height of culm; blades linear, 1–5 mm wide, 3–30 cm long, thinly herbaceous, soft, gradually acuminate at apex; sheaths tinged with brownish purple, ventral side hyaline, basal sheaths

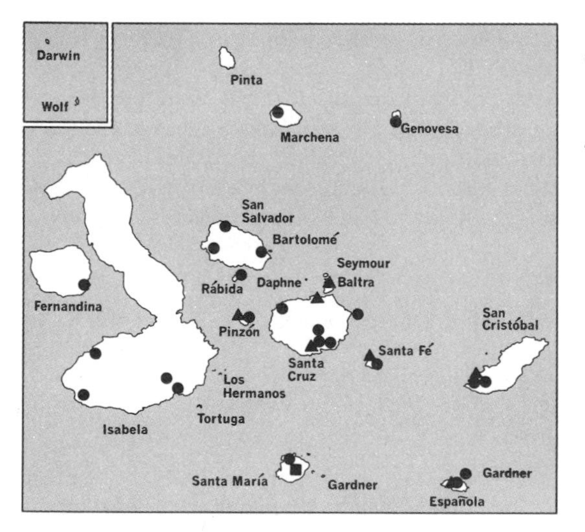

● *Cyperus confertus*
■ *Cyperus distans*
▲ *Cyperus elegans* subsp. *rubiginosus*

bladeless, dark brown-purple; inflorescence umbel-like, simple, a congested head in poorly developed individuals; involucral bracts 2–4, elongate, leaflike, 2 or 3 times as long as inflorescence; rays 2–5, patent, 1–8 cm long, slender; prophylls about 5 mm long, straw-colored, purplish-tinged, orifice shallowly 2-toothed; spikes globose-ovoid, 8–20 spikelets crowded on abbreviated axis, 10–15 mm long and as broad, straw-colored or light stramineous-brown; spikelets divergent, jointed at base, lance-oblong, 6–10 mm long, 2–2.2 mm wide, compressed, straw-colored, 6–18-flowered; rachilla only very narrowly winged or with extremely acute edges; glumes ovate, 2.2–2.5 mm long, thickly membranous or thinly herbaceous, straw-colored or light yellow-brownish, 3- or 4-nerved on both sides, gradually narrowed from above base toward a more or less 3-dentate apex, costa broad, green, projected beyond apex of glumes into a straight awn as much as 0.3 mm long; achene obovate, 1–1.2 mm long, blackish, puncticulate, tapering to base, contracted to mucronulate apex; stigma lobes 3.

Tropical America from the West Indies to the Venezuelan and Colombian coasts; occurring widely throughout the Galápagos Islands in open wet places along sandy shores and on lava beds to 300 m. Specimens examined: Española, Fernandina, Gardner (near Española), Genovesa, Isabela, Las Plazas, Marchena, Pinzón, Rábida, San Cristóbal, San Salvador, Santa Cruz, Santa Fé, Santa María.

Cyperus distans L. f., Suppl. Sp. Pl. 103. 1781

Perennial with short-creeping rhizomes; culms solitary, 30–70 cm tall, relatively stout in well-developed individuals, 3-sided, smooth, ligneous-thickened at base; leaves basal, several to a culm; blades broadly linear, 3–10 mm wide, equaling culm, flattish or slightly recurved, scabrous on margins, gradually long-acuminate at apex; sheaths brownish or purplish brown, eventually slightly disintegrated into fibers; inflorescence a large decompound umbel-like corymb, loose to subdense, 10–20 cm long and as wide; involucral bracts 4–6, leaflike, patent, lower 2 or 3 exceeding inflorescence; rays 5–10, patent, slender, 5–17 cm long; prophylls tubular, truncate at adaxial apex, briefly recurved and bicuspidate at abaxial apex; raylets 3–5 in a secondary corymb; bracteoles equaling secondary corymb; spikes broadly ovoid, 4 cm long, often divided at base, loosely bearing many spikelets; spikelets linear, subterete, 1.5–2 cm long, loosely 10–30-flowered, sanguineous, first obliquely patent, then spreading and eventually reflexed; rachilla flexuous, slender, winged, the wings hyaline and caducous; glumes elliptic, 1.5 mm long, suberect, membranous, hyaline-margined, sanguineous, 3–5-nerved on blunt keel, rounded to obtuse at apex; achenes oblong-cylindrical, 1–1.2 mm long, obtusely trigonous; style 3-cleft.

A pantropical sedge occurring in wet grasslands and along waysides. Specimens examined: Santa María.

Cyperus elegans subsp. **rubiginosus** (Hook. f.) Eliass., Svensk Bot. Tidskr. 59: 475, *f. 21.* 1965

Cyperus rubiginosus Hook. f., Trans. Linn. Soc. Lond. 20: 178. 1847.
Mariscus cornutus Anderss., Kongl. Svensk. Vet.-Akad. Handl. 1853: 151. 1855.
Cyperus cornutus Anderss., Svensk. Freg. Eugen. Res. Bot. 2: 53, *t. 13, f. 1.* 1861.
Cyperus rubiginosus var. *cornutus* Robins., Proc. Amer. Acad. 38: 128. 1902.

Perennial sedge tufted from very short rhizomes; roots purplish, slightly swollen, 0.5–1 mm thick; culms rigid, 6–70 cm tall, obtusely trigonous or nearly terete,

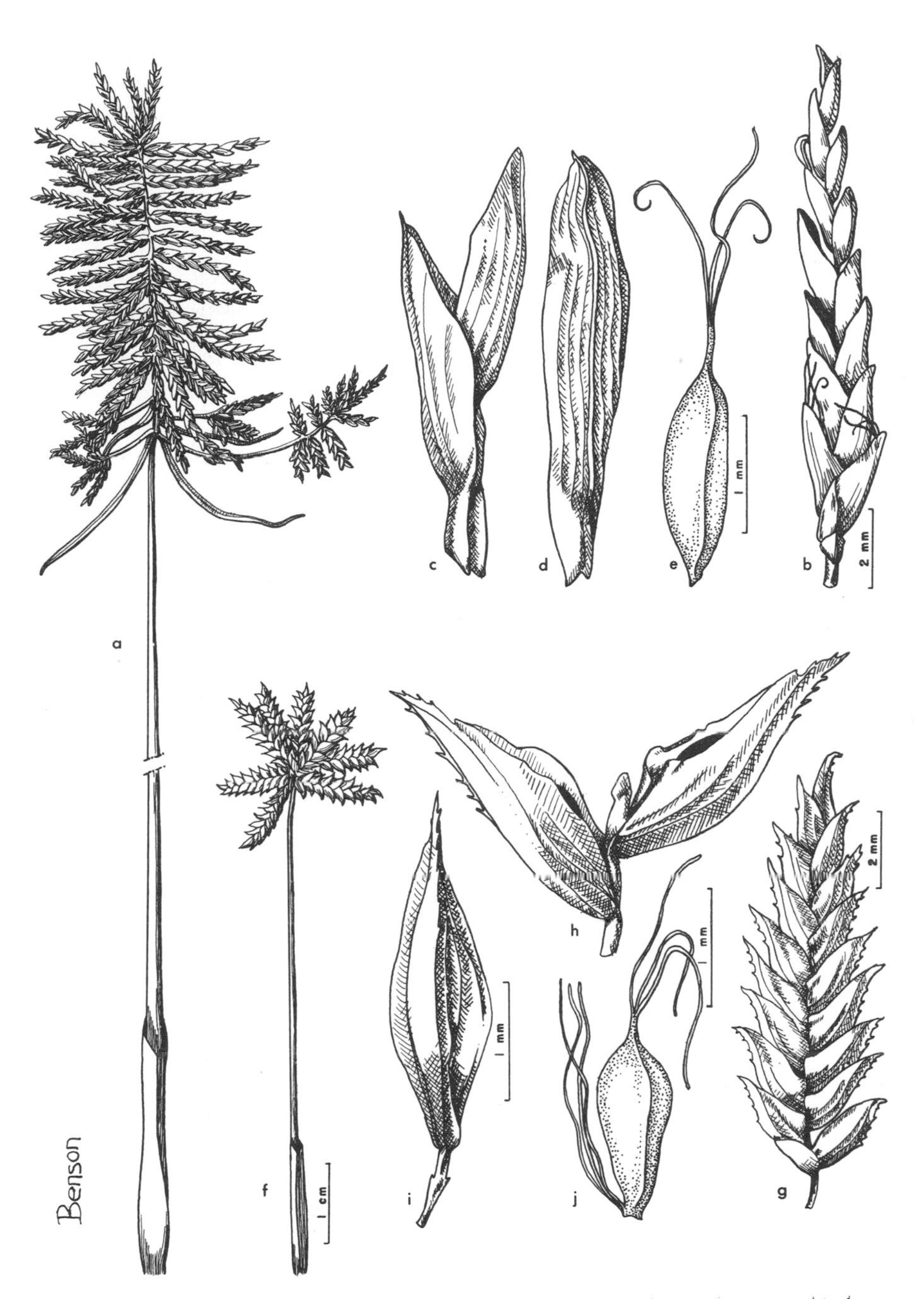

Fig. 227. *Cyperus grandifolius* (*a–e*). *C. elegans* subsp. *rubiginosus* (*f–j*).

smooth throughout, striate; leaves a few to a culm, essentially radical; blades narrowly linear, 1.5–4 mm wide, shorter than to sometimes exceeding culms, convolute, scabrous on margins, obtusely keeled, striate, gradually long-acuminate at apex; sheaths nearly cylindrical, 1–15 cm long, brownish red or sanguineous, many-nerved, weakly septate-nodose; inflorescence umbelliform, simple or in part compound; involucral bracts 3 or 4, leaflike, longer ones 2–5 times as long as inflorescence; rays 2–6, patent, very unequal in length, to 10 cm long; raylets when present to 2 cm long; prophylls tubular, sanguineous, 2-costate, truncated at orifice; spikelets 5–12 to a spike, capitate on very short axis, lance-oblong to ovate-oblong, 6–13 mm long, 3 mm wide, 8–22-flowered; rachilla hardly winged; glumes broadly ovate, 2(2.7) mm long, chartaceous, reddish brown, conspicuously 2-nerved on each side, keel prominent, greenish, serrulate-scabrous toward apex, projecting beyond glume apex into short subulate awn about 0.3 mm long; achene narrowly ovate, 3-sided, 1.2–1.5 mm long, blackish at maturity, cuneate-attenuate at base, contracted to an apiculate apex; style trifid.

Endemic; sandy, muddy, or rocky margins of both brackish and fresh-water pools, and among lava slabs along temporary rills throughout the archipelago. Specimens examined: Baltra, Española, Pinzón, San Cristóbal, Santa Cruz, Santa Fé.

See Fig. 227.

Cyperus esculentus L., Sp. Pl. 45. 1753

Cyperus strigosus sensu Hook. f., Trans. Linn. Soc. Lond. 20: 177. 1847, and sensu Robins., Proc. Amer. Acad. 38: 128. 1902, non L.

Slender perennial with fibrous roots; stolons filiform, elongate, covered with light brown scales, terminated by small tuber; culms solitary, 10–50 cm tall, slender, trigonous, smooth, thickened at base with small cormlike enlargement clothed with brown fibers; leaves radical, several to a culm; blades narrowly linear, 3–6 mm wide, shorter than to surpassing culms, flattish or slightly recurved; sheaths red-brown, disintegrating into fibers; inflorescence a loose umbel-like corymb, simple or in part compound, 4–7 cm long and wide; involucral bracts 3–6, leaflike, patent, lower 1 or 2 surpassing inflorescence; rays slender, 5–8, patent, 1–4 cm long; spike ovoid, loosely bearing 5–14 spikelets; spikelets linear-oblong to oblong, 5–12 mm long, 2 mm wide, slightly compressed, 8–16-flowered, yellow-brown or light brown; rachilla winged; glumes suberect, ovate or ovate-elliptic, 1.6–2.2 mm long, membranous, yellow-brown or reddish brown, pale hyaline on upper margins, obtuse at apex, 7-nerved including midvein, keel obtuse; achenes obovate or oblong-obovate, 1.5–1.75 mm long, obtusely trigonous, grayish brown, shiny, minutely puncticulate; style 3-cleft.

Widespread in tropical and warm regions of the world; on sandy seashore, farms, and open sandy grassland. Specimens examined: Isabela, San Cristóbal, Santa Cruz, Santa María.

Cyperus grandifolius Anderss., Kongl. Svensk. Vet.-Akad. Handl. 1853: 157. 1855

Cyperus galapagensis Caruel, Atti Real. Accad. Lincei (Roma), Ser. 4ª, 5: 621. 1889.

Large perennial with decumbent woody rhizome, 1–2 cm thick; roots stout, brownish, 1 mm thick; culm solitary, 75–120 cm tall, 4–6 mm thick below middle, trigonous, smooth throughout; leaves radical, many, blade broadly linear, 7–15 mm wide,

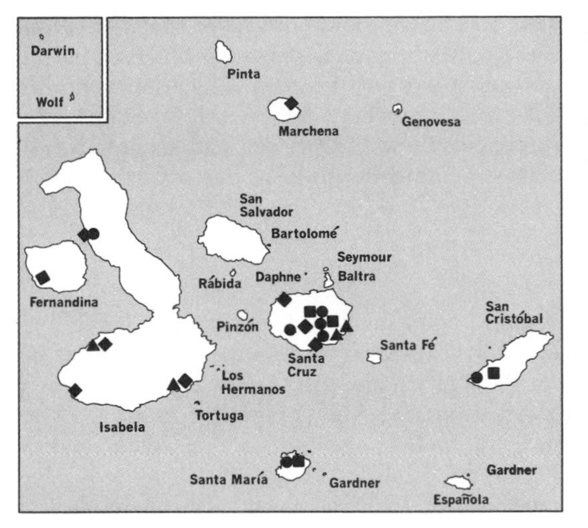

● *Cyperus esculentus*
■ *Cyperus grandifolius*
▲ *Cyperus laevigatus*
◆ *Cyperus ligularis*

shorter than or equaling culm, 1-costate, flattish, scaberulous on margins, gradually long-acuminate at apex; sheaths to 15 cm long, ventrally hyaline and fulvous, yellow-brown above, dark brown below, truncate at orifice; inflorescence a large, simple or compound umbel-like corymb 15–20 cm long and wide; involucral bracts 7–10, leaflike, longest to 60 cm long, 15 mm wide; rays 8–18, to 18 cm long, longer ones relatively equal, raylets when present to 3 cm long; prophylls to 3 cm long, tubular, dark brown, obliquely truncate at bi-aristate orifice; spikes cylindrical, dense, 3.5–5 cm long, 1–2.5 cm across, more or less crowded with spikelets; axis sparsely hispidulous on acute angles; spikelets divaricate, linear, slightly compressed, 10–18 mm long, 1.5 mm wide, 6–18-flowered, yellow-brown; rachilla winged with pair of narrow, white-hyaline wings; glumes loosely disposed, erect, ovate-oblong to elliptic-oblong, 3 mm long, herbaceous, yellow-brownish, 7–9-nerved, obtuse at mucronulate apex, keel smooth or sparsely spinulose; achenes oblong or obovate-oblong, 3-sided, 1.5–2 mm long, 0.5 mm wide, dark brown, puncticulate, contracted at both ends, apiculate at apex; style trifid.

Endemic; a large umbrella-shaped sedge in swampy habitats. Specimens examined: San Cristóbal, Santa Cruz, Santa María.

See Fig. 227.

Cyperus laevigatus L., Mant. Pl. 2: 179. 1771

Juncellus laevigatus Clarke, Fl. Brit. Ind. 6: 596. 1893.

Leafless creeping perennial; rhizome creeping, short or elongate, clothed with brownish scales; culms arranged in single row along rhizome, distant or close, 10–45 cm tall, 3-sided, smooth, stoutish, clothed at base with a few bladeless sheaths only; sheaths 2 or 3, brown or dark brown, obliquely truncate at orifice, uppermost short-bladed or mucronate at orifice, blade subulate, convolute, glaucous-green; inflorescence pseudolateral, capitate with 1–12 sessile spikelets; bracts 2, the upper erect, continuing culm, 3–8 cm long, much exceeding inflorescence, the lower short, patent; spikelets lance-oblong, slightly swollen, 7–12 mm long, 2 mm thick, 12–24-flowered, subacute at apex, straw-colored; rachilla tetragonal, nonwinged;

glumes closely disposed, broadly ovate, 2–2.25 mm long, obtuse or mucronulate at apex, herbaceous, straw-colored, occasionally ferruginous-striate, 3–5-nerved on the dorsally depressed keel; achenes obovate to ovate, 1.5 mm long, dorsiventrally concave-convex, obtuse at tip, grayish brown, puncticulate; style short, bifid.

Widespread in tropical and warm regions; in wet sandy soil and shallow water, especially in brackish habitats. Specimens examined: Isabela, Santa Cruz.

Cyperus ligularis L., Amoen. Acad. 5: 391. 1760

Mariscus ligularis Urban, Symb. Antill. 2[1]: 165. 1900.

Robust perennial with very short rhizomes; culm solitary, 30–80 cm tall, 3.5–5 mm thick, obtusely trigonous, smooth, minutely papillose; leaves many, densely tufted at base of culm; blades broadly linear, 8–12 mm wide, equaling or surpassing culm, 1-costate, flattish, subcoriaceous, glaucous-green, septate-nodulose, serrulate-scabrous on margins and midvein, apex gradually long-acuminate; sheaths reddish brown or purplish brown, eventually loosely disintegrated into fibers; inflorescence a dense, compound umbel-like corymb, 8–13 cm long and as wide; involucral bracts 5–8, leaflike, patent, much longer than inflorescence, longest 4–5 times as long as inflorescence; rays 7–12, patent, longer ones relatively equal in length, 2–10 cm long; prophylls tubular, purplish brown, truncate at orifice; spikes nearly sessile or short-peduncled, congested, the central one oblong-cylindrical, 1.5–2.5 cm long, 1–1.2 cm broad, the lateral subglobose, to 1.5 cm long, divaricate; spikelets divaricate, lance-oblong, subterete, 4–6 mm long, 2–2.7 mm wide, 2–6-flowered; rachilla broadly winged; glumes oval or ovate-oval, 2.5–2.7 mm long, slightly incurved, membranous, pale brown and occasionally tinged with red-brown, blunt at apex, 9-nerved, obtuse on keel; achene obovate, 3-sided, 1.5 mm long, brown, minutely papillose; style trifid.

Occurring widely in the tropics of Africa and America, and introduced into Micronesia; in wet sand of seashore and wet open grassland near the coast. Specimens examined: Fernandina, Isabela, Marchena, Santa Cruz.

Cyperus polystachyos subsp. **holosericeus** (Link) T. Koyama, Madroño 20: 253. 1970

Cyperus holosericeus Link, Hort. Berol. 1: 317. 1827.
Cyperus microdontus Torr., Ann. Lyc. N. Y. 3: 255. 1836.
Cyperus gatesii Torr., loc. cit.
Cyperus microdontus var. *texensis* Torr., op. cit. 430.
Cyperus fugax Liebm., Vidensk. Selsk. Skr. Kjøb. ser. 5, 196. 1851.
Cyperus inconspicuus Liebm., op. cit. 197.
Cyperus liebmannii Steud., Syn. Pl. Glumac. 2, 7. 1855.
Cyperus texensis Steud., op. cit. 9.
Cyperus polystachyos var. *leptostachyus* Boeck., Linnaea 35: 478. 1868.
Pycreus polystachyos var. *laxiflorus* Clarke in Urban, Symb. Antill. 2[1]: 17, 1900.
Cyperus polystachyos var. *leptostachyus* f. *inconspicuus* Kükenth. in Engler, Pflanzenr. IV[20], Heft 101: 376. 1936.
Cyperus polystachyos var. *leptostachyus* f. *fugax* Kükenth., loc. cit.
Cyperus polystachyos var. *texensis* Fern., Rhodora 41: 530. 1939.

Slender annual more or less tufted, with fibrous roots; culms slender, 4–25 cm tall, smooth; leaves few to a culm, all radical; blades shorter than to equaling culm, narrowly linear, 0.5–2 mm wide, flaccid, scaberulous on margins; sheaths brownish or purplish brown; inflorescence an open, umbel-like corymb, simple or subcompound, 3–9 cm long and wide; involucral bracts 2–4, leaflike, patent, longest 2–4 times as long as inflorescence; rays 3–8, slender, patent, unequal in length; prophylls tubu-

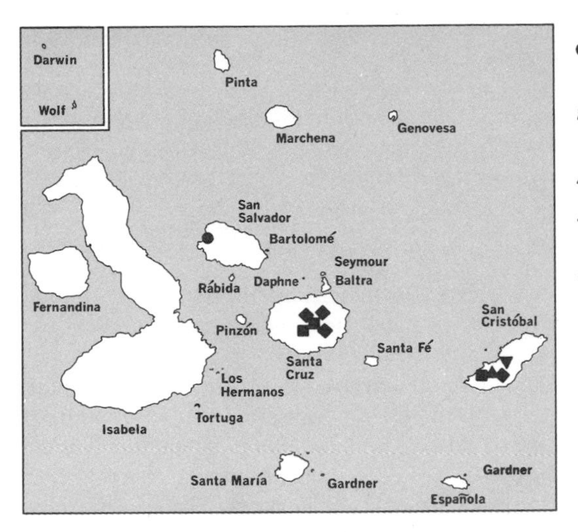

● *Cyperus polystachyos* subsp. *holosericeus*

■ *Cyperus rivularis* subsp. *lagunetto*

▲ *Cyperus rotundus*

◆ *Cyperus virens* subsp. *drummondii*

▼ *Eleocharis atropurpurea*

lar, 0.5–1 cm long, truncate at bicuspidate orifice, purplish toward base; spikes with 3–15 radiate spikelets on an abbreviated axis; spikelets linear, compressed, 6–23 mm long, 1.2–1.8 mm wide, yellow-brown, patent; rachilla flexuose, narrowly winged; glumes ovate, 1.2–1.5(2) mm long, strongly folded, membranous, contracted to muticous apex, the keel green, 3-nerved; achenes obovate, bilaterally compressed, lenticular, contracted at both ends, dark brown, puncticulate; style bifid.

Widespread from the eastern coastal United States through Central America and the West Indies to Venezuela and Ecuador; in open wet sandy places, especially near the shore. Specimens examined: San Salvador.

Cyperus rivularis subsp. **lagunetto** (Steud.) Kükenth. in Engler, Pflanzenr. IV[20], Heft 101: 383. 1936

Cyperus lagunetto Steud., Syn. Pl. Glumac. 2, 5. 1855.
Cyperus argenteus Clarke, Jour. Linn. Soc. 21: 64. 1884.
Pycreus lagunetto Clarke, Bot. Jahrb. 30, Beibl. 68: 8. 1901.
Cyperus rivularis var. *lagunetto* O'Neil, Rhodora 44: 86. 1942.
Cyperus tristachyus sensu Robins., Proc. Amer. Acad. 38: 129. 1902, non Boeck.

Annual with fibrous roots, tufted in small or moderate-sized clumps; culms slender, 8–30 cm tall, few-leaved at base; leaves narrowly linear to setaceous, 1–2 mm wide, soft, exceeding culms; sheaths brownish; inflorescence an open, umbel-like corymb, 3–6 cm long and wide; involucral bracts 3 or 4, leaflike, patent, unequal in length, longer ones exceeding inflorescence; rays 2–5, slender, patent, 1–4 cm long; spikes somewhat loose, bearing 3–8 spikelets; spikelets ovate to ovate-oblong, compressed, 7–12 mm long, 2.5 mm wide, dark chestnut brown or pale green and densely variegated with dark brown, 12–24-flowered; rachilla straight or nearly so, wingless; glumes ovate, strongly folded, 1.8–2 mm long, thickly membranous, pale brownish and sanguineous-tinged, contracted to subobtuse apex, neither sulcate nor nerved on either side, keel green, prominent, 3-nerved; achenes obovate to nearly ovate-orbicular, 1 mm long, laterally lenticular, dark brown, somewhat shiny, the apex contracted to short-pointed; style 2-cleft; stamens 2 or 3.

In wet grassland, from Central America through the Andean region of Venezuela and Colombia south to Chile and Argentina. Specimens examined: Santa Cruz, San Cristóbal.

The subspecies *lagunetto,* to which all Galápagean and continental South American plants belong, differs from the North American subspecies *rivularis* in having relatively broader achenes and spikelets and taller and less densely tufted culms.

Cyperus rotundus L., Sp. Pl. 45. 1753

Relatively small perennial with fibrous roots; rhizome elongate, slender but rigid, covered with brown scales and terminated by small tuber; culm solitary, slender, smooth, 7–30 cm tall, ligneously thickened at base with small cormlike enlargement; leaves few to a culm, radical; blades much shorter than culm, 2–4 mm wide, patent or slightly recurved; sheaths eventually disintegrating into brown fibers; inflorescence a loose, umbel-like corymb, 4–8 cm long, 2–6 cm wide, simple or in part compound; involucral bracts 1 or 2, leaflike, equaling inflorescence or shorter; rays 1–5, patent, slender, unequal in length, longest to 6 cm long; spikes 3–4 cm long, loosely bearing 3–8 spikelets; spikelets linear, 1.5–3 cm long, 2–2.5 mm wide, somewhat compressed, deeply red-brown, slightly shiny, 20–30-flowered; rachilla slender, winged; glumes nearly erect, loosely disposed, oblong-ovate or oblong-elliptic, 3–3.5 mm long, deep red-brown or sanguineous, membranous, scarcely veined but closely 3- or 5-nerved on obtuse keel, midvein greenish; achenes oblong, trigonous, 0.8–1 mm long; style elongate, trifid.

A cosmopolitan weedy sedge, growing in farmlands, grasslands, and sandy or gravelly shores, riverbanks, and waste places. Specimen examined: San Cristóbal.

Since the specimens cited by Hooker f. (1847, 177) and by Andersson (1855, 153) very closely resemble *C. esculentus,* our specimen (*Wiggins & Porter 409* [CAS]) appears to be the first documentation of true *C. rotundus* from the Galápagos Islands. The differentiation between the 2 species is extremely weak and is based almost solely on the glumes. The pale brown or straw-colored glumes of *C. esculentus* show 3 conspicuous and relatively evenly located nerves on each side in addition to a slender midvein, whereas in *C. rotundus* the deep red-brown or sanguineous glumes are essentially nerveless on both sides and their 3 (sometimes 5) nerves are located close together on the relatively prominent keel. With respect to the vegetative parts, *C. esculentus* tends to be larger than *C. rotundus,* and its well-elongated blades surpass the inflorescences. As a rule, in *C. rotundus* the leaves are much shorter than the flowering culms. According to Kükenthal, the underground runner of *C. rotundus* is lignescent and more rigid than that of *C. esculentus,* which is not persistent.

Cyperus virens subsp. **drummondii** (Torr. & Hook.) T. Koyama, Madroño 20: 254. 1970

Cyperus drummondii Torr. & Hook., Ann. Lyc. N. Y. 3: 437. 1836.
Cyperus virens var. *drummondii* Kükenth. in Engler, Pflanzenr. IV[20], Heft 101: 181. 1936.
Cyperus surinamensis sensu Anderss., Kongl. Svensk. Vet.-Akad. Handl. 1853: 153. 1855, sensu Robins., Proc. Amer. Acad. 38: 129. 1902, and sensu Eliass., Svensk Bot. Tidskr. 59: 478. 1965, non Rottb.

Perennial with slightly thickened, short rhizomes; culms solitary or 2 or 3 together, 30–60 cm tall, 1.5–3 mm thick, acutely triquetrous, usually scabrid on angles, septate-nodulose below middle; leaves radical, few; blades linear, equaling or slightly exceeding culm, 3–6 mm wide, soft, herbaceous, more or less septate-nodulose beneath, gradually long-acuminate at apex; sheaths to 20 cm long, reddish brown or purplish brown, septate-nodose; inflorescence a congested, simple umbel, 3–4 cm tall, 3–5 cm wide; involucral bracts 3 or 4, leaflike, patent, longer than inflorescence, longest to 20 cm long, 4 mm wide; rays short, 7–25 mm long; spikelets many, congested in glomerules 8–12 mm across, oblong or lance-oblong, 7–10 mm long, 2–2.5 mm wide, cinereous-green, ferruginous- or fulvous-tinged, 16–22-flowered; rachilla straight or nearly so, wingless; glumes somewhat loosely disposed, ovate or ovate-elliptic, boat-shaped, 1.2–1.5 mm long, thinly coriaceous, cellulate, pale green and ferruginous- or fulvous-tinged on both sides of midvein, contracted to mucronulate apex, keel broadly obtuse, green, obsoletely and closely 3-nerved; achenes three-fourths to four-fifths as long as glume, ovate-oblong or narrowly elliptic, 1.25 mm long, 3-sided, gray-brown, contracted at the base, subacute at the apex; the style 3-cleft.

In wet, open grassland from the southern United States through the West Indies and Central America to Brazil and Ecuador. Specimens examined: San Cristóbal, Santa Cruz.

Eleocharis R. Br., Prodr. Fl. Nov. Holl. 224. 1810

Small or medium-sized annual or perennial rushlike sedges, sometimes with rhizomes or stolons; culms 3-sided or terete, sometimes septate-nodose, leafless, clothed at base only with a few bladeless sheaths. Inflorescence a single terminal spikelet. Spikelet consisting of few to many glumes imbricate on continuous simple axis, each glume bearing axillary flower; flowers hermaphroditic, with pistil and 2 or 3 stamens; hypogynous bristles often present, 4–6(8) to a flower, frequently retrorsely spinulose-scabrous. Achenes trigonous or lenticular, the style base spongy, thickened, persistent at apex of achene. Style bi- or trifid.

About 120 species in wet habitats in most of the world.

Spikelet cylindrical, scarcely wider than culms; glumes herbaceous:
  Glumes loosely imbricate, 4–5 mm long; achenes cancellate with 13–16 rows of cells .... *E. fistulosa*
  Glumes densely imbricate, 2.7–3 mm long; achenes cancellate with 22–24 rows of cells .... *E. mutata*
Spikelet ellipsoid, ovoid, or oblong, clearly wider than culms; glumes membranous:
  Plant perennial with decumbent rhizomes or elongated stolons:
    Plant stoloniferous; culms filiform, rigid, 0.5–0.7 mm across, striate; sheaths obliquely truncate at orifice with a hyaline appendage; achenes black and shiny at maturity; hypogynous bristles 7 or 8, shorter than or equaling achene....... *E. maculosa*
    Plant with decumbent rhizome; culms septate, soft, 1–2.5 mm across, not striate; sheaths transversely truncate at orifice; achenes olive-colored or light brown at maturity; hypogynous bristles 4–6, exceeding achene.............. *E. nodulosa*
  Plant with fibrous roots only, without rhizomes or stolons:
    Orifice of sheaths thinly herbaceous, obliquely truncate with a short, subulate notch; achenes black and shiny at maturity, 0.5 mm long; hypogynous bristles 0–4; culms capillary, about 0.3 mm broad...................... *E. atropurpurea*
    Orifice of sheaths hyaline, fugaceous, irregularly bilobed; achenes olive-colored at maturity, 0.8–1 mm long; hypogynous bristles 7 or 8; culms slender, 0.7–1 mm broad .... *E. sellowiana*

Eleocharis atropurpurea (Retz.) Presl, Rel. Haenk. 1: 196. 1828
Scirpus atropurpureus Retz., Obs. Bot. Fasc. 5, 14. 1789.

Small slender annual with fibrous roots, densely tufted in large clumps; culms capillary, 3–17 cm tall, about 0.3 mm broad, often slightly arcuate, obtusely trigonous, sulcate; sheaths 3 or 4 to a culm, membranous, 5–25 mm long, the lower deep purplish or red-brown, the uppermost pale brown or straw-brown, orifice obliquely truncate with a short, oblique notch; spikes ovoid to ellipsoid, terete, light brown or reddish brown, 2–8 mm long, 1.5–2.25 mm broad, apex subobtuse; glumes densely imbricate, ovate, 1–1.2 mm long, membranous, contracted to subobtuse apex, pale, deeply sanguineous-brown on both sides of green midvein; achenes obovate to obovate-obdeltoid, 0.5 mm long, biconvex, black-brown, shiny, smooth; base of style depressed-conical, about one-fifth as wide as body of achene, olivaceous; style bifid; hypogynous bristles 0–4, pale, shorter than achene, sparsely and retrorsely spinulose.

Occurring relatively sporadically in tropical and warm regions of the world; on wet, open margins of swamps and pools of both fresh and saline habitats. In the Galápagos Islands known only from Isla San Cristóbal.

Eleocharis fistulosa (Poir.) Link in Spreng., Jahrb. Gewächsk. 1, Heft 3: 78. 1820
Scirpus fistulosus Poir., Encycl. Méth. 6: 749. 1804.

Perennial, tufted with creeping, long, stolons; culms 30–80 cm tall, 3–4 mm in diameter, acutely triquetrous, yellowish green; sheaths few, membranous, 5–8 cm long, lower ones brown, upper ones pale brown, obliquely truncate at orifice; spikes cylindrical or oblong-cylindrical, 2–3 cm long, 3–4 mm in diameter, acutish at apex, yellow-green or stramineous-greenish, many-flowered; glumes rather loosely imbricate, broadly ovate, 4–5 mm long, lowest 2 empty, remainder floriferous, herbaceous, light stramineous-greenish and obscurely striate in median portion, margins narrowly scarious-hyaline and pale brownish; achenes obovate or broadly obovate, compressed-trigonous, 1.5–2 mm long, stramineous-brown, shiny, cancellate with

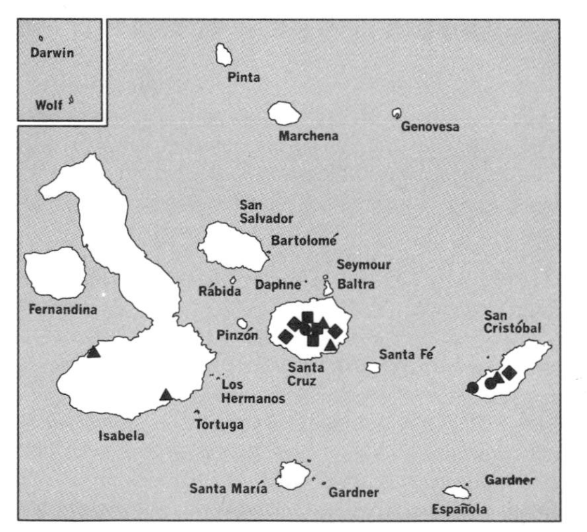

● *Eleocharis fistulosa*
■ *Eleocharis maculosa*
▲ *Eleocharis mutata*
◆ *Eleocharis nodulosa*

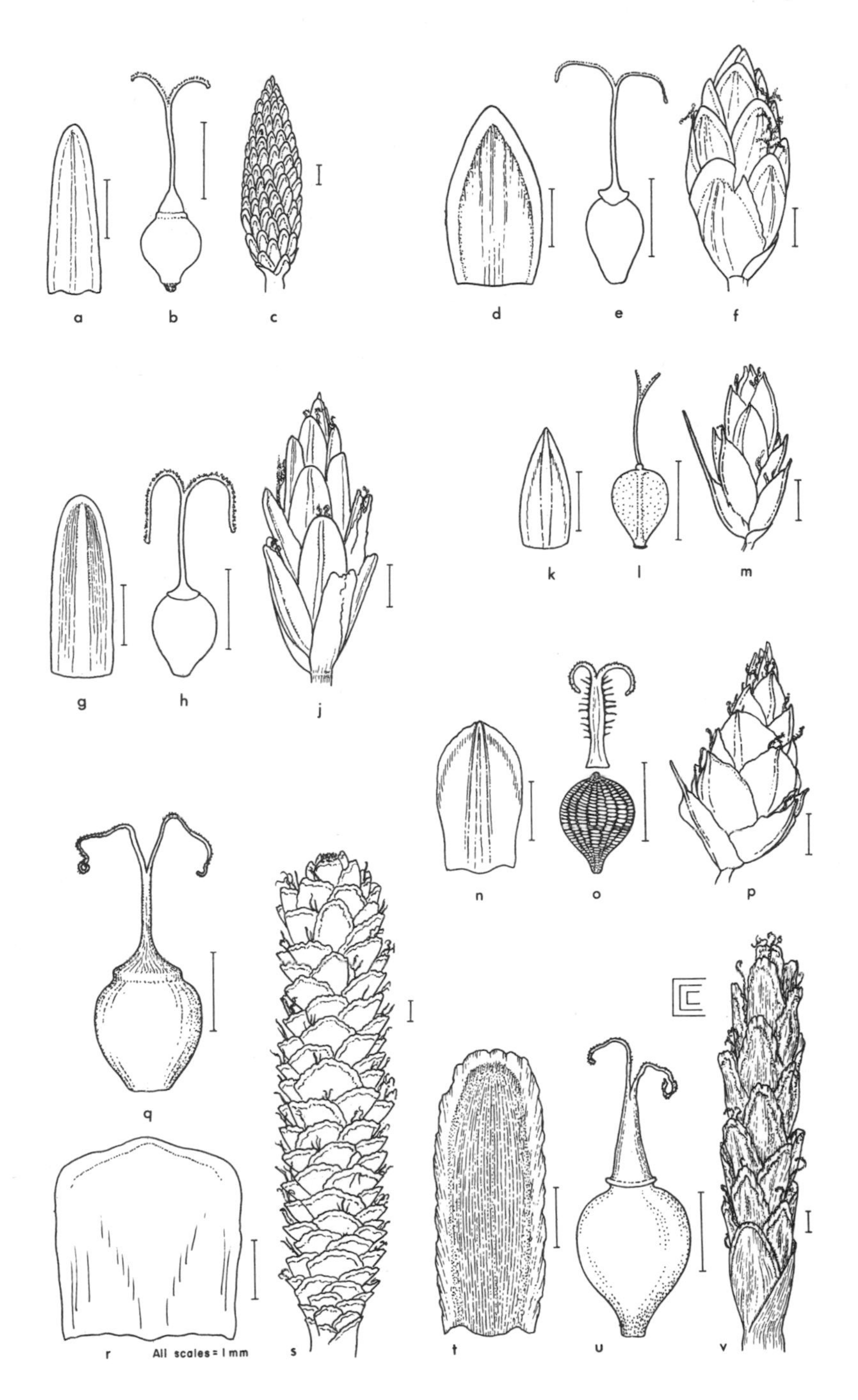

Fig. 228. *Eleocharis nodulosa* (*a–c*). *E. maculosa* (*d–f*). *E. sellowiana* (*g–j*). *Bulbostylis hirtella* (*k–m*). *Fimbristylis dichotoma* (*n–p*). *Eleocharis mutata* (*q–s*). *E. fistulosa* (*t–v*). (Scale lines 1 mm in each case.)

13–16 rows of transversely oblong cells; style base ovate-deltoid, about 0.5 mm long; style trifid; hypogynous bristles 6 or 7, 1 very short, rest attaining height of style base, stoutish, smoothish except for the sparsely retrorsely spinulose apex.

Occurring rather sporadically in swampy places of the Old and New World tropics. Specimens examined: San Cristóbal, Santa Cruz.

See Fig. 228.

Eleocharis maculosa (M. Vahl) Roem. & Schult., Syst. Veg. 2: 1954. 1817

Scirpus maculosus M. Vahl, Enum. Pl. 2: 247. 1805.

Densely tufted perennial with long-creeping, slender stolons; culms slender, stiffish, 7–30 cm tall, 0.5–0.7 mm broad, obtusely trigonous, sulcate; sheaths 2 or 3 to a culm, the lower ones red-brown or reddish purplish, the uppermost to 4 cm long, pale greenish above, reddish-brown below, reddish-striate, orifice obliquely truncate, exceeded by white-hyaline, rugose ligules; spikes ellipsoid or ovoid-ellipsoid, 5–12 mm long, 2–3.5 mm wide, acute at both ends, deeply red-brown; glumes tightly imbricate, elliptical or ovate, 2.5–3.5 mm long, membranous, deeply sanguineous-brown and slightly lustrous on both sides, finely striate, pale-hyaline on margins, costa 1-nerved, ending below the obtuse or rounded hyaline apex; achenes obovate, 1 mm long excluding style base, biconvex, contracted at apex, attenuate at base, surface roughened, shiny black at maturity; style base depressed-conical, two-thirds as wide as body of achene; style 2-cleft; hypogynous bristles 7 or 8, unequal in length, shorter than to nearly as long as achene, rust-brown, minutely retrorse-scaberulous with many spinules.

Sporadically recorded from wet and swampy places in tropical South America; in the Galápagos Islands known only from Isla Santa Cruz.

See Fig. 228.

Eleocharis mutata (L.) Roem. & Schult., Syst. Veg. 2: 155. 1817

Scirpus mutatus L., Amoen. Acad. 5: 391. 1759.

Perennial, tufted from short rhizomes; culms coarse, acutely triangular, 40–90 cm tall, 4–7 mm broad, smooth; sheaths 3 or 4 to a culm, membranous, to 35 cm long, straw-colored or lower ones brownish, obliquely truncate and pointed at apex; spike cylindrical, 2–5 cm long, 4–6 mm thick, subterete, obtuse at apex, stramineous; glumes densely imbricate in many rows, ovate-orbicular to orbicular, shallowly boat-shaped, 2.7–3 mm long, 2.5–2.75 mm wide, thinly herbaceous to thickly membranous, broadly hyaline on margins, straw-colored and hardly veined on median portion, semitranslucent and finely brown-lineolate on margins; achenes broadly obovate, compressed-trigonous, 1.5–2 mm long excluding style base, 1.2–1.5 mm wide, orange-brown, shiny, cancellate with 22–24 rows of shallow transversely oblong cells; style base depressed-deltoid, 0.25–0.5 mm long, with annular thickening at base; style 3-cleft; hypogynous bristles 6, equaling or slightly exceeding achene, brown, retrorsely scabrous with coarse but soft spinules.

Occurring in tropical America, the West Indies, Central America, and South America from Venezuela and the Guianas to Brazil and Paraguay; in shallow pools of fresh water, occasionally in saline habitats. Specimens examined: Isabela, San Cristóbal, Santa Cruz.

See Fig. 228.

Eleocharis nodulosa (Roth) Schult., Mant. Pl. 2: 87. 1824

Scirpus nodulosus Roth, Nov. Pl. Sp. 29. 1821.

Perennial with decumbent rhizomes; culms tufted, 20–80 cm tall, 1–2.5 mm broad, terete, striate, transversely septate; sheaths 2 or 3 to a culm, uppermost one 8–13 cm long, stramineous and tinged with reddish brown or reddish purple toward base, transversely truncate at orifice, lower sheaths to 3 cm long, deeply reddish brown or purplish brown, slightly obliquely truncate at orifice; spike lance-oblong, terete, 0.8–1.5 cm long, 2–4 mm wide, densely many-flowered, subacute at apex, reddish brown; glumes ovate, 2–2.5 mm long, thinly membranous, subacute or subobtuse at apex, reddish brown and whitish on hyaline margin, the keel 1-nerved; achenes broadly obovate, 1 mm long, biconvex, olive brown or chestnut brown, finely pitted-reticulate; style base deltoid, flattened, about half to two-thirds as wide as body of achene, about 0.3 mm long; style 2-cleft; hypogynous bristles 4–6, equaling or slightly exceeding body of achene, brownish, minutely retrorsely spinulose.

Relatively frequent in marshy places from the southern United States through the West Indies and Central America to Uruguay and northern Argentina. Specimens examined: San Cristóbal, Santa Cruz.

See Fig. 228.

Eleocharis sellowiana Kunth, Enum. Pl. 2: 149. 1837

Eleocharis galapagensis Svens., Rhodora 31: 233. 1929.

Annual; culms densely tufted with fibrous roots, (4)8–30(40) cm tall, 0.6–1 mm broad, relatively soft, striate; sheaths 2 or 3 to a culm, membranous, 1–6 cm long, reddish brown or straw-colored, apex white-hyaline, fugacious, irregularly bilobed; spikes elliptical or lanceolate, acute at tip, 5–10 mm long, 1.5–3 mm wide, fuscous or pale brown, rather loosely many-flowered; glumes oblong, 2.5–2.75 mm long, obtuse at apex, membranous, fuscous or pale brownish, white-hyaline on margins, the keel green, finely 3-nerved, ending below glume apex; achenes obovate or broadly so, 0.8–1 mm long, lenticular, olivaceous, finely pitted-reticulate; style base small, conical, one-fourth as wide as achene, 0.25 mm long, stramineous-green; style

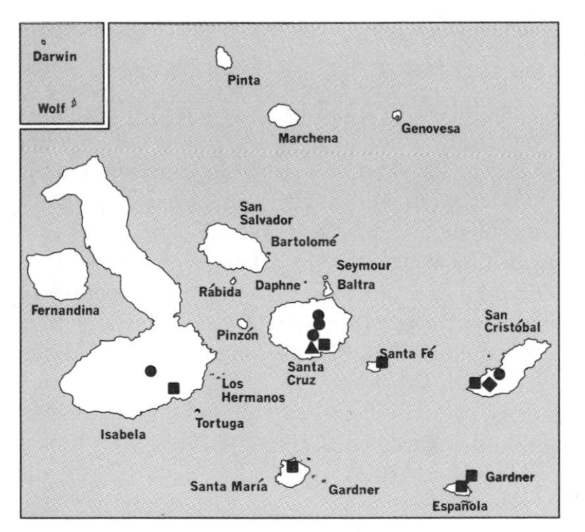

● *Eleocharis sellowiana*
■ *Fimbristylis dichotoma*
▲ *Fimbristylis littoralis*
◆ *Hemicarpha micrantha*

bifid; hypogynous bristles 7 or 8, whitish, shorter than achene, densely retrorse-spinulose.

Relatively sparsely scattered from the central United States to Uruguay; in swampy places and along dried-out margins of pools. Specimens examined: Isabela, San Cristóbal, Santa Cruz.

See Fig. 228.

Fimbristylis M. Vahl, Enum. Pl. 2: 285. 1806

Small or medium-sized annual or perennial sedges. Culms without nodes, leafy only at base. Leaves narrowly linear, flattish, obscurely costate, sometimes folded and obscurely 2-ranked; sheaths without ligules or with a fringe of pubescence at orifice. Inflorescence an umbel-like corymb or a congested head, subtended by involucral bracts, simple or compound with few to numerous spikelets. Spikelets bisexual, ovoid to globose or ellipsoid; glumes many, imbricate or nearly 2-ranked on a continuous simple axis, each glume bearing an axillary hermaphroditic flower. Flowers of pistil and 2 or 3 stamens; hypogynous bristles none. Achenes trigonous or dorsiventrally lenticular; styles dropping from achene after anthesis, filiform or flattened, the flattened form usually fimbriate-margined. Stigmas 2 or 3.

About 150 species, chiefly in tropical and temperate Eurasia, with some in tropical Africa and a few in tropical and temperate North and South America. The 2 species recorded from the Galápagos Islands are apparently of Old World origin. They occur now all over the world, owing possibly to the traffic in rice and other agricultural products.

Leaves dorsiventrally flat; spikelets ovoid, 5–7 mm long; achenes lenticular; styles flattened, fimbriate, bifid . . . . *F. dichotoma*
Leaves laterally compressed; spikelets globular, 2.5–3.5 mm long; achenes trigonous; style filiform, smooth, trifid . . . . *F. littoralis*

Fimbristylis dichotoma (L.) M. Vahl, Enum. Pl. 2: 287. 1806

Scirpus dichotomus L., Sp. Pl. 50. 1753.
Scirpus annuus All., Fl. Pedemont. 2: 277, *t. 88, f. 5.* 1785.
Scirpus diphyllus Retz., Obs. Bot. 5, 15. 1789.
Fimbristylis diphylla M. Vahl, Enum. Pl. 2: 289. 1806.
Fimbristylis annua Roem. & Schult., Syst. Veg. 2: 95. 1817.

Short rhizome sometimes present; culms tufted in small to large clumps, flattened, subtrigonous, 15–60 cm tall, smooth, glabrous or sparsely pilose, especially toward apex, leaves borne only at base; leaves shorter than or sometimes equaling culms; blades narrowly linear, 2–5 mm wide, flattish and slightly incurved, cinereous-green, glabrous or sparsely hairy, scaberulous toward blunt apex; sheaths pubescent or glabrescent, later becoming brownish, the orifice with a fringe of pubescence; inflorescence an open umbel-like corymb, 3–6 cm long and wide, compound or simple; involucral bracts 2–5, leaflike, shorter than or sometimes surpassing inflorescence; rays few to several, very unequal in length, the longest to 6 cm long; spikelets solitary or in a group of 2 or 3, ovoid, terete, 5–7 mm long, 2.5–3 mm broad, abruptly subacute at apex, densely many-flowered; glumes imbricate, appressed, broadly ovate, 2–3 mm long, thinly chartaceous, tinged with fuscous brown or chestnut brown, faintly few-veined on both sides, especially toward base, the keel obscurely 3-nerved, ending in mucro at apex of glume; achenes obovate or broadly so, biconvex, 0.8–1.2 mm long, cream- or straw-colored, cancellate with 7–15 rows of transversely oblong cells, short-stipitate at base, sometimes sparsely verruculose; style 2–2.5 mm long, flattened, fimbriate, falling from achene after anthesis, bifid at apex.

Widespread in the tropical and temperate regions of the world; in wet, open grassland and along margins of cultivated fields. Specimens examined: Española, Gardner (near Española), Isabela, San Cristóbal, Santa Cruz, Santa Fé, Santa María.

Fimbristylis dichotoma is definitely the most variable species of the genus, and the classification of its subspecific taxa requires study of the group on a worldwide basis. All of the specimens I examined from the Galápagos Islands represent a single type, a type quite common throughout the tropics and frequently referred to *F. annua*. It is characterized by relatively slender habit, less compound inflorescences, soft leaves, and bracts with slightly hairy margins. However, some of the Galápagos specimens appear to be perennial, though they do not show a conspicuous rhizome like that of *F. floribunda*.

See Fig. 228.

Fimbristylis littoralis Gaud. in Freyc., Voy. Bot. 413. 1826

Fimbristylis miliacea sensu auct., and sensu Svens., Proc. Calif. Acad. Sci. IV, 22: 190. 1939, non M. Vahl.

Culms densely tufted with fibrous roots only, 10–60 cm tall, strongly compressed, subtetragonous, smooth, clothed at base with pale brownish sheaths, these 2 or 3 to a culm, bladeless or nearly so, orifice obliquely truncate, acuminate; leaves bilaterally compressed, 2-ranked; blades linear, falcate, 1.5–2.5 mm wide, fresh green, gradually attenuate to long-acuminate apex, glabrous; leaf sheaths laterally compressed, broadly hyaline-margined, pale brown; inflorescence a decompound umbel-like corymb with numerous spikelets, 5–8 cm long and broad; involucral bracts 2–4, setaceous, shorter than inflorescence; rays many, weakly scabrous, to 6 cm long; bractlets setaceous; raylets slender, scaberulous; spikelets solitary at apex of raylets, globose or ovoid-globose, 2.5–4 mm long, 1.5–2.5 mm broad, rusty brown, densely many-flowered, apex obtuse or rounded; glumes imbricate, ovate or broadly ovate, 1 mm long, membranous, obtuse at apex, reddish brown or rusty brown on both sides, broadly white-hyaline on margins, keel 3-nerved, ending below apex of glume; achenes obovate, trigonous, 0.6 mm long, cream-colored, umbonate, cancellate with a few rows of transversely oblong cells, sessile; styles 1.5 mm long, filiform, smooth, falling from achene after anthesis, trifid at apex.

A pantropic sedge occurring in rice fields and other wet habitats. In the Galápagos Islands known only from a single collection from Isla Santa Cruz.

This sedge usually passes under the name *F. miliacea* M. Vahl, a name based on a Linnaean binomial, *Scirpus miliaceus*, whose type collection is a mixture of this species and a closely allied plant usually classified as *F. quinquangularis*. Because of this situation, there are two different views in the typification of *S. miliaceus* as expressed by Blake (1954, 217) and Kern (1954, 246), respectively. I follow Blake's opinion and use the name *F. littoralis* for this species, as I explained in my treatment in an earlier publication (Koyama, Jour. Fac. Sci. Univ. Tokyo III, 8: 109. 1961).

Hemicarpha Nees, Edinb. New Phil. Jour. 17: 263. 1834

Small annual sedges tufted from fibrous roots. Culms not nodose, leafy only at base. Leaves usually 2 to a culm. Inflorescence a head with 1 to several sessile spikelets, subtended by involucral bracts. Spikelets with numerous glumes closely imbricate on a continuous indeterminate rachilla. Each glume subtending a small hyaline inner scale and a hermaphroditic flower, the inner scale inserted between flower and rachilla. Stamens 1 or 2 to a flower. Achenes obovate-oblong, nearly terete. Stigmas 2.

About 3 species in North and South America.

Hemicarpha micrantha (M. Vahl) Britt., Bull. Torrey Bot. Club 15: 104. 1888

Scirpus micranthus M. Vahl, Enum. Pl. 2: 254. 1806.
Hemicarpha subsquarrosa Nees in Mart., Fl. Bras. 2^1: 61, *t. 4, f. I.* 1842.

Small slender annual, densely tufted from fibrous roots; culms slender, (2)5–20 cm tall, about 0.3 mm broad, obtusely trigonous, smooth, 2- or 3-leaved at base; leaves much shorter than culms, often reduced to almost bladeless sheaths; blades narrowly linear, 0.3–0.5 mm wide, flattish, smooth, somewhat obtuse at apex, 0.5–10 cm long; sheaths 5–25 mm long, reddish brown or light brown, obliquely truncate at orifice; inflorescence a head with 1–4 spikelets, pseudolateral; involucral bracts usually 2, the lower erect, 1–4 cm long, culmlike, the upper divergent to reflexed, 0.5–1 cm long; spikelets ovoid, terete, 3–4 mm long, 1.5–2 mm wide, subobtuse at apex, brownish green, densely many-flowered; glumes numerous, densely imbricate, rhombic-obovate, acute at apex, shallowly boat-shaped, 0.7–1 mm long, herbaceous, light green and variegated or tinged with chestnut brown on both sides, the keel green, ending in a short, erect mucro; inner scale inserted between achene and rachilla, hyaline, oblanceolate; achene obovate-oblong, 0.5–0.6 mm long, thickly plano-convex to nearly terete, cuneate at base, contracted to apiculate apex, brown, puncticulate; stigmas 2.

In moist sandy soil, from the United States south to tropical South America; in the Galápagos Islands known only from a single collection from San Cristóbal.

Rhynchospora M. Vahl, Enum. Pl. 2: 229. 1806, nom. conserv.

Perennial or occasionally annual sedges of various sizes. Culms scapelike or nodose and leafy. Leaves linear, sheath without ligule. Inflorescences terminal and lateral, corymbose, paniculate, or congested in a head. Spikelets lanceolate, ovate, or elliptic, terete or slightly compressed, brown or white, consisting of few to many glumes distichously disposed or somewhat imbricate on a continuous, simple axis; 1 to several hermaphroditic and fruiting flowers borne in axils of lower glumes; 1 to few staminate flowers in axils of upper glumes, uppermost glumes occasionally empty. Glumes 1-costate. Flowers bisexual or staminate owing to abortive pistil; hypogynous bristles 6 or rarely none, upwardly or retrorsely scabrous or smooth. Achenes lenticular, crowned with a persistent conical style base. Styles 2-cleft or undivided.

About 250 species ranging from tropical to subarctic regions of the world, with the greatest concentration of species in tropical America.

Inflorescence of terminal and lateral corymbs; lateral corymbs subtended by a sheathing green bract; spikelets brown or straw-colored:

Achenes 3–3.5 mm long; style base elongate-conical, about as long as body of achene; style scarcely divided at apex; robust sedge with large, open corymbs; leaves 8–20 mm wide . . . . *R. corymbosa*

Achenes less than 2 mm long; style base depressed or broadly conical, less than half as long as body of achene; style 2-cleft at apex; slender or medium-sized sedges with subdense, small corymbs; leaves 1.5–5 mm wide:

Corymbs borne along nearly whole length of culms; spikelets 1–1.5 mm long; achenes 0.5–0.7 mm long . . . . *R. micrantha*

Corymbs borne only on upper part of culms; spikelets 3–5 mm long; achenes 1–2 mm long:

Culms 50–90 cm tall; leaves 1.5–3 mm wide; spikelets rusty brown; achenes with 6 or 7 hypogynous bristles. . . . . *R. rugosa*

Culms 10–30 cm tall; leaves 1–2 mm wide; spikelets straw-colored or yellow-brown; achenes without hypogynous bristles. . . . . *R. tenuis*

Inflorescence a single terminal head subtended by involucral leaves, the basal portion of which is often snow-white; spikelets snow-white:

Leaf blades 1–2 mm broad; culms 8–30 cm tall; rhizome decumbent; inflorescence with 3–5 spikelets . . . . *R. nervosa* subsp. *nervosa*

Leaf blades 2–4 mm broad; culms 25–60 cm tall; rhizome short, suberect; inflorescence with 4–10 spikelets . . . . *R. nervosa* subsp. *ciliata*

Rhynchospora corymbosa (L.) Britt., Trans. N. Y. Acad. Sci. 11: 84. 1892

Scirpus corymbosus L., Cent. Pl. 2: 7. 1756.
Rhynchospora aurea M. Vahl, Enum. Pl. 2: 229. 1806.

Culm arising from short but thick rhizome, 80–120 cm tall, 3-sided, with several nodes, these leaf-bearing, scaberulous on angles on upper part; leaves many, aggregated at base of culm and several leaves on upper culm; blades broadly linear, 8–20 mm wide, flattish, herbaceous or thinly coriaceous, gradually long-acuminate at apex; sheaths elongate, the basal straw-brown or light brown; corymbs 2–4, on upper parts of culm, compound or decompound, rather contiguous, 10–15 cm long and broad; bracts leaflike, shorter than to much longer than corymb, sheathing at base; corymb rays many, the longer to 12 cm long; final raylets terminated by cluster of 2–5 spikelets, these lanceolate or elliptic, terete, 7–10 mm long, rusty to orange-

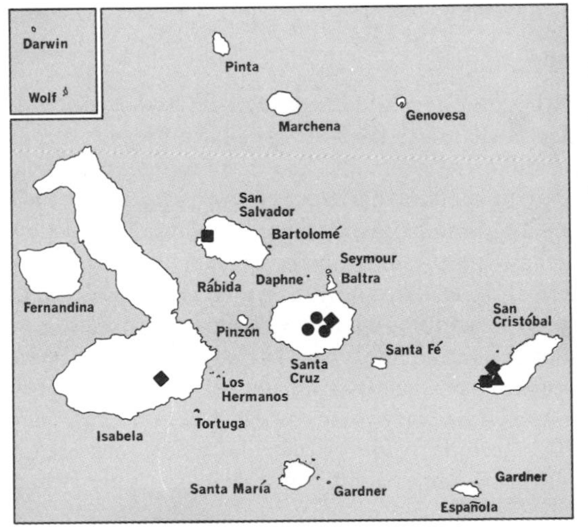

● *Rhynchospora corymbosa*
■ *Rhynchospora micrantha*
▲ *Rhynchospora nervosa* subsp. *nervosa*
◆ *Rhynchospora nervosa* subsp. *ciliata*

brown, bearing 1 fruit; glumes obscurely 2-ranked, the lower ovate, the upper oblong-ovate, 2.5–6 mm long, membranous, light brown, 1-nerved, acute at muticous apex; achenes obdeltoid-obovate, 3–3.5 mm long, yellow-brown, dull, finely wrinkled transversely in median portion, coarsely undulate-rugose toward margins; style base elongate-conical, as long as or slightly longer than achene body, furrowed in median portion; style elongate, scarcely divided at apex; hypogynous bristles 6, 4–5 mm long, brown, upwardly scabrid.

A pantropical sedge growing in swampy habitats; in the Galápagos Islands known only from Isla Santa Cruz.

See Fig. 229.

Rhynchospora micrantha M. Vahl, Enum. Pl. 2: 231. 1805

Small annual with fibrous roots; culms tufted in small clumps, setaceous, 5–30 cm tall, striate, smooth; leaves basal, a few to a culm; blades narrowly linear, shorter than to equaling culm, 1–2 mm wide, soft, light green; sheaths light brownish; corymbs 3–5, compound, open, remotely borne on culms except at extreme base; bract of lateral corymb leaflike, surpassing inflorescence, the sheathing base 4–8 mm long; bract of terminal corymb setaceous, to 2 cm long, hardly sheathing; corymb-rays spreading, 0.6–1 cm long, capillary, each terminated by a headlike cluster of spikelets; spikelets broadly ovoid, 1–1.5 mm long, 2- or 3-flowered, the lower 1 or 2 flowers maturing achenes; glumes few, lance-ovate, the fruit-bearing about 1 mm long, membranous, pale, brown-lineolate, costa green, ending in mucro; achenes obovate-orbicular, 0.5–0.7 mm long, biconvex, straw-brown, marginate on edges, coarsely undulate-rugose transversely, short-stiped; style base depressed-conical, basal lobes decurrent on upper edges of achene; style bifid; hypogynous bristles none.

Occurring abundantly in open sandy grassland and savannas of tropical America from southern Mexico and West Indies southward to Brazil. Specimens examined: San Cristóbal, San Salvador.

See Fig. 229.

Rhynchospora nervosa (M. Vahl) Boeck., Vidensk. Meddel. Kjøb. 1869: 143. 1869, subsp. **nervosa**

Dichromena nervosa M. Vahl, Enum. Pl. 2: 241. 1806.

Rhizome obliquely ascending, clothed with yellowish brown fibers, occasionally with elongate stolons; culms slender, 8–30 cm tall, glabrous, bearing leaves only at base, base thickened into small cormlike enlargement; leaves many; blades linear, 1–2 mm broad, shorter than to equaling culm, herbaceous, soft, gradually long-acuminate at apex, glabrous or sparsely pilose beneath; sheaths yellowish or yellow-brown, finally disintegrating into yellow-brown fibers; inflorescence a terminal head with 3–5 spikelets; involucral bracts 3–7, leaflike, very unequal in length, longest to 7 cm long, shortest 1–1.5 cm long, spreading, ciliate and pilose with long hairs toward base, this usually snow-white or occasionally pale, slightly dilated; spikelets lanceolate, terete, 6–8 mm long, acute at apex, snow-white, lower 4–7 flowers fruit-bearing, upper 3–5 staminate; glumes ovate to lanceolate-ovate, 3–5 mm long, membranous, white, subacute at apex, 1-costate, scabrid on upper costa; achenes broadly to orbicular-obovate, biconvex, 1–1.3 mm long excluding style base, rusty brown,

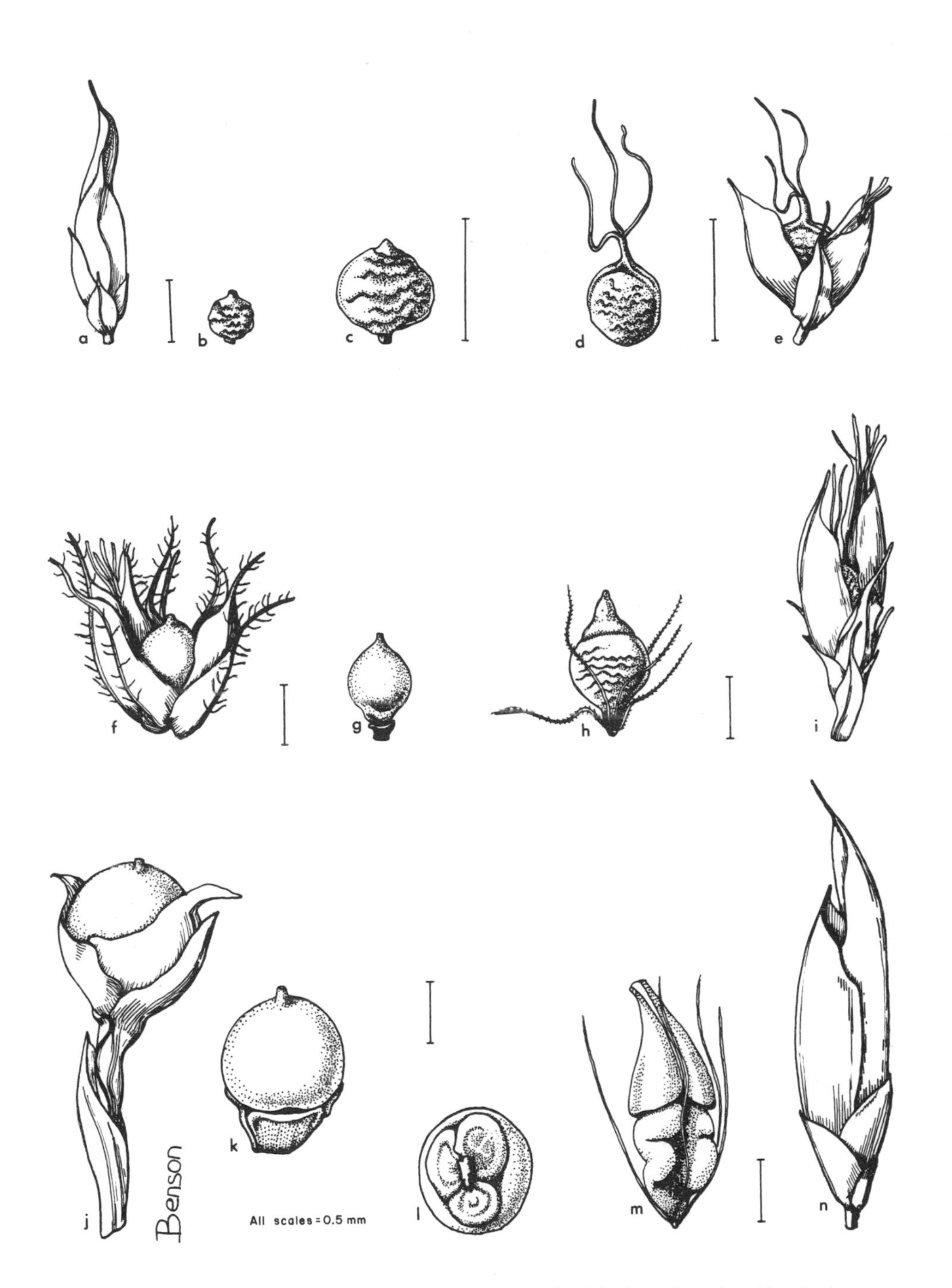

Fig. 229. *Rhynchospora tenuis* (*a–c*). *R. micrantha* (*d, e*). *Scleria hirtella* (*f, g*). *Rhynchospora rugosa* (*h, i*). *Scleria pterota* (*j–l*). *Rhynchospora corymbosa* (*m, n*).

transversely wrinkled; style base broadly conical, 0.4–0.5 mm long; style filiform, 2-cleft at apex; hypogynous bristles none.

In wet grassland of tropical America from the West Indies and Central America south to Brazil; in the Galápagos Islands known only from Isla San Cristóbal.

Rhynochospora nervosa subsp. **ciliata** (M. Vahl) T. Koyama, Madroño 20: 254. 1970

Dichromena ciliata M. Vahl, Enum. Pl. 2: 246. 1806.
Rhynchospora ciliata Kükenth., Bot. Jahrb. 56, Beibl. 125: 16. 1921.
Rhynchospora nervosa var. *ciliata* Kükenth., Bot. Jahrb. 75: 295. 1951.

Rhizome short, not stoloniferous; leaves radical and 1 or 2 above on culm; blades 2–4 mm broad, shorter than to surpassing culms, margins ciliate with long hairs; sheaths straw-colored or light brown; culms to 60 cm tall; head with 4–10 spikelets. Characters otherwise as in subsp. *nervosa*.

Common in wet meadows of tropical America from Central America and the West Indies to southern Brazil. Specimens examined: Isabela, San Cristóbal, Santa Cruz.

Rhynchospora rugosa (M. Vahl) Gale, Rhodora 46: 275, *t. 835, f. 1A–B.* 1944

Schoenus rugosus M. Vahl, Eclog. Amer. 2: 5. 1798.
Rhynochospora glauca M. Vahl, Enum. Pl. 2: 233. 1806.

Culms tufted from short rhizome, slender but rigid, 30–70 cm tall, 3-sided, smooth, 2 or 3 nodes per culm; leaves basal and 2 or 3 on upper culm (at nodes); blades narrowly linear, 2–4 mm wide, shorter than culm, folded, thinly coriaceous, gradually long-acuminate at apex, scaberulous on margins; sheaths of cauline leaves yellowish green, basal sheaths brownish and eventually disintegrated into fibers; corymbs 2–4, remotely located on upper part of culm, 2–5 cm long, 2–3 cm wide, rather loose to moderately dense; rays unequal in length, filiform, obliquely ascending, final raylets terminated by single spikelet or a cluster of 2 or 3 spikelets; spikelets ovate-oblong, 3–4.5 mm long, subcompressed, 2- or 3-flowered, bearing 1 or 2 fruits, rusty brown or orange-brown; glumes ovate or broadly so, membranous, 1-nerved, acute and mucronate at apex; achenes broadly obovate, biconvex, 2 mm long, yellow-brown, finely transversely rugulose; style base depressed-conical, half as long as body of achene; style 2-cleft; hypogynous bristles 6 or 7, slightly shorter than achene, reddish brown, upwardly scabrid.

A pantropical sedge occurring in marshy places. Specimens examined: Santa Cruz.

See Fig. 229.

Rhynchospora tenuis Link in Spreng., Jahrb. Gewächsk. 1, Heft 3: 76. 1820

Rhizome short and knotty, covered with brown fibers; culms densely tufted, slender, rigid, 10–25 cm tall, trigonous, smooth, 1-furrowed on each side; leaves radical and 1 or 2 at upper nodes on culms; blades shorter than culm, narrowly linear, 1–2 mm wide, canaliculate at base, gradually long-acuminate at apex; sheaths 1–4 cm long, straw-colored to grayish brown; inflorescence of 2 or 3 corymbs remotely borne on upper half of culm; terminal corymb compound, 1–2.5 cm wide, 1–1.5 cm high, bearing many rather crowded spikelets; bracts 2 or 3, surpassing corymb; lateral corymbs 7–10 mm wide, 6–8 mm high, dense, the peduncle 1.5–2.5 cm long, en-

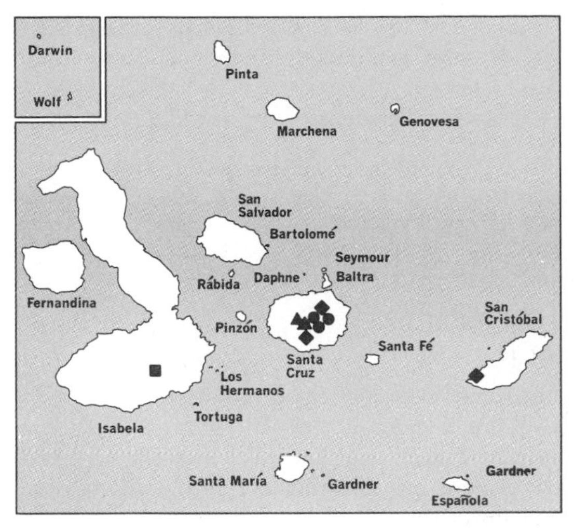

● *Rhynchospora rugosa*

■ *Rhynchospora tenuis*

▲ *Scleria hirtella*

◆ *Scleria pterota*

closed in sheath of bract most of the length; bract elongate, leaflike, to 10 cm long, surpassing inflorescence, sheath 1–1.5 cm long; spikelets in fascicles of 2 to 4, nearly sessile, lance-oblong, 4–5 mm long, subterete, 3- or 4-flowered but only lowest flower maturing; glumes 6 or 7, ovate or ovate-elliptic, membranous, straw-brown and brown-lineolate, whitish on broad hyaline margin, lowest 2 glumes empty and mucronate, others muticous; achenes obovate-orbicular, 1 mm long, biconvex, yellowish brown and later becoming dark brown, coarsely transversely undulate-rugose; style base depressed-deltoid, 0.2 mm long, inversely emarginate at base with lobes decurrent on upper edges of achene; style 2-cleft; hypogynous bristles none.

Occurring abundantly in sandy grassland and savannas in tropical America from Mexico through the West Indies to northern Argentina; in the Galápagos Islands known from a single collection from Isabela.

See Fig. 229.

Scleria Bergius, Kongl. Vet. Akad. Handl. 26: 142. 1765

Perennial or rarely annual sedges, often with ligneous rhizomes. Culms 3-sided, often scabrous on angles, aphyllopodic, clothed at base with a few blade-bearing sheaths. Leaves of normal kind borne on middle portion of culms, all spaced or aggregated at nodes of culms; blades linear, 3- or 5-costate; leaf sheaths often winged, ventral side usually extending into a contra-ligule beyond orifice. Inflorescence of terminal and compound panicles subtended by leaflike bract (in Sect. *Hypoporum* often of a single terminal spike). Fruit-bearing cymelets and staminate spikelets intermingled, or fruit-bearing cymelets on lower portion of branches and staminate spikelets on upper. Plants monoecious, with fruit-bearing cymelets and staminate spikelets on the same inflorescence. Cymelets pistillate and simple or bisexual and compound; pistillate cymelets consisting of a few empty glumes imbricate on an axis terminated by a single terminal fructification; in bisexual cymelets an empty glume immediately below the fructification bearing an axillary staminate spikelet. Fructification of a globose or trigonous bony achene exposed over a discoid or cuplike hypogynium (this not conspicuous in Sect. *Hypoporum*). Staminate spikelets with more or less distichously disposed staminate glumes, each subtending a monandrous or diandrous staminate flower.

About 150 species in tropical and subtropical regions of the world.

Fructifications stipitate, without conspicuous hypogynium; inflorescence spicate, simple; contra-ligule minute or none. . . . . *S. hirtella*
Fructifications sessile, with conspicuous discoid hypogynium; inflorescence paniculate; contra-ligule well developed . . . . *S. pterota*

Scleria hirtella Sw., Prodr. Veg. Ind. Occ. 19. 1788

Hypoporium hirtellum Nees [Linnaea 9: 303. 1834, invalid comb.], Fl. Bras. 2[1]: 303. 1842.

Perennial with horizontal creeping rhizomes 3–4 mm thick; culms arranged in a row along rhizome at intervals of 2–15 mm, slender, acutely triquetrous, 20–60 cm tall, erect, glabrous or pubescent toward apex; basal sheaths bladeless or short-bladed, 1–4 cm long, brown to dark brown; leaves rather crowded, alternate on middle portion of culms, shorter than inflorescence; blades narrowly linear, 5–15 cm long, 2–5 mm wide, flattish, soft, pubescent or glabrescent; leaf sheaths 2–5 cm long, hirsute; contra-ligule minute or absent; inflorescence spikelike, bearing 4–9 glomerules on a simple axis, 5–12 cm long; bracts setaceous, the lower 2–3 times as long as glomerules; glomerules remote, more or less nodding, 5–6 mm high, 5–8 mm wide; bractlets scalelike or subulate, pubescent; cymelets bisexual or unisexual, 5 mm long; glumes ovate-lanceolate to lanceolate, 3–4 mm long, thickly membranous, deep red-brown, pilose with fuscous tubercle-based hairs, midvein projected beyond glume apex into slightly recurved awn; achenes obovoid-globose to globose, obscurely trigonous, 1–1.75 mm long, bony, whitish, shiny, mucronate at apex, cuneate at stipelike base, disk not developing.

The southern United States through Central America and the West Indies to northern Argentina and Chile; also in tropical Africa; common in wet grasslands, savannas, and barrens. Specimens examined: Santa Cruz.

See Fig. 229.

Scleria pterota Presl, Oken Isis 21: 268. 1828

Rhizome creeping, ligneous, 4–7 mm thick, clothed with purple-fuscous scales; culms solitary or 2 or 3 together, 50–100 cm tall, acutely triquetrous, scabrid on angles; basal sheaths bladeless or short-bladed, 2–14 cm long, dark purple or purple-brown; leaves cauline, distantly disposed; blades broadly linear, 15–40 cm long, 7–15 mm wide, herbaceous, 3-costate, glabrous, scabrous on margins, subabruptly narrowed to subobtuse or subacute apex; sheaths 8–13 cm long, narrowly 3-winged, the contra-ligule deltoid, 3–9 mm long, pilose on reddish brown, cartilaginous margin; panicles 3 or 4, pyramidal, 5–15 cm high, loosely flowered, the rachis and branches puberulent or glabrescent, often purplish; bracts of lateral panicles leaflike, equaling or slightly surpassing terminal panicle; bractlets setaceous; staminate spikelets and pistillate cymelets intermingled, greenish brown to deep purple; pistillate glumes ovate-orbicular, 2.5–3.5 mm long, herbaceous, 1-costate, acuminate at often recurved apex, minutely ciliate on margins; achenes depressed obovoid-globose, 1.5–2.5 mm long and broad, bony, whitish or tinged with yellow-brown; hypogynium 3-lobed, lobes rounded, ciliate or glabrous; stigmas 3.

In wet woods and swampy grasslands from the West Indies and tropical and subtropical South America to northern Argentina. Specimens examined: San Cristóbal, Santa Cruz.

See Fig. 229.

GRAMINEAE. GRASS FAMILY*

Herbs or rarely woody plants with terete or somewhat flattened, hollow or solid stems (culms), these closed at the nodes, and 2-ranked usually parallel-veined leaves, these consisting of an open or closed sheath enveloping the culm, a usually flat blade, and between the two, on the inside, a membranous hyaline or hairy appendage (ligule), this rarely obsolete. Flowers perfect (rarely unisexual); perianth reduced to 2(3) small hyaline appendages (lodicules), these sometimes not developed; stamens usually 3 (rarely 6) on slender filaments, the anthers usually deeply sagittate; pistil 1, 1-celled and bearing a single ovule, stigmas 2(3), usually feathery, sessile on ovary or borne on styles of varying lengths; fruit normally a caryopsis, in which seed is adherent to pericarp, but in some genera (e.g. *Sporobolus*) the seed may be free. Flowers borne in spikelets consisting of a shortened axis (rachilla) and 2 to many 2-ranked bracts, lowest 2 (glumes, rarely 1 or both obsolete) empty, the 1 or more succeeding (lemmas) subtending and often partially enclosing a second 2-nerved bract (palea) that bears in its axil, on side toward lemma, the single flower; lemma, palea, and included flower constitute the floret. Spikelets normally aggregated into spikes, racemes, or panicles at end of culm or its branches.

The fourth largest family of flowering plants, comprising more than 500 genera and perhaps 6,000 species. In the present treatment, 55 species and 2 varieties are recognized as occurring in the Galápagos Islands; many of these appear to be endemic.† Doubtful and excluded species are discussed briefly following the description of the genus *Uniola*; included in these exclusions (20 species in 13 genera) are the genera *Ammophila, Chusquea, Distichlis,* and *Poa*.

The 26 genera of the Gramineae follow the key in alphabetical order. Our systematic arrangement would place these genera in three subfamilies as follows: Subfamily Festucoideae, Tribe Aveneae, *Trisetum*; Subfamily Eragrostoideae, Tribe Chlorideae, *Aristida, Eragrostis, Muhlenbergia, Sporobolus, Leptochloa, Trichoneura, Eleusine, Dactyloctenium, Uniola, Cynodon, Chloris, Bouteloua*; Subfam-

* Contributed by John R. Reeder and Charlotte G. Reeder, who add the following: "We are grateful to the curators of the following herbaria, who kindly lent specimens and/or made their facilities available for these studies: Gray Herbarium of Harvard University; Brooklyn Botanic Garden; Royal Botanic Gardens, Kew, England; and the United States National Herbarium. We owe special thanks also to W. D. Clayton, of Kew, who courteously searched the collections at that institution for critical material; our associate D. N. Singh, who made skillful preparations of the epidermis and leaf transections of numerous sterile specimens; and I. L. Wiggins, who arranged to have sent for our studies all of the Galápagos grass material at the California Academy of Sciences and at Stanford University."

† The most extensive collection of grasses to come out of the Galápagos Islands was made by J. T. Howell, who collected there from April 15 to June 15, 1932. These grasses were identified by A. S. Hitchcock and Agnes Chase. Except for the three new species and one variety (cf. Hitchcock 1935, 295–300) that emerged, these determinations were never published.

ily Panicoideae, Tribe Paniceae, *Panicum, Ichnanthus, Echinochloa, Oplismenus, Paspalum, Eriochloa, Setaria, Pennisetum, Cenchrus, Anthephora, Digitaria, Stenotaphrum,* Tribe Andropogoneae, *Coix.*

Spikelets unisexual, both kinds of florets in same inflorescence; pistillate spikelets permanently enclosed in a very hard pearly white or drab beadlike involucre from whose orifice the slender axis of staminate portion of inflorescence protrudes........*Coix*
Spikelets alike, all perfect; inflorescences various:
 Inflorescence of several to many readily deciduous, indurated, often spiny burs borne in spikelike racemes:
  "Burs" not spiny, consisting of usually 4 spikelets with their first glumes indurated, united at base and facing outward, these as long as spikelet........*Anthephora*
  "Burs" spiny, consisting of an involucre of more or less coalesced sterile branchlets, this enclosing 1 or more spikelets; glumes membranous, shorter than lemmas ....*Cenchrus*
 Inflorescence not as above, but spikelets sometimes subtended by 1 or more, usually scabrous bristles:
  Spikelets subtended by 1 or more bristles:
   Bristles falling attached to spikelets at maturity....................*Pennisetum*
   Bristles persistent, spikelets deciduous..............................*Setaria*
  Spikelets not subtended by bristles:
   Spikelets more or less dorsally compressed, 2-flowered (but often appearing 1-flowered); fertile lemma distinctly firmer than glumes, often hard and shiny:
    First glume wanting or much reduced:
     Rachilla thickened below second glume into ringlike or beadlike callus...... ....*Eriochloa*
     Rachilla not thickened:
      Fertile lemma cartilaginous, the margins hyaline, not enrolled; first glume usually present, but minute...........................*Digitaria*
      Fertile lemma strongly indurate, the margins enrolled...........*Paspalum*
    First glume prominent:
     Glumes awned, awn of first glume the longer, usually twice or more the length of spikelet ....*Oplismenus*
     Glumes awnless, or if bearing awns, these shorter than glumes:
      Fertile lemma with lateral appendages or excavations at base; first glume sometimes attenuate or awn-tipped.....................*Ichnanthus*
      Fertile lemma with neither appendages nor excavations:
       Spikelets sunken in thick corky rachis; rachis disarticulating at maturity ....*Stenotaphrum*
       Spikelets not sunken; rachis not disarticulating:
        Lemma pointed, the margins not enrolled over palea at apex, tip of palea free ....*Echinochloa*
        Lemma usually blunt at apex, the margins enrolled over palea throughout ....*Panicum*
   Spikelets terete or laterally compressed, 1- to several-flowered; lemmas and glumes similar in texture:
    Spikelets sessile or short-pedicellate on 1 side of continuous rachis:
     Racemes digitate at summit of culm, often with 1 to several racemes below:
      Spikelets 1-flowered, with no rudimentary florets above fertile one.*Cynodon*
      Spikelets 2- or several-flowered or, if with only 1 functional floret, with 1 or more rudimentary florets above:
       Spikelets with 1 functional floret, and 1 or more modified florets above ....*Chloris*
       Spikelets with several similar florets, the upper sometimes reduced:

Rachis extending beyond uppermost spikelet . . . . *Dactyloctenium*
Rachis not prolonged . . . . *Eleusine*
Racemes not digitate, racemose along an elongate axis:
Spikelets with 1 functional floret and 1 or more usually greatly modified florets above . . . . *Bouteloua*
Spikelets with several similar florets, the upper sometimes reduced:
Lemma margins not conspicuously stiff-ciliate; ligule a ciliate membrane . . . . *Leptochloa*
Lemma margins ciliate with stiff hairs; ligule hyaline, often erose but not ciliate . . . . *Trichoneura*
Spikelets in open or contracted panicles, never in 1-sided racemes:
Spikelets 1-flowered:
Lemmas with 3-branched awn at apex (rarely lateral branches obsolete); caryopsis slender, tightly enclosed in the convolute lemma. . . . *Aristida*
Lemmas awnless or with single awn; caryopsis rather plump, falling freely or easily removed from floret:
Lemma 3-nerved, awned . . . . *Muhlenbergia*
Lemma 1-nerved, awnless . . . . *Sporobolus*
Spikelets 2- to several-flowered:
Lemmas 5-nerved, midnerve extended as geniculate awn arising from below apex . . . . *Trisetum*
Lemmas 3-nerved, awnless:
Spikelets with 1–4 empty lemmas below fertile florets; lemmas firm . *Uniola*
Spikelets with no sterile lemmas below fertile florets; lemmas membranous . . . . *Eragrostis*

Anthephora Schreb., Beschr. Gräser 2: 105, *pl. 44.* 1810

Annuals (ours) or perennials with flat blades and spikelike racemes of readily deciduous burs. Spikelets subsessile, several together and permanently fused at their bases into an indurated fascicle or bur, these sessile or nearly so on a slender axis; fascicles falling entire, the seeds germinating within them. Spikelets arranged so that first glumes face outward; spikelets toward axis usually sterile and represented by glumes only, others fertile and perfecting fruit. First glume broad, indurate, and several-nerved; second narrow, membranous, attenuate-tipped. Sterile lemma about equal to the fertile, membranous; fertile lemma somewhat indurate, membranous near tip, margins clasping palea, but not enrolled. $x = 9$.

About 20 species, mostly in tropical Africa; only one native to the Western Hemisphere.*

Anthephora hermaphrodita (L.) Kuntze, Rev. Gen. Pl. 2: 759. 1891

Tripsacum hermaphrodita L., Syst. Nat. ed. 10, 2: 1261. 1759.
Anthephora elegans Schreb., Beschr. Gräser 2: 105, *pl. 44.* 1810.
Anthephora cuspidata Anderss., Kongl. Svensk. Vet.-Akad. Handl. 1853: 141. 1855.

Annual; culms glabrous, erect or ascending from geniculate or decumbent base and usually branching and rooting from base and lower nodes, 15–50 cm long; sheaths

* Although Schreber, who described the genus, considered it to be allied to *Cenchrus,* Bentham (1882, 29f) suggested a relationship with members of the Zoysieae. This disposition has been followed by most American authors. It is now clear, however, that the affinities of *Anthephora* are with the Paniceae, and that it is probably related most closely to *Cenchrus,* as the early authors thought. (See Reeder 1960, 211–18.)

Fig. 230. *Anthephora hermaphrodita*.

shorter than internodes, glabrous or papillose-hirsute; ligule membranous, brownish, 2–3 mm long; blades flat, 5–20 cm long, 3–8 mm wide, glabrous or more or less pubescent; racemes 5–10 cm long, erect; burs 5–7 mm long, of 4 spikelets; first glume acute to acuminate, margins minutely scabrous; caryopsis as in *Panicum*, dorsally flattened and with punctate hilum, the embryo about two-thirds its length.

A common weedy species occurring throughout tropical America from sea level to about 2,000 m. Specimens examined: Fernandina, Isabela, San Cristóbal, San Salvador, Santa Cruz, Santa María.

Examination of type material of *A. cuspidata* Anderss. (fragment, US), indicates that this specimen is well within the range of variability of *A. hermaphrodita*. It appears to represent a depauperate plant.

See Fig. 230.

Aristida L., Sp. Pl. 82. 1753; Gen. Pl. ed. 5, 35. 1754

Annual or perennial grasses with narrow, flat or involute blades, ciliate ligules, and open, contracted, or sometimes spikelike panicles. Spikelets 1-flowered, disarticulating obliquely above glumes, floret with minutely bearded callus. Glumes membranous, equal or unequal, usually exceeding floret, acute, acuminate, awn-pointed, or awned. Lemma terete, becoming indurated at maturity, 1–3-nerved, nerves converging toward apex and each, or only the central, produced into an awn, or upper part of lemma developing into 3-nerved twisted column, each nerve emerging as awn at apex. $x = 11$.

Perhaps 300 species widely distributed in temperate and tropical regions of both hemispheres.*

Inflorescence densely contracted, spikelike, or occasionally slightly interrupted; glumes more or less scabrous to scabrous-hispid on surface:
- Plant perennial; sheaths glabrous or scabrous, rarely with a very short pubescence; awn column 4 mm long or more . . . . *A. subspicata*
- Plant annual; sheaths conspicuously silky-villous with white spreading hairs, especially above; awn column 2 mm long or less . . . . *A. villosa*

Inflorescence more or less open or lax; glumes smooth except on keel (in *A. divulsa* glumes sometimes minutely scaberulous in lines, especially near tip):
- Plant perennial, erect; panicle branches, at least the lower, naked for about 1 cm . . . . *A. divulsa*
- Plant annual, weak and somewhat decumbent at base; panicle branches mostly spikelet-bearing from base . . . . *A. repens*

---

* All of the species found in the Galápagos Islands belong to Sect. *Pseudarthratherum* Chiovenda, in which the summit of the column is articulated with the awns, which are of about equal length. Moreover, all of the species in the islands have awned glumes. The Galápagos species are obviously closely related, and the spikelets are so similar that they are of limited usefulness for taxonomic purposes. The most diagnostic characters are presence or absence of scabrosity on the glume surfaces, and the length of the awn column. Henrard, in his monograph of the genus *Aristida* (1929), listed five species as occurring in the Galápagos. In the present treatment, four are recognized, all of which appear to be endemic. More thorough collecting and field study could provide data that might improve the taxonomy. Cytological information would be highly desirable, as would population samples that might clear up the extent of suspected hybridization.

Aristida divulsa Anderss., Kongl. Svensk. Vet.-Akad. Handl. 1853: 143. 1855

Perennial; culms erect, terete, densely cespitose, glabrous or minutely scaberulous, 40–60 cm tall, simple or sparingly branched from middle nodes; sheaths shorter than internodes, glabrous or minutely scaberulous under lens; ligule a row of white hairs about 1 mm long; blades flat at base but becoming involute toward tip, sometimes involute throughout, 10–20 cm long, about 1 mm wide, scaberulous below, scaberulous and minutely puberulent above, margins scabrous; panicle 15–20 cm long, somewhat open when mature, axis, branches, and pedicels scabrous, branches (at least the lower) naked at base often for a distance of 1 cm or more; glumes glabrous or very minutely scaberulous in lines, especially near apex, keel scabrous, first glume 6–7 mm long including the scabrous awn, this 1–2 mm long, second glume slightly longer, similar awn often from between 2 short teeth; lemma purple at maturity, punctulate, distinctly scabrous above, narrowed gradually into strongly twisted column, body of lemma including callus about 3.5 mm long, column 2–3 mm, callus white-hairy, 1 mm long; awns minutely scabrous, about equal, 10–12 mm long.

Relatively infrequent and apparently endemic. Specimens examined: Isabela, Marchena, Pinta, San Cristóbal, San Salvador, Santa Cruz.

The collection from Santa Cruz (*Stewart 1251*) at the California Academy of Sciences is a sterile specimen; the material at Gray Herbarium has a fragmentary inflorescence. It was identified by Stewart as *Chloris anisopoda* Robins., and distributed under this name. That it is a species of *Aristida* is shown by our studies of the leaf anatomy. The open inflorescence with branches naked at the base, and the glabrous glumes, leave little doubt that it is *A. divulsa*.

Aristida repens Trin., Mém. Acad. St. Petersb. VI, 1: 87. 1830

Annual; culms terete, glabrous or minutely scaberulous, much branched, ascending from somewhat decumbent base, (20)30–40(60) cm tall; sheaths shorter than internodes, slightly striate, glabrous, somewhat compressed; ligule a row of white

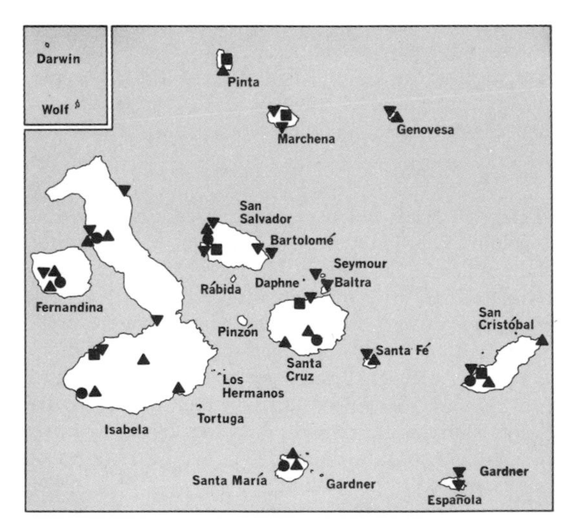

● *Anthephora hermaphrodita*
■ *Aristida divulsa*
▲ *Aristida repens*
▼ *Aristida subspicata*

hairs 1–2 mm long, sometimes with hairs twice this length on either side; blades flat, becoming involute, mostly 8–15 cm long, 1–2 mm wide, glabrous below, densely short-puberulent above, margin smooth or minutely scabrous; panicle 10–15 cm long, contracted but rather lax, the axis, branches, and pedicels scabrous, branches mostly spikelet-bearing from base; glumes about equal, 5–7 mm long, awn-pointed or with an awn nearly 1 mm long, smooth and shining, often purplish, the first scabrous on keel, the second glabrous on keel or scabrous near tip only; lemma pale or slightly purplish, mottled with black spots, glabrous and punctulate below, scabrous above, body of lemma including callus about 3.5–4 mm long, callus white-hairy, about 1 mm long, column 1–2 (rarely 0.5) mm long, loosely to tightly twisted; awns minutely scabrous, about equal, 7–10 mm long.

Apparently endemic; a weedy species found from sea level to about 520 m on most of the islands. Specimens examined: Fernandina, Genovesa, Isabela, Pinta, San Cristóbal, San Salvador, Santa Cruz, Santa Fé, Santa María.

Stewart 1252 (CAS) was reported by him as *Chloris anisopoda* Robins. All of his other collections (*1203, 1205, 1206, 1209, 1216, 1223*) appear in his publication under *A. subspicata* Trin. & Rupr. It is of interest that the only two specimens of the above numbers (*Stewart 1206* and *1209*) at the U.S. National Herbarium have been annotated as *A. repens* by the late Mrs. Agnes Chase.

Aristida subspicata Trin. & Rupr., Mém. Acad. St. Petersb. VI, Sci. Nat. 5(1): 125. 1842

Aristida caudata Anderss., Kongl. Svensk. Vet.-Akad. Handl. 1853: 144. 1855.
Aristida compacta Anderss., op. cit. 145.
Stipa rostrata Anderss., op. cit. 142.

Perennial, sometimes flowering first year and appearing annual; culms densely cespitose, terete, simple or sparingly branched, glabrous or more or less scabrous, especially below nodes, 30–50(70) cm tall; sheaths shorter than internodes, glabrous or somewhat scabrous; ligule a row of hairs about 1.5 mm long, sometimes with longer hairs on either side; blades flat to involute, to 20 cm long, 1–2 mm wide, smooth or somewhat scabrous below, scabrous and densely puberulent above, margins scabrous; panicle dense, spikelike, sometimes slightly interrupted at base, often partly included in upper sheath, 6–10 cm long, the axis, branches, and pedicels scabrous; spikelets congested, subsessile; glumes unequal, more or less scabrous on surface, the first (5.5)6–7(8) mm long including awn, scabrous on keel, awn 2–3 mm long, the second somewhat narrower and longer, keel scabrous upward, awn 1–2 mm long from bifid apex; lemma pale, mottled, punctulate below, scabrous upward, body of lemma including callus about 4 mm long, tapering into 4–6 mm long tightly twisted column, callus hairy, 1 mm long; awns minutely scabrous, central one 10–12(15) mm long, laterals often shorter.

Endemic; this species has been collected on (and specimens examined from) most of the islands and is to be expected on all of them. It is apparently the most abundant and distinctive grass in the Galápagos Islands, occurring from sea level to at least 310 m.

Type fragments of *A. compacta* Anderss. and *A. caudata* Anderss. were examined at the U.S. National Herbarium. Both seem to fall within the range of variability of *A. subspicata* as we understand it. It is of interest that Hitchcock and Chase have indicated (on the sheets) that they also consider both of the Andersson "species"

Fig. 231. *Aristida subspicata.*
(Drawn by Charlotte Reeder, inked by Jeanne R. Janish.)

to be synonyms of *A. subspicata*. Some eight specimens (*Eliasson 1318*; *Howell 8673, 9532, 9705B*; *Snodgrass & Heller 273, 776*; *Stewart 1211, 1220*) are atypical and may represent hybrids between *A. subspicata* and *A. repens*. They have the scabrous glumes of the former species, but the inflorescence is more as in *A. repens*.

Howell 9846, at both CAS and US, has only a single awn developed. It is similar to the type of *Stipa rostrata* Anderss: but although the spikelets suggest a species of *Stipa*, our studies of the leaf anatomy show that it is clearly an *Aristida*. The general habit is that of a typical specimen of *A. subspicata*.

See Fig. 231.

Aristida villosa Robins. & Greenm., Amer. Jour. Sci. III, 50: 144. 1895

Annual; culms terete, scabrous and more or less puberulent, especially below nodes, much branched, ascending from somewhat decumbent base, to 40 cm tall; sheaths shorter than internodes, conspicuously silky-villous, especially above, with white spreading hairs, these typically from papillose base; ligule a row of white hairs about 1.5 mm long, these often fused part of their length, longer weak hairs commonly on either side; blades flat, becoming involute, 5–8 cm long, puberulent-hispid on both surfaces, more densely so above, margins scabrous; panicle contracted, spikelike, base often included in upper sheath, 4–7 cm long, the axis, branches, and pedicels puberulent; spikelets congested, subsessile; glumes unequal, puberulent, the first about 6 mm long including awn, this 2 mm long, the second 7–8 mm long, awn 1.5–2 mm long; lemma 4.5–5 mm long including callus and beak, callus white-hairy, 1 mm long, beak short (1–1.5 mm), scarcely twisted, body of lemma at maturity purplish mottled with black, smooth below and scabrous above; awns scaberulous, central one 10–12 mm long, laterals slightly shorter.

Endemic; apparently of infrequent occurrence, but perhaps more abundant and overlooked because of its superficial resemblance to the common *A. subspicata*. Specimens examined: Baltra, Pinzón, Rábida, San Salvador, Santa Fé, Santa María, Seymour.

Hitchcock and Chase apparently did not recognize *A. villosa*. A note with the

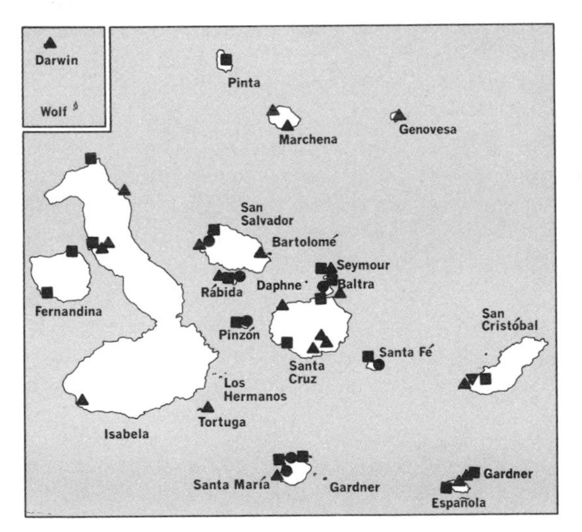

- ● *Aristida villosa*
- ■ *Bouteloua disticha*
- ▼ *Cenchrus echinatus*
- ▲ *Cenchrus platyacanthus*

photograph and fragment of the type of *A. villosa* (*Baur 337*), in Mrs. Chase's script, reads: "Specimen has general appearance of small plants of *A. subspicata* Trin., but foliage is villous and awns are not quite as long." The typical specimens of *A. villosa* are densely silky-villous on the sheaths, the hairs arising from a papillose base. The blades are puberulent-hispid on both surfaces, more densely so below. Among the specimens examined, these characters are perhaps best exemplified by *Howell 9743*. Some of the others (*Howell 9306, 9693*; *Snodgrass & Heller 566, 581*) are much less hairy, but are villous in the throat and along the sheath margins. They are placed here provisionally, but may be the result of hybridization between this species and *A. subspicata.*

Bouteloua Lag., Var. Cienc. Lit. & Art. 2: 134. 1805;
emend. Beauv., Ess. Agrost. 40. 1812

Annual (ours) or perennial grasses, the inflorescence of few to many spicate branches racemosely arranged on a common axis; spikelet disarticulating above glumes, the fertile floret falling with rudiment attached. Spikelets with 1 functional floret and 1 or more rudimentary florets above, these subsessile, arranged in 2 rows along 1 side of a continuous rachis. Glumes unequal, 1-nerved, the first shorter and narrower. Lemma equaling second glume or a little longer, 3-nerved, nerves extending into awns or mucros, internerves often extending into lobes or teeth; rudiments various but commonly 3-awned, awns usually longer than those of fertile lemma. $x = 10$.

About 40 species restricted to the Western Hemisphere, with greatest development in arid regions of the southwestern United States and northern Mexico.

Bouteloua disticha (HBK.) Benth., Jour. Linn. Soc. Bot. 19: 105. 1881

Polyodon distichum HBK., Nov. Gen. & Sp. Pl. 1: 175, *t. 55.* 1816.
Eutriana pilosa Hook. f., Trans. Linn. Soc. Lond. 20: 173. 1847.
Leptochloa hirta Nees ex Steud., Syn. Pl. Glum. 1: 209. 1854.
Bouteloua pilosa (Hook. f.) Benth. ex S. Wats., Proc. Amer. Acad. 18: 179. 1883.*

Annual; culms terete, glabrous, much branched, often decumbent and rooting from nodes, 25–80 cm long or more; sheaths shorter than internodes, glabrous or more or less papillose-pilose; ligule a fringe of hairs about 0.5 mm long; blades flat, elongate, mostly 2–5 mm wide, glabrous or somewhat pilose, margins scabrous; inflorescence with 15–30 branches, these mostly bearing 3–5 spikelets crowded on basal third of the flattened scabrous rachis; spikelets pale to purplish, 5–7 mm long excluding awns; first glume setaceous, 3–4 mm long, second lanceolate, scabrous, 4.5–6 mm long, apiculate or very short-awned from shallowly notched apex; lemma smooth, about equal to second glume, 3-nerved, nerves terminating in short awns; palea equaling lemma, the 2 nerves terminating in short awn points; rudiment including the strong scabrous central awn to 10 mm long, the body from greatly reduced to about 4 mm long, 2 short lateral awns often present (rudiment rarely lacking or reduced to scabrous awn about equaling lemma); caryopsis broadly fusiform, flattened, about 2.5 mm long, hilum punctiform, embryo large, covering about four-fifths of adaxial surface.

A weedy species in tropical and subtropical areas, from southern Mexico through

* For more complete synonymy, see F. W. Gould & Z. J. Kapadia (1964, 188).

Fig. 232. *Bouteloua disticha.*
(Drawn by Charlotte Reeder, inked by Jeanne R. Janish.)

Central America to Ecuador and Peru. Specimens examined: Baltra, Española, Fernandina, Gardner (near Española), Isabela, Pinta, Pinzón, Rábida, San Cristóbal, San Salvador, Santa Cruz, Santa Fé, Santa María, Seymour.

Except for the paper by Gould and Kapadia, all references to this species in the Galápagos have been either by the name *Eutriana pilosa* Hook. f. or by the name *Bouteloua pilosa* (Hook. f.) Benth. Hitchcock (1927, 418) states under *B. pilosa*: "This species is probably a form of *B. disticha.*" Gould and Kapadia have considered the two as taxonomic synonyms, a disposition we are following here.

Although Andersson (1861) lists *Leptochloa hirta* Nees ex Steud. as a synonym of *L. hookeri* Anderss. [= *Trichoneura lindleyana* (Kunth) Ekman], Ekman (1912,

9) points out that Steudel's description of *L. hirta* does not suggest a species of *Trichoneura*, but applies reasonably well to *Eutriana pilosa* Hook. f. [= *Bouteloua disticha* (HBK.) Benth.]. See our more detailed discussion under *Trichoneura*.
See Fig. 232.

Cenchrus L., Sp. Pl. 1049. 1753; Gen. Pl. ed. 5, 470. 1754

Annuals or perennials with flat blades and spikelike racemes of burs, these readily deciduous. Spikelets lanceolate, sessile, 1 to several together, permanently enclosed in a bristly or spiny involucre or bur composed of more or less coalesced sterile branches; burs sessile or nearly so on a slender axis, its apex produced into a short point beyond uppermost bur, burs falling entire, the grains germinating within them. Glumes membranous, first 1-nerved, narrow, sometimes wanting, second 3–5-nerved, usually shorter than spikelet. Sterile lemma about equal to the fertile, membranous, 3–5-nerved, enclosing well-developed palea and often a staminate flower; fertile lemma indurate, membranous near tip, margins clasping palea but not enrolled. $x = 17$ (American species).
About 25 species in warm regions of the world.*

Bur consisting of a single whorl of united flattened spines, with 1 or more whorls of smaller bristles below; spikelets 3–6 per bur. . . . . *C. echinatus*
Bur consisting of several whorls of united flattened spines, without smaller bristles below; spikelets normally 1 per bur. . . . . *C. platyacanthus*

Cenchrus echinatus L., Sp. Pl. 1050. 1753

Annual; culms somewhat compressed, glabrous, or scabrous below inflorescence, ascending from geniculate or decumbent base and usually branching and rooting from base and lower nodes, 25–60(100) cm long; sheaths usually overlapping, glabrous or more or less pubescent, especially on margins toward summit, loose, somewhat keeled; ligule ciliate, about 1 mm long; blades flat, 6–20 cm long, 3–8 mm wide, scabrous or sparsely pilose on upper surface; inflorescence dense, spikelike, 3–10 cm long, often rather long-exserted in age, axis scabrous; burs truncate at base, body 4–7 mm high, as broad or broader, pubescent, outer slender bristles numerous, half as long as body or less, inner bristles stout, usually broadened at base and equaling lobes of body or shorter; lobes of body commonly 10, often pilose, ciliate, erect or bent inward, sometimes 1 or 2 lobes inflexed; spikelets 3–6 in each bur, lanceolate, acuminate, about equaling lobes or shorter, 4.5–6 mm long; first glume narrow, 1-nerved, second two-thirds to three-fourths as long as subequal sterile and fertile lemmas.
From the southern United States through the West Indies and Central America to Argentina; found (perhaps introduced) on most of the Pacific islands, in the Philippines, and in Australia, and sparingly introduced into Africa and the Middle East; in the Galápagos in fields and waste places, especially in sandy soil, to about 1,050 m. Specimens examined: San Cristóbal.
This species has not been reported previously from the Galápagos Islands. The Howell specimen (*8537* [CAS, GH, US]) was determined as *C. echinatus* by Mrs.

* One species, *C. distichophyllus,* has been excluded from the flora; it is discussed briefly in the pages concluding the Gramineae.

Chase. Both specimens examined have been annotated as this species by D. G. DeLisle, but they are not listed in his revision of the genus *Cenchrus* (1963, 282–86). *Snodgrass & Heller 520* (DS, GH) was reported as "*Cenchrus* sp." by both Robinson and Stewart. Curiously the original label bears the name "*Cenchrus agrimonoides* [sic] Trin.," a quite different species, whose type is from Hawaii.

Cenchrus platyacanthus Anderss., Kongl. Svensk. Vet.-Akad. Handl. 1853: 139. 1855
Cenchrus granularis Anderss., op. cit. 140, non L. 1771.

Annual; culms glabrous, weak, erect or ascending from geniculate base, rooting at nodes, (20)30–60(80) cm long; sheaths mostly shorter than internodes, rather loose, glabrous or somewhat pilose, especially along margins above; ligule a row of stiff

Fig. 233. *Cenchrus platyacanthus.*
(Drawn by Charlotte Reeder, inked by Jeanne R. Janish.)

hairs 1–2 mm long; blades elongate, 2–8 mm wide, glabrous below, scabrous and often more or less pilose above; inflorescence very dense, spikelike, mostly 4–6 cm long, the sharply angled axis minutely scabrous, internodes very short, seldom exceeding 1.5 mm in length; burs subsessile, stramineous to purple, glabrous or puberulent, 3–6 mm long, the spines broad and flattened, retrorsely barbed near tip; spikelet usually 1 per bur, 4–5 mm long; first glume usually present, one-third the length of spikelet or longer, second distinctly shorter than lemmas; sterile lemma 3.5–5 mm long, enclosing well-developed palea and often a staminate flower; fertile lemma equal to the sterile, slightly indurate.

Apparently endemic; often abundant, especially in sandy soil near the shore. Collected from (and specimens examined from) almost all the islands.

Many collections are cited in the literature under the name *C. granularis*. Hitchcock and Chase considered this to be a synonym of *C. platyacanthus* at the time they named the Howell specimens. DeLisle, in his revision of *Cenchrus* (op. cit.), states, "It appears that Andersson was dealing with two extremes of the same taxon, and since *C. granularis* is a later homonym, *C. platyacanthus* remains the valid name for this species."

See Fig. 233 and color plate 78.

Chloris Sw., Prodr. Veg. Ind. Occ. 25. 1788

Annual (ours) or perennial grasses with flat or folded blades and 2 to many digitate, verticillate, or subracemose racemes. Spikelets with 1 perfect floret and 1 or more rudimentary florets above, these subsessile and borne in 2 rows on 1 side of a continuous rachis; spikelets disarticulating above glumes, fertile floret falling with rudiment attached. Glumes narrow, 1(3)-nerved, the second a bit longer than first. Fertile lemma 3-nerved, awnless or awned from just below tip, margins usually ciliate with hairs increasing in length toward tip, callus more or less bearded on sides. $x = 10$.

About 70 species in warm regions of the world.

Rudiment broad, truncate, conspicuous; fertile lemma "hump-backed," broadest above middle, margins above long-ciliate with hairs to 3 mm long............*C. virgata*
Rudiment narrow, acute; fertile lemma narrow, broadest at or below middle:
  Culms terete; fertile floret very narrow, purplish, 4–5 mm long; rudiment similar to fertile floret except for its smaller size, the lemma with bearded callus and ciliate upper margins .... *C. mollis*
  Culms flattened, the sheaths compressed-keeled; fertile floret broader, stramineous or slightly brownish, sometimes flecked with purple, 2–3 mm long; rudiment completely glabrous ....*C. radiata*

Chloris mollis (Nees) Swallen, North Amer. Flora 17: 596. 1939

Gymnopogon mollis Nees, Agrost. Bras. 427. 1829.
Gymnopogon rupestre Ridley, Jour. Linn. Soc. Bot. 27: 73. 1890.
Chloris anisopoda Robins., Proc. Amer. Acad. 38: 118. 1902.
Chloris angustiflora Aresch., Svensk. Freg. Eugen. Res. Bot. 3: 118. 1910.
Chloris rupestris (Ridley) Hitchc., Misc. Publ. U.S. Dept. Agric. 243: 126. 1936.

Annual; culms terete, glabrous, erect or somewhat geniculate, 20–100 cm tall; sheaths shorter than internodes, glabrous or sparsely pubescent, margins papillose-ciliate; ligule a row of hairs about 0.5 mm long; blades flat, 5–15 cm long, 2–6 mm wide, scabrous and sometimes sparsely pilose on upper surface; racemes mostly

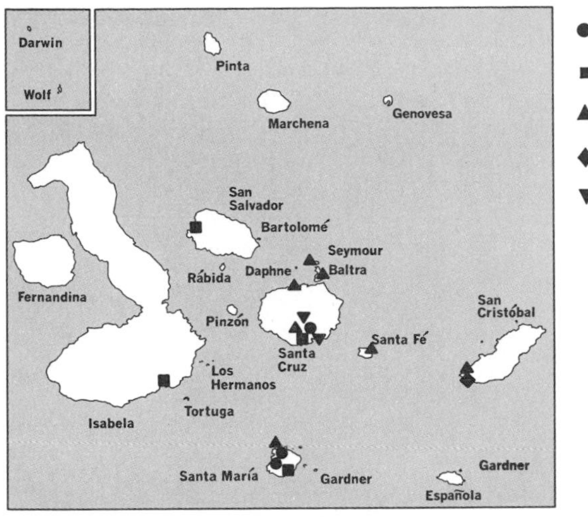

● *Chloris mollis*
■ *Chloris radiata*
▲ *Chloris virgata*
◆ *Coix lacryma-jobi*
▼ *Cynodon dactylon*

5–15, 5–10 cm long, somewhat flexuous, ascending or appressed, aggregate on an axis to 5 cm long; spikelets very narrow; glumes acuminate, awn-pointed, scabrous on keel, the first 3–4 mm long, the second 4–5 mm long; fertile floret slender, glabrous except for callus and ciliate upper margins, purplish when mature, 4–5 mm long, scabrous awn 4–6 mm long; rudiment 1–2 mm long, similar to fertile lemma except for its small size, scabrous awn about 4 mm long.

The West Indies and Guatemala south to Venezuela, Colombia, Ecuador, and Peru. Specimens examined: Santa María.

Reported (as *C. anisopoda*) from Santa María by Robinson, Stewart, and Svenson, and from Santa Cruz by Stewart. The type of *C. anisopoda* is *Baur 333* (holotype: GH!, photo and fragment, US!) from Santa María. *Stewart 1194* (CAS, US) was reported by him as *Stipa rostrata* Anderss. with the following note: "evidently a younger specimen than the one described by Andersson." Stewart's report of the occurrence of *C. anisopoda* on Santa Cruz is based on his collections *1251* and *1252*, both of which are poor and very overmature specimens. The first of these is actually *Aristida divulsa* Anderss. (q.v.), and the second *A. repens* Trin. (q.v.).

Chloris radiata (L.) Sw., Prodr. Veg. Ind. Occ. 26. 1788
Agrostis radiata L., Syst. Nat. ed. 10, 2: 873. 1759.

Annual; culms distinctly flattened, branching, erect or ascending from decumbent base, 15–60 cm tall; sheaths compressed-keeled, overlapping, glabrous or somewhat papillose-pilose, especially on upper margins; blades flat or folded, mostly 5–10 cm long, to 6 mm wide, scabrous and more or less papillose-pilose, especially on lower surface; ligule a ciliate membrane scarcely 0.5 mm long; racemes 5–15, slender, straight or flexuous, 4–8 cm long, subdigitate on an axis usually not more than 1 cm long; spikelets appressed, pale or slightly brownish; glumes narrow, 1-nerved, acuminate to awn-pointed, the first 1.5–2 mm long, the second 2.5–3.5 mm long, glabrous or scabrous on keel; fertile floret 2.5–3 mm long, brownish when mature, glabrous except for bearded callus and stiff-ciliate upper margins, often somewhat scabrous on keel near tip, awn scabrous, to 12 mm long; rudiment glabrous, scarcely 1 mm long, scabrous awn 4–6 mm long.

Fig. 234. *Chloris virgata.*
(Reproduced by permission of U.S. National Herbarium.)

From the West Indies and Mexico to Paraguay. Specimens examined: Isabela, San Salvador, Santa Cruz, Santa María.

Chloris virgata Sw., Flora Ind. Occ. 203. 1797

Chloris elegans HBK., Nov. Gen. & Sp. Pl. 1: 166, *pl. 49.* 1816.

Annual; culms terete, glabrous, erect or ascending from a usually decumbent base, often rooting at lower nodes; sheaths compressed-keeled, glabrous; ligule membranous, ciliolate, about 0.5 mm long; blades flat, to 25 cm long, 3–7 mm wide, scabrous or nearly smooth on both surfaces, sometimes more or less papillose-pilose on upper surface near base, margins scabrous; racemes (3)5–10(15), digitate or nearly so, (3)5–6(9) cm long, flexuous, erect or spreading; spikelets 3–3.5 mm long on very short pedicels; glumes narrow, 1-nerved, scabrous on keel, the first 1.5–2.5 mm long, the second 2.5–3.5 mm long and bearing an awn about 1 mm long; fertile floret 3–3.5 mm long, the lemma "hump-backed" (i.e. keel bowed out above middle), appressed-pubescent on keel and lower margins, long-ciliate on margins near apex, hairs spreading, to 3 mm long, awn rather stout, scabrous, 6–10 mm long; rudiment narrow, truncate at apex, glabrous, 2–2.5 mm long, awn about 6 mm long.

A widely distributed weedy species, occurring in the central and southwestern United States through the West Indies and Mexico to Argentina; also in the Old World. Specimens examined: Baltra, San Cristóbal, Santa Cruz, Santa Fé, Santa María, Seymour.

Reported as *C. elegans* from Seymour and Santa María by both Robinson and Stewart. Stewart reported his *1254* (CAS) from San Cristóbal as *Dactyloctenium aegyptium*.

See Fig. 234.

Coix L., Sp. Pl. 972. 1753; Gen. Pl. ed. 5, 419. 1754

Annuals with branching culms and broad flat blades, inflorescences numerous on stout peduncles clustered in axils of leaves. Spikelets unisexual; staminate spikelets 2-flowered, in 2's and 3's on a slender continuous rachis, glumes membranous, lanceolate, obscurely nerved, lemmas and paleas hyaline, stamens 3; pistillate spikelets 3 together, 1 fertile, other 2 sterile and reduced to narrow tubular glumes. Glumes of fertile spikelet several-nerved, hyaline below, chartaceous in upper pointed part, the first glume broad, enfolding spikelet, the second narrower; sterile lemma similar but a bit narrower; fertile lemma and palea hyaline. Inflorescence of an ovate, oval, or somewhat cylindrical, pearly white or drab beadlike, very hard involucre (much-modified sheathing bract) containing pistillate portion of inflorescence, points of pistillate spikelets and slender axis of staminate portion of inflorescence protruding from orifice at apex, staminate portion to 6 cm long, soon deciduous. $x = 9$.

Some 3 or 4 species native in tropical Asia, with one widely distributed in the tropical regions of the world.

Coix lacryma-jobi L., Sp. Pl. 972. 1753

Coix lacryma L., Syst. Nat. ed. 10, 1261. 1759.

Annual; culms erect, stout, much-branched upward, 1–3 m tall; sheaths glabrous; ligule membranous, ciliate, about 1 mm long; blades glabrous, narrowly lanceolate,

Fig. 235. *Coix lacryma-jobi.*

often cordate at base, acute, 10–60 cm long (the lower sometimes to 120 cm), 2–5 cm wide, margins serrate-scabrous; staminate racemes 1–6 cm long, glabrous; spikelets 8–10 mm long, first glume winged on keels; false fruits ovoid-globose, 6–12 mm long, hard and shiny at maturity.

Introduced from the Old World and found as an escape in moist places throughout the tropics. Specimen examined: San Cristóbal [*Schimpff 143* (CAS)].

Apparently this species has not been reported previously from the Galápagos. It is cultivated for ornament and sometimes as a cereal. The false fruits are used for beads, frequently in rosaries.

See Fig. 235.

Cynodon L. C. Rich. in Pers., Syn. Pl. 1: 85. 1805

Low rhizomatous and stoloniferous perennial grasses with short blades and few to many slender digitate racemes. Spikelets strongly laterally compressed, 1-flowered, the rachilla prolonged beyond base of lemma as naked stipe or sometimes bearing a rudimentary floret, the spikelets borne on very short pedicels in 2 rows on 1 side of narrow triangular rachis. Glumes subequal, shorter than lemma, 1-nerved, keels scabrous. Lemma acute, awnless, 3-nerved; palea narrow, acute, as long as lemma. $x = 9$.

Perhaps 10 species, with one widely distributed in the warmer regions of the world.

Cynodon dactylon (L.) Pers., Syn. Pl. 1: 85. 1805

Panicum dactylon L., Sp. Pl. 58. 1753.

Perennial with strong rhizomes and widely creeping culms or stolons; culms wiry, glabrous, the floriferous branches usually 10–30 cm tall; sheaths usually overlapping, glabrous or pilose in throat; ligule membranous, ciliolate, less than 0.5 mm long, often with row of hairs more than 1 mm long behind it; blades flat, 2–4 mm wide, short and spreading on stolons, or rarely to 15–20 cm long, scabrous and sometimes sparsely pilose, margins scabrous; racemes 4–7, slender, 2–7 cm long; spikelets 2–3 mm long; lemma appressed-pubescent on nerves.

Common at lower altitudes throughout the warmer regions of the world. Specimens examined: Santa Cruz, Santa María.

Previously unreported from the Galápagos Islands, but to be expected on any of them. Frequently used for lawns and pastures.

See Fig. 236.

Dactyloctenium Willd., Enum. Pl. Hort. Berol. 1029. 1809

Annual or perennial grasses with flat blades and 2 to several thick spikes digitate and widely spreading at summit of culms. Spikelets 3–5-flowered, strongly laterally flattened, sessile and closely imbricate in 2 rows along 1 side of the rather narrow rachis, the end naked and produced as a point beyond spikelets; rachilla disarticulating above first glume and between florets. Glumes broad, somewhat unequal, 1-nerved, the first acute, the second mucronate or with a stout somewhat falcate awn to 1 mm long. Lemmas rather firm, broad, acute or short-awned, 3-nerved, the

Fig. 236. *Cynodon dactylon.*
(Reproduced by permission of U.S. National Herbarium.)

lateral nerves indistinct; palea about as long as lemma. Seed somewhat flattened, keystone-shaped, prominently ridged, enclosed in thin pericarp. x = 10, 12?

About 10 species in the warm regions of the Old World; one introduced into the Western Hemisphere.

Dactyloctenium aegyptium (L.) Beauv., Ess. Agrost. 15, 159. 1812

Cynosurus aegyptius L., Sp. Pl. 72. 1753.
Eleusine aegyptiaca (L.) Desf., Fl. Atlant. 1: 85. 1798.

Annual; culms branching, radiate-spreading, rooting at nodes, often forming mats, the ascending flowering portion 20–40 cm long; sheaths mostly overlapping; ligule membranous, ciliolate, about 1 mm long; blades flat, 2–5 mm wide, glabrous or more or less pilose; spikes mostly 2–5, 2–4 cm long; spikelets crowded, about 3 mm long; glumes subequal, the second slightly longer and bearing an awn 1–2.5 mm long; lemmas 2.5–3.5 mm long.

A common weed of open ground and waste places in the tropics and subtropics of the world; native to the Old World but introduced and widespread in the Americas. Specimens examined: Española, San Cristóbal, Santa Cruz, Santa María.

Robinson listed this plant under the name *Eleusine aegyptica.*

See Fig. 237.

Digitaria Heist. ex Haller, Hist. Stirp. Helvet. 2: 244. 1768

Annual or perennial, often weedy grasses with membranous or hyaline ligules and slender racemes digitate or racemose on a short axis. Spikelets 2-flowered (usually appearing 1-flowered), lanceolate or elliptic, nearly plano-convex, in 2's or 3's, rarely solitary, subsessile or short-pedicelled, alternate in 2 rows on 1 side of a 3-angled, winged or wingless rachis, back of fertile lemma turned toward it. Glumes unequal, membranous, the first minute or wanting, the second as long as spikelet or shorter. Sterile lemma as long as spikelet, membranous; fertile lemma cartilaginous, margins hyaline, not enrolled. x = 9.

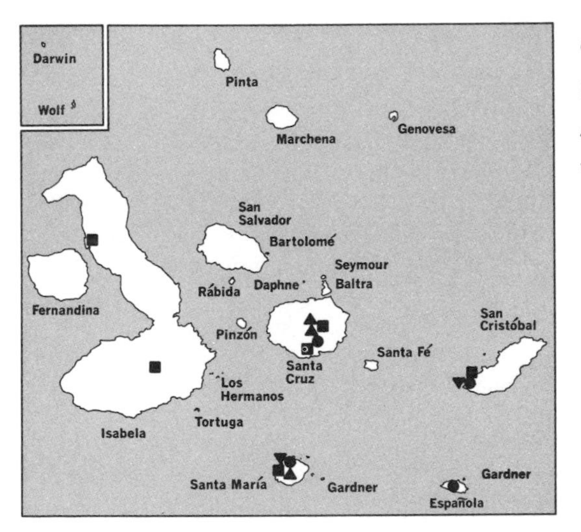

● *Dactyloctenium aegyptium*
■ *Digitaria adscendens*
▲ *Digitaria horizontalis*
▼ *Echinochloa colonum*

Perhaps 300 species or more, widely distributed in the temperate and tropical regions of the world.*

Spikelets about 3 mm long; rachis about 1 mm wide .... *D. adscendens*
Spikelets about 2 mm long; rachis about 0.5 mm wide, bearing few to many spreading weak hairs to 5 mm long .... *D. horizontalis*

Digitaria adscendens (HBK.) Henrard, Blumea 1: 92. 1934

Panicum adscendens HBK., Nov. Gen. & Sp. Pl. 1: 97. 1815.
Digitaria marginata Link, Hort. Berol. 1: 102. 1821.
Digitaria fimbriata Link, Hort. Berol. 1: 226. 1827.
Syntherisma fimbriata (Link) Nash, Bull. Torrey Bot. Club 25: 302. 1898.
Syntherisma marginata (Link) Nash, North Amer. Flora 17: 154. 1912.

Annual; culms glabrous, weak and spreading, to 1 m long, profusely branching below, rooting at nodes; sheaths, at least the lower, papillose-pilose; ligule membranous, 2–3 mm long; blades flat, elongate, 3–8 mm wide, scabrous, especially on margins, rarely more or less pubescent or pilose; racemes (2)3–7, digitate or subdigitate on an axis to 3 cm long; rachis about 1 mm wide, winged, the wings slightly broader than midrib, glabrous except for scabrous margins; spikelets lanceolate-elliptic, about 3 mm long, borne in pairs on scabrous, angled pedicels of unequal length, shorter pedicel about 1 mm long, the longer 3–4 mm long; first glume small, deltoid, about 0.5 mm long, second narrow, more than half the length of spikelet, 3-nerved, ciliate; sterile lemma strongly 7-nerved, 2 lateral nerves close together, appressed-pubescent along margins, the hairs often spreading at maturity; fertile lemma pale or plumbeous.

A widely distributed weedy species of the warmer regions of the world. Specimens examined: Isabela, San Cristóbal, Santa Cruz, Santa María.

Frequently confused with *D. sanguinalis* (L.) Scop. (cf. J. E. Ebinger 1962, 248–53), and so reported from San Cristóbal and Santa María by Stewart. Reported as *Panicum sanguinale* from San Cristóbal by Robinson.

Digitaria horizontalis Willd., Enum. Pl. Hort. Berol. 92. 1809

Milium digitatum Sw., Prodr. Veg. Ind. Occ. 24. 1788.
Panicum horizontale G. F. Meyer, Prim. Fl. Esseq. 54. 1818.
Syntherisma digitata Hitchc., Contr. U. S. Nat. Herb. 12: 142. 1908.
Digitaria digitata Urban, Symb. Antill. 8: 24. 1920, non Buse. 1854.

Annual; culms glabrous, weak and spreading, to 1 m long, profusely branching and rooting at nodes; sheaths papillose-pilose; ligule membranous, erose, about 1 mm long; blades flat, pubescent and often velvety, to 15 cm long or more, 3–8 mm wide, scabrous on margins; racemes mostly 5–10(15), lax and spreading, to 12 cm long, subdigitate or racemose on a short axis; rachis about 0.5 mm wide, narrowly winged, and bearing few to numerous slender spreading white hairs to 5 mm long; spikelets about 2 mm long, in pairs, one nearly sessile, the other on pedicel 1.5–2 mm long; first glume minute or wanting, second half as long as spikelet or longer, 3-nerved, appressed-pubescent along margins; sterile lemma more or less silky pubescent on margins and lateral internerves, the hairs sometimes spreading at maturity; fertile lemma pale or plumbeous.

* One species, *D. serotina,* has been excluded from the flora; it is discussed briefly in the pages concluding the Gramineae.

Fig. 237. *Dactyloctenium aegyptium.*
(Reproduced by permission of U.S. National Herbarium.)

A common weed of fields and waste places in the tropics of the world. Specimens examined: Santa Cruz, Santa María.

We can find no reference to this species in the literature concerned with the Galápagos Islands. The Edmonstone specimen (*s.n.* [K]) was reported as *D. serotina* by Hooker, as *Paspalum serotinum* by Andersson, and as *Panicum serotinum* by both Robinson and Stewart. Through the courtesy of W. D. Clayton, we were able to borrow the specimen, which proves to be *D. horizontalis.* An annotation by O. Stapf on the sheet reads: "*Digitaria digitata.*" Stapf's determination is correct: *D. digitata* (based on *Milium digitatum* Sw.) is a synonym of *D. horizontalis.*

See Fig. 238.

Fig. 238. *Digitaria horizontalis.*
(Reproduced by permission of U.S. National Herbarium.)

Echinochloa Beauv., Ess. Agrost. 53, *pl. 11, fig. 2.* 1812

Slender to coarse annuals or perennials with compressed sheaths and few to many densely flowered racemes distant or approximate on main axis. Spikelets 2-flowered but often appearing 1-flowered, more or less hispid, plano-convex, subsessile, in pairs or irregular clusters along 1 side of panicle branches. First glume about one-third to half the length of spikelet, acute or mucronate; second glume and sterile lemma equal, the latter mucronate to long-awned, enclosing a membranous palea and sometimes a staminate flower. Fertile lemma indurate, slightly crested at apex, smooth and shining, margins enrolled, tip of palea usually free. x = 9.

About 20 species in warm regions of the world, some of them widely distributed and of weedy habit.

Fig. 239. *Echinochloa colonum.*

Echinochloa colonum (L.) Link, Hort. Berol. 2: 209. 1833

Panicum colonum L., Syst. Nat. ed. 10, 2: 870. 1759.
Oplismenus colonus HBK., Nov. Gen. & Sp. Pl. 1: 108. 1815.

Annual; culms glabrous, erect or decumbent, usually much branched at base, 20–80 cm long; sheaths glabrous, shorter than internodes, rather loose, somewhat keeled toward summit; ligule wanting; blades rather lax, 6–15 cm long, 3–8 mm wide, glabrous to sparsely pubescent, margins scabrous; panicle 5–15 cm long, axis and branches scabrous to puberulent; racemes simple, single or occasionally 2 together, 1–2 cm long; spikelets 2–3 mm long, crowded, glumes and sterile lemma glabrous to scabrous-puberulent, nerves scabrous-hispid; first glume one-third to half as long as spikelet, 3-nerved; second glume about equal to sterile lemma, mucronate, 5-nerved; sterile lemma slightly exceeding second glume, mucronate or awn-pointed, flattened on back, enclosing a hyaline palea of equal length; fertile lemma rounded on back, smooth and shining, obscurely 5-nerved, short-acuminate, margins enrolled below, apex of palea not enclosed.

A common weed of ditches and moist ground in warmer regions of the world; introduced in the Americas, where it occurs from the southern United States to Argentina. Specimens examined: San Cristóbal, Santa María.

Reported (as *Panicum colonum*) from Santa María by Hooker, Robinson, and Stewart, and (as *Oplismenus colonus*) from the same island by Andersson. These accounts are all based on 2 collections, one by Darwin, the other by Andersson. Stewart reported his *1288* (CAS) as *Panicum geminatum* Forsk., and his *1300* (GH, US) as *P. multiculmum* Anderss. The specimen of the latter collection at US has been annotated as *E. colonum* by A. S. Hitchcock.

See Fig. 239.

Eleusine Gaertn., Fruct. et Sem. Pl. 1: 7, *pl. 1, fig. 11.* 1788

Annuals (ours) or perennial grasses with flat or folded blades and 2 to several rather stout spikes digitate at summit of culms, sometimes with 1 or 2 additional spikes a short distance below. Spikelets few- to several-flowered, strongly laterally compressed, sessile and closely imbricate in 2 rows along 1 side of a narrowly winged rachis that is spikelet-bearing to the end; rachilla disarticulating above glumes and between florets. Glumes shorter than lowermost floret, the first narrower and shorter than the second. Lemma keeled, 3-nerved, lateral nerves near midrib. Seed prominently ridged, free inside the thin pericarp. $x = 9, 8$.

Some 7 or 8 species in warm regions, chiefly in the Old World, with 1 species in South America and another (*E. indica*) widely distributed throughout the world.

Eleusine indica (L.) Gaertn., Fruct. et Sem. Pl. 1: 8. 1788

Cynosurus indicus L., Sp. Pl. 72. 1753.

Annual; culms glabrous, in spreading clumps, branching below, often decumbent, 15–70(100) cm tall; sheaths compressed-keeled, mostly overlapping, often papillose-pilose along margins, otherwise glabrous; ligule membranous-ciliolate, scarcely 1 mm long; blades elongate, 3–8 mm wide, flat or folded, glabrous or somewhat pilose above, margins scabrous; spikes 2 to several, (3)5–10(15) cm long, stiffly ascending; spikelets 3–8-flowered; glumes unequal, smooth, keeled, the first about 2

Fig. 240. *Eleusine indica.*
(Reproduced by permission of U.S. National Herbarium.)

mm long, the second 3 mm long, both scabrous on the keels; lemmas smooth except for scabrous keel, about 3 mm long, acute; palea a little shorter than lemma; seed reddish brown, about 1 mm long, prominently ridged, ridges radiating outward from embryo region.

A common weed of temperate and tropical regions in the Old and New Worlds; introduced into the Americas. Specimens examined: Isabela, San Cristóbal, San Salvador, Santa Cruz, Santa María.

See Fig. 240.

Eragrostis Host, Icon. Gram. Austr. 4: 14, *pl. 24.* 1809

Annual or perennial grasses with ciliate ligules and open or contracted, sometimes spikelike panicles. Spikelets laterally compressed, few- to many-flowered, the rachilla disarticulating above glumes and between florets, or continuous, lemmas deciduous, the paleas persistent. Glumes somewhat unequal, acute or acuminate, 1-nerved, or the second only rarely 3-nerved. Lemmas acute or acuminate, keeled or rounded on back, 3-nerved, nerves prominent; palea 2-nerved, keels sometimes ciliate. $x = 10$.

Perhaps 300 species in the warm regions of the world; especially abundant in tropical and southern Africa.*

Inflorescence spikelike; palea prominently ciliate on keels. . . . . *E. ciliaris*
Inflorescence an open panicle; palea not conspicuously ciliate on keels:
  Spikelets about 2.5 mm wide; lemmas with prominent glandular depressions on keels. . . . . *E. cilianensis*
  Spikelets about 1.5 mm wide; lemmas without glandular depressions. . . . . . *E. mexicana*

Eragrostis cilianensis (All.) Lutati, Malpighia 18: 386. 1904

Poa cilianensis All., Flora Pedem. 2: 246. 1785.
Poa megastachya Koel., Descr. Gram. 181. 1802.
Eragrostis major Host, Icon. Gram. Austr. 4: 14, *pl. 24.* 1809.
Eragrostis megastachya Link, Hort. Berol. 1: 187. 1827.

Annual; culms tufted, erect to prostrate, branching, often geniculate, 10–60 cm tall, usually with ring of glands below nodes; sheaths keeled, usually with rather prominent glandular depression on keel, more or less pilose on margins above; ligule a row of white hairs about 0.5 mm long; blades flat, linear to linear-lanceolate, mostly 3–15 cm long, 2–7 mm wide, usually with prominent glandular pits on margins and midrib below; panicles 5–20 cm long, densely flowered, open or condensed, branches spikelet-bearing to base; spikelets 10–40-flowered, short-pedicellate; glumes subequal, about 2 mm long, keels more or less glandular-pitted; lemmas closely imbricate, about 2.5 mm long, the keels scabrous above and usually with some prominent glandular pits; palea about two-thirds as long as lemma, minutely ciliate on keels; caryopsis ellipsoidal to subspherical, reddish brown, 0.5–0.8 mm long.

A weed of roadsides, waste places, and cultivated fields; introduced from Europe and now common in the Western Hemisphere from Canada to Argentina. Speci-

* Two species, *E. bahiensis* and *E. pilosa,* have been excluded from the flora; they are discussed briefly in the pages concluding the Gramineae.

Fig. 241. *Eragrostis cilianensis.*
(Reproduced by permission of U.S. National Herbarium.)

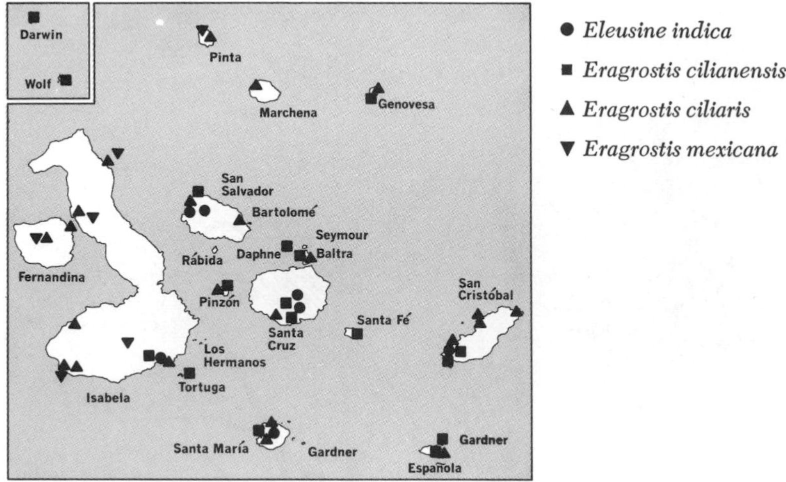

mens examined: Baltra, Daphne, Darwin, Española, Gardner (near Española), Genovesa, Isabela, Pinzón, San Cristóbal, San Salvador, Santa Cruz, Santa Fé, Santa María, Tortuga, Wolf.
See Fig. 241.

Eragrostis ciliaris (L.) R. Br. in Tuckey, Narr. Exp. Congo App. 478. 1818
Poa ciliaris L., Syst. Nat. ed. 10, 2: 875. 1759.

Annual; culms slender, branching, erect to decumbent-spreading, mostly 10–40 cm tall; sheaths glabrous to densely papillose-pilose and with tuft of long hairs at throat; ligule ciliate, less than 0.5 mm long; blades flat to subinvolute, acuminate or attenuate, mostly less than 10 cm long, rarely to 5 mm wide, glabrous to sparsely pilose; panicles dense, spikelike, interrupted toward base, mostly 5–10 cm long; spikelets mostly 2–3 mm long, 6–12-flowered; glumes subequal, glabrous, about 1 mm long; lemmas glabrous, 1–1.5 mm long, midnerve slightly excurrent; palea slightly shorter than lemma, keels conspicuously papillose-ciliate, the stiff, spreading hairs about 0.5 mm long.

A common weedy species in the tropics and subtropics of the world. Specimens examined: Baltra, Española, Fernandina, Genovesa, Isabela, Marchena, Pinta, Pinzón, San Cristóbal, San Salvador, Santa Cruz, Santa María.

Eragrostis mexicana (Hornem.) Link, Hort. Berol. 1: 190. 1827
Poa mexicana Hornem., Hort. Hafn. 2: 953. 1815.

Annual; culms glabrous, erect or ascending from a geniculate or decumbent base, 15–50 cm tall, branching at base; sheaths pilose in throat and often along upper margins; ligule a row of hairs about 0.5 mm long; blades flat to somewhat involute, glabrous or sometimes papillose-pilose, scabrous above, 5–10 cm long, 3–6 mm wide; panicles open, oblong to pyramidal, loosely flowered, 5–15 cm long, the branchlets and pedicels slender, subcapillary, scabrous; spikelets plumbeous, often tinged with red or purple, 4–6(9) mm long, 1.6–2 mm wide, 6–12-flowered; glumes subequal, 1.5–2 mm long, the second usually slightly exceeding the first; lemmas about 2 mm long, acute to subobtuse, often scaberulous, especially on upper half; palea usually slightly shorter than lemma, ciliolate on keels.

From the southwestern United States through Mexico and Central America to Venezuela and Brazil; in fields, along roadsides, and in waste places, often on sandy soil. Specimens examined: Fernandina, Isabela, Pinta.

Not reported previously from the Galápagos Islands. *Stewart 1262* (CAS, GH, US) and *Snodgrass & Heller 110* (DS, GH) were reported as *E. bahiensis* Roem. & Schult. Hitchcock determined *Howell 9571* (CAS, US) as *E. neomexicana* Vasey, and this name appears on the sheets. In 1968, however, we found both *Stewart 1262* and *Howell 9571* in the *E. mexicana* folder at US, presumably placed there by Mrs. Chase. Both *Snodgrass & Heller 811* (DS) and *Stewart 1184* (CAS, GH) were reported erroneously as *Sporobolus domingensis* (Trin.) Kunth.

Eriochloa HBK., Nov. Gen. & Sp. Pl. 1: 94. 1815

Annuals or perennials with terminal panicles consisting of few to many racemes racemose along main axis. Spikelets 2-flowered but appearing 1-flowered, dorsally compressed, more or less pubescent, solitary or in pairs, short-pedicelled or subsessile, in 2 rows on 1 side of a narrow, usually hairy rachis, back of fertile lemma turned away from it; rachilla below second glume thickened into ringlike or beadlike callus. First glume wanting or reduced to a minute sheath surrounding callus and adnate to it; second glume and sterile lemma subequal, or glume somewhat longer, membranous, acute or acuminate. Fertile lemma indurate, minutely papillose-rugose, margins slightly enrolled, apex mucronate or short-awned. $x = 9$.

About 20 species, in warm regions of the world.*

Eriochloa pacifica Mez, Bot. Jahrb. 56, Beibl. 125: 11. 1921

Annual; culms erect or spreading, branching, glabrous or more or less pubescent, especially below nodes and inflorescence, as much as 1 m tall; sheaths glabrous to somewhat pubescent, usually ciliate on margins and pubescent on collar; ligule a row of hairs about 0.5 mm long; blades flat, narrowly lanceolate, mostly 5–10 cm long, more or less pilose, especially on upper surface, margins scaberulous; inflorescence 5–10 cm long, of 4–10 or more ascending or spreading racemes about 2 cm long; axis and branches densely pubescent; pedicels 0.5–1 mm long, bearing stiff hairs, those at summit of pedicel sometimes nearly as long as spikelet; spikelet 5–6 mm long, densely pubescent on lower two-thirds, hairs often appressed; first glume lacking; second glume as much as 6 mm long, extending into an awn about 1 mm long; sterile lemma subulate-pointed, about 1 mm shorter than glume; fertile lemma about 2 mm long, obtuse at apex, mucronate or minutely awn-pointed.

Found at low elevations, Peru and Ecuador. Specimens examined: San Cristóbal, Santa Cruz.

In his Grasses of Ecuador, Peru, and Bolivia, Hitchcock (Contr. U.S. Nat. Herb. 24: 429. 1927) states that *E. pacifica* occurs in the Galápagos, but he cites no

* Two species, *E. distachya* and *E. punctata,* have been excluded from the flora; they are discussed briefly in the pages concluding the Gramineae.

Fig. 242. *Ichnanthus nemorosus* (*a–f*). *Eriochloa pacifica* (*g–m*). (Drawn by Charlotte Reeder, inked by Jeanne R. Janish.)

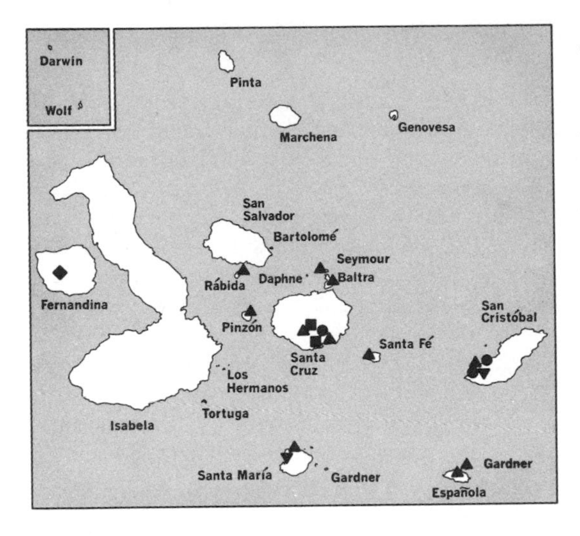

● *Eriochloa pacifica*
■ *Ichnanthus nemorosus*
▲ *Leptochloa filiformis*
▼ *Leptochloa virgata*
◆ *Muhlenbergia microsperma*

specimens. This is apparently the only reference to this species in these islands.

Snodgrass & Heller 547 (US, GH) was reported as *E. distachya* HBK. by Robinson. Stewart reported his *1274* (CAS, GH, US) as *E. punctata* (L.) Desv. However, the Stewart specimen at US has been annotated as *E. pacifica* by A. S. Hitchcock, who also determined *Howell 8598* as this species. Neither *E. distachya* nor *E. punctata* occurs in the Galápagos Islands (see list of doubtful and excluded species).

See Fig. 242.

Ichnanthus Beauv., Ess. Agrost. 56, *pl. 12, fig. 1.* 1812

Weak-stemmed, rather broad-leaved grasses with short-pedicelled spikelets in open or somewhat contracted panicles. Spikelets 2-flowered but often appearing 1-flowered, more or less laterally compressed, the glumes and sterile lemma strongly nerved. First glume acuminate, often as long as spikelet; second glume and sterile lemma equal or subequal, the lemma enclosing a palea and sometimes a staminate flower. Fertile lemma dorsally compressed, chartaceous-indurate, short-stipitate and bearing on either side membranous appendages adnate to base of lemma, or these reduced or indicated by minute excavations only (our species has excavations only). $x = 10$.

Some 40 to 50 species mostly in tropical America but a few in the tropics of the Old World.*

* The most comprehensive treatment is the revision given by Hitchcock & Chase (1920, 1–12). This is not completely satisfactory, however, and several of the species seem poorly differentiated. More field work, supplemented by cytological studies, might provide data for an improved taxonomy. The specimens examined seem to fit most closely the Hitchcock & Chase concept of *I. nemorosus*, but this species may deserve no more than varietal status under *I. pallens*. One of them—*Fournier 239* (RM)—is somewhat taller, has larger leaves and panicles, and has spikelets tending to be slightly larger than described for the species by Hitchcock & Chase, but the attenuate first glume suggests that it belongs here.

Ichnanthus nemorosus (Sw.) Doell in Mart., Fl. Bras. 2²: 289. 1877

Panicum nemorosum Sw., Prodr. Veg. Ind. Occ. 22. 1788.

Culms spreading or creeping, much branched, rooting at nodes, glabrous or somewhat pubescent, nodes sometimes pilose, fertile shoots decumbent or rising 15–30 cm; sheaths shorter than internodes, more or less pilose or nearly glabrous, margins and collar villous; blades thin, ovate-lanceolate, 3–7 cm long, 1–2 cm wide, clasping at the usually asymmetric base, more or less hispid and scaberulous on both surfaces; panicles terminal and axillary, 2–8 cm long, the few branches spreading or appressed; axis, branches, and short pedicels angled, scabrous, branches villous at base; spikelets 2.5–3(3.5) mm long on pedicels 1–2 mm long or less; first glume 3-nerved, attenuate or awn-tipped, as long as or slightly exceeding the second, glabrous or with a few long hairs on margin, keel scabrous; second glume and sterile lemma 5-nerved, glabrous, the glume a little longer, acuminate, scabrous on keel; sterile lemma smooth on keel, acute or rounded at apex, palea well developed and often bearing staminate flower; fertile lemma about 2 mm long, yellowish, smooth and shining, scars at base about 0.3 mm long, margins enrolled.

In shady habitats, from the West Indies and Mexico through Central America to Ecuador. Specimens examined: Santa Cruz, the first records of this species from the Galápagos.

Since all the collections are from the same island, this species may be a recent introduction.

See Fig. 242.

Leptochloa Beauv., Ess. Agrost. 71, 166, *pl. 15, fig. 1.* 1812

Annual or perennial grasses with flat blades, membranous but usually ciliate ligules, and numerous racemes borne on a common, usually elongate, axis. Spikelets laterally compressed, 2- to several-flowered, borne on short pedicels in 2 rows along 1 side of a narrow rachis, spikelets often rather distant; rachilla disarticulating above glumes and between florets. Glumes usually shorter than lowermost lemma, 1-nerved, awnless or mucronate. Lemmas obtuse or acute, sometimes 2-toothed and mucronate or short-awned from between teeth, 3-nerved, nerves sometimes pubescent. $x = 10$.

Some 15 to 20 species in the warm regions of the world.

Inflorescence often half the height of plant or more; lemmas appressed-pubescent on nerves; plant annual . . . . *L. filiformis*

Inflorescence 10–15 cm long, much shorter than culm; lemmas sparsely pubescent on margins; plant perennial . . . . *L. virgata*

Leptochloa filiformis (Lam.) Beauv., Ess. Agrost. 71, 161, 166. 1812

Festuca filiformis Lam., Tabl. Encycl. Méth. 1: 191. 1791.
Eleusine mucronata Michx., Flora Bor. Amer. 1: 65. 1803.
Leptochloa mucronata Kunth, Rév. Gram. 1: 91. 1829.

Annual; culms slender to somewhat coarse, branching, erect, or ascending from geniculate base, 15–100 cm tall; sheaths glabrous or more or less papillose-pilose with long weak hairs; ligule a ciliate membrane 1–2 mm long; blades elongate, 3–8 mm wide, glabrous or sparsely papillose-pilose; inflorescence often half the length of culm, with several to many slender spreading to reflexed branches (1)2–8(15) cm

Fig. 243. *Leptochloa filiformis*.
(Reproduced by permission of U.S. National Herbarium.)

long; spikelets 2- or 3-flowered, 2–3 mm long; glumes 1-nerved, narrow, acuminate, subequal, scabrous on keel, nearly as long as spikelet; lemmas about 1.5 mm long, awnless, appressed-pubescent on nerves.

A common weed of fields and waste places at low altitudes, from southeastern United States to Argentina. Specimens examined: Baltra, Española, Gardner (near Española), Pinzón, Rábida, San Cristóbal, Santa Cruz, Santa Fé, Santa María, Seymour.

See Fig. 243.

Leptochloa virgata (L.) Beauv., Ess. Agrost. 71, 161, 166. 1812

Cynosurus virgatus L., Syst. Nat. ed. 10, 2: 876. 1759.

Perennial; culms cespitose, usually erect, several-noded, 50–100 cm tall; sheaths shorter than internodes, glabrous or nearly so; ligule 0.5 mm long or less, ciliate; blades flat, elongate, 3–10 mm wide, glabrous, scabrous on margins; inflorescence drooping, 10–15 cm long, branches lax, 8–12 cm long (rarely longer); spikelets 3–5-flowered, 2.5–4 mm long; glumes similar, scabrous on keel, first about 1.5 mm long, second about 2 mm long; lemmas mostly 1.5–2 mm long, sparsely pubescent on margins, tip blunt, awnless or with awn to 2 mm long.

A common, often weedy species, from the southern United States and the West Indies to Argentina. Specimens examined: San Cristóbal, Santa María.

Muhlenbergia Schreb. ex Gmel. in L., Syst. Nat. ed. 13, 2: 171. 1791

Annual or perennial grasses with membranous ligules and open, contracted, or sometimes spikelike panicles. Spikelets 1-flowered (rarely some 2-flowered), disarticulating above glumes, floret falling free. Glumes very short to as long as floret, 1-nerved, obtuse to acuminate, awned or awnless. Lemma membranous, 3-nerved, with very short, usually minutely pilose callus, apex acute, mucronate, or extending into straight or flexuous awn. $x = 10$ (rarely 9).

About 190 species, mostly in temperate and warm regions of the Western Hemisphere, with a few in Asia.

Muhlenbergia microsperma (DC.) Kunth, Rév. Gram. 1: 64. 1829

Trichochloa microsperma DC., Cat. Pl. Hort. Monsp. 151. 1813.
Podosaemum debile HBK., Nov. Gen. & Sp. Pl. 1: 128. 1815.
Muhlenbergia debilis (HBK.) Kunth, Rév. Gram. 1: 63. 1829.
Muhlenbergia reamosissima Vasey, Bull. Torrey Bot. Club 13: 231. 1886.

Annual; culms slender, scaberulous at least below nodes, branching at base and often at nodes above, 10–50 cm long; sheaths mostly shorter than internodes, more or less scaberulous; ligule hyaline, 1.5–2 mm long; blades flat or loosely involute, more or less scabrous on both surfaces, mostly less than 5 cm long, 1–3 mm wide; panicles numerous, terminal and axillary, usually purplish, narrow, 5–20 cm long, the axis, branches, and pedicels scabrous; glumes subequal, broad, awnless, usually obtuse, second about 1 mm long; lemma narrow, 2–4 mm long, scabrous, especially on nerves, callus short-pubescent, apex with a slender, straight or slightly flexuous awn 1–3 cm long. Cleistogamous spikelets often developed in lower sheaths.

In rocky places and on dry open ground, from the southwestern United States to Guatemala; also in Ecuador and Peru. Specimen examined: Fernandina (*Eliasson*

Fig. 244. *Muhlenbergia microsperma.*

1689), the first report of any species of *Muhlenbergia* from the Galápagos Islands. See Fig. 244.

Oplismenus Beauv., Fl. Owar. 2: 14, *pl. 58, fig. 1.* 1809

Freely branching, creeping annuals or perennials of shady habitats with flat lanceolate blades and inflorescences of few to several alternate spikelike racemes racemose on main axis, racemes sometimes very short and spikelets crowded and appearing fascicled, or racemes almost totally absent and spikelets solitary or in 2's or 3's along axis. Spikelets 2-flowered but often appearing 1-flowered, terete or somewhat laterally compressed, subsessile, solitary or in pairs, in 2 rows, crowded or approximate on 1 side of a narrow, scabrous or hairy rachis, back of fertile lemma turned toward it. Glumes subequal, awned, awn of first glume considerably longer than that of second, often as long as or exceeding length of spikelet. Sterile lemma empty or rarely enclosing a staminate flower, longer than glumes and fertile floret, awnless or short-awned, awn rarely to 2 mm long; sterile palea absent or when present narrow, hyaline, ciliate at apex. Fertile lemma chartaceous-indurate, pale, smooth and shining, acute, usually minutely crested, sometimes short-awned. x = 9.

About 10 species in the tropics of the world.

Oplismenus setarius (Lam.) Roem. & Schult., Syst. Veg. 2: 481. 1817

Panicum setarium Lam., Tabl. Encycl. Méth. 1: 170. 1791.

Culms slender, creeping, freely branching, ascending flowering branches 10–30 cm tall; sheaths mostly shorter than internodes, usually glabrous, sometimes more or less pubescent or even papillose-pilose, margins and collar pubescent-ciliate; ligule membranous, truncate, erose, ciliate, hairs often equaling or even exceeding membrane; blades lanceolate, narrowed at base, 2–4 cm long (rarely to 8 cm), 5–10 mm wide (rarely to 15 mm), margins usually undulate, glabrous or scabrous to variously pubescent, hairs sometimes papillose-based; racemes mostly 4–6, distant, rachis 5 mm long or less, bearing 2–10 spikelets (rarely only 1), these often ap-

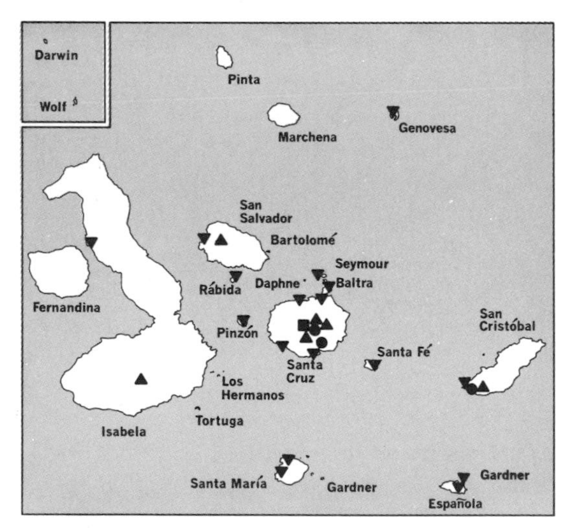

● *Oplismenus setarius*

■ *Panicum arundinariae*

▲ *Panicum dichotomiflorum*

▼ *Panicum fasciculatum*

Fig. 245. *Oplismenus setarius.*
(Reproduced by permission of U.S. National Herbarium.)

pearing fascicled; spikelets terete or somewhat laterally compressed, about 2.5 mm long excluding awns, sparsely to rather densely pubescent; awn on first glume 4–8 mm long.

In shady habitats, usually at low elevations, from the southern United States and the West Indies to Argentina. Species examined: San Cristóbal, Santa Cruz.

See Fig. 245.

Panicum L., Sp. Pl. 55. 1753; Gen. Pl. ed. 5, 29. 1754

Annuals or perennials of various habit. Spikelets 2-flowered but often appearing 1-flowered, more or less dorsally compressed, in open or compact panicles, rarely in racemes, back of fertile lemma turned toward rachis. Glumes 2, membranous, usually unequal, the first often minute, the second usually as long as spikelet (rarely first glume as long as spikelet or second somewhat shorter). Sterile lemma membranous or rarely slightly indurate, usually enclosing a membranous or hyaline palea and sometimes a staminate flower. Fertile lemma chartaceous to strongly indurate, margins enrolled over a palea of same texture (rarely margins merely clasping palea and not strongly enrolled). $x = 9, 10$.

About 500 species, principally in the tropics and subtropics of the world, with a few in temperate regions.*

Spikelets short-pedicelled along 1 side of inflorescence branches, forming more or less 1-sided racemes:
  Spikelets 1–1.5 mm long; sterile palea subrigid at maturity; fertile lemma smooth and shining . . . . *P. laxum*
  Spikelets 2–3 mm long; sterile palea hyaline or membranous; fertile lemma transversely rugose:
    Nodes densely bearded; first glume not clasping at base; fertile lemma minutely transversely rugose . . . . *P. purpurascens*
    Nodes smooth or appressed-pubescent; first glume clasping at base; fertile lemma strongly transversely rugose:
      Panicle branches ascending or spreading; first glume acute; second glume and sterile lemma more or less transversely wrinkled between nerves . . . . *P. fasciculatum*
      Panicle branches short, closely appressed; first glume truncate or obtuse; second glume and sterile lemma smooth between nerves . . . . *P. geminatum*
Spikelets in open or contracted panicles, but not in 1-sided racemes:
  First glume half or more the length of spikelet:
    Glumes subequal, obtuse, slightly shorter than spikelet . . . . *P. glutinosum*
    Glumes unequal, acuminate, first glume half to three-fourths the length of spikelet, second about as long as sterile lemma . . . . *P. hirticaule*
  First glume not more than one-third the length of spikelet:
    Fertile lemma transversely rugose; culms robust, 1–2.5 m tall, cespitose in large clumps . . . . *P. maximum*
    Fertile lemma not transversely rugose:
      First glume narrow, not clasping; fertile lemma sparsely appressed-pilose with short white hairs . . . . *P. arundinariae*
      First glume broad and clasping; fertile lemma smooth and shining . . . . *P. dichotomiflorum*

* Two species, *P. molle* and *P. serotinum*, have been excluded from the flora; they are discussed briefly in the pages concluding the Gramineae.

Panicum arundinariae Trin. ex Fourn., Mex. Pl. 2: 25. 1886

Panicum virgultorum Hack., Oesterr. Bot. Zeitschr. 51: 369. 1901.

Culms clambering or straggling, freely branching, to 1 m or more long, glabrous or more or less pubescent, especially on younger growth, nodes pubescent; sheaths mostly shorter than internodes, glabrous or more or less pubescent, hairs sometimes papillose-based, margins and collar pubescent-ciliate; ligule very short, ciliate; blades lanceolate, 4–10 cm long, 6–10 mm wide, more or less papillose-hispid on both surfaces, margins scabrous; panicles 3–8 cm long, the few solitary branches densely flowered but usually naked toward base, the axis, branches, and short pedicels scabrous and often somewhat pubescent; spikelets glabrous, about 1.8 mm long; first glume narrow, less than one-fourth the length of spikelet, not clasping; second glume and sterile lemma subequal, both 5-nerved; fertile lemma shining when mature, sparsely appressed-pilose with short white hairs.

From southern Mexico and Guatemala through Central America to Ecuador. Specimens examined: Santa Cruz (*Howell 9242*), apparently the only record of *P. arundinariae* from the Galápagos Islands.

Panicum dichotomiflorum Michx., Flora Bor. Amer. 1: 48. 1803

Panicum aquaticum Poir. in Lam., Encycl. Méth. Suppl. 4: 28. 1816.

Culms usually freely branching, often somewhat succulent, erect or more often ascending or spreading from a geniculate base, glabrous; sheaths often compressed, usually longer than internodes, glabrous; ligule a ciliate membrane 1–2 mm long; blades elongate, 3–10 mm wide, flat or folded, glabrous or sparsely pilose above; panicle open, 10–15 cm long, branches rather stiffly ascending, short branchlets appressed, bearing short-pedicelled, often somewhat crowded spikelets, the axis and pedicels angled, scabrous; spikelets glabrous, mostly 2.5–3 mm long; first glume thin, truncate, rounded, or subacute, broad and clasping, one-fifth to one-fourth the length of spikelet; second glume and sterile lemma subequal, rather prominently nerved, pointed beyond "fruit"; fertile lemma smooth and shining.

In moist ground along margins of ponds and streams, from the United States to Argentina. Specimens examined: Isabela, San Cristóbal, San Salvador, Santa Cruz.

The name *P. dichotomiflorum* does not appear in the literature concerned with the grass flora of the Galápagos. Svenson reported a collection from Santa Cruz as *P. aquaticum*, crediting the determination to Mrs. Chase.

The history of the application of the names *P. dichotomiflorum* and *P. aquaticum* is an interesting one. Hitchcock and Chase, in their monograph of *Panicum* (1910) listed *P. aquaticum* as a synonym of *P. dichotomiflorum*, giving the range of the latter as the United States, Mexico, the West Indies, and South America to Uruguay. Five years later, in the *North American Flora*, Hitchcock recognized both species, indicating that *P. dichotomiflorum* is an annual, ranging from Maine to California in the United States through Bermuda, the Bahamas, the West Indies, and Panama, and describing *P. aquaticum* as a perennial, having a distribution in the West Indies, Central America, and tropical South America. Workers at the U. S. National Herbarium seem consistently to have recognized both species since that time, but with changing concepts of their distributions. Thus in the *Manual of Grasses of the West Indies* (1936), we find that the range given for *P. aquaticum* is much the same, but *P. dichotomiflorum* is said to occur from the "United States to Argentina." By 1936, it appears that Hitchcock and Chase had concluded that the two species were sympatric from the West Indies and Mexico southward. The distinctions they list as separating the two taxa seem extremely slight: *P. aquaticum* is perennial, has panicles "similar to those of *P. dichotomiflorum*, but averaging smaller," and has spikelets averaging somewhat larger.

Our study of material of this complex from various parts of the Americas suggests that we are dealing with one wide-ranging polymorphic species. Measurements of the spikelets of South American plants with smaller panicles did not reveal consistently larger spikelets. None of the plants we examined seemed definitely to be perennial. Species with wide distribution often, of course, have many races. Future biosystematic studies of plants within this complex may yield information that will permit meaningful groupings at the varietal or even the specific level. In the light of our present knowledge, however, any attempt to recognize *P. aquaticum* seems only to compound the problem of identification.

Panicum fasciculatum Sw., Prodr. Veg. Ind. Occ. 22. 1788

Panicum fuscum Sw., op. cit. 23.
Panicum multiculmum Anderss., Kongl. Svensk. Vet.-Akad. Handl. 1853: 133. 1855.

Annual; culms slender, erect or decumbent, often much branched, glabrous to more or less pubescent or papillose-hispid, (15)20–60(100) cm long, nodes appressed-pubescent; sheaths papillose or papillose-hispid, especially on margins, sometimes more or less soft-puberulent; blades 5–25 cm long, 0.5–2 cm wide, glabrous to puberulent, or sometimes with scattered, rather stiff hairs; inflorescence 3–15 cm long, often short-exserted until maturity, consisting of a series of spikelike racemes arranged along a scabrous, sometimes pilose axis, the racemes ascending or spreading; spikelets glabrous, 2–2.5(3) mm long, bronze to mahogany-colored, obovate, turgid, abruptly short-pointed; first glume clasping, about one-third the length of spikelet, subacute; second glume and sterile lemma subequal, prominently nerved, faintly to strongly transversely wrinkled between nerves; fertile lemma pale, strongly transversely rugose.

In moist open ground, often a weed in fields and waste places, from the southern United States, the West Indies, and Mexico through Central America to Ecuador and Brazil. Specimens examined: Baltra, Española, Gardner (near Española), Genovesa, Isabela, Pinzón, Rábida, San Cristóbal, San Salvador, Santa Cruz, Santa Fé, Santa María, Seymour.

Panicum multiculmum Anderss., described from Isla Santa María, was characterized as having cespitose, strongly branching culms, short, villous leaves, and a condensed panicle. Robinson recognized the species, as did Stewart, both listing it from a number of islands. It has never been reported from elsewhere than in the Galápagos. The rather ample collections of this complex made by Howell in 1932 from nine islands were all identified by Hitchcock and Chase as *P. multiculmum*; no specimen was named *P. fasciculatum*! This seems curious inasmuch as some of Howell's plants are large, are only slightly branched, and have leaves that are glabrous or essentially so. In their monograph of *Panicum* (1910), however, Hitchcock and Chase list *P. fasciculatum* from the Galápagos, citing Agassiz in 1891. Following their discussion of *P. fasciculatum chartaginense* [= *P. fasciculatum* var. *reticulatum* (Torr.) Beal], they state: "A closely allied species, *P. multiculmum* Anderss., from the Galápagos Islands, has been referred to *P. chartaginense* Swartz by Grisebach." They cite no specimens, nor do they otherwise discuss *P. multiculmum* in their monograph.

In our opinion, *P. multiculmum* is best treated as a synonym of the wide-ranging and variable *P. fasciculatum*. True, some of the specimens from the Galápagos are rather small and profusely branched, and though in many cases their foliage is rather densely pubescent, in others it is essentially glabrous; the amount of pubescence on the leaves of both large and small plants varies considerably. The size of the spikelet in *P. fasciculatum* is also variable, ranging from 2 to 3 mm in length. We have been unable to find meaningful correlations among size of spikelets, indument on the foliage, or degree of branching of the culms. The characters mentioned by Andersson as separating his *P. multiculmum* from *P. fasciculatum* appear to represent extremes within the normal variation of a single, somewhat polymorphic species. Biosystematic studies with plants of this complex can probably give more definitive answers regarding the validity of the taxon *P. multiculmum*, but for the present it does not seem useful to give it even varietal status.

Panicum geminatum Forsk., Flora Aegypt.-Arab. 18. 1775

Panicum fluitans Retz., Obs. Bot. 3: 8. 1783.
Paspalidium geminatum Stapf in Prain, Flora Trop. Africa 9: 583. 1920.

Perennial, glabrous throughout; culms cespitose, spreading from decumbent base, 25–80 cm tall, somewhat succulent at least toward base; sheaths longer than internodes; ligule membranous at base, fimbriate-ciliate, about 1 mm long; blades elongate, flat or loosely involute, 3–6 mm wide, somewhat scabrous on upper surface; inflorescence short-exserted or partially included in upper sheath, 12–30 cm long, with 10–15 or more short, erect or narrowly ascending racemes, the lower 2.5–3 cm long, those above gradually shorter; spikelets 2–2.5 mm long, about 1.5 mm wide, turgid, abruptly pointed, borne on short pedicels in 2 rows on 1 side of the angled rachis; first glume about one-third as long as spikelet, truncate or obtuse, base clasping; second glume distinctly shorter than spikelet; sterile lemma slightly exceeding the fertile, enclosing a hyaline palea; fertile lemma strongly transversely rugose.

In moist ground and ditches, along margins of lakes and ponds at low altitudes in the tropical regions of the world. Specimens examined: Española, Pinzón, San Cristóbal, Santa Cruz, Santa María.

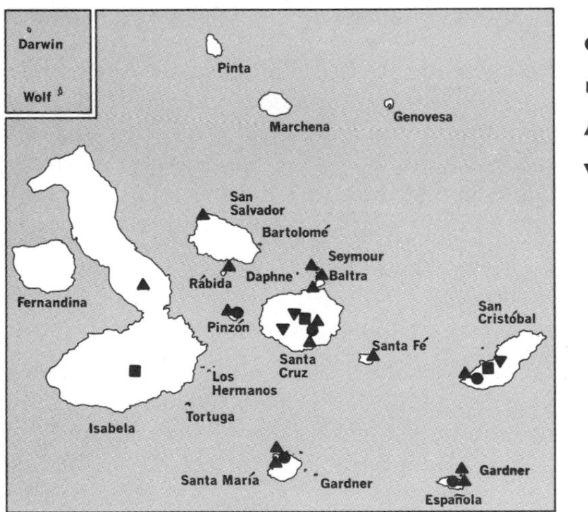

● *Panicum geminatum*
■ *Panicum glutinosum*
▲ *Panicum hirticaule*
▼ *Panicum laxum*

Curiously, the name *P. fluitans* was not listed (either as a valid name or as a synonym) by Hitchcock and Chase in their 1910 monograph of *Panicum* or in subsequent writings. This name has been associated with that of *P. geminatum* in the literature, and most agrostologists seem to consider the two to be synonyms.

Panicum glutinosum Sw., Prodr. Veg. Ind. Occ. 24. 1788

Perennial, somewhat glaucous; culms erect or decumbent at base, often rooting at lower nodes or somewhat stoloniferous, glabrous, to 2 m long; sheaths mostly longer than internodes, glabrous or sometimes pilose; ligule a minute, ciliate membrane; blades elongate-lanceolate, 5–25(50) cm long, 12–24 mm wide, acuminate, narrowed toward base, glabrous or sparingly pilose, but always the adaxial surface papillose-pilose just above ligule; panicle 15–30 cm long, about as wide, lower branches verticillate; spikelets long-pedicelled, 2.8–3.4 mm long, turgid, obtuse, viscid; glumes subequal, slightly shorter than spikelet; sterile and fertile lemmas about equal, the latter pale, smooth and shining.

A common forest grass of tropical America; Mexico and the West Indies, south to Argentina. Specimens examined: Isabela, San Cristóbal, Santa Cruz.

There is no previous report of this species from the Galápagos Islands.

Panicum hirticaule Presl, Rel. Haenk. 1: 308. 1830

Panicum hirticaulon var. *glabrescens* Anderss., Kongl. Svensk. Vet.-Akad. Handl. 1853: 135. 1855.
Panicum hirticaulon var. *majus* Anderss., loc. cit.
Panicum hirticaulon var. *minus* Anderss., loc. cit.

Annual; culms erect or ascending, sparingly branched, 15–70 cm tall, usually papillose-hispid, especially below nodes, sometimes glabrous, the nodes pubescent or hispid; sheaths conspicuously papillose-hispid with spreading hairs, rarely glabrous; ligule a row of hairs 0.5 mm long or less; blades 5–20 cm long, 4–12 mm wide, often cordate at base, glabrous or somewhat hispid, margins scabrous and sometimes papillose-hispid-ciliate toward base; panicles open, 5–15 cm long, slender, scabrous, ascending branches naked for one-third to half their length; spikelets about 3 mm long, lanceolate, acuminate, typically reddish brown; first glume one-half to three-fourths the length of spikelet, acuminate, clasping at base, second glume and sterile lemma subequal, acuminate, strongly nerved; fertile lemma 2 mm long or less, smooth and shining.

In rocky or sandy soil, from the southwestern United States and the West Indies through Mexico, Central America, and western South America to Argentina. Specimens examined: Baltra, Española, Gardner (near Española), Isabela, Pinzón, Rábida, San Cristóbal, San Salvador, Santa Cruz, Santa Fé, Santa María, Seymour.

The three varieties of *P. hirticaule* described from the Galápagos by Andersson appear to be only minor variants and scarcely worthy of taxonomic recognition.

Panicum laxum Sw., Prodr. Veg. Ind. Occ. 23. 1788

Panicum boliviense Hack., Rep. Spec. Nov. Regni Veg. 11: 19. 1912.

Culms glabrous, geniculate-ascending or decumbent and rooting at nodes, branching, 20 cm to 1 m or more long; sheaths mostly shorter than internodes, glabrous or more or less papillose-hispid near summit, margins usually densely ciliate; ligule a ciliate membrane about 0.5 mm long; blades 5–15(20) cm long, 5–15(20) mm wide, glabrous or sometimes scabrous or even sparsely pilose, base narrow or more or less rounded or subcordate, margins often more or less scabrous; inflorescence densely flowered, (5)10–20(35) cm long, composed of several to many spreading or ascending branches, these sometimes pilose in their axils; branchlets secund on lower side of branches, the lower bearing 10 or more spikelets; spikelets 1–1.5(1.8) mm long, turgid; first glume clasping, about half the length of spikelet, scabrous on keel; second glume and sterile lemma equal or subequal, scabrous on keels above, lemma somewhat inflated and subtending a membranous palea; fertile lemma slightly shorter, weakly indurate.

In ditches, on banks of streams, along pond margins, and in moist fields and woodlands, Mexico and the West Indies to Argentina. Specimens examined: San Cristóbal, Santa Cruz. Previously unreported from the Galápagos Islands.

Svenson's *237* (BKL, US) was reported by him as *P. boliviense* Hack., the determination having been supplied by Agnes Chase. In the original description of *P. boliviense*, Hackel cited a single collection (*Buchtien 2501*) from Bolivia. According to him, the new species was closely related to *P. laxum*, but the panicles were more loosely flowered and the glumes were somewhat pointed rather than blunt. Hitchcock and Chase, and later Swallen, have attempted to recognize *P. boliviense* in their publications. The key character they use is "blades somewhat cordate at base" in *P. boliviense*, in contrast to "blades narrowed toward the base" in *P. laxum*. In their descriptions, however, one finds that the leaves of *P. boliviense* are described as "7–20 mm wide, cordate at base," and those of *P. laxum* as "5–10, rarely to 15 mm wide, rounded or subcordate at base." The spikelet size is usually given as 1–1.5 mm long in *P. laxum* and 1.5–1.6 mm long in *P. boliviense*. These authors suggest that in the latter taxon the panicles are more loosely flowered. The same range and habitats are listed for both "species." We have failed to find good correlations among the characters listed. *P. boliviense* appears to be only one of many variants of the wide-ranging *P. laxum*. In our opinion it is undeserving of even varietal status.

Panicum maximum Jacq., Coll. Bot. 1: 76. 1786

Perennial; culms robust, cespitose in large clumps, 1–2.5 m tall, the nodes usually densely hirsute; sheaths shorter than internodes, papillose-hirsute to nearly glabrous, usually densely pubescent on collar; ligule membranous, ciliate, about 1.5 mm long, but with dense row of stiff hairs 4–6 mm long behind it; blades elongate, 1–2(3.5) cm wide, margins very scabrous; panicles densely flowered, finally long-exserted, 20–50 cm long, usually about one-third as wide, branches ascending or spreading, somewhat drooping at maturity, naked at base, the lower in whorls; spikelets glabrous, turgid, often shining, 3–3.5 mm long, about 1.2 mm wide; first glume about one-third the length of spikelet, clasping at base; second glume and sterile lemma subequal, the latter enclosing a well-developed palea and usually a staminate flower; fertile lemma slightly shorter, whitish, transversely rugose.

Probably the most important cultivated forage grass of tropical America; introduced long ago from Africa and now escaped and common in fields and waste places at lower elevations. Specimens examined: Santa Cruz.

The three collections examined, all from essentially the same locality, are the only records of this species from the Galápagos Islands. They were gathered in 1964 and 1966, and may represent a recent introduction.

Fig. 246. *Panicum purpurascens.*
(Reproduced by permission of U.S. National Herbarium.)

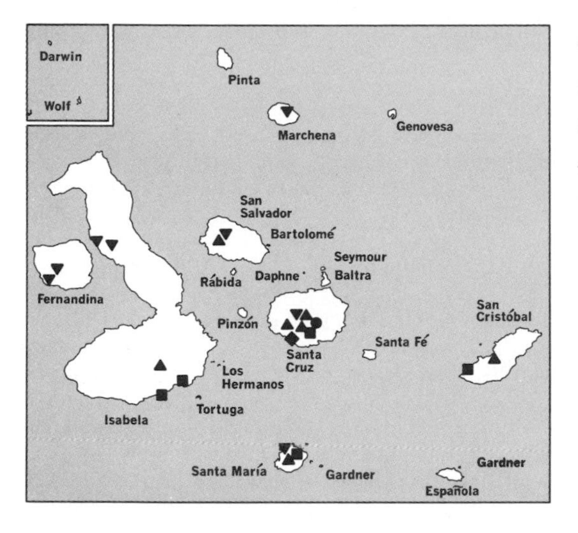

● *Panicum maximum*
■ *Panicum purpurascens*
▲ *Paspalum conjugatum*
◆ *Paspalum distichum*
▼ *Paspalum galapageium* var. *galapageium*

Panicum purpurascens Raddi, Agrost. Bras. 47. 1823

Panicum barbinode Trin., Mém. Acad. St. Petersb. VI, Sci. Nat. 1: 256. 1834.

Perennial; culms decumbent and rooting at base, 2–5 m long, the nodes densely villous; sheaths mostly longer than internodes, papillose or papillose-pilose, densely short-puberulent on collar; ligule a row of hairs about 1 mm long; blades 10–30 cm long, 10–15 mm wide, acuminate, glabrous on both surfaces, margins scabrous; inflorescences 10–25 cm long, about half as broad, lowermost branches compound, upper simple, branchlets and upper branches bearing spikelets solitary or in 2's or 3's on 1 side of a flattened rachis, this, like main axis, angled, scabrous on angles; spikelets glabrous, pointed, about 3 mm long; first glume acute, about 1 mm long, not clasping at base; second glume and sterile lemma equal, the latter subtending a hyaline palea; fertile lemma slightly shorter, obtuse, minutely transversely rugose.

In moist or wet ground, sometimes growing in water, throughout tropical and subtropical America, usually at low altitudes; cultivated for forage. Specimens examined: Isabela, San Cristóbal, Santa Cruz, Santa María.

The name *P. purpurascens* does not appear in the literature on the Galápagos Islands.

See Fig. 246.

Paspalum L., Syst. Nat. ed. 10, 2: 855. 1759

Annual or perennial grasses of various habit, with 1 to many spikelike racemes, solitary, paired, or few to many on a common axis. Spikelets 2-flowered, but usually appearing 1-flowered, plano-convex, or sometimes unequally bi-convex or even concavo-convex, subsessile or short-pedicelled, solitary or in pairs, in 2 rows (sometimes appearing as 4 rows) on 1 side of a narrow or winged rachis, back of fertile lemma turned toward rachis. First glume typically wanting (occasionally developed in some or all spikelets); second glume and lower (sterile) lemma usually similar, membranous (sterile lemma indurate in some cases). Fertile lemma and palea chartaceous-indurate, the margins of lemma enrolled at maturity. x = 10.

Perhaps 400 species in tropical and warm regions of the world; especially abundant in South America.*

Racemes 2, conjugate or nearly so at summit of culm, rarely a third one below:
  Spikelets about 1.5 mm long, ovate, margins long-silky . . . . *P. conjugatum*
  Spikelets 2.5–4.5 mm long, ellipsoid, glabrous or essentially so on margins:
    Spikelets relatively turgid; second glume appressed-pubescent on back . . *P. distichum*
    Spikelets flattened; second glume glabrous . . . . *P. vaginatum*
Racemes 2 to many on an elongate axis, not conjugate:
  Rachis foliaceous, broad and winged; spikelets whitish . . . . *P. penicillatum*
  Rachis neither foliaceous nor winged:
    Inflorescence of numerous spreading or somewhat drooping racemes; spikelets 1.5 mm long or less . . . . *P. paniculatum*
    Inflorescence of 2–6 racemes; spikelets mostly larger (except in *P. galapageium* var. *minoratum*):
      Spikelets 4–5 mm long, tinged with purple-brown . . . . *P. redundans*
      Spikelets not more than 2.8 mm long, pale or obscurely speckled with brown:
        Blades densely softly canescent; spikelets 2–2.5(2.8) mm long . . . . *P. galapageium* var. *galapageium*
        Blades less densely canescent to glabrescent; spikelets 1.5–1.8 mm long . . . . *P. galapageium* var. *minoratum*

Paspalum conjugatum Bergius, Acta Helv. Phys. Math. 7: 129, *pl. 8.* 1762

Perennial; extensively creeping with leafy, compressed, wiry stolons to 2 m long, upright flowering branches 20–60 cm tall, simple or sparingly branching, glabrous or nodes often pubescent, those of stolons usually conspicuously pilose; sheaths loose, compressed, margins ciliate, at least on upper half, often pubescent on collar, otherwise glabrous; ligule membranous, about 0.5 mm long and with row of white hairs back of it, these to 2 mm long; blades flat, thin, 8–12 cm long, 5–15 mm wide, usually glabrous below and sparsely pubescent to glabrous above, margins scabrous or ciliate-hispid; racemes in pairs, rarely a third one below, widely divergent, 8–12 cm long, rachis flattened; spikelets solitary, 1.5–1.7 mm long, ovate, light yellow, conspicuously silky-ciliate on margins, otherwise glabrous; second glume and sterile lemma equal; fertile lemma pale, not strongly indurate.

A common weed throughout the tropics of the world. Specimens examined: Isabela, San Cristóbal, San Salvador, Santa Cruz, Santa María.

See Fig. 247.

Paspalum distichum L., Syst. Nat. ed. 10, 2: 855. 1759

Perennial with slender rhizomes and extensively creeping stolons, nodes of these bearded, flowering branches erect to ascending, 6–50 cm tall; sheaths keeled, commonly pilose on margins toward summit; ligule membranous, about 0.5 mm long; blades flat, 3–12 cm long, 2–6 mm wide, acuminate, ciliate at rounded base, sometimes pubescent adaxially; racemes usually 2, 1.5–7 cm long, erect to reflexed, commonly incurved; rachis minutely scabrous on margins, 1–1.5 mm wide, slightly pedunculate in one raceme, or sometimes in both; spikelets 2.5–3 mm long, usually

* Three species, *P. longepedunculatum*, *P. scrobiculatum*, and *P. serotinum*, have been excluded from the flora; they are discussed briefly in the pages concluding the Gramineae.

Fig. 247. *Paspalum conjugatum.*
(Reproduced by permission of U.S. National Herbarium.)

solitary, elliptic, abruptly acute; first glume often developed; second glume and sterile lemma equal, 3–5-nerved, glume more or less appressed-pubescent on back, lemma glabrous; fertile lemma slightly shorter than spikelet, pale, smooth and shining.

Around ponds and along streams and ditches, from the southern United States and the West Indies to Argentina; also in the Eastern Hemisphere. Specimen examined: Santa Cruz (*Svenson 240*).

Baur 334 (GH), which was the basis for the report of *P. distichum* from Isla San Salvador by Robinson and Stewart, is a sterile specimen. Our studies of the leaf anatomy suggest that it is actually *P. vaginatum* (q.v.).

Paspalum galapageium Chase, Proc. Calif. Acad. Sci. IV, 21: 297. 1935, var. **galapageium**

Paspalum canescens Anderss., Kongl. Svensk. Vet.-Akad. Handl. 1853: 132. 1855, non Nees 1826.

Perennial; culms simple, 40–80 cm tall, leafy to or almost to summit; sheaths mostly overlapping, keeled, glabrous to finely canescent toward summit, margins often ciliate, basal sheaths sometimes partially enclosing a few to several few-flowered racemes of cleistogamous spikelets; ligule membranous, fragile, 2–4 mm long; blades flat, mostly 15–25 cm long, 3–8 mm wide, densely softly canescent on both surfaces, the pale midnerve prominent beneath; racemes (2)4–6(8), ascending to spreading, 3–6.5 cm long, rather distant on a slender axis; rachis 0.5–0.8 mm wide, scabrous; spikelets elliptic to obovate, 2–2.5(2.8) mm long, in pairs on short, flat, scabrous pedicels; first glume occasionally developed; second glume and sterile lemma pale, often speckled with purple or brown, the glume slightly shorter; fertile lemma pale, often yellowish, smooth and shining.

Apparently endemic. Specimens examined: Fernandina, Isabela, Marchena, San Salvador, Santa Cruz, Santa María.

Mrs. Chase described *P. galapageium* to accommodate the material that had been included under *P. canescens* Anderss. She states: "Since that name is preoccupied, Andersson's specimen is incomplete, and his description inadequate, *Paspalum galapageium* is based on a new type instead of on *P. canescens* Anderss."

Paspalum galapageium var. **minoratum** Chase, Proc. Calif. Acad. Sci. IV, 21: 299. 1935

Differs from var. *galapageium* in having less densely canescent to glabrescent blades, and smaller (1.5–1.8 mm long) spikelets. In the protologue, Mrs. Chase points out that there is no sharp line by which these two varieties may be separated. One specimen (*Stewart 1322*) from Isla San Cristóbal has spikelets 2 mm long and glabrescent blades, and is therefore intermediate.

Endemic. Specimens examined: Fernandina, Isabela, San Cristóbal, Santa Cruz, Santa María.

Paspalum paniculatum L., Syst. Nat. ed. 10, 2: 855. 1759

Perennial; culms cespitose, suberect or ascending, 50–100 cm tall, glabrous or nodes often more or less pubescent with stiff ascending hairs; sheaths usually longer than

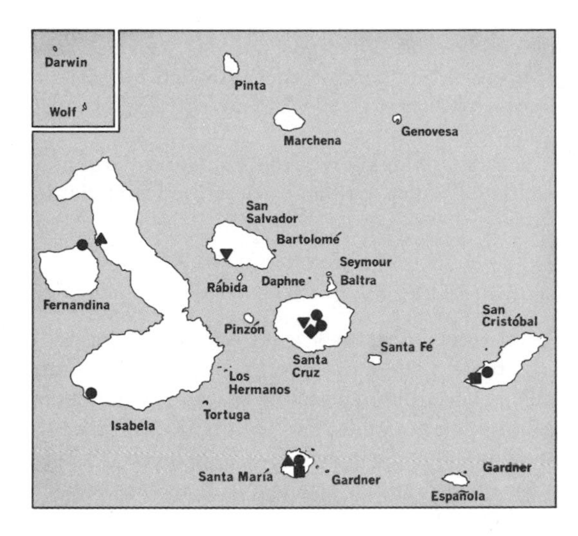

● *Paspalum galapageium* var. *minoratum*

■ *Paspalum paniculatum*

▲ *Paspalum penicillatum*

◆ *Paspalum redundans*

▼ *Paspalum vaginatum*

internodes, keeled, papillose-hispid to papillose-pilose, the hairs to 5 mm long; ligule about 0.5 mm long, membranous and bearing behind it a dense row of white hairs to 12 mm long; blades flat, spreading, 10–40 cm long, 10–25 mm wide, from coarsely hispid on both surfaces with a tuft of long white hairs at base to scabrous or sometimes glabrous except at base and along margin, the midnerve prominent beneath; panicle 10–20 cm long, of numerous spreading to somewhat drooping racemes, lowermost as much as 10 cm long; spikelets in pairs, subhemispheric, 1.3–1.5 mm long, on slender pedicels, crowded along slender angled rachis; glume and sterile lemma equal, 5-nerved, lateral pair of nerves contiguous, glume loosely pubescent with delicate hairs, sterile lemma with similar hairs along margin, sometimes throughout; fertile lemma pale.

In moist open ground and on brushy slopes, often a weed in cultivated ground and along roadsides; Mexico and the West Indies to Argentina, West Africa, the Society Islands, Australia, and New Guinea. Specimens examined: San Cristóbal, Santa María.

The two specimens examined, both collected by Edmonstone, are mounted on the same sheet and have been annotated as *P. paniculatum* by General Munro and by Mrs. Chase. There are no data indicating when they were collected. However, Hooker (1847, 238) indicates that Edmonstone was in the Galápagos Islands in the winter of 1845. It seems curious that this rather aggressive species has not been picked up by others.

Paspalum penicillatum Hook. f., Trans. Linn. Soc. Lond. 20: 171. 1847

Annual; culms glabrous, freely branching, erect or ascending from decumbent base, 15–40 cm tall; sheaths shorter than internodes, keeled, glabrous or sparingly pilose, hairs sometimes papillose-based, margins hyaline; ligule a hyaline membrane about 1 mm long; blades thin, flat or folded, 3–10 cm long, 4–10 mm wide, glabrous above, more or less densely pilose below, margins scabrous; racemes 5 to many, sometimes in pairs, 10–20 mm long, rather distant on a slender angled axis; rachis membranous, green, about 1 mm wide, produced beyond uppermost spikelet, margins scaberulous; spikelets solitary, oblong-elliptic, whitish, somewhat crowded, about 1.7 mm long; first glume wanting; second glume and sterile lemma equal, glabrous; fertile lemma whitish, smooth and shining.

A species of shaded or semishaded habitats; also known from Ecuador, Peru, and Bolivia. Specimens examined: Isabela, Santa María. Originally described from Santa María.

Paspalum redundans Chase, Proc. Calif. Acad. Sci. IV, 21: 300. 1935

Perennial; culms compressed, in dense tufts, erect to ascending, 70–90 cm tall, leafy to or almost to summit; sheaths mostly overlapping, pilose toward summit, or nearly glabrous, basal sheaths usually partially enclosing a few to several short racemes of cleistogamous spikelets; ligule membranous, fragile, about 2 mm long; blades flat, 5–20 cm long, 3–7 mm wide (the uppermost much reduced), finely papillose-pilose on both surfaces, pale midnerve rather prominent beneath; racemes of terminal panicle 2–4, erect to ascending, 2.5–6 cm long, on a slender axis; rachis flexuous, 1–1.3 mm wide, scabrous and usually with a few long hairs at base; spikelets obovate, turgid, 4–5 mm long, in pairs on short scabrous pedicels; glume and sterile lemma 5-nerved, glabrous, rather firm, tinged with purplish brown, sterile lemma equaling spikelet, glume slightly shorter and exposing tip of fertile floret; fertile lemma yellowish, smooth and shining.

Endemic. Specimens examined: Santa Cruz.

Paspalum vaginatum Sw., Prodr. Veg. Ind. Occ. 21. 1788

Perennial with long creeping rhizomes and slender wiry stolons; flowering culms 8–60 cm tall, glabrous; sheaths longer than internodes, overlapping, often keeled, the margins hyaline, glabrous except for a few weak hairs near ligule; ligule membranous, about 0.5 mm long; blades 2.5–15 cm long, 3–8 mm wide at base, narrower than summit of sheath, more or less involute; racemes 2(3–5), conjugate or closely approximate at first, often spreading or reflexed at maturity; rachis 1–2 mm wide, 3-angled, often distinctly zigzag, especially toward end; spikelets solitary, imbricate, elliptic, glabrous, 3.5–4 mm long; first glume wanting or rarely slightly developed; second glume and sterile lemma equal, thin, weakly 5-nerved, but midnerve of second glume often suppressed, sterile lemma often transversely undulate; fertile lemma 2.5–3 mm long, slightly concave-convex, with tuft of short, stiff hairs at apex, clasping palea for only about two-thirds of its length.

Along sandy lake shores and seacoasts, in the tropics and subtropics of the world. Specimens examined: San Salvador, Santa Cruz.

The specimen from San Salvador (*Baur 334* [GH]) is sterile. It was reported by Robinson & Greenman as *P. vaginatum,* but a note on the sheet states: "I think there can be no doubt that this is *P. distichum* L. [B.L.R. May 1901]." Another note on the same sheet by A. S. Hitchcock reads: "I think it is *Pasp. vaginatum.* ASH 11-18-15." We have examined the leaf anatomy in comparison with known specimens of both species, and there seems little doubt that Baur's specimen represents *P. vaginatum.* For a detailed comparison of the leaf histology in these two species, see Anna M. Türpe (1966, 96–98, 188–90, *figs. 19, 72*).

Pennisetum L. C. Rich. in Pers., Syn. Pl. 1: 72. 1805

Annuals or perennials with usually flat blades and dense spikelike panicles. Spikelets 2-flowered, lanceolate, sessile or short-pediceled, solitary or 2 or 3 together, surrounded by an involucre of bristles (sterile branchlets), the fascicles subsessile or short-peduncled, usually crowded on common axis and falling entire with spikelets enclosed. Glumes shorter than spikelet, unequal, the first smaller, thin, usually nerveless, the second 1-nerved. Sterile lemma few- to several-nerved, acute or awn-pointed, enclosing a palea and often a staminate flower; fertile lemma equaling sterile or shorter, smooth, chartaceous, rounded on back, margins flat, thin, enclosing a similar palea. $x = 9$ (perhaps others).

About 130 species in warm regions of the world; especially abundant in Africa.

Culms slender; inflorescence slender; bristles sparse, whitish, shorter than to scarcely equaling spikelet, and more or less clustered on either side of it. . . . . *P. pauperum*
Culms stout (to 2 cm); inflorescence 1 cm or more thick; bristles dense, yellowish, much exceeding spikelet . . . . *P. purpureum*

Pennisetum pauperum Nees ex Steud., Syn. Pl. Glum. 1: 102. 1854

Amphochaeta exaltata Anderss., Kongl. Svensk. Vet.-Akad. Handl. 1853: 137. 1855.
Pennisetum exaltatum Hook. f. & Jacks., Index Kew. 1: 112. 1893.

Perennial in dense clumps to 2 m tall (fide Stewart); culms slender, firm, glabrous, simple or branching at base; sheaths glabrous, distinctly shorter than internodes; ligule a row of whitish hairs 1.5–2 mm long; blades 10–20(30) cm long, 3–4 mm wide, attenuate, abruptly narrowed into a petiole-like base, glabrous below, more or less scabrous or pubescent above, becoming involute; inflorescences terminal and from the upper sheaths, often included at base, slender, 4–6 cm long; rachis slender, angled, minutely puberulent; the very short pedicels similar; spikelets lanceolate, 4–6 mm long, whitish to tawny, subtended by a few whitish bristles, these shorter than or scarcely exceeding spikelet, and more or less clustered on either side of it; first glume rounded or subacute, about 1 mm long; second glume similar,

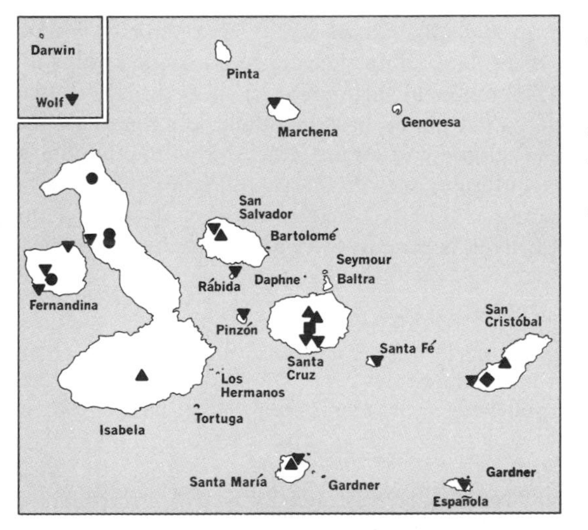

● *Pennisetum pauperum*
■ *Pennisetum purpureum*
▲ *Setaria geniculata*
▼ *Setaria setosa*
◆ *Setaria vulpiseta*

about twice as long; sterile and fertile lemmas subequal, the sterile scabrous and sometimes slightly puberulent on upper half, otherwise glabrous; fertile lemmas glabrous, scarcely indurate, margins hyaline, not enrolled.

Endemic; known only from high elevations in lava and cinder beds on Fernandina and Isabela. Svenson reported this as common around Academy Bay, but he collected no specimens.

See Fig. 248.

Pennisetum purpureum Schum., Beskr. Guin. Pl. 64. 1827

Robust perennial, often forming large bamboo-like clumps; culms erect, 2–6 m tall, to 2 cm broad at base, branching upward, frequently pruinose or glaucous, sometimes pubescent below inflorescence; sheaths terete or the lower compressed, usually glabrous, but sometimes papillose-hirsute toward apex; ligule a row of hairs to 3–4 mm long; blades attenuate, 30–120 cm long, to 3 cm or more wide, more or less pubescent above, especially toward base, scabrous on margins; inflorescence dense, cylindric, erect, 8–30 cm long, usually yellow or tinged with brown, sometimes purplish, rachis slender or rather stout, densely pubescent or pilose; bristles numerous, exceeding spikelet, of unequal length, scabrous or sometimes nearly smooth; spikelets lanceolate, about 5 mm long, solitary within involucre, or sometimes in clusters of 2–5 in lower part of inflorescence; first glume absent or very small, second thin, about 2 mm long; sterile lemma prominently nerved, slightly shorter than the fertile, this smooth below, very minutely scabrous upward.

A native of Africa, introduced as a forage crop in several parts of the American tropics; sometimes called "elephant grass" or "Napier grass." Specimen examined: Santa Cruz (*Colinvaux 462*).

There is no previous report of this species from the Galápagos. The specimen examined is somewhat fragmentary. Our description has been adapted from Stapf & Hubbard in Prain (1934, 1016–17).

Setaria Beauv., Ess. Agrost. 51, *pl. 13, fig. 3.* 1812, nom. conserv.

Annuals or perennials with narrow, usually spikelike or rarely open panicles. Spikelets dorsally compressed, elliptic or lanceolate, solitary or clustered, some or all subtended by 1 to several bristles (sterile branches), articulate on short discoid-tipped pedicels and falling free from bristles. First glume broad, usually less than half the length of spikelet, 3–5-nerved; second glume and sterile lemma equal or the glume shorter, 5–7-nerved. Fertile lemma indurate, smooth or prominently transversely rugose, margins enrolled. $x = 9$.

About 100 species in the world's warm regions; many are widely distributed weeds.*

Bristles below each spikelet 5 or more . . . . *S. geniculata*
Bristles below each spikelet 1 or 2:
  Inflorescence usually no more than 1 cm thick; bristles not conspicuously yellow; culms slender; blades 10 mm wide or less, pubescent . . . . *S. setosa*
  Inflorescence 4–5 cm thick; bristles yellowish or brownish; culms stout; blades to 3 cm wide, scabrous . . . . *S. vulpiseta*

* Three species, *S. antillarium, S. floriana,* and *S. rottleri,* have been excluded from the flora; they are discussed briefly in the pages concluding the Gramineae.

Fig. 248. *Pennisetum pauperum* (*a–e*). *Setaria setosa* (*f–l*).
(Drawn by Charlotte Reeder, inked by Jeanne R. Janish.)

Setaria geniculata (Lam.) Beauv., Ess. Agrost. 51: 178. 1812

Panicum geniculatum Lam., Tabl. Encycl. Méth. 4: 727 (err. typ. 737). 1798.
Chaetochloa geniculata Millsp. & Chase, Field Mus. Bot. 3: 37. 1903.

Perennial; culms glabrous, erect or geniculate-spreading from short, knotty rhizomes, to 1 m tall, branching at lower nodes; sheaths keeled, glabrous or scabrous toward summit; ligule 0.5–1 mm long, densely ciliate; blades flat, 5–15(20) cm long, 4–6(8) mm wide, scabrous, often more or less villous on upper surface toward base; inflorescence spikelike, yellow, purple, or greenish, (2)4–6(10) cm long, the axis densely and softly pubescent; bristles below each spikelet 5 or more, mostly 5–10 mm long, antrorsely scabrous; spikelets ovoid, 2–2.5(3) mm long; first glume 3-nerved, about one-third as long as spikelet, second glume 5-nerved, half to two-thirds as long as spikelet; sterile lemma equaling fertile one, with a well-developed palea and sometimes a staminate flower; fertile lemma strongly transversely rugose.

In swamps and fields, along roadsides, and common in cultivated ground, at low and medium elevations from eastern and southern United States to Argentina. Specimens examined: Isabela, San Cristóbal, San Salvador, Santa Cruz, Santa María.

Setaria setosa (Sw.) Beauv., Ess. Agrost. 51, 171, 178. 1812

Panicum setosum Sw., Prodr. Veg. Ind. Occ. 22. 1788.
Panicum caudatum Lam., Tabl. Encycl. Méth. 1: 171. 1791.
Setaria caudata Roem. & Schult., Syst. Veg. 2: 495. 1817.
Chaetochloa setosa Scribn., U. S. Dept. Agric. Div. Agrost. Bull. 4: 39. 1897.

Perennial; culms erect, spreading, or sometimes decumbent at base, often wiry, scabrous or pubescent below inflorescence, otherwise usually glabrous; sheaths glabrous or more or less pubescent, mostly shorter than internodes; ligule 1–2 mm long, membranous at base, long-ciliate; blades flat or folded, 10–15(25) cm long, usually 5–10 mm wide, more or less pubescent on both surfaces; inflorescence terminal, spikelike, 4–10 cm long, the slender axis villous with white hairs to 1 mm long; bristles mostly 1 below each spikelet, flexuous, antrorsely scabrous, mostly 5–10 mm long; spikelet about 2 mm long, turgid; first glume about half as long as spikelet, 3-nerved; second glume three-fourths as long as spikelet, 5-nerved; sterile and fertile lemmas subequal, the sterile with a well-developed palea; fertile lemma pale, finely but distinctly transversely rugulose.

In dry woods and on rocky hills at low altitudes, the West Indies to Colombia. Specimens examined: Española, Fernandina, Isabela, Marchena, Pinzón, Rábida, San Cristóbal, San Salvador, Santa Cruz, Santa Fé, Santa María, Wolf.

Stewart listed his numbers *1293* to *1297* under *S. setosa.* This is an error. These collections are actually *Panicum hirticaule* Presl, and were correctly reported as such by Stewart. The *S. setosa* numbers read *1093* to *1097*. Stewart's *1093* and *1094* actually represent *S. geniculata*, not *S. setosa* as reported by him. It should be noted, however, that a specimen of *Stewart 1094* at GH is a mixture and consists of plants of both *S. setosa* and *S. geniculata.*

See Fig. 248.

Setaria vulpiseta (Lam.) Roem. & Schult., Syst. Veg. 2: 495. 1817

Panicum vulpisetum Lam., Encycl. Méth. 4: 735 (err. typ. 745). 1798.
Chaetochloa vulpiseta Hitchc. & Chase, Contr. U. S. Nat. Herb. 18: 350. 1917.

Perennial; culms in large tufts, often decumbent at base, to 2 m tall, glabrous, or scabrous below inflorescence; sheaths keeled, glabrous to rather densely hirsute, the collar bearing a ring of stiff hairs 2–4 mm long; ligule about 2 mm long, membranous below but ciliate with long, stiff hairs; blades to 50 cm long, 2–3.5 cm wide, scabrous, narrowed from middle toward both ends, base often somewhat petiolate; inflorescence cylindric, 15–30 cm long, to 6 cm wide including bristles, somewhat narrowed toward apex, the axis angled, scabrous, and densely villous, branches slender, ascending or spreading to 3 cm long; bristles 1 or 2 below each spikelet 15–25 mm long, yellowish or brownish, antrorsely scabrous; spikelets 2–2.5 mm long, turgid; first glume about half as long as spikelet, 3-nerved; second glume about two-thirds as long as spikelet, 7-nerved; sterile and fertile lemmas subequal, sterile one enclosing a well-developed palea; fertile lemma strongly transversely rugose.

On open ground and brushy slopes, the West Indies and Mexico south to Argentina. Specimen examined: San Cristóbal (*Schimpff 181*).

Sporobolus R. Br., Prodr. Flora Nov. Holl. 1: 169. 1810

Annual or perennial grasses with narrow, flat, folded, or involute blades, ciliate ligules, and open or contracted, sometimes spikelike panicles. Spikelets 1-flowered, disarticulating above glumes. Glumes awnless, equal and shorter than floret, or remarkably unequal in length, the second often as long as or longer than floret. Lemma membranous, 1-nerved, awnless, acute to obtuse; palea of same texture, as long as or sometimes longer than lemma, frequently splitting between the 2 nerves as fruit matures. Fruit readily falling from floret at maturity, pericarp free from seed. $x = 6, 9, 10$.

About 150 species in temperate and tropical regions of the world.*

Plant extensively creeping, with tough wiry culms and hard scaly rhizomes; leaves often conspicuously distichous . . . . S. *virginicus*

Plant not creeping; culms in dense or loose tufts; leaves not conspicuously distichous:

Glumes subequal, mostly less than half as long as floret; panicle narrow, usually spikelike, mostly 20–40 cm long . . . . S. *indicus*

Glumes very unequal, the second about as long as floret; panicle 8 cm long or less, often becoming pyramidal at maturity, the lower branches whorled . . . . . . S. *pyramidatus*

Sporobolus indicus (L.) R. Br., Prodr. Flora Nov. Holl. 1: 170. 1810

Agrostis indica L., Sp. Pl. 63. 1753.
Agrostis compressa Poir. in Lam., Encycl. Méth. Bot. Suppl. 1: 258. 1810, non Willd. 1790.
Axonopus poiretii Roem. & Schult., Syst. Veg. 2: 318. 1817. Based on *Agrostis compressa* Poir.
Vilfa berteroana Trin., Mém. Acad. St. Petersb. VI, Sci. Nat. 1–2: 100. 1840.
Sporobolus berteroanus Hitchc. & Chase, Contr. U.S. Nat. Herb. 18: 370. 1917.
Sporobolus poiretii Hitchc., Bartonia 14: 32. 1932.

Perennial; culms slender, in erect clumps 30–100 cm tall, glabrous; sheaths glabrous, the lower loose and shiny; ligule a row of hairs about 0.5 mm long; blades flat or folded, sometimes becoming loosely involute, elongate, 3–5 mm wide, attenuate, glabrous; panicles spikelike, mostly 20–40 cm long, very densely flowered, often interrupted toward base; spikelets (1.8)2–2.3(2.5) mm long; glumes subequal, about half as long as spikelet, obtuse or subacute, glabrous; lemma and palea subequal, the palea usually slightly shorter; anthers 0.5–0.7 mm long; fruit

* One species, S. *domingensis*, has been excluded from the flora; it is discussed briefly in the pages concluding the Gramineae.

1–1.2 mm long, rectangular in outline, reddish brown, frequently afflicted with smut.

In fields, along roadsides, and in waste places, the southeastern United States, Mexico, and the West Indies to Paraguay. Specimens examined: Isabela, San Cristóbal, Santa Cruz, Santa María.

The Howell specimens—*8990* (CAS, GH, US) and *9324* (CAS, GH, US)—were determined by Hitchcock & Chase as *S. poiretii,* which we are treating as a synonym of *S. indicus* (see W. D. Clayton 1965, 287–93). *Snodgrass & Heller 862* was reported by Robinson as *S. domingensis* (Trin.) Kunth (see list of excluded species, at conclusion of the Gramineae).

Sporobolus pyramidatus (Lam.) Hitchc., U. S. Dept. Agric. Misc. Publ. 243: 84, *fig. 48*. 1936

Agrostis pyramidata Lam., Tabl. Encycl. Méth. 1: 161. 1791.
Vilfa arguta Nees, Agrost. Bras. 395. 1829.
Sporobolus argutus Kunth, Rév. Gram. 1: Suppl. XVII. 1830.

Perennial; culms tufted, erect or somewhat spreading, 10–40 cm tall, glabrous; sheaths shorter than internodes, pilose in throat, otherwise usually glabrous, but sometimes papillose-pilose on margins; ligule a row of hairs 0.5–0.7 mm long; blades flat, 3–10(15) cm long, 2–4 mm wide, glabrous or upper surface with scattered

Fig. 249. *Sporobolus pyramidatus.*

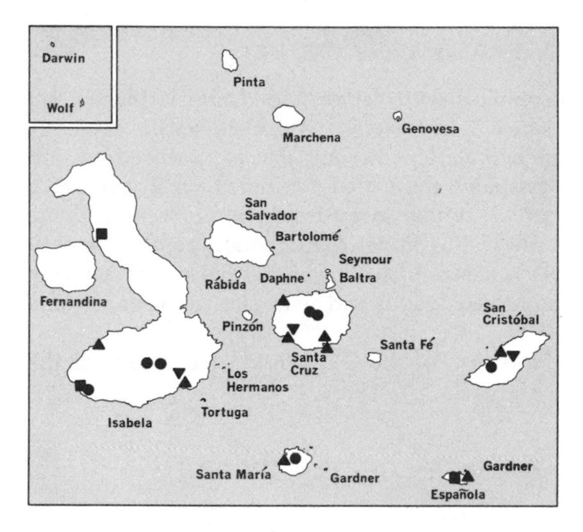

● *Sporobolus indicus*
■ *Sporobolus pyramidatus*
▲ *Sporobolus virginicus*
▼ *Stenotaphrum secundatum*

long hairs, especially toward base, margins scabrous; panicles 3–8 cm long, at first contracted but becoming narrowly pyramidal, branches ascending or finally spreading, the lowermost whorled, at least some of them naked at base; spikelets about 1.5–1.8 mm long, short-pedicellate, appressed; first glume half (or less) the length of spikelet; second glume as long as spikelet or slightly shorter; lemma and palea subequal, anthers 0.3–0.4 mm long, fruit more or less rectangular in outline, reddish brown, 1 mm long or less, the pericarp loose.

In open sandy or alkaline soil, from the southern United States, the West Indies, and Mexico to Argentina. Specimens examined: Española, Isabela.

This name does not appear in the literature concerned with the Galápagos Islands. *Snodgrass & Heller 140, 717* (DS, GH) were reported as *S. domingensis* (Trin.) Kunth by Robinson. This was repeated by Stewart, who also cited his own specimens under this name (see list of doubtful and excluded species). *Wiggins & Porter 488* (CAS, RM) has a somewhat unusual base (perhaps owing to having become partially buried) but seems certainly to represent this species.

See Fig. 249.

Sporobolus virginicus (L.) Kunth, Rév. Gram. 1: 67. 1829

Agrostis virginica L., Sp. Pl. 63. 1753.

Extensively creeping, tough wiry perennial with hard scaly rhizomes; culms glabrous and shiny, much branched, the upright flowering portion mostly 10–40 cm tall; sheaths overlapping, pilose at throat; ligule a very short row of hairs; blades firm, involute-pointed, often conspicuously distichous, 3–8(12) cm long, 2–5 mm wide, glabrous or with a few long hairs on upper surface near base; panicle spikelike, dense, 2–10(15) cm long, often included at base; spikelets 2–2.5 mm long; glumes subequal, the second as long as floret, the first a little shorter; anthers 1.2–1.5 mm long.

In salt marshes and along sandy seacoasts throughout the warm regions of the world. (See W. D. Clayton, Kew Bull. 19: 295f. 1965.) Specimens examined: Isabela, San Cristóbal, Santa Cruz, Santa María, Española.

See color plate 26.

Stenotaphrum Trin., Fund. Agrost. 175. 1820

Coarse, branching, stoloniferous perennials with flat or folded obtuse blades and solitary, terminal and axillary thickened false racemes. Spikelets sessile or nearly so, solitary in 2 rows on 1 side of a short flattened rachis that is produced beyond upper spikelet, the short racemes partially embedded (in ours) on 2 sides of a thickened corky axis. First glume small, turned away from rachis; second glume membranous, equaling lemmas. Sterile lemma firmer than glumes, pointed, enclosing a palea of equal length and often a staminate flower; fertile lemma scarcely firmer than the sterile, of equal shape and length and with similar palea, lemma margins thin, not enrolled. x = 9.

Some 7 or 8 species, mostly in the tropics of the Old World; one species in the Western Hemisphere.*

Stenotaphrum secundatum (Walt.) Kuntze, Rev. Gen. Pl. 2: 794. 1891

Ischaemum secundatum Walt., Flora Carol. 249. 1788.
Stenotaphrum americanum Schrank, Pl. Rar. Hort. Acad. Monac. 2(10): *pl. 98.* 1822.

Perennial; culms branching, glabrous, erect from usually arching stolons, to 50 cm tall; sheaths compressed, keeled, glabrous, sometimes ciliate on margins; ligule a row of hairs about 1 mm long; blades flat or folded, obtuse, to 15 cm long, 12 mm wide, glabrous, the margins scabrous near tip; inflorescences terminal and axillary, 4–10 cm long, straight or somewhat curved, the thick rachis 3–4 mm wide; spikelets 4–5 mm long, shortly acuminate, glabrous; first glume obtuse, truncate, or sometimes emarginate, at least the margin hyaline; remainder of spikelet as in the generic description.

At low altitudes, especially near the coast, from the southern United States to Argentina. Specimens examined: San Cristóbal, Santa Cruz, Isabela.

See Fig. 250.

Trichoneura Anderss., Kongl. Svensk. Vet.-Akad. Handl. 1853: 148. 1855

Annual or perennial grasses with usually flat blades and few to several spicate branches racemosely arranged on a common axis, the lower part of inflorescence often with a single spikelet borne at each rachis joint. Spikelets 4- to several-flowered, laterally compressed, borne on very short pedicels in 2 rows along 1 side of a slender, continuous rachis; rachilla disarticulating above glumes and between florets, the break occurring at a sharp angle a short distance below floret, thus leaving a sharp-pointed callus attached to its base. Glumes longer than lowest floret, often equaling or exceeding upper floret, 1-nerved, or frequently upper glume with 1 or 2 additional nerves, these sometimes as prominent as central nerve. Lemma bidentate or shortly 2-lobed, 3-nerved, lateral nerves near margin, midnerve usually excurrent as short awn, margins ciliate with stiff hairs; palea flat, 2-nerved, nerves near margins. x = 10.

About 8 species in Africa and America.

Although the genus *Trichoneura* is predominantly African, it is based on *T.*

* One species, *S. glabrum,* has been excluded from the flora; it is discussed briefly in the pages concluding the Gramineae.

Fig. 250. *Stenotaphrum secundatum.*
(Reproduced by permission of U.S. National Herbarium.)

lindleyana, which appears to be endemic to the Galápagos Islands. We recognize only a single species, differentiable into 2 varieties, as occurring in the archipelago.

Sheaths papillose-pilose, usually densely so . . . . *T. lindleyana* var. *lindleyana*
Sheaths essentially glabrous . . . . *T. lindleyana* var. *albemarlensis*

Trichoneura lindleyana (Kunth) Ekman, Arkiv för Bot. 11(9): 9. 1912, var. **lindleyana**

Leptochloa lindleyana Kunth, Enum. Pl. 1: 525. 1833.
Calamagrostis pumila Hook. f., Trans. Linn. Soc. Lond. 20: 176. 1847.
Trichoneura hookeri Anderss., Kongl. Svensk. Vet.-Akad. Handl. 1853: 148. 1855.
Leptochloa hookeri Anderss., Svensk. Freg. Eugen. Res. Bot. 2: 51. 1861.

Annual; culms erect, or ascending from decumbent base, much branched, 10–30 cm tall; sheaths mostly shorter than internodes, densely papillose-pilose; ligule hyaline, erose, 0.5–1 mm long; blades flat or folded, or sometimes weakly involute, 2–4 cm long, about 2 mm wide, more or less densely pilose on both surfaces, margins scabrous; inflorescence 5–7 cm long, simple below and occasionally throughout, but usually with ascending or spreading raceme-like branches above, these 1–2 cm long, the axis angled, scabrous; spikelets mostly 5–7 mm long, 4–8-flowered; glumes subequal, often as long as spikelet, narrow, acuminate, rather firm, scabrous, 1-nerved or (especially the second) with 1 or 2 additional nerves; lowermost lemma 3–3.5 mm long, appressed-puberulent with stiff white hairs over the back, margins stiff-ciliate with hairs to 1 mm long, apex 2-lobed and with short awn only slightly exceeding lobes, the callus minutely bearded; upper florets similar to lowermost, but decreasing in size upward; caryopsis dorsally flattened, 1.5–2 mm long, embryo one-third or more its length, hilum punctiform.

Endemic; among lava boulders, in cracks in flows, and on cinder fields. Specimens examined: Baltra, Fernandina, Gardner (near Española), Genovesa, Isabela, Marchena, Pinta, Pinzón, Rábida, San Salvador, Santa Cruz, Santa María, Seymour. See Fig. 251.

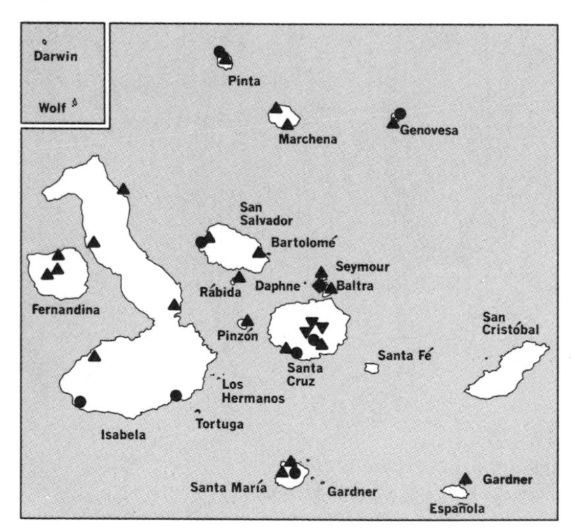

▲ *Trichoneura lindleyana* var. *lindleyana*

● *Trichoneura lindleyana* var. *albemarlensis*

▼ *Trisetum howellii*

◆ *Uniola pittieri*

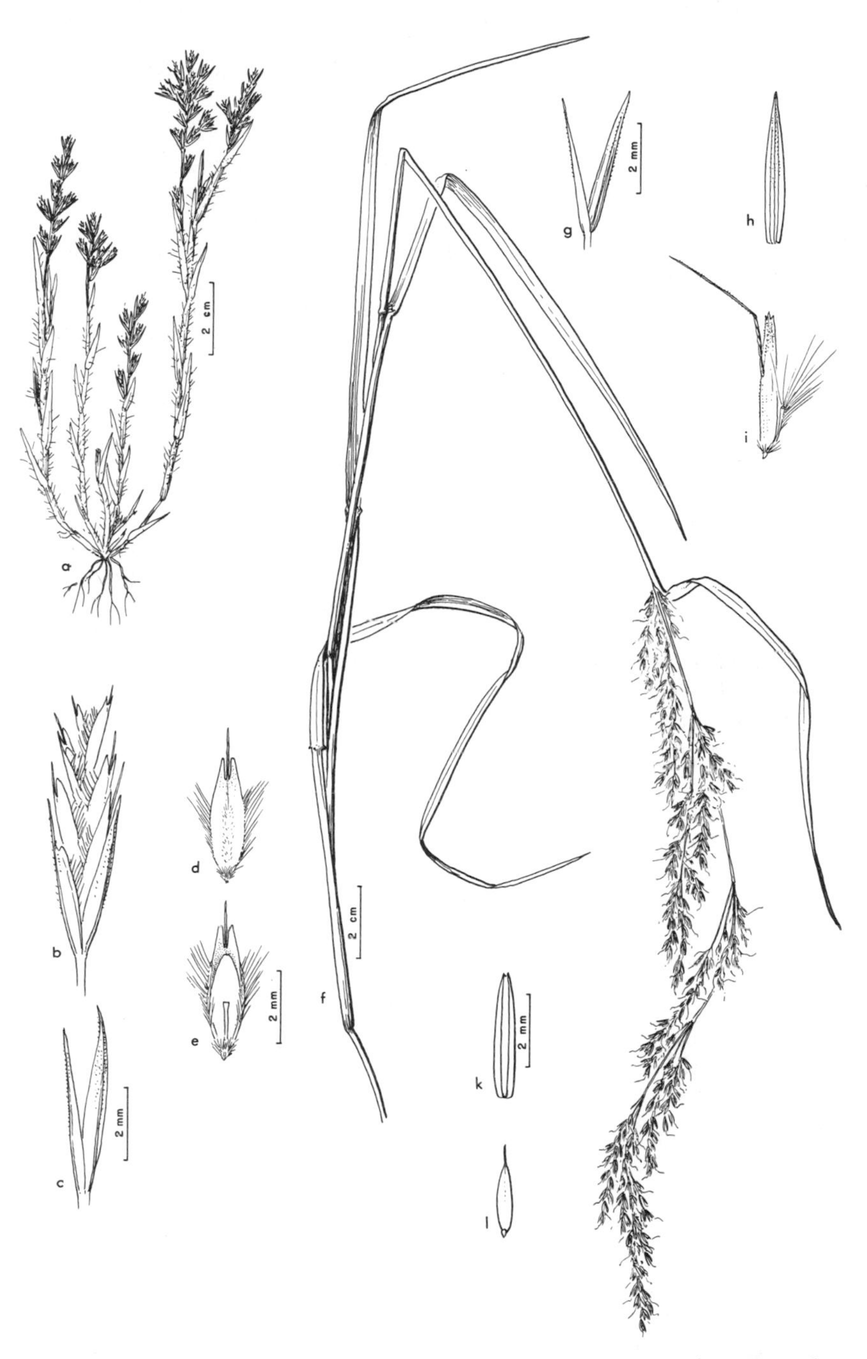

Fig. 251. *Trichoneura lindleyana* var. *lindleyana* (*a–e*). *Trisetum howellii* (*f–l*).
(Drawn by Charlotte Reeder, inked by Jeanne R. Janish.)

Trichoneura lindleyana var. **albemarlensis** (Robins. & Greenm.) J. & C. Reeder, Madroño, 20: 253. 1970

Leptochloa albemarlensis Robins. & Greenm., Amer. Jour. Sci. III, 50: 145. 1895.

Differs from var. *lindleyana* in having essentially glabrous foliage and perhaps more consistently 3-nerved second glumes. These characters are by no means constant. It is a rare specimen that has completely glabrous foliage; usually the sheaths, at least, are sparingly pilose, and the hairs are often papillose-based. Though the 3-nerved character of the glumes is quite obvious in many specimens, some glumes in the same inflorescence may bear only a single nerve. The florets in var. *albemarlensis* are indistinguishable in size or other features from those of var. *lindleyana.*

Endemic; along trails, in cracks in lava, and on cinder fields, mainly in Arid and Transition zones. Specimens examined: Genovesa, Isabela, Pinta, San Salvador, Santa Cruz, Santa María.

Macrae was in the Galápagos on one occasion—from March 28 to April 2, 1825. His collections apparently all came from Isla Isabela (Albemarle), and consisted of 41 unnumbered gatherings, about half of which were described as new. Of these 41 collections, only seven seem to have been grasses. *Calamagrostis pumila* Hook. f. is based on: "Albemarle Island, *Mr. Macrae.*" In the original description of *Leptochloa lindleyana,* Kunth cites no specimen, but gives, "Insula Albemarle, una ex Gallapogos." A Macrae specimen (fragment "ex Herb. Berlin" at US) bore on the label "*Leptochloa Lindleyana* Kunth," along with a notation that it was from Albemarle Island. There is a more complete specimen at the Gray Herbarium. There seems little doubt that this is the material upon which Kunth based his description. Moreover, it is possible that this represents a part of the same collection on which Hooker's *Calamagrostis pumila* was based.

Andersson (1861) lists *L. hirta* Nees ex Steud. as a synonym of *L. hookeri* Anderss. (= *Trichoneura lindleyana*). Robinson likewise cites *L. hirta* as a synonym of *L. lindleyana* (= *T. lindleyana*). Ekman (1912, 9), however, points out that Steudel's description of *L. hirta* does not suggest a species of *Trichoneura,* but applies reasonably well to *Eutriana pilosa* Hook. f. [= *Bouteloua disticha* (HBK.) Benth.]. We agree with Ekman. Certainly "spiculis subbifloris superioribus in setam subuliformen simplicem excurrentibus" does not seem to apply to a species of *Trichoneura.* Steudel cites no specimen, merely stating that the plant is from Albemarle Island. It seems significant that he indicates that Nees considered the plant to be a species of *Dineb[r]a,* a genus to which various species of *Bouteloua* have been referred at various times. Actually, *Dinebra* is a monotypic genus of the Old World.

Trisetum Pers., Syn. Pl. 1: 97. 1805

Slender or sometimes tall and rather coarse perennial grasses with usually flat, narrow blades and open or narrow, sometimes rather dense and spikelike panicles. Spikelets 2- or 3-flowered, laterally compressed, disarticulating above glumes and between florets, rachilla usually villous and prolonged beyond base of upper floret. Glumes equal or the first shorter, 1 or both exceeding lower floret. Lemmas usually short-bearded at base, 2-toothed, bearing a usually geniculate and twisted awn from below apex. $x = 7$.

About 75 species in the temperate and cool regions of the world.

Trisetum howellii Hitchc., Proc. Calif. Acad. Sci. IV, 21: 296. 1935

Annual; culms loosely cespitose, erect, weak, slender, glabrous, 3–5-noded, to 70 cm or more tall; sheaths striate, glabrous, or the lower somewhat pubescent; ligule hyaline, erose, 1–2 mm long; blades elongate, flat, lax, glabrous or slightly scaberulous above, 2–4 mm wide; panicle narrow, erect or nodding, mostly 15–20 cm long, the axis slender, scaberulous or nearly glabrous, the slender, loosely flowered branches weakly appressed, to 8 cm long, spikelet-bearing nearly to base; spikelets 5–6 mm long, 2-flowered; glumes subequal, narrow, 4–5 mm long, acuminate and often with short aristate tip, keels more or less scabrous, the first 1-nerved, the second slightly broader, strongly 3-nerved; lemmas lanceolate, faintly 5-nerved, glabrous but under a lens minutely papillose-roughened, the lower one 4–5 mm long, callus minutely bearded with hairs about 0.5 mm long; awn 5–6 mm long, geniculate, attached about 1 mm below tip of lemma; rachilla conspicuously hairy, hairs at summit of each segment much longer than rest and to 1 mm in length; upper floret similar to lower, but smaller, the rachilla extending beyond its base about 1 mm and bearing hairs like those on rachilla joint below.

Endemic; known only from Mount Crocker, Santa Cruz.

This species is apparently most closely related to *T. deyeuxioides* (HBK.) Kunth, from which it differs most noticeably in possessing a distinctly 3-nerved second glume.

See Fig. 251.

Uniola L., Sp. Pl. 71. 1753; Gen. Pl. ed. 5, 32. 1754

Coarse perennial stoloniferous or rhizomatous beach grasses with firm attenuate leaf blades and relatively large spikelets crowded in erect or drooping panicles. Spikelets 5- to many-flowered, strongly laterally compressed, disarticulating below glumes and falling entire. Glumes subequal, shorter than lemmas, 3-nerved, keeled. Lemmas (the lower 2 to 6 empty) acute or slightly truncate and mucronate, somewhat indurate, the nerves often obscure; palea usually slightly shorter than lemma, strongly 2-keeled. Caryopsis elongate-deltoid in face view, embryo about one-third its length. x = 10.

Only 2 species, confined to the seacoasts of the Americas from Virginia through the West Indies to Ecuador.*

Uniola pittieri Hack., Oesterr. Bot. Zeitschr. 52: 309. 1902

Stout, glabrous, extensively stoloniferous perennial; culms in large clumps, to 2 m tall; sheaths overlapping below, shorter than internodes above, glabrous, margins ciliate; ligule a row of hairs about 1 mm long; blades elongate, firm, attenuate into a fine tip, flat, becoming involute in drying, glabrous or somewhat scabrous above; panicles contracted, 20–30 cm long, interrupted below, the axis, branches, and very short pedicels scaberulous-puberulent; spikelets many-flowered, 1–2 cm long, 8–10 mm wide; glumes about as long as lemmas, less firm, ciliate on keel, the 2–6 sterile lemmas somewhat transitional in size and texture; lemmas acute to slightly mucron-

* See H. O. Yates (1966).

Fig. 252. *Uniola pittieri.*

ate, ciliolate on keel and margins, otherwise glabrous, hairs on margin somewhat longer near base; palea shorter than lemma, keels ciliate.

On sandy beaches, from Mexico to Ecuador. Specimen examined: Baltra.

The collection examined (*Stewart 1195* [CAS]) is the only record of this species from the Galápagos Islands known to us. It was reported by Stewart as *Ammophila arenaria* (L.) Link. The specimen is sterile, but has stout culms with long internodes; the base of the plant bears only a superficial resemblance to an *Ammophila*. The ligule is a row of short hairs; in *Ammophila* the ligule is a membrane. Study of leaf epidermis and transection reveals that these are of the Eragrostoid-Chloridoid type; *Ammophila* is Festucoid. Comparison of our anatomical preparations with those from a flowering specimen of *U. pittieri* from Panama (*Duke 4217* [RM]) reveals a very close correspondence. Moreover, the Stewart specimen matches rather closely the vegetative portion of the illustration reproduced here.

See Fig. 252.

GRAMINEAE SPECIES EXCLUDED FROM THE FLORA

Ammophila arenaria (L.) Link, Hort. Berol. 1: 105. 1827

Reported by Stewart on the basis of a sterile specimen (*1195*) collected by him on Isla Baltra. (See our note under *Uniola pittieri* Hack.)

Cenchrus distichophyllus Griseb., Cat. Pl. Cub. 234. 1866

Reported by Robinson on the basis of a sterile specimen (*Baur 335*) collected on Isla Isabela and by Stewart on the basis of two collections of sterile material made by him: one (*1236*) from Isla San Cristóbal, the other (*1235*) from Española. We have examined the leaf anatomy of this material, and find that these three collections are all *Sporobolus virginicus* (L.) Kunth. Both of the Stewart specimens at US have been so annotated by Hitchcock.

Chusquea sp.

Reported by Robinson on the basis of a Snodgrass & Heller collection (*230*) made on Isla Isabela and consisting only of dry, leafless culms. Robinson states "recognizable from the peculiar mode of branching." Stewart examined the material later and suggested that it was probably *Pennisetum exaltatum* (Anderss.) Hook. f. & Jacks. (= *P. pauperum* Nees ex Steud.). Stewart is correct. The leafless culms of the specimen are very like those on some flowering specimens of *P. pauperum*, such as *Stewart 1323,* which also shows the same type of branching.

Digitaria serotina (Walt.) Michx., Flora Bor. Amer. 1: 46. 1803

Reported from Isla Santa María by Hooker, who cites a collection of Edmonstone. Andersson, Robinson, and Stewart refer to the same collection, Andersson under the name *Paspalum serotinum* (Walt.) Flügge, Robinson and Stewart under the name *Panicum serotinum* Trin. (Robinson apparently believed that *Panicum serotinum* was a synonym of *D. serotina.*)

Dr. W. D. Clayton of Kew has very kindly searched through the collections of *Digitaria* at that institution and reports finding only one collected by Edmonstone in the Galápagos, and this from Isla Santa María (Charles Island)! We have borrowed this sheet, which proves to be *D. horizontalis* Willd. (q.v.).

Distichlis sp.

Reported by Rose (Contr. U. S. Nat. Herb. 1: 138. 1892), who cites an Agassiz collection from Isla San Cristóbal. We have examined the leaf anatomy of this sterile specimen (at the Gray Herbarium), and find that it is actually *Sporobolus virginicus* (L.) Kunth. Robinson also referred this collection correctly to *S. virginicus.*

Eragrostis bahiensis Schrad. ex Schult., Mant. 2: 318. 1824

Reported by Robinson, who cites *Snodgrass & Heller 110* from Isla Isabela. Stewart lists the same collection, and also one of his own (*1262*) from Isla Pinta. We have determined both of these as *E. mexicana* (q.v.). There is no specimen of *Snodgrass & Heller 110* at US, but in March 1968 we found *Stewart 1262* in the *E. mexicana* folder at that institution, presumably placed there by either Hitchcock or Chase.

Eragrostis pilosa (L.) Beauv., Ess. Agrost. 71, 162, 175. 1812

Reported by Hooker (as *Poa pilosa* L.) on the basis of a Darwin collection from Isla San Salvador. Robinson and Stewart list the same Darwin collection, but under *E. pilosa,* and Robinson also cites "Galápagos Islands: *Habel s.n.*" There are no other reports of this species. We have not seen the Darwin specimen, nor have we seen any others from the Galápagos that we would determine as *E. pilosa.* The closely related *E. mexicana* (Hornem.) Link is known to occur in these islands, and the Darwin and Habel plants may represent this species.

Eriochloa distachya HBK., Nov. Gen. & Sp. Pl. 1: 95, *pl. 30.* 1815

Reported by Robinson from Isla San Cristóbal on the basis of *Snodgrass & Heller 547*. This is repeated by Stewart, who cites the same specimen. The plant is *E. pacifica* Mez (q.v.).

Eriochloa punctata (L.) Desv. ex Hamilton, Prodr. Plant. Ind. Occ. 5. 1825

Stewart reports this species from Isla San Cristóbal, and cites his own collection (*1274*). This plant too is *E. pacifica* Mez, and the specimen at US has been correctly so annotated by Hitchcock.

Panicum molle Sw., Prodr. Veg. Ind. Occ. 22. 1788

Caruel (1889, 621) reported this species from Isla San Cristóbal, and cites a collection by Chierchia in March 1884. This was repeated by Robinson and by Stewart, but neither cites any additional collections. We have not seen the Chierchia material, but it seems likely that it represents *P. purpurascens* Raddi or *P. fasciculatum* Sw.

Panicum serotinum sensu Robins. 1902, non Trin. 1834

See *Digitaria serotina* in this list.

Paspalum longepedunculatum Le Conte, Jour. Phys. Chim. 91: 284. 1820

The specimens reported under this name have been shown by Chase to be *P. galapageium* Chase, or its var. *minoratum* Chase, which are apparently endemic to the Galápagos. *P. longepedunculatum* is not found outside of the continental United States.

Paspalum scrobiculatum L., Mant. Pl. 1: 29. 1767

This was reported originally by Caruel (1889, 621), who cited a specimen collected on Isla San Cristóbal by Chierchia in 1884. This was repeated by Robinson and by Stewart, but they cite no additional collections. In fact, Robinson states: "It is remarkable, if Prof. Caruel was correct in his identification of this gerontogeous grass, that no trace of it has been found in the Galápagos Archipelago by any other collector." Robinson's words are quite appropriate today. We have not seen the Chierchia gathering, nor any other that we would determine as *P. scrobiculatum*. It seems likely that the specimen in question was a large-fruited form of *P. galapageium* Chase, or perhaps the less common *P. redundans* Chase.

Paspalum serotinum (Walt.) Flügge, Monogr. Pasp. 145. 1810

See *Digitaria serotina* in this list.

Poa pilosa L., Sp. Pl. 68. 1753

See *Eragrostis pilosa* in this list.

Setaria antillarium (Poir.) Kunth, Rév. Gram. 1: 46. 1829

Reported by both Andersson and Robinson from Isla San Cristóbal. The specimens are *Setaria setosa* (Sw.) Beauv. According to Hitchcock (1936, 359f), *S. antillarium* is a synonym of *Pennisetum antillarium* (Poir.) Desv. [= *P. hordeoides* (Lam.) Steud.], an African species.

Setaria floriana Anderss., Kongl. Svensk. Vet.-Akad. Handl. 1853: 138. 1855

Based on an Andersson collection from Isla Santa María. We have not seen the specimen, but judging from the description, and the fact that the author compares it to *S. glauca* and *S. imberbis*, it seems likely that Andersson had a plant of *S. geniculata* (Lam.) Beauv. (q.v.). *Setaria imberbis* (Poir.) Roem. & Schult. is generally considered to be a synonym of *S. geniculata*.

Setaria rottleri Spreng., Syst. Veg. 1: 304. 1825

Hooker cites a Macrae specimen from Isla Isabela and this is repeated by Andersson. A Macrae specimen from Isabela (borrowed from K) proved to be S. *setosa* (Sw.) Beauv. According to Rominger (1962, 86–88), *S. rottleri* is a synonym of *S. adhaerans* (Forsk.) Chiov.

Sporobolus domingensis (Trin.) Kunth, Rév. Gram. 1: Suppl. XVII. 1830

A number of collections were reported under this name by both Robinson and Stewart. *Snodgrass & Heller 811* and *Stewart 1184* proved to be *Eragrostis mexicana* (Hornem.) Link. *Snodgrass & Heller 862* is *Sporobolus indicus* (L.) R. Br. All of the others are *Sporobolus pyramidatus* (Lam.) Hitchc. Apparently *S. domingensis* does not occur south of the West Indies and Mexico, since we found no specimens of this species in the South American folders at US. Svenson (1946, 429) reports *S. pyramidatus* from the region of Guayaquil, Ecuador. He comments that inasmuch as Hitchcock gives the distribution as "Southwestern United States to Argentina," it is strange that it has not been reported from the Galápagos Islands.

Stenotaphrum glabrum Trin., Fund. Agrost. 176. 1820

Reported by Robinson from Isla Isabela on the basis of *Snodgrass & Heller 57* and *873*, and from San Cristóbal on the basis of *Baur 325*. The specimens are *S. secundatum* (Walt.) Kuntze. According to Bor (*Grasses of Burma, Ceylon, India, and Pakistan* 366. 1960), *S. glabrum* [= *S. dimidiatum* (L.) Brongn.] is an Old World species confined to India and East Africa.

HYPOXIDACEAE. STAR-GRASS FAMILY*

Acaulescent herbs, growing from small monocarpic corm or large, hard, polycarpic, tuberous rhizome, herbage nearly always pubescent, trichomes setose, or sometimes stellate, never glandular. Stem bases ensheathed by persistent membranous or fibrous remnants of old leaves. Leaves filiform to lanceolate, rarely cuneate, sessile or petiolate, usually prominently veined, sometimes plicate, persistent. Scapes elongate to almost absent; bracts linear-setaceous to lanceolate, persistent. Flowers regular, bisexual, perigynous, solitary, few and corymbose or racemose, or sometimes many and densely capitate, usually yellow, sometimes white, rarely red. Perianth segments (4)6, usually in 2 subequal whorls of 3 each, spreading, oblong to lanceolate or elliptic, segments of exterior whorl often somewhat more acute than those of inner whorl, dorsally green and pubescent; tube none, shortly elongate and funnelform, or elongate and filiform. Stamens (3)6, inserted in 1 whorl within or at mouth of tube; filaments short, filiform or thickened; anthers 2-loculed, linear to oblong, basifixed or versatile, dehiscent longitudinally, introrse or extrorse. Gynoecium 3-carpeled; ovary inferior, (2)3-loculed; ovules few to many per locule, superposed in 2 rows, anatropous, placentation axile or rarely parietal; style simple or trifid; stigmas (2)3, united by growth or free. Fruit a capsule, usually crowned by the

* Contributed by Duncan M. Porter.

persistent perianth, circumscissilely or loculicidally dehiscent, or sometimes indehiscent and fleshy; sometimes spuriously 1-loculed by early disappearance of septa. Seeds small, turgid, dark; testa thick, crustaceous, hilum prominent; endosperm fleshy; embryo straight, axial.

A family of 7 genera and perhaps 125 species, occurring mainly in the Southern Hemisphere and in tropical Asia.

Hypoxis L., Syst. Nat. ed. 10, 2: 986. 1759

Small, acaulescent herbs arising from corms accompanied by somewhat fleshy bulbous or tuberous vertical rhizomes. Leaves grasslike, linear-lanceolate to nearly filiform, flat or terete, usually pubescent. Scapes simple or branched, slender, 1- to several-flowered, leafless or with single sheathing leaf, usually slightly pubescent above, often becoming glabrate below; pedicels usually very short; bracts usually shorter than pedicels. Flowers solitary or few in a short raceme, yellow or white. Perianth segments (4)6, nearly equal, glabrous and yellow to whitish within, usually green and pubescent without, connivent at least following anthesis, persistent, tube completely adnate to ovary. Stamens (4)6, inserted at base of perianth segments; anthers oblong to linear, versatile or basifixed, more or less lobed basally. Ovary 2- or 3-loculed, usually pubescent when young; ovules numerous. Style short; stigmas 2 or 3, erect, connate or free, oblong or linear, externally papillose. Capsule globose to linear, usually 3-lobed, becoming nearly glabrate at maturity; crowned by persistent perianth, which usually falls at dehiscence of capsule, carrying away top of latter. Seeds globose to ellipsoid; testa tuberculate, pebbled, or sculptured; hilum prominent, hooked.

Perhaps 90 species in temperate and tropical regions of the world, but most common in tropical America, southern Africa, and Australia.

Hypoxis decumbens L., Pl. Jam. Pugill. 11. 1759

Anthericum sessile Mill., Icon. *t.* 39, *f.* 2. 1760.
Hypoxis caricifolia Salisb., Prodr. 248. 1796.
Hypoxis pusilla HBK., Nov. Gen. & Sp. Pl. 1: 228. 1816.
Hypoxis elongata HBK., op. cit. 229.
Hypoxis gracilis Lehm. ex Schult. & Schult., Syst. Veg. 7: 764. 1830.

Scapose herb 1–3 dm high; corm ellipsoid to globose, 1–2 cm long, leaf sheaths dark and membranous, rhizomes tuberous, elongate, brown; leaves linear-lanceolate, becoming falcate, flat, flaccid, sparsely villous, especially on margins, to almost glabrous, 5–30 cm long, 2–7 mm wide; scapes simple, 1- or 2(4)-flowered, filiform, loosely ascending, sparsely villous to glabrous, 1–20 cm long; bracts linear-lanceolate, sparsely villous to glabrous, margins scarious, acuminate, 5–10 mm long; pedicels 2–6 mm long, shorter than bracts; perianth segments 6, ovate-lanceolate, acute, yellow, dorsally villous, 4–6 mm long, 1.5–3 mm wide, the tube pinched inward below perianth; stamens half as long as perianth segments; anthers lanceolate, sagittate basally, 1.5 mm long; filaments equal, filiform, 2 mm long; ovary obovoid, villous, 4 mm long, 2 mm wide; style conical, pubescent or glabrous, about 3 mm high; stigmas connate; capsule oblong-ellipsoid, villous to glabrate, 7–12 mm long, 2–3 mm wide; mature seeds black, ellipsoid, covered with low, rounded, scarcely confluent tubercles, about 1 mm broad.

Fig. 253. *Hypoxis decumbens* (*a–d*). *Sisyrinchium macrocephalum* (*e–i*).

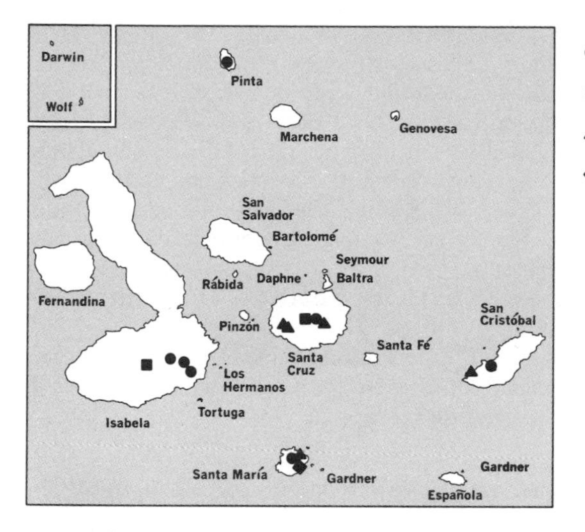

Mainly in open grassy areas from Mexico through the West Indies and Central America to Paraguay. Specimens examined: Isabela, San Cristóbal, Santa Cruz; reported also from Pinta and Santa María.

See Fig. 253.

IRIDACEAE. Iris Family*

Perennial herbs from rhizomes, bulbs, or corms, rarely with a cluster of tuberous-thickened roots or fibrous-rooted annuals. Leaves narrow, sessile, 2-ranked, usually equitant, mostly firm and persistent. Flowers bisexual, regular or irregular, small or large, often showy, solitary or in clusters from spathelike bracts. Perianth of 6 equal or unequal petaloid segments in 2 whorls, all usually connate basally, tube adnate to ovary, convolute, withering-persistent. Stamens 3; filaments filiform and free, or partially connate, inserted on perianth opposite outer segments; anthers 2-loculed, extrorse. Gynoecium 3-carpeled, syncarpous; ovary inferior, usually 3-loculed; ovules usually many per locule, superposed, anatropous, placentation axile; styles 3, free, entire or parted, sometimes petaloid; stigmas terminal or adaxial, opposite ovary locules. Fruit a 3-loculed, loculicidal capsule dehiscing by 3 persistent valves. Seeds many, in 1 or 2 rows per locule; testa thin, membranous; endosperm copious, fleshy or horny; embryo small, straight.

Some 63 genera and about 1,500 species, occurring throughout the world except in the coldest regions; the center of distribution is in southern Africa.

Sisyrinchium L., Sp. Pl. 954. 1753; Gen. Pl. ed. 5, 409. 1754

Perennial or annual, more or less cespitose herbs; rhizomes usually short or absent, roots fibrous or often thickened and fleshy. Leaves linear or narrowly ensiform, mostly in an obvious basal cluster, this sometimes lacking and cauline leaves then more numerous and well-developed. Spathes terminal on simple or branched stems,

* Contributed by Duncan M. Porter.

sometimes appearing lateral when a pseudoterminal cauline leaf is present; 1- to several-flowered, flower clusters solitary or fasciculate, rarely spicate or paniculate. Flowers small, rotate; peduncles flattened or terete, leafy or naked, often winged. Perianth segments oblong, subequal, spreading above base, connate basally. Stamens inserted at base of perianth; filaments from partially to almost completely connate into a glandular, villous, or glabrous tube; anthers from small and approximate to large and divergent, erect or versatile. Ovary 3-loculed, oblong, turbinate, or globose; ovules many per locule; style at least as long as staminal tube, subulate, style branches from minute to very long, slender, linear, alternate with stamens; stigmas terminal. Capsule subglobose or turbinate, exserted from spathe. Seeds minute.

Temperate and tropical North and South America from Canada and Greenland to Argentina, possibly indigenous to Ireland; 60 or more species, several of which have been introduced into the Old World as garden escapes.*

Sisyrinchium macrocephalum Graham, Edinb. New Phil. Jour. 1833: 176. 1833

Moraea alata M. Vahl, Enum. Pl. 2: 154. 1805, non *Sisyrinchium alatum* Hook., Icon. 3: 219. 1840.
Sisyrinchium altissimum Ten., Atti 3a. Riun. Soc. Bot. Ital. 504. 1841.
Sisyrinchium giganteum Ten., Cat. Orto Bot. Napoli 96. 1845.
Sisyrinchium marginatum Klatt, Linnaea 31: 83. 1861–62.
Sisyrinchium grande Baker, Bull. Herb. Boiss. ser. 2, 3: 1106. 1903.

Coarse, vigorous perennial, forming clumps 3–5 dm high; rhizome short or absent; roots coarse and fibrous; leaves many, crowded at stem base in a conspicuous cluster, flattened, ensiform, acute, glabrous, the margins thin and scarious, numerous ribs prominent, becoming fibrous in age, 4–9 mm wide, to 30 cm long; stem simple, glabrous, ridged, winged, surpassing leaves, 4–7 mm wide, to 35 cm high, terminated by an erect cauline leaf about 10 cm long appearing to be a continuation of stem that subtends the pseudolateral fasciculate clusters of spathes; peduncles leafy, flattened, winged; spathes overtopping the leafy bracts, herbaceous, ribbed, sessile, 3- or 4-flowered, becoming fibrous in age; inflorescence repeatedly branched, congested; pedicels glabrous, angular; perianth segments yellow, veins brown, glabrous; filaments connate basally for 1 mm, free and diverging above, glabrous, 3 mm long; ovary oblong, glabrous, 2.5 mm high; capsule subglobose, glabrous, 5–6 mm in diameter; seeds black, subglobose, reticulate-foveolate, micropylar pit present, 1.5 mm in diameter.

In swamps, meadows, and sand, and on rocky hillsides; a rather variable species occurring from Colombia to Argentina. Specimens examined: Isabela, Santa Cruz.

According to Johnston (1938), the correct name for this plant may prove to be *S. palmifolium* L.

See Fig. 253.

LEMNACEAE. Duckweed Family†

Floating or submerged perennial plants without roots or with unbranched, slender roots, each root bearing a cylindrical rootcap, the plant body a small, oval, oblong,

* "The species are often difficult of segregation, and the whole genus is badly in need of critical study" (Standley & Steyermark 1952, 172).

† Contributed by Ira L. Wiggins.

flat or subglobose thallus without leaves. Reproducing by overwintering buds that sink to the bottom, and by new plants growing from lateral pouches near base of thallus, or by seeds produced in same pouches as vegetative offshoots. Flowers unisexual, monoecious, initially in a membranous cup or spathe, or flower naked. Staminate flowers solitary or in pairs, consisting of 1(2) stamens; anthers 1- or 2-celled, sessile or on filiform or fusiform filaments. Pistillate flowers solitary, of a simple, unilocular, 1–7-ovuled, sessile ovary; style and stigma 1, simple. Fruit a minute utricle. Seeds with scarcely any endosperm or none.

Four genera and 25–30 species distributed throughout most of the temperate and tropical zones of the world in quiet pools, ditches, slow-flowing streams, and brackish estuaries.

Lemna L., Sp. Pl. 970. 1753; Gen. Pl. ed. 5, 417. 1754

Diminutive floating aquatic plants, or sometimes growing on wet surfaces when stranded. Thallus oval to obovate, flat, convex, or slightly concave above, usually markedly convex beneath, with a meristematic pouch forming a slit in each side about opposite the single root, these pouches producing new thalli or flowers or both. Vegetative buds usually disarticulating soon after development to form independent plants, sometimes remaining attached in chains or colonies of several plants. Staminate flowers 2 in each spathe, consisting of a single stamen each. Pistillate flowers nearly always solitary in a spathe, consisting of a single pistil. Ovary 1–3-ovuled. Seeds usually finely ribbed and normally bearing an operculum.

About a dozen species, widely distributed in all regions except the arctic and subarctic zones.

Lemna aequinoctialis Welw., Ann. Cons. Ultramarino 56: 545. 1858

Lemna angolensis Welw. ex Hegelmaier, Jour. Bot. 3: 112. 1865.

Fronds solitary or in groups of 2–7, the thalli slightly asymmetrically obovate to subelliptic, (1)2(2.5) mm long, (0.6)1(2) mm wide, plane or slightly concave on upper surface, slightly convex beneath, about 0.2 mm thick, with a single series of air chambers in thinner portion, or part of thallus with 2 series of such chambers, the frond delicately 3-nerved, often faintly carinate and bearing a few minute papillae above midvein, a more prominent papilla at apex, and sometimes 1 near proximal end of frond; root solitary, very slender, often to 2 cm long, emerging from a short, slightly winged sheath, bearing cylindrical rootcap 1–1.5 mm long, this sometimes slightly swollen in part of its length; reproductive pouches about opposite root or slightly posterior to it, often giving rise to flowers in 1 pouch and to vegetative buds in the other, but sometimes both flowers and vegetative bud in the same pouch at one time, at which time the vegetative bud nearly always situated dorsally to flower; spathe short, roughly reniform, open on 1 side, about 0.1–0.2 mm long; stamens exserted about 0.2–0.6 mm at anthesis; anthers about 0.1 mm thick; pistil about 0.3–0.4 mm long at fertilization, the slender style to 0.2 mm long; ovule 1 in each ovary, orthotropous but slightly oblique; fruit obovoid, brownish, 0.5–0.8 mm long, about 0.4 mm wide, style more or less persistent, dark brown; seeds slightly included within utricle, oblong or narrowly obovate, about 0.45–0.55 mm long, 0.35–0.40 mm wide, with a low, dark brown operculum, with 12–18 delicate ribs and microscopic striae transverse to ribs.

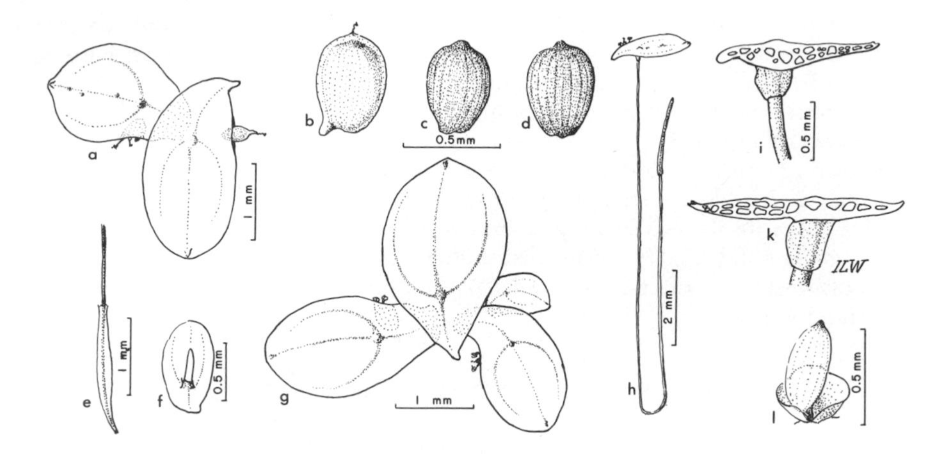

Fig. 254. *Lemna aequinoctialis*: *a, g*, plants, upper surface; *b*, fruit; *c, d*, seeds; *e*, root cap; *f*, young plant, ventral surface; *h*, whole plant, root doubled back; *i, k*, thallus in longisection; *l*, pistillate flower in spathe.

In fresh water, from southern Arizona to Argentina, and in Africa and the Philippine Islands; known with certainty from several of the larger islands in the Galápagos archipelago. Specimens examined: San Cristóbal, Santa Cruz, Santa María.

Stewart (1911, 45) reported *L. minor* L. from Isabela and San Cristóbal, and Robinson (1902, 130) listed "*Lemna* sp." from Santa María on the basis of a collection by Wolf. Svenson (1935, 226) collected material on Santa Cruz that he identified as *L. minor* Philippi, and stated that its small seeds and fronds readily separated it from *L. perpusilla* Torr. and from *L. minor* L. But Giardelli (1959, 588) cited Svenson's collection as *L. aequinoctialis* Welw.

Careful examination of two specimens, both of which bore flowers and fruit in profusion, convinces me that Dr. Giardelli was correct in her identification of Svenson's collections, for Hendrickson's collection came from the same locality as Svenson's. The Svenson material I saw was less copious than the Hendrickson and Wiggins material, but it too was flowering and is obviously the same as the other two collections. The veins in these plants are delicate and can be overlooked unless the light is adjusted carefully. With strong back-lighting so that the rays pass through the frond, the 3 nerves are obvious in mature fronds, though they may be obscure in young plants.

See Fig. 254.

MUSACEAE. BANANA FAMILY*

Large herb with treelike appearance and with a stout, unbranched stem surrounded by sheathing petioles. Leaves large, alternate; blade entire or lacerate, pinnately veined. Inflorescence a spike, panicle, or rarely a head, subtended by spathelike bracts. Flowers bisexual or unisexual and plants monoecious with pistillate flowers toward base of inflorescence, staminate toward tip; perianth of 2 whorls of 2 or 3

* Contributed by Duncan M. Porter.

unequal segments. Stamens 6, 1 usually reduced to staminode. Pistil 1; ovary inferior, 3-locular. Fruit a 3-loculed capsule or an elongate berry. Seeds often arillate. Endosperm present.

Five genera and about 150 species widely distributed in moist tropics.

Musa L., Sp. Pl. 1043. 1753; Gen. Pl. ed. 5, 466. 1754

Leaf blades entire at first, soon lacerate to midrib. Flowers hermaphroditically unisexual, in clusters, each cluster subtended by spathelike reddish to purplish bract. Calyx splitting down 1 side, 3–5-toothed (or sometimes entire). Petal 1, opposite calyx, simple or 3-toothed. Fruit a berry, pulpy or dry. Seeds present, with mealy endosperm, or abortive.

About 40 species and probably over 200 cultivated hybrids, native in tropical Asia, but cultivated forms now pantropical.

Calyx nearly equaling petal; leaves glaucous beneath; fruit ellipsoid, 8–12 cm long, acuminate .... *M. acuminata*
Calyx half as long as petal; leaves not glaucous; fruit cylindrical, 15–30 cm long, blunt, not acuminate .... *M. × paradisiaca*

Musa acuminata Colla, Mem. Mus. 66. 1820

Musa simiarum S. Kurz, Jour. Agric. Soc. India 14: 297. 1867.
Musa rumphiana S. Kurz, op. cit. n.s. 5: 164. 1878.
Musa corniculata S. Kurz, op. cit. n.s. 5: 166. 1878.

Plant strongly rhizomatous; trunk to 6 m tall; leaf blades about 2 m long; pulp of fruit usually reddish to red-orange.

Native to Java, Malaysia, and New Guinea; widely cultivated in all tropics. Commonly called "plantain," and some cultivars produce rather mealy fruit normally cooked before consumption.

Musa × paradisiaca L., Sp. Pl. 1043. 1753

Plant strongly rhizomatous or stoloniferous; trunk to 8 m tall; leaf blade 2–2.5 m long, about 60 cm wide; pulp of berry white to pale cream.

Perhaps originating in India; clones of this hybrid are pantropical as a cultivated fruit tree. Banana.

These Malaysian taxa are commonly cultivated in the Galápagos Islands (Koford, 1966). Both will persist after cultivation has ceased, and in the absence of habitations may appear to be a part of the spontaneous flora. There is no conclusive evidence that either has actually become naturalized and established independent of human effort.

See Fig. 255.

NAJADACEAE. Water-Nymph Family*

Submerged, aquatic plants in fresh and brackish water of ponds, sluggish streams, and estuaries, with slender stems openly and often diffusely branched, or condensed

* Contributed by Ira L. Wiggins.

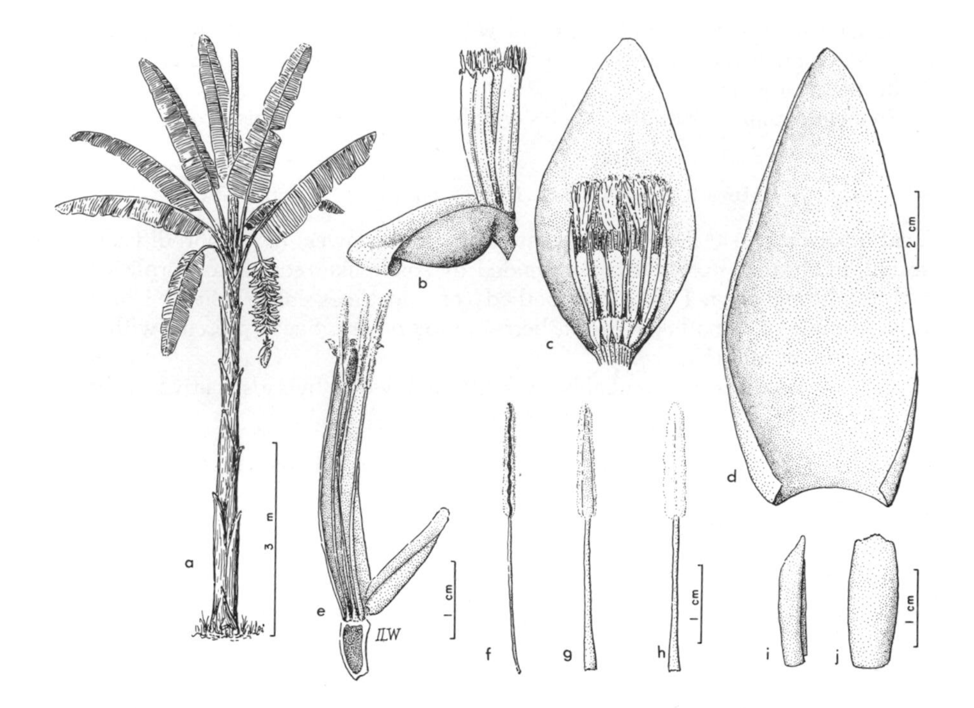

Fig. 255. *Musa* × *paradisiaca*.

and compactly much branched. Leaves linear to narrowly oblong, minutely to coarsely toothed, opposite but each pair consisting of a lower and an upper leaf on opposite sides of stem, or leaves in whorls or appearing fascicled owing to the short internodes and crowding; leaf bases dilated to form conspicuous sheaths, these varying in size and shape, minute hyaline scales often present within sheath. Flowers monoecious or dioecious, solitary or fascicled in axils of sheaths. Staminate flowers each consisting of a single stamen in a perianth-like spathe; anthers 1- or 4-celled, at first sessile, short-stalked at dehiscence. Pistillate flowers naked, each consisting of a solitary ovary; style short; stigmas 2–4, linear and somewhat roughened, often accompanied by sterile spinelike processes. Fruit a nutlet, within a loose, membranous jacket. Seeds ovoid to elliptic.

A monogeneric family of wide distribution.

Najas L., Sp. Pl. 1015. 1753; Gen. Pl. ed. 5, 445. 1754

Characters of the family.

Leaves delicately fine-toothed, 0.5–1 mm wide; internodes and backs of leaves smooth, not spiny; seeds narrowly ellipsoid, about one-third as wide as long; plants monoecious . . . . *N. guadalupensis*

Leaves coarsely toothed, 1–3 mm wide; internodes and backs of leaves usually coarsely spiny; seeds broadly ovoid, nearly as wide as long; plants dioecious . . . . . *N. marina*

Najas guadalupensis (Spreng.) Morong, Mem. Torrey Bot. Club 3: 60. 1893

Caulinia guadalupensis Spreng., Syst. 1: 20. 1826.
Najas flexilis var. ?*fusiformis* A. W. Chapm., Fl. So. U.S. 444. 1860.
Najas flexilis var. *guadalupensis* A. Braun, Jour. Bot. Brit. & For. 2: 274. 1864.
Najas microdon A. Braun, Sitzungsb. Ges. Naturf. Fr. Berlin (June 16) 17. 1868.
Najas microdon var. *guadalupensis* A. Braun, loc. cit.

Stems very slender, about 0.5 mm thick or less, 3–6 dm long; leaves narrowly linear, flat or slightly crispate, 0.5–1 mm wide, 1–2.5 cm long, acute to acuminate and with 1–4 short, 1-celled spines at apex, marginal teeth minute, 1-celled, brownish, 10–18 along each side (sometimes more numerous or lacking); shoulders at base of leaf oblique or rounded, not auriculate, usually with 4–7 teeth like those of upper margins arranged above middle of sheath, but sometimes lacking; sheaths about 2–3 mm long, enfolding stem; plants monoecious; staminate flowers 2–3 mm long, anthers 4-celled, though often broader than thick; pistillate flowers 2–3 mm long; ovary with 2 or 3 roughened stigmas and often 1 or 2 sterile, smooth, spiny processes about equaling or slightly exceeding stigmas; fruit and seed narrowly ellipsoid, seed 0.5–0.6 mm in diameter, 1.4–1.6 mm long, tapering toward each end, with about 20 rows of minute reticulations from base to apex, the areoles squarish.

Widely distributed in fresh-water pools, ditches, and sluggish streams, central California into South America. Specimen examined: Santa María.

It is possible that the specimen examined (*Howell 8896* [CAS]) is *N. flexilis*, for the anthers were quite immature and at the stage present in the small sample soaked up were distinctly thinner than wide. However, mature seeds were a good match with those taken from authentic material of *N. guadalupensis*. This is the first collection of *N. guadalupensis* (or *N. flexilis*, if it so proves to be) from the Galápagos Islands of which I am aware.

Najas marina L., Sp. Pl. 1015. 1753

Stems slender, 0.5–1.2 mm thick, 2–8 dm long, considerably branched above, rooting at lower nodes, internodes 1–5 cm long, irregularly armed with deltoid-trian-

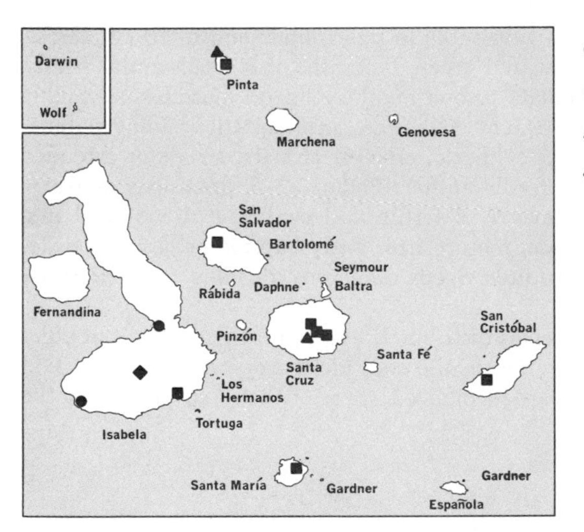

● *Najas marina*
■ *Epidendrum spicatum*
▲ *Govenia utriculata*
◆ *Habenaria alata*

gular teeth 0.5–2 mm long; leaves rigid, thick, linear-oblong, 3–6 mm wide, 2–4.5 cm long, coarsely 5–10-toothed on each margin, the teenth spreading, narrowly deltoid, 1.5–2.5 mm wide at base, tipped with sharp, brownish, rigid spine 0.2–0.4 mm long and slightly curved forward, midrib sometimes bearing spinose processes similar to marginal teeth; sheath at base of each leaf with rounded shoulders, about 3 mm deep, slightly and shallowly dentate; staminate flowers 3–4 mm long, spathe flasklike; anthers 4-locular; pistillate flowers naked, stigmas 2(3), linear; fruit ovoid, 2–3 mm wide, 3–3.5 mm long, tawny, minutely reticulate-rugulose, the areoles slightly longer than wide.

In ponds, slow-moving streams, and quiet estuaries, usually in more or less saline water; cosmopolitan in the temperate and tropical parts of the world. Specimens examined: Isabela.

See Fig. 256.

ORCHIDACEAE. Orchid Family*

Plants perennial, herbs, vines, or shrublike, exceedingly variable in habit and habitat preference; terrestrial, lithophytic, epiphytic, rarely semiaquatic or subterranean. Most commonly autophytic, occasionally saprophytic, never parasitic. Roots adventitious, fibrous, fleshy, rarely tuberous, single, fascicled or scattered along rhizome or stem. Rhizome or primary stem commonly much reduced so that plants appear cespitose. Stems usually rounded or laterally compressed, occasionally angular, commonly elongated or reduced in size, often modified into pseudobulbs and rarely cormlike. Developmentally either sympodial (i.e. with a determinate apical yearly growth and a lateral bud for the following year's shoot) or monopodial (i.e. with an indeterminate apical growth year after year). Leaves 1 to many, radical or cauline, alternate, opposite or whorled, rarely absent, commonly entire in outline, occasionally digitately lobed or sagittate; in origin convolute or conduplicate, the latter often modified into a terete or equitant shape. Inflorescence a 1- to many-flowered spike, raceme, or panicle, occasionally a pseudoumbel, usually supported by a peduncle of varying length. Flowers gynandrous, zygomorphic, hermaphroditic, rarely monoecious or dioecious, variable in size from 2 mm to 20 cm across, and in colors, mostly purple, yellow, and green. Perianths of 3 sepals and 3 petals, all free from, connate with, or adnate to one another; lip or ventral petal highly modified in form and usually of different coloration, an adaptation aiding pollination. Column or gynandrium fleshy, cylindric, erect or arcuate, sessile or extended into a foot, and capped by single fertile anther. Pollinia 2–8, granulous or waxy, with or without a viscidium. Stigmas 3, 2 fertile and confluent, 1 modified into rostellum. Ovary inferior, pedicellate; mature fruit a dry capsule or fleshy pod dehiscing through 1–6 longitudinal sutures. Seeds numerous, dustlike, without endosperm.

Some 800 genera and 35,000 species distributed everywhere except in the polar regions, with the greatest concentration in the cool moist tropical belt of the Old and New Worlds. Five subfamilies are recognized, three of these represented in the Galápagos Islands.

* Contributed by Leslie A. Garay.

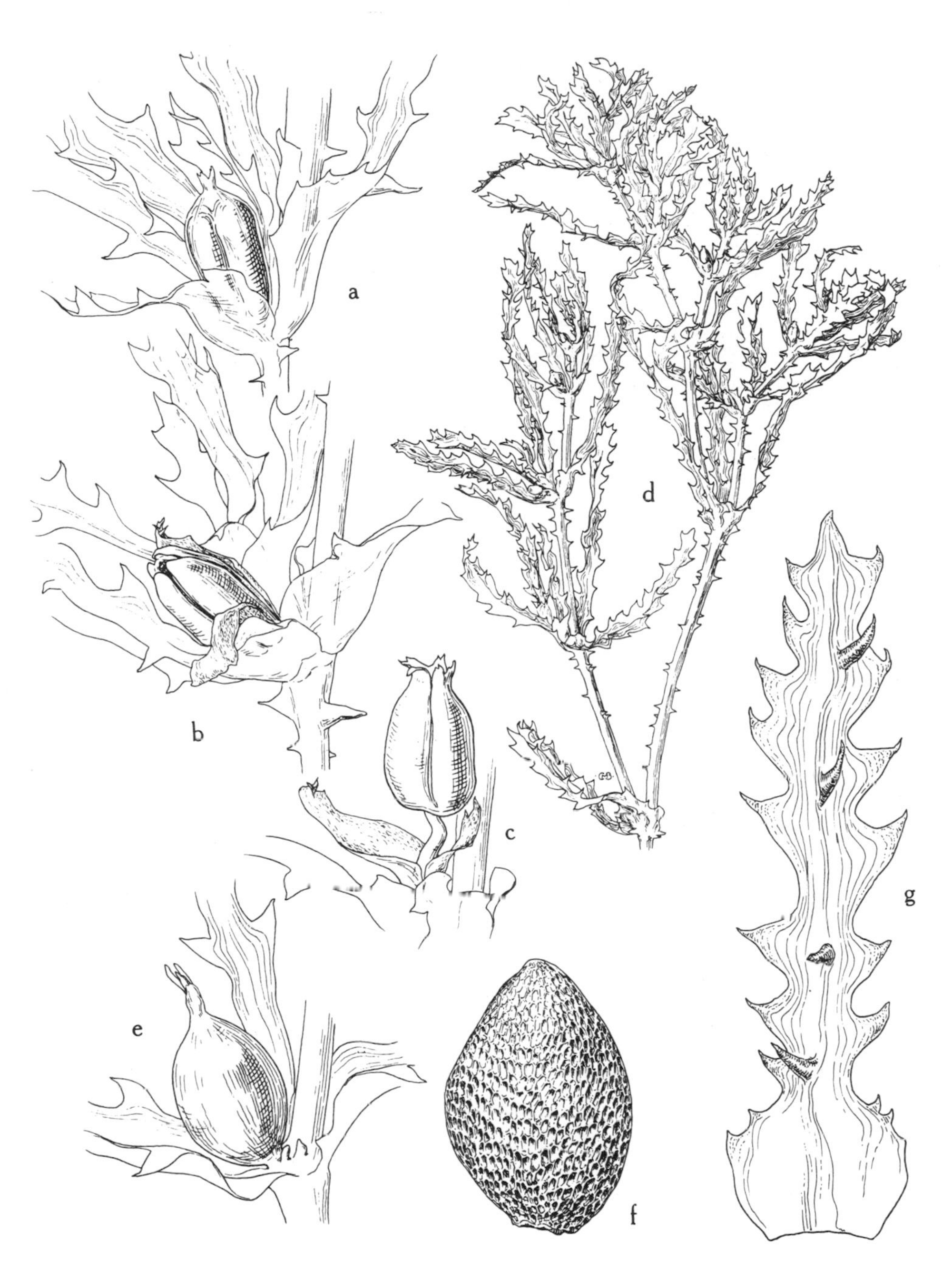

Fig. 256. *Najas marina*: *a–c*, staminate flowers in axil of leaf; *d*, habit; *e*, mature pistillate flower; *f*, seed; *g*, leaf blade.
(Reproduced by permission of the Regents of the University of California.)

Flowers not resupinate, i.e. lip uppermost:
  Petals adnate to side of column above its base; lip minute, much smaller than other floral segments .... *Ponthieva*
  Petals free from column; lip more or less same size as other floral segments or larger:
    Main stem with superimposed branches; leaves plicate; flowers in compact panicle .... *Tropidia*
    Main stem not branched; leaves conduplicate, fleshy; flowers in raceme:
      Sepals free, not forming a cup; column without a foot............... *Cranichis*
      Sepals basally connate to form shallow cup; column with distinct foot... *Prescottia*
Flowers resupinate, i.e. lip lowermost:
  Inflorescence lateral, arising from base of a pseudobulb:
    Pseudobulb 2-leaved, leaves well developed, long-petiolate; rhizome none... *Govenia*
    Pseudobulb with single bractlike leaf or leafless; rhizome elongate, well developed .... *Ionopsis*
  Inflorescence terminal:
    Stem pseudobulbous; leaves 2–5 in cluster............................ *Liparis*
    Stem without pseudobulbs; leaves several to many, distichous:
      Plants epiphytic; lip adnate to column, without spur or saccate base... *Epidendrum*
      Plants terrestrial; lip free from column, its base spurred or saccate:
        Stems rhizomatous, rooted at nodes; lip saccate; anther free, dorsal, parallel with column .... *Erythrodes*
        Stems with cespitose fleshy tubers; lip spurred; anther terminal, completely fused with column .... *Habenaria*

Cranichis Sw., Nov. Gen. Sp. Pl. Prodr. 120. 1788

Terrestrial herbs, cespitose. Roots fascicled, villous-hairy. Leaves one or two, basal, fleshy, commonly with a distinct petiole, suberect or arcuately spreading. Peduncle erect with a few appressed sheaths, racemose above. Flowers rather small, not resupinate. Sepals free, subequal, more or less spreading. Petals much narrower, often ciliolate on margin. Lip uppermost, rather fleshy, cochleate, simple or lobed, often with conspicuously marked or colored reticulate veins. Column short to very short, fleshy, cylindric. Anther dorsal. Pollinia 2, both pulverulent. Rostellum well developed.

Some 30 species distributed in the American tropics and subtropics.

Cranichis schlimii Rchb. f., Linnaea 41: 19. 1876

Terrestrial, erect plant to 50 cm tall; roots fleshy, fasciculate, hirsute; leaves 1 or 2, fleshy, elliptic or ovate-elliptic, acute or subacuminate, cuneate to subcordate at base, tapering into a slender, occasionally quite long petiole, including the petiole to 17 cm long, 4.5 cm wide; stem erect, glabrous below, finely puberulent above, remotely 3- or 4-sheathed, subdensely racemose at apex, to 50 cm long; flowers rather small, white with green nerves; sepals spreading, similar, elliptic or ovate-elliptic, acute, sparsely hirsute dorsally toward base, to 4 mm long; petals spreading, linear-oblong to linear-oblanceolate, obtuse, to 3.5 mm long; lip fleshy, distinctly concave, cucullate, suborbicular in outline when expanded, disk to 3.5 mm long with 3 prominent green nerves, each anastomosing transversely; column fleshy, cylindric; ovary pedicellate, cylindric, glandular-pubescent.

Widespread in moist forests but rather uncommon in Colombia, Ecuador, and

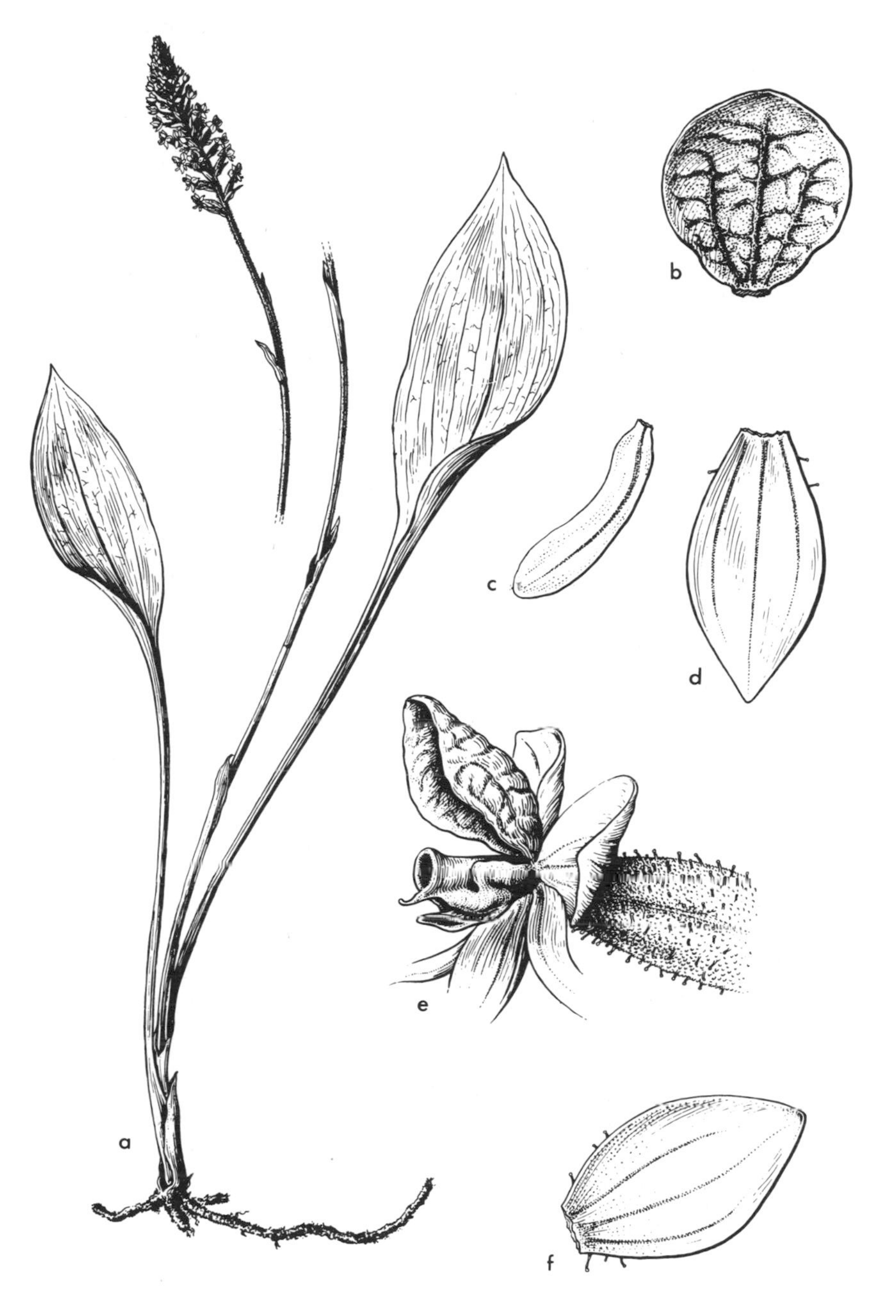

Fig. 257. *Cranichis schlimii*: *b*, upper lip; *c*, petal; *d, f*, sepals; *e*, flower, lateral view.

Peru. Specimens examined: Santa María, a new record for the Galápagos Islands, based on *D. Weber 249B*.
See Fig. 257.

Epidendrum L., Sp. Pl. 952. 1753; Gen. Pl. ed. 5, 408. 1754

Epiphytic or lithophytic plants, exceedingly variable in both habit and flower structure. Roots fleshy, glabrous. Primary stem or rhizome either well-developed or wanting. Secondary stem erect, ascending, rarely pendent, often pseudobulbous. Leaves 1 to many, fleshy, occasionally leathery or coriaceous, alternate, distichous or imbricate, entire, conduplicate, rarely equitant. Inflorescence normally terminal or subterminal, capitate, racemose, paniculate, or reduced to a single flower. Flowers fleshy, from minute to showy, variously colored. Sepals and petals free, ringent, patent or reflexed. Lip conspicuous, either free from, or variously adnate to, column, simple or variously lobed and dissected, with or without calli. Column fleshy, cylindric, from free to fully adnate to lip, without columnar foot. Anther terminal, incumbent. Pollinia 4, waxy, held together and attached to rostellum by a viscidium. Ovary cylindric.

The largest genus in the New World, with over 1,000 species occurring in all subtropical and tropical regions.

Epidendrum spicatum Hook. f., Trans. Linn. Soc. Lond. 20: 180. 1847

Epiphytic on trunks and branches of trees; rhizome short, ascending; roots well developed, glabrous; stems from an ascending base, erect, unbranched, somewhat flexuous, laterally compressed; basal third completely enclosed in leafless, infundibuliform sheaths, upper two-thirds distichously leaf-bearing, to 50 cm tall; leaves fleshy, deciduous, erect, lanceolate or ovate-lanceolate, acute, tapering into a compressed sheathing base, to 15 cm long, 3 cm broad; inflorescence a many-flowered raceme, rarely subpaniculate, commonly shorter than subtending leaves; bracts conspicuous, linear-oblong, dorsally carinate, shorter than pedicellate ovaries; flowers rather thin in texture; dorsal sepals obovate, concave, abruptly acuminate, to 12 mm long, 5 mm wide; lateral sepals obliquely obovate, somewhat concave, acute, as long as dorsal sepal; petals linear to linear-spatulate, obtuse or subacute, to 11 mm long, 2 mm wide; lip completely adnate to column, 3-lobed; lateral lobes rotund with erose or erose-dentate margins, midlobe trapezoid, emarginate to bilobed in front with small triangulate apicule in middle; disk with 2 calli at the base and a median keel; whole lip 5 mm long, 10 mm wide; column fleshy, arcuate-clavate, 8 mm long; ovary pedicellate, 8–10 mm long.

Endemic. Specimens examined: Isabela, Pinta, San Cristóbal, San Salvador, Santa Cruz, Santa María.

Flowers throughout the year.

See Fig. 258.

Erythrodes Bl., Bijdr. 8: 410. 1825

Terrestrial herbs with decumbent rhizomes rooted at nodes. Roots simple, fleshy. Leaves fleshy, few to several, with distinct petioles dilated into an infundibuliform

Fig. 258. *Epidendrum spicatum*; *b*, bud; *c*, flower, face view.

sheath. Inflorescence simple, usually with distinct peduncle, spicate or racemose above. Flowers commonly small, rather fleshy. Sepals free, concave. Petals partially connivent with dorsal sepal to form a hood. Lip lowermost, 3-lobed, rarely entire, basally produced into distinct spur or spurlike sac. Column short, cylindric. Anther dorsal, parallel with column. Pollinia 2, pulverulent. Rostellum well developed, bifid.

About 100 species native to the tropics and subtropics of America, Asia, Malaysia, and Oceania.

Erythrodes weberiana Garay, Orch. Rev. 78: 114, *fig. 58/70.* 1970

Terrestrial, to 30 cm tall; rhizome decumbent, fleshy, stemlike, rooted at nodes; roots fleshy, villous; stem suberect or ascending, leafy; leaves fleshy, ovate to ovate-lanceolate, somewhat oblique, acute or subacuminate above, abruptly contracted at rounded base into a slender petiole at apex of amplexicaul sheath, blade to 5 cm long, 2 cm wide; peduncle slender, minutely puberulent, terminated by a rather laxly flowered, cylindric spike; bracts narrowly ovate-lanceolate, acuminate, to 5 mm long; flowers white, small; dorsal sepal ovate, concave, acute, sparsely glandular-pubescent dorsally toward base, to 3 mm long; lateral sepals obliquely oblong-ligulate, obtuse, sparsely glandular-pubescent dorsally, to 3 mm long; petals linear-oblong, falcate, truncate at apex, partially adnate to dorsal sepal to form a hood, somewhat shorter than dorsal sepal; lip navicular, basally extended into a deep, conical sac, at apex provided with divaricate, subquadrate, puberulent, reflexed lobes, to 3 mm long; column short; rostellum prominent; ovary cylindric, sparsely glandular-pubescent.

Endemic; in shaded, moist soil at about 300 m on Isabela. Known only from the type collection (*D. Weber 211A*).

See Fig. 259.

Govenia Lindl. ex Lodd., Bot. Cab. 18: *t. 1709.* 1831

Terrestrial, erect plants. Roots fibrous. Stems pseudobulbous, enclosed in close-fitting, tubular sheaths. Pseudobulb 2-leaved; leaves long-petiolate, plicate, well-developed, thin in texture. Inflorescence lateral from base of pseudobulb, erect, racemose above, few- to many-flowered. Flowers rather delicate, often fragrant, medium in size. Sepals and petals free, glabrous; dorsal sepal incurved; lateral sepals falcate, deflexed; petals similar to lateral sepals. Lip articulate with column-foot, simple, reflexed, fleshy. Column arcuate, fleshy, with short but distinct columnar foot. Anther terminal, incumbent. Pollinia 4, waxy, compressed. Ovary ellipsoidal.

About 12 species; known from Florida, Central America, South America, and the West Indies.

Govenia utriculata (Sw.) Lindl., Bot. Reg. 25: Misc. p. 46. 1839

Limodorum utriculatum Sw., Prodr. 119. 1788.
Cymbidium utriculatum Sw., Nov. Act. Upsal. 6: 75. 1799.
Govenia gardneri Hook., Bot. Mag. 65: *t. 3660.* 1839.
Govenia boliviensis Rolfe, Mem. Torrey Bot. Club 4: 263. 1895.
Govenia ernstii Schltr., Rep. Spec. Nov. Regni Veg. Beih. 6: 43. 1919.
Govenia sodiroi Schltr., op. cit. 8: 91. 1921.
Govenia powellii Schltr., op. cit. 17: 51. 1922.

Terrestrial herb; roots fleshy, hirsute; pseudobulb ovoid, completely enclosed by 2–4, close-fitting tubular sheaths, 2-leaved; leaves large, broadly elliptic or obovate,

Fig. 259. *Erythrodes weberiana*: *a*, flower, *c*, pollinium.

acute or subacuminate, long-petiolate, to 60 cm long, 14 cm wide; peduncle erect, shorter than subtending leaves, to 50 cm long; raceme rather short, lax, many-flowered, to 15 cm long; bracts scalelike, acute, shorter than ovaries; flowers white or pale cream-colored, fleshy; dorsal sepal elliptic or obovate-oblanceolate, acute, to 25 mm long, 8 mm wide; lateral sepals oblique, ovate-falcate, acute, to 15 mm long, 5 mm wide; petals falcate, ovate-lanceolate, acute, much broader than sepals, to 20 mm long, 10 mm wide; lip sigmoid-arcuate, recurved with short claw, ovate to ovate-elliptic, acute or obtuse, deeply canaliculate, to 12 mm long, 8 mm wide; column arcuate; ovary pedicellate, cylindric, to 30 mm long.

Locally abundant in Central America from Mexico to Panama, and in South America from Venezuela to Bolivia, Argentina, and Brazil; also in Florida and the West Indies. Specimens examined: Pinta, Santa Cruz.

The sterile specimens reported by Hemsley (1900, 177), Stewart (1911, 47), and Schlechter (1921, 165) as *Eulophia* sp. are referable to this species.

See Fig. 260.

Habenaria Willd., Sp. Pl. 4: 44. 1805

Terrestrial, rarely semiaquatic herbs with fleshy tubers and fibrous roots. Stem erect, simple with either basal or cauline leaves terminating in a simple, unbranched inflorescence. Leaves entire, rather thin in texture, conduplicate, often reduced in size and bractlike. Inflorescence a raceme or spike, loosely or densely flowered, rarely reduced to a single flower. Bracts commonly well-developed and equal in length with pedicellate ovary, occasionally foliaceous. Flowers variable in size, rarely showy. Sepals free; dorsal sepal often forming a hood over the column; lateral sepals spreading or reflexed. Petals either simple or bifid, rarely comose, connivent with dorsal sepal. Lip undivided or more commonly 3-lobed, occasionally many-parted, produced at base into nectariferous spur. Column short, fleshy and cylindric, firmly fused with anther. Anther 2-celled with prominent anther canals. Stigmas 2, confluent. Rostellum absent. Pollinia granular on distinct caudicles.

A polymorphic genus of approximately 600 species distributed throughout the subtropical and tropical regions of the world.

Stem plain; ovary winged; lip simple . . . . *H. alata*
Stem maculose; ovary wingless; lip 3-lobed . . . . *H. monorrhiza*

Habenaria alata Hook., Exot. Fl. 3: *t. 169*. 1826

Orchis foliosa Spreng., Syst. Veg. 3: 688. 1826.
Habenaria brachyceras Lindl., Gen. & Sp. Orch. Pl. 315. 1835.
Habenaria bidentata Poeppig, ex Steud., Nomencl. ed. 2, 1: 716. 1841.
Habenaria stricta Rich. & Gal., Ann. Sci. Nat. ser. 3, 3: 29. 1845.
Habenaria triptera Rchb. f., Linnaea 22: 814. 1849.
Habenaria platantheroides Schltr., Beih. Bot. Centralb. 36^2: 372. 1918.

Terrestrial, erect plant up to 70 cm tall; roots many, fibrous; tubers 1 or 2, subspherical; stem leafy, without maculations; leaves lanceolate, conduplicate, distinctly keeled on back, acuminate, sheathing at base, to 15 cm long, 2.5 cm wide, progressively decreasing upward, then becoming bractlike; inflorescence racemose, sublax, many-flowered, to 20 cm long; bracts foliaceous, ovate-lanceolate, usually longer than ovaries; flowers fleshy, ringent, greenish or greenish yellow; dorsal sepal ovate, concave, apiculate, with minutely papillose margins, dorsally keeled, to 10

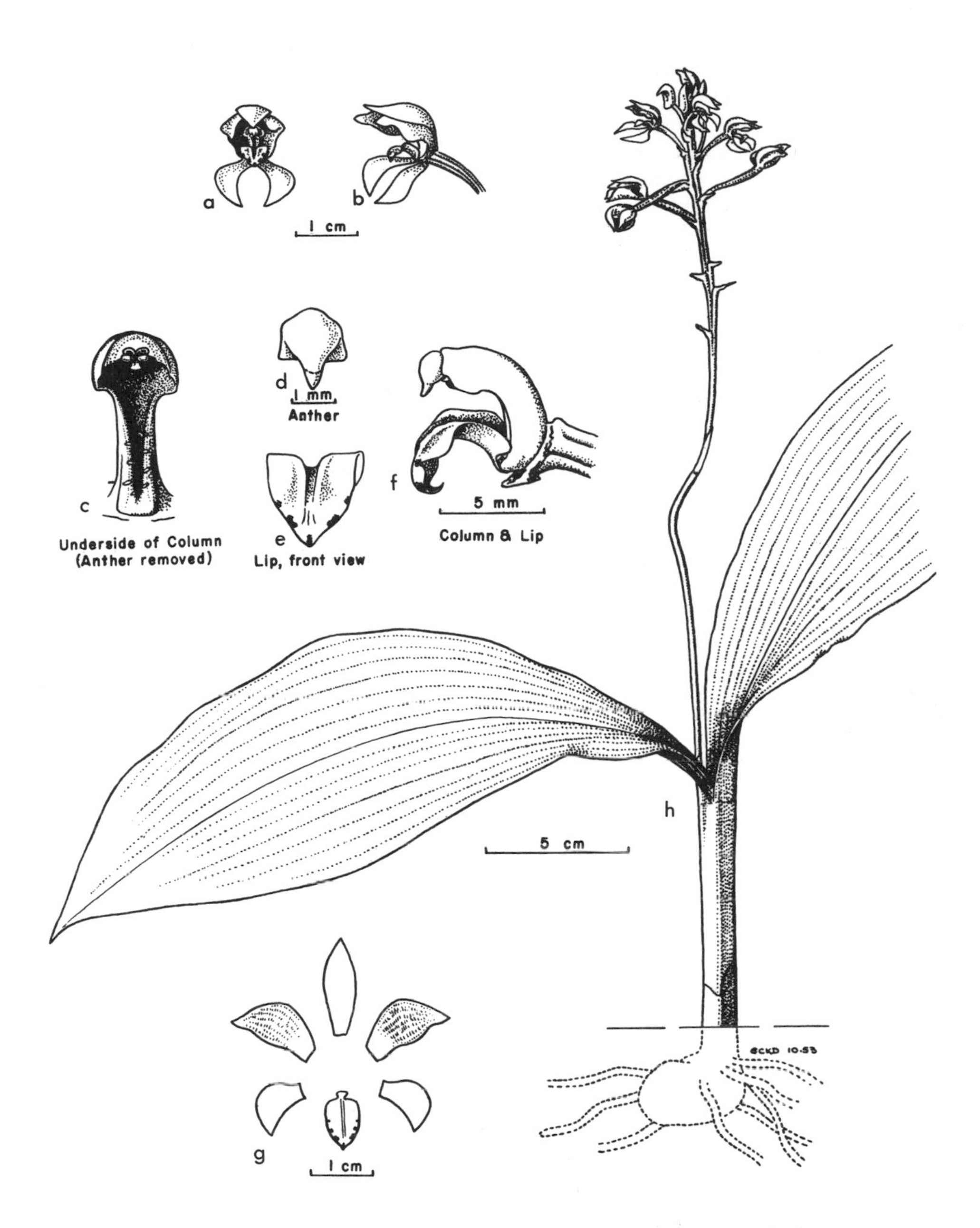

Fig. 260. *Govenia utriculata*: *a, b*, flower, face and side; *c*, underside of column; *d*, anther; *e*, lip, front view; *f*, column and lip, side view; *g*, exploded perianth.

mm long, 5 mm wide; petals linear-lanceolate, acute, slightly dilated toward anterior margin at base, light green grading to white at base, to 10 mm long, 2.5 mm wide; lip fleshy, porrect, linear-lanceolate, obtuse, simple or rarely sublobulate at base, light green, grading to white toward base, as long as petals; spur filiform, subclavate, as long as ovary, to 16 mm long; ovary prominently winged.

Widespread but uncommon in the Caribbean region, including the Antilles and Mexico, and from Central America to Ecuador and Bolivia. Specimen examined: Isabela, a new record for the Galápagos Islands.

See Fig. 261.

Habenaria monorrhiza (Sw.) Rchb. f., Ber. Deutsch. Bot. Gesellsch. 3: 274. 1885

Orchis monorrhiza Sw., Prodr. 118. 1788.
Orchis setacea Jacq., Enum. Pl. Carib. 28. 1760, non *Habenaria setacea* Lindl. 1835.
Habenaria brachyceratitis Willd., Sp. Pl. 4: 44. 1805.
Habenaria speciosa Poeppig & Endl., Nov. Gen. & Sp. Pl. 1: 44. 1835.
Habenaria maculosa Lindl., Gen. & Sp. Orch. Pl. 309. 1835.
Habenaria sodiroi Schltr., Rep. Spec. Nov. Regni Veg. 14: 115. 1915.

Terrestrial, erect plant, to 12 dm tall; roots many, fibrous; tuber ovoid, glabrous; stem leafy, distinctly dark-maculate; leaves rather soft in texture, ovate, ovate-lanceolate, or oblong-lanceolate, acute or subacuminate, sheathing at base, to 15 cm long, 4 cm wide, decreasing in size toward inflorescence; inflorescence racemose, variable in size, to 25 cm long; flowers rather delicate, white; dorsal sepal ovate to elliptic, acute or obtuse, deeply concave, to 14 mm long; lateral sepals oblique, similar in shape to dorsal sepal but slightly longer and narrower; petals deeply bipartite, posterior segment ovate or oblong-ligulate, anterior segment filiform, to 8 mm long; lip distinctly 3-lobed, midlobe oblong-ligulate, obtuse, equal to or shorter than side lobes; lateral lobes divergent, filiform, often reflexed; all segments glabrous; spur filiform, longer than ovary, to 24 mm long; ovary pedicellate, cylindric.

Common and locally abundant throughout the American tropics. Specimens examined: Isabela, Santa Cruz.

See color plate 64.

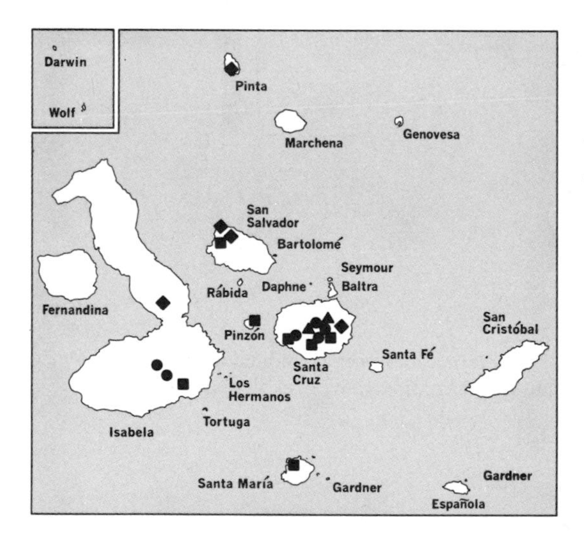

● *Habenaria monorrhiza*
■ *Ionopsis utricularioides*
▲ *Liparis nervosa*
◆ *Ponthieva maculata*

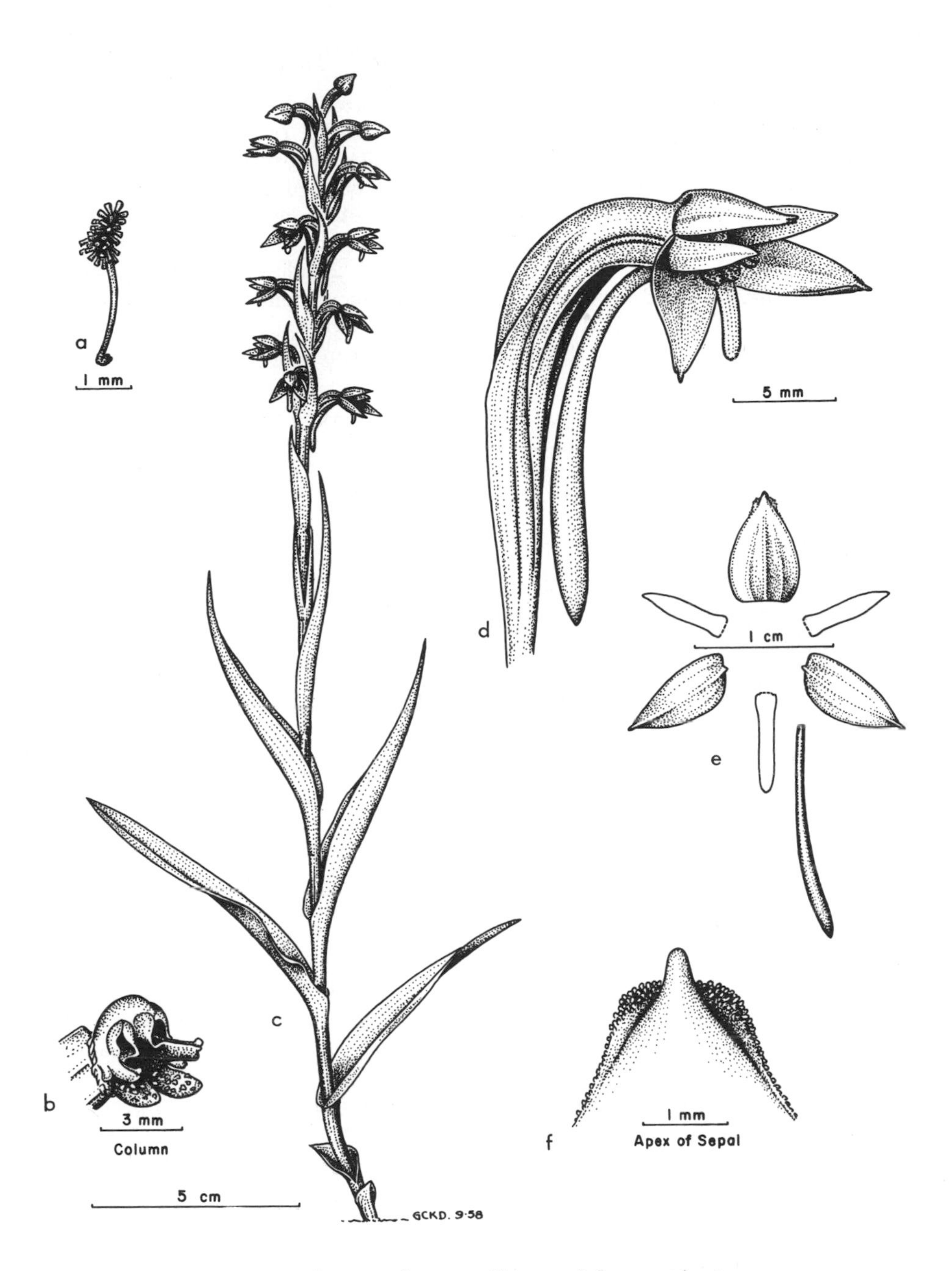

Fig. 261. *Habenaria alata*: *a*, pollinium; *d*, flower, side view; *e*, perianth and spur, exploded pattern.

Ionopsis HBK., Nov. Gen. & Sp. Pl. 1: 348. 1816

Epiphytic plants with a more or less developed rhizome. Roots fleshy, glabrous. Stem pseudobulbous, concealed by leaf-bearing sheaths. Leaves coriaceous, distichous. Inflorescence lateral from base of pseudobulbs, long-pedunculate, racemose or paniculate, lax, few- to many-flowered. Bracts inconspicuous, much shorter than ovaries. Flowers delicate, mostly showy. Sepals and petals similar, connivent; lateral sepals connate at base, forming a gibbous mentum. Lip much larger than other floral segments, attached to base of column; disk with 2 calli. Column fleshy, short, slightly arcuate, provided with distinct foot. Anther incumbent, 2-celled. Pollinia 2, waxy, on distinct linear stipe. Ovary pedicellate, cylindric.

About 12 species, extending throughout the American tropics and subtropics.

Ionopsis utricularioides (Sw.) Lindl., Coll. Bot. *t. 39A.* 1826

Epidendrum utricularioides Sw., Prodr. 122. 1788.
Dendrobium utricularioides Sw., Nov. Act. Upsal. 6: 83. 1799.
Ionopsis pulchella HBK., Nov. Gen. & Sp. Pl. 1: 348. 1816.
Iantha pallidiflora Hook., Exot. Fl. 2: *t. 113.* 1824.
Cybelion pallidiflorum Spreng., Syst. Veg. 3: 721. 1826.
Cybelion pulchellum Spreng., loc. cit.
Cybelion utriculariae Spreng., loc. cit.
Epidendrum crenatum Vell., Fl. Flum. Icon. 9: *t. 6.* 1827.
Ionopsis tenera Lindl., Bot. Reg. 22: *t. 1904.* 1836.
Ionopsis pallidiflora Lindl., op. cit. sub *t. 1904.* 1836.
Ionopsis paniculata Lindl., loc. cit.
Cybelion tenerum Steud., Nomencl. ed. 2, 1: 458. 1840.
Ionopsis gardneri Lindl. in Paxt., Fl. Gard. 2: 13. 1851.
Ionopsis zonalis Lindl., loc. cit.
Epidendrum calcaratum Sesse & Moç., Fl. Mex. ed. 2, 201. 1894.

Epiphytic plant with elongate, filiform, glabrous roots; rhizome ascending, densely covered with roots, terminating in small ovoid pseudobulb; leaves coriaceous or fleshy, narrowly oblong or linear-ligulate, abruptly acuminate and apiculate, canaliculate at base, to 16 cm long, 1.5 cm wide; inflorescence lax, many-flowered, either a simple raceme or a well-developed panicle, to 60 cm tall; bracts small, triangular to lanceolate, much shorter than ovaries; flowers delicate, white, pale rose, or white with lilac stripes, scented; sepals oblong-ligulate to ovate-lanceolate, acute; lateral sepals slightly connate basally, gibbous, to 6 mm long, 2 mm wide; petals ovate-oblong, acute or abruptly acuminate, to 6 mm long, 3.5 mm wide; lip from a cuneate base abruptly dilated into a suborbicular lamina, deeply bilobed in front; the cuneate base often biauriculate with rounded lobes; disk biauriculate at base, to 12 mm long and wide; column short, fleshy, 2 mm long; ovary pedicellate, cylindric, to 12 mm long.

Locally abundant throughout the American tropics and subtropics. Specimens examined: Isabela, Pinzón, San Salvador, Santa Cruz, Santa María.

A distribution map of this species was published by Svenson (1946, 410).

See Fig. 262.

Liparis L. C. Rich., Mém. Mus. Paris 4: 43. 1818

Terrestrial or epiphytic herbs, often with well-developed rhizome. Roots fleshy. Stem often pseudobulbous, leaf-bearing. Leaves thin in texture, conduplicate or plicate, basal or cauline, sessile or petiolate. Inflorescence a lax, few- to many-flowered raceme. Flowers variable in size, minute to showy, rather delicate in texture.

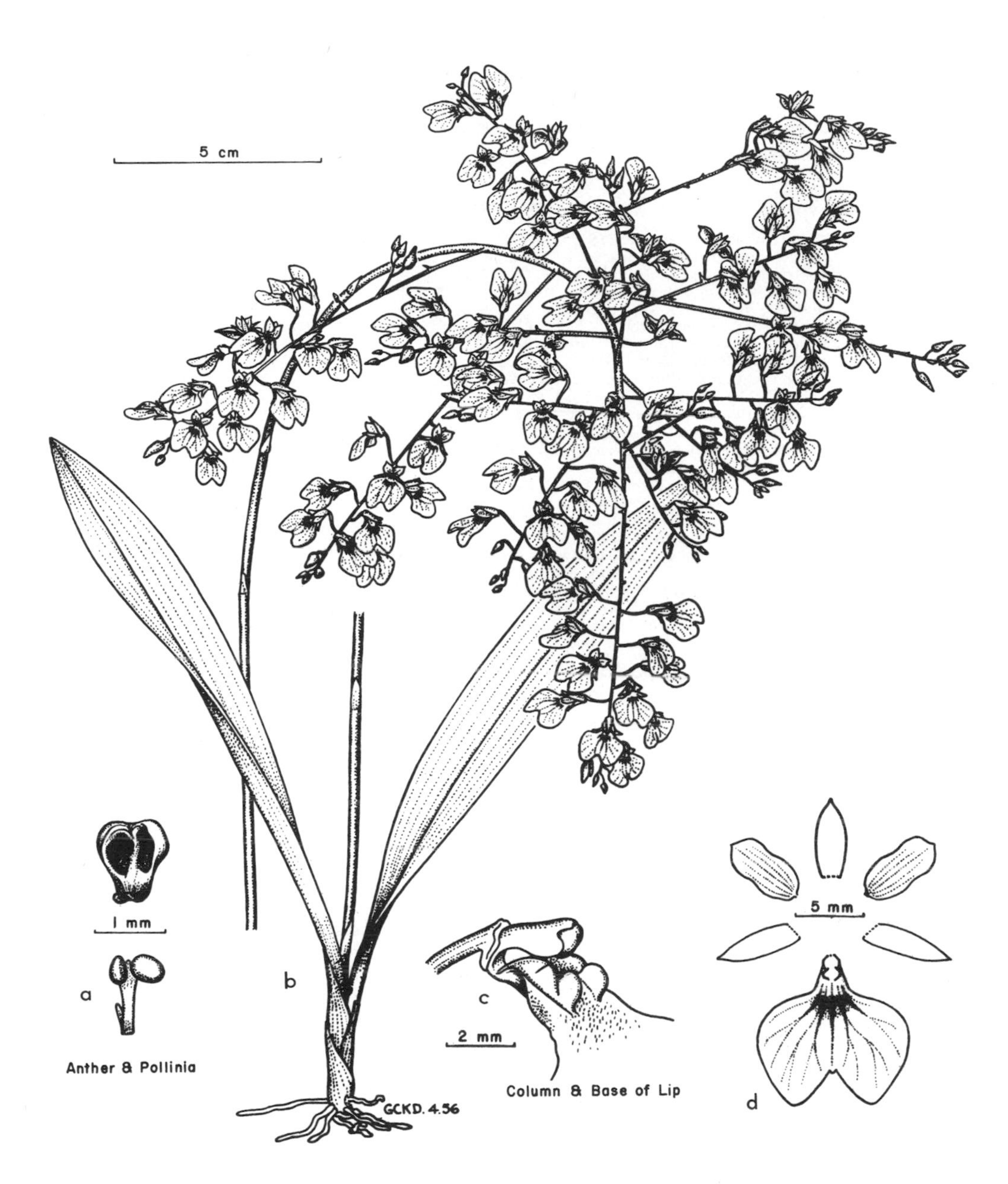

Fig. 262. *Ionopsis utricularioides*: *d*, perianth parts separated.

Sepals and petals free, spreading, occasionally reflexed with revolute margins. Lip articulate with column, porrect or reflexed, simple or lobed, occasionally laciniate. Column arcuate-clavate, narrowly marginate or alate, without columnar foot. Anther incumbent. Pollinia 4, waxy, ovoid, without stipe. Pedicellate ovary cylindric.

A cosmopolitan genus of some 200 species, dispersed in the temperate and tropical regions of the world.

Liparis nervosa (Thunb.) Lindl., Gen. & Sp. Orch. Pl. 26. 1830

Ophrys nervosa Thunb., Fl. Jap. 27. 1784.
Epidendrum nervosum Thunb., Trans. Linn. Soc. Lond. 2: 327. 1794.
Cymbidium nervosum Sw., Nov. Act. Upsal. 6: 76. 1799.
Malaxis nervosa Sw., Kongl. Vetensk. Acad. Handl. 21: 235. 1800.
Malaxis odorata Willd., Sp. Pl. 4: 91. 1805.
Malaxis lancifolia Sm. in Rees, Cyclop. 22: 7. 1814.
Empusa paradoxa Lindl., Bot. Reg. 10: sub *t. 825*. 1824.
Cymbidium bituberculatum Hook., Exot. Fl. 2: *t. 116*. 1824.
Liparis bituberculata Lindl., Bot. Reg. 11: sub *t. 882*. 1825.
Bletia bicallosa D. Don, Prodr. Fl. Nepal. 30. 1825.
Limodorum bicallosum Hamilt. ex Don, loc. cit., in synonymy.
Liparis elata Lindl., Bot. Reg. 14: *t. 1175*. 1828.
Liparis odorata Lindl., Gen. & Sp. Orch. Pl. 26. 1830.
Liparis olivacea Lindl., op. cit. 27.
Liparis guineensis Lindl., Bot. Reg. 20: *t. 1671*. 1834.
Dituilis nepalensis Raf., Fl. Tellur. 4: 49. 1836.
Sturmia bituberculata Rchb. f., Bonpl. 2: 22. 1854.
Sturmia nervosa Rchb. f., op. cit. 3: 250. 1855.
Liparis elata var. *purpurascens* Regel, Ind. Sem. Hort. Petrop. 18. 1856.
Liparis paradoxa Rchb. f. in Walp., Ann. Bot. Syst. 6: 218. 1861.
Liparis odontostoma Rchb. f., Linnaea 41: 97. 1876.
Liparis elata var. *inundata* Barb., Rodr., Gen. & Sp. Orch. Nov. 1: 36. 1877.
Liparis formosana Rchb. f., Gard. Chron. n.s. 13: 394. 1880.
Liparis eggersii Rchb. f., Ber. Deutsch. Bot. Gesellsch. 3: 278. 1885.
Liparis lutea Ridley, Jour. Linn. Soc. Bot. 21: 458. 1885.
Liparis bicornis Ridley, loc. cit.
Liparis elata var. *latifolia* Ridley, op. cit. 22: 260. 1886.
Liparis elata var. *rufina* Ridley, loc. cit.
Liparis bituberculata var. *formosana* Ridley, op. cit. 263.
Liparis cornicaulis Makino, Ill. Fl. Jap. 1, No. 8: 1, *t. 47*. 1891.
Leptorchis bicornis Kuntze, Rev. Gen. Pl. 2: 671. 1891.
Leptorchis bituberculata Kuntze, loc. cit.
Leptorchis eggersii Kuntze, loc. cit.
Leptorchis elata Kuntze, loc. cit.
Leptorchis guineensis Kuntze, loc. cit.
Leptorchis lutea Kuntze, loc. cit.
Leptorchis nervosa Kuntze, loc. cit.
Leptorchis odontostoma Kuntze, loc. cit.
Leptorchis odorata Kuntze, loc. cit.
Leptorchis olivacea Kuntze, loc. cit.
Liparis bambusaefolia Makino, Tokyo Bot. Mag. 6: 48. 1892.
Liparis elata var. *longifolia* Cogn. in Mart., Fl. Bras. 3^4: 287. 1895.
Liparis rufina Rchb. f. ex Rolfe in Thiselt.-Dyer, Fl. Trop. Afr. 7: 19. 1897.
Liparis melanoglossa Schltr., Rep. Spec. Nov. Regni Veg. Beih. 1: 184. 1911.
Liparis perrieri Schltr., Ann. Mus. Col. Mars. ser. 3, 1: 21. 1913.
Liparis guamensis Ames, Phil. Jour. Sci. 9: 11. 1914.
Liparis nyassa Schltr., Engl. Bot. Jahrb. 53: 560. 1915.
Liparis bicallosa Schltr., Rep. Spec. Nov. Regni Veg. Beih. 4: 196. 1919.
Tribrachia racemosa Lindl. ex Schltr., loc. cit., in synonymy.
Liparis hachijoensis Nakai, Tokyo Bot. Mag. 35: 97. 1921.
Liparis odorata var. *longiscapa* Rolfe ex Dow., Kew Bull. 370. 1925.
Liparis siamensis Rolfe ex Dow., op. cit. 371.
Liparis odorata var. *intacta* Kerr, Kew Bull. 216. 1927.
Liparis longiscapa Gagn. & Guill. in Lecomte & Humbert, Fl. Gen. Indo-Chine 6: 182. 1932.
Liparis bicallosa var. *hachijoensis* Kitamura, Acta Phytotax. Geobot. 22: 66. 1966.*

Terrestrial or lithophytic, occasionally epiphytic herb, to 70 cm tall; pseudobulbs conical, completely enveloped in leaf sheaths, 3–4 cm long; leaves 2–5, thin, plicate,

* In the American botanical literature this species is listed always as *L. elata* Lindl.

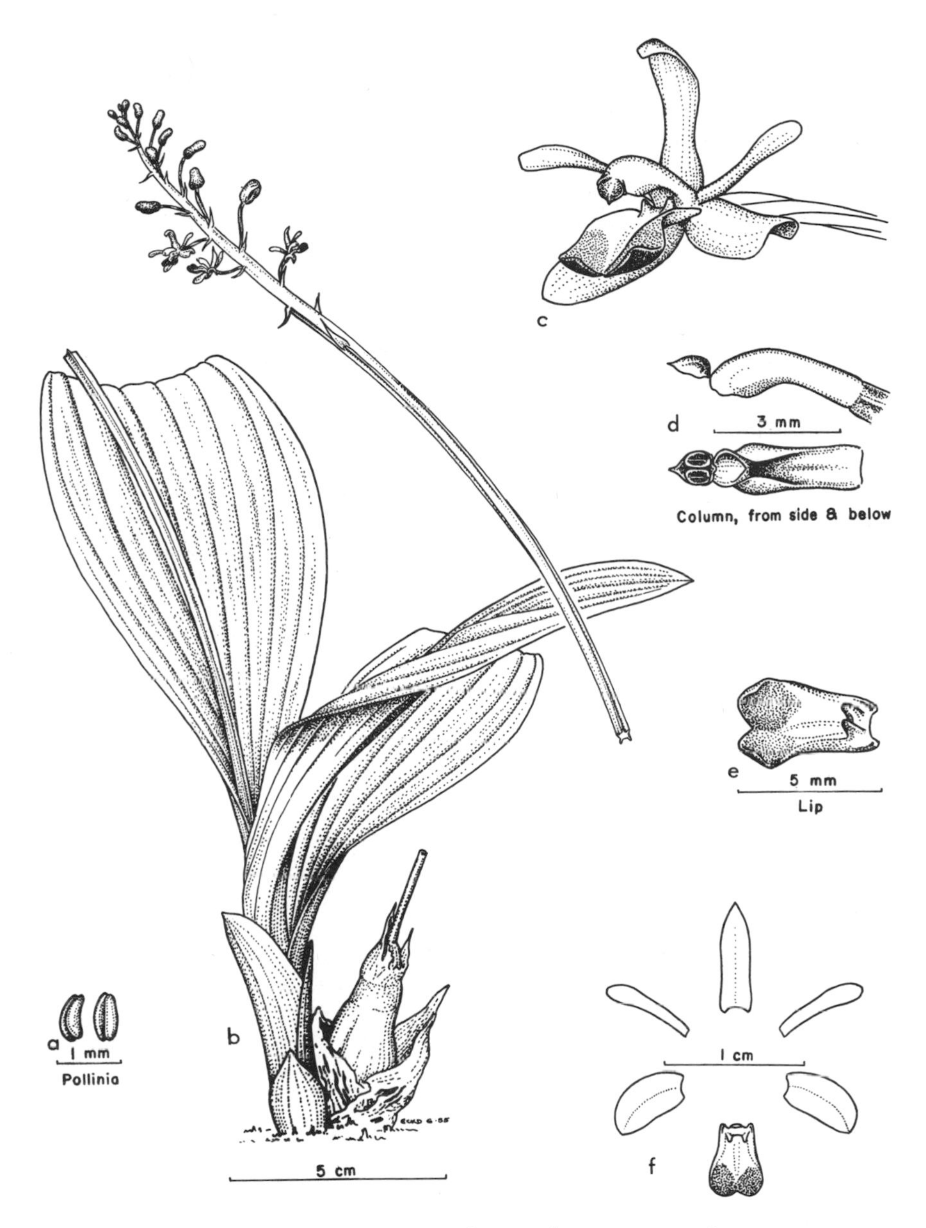

Fig. 263. *Liparis nervosa*: *f*, perianth parts separated.

ovate, ovate-elliptic, or ovate-lanceolate, acute or subacuminate, sheathing stem below, to 30 cm long, 10 cm wide; scape erect, much surpassing leaves, glabrous, terminating in lax, many-flowered raceme to 70 cm long; bracts erect, spreading or reflexed, usually shorter than ovaries; flowers rather delicate, opening in succession; sepals and petals reflexed; dorsal sepal oblong-ligulate to elliptic-oblong, acute or obtuse, margins revolute, to 8 mm long, 3 mm wide; lateral sepals falcate-ovate, acute or obtuse, margins revolute, to 7 mm long, 5 mm wide; petals linear-oblanceolate or linear-spatulate, to 7 mm long, 3 mm wide; lip cuneate-flabellate, bilobed in front, strongly recurved with 2 triangular obtuse calli at base; column fleshy, somewhat arcuate, thin-margined in front; ovary pedicellate, cylindric, to 1.5 cm long.

Variable and widespread in humid regions of the Old and New World tropics and subtropics. Specimens examined: Santa Cruz.

The extensive list of synonyms is based on an examination of all holotypes as well as other authentic materials. The fact that we are dealing with a species exceedingly variable in both habit and floral structures is evident from the following summary of selected taxonomic treatments and opinions.

Although *L. bituberculata* Lindl. is accredited to India and Nepal, Reichenbach not only reports it from Venezuela (Bonpl. 2: 22. 1854), but also considers it to be conspecific with *L. elata* Lindl. from Brazil. Lindley expressed the same opinion in his *Genera and Species of Orchidaceous Plants.*

Ridley described *L. elata* var. *rufina* from Africa. This variety was elevated to specific status by Rolfe with the comment, "It is also near the Indian *L. bituberculata* Lindl." Moreover, Rolfe considered Wilford's material cited by Ridley as *L. guineensis* Lindl. to be referable to *L. rufina* Rchb. f. ex Rolfe. These latter two species are kept apart by Rolfe because of the shape of the leaves—ovate-lanceolate versus broadly elliptic—even though both extremes are to be found within a single population.

Thwaites (1864, 295) considers *L. odorata* Lindl. to be the same as *L. elata* Lindl., while King and Pantling (Ann. Roy. Bot. Gard. Calcutta 8: 27. 1898) cite *L. odorata* Lindl. in synonymy of *L. paradoxa* Rchb. f. On the other hand, Hooker (cf. Trimen, Fl. Ceylon 4: 145. 1898) unites *L. paradoxa* Rchb. f. and *L. odorata* Lindl. with *L. nervosa* (Thunb.) Lindl. *L. formosana* Rchb. f. has been reduced by Ridley to a variety of *L. bituberculata* Lindl. Rolfe (Jour. Linn. Soc. Bot. 36: 7. 1903), however, states that *L. formosana* Rchb. f. is conspecific with *L. nervosa* (Thunb.) Lindl.

It is obvious that the correct identification of this species has been hampered greatly by the local aspect of each floristic study.

See Fig. 263.

Ponthieva R. Br. in Ait., Hort. Kew ed. 2, 5: 199. 1813

Terrestrial or epiphytic herbs, glabrous or hairy throughout. Roots fascicled, villous. Leaves mostly basal, sessile or petioled, erect or patent. Peduncle erect, more or less developed, usually elongate, terminated by lax, few- to many-flowered raceme. Flowers not resupinate, mostly inconspicuous, delicate in texture. Sepals free, spreading; lateral sepals occasionally connate. Petals always asymmetrical, adnate well above base to side of column. Lip small, fleshy, upper in position, adnate to column. Column fleshy, cylindric, dilated above. Anther incumbent. Pollinia 2, pulverulent.

Some 30 species distributed in the American tropics and subtropics.

Ponthieva maculata Lindl., Ann. & Mag. Nat. Hist. 15: 385. 1845

Ponthieva formosa Schltr., Rep. Spec. Nov. Regni Veg. Beih. 19: 12. 1923.
Ponthieva brenesii Schltr., op. cit. 165.

Epiphytic, rarely terrestrial herb, villous throughout; roots fasciculate, fleshy, densely covered with soft hairs; leaves basal, 1–4, soft in texture, erect to suberect, conduplicate, lanceolate, elliptic-lanceolate, or oblanceolate, acute to sharply acuminate, sessile or subpetiolate, to 30 cm long, 5 cm wide; scape commonly single, rarely 2, erect or slightly arcuate, usually exceeding leaves in length; peduncle terete with single sheath in middle, as long as the laxly-flowered raceme; bracts erect, cucullate, acute or shortly acuminate, about one-third to half the length of ovaries; flowers rather large for the genus, yellow-green with brownish veins and spots; dorsal sepal elliptic, concave, acute, 10–15 mm long; lateral sepals broadly elliptic, slightly oblique, acute or obtuse, 10–16 mm long; petals with short claw, dolabriform, obtuse, 8–13 mm long; lip very fleshy, reddish green in color, spoon-shaped, shortly acuminate, 4–5 mm long, 3–4 mm wide; column cylindric-clavate, 4–5 mm long; ovary pedicellate, cylindric, to 2 cm long.

Rather uncommon throughout Central America from Mexico to Panama; also in Venezuela, Colombia, and Ecuador. Specimens examined: Isabela, Pinta, San Salvador, Santa Cruz.

See Fig. 264.

Prescottia Lindl. in Hook., Exot. Fl. 2: *t. 115.* 1824

Terrestrial herbs, glabrous or rarely pubescent throughout. Roots fleshy, often tuberous, fascicled, villous. Leaves fleshy, rosulate, rarely cauline, sessile or with well-developed petioles. Inflorescence erect, terminal. Peduncle mostly covered with

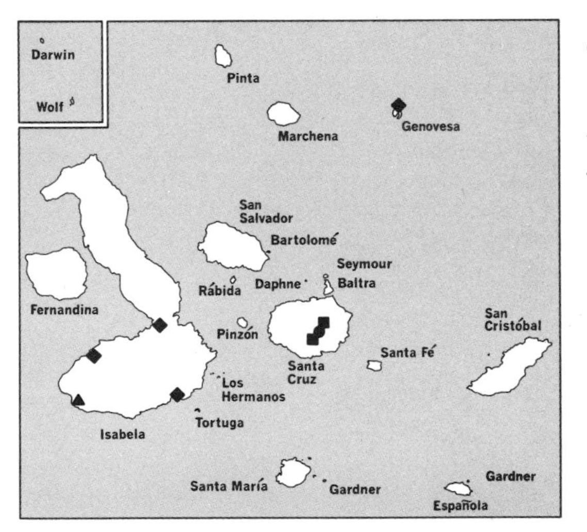

● *Prescottia oligantha*
■ *Tropidia polystachya*
▲ *Potamogeton pectinatus*
◆ *Ruppia maritima*

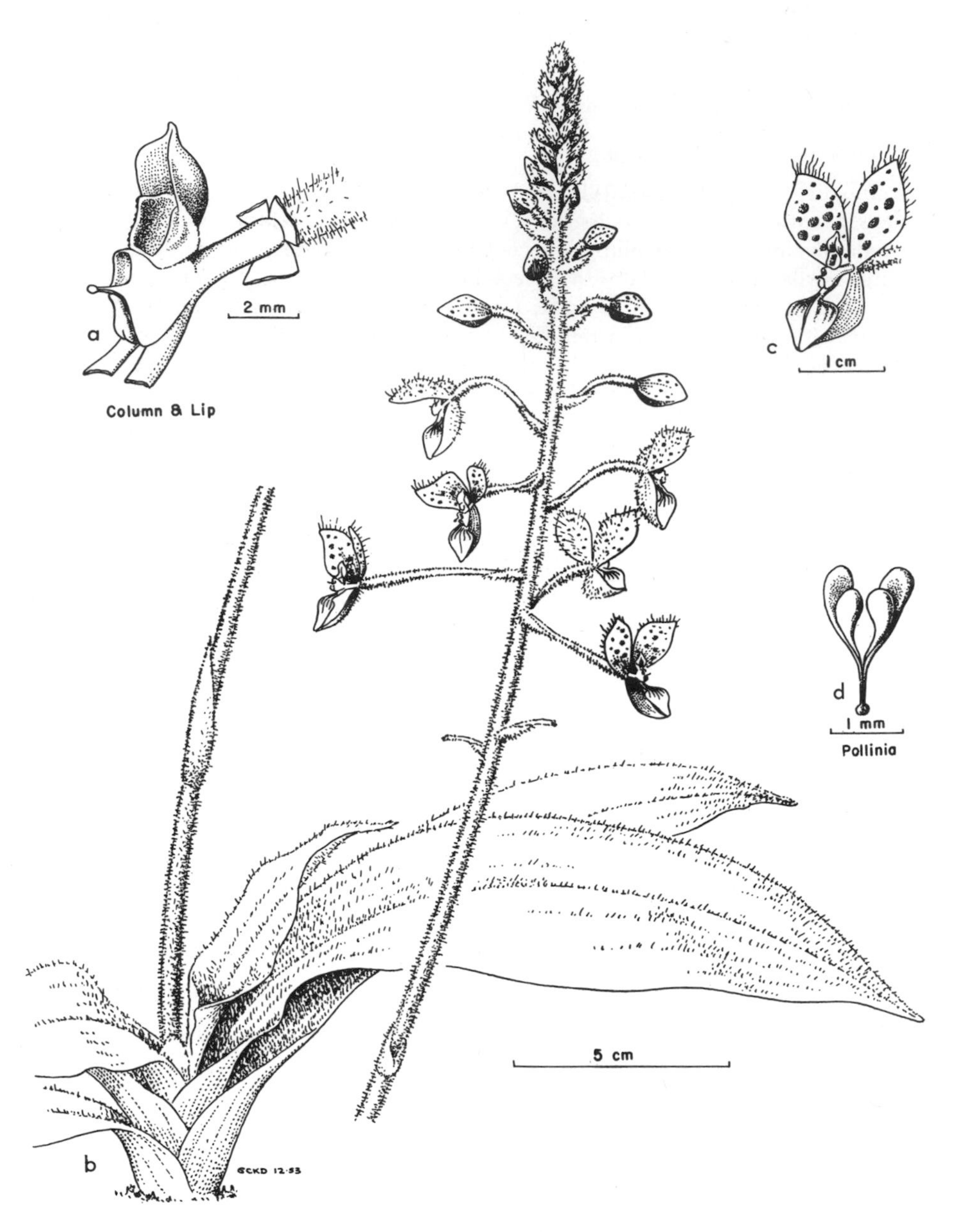

Fig. 264. *Ponthieva maculata.*

bractlike sheaths. Spike commonly dense, many-flowered, unbranched. Flowers minute, nonresupinate, i.e. lip uppermost. Sepals thin, basally connate to form a short cup, spreading. Petals narrower, thin, adnate to sepaline cup. Lip fleshy, galeate or cochleate, basally auriculate, the claw adnate to sepaline cup. Column short, fleshy, winged above, basally produced into short foot, this confluent with claw of lip. Pollinia 4, granular; capsule ovoid.

About 25 species distributed throughout the American tropics and subtropics, with the greatest concentration in Brazil.

Prescottia oligantha (Sw.) Lindl., Gen. & Sp. Orch. Pl. 454. 1840

Cranichis oligantha Sw., Prodr. Veg. Ind. Occ. 120. 1788.
Cranichis micrantha Spreng., Syst. Veg. 3: 700. 1826.
Serapias pumila Vell., Fl. Flum. 9: *t. 55.* 1827.
Decaisnea densiflora Brongn. in Duper., Bot. Voy. Coquill. 192. 1829.
Prescottia micrantha Lindl., Bot. Reg. 22: sub *t. 1916.* 1836.
Prescottia tenuis Lindl., Gen. & Sp. Orch. Pl. 454. 1840.
Prescottia densiflora Lindl., op. cit. 455.
Prescottia corcovadensis Rchb. f., Linnaea 22: 814. 1849.
Prescottia myosurus Rchb. f. ex Griseb., Fl. Br. W. Ind. 640. 1864.
Prescottia viacola Barb., Rodr., Gen. & Sp. Orch. Nov. 2: 279. 1882.
Prescottia viacola var. *polyphylla* Cogn. in Mart., Fl. Bras. 3^6: 549. 1906.
Prescottia panamensis Schltr., Rep. Spec. Nov. Regni Veg. 16: 357. 1920.
Prescottia polysphaera Schltr., loc. cit.
Prescottia filiformis Schltr., Rep. Spec. Nov. Regni Veg. Beih. 7: 50. 1920.
Prescottia gracilis Schltr., op. cit. 51.

Terrestrial, erect plant to 40 cm tall; roots fasciculate, fleshy, densely covered with soft hairs; leaves basal, rosulate, 1–5, ovate-oblong to obovate, suborbicular or elliptic, acute or obtuse, petiolate or subsessile, to 4 cm long, 3 cm wide; scape erect, glabrous, covered with lamellate, acute bracts, densely many-flowered above; spike commonly short, cylindric, occasionally elongate, slender, to 15 cm long; bracts ovate-lanceolate, long-acuminate, to 12 mm long, equaling or exceeding ovaries; flowers minute, nonresupinate, with dorsal sepal and petals reflexed; dorsal sepal ovate, obtuse, concave, to 2 mm long, 1 mm wide; lateral sepals connate at base to form a gibbous mentum, triangular-ovate, acute, somewhat concave, 2 mm long, 1.5 mm wide; petals linear or linear-spatulate, obtuse, 1.5 mm long, 0.5 mm wide; lip fleshy, subrotund, saccate with an auriculate base and apiculate apex, to 2 mm long and wide; column short, fleshy, laterally winged at apex; ovary ovoid to cylindric, to 5 mm long.

Widespread and locally common in Florida, the Caribbean region, Mexico, and Central America south to Brazil, Argentina, and Peru. Specimen examined: Santa Cruz, a new record for the Galápagos Islands.

See Fig. 265.

Tropidia Lindl., Bot. Reg. 19: sub *t. 1618.* 1833

Terrestrial, erect, simple or branching herbs. Roots fibrous, fasciculate or along short rhizome. Leaves few, thin in texture, plicate, usually tapering into sheathing base. Inflorescence terminal or occasionally from axils of upper leaves, racemose or paniculate, rarely glomerate, sessile or pedunculate, few- to many-flowered. Flowers small, rather thin in texture. Sepals and petals ringent; lateral sepals slightly connate at base forming a mentum. Lip sessile, parallel with column. Column short, fleshy, with short foot. Anther dorsal, erect. Pollinia 2, pulverulent.

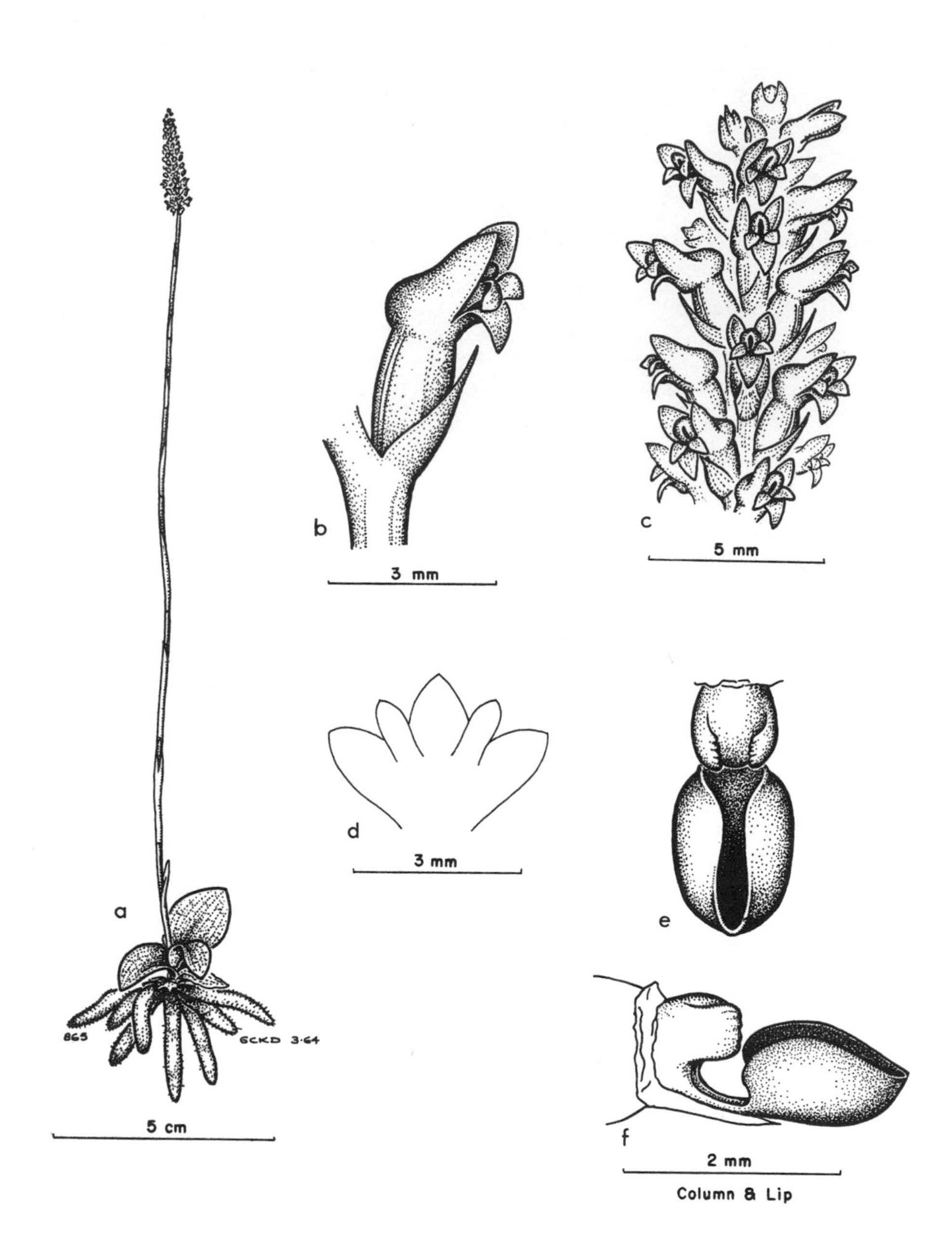

Fig. 265. *Prescottia oligantha*: *d*, upper part of perianth; *e*, *f*, lower lip and column from front and side.

About 30 species native to most tropical regions in India, Ceylon, Malaysia, New Caledonia, and Fiji, with 1 species in Florida, Central America, the West Indies, and the Galápagos Islands.

Tropidia polystachya (Sw.) Ames, Orchid. 2: 262. 1908

Serapias polystachya Sw., Prodr. 119. 1788.
Neottia polystachya Sw., Flora Ind. Occ. 3: 1415. 1799.
Chloidia vernalis Lindl., Gen. & Sp. Orch. Pl. 484. 1840.
Polystachya membranacea A. Rich. in Sagra, Ist. Isla Cuba, Hist. Nat. 11^2: 248. 1853.
Macrostylis vernalis Rchb. f., Bonpl. 2: 11. 1854.
Chloidia polystachya Rchb. f. in Walp., Ann. Bot. Syst. 6: 644. 1863.
Corymborchis polystachya Kuntze, Rev. Gen. Pl. 2: 658. 1891.
Tropidia eatoni Ames, Contr. Orch. S. Florida 14. 1904.

Terrestrial, erect or ascending, occasionally branching, leafy plant, glabrous throughout, to 60 cm tall; roots fasciculate, fibrous, hirsute; stem slender, wiry, to 4-leaved above, covered with scarious sheath below; leaves lanceolate, elliptic, or ovate-lanceolate, acute or long-acuminate, subplicate, to 30 cm long, 6 cm wide; inflorescence a lax, many-flowered panicle with rather short branches; peduncle terete, slender, to 15 cm long; bracts linear-lanceolate to ovate-lanceolate, much shorter than ovaries; flowers fleshy, rather small, green with purplish suffusion; dorsal sepal concave, ovate, acute or obtuse, to 7 mm long, 2.5 mm wide; laterals slightly oblique, similar in shape and size to dorsal sepal; petals linear-oblong to ovate, subacute or obtuse, occasionally truncate-emarginate, as long as sepals; lip fleshy, sessile, ovate-oblong, concave, obtuse, saccate at base; disk sparsely pubescent in middle, laterally with 2 falcate-incurved calli, to 6 mm long; column cylindric, to 4 mm long; ovary pedicellate, to 15 mm long.

Locally abundant in Florida, Mexico, Guatemala, Costa Rica, and the West Indies. Specimens examined: Santa Cruz.

The collections reported by Christophersen (1932, 71) and Svenson (1935, 227) as *Corymborchis* sp. are undoubtedly referable here. It is of special interest to note that the genus *Tropidia* does not occur on the mainland of South America. Although the material from the Galápagos Islands clearly shows long isolation—more spreading habit, longer and thinner leaves, occasional axillary inflorescences—the differences are not sufficient for the recognition of a separate species. A distribution map of the genus is published by Garay in Proc. 4th World Orch. Conf. 184. 1964.

See Fig. 266.

POTAMOGETONACEAE. Pondweed Family*

Perennial, rhizomatous herbs growing in fresh water, with submerged or floating leaves or both. Stems jointed, leafy, fibrous-rooted at lower nodes. Leaves alternate or opposite, in 2 ranks, sessile and capillary or petiolate and floating, both types present on the same plant in some species, in which case the submerged leaves usually linear, the floating leaves thicker and broader, sheathing at base, the sheath free or adnate to petiole. Flowers perfect or monoecious, in axillary clusters or on axillary, sessile or pedunculate spikes, base of peduncle surrounded by the sheath. Bracts none. Perianth of 4–6 valvate, rounded, short-clawed segments or these cup-shaped, or wanting. Stamens 1–4, inserted on claws of the perianth segments; an-

* Contributed by Ira L. Wiggins.

Fig. 266. *Tropidia polystachya*: *b*, column; *c*, flower, exploded, lower lip below.

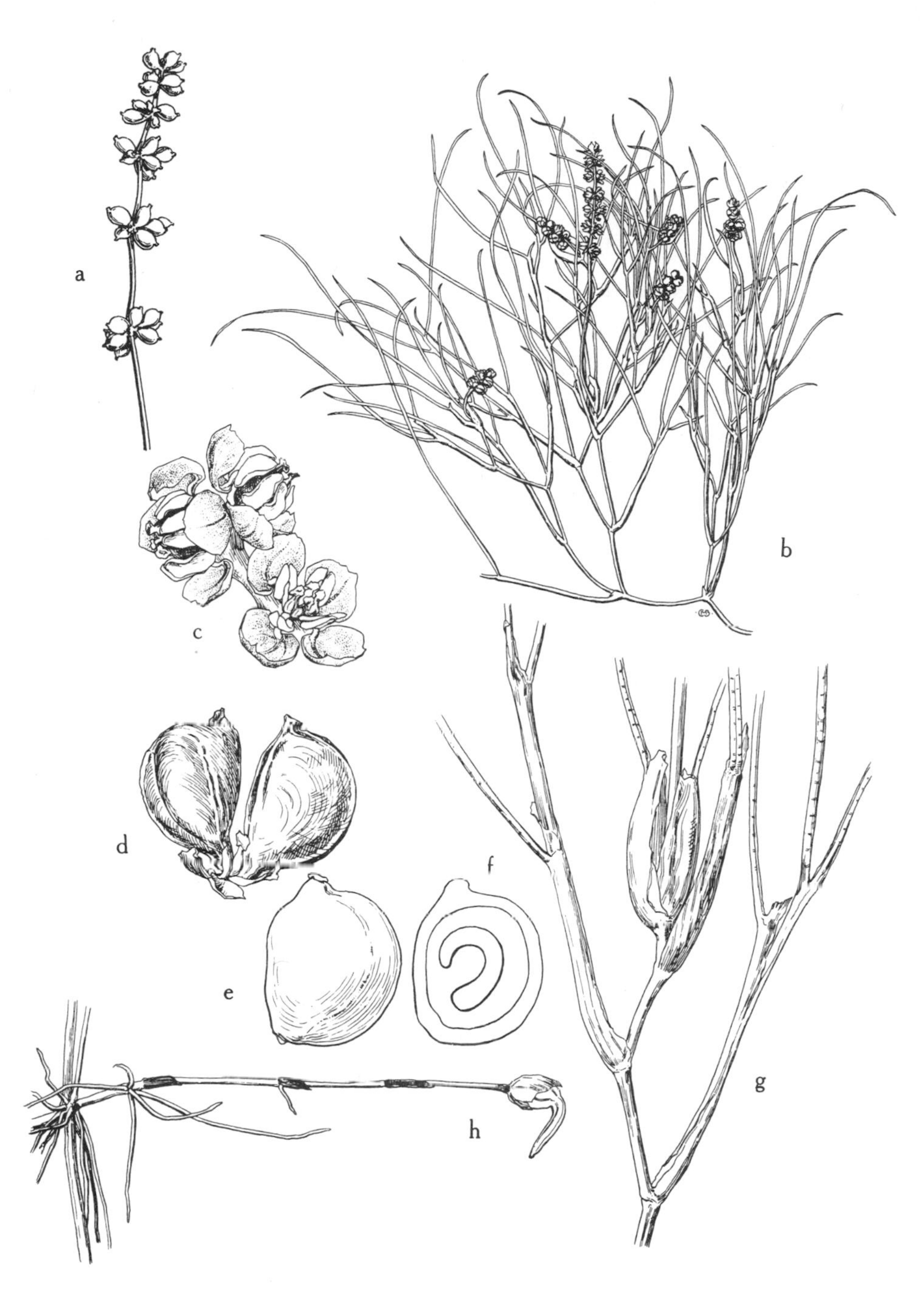

Fig. 267. *Potamogeton pectinatus*: *a*, spike with mature achenes; *b*, habit; *c*, flowers; *d–f*, achenes; *g*, stipules sheathing stem; *h*, rhizome bearing winter corm. (Reproduced by permission of the Regents of the University of California.)

thers 1- or 2-celled, sessile, extrorse. Pistils 1–4-carpeled, carpels 1-celled, 1-ovuled, distinct when more than 1. Stigmas sessile or on short styles. Fruit drupelike, sessile or stipitate, more or less compressed, with bony endocarp. Seed 1; endosperm none; embryo curved to uncinate.

A single genus, worldwide in distribution.

Potamogeton L., Sp. Pl. 126. 1753; Gen. Pl. ed. 5, 61. 1754

Characters of the family. About 90 species.

Potamogeton pectinatus L., Sp. Pl. 127. 1753

Potamogeton vaginatus Turcz., Bull. Soc. Imp. Nat. Moscow 27[2]: 66. 1854.
Potamogeton columbianus Suksd., Deutsch. Bot. Monatss. 19: 92. 1901.

Totally submerged plant from densely matted, slender rhizomes 1–1.5 mm thick, these bearing fleshy winter corms 3–5 mm thick; stems slender, repeatedly dichotomously branched, 0.8–1.2 mm thick, to 2 or 3 m long; leaf blades linear, 3–35 cm long, 0.6–2.5 mm wide, appearing to originate at apex of sheath formed by stipules, these adnate to margins of leaf base, free, or joined to each other part or all the way to apex of sheath, free from leaf at apex and forming ligule 3–10 mm long, this hyaline, irregularly dentate or lacerate, notched or lobed and rounded at apex; sheaths surrounding inflorescences slightly to markedly inflated; leaf blades 3–5-veined, midvein usually slightly to markedly off-center toward 1 margin, cross veins distinct to obscure, irregularly or regularly arranged; tips of blades acute, attenuate or obtuse, midvein entering apiculation when this is present; peduncles filiform, 5–25 mm long; spikes of 2–6 interrupted whorls, axis usually becoming lax and more or less pendent at maturity, not swollen toward apex; nutlets plump, 2.5–5 mm long, obliquely ovoid, smooth and convex on sides, back scarcely keeled, usually rounded, finely striate to smooth, tipped with short, blunt or flattened beak about 0.5 mm long, erect or recurved.

Floating below the surface in fresh, brackish, or salt water in streams, ponds, and estuaries, and occasionally in protected coves in sea water, often in extensive masses over many acres; cosmopolitan in distribution from cool temperate to tropical zones. In the Galápagos Islands known only from Isla Isabela, but probably present elsewhere in the archipelago.

The specimens examined have some leaves 0.6 mm wide, with 3 nerves, and others to 1.4 mm wide, with 5 nerves, on the same plant. The stipules and venation of the widest leaves closely approach those characteristics in the related *P. latifolius*. Many names have been applied to this species complex. Ascherson and Graebner (1907, 121–26) recognize 11 variants below the species level, under most of which they list several synonyms.

See Fig. 267.

RUPPIACEAE. DITCH-GRASS FAMILY*

Aquatic perennial herbs of brackish and saline ponds and marshes, with slender to filiform, profusely branching stems. Leaves alternate or opposite, linear, sheathing

* Contributed by Ira L. Wiggins.

Fig. 268. *Ruppia maritima*: *a, b*, variations in habit; *c*, peduncle bearing 2 flowers, each of 2 anthers, 4 pistils; *d*, 2 flowers, after fertilization; *e*, elongation of pedicels; *f*, mature nutlets; *g*, stipular sheaths; *h*, variation in habit; *i*, serrate leaf tip. (Reproduced by permission of the Regents of the University of California.)

at base. Flowers enclosed in spathelike sheaths on a considerably coiled and folded or straight peduncle, this straightening to lift flowers to surface of water just before anthesis and recoiling or not after pollination. Perianth none. Anthers 2, bilocular. Pistils 4, subsessile at first, later distinctly stalked. Stigmas peltate, sessile. Fruit a nutlet.

A single genus, widely distributed in all continents in temperate to tropical areas.

RUPPIACEAE. DITCH-GRASS FAMILY*

Characters of the family. One or possibly 2 species.

Ruppia maritima L., Sp. Pl. 127. 1753

Ruppia didyma Sw. & Wikstr., Kongl. Svenska Vetenskapsakad. Handl. 427. 1825.
Ruppia marina E. Fries, Semma Veg. Skand. 68. 1846.
Potamogeton filifolius Philippi, Reise in de Atacama 357. 1860.
Ruppia andina Philippi, Anal. Univ. Santiago 91: 525. 1896.
Ruppia pectinata Rydb., Mem. N.Y. Bot. Gard. 1: 18. 1896.
Ruppia filifolia Skottsb., Kongl. Svensk. Vet.-Akad. Handl. 56: 171. 1916.

Plant 6–10 dm long, fully submerged but floating up toward or to the surface; leaves usually alternate, narrowly linear to capillary, 0.5 mm wide or less, 2–10 cm long, with membranous stipular sheath 5–10 mm long, free part of sheath very short or lacking; flowers on slender, axillary peduncles, these at first sometimes short and enclosed by the spathelike basal leaf sheath, each peduncle usually bearing 2 flowers, each of these of 2 bilocular anthers and 4 sessile or subsessile, free pistils; peduncles elongating at anthesis; each pistil developing a slender stalk or pedicel 4–25 mm long in fruit, the fruits of the 2 flowers in a terminal and a subterminal umbellike group; nutlets ovoid, obliquely pointed, 1.5–2 mm long, smooth, terete, at length tawny or black.

In brackish or saline waters along coasts and inland, in tropical to cool temperate zones throughout the world; in the Galápagos Islands known only from Isabela and Genovesa, but to be expected elsewhere.

Svenson (1935, 221) accepted Skottsberg's *R. filifolia,* and placed his collection taken on Isla Genovesa in that "species." Setchell (1946) presented convincing arguments against recognizing more than 2 species to encompass the extremely variable populations constituting this genus and family. I prefer to follow his recommendations and place all Galápagos material in *R. maritima* L.

See Fig. 268.

Glossary of Botanical Terms

Glossary of Botanical Terms

Abaxial. On the side (of an organ) away from the axis, as the lower face of a leaf; dorsal; opposite of *adaxial.*

Aberrant. Departing from the normal.

Abortion. Imperfect development, or non-development, of an organ.

Abortive. Defective or barren; sterile.

Abscissing. Naturally separating, as flowers, fruit, or leaves, from the parent plant at a special separation layer.

Acaulescent. Stemless; having the stem very short and often underground, thus appearing stemless, as shrubs or bushes; opposite of *caulescent.*

Accessory. Additional; extra; constituting an appendage, as a branch arising from an adventitious bud (as contrasted with axillary or terminal branches).

Accrescent. Increasing in size with age, as the calyces of *Physalis* and other flowering plants.

Accumbent. Of a pair of cotyledons, lying face to face but curved back down the hypocotyl such that an edge of each is turned toward the hypocotyl (the cotyledons thus rotated 90° from the *incumbent* orientation).

Acerose. Needle-shaped.

Achene. A small, dry, indehiscent, 1-seeded fruit with a thin, tight pericarp, as the individual fruit of a composite flower or of the Amaranthaceae.

Acicular. Needle-shaped; slender, terete, sharp-pointed.

Acropetal. Beginning development at or near the base of an axis and progressing toward the apex; from the base upward; opposite of *basipetal.*

Acroscopic. Toward the apex or summit; opposite of *basiscopic.*

Actinomorphic. Symmetrical, capable of bisection through 2 or more planes into similar halves; of a floral whorl, having the parts all of the same shape and size; opposite of *zygomorphic, irregular.*

Aculeate. Armed with prickles or short spines, as the stem of a rose or of *Zanthoxylum fagara.*

Acuminate. Tapering gradually to an acute apex, with the sides concave along the taper. Cf. *acute.*

Acute. Sharp; tapering to a point, but with the sides straight along the taper. Cf. *acuminate.*

Adaxial. On the side (of an organ) toward the axis, as the upper face of a leaf; ventral; opposite of *abaxial.*

Adnate. United with another part of a different kind or cycle, as the stamens to the corolla. Cf. *free.*

Adnation. The growth process of becoming adnate.

Aduncate, aduncous. Bent or crooked, like a hook.

Adventitious. Arising in irregular places or in an irregular order, as buds near a wound or along an internode.

Adventive. Imperfectly naturalized.

Aerial. Growing or living in the air, as contrasted with subterranean or submerged.

Aestivation. The arrangement of the perianth parts in the bud of a flower.

Agglomerate. Heaped or crowded into a dense cluster, but not cohering.

Agglutinate. Glued together, as the pollen grains in the Asclepiadaceae and Orchidaceae.

Aggregate. Agglomerate.

Aggregation. A group or mass composed of distinct parts, as a head.

Alate. Winged; bearing a wing or wings.

Alternate. Arranged singly at regular intervals along an axis and alternating in orientation to the axis, as leaves; not in whorls or pairs. Cf. *opposite.*

Alternation of generations. Reproduction involving 2 dissimilar phases, as the alternation of the gametophyte (prothallus) generation with the sporophyte generation in the life history of ferns.

Alveolate. Pitted, like a honeycomb; faveolate; favose; foveolate.

Ament. A compact spike of unisexual flowers, each flower usually subtended by a bract, the whole inflorescence frequently deciduous; a catkin.

Amorphous. Shapeless or of indefinite form.

Amphicarpous. Bearing 2 kinds of fruit, the 2 differing in character or in time of ripening.

Amphitropous. Having the ovule itself curved almost 180° so that its micropylar end approaches, but not closely, its funicular end. Cf. *campylotropous, anatropous, hemianatropous, apotropous, epitropous, orthotropous.*

Amplexicaul(e). Stem-clasping; with the base of the leaf dilated and embracing the stem.

Ampliate. Enlarged; expanded; dilated.

Anadromous. Having the lowest superior element (segment, nerve, vein branch) of a pinna nearer the rachis than the lowest inferior one; opposite of *catadromous.*

Anastomosing. Forming an anastomosis.

Anastomosis. A network formed by the rejoining of veins after forking; the process of forming such a network.

Anatropous. Having the funiculus extended into a raphe fused dorsally halfway about the ovule so that the ovule is fully inverted but uncurved and the micropyle is close to and above the placental attachment of the funiculus. Cf. *hemianatropous, apotropous, epitropous, orthotropous, amphitropous, campylotropous.*

Androecium. The stamens considered as a unit or system of a flower. Cf. *gynoecium.*

Androgynophore. A stalk bearing both the stamens and the pistil; a prolongation of the apical part of the receptacle, bearing stamens and pistil above the level of attachment of the outer floral envelope. Cf. *androphore, gynophore.*

Androgynous. Composed of both staminate and pistillate flowers, the staminate at the apex; used chiefly in the Cyperaceae.

Androphore. A stalk bearing an androecium. Cf. *androgynophore, gynophore.*

Anemophilous. Pollinated by wind; opposite of *entomophilous.*

Anfractuose. Full of windings and intricate turnings; tortuous.

Angiospermous. Having the seeds borne within a pericarp, as the Angiospermae.

Angulate. Angular, as contrasted with curved or rounded.

Anisopolar. Differing in some way at opposite ends of the main axis.

Annual. Of only one year's duration.

Annular. In the form of a ring; arranged in a circle.

Annulate. Furnished with or composed of rings.

Annulus. A rim of thickened cells around a fern sporangium functioning as an elastic mechanism in shedding spores.

Anterior. On the side away from the axis; toward a subtending bract or leaf; abaxial; opposite of *posterior.*

Anther. The pollen-producing part of a stamen, including or bearing a pollen sac or sacs.

Anther connective. The tissue between the 2 lobes of an anther and distal to the filament of the stamen.

Antheridium (pl. **-ia**). The male organ in spore-bearing plants; a structure producing antherozoids.

Antheriferous. Bearing an anther or anthers.

Antherozoid. One of the motile male sex cells produced in antheridia, moving by pulsations of cilia or flagella.

Anther sac. One of the cavities in an anther containing the pollen grains; a pollen sac; a theca.

Anthesis. The process or the time of opening of a flower bud and the expansion of the flower parts.

Anthocarp. A fruit formed by the fusion of part or all of the floral parts with the fruit itself, as in the Nyctaginaceae.

Anthocyanic. Having a blue or bluish color, the color imparted to the flower by anthocyanin in the tissue.

Anthocyanin. A pigment dissolved in the fluid of some plant cells that is red in acid, blue in basic cell sap.

Antipodal. At the opposite end; at the apex; terminal.

Antrorse. Directed upward or toward the apex; opposite of *retrorse.* Cf. *extrorse, introrse.*

Apetalous. Lacking petals.

Apex. The uppermost point; vertex; tip.

Aphyllopodic, aphyllopodous. Lacking a basal rosette of leaves, where such a rosette is normally to be expected.

Aphyllous. Lacking leaves.

Apical. At the apex or tip of an organ, part, or plant; opposite of *basal.*

Apiculate. Terminating in an apiculus.

Apiculation. An apiculus; the growth pattern culminating in an apiculus.

Apicule. Apiculus.

Apiculus (pl. **-li**). A short, sharp, somewhat flexible point, but not a spine.

Apocarpous. Having the carpels separated

and free, not united; opposite of *syncarpous.*

Apopetalous. Having the petals mutually separate, as in the Polypetalae.

Apotropous. Having the ovule anatropous and the raphe ventral, so that the micropyle is below the placental attachment of the funiculus. Cf. *anatropous, hemianatropous, epitropous, orthotropous, amphitropous, campylotropous.*

Appendage. An addition; an extra part; a part added or attached to another, as the marginal expansion of glands on the cyathia in *Chamaesyce.*

Appendiculate. Furnished with an appendage or appendages.

Appressed. Flattened against underlying or adjacent tissues; pressed down or against.

Approximate. Close together but not united, as leaves along a stem; opposite of *distant.*

Aquatic. Living in water.

Arachnoid. Cobwebby, with soft, entangled hairs; spider-like.

Arboreal. Dwelling or growing on trees, but not necessarily parasitic.

Arborescent. Treelike in habit, as some shrubs.

Archegonium (pl. **-ia**). The female organ in cryptogams (ferns and lycopods) that contains an egg cell.

Arcuate. Curved; bowed; bent in an arc.

Areolate. Marked out into small spaces; reticulate.

Areole (pl. **areoles, areolae**). A space surrounded by anastomosing veins; a pit or spot bearing hairs, glochids, or spines, or all three, as in many cacti.

Argillaceous. Clay-colored; growing in clay; of the consistency of clay.

Aril. An extra covering of part or all of a seed, which is an outgrowth from the hilum and which may be pulpy and juicy, or dry and bony.

Arillate. Bearing an aril.

Aristate. Having an elongated, narrow appendage at the apex; bearing a terminal bristle or awn.

Armature. A covering of protective spines, barbs, hooks, or prickles.

Armed. Furnished with 1 or more spines, barbs, etc.

Aromatic. Having a spicy smell.

Arrested bud. A bud that reaches an intermediate stage of development, then becomes inactive, as a bud buried in wound tissue.

Article. A 1- or several-seeded portion of a multi-seeded fruit, especially in the Leguminosae, separated from others by a constriction or joint. Cf. *coccus.*

Articulate(d). Jointed; with zones or nodes for natural or regular separation of parts.

Articulation. A joint; the manner or pattern of jointing.

Ascending. Growing upward; standing obliquely upward.

Asexual. Without sex; reproducing without the union of gametes.

Assurgent. Ascending; rising or curving obliquely upward.

Atomiferous. Bearing microscopic or very small structures, such as glands; particle-bearing.

Atropurpureous. Black-purple; very dark purple.

Attenuate. Drawn out; having a long, gradual taper at the base or apex.

Auricle. An ear-shaped appendage or part, as at the base of the leaves or petals of some plants.

Auricular, auriculate. Having an earlike lobe or appendage; auricled.

Autogamous. Reproducing by pollination with its own pollen.

Autophyte. A plant not dependent on humus, as opposed to a saprophyte or parasite; a self-supporting plant that manufactures its own food.

Autophytic. Pertaining to or of the nature of an autophyte.

Autotrophic. Capable of self-nourishment, as plants that manufacture their own food from raw materials, as contrasted with parasitic and saprophytic plants.

Awl-shaped. Linear and tapering to a fine point.

Awn. A bristle-shaped appendage.

Axil. The upper angle formed by a petiole and the stem, or by a peduncle and the stem. Cf. *subtending.*

Axile. Belonging to the axis (the main stem, line of growth in an organ, etc.); central in position, as axile placentation.

Axillary. Situated in, or pertaining to, an axil.

Axis. The main or central line of development of a plant or of part of a plant; the main stem.

Baccate. Berry-like; pulpy or fleshy.

Banner. The uppermost petal of a papilionaceous corolla; a standard; a vexillum.

Barbate. Bearded, especially with long, stiff hairs.

Barbed. Having terminal or lateral spine-like hooks bent sharply backward.

Barbellate. Finely barbed.

Barbulate. Finely bearded.
Basal. At the base of an organ, part, or plant; opposite of *apical.* Cf. *cauline.*
Basifixed. Attached at the base.
Basinerved, basinervous. Veined from the base, as contrasted with a pinnately nerved leaf, in which lateral nerves or veins are arranged at successive levels along the sides of the midvein.
Basionym. The original epithet, retained when transferred to a new position.
Basipetal. Developing from the apex downward toward the base, as the order in which buds open in a determinate inflorescence; from the top down; opposite of *acropetal.*
Basiscopic. Toward the base; opposite of *acroscopic.*
Beak. A long, substantial point; the firm prolongation of the tip of a fruit or pistil.
Beard. A zone of hairs in an otherwise glabrous structure, or a zone of longer hairs among shorter ones; a long awn or bristle-like hair.
Berry. A pulpy, indehiscent, few- to many-seeded fruit arising from a single pistil but containing 2 or more seeds (or 1 by abortion).
Bi-. In combination, 2, twice, or doubled, as bilobed for 2-lobed.
Bicolorous. Two-colored, as a fern with red juvenile and green mature leaves, or a corolla displaying 2 colors.
Bidentate. Having 2 teeth.
Biennial. A plant having a 2-year life cycle, usually growing the first year, flowering and fruiting the second.
Bifid. Two-cleft; cleft at least halfway to the base or midrib into 2 lobes, teeth, etc.
Bifurcate. Two-cleft, as the tips of some petals and leaves.
Bilateral. Arranged on opposite sides.
Binate. In pairs; twinned; geminate.
Binomial. The taxonomic name of a species, consisting of genus and species names, as *Bursera graveolens.*
Biotype. A plant of a biotically but not genetically stable population, as a plant that grows consistently in saline soil, i.e. a plant conditioned by its habitat, rather than its heredity, to a consistent growth pattern. Cf. *genotype.*
Biparous. Bearing 2; having parts produced in pairs. Cf. *uniparous.*
Biseriate. In 2 cycles, rows, or whorls, as with 1 whorl of sepals and another of petals.
Bisexual. Having both stamens and pistil(s) in the same flower; perfect.
Bitypic. Of 2 kinds or types.
Bladder. An inflated structure resembling the bladder of an animal.
Bladdery. Inflated; with a thin-walled sac. Cf. *utricular, vesicular.*
Blade. The broad part of a leaf.
Blister. A bladder-like elevation of the epidermis, caused by a burn, bruise, infection, etc.
Bloom. A whitish powdery and glaucous covering of the surface, often of a waxy nature.
Bole. The main trunk of a tree, often with few or no lateral branches attached.
Bony. Dense and hard-textured, as the coats of *Opuntia* seeds.
Boss. A knoblike or rounded protuberance; an umbo.
Bract. A scale-like leaf; a reduced leaf subtending a flower.
Bracteate. Having bracts.
Bracteolate. Furnished with bracteoles or bractlets.
Bracteole. A secondary bract; a bractlet.
Bractlet. A bract borne on a petiole, pedicel, or other secondary axis.
Branch. A primary division of the main axis of a plant.
Branchlet. A division or subdivision of a branch; a twig.
Bristle. A stiff, strong hair or similar body.
Bristly. Bearing bristles.
Broomlike. Bearing a bundle of bristles or firm, stiff hairs at the end of an axis.
Brunnescent. Becoming brown in time.
Bud. The rudimentary state of a stem or branch; an unexpanded flower.
Bud scale. A leaf that is greatly modified and reduced to form a papery, leathery, or horny protective covering around a bud.
Bulb. The thickened, chiefly underground base of some plants, capable of reproducing the plant vegetatively and made up of appressed scales or plates on a short axis.
Bulbiferous. Bearing bulbs.
Bulbil. A small bulb, usually axillary in position, and often arising on a main bulb as a secondary structure.
Bulblet. A small bulb produced in an unusual place, as in an inflorescence.
Bulbous. Pertaining to or of the nature of a bulb; rotund.
Bullate. Blistered or swollen; having a mound or mounds on the epidermis.
Bundle. A strand of specialized tissue, variously modified, as a vascular bundle.
Bush. A low, thick, densely branched shrub.

Buttress. The broadened base of a tree trunk or a thickened vertical part of it.

Caducous. Falling early or prematurely; of short duration or life. Cf. *persistent.*
Caespitose. Cespitose.
Calcarate. Furnished with or produced into a spur.
Callose. Hard and thick in texture; bearing callosities.
Callosity. A thickened or raised area; an area of tissue firmer than the surrounding tissue.
Callous. Having the texture of a callus.
Callus (pl. **-i**). A hard protuberance; in the grasses, the swelling at the point of insertion of the lemma or palea; a roll of new tissue developed around a wound, or just interior to the bark on a cutting, sealing off the wound.
Calycine. Belonging or pertaining to a calyx; calyx-like.
Calyculate. Bearing a calyx or a part similar to a calyx; calyx-like; having an epicalyx.
Calyptra. A lid or cap; the cap of a moss capsule; a lid over a bud.
Calyptrate. Furnished with or of the nature of a calyptra.
Calyx (pl. **-yces**). The outer whorl of the perianth or floral envelope; the sepals, whether distinct or connate, but considered collectively.
Calyx cup. A low calyx tube.
Calyx lobe. One of the free, projecting parts of a calyx.
Calyx tube. The tube or cup of a gamosepalous calyx; sometimes applied to the hypanthium.
Cambium. A layer, usually regarded as one cell thick, of persistent meristematic (vascular) tissue; or a persistent meristematic layer that gives rise to secondary wood and secondary phloem.
Campanulate. Bell-shaped.
Campylotropous. Amphitropous but with the curvature more marked, so that the micropylar end of the ovule is almost in contact with the hilum. Cf. *amphitropous, anatropous, hemianatropous, apotropous, epitropous, orthotropous.*
Canaliculate. Having a channel or groove.
Cancellate. Latticed; cross-meshed, the meshes usually rectangular and the long axis of the individual meshes mostly at right angles to the axis of the latticed organ or structure; coarsely reticulated. Cf. *clathrate.*
Canescent. Gray-pubescent and hoary.
Canotomentulose. Having an indument of very fine, tangled, wool-like hairs.
Cap. A convex, removable part or covering; a lid. Cf. *calyptra, operculum.*
Capillary. Very slender; hairlike.
Capitate. Borne in heads; formed like a head; in a compact cluster.
Capituliform. Having the form of a capitulum.
Capitulum (pl. **-a**). A dense inflorescence consisting of sessile flowers; a head.
Capsular. Belonging to a capsule; formed like a capsule.
Capsule. A dry, dehiscent fruit resulting from the maturation of a compound (multi-carpeled) ovary.
Carina. A keel; a ridge.
Carinal. Pertaining to or of the nature of a ridge or keel; in aestivation, having the parts of the flower enfolded within a keel.
Carinate. Keeled; having a ridge or line along the dorsal (or ventral) side.
Carneous. Flesh-colored; pale pink.
Carpel. A simple pistil or 1 member of a compound pistil, bearing an ovule or ovules along part of its inner wall.
Carpellate, carpeled, carpous. Having, pertaining to, or of the nature of a carpel or carpels.
Carpophore. The slender prolongation of the axis which in the Umbelliferae supports the pendulous ripe carpels.
Cartilaginous. Tough and firm but not bony; like gristle or cartilage.
Caruncle. A protuberance near the hilum of a seed, as on the seed of *Ricinus communis*; a strophiole.
Carunculate. Possessing a caruncle.
Caryopsis. An achene; usually restricted to the fruit of grasses, which differs from the achene of the Compositae in being derived from a superior ovary.
Castaneous. Chestnut-colored.
Catadromous. Having the lowest inferior element (segment, nerve, vein branch) of a pinna nearer the rachis than the lowest superior one; opposite of *anadromous.*
Cataphyll. The early leaf of a seedling (or cotyledon); a bud scale; a rhizome scale.
Catkin. Ament.
Caudate. Bearing a tail-like appendage.
Caudex (pl. **-ices**). The stem; the main shoot; the persistent base of an otherwise annual herbaceous stem.
Caudicle. The slender stalklike appendage of the pollen masses in orchids.
Caudiform. Having the form of a caudex.
Caulescent. Having an obvious main stem above ground; opposite of *acaulescent.*

Cauliflorous. Producing flowers from old growth.
Cauline. Pertaining or belonging to a stem or axis, as opposed to basal.
Cell. One of the minute vesicles, of various forms, of which living organisms are composed; sometimes, one of the cavities in an ovary.
Cellular, cellulate. Made up of cells; containing small, enclosed spaces of similar shape and size.
Centrifugal. Developing from the center toward the periphery; opposite of *centripetal.* Cf. *determinate.*
Centripetal. Developing from the periphery toward the center; opposite of *centrifugal.* Cf. *indeterminate.*
Ceraceous. With a waxy white or pale yellow covering or bloom.
Cerebriform. Having an irregularly furrowed, brainlike surface or appearance.
Cernuous. Drooping.
Cespitose, caespitose. Matted; growing in dense tufts; growing in small clumps.
Chaff. A thin, small, dry, and membranous scale or bract, as the bracts on the disk of some Compositae; such scales collectively.
Chaffy. Furnished with or having the texture of chaff.
Chalaza. The basal part of an ovule; the area of attachment of an ovule to its funiculus and its integument. Cf. *hilum.*
Channeled. Deeply grooved longitudinally; canaliculate.
Chartaceous. Papery or tissue-like in texture, usually colorless or faintly tinted, not green.
Chlorophyll. The green photosynthetic coloring matter in the cells of plants, an indispensable catalyst in the manufacture of sugars from raw materials under the energy impact of light.
Chlorophyllous. Producing chlorophyll.
Choripetalous. Polypetalous.
Ciliate. Marginally fringed with fine hairs. Cf. *fimbriate.*
Ciliolate. Minutely ciliate.
Cilium (pl. **-ia**). A fine marginal hair.
Cinereous. Ash-colored; light gray.
Cinnamomeous. Cinnamon-colored; yellow- or orange-brown.
Circinate. Coiled from the tip downward, as an unexpanded fern frond.
Circumcinct. Girt about; girdled.
Circumscissile. Opening by a line of dehiscence around a fruit or anther, with the upper part shedding as a lid or cap; dehiscing circumferentially.
Cirrhate, cirrhous, cirrose. Having tendrils or spirally twining organs that resemble and function like tendrils.
Cladophyll. A flattened foliaceous stem that functions like a leaf but arises in the axil of a bractlike, often caducous leaf. Cf. *phyllodium.*
Clambering. Climbing and spreading awkwardly, as across rocks.
Clasping. Enclosing and holding firmly.
Cathrate. Latticed; cross-meshed, the meshes usually rectangular and the long axis of the individual meshes mostly parallel with the axis of the latticed organ or structure; pierced with apertures; arranged in rectangular cells with darkened cross walls, as in some scales of ferns. Cf. *cancellate.*
Clavate. Club-shaped; with the body thicker toward the apex than at the base.
Clavellate. Minutely clavate.
Claw. The slender stalk or petiole-like part of a petal (or of a sepal in rare cases).
Cleft. Divided about to the middle into 2 or more parts. Cf. *divided, -fid.*
Cleistogamous. Self-fertilized in the bud or in small, closed flowers, as some violets, *Dichondra*, etc., often underground.
Clone. A group of individual plants propagated without sexual reproduction from a single individual, as by vegetative growth from fallen stem segments, or by the rise of new plants from stolons.
Coadnate, coadunate. Adnate; attached to a floral organ of a different whorl. Cf. *free.*
Coalescence. The union of parts or organs of the same kind, as sepals into a calyx cup; the process of joining of such parts.
Coalescent, coalesced. Joined together by coalescence.
Coarctate. Crowded together.
Coccus (pl. **cocci**). One of the parts into which a dry, lobed, membranous or leathery fruit with 2 or more 1-seeded and often indehiscent cells splits. Cf. *article.*
Cochleate. Shell-shaped; shaped like a snail shell.
Coenosorus. Having sporangia from different sori continuously covering a large area, as in *Elaphoglossum* or *Doryopteris.*
Coherent. Having 2 or more similar parts touching but not fused.
Collar. An encircling outgrowth or ring of tissue.
Collateral. Located side by side, as pairs of stamens in many of the Labiatae.
Colonial. Given to growing in colonies.
Colony. A distinguishable localized population of a species; a stand or patch of

plants of a given species having common ancestry.

Colporate. Having pollen grains with distinct pores through which the pollen tubes emerge.

Columella. The carpophore of the mericarpous fruits of the Umbelliferae.

Column. The structure formed by the adnation of the stamen, style, and stigma in orchid flowers; the fusion of stamen filaments in the flowers of the Malvaceae.

Columnar. Pertaining to or of the nature of a column.

Coma. A tuft or crown of fine hairs, as on the seed of *Asclepias*; sometimes, a dense crown of leaves, as in some palms.

Commissural. Pertaining to or of the nature of a commissure.

Commissural face, commissural plane. One of the plane surfaces of a spore, pollen grain, carpel, etc., formed by its previous juncture with another such surface at a commissure.

Commissure. The plane of coherence of 2 parts, as the interface of 2 spores or pollen grains derived from the same spore mother cell, or of 2 carpels, as in the Umbelliferae; the place where 2 coherent style branches join.

Comose. Bearing a coma; crowned with fine hairs.

Complete annulus. An annulus completely encircling the rim of the sporangium.

Composite. Apparently simple in structure but made up of several joined or closely approximate parts; compound; the common name for a member of the Compositae.

Compound. Consisting of 2 or more parts. Cf. *simple*.

Compound leaf. A leaf consisting of 2 or more leaflets, as in ferns.

Compound pistil. A pistil formed by the fusion of 2 or more carpels.

Compressed. Flattened, especially laterally.

Concave. Hollowed or rounded inward, like the inside of a bowl.

Concavo-convex. Concave on one side and convex on the other.

Concolorous. Uniform in color; of 1 color, as contrasted with bicolorous, heterochromous.

Conduplicate. With the lateral margins bent abruptly inward toward the axis and touching, as the leaves in some buds. Cf. *induplicate*.

Cone. An elongated aggregation of flowers or fruits forming a detachable spore-bearing or seed-bearing structure.

Confluent. Blending together; merging.

Conform, conformed. Similar in form; closely fitting, as the sheath of a grass leaf.

Congested. Crowded together.

Conglomerate. Densely clustered or heaped together.

Conical. Cone-shaped.

Conjugate. Joined together in pairs. Cf. *jugate*.

Connate. United; congenitally fused; used especially of like organs, as leaves.

Connate-perfoliate. Having opposite sessile leaves that are fused at the base, with the stem axis passing through the fused blades.

Connective. The tissue between 2 pollen sacs of an anther.

Connivent. With the parts coming together but not fused; coherent.

Conspecific. Of the same species.

Contiguous. Touching but not fusing, used regardless of whether the parts are of the same or of different whorls.

Contorted. Twisted; bent or twisted on itself.

Contra-. In combination, against; opposite; as contra-ligule for a phyllary of a composite head just outside a ligule (ray flower) of the inflorescence.

Convex. Rounded outward, like the surface of a dome.

Convolute. Rolled together longitudinally, as a floral bud with its petals (or sepals or both) overlapped successively.

Coralloid. Coral-like.

Cordate. Heart-shaped, with the notch at the base of the structure.

Coriaceous. Leathery in consistency or texture.

Corm. The solid, bulblike, usually underground part of a stem.

Corneous. Of the texture or character of horn.

Corniculate. Furnished with a small horn or horns.

Corolla. The inner or second whorl of the perianth or floral envelope; the petals, whether distinct or connate, but considered collectively.

Corolliform. Resembling a corolla, as the calyx of the flowers of *Polygonum*.

Corolloid. Corolla-like.

Corona. Crown; the petal-like appendage or corolloid structure standing between the corolla and the stamens, as in the flowers of many species of *Asclepias*; a crown.

Coronal. Pertaining to or of the nature of a corona.

Corpusculum. The connective between the

arms of the pollen masses in the flowers of the Asclepiadaceae.

Cortex. The typically parenchymatous layer of tissue external to the vascular tissue and internal to the corky or epidermal tissues in green plants.

Corymb. A broad, more or less flat-topped, indeterminate inflorescence. Cf. *cyme.*

Corymbiform. Corymb-shaped; corymbose.

Corymbose. Arranged like a corymb or in corymbs.

Cosmopolitan. Occurring naturally in all or most parts of the world.

Costa (pl. **-ae**). A rib or vein of a leaf, especially the midvein; the midvein of the ultimate segment of a fern frond.

Costal, costalar. Relating to, originating at, or of the nature of a costa.

Costate. Ribbed; having 1 or more longitudinal nerves.

Costule. The primary vein of a segment of a fern frond.

Cotyledon. A seed leaf; the primary leaf or leaves of an embryo.

Crateriform. Crater-shaped.

Creeping. Running along at or near the ground surface. Cf. *procumbent, repent.*

Crenate. Shallowly round-toothed; scalloped.

Crenation. One of a series of rounded projections, as a scallop on a leaf margin. Cf. *tooth.*

Crenulate. Minutely crenate.

Crested. Bearing elevated ridges or projections on the surface, especially on petals.

Crispate, crisped. Curled; extremely undulate; of hairs, sharply kinked.

Cristate. Crested; having a tasseled margin.

Crown. A corona; also, the part of a stem at the soil surface; a part of a rhizome bearing a large bud.

Cruciate. Cross-shaped, as the petals of the flowers of the Cruciferae.

Cruciferous. Belonging to the Cruciferae.

Crustaceous. Of a hard and brittle texture.

Cryptogam. A plant lacking true seeds but often reproduced as the result of a sexual act.

Crystalline. Made up of or filled with crystals, as many subepidermal cells in stems of *Opuntia* plants that contain calcium oxalate crystals.

Crystalline denticle. One of the minute, toothlike projections of crystallized salt along the leaf margins in mangroves during dry, hot weather; one of the similar silicate projections along ridges on the stems of *Equisetum.*

Cucullate, cuculiform. Hood-shaped; hooded.

Culm. The stem of a grass or sedge, usually hollow except at the swollen nodes.

Cultivar. A plant form that originated in, and is maintained by, cultivation, as contrasted with plants growing in the wild or native state.

Cuneate. Wedge-shaped; with the narrow point at the base of the leaf, petal, or other structure.

Cupulate. Furnished with a cupule.

Cupule. A cup; an involucre of bracts more or less investing a seed or several seeds.

Cupuliform. Cup-shaped.

Cusp. A sharp, rigid point.

Cuspidate. Abruptly narrowed into an elongate, concave-margined, sharp-pointed tip.

Cyathiform. Shaped like a drinking cup; resembling a cyathium.

Cyathium (pl. **-ia**). A cup-shaped involucre containing individually free male flowers and usually 1 female flower, commonly present in the Euphorbiaceae.

Cyclic. With parts arranged in whorls, as contrasted with spiral.

Cylindric, cylindrical. Shaped like a cylinder.

Cyme. A broad, more or less flat-topped, determinate inflorescence. Cf. *corymb.*

Cymelet. A small cyme.

Cymose. Cyme-like; arranged in cymes.

Cymule. A diminutive cyme; a few-flowered cyme.

Cystolith. An intercellular concretion; a structure, usually acicular or elliptic in shape, lying flat on the epidermis of some plants.

Deciduous. Falling at the end of the growing season; opposite of *persistent.* Cf. *caducous.*

Declinate. Bent downward or forward, often with the tip recurved, as the stamens in the flowers of *Tropaeolum.*

Decompound. More than once compound; repeatedly divided.

Decumbent. Lying on the ground, with the tip turned upward.

Decurrent. Extending downward along and adnate to the stem.

Decurved. Curved downward; bent down.

Decussate. With opposite leaves alternating in pairs at successive levels, each level at right angles to the last.

Definite. Constant in number, and occurring in multiples of the petal number, as stamens; opposite of *indefinite.*

Deflexed. Reflexed; bent sharply downward at the point of attachment.
Dehiscence. The method of opening of a bud, a seed pod, or an anther, at maturity.
Dehiscent, dehiscing. Opening regularly by valves, slits, etc., as a capsule or anther; opposite of *indehiscent.*
Deliquescent. Softening or becoming semiliquid and sticking together, as some petals soon after anthesis; absorbing water and becoming liquid.
Deltoid. Broadly triangular; shaped like the Greek letter Δ.
Dentate. With sharp, spreading, usually coarse teeth along the margins.
Dentation. The teeth along a margin, as of a leaf; the pattern of such teeth.
Denticle. A minute tooth.
Denticulate. Minutely dentate; with the margins minutely toothed.
Depauperate. Impoverished; stunted; reduced in stature or function.
Depressed. More or less flattened; pressed down.
Determinate. Of an inflorescence, having the central (terminal) flower opening first, with no subsequent elongation of that axis, as in a cyme; opposite of *indeterminate.*
Developed. At full growth; not rudimentary.
Dextrorse. Twining spirally upward around an axis, from left to right; opposite of *sinistrorse.*
Diad, dyad. A subdivision of a tetrad in meiosis or mitosis, which again divides into a pair of single elements.
Diadelphous. Of stamens, having an androecium consisting of 2 often unequal bundles or clusters; in many legumes represented by 9 stamens attached to a cleft tube, and a tenth separate and distinct.
Diandrous. Having 2 stamens.
Diaphragm. A thin partition; a thin membrane across a space.
Dichasial. Pertaining to or of the nature of a dichasium.
Dichasium (pl. **-ia**). A determinate inflorescence in which the first flower to open is between 2 lateral flowers, forming a false dichotomy, the pedicels of the lateral flowers often elongating, that of the terminal flower not, as in *Galium*; a cyme with 2 lateral axes.
Dichotomous. Successively branched into 2 or more pairs.
Dicotyledonous. Having 2 cotyledons.
Dictyostele. A stele with large, overlapping leaf-gaps, arranged in a cylinder around a central pith. Cf. *protostele, siphonostele, solenostele.*
Dictyostelic. Having the properties of a dictyostele.
Didymous. Occurring in pairs, as the mericarps of the Umbelliferae.
Didynamous. Having 2 pairs of stamens of 2 different lengths, as in *Browallia americana.*
Diffuse. Widely or loosely spreading.
Digestion. An enzymatic action in cells or tissues of plants that change the physicochemical nature of the substance acted upon, as the dissolution of cell walls and contents of a host plant by the haustorial cells of a parasitic plant like *Cuscuta.*
Digitate. Shaped like a hand; compound, with the members arising at the same level, as the leaflets of *Oxalis corymbosa.*
Dilated. Expanded laterally; flattened.
Dimerous. Having all the parts in 2's.
Dimidiate. Halved unequally, as when 1 part of a leaf is lacking or is much smaller than the part on the opposite side of the midrib.
Dimorphic. Occurring in 2 forms, as the fertile and sterile fronds of *Blechnum,* or the short-styled and long-styled flowers in *Nicotiana.*
Dimorphism. The property of being dimorphic.
Dimorphous. Dimorphic.
Dioecious. Having staminate and pistillate flowers on different plants; opposite of *monoecious.*
Diplostemonous. Having the stamens arranged in 2 distinct whorls, 1 exterior to the other.
Dipterous. Two-winged.
Disarticulate(d). Separated at a joint, as when a leaf falls from a stem in autumn.
Disarticulating. Separating joint from joint, or part from part.
Disc, disk. A more or less fleshy process of the receptacle; the coalesced nectaries or staminodes around the base of a pistil; the flattened or conical receptacle in the heads of many Compositae; the flattened, clinging tip of some tendrils.
Disc flowers. The tubular flowers in the center of the head of many Compositae, as distinguished from the strap-shaped ray (marginal) flowers.
Disciform. Flat and circular; shaped like a disc.
Discoid. Having only disc flowers; having a disclike shape, as a stigma.
Discrete. Separate; not coalescent.

Disk. Disc.
Dissected. Divided into numerous segments.
Dissection. The process or pattern of dividing into numerous segments.
Dissepiment. A partition in a compound ovary or fruit, usually formed by the adhesion of the sides of adjacent carpels; a transverse wall in a cavity.
Distal. Remote from the point of attachment; away from the center of a body; opposite of *proximal.*
Distant. Widely removed, as leaves along a stem; opposite of *approximate.*
Distichous. Two-ranked; with the leaves, leaflets, or flowers on opposite sides of a stem or rachis and lying in the same plane; spreading in opposite directions. Cf. *polystichous.*
Distinct. Separate; free from adjacent units.
Diurnal. Opening during daylight hours; opposite of *nocturnal.*
Divaricate. Spreading markedly; extremely divergent.
Divergent. Spreading moderately.
Divided. Separated to or near the base into 2 or more parts. Cf. *cleft, -fid.*
Dolabriform. Hatchet-shaped.
Dorsal. On or of the outer or back surface (of an organ); abaxial.
Dorsifixed. Attached on the back or dorsal surface, as the awn on the lemmas of some grasses.
Dorsiventral. Of an organ, having principally an under and an upper surface, as many leaves.
Dorsum. The upper or inner surface of an appendage or part.
Doubly reverse bent. Bent fully downward, then fully upward, producing 2 reverses in the direction of growth, as the pappus-awn of *Pseudelephantopus spicatus.*
Downy. Covered with short, soft hairs; fuzzy.
Dropsical. Puffy; swollen; turgid.
Drupaceous. Resembling a drupe; bearing drupes.
Drupe. A fleshy, 1-seeded, indehiscent fruit; a stone fruit, as the fruit of *Castela.*
Drupelet. A diminutive drupe.
Duct. An elongated tubular structure occurring frequently in fibrovascular systems, which may contain latex, resin, mucilage, or other fluids.
Duplicate. Double or folded; a twin or copy.
Dwarfed. Small in size or height as compared to its close relatives; stunted.
Dyad. Diad.
Ebenaceous, ebeneous. Black as ebony.
Ebracteate. Lacking bracts.
Ecarunculate. Lacking a caruncle.
Echinate. Bearing stout, blunt prickles, as the fruit of *Luffa astorii.*
Eciliate. Lacking cilia.
Effuse. Very loosely spreading.
Egg cell. The female gamete, consisting of a single cell.
Eglandular. Lacking glands.
Elater. The hygroscopic, ribbon-like appendages on the spores of *Equisetum.*
Eligulate. Lacking a ligule or ligules.
Ellipsoid, ellipsoidal. Solid but with an elliptical outline; having the form of an extended sphere, as a plum. Cf. *oblate.*
Elliptic, elliptical. Ellipse-shaped but essentially 2-dimensional, rounded about equally toward both ends.
Elongate(d). Drawn out in length; extended.
Emarginate. With a shallow notch or sinus at the apex.
Embedded. Enclosed closely in a matrix of tissue.
Embryo. The rudimentary plant within a seed.
Emersed. Extending above the surface of the water.
Endemic. Native; confined to a particular area; when said of a plant recorded from the Galápagos Islands, known from no other area.
Endocarp. The inner layer of a pericarp or fruit wall; opposite of *exocarp.*
Endogenous. Growing throughout the substance of the stem, instead of by superficial layers. Cf. *exogenous.*
Endoglossum. An internal flap or septum, often incomplete or only partially attached to side walls, as in a locule or mericarp in the Malvaceae.
Endophyte. A plant growing wholly or partially within another plant, as the algae in the tissues of a lichen, or the ramifying growth of the feeder branches of a parasitic plant within its host plant. Cf. *epiphyte, saphrophyte.*
Endophytic. Of the character of an endophyte.
Endosperm. The starchy, oily, or horny nutritive tissue associated with the embryo in many flowering plants; a nutritive tissue inside the seed.
Enfolded. Clasped within folds, as of a leaf or sheath.
Enrolled. Clasped within a rolled surface, as a stem within a leaf sheath.
Ensheathed. Surrounded by a sheath.

Ensiform. Sword-shaped.

Entire. Having a continuous margin; not toothed or scalloped.

Entomophilous. Pollinated by insects; opposite of *anemophilous.*

Enwrapped. Enrolled.

Epaleaceous. Lacking a palea.

Epappose. Lacking a pappus.

Ephemeral. Short-lived; evanescent; of flowers, persisting for only a day.

Epicalyx. An involucre resembling an exterior calyx.

Epicarp. The outer layer of the pericarp or mature ovary.

Epidermal. Pertaining to or of the nature of epidermis.

Epidermis. The superficial layer of cells.

Epigynous. Borne on top of or arising from the summit of the gynoecium, or apparently so. Cf. *hypogynous, perigynous.*

Epipetalous. Borne on a petal or on the petals of a corolla.

Epiphyte. A plant growing upon but not parasitizing another plant. Cf. *endophyte.*

Epispore. The external coat around a spore.

Epithet. The specific half of a binomial. Cf. *basionym.*

Epitropous. Anatropous, but with the ovule oriented differently at different stages of development, the raphe away from the central axis when the free part of the funiculus is ascending and toward the central axis when it is descending. Cf. *anatropous, apotropous, hemianatropous, orthotropous, amphitropous, campylotropous.*

Equilateral. Having roughly equal sides or faces.

Equitant. Astride; overlapping in 2 conduplicate ranks with the bases of the leaves astride the stem and each other, as in *Iris* or *Sisyrinchium.*

Erect. Vertical; not spreading or decumbent.

Ericoid. With small narrow leaves, as in *Erica*; resembling *Erica.*

Erose. With an irregularly uneven margin; gnawed or jagged.

Erostrate. Without a beak.

Erosulate. Minutely erose.

Erumpent. Breaking through; erupting.

Escape. A plant now self-sufficient, and reproducing naturally, itself or its progenitors having been sustained by cultivation.

Escaped. Occurring as an escape.

Estipulate, exstipulate. Without stipules.

Eusporangiate. In ferns, having sporangia derived from superficial cells; opposite of *leptosporangiate.*

Evanescent. Soon disappearing; lasting only a short time; ephemeral.

Even-pinnate. Feather-shaped and with the same number of leaflets along each side of the rachis, but with no terminal leaflet at the rachis apex; opposite of *odd-pinnate.*

Exappendiculate. Having no appendages.

Excavate(d). Hollow or with a depression or hole, as if dug out.

Exceeding. Extending beyond or above some other part, as a leaf sheath subtending a (shorter) spike; surpassing; overtopping.

Excentric. Off-center; 1-sided; away from the center; abaxial.

Excrescence. A surficial enlargement or outgrowth, of whatever sort.

Excurrent. Extending beyond the margin or tip of a leaf or leaflet.

Exfoliate. To peel off in sheets or shreds; to flake away.

Exindusiate. Lacking an indusium or indusia.

Exine. The outer layer of the wall of a spore or pollen grain.

Exocarp. The outer layer of a pericarp or fruit wall; opposite of *endocarp.*

Exogenous. Growing by annular layers near the surface. Cf. *endogenous.*

Exospore. An asexual spore formed by the cutting off of a portion of the sporophore through the growth of septa.

Expanded. Fully opened out, as a flower.

Exserted. Projecting beyond the perianth, as stamens projecting from a corolla throat; opposite of *included.*

Exstipulate. Estipulate.

Extra-axillary. Situated other than in an axil, as a flower emerging laterally from a stem.

Extrorse. Facing outward, away from the axis, as anthers that split down the abaxial side; opposite of *introrse.* Cf. *antrorse, retrorse.*

Exudate. Exuded matter.

Exude. To force out through pores or other incisions, as water or other fluids; to seep out.

Falcate. Sickle-shaped; curved like a sickle.

False. Spurious; specious; having a deceptive appearance.

Falsely (2-, 3-, etc.)celled. Of a fruit, appearing to be several-celled, but having the apparent locules all derived from a single carpel by invagination of its walls, or by extrusion from the placenta.

False vein. A thick-walled cell or column

of cells having the appearance of a vein, but which does not transport water or food materials.

Farinaceous. Containing starch; having a surface with a mealy coating, as the leaves of *Chenopodium murale.*

Farinose. Covered with a meal-like powder.

Fasciate(d). Having an abnormally flattened and widened stem or other part, often considerably divided or branched, the parts often appearing to have coalesced.

Fascicle. A close or dense cluster or bundle, as flowers or leaves.

Fascicled, fasciculate. Bearing or arranged in fascicles.

Fastigiate. Having the branches erect and more or less appressed; broomlike.

Faveolate, favose. Marked with a regular, uniform pattern like a honeycomb; having very shallow pits in such a pattern; of a membrane, having small perforations in such a pattern; alveolate; foveolate.

Fecundating. Fertilizing the female gamete.

Fenestrate. Having perforations or transparent areas; window-like.

Ferrugineous, ferruginous. Rust-colored; reddish.

Fertile. Capable of impregnation, reproduction, or germination, as pollen-bearing stamens and seed-producing fruits; productive.

Fertilization. The process of union of two germ cells.

Fibrillose, fibrillous. Of the nature of a small filament or fiber, as a root hair; furnished with or abounding in fine hairs.

Fibrous. Composed of, or resembling, fibers; having much woody tissue.

Fibrous tissue. A tissue formed of elongated, thick-walled cells; woody tissue.

Fibrovascular. Composed of a bundle of woody fiber cells and associated water- and food-conducting elements.

-fid. In combination, cleft at least halfway to the base or midrib, as 10-fid for cleft into 10 lobes, teeth, etc.

Filament. The stalk of a stamen; any very slender or threadlike structure.

Filamentous. Composed of threads.

Filiform. Threadlike; long and slender.

Filmy. Resembling, or composed of, a thin membranous tissue.

Fimbria (pl. **-iae**). A bordering fringe.

Fimbriate. Having a fringed margin, the hairs longer than those on a *ciliate* margin.

Fimbrillate, fimbriolate. Minutely or very finely fimbriate.

Fissile. Friable or splitting.

Fissured. Interrupted or broken by long, narrow openings or cracks, usually parallel.

Fistular, fistulose. Hollow; cylindrical.

Flabellate, flabelliform. Fan-shaped; narrow at the base and much broader at the apex.

Flaccid. Limp; lacking rigidity.

Flagellate. Provided with whiplike propulsive structures on the male sex cells.

Flagellum (pl. **-a**). A long, mobile cilium.

Flange. A ringlike projection; a horizontally spreading rim.

Fleshy. Succulent; pulpy; not thin, dry, or membranaceous.

Flexuous. Having a zigzag or wavy form; curved alternately in opposite directions.

Floccose. Covered with tufts of woolly hairs.

Flocculent. Minutely floccose.

Floral envelope. A structure investing the ovary, formed either by extension of the margin of the receptacle or by coalescence and adnation of the bases of the sepals, petals, and stamens.

Floret. An individual flower, especially of the grasses and the Compositae; a diminutive flower.

Floriferous. Bearing flowers.

Flower. An axis bearing 1 or more pistils or 1 or more stamens, or both, with or without floral envelopes.

Fluted. Furnished with 1 or (usually) more rounded (parallel) grooves.

Foliaceous. Leaflike; bearing leaves.

Foliar. Of or relating to a leaf.

-foliate. In combination, -leaved, as 12-foliate for having 12 leaves.

Foliolate. With separate leaflets.

-foliolate. In combination, -leafleted, as 12-foliolate for having 12 leaflets.

Foliose. Bearing numerous or crowded leaves; resembling a leaf.

Follicle. A dry, dehiscent fruit opening along 1 side only, as the fruits of the Asclepiadaceae.

Follicular. Like a follicle.

Foraminate. Perforated; having holes in an integument or covering.

Forked. Divided at a common point into 2 or a few nearly equal branches.

Foveolate. Faveolate; favose; alveolate.

Free. Not adnate or adherent to other organs, as the veinlets in some ferns that do not anastomose with other veins or veinlets. Cf. *adnate, coadnate.*

Free-living. Living independently of the parent plant, as the gametophytes of ferns and some other structures that live separately from and independent of the sporophyte.

Friable. Easily crumbled or pulverized.

Fringed. Furnished with marginal short, straight or twisted appendages, hairs, processes, etc., in the manner of an unraveled edge of cloth.

Frond. The leaf of a fern; sometimes the leaf of certain palms or the thallus-like stem of *Azolla.*

Fructification. The act or organ(s) of fruiting; a fruiting body, as a pod, drupe, or achene, or a group of these.

Fruit. A ripened ovary (pistil) with its adnate parts of whatever form; a seed-producing organ.

Frutescent. Shrubby or becoming so.

Fruticose. Shrublike or shrubby; woody at the base.

Fugacious. Withering or falling early; fugitive; short-lived.

Fulvous. Tawny; yellowish.

Functionless scales. Scales not bearing spores or other reproductive structures, as the scales at the base of stamens in the flowers of *Cuscuta.*

Funicle. Funiculus.

Funiculate. Having a funiculus.

Funiculus (pl. **-ae**). The stalk (usually slender) on which an ovule is borne.

Funnelform. Funnel-shaped.

Furcate. Forked; divergently parted; cleft; sometimes, bifurcate.

Furcation. A forking; a branching of the axis.

Furf. Minute scales, usually soft and often flaky and easily detached.

Furfuraceous. Scurfy; bearing furf.

Furrowed. Longitudinally grooved or channeled.

Fuscous. Dusky; grayish brown.

Fusiform. Spindle-shaped; swollen in the middle and narrowing gradually toward each end.

Galea. A helmet-shaped corolla lip, calyx, or sepal.

Galeate. Hollow and vaulted; having a galea, as in the corollas of *Castilleja.*

Gamete. A mature germ cell, capable of initiating formation of a new individual by fusion with another gamete.

Gametophyte. The generation that bears sex organs producing gametes, eventually giving rise to the sporophyte; in ferns, the prothallus.

Gamopetalous. Having the corolla parts fused, at least at the base, to form a tube, a cup, or a ring, the whole corolla falling as a unit; sympetalous; opposite of *polypetalous.*

Gamosepalous. Having a calyx composed of connate sepals; symsepalous; opposite of *polysepalous.*

Geminate. In pairs; twinned; binate.

Gemma (pl. **-ae**). The asexual product of some cryptogams, analogous to a leaf-bud; a young plant asexually produced on the margin of a leaf or in an inflorescence.

Gemmiparous. Producing buds or gemmae. Cf. *viviparous.*

Geniculate. Bent abruptly; having kneelike nodes or joints.

Geniculum (pl. **-a**). A kneelike structure; a bend at a node or joint.

Germination. The process of causing growth to ensue.

Gerontogenous. Restricted to, or originating in, the Old World.

Gibbous. Moderately swollen on 1 side, usually basally. Cf. *ventricose.*

Glabrate. Nearly glabrous, bearing only a few hairs.

Glabrescent. Becoming glabrous with age.

Glabrous. Lacking hairs; smooth-surfaced.

Gland. A secreting prominence, pit, or appendage; often applied to a glandlike body that is nonsecretory.

Gland-dotted. Dotted or sprinkled with glands; glanduliferous.

Gland-tipped. Furnished with a gland or glands at the tip.

Glandular. Having glands.

Glandular-puberulent, -pubescent. Having a mixture of glandular and nonglandular hairs.

Glanduliferous. Bearing glands.

Glaucescent. Slightly glaucous.

Glaucous. Covered with a bloom of whitish or bluish substance that is easily rubbed off, as a plum.

Globose. Almost spherical.

Globular. Having the form of a globe or globule.

Glochid. A minute barbed bristle, often growing in tufts, as in the areoles of *Opuntia.*

Glochidiate. Barbed at the tip.

Glochidium (pl. **-ia**). One of the barbed bristles on the massulae of the microsporangia of *Azolla* and some other cryptogams.

Glomerate. Densely clustered.

Glomerulate. Arranged in small compact clusters or glomerules; minutely glomerate.

Glomerule. A compact inflorescence consisting of a headlike, often sessile cyme.
Glume. A small chafflike bract; a sterile bract or bracts at the base of a grass spikelet.
Glutinous. Sticky; gluelike; covered with a sticky exudation.
Granular, granulate, granulose, granulous. Covered, or appearing covered, with small grains; finely mealy.
Gregarious. Growing in company; associated closely; occurring in large numbers in close groupings.
Grooved. Furnished with a long narrow depression or depressions.
Gymnospermous. Having ovules developed "naked" on a cone scale, not enclosed in a capsule or other fruit cover; of or resembling the Gymnospermae.
Gynandrium. A column bearing both a pistil and stamens, both sexes present.
Gynandrous. Having the characteristics of a gynandrium.
Gynobase. An enlargement or prolongation of the receptacle bearing the ovary; a style surmounting a prolongation of the torus between the tips of the carpels, as in many Boraginaceae.
Gynobasic. Having or of the nature of a gynobase.
Gynoecium. The pistil (or pistils) considered as a unit or system of a flower. Cf. *androecium.*
Gynophore. A stalk bearing a gynoecium; a prolongation of the apical part of the receptacle, bearing the ovary above the level of attachment of the outer floral envelope. Cf. *androgynophore, androphore.*
Gynostegium. The staminal crown in the *Asclepias* flower.

Habit. The general appearance of a plant.
Habitat. The normal situation in which a plant lives.
Hair. A single elongate cell or column of cells arising from the epidermis.
Halophilous. Salt-loving; salt-tolerating, as *Salicornia* and *Batis.*
Halophyte. A salt-tolerating plant; a plant of saline soil.
Halophytic. Growing in saline soil; salt-tolerating.
Hastate. Shaped like an arrowhead, but with the basal lobes standing nearly at right angles to the midline of the axis (or midrib).
Haustorium (pl. **-ia**). The food-absorbing sucker of a parasitic plant.
Head. A dense cluster of sessile or nearly sessile flowers on a very short axis or flat to conical receptacle.
Helicoid. Curved or spiraled like a snail-shell.
Helicoid cyme. A sympodial determinate inflorescence whose lateral branches all develop along the same side of the axis and whose main axis is coiled like a helix at the tip.
Hemianatropous. Half-anatropous, the funiculus attached to the integument of the ovule for about half the length of the ovule and the ovule uncurved but almost inverted. Cf. *anatropous, apotropous, epitropous, orthotropous, amphitropous, campylotropous.*
Hemiparasitic. Parasitic, but containing some chlorophyll and capable of photosynthesis, as mistletoe.
Hemispheric, hemispherical. Having the form of half a sphere.
Herb. A plant with no persistent woody stem above ground.
Herbaceous. Not woody; having relatively nonrigid and soft tissue; dying at the end of the growing season; opposite of *ligneous.*
Herbage. The vegetative parts of a plant.
Hermaphrodite. Bisexual; with the stamens and pistils in the same flower.
Hermaphroditic. Having both male and female parts.
Hermaphroditically unisexual. Having both stamens and pistils present, but only 1 of the 2 functioning in a particular flower or inflorescence.
Hesperidium. A superior, polycarpous, syncarpous berry, pulpy within and covered with a leathery rind, as an orange or a lime.
Heterocarpous. Producing more than 1 kind of fruit, as the dehiscent, aerial and the indehiscent, subterranean capsules of some violets; opposite of *homocarpous.*
Heterochlamydeous. Of a flower having calyx and corolla clearly distinct. Cf. *monochlamydeous.*
Heterochromous. Of more than 1 color.
Heterophyllous. Producing more than 1 kind of leaf.
Heterosporous. Producing 2 kinds of spores, as *Azolla* or *Lycopodium*; opposite of *homosporous.*
Heterostylous. Having styles of 2 or more distinct forms or of different lengths; opposite of *homostylous.*
Heterozygous. Producing genetically different zygotes, as peas with smooth and

wrinkled seeds within the same pod; opposite of *homostylous*.

Hilum. The scar on a seed where the funiculus was attached. Cf. *chalaza*.

Hirsute. Bearing long, rather coarse hairs.

Hirsutulous. Minutely or only slightly hirsute.

Hirtellous. Finely hirsute.

Hispid. Having bristly or stiff hairs.

Hispidulous. Minutely hispid.

Hoary. Covered with a close white or whitish pubescence.

Holdfast. A part by which a plant clings to a flat surface.

Holotype. The particular specimen (of perhaps several described or discussed) cited by an author as the type of a new species in his original description of the species. Cf. *syntype*.

Homocarpous. Producing fruits that are all alike; opposite of *heterocarpous*.

Homosporous. Producing only 1 kind of spore; opposite of *heterosporous*.

Homostylous. Having styles all of the same character or length; opposite of *heterostylous*.

Homozygous. Producing like zygotes; opposite of *heterozygous*.

Hood. A hood-shaped upper petal.

Hooked. Furnished with, or in the form of, a hook.

Horned. Furnished with a horn-like appendage or projection.

Hyaline. Translucent to transparent; tissue-like.

Hybrid. A plant produced by pollination of a flower by pollen from a different species or race; a cross-bred offspring, as hybrid seed corn.

Hybridize. To cause to reproduce hybrids; to interbreed.

Hydathode. A water-pore or water-gland; an organ that exudes water.

Hygroscopic. Expanding when taking up moisture from the atmosphere or other sources, shrinking and often changing shape when dry.

Hypanthium. A cuplike or flattened receptacle formed from the fusion of the floral envelope and the androecium, on which the corolla and stamens, and often the calyx, are borne; present in a hypogynous flower.

Hypocotyl. The axis of an embryo below the cotyledons, developing into the radicle upon germination of the seed.

Hypocrateriform. Salverform.

Hypogeous. Growing or remaining below ground.

Hypogynium. The organ supporting the ovary in certain sedges.

Hypogynous. Borne on the receptacle or under the ovary; having a hypanthium; said of stamens or petals but not the ovary. Cf. *epigynous, perigynous*.

Imbricate. Overlapping, like shingles on a roof.

Immersed. Growing wholly under water.

Imparipinnate. Pinnate with an odd terminal leaflet; odd-pinnate.

Imperfect. Having either functional pistil or stamens, but not both; opposite of *perfect*.

Impressed. Bent inward, hollowed or furrowed as if by pressure.

Impressed-venose. Having the veins embedded into the tissue, thus forming a groove in, instead of a ridge on, the surface of the leaf or petal.

Inaperturate. Lacking an aperture.

Incanous. Hoary; quite gray.

Incised. Cut or slashed irregularly and more or less deeply; of a form intermediate between toothed and lobed.

Included. Remaining within a corolla or other structure; opposite of *exserted*.

Incrassate. Thickened, as in the fleshy leaves of *Batis maritima*.

Incumbent. Of a pair of cotyledons, lying face to face but curved back down the hypocotyl such that the back of one is toward the hypocotyl, the back of the other away from it; also said of anthers turned sharply inward. Cf. *accumbent*.

Incurved. Curved or bent inward.

Indefinite. Not constant in number; not occurring in multiples of the petal number; opposite of *definite*.

Indehiscent. Not opening by a regular process; said of a fruit that remains closed, as an achene or a berry; opposite of *dehiscent*.

Indeterminate. Of an inflorescence, having the central (terminal) flower opening last, with considerable growth possible after the first flowers open; opposite of *determinate*.

Indument, indumentum. A dense or heavy pubescent covering.

Induplicate. With the lateral margins bent abruptly inward toward the axis but not touching, as the leaves in some buds. Cf. *conduplicate*.

Induplicate-valvate. Having the carpel margins bent inward so that the edges of adjacent valves are in contact.

Indurate(d). Hardened with age.

Indusial. Pertaining to or of the nature of an indusium.

Indusiate. Having an indusium or indusia.

Indusium (pl. **-ia**). A thin shield or covering overarching clusters of sporangia in ferns.

Inequilateral. Having unequal sides or faces.

Inferior. Beneath; below; at a lower level.

Inferior fruit. A fruit derived from an inferior ovary and thus bearing the withered corolla and calyx or a scar formed by their dehiscence at the end opposite the pedicel.

Inferior ovary. An ovary situated below the calyx segments and other floral parts; opposite of *superior ovary.*

Inflated. Bladdery.

Inflexed. Turned abruptly or bent inward; incurved.

Inflorescence. The disposition of the flowers on an axis; the arrangement of the flowers as they develop and open; often, but less correctly, all the flowers on a branch or axis, considered collectively.

Inframarginal. Within the margin; near but not at the margin; submarginal.

Inframedial. Between the middle and the margin, but nearer the middle.

Infundibular, infundibuliform. Funnel-shaped.

Ingrowth. A structure growing inward, as the placental wings of some capsules having parietal placentation.

Insectivorous. Deriving nourishment from trapped insects.

Inserted. Attached to or growing out of.

Insertion. A part attached to, or growing out of, another.

Integument. The outer envelope of an ovule, becoming the testa; the seed coat.

Intercrossing. Cross-fertilization, i.e. pollination of a stigma of 1 flower by pollen from a different plant; hybridization.

Intergradation. A gradual change from one state to another, as the suffusion of 2 colors from base to apex of a petal; the intermingling of characteristics of 2 genetically different populations along the area of their contact with one another.

Internerve. Intervenous.

Internode. The section of an axis between two nodes.

Interpetiolar. On or of the portions of a stem or branch that lie between adjacent leaves, as the inflorescences on the stems of *Urera*, as contrasted with those borne at the axils of leaves at the immediate bases of petioles.

Interrupted. Having broken symmetry of design; having gaps in a pattern.

Intervenous, intervein. Belonging to those areas of a leaf or petal surface that lie between veins; lying between veins.

Intramarginal. Within and near, but not at, the margin.

Intrapetiolar. Within the petiole; borne between the base of the petiole and the stem (in an axil).

Intrastaminal. Within the axil of a stamen.

Intrastylar. Within the axil of the style.

Introduced. Brought from another region by the agencies of man, whether intentionally or not; not native.

Introduction. An introduced plant.

Introgression. An infiltration of genetic units (genes) into a population from a genetically different population by way of occasional hybridization, usually with some gene flow in each direction and often resulting in intergradation.

Introrse. Facing inward; toward the axis; as anthers that split down the adaxial side; opposite of *extrorse.* Cf. *antrorse, retrorse.*

Invaginated. Ensheathed; folded in so that an outer becomes an inner surface.

Invested. Covered completely; clothed; wrapped within.

Investing. Covering completely; enwrapping.

Investiture. A part of a plant that covers or encloses another.

Involucel. A secondary involucre; a small involucre around the units of a cluster.

Involucral. Pertaining to or of the nature of an involucre.

Involucre. The whorl or whorls of small leaves or bracts (phyllaries) around or immediately beneath a flower cluster.

Involute. Rolled inward or toward the upper side; applied chiefly to a leaf or leaves.

Irregular. Asymmetrical in design; of a flower, having similar parts of a whorl different in size and shape, or both, so that the flower can be divided into mirror halves along only 1 vertical plane; zygomorphic; opposite of *regular, actinomorphic.*

Isomorphic. Of equal form.

Isotype. A typical specimen of a plant species taken from the type series or type locality, but not the type specimen itself.

Joint. An articulation; a node in the culm of a grass.

Jugate. Paired, but in any number of pairs,

as with 2 leaves of similar shape and size at each node.

-jugate. In combination, paired, as 10-jugate for having 10 pairs of florets, leaflets, etc.

Jugum (pl. **juga**). A pair of leaflets; one of the ridges on the fruits of the Umbelliferae.

Juvenile. At a young stage; a plant at a young stage; said of seedlings or very young, immature plants.

Keel. A central dorsal ridge, like the bottom of a boat; a carina; the 2 united petals of a papilionaceous flower.

Keeled. Having a keel.

Keel petal. One of the 2 lower petals in a papilionaceous flower that usually embrace the pistil and stamens, as in *Phaseolus* or *Galactia.*

Knee. An abrupt bend in a stem or in the trunk of a tree.

Knoblike. Having the form of a lump or other rounded protuberance.

Labellum (pl. **-a**). A lip; a perianth part, particularly the lip of an orchid flower; the lower or upper set of sepals or corolla parts of the flowers of the Scrophulariceae and the Labiatae.

Labiate. Lipped; belonging to the Labiatae.

Labyrinthine. Of an intricate texture or composition.

Lacerate. Irregularly cleft or cut; torn along the margin.

Laciniate. Slashed into narrow, irregular, pointed lobes.

Lactescent, lactiferous. Producing latex.

Lacuna (pl. **-ae**). A hole; a cavity; a gap; an air-space in the midst of tissue.

Laesurae. The network pattern of cells in some scales or on the spores of ferns.

Lamella (pl. **-ae**). A thin flat plate or laterally flattened ridge.

Lamellate. Having thin plates; having thin cross partitions.

Lamina (pl. **-ae**). A blade; an expanded part.

Lanate. Woolly, with long, intertangled hairs.

Lance-. In combination, lanceolate, as lance-ovate for a shape intermediate between lanceolate and ovate.

Lanceolate. Lance-shaped; several times longer than wide, with the narrowest end at the apex.

Lanuginous. Woolly or cottony; clothed with interwoven hairs.

Lanulose. Short-woolly.

Lateral. Belonging to or borne on or near the sides; an organ occupying such a position.

Latero-anterior. Arranged at the sides of some axis, but toward its anterior end.

Latero-posterior. As latero-anterior, but toward the posterior end.

Latex. Milky sap.

Lax. Loose; open; spread apart, as the constituents in some composite flowers.

Leaf. The usually green photosynthetic organ of a plant, usually comprising a flat expanded blade or lamina, a stalk or petiole, 1 or more ribs or nerves, and often 2 stipules, and attached to the stem or branch at a node.

Leaf-gap. A narrow, longitudinally oriented break in the cylinder of the stele just above the departure of a leaf-trace, usually filled with thin-walled, unspecialized cells (parenchyma cells).

Leaflet. One of the segments making up a compound leaf; a secondary leaf.

Leaf scar. The scar remaining on a stem, etc., after a leaf has fallen off.

Leaf sheath. The portion of a leaf in grasses, sedges, and some orchids that enwraps the stem for some distance above the node to which it is attached, the sheath usually free from the stem although in close contact with it, and the leaf blade spreading or not beyond the area of sheathing.

Leaf-trace. A strand of vascular tissue running from the stele into the petiole of a leaf.

Legume. A simple fruit, usually dehiscing along both sutures, derived from the maturation of a unicarpellate ovary; the fruit of the bean family (Leguminosae).

Leguminous. Pertaining to a legume or to the Leguminosae; of the nature of a legume.

Lemma. The chaffy structure, usually the lower 1 of 2, subtending an individual floret in the grass family.

Lenticel. A corky spot on young bark that serves as a path of gas exchange between the atmosphere and internal tissues.

Lenticellate. Having, or of the nature of, a lenticel or lenticels.

Lenticular. Lens-shaped; of the shape of a biconvex lens.

Lepidote. Covered with small scurfy scales.

Leptosporangiate. Having sporangia developed from deep-seated cells several layers beneath the epidermis; opposite of *eusporangiate.*

Liana. Any climbing plant that roots in the ground; an assemblage of such plants, sometimes constituting a microhabitat.

Lianous. Vinelike.

Ligneous. Woody; having cells whose walls are thickened and reinforced by lignin; opposite of *herbaceous.*

Lignescent. Becoming woody in time.

Lignin. A substance related physiologically to cellulose, and with it constituting the essential part of woody tissue.

Ligulate. Strap-shaped; having a ligule; also said of some parts of certain orchid flowers.

Ligule. A strap-shaped corolla limb, as in the ray flowers of the Compositae; a projection from the top of the sheath in grass leaves and in some other plants.

Liguliform, liguloid. Strap-shaped; having the form of a ligule.

Limb. The expanded part of a petal or gamopetalous corolla; a branch of a tree or shrub.

Limen. A more or less horizontal ring of tissue extending inward from the inner side of the hypanthium cup in *Passiflora* flowers.

Linear. Long and narrow, with more or less parallel sides.

Lineate. Bearing more or less closely spaced parallel lines or minute ridges.

Lineolate. Marked with fine lines.

Linguiform, lingulate, linguloid. Shaped like or resembling a tongue.

Lip. One of the parts of an unequally divided corolla or calyx, usually 2 in a flower, an upper lip and a lower lip.

Livid. Ashen; pallid; lurid.

Lithophytic. Growing on rocks; rock-loving.

Litoral, littoral. Growing on shores or pertaining to shores; the shoreline habitat.

Lobate. Divided into or bearing lobes.

Lobe. A segment of a dissected organ, especially if rounded; one of the parts of a petal, leaf, calyx, etc., that is divided in the middle.

Lobing. The pattern of lobes along the margin of a leaf, petal, etc.

Lobulate. Divided into small lobes.

Lobule. A minute lobe.

Locular. Having or pertaining to a locule or locules.

Locule. A compartment or "cell" in an ovary, anther, or fruit; a descriptive term with functional rather than morphological significance.

Loculicidal. Dehiscent (into the cavity of a locule) through the dorsal suture, i.e. about midway between the partitions or lines of union of the carpels, permitting egress of mature seeds.

Lodicule. One of the 2 or 3 scales appressed to the base of the ovary in most grass flowers.

Loment. A legume composed of 1-seeded articles; a leguminous fruit, contracted between the seeds, whose 1-seeded segments usually separate from one another but do not dehisce at maturity.

Lunate. Of the shape of a half-moon or crescent.

Lunulate. Minutely lunate.

Lurid. Dingy; livid.

Lutescent. Yellowish or turning yellow.

Lyrate. Pinnatifid, with the terminal lobe larger than the lower lobes; lyre-shaped.

Macrosporangium. Megasporangium.

Macrospore. Megaspore.

Macular, maculate, maculose. Spotted or blotched.

Maculation. The spots or blotches causing a macular condition.

Maculose. Macular.

Malpighiaceous. Of hairs, especially as in the Malpighiaceae, attached at or near the middle, with the 2 points oriented in different, often opposite, directions.

Mammillate. Bearing teat-shaped processes, as the stems of *Brachycereus* or the fruits of some species of *Annona.*

Marcescent. Withering, but persisting on the plant.

Margin. The outer edge or periphery, as of a leaf blade.

Marginal. Along or attached to the outer edge.

Marginate. Having a distinct margin.

Massula (pl. **-ae**). The frothy or hardened mucilage enclosing groups of spores in some heterosporous ferns, as in *Azolla*; in seed plants, groups of coherent pollen grains, as in the Asclepiadaceae and the Orchidaceae.

Mat. Part of a plant or plants so intertwined or tangled as to form a carpetlike layer.

Median. Belonging to the middle.

Medullary. Of or relating to the pith of a plant.

Medullary ray. One of the plates of parenchyma or other cellular tissue radiating from the pith to the cortex in a stem; one of the structures that forms the "silver grain" in many cabinet woods.

Megaphyllous. Having very large leaves.

Megasporangium. A sporangium containing only megaspores.

Megaspore. The larger kind of spore in

heterosporous ferns or fern allies, which usually gives rise to the female gametophyte when the 2 sexes are on separate gametophytes.

Megasporocarp. The sac or vesicle containing the megasporangium and megaspore in *Azolla*.

Megasporophyll. The sporophyll on which a megasporangium develops; in seed plants, the carpel.

Membranaceous, membranous. Thin, rather soft, and more or less translucent and pliable.

Mentum. An extension of the foot of the column in some orchids; a projection extending forward in the flower.

Mericarp. A 1-seeded, indehiscent part of a fruit that falls from the receptacle with the seed still inside, 2 or more mericarps making up the fruit (schizocarp).

Meristem. A growth region that produces new cells, as the growing tip of a stem or root; a cambium zone.

Meristematic. Pertaining to or of the nature of a meristem.

-merous. In combination, -parted, as 10-merous for having 10 parts.

Mesic. Moderately moist; said of a habitat.

Mesocarp. The middle of a pericarp.

Mesophyll. The parenchyma between the epidermal layers of a foliage leaf.

Mesophyte. A plant having moderate water requirements, i.e. intermediate between the xerophytes and the hydrophytes (plants that grow in the water); a plant of moist land.

Mesophytic. Pertaining to or of the nature of a mesophyte.

Microgametophyte. A cell that divides to produce the smaller (usually male) gametes, as the pollen grain of flowering plants; the prothallium that produces antherozoids in ferns having dioecious prothallia.

Microhabitat. A habitat of very restricted spatial extent; a minute habitat.

Micropylar. Of the micropyle.

Micropyle. A minute aperture at the apex of the integument of an ovule through which the pollen tube often enters to reach the egg shell.

Microsporangium (pl. **-ia**). A sporangium containing microspores; a pollen sac.

Microspore. The smaller of 2 spores in a heterosporous cryptogam.

Microsporocarp. The structure from which the microsporangia of *Azolla* arise.

Midnerve, midrib, midvein. The main nerve or rib of a leaf; a continuation of the petiole into the leaf.

Midpetal. Along the main axis of a petal.

Milk, milky sap. Latex of a milky color and consistency found in the stems, leaves, and other parts of some plants as in the Cichoreae of the Compositae, or in the twigs of *Hippomane*, *Croton*, and *Chamaesyce*.

Modified. Altered from the original form or function or both owing to external conditions; so altered to fill a special requirement, as incipient leaves reduced to protective scales around buds.

Monadelphous. Having stamens united by their filaments into a single group, as in some legumes and most Malvaceae.

Monandrous. Having only 1 stamen.

Moniliform. Regularly constricted so as to appear like a string of beads; applied to some hairs and to fruits with constrictions between seeds.

Monocarpic. Bearing fruit but once and dying.

Monocephalous. Bearing a single head.

Monochasial. Having or of the nature of monochasia.

Monochasium (pl. **-ia**). A cymose inflorescence that produces 1 main axis.

Monochlamydeous. Of a flower, having a single kind of perianth envelope, i.e. a calyx but no corolla. Cf. *heterochlamydeous*.

Monocotyledonous. Having 1 cotyledon.

Monoecious. Having staminate and pistillate flowers on the same plant; opposite of *dioecious*.

Monogeneric. Comprising a single genus, as a family or subfamily.

Monolete. Of a fern spore, having a single, surficial line or minute ridge separating 2 more or less flat areas, the line marking the narrow zone where 4 spore faces end, the flat areas marking former areas of contact with 2 neighboring spores, found where 4 spores are radially disposed about the zone-line in a disclike uniaxial tetrad before separation. Cf. *trilete*.

Monopodial. Having a stem that is a single, continuous axis; also said of a rhizome that grows forward without branching.

Monosporic. Producing a single spore.

Monotypic. Comprising a single species, as a genus or section.

Morphological. Having to do with the physical form and structure of a plant, as opposed to its genetic, geographic, etc., characteristics.

Morphological intergradation. The development of a continuous range of sizes or shapes of an organ of a plant, from one part of a plant to another, and to other plants, as the formation of leaves ranging in shape between the ovate leaves of 1 population and the linear leaves of a genetically different population.

Motile. Capable of moving by its own propulsive force, as the male sex cells (antherozoids) of ferns and other cryptogams.

Mucilage. A usually adhesive, sticky vegetable gelatin belonging to the amylose group of hydrocarbons.

Mucilaginous. Slimy or mucilage-like.

Mucro. A short, sharp or spiny tip.

Mucronate. Terminated by a mucro.

Mucronulate. Minutely mucronate.

Multicipital. Having many heads; said of a much-branched rhizome or of the crown of a single, much-branched root.

Multistriate. Many-lined.

Muricate. Rough, with many minute sharp processes on the surface; finely scabrous.

Muriculate. Very finely muricate.

Muticous. Blunt; pointless; awnless.

Mycorhiza, mycorrhiza. Minute strands of fungus that form symbiotic unions with the roots of most plants.

Naked. Lacking pubescence or enveloping, subtending, or sheltering parts; uncovered.

Nascent. Young; immature; in the process of being formed.

Naturalized. Foreign but established and reproducing as if native.

Navicular. Boat-shaped.

Nectariferous. Producing or having nectar.

Nectar scale. A flap of tissue on or under which nectar is borne.

Nectary. A nectar-secreting gland or area on floral or leaf parts.

Nerve. A vein or veinlet in a leaf; a simple, unbranched vein or slender rib.

Netted. Reticulated.

Nigrescent. Blackish or turning black.

Nocturnal. Opening at night and closing in daylight; functioning at night; opposite of *diurnal.*

Nodal. Situated at a node or nodes; pertaining to or of the nature of a node.

Nodding. Bending or swaying gently downward or forward, with the wind.

Node. A joint, often somewhat swollen, where a leaf or whorl of leaves occurs.

Nodose. Knobby; knotty; having nodes.

Nodular. Pertaining to or of the nature of a nodule.

Nodule. A small node or knob.

Nodulose. Finely nodose.

Non-. In combination, not, as nonglandular for lacking glands.

Nucellus. The central and chief part of an ovule containing the embryo sac.

Nucleate. Of or possessing a nucleus, as contrasted with water-conducting cells containing no nuclei, and as contrasted with binucleate, etc.

Nut. An indehiscent, 1-celled, 1-seeded, hard or bony fruit.

Nutlet. A small nut.

Nyctitropic. Having leaves or leaflets that droop or fold together during darkness; assuming the "sleep" position in darkness, as the leaflets of clover.

Ob-. In combination, inversely or oppositely, as obovate for ovate with the narrow end at the base or petiolar attachment.

Obcompressed. Flattened somewhat at right angles to the primary axis or plane, as some fruits; oblate.

Obconic, obconical. Conical, with the apex of the cone at the point of attachment.

Obcordate. Deeply lobed at the apex; heart-shaped but with the notch at the apex instead of the base.

Obdiplostemonous. Having twice as many stamens as petals, with the outer whorl of stamens opposite the petals.

Oblanceolate. Inversely lanceolate.

Oblate. Flattened at the poles, as a tangerine; having the form of a compressed sphere; obcompressed. Cf. *ellipsoid.*

Oblique. Slanted unequally on opposite sides of the midrib or midline.

Oblique annulus. An annulus encircling a sporangium with its plane oblique to the plane of the sporangium, i.e. with its distal part offset to one side of the pole of the sporangium.

Oblong. Longer than broad and with more or less parallel sides. Cf. *elliptic.*

Obovate. Inversely egg-shaped; ovate with the narrow end at the base or petiolar attachment.

Obscure. Inconspicuous; hidden; poorly developed; indistinct, as a faintly evident vein or nerve; uncertain in affinity.

Obsolescent. Becoming rudimentary or vestigial.

Obsolete. Not evident; rudimentary.

Obturate. Having an opening closed by a plug or special structure.

Obtuse. Blunt or rounded at the end.

Obvolute. Overlapping; convolute; having 1 margin of 1 organ overlapping 1 of the opposite organ, the other margin overlapped by that of the opposite organ, as 2 opposed leaf sheaths at a node jointly clasping a stem.

Ocella (pl. **-ae**). An eyelike spot or marking; a circular patch of color, as at the base of the corolla lobe in some species of *Physalis*.

Ochraceous. Ochre-colored.

Ochroleucous. Yellow-white; buff; cream-colored.

Ocrea (pl. **-eae**). A nodal sheath formed by the fusion of 2 stipules, as in many Polygonaceae.

Ocreate. Having sheathing stipules.

Ocreola (pl. **-ae**). A small ocrea, usually on secondary branches.

Odd-pinnate. Feather-shaped, with the same number of leaflets along each side of the rachis, but with a terminal leaflet at the rachis apex; imparipinnate; opposite of *even-pinnate*.

Odoriferous. Yielding an odor, unpleasant or otherwise.

Oil gland. A secretory structure containing or exuding oil, as glands on the surface of the leaves, stems, and achenes in *Porophyllum*.

Olivaceous. Olive-green.

Operculate. Having a cap or cover, as an operculum.

Operculum. A lid or cap formed by the circumscissile dehiscence of a capsule.

Opposite. Arranged in pairs at regular intervals along an axis, and separated by half the circumference of the axis, as leaves; opposed. Cf. *alternate, decussate*.

Orbicular, orbiculate. Like an orb; sphere-like.

Orthotropous. Having the ovule standing erect on its funiculus so that the axes of both form a straight line, with the ovule uncurved and the micropyle at the end of the ovule opposite the funiculus. Cf. *anatropous, hemianatropous, apotropous, epitropous, amphitropous, campylotropous*.

Osseous. Bony.

Outrolled. Having the lateral margins rolled downward, toward the ventral side.

Oval. Broadly elliptical.

Ovary. The ovule-bearing part of the pistil.

Ovate. Egg-shaped, with the broader end at the base.

Overtopping. Exceeding.

Overwintering. Persisting through the winter.

Ovoid. A solid with an ovate outline; egg-shaped.

Ovulate. Bearing ovules.

Ovule. The structure that becomes a seed after the fertilization and development of the egg cell and its attendant structures.

Ovuliferous. Bearing ovules.

Palate. A ridge or inward fold in the throat at the base of the lower lip in some gamopetalous flowers, often more or less closing the throat.

Palatine. Pertaining to, lying near, or having a palate.

Palea (pl. **-eae**), **palet.** The upper of the 2 enclosing bracts of a grass floret.

Paleaceous. Chaffy; furnished with paleae; chafflike in texture.

Palmate. Lobed, divided, or ribbed in a palm-shaped pattern.

Palmatifid. Cut about halfway to the base in a palmate pattern, with the lobes spreading like fingers from the palm of a hand.

Palmatinerved. Having nerves or veins arranged in a palm-and-finger pattern. Cf. *penninerved*.

Pandurate. Fiddle-shaped.

Panicle. An indeterminate inflorescence with pedicellate flowers or florets.

Paniculate. Resembling a panicle; arranged like a panicle or in panicles.

Paniculiform. Having the form of a panicle.

Papilionaceous. Of some flowers, shaped like a butterfly; having a corolla comprising a standard or banner, 2 wing petals, and 2 keel petals (these usually more or less connate along 1 side), as in many legumes.

Papilla (pl. **-ae**). A small, nipple-like protuberance.

Papillate, papillose. Having papillae on the surface.

Papilliform. Having the form of a papilla or papillae.

Pappiferous. Bearing a pappus or pappi.

Pappus (pl. **-i**). The modified outer perianth series in the florets of the Compositae, consisting variously of scales, bristles, etc., that forms a crown at the apex of the achene.

Papyraceous. Paper-like.

Paraphyses. The sterile filaments among the sporangia in ferns.

Parasite. A plant growing on and deriving nourishment from another plant.

Parasitic. Of the nature of a parasite.

Parenchyma. A soft tissue composed of unspecialized, thin-walled, more or less isodiametric cells, such as those in the pith of stems and the mesophyll of leaves. Cf. *sclerenchyma*.

Parenchymatous. Having the nature or function of parenchyma.

Parietal. Borne on the walls of an ovary that is composed of 2 or more carpels fused along their margins, or on the slight intrusion along these edges.

Paripinnate. Pinnate with an even number of leaflets, i.e. lacking a terminal leaflet; even-pinnate.
Parted. Cut or cleft almost to the base.
-parted, -partite. In combination, having parts, of some number and form.
Patelliform. Shaped like a small, shallow dish with a rimmed margin.
Patent. Spreading widely.
Pebbled. Having a surface of the texture of pebbles.
Pectinate. Pinnatifid, but with the segments narrow and set close like the teeth of a comb.
Pedate. Palmately divided, with the lateral divisions 2-cleft.
Pedicel. The stalk of a flower.
Pedicellate, pedicelled. Borne on or having a pedicel.
Peduncle. The stalk of a flower cluster; also, the stalk of a single flower if the flower is a remnant of a cluster.
Pedunculate. Pertaining to or borne upon a peduncle.
Pellucid. Clear; said of an area that is more or less translucent, such as the oil glands in some leaves.
Pellucid-dotted. Bearing a more or less regular pattern of transparent or translucent dots, as a leaf, stem, etc.
Peltate. Target-shaped; shield-shaped; of a leaf, attached to the petiole on its undersurface instead of at the margin, as in *Tropaeolum* and *Hydrocotyle.*
Pendent. Supported from above; suspended.
Pendulous. More or less hanging or declined.
Penicillate. Brushlike; having a terminal tuft of hairs.
Penninerved. Pinnately nerved or veined. Cf. *palmatinerved.*
Pentamerous. Having all the parts in 5's.
Penultimate. The next to last; the ultimate but 1.
Pepo. A 1-celled, many-seeded, inferior fruit with parietal placentation and a more or less pulpy center, as a gourd, pumpkin, or melon.
Perennial. Lasting year after year.
Perfect. Having both functional pistil and stamens; opposite of *imperfect.*
Perfoliate. Having a sessile leaf or bract whose base completely surrounds the stem.
Perforate. Perforated by 1 or more holes, as some leaves.
Perianth. The 2 floral envelopes—corolla and calyx—considered together.
Pericarp. The wall of a ripened ovary or fruit.
Perigynium. Collectively, the setae borne around the base of the ovary in many sedges; any hypogynous disk; any structure surrounding the ovary of a flower.
Perigynous. Having the perianth parts and stamens borne on or arising from the periphery of the ovary but not beneath it; with the calyx lobes, corolla, and stamens arising on or near the rim of the hypanthium.
Perine. The outermost layer of sculpturing on pollen grains.
Peripheral. Surrounding; marginal; pertaining to the edge, margin, periphery, etc.
Perisperm. The stored nutrient material in a seed that lies exterior to or around the embryo; the part of the endosperm outside the embryo.
Perispore. The membrane or case surrounding an individual spore.
Perisporiate. Pertaining to or having a perispore.
Permanent. Long-persistent; lasting a long time.
Persistent, persisting. Long-remaining, as a calyx upon the fruit, or leaves through winter, even when withered; opposite of *deciduous.* Cf. *caducous.*
Personate. Of a 2-lipped flower, having the throat closed by a palate.
Petal. A unit of the inner floral envelope or corolla of a polypetalous flower, usually white or variously colored, not green.
Petaloid. Resembling a petal in shape or color, or both.
Petiolar. Pertaining to or of the nature of a petiole.
Petiolate. Having a petiole.
Petiole. The stalk of a leaf.
Petiolulate. Having a petiolule.
Petiolule. The stalk of a leaflet.
Phanerogam. A seed-bearing, as opposed to spore-bearing, plant; a spermatophyte.
Phloem. The abaxial part of a vascular strand or bundle through which elaborated foods are transported from photosynthetic areas to other areas of the plant body.
Photosynthesis. The synthesis of complex carbon compounds with the aid of light, carried on in the chlorophyll-containing tissues of plants.
Photosynthetic. Relating to or conducting photosynthesis.
Phyllary. A bract, usually of an involucre, as in the Compositae.

Phyllode. A flat expanded petiole that replaces or complements the blade of a foliage leaf.

Phyllodium (pl. **-ia**). A somewhat dilated petiole having the form of and functioning as a leaf blade.

Phylloideous. Having or of the nature of a phyllode or phyllodes; foliaceous; leaflike.

Pilose. Shaggy, with slender, soft hairs.

Pilosulous. Minutely pilose.

Pinna (pl. **-ae**). One of the primary divisions of a pinnate leaf or frond, as of a fern.

Pinnate. Feather-shaped; compound, with the leaflets placed on opposite sides of a midrib or rachis and lying in 1 plane along each side of the midrib.

Pinnatifid. Cut or cleft partway to the midrib in a pinnate pattern.

Pinnatisect. Cut to the midrib in a pinnate pattern.

Pinnipalmate. Of a decompound leaf, having the divisions of the first order arranged in a palmate pattern and each of these bearing pinnately arranged leaflets.

Pinnule. A secondary pinna or leaflet in a pinnately decompound leaf or frond.

Pistil. A unit of the gynoecium consisting of the ovary, the style (sometimes absent), and the stigma, sometimes comprising 2 or more carpels.

Pistillate. Female; having pistils and no functional stamens; pertaining to or of the nature of a pistil.

Pit. The hard endocarp of a drupaceous fruit, as a peach; a stone.

Pith. The spongy center of an exogenous stem, consisting chiefly of parenchyma.

Pitted. Marked with small depressions or hollows, but in no particular pattern.

Placenta (pl. **-ae**). The part of the ovary or carpel where the ovules are attached. The placenta may be linear, and may occur along the outer walls, along the central axis, or at the base of a locule or locules.

Placental. Pertaining to or of the nature of a placenta.

Placentation. The arrangement of the placenta(e) in the carpel or carpels.

Plaited. Plicate.

Plane. Flat; even; level; any 2-dimensional surface, section, or orientation.

Plano-convex. Flat on 1 surface, outwardly bulging or curving on the other.

Plexus. A network of anastomosing nerves or other parts.

Plicate. Folded into plaits, like a fan.

Plicatulate. Finely plicate.

Plumbeous. Lead-colored.

Plumose. Plumelike; like a fluffy feather; having fine hairs arranged in a pinnate pattern.

Plumule. The embryonic leaf or leaf bud of an embryo, borne at the apex of the hypocotyl.

Pluriseriate. In more than 1 series; multiseriate.

Pneumatophore. An air vessel; an intercellular space in the tissue of the Rhizophoraceae.

Pocket. A relatively deep depression or excavation.

Pod. A dry, (usually) dehiscent fruit; an uncritical term.

Pollen. The minute fecundating grains borne in the anther of a stamen, considered collectively; the microgametophytes.

Pollen sac. The microsporangium of seed plants, borne on or part of the anther; an anther sac; a theca.

Pollen tube. The tube emitted by a pollen grain, passing down from the stigma to the ovary and ovules.

Pollination. The transfer of pollen from the androecium to the gynoecium, whether within 1 flower, from flower to flower on the same plant, or from plant to plant.

Polliniferous. Bearing pollen.

Pollinium (pl. **-ia**). A coherent mass of pollen, as in orchids and milkweeds.

Polsterform. Cushion- or pillow-shaped.

Polyadelphous. Having the stamens aggregated in several to many bundles.

Polyandrous. Having more than 1 stamen.

Polycarpellate, polycarpous, polycarpic. Having more than 1 carpel; having a gynoecium forming 2 or more distinct ovaries.

Polyembryony. The product of more than 1 embryo in an ovule.

Polygamodioecious. Of a plant, functionally dioecious but having some flowers of the opposite sex or some bisexual flowers.

Polygamomonoecious. Having both perfect and unisexual flowers borne on the same plant, with at least some of each type functional.

Polygamous. Bearing perfect and unisexual flowers on the same plant.

Polygonal. Many-sided, whether in 2 dimensions or 3; not rounded.

Polymorphous. Extremely variable in form, perhaps encompassing several subspecific taxa.

Polypetalous. Having a corolla composed

of separate petals; opposite of *gamopetalous, sympetalous.*

Polysepalous. Having a calyx composed of separate sepals; opposite of *gamosepalous, symsepalous.*

Polystichous. Having leaves borne in many series in low spirals and spreading in many directions. *Cf. distichous.*

Porate. Pored; porous.

Pore. A minute opening, especially one by which matter passes through the epidermis.

Porose, porous. Pierced with small holes or pores.

Porrect. Directed outward and forward.

Posterior. On the side toward the axis; away from a subtending bract or leaf; adaxial; opposite of *anterior.*

Pouch. A saclike structure; a hollow spur with a rounded bottom.

Precocious. Appearing or developing early, as flowers produced the first season by a perennial plant, or leaves that expand very early in spring and risk frost damage.

Prehensile. Twining and contracting about, as tendrils; also said of flowers whose stamens and stigmas are so arranged that insect pollinators grasping the style and stamens become covered with pollen.

Prickle. A small, weak, spinelike body on the epidermis or bark.

Primary. Principal; the first organ of a series to develop; the main branch or stem in a multistemmed plant, etc.

Prismatic. Prism-shaped.

Process. Any projection, appendage, or part arising beyond or above a surface; said generally of bodies not obviously functional.

Procumbent. Trailing or flat on the ground but without rooting at the nodes. Cf. *repent.*

Produced. Extended in length, area, or volume; sometimes, disproportionately elongate.

Proliferating, proliferous. Bearing many offshoots; dividing excessively into many twigs or branches.

Proliferations. New parts, as cells, buds, etc., rapidly and extensively produced.

Prolonged. Produced; extended.

Prominulent, prominulose, prominulous. Somewhat prominent; subprominent.

Prophyll, prophyllum (pl. **-a**). The bracteole at the base of an individual flower.

Prop root. A root that serves as a prop or support to the plant, as many of the roots of mangroves.

Prostrate. Lying flat on the ground.

Prothallium (pl. **-ia**), **prothallus.** The gametophyte of a fern, usually flattened and thin, which bears sex organs.

Protostele. A simple, primitive stele devoid of pith and leaf gaps, characteristic of most roots and the earliest portions of stems. Cf. *dictyostele, siphonostele, solenostele.*

Protostelic. Having the properties of a protostele.

Proximal. Close to the point of attachment or the center of a body; opposite of *distal.*

Pruinose. Having a bloom that can be rubbed off the surface.

Psammophilous. Sand-loving; said of plants that grow on sand dunes or sandy beaches.

Pseud-, pseudo-. In combination, false, as pseudoumbel for a false umbel.

Pseudanthial. Of a flower, having more than 1 simple axis but simulating a simple flower; having subsidiary flowers.

Pseudobulb. The thickened or bulblike stem of some orchids that is solid and borne above ground.

Puberulent, puberulous. Minutely pubescent.

Pubescence. A pubescent covering.

Pubescent. Covered with soft, straight, short hairs.

Pulp. A moist, usually coherent mass of soft tissue.

Pulverulent. Dusted over with fine, powder-like granules.

Pulvinate, pulviniform. Cushion-shaped; having a pulvinus.

Pulvinus. A minute gland or swollen base of a petiole that is sensitive to touch, vibration, or heat, and that causes rapid folding or drooping of leaves under stimulation, as in *Mimosa pudica.*

Punctate. Bearing translucent or colored dots or depressions; having points of translucent material.

Puncticulate, puncticulous. Minutely punctate.

Punctiform. Reduced to a mere point; in the form of a dot or point.

Pungent. Terminating in a sharp, stiff point or tip; acrid to the taste or smell.

Pustular, pustulate. Having or of the nature of a pustule or pustules; more or less covered with pustules.

Pustule. A small, often distinctively colored and sometimes translucent surface elevation or spot resembling a blister.

Pyramidal. Having the form of pyramid.

Pyrene. A small hard fruit; a nutlet; the stone of a small drupe or drupelet.

Pyriform. Pear-shaped.
Pyxis, pyxidium (pl. **-ides**). A small capsule dehiscing circumscissilely, the upper portion acting as a lid or operculum, as in *Plantago, Portulaca.*

Quadrangular. Four-angled; 4-faced.
Quadrate. Nearly square in form.
Quadridentate. Four-toothed.
Quadripinnate. Four times pinnate.
Quintipinnate. Five times pinnate.

Raceme. A simple, elongated, indeterminate inflorescence of pedicellate flowers.
Racemiform. In the form of a raceme.
Racemose. Having flowers in raceme-like inflorescences.
Rachilla, rhachilla (pl. **-ae**). A diminutive secondary axis or rachis; in grasses and sedges, the axis that bears the florets.
Rachis (pl. **rachises**). An axis bearing flowers or leaflets laterally; in compound leaves, as in ferns, the continuation of the leaf petiole or stipe, to which the pinnae are attached.
Radiate. Spreading from the center; having rays, as in the flowers of the Compositae.
Radical. Arising from the root crown; of leaves, arranged at the base of the stem or in rosettes.
Radicant. Rooting; usually applied to stems or leaves.
Radicle. The embryonic root of a germinating seed.
Ramification. Branching; spreading by division.
Ramose. Branching; having many branches.
Ramulose. Having many branchlets.
Ranked. Arranged in distinct (especially vertical) rows or sets.
Raphe. That part of the funiculus adnate to or covered by the integument of the ovule.
Ray, ray flower. The outer, modified floret of many Compositae, with a straplike extension of the corolla tube; a branch of an umbel; one of the individual spreading arms of a stellate trichome (hair).
Receptacle. The enlarged or elongated end of the stem or flower axis on which some or all of the flower parts are borne; the torus; the flattened or conical expansion at the end of a peduncle on which the florets of the inflorescence in the Compositae are borne; one of certain other specialized bodies, such as the sporocarp of ferns.
Receptacular. Pertaining to or of the nature of a receptacle.
Receptive. At the stage of growth when the stigmatic surface is ready for pollination, often by the secretion of viscid material, or by the expansion of previously closed or inrolled branches, or both.
Reclinate. Turned or bent downward.
Recurved. Curved or bent downward or backward.
Reduced. Altered by reduction.
Reduction. A loss of some parts as compared with those of an ancestral type; diminution in size, vigor, flowering potential, etc.; the halving of the number of chromosomes occurring in somatic cells during meiosis (reduction division) in preparation for establishing gametes.
Reflexed. Abruptly bent or curved downward or backward.
Regular. Of a flower, having all similar parts in a whorl of the same size and shape, and arranged so that the flower can be divided into mirror halves along more than 1 vertical plane; actinomorphic; opposite of *irregular, zygomorphic.*
Relict. A localized plant evidently left over from past geological epochs.
Remote. Widely spaced.
Reniform. Kidney-shaped; bean-shaped.
Repand. Weakly sinuate; with a slightly uneven margin.
Repent. Creeping; prostrate and rooting at the nodes. Cf. *procumbent.*
Replicate. Folded backwards.
Resin. One of a group of oxidized hydrocarbons, constituting the solid portions of turpentine, insoluble in water and produced in many plants, as the pitch in pine trees or the sticky material on many of the Compositae.
Resin duct. A tubular or cylindrical cavity surrounded by epithelial (secretory) cells that exude resin into the cavity, as the latex in the stems of *Sonchus* and *Hippomane.*
Resiniferous. Producing or pertaining to resin.
Resupinate. Twisted 180°, as the flower of many orchids; upside down.
Reticulate(d). Netted; in the form of a network; a meshwork, the meshes of 1 of a number of shapes, as square, rectangular, hexagonal, etc.
Reticulation. A reticulated formation or network; something reticulated.
Reticulum. A network; a net of fibers around the stem or other part of a plant; strands of fibers forming a net, as the vascular strands in a pad of *Opuntia.*
Retinaculum (pl. **-a**). The gland to which 1 or more pollinia are attached in orchid

flowers; in flowers of the Asclepiadaceae, a horny body (the corpusculum) to which pollinia are attached; the funiculus in the capsules of many of the Acanthaceae.

Retrorse. Bent or turned backward or downward, as trichomes or hairs; opposite of *antrorse*. Cf. *extrorse, introrse.*

Retuse. Notched shallowly at an obtuse apex.

Revolute. With the margins rolled under toward the lower (abaxial) side, as a leaf.

Rhachilla. Rachilla.

Rhizoid. A rootlike structure in mosses.

Rhizomatose, Rhizomatous. Producing or of the nature of rhizomes.

Rhizome. An underground stem that gives rise to aerial shoots, distinguished from a root by the presence of nodes, buds, or scalelike leaves.

Rhombate, rhombic. Rhombus-shaped and essentially 2-dimensional; having the outline of an equilateral, oblique-angled figure.

Rhomboid, rhomboidal. Solid but with a rhombate outline.

Rib. A prominent vein or nerve; the primary vein; any riblike process.

Ribbed. With a prominent rib or ribs or riblike processes.

Ringent. Gaping; wide open, as the mouth of the corolla of a mint flower.

Root. The descending axis of a plant, or any of its main branches, mostly developing underground.

Rootcap. A protective cap of parenchyma cells that covers the terminal meristem in most root tips.

Root crown. The uppermost portion of a root that gives rise to the aerial stem of a plant.

Root hair. A unicellular absorptive hair on a root.

Rootlet. A minute root; a root branch.

Rootstock. The main root near its junction with the stem or stems; a rhizome.

Rosette. An arrangement of leaves radiating from a crown or center, usually at ground surface.

Rostellate. Somewhat beaked; bearing a rostellum.

Rostellum. A small beak; a narrow extension of the upper edge of the stigma in some orchid flowers.

Rostrate. Beaked; bearing a rostrum.

Rostrum. Any beak or beaklike process; one of the inner segments of the coronal lobes in the flowers of some Asclepiadaceae.

Rosulate. In the form of a rosette or in rosettes.

Rotate. Wheel-shaped; having a gamopetalous corolla with a flat, circular limb.

Rotund. Rounded in outline; plump.

Rudiment. An undeveloped organ, as a bud; sometimes, an imperfectly developed and functionally useless organ. Cf. *vestige.*

Rudimentary. Of the character of a rudiment.

Rufopilosulous. Shaggy with minute, soft, reddish hairs.

Rufous. Reddish-brown, in any of various shades.

Ruga (pl. **-ae**). A fold or wrinkle in tissue.

Rugose. Wrinkled; having veins that are more or less impressed, producing an uneven surface.

Rugulose. Minutely rugose.

Ruminate. Appearing as if chewed; crumpled, as the surface of a nutmeg.

Runcinate. Sharply incised, with the teeth directed backward.

Runner. A trailing, slender shoot that often roots at the nodes.

Rupture. A break in a structure, as in an indefinitely dehiscent capsule; a breaking out from within.

Rupturing. Bursting irregularly.

Rusty. Rust-colored; ferrugineous.

Sac. A pouchlike part, often containing a fluid.

Saccate. Bag- or sac-shaped; pouchlike.

Sagittate. Shaped like an arrowhead, with prominent basal lobes pointing or curving downward beyond the juncture of blade and petiole.

Salverform. In the shape of a salver; said of a gamopetalous corolla with a slender tube and a rotate limb.

Samara. An indehiscent, winged fruit, as in *Dodonaea.*

Samaroid. Resembling a samara.

Sanguineous. Crimson; blood-colored.

Saprophyte. A plant obtaining food by absorbing dissolved organic material, especially osmotically from the products of organic breakdown or decay (typically humus), but not parasitic, as some plants of the forest floor. Cf. *autophyte, endophyte.*

Saprophytic. Pertaining to or of the nature of a saprophyte.

Sarmentose. Producing long, slender runners.

Saxicolous. Growing on rocks.

Scaberulous, scabridulous. Slightly rough-

ened, as with small, rigid trichomes; minutely scabrous.
Scabrous. Rough, harsh; with low, sharp processes on the surface.
Scale. A small, dry, often scarious, usually appressed leaf, usually vestigial and often adapted to a specialized function. Cf. *bract.*
Scalloped. With the margin cut into a series of semicircular lobes.
Scandent. Climbing or scrambling over rocks and other plants without the use of tendrils; lying on other plants or objects.
Scape. A leafless stalk arising from the ground; a peduncle growing directly from underground roots, bulbs, or bulblike structures.
Scapiform. Resembling a scape.
Scapose. Bearing an inflorescence on a scape.
Scarious. Thin, dry, and membranous; parchment-like, as a leaf or scale.
Scattered. Irregularly and somewhat distantly placed, as leaves along a stem.
Schizocarp. A dry, indehiscent, compound fruit splitting at maturity into 2 or more 1- to several-seeded parts or carpels (mericarps).
Schizocarpous. Producing, resembling, or related to a schizocarp.
Sclerenchyma. Thick-walled cells devoid of protoplasm after the walls are laid down, often associated with the vascular bundles, and increasing the mechanical strength of the organ. Cf. *parenchyma.*
Sclerenchymatous. Composed of or of the nature of sclerenchyma.
Sclerotic. Hardened; stiff and unyielding; containing stone-like inclusions of material in the walls, as the gritty aggregations in the flesh of a pear.
Scorpioid. Having a circinately coiled, determinate inflorescence whose flowers are 2-ranked and borne alternately on opposite sides of the rachis.
Scrambling. Climbing over, as vines or runners over rocks; scandent.
Scrobiculate. Marked by minute or shallow pits or depressions in the surface.
Sculptured. Having the surface formed into 3-dimensional protrusions, ridges, etc., producing the effect of sculpturing.
Scurfy. With fine, scale-like or branlike particles.
Scutate. Shaped like a small shield.
Scutelliform. Platter-shaped.
Secondary. A second-rank or subordinate organ developing after the primary.
Secretory. Promoting secretion of oil or some other fluid, as a gland.
Secund. Directed to 1 side only, usually by torsion; having an inflorescence whose flowers appear to be borne on only 1 side of the axis because the pedicels bend or curve to so orient them; having a similar arrangement of leaves.
Seed. A ripened ovule; the essential parts of the embryo, together with nutritive tissue (the endosperm) if present, contained within the integument of the ovule.
Seedling. A young plant, growing from seed.
Segment. One of the parts of a leaf, petal, calyx, perianth, etc., that is cleft or divided.
Segmentation. The process or pattern of segmenting in a body, as leaf blades into lobes.
Segregate. A species separated from a former, more inclusive species; a member of a segregating generation.
Segregating generation. One of the genetically and morphologically (or physiologically) different strains or races that appear in a large population of hybrids; one of the plants resulting from differently segregating gene combinations.
Semi-. In combination, partly; to some extent; half; incompletely; etc.
Sepal. One of the separate parts of a calyx, usually green.
Sepaline. Pertaining to or belonging to sepals.
Sepaloid. Resembling a sepal but not morphologically of that envelope.
Sepalous. Sepaline.
Septate. Divided by partitions.
Septicidal, septifragal. Dehiscent along partitions (septa).
Septum (pl. **-a**). A partition; a cross wall; a dividing membrane.
Seriate. In series, usually in whorls, cycles, or rows.
Sericeous. Silky, with very fine hairs.
Serotinous. Produced late in the season.
Serrate. Having a saw-toothed margin, with the teeth pointing toward the apex.
Serrulate. Minutely serrate.
Sessile. Without a stalk; sitting directly on the stem, receptacle, etc.
Seta (pl. **-ae**). A bristle; a slender, needle-like process.
Setaceous, setiferous. Bearing setae; covered with bristles.
Setiform. Bristle-shaped.
Setose. Covered with bristles.
Setulose. Minutely setose.

Sexual. Having the structural and functional properties subserving reproduction by the joining of male and female gametes.

Sheath. Any more or less tubular structure surrounding a part, as the basal part of a grass leaf around the culm of the plant.

Sheathing. Surrounding in a sheath-like manner.

Shield. A protective covering, often attached centrally or marginally by a short stalk, as the indusium of many fern sori, or the sporangiophore in *Equisetum.*

Shoot. A stem or branch with its leaves and appendages, but not yet mature.

Shrub. A woody perennial plant with several to many axes arising from rootstock, but no main trunk.

Sieve tube. A long, articulated tube in the phloem composed of relatively large cells containing protoplasm but no nucleus, and with sieve-like areas of pores in the end walls and certain areas on the side walls, such tubes functioning collectively to transport elaborated foods from one part of the plant body to another.

Sigmoid. Shaped like the letter S.

Siliceous, silicious. Composed of or abounding in silica.

Silicle. A broad, short silique.

Silique. The long, slender fruit of the Cruciferae.

Silky. Covered with close-pressed soft and straight pubescence.

Simple. Not compound, as the leaves of most grasses or as an unbranched inflorescence.

Simple-pinnate. Once pinnate, i.e., with a single, simple rachis or midrib, as contrasted with bi-, tripinnate, etc.

Sinistrorse. Twining spirally upward around an axis, from right to left; opposite of *dextrorse.*

Sinuate, sinuous. With the outline of the margin strongly wavy.

Sinus. The space between 2 lobes or divisions of a leaf or petal; a notch.

Siphonostele. A hollow cylindrical stele, with or without pith. Cf. *dictyostele, protostele, solenostele.*

Siphonostelic. Having the properties of a siphonostele.

Solenostele. A vascular tube with widely separated leaf gaps. Cf. *dictyostele, protostele, siphonostele.*

Solenostelic. Having the properties of a solenostele.

Somatic. Of or relating to the vegetative body cells of a plant, as contrasted with the sexually reproductive cells.

Soral. Pertaining to or resembling a sorus.

Sordid. Dirty in tint; applied chiefly to the pappus when of an impure white.

Soriferous. Bearing a sorus or sori.

Sorus (pl. **sori**). A cluster of sporangia in a fern.

Spadix. A spike with closely crowded, usually small flowers, surrounded or subtended by a spathe, as in the Araceae.

Spathaceous. Pertaining to, resembling, or having a spathe.

Spathe. A bract or leaf surrounding or closely subtending a spadix.

Spathiform. Shaped like a spathe.

Spatulate. Spatula- or spoon-shaped; having an oblong, plane structure.

Spermatophyte. Phanerogam.

Spherical. Having the form of a sphere; globose.

Spheroid. An essentially spherical body, sometimes somewhat modified in form, as oblate-spheroid for a sphere flattened at the poles.

Spheroidal. Having the form of a spheroid.

Spicate. Spikelike; arranged in or having spikes.

Spiciform. Shaped like a spike.

Spicular, spiculate, spiculose. Having the surface covered with fine points; having the surface covered with minute needle-like spines or glochids, as the areoles of *Opuntia.*

Spicule. A minute, slender, pointed, usually hard body.

Spike. An unbranched, simple, elongate, indeterminate inflorescence bearing sessile or subsessile flowers.

Spikelet. A secondary spike; a part of a compound inflorescence that is itself a spike; the ultimate cluster of flowers or florets in a grass inflorescence.

Spination. The process or pattern of spine growth on a stem, etc.; a covering of spines.

Spine. A sharp-pointed woody or hardened body, usually a modified branch, petiole, or stipule, or some other part.

Spinescent. Terminated by a sharp tip or spine.

Spinose, spinous. Bearing a spine or spines.

Spinule. A small, slender spine.

Spinulose. Minutely spinose; bearing a spinule or spinules.

Spiral. A coil of nearly uniform diameter as though wound obliquely about an axis, as the tendrils on a grapevine; of leaves on a stem, arranged so that a line drawn from the base of each leaf to that of the one next above would form a coiled path like a spiral staircase.

Spiricle. One of the delicate coiled threads in the surface cells of some seeds and achenes that uncoil and often become mucilaginous when wet.

Spongiose. Ramified with irregular channels and cavities; like a sponge in structure and texture.

Sporangiophore. The stalk of a sporangium.

Sporangium (pl. **-ia**). The spore case; the sac of the minute body containing spores.

Spore. In ferns and other cryptogams, a simple, unicellular reproductive body, or a body containing a mass of nucleated protoplasm but not an embryo.

Sporocarp. A receptacle containing sporangia.

Sporogenous. Producing spores.

Sporophore. The part of a sporophyte that develops spores.

Sporophyll. A spore-bearing leaf.

Sporophyte. In ferns and allied plants, the foliaceous vegetative plant succeeding the gametophyte generation.

Sporulating. Producing spores or spore-containing sporangia, as a fertile fern frond at the time spores are shed.

Spreading. Diverging nearly at right angles; taking an outward direction.

Sprout. A shoot, especially as arising aerially from a root.

Spur. A tubular or saclike projection, often containing a nectary or nectar-secreting area; a contracted lateral-bearing shoot with several to many foliage leaves in a tuft.

Spur branch. A short, slow-growing, lateral branch with greatly shortened internodes, on which leaves or flowers, or both, are borne, usually in crowded clusters.

Spurred. Furnished with, or produced into, a spur or spurs.

Squama (pl. **-ae**). Scale.

Squamate. Scaly.

Squamella (pl. **-ae**). A small squama, or secondary scale; a small bract; a palea.

Squamellate. Having squamellae.

Squamose. Covered with scales.

Squamulose. Minutely squamose.

Squarrose. Having the parts or processes (such as the tips of phyllaries) spreading or recurved at the end.

Stalk. A part of a plant (as a petiole, stipe, or peduncle) that supports another.

Stamen. A unit of the androecium, usually consisting of a filament and an anther, the anther sometimes sessile; the pollen-bearing organ in a flower.

Stamen trace. The strand of vascular cells leading into a stamen, supplying the latter with nutrients and water.

Staminal. Pertaining to or of the nature of a stamen.

Staminal column. A tube formed by fusion of the filaments of the stamens within a flower, as in the flowers of most Malvaceae.

Staminate. Having a male flower or part of a flower; having a stamen or stamens.

Staminode. A structure occupying the position of, and often shaped like, a stamen, but without a fertile anther; a sterile stamen or stamen-like structure.

Staminodium (pl. **-ia**). Staminode.

Standard. The upper, broad, often erect to recurved petal in many leguminous flowers; the erect, usually narrow part of an *Iris* flower; banner; vexillum.

Stele. An axial cylinder of vascular tissues in a stem, branch, or root; in roots, often a solid rod rather than a cylinder. Cf. *dictyostele, protostele, siphonostele, solenostele.*

Stellate. Star-shaped; having hairs that branch in a star-shaped pattern.

Stellulate. Minutely stellate.

Stem. A primary plant axis that supports buds and shoots, as opposed to roots.

Sterile. Barren; not producing spores, pollen grains, or ovules.

Stigma (pl. **stigmas** or **stigmata**). The top or receptive part of the pistil where pollen grains germinate to bring about the fertilization of egg cells in the ovule(s).

Stigmatic. Of or relating to the stigma; having the properties of a stigma.

Stigmatiferous. Bearing a stigma or stigmata.

Stipe. In some seed plants, a stalklike prolongation of the receptacle beneath the ovary; in ferns, the petiole or "leaf-stalk" of a frond.

Stipel. An appendage of a leaflet in a compound leaf analogous to a stipule.

Stipellate. Furnished with or of the nature of a stipel.

Stipitate. Having or of the nature of a stipe.

Stipular, stipulate. Of, resembling, or provided with stipules.

Stipule. An appendage at the base of a petiole or leaf or on each side of its insertion.

Stolon. A runner that roots at the nodes and gives rise to new plants at some nodes or at its tip, or both.
Stoloniferous. Producing stolons.
Stoma (pl. **-ata**). A pore in the leaf epidermis between 2 reniform flanking cells (the guard cells) through which gases are exchanged with the air.
Stomium. A mouth; an opening on the side of a fern sporangium, between liplike cells, through which the spores are shed.
Stone. The hard endocarp of a drupaceous fruit, as a peach; a pit.
Straggly. Having the parts, as leaves or branches, spread out irregularly, seemingly arbitrarily and inefficiently.
Stramineous. Strawlike; straw-colored.
Stria (pl. **-ae**). A linear marking; a minute ridge; one of a broad pattern of lines.
Striate. Having fine, linear lines; marked with striae.
Strict. Upright and straight, slightly if at all branched; having branches closely appressed to the main axis.
Striga (pl. **-ae**). A small, straight, hair-like scale.
Strigillose, strigulose. Minutely or finely strigose.
Strigose. Bearing appressed, sharp, straight hairs, often set on swollen bases.
Strobilus (pl. **-ili**). A cone or conelike structure containing the seeds or spores of 1 or both sexes, as in *Lycopodium* and *Equisetum*; sporophylls set close together, either spirally or in whorls, and considered collectively.
Strophiole. An appendage at the hilum of some seeds; a caruncle.
Stylar. Pertaining to or of the nature of a style.
Style. The elongate, subterminal part of the pistil, bearing the stigma at its apex, lacking in flowers with sessile stigmas.
Styliferous. Bearing a style or styles.
Stylopodium. A disclike or cushion-like enlargement at the base of the style, as in the Umbelliferae.
Sub-. In combination, less than; almost; approaching; etc.; as subbilobate for slightly or shallowly bilobed.
Subcostal. Near the costa or midnerve; almost costal.
Subdidynamous. Obscurely didynamous, the 4 stamens inserted almost equidistant from one another instead of at distinctly different levels in pairs.
Subepidermal. Under or just within the epidermis.
Submarginal. Near, but not at, the margin.
Submedian. Almost median; near, but not at, the midline or midvein.
Submerged. Growing beneath the surface of the water; underwater.
Subshrub. A very small shrub; a fruticose perennial with a woody base and mostly herbaceous upper parts.
Subsidiary wing. The smaller or lower of 2 or more wings.
Subsimple. Almost without branches.
Substrate. The underlying soil, sand, rock, or other material on which or in which a plant grows.
Subtending. Standing below, as a leaf whose axil contains a bud or twig.
Subterminal. Just below the apex; not quite terminal.
Subterranean. Underground; below the surface of the substrate.
Subula. A fine, sharp point; an awl-shaped tip of a leaf, bract, or other structure.
Subulate. Awl-shaped.
Succulent. Juicy; fleshy; having tissues organized to conserve moisture.
Sucker. A vegetative, nonsexual shoot often of subterranean origin.
Suffrutescent. Low and slightly or obscurely woody or shrubby, with a woody base and herbaceous terminal parts.
Suffruticose. Very low and woody; diminutively shrubby.
Sulcate. Grooved lengthwise; furrowed.
Sulcus. A furrow or groove.
Superior. Above; at a higher level.
Superior ovary. An ovary situated above the calyx segments and other floral parts; opposite of *inferior ovary*.
Superposed. Growing or situated vertically or dorsally over another part or organ.
Supramedial. Above the middle.
Surculose. Producing suckers or rapidly growing shoots, especially those that originate from adventitious buds near the base of the plant.
Surficial. On the surface.
Suspended ovule. An ovule hanging from the apex of a locule or from its point of attachment.
Suture. A line or mark of dehiscence; a groove marking a natural division or union.
Swollen. Somewhat inflated, or apparently so.
Symbiosis. The relationship by which 2 or more dissimilar organisms live together, with benefit to 1 or more, as that between many leguminous plants and the nitrogen-fixing bacteria in the nodules on their roots.

Sympatry. The condition in which 2 species or 2 populations of a species co-occur in a given locality without intergradation by hybridization.

Sympetalous. Having the petals united, at least at the base, so as to fall as a unit; gamopetalous; opposite of *polypetalous*.

Sympodial. Having a stem made up of a series of superposed branches arranged so as to appear like a simple axis.

Symsepalous. Having sepals marginally connate, at least at the base, to form a disc, cup, or tube; gamosepalous.

Synandrium. The fused anthers of the stamens in a flower.

Syncarp. A multiple or fleshy aggregate fruit, as the mulberry.

Syncarpous. Having the carpels united in a compound ovary; opposite of *apocarpous*.

Synonym. A superseded or unused name, usually a binomial.

Synonymy. The series of discarded names for identical objects.

Synpetalous. Sympetalous.

Synpodial. Sympodial.

Synsepalous. Symsepalous.

Syntype. One of 2 or more specimens cited in the original description of a species in which the author did not designate a particular specimen of those cited as being the type (holotype).

Tactile. Sensitive to touch.

Talus. The accumulation of rock fragments at the base of a cliff or steep incline, usually sloped and often rather unstable.

Taproot. The primary descending root; more specifically, a prolonged primary root seeking water at lower soil levels.

Taxon (pl. **taxa**). A unit of classification, such as a species, or one of the varieties composing it, or a larger unit, such as a family or genus.

Teeth sheath. A sheath formed by lateral connation of the teeth of leaves, as in *Equisetum*, or of the stipules, as in *Borreria laevis*.

Tendril. A slender twining or clasping cauline or foliar outgrowth, with or without adhesive discs, by which a plant climbs over other plants or objects. Cf. *cirrhate*.

Tepal. A segment or unit of the perianth, otherwise a sepal or petal, where sepals and petals are not clearly differentiated, as in *Polygonum*.

Tepaloid. Resembling or simulating a tepal.

Terete. Circular in cross section.

Terminal. At the tip; belonging to the apex; apical.

Ternate. Arranged in 3's.

Terrestrial. Occurring on land, as contrasted with aquatic or subterranean.

Tessellate. In a checkerboard pattern, as the squarish areas of thin tissue surrounded by thicker areas on the sides of schizocarps of several genera of the Malvaceae.

Testa. The outer coat of a seed. Cf. *integument*.

Tetrad. A group of 4, as the 4 pollen grains derived from a pollen mother cell.

Tetradynamous. An androecium having 4 long and 2 shorter stamens, as in most of the Cruciferae.

Tetragonal, tetragonous. Four-angled, as the stems of many of the Labiatae.

Tetrahedral. Four-sided, as a 3-sided pyramid and its base.

Tetramerous. Having all the parts in 4's.

Thalloid, thallose. Of the nature or shape of a thallus.

Thallus (pl. **-i**). A vegetative body without differentiation into stem and leaf, as the prothallium (gametophyte) of a fern, or the body of some marine algae.

Theca (pl. **-ae**). A sac or case; the pollen sac of the anther.

Thorn. Spine.

Throat. The orifice in a gamopetalous flower or calyx, or the expanding area just below.

Thyrse. A panicle-like cluster of flowers with an indeterminate main axis and determinate lateral axes; a compound panicle.

Thyrsiform. Having the form of a thyrse.

Thyrsoid. Resembling a thyrse.

Tomentose. Densely woolly; covered with matted, soft, wool-like hairs.

Tomentulose. Minutely tomentose.

Tomentum. Closely intermeshed woolly hair.

Tooth. One of a series of small marginal lobes, usually acute at the apex and regularly arranged along the margin of a leaf, calyx, etc. Cf. *crenation*.

Toothed. Having the margin dissected into a series of points; serrate.

Torose, torulose. Twisted or knobby; cylindrical but irregularly swollen and constricted at close intervals like a string of beads.

Torsion. A spiral bending or twisting.

Tortuous. Formed into or growing in repeated turns, bends, or twists; winding in no particular pattern.

Torus. The receptacle of a flower; the part of an axis on which floral organs are borne; a fleshy ring at or near the base of the calyx cup in some flowers.
Trace. A strand of tissue, usually the vascular supply to a leaf, flower, lateral branch, or similar structure.
Trachea (pl. **-eae**). A short, broad, water-conducting cell in a vessel, its ends oblique or square and its end walls lacking or perforated by 1 or more large pores.
Tracheid. In the woody part of a vascular bundle, an elongated cell with tapered ends, secondarily thickened and conducting water upward to the stems and leaves; an individual element in many vascular bundles.
Trailing. Prostrate but not rooting at the nodes.
Transitional. Representing some point genetically intermediate between 2 intergraded populations.
Translator. Sometimes, the retinaculum of a flower of the Asclepiadaceae.
Transverse. Lateral; oriented at 90° to the axis.
Trapeziform. Having 4 sides, no 2 of which are parallel.
Tree. A perennial, woody plant with an evident trunk, variously branched.
Tri-. In combination, 3, as triandrous for having 3 stamens.
Triad. A group of 3.
Trichome, trichoma. A stiff hair or bristle.
Trifid. Three-cleft; cleft at least halfway to the base or midrib into 3 lobes, teeth, etc.
Trigonous. Three-angled or 3-sided.
Trilete. Of a fern spore, having 3 surficial lines radiating from a central apex, thus forming a shallow 3-sided pyramid on 1 side of the spore, the 3 lines and 3 sides marking the juncture of the spore with the 3 adjacent spores in a 4-axial tetrad before separation. Cf. *monolete.*
Trimerous. Having all the parts in 3's, as the flowers of the Liliaceae.
Triplinerved. Trinerved.
Tripterous. Three-winged.
Triquetrous. Triangular in cross section; having 3 salient angles.
Triradiate-ridged. Bearing 3 ridges that radiate from a central point, as a trilete fern spore.
Truncate. Ending abruptly; squared off at the tip; having a base or apex that is nearly straight across.
Tube. Any hollow, elongated body.
Tuber. A subterranean, short, often knobby stem or rootstock capable of vegetative reproduction, even if separated from the parent plant.
Tubercle. A rounded, protruding body; an excrescence; a warty outgrowth.
Tuberculate. Bearing small processes or tubercles; of the nature of a tubercle.
Tuberiferous. Producing tubers.
Tuberoid. A false tuber; resembling a tuber.
Tuberous. Of the nature of a tuber.
Tufa. A porous rock formed of volcanic detritus.
Tufaceous. Derived from volcanic tufa; resembling tufa.
Tuft. One of a group of erect or ascending plants, organs, or other structures that are close together but not joined except perhaps at the base, as closely crowded grass plants, or small groups of glochids in the areoles of certain cacti.
Tufted. Furnished with a tuft or tufts.
Tumefaction. A tumor; swelling.
Tumid. Swollen, puffy.
Tunicate(d). Having concentric enwrapping leaves or layers, as an onion; bulbous.
Turbinate. Top-shaped; inversely conical.
Turbinoid. Solid but with a turbinate outline.
Turgid. Swollen, as by pressure from within; tightly filled.
Twining. Twisting together or about another part or object, more or less spirally.
Type. In taxonomy, the specimen on which the original description was based. Cf. *holotype, syntype.*
Typical. In classification, referring to the originally described or the ordinary variation.

Ultimate. Final; apical; terminal.
Umbel. An indeterminate, often flat-topped inflorescence whose pedicels and their peduncles arise from a common point (at the same level).
Umbellate. Arranged in umbels; of or relating to an umbel or umbels.
Umbellet, umbellule. The ultimate umbel in a compound umbel; a small umbel.
Umbelliform. Umbel-like; umbrella-shaped.
Umbilicate. Depressed in the center; navel-like.
Umbo. A low, stout, conical projection; a boss.
Umbonate. Bearing an umbo or a boss at the center of a part or organ.
Un-. In combination, non-; not.

Unarmed. Having no hard and sharp projections (as spines, spurs, etc.).
Uncinate. Hooked at the tip, like a cat's claw.
Understory. The vegetation on the floor of a forest; beneath the larger trees.
Undulate. Wavy along the margin or surface; sinuous; repand.
Unexpanded. Not yet opened out, as a bud.
Unguiculate. Clawed; narrowed at the base into a claw.
Uni-. In combination, 1, as unilocular for having a single locule.
Unicarpellate, unicarpellary, unicarpeled, unicarpous. Consisting of a single carpel.
Unifoliolate. Of a compound leaf, reduced to a single, terminal leaflet, this distinguishable from a simple leaf by the presence of an articulation at the base of the petiolule.
Unijugate. Of a compound leaf, having only 2 (a pair of) leaflets.
Unilateral. One-sided, as a unilateral inflorescence for an inflorescence with all the flowers along 1 side of the rachis.
Uniparous. Having parts produced singly. Cf. *biparous.*
Unisexual. Either staminate or pistillate; having only 1 sex represented in a flower.
Urceolate. Urn-shaped; hollow and cylindrical or ovoid, and contracted at or below the mouth.
Utricle. A small bladder-like body; a bladdery, 1-seeded, usually indehiscent fruit, as in the Urticaceae.
Utricular. Inflated; bladder-like.

Vaginate. Sheathed.
Valvate. Having or pertaining to valves; opening by valves, as a capsule.
Valve. A segment of a dehiscing capsule; a segment of a calyx meeting another segment in bud but not overlapping; a partially detached flap of an anther wall, as in *Persea.*
Vascular. Furnished with vessels in the conducting tissues of a stele, including both water-conducting tracheae and tracheids and phloem elements; used also with respect to the ramification or extension of such strands into the leaves and floral parts.
Vascular bundle. A strand of cells or cell derivatives transporting water and elaborated food materials.
Vein. A strand of vascular tissue in a leaf blade, typically the midrib or a secondary later branch from the midrib; sometimes called a nerve.
Vein angle. The angle formed between the main vein and the base of a lateral branch vein.
Veinlet. A small vein; a branch of a vein.
Veiny. Liberally veined; of a texture as if veined.
Velutinous. Velvety; bearing a velvet-like indument.
Venation. The arrangement of veins in a leaf, leaflet, sepal, or petal; the mode or pattern of vein distribution in such a structure.
Venose. Having numerous or conspicuous veins.
Ventral. On or of the inner or front surface (of an organ); adaxial.
Ventricose. Markedly swollen on 1 side. Cf. *gibbous.*
Ventro-apical. Oriented or arranged on the upper or inner surface (as of a leaf) at or near the apex.
Ventromarginal. Arranged on the upper or inner surface (as of a leaf) near the margin.
Vermiform. Wormlike; shaped like a worm.
Vernation. The arrangement of young leaves in a bud; the order of the unfolding of young leaves.
Verruciform. Wart-shaped; formed like a wart.
Verrucose. Having wartlike or nodular excrescences on the surface.
Versatile. Attached at or near the middle and usually moving freely; said chiefly of some anthers.
Verticil. A whorl of organs or parts.
Verticillaster. A false whorl, composed of opposed cymes, as in the nodal clusters of flowers in many Labiatae.
Verticillate. In whorls; in verticils.
Vesicle. A small bladdery sac filled with liquid or air, usually on the surface.
Vesicular. Comprising or furnished with a vesicle or vesicles.
Vessel. A duct or articulated tube produced by the absorption of the cross walls at the ends of the tracheae (vessel segments), occurring in the water-conducting part of a vascular bundle in most Angiosperms.
Vestige. An organ or part once functional that has become (either in the life of the individual or in the evolution of the species) subfunctional or nonfunctional; a much reduced remnant of a formerly larger or more obvious structure. Cf. *rudiment.*
Vestigial. Pertaining to or of the nature of a vestige.

Vesture, vestiture. Any surface covering, such as hairs, trichomes, or scales.
Vexillar. Of or relating to a vexillum.
Vexillum. The standard or banner of a papilionaceous flower; the large posterior petal in such a flower.
Villose, villous. Provided with long, soft, nonmatted hairs.
Villosulous. Minutely villose.
Virgate. Wandlike; long, slender, and fairly straight.
Viscid. Glue-like; sticky.
Viscidium. The viscid-coated, sticky disc in certain orchid flowers.
Vitta (pl. **-ae**). An oil tube commonly present in the pericarp of many Umbelliferae.
Viviparous. Sprouting or germinating on the parent plant.
Volatile. Easily passing from the liquid to the gaseous state, as the oils produced in mint plants, *Bursera*, and *Pectis*.

Waif. An introduced plant that is unable to compete with well-established native or naturalized plants and disappears in a few seasons.
Warty. More or less covered with glandular excrescences or hard protuberances.
Wavy. Undulate; sinuous; repand.
Wax. A non-oily but fatty substance occurring either as a thin layer of bloom on the surface of leaves or fruits or as a more massive accumulation in some fruits.
Waxy. Resembling beeswax in color and consistency.
Weed. An aggressive plant that intrudes where it is not wanted.
Weedy. Of the character of a weed.
Whorl. Three or more organs or parts at a node or in a circle around some point on an axis.
Whorled. Arranged in whorls; having a whorl or whorls.
Wing. A thin, often dry and membranous expansion found on various organs, such as fruits, branches, and petioles; one of the 2 intermediate petals in many papilionaceous flowers.
Winged. Furnished with a winglike foliaceous, membranous, or woody expansion of tissue, especially along a stem or on a samara or capsule.
Withering. Becoming dry and sapless and often becoming limp or prostrate, but not necessarily falling away.
Woody. Of the composition or texture of wood.
Woolly. Covered with long, soft, more or less tangled hairs; of the character of wool.

Xeric. Dry; arid; with a low water supply; said of habitats.
Xeromorphic. Having the form of a xerophyte.
Xerophyte. A plant adapted to growing in arid regions.
Xerophytic. Growing in arid soil; tolerant of aridity.
Xylem. A complex tissue in the vascular system of higher plants consisting of vessels, tracheids, or both usually together with wood fibers and parenchyma cells, functioning chiefly in conduction but also in support and storage, and typically constituting the woody element (as of a plant stem).
Xylem ray. A ray similar to a medullary ray but originating in the secondary wood of a perennial stem and radiating outward into the cortex.

Zygomorphic. Irregular, capable of bisection along only 1 plane of symmetry, as flowers capable of division into mirror halves along only 1 vertical plane; opposite of *actinomorphic*.
Zygote. The body formed from the union of a male and a female gamete cell before the resulting fusion cell has divided further.

Literature Cited

Literature Cited

Amshoff, G. Jane H. 1939. Papilionaceae, in A. Pulle, *Flora of Suriname*, Vol. 2^2. J. H. de Bussy, Ltd., Amsterdam. 257 pp.

Andersson, N. J. 1855. Om Galápagos Öarnes Vegetation. Kongl. Svensk. Vet.-Akad. Handl. 1853: 61–256.

———. 1861. Über die Vegetation der Galápagos Inseln. Linnaea 31: 571–631.

Ascherson, P., & P. Graebner. 1907. Potamogetonaceae, in A. Engler, Das Pflanzenreich IV. 11: 1–184.

Bacigalupi, R. 1931. Taxonomic studies in *Cuphea*. Contr. Gray Herb. 95: 3–26.

Badillo, V. M. 1967. Inventorio de las Caricaceae hasta hoy tenidas como válidas. Rev. Fac. Agron. Univ. Venezuela 4^2: 48–66.

Banfield, A. F., C. N. Behre, Jr., & D. St. Clair. 1956. Geology of Isabela (Albemarle) Island, Archipelago de Colón (Galápagos). Bull. Geol. Soc. Amer. 67: 215–34.

Barneby, R. C. 1964. Atlas of North American *Astragalus*. Mem. New York Bot. Gard. 13^2: 597–1188.

———. 1965. Conservation and typification of *Dalea*. Taxon 14: 160–64.

Bartholomew, George A. 1966. Interaction of physiology and behavior under natural conditions, in R. I. Bowman, ed., *The Galápagos*, listed below.

Baur, Georg. 1891. The Galápagos Islands. Amer. Antiq. Soc. Proc. n.s. 7: 418–23.

———. 1897. New observations on the origins of the Galápagos Islands. Amer. Naturalist 31: 661–80; 864–96.

Beebe, William. 1924. *Galapagos World's End*. Putnam. New York. xxi + 443 pp.

———. 1926. *The Arcturus Adventure*. Putnam. New York. xix + 439 pp.

Bentham, George. 1875. Revision of the suborder Mimoseae. Trans. Linn. Soc. Lond. 30: 335–664, *pls. LXVI–LXX*.

———. 1882. Notes on Gramineae. J. Linn. Soc. Bot. 19: 14–134.

Blake, S. F. 1954. The Cyperaceae collected in New Guinea by L. J. Brass, IV. Jour. Arnold Arb. 55: 203–38.

Bowman, R. I. 1960. Report on a biological reconnaissance of the Galápagos Islands. UNESCO, Paris. 65 pp. (mimeo.).

———. 1961. Morphological differentiation in the Galápagos finches. Univ. Calif. Publ. Zool. 58: 1–302.

———. 1963. Evolutionary patterns in Darwin's Finches. Occ. Papers Calif. Acad. Sci. 44: 107–40.

———, ed. 1966. *The Galápagos*. Univ. Calif. Press, Berkeley. xvii + 318 pp.

Britton, N. L., & E. P. Killip. 1936. Mimosaceae and Caesalpinaceae of Colombia. Ann. New York Acad. Sci. 35: 101–208.

Britton, N. L., & J. N. Rose. 1928. Mimosaceae: in North American Flora 23: 1–194.

Brizicky, G. K. 1962. Taxonomic and nomenclatural notes on *Zanthoxylum* and *Glycosmis* (Rutaceae). Jour. Arnold Arb. 43: 80–93.

Bullock, A. A. 1936. Notes on the Mexican species of the genus *Bursera*. Kew Bull. 1936: 346–87.

Burkhart, A. 1940. Materiales para una monografía del género *Prosopis* (Leguminoseae). Darwiniana 4: 57–132, *pls. I–XXIII, figs. 1–15.*

———. 1947. Leguminosas nuevas o críticas, II. Darwiniana 7: 504–40.

Carpenter, Charles C. 1966. Comparative behavior of the Galápagos lava lizards (Tropidurus), in R. I. Bowman, ed., *The Galápagos*, listed above.

Caruel, T. 1889. Contribuzione alla flora delle Galápagos. Rendic. Accad. Lincei 5: 619–25.

Christensen, Carl. 1920. A monograph of the genus *Dryopteris*. Part II, The tropical American bipinnate-decompound species. Dansk Vidensk. Selsk. Skr. ser. 8, 6: 1–132, *figs. 1–29.*

———. 1938. Filicinae, in Frans Verdoorn, ed., *Manual of Pteridology*. Martinus Nijhoff, The Hague. xx + 640 pp.

Christophersen, E. 1932. A collection of plants from the Galápagos. Nyt Mag. Naturvidensk. 70: 67–97.

Chubb, Laurence J. 1933. Geology of Galápagos, Cocos, and Easter islands. Bernice P. Bishop Mus. Bull: 110. 67 pp.

Clayton, W. D. 1965. Studies in the Gramineae: VI. Sporoboleae; the *Sporobolus indicus* complex. Kew Bull. 19: 287–93.

Cronquist, A. 1945. Additional notes on the Simaroubaceae. Brittonia 5: 469–70.

Dall, W. H., & W. H. Ochsner. 1928. Land shells of the Galápagos Islands. Proc. Calif. Acad. Sci. IV, 17: 141–85, *pls. 8, 9.*

Dandy, J. E. 1967. Index of generic names of vascular plants, 1753–1774. Regnum Veg. 51: 5–130.

Dandy, J. E., & E. Milne-Redhead. 1963. The typification of *Hedysarum diphyllum*. Kew Bull. 17: 73–74.

Dandy, J. E., & D. P. Young. 1959. *Oxalis megalorrhiza* Jacq. Bot. Soc. Brit. Isles Proc. 3: 174–75.

Dawson, E. Yale. 1966. Cacti in the Galápagos Islands, with special reference to the relationship with tortoises, in R. I. Bowman, ed., *The Galápagos*, listed above.

Delisle, Donald G. 1963. Taxonomy and distribution of the genus *Cenchrus*. Iowa State Coll. J. Sci. 37: 259–351.

Duke, James A. 1961. Preliminary revision of the genus *Drymaria*. Ann. Missouri Bot. Gard. 48: 173–268.

Eames, Arthur J. 1936. *Morphology of Vascular Plants: Lower Groups*. McGraw-Hill. New York. xviii + 433 pp.

Ebinger, John E. 1962. Validity of the grass species *Digitaria adscendens*. Brittonia 14: 248–53.

Eiten, G. 1963. Taxonomy and regional variation in *Oxalis* section Corniculatae. I. Introduction, keys, and synopsis of the species. Amer. Midl. Naturalist 69: 257–309.

Ekman, E. L. 1912. Ueber die Gramineengattungen *Trichoneura* und *Crossotropis*. Ark. Bot. 11[9]: 1–19, *3 pls.*

Eliasson, Uno. 1965a. Studies in Galápagos plants I. Leguminosae. Svensk. Bot. Tidskr. 59[3]: 345–67.

———. 1965b. Studies in Galápagos plants II. Cyperales. Svensk. Bot. Tidskr. 59[4]: 469–91.

———. 1966. Studies in Galápagos plants III. Centrospermae. Svensk. Bot. Tidskr. 60[3]: 393–439.

———. 1967. Studies in Galápagos plants IV. The genus *Chrysanthellum*. Svensk. Bot. Tidskr. 61[1]: 88–92.

———. 1968a. Studies in Galápagos plants V. Some new records from the archipelago. Svensk. Bot. Tidskr. 62[1]: 243–48, *pls. I, II.*

———. 1968b. Studies in Galápagos plants VI. On the identity of *Calandrinia galapagosa* St. John. Svensk Bot. Tidskr. 62[2]: 365–68, *pl. I.*

———. 1968c. Studies in Galápagos plants VII. On some new or otherwise noteworthy plants from the archipelago. Bot. Notiser 121: 630–40, *figs. 1–5.*

Fassett, N. C. 1951. *Callitriche* in the New World. Rhodora 53: 137–55; 161–82; 185–94; 209–22.

Fawcett, William, & A. B. Rendle. 1920. *Flora of Jamaica.* Vol. IV. British Museum, London. xv + 369 pp.

Giardelli, Maria L. 1959. *Lemna aequinoctialis* Welwitsch nueva para la flora de América y de las Filipinas. Darwiniana 11: 584–90.

Gleason, H. A. 1958. Flora of Panama. Melastomaceae. Ann. Missouri Bot. Gard. 45: 203–4.

Goodspeed, T. H. 1954. *The Genus Nicotiana.* Chronica Botanica Co., Waltham, Mass. xxi + 536 pp.

Gould, F. W., & Z. J. Kapadia. 1964. Biosystematic studies in the *Bouteloua curtipendula* complex II. Taxonomy. Brittonia 16: 182–207.

Graham, S. A. 1964a. The genera of Lythraceae in the southeastern United States. J. Arnold Arb. 45: 235–50.

———. 1964b. The genera of Rhizophoraceae and Combretaceae in the southeastern United States. J. Arnold Arb. 45: 285–301.

Gregory, D. P. 1958. Flora of Panama. Rhizophoraceae. Ann. Missouri Bot. Gard. 45: 136–42.

Hall, Carlotta C., & David B. Lellinger. 1967. A revision of the fern genus *Mildella.* Amer. Fern J. 57: 113–34.

Harling, Gunnar. 1962. On some Compositae endemic to the Galápagos Islands. Acta Horti Berg. 20[3]: 63–120, *pls. I–VIII, figs. 1–25.*

Harris, W. F. 1955. A manual of the spores of New Zealand Pteridophyta. New Zeal. Dept. Sci. Indus. Res. Bull. 116: 1–186.

Hassler, E. 1910. Ex Herbario Hassleriano: Novitates paraguariensis IX. Rep. Spec. Nov. Regni Veg. 9: 1–18.

Heiser, C. B., & Paul G. Smith. 1953. The cultivated *Capsicum* peppers. Econ. Bot. 7: 214–27.

Hemsley, W. B. 1900. The vegetation of the Galápagos Islands. Gard. Chron. *ser. 3,* 27: 177–78.

Hendrickson, John D. 1966. The Galápagos tortoises, *Geochelone* Fitzinger 1875 (*Testudo* Linnaeus 1758 in part), in R. I. Bowman, ed., *The Galápagos,* listed above.

Henrard, J. T. 1929. Monograph of the genus *Aristida*. Meded. Rijks-Herb. 59: 1–137.
Hermann, F. J. 1962. A revision of the genus *Glycine* and its immediate allies. Tech. Bull. U.S.D.A. 1268.
Heyerdahl, T., & A. Skjölsvold. 1956. Archeological evidence of pre-Spanish visits to the Galápagos Islands. Amer. Antiquity (Suppl.) 22^2, pt. 3: 1–71, *1 col. pl., figs. 1–48.*
Hitchcock, A. S. 1920. The North American species of *Ichnanthes*. Contr. U.S. Nat. Herb. 22: 1–12, *pls. 1–9.*
———. 1927. The grasses of Ecuador, Peru and Bolivia. Contr. U.S. Nat. Herb. 24: 291–556.
———. 1935. New species of grasses from the Galápagos Islands. Proc. Calif. Acad. Sci. IV, 21: 295–300.
———. 1936. Manual of the grasses of the West Indies. U.S.D.A. Misc. Circ. 243.
———, & Agnes Chase. 1910. The North American species of *Panicum*. Contr. U.S. Nat. Herb. 15: 1–396, *370 figs.*
Hitchock, C. L. 1932. A monographic study of the genus *Lycium* of the Western Hemisphere. Ann. Missouri Bot. Gard. 19: 179–374, *pls. 12–24, figs. 1, 2.*
———. 1945. The South American species of *Lepidium*. Lilloa 11: 75–134, *figs. 1–44.*
Hooker, J. D. 1847a. An enumeration of the plants of the Galápagos Archipelago, with descriptions of those which are new. Trans. Linn. Soc. Lond. 20: 163–233.
———. 1847b. On the vegetation of the Galápagos Archipelago as compared with that of some other tropical islands of the continent of America. Trans. Linn. Soc. Lond. 20: 235–62.
Hooker, W. J. 1858. *Species Filicum*. William Pamplin, London. Vol. II. 250 pp., *pls. LXXI–CXL.*
Hou, D. 1960. A review of the genus *Rhizophora* with special reference to the Pacific species. Blumea 10: 625–34.
Howell, J. T. 1933a. The genus *Mollugo* in the Galápagos Islands. Proc. Calif. Acad. Sci. IV, 21: 13–23.
———. 1933b. The Cactaceae of the Galápagos Islands. Proc. Calif. Acad. Sci. IV, 21: 41–54.
———. 1933c. The Amaranthaceae of the Galápagos Islands. Proc. Calif. Acad. Sci. IV, 21: 87–116.
———. 1934a. Cacti in the Galápagos Islands. Cact. & Succ. J. (Los Angeles) 5: 515–18; 531–32.
———. 1934b. Concerning names of species of *Cereus* in the Galápagos Islands. Cact. & Succ. J. (Los Angeles) 5: 585–86.
———. 1935. New flowering plants from the Galápagos Islands. Proc. Calif. Acad. Sci. IV, 21: 329–36.
———. 1937. The plant genus *Coldenia* in the Galápagos Islands. Proc. Calif. Acad. Sci. IV, 22: 99–110, *pls. 26, 27.*
———. 1941a. The genus *Scalesia*. Proc. Calif. Acad. Sci. IV, 22: 221–71.
———. 1941b. Hugh Cuming's visit to the Galápagos Islands. Lloydia 4: 291–92.
———. 1942. Up under the equator. Sierra Club Bull. 27: 79–82.
———. 1966. Plantae Galapaganae excerptae. Leafl. West. Bot. 10: 351–52.

———, & D. M. Porter. 1968. The plant genus *Polygala* in the Galápagos Islands. Proc. Calif. Acad. Sci. IV, 32: 581–86.
Hutchinson, J. B., R. A. Silow, & S. G. Stephens. 1947. *The Evolution of Gossypium.* Oxford Univ. Press, London. xi + 161 pp., *pls. 1–8, figs. 1–10.*
Johnston, I. M. 1924. A tentative classification of the South American *Coldenias.* Contr. Gray Herb. 70: 55–61.
———. 1928. Studies in the Boraginaceae, VII. The South American species of *Heliotropium.* Contr. Gray Herb. 81: 3–73.
———. 1936. A study of the Nolanaceae. Contr. Gray Herb. 112: 1–87.
———. 1938a. Notes on some *Astragalus* species of Ecuador and Peru. J. Arnold Arb. 19: 88–96.
———. 1938b. The speices of *Sisyrinchium* in Uruguay, Paraguay and Brazil. J. Arnold Arb. 19: 376–401.
Johnston, Marshall C. 1962. The North American mesquites, *Prosopis* Sect. *Algarobia* (Leguminosae). Brittonia 14: 72–90.
Kern, J. H. 1954. Be cautious with typification! Taxon 3: 246.
Kilipp, E. P. 1938. The American species of Passifloraceae. Field Mus. Nat. Hist. Bot. Ser. 19: 1–613.
Koford, C. B. 1966. Economic resources of the Galápagos Islands, in R. I. Bowman, ed., *The Galápagos,* listed above.
Koyama, Tetsuo. 1961. Classification of the family Cyperaceae (1). J. Fac. Sci. Univ. Tokyo, sect. 3, Bot. 8: 109.
Krapovickas, A. 1954. Estudio de las especies de *Anurus,* sección del genero *Urocarpidium* Ulbr. (Malvaceae). Darwiniana 10: 606–36.
Krukoff, B. A. 1939. The American species of *Erythrina.* Brittonia 3: 205–337.
Lack, David. 1945. The Galápagos Finches (Geospizinae). A study in variation. Occ. Papers Calif. Acad. Sci. 21: vii + 1–151, *pls. 1–4.*
Laruelle, Jacques. 1966. Study of soil sequences on Indefatigable Island, in R. I. Bowman, ed., *The Galápagos,* listed above.
Lawrence, George H. M., A. F. Günther Buchheim, Gilbert S. Daniels, & Hellmut Dolzal. 1968. *B.-P.-H: Botanico-Periodicum-Huntianum.* Hunt Botanical Library, Pittsburgh, Pa. 1063 pp.
Lay, K. K. 1950. The American species of *Triumfetta* L. Ann. Missouri Bot. Gard. 37: 315–95.
Linsley, E. G. 1966. Pollinating insects of the Galápagos Islands, in R. I. Bowman, ed., *The Galápagos,* listed above.
McBirney, A. R., & K. Aoki. 1966. Petrology of the Galápagos Islands, in R. I. Bowman, ed., *The Galápagos,* listed above.
Macbride, J. F. 1937. Flora of Peru. Urticaceae. Field Mus. Nat. Hist. Bot. Ser. 13(2): 331–67.
———. 1943. Flora of Peru. Leguminosae. Field Mus. Nat. Hist. Bot. Ser. 13(3, no. 1): 1–507.
———. 1949a. Flora of Peru. Oxalidaceae. Field Mus. Nat. Hist. Bot. Ser. 13(3, no. 2): 544–608.
———. 1949b. Flora of Peru. Burseraceae. Field Mus. Nat. Hist. Bot. Ser. 13(3, no. 2): 703–17.
———. 1960. Flora of Peru. Boraginaceae. Field Mus. Nat. Hist. Bot. Ser. 13(5, no. 2): 539–609.

———. 1962. Flora of Peru. Solanaceae. Field Mus. Nat. Hist. Bot. Ser. 13(5-B, no. 1): 3–267.

McVaugh, R., & J. Rzedowski. 1965. Synopsis of the genus *Bursera* L. in western Mexico with notes on the material of *Bursera* collected by Sessé & Mociño. Kew Bull. 18: 317–82.

Martius, C. F. P. 1859–63. *Flora Brasiliensis*. Fleischer, Leipzig. 15[1]. 350 pp., *pls. 1–127*.

Mohlenbrock, R. H. 1957. A revision of the genus *Stylosanthes*. Ann. Missouri Bot. Gard. 44: 299–355.

Morton, C. V. 1947. The American species of *Hymenophyllum*, Section *Sphaerocionium*. Contr. U.S. Nat. Herb. 29: 139–201.

———. 1957. Ferns of the Galápagos Islands. Leafl. West. Bot. 8: 188–95.

———. 1963. Some West Indian species of *Thelypteris*. Amer. Fern J. 53: 59–70.

———. 1968. The genera, subgenera, and sections of the Hymenophyllaceae. Contr. U.S. Nat. Herb. 38: 153–214.

———, & David Lellinger. 1966. The Polypodiaceae subfamily Asplenioideae in Venezuela. Mem. New York Bot. Gard. 15: 1–49.

Nelson, Bryan. 1968. *Galápagos: islands of birds*. Longmans, Green, London. xx + 338 pp.

O'Donnell, Charles A. 1941. Revisión de las especies Americanas de *Merremia*. Lilloa 6: 469–554.

Ownbey, Gerald B. 1958. Monograph of the genus *Argemone* for North America and the West Indies. Mem. Torrey Bot. Club 21: 1–159.

Palmer, Clarence E., & Robert L. Pyle. 1966. The climatological setting of the Galápagos, in R. I. Bowman, ed., *The Galápagos*, listed above.

Pennell, F. W. 1946. Reconsideration of the *Bacopa-Herpestris* problem of the Scrophulariaceae. Proc. Acad. Nat. Sci. Phila. 98: 83–98.

———. 1951. The genus *Calceolaria* in Ecuador, Colombia and Venezuela. Proc. Acad. Nat. Sci. Phila. 103: 85–196.

Pierce, John H. 1942a. The nomenclature of balsa wood (*Ochroma*). Trop. Woods 69: 1–2.

———. 1942b. An evaluation of the type material of *Ochroma*, the source of balsa wood. Trop. Woods 70: 20–23.

Porter, Duncan M. 1969a. The flora of the Galápagos Islands. Ann. Missouri Bot. Gard. 55: 173–75.

———. 1969b. *Psidium* (Myrtaceae) in the Galápagos Islands. Ann. Missouri Bot. Gard. 55: 368–71.

———. 1969c. The genus *Dodonaea* (Sapindaceae) in the Galápagos Islands. Occ. Papers Calif. Acad. Sci. 81: 1–4.

Prain, D. 1934. *Flora of Tropical Africa*. Vol. IX. Gramineae. L. Reeve & Co., Ashford, Kent.

Radlkofer, L. 1932. Sapindaceae. II. *Pflanzenreich* IV. 165: 321–640.

Reeder, J. R. 1960. The systematic position of the grass genus *Anthephora*. Trans. Amer. Microscop. Soc. 79: 211–18.

Rendle, A. B. 1934. Linnaean species of *Spermacoce*. J. Bot. Brit. & For. 72: 329–33, *pls. 607, 608*.

———. 1936. *Spermacoce remota* Lam. J. Bot. Brit. & For. 74: 10–12.

Rhodes, O. G. 1962. Flora of Panama. Menispermaceae. Ann. Missouri Bot. Gard. 49: 157–72.

Rick, C. M. 1956. Genetic and systematic studies on accessions of *Lycopersicon* from the Galápagos Islands. Amer. J. Bot. 43: 687–96.

———. 1966. Some plant-animal relations on the Galápagos Islands, in R. I. Bowman, ed., *The Galápagos*, listed above.

———, & R. I. Bowman. 1961. Galápagos tomatoes and tortoises. Evolution 15: 407–17.

Riley, L. A. M. 1925. Critical notes on Galápagos plants. Kew Bull, 1925: 216–31.

Robinson, B. L. 1902. Flora of the Galápagos Islands. Proc. Amer. Acad. 38: 78–270.

———, & J. M. Greenman. 1895. On the flora of the Galápagos Islands as shown by the collections of Dr. Baur. Amer. J. Sci. 50: 135–49.

Robyns, André. 1964. Flora of Panama. Sterculiaceae. Ann. Missouri Bot. Gard. 51: 69–107.

Rominger, James M. 1962. Taxonomy of *Setaria* (Gramineae) in North America. Illinois Biol. Monogr. 29: 1–132.

Rowlee, W. W. 1919. Synopsis of the genus *Ochroma*, with descriptions of new species. J. Wash. Acad. Sci. 9: 157–67.

Rudd, Velva E. 1966. *Acacia cochliacantha* or *Acacia cymbispina* in Mexico? Leafl. West. Bot. 10: 257–62.

Salvoza, F. M. 1936. *Rhizophora.* Nat. & Appl. Sci. Bull. Univ. Philipp. 5: 179–237.

Sauer, Jonathan D. 1964. Revision of *Canavalia.* Brittonia 16: 106–81.

Schofield, Eileen K., & Paul A. Colinvaux. 1969. Fossil *Azolla* from the Galápagos Islands. Bull. Torrey Bot. Club 96: 623–28.

Schubert, Bernice G. 1940. *Desmodium*: Preliminary studies, I. A. *Desmodium procumbens* and related species. B. Miscellaneous notes and records. Contr. Gray Herb. 129: 3–31.

Schultes, R. E., & R. Romero-Castaneda. 1962. Edible fruits of *Solanum* in Colombia. Bot. Mus. Leafl. 19: 235–86.

Senn, Harold A. 1939. The North American species of *Crotalaria.* Rhodora 41: 315–67.

———. 1943. *Crotalaria*, in J. F. Macbride, Flora of Peru. Field Mus. Nat. Hist. Bot. Ser. 13(3, no. 1): 454–58.

Setchell, W. A. 1946. The genus *Ruppia* L. Proc. Calif. Acad. Sci. IV, 25: 469–78.

Slevin, J. R. 1959. The Galápagos Islands. A history of their exploration. Occ. Papers Calif. Acad. Sci. 25: 1–150.

Stafleu, F. A. 1967. *Taxonomic Literature.* Regnum Veg. 52. xx + 556 pp.

Standley, P. C. 1922. Trees and shrubs of Mexico (Fagaceae-Fabaceae). Contr. U.S. Nat. Herb. 23^2: 171–515.

———. 1923. Trees and shrubs of Mexico (Oxalidaceae-Turneraceae). Contr. U.S. Nat. Herb. 23^3: 517–848.

———. 1924. Trees and shrubs of Mexico (Passifloraceae-Scropulariaceae). Contr. U.S. Nat. Herb. 23^4: 849–1312.

Standley, P. C., & J. Steyermark. 1946a. Flora of Guatemala (Oxalidaceae). Fieldiana: Bot. 24^5: 374–85.

———. 1946b. Flora of Guatemala (Rutaceae). Fieldiana: Bot. 24^5: 398–425.

———. 1949. Flora of Guatemala (Trigoniaceae-Saurauiaceae). Fieldiana: Bot. 24[6]: 1–440.

———. 1952a. Flora of Guatemala (Iridaceae). Fieldiana: Bot. 24[3]: 159–78.

———. 1952b. Flora of Guatemala (Utricaceae). Fieldiana: Bot. 24[3]: 396–430.

Standley, P. C., & L. O. Williams. 1961. Flora of Guatemala (Bixaceae). Fieldiana: Bot. 24[7]: 65–67.

Stearn, W. T. 1958. A key to the West Indian mangroves. Kew Bull. 13: 33–37.

Stephens, S. G., & C. M. Rick. Problems of origin, dispersal, and establishment of the Galápagos cottons, in R. I. Bowman, ed., *The Galápagos*, listed above.

Stewart, A. 1911. A botanical survey of the Galápagos Islands. Proc. Calif. Acad. Sci. IV, 1: 7–288.

———. 1915a. Some observations concerning the botanical conditions on the Galápagos Islands. Trans. Wisconsin Acad. Sci. 18: 272–340.

———. 1915b. Notes on the forms of *Castela galapageia*. Amer. J. Bot. 2: 279–88.

Svenson, H. K. 1935. Plants of the Astor Expedition, 1930 (Galápagos and Cocos islands). Amer. J. Bot. 22: 208–77.

———. 1944. The new world species of *Azolla*. Amer. Fern J. 34: 69–84.

———. 1946. Vegetation of the coast of Ecuador and Peru and its relation to that of the Galápagos Islands. Amer. J. Bot. 33: 394–426; 427–98.

Swarth, Harry S. 1931. The avifauna of the Galápagos Islands. Occ. Papers Calif. Acad. Sci. 18: 1–229, *figs. 1–57.*

Swingle, W. T. 1943. The botany of *Citrus* and its wild relatives of the orange family (family Rutaceae, subfamily Aurantioideae), in H. J. Webber & L. D. Batchelor, eds., *The Citrus Industry*. Vol. I. *History, Botany & Breeding*. Univ. Calif. Press, Berkeley. xx + 1028 pp.

Tanaka, T. 1937. Further revision of Rutaceae-Aurantioideae of India and Ceylon. J. Indian Bot. Soc. 16: 227–40.

———. 1954. Species problem in *Citrus*. Hort. Inst. Tokyo Agr. Univ. Tech. Papers 10: 1–152, *pls. 1–3.*

Tenny, F. A. 1928. Costa Rican balsa. Trop. Woods 15: 34–37.

Thwaites, G. H. K. 1864. *Enumeratio plantarum Zeylanicae: an enumeration of the Ceylon plants*. Delau & Co., London. viii + 483 pp.

Trelease, W., & T. G. Yuncker. 1950. *The Piperaceae of Northern South America*. Univ. Ill. Press, Urbana. 2 vols. vii + 674 pp., 838 figs.

Tryon, R. M. 1956. A revision of the American species of *Notholaena*. Contr. Gray Herb. 179: 1–106.

———. 1962. Taxonomic notes. II. *Pityrogramma* (including *Trismeria* and *Anogramma*). Contr. Gray Herb. 189: 52–76.

Türpe, Anna Maria. 1966. Histotaxonomía de las especies Argentinas del género *Paspalum*. Lilloa 32: 35–299, *figs. 1–81.*

Urban, I. 1883. Monographie der Familie der Turneraceen. Jahrb. Bot. Gart. Berlin 2: 1–152.

Usinger, R. L., & Peter D. Ashlock. 1966. Evolution of Orsilline insects in oceanic islands (Hemiptera, Lygaeidae), in R. I. Bowman, ed., *The Galápagos*, listed above.

Van Dyke, E. C. 1953. The *Coleoptera* of the Galapagos Islands. Occ. Papers Calif. Acad. Sci. 22: 1–181, *pls. 1–7.*

Van Meemoen, M. G., et al. 1961. Preliminary revisions of some genera of Maylaysian Papilionaceae I. Reinwardtia 5: 419–56.

Van Ooststroom, S. J. 1940. The Convolvulaceae of Maylaysia, III. Blumea 3: 481–584.

Vercammen-Grandjean, P. H. 1966. Evolutionary problems in mite-host specificity and its relevance to studies of Galápagos organisms, in R. I. Bowman, ed., *The Galápagos*, listed above.

Webber, H. J. 1943. Cultivated varieties of *Citrus*, in H. J. Webber & L. D. Batchelor, eds., *The Citrus Industry*. Vol. I. *History, Botany & Breeding*. Univ. Calif. Press, Berkeley. xx + 1028 pp.

Weber, W. C. 1966. Lichenology and Bryology in the Galápagos Islands, with check lists of the lichens and bryophytes thus far reported, in R. I. Bowman, ed., *The Galápagos*, listed above.

Wheeler, Louis C. 1941. *Euphorbia* subgenus *Chamaesyce* in Canada and the United States exclusive of southern Florida. Rhodora 43: 97–205; 223–86, *pls. 654–68*.

Wiggins, Ira L. 1966. Origins and relationships of the flora of the Galápagos Islands, in R. I. Bowman, ed., *The Galápagos*, listed above.

Wilks, W., ed. 1914. *Journal kept by David Douglas during his travels in North America 1823–1827*. William Wesley & Son, London. 364 pp.

Williams, F. X. 1907. Report on expedition of California Academy of Sciences. Entomological News 18: 260–61.

Williams, Howell. 1966. Geology of the Galápagos Islands, in R. I. Bowman, ed., *The Galápagos*, listed above.

Windler, D. R. 1966. A revision of the genus *Neptunia* (Leguminosae). Austral. J. Bot. 14: 379–420.

Wolf, T. 1879. Ein Besuch der Galápagos Inseln, mit drei Kärtchen. Samml. Vort. deutsche Volk. 1: 259–302.

Woodson, R. E., Jr. 1945. Flora of Panama. Cannaceae. Ann. Missouri Bot. Gard. 32: 74–80.

Yates, H. O. 1966a. Morphology and cytology of *Uniola* (Gramineae). Southw. Naturalist 11: 145–89.

———. 1966b. Revision of grasses traditionally referred to *Uniola*. I, *Uniola* and *Leptochloöpsis*. Southw. Naturalist 11: 372–94.

———. 1966c. Revision of grasses traditionally referred to *Uniola*. II, *Chasmanthium*. Southw. Naturalist 11: 415–55.

Young, D. P. 1958. *Oxalis* in the British flora. Watsonia 4: 51–69.

Index of Botanical Names

The index is to plant names only (the few nonvascular plants mentioned are included); and with the exception of common names for families (e.g. "SEDGE FAMILY"), only taxonomic names are given. All taxonomic levels are indexed: taxa above genera are in capital and small-capital letters; recognized taxa at generic or lower levels are in roman type with initial capitals; synonyms are in italic type with initial capitals; and taxa discussed but occurring elsewhere than in the Galápagos Islands are distinguished by italic page numbers. The second and following pages of continuous discussions are not cited; new discussions on such pages are. Occurrences of names in keys to species (and other taxa within genera) are not cited, since the genus name immediately locates such a key; occurrences of names in keys to the higher taxa *are* cited. Wherever a taxon is illustrated with a line drawing, the page number for the drawing is cited; the page number for a taxon's range map is not. Citations to "color plates" are best referred to the "Detailed Descriptions of Color Plates," on pp. xiii–xx; the color plates themselves follow p. 42.

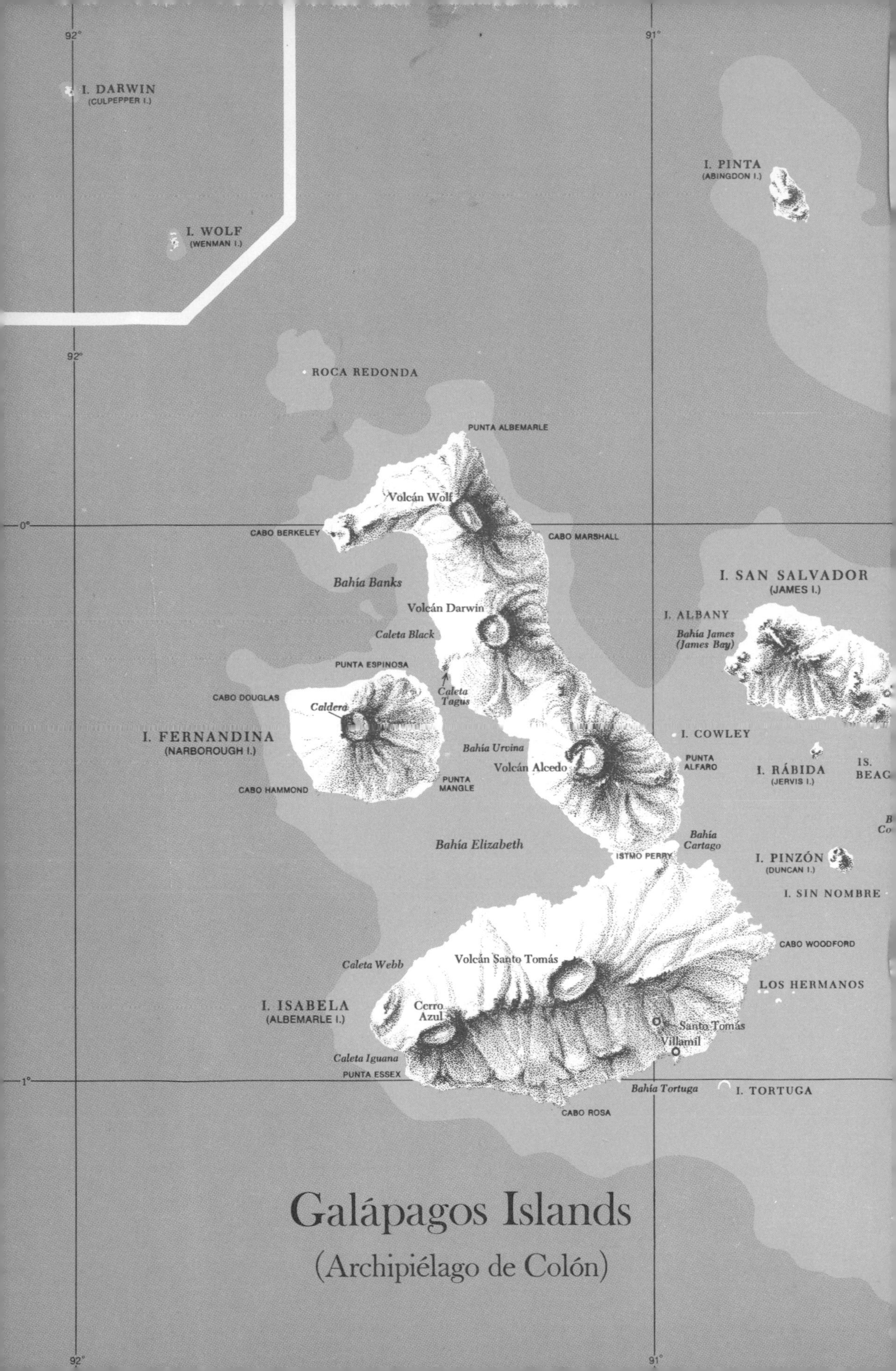
92°
91°
I. DARWIN
(CULPEPPER I.)
I. WOLF
(WENMAN I.)
I. PINTA
(ABINGDON I.)
92°
ROCA REDONDA
PUNTA ALBEMARLE
Volcán Wolf
0°
CABO BERKELEY
CABO MARSHALL
Bahía Banks
I. SAN SALVADOR
(JAMES I.)
Volcán Darwin
I. ALBANY
Bahía James
(James Bay)
Caleta Black
PUNTA ESPINOSA
Caleta
Tagus
CABO DOUGLAS
Caldera
I. FERNANDINA
(NARBOROUGH I.)
I. COWLEY
Bahía Urvina
PUNTA
ALFARO
Volcán Alcedo
I. RÁBIDA
(JERVIS I.)
IS.
BEAG
CABO HAMMOND
PUNTA
MANGLE
Bahía
Cartago
Bahía Elizabeth
ISTMO PERRY
I. PINZÓN
(DUNCAN I.)
I. SIN NOMBRE
CABO WOODFORD
Caleta Webb
Volcán Santo Tomás
LOS HERMANOS
I. ISABELA
(ALBEMARLE I.)
Cerro
Azul
Santo Tomás
Villamil
Caleta Iguana
PUNTA ESSEX
1°
Bahía Tortuga
I. TORTUGA
CABO ROSA
Galápagos Islands
(Archipiélago de Colón)
92°
91°